W9-CQW-170

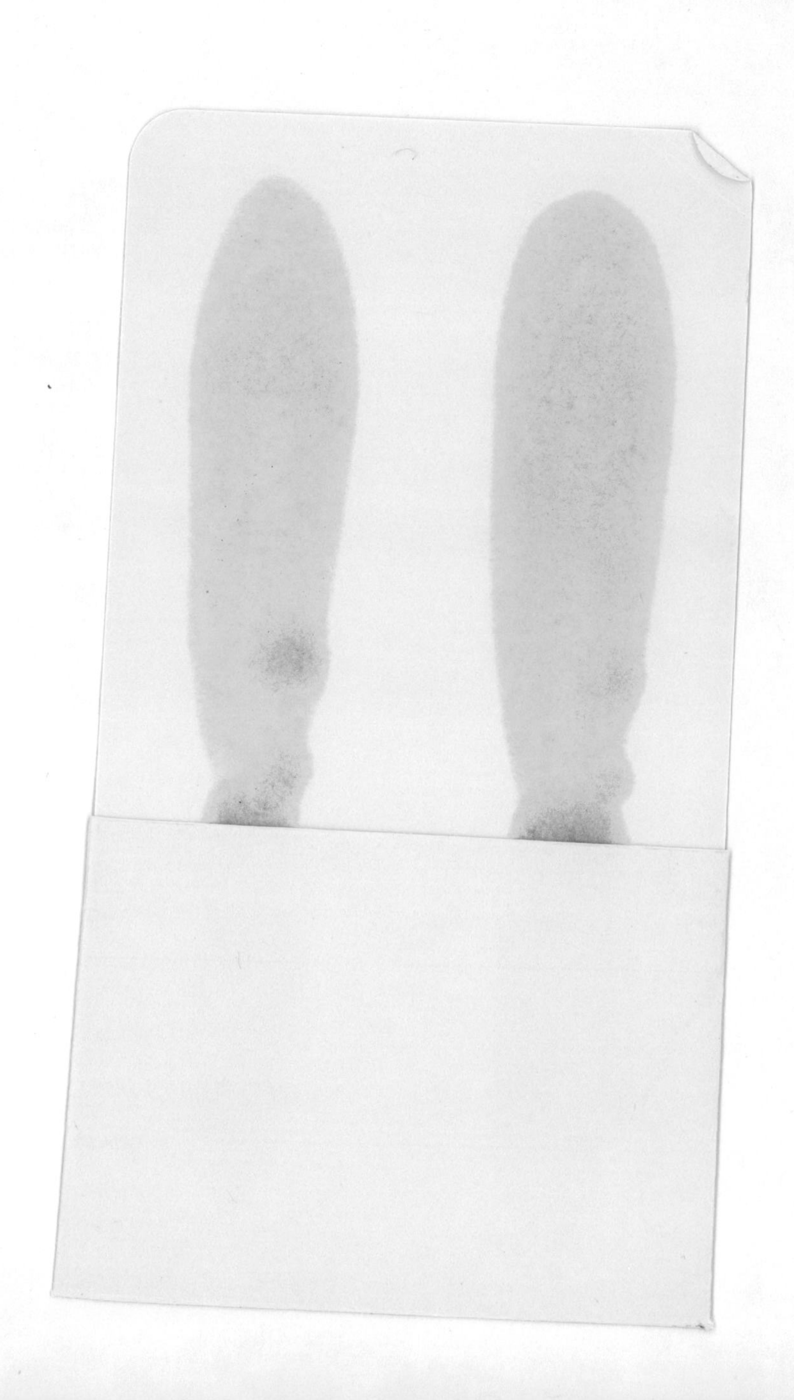

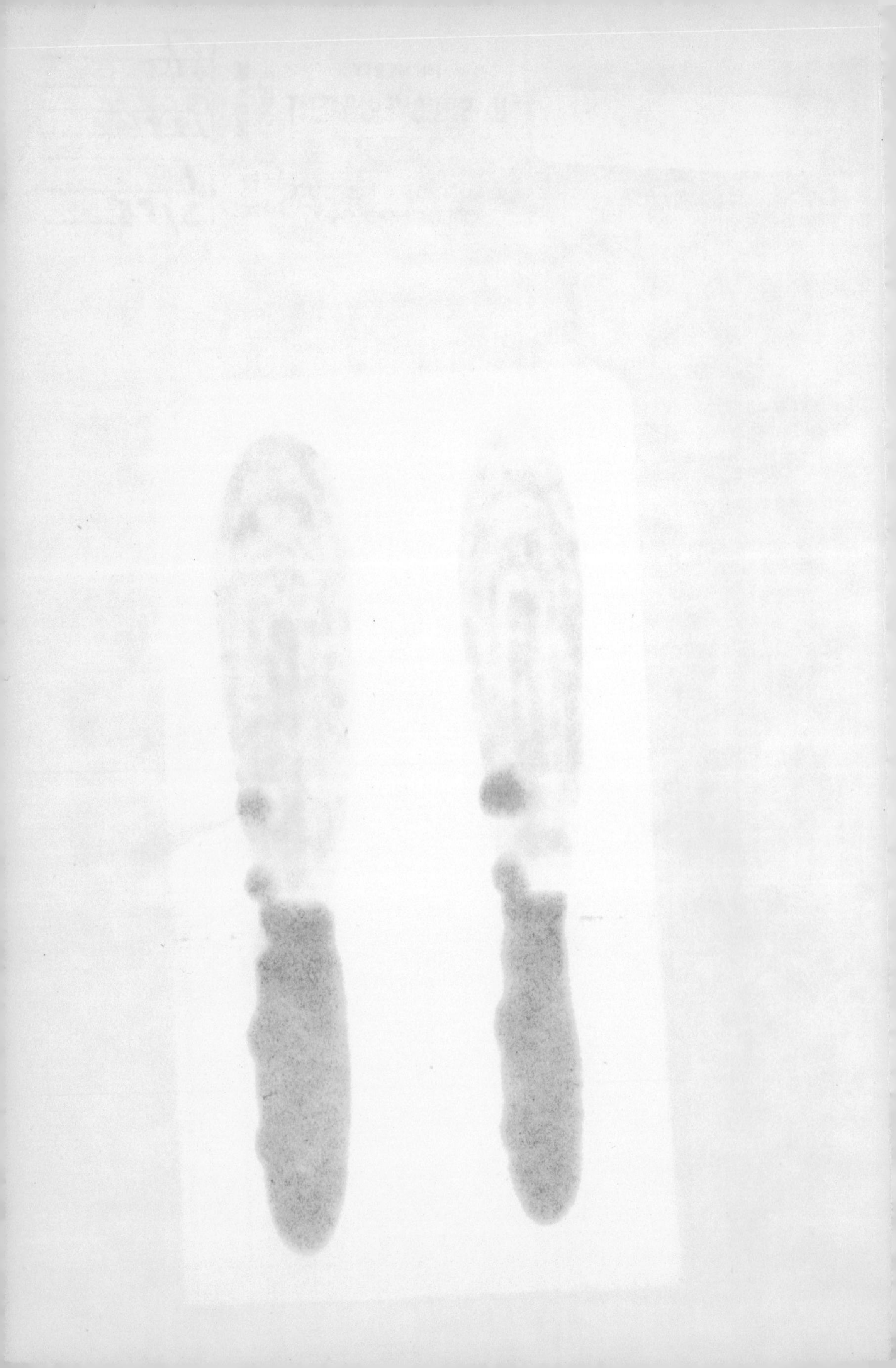

BIOCHEMICAL ENGINEERING FUNDAMENTALS

McGraw-Hill Chemical Engineering Series

BUILDING THE LITERATURE OF A PROFESSION

Fifteen prominent chemical engineers first met in New York more than 60 years ago to plan a continuing literature for their rapidly growing profession. From industry came such pioneer practitioners as Leo H. Baekeland, Arthur D. Little, Charles L. Reese, John V. N. Dorr, M. C. Whitaker, and R. S. McBride. From the universities came such eminent educators as William H. Walker, Alfred H. White, D. D. Jackson, J. H. James, Warren K. Lewis, and Harry A. Curtis. H. C. Parmelee, then editor of *Chemical and Metallurgical Engineering*, served as chairman and was joined subsequently by S. D. Kirkpatrick as consulting editor.

After several meetings, this committee submitted its report to the McGraw-Hill Book Company in September 1925. In the report were detailed specifications for a correlated series of more than a dozen texts and reference books which have since become the McGraw-Hill Series in Chemical Engineering and which became the cornerstone of the chemical engineering curriculum.

From this beginning there has evolved a series of texts surpassing by far the scope and longevity envisioned by the founding Editorial Board. The McGraw-Hill Series in Chemical Engineering stands as a unique historical record of the development of chemical engineering education and practice. In the series one finds the milestones of the subject's evolution: industrial chemistry, stoichiometry, unit operations and processes, thermodynamics, kinetics, and transfer operations.

Chemical engineering is a dynamic profession, and its literature continues to evolve. McGraw-Hill and its consulting editors remain committed to a publishing policy that will serve, and indeed lead, the needs of the chemical engineering profession during the years to come.

THE SERIES

Bailey and Ollis: *Biochemical Engineering Fundamentals*
Bennett and Myers: *Momentum, Heat, and Mass Transfer*
Beveridge and Schechter: *Optimization: Theory and Practice*
Carberry: *Chemical and Catalytic Reaction Engineering*
Churchill: *The Interpretation and Use of Rate Data—The Rate Concept*
Clarke and Davidson: *Manual for Process Engineering Calculations*
Coughanowr and Koppel: *Process Systems Analysis and Control*
Daubert: *Chemical Engineering Thermodynamics*
Fahien: *Fundamentals of Transport Phenomena*
Finlayson: *Nonlinear Analysis in Chemical Engineering*
Gates, Katzer, and Schuit: *Chemistry of Catalytic Processes*
Holland: *Fundamentals of Multicomponent Distillation*
Holland and Liapis: *Computer Methods for Solving Dynamic Separation Problems*
Johnson: *Automatic Process Control*
Johnstone and Thring: *Pilot Plants, Models, and Scale-Up Methods in Chemical Engineering*
Katz, Cornell, Kobayashi, Poettmann, Vary, Elenbaas, and Weinaug: *Handbook of Natural Gas Engineering*
King: *Separation Processes*
Klinzing: *Gas–Solid Transport*
Knudsen and Katz: *Fluid Dynamics and Heat Transfer*
Luyben: *Process Modeling, Simulation, and Control for Chemical Engineers*
McCabe, Smith, J. C., and Harriott: *Unit Operations of Chemical Engineering*
Mickley, Sherwood, and Reed: *Applied Mathematics in Chemical Engineering*
Nelson: *Petroleum Refinery Engineering*
Perry and Chilton (Editors): *Chemical Engineers' Handbook*
Peters: *Elementary Chemical Engineering*
Peters and Timmerhaus: *Plant Design and Economics for Chemical Engineers*
Probstein and Hicks: *Synthetic Fuels*
Ray: *Advanced Process Control*
Reid, Prausnitz, and Sherwood: *The Properties of Gases and Liquids*
Resnick: *Process Analysis and Design for Chemical Engineers*
Satterfield: *Heterogeneous Catalysis in Practice*
Sherwood, Pigford, and Wilke: *Mass Transfer*
Smith, B. D.: *Design of Equilibrium Stage Processes*
Smith, J. M.: *Chemical Engineering Kinetics*
Smith, J. M., and Van Ness: *Introduction to Chemical Engineering Thermodynamics*
Thompson and Ceckler: *Introduction to Chemical Engineering*
Treybal: *Mass Transfer Operations*
Valle-Riestra: *Project Evolution in the Chemical Process Industries*
Van Ness and Abbott: *Classical Thermodynamics of Nonelectrolyte Solutions: With Applications to Phase Equilibria*
Van Winkle: *Distillation*
Volk: *Applied Statistics for Engineers*
Walas: *Reaction Kinetics for Chemical Engineers*
Wei, Russell, and Swartzlander: *The Structure of the Chemical Processing Industries*
Whitwell and Toner: *Conservation of Mass and Energy*

BIOCHEMICAL ENGINEERING FUNDAMENTALS

Second Edition

James E. Bailey
California Institute of Technology

David F. Ollis
North Carolina State University

McGraw-Hill Book Company

New York St. Louis San Francisco Auckland Bogotá Hamburg
Johannesburg London Madrid Mexico Montreal New Delhi
Panama Paris São Paulo Singapore Sydney Tokyo Toronto

This book was set in Times Roman.
The editors were Kiran Verma and Cydney C. Martin.
The production supervisor was Diane Renda;
the cover was designed by John Hite;
project supervision was done by Albert Harrison, Harley Editorial Services.
Halliday Lithograph Corporation was printer and binder.

BIOCHEMICAL ENGINEERING FUNDAMENTALS

34567890 HALHAL 8987

ISBN 0-07-003212-2

Library of Congress Cataloging-in-Publication Data

Bailey, James E. (James Edwin), 1944–
Biochemical engineering fundamentals.

(McGraw-Hill chemical engineering series)
Includes bibliographies and index.
1. Biochemical engineering. I. Ollis, David F.
II. Title. III. Series
TP248.3.B34 1986 660'.63 85-19744
ISBN 0-07-003212-2

To Sean
and
Andrew, Mark, Stephen, and Matthew

CONTENTS

PREFACE

Processing of biological materials and processing using biological agents such as cells, enzymes, or antibodies are the central domains of biochemical engineering. Success in biochemical engineering requires integrated knowledge of governing biological properties and principles and of chemical engineering methodology and strategy. Work at the forefront captures the latest, best information and technology from both areas and accomplishes new syntheses for bioprocess design, operation, analysis, and optimization. Reaching this objective clearly requires years of careful study and practice.

This textbook is intended to start its readers on this challenging and exciting path. Central concepts are defined and explained in the context of process applications. Principles of current bioprocesses for reaction and separation are presented. Special attention is devoted throughout to the central roles of biological properties in facilitating and enabling desired process objectives. Also, process constraints and limitations imposed by sensitivities and instabilities of biological components are highlighted. By focusing on pertinent fundamental principles in the biological and engineering sciences and by repeatedly emphasizing the importance of their syntheses, the text seeks to endow its readers with a strong foundation for future study and practice. Learning fundamental properties and mechanisms on an ongoing basis is absolutely essential for long-term professional viability in a technically vibrant area such as biotechnology.

The book has been written for the first course in biochemical engineering for senior or graduate students in chemical engineering. However, selected portions of the text can provide bases for other courses in chemical, environmental, civil, or food engineering. As in the first edition, the book is presented in a systematic, logical sequence building from the most fundamental biological concepts. It is therefore well suited for self study by industrial practitioners.

To facilitate the book's accessibility for independent reading and to provide required background in a one- or two-term course taken as an elective or introduction, the text includes a self-contained presentation of key concepts from biochemistry, cell biology, enzyme kinetics, and molecular genetics. Clearly, this treatment is intended as an introductory exposure to these topics and not as complete coverage of the life science fundamentals needed by those who will study biochemical engineering in depth or who practice in the field. Further formal or self study in biological fundamentals and practical properties is essential in these cases.

Throughout, we have tried to interweave descriptive material on the life sciences with engineering processes and analytical techniques. The implications of bioscience fundamentals for bioprocess engineering are frequently indicated in sections dealing with biological principles. Treatment of engineering analysis is presented after required descriptive, background material has been covered. Thus, enzyme kinetics and reaction engineering are introduced immediately following description of proteins and other biochemicals, and cell kinetics follows description of metabolic pathway structure, stoichiometry, and regulation.

Text examples and end-of-chapter problems provide the student with opportunities to apply the concepts presented and to broaden understanding of the subject. More than 150 problems, spanning a range of difficulty, require discussions, derivations, and/or calculations by the student.

Compelling motivations for this second edition have come from explosive developments in the biological sciences which provide revolutionary new organisms and materials with tremendous promise for new products and processes. Recombinant DNA and hybridoma technology have stimulated a new biotechnology industry. The text has been expanded and updated to present the materials and methods of gene cloning and expression and cell fusion. New process challenges and strategies for large-scale manufacture of new, ultra-pure protein products are summarized.

Several engineering topics have received greater emphasis in the second edition. This is immediately apparent from the new chapters on separation processes, bioprocess instrumentation and control, and bioprocess economics. Important new topics such as metabolic stoichiometry, multiphase reactor engineering, and animal and plant cell reactor technology have also been integrated into the earlier text.

In addition, the opportunity of preparing a second edition has enabled numerous improvements in organization and presentation of material included in the first edition. This contributes, for example, to more concise yet more informative description of background material, and to a more systematic approach to stoichiometry, kinetics, and bioreactor design. The importance of coalescence and dispersion processes in multiphase reactor contacting exemplifies another area of enhanced presentation.

Cogent and critical comments on the second edition from Michael Shuler, Douglas Lauffenberger, Peter Reilly, Frances Arnold, Donald Kirwan, and Elmer Gaden provided many improvements. Numerous colleagues and current and former students including Dinesh Arora, Ruben Carbonell, Douglas Clark, Kathy Dennis, Jorge Galazzo, L. Gary Leal, Sun Bok Lee, Harold Monbouquette, Mustafa Ozilgen, Steven Peretti, Alex Seressiotis, Robert Siegel, Friedrich Srienc, and Gregory Stephanopoulos contributed ideas, background research, and/or new homework exercises to the second edition. To those who contributed in numerous ways to the first edition, including Peter Reilly, Elmer Gaden, Harold Bungay, Murray Moo-Young, and George Tsao, we again offer our thanks. Of course the authors take full responsibility for any errors, and welcome comments and suggestions from readers.

This book would not exist without the patient, steadfast efforts of April Olson, Kathy Lewis, Heidi Youngkin, Sandra Cantrell, Bessie See, and Kathy Cannady who typed the several drafts. Hundreds of hours of proofreading assistance were generously donated by Doug Axe, Nancy da Silva, Jorge Galazzo, Chris Guske, Justin Ip, Anne McQueen, Kim O'Connor, Steve Peretti, Mike Prairie, Todd Przybycien, Ken Reardon, Jin-Ho Seo, Alex Seressiotis, Jackie Shanks, Friedrich Srienc, and Dane Wittrup. Finally, we would like to extend our heartfelt gratitude to many friends, colleagues, students, and sponsors who have stimulated our development as biochemical engineers in the years since the first edition. They are in many ways the true authors of this book.

James E. Bailey
David F. Ollis

CHAPTER

ONE

A LITTLE MICROBIOLOGY

Small living creatures called *microorganisms* interact in numerous ways with human activities. On the large scale of the biosphere, which consists of all regions of the earth containing life, microorganisms play a primary role in the capture of energy from the sun. Their biological activities also complete critical segments of the cycles of carbon, oxygen, nitrogen, and other elements essential for life. Microbes are also responsible for many human, animal, and plant diseases.

In this text we concentrate primarily on mankind's use of microbes. These versatile biological catalysts have served mankind for milennia. The ancient Greeks credited the god Dionysus with invention of fermentation for wine making, and the "Monument bleu," which dates from 7000 B.C., shows beer brewing in Babylon. Fermented foods such as cheese, bread, yoghurt, and soy have long contributed to mankind's nutrition. Late in the 19th century, the work of Pasteur and Tyndall identified microorganisms as the critical, active agents in prior fermentation practice and initiated the emergence of microbiology as a science. From these beginnings, further work by Buchner, Neuberg, and Weizmann led to processes for production of ethanol, glycerol, and other chemicals in the early 20th century.

In the 1940s complementary developments in biochemistry, microbial genetics, and engineering ushered in the era of antibiotics with tremendous relief to mankind's suffering and mortality. This period marks the birth of *biochemical engineering*, the engineering of processes using catalysts, feedstocks, and/or sorbents of biological origin. Biotechnology began to change from empirical art to predictive, optimized design.

A later generation of fermentation processes produced steroids for birth control and for treatment of arthritis and inflammation. Methods for cultivation of plant and animal cells made possible mass production of vaccines and other useful biological agents. Clearly, mankind's successful harnessing and direction of cellular activities has had many health, social, environmental, and economic impacts on past and contemporary human civilization.

An interwoven fabric of research in molecular biology and microbial genetics has led to fundamental understanding of many of the controls and catalysts involved in complex biochemical syntheses conducted by living cells. On this foundation of basic knowledge, the methods of *recombinant DNA technology* have been erected. It is difficult to imagine the scope and magnitude of the eventual benefits of these marvelous tools. New vaccines and drugs have already been produced, but these are only the beginnings of revolutionary developments to come.

Our challenge in learning biochemical engineering is to understand and analyze the processes of biotechnology so that we can design and operate them in a rational way. To reach this goal, however, a basic working knowledge of cell growth and function is required. These factors and others peculiar to biological systems usually dominate biochemical process engineering. Consider for a moment that a living microorganism may be viewed in an approximate conceptual sense as an expanding chemical reactor which takes in chemical species called *nutrients* from its environment, grows, reproduces, and releases products into its surroundings. In instances such as sewage treatment, consumption of nutrients (here the organic sewage material) is the engineering objective. When microbes are grown for food sources or supplements, it is the mass of microbial matter produced which is desired. For a sewage-treatment process, on the other hand, this microbial matter produced by nutrient consumption constitutes an undesirable solid waste, and its amount should be minimized. Finally, the products formed and released during cellular activity are of major concern in many industrial and natural contexts, including penicillin and ethanol manufacture. The relative rates of nutrient utilization, growth, and release of products depend strongly on the type of cells involved and on the temperature, composition, and motion of their environment. Understanding these interactions requires a foundation built upon biochemistry, biophysics, and cell biology. Since study of these subjects is not traditionally included in engineering education, a substantial portion of our efforts must be dedicated to them.

Whenever possible we shall extend our study of biological processes beyond qualitative understanding to determine quantitative mathematical representations. These mathematical models will often be extremely oversimplified and idealized, since even a single microorganism is a very complicated system. Nevertheless, basic concepts in microbiology will serve as a guide in formulating models and checking their validity, just as basic knowledge in fluid mechanics is useful when correlating the friction factor with the Reynolds number.

1.1 BIOPHYSICS AND THE CELL DOCTRINE

Microbiology is the study of living organisms too small to be seen clearly by the naked eye. As a rough rule of thumb, most microorganisms have a diameter of 0.1 mm or less. Present knowledge indicates that even the simplest microorganism houses chemical reactors, information and control systems, and mass-transfer operations of amazing sophistication, efficiency, and organization. These conclusions have been reached in numerous experimental studies involving methods adapted from the physical sciences. Since this approach has proved so fruitful, the applicability of the principles of chemistry and physics to biological systems is now a widespread working hypothesis within the life sciences. The term *biophysics* is sometimes used to indicate explicitly the union of the biological and physical sciences.

A development critical to the understanding of living systems started in 1838, when Schleiden and Schwann first proposed the *cell theory*. This theory stated that all living systems are composed of cells and their products. Thus, the concept of a basic module, or building block, for life emerged. This notion of a common denominator permits an important decomposition in the analysis of living systems: first the component parts, the cells, can be studied, and then this knowledge is used to try to understand the complete organism.

The value of this decomposition rests on the fact that cells from a wide variety of organisms share many common features in their structure and function. In many instances this permits successful extrapolation of knowledge gained from experiments on cells from one organism to cells of other types. This existence of common cellular characteristics also simplifies our task of learning how microorganisms behave. By concentrating on the apparently universal features of cellular function, a basic framework for understanding all living systems can be established.

We should not leave this section with the impression that all cells are alike, however. Muscle cells are clearly different from those found in the eye or brain. Equally, there are many different types of single-celled organisms. These in turn can be classified in terms of the two major types of cellular organization described next.

1.2 THE STRUCTURE OF CELLS

Observations with the electron microscope have revealed two markedly different kinds of cells. Although still linked by certain common features, these two classes are sufficiently distinct in their organization and function to warrant individual consideration here. So far as is known today, all cells belong to one of these groups.

1.2.1 Procaryotic Cells

Procaryotic cells, or *procaryotes*, do not contain a membrane-enclosed nucleus. Procaryotes are relatively small and simple cells. They usually exist alone, not

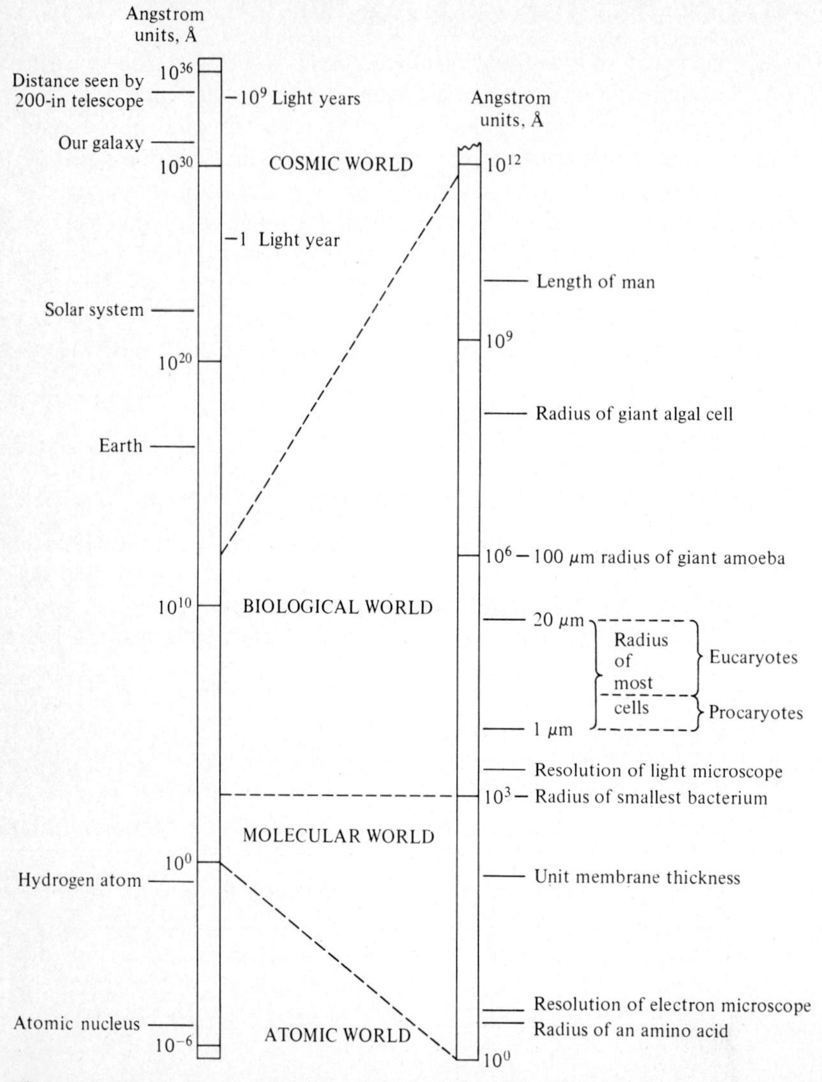

Figure 1.1 Characteristic dimensions of the universe. The biological world encompasses a broad spectrum of sizes. *(From "Cell Structure and Function," 2d ed., p. 35, by Ariel G. Loewy and Philip Siekevitz. Copyright © 1963, 1969 by Holt, Rinehart and Winston Inc. Reprinted by permission of Holt, Rinehart and Winston.)*

associated with other cells. The typical dimension of these cells, which may be spherical, rodlike, or spiral, is from 0.5 to 3 μm.† In order to gain a qualitative feel for such dimensions, it is instructive to compare the relative sizes of cells with other components of the universe. As Fig. 1.1 reveals, the size of a procaryote

† 1 m (meter) = 10^3 mm (millimeter) = 10^6 μm (micrometer, formerly known as micron) = 10^9 nm (nanometer) = 10^{10} Å (angstrom units).

relative to a man is approximately equal to the size of a man relative to the earth and less than the size of the hydrogen atom compared with that of a cell. These size relationships are very significant considerations when the details of cell function are investigated, as we shall see later. The volume of procaryotes is on the order of 10^{-12} ml per cell, of which 50 to 80 percent is water. As a rough estimate, the mass of a single procaryote is 10^{-12} g.

Microorganisms of this type grow rapidly and are widespread in the biosphere. Some, for example, can double in size, mass, and number in 20 min. Typically, procaryotes are biochemically versatile; i.e., they often can accept a wide variety of nutrients and further are capable of selecting the best nutrient from among several available in their environment. This feature and others to be recounted later make procaryotic cells adaptable to a wide range of environments. Since procaryotes usually exist as isolated single-celled organisms, they have little means of controlling their surroundings. Therefore the nutrient flexibility they exhibit is an essential characteristic for their survival. The rapid growth and biochemical versatility of procaryotes make them obvious choices for biological research and biochemical processing.

In Fig. 1.2 the basic features of a procaryotic cell are illustrated. The cell is surrounded by a rigid *wall*, approximately 200 Å thick. This wall lends structural strength to the cell to preserve its integrity in a wide variety of external surroundings. Immediately inside this wall is the *cell membrane*, which typically has a thickness of about 70 Å. This membrane has a general structure common to membranes found in all cells. It is sometimes called a *plasma membrane*. These membranes play a critical role: they largely determine which chemical species can be transferred between the cell and its environment as well as the net rate of such transfer. Within the cell is a large, ill-defined region called the *nuclear zone*, which is the dominant control center for cell operation. The grainy dark spots apparent

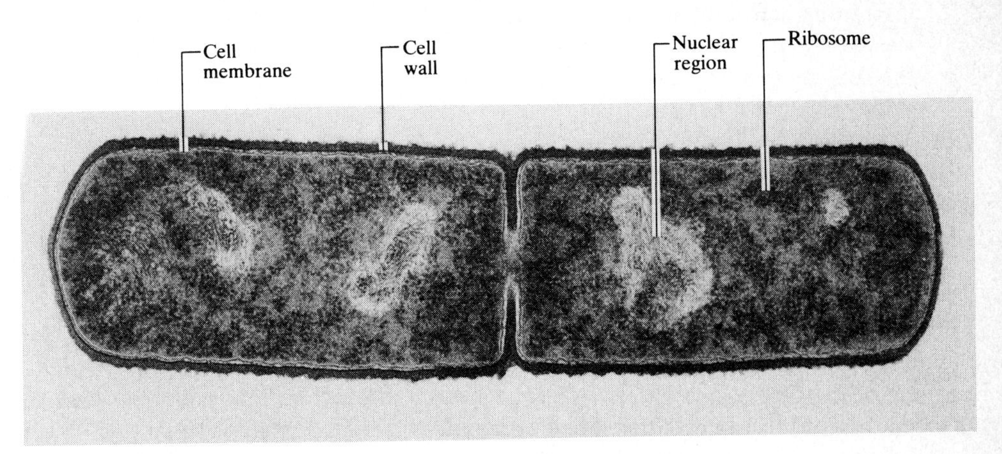

Figure 1.2 Electron micrograph of a procaryote, *Bacillus subtilis*. This soil bacterium, shown here near completion of cell division, is used commercially to make several biological catalysts and antibiotics. Typical cell dimensions are around 1 μm diameter and 2 μm length. *B. subtilis* is also an important host organism for recombinant DNA. *(Courtesy of Antoinette Ryter.)*

in the cell interior are the *ribosomes*, the sites of important biochemical reactions. The *cytoplasm* is the fluid material occupying the remainder of the cell. Not evident here but visible in some photographs are clear, bubblelike regions called *storage granules*. We shall explore the composition and function of these structures within the procaryotic cell in greater detail after establishing the necessary background and defining some terms.

While sharing many common structural and biochemical features, procaryotes exhibit considerable diversity. The blue-green algae, for example, contain membranes which capture light energy for *photosynthesis*. This complex process uses light energy from the sun, provides the cells with organic molecules suitable for its reactions, and releases oxygen into the atmosphere.

1.2.2 Eucaryotic Cells

Eucaryotic cells, or *eucaryotes*, make up the other major class of cell types. Eucaryotes may be defined most concisely as cells which possess a membrane-enclosed nucleus. As a rule these cells are 1000 to 10,000 times larger than procaryotes. All cells of higher organisms belong to this family. In order to meet the many specialized needs of animals, for example, eucaryotic cells exist in many different forms. By coexisting and interacting in a cooperative manner in a higher organism, these cells can avoid the necessity for biochemical flexibility and adaptability so essential to procaryotes. Many important microbial species are also eucaryotes. In the next section we shall see several examples of eucaryotes which exist as single-celled organisms.

The internal structure of eucaryotes is considerably more complex than that of procaryotic cells, as can be seen in Figs. 1.3 and 1.4. Here there is a substantial degree of spatial organization and differentiation. The internal region is divided into a number of distinct compartments, which we shall explore in greater detail later; they have special structures and functions for conducting the activities of the cell. At this point we shall only consider the general features of eucaryotic cells.

The cell is surrounded by a plasma, or unit, membrane similar to that found in procaryotes. On the exterior surface of this membrane may be a cell coat, or wall. The nature of the outer covering depends on the particular cell. For example, cells of higher animals usually have a thin cell coat. The specific adhesive properties of this coat are important in binding like cells to form specialized tissues and organs such as the liver. Plant cells, on the other hand, are often enclosed in a very strong, thick wall. Lumber consists for the most part of the walls of dead tree cells.

Important to the internal specialization of eucaryotic cells is the presence of unit membranes within the cell. A complex, convoluted membrane system, called the *endoplasmic reticulum*, leads from the cell membrane into the cell. The *nucleus* here is surrounded by a porous membrane. *Ribosomes*, biochemical reaction sites seen before in procaryotes, are embedded in the surface of much of the endoplasmic reticulum. (Ribosomes in procaryotes are smaller, however.)

A major function of the *nucleus* is to control the catalytic activity at the

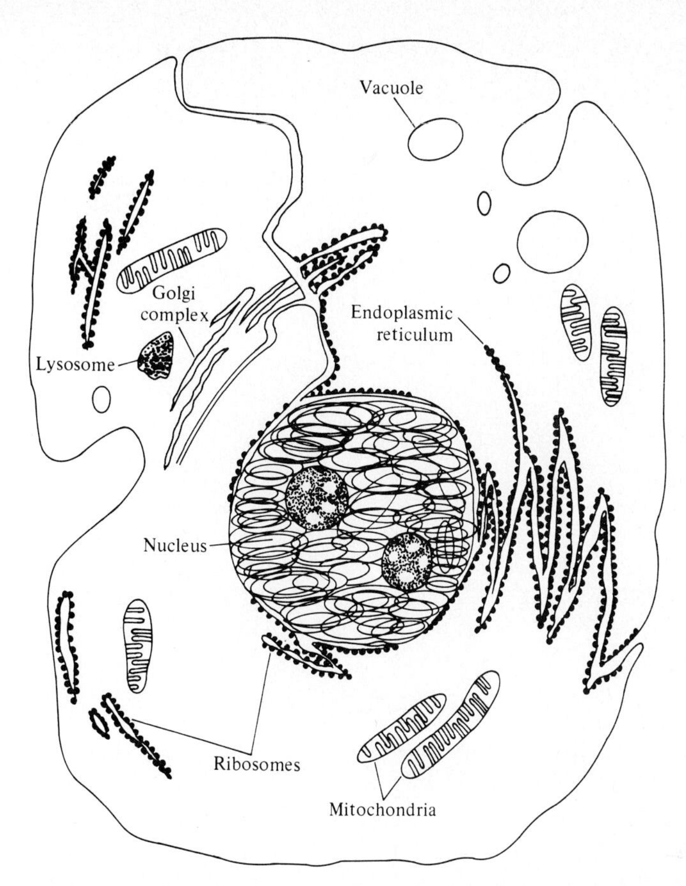

Figure 1.3 A typical eucaryotic cell. Such a typical cell is an imaginary construct, for there are wide variations between different eucaryotes. Still, many of these cells share common features and components, making the typical eucaryote a convenient and useful concept.

ribosomes. Not only are the reaction rates regulated, but the particular reactions which occur are determined by chemical messengers manufactured in the nucleus.

The nucleus is one of several interior regions surrounded by unit membranes. These specialized membrane-enclosed domains are known collectively as *organelles.* Catalyzing reactions whose products are major energy supplies of the cell, the *mitochondria* are organelles with an extremely specialized and organized internal structure. They are found in all eucaryotic cells which utilize oxygen in the process of energy generation. In *phototrophic cells,* which are those using light as a primary energy source, the *chloroplast* (see Fig. 1.5) is the organelle serving as the major cell powerhouse. Chloroplasts and mitochondria are the sites of many other important biochemical reactions in addition to their role in energy production.

The Golgi complex, lysosomes, and vacuoles are the remaining organelles illustrated in Figs. 1.3 to 1.5. In general, they serve to isolate chemical reactions or certain chemical compounds from the cytoplasm. This isolation is desirable

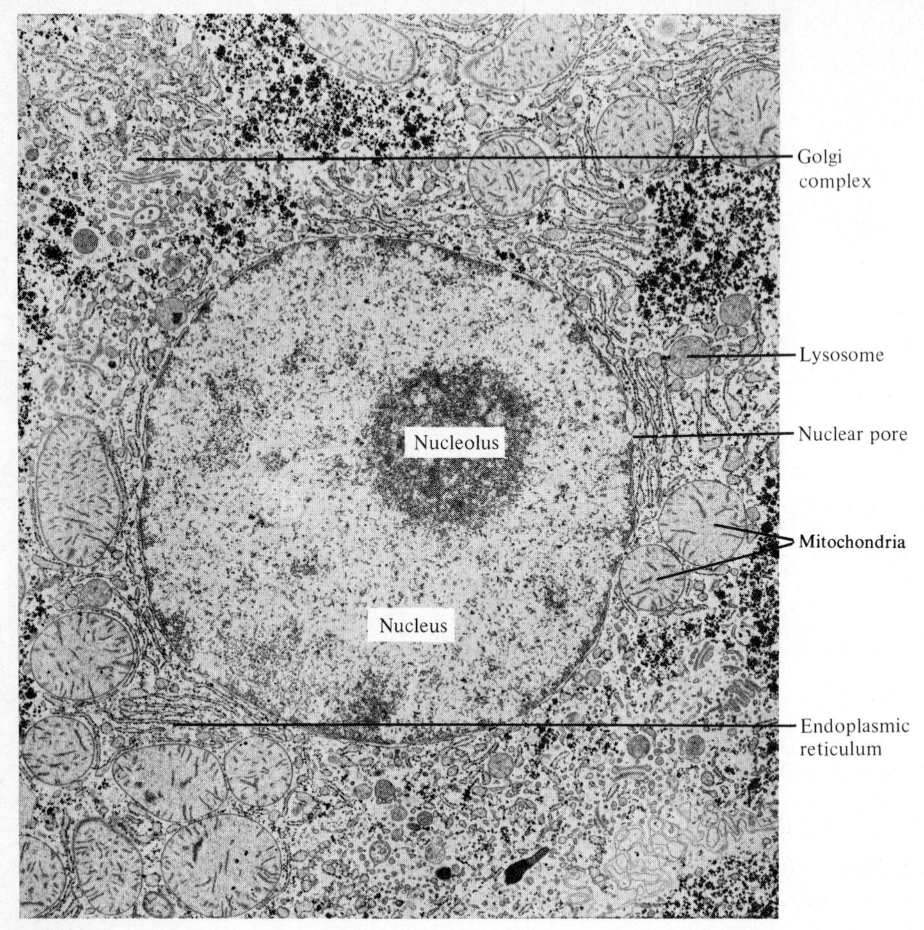

Figure 1.4 Electron micrograph of a rat liver cell ($\times$ 11,000). *(Courtesy of George E. Palade, Yale University.)*

either from the standpoint of reaction efficiency or protection of other cell components from the contents of the organelle.

The discovery of similar organelles in many different eucaryotes allows a refinement of the major working advantages of the cell doctrine. The activities of the cell itself can now be decomposed conceptually into the activities of its component organelles, which in turn can be studied in isolation. In the absence of contrary evidence, similar organelles are assumed to perform similar operations and functions, regardless of the type of cell in which they are found.

Determination of the chemical composition, structure, and biochemical activities of organelles is a major goal of cell research. Much of the present knowledge of cell biochemistry came from investigation of these questions. Consequently we shall briefly examine the centrifugation techniques widely employed to isolate components of the cell.

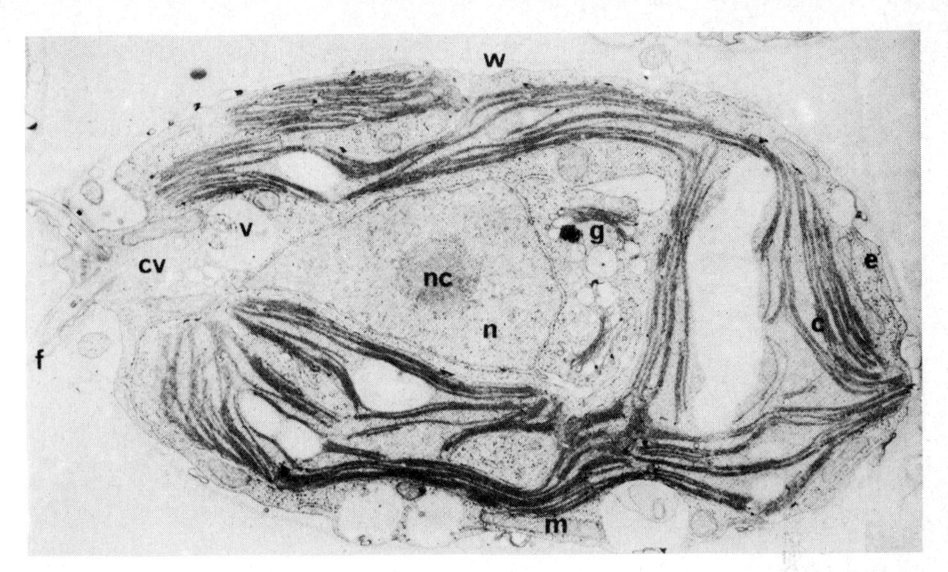

Figure 1.5 Electron micrograph of the eucaryotic alga, *Chamydonomas reinhardii* (× 13,000). Visible are the chloroplasts (*c*), wall (*w*), nucleus (*n*) and nucleolus (*nc*), vacuoles (*v*), and the Golgi complex (*g*). *(Reprinted from U. R. Goodenough and K. R. Porter, J. Cell Biol., vol. 38, p. 403, 1968.)*

1.2.3 Cell Fractionation

A major problem in analyzing the characteristics of a particular organelle from a given type of cell is obtaining a sufficient quantity of the organelle for subsequent biochemical analysis. Typically this requires that a large number of organelles be isolated from a large number of cells, or a *cell population*. Let us follow a common procedure for this purpose: First a cell suspension is homogenized in a special solution using either a rotating pestle within a tube or ultrasonic sound. Here an attempt is made to break the cells apart without significantly disturbing or disrupting the organelles within. Fractionation of the resulting suspension, which now ideally contains a variety of isolated intact organelles, is the next step.

As process engineers, we know that any separation process is based upon exploitation of differences in the physical and/or chemical properties of the components to be isolated. The standard centrifugation techniques for fractionating cell organelles rely upon physical characteristics: size, shape, and density. A rudimentary analysis of centrifugation is presented in the following example.

Example 1.1: Analysis of particle motion in a centrifuge Suppose that a spherical particle of radius R and density ρ_p is placed in a centrifuge tube containing fluid medium of density ρ_f and viscosity μ_c. If this tube is then placed in a centrifuge and spun at angular velocity ω (see Fig. 1E1.1), we may calculate the particle motion employing the following force balance (what approximations have been invoked here?):

$$\text{Drag force on particle} = \text{buoyancy force}$$

$$6\pi\mu_c R u_r = \frac{4\pi R^3}{3} G(\rho_p - \rho_f) \tag{1E1.1}$$

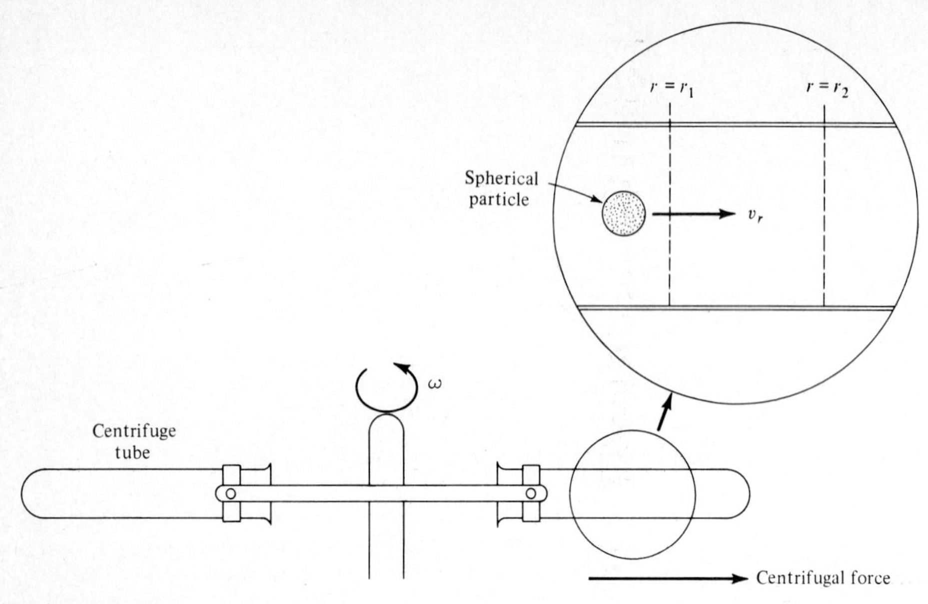

Figure 1E1.1 When a centrifuge is spun at high speed, particles suspended in the centrifuge tubes move away from the centrifuge axis. Since the rate of movement of these particles depends on their size, shape, and density, particles differing in these properties can be separated in a centrifuge.

where u_r is the particle velocity in the r direction

$$u_r = \frac{dr}{dt} \tag{1E1.2}$$

and G is the acceleration due to centrifugal forces

$$G = \omega^2 r \tag{1E1.3}$$

Stokes' law has been used in Eq. (1E1.1) to express the drag force since particle velocities (and therefore particle Reynolds numbers) are usually very low in this situation. The usual gravitational-force term does not appear in Eq. (1E1.1) because the r direction in Fig. 1E1.1 is normal to the gravity force. Rotation of the centrifuge at high speed, however, produces an acceleration G usually many times larger than the acceleration of gravity g; for example, $G = 600\text{–}600{,}000\ g$.

Integration of this expression gives the time required for movement of the particle from position r_1 to r_2:

$$t = \frac{9}{2} \frac{\mu_c}{\omega^2 R^2(\rho_p - \rho_f)} \ln \frac{r_2}{r_1} \tag{1E1.4}$$

Spheres with different sizes and/or densities will take different times to traverse the same distance in the centrifuge tube. This is the basis for the method of *differential centrifugation.* Since the larger particles such as nuclei and unbroken cells sediment more rapidly, they can be collected as a precipitate by spinning the suspension for a limited time at relatively low velocities. The supernatant suspension is then subjected to additional centrifugation at higher rotor speeds for a short time, and another precipitate containing mitochondria is isolated. By continuing this procedure a series of cell fractions can be obtained. The overall process is illustrated schematically in Fig. 1E1.2.

More sophisticated centrifugation methods employ liquid media with density gradients along the centrifuge tube. These techniques are also applicable for continued subdivision and fractionation

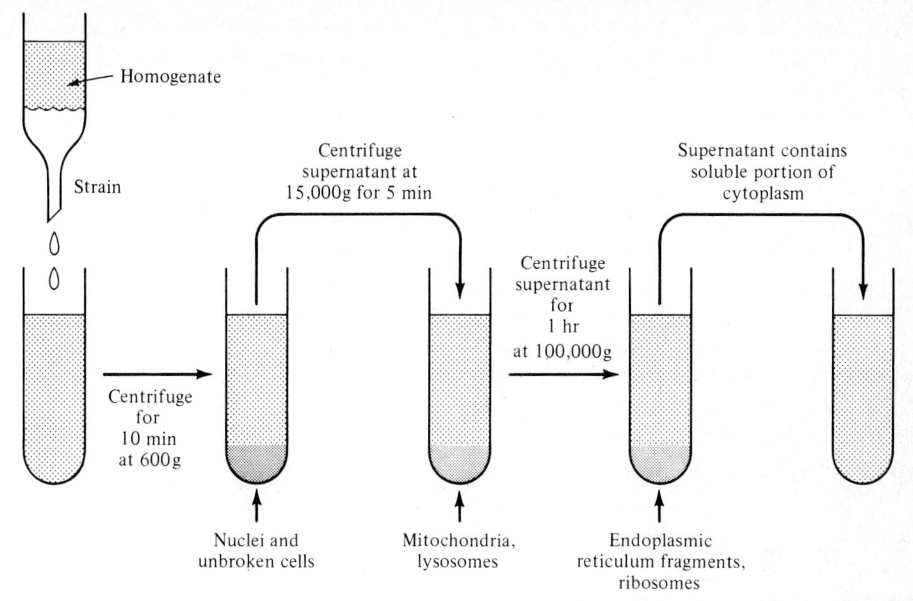

Figure 1E1.2 The steps in a typical differential centrifugation separation of cell constituents. Smaller components are isolated as the process proceeds.

of smaller cell constituents such as particular types of macromolecules. Distinctions in chemical properties, also very valuable in such fine-scale separations, will be investigated in greater detail in Chap. 11.

There are several limitations in the application and interpretation of centrifugal cell-fractionation results. For an excellent summary the text of Mahler and Cordes† should be consulted. One difficulty, however, will plague us at almost every turn in investigating and utilizing microorganisms. In order to obtain a sufficient quantity of cells, organelles, biological molecules, or the like for analysis we are compelled to use a *population*, or a large number, of individual objects. It is common to assume that this population is *homogeneous*, i.e., that all its members are alike. In such a case the population serves only to amplify the characteristics of the individual so that it can be more conveniently observed.

Usually, however, the members of the population are different; the population is *heterogeneous*. For example, a growing cell population typically contains a mixture of old and young, bigger and smaller cells, often with different biochemical compositions and activities. On a finer scale, similar organelles such as mitochondria within a single cell are generally different in some respects. Consequently, a cell fraction containing mitochondria, for example, is a heterogeneous population. When such a mixture of different components is analyzed,

† H. R. Mahler and E. H. Cordes, *Biological Chemistry*, 2d ed., Harper & Row, Publishers, Incorporated, New York, 1971.

properties representing some kind of average over the cell population are obtained. Therefore, the measured properties will depend upon the makeup of the population.

1.3 IMPORTANT CELL TYPES

In this section we shall briefly review the classification of the kingdom of *protists*, which consists of all living things with a very simple biological organization compared to plants and animals. All unicellular (single-celled) organisms belong to this kingdom, and organisms containing multiple cells which are all of the same type are also classed as protists. Plants and animals, on the other hand, are distinguished by a diversity of cell types. A classification of plant and animal cells that can be grown on solid or in liquid nutrient media is included at the end of this section.

Table 1.1 shows a breakdown of the protist kingdom into groupings convenient for our purposes. These classifications show differences in several characteristics including the following: energy and nutritional requirements, growth and product-release rates, method of reproduction, and capability and means of motion. All these factors are of great practical importance in applications. Also significant in classification are differences in the *morphology*, or the physical form and structure, between these various types of organisms. The morphology of a microorganism has an influence on the rate of nutrient mass transfer to it and also can profoundly affect the fluid mechanics of a suspension containing the organism. Obviously then we must examine each group in Table 1.1 individually.

Taxonomy is the art of biological classification. The basic unit in this classification scheme is the *species* which is characterized by a high degree of similarity

Table 1.1 Classifications of microorganisms belonging to the kingdom of protists

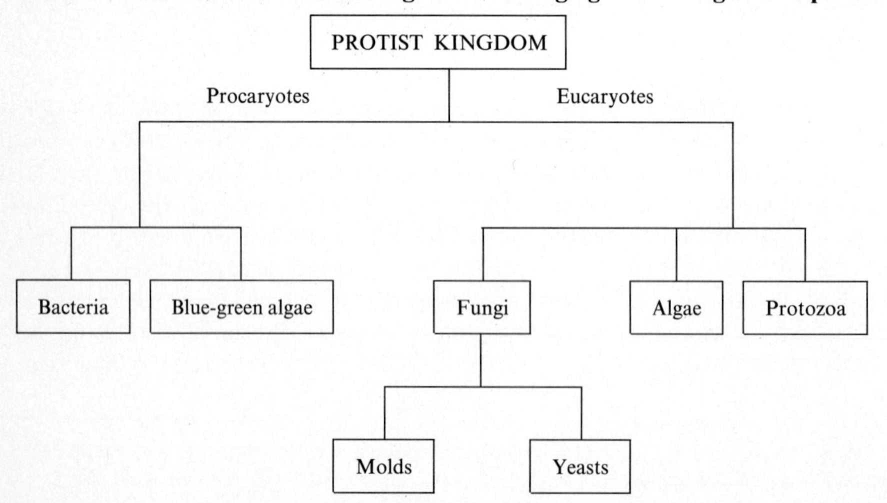

in physical and biochemical properties and significant differences from the properties of related organisms. Biological species are assigned a two-word latinized name in which the capitalized first word is the *genus* or generic name and the second word is the specific name, often a descriptive term. For example, an extensively studied bacterium found in the human intestine has the name *Escherichia* (generic name) *coli* (specific, descriptive name). The species name is italicized, and, if the generic name is clear from context, it is usually abbreviated to its first letter: *E. coli.*

In order to organize the cataloging of species and genera, a hierarchical system of taxonomy has been developed in which related genera first are grouped into families, then related families collected in orders, orders in turn organized into classes, next classes gathered in divisions or phyla, and finally phyla grouped into kingdoms. In Table 1.1, for example, protist designates a kingdom, fungi constitute a division, and yeasts belong to a common class. Often, gradations in properties between microorganisms are sufficiently smooth that detailed classification becomes somewhat artificial and arbitrary, especially for bacteria and yeasts.

1.3.1 Bacteria

As seen earlier in our discussion of procaryotes, bacteria are relatively small organisms usually enclosed by rigid walls. In many species the outer surface of the cell wall is covered with a slimy, gummy coating called a *capsule* or *slime layer*. Bacteria are typically unicellular, but they may exist in three basic morphological forms (Fig. 1.6). Most cannot utilize light energy, are capable of motion

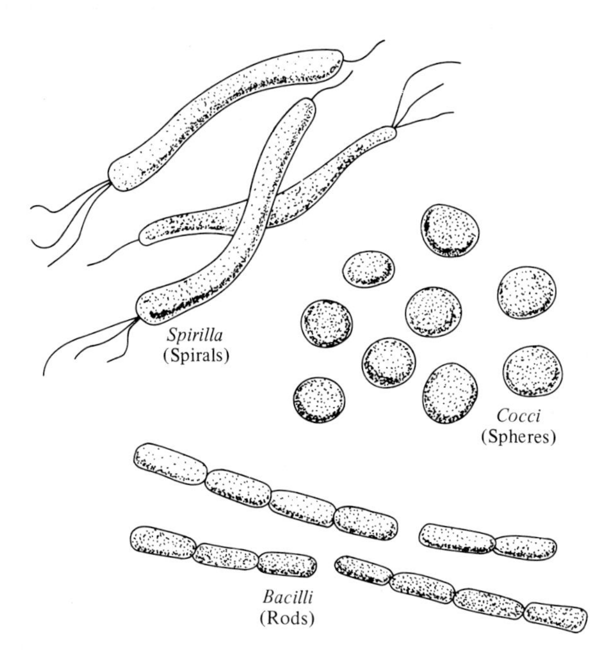

Figure 1.6 The three forms of bacteria.

Table 1.2 Some major groups of bacterial species and some of their distinguishing characteristics

Group	Dominant morphological form	Some nutritional habits	Common habitat	Most members require O_2? (aerobic?)	Photo-synthetic?	Forms endospores?	Gram reaction
Acetic acid bacteria (*Acetobacter*, *Gluconobacter*)	Rods; some *Acetobacter* species form extensive slime layers	Often consume alcohol; acid-tolerant	Decaying plants	Yes	No	No	Negative
Bacillus	Rods	Versatile; exists on many different nutrients	Soil	Yes	No	Yes	Positive
Clostridium	Rods	Many varieties have special nutrient requirements	Soil	Most species cannot tolerate O_2	No	Yes	Positive
Corynebacterium	Irregular form; reproduction not by binary fission; often nonmotile	Simple requirements	Soil, human body	No, but use O_2 if present	No	No	Positive

Enteric or coliform bacteria (*E. coli* is one)	Rods	Simple organic compounds	Some naturally reside in intestine of higher animals	No, but can use O_2 if present	No	No	Negative
Lactic acid (*Lactobacillus, Streptococcus, Leuconostoc*)	Rods or cocci	Lactic acid is a major end product of nutrient utilization; species acid-tolerant	Plants	No	No	No	Positive
Pseudomonas	Rods	Some extremely versatile and live on very wide range of nutrients	Soil, water	Yes	No	No	Negative
Rhizobium	Rods	Fixes nitrogen in association with legumes	Soil; in nodules of leguminous plants	Yes	No	No	Negative
Rhodospirillum	Rods, spirals	Can fix N_2 or produce H_2	Specialized aqueous environments	No	Yes	No	Negative
Zymomonas	Rods	Ferments glucose to ethanol	Soil	No, but can tolerate some O_2	No	No	Negative

(motile), and reproduce by division into two daughter cells (binary fission). Still, many exceptions to each of these rules are known.

There are many subdivisions of bacteria: some of the general groups of bacteria and some of their distinguishing characteristics are given in Table 1.2. The column labeled "Gram reaction" refers to the response of the bacteria to a relatively straightforward and rapid staining test. Cells are first stained with the dye crystal violet, then treated with an iodine solution and washed in alcohol. Bacteria retaining the blue crystal-violet color after this process are called *gram-positive*; loss of color indicates a *gram-negative* species. Many characteristics of bacteria correlate very well with this test, which also indicates basic differences in cell-wall structure.

Whether or not oxygen is supplied to the cells is especially important in commercial exploitation of microorganisms (Chaps. 8, 12, and 14). In an *aerobic* process, oxygen is provided, usually as air, for use by the microorganisms. Manufacture of vinegar, some antibiotics, and animal-feed supplements are among the important microbial applications which employ aeration. The sparing solubility of oxygen in the aqueous media typical of these systems has major implications in process design (Chap. 8). The protists function without oxygen in an *anaerobic* process such as production of some alcohols or digestion of organic wastes.

Especially important in commerical utilization and control of bacteria is their ability to form *endospores* under adverse conditions. Spores are dormant forms of the cell, capable of resisting heat, radiation, and poisonous chemicals. When the spores are returned to surroundings suitable for cell function, they can germinate to give normal, functioning cells. This normal, biologically active cell state is often called the *vegetative form* in order to distinguish it from the spore form. As Table 1.2 indicates, there are two major groups of sporeforming bacteria. The aerobic *Bacillus* species are extremely widespread and adaptable. Several *Clostridium* species, which normally function under anaerobic conditions, die in the presence of oxygen in the vegetative state but form spores unaffected by oxygen. Some bacteria whose vegetative forms are rapidly killed at 45°C can form spores which survive boiling in water for several hours. Therefore, when we are attempting to kill microorganisms by heating (*heat sterilization*) the spore-forming capability of bacteria demands use of higher temperatures, typically achieved by boiling under pressure in an autoclave to give $T > 120°C$.

The blue-green algae will not be discussed here since they are not of great commerical significance. They are important, however, in the overall operation of natural aquatic systems since they participate in the nitrogen cycle (Chap. 14).

1.3.2 Yeasts

Yeasts form one of the important subgroups of fungi. Fungi, like bacteria, are widespread in nature although they usually live in the soil and in regions of lower relative humidity than bacteria. They are unable to extract energy from sunlight and usually are free-living. Although most fungi have a relatively complex morphology, yeasts are distinguished by their usual existence as single, small cells from 5 to 30 μm long and from 1 to 5 μm wide.

The various paths of reproduction of yeasts are asexual (budding and fission) as shown in Fig. 1.7, and sexual. In budding, a small offspring cell begins to grow on the side of the original cell; physical separation of mature offspring from the parent may not be immediate, and formation of clumps of yeast cells involving

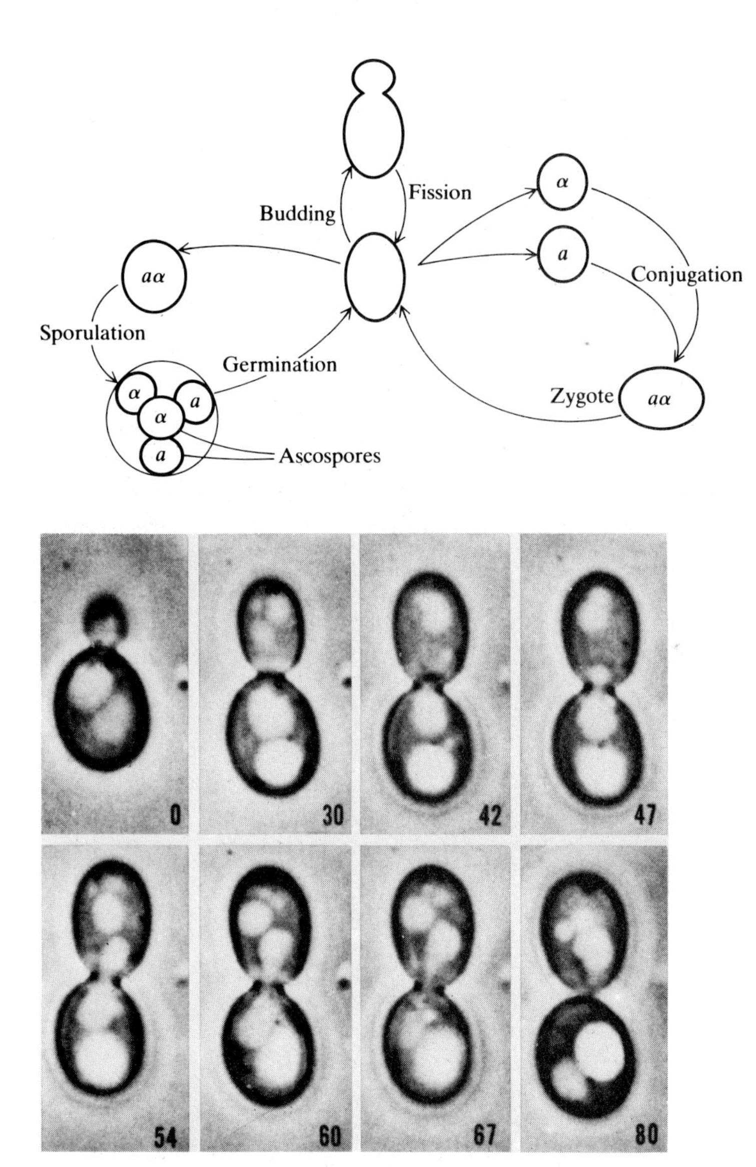

Figure 1.7 Reproduction of yeast by asexual budding is shown in the lower series of photographs. Numbers denote elapsed time in minutes. As illustrated in the upper sketch, sexual reproduction also occurs in the yeast life cycle. *(Photographs courtesy of C. F. Robinow.)*

several generations is possible. Fission occurs by division of the cell into two new cells. Sexual reproduction occurs by conjugation of two *haploid* cells (each having a single set of chromosomes) with dissolution of the adjoining wall to form a *diploid* cell (two sets of chromosomes per cell) zygote. The nucleus in the diploid cell may undergo one or several divisions and form *ascospores*; each of these eventually becomes an individual new haploid cell, which may then undergo subsequent reproduction by budding, fission, or sexual fusion again. The ascospores, which here are a normal stage in the reproductive cycle of these organisms, should not be confused with the endospores, discussed above, which are a defense mechanism against hostile environments.

In the production of alcoholic beverages, yeasts are the only important industrial microbes. In addition to supplying the consumer market for beer and wine, anaerobic yeast activities produce industrial alcohol and glycerol. The yeasts themselves are also grown for baking purposes and as protein supplements to animal feed (Chap. 12).

1.3.3 Molds

Molds are higher fungi with a vegetative structure called a *mycelium*. As illustrated in Fig. 1.8, the mycelium is a highly branched system of tubes. Within these enclosing tubes is a mobile mass of cytoplasm containing many nuclei. The mycelium may consist of more than one cell of related types. The long, thin

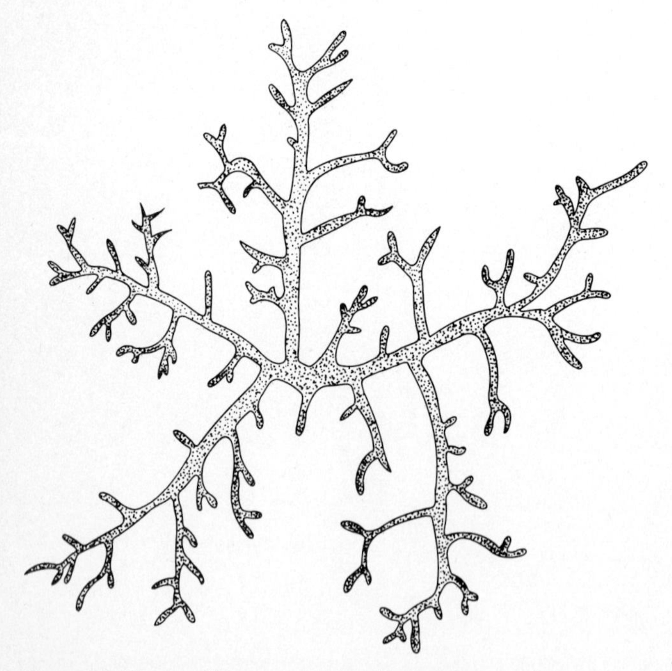

Figure 1.8 The mycelial structure of molds. A dense mycelium can cause conditions near its center to differ considerably from those at the outer extremities.

filaments of cells within the mycelium are called *hyphae*. In some cases the mycelium may be very dense. This property, coupled with molds' oxygen-supply requirements for normal function, can cause complexities in their cultivation, since the mycelium can represent a substantial mass-transfer resistance. This problem and the unusual flow properties of suspensions of mycelia will be explored in further detail in Chaps. 4 and 8.

Molds, like yeasts, do not contain chlorophyll, and they are generally nonmotile. Reproduction, which may be sexual or asexual, is typically accomplished by means of spores. Spore properties form an important component in fungi classification.

The most important classes of molds industrially are *Aspergillus* and *Penicillium* (Fig. 1.9). Major useful products of these organisms include antibiotics (biochemical compounds which kill certain microorganisms or inhibit their growth), organic acids, and biological catalysts.

The strain *Aspergillus niger* normally produces oxalic acid (HO_2CCO_2H). Limitation of both phosphate nutrient and certain metals such as copper, iron, and manganese results in a predominant yield of citric acid [$HOOCCH_2COH(COOH)CH_2COOH$]. This limitation method is the basis for the commercial biochemical citric acid process. Thus *A. niger* is an interesting example differentiating approaches to biochemical-reactor design and optimization from those of nonbiological reactors: a much greater selectivity can sometimes be achieved in the biological system by minor alteration of feed composition to the reactor.

This example (as well as that of penicillin below) should motivate us to learn

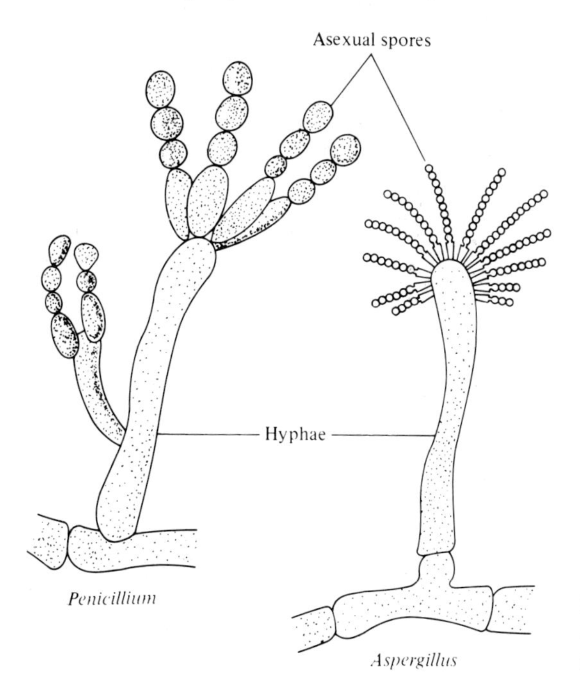

Figure 1.9 The hyphae of *Aspergillus* and *Penicillium*, two industrially important varieties of molds.

the essentials of cell structure, metabolism, and function which are woven into the following chapters. Without this background in cellular processes and characteristics, our skills as process engineers, which are well suited to design and analysis of many aspects of biochemical processes, could be wasted because key biological features of the system would be ignored.

Penicillin production offers an example of a second fundamental difference between microbiological and nonbiological reactors. Major improvement in production of penicillin has arisen by use of ultraviolet irradiation of *Penicillium* spores to produce *mutants* of the original *Penicillium* strain (Fig. 1.10). Cell mutation by various techniques may result in orders-of-magnitude improvements in desired yield. Recombinant DNA technology, summarized in Chap. 6, provides carefully controlled and novel genetic alterations in certain organisms. This approach has made possible manufacture of valuable animal products (proteins) in simple bacteria. These examples indicate the central importance of genetics to the biochemical engineer. This subject occupies a large fraction of the present efforts in both university and industrial biological research. The practical importance of designing genetic modifications (or, in other situations, of avoiding such changes) emphasizes the need for close cooperation between engineers, biologists, and biochemists in biochemical process design and evaluation. The history of penicillin production is in itself a story of joint development of new techniques, includ-

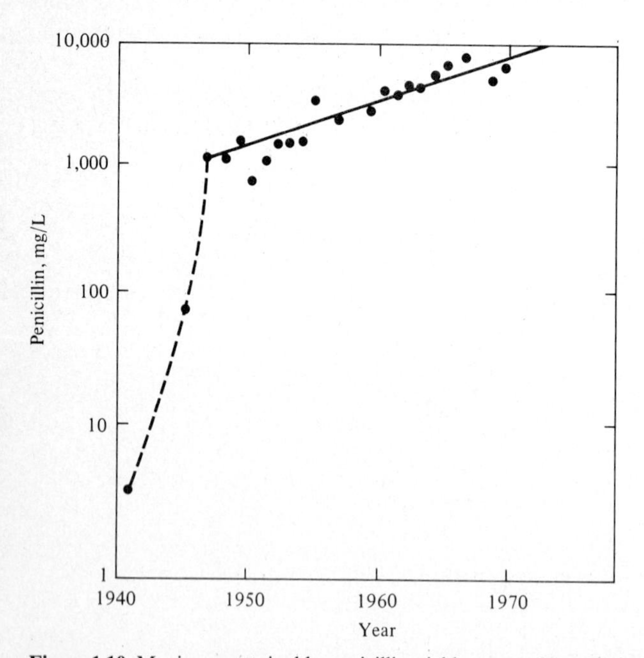

Figure 1.10 Maximum attainable penicillin yields over a 30-yr period. Development of special mold strains by mutation has produced an exponential increase in yields for the past 25 years. A similar trend also holds for yields of the antibiotic streptomycin. *[Reprinted by permission of A. L. Demain. Overproduction of Microbial Metabolites due to Alteration of Regulation, In T. K. Ghose and A. Fiechter (eds.), "Advances in Biochemical Engineering 1," p. 129, Springer-Verlag, New York, 1971.]*

ing deep-submerged production, solvent extraction of a delicate product on a large scale, air-sterilization procedure for high volumetric flow rates, and isolation of mutants with high penicillin yields.

Before leaving bacteria and fungi, we should mention the *actinomycetes*, a group of microorganisms with some properties of both fungi and bacteria. These organisms are extremely important for antibiotic manufacture. Although formally classified as bacteria, actinomycetes resemble fungi in their formation of long, highly branched hyphae. Also, design of antibiotic production processes utilizing actinomycetes is very similar to those involving molds. One difference, however, is the susceptibility of actinomycetes to infection and disease by viruses which also can attack bacteria. These agents will be examined briefly later in Chap. 6.

1.3.4 Algae and Protozoa

These relatively large eucaryotes have sophisticated and highly organized structures. For example, *Euglena* has flagella for locomotion, lacks a rigid wall, and has an eyespot sensitive to light. The cell, guided by the eyespot, moves in response to stimulus by illumination—clearly a valuable asset since most algae require energy in the form of light. Many diatoms (another kind of algae) have exterior skeletons of complex architecture which are impregnated with silica. These skeletons are widely employed as filter aids in industry.

Considerable commercial interest in algae is concentrated on their possible exploitation as foodstuffs and food supplements. In Japan, several processes for algae food cultivation are in operation today. Also important in Asia is use of seaweed in the human diet. While not microorganisms, many seaweeds are actually multicellular algae. Like the simpler blue-green algae, eucaryotic algae serve a vital function in the cycles of matter on earth (Chap. 14).

Just as algae may be viewed as primitive plants, *protozoa*, which cannot exploit sunlight's energy, are in a sense primitive animals. The habitats, morphology, and activities of protozoa span a broad spectrum. For example, some trypanosomes carry serious disease, including African sleeping sickness. The *Trichonympha* inhabit the intestines of termites and assist them in digesting wood. While the amoeba has a changing, amorphous shape, the heliozoa have an internal skeleton and definite form.

Although protozoa are not now employed for industrial manufacture of either cells or products, their activities are significant among the microorganisms which participate in biological waste treatment (Chap. 14). These processes, widely employed in urban communities and large industrial plants throughout the world, are suprisingly complicated from a microbiological viewpoint. Since a complex mixture of different nutrients and microbes are present in sewage or industrial wastes, a correspondingly large collection of different protists are present and indeed necessary in treatment operations. These diverse organisms compete for nutrients, devour each other, and interact in numerous ways characteristic of a small-scale ecological system. A survey and analysis of interactions between different species will be considered in Chap. 13.

1.3.5 Animal and Plant Cells

Many vaccines and other useful biochemicals are produced by growth of animal cells in process reactors; i.e., by cell propagation outside of the whole animal. Improvements in cultivation techniques for these tissue-derived cells and emerging methods for genetic manipulation of animal and plant cells offer great potential for expanded commercial utilization of these higher cells. The reactors in which "tissue" cultures may be propagated may be quite similar to microbial reactors, admitting a unified treatment of cell kinetics and biochemical reactors in Chaps. 7 and 9, respectively. We next consider a very condensed summary of some of the important lines of higher cells which can be propagated "in culture"—that is, in a process device apart from the animal or plant of origin.

When a piece of animal tissue, perhaps after disruption to break the cells apart, is placed in appropriate nutrient liquid, many cell types, such as blood cells, die within a few days, weeks, or months. Other cells multiply and are called *primary cell lines*. Often these cells can be "passaged" by transfer to fresh medium after which further cell multiplication occurs giving a *secondary cell line*. Some secondary cells can be passaged apparently indefinitely; these cells are then dubbed an *established*, *permanent*, or *stable cell line*.

Many cell lines have been developed from the epithelial tissues (skin and tissues which cover organs and line body cavities), connective tissues, and blood and lymph of several animals including man, hamster, monkey, and mouse. Figure 1.11 shows LA-9 cells, a line derived from mouse fibroblasts, growing attached to a solid surface. A sampling of cell lines and their sources and names are listed in Table 1.3. As noted there, several cell lines are derived from malignant growths called *carcinomas* arising from various tissues. Malignant growths of blood and lymph tissues are usually designated *leukemias*.

Some cultured tissue cells can be grown suspended in liquid, but growth of most cell lines requires attachment to a solid surface. This anchorage dependence poses stringent restrictions on scale-up for production of vaccines and other biological products from animal cell culture. Microcarrier culture techniques, which we will investigate in Chap. 9, have greatly increased the volumetric productivity of reactors for anchorage-dependent culture cells.

It is also possible to grow certain plant cells in culture, either as a *callus* (a lump of undifferentiated plant tissue growing on solid nutrient medium) or as aggregated cells in suspension. Since plants produce many commercially important compounds including perfumes, dyes, medicinals, and opiates, there is significant potential for future applications of plant cell culture. Cultured plant cells can also catalyze highly specific useful transformations. Cultured plant cells have several potential agricultural applications including whole plant regeneration. Greater knowledge of plant biology and of requirements and constraints in plant cell culture are needed before significant commercialization of this approach occurs. After noting that tissue cells from insects and other invertebrates can be grown in culture, we will confine discussion in later chapters of eucaryote tissue cell culture to animal and plant cells.

Figure 1.11 Scanning electron micrograph of LA-9 cells grown in culture on a solid surface. These cells, derived from mouse fibroblasts, are approximately 15 μm wide and 75–90 μm long. *(Courtesy of Jean-Paul Revel.)*

Table 1.3 A sampling of standard animal cell lines and their origins

Cell line designation	Animal	Tissue
HeLa (CCL 2)†	Human	Cervix carcinoma
HLM	Human	Fetal liver
FS-4	Human	Foreskin fibroblast
MK2	Monkey	Kidney
CHO (CCL 61)	Chinese hamster	Ovary
L-M (CCL 2)	Mouse	Connective tissue

† The CCL designation is used by the Cell Culture Repository, American Type Culture Collection, 12301 Parklawn Drive, Rockville, Maryland 20852 USA, which maintains and supplies these lines.

1.4 A PERSPECTIVE FOR FURTHER STUDY

In our brief sojourn through cell structure and classification in this chapter, we have seen the basic importance of the cell doctrine. The importance of basic biology for understanding biochemical process systems has been emphasized. In the next few chapters, our attention will continue to be concentrated on fundamental cell biology.

Chapter 2 reviews the chemicals which the cell must synthesize for survival and reproduction; the catalysts used to facilitate reactions within the cell are examined in Chaps. 3 and 4. Reaction sequences necessary for cellular function are then studied in Chap. 5, with control of these reactions and genetics the primary topics of Chap. 6. After investigating the kinetics of microbial systems in Chap. 7, we direct our remaining efforts to analysis, design, control, and optimization of biological process systems.

PROBLEMS

1.1 Microbiologists Read a biography of one of the early influential men in microbiology, e.g., Robert Hooke, Anton van Leeuwenhoek, Lazzaro Spallanzani, Louis Pasteur, Walter Reed, D. Iwanowski, P. Rous, Theodor Schwann, or M. J. Schleiden. Prepare a short summary indicating what technical and social obstacles were or were not overcome, what technical achievements did or did not result from careful quantitation, and the place of induction and deduction in the studies of these men.

1.2 Experimental microbiology Many techniques in microbiology are simple and relatively well established but unfamiliar to those who have completed general, organic, and physical chemistry. As observation and measurement underlie any useful analysis, some appreciation of techniques and their accuracy is indispensable. Take a microbiology laboratory course in parallel or following this course if you have not previously done so. Lacking the opportunity, read carefully through a short laboratory paperback on this topic as you follow the chapters of this text, e.g., Ref. [7]. As you do this exercise, remember Claude Bernard: "to experiment without a preconceived idea is to wander aimlessly." For the laboratory course or paperback, summarize the purpose(s) of each experiment. For each such experimental setup, what other information could you obtain?

1.3 Observation Read a brief account of microscope techniques including dark-field, phase-contrast, fluorescence, and electron microscopy. Develop a list of relative advantages and limitations for each technique.

1.4 Definitions Define the following terms and when there is more than one, compare and contrast their general characteristics with those of others in the same group:

(*a*) Procaryotes, eucaryotes
(*b*) Cell wall, plasma membrane, endoplasmic reticulum
(*c*) Cytoplasm
(*d*) Nucleus, nuclear zone
(*e*) Ribosome, mitochondria, chloroplast
(*f*) Morphology
(*g*) Spirilla, cocci, bacilli
(*h*) Budding, sexual fusion, fission, sporulation
(*i*) Protozoa, algae, mycelia, amoeba

1.5 Identification and classification (*a*) Sketch the diagram showing the kingdom of the protists (from memory).

(*b*) The taxonomy of microbial species is based largely on visual observation with the optical microscope. Using either Bergey's *Manual of Determinative Bacteriology* or *A Guide to the Identification of the Genera of Bacteria*, by Skerman, locate the organisms *Escherichia coli*, *Staphylococcus aureus*, *Bacillus cereus*, and *Spirillum serpens*. For any two, list the distinguishing characteristics which would lead to ultimate identification. Begin most generally, passing from family through subgroups (tribe, genus) to species.

1.6 "The view from the ground floor" Pick a microbial topic of interest (beer fermentation, antibiotic production, yeast growth, wastewater treatment, soil microbiology, vaccines, pickling, cheese manufacture, lake ecology, saltwater microbiology, etc.). Read a descriptive account of the topic in a text such as the Kirk-Othmer *Encyclopedia of Chemical Technology*, etc. Sketch a process or natural flow scheme indicating important microbial species, food sources, and waste or exit streams. Each week during the course, add one or two pages to the description indicating the (lack of) importance of the chapter topic to the process. At the end of the course, prepare a short paper on your topic and present it to the class.

1.7 Industrial microbiology A brief description of the history of industrial microbiology and some suggestions of developments to come is accessible in "Industrial Microbiology," A. L. Demain and N. A. Solomon, *Scientific American*, **245**, 67–75, 1981. Read the article, and prepare a one-page outline of the historical development of industrial microbiology, highlighting dates, products, and/or processes and microbial species.

1.8 Protozoan motility You are observing the motion of spherical protozoa through the microscope and can estimate the sizes of the organisms and their speeds in body lengths per unit time.

(*a*) For the range of body sizes and speeds shown below, compute the Reynolds number of the flow around the organism. Assume that the fluid is water at 20°C and that the flow around the organism is relatively undisturbed by any spines, cilia, or slime.

Sizes: 10 μm, 50 μm, 100 μm

Speeds: 10 body lengths per second
1 body length per second
0.1 body length per second

What flow regime characterizes these cellular motions?

(*b*) Methyl cellulose solution is frequently employed in microscopy to slow motile species and render them more convenient for observation. Assuming that the organism's energetic commitment to motion is constant, what effect on velocity would you expect from a 10^4 increase in viscosity of the fluid surrounding the cells?

1.9 Plating of *E. coli* When growing *E. coli* for use in experiments it is desirable to have a genetically homogeneous population. This is usually achieved by diluting the source broth to a concentration which results in distinct colonies originating from single cells when spread on an agar plate. A single colony is then used to grow a population.

(*a*) Given a circular plate of agar with a radius of 4 cm, 1 mL of a broth of 10^{10} cells/liter, and 1 liter of sterile broth for dilutions, what is the proper dilution of bacterial broth required to give about 100 distinct colonies on a plate? (It takes 5 mL of solution to properly cover a 4 cm radius plate.)

(*b*) How would you get colonies from individual cells if you did not know the concentration of cells in the broth?

1.10 Centrifugal separation Consider a dilute suspension of particles of type A and type B. The initially uniform suspension is spun in a centrifuge at an angular velocity ω for a time t. In terms of the initial fraction of type A particles, f_0, find an expression for the final fraction of type A particles in the supernatant suspension. Assume that t is sufficiently small that some particles of both types remain in suspension.

1.11 Centrifugation in an angled rotor Consider the "angled" centrifuge shown below. Given angular velocity ω, particle (spherical) radius R, particle density ρ_p, fluid density ρ_f and viscosity μ_f, and angle from the vertical θ, find the time for the particle to move from distance $r = r_1$ to $r = r_2$. Note that the glass wall applies a normal force to the particle.

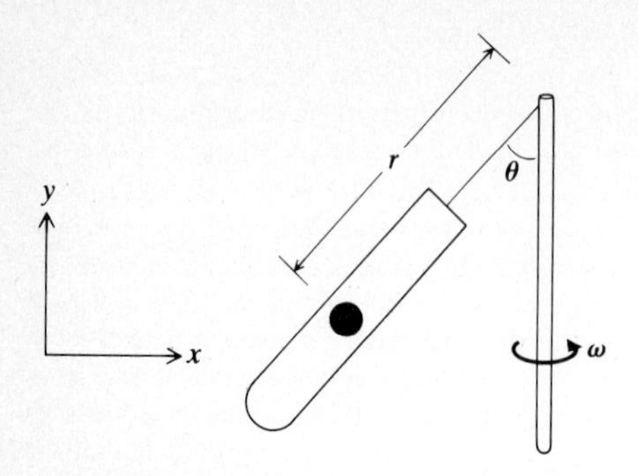

Figure 1P11.1

REFERENCES

1. W. R. Sistrom, *Microbial Life*, 2d ed., Holt, Rinehart, and Winston, Inc., N.Y., 1969. The first three chapters of this introductory microbiology text cover most of the topics in this chapter. Also included is material on metabolism, growth, and genetics, which are treated later in this text.
2. M. Frobisher, *Fundamentals of Microbiology*, 8th ed., W. B. Saunders Company, Phila., 1968. A descriptive presentation of microbial life forms including cell classification, viruses, sterilization, immunology, and microbial applications. The ever-present connections between microorganism and man are emphasized.
3. Michael J. Pelczar, Jr., Roger D. Reid, and E. C. S. Chan, *Microbiology*, 4th ed., McGraw-Hill, New York, 1977. A more advanced text. Includes more detail on fungi, algae, protozoa, and viruses. Also extended coverage of environmental and industrial microbiology and on physical and chemical methods for control of microbial proliferation.
4. R. Y. Stanier, M. Doudoroff, and E. A. Adelberg, *The Microbial World*, 4th ed., Prentice-Hall, Inc., Englewood Cliffs. N.J., 1975. Another relatively advanced text; very well written. Covers microbiology history, classes of microbes, symbiosis, and disease, as well as microbial metabolism. Genetics at the molecular level, including mutation and regulation, is presented in detail.
5. "The Living Cell," readings from *Scientific American*, W. H. Freeman and Company, San Francisco, 1965. Reprints of *Scientific American* articles including levels of cell complexity, energetics, synthesis, division and differentiation, and special activities such as communication, stimulation, and muscle action. While some of the material is now somewhat outdated, this wide-ranging and profusely illustrated collection is still very worthwhile reading.
6. John Paul, *Cell and Tissue Culture*, 5th ed., Churchill Livingstone, Edinburgh, 1975. An excellent, comprehensive introductory monograph which includes animal cell lines, cell culture principles and techniques, and biological sciences applications.

Problems

7. K. T. Crabtree and R. D. Hinsdill: *Fundamental Experiments in Microbiology*. W. B. Saunders Co., Philadelphia, 1974.

CHAPTER

TWO

CHEMICALS OF LIFE

The organism must synthesize all the chemicals needed to operate, maintain, and reproduce the cell. In the following chapters we investigate the kinetics, energetics, and control of the major biochemical pathways for such syntheses. A necessary prerequisite for such studies is familiarity with the reactants, products, catalysts, and chemical controllers which participate in reaction networks of the cell.

The present chapter is concerned with the predominant cell polymeric chemicals and the smaller monomer units from which the larger polymers are derived. The four main classes of polymeric cell compounds are the fats and lipids, the polysaccharides (cellulose, starch, etc.), the information-encoded polydeoxyribonucleic and polyribonucleic acids (DNA, RNA), and proteins. The physicochemical properties of these compounds are important both in understanding cellular function and in rationally designing processes incorporating living cells.

The various biological polymers may be usefully regarded as being either repetitive or nonrepetitive in structure. *Repetitive biological polymers* contain one kind of monomeric subunit: distinctions between different types of the same polymers are primarily due to differences in molecular weight and the degree of branching of the polymer chains. The major function of repetitive polymers in the cell is to provide *structures* with the desired mechanical strength, chemical inertness, and permeability. Repetitive polymers also provide a means of *nutrient storage*. In the latter function, for example, a 1 *M* glucose solution can be stored as the polymer glycogen, a cell polysaccharide reserve, with a concurrent reduction in molarity by a factor of 10,000 or greater. As cells may need to store excess food supplies without seriously disrupting the intracellular osmotic pressure, polymer is a useful form for commodity storage.

Nonrepetitive polymers may contain from several up to 20 *different* monomer species. Further, each of these biological polymers has a fixed molecular weight and monomer composition, and the monomers are linked together in a fixed, genetically determined sequence.

The elemental pool from which the polymers are constructed is exemplified by the *E. coli* composition given in Table 2.1. The predominant elements (hydrogen, oxygen, nitrogen, and carbon) form chemical bonds by completing their outer shells with one, two, three, and four electrons, respectively. They are the lightest elements in the periodic table with such properties, and (except for hydrogen) they can form multiple bonds as well. The variety of chemicals which can be assembled from these four elements include, if we add a little phosphorus and sulfur, the four major biopolymer classes.

In addition to variety, the biochemical compounds assembled from these elements are quite stable, reacting only slowly with each other, water, and other cellular compounds. Chemical reactions involving such compounds are catalyzed by biological catalysts: proteins which are called *enzymes* (recall that a catalyst is a substance which allows an increase in a reaction rate without itself undergoing a permanent change). Consequently, by controlling both the number and type of enzymes which the cell contains, the cell regulates both the type and rate of chemical reactions which occur within it. Details of these control mechanisms are considered in Chap. 6.

While phosphorus and sulfur occur in the organic matter of all living things, they are present in relatively small amounts. The ionized forms of sodium, potassium, magnesium, calcium, and chlorine are always present, and trace amounts of manganese, iron, cobalt, copper, and zinc are necessary for proper activation of certain enzymes. Some organisms also require miniscule amounts of boron, aluminum, vanadium, molybdenum, iodine, silicon, fluorine, and tin. Thus, at least 24 different elements are necessary for life.

Table 2.1 The elemental composition of *E. coli*

Element	Percentage of dry weight	Element	Percentage of dry weight
Carbon	50	Sodium	1
Oxygen	20	Calcium	0.5
Nitrogen	14	Magnesium	0.5
Hydrogen	8	Chlorine	0.5
Phosphorus	3	Iron	0.2
Sulfur	1	All others	~0.3
Potassium	1		

Data for *E. coli* assembled by S. E. Luria, in I. C. Gunsalus and R. Y. Stanier (eds.), "*The Bacteria*," vol. 1, chap. 1, Academic Press Inc., New York, 1960.

The solvent within which cells live is, of course, water. In addition to its relatively unusual properties (a high heat of vaporization, a high dielectric constant, the ability to ionize into acid and base, and propensity for hydrogen bonding), water is an extremely important reactant which participates in many enzyme-catalyzed reactions. Also, the properties which biopolymers exhibit depend strongly on the properties of the solvent within which they are placed: this fact provides the means of many separation process designs. The biological fitness of water and other common cell chemicals is discussed by Blum [12].[†]

2.1 LIPIDS

By definition, lipids are biological compounds which are soluble in nonpolar solvents such as benzene, chloroform, and ether, and are practically insoluble in water. Consequently, these molecules are diverse in their chemical structure and their biological function. Their relative insolubility leads to their presence predominantly in the nonaqueous biological phases, especially the plasma and organelle membranes. Fats, which simply serve as polymeric biological fuel storage, are lipids, as are several important mediators of biological activity. Lipids also constitute portions of more complex molecules such as lipoproteins and liposaccharides, which again appear predominantly in biological membranes of cells and the external walls of some viruses.

2.1.1 Fatty Acids and Related Lipids

Saturated fatty acids are relatively simple lipids with the general formula:

$$CH_3{-}(CH_2)_n{-}C\begin{matrix}{\diagup\!\!\diagup}O\\ \diagdown OH\end{matrix}$$

The hydrocarbon chain is constructed from identical two-carbon monomers, so that fatty acids may be regarded as noninformational biopolymers with a terminal carboxylic group. The value of n is typically between 12 and 20 (even numbers) in biological systems.

Unsaturated fatty acids are formed upon replacement of a saturated (—C—C—) bond by a double bond (—C=C—). For example, oleic acid is the unsaturated counterpart of stearic acid ($n = 16$).

$CH_3{-}(CH_2)_{16}{-}COOH$	$CH_3{-}(CH_2)_7{-}HC{=}CH{-}(CH_2)_7{-}COOH$
Stearic acid	Oleic acid

The hydrocarbon chain is nearly insoluble in water, but the carboxyl group is very hydrophilic. When a fatty acid is placed at an air–water interface, a small amount of the acid forms an oriented monolayer, with the polar group hydrated

[†] Numbers in brackets indicate the reference listed at the end of the chapter.

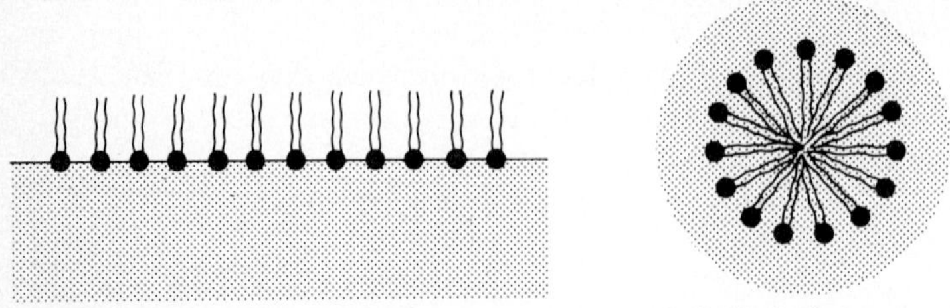

Lipid monolayer at an air-water interface　　Lipid micelle in water

Figure 2.1 Some stable configurations of fatty acids in water.

in the water and the hydrocarbon tails on the air side (Fig. 2.1). The same phenomenon occurs in the action of soaps, which are fatty acid salts. The soap-monolayer formation greatly lowers the air–water interfacial tension, and the ability of the solution to wet and cleanse confined regions is greatly increased.

$$Na^{+}\ {}^{-}O-\overset{\overset{\displaystyle O}{\|}}{C}-(CH_2)_7-HC=CH-(CH_2)_7-CH_3$$

A soap: sodium oleate

These hydrophilic-hydrophobic lipid molecules possess very small solubilities: elevation of the solution concentration above the monomolecular solubility limit results in the condensation of excess solutes into larger ordered structures called *micelles* (Fig. 2.1). This spontaneous process occurs because the overall free energy of the resultant (micelle plus solution) mixture is lower than that of the original solution. The structure of the micelle is dictated by the favorable increase in the number of hydrophobic-hydrophobic and hydrophilic-hydrophilic contacts and concurrent diminution of hydrophilic-hydrophobic associations. Such interactions between hydrophilic and hydrophobic portions of the *same* biopolymer are also known to favor the folding of such polymer chains into a single preferred configuration. This behavior of DNA and proteins will be discussed shortly.

Important as reservoirs of fuel, *fats* are esters formed by condensation of fatty acids with glycerol.

$$\begin{array}{lcl}
CH_2OH & HO-OC(CH_2)_{n_1}-CH_3 & & CH_2O-\overset{\overset{O}{\|}}{C}-(CH_2)_{n_1}-CH_3 \\
| & + & & | \\
CHOH\ + & HO-OC(CH_2)_{n_2}-CH_3 & \xrightarrow{-3H_2O} & CHO-\overset{\overset{O}{\|}}{C}-(CH_2)_{n_2}-CH_3 \\
| & + & & | \\
CH_2OH & HO-OC(CH_2)_{n_3}-CH_3 & & CH_2O-\overset{\overset{O}{\|}}{C}-(CH_2)_{n_3}-CH_3 \\
\text{Glycerol} & \text{Fatty acids} & & \text{A fat}
\end{array}$$

Fats and other lipids discussed in this subsection are hydrolyzed to glycerol and soap by heating in alkaline solution, the historical method for making soap from animal fats. The reverse of the fat synthesis reaction shown above is catalyzed by fat-splitting enzymes at body temperatures in the digestive tract of animals: microbes also secrete such enzymes to hydrolyze particulate fats into smaller fragments, which can then be taken in through the cell membrane.

Closely related to the fats in structure but not function are the phosphoglycerides. In these molecules, phosphoric acid replaces a fatty acid esterified to one end of glycerol. The result is again a molecule with strongly hydrophilic and hydrophobic portions; thus micelle formation is again observable at sufficiently large phosphoglyceride concentrations. A flat double-molecular layer structure may be formed across a small aperture in a sheet submerged in a phospholipid solution (Fig. 2.2*a*). The resulting planar lipid bilayer has a thickness of about 70 Å (7×10^{-7} cm). Biological plasma membranes typically contain appreciable concentrations of phospholipids and other polar lipids. Also, plasma membranes show an apparent molecular bilayer (Fig. 2.2*b*) of thickness similar to the spontaneously formed phosphoglyceride double layer in Fig. 2.2*a*. Consequently, it appears that the bilayer lipid membranes might serve as convenient synthetic systems for fundamental characterization of thin membrane processes.

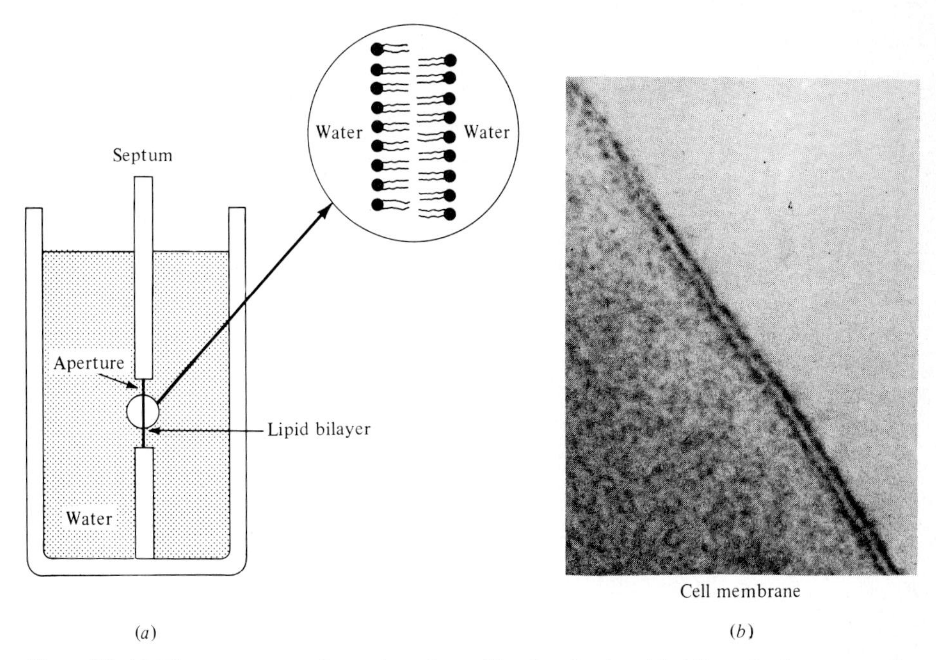

Figure 2.2 (*a*) The spontaneous formation of a stable phosphoglyceride bilayer in the aperture between two compartments filled with water and lipid. (*b*) This structure strongly resembles the bilayer appearance of cell membranes in electron micrographs. *[Electron micrograph reprinted by permission from J. B. Robertson, Membrane Models: Theoretical and Real, in "The Nervous System, vol. 1: The Basic Neurosciences," D. B. Tower (ed.), p. 43, Raven Press, New York.]*

Several physical properties of lipid bilayer membranes are similar to those of plasma membranes. Both lipid and plasma membranes have high passive electrical resistance and capacitance. The resultant impermeability of natural membranes to such highly charged species as phosphorylated compounds is largely a result of this property. The membrane thereby allows the cell to contain a reservoir of charged nutrients and metabolic intermediates, as well as maintaining a considerable difference between the internal and external concentrations of small cations such as H^+, K^+, and Na^+.

Other membrane components and their influences on material exchange between the cell environment and interior will be considered in Secs. 2.5 and 5.7. These barriers and passages for specific biochemicals determine which enter, are confined, and leave the catalytic reaction network housed inside the cell. These mass transport regulation functions are critical for life, and they are obviously of major importance in process applications employing cells as catalysts.

An intriguing similarity between bilayer lipid membranes and plasma membranes is their ability to be modified in their selective ion permeabilities by the addition of small amounts of various substances. In particular, several antibiotics and other cation-complexing molecules have been found to markedly increase passive ion transport in both types of membranes. In more complex processes, the cell walls of viable organisms can be rendered leaky by mild chemical or heat treatment. This has been used advantageously in the microbial production of metabolic intermediates and in the treatment of cells to decrease their nucleic acid content before use as animal feed.

2.1.2 Fat-soluble Vitamins, Steroids, and Other Lipids

A *vitamin* is an organic substance which is required in trace amounts for normal cell function. The vitamins which cannot be synthesized internally by an organism are termed *essential vitamins*: in their absence in the external medium, the cell cannot survive. (This fact has been used advantageously by growing microbes in test media as a probe for the presence or absence of a particular vitamin.) The water-soluble vitamins such as vitamin C (ascorbic acid) are not lipids by definition. However, vitamins A, E, K, and D are water-insoluble and dissolve in organic solvents. Consequently these vitamins are classified as lipids. The ultimate role of the lipid-soluble vitamins appears obscure with the exception of vitamin A which is necessary to prevent night blindness in humans.

Interest in vitamin supply from microbial and other food derives largely from the fact that the water-soluble vitamins thiamine, riboflavin, niacin (nicotinic acid), pantothenic acid, biotin, folic acid, and choline and the lipid vitamins A, D, E, and K are all essential (or probably essential) for children and/or adults. Many microorganisms can synthesize a number of these compounds. Yeast, for example, provides the precursor ergosterol, which is converted by sunlight to vitamin D_2 (calciferol). The fat-soluble vitamin K is synthesized by microbes in animal and human digestive tracts, an excellent example of mutually assisting populations (commensalism, considered further in Chap. 13). Several water-soluble vitamins are known to be necessary for activity of specific enzymes.

Steroids are a class of lipid biochemicals with the general structure shown in Fig. 2.3*a*. Of these, a subgroup (hormones) constitutes some of the extremely potent controllers of biological reaction rates: hormones may be effective at levels of 10^{-8} M in human tissue. Microbes are currently used to carry out relatively minor transformations of such active steroids to yield more valuable products. For example, progesterone can be converted into *cortisone* in a two-step process (microbial, then chemical) (Fig. 2.3*b*). Further examples appear in

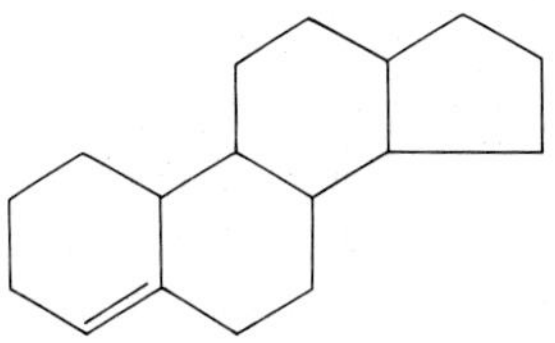

(*a*) General steroid base: perhydroxycyclopentano phenanthrene

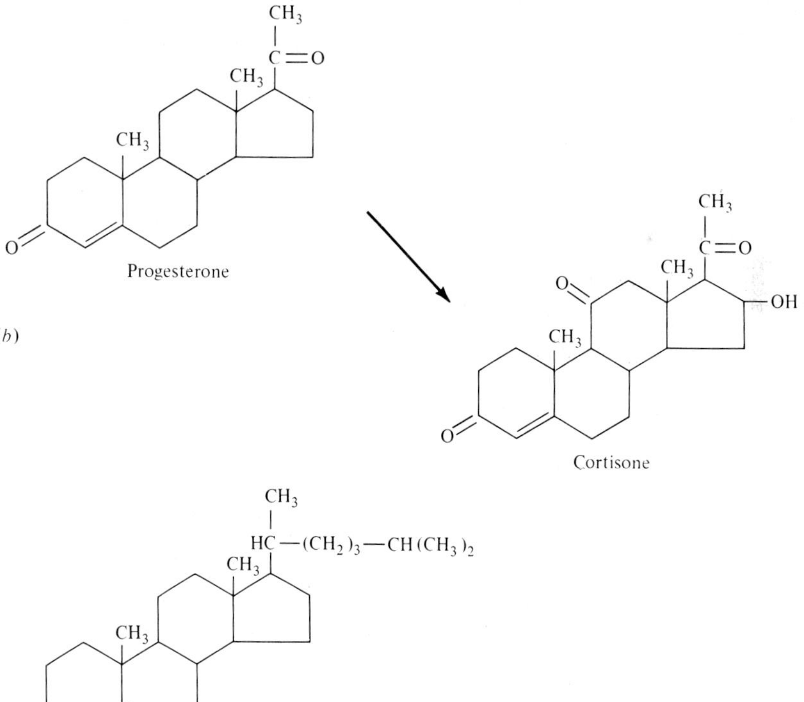

Progesterone

Cortisone

(*b*)

Cholesterol

(*c*)

Figure 2.3 Some examples of steroid structure.

Table 12.15. Evidently, the complexity of the reactant is such that only the action of an enzyme, produced by perhaps only one or several kinds of microbes, carries out the reaction with a useful *selectivity* (minimal side-product generation). The familiar steroid *cholesterol* (Fig. 2.3c) occurs almost exclusively in membranes of animal tissues. Related sterol compounds have been shown to alter cell plasma-membrane permeabilities.

An important food-storage polymer for some bacteria, including *Alcaligenes eutrophus*, is poly-β-hydroxybutyric acid (PHB). The repeating unit is

$$-\underset{\displaystyle |}{\overset{\displaystyle CH_3}{}}\!\!\!\!\!CH-CH_2-\overset{\displaystyle O}{\overset{\|}{C}}-O-$$

The polymer occurs as granules within the cells. In the absence of sufficient food supply, the cell depolymerizes this reserve to yield the soluble, easily metabolized β-hydroxybutyric acid. PHB is a possible candidate for large-scale manufacture because it is biodegradable and has properties adaptable to packaging.

2.2 SUGARS AND POLYSACCHARIDES

The *carbohydrates* are organic compounds with the general formula $(CH_2O)_n$, where $n \geq 3$. These compounds are found in all animal, plant, and microbial cells; the higher-molecular-weight polymers serve both structural and storage functions. The formula $(CH_2O)_n$ is sufficiently accurate to be useful in calculating overall elemental balances and energy release in cellular reactions.

In the biosphere, carbohydrate matter (including starches and cellulose) exceeds the combined amount of all other organic compounds. When photosynthesis occurs, carbon dioxide is converted to simple sugars C_3 to C_9 in reactions driven by the incident sunlight (considered further in Chap. 5, bioenergetics). These sugars are then polymerized into forms suitable for structure (cellulose) or sugar storage (starches). By these processes, radiant solar energy is stored in chemical form for subsequent utilization. The magnitude of this energy transformation is estimated to be 10^{18} kcal per year, corresponding to storage of 0.1 percent of the annual incident radiant energy. Much of the annual 10^{18} kcal stored is of course ultimately released in subsequent oxidation (largely cellular respiration) to carbon dioxide.

2.2.1 D-Glucose and Other Monosaccharides

Monosaccharides, or *simple sugars*, are the smallest carbohydrates. Containing from three to nine carbon atoms, monosaccharides serve as the monomeric blocks for noninformational biopolymers with molecular weights ranging into the millions.

D(+)-Glucose, the optical isomer which rotates polarized light in the + direction, is a polyhydroxyalcohol derivative like other simple sugars (Table 2.2).

Table 2.2 Monosaccharides commonly found in biological systems

	Aldoses (aldehyde derivatives; prefix *aldo*-)			Ketoses (ketone derivatives; prefix *keto*-)
Triose (three-carbon)		CHO HCOH CH_2OH D-Glyceraldehyde		CH_2OH C=O CH_2OH Dihydroxyacetone
Pentose (five-carbon)		CHO HCOH HCOH HCOH CH_2OH D-Ribose		CH_2OH C=O HCOH HCOH CH_2OH D-Ribulose
Hexose (six-carbon)	CHO HCOH HOCH HCOH HCOH CH_2OH D-Glucose	CHO HOCH HOCH HCOH HCOH CH_2OH D-Mannose	CHO HCOH HOCH HOCH HCOH CH_2OH D-Galactose	CH_2OH C=O HOCH HCOH HCOH CH_2OH D-Fructose

Although D-glucose is by far the most common monosaccharide, other simple sugars are also found in living organisms (Table 2.2). These common sugars are all either aldehyde or ketone derivatives. In sugar nomenclature, prefixes indicating these functional groups are often combined with a name fixing the length of the carbon chain. Thus, glucose is an aldohexose; the notation D referring to a particular optical isomer occurring almost exclusively in living systems (see optical activity, Sec. 2.4).

In solution D-glucose is present largely as a ring structure, *pyranose*, which results from reaction of the C-1 aldehyde in glucose with the C-5 hydroxyl (note the standard numbering scheme for six carbons and the α,β labels for the position of the —OH group on the number 1 carbon).

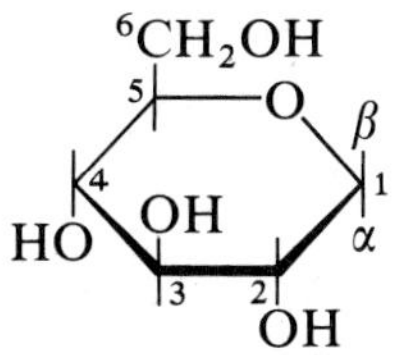

$^6CH_2OH$
5 O β
4 1
HO OH α
3 2
OH

The five-membered sugars D-ribose and deoxyribose are major components of the nucleic acid monomers of DNA and RNA and other biochemicals to be discussed shortly.

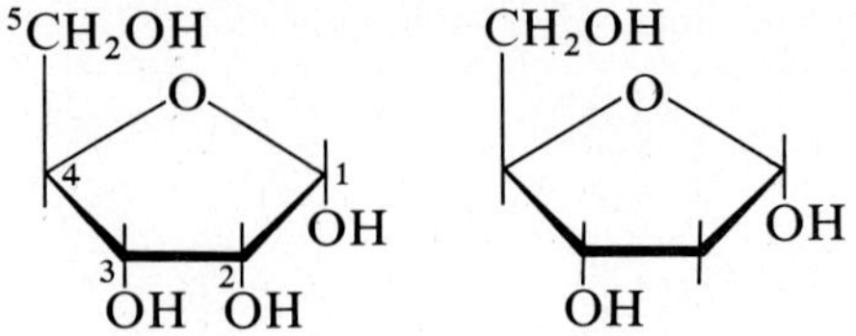

D-Ribose

Deoxyribose

2.2.2 Disaccharides to Polysaccharides

Because the ringed form of many simple sugars predominates in solution, they do not exhibit the characteristic reactions of aldehydes or ketones. In the D-glucopyranose ring above, the —OH attached to position 1 is relatively reactive. As shown below, this hydroxyl group, here attached in the α-position, can condense with an —OH on the 4 carbon of another sugar to eliminate a water molecule and form an α-1,4-*glycosidic bond*:

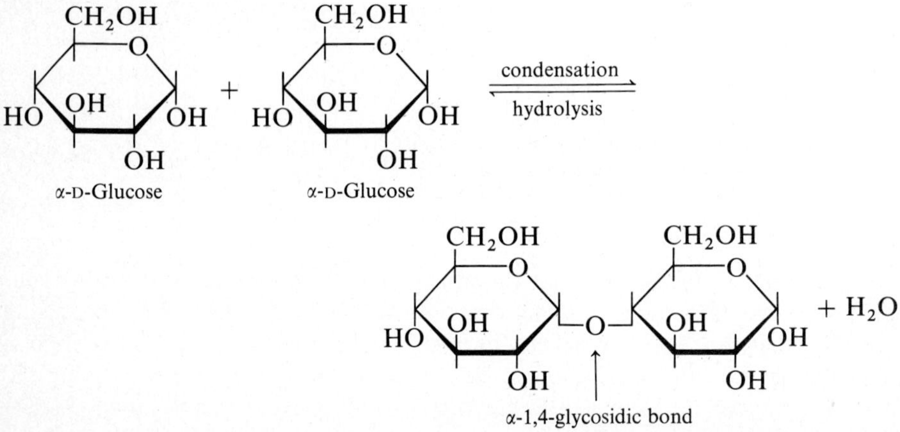

α-D-Glucose + α-D-Glucose (condensation / hydrolysis) α-Maltose $+ H_2O$

α-1,4-glycosidic bond

α-Maltose

The condensation product of two monosaccharides is a *disaccharide*. In addition to maltose, which is formed from two D-glucose molecules, the following disaccharides are relatively common:

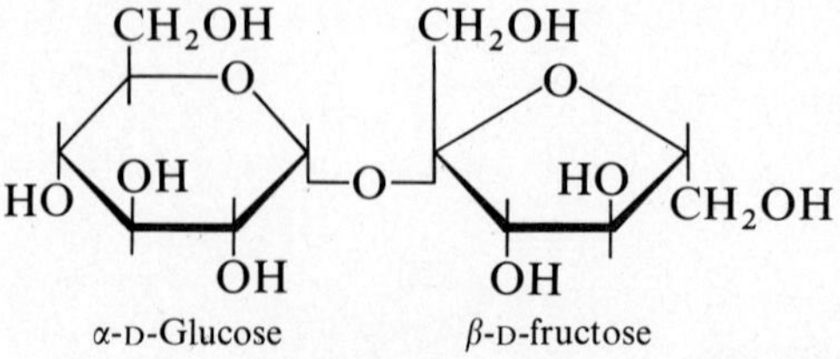

α-D-Glucose β-D-fructose

Sucrose

β-D-galactose β-D-glucose

Lactose

Table sugar is *sucrose*, a major foodstuff which is found in all photosynthetic plants. Among all disaccharides, sucrose is the easiest to hydrolyze: the resulting mixture of glucose and fructose monosaccharides is called *invert sugar*. Found only in milk, *lactose* is a relatively rare but important disaccharide. Since some people are lactose-intolerant and therefore cannot digest milk, enzyme processes to hydrolyze milk lactose are currently under development.

Continued polymerization of glucose can occur by formation of new 1,4-glycosidic bonds. *Amylose* is a straight-chain polymer of glucose subunits with molecular weight which may vary from several thousand to half a million:

α-1,4-glycosidic linkages

Portion of an amylose chain (—OH groups omitted for clarity)

Amylose typically constitutes about 20 percent of *starch*, the reserve food in plants, although this percentage varies widely. Granules of starch are large enough to be seen in many plant cells examined through a microscope.

While the amylose fraction of starch consists of straight-chain, water-insoluble polymers,[†] the bulk of starch is *amylopectin*. Amylopectin, also D-glucose polymer, is distinguished by a substantial amount of branching. Branches occur from the ends of amylose segments averaging 25 glucose units in length. Such structures arise when condensation occurs between the glycosidic —OH on one

[†] M. M. Green, G. Blankenhorn, and H. Hart, "Textbook Errors. 123; Which Starch Fraction Is Water-Soluble, Amylose or Amylopectin?," *J. Chem. Educ.*, **B52**: 729, 1975.

chain and the 6 carbon on another glucose:

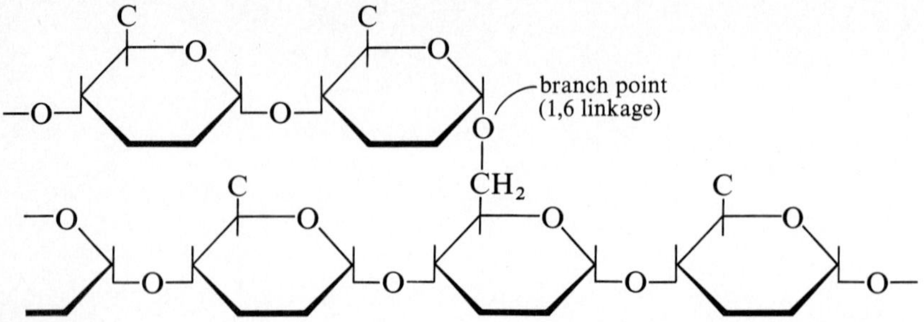

Amylopectin molecules are typically larger than amylose, with molecular weights ranging up to 1 to 2 million. Amylopectin is soluble in water, and it can form gels by absorbing water. After partial hydrolysis of starch, by acid or certain enzymes, many branched remnants of amylopectin called *dextrins* remain. Dextrins are used as thickeners and in pastes. Naturally, glucose, maltose, and other relatively small sugars are also obtained by partial hydrolysis. Corn syrup is derived from corn starch in this manner.

The glucose reservoirs in animals, especially numerous in liver and muscle cells, are granules of *glycogen*, a polymer which resembles amylopectin in that it is highly branched. Here, the degree of branching is greater as there are typically only 12 glucose units in the straight-chain segments between branches. Glycogen molecular weights of 5 million and greater are not uncommon. Glycogen also serves as an energy reserve for some microorganisms, including the enteric bacteria.

2.2.3 Cellulose

Cellulose, a major structural component of all plant cells from algae to trees, is the most abundant organic compound on earth. Cotton and wood are two common examples of materials rich in cellulose. Estimates place the amount of cellulose formed in the biosphere at 10^{11} tons per year. Each cellulose molecule is a long, unbranched chain of D-glucose subunits with a molecular weight ranging from 50,000 to over 1 million.

Although the glycosidic linkage in cellulose occurs between the 1 and 4 carbons of successive glucose units, the subunits are bonded differently than in amylose (compare the following structural formula with the preceding one for amylose). This difference in structure is significant. While many microorganisms, plants, and animals possess the necessary enzymes to break (hydrolyze) the α-1,4-glycosidic bonds which are found in starch or glycogen, few living creatures can hydrolyze the β-1,4 bonds of cellulose. One of the common products of enzymatic cellulose hydrolysis is *cellobiose*, a dimer of two glucose units joined by a β-1,4-glycosidic linkage.

The resistance of cellulose to natural and process degradation derives more from the crystalline structure of cellulose and upon its biological "packaging"

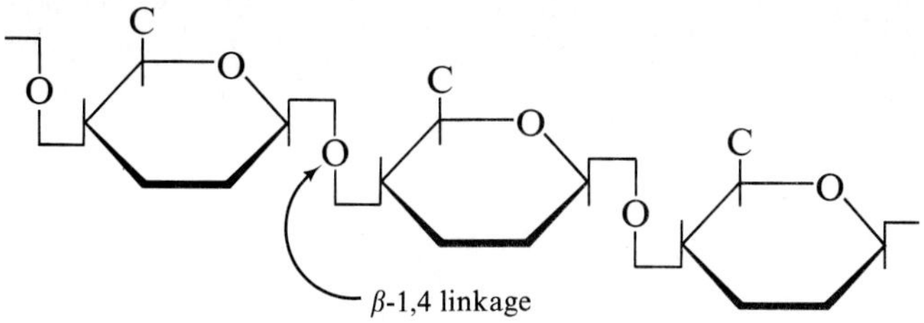

β-1,4 linkage
The glucose chain of cellulose

than on its use of β-1,4-glycosidic bonds. As shown in Fig. 2.4, intrachain hydrogen bonding occurs between the C-3 hydroxyl and the oxygen in the pyranose ring. Occasional interchain hydrogen bonding is also present. This hydrogen bonding makes cellulose chains combine to give crystallites, larger structures which are visible in the electron microscope.

Several different models for the crystalline structure of cellulose have been proposed; the essential concepts for our purposes can be gleaned from the schematic view given in Fig. 2.5. Most of the cellulose is organized into highly ordered *crystalline regions*, in which cellulose chains or *fibrils* are so tightly packed that even water molecules scarcely penetrate. Cellulose is, accordingly, water insoluble. Less ordered portions of the assembly, called *amorphous regions*, comprise typically about 15 percent of the cellulose microstructure. The amorphous regions are easily hydrolyzed by, for example, acids; the crystalline regions on the other hand are much more difficult to decompose.

The cellulose fibrils are clustered in *microfibrils* which are often depicted in cross section as in Fig. 2.5*a*. Here the solid lines denote the planes of the glucose building blocks, and the broken lines represent orientations of another important group of polysaccharides called *hemicellulose*. Hemicellulose molecules are found

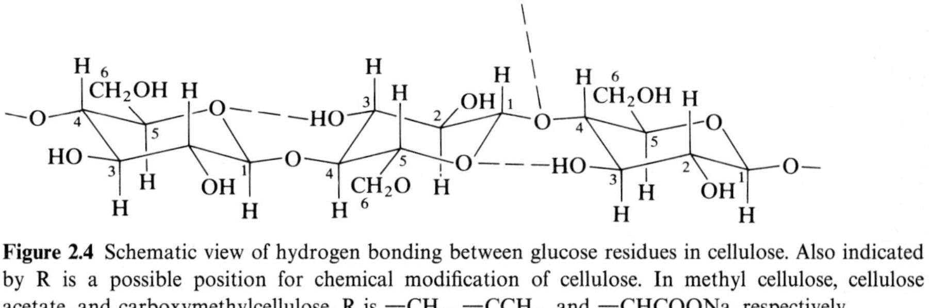

Figure 2.4 Schematic view of hydrogen bonding between glucose residues in cellulose. Also indicated by R is a possible position for chemical modification of cellulose. In methyl cellulose, cellulose acetate, and carboxymethylcellulose, R is $-CH_3$, $-\underset{\underset{O}{\|}}{C}CH_3$, and $-CHCOONa$, respectively.

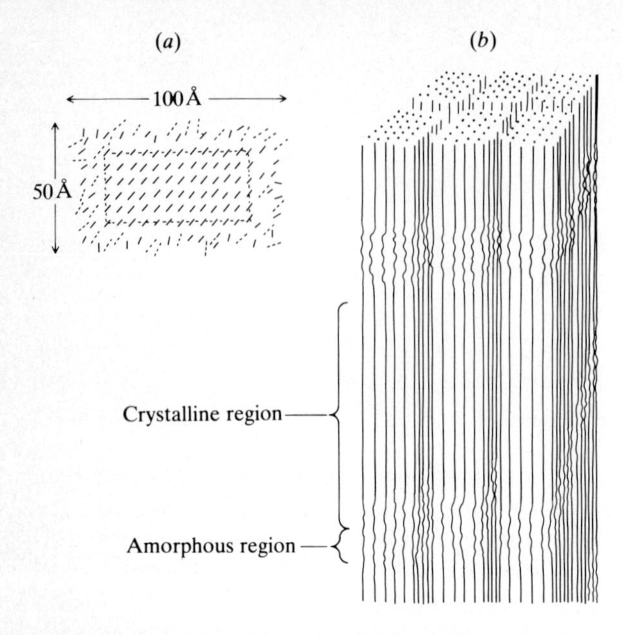

Figure 2.5 Diagram of the structure of cellulose microfibrils. *(Adapted from K. Mülethaner, Ann. Rev. Plant Phys., vol. 18, p. 1, 1967.)*

in wood and other cellulosic materials surrounding clusters of microfibrils, and these structures are in turn encased in an extensively cross-linked coating of *lignin.*

Because of this packaging, the term *lignocellulose* is often used to describe these materials. In Table 2.3, we can see that many types of biomass and cellulose waste materials contain significant amounts of lignocellulose and that the relative proportions of cellulose, hemicellulose, and lignin vary considerably. These composition data remind us that, in addition to serving as a renewable source of cellulose feedstock, these biomass resources also contain large quantities of other potentially useful raw materials. For this reason, we shall examine further the properties of hemicellulose and lignin.

Hemicelluloses, which vary in specific composition among different plants, are short, branched polymers of pentoses (xylose and arabinose) and some hexoses (glucose, galactose, and mannose). These units, which typically contain a

Table 2.3 Distribution (wt %) of cellulose, hemicellulose, and lignin in biomass and waste resources

Material	% cellulose	% hemicellulose	% lignin
Hardwoods stems	40–55	24–40	18–25
Softwoods stems	45–50	25–35	25–35
Grasses	25–40	25–50	10–30
Leaves	15–20	80–85	~0
Cotton seed hairs	80–95	5–20	~0
Newspaper	40–55	25–40	18–30
Waste papers from chemical pulps	60–70	10–20	5–10

large number of acetyl groups, are linked by 1,3-, 1,6-, and 1,4-glycosidic bonds. Many of these features are illustrated in the following representation of hardwood hemicellulose:

```
┌X—X—X—X—X—X—X—X—X—X┐
 3 2   2 3   3 3 2 3
└A A   α A   A A A A┘80–270
       1
       GA
       4
       M
```

X: xylose
A: acetyl group
GA: glucuronic acid
M: methoxy group
—: β-1,4 glycosidic bond

Here the numbers denote the carbon numbers of various bonds.

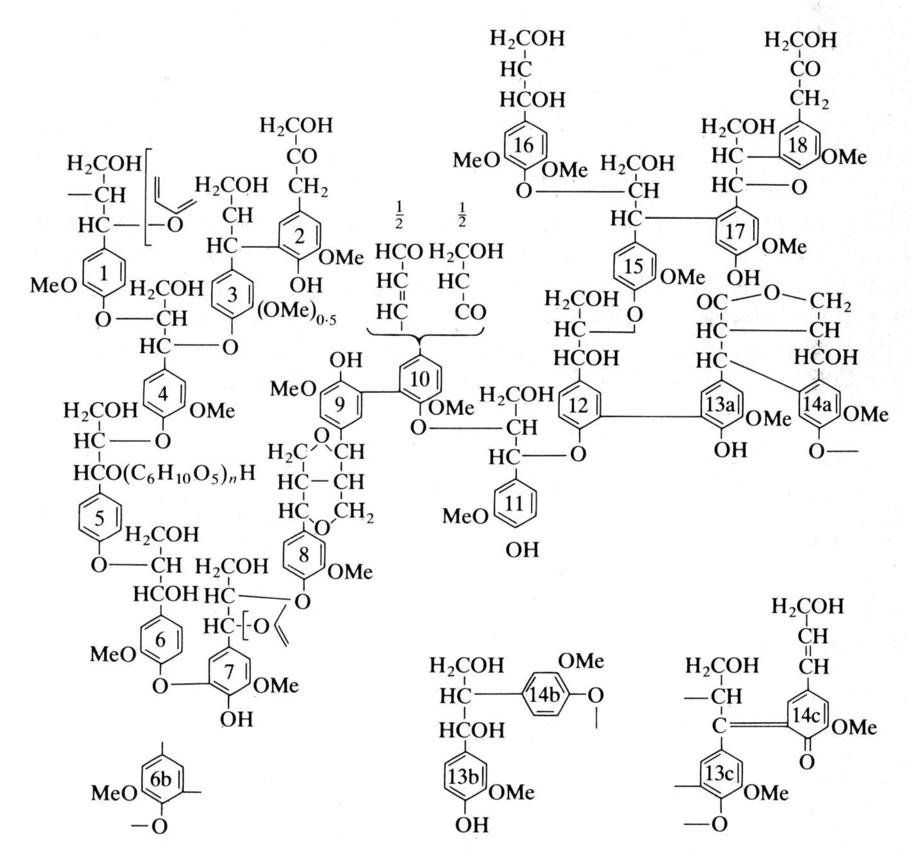

Figure 2.6 Schematic two-dimensional illustration of components and bonds found in spruce lignin. *(Reprinted by permission from K. Freudenberg, "Lignin: Its Constitution and Formation from p-Hydroxycinnamyl Alcohols," Science, vol. 148, p. 595. Copyright 1965 by the American Association for the Advancement of Science.)*

Glucuronic acid has not been mentioned before. It is one member of the family of *uronic acids* which are obtained by specific oxidation of one of the primary alcohol groups of a monosaccharide. In particular, glucuronic acid is one in which a carboxyl group occupies the 6 position.

We can hydrolyze hemicellulose to soluble products relatively easily. This can be achieved, for example, with dilute sulfuric acid (0.05 to 3 percent) or even with hot water. Much more recalcitrant, however, is the lignin casing which encloses the polysaccharide components of biomass.

Lignin is a polyphenolic material of irregular composition. In the model structure for spruce lignin in Fig. 2.6, we get a clear impression of the chemical complexity and heterogeneity of this material. This random arrangement of different building blocks strongly resists chemical and enzymatic attack. We shall briefly examine some of the available techniques in Chap. 4. In closing this overview of biomass chemistry and structure, we should note that lignin fragments are themselves potentially interesting chemicals and feedstocks.

2.3 FROM NUCLEOTIDES TO RNA AND DNA

The *informational* biopolymer DNA (deoxyribonucleic acid) contains all the cell's hereditary information. When cells divide, each daughter cell receives at least one complete copy of its parent's DNA, which allows offspring to resemble parent in form and operation. The information which DNA possesses is found in the *sequence* of the subunit nucleotides along the polymer chain. The mechanism of information transfer and the timing of DNA replication in the cell cycle are discussed in Chap. 6, along with other aspects of cellular controls. The subunit chemistry, the structure of the DNA polymer, and the information coded in the subunit sequence are now considered.

2.3.1 Building Blocks, an Energy Carrier, and Coenzymes

In addition to their presence in nucleic acids, nucleotides and their derivatives are of considerable biological interest on their own. Three components make up all nucleotides: (1) phosphoric acid, (2) ribose or deoxyribose (five-carbon sugars), and (3) a nitrogenous base, usually derived from either purine or pyrimidine.

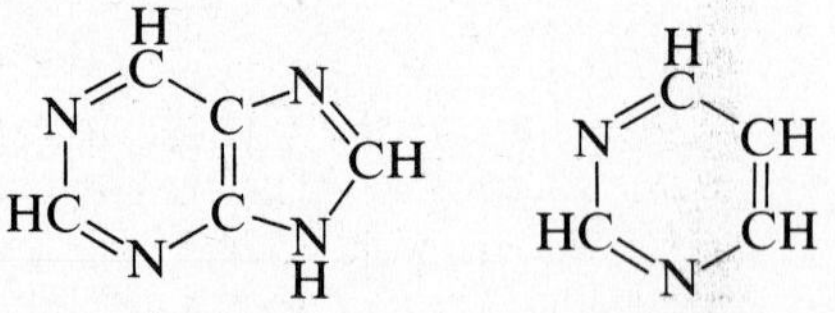

Purine

Pyrimidine

These three components are joined in a similar fashion in two nucleotide types (Fig. 2.7) which are distinguished by the five-carbon sugar involved. Ribonucleic acid (RNA) is a polymer of ribose-containing nucleotides, while deoxyribose is the sugar component of nucleotides making up the biopolymer

Deoxyribonucleotides
(general structure)

Base

OH
5′
HO—P—O—CH_2 O
O
1′
H H
H H
3′ 2′
OH H

Phosphoric
acid

Sugar Base

Ribonucleotides
(general structure)

Base

OH
HO—P—O—CH_2 O
O
H H
H H
OH OH

(*a*)

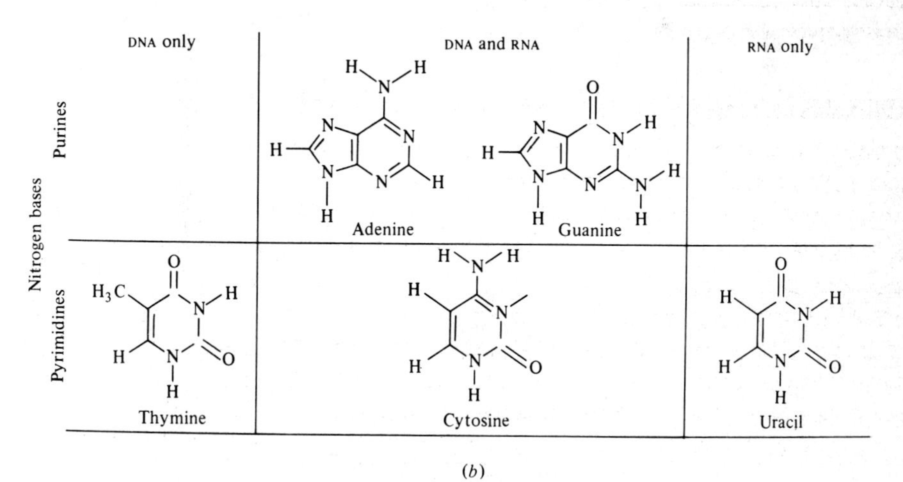

Figure 2.7 (*a*) The general structure of ribonucleotides and deoxyribonucleotides. (*b*) The five nitrogenous bases found in DNA and RNA.

deoxyribonucelic acid (DNA). Also shown in Fig. 2.7 are the five nitrogenous bases found in DNA and RNA nucleotide components. Three of the bases, adenine (A), guanine (G), and cytosine (C), are common to both types of nucleic acids. Thymine (T) is found only in DNA, while uracil (U), a closely related pyrimidine base, is unique to RNA. Both series of nucleotides are strong acids because of their phosphoric acid groups.

Removal of the phosphate from the 5′-carbon of a nucleotide gives the corresponding nucleoside. As indicated in Table 2.4, the nomenclature for nucleosides and nucleotides derives from the names of the corresponding nitrogenous bases. We should note that alternative designations are sometimes used for the nucleotides. For example, adenylate can also be called adenosine-5′-monophosphate. It is the latter nomenclature that is typically extended to describe derivatives with diphosphate or triphosphate groups esterified to the nucleoside 5′-hydroxyl; for example, adenosine 5′-triphosphate (ATP).

Of particular biological significance is the nucleoside *adenosine*, made with ribose and adenine. Sketched in Fig. 2.8, along with its important derivatives, is *adenosine monophosphate* (AMP). One or two additional phosphoric acid groups can condense with AMP to yield ADP (adenosine diphosphate) and ATP, respectively. The phosphodiester bonds connecting the phosphate groups have especially useful free energies of hydrolysis. For example, the reaction of ATP to yield ADP and phosphate is accompanied by a standard Gibbs free-energy change of -7.3 kcal/mol at 37°F and pH 7 [recall that pH = $-\log$ (molar H^+ concentration)].

Although we are accustomed to thinking primarily in terms of thermal energy, the cell is essentially an isothermal system where *chemical* energy transformations are the rule. As examined in considerable additional detail in Chap. 5, ATP is the major currency of chemical energy in all cells. That is, ATP is the means by which the cell temporarily stores the energy derived from nutrients or sunlight for subsequent use such as biosynthesis of polymers, transport of materials through membranes, and cell motion. While adenosine phosphates are the predominant forms of energy carriers, the diphosphate and triphosphate derivatives of the other nucleosides also serve related functions in the cell's chemistry.

Table 2.4 Nomenclature of nucleosides and nucleotides. As indicated in the last row, the prefix deoxy- is used to identify those molecules containing deoxyribose.

Base	Nucleoside	Nucleotide
Adenine (A)	Adenosine	Adenylate (AMP)
Cytosine (C)	Cytidine	Cytidylate (CMP)
Guanine (G)	Guanosine	Guanylate (GMP)
Uracil (U)	Uridine	Uridylate (UMP)
Thymine (T)	Deoxythymidine	Deoxythymidylate (dTMP)

The cyclic form of AMP, so called because of an intramolecular ring involving the phosphate group (Fig. 2.8), serves as a regulator in many cellular reactions, including those that form sugar and fat-storage polymers. An inadequacy of cyclic AMP in tissue is related to one kind of cancer, a condition of relatively uncontrolled cell growth.

In addition to providing components for nucleic acids, the adenine-ribose monophosphate is a major portion of the *coenzymes* shown in Fig. 2.9. Enzyme kinetics are considered in the following chapter: it suffices here to note that the *coenzyme* is the additional organic moiety which is necessary for the activation of certain enzymes to the catalytically useful form. As essentially all reactions within

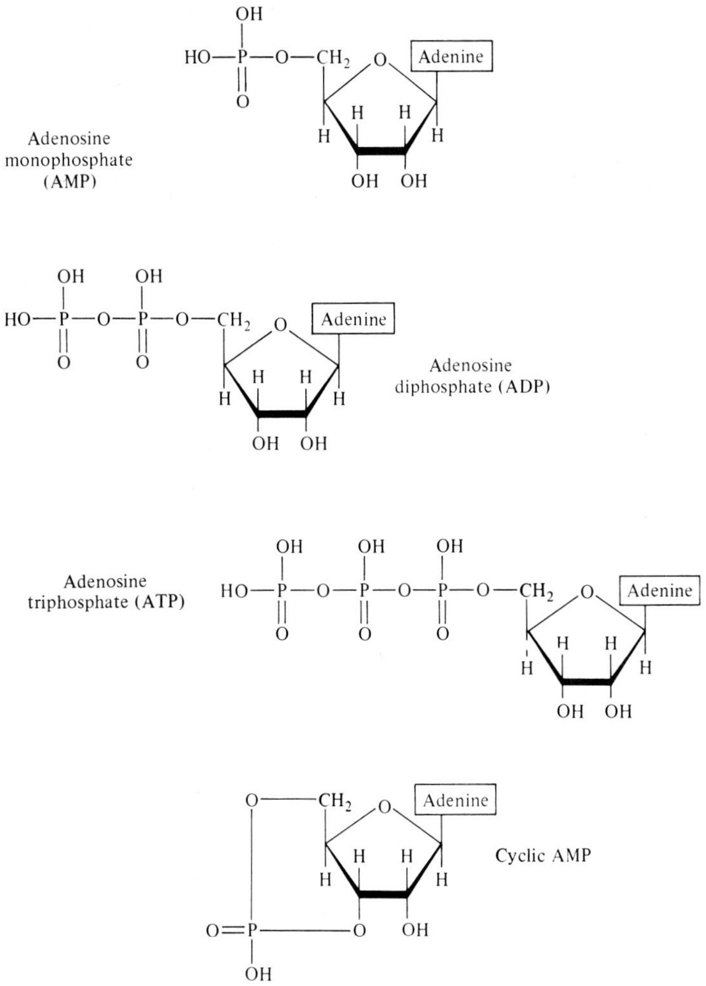

Adenosine monophosphate (AMP)

Adenosine diphosphate (ADP)

Adenosine triphosphate (ATP)

Cyclic AMP

Figure 2.8 The phosphates of adenosine. AMP, ADP, and ATP are important in cellular energy transfer processes, and cyclic AMP serves a regulatory function.

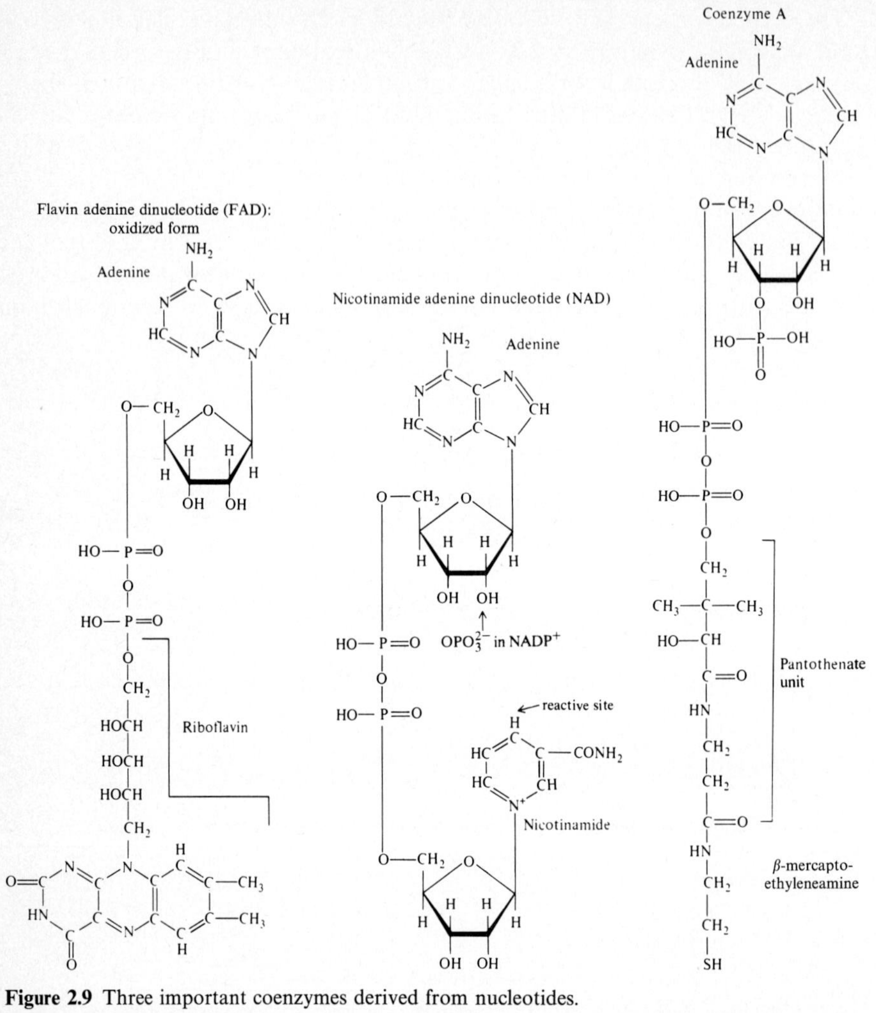

Figure 2.9 Three important coenzymes derived from nucleotides.

the cell are catalyzed by enzymes, variations of the coenzyme level provide a convenient short-term method for cellular regulation of *active* enzyme and thus of the rate of certain intracellular reactions.

2.3.2 Biological Information Storage: DNA and RNA

As with the polysaccharides, the polynucleotides are formed by condensation of its monomers. For both RNA and DNA, the nucleotides are connected between the 3′ and 5′ carbons of successive sugar rings. Figure 2.10 illustrates a possible sequence in a trinucleotide. The DNA molecules in cells are enormously larger: all the essential hereditary information in procaryotes is contained in one DNA molecule with a molecular weight of the order of 2×10^9. In eucaryotes, the

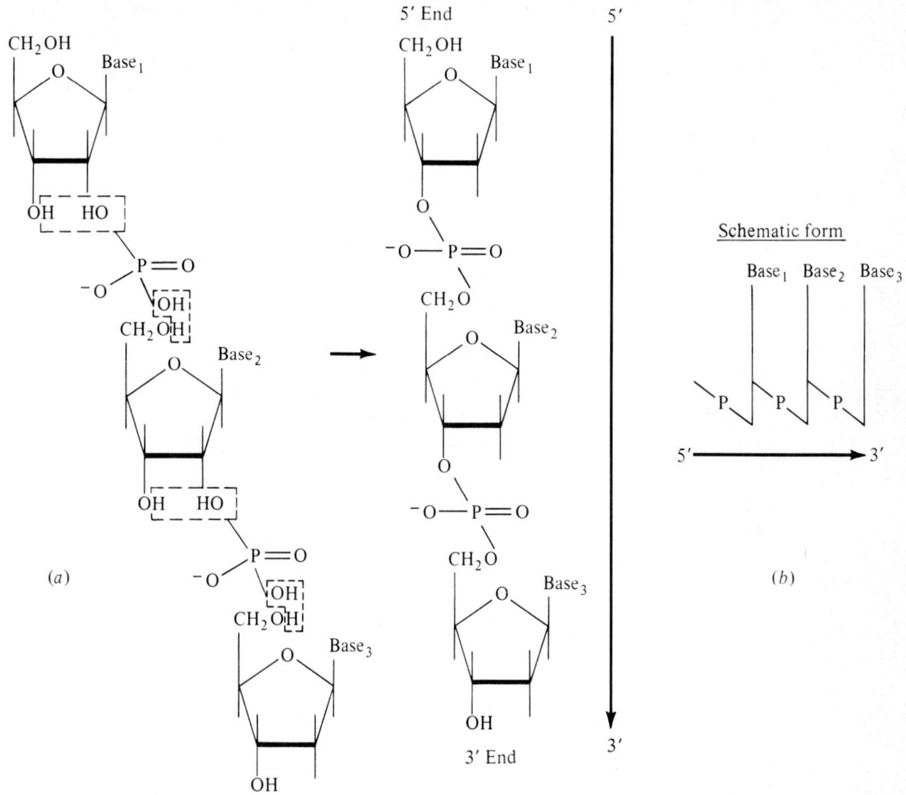

Figure 2.10 (*a*) Condensation of several nucleotides to form a chain linked by phosphodiester bonds. (*b*) Another schematic for a nucleotide chain.

nucleus may contain several large DNA molecules. The negative charges on DNA are balanced by divalent ions (procaryotes) or basic amino acids (eucaryotes).

It is important to notice in Fig. 2.10*a* that the sequence of nucelotides has a direction or *polarity*, with one free 5′-end and the other terminus possessing a 3′-end not coupled to another nucleotide. As indicated in Fig. 2.10*b*, it is conventional to write the nucleotide sequence in the 5′ to 3′ direction. More abbreviated notations are often used. Suppose in the example in Fig. 2.10*b*, which contains a phosphate group esterified to the 5′-hydroxyl, that $Base_1$ is cytosine, $Base_2$ is adenine, and $Base_3$ is thymine. Then the same nucleotide sequence information is often written pCpApT, or, even more briefly, CAT.

As James Watson and Francis Crick deduced in 1953, the DNA molecule consists of two polynucleotide chains coiled into a double helix (Fig. 2.11). The regular helical backbone of the molecule is composed of sugar and phosphate units. In the interior of the double helix are the purine and pyrimidine bases. It is the *sequence* of the four different nitrogenous bases along the polynucleotide

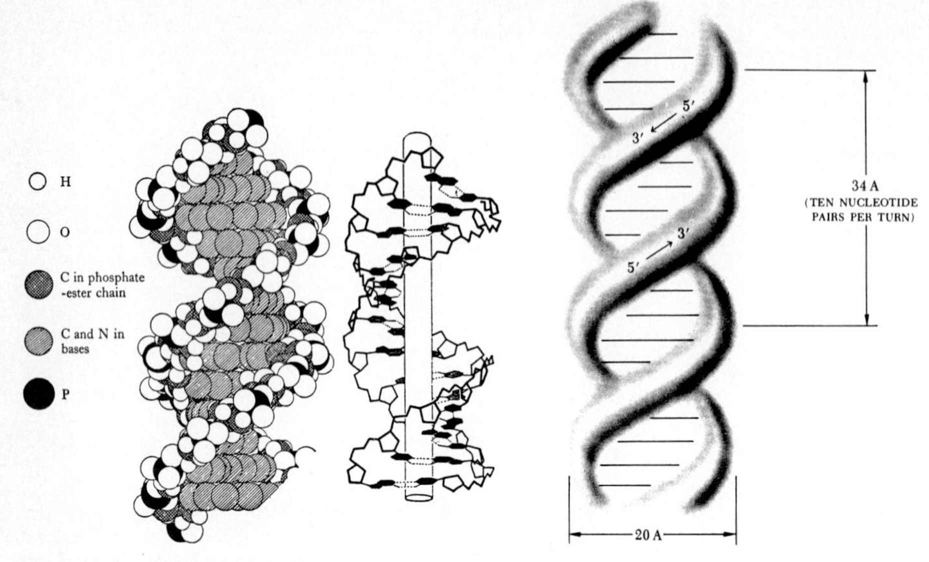

Figure 2.11 Several views of the double-helical structure of DNA. These diagrams show the classical Watson-Crick structure [7], also called B-DNA, a right-handed helix in which the planes of the base pairs are perpendicular to the helix axis. The sketch on the right reveals some geometrical parameters of this structure. *(Left-hand and center diagrams reprinted with permission from M. Yudkin and R. Offord, "A Guidebook to Biochemistry," p. 52, Cambridge University Press, London, 1971.)*

chains which carries genetic information—namely the instructions for ribonucleic acid and protein synthesis. We shall investigate the mechanisms for control and utilization of the genetic information in DNA in Chap. 6.

The nitrogenous base at a particular position on one strand is paired with the adjacent base on the other strand in a particular and precise fashion: adenine on one strand pairs with thymine on the other. Guanine-cytosine is the second *base pair* found in DNA. Figure 2.12 indictates the remarkable geometric similarity of the two base pair configurations and illustrates the presence of two and three hydrogen bonds in the A—T and C—G base pairs, respectively.

The two strands of DNA have opposite polarity: they run $5' \rightarrow 3'$ in opposite directions. Combining this information with the base pairing rules, we can infer from the nucleotide sequence of a single deoxyribonucleotide chain, for example

5′ CGAATCGTA 3′

that the corresponding fragment of an intact, double-stranded DNA molecule looks like

5′ CGAATCGTA 3′
3′ GCTTAGCAT 5′

Similarly, if we have been given only the sequence

5′ TACGATTCG 3′

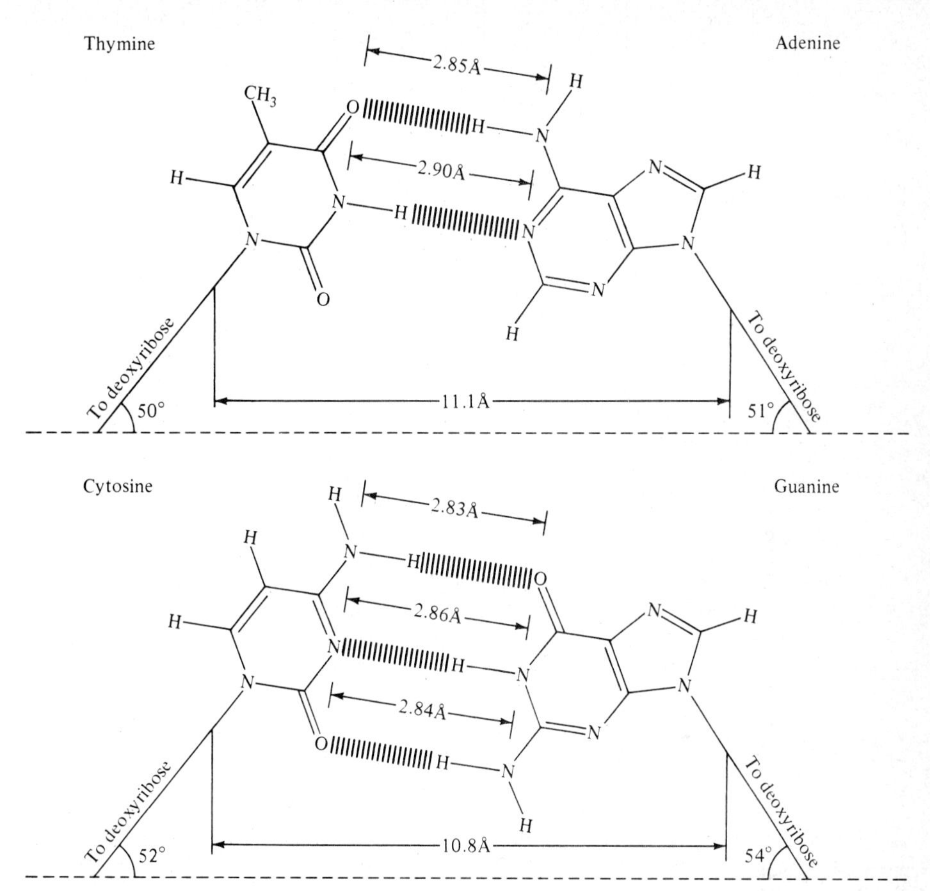

Figure 2.12 The complementary structures of the adenine-thymine and guanine-cytosine pairs provide for hydrogen bonding while maintaining very similar geometries. This feature permits base pairing within the DNA molecule double helix. *(From "Cell Structure and Function," 2d ed., p. 141, by Ariel G. Loewy and Philip Siekevitz. Copyright © 1963, 1969 by Holt, Rinehart, and Winston. Reprinted by permission of Holt, Rinehart, and Winston.)*

we would also have deduced the same double-stranded structure. Thus, in an informational sense, knowledge of one strand's sequence implies the sequence of the complementary strand. Each strand is a *template* for the other.

This feature of DNA provides the directions for synthesis of daughter DNA from a parent DNA molecule, a process called DNA *replication.* If the two complementary DNA strands are separated and double helices are constructed from each strand following the base-pairing rules, the end products are two new molecules, each identical to the original double-stranded DNA and each containing one new strand and one old strand (Fig. 2.13). Therefore, base pairing provides a chemical reader for the biological message coded in the DNA nucleotide sequence. The overall mechanism depicted in Fig. 2.13 was successfully proved in an experiment in which *E. coli* was grown first in a medium containing ^{15}N

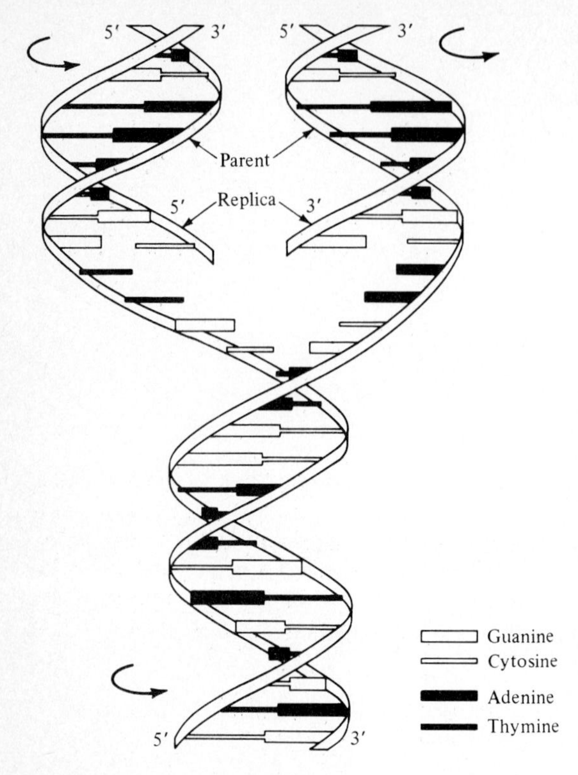

Figure 2.13 A simplified diagram of DNA replication. As the parent strands separate, complementary strands are added to each parent, resulting in two daughter molecules identical to the parent. Notice that each daughter molecule contains one strand from the parent. *(From "Cell Structure and Function," 2d ed., p. 145, by Ariel G. Loewy and Philip Siekevitz. Copyright © 1963, 1969 by Holt, Rinehart, and Winston. Reprinted by permission of Holt, Rinehart, and Winston.)*

nitrogen, then in a medium with only the normal ^{14}N isotope. DNA containing the different isotopes could be separated in an ultracentrifuge. Ultracentrifugation and the use of isotopic labels, including radioisotopes, are major tools in contemporary biochemistry.

Another important feature of DNA structure is the hydrogen bonding which occurs between the base pairs (See Fig. 2.12). This helps to stabilize the double-stranded configuration, but not in a permanent or irreversible fashion. The possibility of unwinding or separating the two strands is crucial to the biological function of DNA as the replication sketch in Fig. 2.13 illustrates. In further investigation of molecular genetics in Chap. 6, we will encounter other central roles for single-stranded DNA segments.

Separation of double-stranded DNA by heat is an important method in DNA characterization. Because AT base pairs involve two hydrogen bonds and GC base pairs have three, AT-rich regions of DNA melt (i.e., the two strands separate) before GC-rich regions. The melting process is readily monitored by following absorbance of the DNA solution at 260 nm: single-stranded DNA absorbs more strongly, so that DNA melting is measured as an increase in overall absorbance. The *melting temperature* T_m is the temperature at which the absorbance is midway between the fully double-stranded and completely melted limits. As expected, the T_m value for DNA correlates with its GC content. For example, *E. coli* DNA, which is 50 percent GC, has a T_m of 69°C, while DNA from *Pseudomonas aeruginosa*, 68 percent GC pairs, gives $T_m = 76°C$. GC content of different organisms span a wide range from 23 to 75 percent. Base composition is one parameter sometimes used in characterizing species.

If a solution of melted DNA is cooled, the separated complementary strands will *anneal* to reform the double helix. Similarly, if two different single-stranded segments of DNA have complementary base pair sequences, these will *hybridize* to form a double-stranded segment. Hybridization is a critical experimental technique in biochemistry and in recombinant DNA technology (Chap. 6). Before leaving this brief introduction to the physical chemistry of DNA, we should note that double-stranded DNA can be melted or denatured by addition of alkali or acid to ionize the bases. Also, double-stranded DNA in solution behaves hydrodynamically like a rigid rod, while single-stranded DNA acts like a randomly coiled polymer.

As mentioned earlier, all of the DNA required for essential functions—growth and multiplication of cell numbers—in a bacterium like *E. coli* is contained in a single DNA molecule which is a closed circle. This molecule, which exists naturally in a highly supercoiled form, is evident in electron micrographs of bacteria as a distinct but somewhat amorphous nuclear region (recall Fig. 1.2). This large, essential DNA is called the *chromosome* of the bacterium; this chromosome is 4.7 million, or 4700 kb, base pairs in circumference (kb denotes kilobases or 1000 nucleotides†).

The DNA in the membrane-enclosed nucleus of eucaryotes is typically divided among several chromosomes, which may be much larger structures than the procaryote chromosome. Eucaryotic chromosomes also contain small basic proteins called *histones* which account for roughly 50 percent of the chromosomes' mass. The nucleoprotein material in the eucaryotic chromosome is called *chromatin*. Eucaryotic cells also contain separate and smaller DNA molecules in their chloroplast and mitochondria organelles.

It is very important to recognize that cells may contain additional DNA molecules. *Plasmids* are relatively small DNAs which are found in many bacteria and eucaryotes. Bacterial plasmids, for example, are circular DNA molecules with sizes ranging from about 4 to 50 kb. Plasmids in nature endow their host cells with useful but nonessential functions such as antibiotic resistance. In the biochemistry laboratory and the production process, plasmids assume a central role: they are important vehicles of recombinant DNA technology. As we shall examine further in Chap. 6, manipulation of plasmids in the test tube and subsequent introduction of the recombinant plasmids into living cells genetically reprograms the cells to produce new compounds or grow more efficiently.

Other small DNA molecules may be inserted into living cells by viruses. *Bacteriophage*, or often called simply *phage*, are small viruses which infect bacteria. These can cause problems in large-scale processes utilizing pure bacterial

† Lengths in kilobases may be converted to linear contour lengths using the factor 0.34 μm/kb. Also, for a double-stranded DNA, 1 kb equals approximately 660 kdal mass.

cultures. On the other hand, studies of phage have made essential contributions to molecular genetics, and phage also are used to maintain stable working collections, or *libraries*, of DNA segments. The phage λ of *E. coli* contains a single circular double-stranded DNA molecule 48.6 kb in length. The DNA in T2 phage of *E. coli* is 166 kb long.

The function of DNA is to store instructions for synthesis of RNA molecules of specific nucleotide sequence and length, some of which then coordinate synthesis of a variety of specific proteins. *A segment of DNA coding for an RNA molecule's sequence is called a gene.* We consider different RNA species briefly next, and then turn to proteins and their building blocks.

There are three distinct classes of ribonucleic acid (RNAs) in all cells and a fourth group found in some viruses. RNAs are comprised of ribonucleotides in which ribose rather than deoxyribose is the sugar component. Like DNA, the nucleotides in an RNA molecule are arranged in a definite sequence. In fact, the RNA sequences are constructed using information contained in segments of DNA. Again, base pairing is the central device in transmission of information from DNA to RNA. The base pairing rules are similar to those within double-stranded DNA, except in RNAs uracil substitutes for thymine in base pairing with adenine:

DNA (template)	RNA (being assembled)
T	A
A	U
C	G
G	C

The various RNAs which participate in normal cell function serve the purpose of reading and implementing the genetic instructions of DNA. *Messenger* RNA (mRNA) is complementary to a base sequence from a gene in DNA. Each mRNA molecule carries a message from DNA to another part of the cell's biochemical apparatus. Since the length of these messages varies considerably, so do the size of mRNAs: a chain of 10^3–10^4 nucleotides in length is typical. RNAs are typically single-stranded. However, intermolecular and intramolecular base pairing of RNA nucleotide sequences often play a role in the structure or function of the RNA.

The message from mRNA is read in the *ribosome* (Figs 1.2 and 1.3). Up to 65 percent of the ribosome is *ribosomal* RNA (rRNA), which in turn can be separated in a centrifuge into several RNA species. For example, the *E. coli* ribosome contains three different rRNAs denoted 23S, 16S, and 5S and with characteristic chain lengths of 3×10^3, 1.5×10^3, and 10^2 nucleotides, respectively. Ribosomes found in the cytoplasm of eucaryotic cells, on the other hand, contain 28S, 18S, 7S, and 5S rRNAs. Different still are the ribosomes found in mitochondria and chloroplasts.

Table 2.5 Properties of *E. coli* RNAs†

Type	Sedimentation coefficient	mol wt	No. of nucleotide residues	Percent of total cell RNA
mRNA	6S–25S	25,000–1,000,000	75–3000	~2
tRNA	~4S	23,000–30,000	75–90	16
rRNA	5S	~35,000	~100	82
	16S	~550,000	~15000	
	23S	~1,100,000	~3100	

† From A. L. Lehninger, *Biochemistry*, 2d ed., table 12-3, Worth Publishers, Inc. New York, 1975.

Transfer RNA species (tRNAs), the smallest type, with only 70 to 95 nucleotide components, are found in the cell's cytoplasm and assist in the translation of the genetic code at the ribosome. Table 2.5 characterizes the various RNAs in the *E. coli* bacterium. The nucleotide sequences and thereby the structures and functions of the various rRNAs and tRNAs found in the cell are, like mRNA sequences, dictated by nucleotide sequences of the corresponding genes in the cell's DNA.

The end result of this intricate information transmitting and translating system is a protein molecule. Proteins are the tangible biochemical expression of the information and instructions carried by DNA. In the terminology of classical process control, they are the final control elements which implement the DNA controller messages to the cellular process. The dynamics and properties of these controllers are considered next and in several later chapters.

2.4 AMINO ACIDS INTO PROTEINS

Proteins are the most abundant organic molecules within the cell: typically 30 to 70 percent of the cell's dry weight is protein. All proteins contain the four most prevalent biological elements: carbon, hydrogen, nitrogen, and oxygen. Average weight percentages of these elements in proteins are C (50 percent), H (7 percent), O (23 percent), and N (16 percent). In addition, sulfur (up to 3 percent) contributes to the three-dimensional stabilization of almost all proteins by the formation of disulfide (S—S) bonds between sulfur atoms at different locations along the polymer chain. The size of these nonrepetitive biopolymers varies considerably, with molecular weights ranging from 6000 to over 1 million. The two major types of protein configuration, fibrous and globular, are shown in Fig. 2.14.

In view of the great diversity of biological functions which proteins can serve (Table 2.6), the prevalence of proteins within the cell is not surprising. A critical role of many proteins is catalysis. Protein catalysts called *enzymes* determine the rate of chemical reactions occurring within the cell. Enzymes are found in a

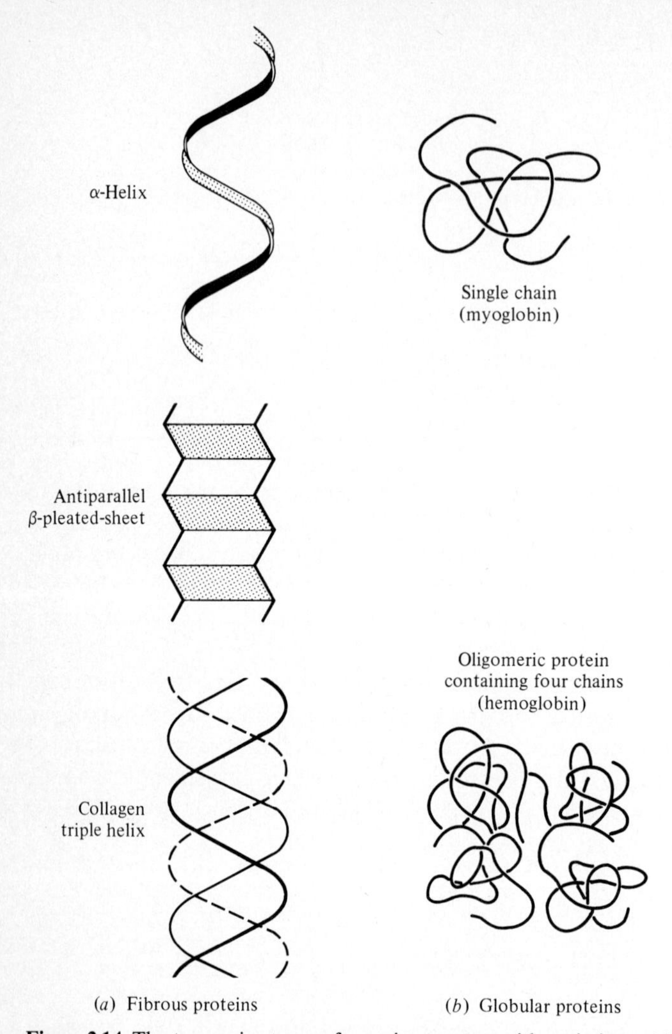

Figure 2.14 The two major types of protein structure with variations: (*a*) fibrous and (*b*) globular.

variety of cell locales: some are suspended in the cytoplasm and thus dispersed throughout the cell interior. Some are localized by attachment to membranes or association with larger molecular assemblies. Other membrane-associated proteins called *permeases* aid in transport of specific nutrients into the cell.

Still different proteins contribute to the structure of cell membranes. Others can serve motive functions. Many single-celled organisms possess small, hairlike appendages called *flagella* whose motion serves to propel the cell; the flagella are driven by contractile proteins. Bacterial protein appendages called *pili* are implicated in the initial attachment of pathogenic bacteria to susceptible tissues.

Proteins are isolated, purified, and characterized by several different types of physical and chemical techniques. Protein separation methods (Chap. 11) are

Table 2.6 The diverse biological functions of proteins

Protein	Occurrence or function
Enzymes (biological catalysts):	
Glucose isomerase	Isomerizes glucose to fructose
Trypsin	Hydrolyzes some peptides
Alcohol dehydrogenase	Oxidizes alcohols to aldehydes
RNA polymerase	Catalyzes RNA synthesis
Regulatory proteins:	
lac repressor	Controls RNA synthesis
Catabolite activator protein	Catabolite repression of RNA synthesis
Interferons	Induce virus resistance
Insulin	Regulates glucose metabolism
Bovine growth hormone	Stimulates growth and lactation
Transport proteins:	
Lactose permease	Transports lactose across cell membrane
Myoglobin	Transports O_2 in muscle
Hemoglobin	Transports O_2 in blood
Protective proteins in vertebrate blood:	
Antibodies	Form complexes with foreign molecules
Thrombin	Component of blood clotting mechanism
Toxins:	
Bacillus thuringiensis toxin	Kills insects
E. coli ST toxin	Causes disease in pigs
Clostridium botulinum toxin	Causes bacterial food poisoning
Storage proteins:	
Ovalbumin	Egg-white protein
Casein	Milk protein
Zein	Seed protein of corn
Contractile proteins:	
Dynein	Cilia and flagella
Structural proteins:	
Collagen	Cartilage, tendons
Gylcoproteins	Cell walls and coats
Elastin	Ligaments

based on differences in molecular properties, which in turn are partly due to the characteristics of the constituent amino acids, our next topic.

2.4.1 Amino Acid Building Blocks and Polypeptides

The monomeric building blocks of polypeptides and proteins are the α-amino acids, whose general structure is

$$\begin{array}{c} \text{H} \\ | \\ \text{H}_2\text{N}-\text{C}-\text{COOH} \\ | \\ \text{R} \end{array}$$

Thus, amino acids are differentiated according to the R group attached to the α carbon (the one closest to the —COOH group). Since there are generally four different groups attached to this carbon (the exception being glycine, for which R = H), it is asymmetrical.

Many organic compounds of biological importance, sugars and amino acids included, are optically active: they possess at least one asymmetric carbon and therefore occur in two isomeric forms, as shown below for the amino acids:

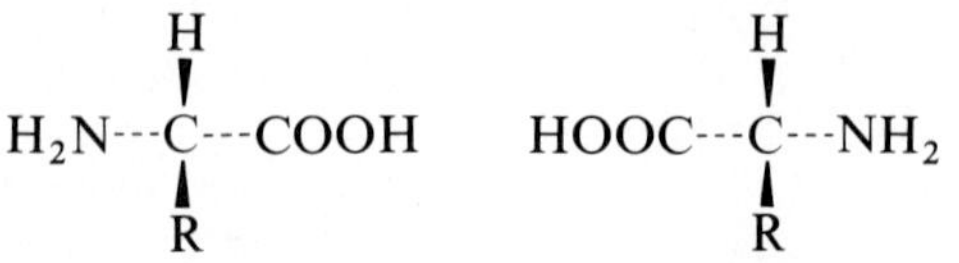

$$\begin{array}{ccc} \text{H} & & \text{H} \\ \text{H}_2\text{N}\cdots\text{C}\cdots\text{COOH} & \qquad & \text{HOOC}\cdots\text{C}\cdots\text{NH}_2 \\ \text{R} & & \text{R} \\ \text{L form} & & \text{D form} \end{array}$$

Solutions of a pure isomer will rotate plane polarized light either to the right (dextro- or *d*-) or left (levo- or *l*-). This isomerism is extraordinarily important since viable organisms can usually utilize only one form, lacking the isomerization catalysts needed to interconvert the two. The enzyme catalysts themselves are typically capable of catalyzing reactions involving one of the two forms only. This property has been used commercially to resolve a racemic mixture of acylamino acids, converting one isomer only, so that the final solution consists of two considerably different and hence more easily separable components. Additional details on this process follow in Chap. 4. As direct physical resolution of optical isomers is expensive and slow, such microbial and enzymatic (as well as other chemical) aids for resolution may become more important.

It is intriguing to note that only the L-amino acid isomers are found in most proteins. Appearance of D-amino acids in nature is rare: they are found in the cell walls of some microorganisms and also in some antibiotics.

The acid (—COOH) and base (—NH_2) groups of amino acids can ionize in aqueous solution. The amino acid is positively charged (cation) at low pH and negatively charged (anion) at high pH. At an intermediate pH value, the amino acid acts as a dipolar ion, or zwitterion, which has no net charge. This pH is the *isoelectric point*, which varies according to the R group involved (Table 2.7). An amino acid at its isoelectric point will not migrate under the influence of an applied electric field; it also exhibits a minimum in its solubility. These properties of amino acids allow utilization of such separation techniques as ion exchange, electrodialysis, and electrophoresis for mixture resolution (Chap. 11).

Illustrated in Fig. 2.15 are the 20 amino acids commonly found in proteins. In addition to the carboxyl and amine groups possessed by all amino acids (except proline), the R groups of some acids can ionize. Several of the acids possess nonpolar R groups which are hydrophobic; other R groups are hydrophilic in character. The nature of these side chains is important in determining both the ultimate role played by the protein and the structure of the protein itself, as discussed in the next section.

Simple proteins are polymers formed by condensation of amino acids. In protein formation, the condensation reaction occurs between the amino group of one amino acid and the carboxyl group of another, forming a *peptide bond*:

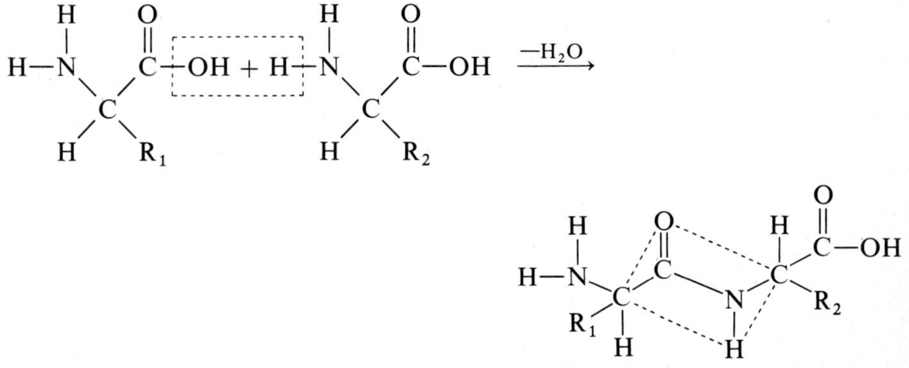

```
  H     O              H     O
  |     ||   ┌──────────┐ |     ||
H—N     C—┼OH + H—┼N     C—OH   —H2O→
   \   /  └──────────┘   \   /
     C                     C
   /   \                 /   \
  H     R1              H     R2

                                                 O
                          H        O           H ||
                          |        ||          | C—OH
                        H—N        C           C
                            \    /   \      /    \
                              C         N          R2
                           R1   |       |
                                H       H
```

$$\xrightarrow{-H_2O}$$

Since the peptide bond has some double-bond character, the six atoms in the dashed-box region lie in a plane. Note that every amino acid links with the next via a peptide bond, thus suggesting that a single catalyst (enzyme) can join all the subunits provided that some other mechanism orders the subunits in advance of peptide-bond formation (Chap. 6).

The amino acid fragment remaining after peptide-bond formation is the *residue*, which is designated with a *-yl* ending, e.g., glycine, glycyl; alanine, alanyl. The list of a sequence of residues in an oligopeptide is begun at the terminus containing the free amino group.

Polypeptides are short condensation chains of amino acids (Fig. 2.16). A little reflection suggests that as the length of the chain increases, the physicochemical characteristics of the polymer will be increasingly dominated by the R groups of the residues while the amino and carboxyl groups on the ends will shrink in importance. By convention, the term polypeptide is reserved for these relatively short chains. These molecules have considerable biological significance. Many hormones such as insulin, growth hormone, and somatostatin are polypeptides.

Larger chains are called *proteins*, with the diffuse dividing line between these two classifications ranging from 50 to 100 amino acid residues. Since the average molecular weight of a residue is about 120, proteins have molecular weights larger than about 10,000, and some protein molecular weights exceed 1 million.

Determination of the amino acid content of a particular protein or a protein mixture can be accomplished in an automated analyzer. Total hydrolysis of the protein can be achieved by heating it in 6 N HCl for 10 to 24 h at 100 to 120°C. All the amino acids with the exceptions of tryptophan, glutamic acid, and aspartic acid are recovered intact in the hydrochloric form. These can then be separated and measured. Alkali hydrolysis can be used to estimate the tryptophan content. The result of such experiments on the proteins from *E. coli*, an intestinal bacterium, are listed in Table 2.8. From these and other data, it has been found that *not* all 20 amino acids in Fig. 2.15 are found in all proteins. Amino acids do not occur in equimolar amounts in any known protein. For a

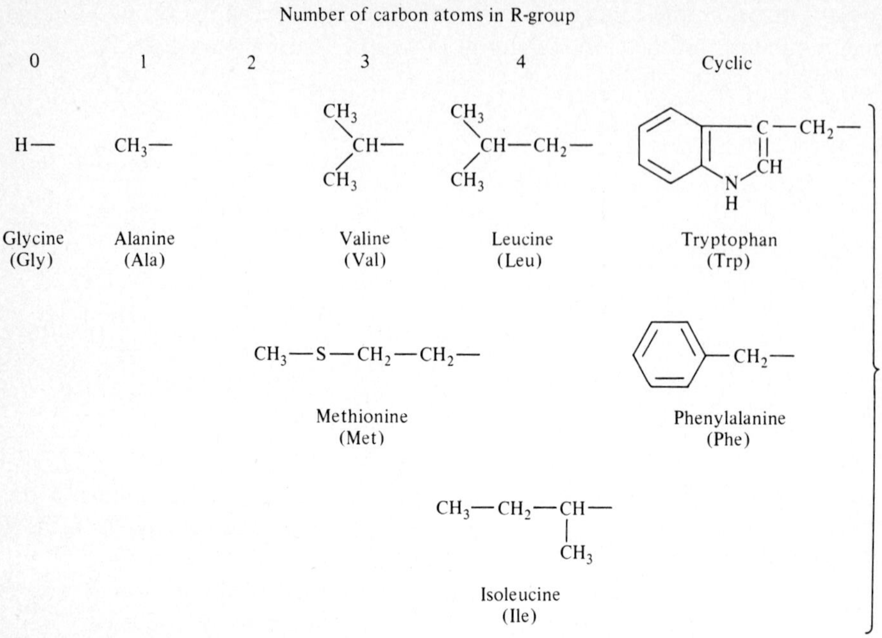

Number of carbon atoms in R-group

0 1 2 3 4 Cyclic

H— Glycine (Gly)

CH_3— Alanine (Ala)

$(CH_3)_2CH$— Valine (Val)

$(CH_3)_2CH—CH_2$— Leucine (Leu)

C—CH_2— (indole ring, CH, N, H) Tryptophan (Trp)

$CH_3—S—CH_2—CH_2$— Methionine (Met)

(benzene ring)—CH_2— Phenylalanine (Phe)

$CH_3—CH_2—CH(CH_3)$— Isoleucine (Ile)

Hydrophobic

Figure 2.15 The 20 amino acids found in proteins.

Table 2.7 pK (= $-\log K$) values of terminal amino, terminal carboxyl, and R groups for several amino acids[†]

Amino acid	pK_1 α-COOH	pK_2 α-NH_3	pK_R R group
Glycine	2.34	9.6	
Alanine	2.34	9.69	
Leucine	2.36	9.60	
Serine	2.21	9.15	
Threonine	2.63	10.43	
Glutamine	2.17	9.13	
Aspartic acid	2.09	9.82	3.86
Glutamic acid	2.19	9.67	4.25
Histidine	1.82	9.17	6.0
Cysteine	1.71	10.78	8.33
Tyrosine	2.20	9.11	10.07
Lysine	2.18	8.95	10.53
Arginine	2.17	9.04	12.48

[†] From A. L. Lehninger, *Biochemistry*, 2d ed., table 4-2. Worth Publishers, Inc., New York, 1975.

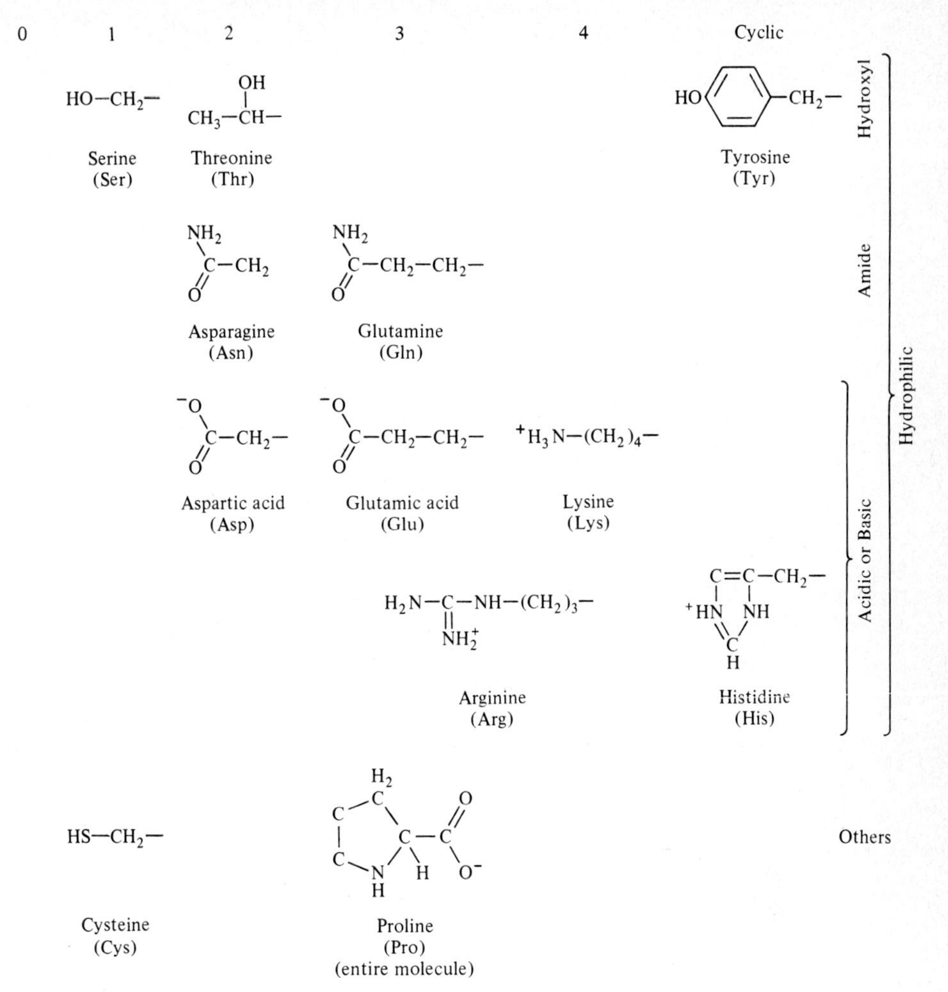

Figure 2.15 (*continued*)

given protein, however, the relative amounts of the various amino acids are fixed.

Amino acids are not the only constituents of all proteins. Many proteins contain other organic or even inorganic components. The other part of these *conjugated proteins* is a *prosthetic group*. If only amino acids are present, the term *simple protein*, already introduced above, is sometimes used. *Hemoglobin*, the oxygen-carrying molecule in red blood cells, is a familiar example of a conjugated protein: it has four heme groups, which are organometallic complexes containing iron. A related smaller molecule, *myoglobin*, has one heme group. Each molecule of the enzyme L-amino acid oxidase, which catalyzes deamination of a number of L-amino acids, contains two FAD (Fig. 2.9) units. As discussed above, the prosthetic portion of ribosomes is RNA.

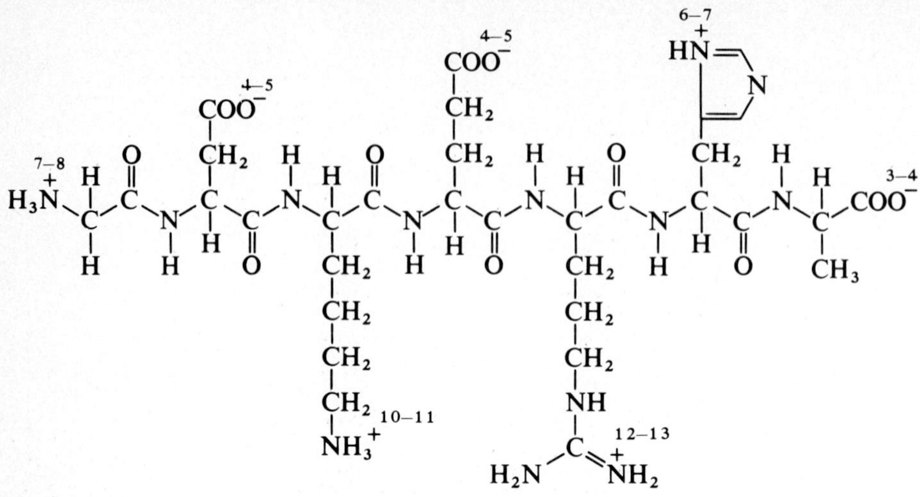

Gly-Asp-Lys-Glu-Arg-His-Ala

Figure 2.16 A hypothetical polypeptide, showing the ionizing groups found among the amino acids. Notice that the listing sequence for the amino acid residues starts at the N-terminal end. Numbers near each ionizing group denote the corresponding p*K* value.

2.4.2 Protein Structure

The many specific tasks served by proteins (Table 2.6) are in large part due to the variety of forms which proteins may take. The structure of many proteins is conveniently described in three levels. A fourth structural level must be considered if the protein consists of more than a single chain. As seen from Table 2.9, each level of structure is determined by different factors, hence the great range of protein structures and functions.

Table 2.8 Proportions of amino acids contained in the proteins of *E. coli*†

Amino acid	Relative frequency (Ala = 100)	Amino acid	Relative frequency (Ala = 100)
Ala	100	Thr	35
Glx(Gln + Glu)	83	Pro	35
Asx(Asn + Asp)	76	Ile	34
Leu	60	Met	29
Gly	60	Phe	25
Lys	54	Tyr	17
Ser	46	Cys	14
Val	46	Trp	8
Arg	41	His	5

† From A. L. Lehninger, *Biochemistry*, 2d ed., table 5.3, Worth Publishers, Inc., New York, 1975.

Table 2.9 Protein structure

1. Primary	Amino acid sequence, joined through peptide bonds
2. Secondary	Manner of extension of polymer chain, due largely to hydrogen bonding between residues not widely separated along chain
3. Tertiary	Folding, bending of polymer chain, induced by hydrogen, salt, and covalent disulfide bonds as well as hydrophobic and hydrophilic interactions
4. Quaternary	How different polypeptide chains fit together; structure stabilized by same forces as tertiary structure

2.4.3 Primary Structure

The *primary structure* of a protein is its sequence of amino acid residues. The first protein structure completely determined was that of insulin (Sanger and coworkers in 1955). It is now generally recognized that every protein has not only a definite amino acid content but also a *unique sequence* in which the residues are connected. As we shall explore further in Sec. 6.1, this sequence is prescribed by a sequence of nucleotides in DNA.

While a detailed discussion of amino acid sequencing would take us too far from our major objective, we should indicate the general concept employed. By a variety of techniques, the protein is broken into different polypeptide fragments. For example, enzymes are available which break bonds between specific residues within the protein. The resulting relatively short polypeptide chains can be sequenced using reactions developed by Sanger and others. The fragmental sequences, some of which partially overlap, are then shuffled and sorted until a consistent overall pattern is obtained and the complete amino acid sequence deciphered. Automated spinning cup analyzers can determine an amino acid sequence up to 100 residues in length; about two hours are required per amino acid. Recent gas phase protein sequencing technology has greatly reduced the amount of protein sample required to obtain sequence information. For example, to obtain the complete amino acid sequence for a 50 residue polypeptide, 50–100 picomoles are required for the gas-phase sequencer, while 1–5 nanomoles of the protein would be needed for spinning-cup analysis. Also, as discussed further in Chap. 6, protein sequences may now be deduced from experimentally determined nucleotide sequences of the corresponding gene.

The sequence of amino acid residues in many polypeptides and proteins is now known. So far no repeating pattern of residues has been found, although the available evidence suggests some patterns may exist in fibrous proteins. As an example, Fig. 2.17 gives the amino acid sequence of an enzyme called lysozyme found in egg white. As mentioned previously, the sequence is conventionally numbered starting with the amino-terminal residue, here lysine, and proceeding to the carboxyl terminus, occupied in this enzyme by leucine.

The various side chains of the amino acid residues interact with each other and with the immediate protein environment to determine the geometrical configuration of the protein molecule. The folding of the polypeptide chain into a precise three-dimensional structure often yields a complex, convoluted form as

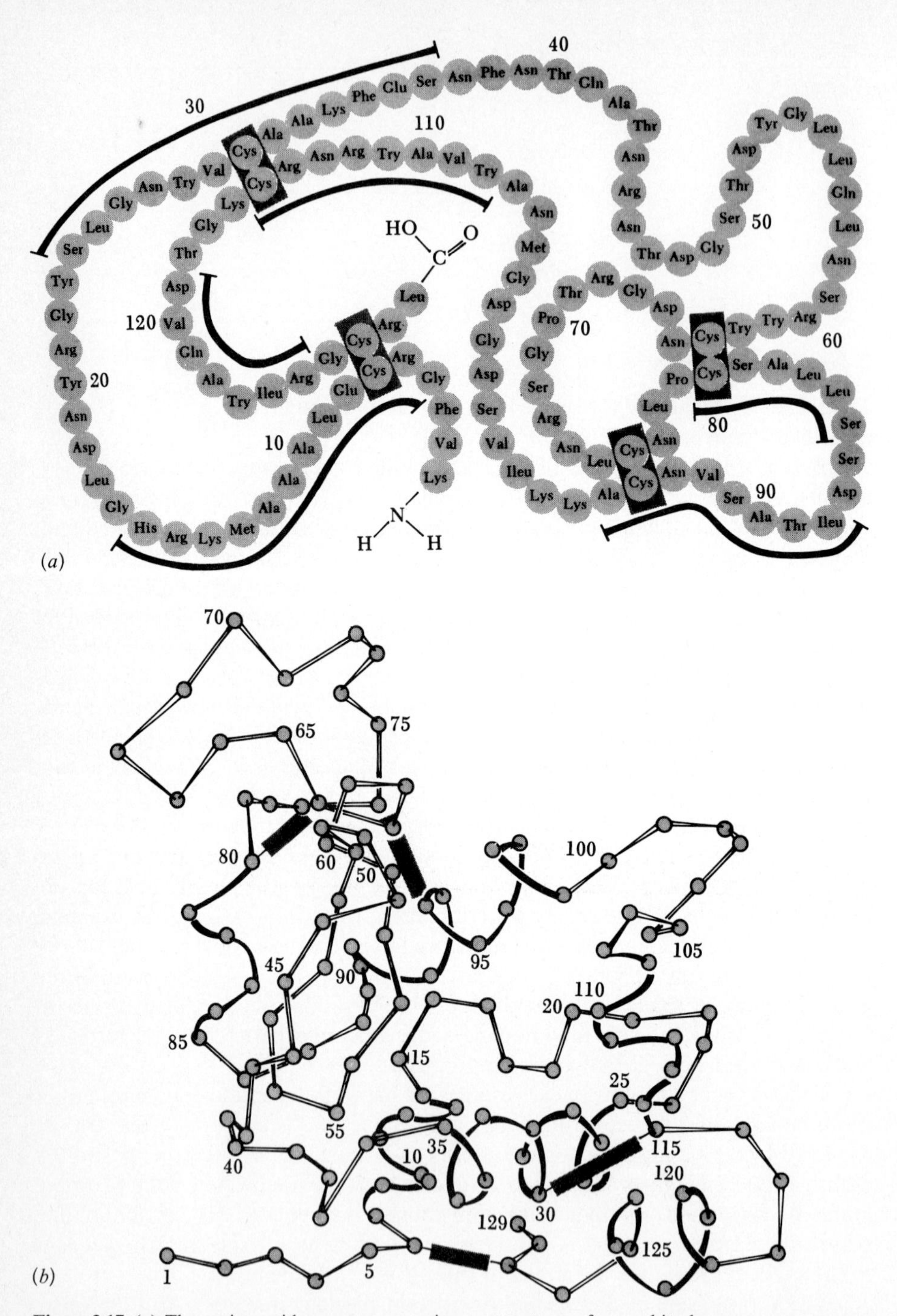

Figure 2.17 (*a*) The amino acid sequence, or primary structure, of egg white lysozyme, an enzyme. (*b*) The corresponding three-dimensional molecular structure (tertiary structure) of crystalline lysozyme. *(By permission of C. C. F. Blake and the editor of Nature. From "Structure of Hen Egg-White Lysozyme," Nature, vol. 206, p. 757, 1965.)*

seen in Fig. 2.17*b* for the lysozyme molecule. This elaborate geometrical structure is dictated primarily by the protein's amino acid sequence.

When this fact is coupled with knowledge of the cell's coding system for the amino acid sequence (Sec. 6.1), the significance of primary protein structure emerges. It is the link between the central DNA controller of the cell and the elaborate, highly specific protein molecules which mediate and regulate the diverse biochemical activities essential for growth and survival.

2.4.4 Three-Dimensional Conformation: Secondary and Tertiary Structure

Secondary structure of proteins refers to the structural configuration of the amino acid residues which are close neighbors in the linear amino acid sequence. We recall that the partial double-bond character of the peptide bond holds its neighbors in a plane. Consequently, rotation is possible only about two of every three bonds along the protein backbone (Fig. 2.18). This restricts the possible shapes which a short segment of the chain can assume.

As already sketched in Fig. 2.14, there are two general forms of secondary structure: helices and sheets. The configuration believed to occur in hair, wool, and some other fibrous proteins arises because of hydrogen bonding between atoms in closely neighboring residues. If the protein chain is coiled, hydrogen bonding can occur between the —C=O group of one residue and the —NH group of its neighbor four units down the chain. A configuration called the *α-helix* is the only one which allows this hydrogen bonding and also involves no deformation of bonds along the molecular backbone. Collagen, the most abundant protein in higher animals, is thought to contain three such α-helices, which are intertwined in a *superhelix*. Rigid and relatively stretch-resistant, collagen is a biological cable. Found in skin, tendons, the cornea of the eye, and numerous other parts of the body, collagen is important in enzyme and cell immobilization and in formulation of biocompatible materials.

Because of the relative weakness of individual hydrogen bonds, the α-helical structure is easily disturbed. Proteins can lose this structure in aqueous solution because of competition of water molecules for hydrogen bonds. If wool is

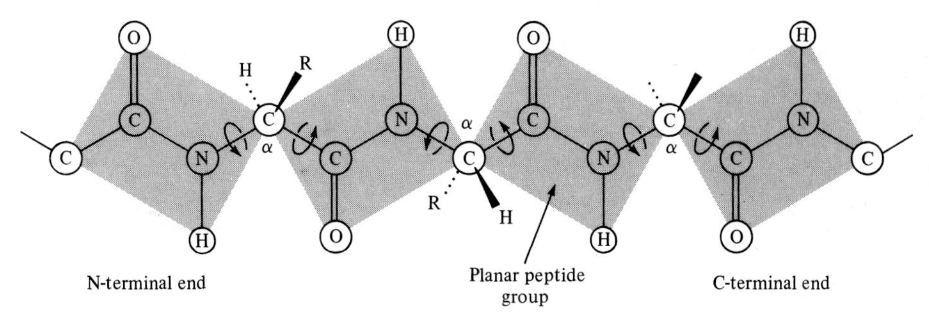

Figure 2.18 The planar nature of the peptide bond restricts rotation in the peptide chain. *(From A. L. Lehninger, "Biochemistry," 2d ed., p. 128, Worth Publishers, New York, 1975.)*

steamed and stretched, it assumes a different, *pleated-sheet structure.* This arrangement is the naturally stable one in silk fibers. The pleated-sheet configuration is also stabilized by hydrogen bonds which exist between neighboring parallel chains. While the parallel sheets are flexible, they are very resistant to stretching.

Achievement of the secondary structure of fibrous proteins is of importance in the food industry. While plant products such as soybeans are valuable sources of essential amino acids, they lack the consistency and texture of meat. Consequently, in the preparation of "textured protein," globular proteins in solution are spun into a more extended, fibrous structure.

The complete three-dimensional configurations, or *tertiary structures*, of a few proteins have been determined by painstaking and laborious experiments. After crystals of pure protein have been prepared, x-ray crystallography is employed. Each atom is a scattering center for x-rays; the crystal diffraction pattern contains information on structural details down to 2 Å resolution. In order to obtain this detail, however, it was necessary in the case of myoglobin (Fig. 2.19) to calculate 10,000 Fourier transforms, a task which had to wait for the advent of the high-speed digital computer.

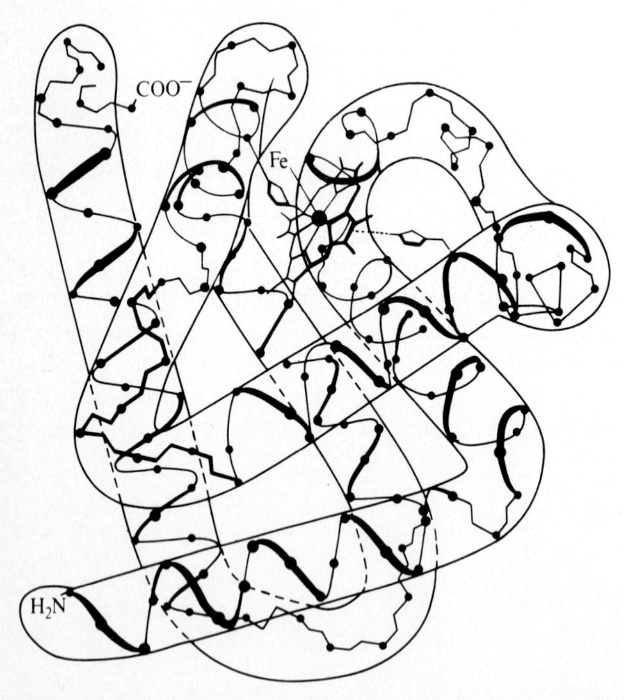

Figure 2.19 The three-dimensional structure of myoglobin. Dots indicate the α-carbons of the 121 amino acid residues in this oxygen-carrying protein. A single heme group, with its central iron atom larger and labeled, is shown in the upper central portion of the molecule. *(Reprinted from N. A. Edwards and K. A. Hassall, "Cellular Biochemistry & Physiology," p. 57, McGraw-Hill Publishing Company Ltd., London, 1971.)*

As the diagrams in Figs. 2.17*b* and 2.19 illustrate, protein tertiary structure, especially in these globular proteins, is exquisitely complex. Subdomains of helical secondary structure are clearly evident in these examples, and the lysozyme molecule also contains regions of sheet secondary structure. How are these convoluted protein architectures stabilized and what purpose is served by a class of molecules with such widely diverse, elegant shapes?

Interactions between R groups widely separated along the protein chains determine how the chain is folded or bent to form the compact configuration typical of globular proteins. Several weak interactions, including ionic effects, hydrogen bonds, and hydrophobic interactions between nonpolar R groups (Fig. 2.20), are prime determinants of protein three-dimensional structure. Like micelle formation by lipids, examined earlier, many globular proteins have their hydrophobic residues concentrated in the molecule's interior with relatively hydrophilic groups on the outside of the molecule. This configuration, sometimes called the *oil-drop model* of protein structure, is presumably the most stable in the protein's natural aqueous environment. (What differences in structure would you expect for a protein embedded in a cell membrane?)

It is *extremely* important at many points of our journey into biochemical engineering to realize that these weak interactions are easily disrupted by changes of several different kinds in the environment of the protein. In biological systems, hydrogen bond energies range from 3 to 7 kcal/mol, and 5 kcal/mol is typical of ionic bond energies. Modest heating can disrupt several of these stabilizing interactions. Changes in pH, ionic strength, physical forces, and addition of

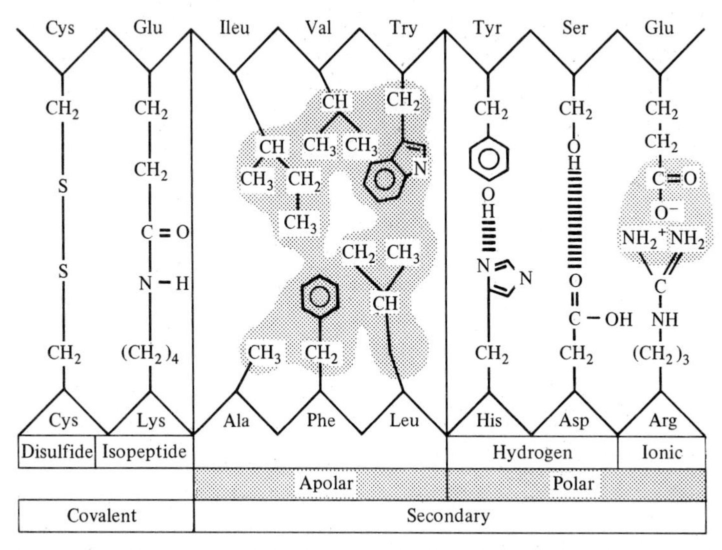

Figure 2.20 A summary of the bonds and interactions which stabilize protein molecular structure. *(From "Cell Structure and Function," p. 210 by Ariel G. Loewy and Philip Siekevitz, Copyright © 1963, 1969 by Holt, Rinehart, and Winston, Inc. Reprinted by permission of Holt, Rinehart, and Winston.)*

organics are other obvious candidates for disturbance of the natural tertiary structure of a protein.

Also important in protein conformational architecture are covalent bonds which often exist between two cysteinyl residues. By eliminating two hydrogen atoms, a disulfide bond is formed between two —SH groups:

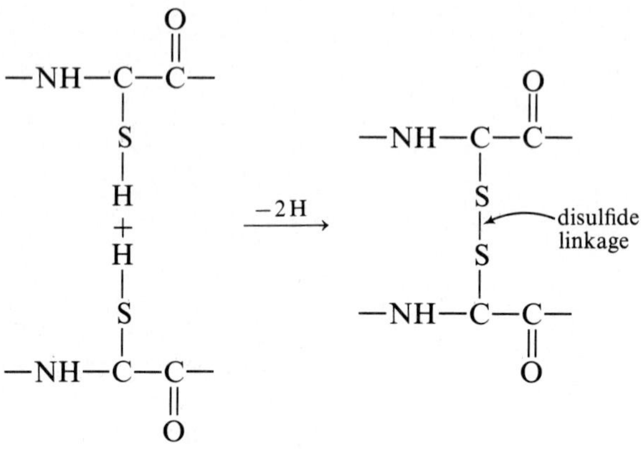

```
            O
            ‖
   —NH—C—C—                         O
       |                            ‖
       S                   —NH—C—C—
       |                       |
       H         −2H           S
       +       ———→            |←——disulfide
       H                       S    linkage
       |                       |
       S                   —NH—C—C—
       |                           ‖
   —NH—C—C—                        O
           ‖
           O
```

Such bonds are found as cross-links within a polypeptide chain, and they sometimes hold two otherwise separate chains: insulin is actually two subchains of 21 and 30 residues which are linked by disulfide bonds (Fig. 2.21). There is also an internal cross-link in the shorter subchain. These covalent bonds are more resistant to thermal disruption than the weak interactions mentioned above. However, disulfide bonds can be reduced by an excess of a sulfhydryl compound such as β-mercaptoethanol.

Protein conformation is a major determinant of protein biological activity. There is considerable evidence that in many cases a definite three-dimensional shape is essential for the protein to work properly. This principle is embodied, for example, in the *lock-and-key model* for enzyme specificity. This simple model suggests why the proteins which catalyze biochemical reactions are often highly specific. That is, only definite reactants, or *substrates*, are converted by a particular enzyme. The lock-and-key model (Fig. 2.22) holds that the enzyme has a specific site (the "lock") which is a geometrical complement of the substrate (the "key") and that only substrates with the proper complementary shape can bind to the enzyme so that catalysis occurs. The available information on tertiary enzyme structure has not only confirmed this hypothesis but also has provided insights into the catalytic action of enzymes, a topic we consider further in the next chapter. The direct connection between protein geometry and highly specific interactions with other reagents is also well established for permeases, hormones, antibodies, and other proteins.

When protein is exposed to conditions sufficiently different from its normal biological environment, a structural change, again called *denaturation*, typically leaves the protein unable to serve its normal function (Fig. 2.23). Relatively small changes in solution temperature or pH, for example, may cause denaturation,

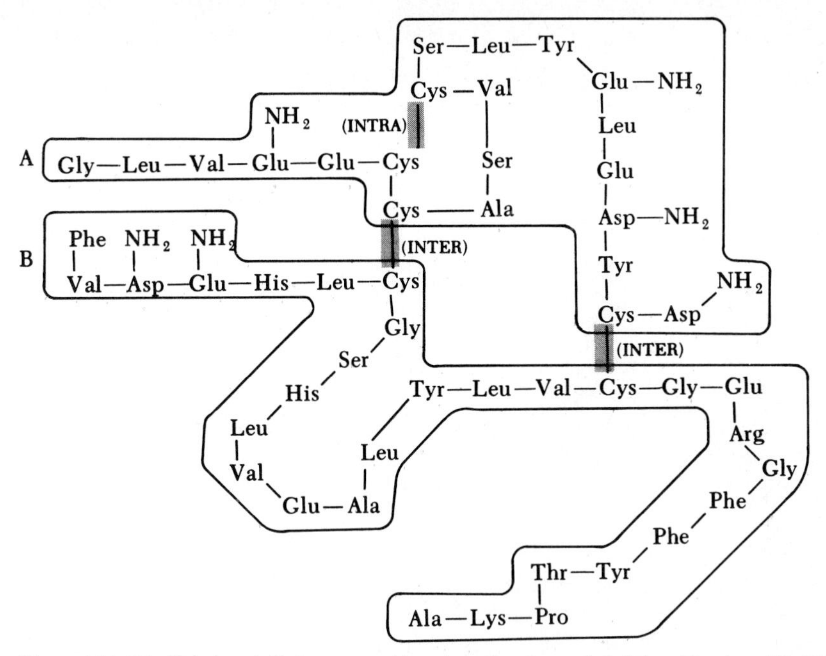

Figure 2.21 Disulfide bonds link two peptide subchains, denoted A (21 residues) and B (30 residues), in the beef (shown here) and other species' insulin molecules. Note also the intrapeptide disulfide linkage between two residues of the shorter subchain. *(Reprinted from F. J. Reithel, "Concepts in Biochemistry, p. 237, McGraw-Hill Book Company, New York, 1967.)*

which usually occurs without severing covalent bonds. If a heated dilute solution of denatured protein is slowly cooled back to its normal biological temperature, the reverse process or, *renaturation*, with restoration of protein function, often occurs. As with other transitions, higher temperatures favor the state of greater entropy (disorder) hence extendedness; cooling favors the greatest number of favorable interactions leading to compactness and renaturation.

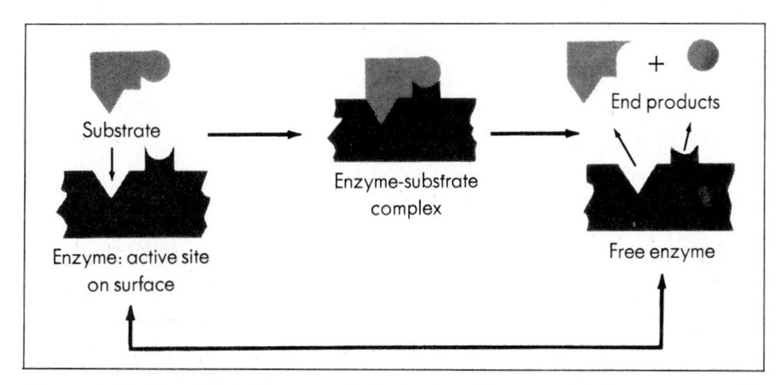

Figure 2.22 Simplified diagram of the lock-and-key model for enzyme action. Here the shape of the catalytically active site (the lock) is complementary to the shape of the substrate key. *(Reprinted from M. J. Pelczar, Jr. and R. D. Reid, "Microbiology," 3d ed., p. 158, McGraw-Hill Book Company, New York, 1972.)*

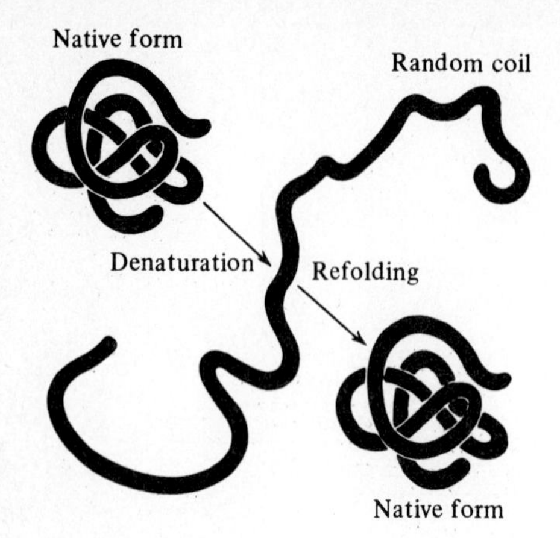

Figure 2.23 After a protein has been denatured by exposure to an adverse environment, it will often return to the native, biologically active conformation following restoration of suitable conditions. This result indicates that primary protein structure determines secondary and tertiary structure. *(From A. L. Lehninger, "Biochemistry," 2d ed., p. 62, Worth Publishers, New York, 1975.)*

The configuration assumed by a protein, and thus which determines its properties, is the one which minimizes the molecule's free energy. While we cannot yet utilize this concept to predict protein structure and function knowing only its amino acid sequence, this perspective is helpful in thinking about how proteins work under process conditions (which often differ significantly from the protein's native environment) or when their amino acid sequence is altered. Equipped with this conceptual outlook, we will not be surprised to find, for example, that an enzyme attached to a solid surface has less catalytic activity than it does in solution.

After placing so much emphasis on the connection between protein structure and function, we should remember that most of the forces which stabilize protein structure are weak, allowing the molecule substantial opportunities to "flex" about a characteristic average geometry. Just as the biological function of DNA depends on the separability of its two strands under isothermal, intracellular conditions, the ability to flex slightly has been implicated in natural function of several proteins. From a process perspective, the relative ease of thermally or chemically inducing denaturation of many proteins reminds us that the biochemical engineer may operate enzyme and cellular processes only over a fairly narrow range of, for example, pH, temperature, and ionic strength. This knowledge is also useful in other subjects such as protein recovery (Chap. 11) and sterilizer design (Chaps. 7 and 9).

2.4.5 Quaternary Structure and Biological Regulation

Proteins may consist of more than one polypeptide chain, hemoglobin being perhaps the best known example. How these chains fit together in the molecule is the *quaternary structure.* As Table 2.10 indicates, a great many proteins, especially enzymes (usually indicated by the *-ase* suffix), are oligomeric and consequently possess quaternary structure. The forces stabilizing quaternary structure are believed to be the same as those which produce tertiary structure. While disulfide bonds are sometimes present, as in insulin, most of the proteins in

Table 2.10 Characteristics of several oligomeric proteins†

Protein	Molecular weight	No. of chains	No. of —S—S—bonds
Insulin	5,798	1 + 1	3
Ribonuclease	13,683	1	4
Lysozyme	14,400	1	5
Myoglobin	17,000	1	0
Papain	20,900	1	3
Trypsin	23,800	1	6
Chymotrypsin	24,500	3	5
Carboxypeptidase	34,300	1	0
Hexokinase	45,000	2	0
Taka-amylase	52,000	1	4
Bovine serum albumin	66,500	1	17
Yeast enolase	67,000	2	0
Hemoglobin	68,000	2 + 2	0
Liver alcohol dehydrogenase	78,000	2	0
Alkaline phosphatase	80,000	2	4
Hemerythrin	107,000	8	0
Glyceraldehyde-phosphate dehydrogenase	140,000	4	0
Lactic dehydrogenase	140,000	4	0
γ-Globulin	140,000	2 + 2	25
Yeast alcohol dehydrogenase	150,000	4	0
Tryptophan synthetase	159,000	2 + 2	
Aldolase	160,000	4(?)	0
Phosphorylase *b*	185,000	2	
Threonine deaminase (*Salmonella*)	194,000	4	
Fumarase	200,000	4	0
Tryptophanase	220,000	8	4
Formyltetrahydrofolate synthetase	230,000	4	
Aspartate transcarbamylase	310,000	4 + 4	0
Glutamic dehydrogenase	316,000	6	0
Fibrinogen	330,000	2 + 2 + 2	
Phosphorylase *a*	370,000	4	
Myosin	500,000	2 + 3	0
β-Galactosidase	540,000	4	
Ribulose diphosphate carboxylase	557,000	24	

† From *Cell Structure and Function*, 2d ed., p. 221, by Ariel G. Loewy and Philip Siekevitz. Copyright © 1963, 1969 by Holt, Rinehart, and Winston. Reprinted by permission of Holt, Rinehart, and Winston.

Table 2.10 are held together by relatively weak interactions. Many oligomeric proteins are known to be *self-condensing*; e.g., separated hemoglobin α chains and β chains in solution rapidly reunite to form intact hemoglobin molecules. This feature is of great significance, for it indicates that at least in some cases the one-dimensional biochemical coding from DNA which specifies primary protein structure ultimately determines all higher structural levels and hence the specific biological role of the protein.

Evidence to date suggests that the subunit makeup of some protein molecules is especially suited for at least two important biological functions: control of the catalytic activity of enzymes, and flexibility in construction of related but different molecules from the same collection of subunits. Different proteins known as *isozymes* or *isoenzymes* demonstrate the latter point. Isozymes are different molecular forms of an enzyme which catalyze the same reaction within the same species. While the existence of isozymes might seem a wasteful duplication of effort, the availability of parallel but different catalytic processes provides an essential ingredient in some biochemical control systems (Chaps. 3 and 5). Several isozymes are known to be such oligomeric proteins; in one instance isoenzymes are composed of five subunits, only two of which are distinct. This might be viewed as an economical device since five different proteins may be constructed from the two polypeptide components.

2.5 HYBRID BIOCHEMICALS

Many biologically and commercially important biochemicals do not fit clearly into any of the chemical classes described above. These diverse compounds are hybrids which contain various combinations of lipid, sugar, and amino acid building blocks. As indicated in Table 2.11, hybrid biochemicals provide several different significant functions. In the next subsection, we consider in more detail the chemistry and organization of the outer surfaces of different types of cells. Here several hybrid compounds as well as new polysugars and proteins are encountered.

Table 2.11 Some examples of the nomenclature, subunit composition, and biological function of hybrid biochemicals

Name	Building blocks	Location, function
Peptidoglycans	Crosslinked disaccharide and peptides	Bacterial cell walls
Proteoglycans	95% carbohydrate, 5% protein	Connective tissue
Glycoproteins	1–30% carbohydrate, remainder protein	Diverse; includes antibodies, eucaryotic cell outer surface components
Glycolipids	Lipids which contain from one to seven sugar residues	Membrane components
Lipoprotein	1–50% protein, remainder lipid in plasma lipoproteins	Membrane, blood plasma components
Lipopolysaccharides	Lipid plus a highly variable oligosaccharide chain	Outer portion of gram-negative bacterial envelope

2.5.1 Cell Envelopes: Peptidoglycan and Lipopolysaccharides

We have already seen in our perusal of lipids that cell membranes play vital roles in regulating component fluxes into and out of the cell interior. Other parts of the outer surfaces of microorganisms and tissue cells are also extremely important. From a natural biological viewpoint, microorganisms, such as bacteria that live in environments much more diverse and much less controlled than cells in liver tissue in an animal, must possess greater physical rigidity and resistance to bursting under large osmotic pressure. Accordingly, bacteria have much different outer envelopes than animal cells.

Looking at cell envelopes from a biochemical process engineering perspective, we find several points of interest. The properties of outer cell surfaces determine the cells' tendency to adhere to each other or to walls of reactors, pipes, and separators. As we shall develop in the following chapters, such phenomena have important influences in formulation of immobilized cell catalysts, operation of continuous microbial reactors, and cell separation from liquid. The chemical and mechanical characteristics of cell envelopes determine the cells' resistance to disruption by physical, enzymatic, and chemical methods, essential in recovery of intracellular products.

As indicated in part by their opposite responses to gram staining, the envelopes of gram-positive and gram-negative cells are very different (Fig. 2.24). In each case, the envelope is comprised of several layers, but their disposition, thicknesses, and composition are quite distinctive. Not shown in these sketches are a variety of proteins embedded in and on the surfaces of the cell membranes; these will be one focus of our discussion of membrane transport in Chap. 5.

While the envelope of a gram-positive bacterium has a single membrane, there are two membranes of this type in gram-negative cells. The outer and cytoplasmic membranes in gram-negative bacteria enclose a region called the *periplasmic space* or *periplasm.* The periplasmic space, which may include from 20 to 40 percent of the total cell mass, contains a number of enzymes and sugar and amino acid binding proteins.

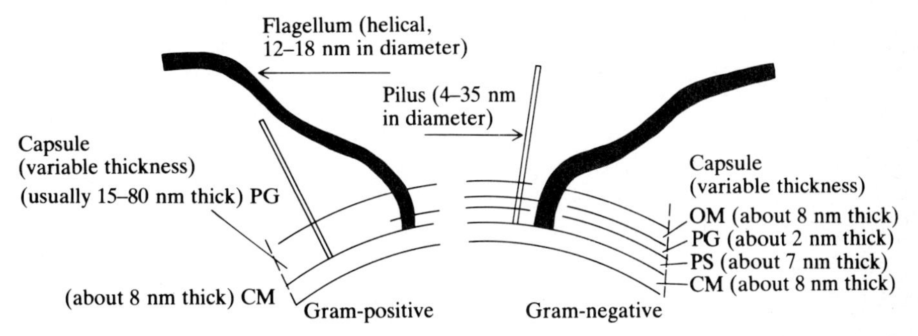

Figure 2.24 Schematic diagrams of the cell envelope structure of gram-positive and gram-negative bacteria (CM cytoplasmic membrane, PG peptidoglycan, PS periplasmic space, OM outer membrane). *(Adapted from R. Y. Stainer, E. A. Adelberg, J. L. Ingraham, and M. L. Wheelis, "Introduction to the Microbial World," p. 128, Prentice-Hall, Englewood Cliffs, N.J., 1979.)*

Adjacent to the outer surface of the cytoplasmic membrane in both types of bacteria is a layer of *peptidoglycan*. The building blocks of peptidoglycan are a β-1,4 linked disaccharide of N-acetylmuramic acid (NAM) and N-acetylglucosamine (NAG), a bridging pentapeptide comprised entirely of glycyl residues, and a tetrapeptide. This last peptide varies from species to species; in *Staphylococcus aureus*, for example, the tetrapeptide component consists of L-alanine, D-glutamine, L-lysine, and D-alanine (Fig. 2.25*a*; note the occurrence here of rare D-amino acids). These units are extensively cross-linked to give an enormous single macromolecule which surrounds the entire cell (Fig. 2.25*b*).

Lysozyme, the enzyme examined in Fig. 2.17, is an effective antibacterial agent because it catalyzes hydrolysis of the β-1,4 glycosidic linkage between the NAM and NAG components of peptidoglycan. This causes breakdown and removal of peptidoglycan from the bacterium which results in cell bursting or *lysis* in natural hypotonic solutions. Living bacteria cells without their peptidoglycan walls can be maintained in the laboratory in isotonic media. Under these conditions, lysozyme treatment yields cells which have lost their normal shapes and become spherical. These forms are called *spheroplasts*. If it is known that the cell wall is completely removed, the denuded cells are designated *protoplasts*.

Lipopolysaccharides (LPS) are important components of the outer portion of the outer membrane of gram-negative bacteria. Some of these compounds are called *endotoxins*: they produce powerful toxic effects in animals. LPS toxicity is one reason why an infection by *E. coli* in the bloodstream is extremely dangerous. Also, stringent removal of cell envelope endotoxins is a major requirement in purifying proteins sythesized using genetically engineered *E. coli*.

Outer membrane LPS molecules have three different regions: (1) a lipid-A component which consists of six unsaturated fatty acid chains, which extend into the membrane, joined to a diglucosamine head, (2) a core oligosaccharide region containing ten sugars, some of them rare, and (3) an O side chain comprised of many repeating tetrasaccharide building blocks. The core region and O side chain are extended outward, away from the cell. Thus, it is the external O side chains which interact with the immune system of an infected animal. As part of their anti-immune system defenses, bacteria can mutate rapidly to change the O chain structure.

This is not yet the end of the story of envelope layers in bacteria. Beyond the outer membrane of many species is a *capsule* or *slime layer* which consists of polysaccharide. In one strain of pneumococci, bacteria that cause pneumonia, the capsule is composed of alternating glucose and glucuronate residues. Mutants without this polysaccharide capsule are not pathogenic. The capsule polysaccharide of some bacteria dissolves in the surrounding media. Production of extracellular polysaccharides is an important commercial process which is complicated considerably by the non-Newtonian flow properties of the medium (see Chap. 8). Slime layers also contribute to bacterial flocculation, an important feature in the activated sludge wastewater treatment process.

Turning now to cell envelopes of yeast cells, the cytoplasmic membrane consists of lipids, protein, and polysaccharide containing mannose. Moving outward

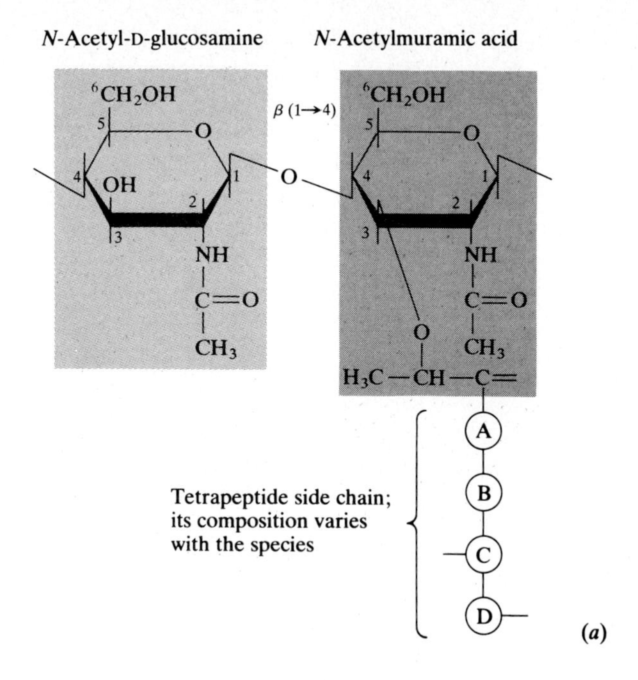

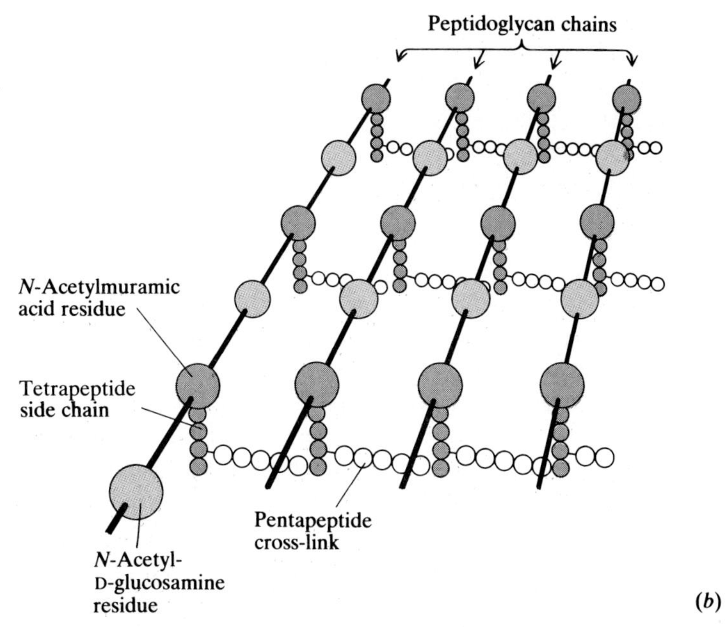

Figure 2.25 (*a*) Structure of the building block for peptidoglycan; (*b*) schematic illustration of highly cross-linked peptidoglycan in the cell wall of *Staphlococcus aureus*, a gram-positive bacterium (*From A. L. Lehninger, "Principles of Biochemistry," p. 293, Worth Publishers, N.Y., 1982.)*

from this membrane through the periplasmic space toward the cell exterior, we find next a cell wall. In baker's yeast, *Saccharomyces cerevisiae*, the wall contains 6 to 8 percent protein including several enzymes, and about 30 percent each, by weight, of *glucan*, D-glucose residues joined with β-1,6 linkages and frequent β-1,3 cross-links, and *mannan*, whose D-mannose building blocks are α-1,6 linked with α-1,2 branches. Protoplasts of *S. cerevisiae* may consequently be prepared using a glucan hydrolyzing enzyme such as a 1,3 glucanase. This is a common step in protocols for introduction of recombinant DNA molecules into yeast hosts.

The cell walls of yeasts and many molds contain *chitin*, a macromolecule comprised of N-acetyl glucosamine building blocks linked with β-1,4 glycosidic bonds.

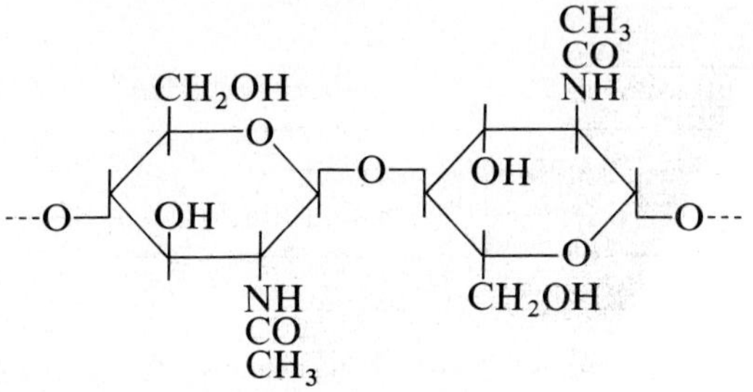

Chitin

Animal cells, because they normally live in a well-controlled isotonic environment, do not possess cell walls. In their plasma membranes, in addition to phospholipids and proteins, we find 2 to 10 percent carbohydrate. These sugar components, which are found on the external membrane surface in all mammalian cells examined to date, are combined with lipids and proteins in the form of *glycolipids* and *glycoproteins*, respectively.

2.5.2 Antibodies and Other Glycoproteins

Proteins with covalently coupled monosaccharides or short oligosaccharide chains, the *glycoproteins*, are found in considerable variety in and around eucaryotic cells. Several enzymes such as glucose oxidase produced by the mold *Aspergillus niger* are glycoproteins. Collagen, the biological coaxial cable mentioned earlier, consists of glycosylated protein. Some interferons, potent antiviral compounds, are glycoproteins. In fact, most eucaryotic proteins which are in contact with or are secreted into the cell environment are glycosylated. Several glycoproteins are currently or likely soon will be important commercial products. We shall examine their biosynthesis and transport properties in Chaps. 5 and 6.

Antibodies, primary agents in immunological defenses of vertebrates, are glycoproteins (Fig. 2.26). Cell fusion techniques discussed in Chap. 6 now enable production of homogeneous antibody molecules in animals or in bioreactors. This important technology signals greater future commercial importance of antibodies for diagnosis, drug delivery, and bioseparations. Accordingly, it is appropriate to examine briefly the origin, structure, and function of antibodies.

(a)

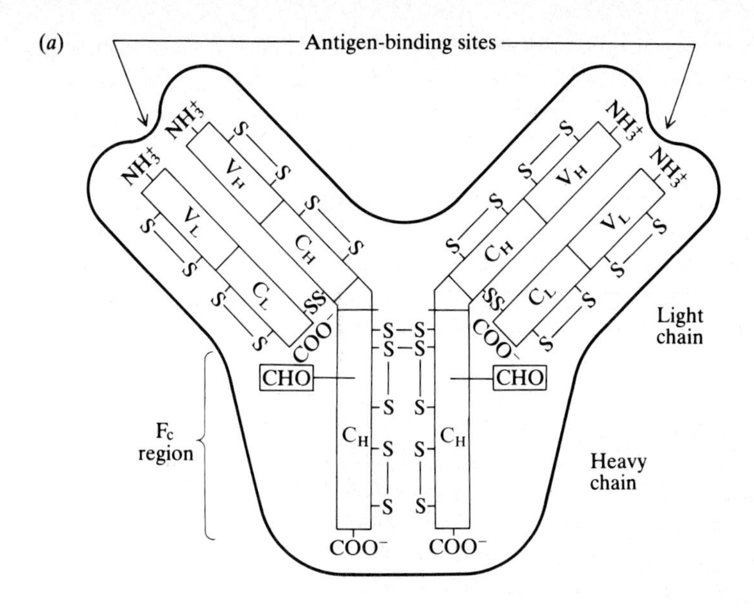

(b)

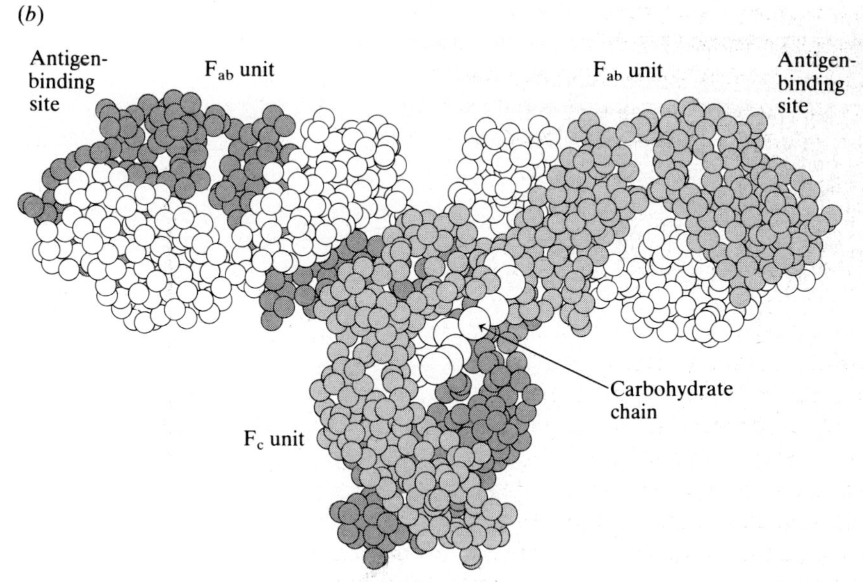

Figure 2.26 (*a*) Schematic diagram of the structure of an immunoglobulin. C and V denote constant and variable regions, respectively, while subscripts identify heavy (*H*) and light (*L*) chains. Disulfide bonds are shown, as is the attachment of carbohydrate groups (CHO) to the heavy chains. The stem region (Fc) and the antigen-binding sites are found at the carboxyl- and amino-termini, respectively, of the heavy chains. Antigen binding specificity and affinity depend on the variable regions of both the heavy and light chains. (*b*) Diagram of the molecular structure of immunoglobulin G, or IgG, an antibody. *(Reprinted by permission from E. W. Silverton, M. A. Navia, and D. R. Davies, Proc. Natl. Acad. Sci., vol. 74, p. 5142, 1977.)*

As part of the vertebrate immune system, *B cells*, one of two classes of lymphoid cells found in the body, differentiate in the presence of a foreign substance, virus, or cell (the *antigen*) to form plasma cells which secrete antibodies. These antibodies bind specifically to the antigen and closely related molecules, serving to coagulate the invading species for subsequent capture and removal.

Antibodies are members of a class of proteins called *immunoglobulins* which have the common structural features indicated in Fig. 2.26. The immunoglobulin molecule consists of two identical, longer, "heavy" chains linked together by disulfide bonds and noncovalent interactions to two identical, shorter, "light" chains. For example, in the most common class of immunoglobulins, designated IgG, the molecular weights of the light and heavy chains are 23,000 and 53,000, respectively. The carboxyl-terminal portion of both types of chains, nearly the same for antibodies of one class, is called the *constant region*. The *Fc* region, the stem of the antibody, is comprised of the carboxyl-terminal ends of the heavy chain constant regions.

The primary differentiation within a class of antibodies occurs at the amino-terminal ends of the chains, named the *variable regions*. The antigen-binding site of the antibody is formed by the variable regions of the light and heavy chains. Like enzyme catalytic activities, antigen binding sites in different antibodies may differ significantly in their specificity and affinity for antigens.

2.6 THE HIERARCHY OF CELLULAR ORGANIZATION

In the previous sections we have reviewed the major smaller biochemicals and the biopolymers constructed from them. Although the relationships between chemical structure and cellular functions have already been emphasized, it will provide a useful perspective to reconsider the dynamic nature and spatial inhomogeneity of the structures within which such functions occur. As viable cells grow, the biopolymers of this chapter must be repeatedly synthesized by the cells. Usually, the nutrient medium of the cell consists of entities such as sugar, carbon dioxide, a few amino acids, water, and some inorganic ions, but typically significant amounts of the biopolymer starting materials are absent. Thus, from these simplest of precursors, the cell must synthesize the remainder of the needed amino acids, nucleic acids, lipids, proteins, etc. The energy-consuming processes inherent in precursor synthesis and biopolymer formation are considered in Chap. 5.

There are many *supramolecular* assemblies within the cell. We have already seen that the cell membrane is a grand collection of a range of molecular varieties. The ribosomes are a separable combination of several different proteins and nucleic acids. In many cases enzymes catalyzing a sequence of chemical reactions are maintained in close proximity, presumably to maximize utilization of the reaction intermediates. The organelles such as mitochondria and chloroplasts represent the final level of organizational sophistication before the cell itself. The

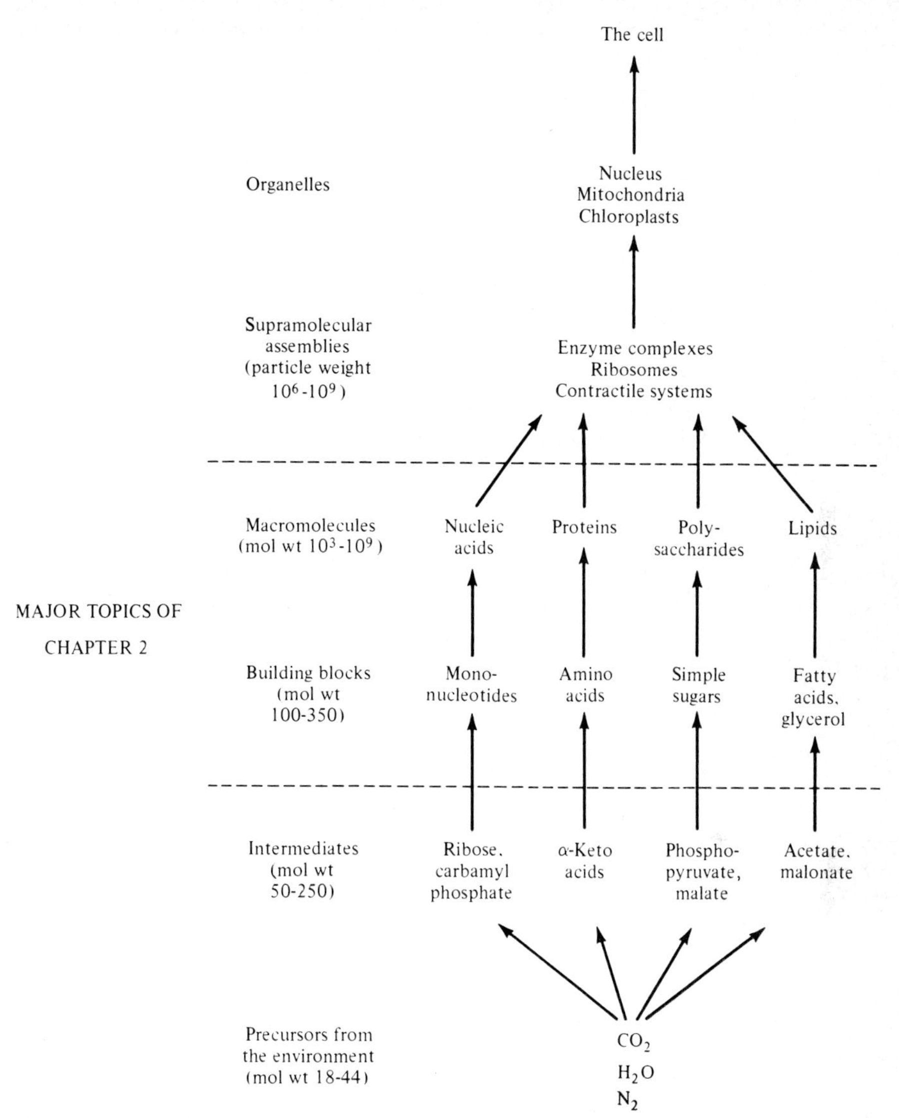

Figure 2.27 Ascending order of increasing complexity and organization of the biomolecules described in this chapter. *(From A. L. Lehninger, "Biochemistry," 1st ed., p. 19, Worth Publishers, N.Y. 1970.)*

different levels of complexity, from elemental components to the cell, are presented in Fig. 2.27.

The locations of the various biochemicals of this chapter within a procaryotic microorganism are summarized in Fig. 2.28, and the overall chemical composition of *E. coli* is outlined in Table 2.12. Small molecules like amino acids and simple sugars are dispersed throughout the cytoplasm, as are some larger molecules including certain enzymes and tRNA. Other biopolymers are localized within the cell or near a surface such as the cell membrane.

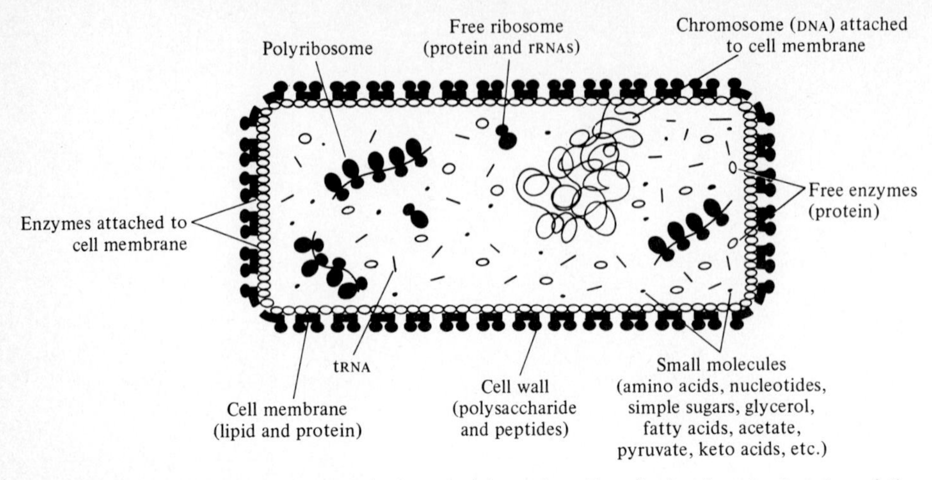

Figure 2.28 A schematic diagram of the intestinal bacterium *E. coli*, showing the location of the biomolecules described in this chapter.

Table 2.12 Composition of a rapidly growing *E. coli* cell†

Component	Percent of total cell weight	Average mol wt	Approximate no. per cell	No. of different kinds
H_2O	70	18	4×10^{10}	1
Inorganic ions (Na^+, K^+, Mg^{2+}, Ca^{2+}, Fe^{2+}, Cl^-, PO_4^{4-}, SO_4^{2-}, etc.)	1	40	2.5×10^8	20
Carbohydrates and precursors	3	150	2×10^8	200
Amino acids and precursors	0.4	120	3×10^7	100
Nucleotides and precursors	0.4	300	1.2×10^7	200
Lipids and precursors	2	750	2.5×10^7	50
Other small molecules‡	0.2	150	1.5×10^7	200
Proteins	15	40,000	10^6	2000–3000
Nucleic acids:				
DNA	1	2.5×10^9	4	1
RNA	6			
16S rRNA	...	500,000	3×10^4	1
23S rRNA	...	1,000,000	3×10^4	1
tRNA	...	25,000	4×10^5	40
mRNA	...	1,000,000	10^3	1000

† J. D. Watson, *Molecular Biology of the Gene*, table 3.3, W. A. Benjamin, Inc., New York, 1970.
‡ Heme, quinones, breakdown products of food molecules, etc.

The ions, nutrients, intermediates, biopolymers and other constituents of the cell are linked together through reaction networks (Chaps. 5 and 6). Individual reactions within such networks are catalyzed by enzymes under the mild conditions conducive to maintenance of the native protein forms discussed earlier. Membranes, catalyst complexes, and organelles (when present) are assembled by the combination of energy-consuming reactions (Chap. 5) and the natural (spontaneous) tendency of their constituent molecules to form spatially inhomogeneous structures rather than a uniform solute sea.

Chapters 3 and 4 examine the kinetics of enzyme-catalyzed reactions and enzyme applications. Next, Chaps. 5 through 7 consider energy flows and mass balances in cellular processes, genetics and the control of cellular reaction networks, and finally mathematical representations of cell-population kinetics. This collection is the prelude to biochemical-reactor analysis, a major theme of the balance of the text.

PROBLEMS

2.1 Chemical structure (*a*) Write down the specific or general chemical structure, as appropriate, for the following substances. Comment on the (possible) functional importance of various groups on the molecule.

1. Fat, phospholipid, micelle, vitamin A
2. D(+)-glucose, pyranose, lactose, starch, amylopectin, cellulose
3. Ribonucleotide, adenosine monophosphate (AMP), adenosine triphosphate (ATP), nicotinamide adenine dinucleotide (NAD), deoxyribonucleic acid (DNA)
4. Amino acid, lysine, tripeptide, protein
5. Globular and fibrous proteins

(*b*) Sketch the hierarchy of protein configuration and indicate the important features of each level of protein configuration (primary through quaternary).

2.2 Spectrophotometry Measurements of the amount of light of a specific wavelength absorbed by a liquid solution are conducted with a spectrophotometer, an instrument of basic importance in biochemical analyses. Spectrophotometric measurements are used to determine solute concentrations and to identify solutes through their absorption spectra. Interpretation of the former class of spectrophotometric data is based on the Beer-Lambert law

$$A = \log \frac{I_0}{I} = a_m bc$$

where A = absorbancy, or optical density (OD)
I_0 = intensity of incident light
I = intensity of transmitted light
a_m = molar extinction coefficient
b = path length
c = molar concentration of solute

Usually the incident-light wavelength is adjusted to the value λ_{max} at which the solute of interest absorbs most strongly. Table 2P2.1 gives λ_{max} values for a few substances of biological interest and the corresponding molar extinction coefficients (for 1-cm light path and pH 7). Additional data are available in handbooks; notice that many of the λ_{max} values lie in the ultraviolet range.

Table 2P2.1

Compound	λ_{max}, nm	$a_m \times 10^{-3}$, L/(cm·mol)
Tryptophan	218	33.5
ATP	259	15.4
NAD^+	259	18
Riboflavin	260	37
NADH	339	6.22
FAD	450	11.3

(*a*) An ATP solution in a 1-cm cuvette transmits 70 percent of the incident light at 259 nm. Calculate the concentration of ATP. What is the optical density of a 5×10^{-5} *M* ATP solution?

(*b*) Suppose two compounds A and B have the molar extinction coefficients given in Table 2P2.2. If a solution containing both A and B has an OD (1-cm cuvette) of 0.35 at 340 nm and an OD of 0.220 at 410 nm, calculate the A and B concentrations.

Table 2P2.2

Wavelength, mμ	A, L/(cm·mol)	B, L/(cm·mol)
340	14,000	7100
410	2,900	6600

2.3 Molecular weights Sedimentation and diffusion measurements are often used to provide molecular-weight estimates based on the Svedberg equation

$$M = \frac{RTs}{\mathscr{D}(1 - \bar{v}\rho)}$$

where R = gas constant
T = absolute temperature
s = sedimentation coefficient, s
$\mathscr{D}$ = diffusion coefficient, cm^2/s
$\bar{v}$ = partial specific volume of substance, cm^3/g
ρ = solution density, g/cm^3

Verify the molecular weights of the substances in Table 2P3.1 (T = 20°C). Assume ρ = 1.0.

Table 2P3.1

Substance	$s \times 10^{13}$	$\mathscr{D} \times 10^7$	$\bar{v}$	M
Lysozyme	1.91	11.2	0.703	14,400
Fibrinogen	7.9	2.02	0.706	330,000
Bushy stunt virus	132	1.15	0.74	10,700,000

2.4 pH and buffering capacity of polymer (*a*) Calculate the pH of a 0.2 M solution of serine; serine·HCl; potassium serinate.

(*b*) Sketch the titration profile for addition of the strong base KOH to the serine·HCl solution; use scales of pH vs. moles KOH added per mole serine.

(*c*) Repeat part (*a*) for the polypeptides serine_5, serine_{20}, and serine_{100} for solutions of the same weight percent solute as 0.2 M serine·HCl. Assume the dissociation constant is independent of the degree of polymerization.

(*d*) Repeat part (*c*) for lysine·HCl, assuming that protonation of the amino NH_2 has an association equilibrium constant $K_{eq} = 8.91 \times 10^{-6}$ L/mol and the side-chain amino acid an association constant of 2.95×10^{-11} L/mol.

(*e*) Estimate the pH of a 1 wt% solution of cellulose. Repeat for starch (take molecular weight to be very large, $>10^6$). What influence do storage polymers in general have on cell pH?

2.5 Amino acid separations (*a*) Answer the following statements *true* or *false*; justify your responses:

1. The pK of a simple amino acid is given by

$$pI = \tfrac{1}{2}(pK_{COOH} + pK_{NH_2})$$

2. Amino acids are neutral between pK_{COOH} and pK_{NH_2}.

(*b*) In amino acid chromatography, useful chromatographic media are sulfonic acid derivatives of polystyrene. The basic components of amino acids interact with the negative sulfonic acid groups, and the hydrophobic portions interact favorably with the aromatic structure of the polymer. At pH values of 3, 7, and 10, give the (approximate) sequence of amino acids which would elute (appear in order) from a column containing:

1. Only sulfonic acid groups on inert support
2. Only polystyrene
3. Equal quantities of accessible sulfonic acid and polystyrene

(*c*) On the same chromatography matrices as in part (*b*) above, would you expect hydrophobic or hydrophilic interactions to be relatively more important in separating mixtures of polypeptides? Of proteins?

2.6 Macromolecular physical chemistry From your own knowledge, a chemistry text, etc., describe the general reactions and/or physical conformation changes involved in:

(*a*) Frying an egg, burning toast, curdling milk, setting Jell-O, blowing a soap bubble, crystallizing a protein, tanning of leather, whipping meringue.

(*b*) In the development of substitute foods from microbial or other sources, e.g., single-cell protein, soyburgers, texture and rheology (flow response to stress) are clearly important parameters. Outline how you would organize a synthetic-food testing laboratory. (What backgrounds would you look for in personnel; what kinds of tests would you organize?)

2.7 Mass balances: stoichiometry The major elemental demands of a growing cell are carbon, oxygen, hydrogen, and nitrogen. Assuming the carbon source to be characterized by (CH_2O) and nitrogen by NH_4^+

$$\alpha CH_2O + \beta NH_4^+ + \xi O_2 \longrightarrow \underset{\text{New cellular material}}{C_aH_bO_cN_d} + \eta CO_2 + \gamma H_2O$$

(*a*) Write down the restriction among the coefficients $\alpha, \beta, \ldots$ from consideration of elemental conservation. Given the "composition" of a cell, what additional information is needed to solve for the coefficient values?

(*b*) Repeat part (*a*) for photosynthetic systems:

$$\alpha CO_2 + \beta NH_4^+ + \gamma H_2O \longrightarrow C_aH_bO_cN_d + \epsilon O_2$$

(*c*) Describe an explicit series of experiments by which you would evaluate the stoichiometric formula of the cell (a, b, c, d).

2.8 Stoichiometry: aerobic vs. anaerobic species In order to calculate conveniently nutrient consumption, aeration, and heat-transfer rates (Chaps. 5, 8), a general stoichiometric equation can be formulated by assigning to the cell an apparent molecular formula.

(*a*) Calculate the cell formula for the growth of yeast on sugar when it is observed to give the following balance (assume 1 mol of cell is produced):

$$100 \text{ g } C_6H_{12}O_6 + 5.1 \text{ g } NH_3 + 46.63 \text{ g } O_2 \longrightarrow 43.23 \text{ g cell mass} + 41.08 \text{ g } H_2O + 67.42 \text{ g } CO_2$$

(*b*) If the above reaction produces 0.30017 "mol" of cell, what cell formula results?

(*c*) Anaerobic fermentations typically produce a variety of partially oxygenated compounds in addition to cell mass. Keeping the cell "molecular formula" of part (*b*) above, calculate the unknown coefficients for the following typical equation (coefficients are mole quantities, not mass):

0.55556 glucose + α ammonia
→ β cell mass
+ 0.05697 glycerol
+ γ butanol, (a typical alcoholic product)
+ ε succinic acid, (a typical acid product)
+ 0.01164 water + 1.03076 carbon dioxide
+ 1.00380 ethanol

Kinetically, the rates of processes (*a*) and (*c*) can conveniently be controlled by changing the aeration rate, giving yeast and ethanol at any intermediate desired level (*Vienna process*) (J. B. Harrison, *Adv. Ind. Microbiol.*, **10,** 129, 1971).

2.9 Plasma and bilayer lipid membranes Tien summarized certain properties of natural and bilayer lipid membranes as shown in Table 2P9.1.

Table 2P9.1†

	Natural membranes	Lipid bilayers
Thickness, Å:		
Electron microscopy	40–130	60–90
X-ray diffraction	40–85	
Optical methods	....	40–80
Capacitance technique	30–150	40–130
Resistance, $\Omega \cdot cm^2$	10^2–10^5	10^3–10^9
Breakdown voltage	100	100–550
Capacitance, $\mu F/cm^2$	0.5–1.3	0.4–1.3

† H. T. Tien, Bilayer Lipid Membranes: An Experimental Model for Biological Membranes, in M. L. Hair (ed.), *Chemistry of Biosurfaces*, vol. 1, p. 239, Marcel Dekker, New York, 1971.

(*a*) Why do these membranes have a common range of thicknesses? (Be as specific as possible.)

(*b*) What cell functions are (possibly) served by membrane resistance and capacitance?

(*c*) Compare the breakdown field of the order of 1 V/Å with the ionization potential per length of any C_{10} or higher hydrocarbon. Comment.

2.10 Concerted-transition model for hemoglobin O_2 binding The concerted-transition model envisions an inactive form of protein T_0 in equilibrium with an active form R_0:

$$R_0 \underset{}{\overset{L}{\rightleftharpoons}} T_0 \qquad \text{equilibrium constant } L$$

In turn, the active form, which in the case of hemoglobin has four subunits, may bind one molecule of substrate S per subunit:

$$R_0 \underset{-S}{\overset{+S}{\rightleftharpoons}} R_1 \underset{-S}{\overset{+S}{\rightleftharpoons}} R_2 \underset{-S}{\overset{+S}{\rightleftharpoons}} R_3 \underset{-S}{\overset{+S}{\rightleftharpoons}} R_4$$

Let K_{DS} denote the dissociation constant for S binding in each step above

$$K_{DS} = \frac{[\text{S}][\text{unbound sites}]}{[\text{bound sites}]}$$

and evaluate the ratio

$$y = \frac{\text{concentration of S-occupied subunits}}{\text{total subunit concentration}}$$

in terms of s/K_{DS} and L. Plot y vs. s/K_{DS} for $L = 9000$. Look up a graph of hemoglobin-bound O_2 vs. O_2 partial pressure in a biochemistry text, and compare it with your plot. Read further in the reference you have found about cooperative phenomena in biochemistry.

2.11 Multiunit enzymes: mistake frequency Multiunit proteins, e.g., some allosteric enzymes, may seem to be more complicated structures than single-chain proteins, yet the percentage of mistakes in the final structures may be fewer.

(*a*) Consider the synthesis of two proteins, each with 850 amino acid residues. One is a single chain; the second is a three-unit structure of 200, 300, and 350 residues. If the probability of error is the same for each residue and is equal to 5×10^{-9} per residue, evaluate the fraction of complete structures with one or more errors when the presence of two mistakes in any chain prevents it from achieving (or being incorporated in) an active structure.

Convincing as part (*a*) may be, it has also been observed that (1) small portions of some enzymes may be removed without loss of activity and that (2) the same function may be accomplished by a considerable variety of protein sequences (see, for example, different cytochrome *c* structures on p. 136 of A. L. Lehninger, *Principles of Biochemistry*, Worth Publishers, Inc., New York, 1982.) What other function may be served in the cell by these replaceable or removable residues? (Look back at the internal cell structure.)

2.12 Mass flows Analysis of correct mass and energy balances in cellular processes requires a careful appreciation of the *sequence* of cellular chemical events in time and in space. (More detailed examples of various metabolic paths appear in subsequent chapters.) For the major biopolymer and higher structures of this chapter, form composite sketches of Figs. 2.27 and 2.28 which follow, by arrows, the time-space movement involved in cell-material synthesis from precursors, for example, O_2, H_2O, NH_4^+ and glucose. Comment where appropriate on the (lack of) restriction in space for particular species.

2.13 Fitness of biochemicals This chapter has surveyed the major existing biochemicals and biopolymers in microbial systems. For an interesting account of *why* these particular biochemicals may have arisen, given the history of the earth's development, read Ref. [12]. Summarize Blum's major arguments relating to the fitness of water, glucose, and ATP for their biological roles.

2.14 Gram staining The Gram stain is the most widely used staining procedure in bacteriology. Transcribe from a manual of microbiology a standard Gram-staining technique. Discuss both basic aspects of the stain procedure and the mechanism of the Gram reaction.

2.15 Intracellular concentrations Calculate the molar concentrations of DNA, the different RNA forms, amino acids and precursors, and lipids and precursors in a single cell of *E. coli*. Estimate the total molar concentration in *E. coli*, and comment on the applicability of ideal, dilute solution thermodynamics to the *E. coli* cytoplasm.

2.16 Enzyme structure The three catalytic groups in the 307 amino acid enzyme carboxypeptidase are arginine 145, tyrosine 248, and glutamic acid 270 (numbers denote position relative to the amino-terminal residue). Calculate the relative positions of these three residues assuming that the molecule has only straight, α-helix structure. What structural alternatives are used in the real molecule to bring these three residues together, within a few tenths of a nanometer, to effect catalysis?

2.17 Protein folding thermodynamics Consider a partially folded protein in aqueous phase which contains an unfolded sequence with Ser, Thr, Asn, and two nonpolar residues. For each of the three scenarios below, estimate ΔG (kcal/mol protein molecules) and determine in which situations folding will take place.

(*a*) All polar groups in the sequence form hydrogen bonds.

(*b*) All but the side-chain polar groups in the sequence form internal hydrogen bonds.

(*c*) No polar groups in the sequence form internal hydrogen bonds.

Assume that (i) folding of a hydrophobic residue from neutral pH aqueous medium to nonpolar protein interior gives $\Delta G \simeq -4$ kcal/mol residue and (ii) $\Delta G \simeq -5$ kcal/mol polar groups for hydrogen bond formation between any two unbonded polar groups. Remember that unfolded residues may hydrogen bond with water.

2.18 Protein sequencing Selective hydrolyses of protein chains gives fragments whose sequences can sometimes be used to deduce the amino acid sequence of the parent protein. The following table lists agents which hydrolyze specific peptide bonds.

Peptide bond:

```
   H      ⁞  H   R2
   |      ⁞  |   |
 —C——C—⁞—N——C—
   |   ||  ⁞     |
   R1  O  ⁞      H
```

Table 2P18.1

Agent	R_1 required	R_2 required
Trypsin	Lys, Arg	Any
Chymotrypsin	Phe, Trp, Tyr	Any
Pepsin	Any	Phe, Trp, Try, Leu, Asp, Glu
Cyanogen bromide	Met	Any

An important adjunct in sequencing studies is the Sanger reaction, in which 1-fluoro-2, 4-dinitrobenzene (FDNB) reacts with the NH_2-terminal amino acid residue to give the corresponding DNP-amino acid.

For the experimental data listed below for the A chain of beef insulin, deduce its primary structure.

Table 2P18.2

Treatment	Result
Partial acid hydrolysis	Peptide fragments with the following sequences: Gly-Ile-Val-Glu-Glu, Glu-Glu-Cys, Glu-Asp-Tyr, Ser-Val-Cys, Ser-Leu-Tyr, Glu-Cys-Cys, Glu-Leu-Glu, Cys-Asp, Leu-Tyr-Glu, Cys-Cys-Ala, Tyr-Cys
Pepsin	A peptide which when hydrolyzed gave fragments Ser-Val-Cys, Ser-Leu
Sanger reaction	DNP-glycine
Analysis to determine terminal —COOH group	Aspartate

REFERENCES

1. L. Stryer, *Biochemistry*, 2d ed., W. H. Freeman and Company, San Francisco, 1981. An advanced yet highly readable text presenting biochemical principles in the context of recent molecular biology.
2. M. Yudkin and R. Offord, *A Guidebook to Biochemistry*, Cambridge University Press, London, 1971. A good introductory text which describes the essentials of biochemistry with a minimum amount of detail.
3. A. G. Loewy and P. Siekevitz, *Cell Structure and Function*, Holt, Rinehart and Winston, Inc., New York, 1969. An excellent introductory short course in biochemistry is found in Part Three. Emphasis on the experimental verification of biological theories and models is an especially strong feature.
4. A. L. Lehninger, *Biochemistry*, 2d ed., Worth Publishers, New York, 1975. An excellent, more advanced biochemistry text. Continued reference to examples from cell and human physiology helps the reader in appreciating, remembering, and organizing the material.
5. A. L. Lehninger, *Principles of Biochemistry*, Worth Publishers, New York, 1982. A recent text written for undergraduates, in the same style as Lehninger's previous biochemistry book and with a more modern, molecular biology perspective.
6. H. R. Mahler and E. H. Cordes, *Biological Chemistry*, Harper & Row, Publishers, Incorporated, New York, 1971. Greater chemical and mathematical detail than Lehninger; the sections on experimental techniques in biochemistry and microbiology go further in modeling and analysis in a transport-phenomena vein than most other texts.
7. J. D. Watson, *Molecular Biology of the Gene*, 3d ed., W. A. Benjamin, Inc., Menlo Park, California, 1976. Although molecular genetics is the main theme, this well-written classic begins with a concise presentation of biochemical principles. The significance of weak interactions and molecular dynamics in biochemical structure and function is presented lucidly.
8. W. B. Wood, J. H. Wilson, R. M. Benbow, and L. E. Hood, *Biochemistry: A Problems Approach*, 2d ed., Benjamin/Cummings Publishing Company, Menlo Park, California, 1981. Concise presentation of biochemistry principles illustrated and amplified by problems (and answers).
9. R. M. Dowben, *General Physiology: A Molecular Approach*, Harper & Row, Publishers, Incorporated, New York, 1969. A book on cell physiology drawing upon fundamental principles of physical chemistry: challenging but enjoyable reading.
10. R. W. Dickerson and I. Geis, *The Structure and Action of Proteins*, Harper & Row, Publishers, Incorporated, New York, 1969. A joint project of a biologist and a science artist, this monograph provides a refreshing survey of the various protein functions with special emphasis on relations with protein structure. Most useful with one of the above references.
11. J. D. Watson, *The Double Helix*, Atheneum, New York, 1968. The story of the discovery of DNA structure, a turning point in biology. Also a candid glimpse at the process of discovery in modern research.
12. H. F. Blum, *Time's Arrow and Evolution*, 3d ed., Princeton University Press, Princeton, N.J., 1968. An inquiry into the biological fitness of elements and chemicals from an evolutionary viewpoint.
13. Fukui and Ishida, *Microbial Production of Amino Acids*, Kodansha Ltd., Tokyo, and John Wiley and Sons, Inc., New York, 1972. An interesting survey of devices to release metabolic intermediates, in this case amino acids.

CHAPTER

THREE

THE KINETICS OF ENZYME-CATALYZED REACTIONS

In the previous chapter we learned that there are many chemical compounds within the cell. How are they manufactured and combined at sufficient reaction rates under relatively mild temperatures and pressures? How does the cell select exactly which reactants will be combined and which molecules will be decomposed? The answer is catalysis by enzymes, which, as we already know, are globular proteins.

A breakthrough in enzymology occurred in 1897 when Büchner first extracted active enzymes from living cells. Büchner's work provided two major contributions: first, it showed that catalysts at work in a living organism could also function completely independently of any life process. As we shall explore in the next chapter, isolated enzymes enjoy a wide spectrum of applications today. Also, efforts to isolate and purify individual enzymes were initiated by Büchner's discoveries. The first successful isolation of a pure enzyme was achieved in 1926 by Sumner. Studies of the isolated enzyme revealed it to be protein, a property now well established for enzymes in general.

Since Sumner's time the number of known enzymes has increased rapidly, and the current total is well in excess of 1500. Based upon the known genetic content of even simple organisms like the *E. coli* bacterium, it is likely that many new enzymes will be identified and characterized in the future. For example, the single DNA molecule which comprises the *E. coli* chromosome contains enough information to code for 3000 to 4500 different proteins.

Since we shall have an occasion to mention several enzymes in this chapter and later, a few words on enzyme nomenclature are in order. Unfortunately, there is no scheme which is consistently applied to all enzymes. In most cases, however, enzyme nomenclature derives from what the enzyme *does* rather than what the enzyme *is*: the suffix *-ase* is added either to the substrate name (urease is the enzyme which catalyzes urea decomposition) or to the reaction which is catalyzed (alcohol dehydrogenase catalyzes the oxidative dehydrogenation of an alcohol). Exceptions to this nomenclature system are long familiar enzymes such

Table 3.1 Enzyme Commission classification system for enzymes (class names, Enzyme Commission type numbers, and type of reactions catalyzed)†

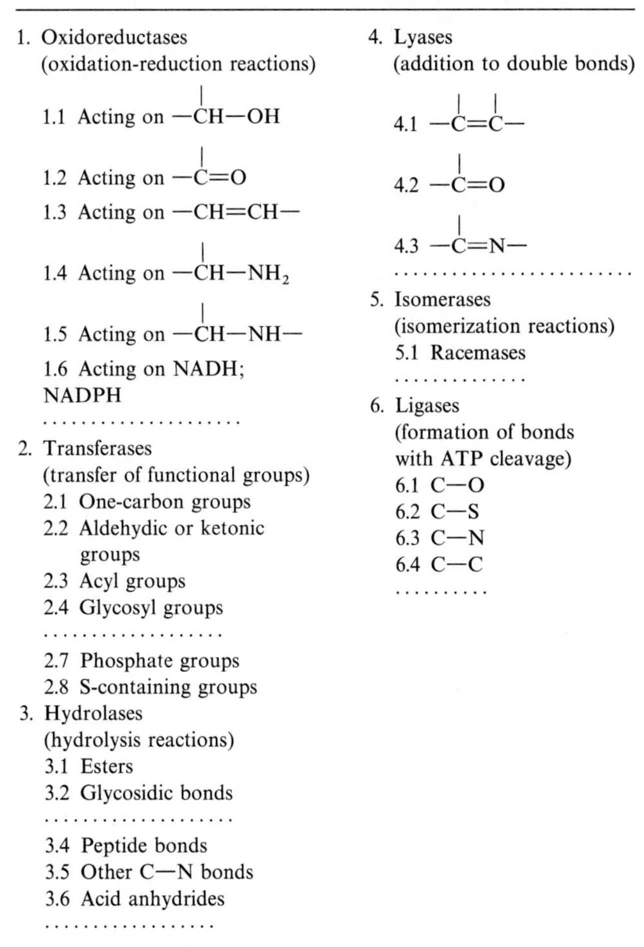

1. Oxidoreductases
   (oxidation-reduction reactions)
   1.1 Acting on $-\overset{|}{C}H-OH$
   1.2 Acting on $-\overset{|}{C}=O$
   1.3 Acting on $-CH=CH-$
   1.4 Acting on $-\overset{|}{C}H-NH_2$
   1.5 Acting on $-\overset{|}{C}H-NH-$
   1.6 Acting on NADH; NADPH
   . . . .
2. Transferases
   (transfer of functional groups)
   2.1 One-carbon groups
   2.2 Aldehydic or ketonic groups
   2.3 Acyl groups
   2.4 Glycosyl groups
   . . . .
   2.7 Phosphate groups
   2.8 S-containing groups
3. Hydrolases
   (hydrolysis reactions)
   3.1 Esters
   3.2 Glycosidic bonds
   . . . .
   3.4 Peptide bonds
   3.5 Other C—N bonds
   3.6 Acid anhydrides
   . . . .
4. Lyases
   (addition to double bonds)
   4.1 $-\overset{|}{C}=\overset{|}{C}-$
   4.2 $-\overset{|}{C}=O$
   4.3 $-\overset{|}{C}=N-$
   . . . .
5. Isomerases
   (isomerization reactions)
   5.1 Racemases
   . . . .
6. Ligases
   (formation of bonds with ATP cleavage)
   6.1 C—O
   6.2 C—S
   6.3 C—N
   6.4 C—C
   . . . .

† A. L. Lehninger, *Biochemistry*, 2d ed., table 8-1, Worth Publishers, Inc., New York, 1975.

as pepsin and trypsin in the human digestive tract, rennin (used in cheese making), and the "old yellow" enzyme, which causes browning of sliced apples.

There are six major classes of reactions which enzymes catalyze. These units form the basis for the Enzyme Commission (EC) system for classifying and assigning index numbers to all enzymes (Table 3.1). Although the common enzyme nomenclature established by past use and tradition is often employed instead of the "official" names, the EC system provides a convenient tabulation and organization of the variety of functions served by enzymes.

To be sure there is no confusion, let us remember what a catalyst is. A catalyst is a substance which increases the rate of a chemical reaction without undergoing a permanent chemical change. While a catalyst influences the *rate* of a chemical reaction, it does not affect reaction equilibrium (Fig. 3.1). Equilibrium concentrations can be calculated using only the thermodynamic properties of the substrates (remember that substrate is the biochemist's term for what we usually call a reactant) and products. Reaction kinetics, however, involve molecular

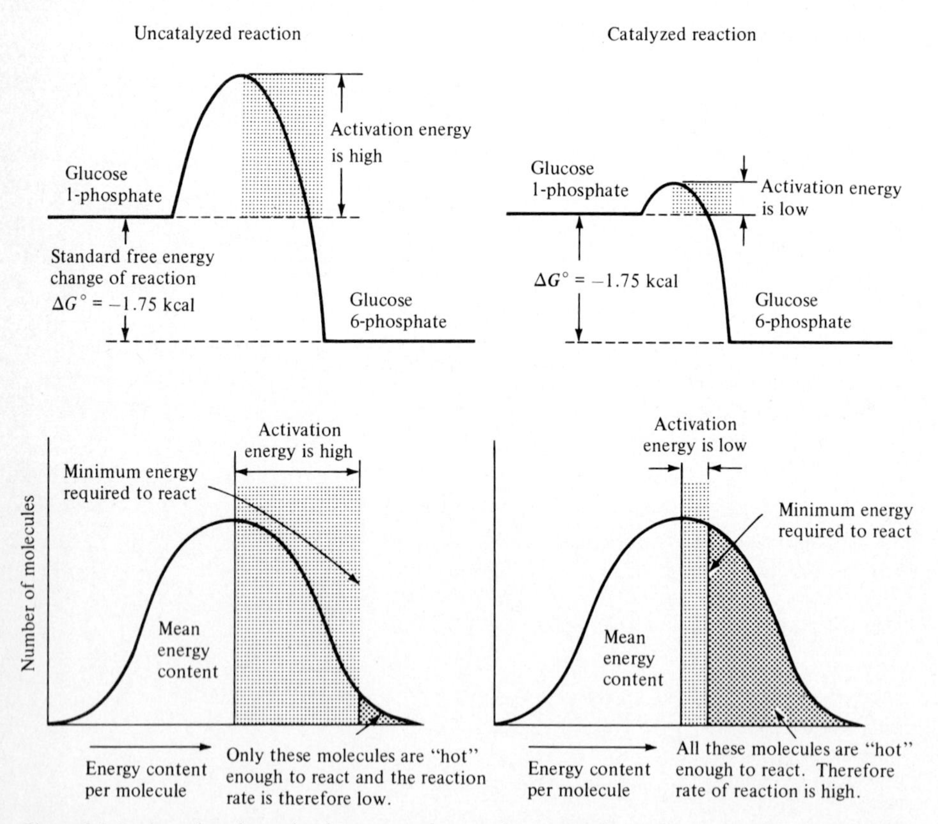

Figure 3.1 By lowering the activation energy for reaction, a catalyst makes it possible for substrate molecules with smaller internal energies to react. *(Reprinted with permission from A. L. Lehninger, "Bioenergetics," 2d ed. p. 35, W. A. Benjamin, Inc., Menlo Park, CA, 1971.)*

dynamics and are presently impossible to predict accurately without experimental data.

For design and analysis of a reacting system, we must have a mathematical formula which gives the *reaction rate* (moles reacted per unit time per unit volume) in terms of composition, temperature, and pressure of the reaction mixture. If you have studied catalyzed reactions before, you have seen a strategy for obtaining a reasonable reaction rate expression. First, an educated guess is made about the elementary reactions which occur at the molecular scale. Then, invoking certain approximations concerning the dynamics of one or more reactive intermediates, an expression for the overall reaction rate is derived by some simple mathematical manipulations. We shall follow precisely this approach in this chapter's quest for rate expressions for enzyme-catalyzed reactions.

The similarities between synthetic catalysts and enzymes go beyond the use of a common technique for modeling reaction kinetics. The rate expressions eventually obtained for both types of catalysts are very similar and sometimes of identical forms. This arises because, in both cases, it is known that the reacting molecules form some sort of complex with the catalyst. This general phenomenon in catalysis accounts for such similarity in rate expressions. We shall have more to say on this shortly.

Although some have already been mentioned, it is important that we remember the differences between enzymes and synthetic catalysts. Most synthetic catalysts are not *specific*; i.e. they will catalyze similar reactions involving many different kinds of reactants. While some enzymes are not very specific, many will catalyze only one reaction involving only certain substrates. Usually the degree of *specificity* of an enzyme is related to its biological role. Obviously it is not desirable for an enzyme with the task of hydrolyzing proteins into smaller peptides and amino acids to be highly specific. On the other hand, an enzyme catalyzing an isomerization of one particular compound must be very discriminating. As mentioned earlier in Sec. 2.4.4, enzyme specificity is thought to be a consequence of its elaborate three-dimensional conformation which allows formation of the *active site* responsible for the catalytic ability of the enzyme.

Another distinguishing characteristic of enzymes is their frequent need for *cofactors*. A cofactor is a nonprotein compound which combines with an otherwise inactive protein (the *apoenzyme*) to give a catalytically active complex. While this complex is often called a *holoenzyme* by biochemists, we shall often simply call it an *enzyme*. Two distinct varieties of cofactors exist. The simplest cofactors are *metal ions* (Table 3.2). On the other hand, a complex organic molecule called a *coenzyme* may serve as a cofactor. We have already encountered the coenzymes NAD, FAD, and coenzyme A (CoA) in the last chapter, and ATP serves as a cofactor in some instances. Often these cofactors are bound relatively loosely to the enzyme, and there is equilibrium between enzyme, apoenzyme, and cofactor. In the strict sense of the word, tightly bound nonprotein structures such as the heme group found in cytochrome *c* are coenzymes. However, the name prosthetic group, introduced previously, is usually reserved for such irreversibly attached groups.

Table 3.2 Some enzymes containing or requiring metal ions as cofactors

Ca^{2+}	α-Amylase (swine pancreas)
	Collagenase
	Lipase
	Micrococcal nuclease
Co^{2+}	Glucose isomerase (*Bacillus coagulans*) (also requires Mg^{2+})
$Cu^{2+}(Cu^{+})$	Galactose oxidase
	Tryosinase
Fe^{2+} or Fe^{3+}	Catalase
	Cytochromes
	Peroxidase
Mg^{2+}	Deoxyribonuclease (bovine pancreas)
Mn^{2+}	Arginase
Na^{+}	Plasma membrane ATPase (also requires K^{+} and Mg^{2+})
Zn^{2+}	Alcohol dehydrogenase
	Alkaline phosphatase
	Carboxypeptidase

The listing of enzymes in Table 3.2 introduces a very important point that will recur frequently in our considerations of enzyme production and application. Notice in some of the listings that a particular enzyme source is mentioned—for example, glucose isomerase from the bacterium *Bacillus coagulans* requires a cobalt cofactor for maximum activity. Specification of the source is essential in many cases to define sufficiently the particular enzyme being discussed. Other organisms, for example, synthesize enzymes that also catalyze the isomerization of glucose to fructose and hence are called also glucose isomerases. *Enzymes with the same name but obtained from different organisms often have different amino acid sequences and hence different properties and catalytic activities.* While the glucose isomerase from *B. coagulans* requires Co^{2+}, the glucose isomerase from a mutant of this organism does not require cobalt at pH greater than 8. We must be aware that an enzyme name by itself may not specify a particular protein with a corresponding particular set of catalytic properties and process requirements. Complete identification of the source organism eliminates this possible ambiguity.

Both synthetic and biological catalysts can gradually lose activity as they participate in chemical reactions. However, enzymes are in general far more fragile. While their complicated, contorted shapes in space often endow enzymes with unusual specificity and activity, it is relatively easy to disturb the native conformation and destroy the enzyme's catalytic properties. We will examine different modes of enzyme deactivation and their kinetic representation in Sec. 3.7.

It is often asserted that enzymes are more active, i.e., allow reactions to go faster, than nonbiological catalysts. A common measure of activity is the *turnover number*, which is the net number of substrate molecules reacted per catalyst site per unit time. To permit some comparison of relative activities, turnover numbers

Table 3.3 Some turnover numbers for enzyme- and solid-catalyzed reactions
N = molecules per site per second†

Enzyme	Reaction	Range of reported N values at 0–37°C	
Ribonuclease	Transfer phosphate of a polynucleotide	$2\text{–}2 \times 10^3$	
Trypsin	Hydrolysis of peptides	$3 \times 10^{-3}\text{–}1 \times 10^2$	
Papain	Hydrolysis of peptides	$8 \times 10^{-2}\text{–}1 \times 10^1$	
Bromelain	Hydrolysis of peptides	$4 \times 10^{-3}\text{–}5 \times 10^{-1}$	
Carbonic anhydrase	Reversible hydration of carbonyl compounds	$8 \times 10^{-1}\text{–}6 \times 10^5$	
Fumarate hydratase	L-Malate $\rightleftharpoons$ fumarate + H_2O	1×10^3 (forward)	
		3×10^3 (reverse)	
Solid catalyst	**Reaction**	**N**	**Temp., °C**
SiO_2–Al_2O_3 (impregnated)	Cumene cracking	3×10^{-8}	25
		2×10^4	420
Decationized zeolite	Cumene cracking	$\sim 10^3$	25
		$\sim 10^8$	325
V_2O_3	Cyclohexene dehydrogenation	7×10^{-11}	25
		10^2	350
Treated Cu_3Au	HCO_2H dehydrogenation	2×10^7	25
		3×10^{10}	327
$AlCl_3$–Al_2O_3	n-Hexane isomerization (liquid)	1×10^{-2}	25
		1.5×10^{-2}	60

† R. W. Maatman, "Enzyme and Solid Catalyst Efficiencies and Solid Catalyst Site Densities," *Catal. Rev.*, **8**: 1, 1973.

for several reactions are listed in Table 3.3, which shows that at the ambient temperatures where enzymes are most active they are able to catalyze reactions faster than the majority of artificial catalysts. When the reaction temperature is increased, however, solid catalysts may become as active as enzymes. Unfortunately, enzyme activity does not increase continuously as the temperature is raised. Instead, the enzyme usually loses activity at quite a low temperature, often only slightly above that at which it is typically found.

A unique aspect of enzyme catalysis is its susceptibility to control by small molecules. Several enzymes are "turned off" by the presence of another chemical compound, often the end product of a sequence of reactions in which the regulated enzyme participates. This feature of some enzymes is essential for normal cell function. Some aspects of the kinetics of such enzymes will be explored in Sec. 3.5, and in Chap. 6 we shall learn how altering the normal channels of cellular control can improve industrial biological processes tremendously.

Before beginning our efforts to model the kinetics of enzyme-catalyzed reactions, we must review the available experimental evidence on molecular events which actually occur during reaction. With this foundation, we shall be able to make reasonable hypotheses about the reaction sequence and apply them to derive useful reaction rate expressions.

3.1 THE ENZYME-SUBSTRATE COMPLEX AND ENZYME ACTION

There is no single theory currently available which accounts for the unusual specificity and activity of enzyme catalysis. However, there are a number of plausible ideas supported by experimental evidence for a few specific enzymes. Probably, then, all or some collection of these phenomena acting together combine to give enzymes their special properties. In this section we shall outline some of these concepts. Our review will necessarily be brief, and the interested reader should consult the references for further details. Since all the theories mentioned here are at best partial successes, we must be wary of attempting to synthesize a single theory of enzyme activity.

Verified by numerous experimental investigations involving such diverse techniques as x-ray crystallography, spectroscopy, and electron-spin resonance is the existence of a *substrate-enzyme complex*. The substrate binds to a specific region of the enzyme called the *active site*, where reaction occurs and products are released. Binding to create the complex is sometimes due to the type of weak attractive forces outlined in Sec. 2.4.3 although covalent attachments are known for some cases. As shown schematically in Figs. 2.24 and 3.2, the complex is formed when the substrate key joins with the enzyme lock. Especially evident in Fig. 3.2 are the hydrogen bonds which form between the substrate and groups widely separated in the amino acid chain of the enzyme.

This example also nicely illustrates the notion of an active site. The protein molecule is folded in such a way that a group of reactive amino acid side chains in the enzyme presents a very specific site to the substrate. The reactive groups encountered in enzymes include the R group of Asp, Cys, Glu, His, Lys, Met, Ser, Thr, and the end amino and carboxyl functions. Since the number of such groups near the substrate is typically 20 (far less than the total number of amino acid residues present), only a small fraction of the enzyme is believed to participate directly in the enzyme's active site. Large enzymes may have more than one active site. Many of the remaining amino acids determine the folding along a chain of amino acids (secondary structure) and the placement of one part of a folded chain next to another (tertiary structure), which help create the active site itself (Fig. 3.4).

While some of the ideas described below are still somewhat controversial, we should emphasize that the notions of active sites and the enzyme-substrate complex are universally accepted and form the starting point for most theories of enzyme action. These concepts will also be the cornerstone of our analysis of enzyme kinetics.

Two different aspects of current thinking on enzyme activity are shown schematically in Fig. 3.3. Enzymes can hold substrates so that their reactive regions are close to each other and to the enzyme's catalytic groups. This feature, which quite logically can accelerate a chemical reaction, is known as the *proximity effect*. Consider now that the two substrates are not spherically symmetrical molecules. Consequently, reaction will occur only when the molecules come together

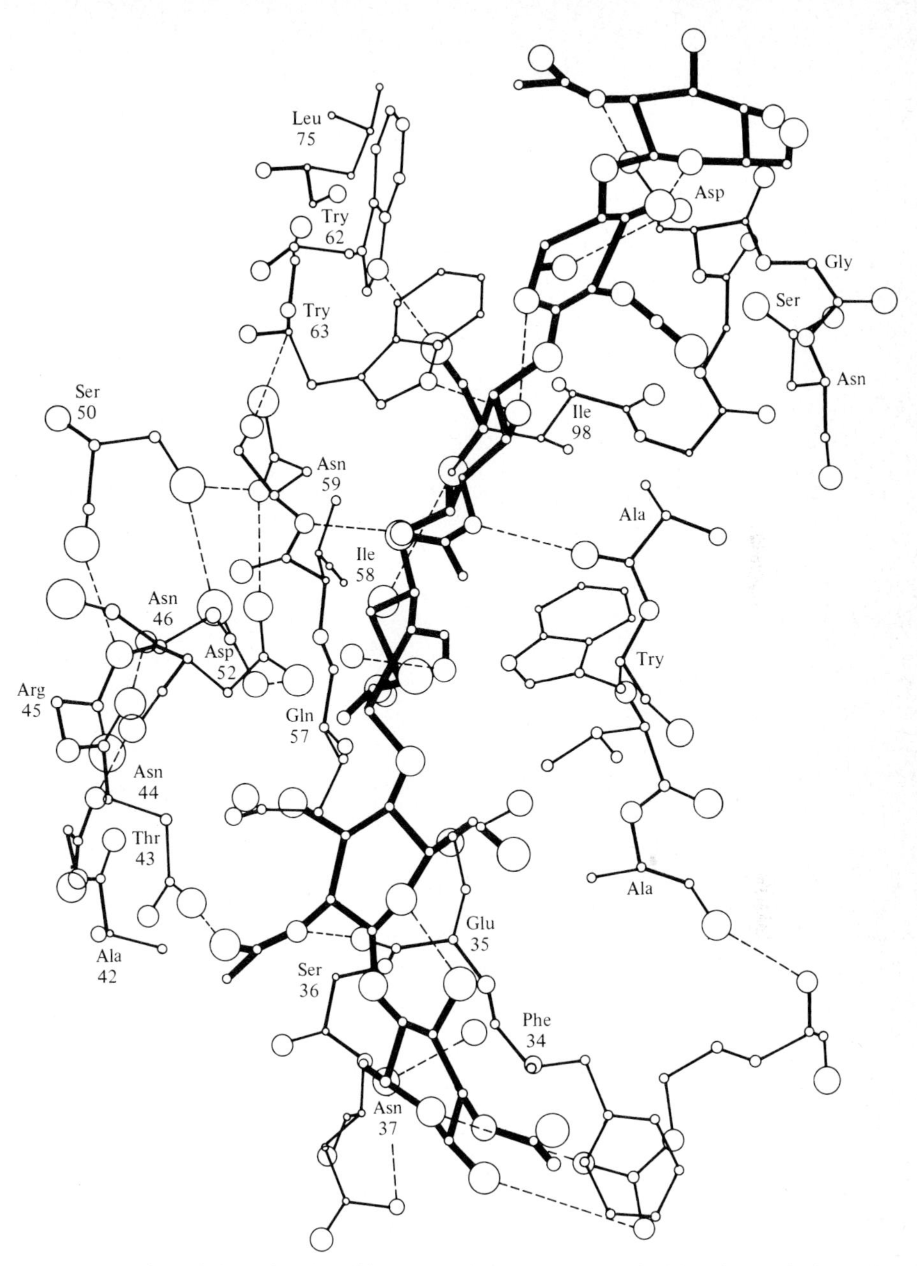

Figure 3.2 A view of the active site of lysozyme, showing a hexasaccharide substrate in heavy lines. Larger and smaller circles denote oxygen and nitrogen atoms, respectively, and hydrogen bonds are indicated by dashed lines. *(Reprinted with permission from M. Yudkin and R. Offord, "A Guidebook to Biochemistry," p. 48, Cambridge University Press, London, 1971).*

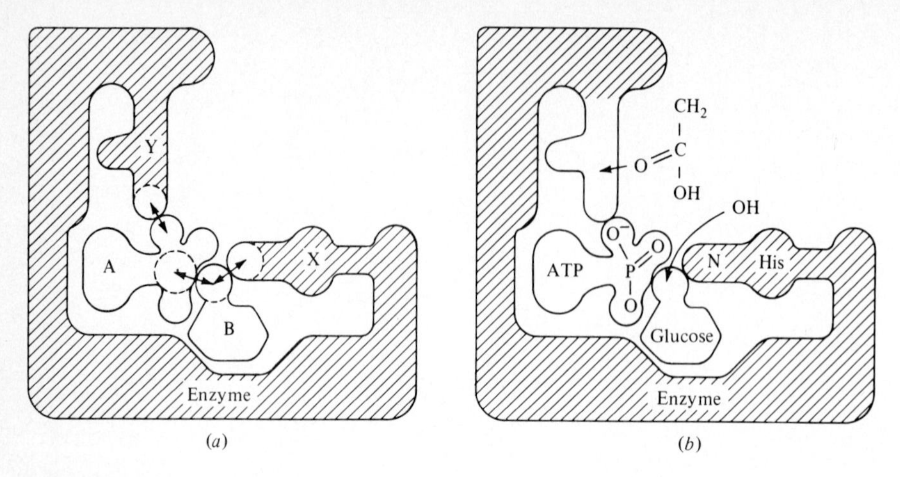

Figure 3.3 An enzyme may accelerate reaction by holding two substrates close to each other (proximity effect) and at an advantageous angle (orientation effect). *(Reprinted by permission of D. E. Koshland.)*

at the proper orientation so that the reactive atoms or groups are in close juxtaposition. Enzymes are believed to bind substrates in especially favorable positions, thereby contributing an *orientation effect*, which accelerates the rate of reaction. Also called *orbital steering*, this phenomenon has qualitative merit as a contributing factor to enzyme catalysis. The quantitative magnitude of its effect, however, is still difficult to assess in general.

Before a brief foray into the chemistry of enzyme action, we should mention one other hypothesis related to enzyme geometry. It is known that for some enzymes the binding of substrate causes the shape of the enzyme to change slightly. Studies of the three-dimensional structure of the enzymes lysozyme and carboxypeptidase A with and without substrates have shown a change in enzyme conformation upon addition of the substrate. This *induced fit* of enzyme and substrate may add to the catalytic process. There are also more sophisticated extensions of the induced-fit model, in which a number of different intermediate enzyme-substrate complexes are formed as the reaction progresses. A slightly elastic and flexible enzyme molecule would have the ability to make delicate adjustments in the position of its catalytic groups to hasten the transformation of each intermediate. Since direct measurements of the influence of substrate binding on enzyme structure are available only for a few enzymes, it is possible that a substrate-induced change in active-site configuration may be a general characteristic of enzyme catalysis.

Catalytic processes well known to the organic chemist also appear to be at work in some enzymes. One of these is *general acid-base catalysis*, where the catalyst accepts or donates protons somewhere in the overall catalytic process. In one of the few enzymes for which a reasonable complete catalytic sequence is proposed (Fig. 3.4), this mode of catalysis is present. *Chymotrypsin*, derived from the pancreas, is a *proteolytic* (protein-hydrolyzing) enzyme with specificity for peptide bonds where the carbonyl side is a tyrosine, tryptophan, or phenylalanine residue. Water is believed to serve as a proton-transfer agent in both the general acid- and general base-catalyzed portions of the chymotrypsin mechanism. Several reactions important to cellular chemistry, including carbonyl addition and ester hydrolysis, are in principle amenable to general acid-base catalysis.

Participating in enzyme catalysis may be a number of other phenomena such as covalent catalysis, strain, electrostatic catalysis, multifunctional catalysis, and solvent effects (recall the oil-drop structure from Chap. 2). Details on these mechanisms available in the references reveal that, like the notions reviewed here, they are probably involved in some enzyme-catalyzed reactions. From the possible involvement of these factors and others, it is not surprising that no one has yet devised a simple general scheme for assessing their combined influence and relative importance. Fortunately, armed only with the basic idea of an enzyme-substrate complex as an essential reaction intermediate, we shall be able to formulate useful rate expressions for enzyme-catalyzed reactions.

3.2 SIMPLE ENZYME KINETICS WITH ONE AND TWO SUBSTRATES

Our objective in this section is to develop suitable mathematical expressions for the rates of enzyme-catalyzed reactions. Naturally, the crucial test of a reaction rate equation is comparison with experimentally determined rates. Since there are some pitfalls awaiting the unwary, we should first briefly outline some of the experimental methods employed to measure reaction rates.

Let us be quite clear on what we mean by a reaction rate. If the reaction is

$$\mathrm{S} \longrightarrow \mathrm{P}$$

the reaction rate v, in the quasi-steady state approximation (Sec. 3.2.1), is defined by

$$v = -\frac{ds}{dt} = \frac{dp}{dt} \tag{3.1}$$

where lowercase letters denote molar concentration. The dimensions of v are consequently moles per unit volume per unit time. The reaction rate is an *intensive* quantity, defined at each point in the reaction mixture. Consequently, the rate will vary with position in a mixture if concentrations or other intensive state variables change from point to point. In experimental kinetic studies, well-mixed reactors are often used so that the reaction rate is spatially uniform.

As usual in our engineering modeling work, the term *point* in the rate definition is not used in a strict geometrical sense. Instead, a point is a volume large enough to contain many molecules but very small relative to the entire reacting

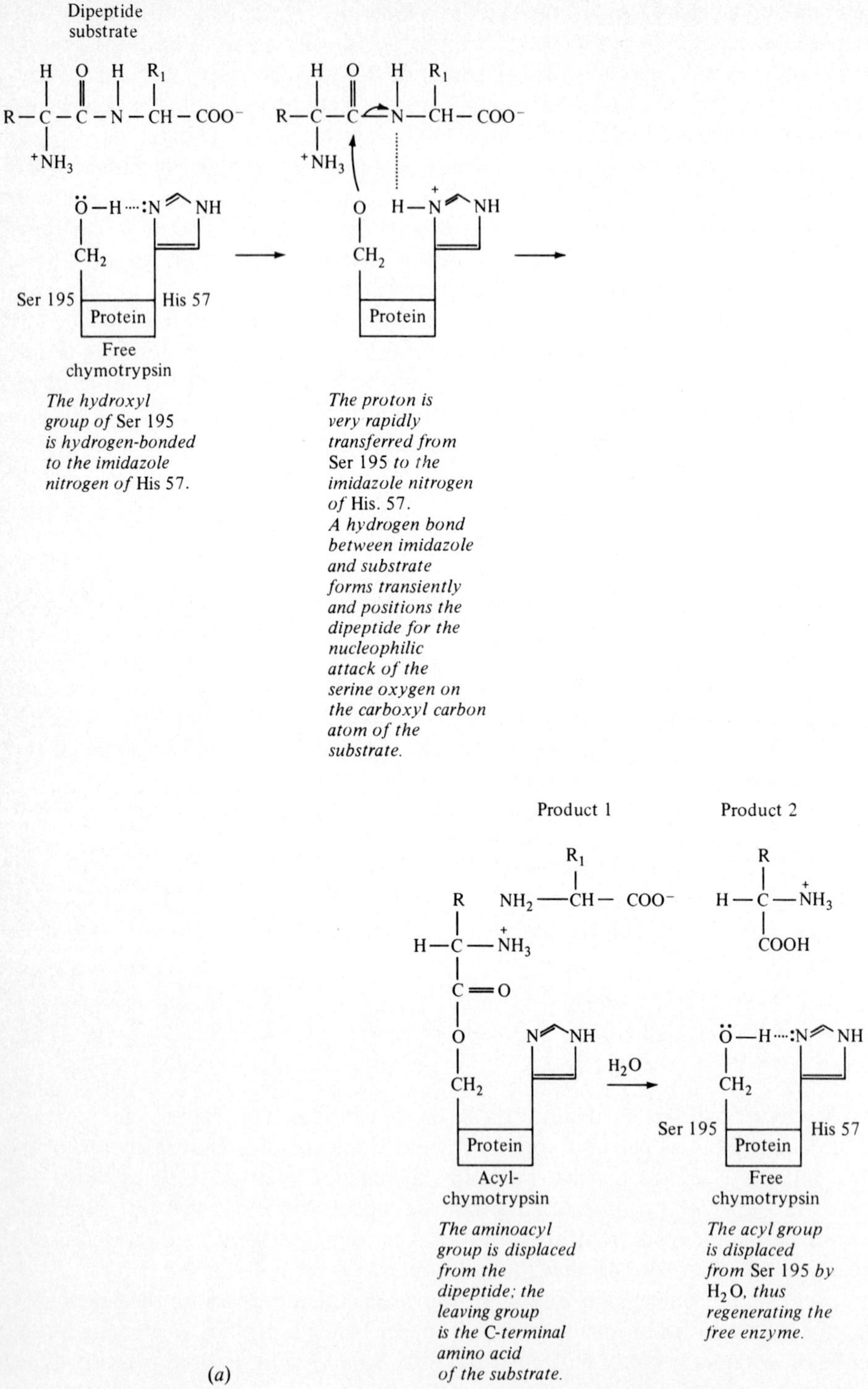
Dipeptide substrate
Ser 195
His 57
Protein
Free chymotrypsin
The hydroxyl group of Ser 195 is hydrogen-bonded to the imidazole nitrogen of His 57.
Protein
The proton is very rapidly transferred from Ser 195 to the imidazole nitrogen of His. 57. A hydrogen bond between imidazole and substrate forms transiently and positions the dipeptide for the nucleophilic attack of the serine oxygen on the carboxyl carbon atom of the substrate.
Product 1
Product 2
H_2O
Protein
Acyl-chymotrypsin
The aminoacyl group is displaced from the dipeptide; the leaving group is the C-terminal amino acid of the substrate.
Ser 195
His 57
Protein
Free chymotrypsin
The acyl group is displaced from Ser 195 by H_2O, thus regenerating the free enzyme.

(*a*)

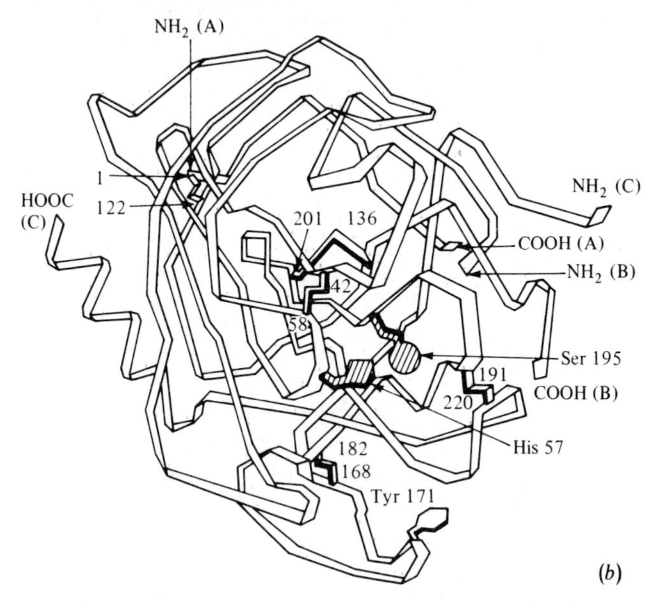

Figure 3.4 (*a*) A proposed reaction sequence for chymotrypsin action on a peptide bond, and (*b*) structure of chymotrypsin. *(From A. L. Lehninger, "Biochemistry," 2d ed., pp. 229, 231. Worth Publishers, New York, 1975. Structure redrawn from B. W. Matthews, et al., "Three-Dimensional Structure of Tosyl-α-Chymotrypsin," Nature, vol. 214, p. 652, 1976.)*

system. It is very important to remember this seemingly minor detail when applying engineering analysis to biological systems. Since cells may contain only a few molecules of a given compound A_1, the concepts of A_1 concentration and reaction rate at a point are potentially erroneous idealizations in modeling molecular processes of an isolated single cell. We shall return to this theme in our consideration of cell kinetics in Chap. 7.

A typical experiment to measure *enzyme kinetics* (shorthand terminology for *kinetics of reactions catalyzed by enzymes*) might proceed as follows. At time zero, solutions of substrate and of an appropriate purified enzyme are mixed in a well-stirred, closed isothermal vessel containing a buffer solution to control pH. The substrate and/or product concentrations are monitored at various later times. Techniques for following the time course of the reaction may include spectrophotometric, manometric, electrode, polarimetric, and sampling-analysis methods. Typically, only *initial-rate* data are used. Since the reaction conditions including enzyme and substrate concentrations are known best at time zero, the initial slope of the substrate or product concentration vs. time curve is estimated from the data:

$$\left.\frac{dp}{dt}\right|_{t=0} = -\left.\frac{ds}{dt}\right|_{t=0} = v\Big|_{t=0} \qquad \text{where } e = e_0,\ s = s_0, \text{ and } p = 0 \tag{3.2}$$

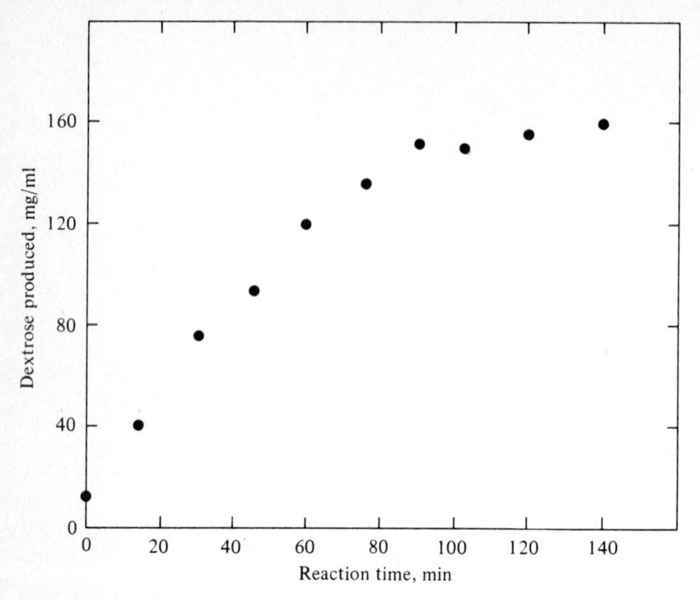

Figure 3.5 Batch reactor data for hydrolysis of 30 percent starch by glucoamylase (60°C, $e_0 = 11{,}600$ units, reactor volume = 1 L).

Figure 3.5 shows actual data for such an experiment. Notice that some product is present initially, indicating that the reported zero reaction time is not the actual time when the reaction started. This is one of the inherent difficulties of the initial-rate method, and others are described in texts on chemical kinetics. Still, initial concentrations as well as enzymatic activity are reasonably reproducible, and these considerations favor the initial-rate approach.

The problem of reproducible enzyme activity is an important one which also has serious implications in design of reactors employing isolated enzymes. Since we have already mentioned that proteins may denature when removed from their native biological surroundings, it should not surprise us that an isolated enzyme in a "strange" aqueous environment can gradually lose its catalytic activity (Fig. 3.6). Although this gradual deactivation is known to occur for many enzymes, the phenomenon is scarcely mentioned in many treatments of enzyme kinetics. While perhaps not important *in vivo* (in the intact living organism), where enzyme synthesis compensates for any loss of previously active enzymes, enzyme deactivation *in vitro* (removed from a living cell) cannot be overlooked in either kinetic studies or enzyme reactor engineering. This topic will surface again in Sec. 3.7.

We see in the caption of Fig. 3.5 another typical feature of much enzyme kinetic information. Notice that the amount of enzyme catalyst used in this experiment is given in terms of "units." What are these mysterious units, and why isn't the catalyst quantity stated in more familiar, less ambiguous terms such as moles or mass?

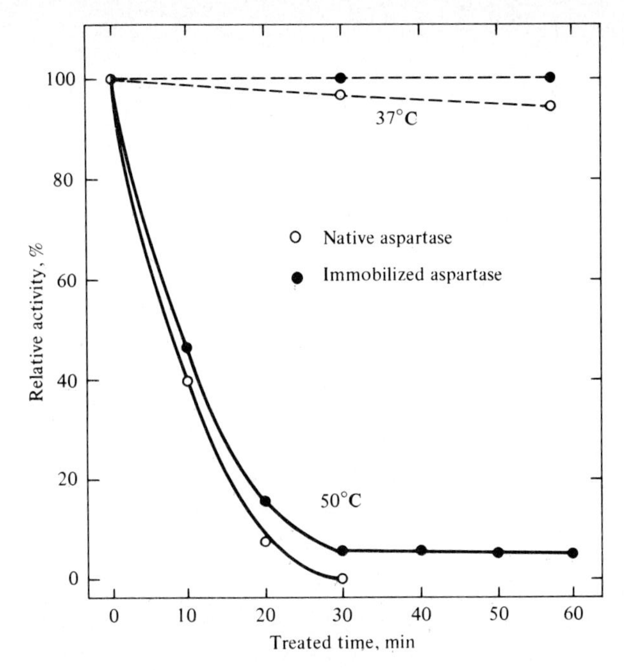

Figure 3.6 Loss of activity in solution. At 37°C, aspartase in solution loses more than 5 percent of its natural activity within 1 h. *(Reprinted from T. Tosa et al., "Continuous Production of* L-*Aspartic Acid by Immobilized Asparatase," Biotech. Bioeng., vol. 15, p. 68, 1973.)*

Let's first answer the second part of the question: the reason for not using, say, "mass of enzyme" is that this is rarely known for enzyme preparations, most of which are (primarily) protein mixtures in which the particular enzyme of interest is one of several components. The fraction of this preparation which consists of the enzyme we are studying is often unknown, however, and the composition of this mixture can vary from batch to batch obtained from its manufacturer. To circumvent this difficulty, we more often find enzyme content specified in terms of units of catalytic activity per mass of the enzyme preparation.

Units or *units of activity* designate the amount of enzyme which gives a certain amount of catalytic activity under a prescribed set of standard conditions for that particular enzyme. For example, in the starch hydrolysis experiment of Fig. 3.5, one unit of activity of glucoamylase is defined as the amount of enzyme which produces 1 μmol of glucose per minute in a 4% solution of Lintner starch at pH 4.5 and 60°C. Thus, we can see that units of activity will be defined differently for different enzymes, and may have different definitions for the same enzyme studied in different contexts. Obvious pitfalls lurk here for anyone who does not carefully check the definition of activity units when interpreting or utilizing enzyme kinetic data. Greater clarity is possible, of course, when activity data is available for highly purified enzyme.

3.2.1 Michaelis-Menten Kinetics

Now let us suppose that armed with a good set of experimental rate data (Fig. 3.7) we face the task of representing it mathematically. From information of the type shown in Fig. 3.7, the following qualitative features often emerge:

1. The rate of reaction is first order in concentration of substrate at relatively low values of concentration. [Recall that if $v = (\text{const})(s^n)$, n is the order of the reaction rate.]
2. As the substrate concentration is continually increased, the reaction order in substrate diminishes continuously from one to zero.
3. The rate of reaction is proportional to the total amount of enzyme present.

Henri observed this behavior, and in 1902 propsed the rate equation

$$v = \frac{v_{\max} s}{K_m + s} \qquad \text{where } v_{\max} = \alpha e_0 \tag{3.3}$$

which exhibits all three features listed above. Notice that $v = v_{\max}/2$ when s is equal to K_m. To avoid confusion of the type found in some of the literature, we should strongly emphasize that s is the concentration of free substrate in the reaction mixure, while e_0 is the concentration of the total amount of enzyme present in both the free and combined forms (see below).

Although Henri provided a theoretical explanation for Eq. (3.3) using a hypothesized reaction mechanism, his derivation and the similar one offered in 1913 by Michaelis and Menten are now recognized as not rigorous in general. However, the general methodology of the Michaelis-Menten treatment will find repeated useful (although still not generally rigorously justified) application in derivation of more complicated kinetic models later in this chapter. Consequently, we shall provide a brief summary of their development before proceeding to others.

As a starting point, it is assumed that the enzyme E and substrate S combine to form a complex ES, which then dissociates into product P and free (or uncombined) enzyme E:

$$\text{S} + \text{E} \underset{k_{-1}}{\overset{k_1}{\rightleftharpoons}} \text{ES} \tag{3.4a}$$

$$\text{ES} \xrightarrow{k_2} \text{P} + \text{E} \tag{3.4b}$$

This mechanism includes the intermediate complex discussed above, as well as regeneration of catalyst in its original form upon completion of the reaction sequence. While perhaps considerably oversimplified, Eqs. (3.4) are certainly reasonable.

Henri and Michaelis and Menten assumed that reaction (3.4a) is in equilibrium which, in conjunction with the mass-action law for the kinetics of

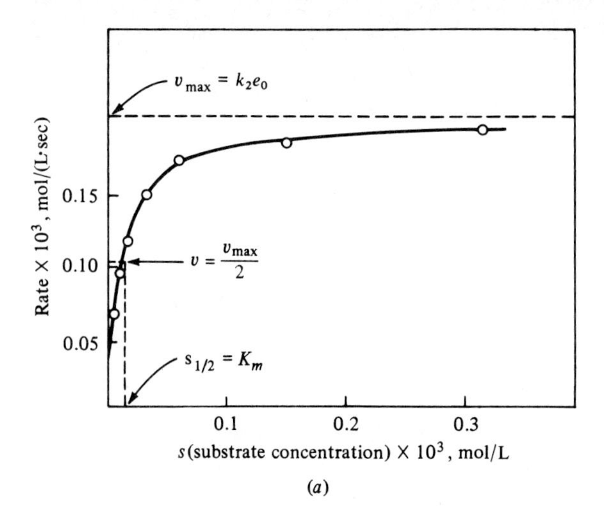

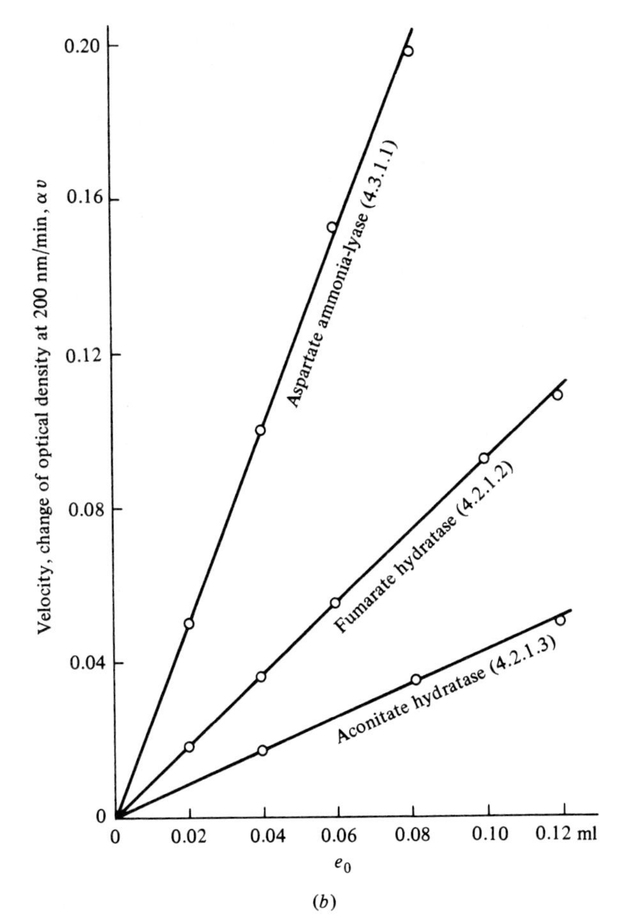

Figure 3.7 Kinetic data for enzyme-catalyzed reactions. (*a*) Enzyme concentration is held constant when studying substrate concentration dependence; (*b*) the converse holds for investigation on the influence of enzyme concentration. *[(a) Reprinted from K. J. Laidler, "The Chemical Kinetics of Enzyme Action," p. 64, The Clarendon Press, Oxford, 1958, Data of L. Oullet, K. J. Laidler, M. F. Morales, "Molecular Kinetics of Muscle Adenosine Triphosphate," Arch. Biochem. Biophys., vol. 39, p. 37, 1942; (b) Reprinted by permission from M. Dixon and E. C. Webb, "Enzymes," 2d ed., p. 55, Academic Press, Inc., New York, 1964.]*

molecular events, gives

$$\frac{se}{(es)} = \frac{k_{-1}}{k_1} = K_m = \text{dissociation constant} \tag{3.5}$$

Here, s, e, and (es) denote the concentrations of S, E and ES, respectively. Decomposition of the complex to product and free enzyme is assumed irreversible:

$$v = \frac{dp}{dt} = k_2(es) \tag{3.6}$$

Since all enzyme present is either free or complexed, we also have

$$e + (es) = e_0 \tag{3.7}$$

where e_0 is the total concentration of enzyme in the system. This is known from the amount of enzyme initially charged into the reactor. Equation (3.3) with v_{max} equal to $k_2 e_0$ can now be obtained by eliminating (es) and e from the three previous equations. We should note here that a reaction described by Eq. (3.3) is commonly referred to as having *Michaelis-Menten kinetics*, although certainly other investigators made equal contributions to the development and justifcation of this kinetic form. The parameter v_{max} is called the *maximum* or *limiting velocity*, and K_m is known as the *Michaelis constant*. While the Michaelis-Menten equation successfully describes the kinetics of many enzyme-catalyzed reactions, it is not universally valid. We shall explore the extensions and modifications necessary for certain enzymes and reaction conditions in later sections and in the problems.

Briggs and Haldane have provided the derivation of Eq. (3.3) which later kinetic studies and mathematical analyses have shown to be the most general. For reaction in a well-mixed closed vessel we can write the following mass balances for substrate and the ES complex:

$$v = -\frac{ds}{dt} = k_1 se - k_{-1}(es)$$

and

$$\frac{d(es)}{dt} = k_1 se - (k_{-1} + k_2)(es)$$

Using Eq. (3.7) in the previous two equations gives a closed set: two simultaneous ordinary differential equations in two unknowns, s and (es). The appropriate initial conditions are, of course,

$$s(0) = s_0 \qquad (es)(0) = 0$$

These equations cannot be solved analytically but they can be readily integrated on a computer to find the concentrations of S, E, ES, and P as functions of time. In calculated results illustrated in Fig. 3.8, it is evident that to a good approximation

$$\frac{d(es)}{dt} = 0 \tag{3.8}$$

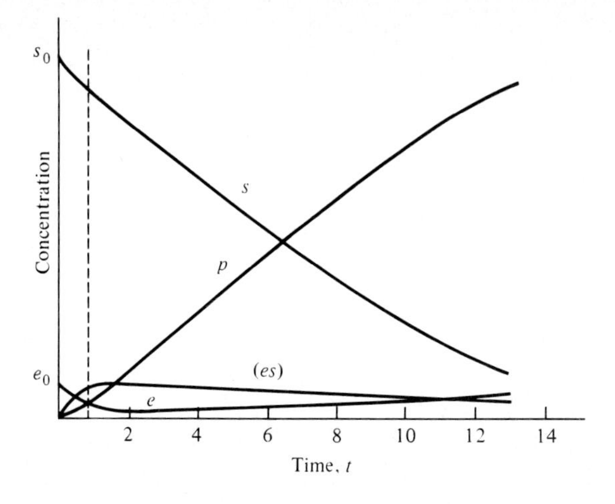

Figure 3.8 The time course of the reaction

$$E + S \underset{k_{-1}}{\overset{k_{+1}}{\rightleftharpoons}} ES \xrightarrow{k_{+2}} P + E$$

with $k_{+1} \cong k_{-1} \cong k_{+2}$.

after a brief start-up period. In the key step in their analysis, Briggs and Haldane assumed condition (3.8) to be true. This assumption, commonly called the *quasi-steady-state approximation*, can be proved valid for the present case of enzyme-catalyzed reaction in a closed system provided that e_0/s_0 is sufficiently small. If the initial substrate concentration is not large compared with the total enzyme concentration, the assumption may break down. In many instances, however, the amount of catalyst present is considerably less than the amount of reactant, so that (3.8) is an excellent approximation after start-up.

Proceeding from Eq. (3.8). as Briggs and Haldane did, it is possible to use Eq. (3.7) to eliminate e and (es) from the problem, leaving

$$v = -\frac{ds}{dt} = \frac{k_2 e_0 s}{[(k_{-1} + k_2)/k_1] + s} \tag{3.9}$$

Consequently, the Michaelis-Menten form results, where Eq. (3.9) shows now

$$v_{\max} = k_2 e_0 \tag{3.10}$$

and

$$K_m = \frac{k_{-1} + k_2}{k_1} \tag{3.11}$$

Notice that K_m no longer can be interpreted physically as a dissociation constant.

With the Michaelis-Menten rate expression at hand, the time course of the reaction can now be determined analytically by integrating

$$\frac{ds}{dt} = \frac{-v_{\max} s}{K_m + s} \qquad \text{with } s(0) = s_0 \tag{3.12}$$

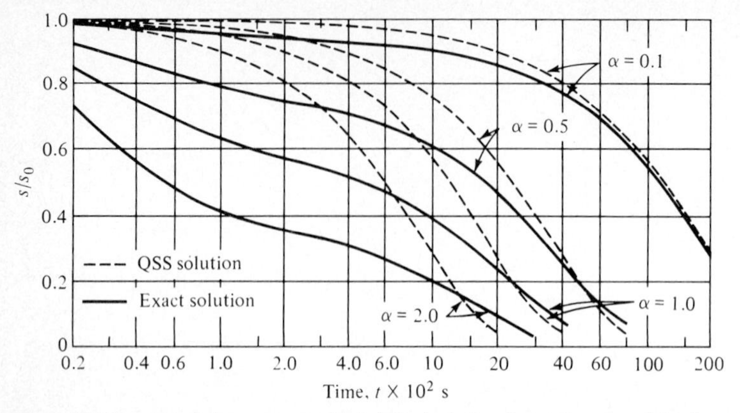

Figure 3.9 Computed time course of batch hydrolysis of acetyl L-phenylalanine ether by chymotrypsin. Considerable discrepancies between the exact solution and the quasi-steady-state solution arise when e_0/s_0 $(=\alpha)$ is not sufficiently small. *(Reprinted by permission from H. C. Lim, "On Kinetic Behavior at High Enzyme Concentrations," Am. Inst. Chem. Eng. J., Vol. 19, p. 659, 1973.)*

to obtain

$$v_{\max}t = s_0 - s + K_m \ln \frac{s_0}{s} \tag{3.13}$$

Naturally this equation is most easily exploited indirectly by computing the reaction time t required to reach various substrate concentrations, rather than vice versa.

It is instructive to compare the behavior predicted by Eq. (3.13) with that obtained without the quasi-steady-state approximation. As Fig. 3.9 reveals, the deviation is significant when the total enzyme concentration approaches s_0. Consequently the Michaelis-Menten equation should not be used in such cases. As another example, we show the dependence of esterase activity on enzyme concentration in Fig. 3.10. The linear dependence predicted by the Michaelis-Menten model is followed initially but does not hold for large enzyme concentrations. Remember that the slope at the linear portion is equal to $k_2 s/(K_m + s)$.

We should consider whether or not the quasi-steady-state approximation can be justified in some cases of large e_0/s_0, perhaps with different rate constants. This could be done if the ES complex tended to disappear much faster than it was formed, i.e., if K_m were very large relative to s_0. Since the Michaelis constant is quite small, with typical values in the range 10^{-2} to 10^{-5}, this case does not arise in many situations.

Consequently, if the enzyme concentration is comparable to s_0, we usually have no proper justification for simplifying the kinetic model with the quasi-steady-state approximation. Situations with relatively large enzyme concentration can occur in several instances. If an enzyme reactor is operated under conditions where most of the substrate is converted, s falls to a value comparable to e_0 and the Michaelis-Menten equation may be inappropriate for the final

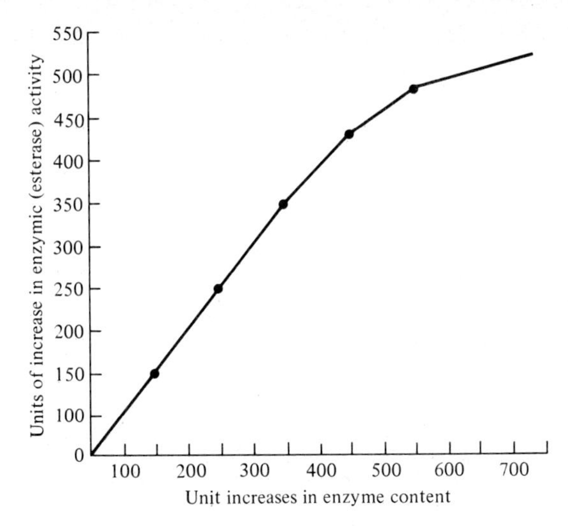

Figure 3.10 These data on esterase activity show deviations from Michaelis-Menten kinetics at large values of initial enzyme content. *(Reprinted by permission from M. Frobisher, "Fundamentals of Microbiology," 8th ed., p. 61, W. B. Saunders Co., Philadelphia, 1968.)*

stages of reaction. Fortunately, the reaction has become slow here, and this situation is consequently seldom of interest. Also in the case of enzyme reactions at interfaces, which we shall investigate in Sec. 4.4, the substrate concentration in the neighborhood of the enzyme may be quite small. Here, the rate of substrate transport to the interface must also be considered. Possible pitfalls in the Michaelis-Menten model have come to light only very recently, and consequently it is still commonly employed for analyzing and designing enzyme reactors. Although they lack rigorous justification in some situations, the equilibrium assumption and the Michaelis-Menten rate equation have proved to be valuable working tools. We shall often employ them in this spirit.

While on the subject of simple Michaelis-Menten kinetics, we should emphasize that exactly the same mathematical form is widely used for expressing the rates of many solid-catalyzed reactions. In chemical engineering practice, kinetics described by Eq. (3.3) is called *Langmuir-Hinshelwood* or *Hougen-Watson kinetics.* Since reactions involving synthetic solid catalysts are widespread throughout the chemical and petroleum industries, a large effort has been devoted to the design and analysis of catalytic reactors. Much of this work has employed Langmuir-Hinshelwood kinetics and is therefore directly applicable to enzyme-catalyzed reactions described by Michaelis-Menten kinetics. In the remainder of our studies, we shall continue to observe analogies, similarities, and identities between classical chemical engineering and biochemical technology. Also, of course, we shall encounter many novel features of biological processes.

3.2.2 Evaluation of Parameters in the Michaelis-Menten Equation

The Michaelis-Menten equation in its original form [Eq. (3.3)] is not well suited for estimation of the kinetic parameters $v_{\max}$ and K_m. As Fig. 3.7 shows, it is quite difficult to estimate $v_{\max}$ accurately from a plot of v vs. s. By rearranging Eq. (3.3) we can derive the following options for data plotting and graphical parameter evaluation:

$$\frac{1}{v} = \frac{1}{v_{\max}} + \frac{K_m}{v_{\max}} \frac{1}{s} \tag{3.14}$$

$$\frac{s}{v} = \frac{K_m}{v_{\max}} + \frac{1}{v_{\max}} s \tag{3.15}$$

$$v = v_{\max} - K_m \frac{v}{s} \tag{3.16}$$

Each equation suggests an appropriate linear plot. In evaluation of the kinetic parameters of the model using such plots, however, several points should be noted. Plotting Eq. (3.14) as $1/v$ vs. $1/s$ (known as a *Lineweaver-Burk plot*) cleanly separates dependent and independent variables (Fig. 3.11*a*). The most accurately known rate values, near $v_{\max}$, will tend to be clustered near the origin, while those rate values which are least accurately measured will be far from the origin and will tend most strongly to determine the slope $K_m/v_{\max}$. Thus, a linear least-squares fitting should not be used with such a plot. The second equation, (3.15), tends to spread out the data points for higher values of v so that the slope $1/v_{\max}$ can be determined accurately. The intercept often occurs quite close to the origin, so that accurate measure of K_m by this method is subject to large errors. The third method uses the *Eadie-Hofstee plot*: v is graphed against v/s (Fig. 3.11*b*). Both coordinates contain the measured variable v, which is subject to the largest errors.

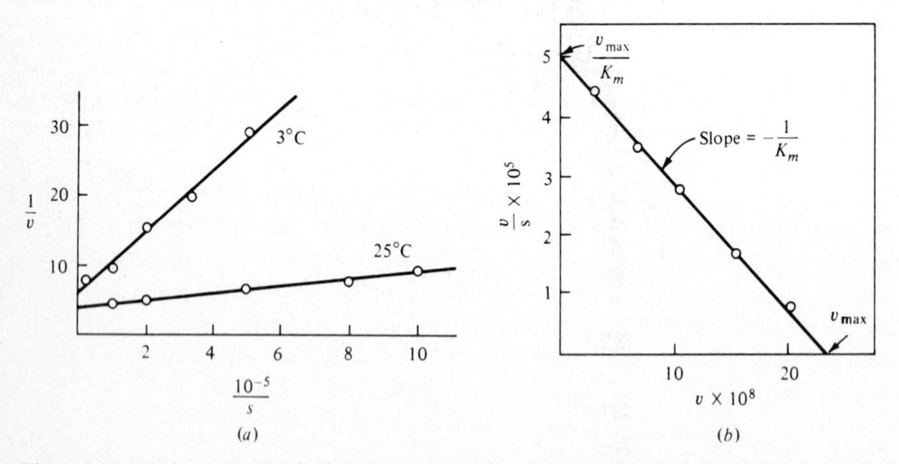

Figure 3.11 (*a*) Lineweaver-Burk plot of experimental data for pepsin. (*b*) Eadie-Hofstee plot of data for hydrolysis of methyl hydrocinnamate catalyzed by chymotrypsin. *(Reprinted by permission from K. J. Laidler, "The Chemical Kinetics of Enzyme Action," pp. 65–66, The Clarendon Press, Oxford, 1958.)*

Table 3.4 Kinetic parameters for several enzymes[†]

Enzyme	Substrate	Temp, °C	pH	k_2, s^{-1}	$1/K_m$, M^{-1}
Pepsin	Carbobenzoxy-L-glutamyl-L-tyrosine ethyl ester	31.6	4.0	0.00108	530
	Carbobenzoxy-L-glutamyl-L-tyrosine	31.6	4.0	0.00141	560
Trypsin	Benzoyl-L-argininamide	25.5	7.8	27.0	480
	Chymotrypsinogen	19.6	7.5	2,900	<770
	Sturin	24.5	7.5	13,100	400
	Benzoyl-L-arginine ester	25.0	8.0	26.7	12,500
Chymotrypsin	Methyl hydrocinnamate	25.0	7.8	0.026	256
	Methyl *dl*-α-chloro-β-phenylpropionate	25.0	7.8	0.135	83.3
	Methyl *d*-β-phenyllactate	25.0	7.8	0.139	28.6
	Methyl-*l*-β-phenyllactate	25.0	7.8	1.38	100
	Benzoyl-L-phenylalanine methyl ester	25.0	7.8	51.0	217
	Acetyl-L-tryptophan ethyl ester	25.0	7.8	30.7	588
	Acetyl-L-tyrosine ethyl ester	25.0	7.8	193.0	31.2
	Benzoyl-L-*o*-nitrotyrosine ethyl ester	25.0	7.8	3.27	90.9
	Benzoyl-L-tyrosine ethyl ester	25.0	7.8	78.0	250
	Benzoyl-L-phenylalanine ethyl ester	25.0	7.8	37.4	167
	Benzoyl-L-methionine ethyl ester	25.0	7.8	0.77	1,250
	Benzoyl-L-tyrosinamide	25.0	7.8	0.625	23.8
	Acetyl-L-tyrosinamide	25.0	7.8	0.279	12.3
Carboxypeptidase	Carbobenzoxyglycyl-L-tryptophan	25.0	7.5	89	196
		25.0	8.2	94	164
	Carbobenzoxyglycyl-L-phenylalanine	25.0	7.5	181	154
	Carbobenzoxyglycyl-L-leucine	25.0	7.5	10.6	37
Adenosine triphosphatase	ATP	25.0	7.0	104	79,000
Urease	Urea	20.8	7.1	20,000	250
		20.8	8.0	30,800	256

[†] K. J. Laidler, *The Chemical Kinetics of Enzyme Action*, p. 67, Oxford University Press, London, 1958.

These considerations suggest the following strategy for evaluating v_{max} and K_m: determine v_{max} from a plot of Eq. (3.14) (find intercept accurately) or Eq. (3.15) (find slope accurately). Then return to a graph of v vs. s and find $s_{1/2}$, the substrate concentration where v is equal to $v_{max}/2$. Recalling the comment following Eq. (3.3), we see that K_m is equal in magnitude and dimension to $s_{1/2}$.

It is important to have a feel for the magnitudes of these kinetic parameters. Table 3.4 shows the range of values encountered for several different enzymes. While k_2 varies enormously, note that K_m has typical values of 2-10 $\times 10^{-3}$ M. Almost all the experiments reported were performed at moderate temperatures and pH values. The exception is pepsin, which has the task of hydrolyzing proteins in the acid environment of the stomach. Consequently, the enzyme has the greatest activity under the acidic conditions employed in the experimental determination of its kinetic parameters. Models for representing the effects of pH and temperature on enzyme kinetics will be explored in Sec. 3.6.

3.2.3 Kinetics for Reversible Reactions, Two-Substrate Reactions, and Cofactor Activation

Many reactions catalyzed by enzymes, such as biopolymer hydrolyses, have equilibria that greatly favor the products so that the reactions may be considered irreversible under most circumstances. In other cases, such as conversion of glucose to fructose by the enzyme glucose isomerase, equilibrium conditions are frequently approached, necessitating consideration of the reverse reaction. In the simplest model for reversible enzyme kinetics, we begin with the model reaction sequence

$$\mathrm{S} + \mathrm{E} \underset{k_{-1}}{\overset{k_1}{\rightleftharpoons}} \mathrm{ES} \qquad (3.17a)$$

$$\mathrm{ES} \underset{k_{-2}}{\overset{k_2}{\rightleftharpoons}} \mathrm{P} + \mathrm{E} \qquad (3.17b)$$

which is the same as the Henri/Michaelis-Menten sequence in Eq. (3.4) except that now formation of ES complex from product combination with the free enzyme E is included in Eq. (3.17b).

Writing the material balances on S, ES, and P in a closed, well-mixed vessel as before, taking advantage of the mass balance on different enzyme forms in Eq. (3.7), and invoking the quasi-steady-state approximation Eq. (3.8), we readily obtain

$$v = -\frac{ds}{dt} = \frac{dp}{dt} = \frac{(v_s/K_s)s - (v_p/K_p)p}{1 + s/K_s + p/K_p} \qquad (3.18)$$

Here, v_s and K_s are identical to v_{max} and K_m in Eqs. (3.10) and (3.11), respectively, and

$$v_p = \frac{k_{-1}v_s}{k_2} \qquad K_p = \frac{k_1 K_s}{k_{-2}} \qquad (3.19)$$

The vast majority of enzyme-catalyzed reactions involve at least two substrates. In many, however, one of the substrates is water, the concentration of which is essentially constant and typically 1000 or more times larger than the concentration of the other substrates. As our analysis in this section will reveal, in such a case the reaction may be treated as in the preceding section, where S is the only substrate considered. Moreover, the kinetic models developed here can be applied in some instances to explain the influence of cofactors on enzyme reaction rates.

In many two-substrate reactions it appears that a ternary complex may be formed with both substrates attached to the enzyme. One possible sequence in this situation is

$$\begin{array}{llll} & & & \text{Dissociation equilibrium constant} \\ E + S_1 & \rightleftharpoons & ES_1 & K_1 \\ E + S_2 & \rightleftharpoons & ES_2 & K_2 \\ ES_1 + S_2 & \rightleftharpoons & ES_1S_2 & K_{12} \\ ES_2 + S_1 & \rightleftharpoons & ES_1S_2 & K_{21} \\ ES_1S_2 & \xrightarrow{k} & P + E & \end{array} \tag{3.20}$$

so that

$$v = k(es_1s_2). \tag{3.21}$$

As before, lowercase symbols are used to denote concentration, and a lowercase symbol in parentheses is the concentation of one species, often a complex. Assuming equilibria in the first four reactions in (3.20) leads to

$$v = \frac{ke_0}{1 + \dfrac{K_{21}}{s_1} + \dfrac{K_{12}}{s_2} + \dfrac{\frac{1}{2}(K_2K_{21} + K_1K_{12})}{s_1s_2}} \tag{3.22}$$

Derivation of this expression is straightforward so long as we remember that the total concentration of enzyme e_0 must be equal to the sum of the concentration of free enzyme e plus the concentrations of the *three* complexes ES_1, ES_2 and ES_1S_2. As for the one-substrate situation, an analysis based on the quasi-steady-state approximation can be undertaken. This leads in the general case to a rather unwieldy equation with too many parameters for practical application. For some parameter values, that equation can be well approximated by Eq. (3.22).

Equation (3.22) can be shortened slightly by noting that equilibrium requires

$$K_1K_{12} = K_2K_{21} \tag{3.23}$$

Rearrangement of (3.22) results in the now familiar form

$$v = \frac{v^*_{max}s_1}{K^*_1 + s_1} \tag{3.24}$$

where

$$v_{\max}^* = \frac{ke_0 s_2}{s_2 + K_{12}} \tag{3.25}$$

and

$$K_1^* = \frac{K_{21}s_2 + K_1 K_{12}}{s_2 + K_{12}} \tag{3.26}$$

From the previous three equations it is apparent that if s_2 is held constant and s_1 is varied, the reaction will follow Michaelis-Menten kinetics. However, Eqs. (3.25) and (3.26) show that the apparent maximum rate and Michaelis constant are functions of S_2 concentration.

Assuming that the above sequence and the rate expression for a two-substrate reaction are correct, we can verify the original Michaelis-Menten form (3.3) when one substrate is in great excess. In this case, $v_{\max}^*$ becomes $= ke_0$, while K_1^* approaches the constant value K_{21}. Therefore, the two-substrate reaction can be treated as though s_1 were the only substrate when $s_2 \gg K_{12}$.

The participation of a cofactor, whether metal ion or coenzyme, in a one-substrate enzymatic reaction (or a two-substrate reaction with $s_2 \gg K_{12}$) can be modeled as in Fig. 3.12. Since this situation resembles so closely the two-substrate mechanism just analyzed, the details of the derivation based on the equilibrium assumption are not necessary. Assuming that substrate complexes only with the apoenzyme-coenzyme complex, the final result is

$$v = \frac{ke_0 cs}{cs + K_s(c + K_c)} \tag{3.27}$$

where c is the cofactor concentration. If substrate concentration s is considered fixed, this rate expression shows a Michaelis-Menten dependence on cofactor

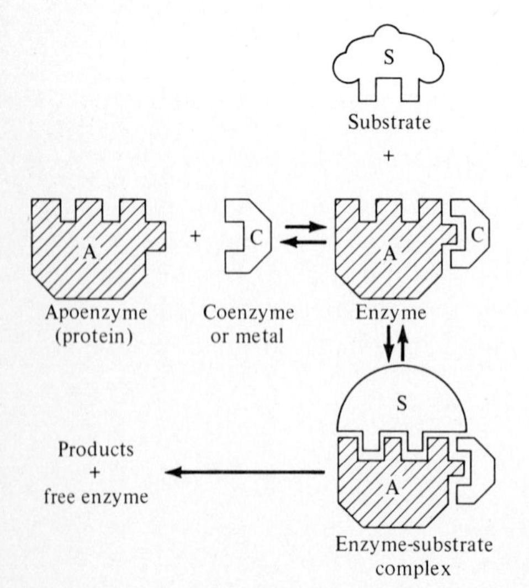

Figure 3.12 Schematic illustration of a plausible mechanism for enzyme catalysis requiring a cofactor.

concentration c. Thus, if there is little cofactor present ($c \ll K_c$), the reaction velocity is first order in c. On the other hand, for $c \gg K_c$, the single-substrate equation is recovered, and the rate is independent of cofactor concentration.

If s_1 and s_2 or c and s_1 must bind to e in an obligatory order, the appropriate kinetic equation is obtained by letting K_{ij} of the forbidden reaction approach infinity.

3.3 Determination of Elementary Step Rate Constants

In Sec. 3.2.2, we explored convenient plotting techniques for determining the parameters v_{max} and K_m which appear in the Michaelis-Menten rate equation. As the Briggs-Haldane derivation reveals, however, these parameters depend on the elementary step rate constants k_1, k_{-1}, and k_2. Examination of the specific relationships in Eqs. (3.10) and (3.11) shows that knowledge of K_m and v_{max} is not sufficient to determine the elementary rate constants. An additional, independent relationship between the elementary step rate constants and experimental data is required.

These more fundamental kinetic parameters are of interest for at least two reasons: (1) They provide a better picture of what occurs during the process of enzyme catalysis. We can learn, for example, how rapidly substrate combines with free enzyme and how this rate compares with the reverse process of ES complex decomposition. (2) We saw earlier that the simplification leading to the Michaelis-Menten equation is rigorously justified only when the total enzyme concentration is relatively small. In cases where this is not true, a careful treatment of enzyme reaction kinetics would involve integration of the mass balances for s, p, and (es). Such an analysis cannot be undertaken, however, unless the elementary step rate constants k_1, k_{-1}, and k_2 are known. With these motivations, we consider experimental techniques for determining these parameters and some of the results of such measurements.

In the *pre-steady-state kinetics* method, attention is focused on the short period ($\sim$1 s or less) just after the reaction is initiated when the quasi-steady-state approximation does not apply. During this interval, substrate concentration changes little, allowing approximate solution of the mass balances for (es) and p. Comparison of the $p(t)$ equation so obtained with experimental data gives, in conjunction with measured values for v_{max} and K_m and their definitions, all the information needed to estimate k_1, k_{-1}, and k_2. The stopped-flow method perfected by Britton Chance and colleagues is the central experimental component of this method. After rapid mixing of substrate and enzyme solutions in an absorption cell to initiate the reaction, product composition changes on a time scale down to milliseconds are monitored by a spectrophotometer.

Relaxation methods, developed and widely applied by Eigen and coworkers, can explore reactions which occur on a time scale of nanoseconds. Although there are several variations of the method, all are based on the same principles: an external perturbation of some reaction condition such as temperature, pressure, or electric field density is applied to a reaction mixture at equilibrium (or steady state). The resulting response of the reaction system is then monitored continuously. As illustrated schematically in Fig. 3.13, the response following a sudden step change in reaction conditions is a transition to a new, nearly equilibrium or steady state.

3.3.1 Relaxation Kinetics

Here we shall explore the step-change relaxation method. The theory and practice of relaxation methods has also been developed for oscillatory perturbations in reaction conditions. Let us consider for the moment only the equilibrium between substrate, enzyme and enzyme-substrate complex:

$$\mathrm{E} + \mathrm{S} \underset{k_{-1}}{\overset{k_1}{\rightleftharpoons}} \mathrm{ES}$$

For this reaction, we know that

$$s + (es) = s_0$$

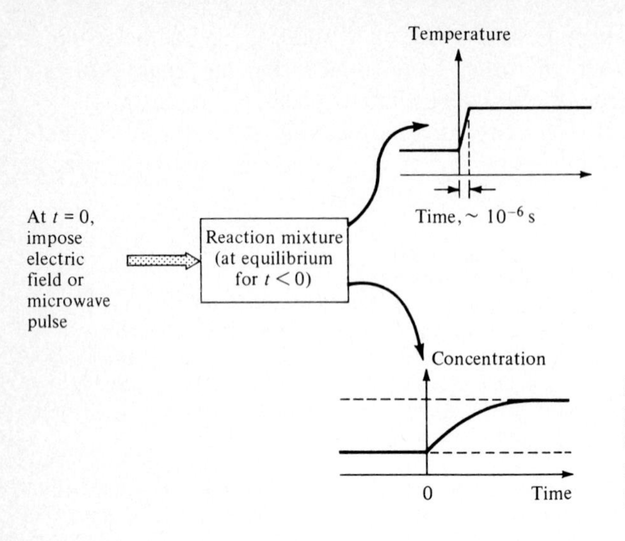

Figure 3.13 In a relaxation experiment, a small step change in reaction conditions, e.g., in temperature, causes a transient approach to a new steady state.

and

$$e + (es) = e_0$$

so that in a well-mixed batch reactor, the mass balance for s

$$\frac{ds}{dt} = k_{-1}(es) - k_1 se \tag{3.28}$$

becomes

$$\frac{ds}{dt} = k_{-1}(s_0 - s) - k_1 s(e_0 + s - s_0) \equiv f(s) \tag{3.29}$$

If we let s^* denote the s concentration at equilibrium *after* the step change in reaction conditions, s^* can be determined by solving

$$f(s^*) = 0 \tag{3.30}$$

where $f(s)$ is the right-hand side of Eq. (3.29) and is equal to ds/dt. In any relaxation experiment s will be close to s^*, and so we can approximate $f(s)$ by the first terms in its Taylor expansion

$$f(s) \approx \underbrace{f(s^*)}_{\text{zero because of Eq. (3.30)}} + \frac{df(s^*)}{ds}(s - s^*) + \text{terms of order } (s - s^*)^2 \text{ and higher} \tag{3.31}$$

Letting χ be the deviation from the equilibrium concentration

$$\chi = s - s^* \tag{3.32}$$

and remembering that s^* is time-invariant, we can combine Eqs. (3.29) to (3.32) to obtain the *linearized* mass balance [squared and higher-order terms neglected in (3.31)]

$$\frac{d\chi}{dt} = -[k_{-1} + k_1(e_0 - s_0 + 2s^*)]\chi \tag{3.33}$$

Table 3.5 Rate constants for several elementary substrate-enzyme reactions†‡

Protein or enzyme	Substrate	k_1, $(M \cdot s)^{-1}$	k_{-1}, s^{-1}
Fumarase	Fumarate	$> 10^9$	$>4.5 \times 10^4$
	L-Malate	$> 10^9$	$>4 \times 10^4$
Acetylcholinesterase	Acetylcholine	$> 10^9$	
Urease	Urea	$>5 \times 10^6$	
Myosin ATPase	ATP	$>8 \times 10^6$	
Hexokinase	Glucose	3.7×10^6	1.5×10^3
	MgATP	$>4 \times 10^6$	$>6 \times 10^2$
β-Amylase	Amylose	$>5.8 \times 10^7$	
Liver alcohol dehydrogenase	NAD	5.3×10^5	74
	NADH	1.1×10^7	3.1
Liver alcohol dehydrogenase-NAD	C_2H_5OH	$>1.2 \times 10^4$	>74
	CH_3CHO	$>3.7 \times 10^5$	>3.1
Malate dehydrogenase	NADH	6.8×10^8	2.4×10^2
Old yellow enzyme	FMN	1.5×10^6	$\sim 10^{-4}$
Catalase	H_2O_2	5×10^6	
Catalase-H_2O_2	H_2O_2	1.5×10^7	
Peroxidase	H_2O_2	9×10^6	<1.4
Peroxidase-H_2O_2(II)	Hydroquinone	2.5×10^6	
Peroxidase	Cytochrome *c*	1.2×10^8	
Glutamic-aspartic transaminase, aldehydic	Glutamate	3.3×10^7	2.8×10^3
	Aspartate	$> 10^7$	$>5 \times 10^3$
	NH_2OH	3.7×10^6	38
	Ketoglutarate, oxalacetate	$>5 \times 10^8$	$>5 \times 10^4$
Aminic	Oxalacetate	7×10^7	1.4×10^2
	Ketoglutarate	2.1×10^7	70
BSA	NR′	2×10^6	35
	NSR′	3.5×10^5	2.5
Anti-R antibodies§	NR′	2.2×10^7	50
Anti-DNP antibodies§	DNP-lysine	8.1×10^7	1.1
$Hb(O_2)_3$	O_2	2×10^7	36
Globin	Carboxyheme	7×10^7	

† H. R. Mahler and E. H. Cordes, *Biological Chemistry*, 2d ed., p. 322, Harper & Row, Publishers, Incorporated, New York, 1971.

‡ FMN, flavin mononucleotide; BSA, bovine serum albumin; NR′, 1-naphthol-4-(4′-azobenzene azophenyl)arsonic acid; NSR′, [naphthol-3-sulfonic acid 4-(4′-azobenzene azo)]phenylarsonic acid; DNP, 2,4-dinitrophenyl; Hb, hemoglobin.

§ Rabbit γ-globulin.

If the system is initially at the old equilibrium corresponding to conditions before the step perturbation

$$\chi(0) = \Delta\chi_0 \tag{3.34}$$

and consequently

$$\chi(t) = \Delta\chi_0 e^{-t/\tau} \tag{3.35}$$

where

$$\frac{1}{\tau} = k_{-1} + k_1(e_0 - s_0 + 2s^*) = k_{-1} + k_1(e^* + s^*) \tag{3.36}$$

In Eq. (3.36), e^* represents the concentration of uncomplexed enzyme at the final equilibrium state. From Eq. 3.35, it is clear that τ can be determined from the slope of a semilog plot of $\chi(t)/\Delta\chi_0$. Alternatively, τ is the time required for the deviation $\chi(t)$ to decay to 37 percent of its initial value. After τ, χ^*, and e^* (or e_0 and s_0) have been measured, Eq. (3.36) provides a relationship between k_1 and k_{-1} which is independent of the equilibrium equation. Consequently, both k_1 and k_{-2} can be calculated.

This technique has enjoyed wide application in chemical kinetics research, including investigations of enzyme catalysis which we now review.

3.3.2 Some Results of Transient-Kinetics Investigation

Listed in Table 3.5 are k_1 and k_{-1} values for a variety of enzymes and substrates. These numbers were obtained by the methods just outlined as well as other specialized approaches described in the references. Examining the rate constant k_1 for combination of enzyme and substrate, we see that measured values are all very near the theoretical maximum, since the absolute bimolecular-collision rate constant in solution is 10^9 to 10^{11} $[(\text{mol-s})/\text{L}]^{-1}$. On the other hand, the reverse process is relatively slow and exhibits considerably wider variation from reaction to reaction. The same comments apply to decomposition of the intermediate complex into product and free enzyme (see k_2 values in Table 3.4).

Another interesting feature of enzyme catalysts which has been revealed by pre-steady-state and relaxation studies is the presence of more than one simple reaction intermediate ES. In some cases the intermediate complex passes through a series of different configurations which may be regarded as isomerizations. Steady-state experiments are incapable of distinguishing whether more than one reaction intermediate exists when the dissociation rate (to products) of the last such intermediate complex is slow compared with that of earlier complexes. Transient relaxation studies provide a valuable probe for such situations.

3.4 OTHER PATTERNS OF SUBSTRATE CONCENTRATION DEPENDENCE

Not all rates of enzyme-catalyzed reactions follow Michaelis-Menten kinetics as described in Sec. 3.2. In this section we shall examine the more common forms of deviation. The first kind of unusual behavior, often associated with regulatory enzymes, can be explained by reaction mechanisms where more than one substrate molecule can bind to the enzyme. In the second case, we shall consider the possibility that the substrate is actually a mixture of different reacting species with different kinetic properties. This analysis will reveal some of the problems encountered when working with impure or poorly defined substrate mixtures.

3.4.1 Substrate Activation and Inhibition

The sigmoidal dependence of reaction rate on substrate concentration exhibited by some enzymes is evidence of a type of activation effect (Fig. 3.14). At low substrate concentrations, the bonding of one substrate molecule enhances the binding of the next one (mathematically stated, the increment in v resulting from an increment in s, dv/ds, is increasing). Such behavior can be modeled by assuming a concerted transition of protein subunits: we assume that the enzyme, which is probably oligomeric, has multiple binding sites for substrate and that the first substrate molecule bound to the enzyme alters the enzyme's structure so that the remaining sites have a stronger affinity for the substrate. Since the mathematical analysis for this model parallels the hemoglobin model outlined in Prob. 2.10, we need not pursue it here.

Sometimes when a large amount of substrate is present, the enzyme-catalyzed reaction is diminished by the excess substrate (Fig. 3.15). This phenomenon is called *substrate inhibition*. We should note from Fig. 3.15 that the reaction rate v passes through a maximum as substrate concentration is increased. When s is larger than $s(v = v_{max})$, a decrease in substrate concentration causes an increase in reaction rate. This *autocatalytic* feature of this type of kinetics can have important implications on the behavior of biochemical reactors.

The quantitative relationship between substrate concentration and reaction rate for substrate-inhibited reactions can be modeled quite nicely using the Michaelis-Menten approach. In the assumed mechanism here, however, a second substrate molecule can bind to the enzyme. When S joins the ES complex, an unreactive intermediate results:

Reaction steps at equilibrium:

Dissociation constant

$$\mathrm{E} + \mathrm{S} \underset{k_{-1}}{\overset{k_1}{\rightleftharpoons}} \mathrm{ES} \qquad K_1\left(\text{for example, } \frac{k_{-1}}{k_1}\right) \tag{3.37}$$

$$\mathrm{ES} + \mathrm{S} \rightleftharpoons \mathrm{ES}_2 \qquad K_2$$

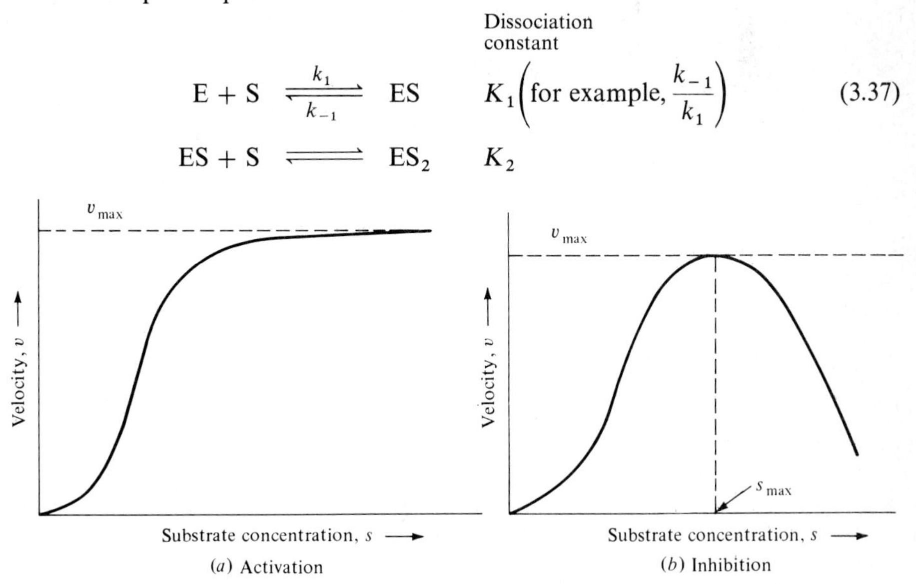

Figure 3.14 Substrate activation (*a*) and inhibition (*b*).

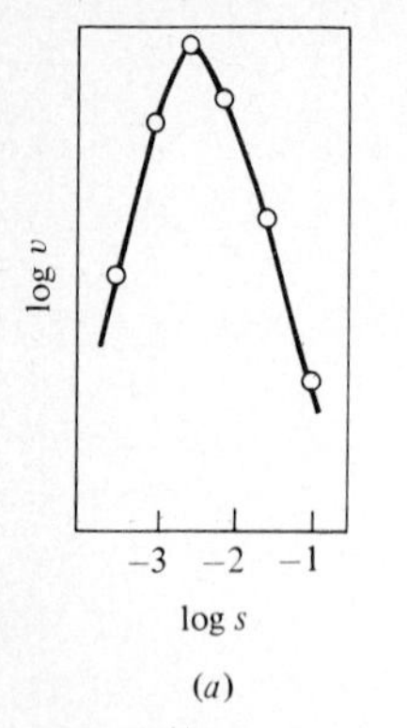

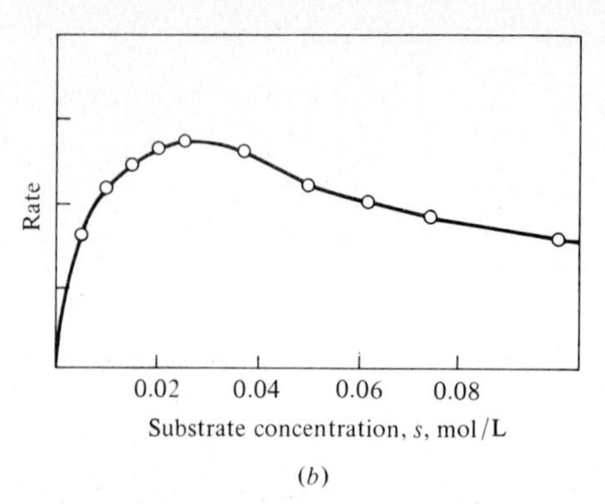

Figure 3.15 Experimental evidence of substrate inhibition. *(K. J. Laidler, "The Chemical Kinetics of Enzyme Action," p. 71, The Clarendon Press, Oxford, 1958. (a) Data of K.-D. Augustinsson, "Substrate Concentration and Specificity of Choline Ester Splitting Enzymes," Arch. Biochem. vol. 23, 111, 1949; (b) Data of R. Lumry, E. L. Smith and R. R. Glantz, "Kinetics of Carboxypeptidase Action. I. Effect of Various Extrinsic Factors on Kinetic Parameters," J. Am. Chem. Soc., vol. 73, p. 4330, 1951.)*

Slow step:
$$\mathrm{ES} \xrightarrow{k} \mathrm{E + P} \tag{3.38}$$

When the two dissociation-equilibrium relationships and an enzyme balance analogous to Eq. (3.7) are used, straightforward algebraic manipulations yield

$$v = \frac{ke_0}{1 + K_1/s + s/K_2} \tag{3.39}$$

It is not difficult to deduce the necessary parameters in this equation from experimental data. First, k can be determined by varying e_0. Next, by plotting $1/v$ against s, a straight line with slope $1/K_2ke_0$ is obtained at high substrate concentrations (Fig. 3.16). Finally, we can evaluate K_1 from Eq. (3.40) where s_{max} satisfies $dv/ds = 0$ using Eq. (3.39)

$$s_{max} = \sqrt{K_1K_2} \tag{3.40}$$

3.4.2 Multiple Substrates Reacting on a Single Enzyme

Many enzymes can utilize more than a single substrate. In such cases, the various reactants compete with each other for the active site, which is here presumed to be the same for each reaction. The most obvious examples of such enzymes are hydrolytic depolymerases, which act on many identical bonds, often regardless of the length of the polymer segment. When the substrate takes the form of a dimer, trimer, or oligomer, the kinetic parameters characterizing the reaction may change from the values applicable to large polymeric units. An example of such a system is lysozyme, the enzyme found in egg whites, mucus, and tears. Lysozyme hydrolyses complex mucopolysaccharides, including the murein cell wall, thereby killing gram-positive bacteria. The previous comments are reflected in Table 3.6,

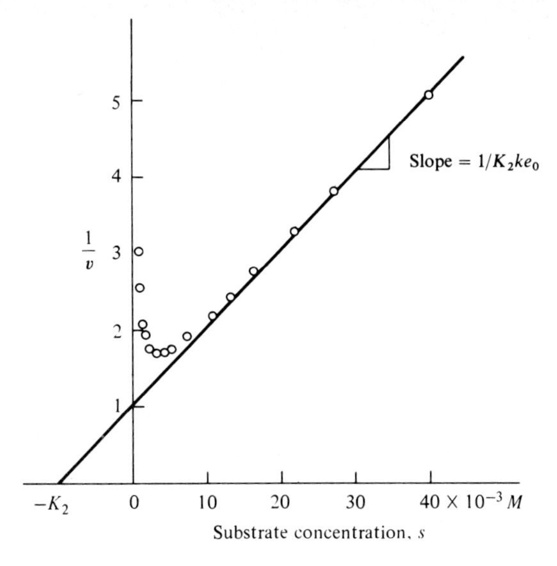

Figure 3.16 K_2 can be determined from the slope of a $1/v$ vs. s plot. Data for hydrolysis of ethyl butyrate by sheep liver carboxylesterase. *(Reprinted by permission from M. Dixon and E. C. Webb, "Enzymes," 2d ed., p. 78, Academic Press, Inc., New York, 1964.)*

which summarizes the measured rate and dissociation constants for each lysozyme substrate when it is the only substrate present. The parameter variations are seen to be greatest with the monomer and oligomers.

In the above examples, as in the following one, each condensation linkage in the oligomer or polymer can be viewed conceptually as a separate substrate. Also, a polymeric substrate like starch is not really a single substrate in the normal sense, for the polymer is a mixture of long chains of different lengths, which may also contain various monomer-monomer links (Sec. 2.2.2). Shown in

Table 3.6 Kinetic parameters for lysozyme action on bacterial cell wall (GlcNAc-MurNAc copolymer)†

GlcNAc = N-acetylglucosamine, MurNAc = N-acetylmurein

	Reaction	Type
1	$S_i = (\text{GlcNAc-MurNAc})_i = i$ unit oligomer	
	$E + S_i \underset{}{\overset{K_{ij}}{\rightleftharpoons}} C_{ij} \quad j = 0, \ldots, i+1$	Association
2	$C_{ij} \xrightarrow{k_c} A_j + S_{i-j} \quad j = 1, \ldots, i-1$	Hydrolysis of complex
3	$C_{ii} \xrightarrow{k_v} A_i + H_2O$	Dissociation of intermediate ($j = i$)
4	$A_i(+H_2O) \xrightarrow{k_H} E + S_i$	Hydrolysis of intermediate
5	$A_i + S_j \xrightarrow{k_T} E + S_{i+j}$	Transglycosylation to yield $i + j$ unit

$K_{11} = 0.5\ M^{-1}$ $\quad K_{21} = 6.3 \times 10^{-4}\ M^{-1}$ $\quad K_{22} = 2000\ M^{-1}$ $\quad K_{32} = 3.5 \times 10^4\ M^{-1}$

$k_c = 1.75\ s^{-1}$ $\quad k_v = 100\ s^{-1}$ $\quad k_T = 2.77 \times 10^5 (M \cdot s)^{-1}$

† After D. M. Chipman, A. Kinetic Analysis of the Reaction of Lysozyme with Oligosaccharides from Bacterial Cell Walls, *Biochemistry*, **10**: 1714, 1971.

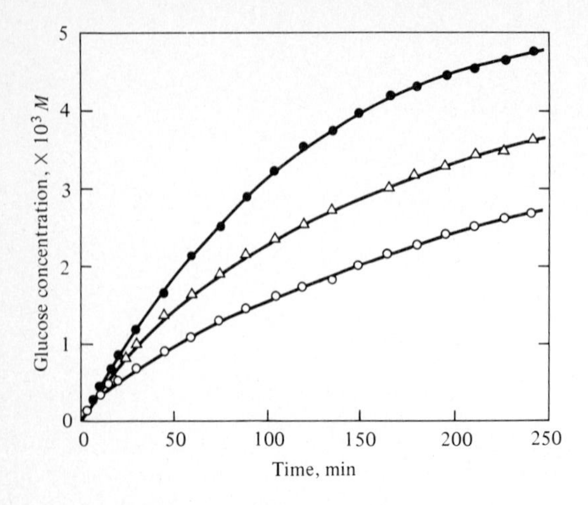

Figure 3.17 Reaction-time curves for hydrolysis of different amylose fractions using amyloglucosidase (25°C, pH 4.6, $s_0 = 0.1$ g/100 mL, $e_0 = 6.8 \times 10^{-8}$ M): ○ mol wt ≈ 1,650,000, △ mol wt ≈ 1,100,000, ● mol wt ≈ 360,000. *[(Reprinted by permission from K. Hiromi et al., "A Kinetic Method for the Determination of Number-Average Molecular Weight of Linear High Polymer by Using an Exo-Enzyme," J. Biochem. (Tokyo) vol. 60, p. 439, 1966.]*

Fig. 3.17 are the differences in the course of enzymatic hydrolysis of several starches. The kinetic behavior of a better-defined system is illustated in Fig. 3.18. In commercial processes for thinning of starch solutions by amylases, these reactions are important. Similar effects are expected in analogous reactions, e.g., hydrolysis of proteins with proteolytic enzymes. Before proceeding to a quantitative treatment of competition between different substrates, we should point out that other phenomena may be involved in the above examples, including variations in physical size and state and differences in the number of reactive bonds in the diverse substrates in the reaction mixture.

Some hydrolytic enzymes exhibit selectivity for different regions of a biopolymer chain. An *exo*-enzyme cleaves the chain at or near the chain end, while *endo*-hydrolases catalyze hydrolysis of internal bonds in the chain. In kinetic descriptions of such systems, it is important to consider both types of enzyme action and the concentrations of internal and end linkages in the biopolymer mixture. We will encounter such an approach in our examination of enzymatic hydrolysis of cellulose in the next chapter.

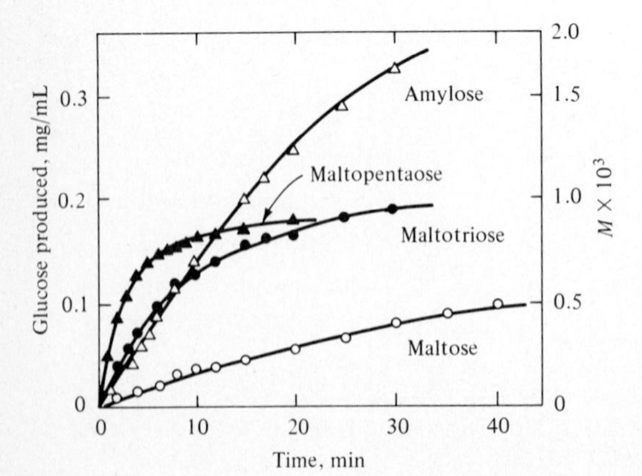

Figure 3.18 Reaction-time curves for glucose oligomers and polymers (15°C, pH 5.15, $s_0 = 0.04$ M, $e_0 = 2.82 \times 10^{-7}$ M). The enzyme is amyloglucosidase. *[S. Ono et al., "Kinetic Studies of Gluc-amylase," J. Biochem. (Tokyo), vol. 55, p. 315, 1964.]*

To gain a perspective of some of the difficulties involved in kinetic description of reactions of mixtures, we consider next a simple illustrative example. Suppose that the sequence below describes the reactions of two different substrates catalyzed by one enzyme:

$$\begin{array}{llll} & & & \text{Dissociation constant} \\ E + S_1 & \rightleftharpoons & ES_1 & K_1 \\ E + S_2 & \rightleftharpoons & ES_2 & K_2 \end{array} \tag{3.41}$$

$$\text{Slow steps:} \quad ES_1 \xrightarrow{k_1} E + P_1 \qquad ES_2 \xrightarrow{k_2} E + P_2 \tag{3.42}$$

Here, each substrate binds a certain fraction of the enzyme present. When we employ the overall enzyme balance and equilibria, as in our earlier analysis, the rate expressions are:

$$-\frac{ds_1}{dt} = v_1 = \frac{k_1 e_0 s_1/K_1}{1 + s_1/K_1 + s_2/K_2} \tag{3.43}$$

and

$$-\frac{ds_2}{dt} = v_2 = \frac{k_2 e_0 s_2/K_2}{1 + s_1/K_1 + s_2/K_2} \tag{3.44}$$

Equations (3.43) and (3.44) indicate that if two reactions are catalyzed by the same enzyme, the individual velocities are slower in the presence of both substrates than in the absence of one of the substrates. This fact has been used advantageously to determine whether the same enzyme in an undefined sample acts upon both substrates or each substrate's reaction is catalyzed by a separate enzyme.

Often in experiments and process operation, it is neither possible nor practical to measure and monitor the concentrations of all species present in the reaction mixture. For example, in Fig. 3.18, the plotted glucose production is actually the net overall effect of many simultaneous reactions, including hydrolysis of maltose, maltotriose, and so forth. Stated in other terms, the overall reaction considered in the kinetic analysis is actually

$$\text{Glucose polymers} \xrightarrow{\text{amyloglucosidase}} \text{glucose}$$

Thus, all the species containing glucose subunits have been *lumped* together into one imaginary species whose concentration is relatively easy to measure. Lumping of this type is extremely widespread in all branches of chemical kinetics. Consider, for example, catalytic cracking of gas oil, which is often usefully represented as a three-species reaction

$$\text{Gas oil} \rightarrow \text{gasoline} \rightarrow \text{gases}$$

when designing or analyzing the reactor although each "species" is clearly a mixture of compounds. Although not often explicitly recognized in the literature of the life sciences and biotechnology, lumping is a pervasive practice when investigating, modeling, or designing biological processes. In this book, we shall watch carefully for this and other simplifying assumptions employed, usually implicitly, in describing biological systems. An unjustified assumption can cause a breakdown in the analysis and the design.

One possible pitfall of lumping can be illustrated with the present example. If s_T is the total concentration of all substrates present

$$s_T = s_1 + s_2 \tag{3.45}$$

then the overall rate of disappearance of s_T is

$$v_T = -\frac{ds_T}{dt} = -\left(\frac{ds_1}{dt} + \frac{ds_2}{dt}\right) = -\frac{e_0(k_1 s_1/K_1 + k_2 s_2/K_2)}{1 + s_1/K_1 + s_2/K_2} \tag{3.46}$$

By varying the relative values of s_1 and s_2, this rate can be changed. For example, if the total substrate concentration has a specified value s_{T0}, the overall rate v_T does *not* in general have a unique value. Instead, it may lie anywhere in the range indicated below, depending on the s_1/s_2 ratio:

$$v_T\Big|_{\substack{s_1 = s_{T0} \\ s_2 = 0}} = \frac{e_0 k_1 s_{T0}}{K_1 + s_{T0}} \quad \xrightarrow[\text{increase } s_2]{\text{decrease } s_1} \quad v_T\Big|_{\substack{s_1 = 0 \\ s_2 = s_{T0}}} = \frac{e_0 k_2 s_{T0}}{K_2 + s_{T0}} \tag{3.47}$$

Clearly, then, the values of v_{max} and K_m we would measure in a kinetic experiment would depend on the detailed composition (s_1/s_2 ratio) as well as the total substrate concentration. Equally, the performance of an enzyme reactor fed a mixture of S_1 and S_2 will change if their relative amounts do, even though the total feed reactant concentration is maintained constant.

3.5 MODULATION AND REGULATION OF ENZYMATIC ACTIVITY

Chemical species other than the substrate can combine with enzymes to alter or modulate their catalytic activity. Such substances, called *modulators* or *effectors*, may be normal constituents of the cell. In other cases they enter from the cell's environment or act on isolated enzymes. Although most of our attention in this section will be concentrated on *inhibition*, where the modulator decreases activity, cases of enzyme *activation* by effectors are also known.

The combination of an enzyme with an effector is itself a chemical reaction and may therefore be fully reversible, partially reversible, or essentially irreversible. Known examples of irreversible inhibitors include *poisons* such as cyanide ions, which deactivate xanthine oxidase, and a group of chemicals collectively termed *nerve gases*, which deactivate cholinesterases (enzymes which are an integral part of nerve transmission and thus of motor ability). If the inhibitor acts irreversibly, the Michaelis-Menten approach to inhibitor-influenced kinetics cannot be used since this method assumes equilibrium between the free and complexed forms. Often, irreversible inhibition increases with time as more and more enzyme molecules are gradually deactivated. Other cases, more difficult to detect, involve the partial deactivation of enzyme. In such an instance, the inhibited enzyme retains catalytic activity although at a level reduced from the pure form.

Section 2.6 mentioned the remarkable chemical capabilities of most microorganisms which, supplied with only a few relatively simple precursors, manufacture a vast array of complex molecules. In order to perform this feat, it is necessary for the supply of precursors to be proportioned very efficiently among the many synthesis routes leading to many end products. Optimum utilization of the available chemical raw materials in most instances requires that only the necessary amount of any end product be manufactured. If enough of one monomer is present, for example, its synthesis should be curtailed and attention devoted to making other compounds which are in relatively short supply.

Reversible modulation of enzyme activity is one control mechanism employed by the cell to achieve efficient use of nutrients. (The other major control device is discussed in Sec. 6.1.) The most intriguing examples of enzyme regulation involve interconnected networks of reactions with several control loops, but their analysis must wait until we have studied the basic chemistry of cellular operation in Chap. 5. For present purposes, we shall employ control of a single sequence of reactions as an example. Figure 3.19 shows a five-step sequence for the biosynthesis of the amino acid L-isoleucine. Regulation of this sequence is achieved by *feedback inhibition*: the final product, L-isoleucine, inhibits the activity of the first enzyme in the path. Thus, if the final product begins to build up, the biosynthesis process will be stopped.

Since the reactions catalyzed by enzymes E_2 through E_5 are essentially at equilibrium and the first reaction is "irreversible" under cellular conditions, the response of this control device is especially fast. Indeed, it is a general property of most regulatory enzymes that they catalyze "irreversible" reactions. Also, it should be obvious from the biological context that such "natural" enzyme modulation must be reversible. For example, if L-isoleucine is depleted by its use in protein synthesis, inhibition of enzyme E_1 must be quickly relaxed so that the required supply of the amino acid is restored.

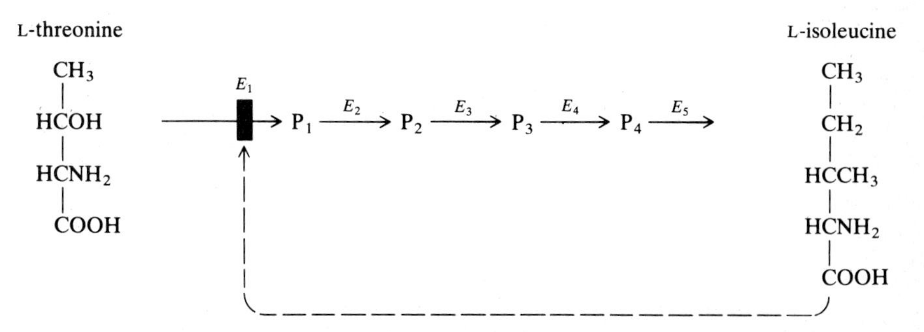

Figure 3.19 In this feedback-inhibition system, the activity of enzyme E_1 (L-threonine deaminase) is reduced by the presence of the end product, L-isoleucine. ($P_1 = \alpha$-ketobutyrate, $P_2 = \alpha$-acetohydroxybutyrate, $P_3 = \alpha,\beta$-dihydroxy-β-methylvalerate, $P_4 = \alpha$-keto-β-methylvalerate.)

Table 3.7 Partial classification of reversible inhibitors[†]

	Type	Description	Result
I*a*	Fully competitive	Inhibitor adsorbs at substrate binding site	Increase in apparent value of K_m
b	Partially competitive	Inhibitor and substrate combine with different groups; inhibitor binding affects substrate binding	Increase in apparent value of K_m
II*a*	Noncompetitive	Inhibitor binding does not affect ES affinity, but ternary EIS (enzyme-inhibitor-substrate) complex does not decompose into products	No change in K_m, decrease of v_{max}
b	Noncompetitive	Same as II*a* except that EIS decomposes into product at a finite rate different from that of ES	No change in K_m, decrease of v_{max}
III	Mixed inhibitor		Affects both K_m and v_{max}

[†] M. Dixon and E. C. Webb, *Enzymes*, 2d ed., table VIII.1, Academic Press, Inc., New York, 1964.

In the case of reversible inhibition, the approaches described in Sec. 3.2 prove quite fruitful. Since many isolated enzyme-substrate systems exhibit Michaelis-Menten kinetics, it is traditional to classify inhibitors by their influence on the Michaelis-Menten equation parameters v_{max} and K_m.

Reversible inhibitors are termed *competitive* if their presence increases the value of K_m but does not alter v_{max}. The effect of such inhibitors can be countered or reversed by increasing the substrate concentration. On the other hand, by rendering the enzyme or the enzyme-substrate complex inactive, a *noncompetitive* inhibitor decreases the v_{max} of the enzyme but does not alter the K_m value. Consequently, increasing the substrate concentration to any level cannot produce as great a reaction rate as possible with the uninhibited enzyme. Common noncompetitive inhibitors are heavy-metal ions, which combine reversibly with the sulfhydryl (—SH) group of cysteine residues.

Several different combinations and variations on the two basic types of reversible inhibitors are known. Some of these are listed in Table 3.7. Experimental discrimination between these possibilities will be discussed shortly. First, however, we shall briefly review some of the current theories and available data on the mechanisms of modulator action.

3.5.1 The Mechanisms of Reversible Enzyme Modulation

Many known competitive inhibitors, called *substrate analogs*, bear close relationships to the normal substrates. Thus it is thought that these inhibitors have the key to fit into the enzyme active site, or lock, but that the key is not quite right so the lock does not work; i.e., no chemical reaction results. For example, consider the inhibition of succinic acid dehydrogenation by malonic acid:

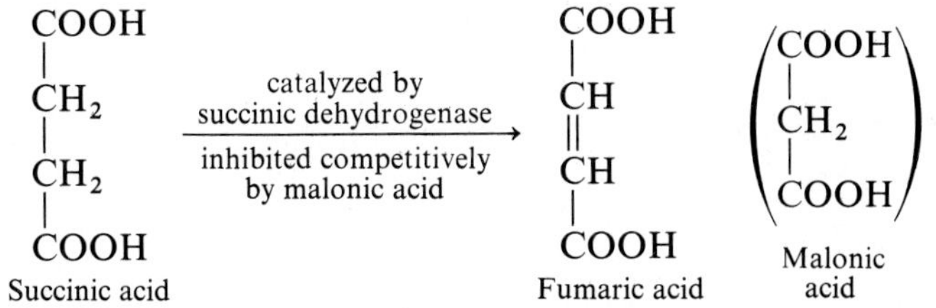

$$\begin{array}{c} COOH \\ | \\ CH_2 \\ | \\ CH_2 \\ | \\ COOH \\ \text{Succinic acid} \end{array} \xrightarrow[\text{inhibited competitively by malonic acid}]{\text{catalyzed by succinic dehydrogenase}} \begin{array}{c} COOH \\ | \\ CH \\ \| \\ CH \\ | \\ COOH \\ \text{Fumaric acid} \end{array} \qquad \left(\begin{array}{c} COOH \\ | \\ CH_2 \\ | \\ COOH \end{array}\right) \begin{array}{c} \\ \\ \\ \text{Malonic acid} \end{array}$$

The malonic acid can complex with succinic dehydrogenase, but it does not react.

The action of one of the sulfa drugs, sulfanilamide, is due to its effect as a competitive inhibitor. Sulfanilamide is very similar in structure to *p*-aminobenzoic acid, an important vitamin for many bacteria. By inhibiting the enzyme which causes *p*-aminobenzoic acid to react to give folic acid, the sulfa drug can block the biochemical machinery of the bacterium and kill it.

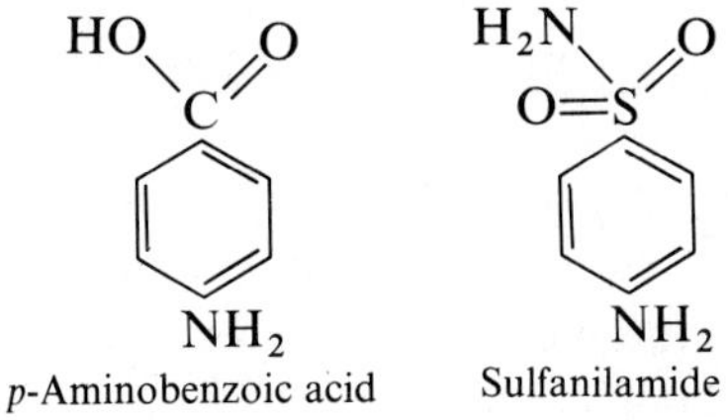

HO–C(=O)–C$_6$H$_4$–NH_2 H_2N–S(=O)$_2$–C$_6$H$_4$–NH_2

p-Aminobenzoic acid   Sulfanilamide

Another mechanism, called *allosteric control*, yields behavior typical of noncompetitive inhibition and is thought to be the dominant mechanism for noncompetitive inhibition and activation. The name allosteric (other shape) was originally coined for this mechanism because many effectors of enzymic activity have structures much different from the substrate. From this fact it has been concluded that effectors work by binding at specific regulatory sites distinct from the sites which catalyze substrate reactions. An enzyme which possesses sites for modulation as well as catalysis has consequently been named an *allosteric enzyme.*

Allosteric control may either inhibit (reduce) or activate (increase) the catalytic ability of the enzyme. One schematic view of this process is depicted in Fig. 3.20. Notice that the enzyme here is visualized as having two pieces. Many allosteric enzymes are known to be oligomeric.

Experimental studies on one oligomeric allosteric enzyme, aspartyl transcarbamoylase (ATCase), have provided the most dramatic evidence to date in support of the allostery theory of enzyme activity control. The enzyme was physically separated into two subunits. It was found that the larger subunit, which is catalytically active, is not influenced by CTP (cytidine triphosphate), an inhibitor of the intact ATCase enzyme. The smaller subunit, on the other hand, has no catalytic activity but does bind CTP.

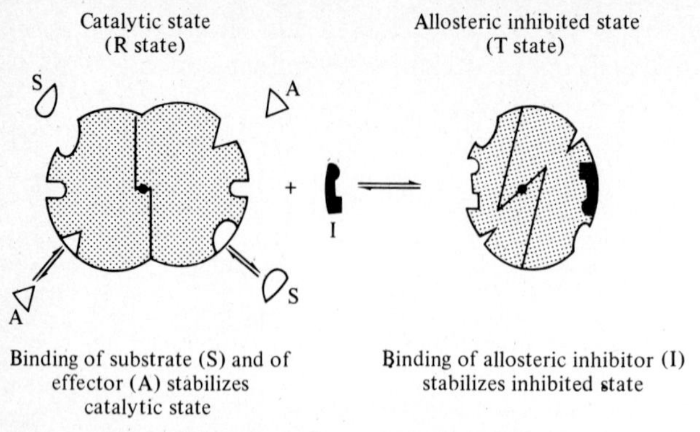

Figure 3.20 In the symmetry model for allosteric control of enzyme activity, binding of an activator A and substrate S gives a catalytically active R enzyme configuration, while binding of an inhibitor changes all subunits in the oligomeric protein molecule to an inactive T state. *(From "Cell Structure and Function," 2d ed., p. 265, by Ariel G. Loewy and Philip Siekevitz. Copyright © 1963, 1969 by Holt, Rinehart and Winston, Inc. Reprinted by permission of Holt, Rinehart and Winston.)*

In the absence of detailed information of the type just mentioned for ATCase, we cannot be certain that an effector does not bind at the active site. All that matters from the kinetic viewpoint, however, is that substrate affinity, unbound enzyme concentration, and/or the rate of complex breakdown are changed by presence of the modulator. In the following section, we shall seek mathematical formulas suitable for representing these effects.

3.5.2 Analysis of Reversible Modulator Effects on Enzyme Kinetics

A major contribution of the Michaelis-Menten approach to enzyme kinetics is accounting quantitatively for the influence of modulators. To begin our analysis, we shall assume that the following sequence reasonably approximates the interactions of a (totally) *competitive* inhibitor and substrate with enzyme. Here E and S have the usual significance, and I denotes the inhibitor. Also, EI is an enzyme-inhibitor complex.

Reaction steps at equilibrium:

Dissociation constant

$$\mathrm{E} + \mathrm{S} \rightleftharpoons \mathrm{ES} \qquad K_s \tag{3.48}$$

$$\mathrm{E} + \mathrm{I} \rightleftharpoons \mathrm{EI} \qquad K_i$$

Slow step:

$$\mathrm{ES} \xrightarrow{k} \mathrm{E} + \mathrm{P} \tag{3.49}$$

In this case binding of substrate and inhibitor to enzyme are mutually exclusive. Because some enzyme is bound in the EI complex, not all the enzyme is available for catalyzing substrate conversion, so the reaction rate is lowered by the inhibitor.

Making use of the equilibrium relationships indicated in Eq. (3.48) and a total mass balance on enzyme

$$e + (es) + (ei) = e_0 \tag{3.50}$$

we can write the reaction rate in terms of *total* enyme concentration e_0 and *free* substrate and inhibitor concentrations (s and i, respectively):

$$v = \frac{ke_0 s}{s + K_s(1 + i/K_i)} \qquad \textit{(totally) competitive inhibition} \tag{3.51}$$

Comparing this with the original Michaelis-Menten form shows that v_{max} is unaffected, but the apparent Michaelis constant K_m^{app}, here

$$K_m^{app} = K_s\left(1 + \frac{i}{K_i}\right) \tag{3.52}$$

is increased by the presence of the competitive inhibitor. Stated in different terms, the rate reduction caused by a competitive inhibitor can be completely offset by increasing the substrate concentration sufficiently; the maximum possible reaction velocity is not affected by the competitive inhibitor.

We can obtain a useful model for (totally) *noncompetitive* inhibition by adding the following equilibrium steps to the reaction sequence given in Eqs. (3.48) and (3.49) above:

			Dissociation constant	
$EI + S$	$\rightleftharpoons$	EIS	K_s	(3.53*a*)
$ES + I$	$\rightleftharpoons$	EIS	K_i	(3.53*b*)

In this situation, inhibitor and substrate can simultaneously bind to the enzyme, forming the ternary complex designated EIS. We assume in this simplest noncompetitive inhibition model that binding of either inhibitor or substrate does not influence the affinity of either species to complex with the enzyme. Consequently, K_s and K_i in Eq. (3.53) are identical to the corresponding dissociation constants in Eq. (3.48). Also, it is assumed here that the EIS complex does not react to give product P.

Using the now familiar procedure of writing the complex concentrations in terms of e_0, s, and i, we obtain the following rate expression:

$$v = \frac{ske_0/(1 + i/K_i)}{s + K_s} \qquad \text{(totally) noncompetitive inhibition} \tag{3.54}$$

Now the Michaelis constant is unaffected, but the maximum reaction velocity which can be obtained is reduced to

$$v_{max}^{app} = \frac{ke}{1 + i/K_i} \tag{3.55}$$

In the presence of a noncompetitive inhibitor, no amount of substrate addition to the reaction mixture can provide the maximum reaction rate which is possible without the inhibitor.

Competitive and noncompetitive inhibitions are easy to distinguish in a Lineweaver-Burk plot. In the competitive inhibitor case, the intercept on the $1/s$ axis is increased and the intercept on the $1/v$ axis is unchanged by the addition of the inhibitor. Conversely, with a noncompetitive inhibitor, only the $1/v$-axis intercept is increased. As an example, examine the data in Fig. 3.21 for starch hydrolysis. Comparing the data plots for different inhibitors with the no inhibitor line, we see that α-dextrin is a competitive inhibitor, while noncompetitive inhibition is caused by both maltose and limit dextrin.

Already mentioned in Table 3.7 are the possibilities of other inhibitor effects on enzyme kinetics. We shall summarize next very briefly how these cases can be described by modification of the inhibitor models already described. In *partially competitive inhibition* (case I*b* in Table 3.7), the inhibitor and substrate combine with different groups of the enzyme, and the inhibitor affects the enzyme affinity for substrate. This case can be described by adding to the reaction sequence

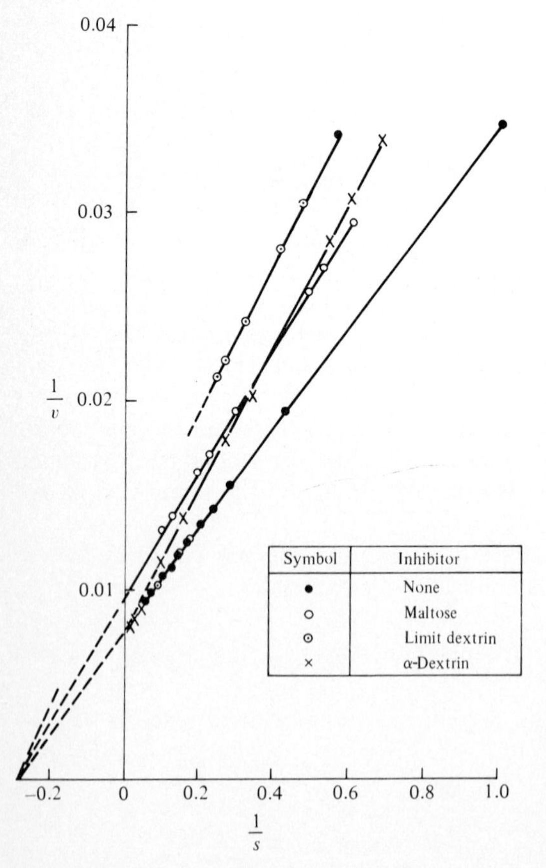

Figure 3.21 Competitive (α-dextrin) and noncompetitive (maltose, limit dextrin) inhibition of α-amylase. *(Reprinted by permission from S. Aiba, A. E. Humphrey, and N. Millis, "Biochemical Engineering," 2d ed., p. 101, University of Tokyo Press, Tokyo, 1974.)*

in Eqs. (3.48) and (3.49) the following reactions which are assumed to be in equilibrium:

$$\begin{array}{lcll} & & & \text{Dissociation constant} \\ \mathrm{EI + S} & \rightleftharpoons & \mathrm{EIS} & K_{is} \\ \mathrm{ES + I} & \rightleftharpoons & \mathrm{EIS} & K_{si} \end{array} \tag{3.56}$$

and the slow step:

$$\mathrm{EIS} \xrightarrow{k} \mathrm{EI + P} \tag{3.57}$$

(Note that it can be shown $K_{is} = K_s K_{si}/K_i$.)

In another *noncompetitive inhibition* situation (case II*b*), the noncompetitive inhibition model above is altered to take into account formation of product from the EIS complex, although with a smaller rate than ES complex reacts:

$$\mathrm{EIS} \xrightarrow{k^i} \mathrm{EI + P} \tag{3.58}$$

One *mixed inhibitor* model which combines pure noncompetitive with partially competitive inhibition is obtained by deleting reaction (3.57) from the partial competition model above. The results for all of these cases may be written in Michaelis-Menten form

$$v = \frac{v_{\max}^{\text{app}} s}{s + K_m^{\text{app}}} \tag{3.59}$$

with apparent parameters $v_{\max}^{\text{app}}$ and K_m^{app} that depend on inhibitor concentration and on various equilibrium and reaction rate parameters (see Table 3.8). Figure 3.22 shows schematically how various types of inhibition influence different forms of kinetic data plots.

Table 3.8 Intercepts of reciprocal (Lineweaver-Burk) plots in presence of inhibitor

	Type		
I*a*	Purely competitive	$\dfrac{1}{v_{\max}}$	$\dfrac{1}{K_s(1 + i/K_i)}$
I*b*	Partially competitive	$\dfrac{1}{v_{\max}}$	$\dfrac{1 + iK_s/K_iK_{is}}{K_s(1 + i/K_i)}$
II*a*	Purely noncompetitive	$\dfrac{1 + i/K_i}{v_{\max}}$	$\dfrac{1}{K_s}$
II*b*	Partially noncompetitive	$\dfrac{1 + i/K_i}{v_{\max} + k^i e_0/K_i}$	$\dfrac{1}{K_s}$
I*b* and II*a*	Mixed	$\dfrac{1 + iK_s/K_iK_s}{v_{\max}}$	$\dfrac{1 + iK_s/K_iK_{is}}{K_s(1 + i/K_i)}$

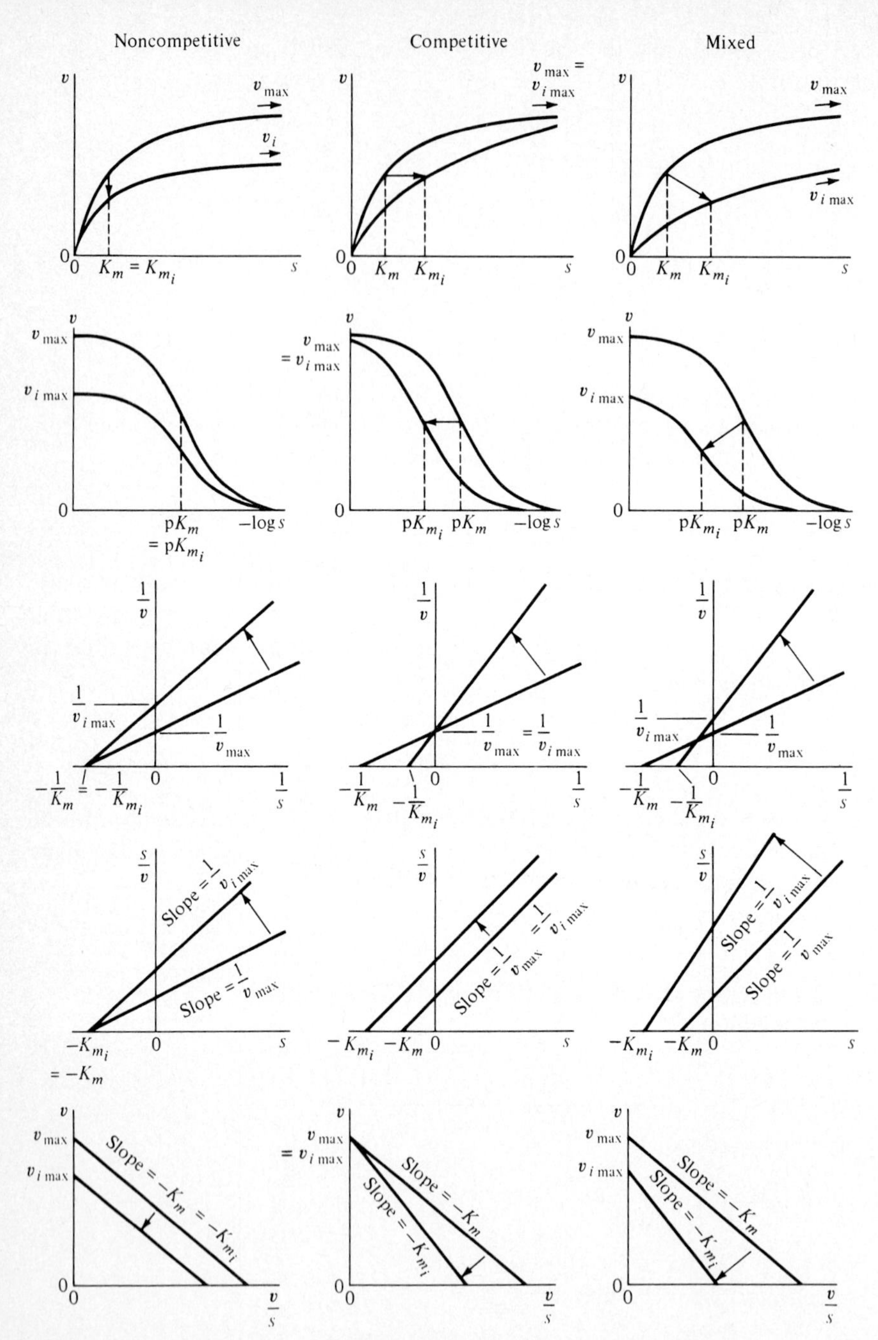

Figure 3.22 Effects of various types of inhibition as reflected in different rate plots. Subscripts i here denote the presence of inhibitor. *(Reprinted by permission from M. Dixon and E. C. Webb, "Enzymes," 2d ed., p. 326, Academic Press Inc., New York, 1964.)*

One form of enzyme activation by an effector can be described by the case II*b* model above in which k^i is greater than k. In physical terms, this means that the substrate complexed with EI reacts to give product more rapidly than the ES complex. Thus, species I serves to *activate* the enzyme.

3.6 OTHER INFLUENCES ON ENZYME ACTIVITY

Before proceeding, it may be helpful to recall the major objective of this chapter: to be able to represent the rate of an enzyme-catalyzed reaction mathematically. Without suitable rate expressions, we cannot design reactors or experiments employing isolated enzymes. Moreover, when we reach the subject of cell growth kinetics, we shall discover that many aspects of enzyme kinetics can be applied. Therefore a thorough exploration of the variables which affect enzyme catalysis and a quantitative analysis of their influence are essential.

We have already explored how different chemical compounds which bind with enzymes can influence the rate of enzyme-catalyzed reactions. Many other factors can influence the catalytic activity of enzymes, presumably by affecting the enzyme's structural or chemical state. Included among these factors are:

1. pH
2. Temperature
3. Fluid forces (hydrodynamic forces, hydrostatic pressure and interfacial tension)
4. Chemical agents (such as alcohol, urea and hydrogen peroxide)
5. Irradiation (light, sound, ionizing radiation)

Sometimes the change in catalytic activity caused by a shift in pH, for example, is reversed by returning to original reaction conditions. This situation is in a sense analogous to the reversible-inhibition cases considered above. The equilibrium (or quasi-steady-state) conditions prevailing in the original environment are merely shifted slightly by a small change in one of the factors above. In general, the amount of change in the environment from the enzyme's native biological habitat must be relatively small (or brief), or deactivation of the enzyme will likely occur. Many of the variables listed above will be important in our discussion of enzyme deactivation in the next section. Here we concentrate on "reversible" effects of pH and temperature on enzyme catalytic activity.

We should point out that the borderline between "reversible" and "irreversible" deactivation of proteins is often vague. For example, an enzyme exposed to an elevated temperature for a short time may exhibit its original activity when returned to a natural working temperature. On the other hand, if the protein is exposed longer to the elevated temperature or to a higher temperature for the same time, only partial activity may remain when the temperature is lowered. All of this makes good intuitive sense if we remember the structure-function connection for proteins and if we imagine the molecular dynamics of the protein and rupture of some of its weak bonds as the environment is changed.

3.6.1 The Effect of pH on Enzyme Kinetics in Solution

Figure 2.15 lists the various amino acids from which all proteins are constructed. These biochemical units possess basic, neutral, or acidic groups. Consequently, the intact enzyme may contain both positively or negatively charged groups at any given pH. Such ionizable groups are often apparently part of the active site since acid- and base-type catalytic action has been linked closely to several enzyme mechanisms. For the appropriate acid or base catalysis to be possible, the ionizable groups in the active site must often each possess a particular charge; i.e., the catalytically active enzyme exists in only *one* particular ionization state. Thus, the catalytically active enzyme may be a large or small fraction of the total enzyme present, depending upon the pH. Figure 3.23 illustrates the influence of pH on several enzymes. In several cases shown there, catalytic activity of the enzyme passes through a maximum (at the optimum pH) as pH is increased.

We can obtain a useful form for representing such pH effects on enzyme kinetics using the following simple model of the active site ionization state:

$$E \underset{+H^+ \atop K_1}{\overset{-H^+}{\rightleftharpoons}} E^- \underset{+H^+ \atop K_2}{\overset{-H^+}{\rightleftharpoons}} E^{2-} \tag{3.60}$$

In these acid-base reactions, E^- denotes the active enzyme form while E and E^{2-} are inactive forms obtained by protonation and deprotonation of the active site of E, respectively. K_1 and K_2 are equilibrium constants for the indicated reactions. Further ionizations away from the E^- state of the enzyme are not

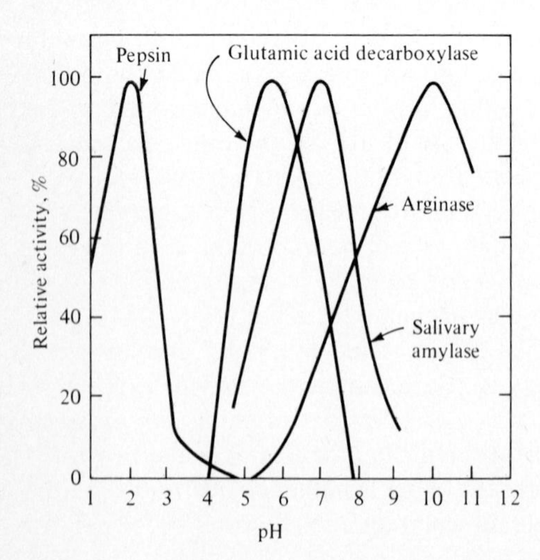

Figure 3.23 Enzyme activity is strongly pH dependent, and the range of maximum activity differs greatly depending on the enzyme involved. *(Redrawn with permission from J. S. Fruton and S. Simmonds, "General Bichemistry," p. 260, John Wiley & Sons, Inc., New York, 1953.)*

considered since the first ionization is assumed here to completely eliminate enzyme activity.

After writing the equilibrium relations for the two ionizations

$$\frac{h^+e^-}{e} = K_1 \qquad \frac{h^+e^{2-}}{e^-} = K_2, \tag{3.61}$$

we can determine the fraction of total enzyme present which is active. When we let e_0 denote the total enzyme concentration,

$$e_0 = e + e^- + e^{2-} \tag{3.62}$$

the active fraction y^- is e^-/e_0 and is given by

$$y^- = \frac{1}{1 + h^+/K_1 + K_2/h^+} \tag{3.63}$$

This is one of the *Michaelis pH functions*. The other two, y and y^{2-}, give the fraction of enzyme in the acid and base forms, respectively.

The y^- function depends upon pH in a manner quite similar to the enzyme activity-pH data shown in Fig. 3.23. There is a single maximum of y^- with respect to pH which occurs at the value

$$h^+_{\text{optimum}} = \sqrt{K_1 K_2} \qquad \text{or} \qquad (\text{pH})_{\text{optimum}} = \tfrac{1}{2}(\text{p}K_1 + \text{p}K_2) \tag{3.64}$$

where $\text{p}K_i$ is defined as $-\log K_i$. The y^- function declines smoothly and symmetrically as pH is varied away from the optimum pH.

Protonation and deprotonation are very rapid processes compared with most reaction rates in solution. Therefore, it can be assumed that the fraction of enzyme in the active ionization state is y^- even when the enzyme is serving as a catalyst. Consequently, the influence on the maximum reaction velocity v_{max} is obtained by replacing the total enzyme concentration e_0 with the total active form concentration $e_0 y^-$:

$$v_{\text{max}} = k e_0 y^- = \frac{k e_0}{1 + h^+/K_1 + K_2/h^+} \tag{3.65}$$

Using this relationship, enzyme activity data may be used to evaluate the parameters K_1 and K_2. The pH of maximal activity is related to K_1 and K_2 by Eq. (3.64). A measured value of enzyme activity at a different pH gives an independent relationship for K_1 and K_2, allowing determination of these parameters.

According to an extension of the Michaelis-Menten treatment of the simplest enzyme-catalyzed reaction sequence given in Eq. 3.4, pH can also affect the Michaelis constant K_m. However, if the substrate does not have different ionization states with different affinities for free enzyme and if formation of the enzyme-substrate complex does not influence K_1 and K_2, then the analysis shows K_m is independent of pH. In any event, pH effects on K_m are usually relatively insignificant. In practice, Eq. (3.65) alone is generally used to represent the dependence of enzyme-catalyzed reaction rates on pH.

We should always remember that the theory just described may not apply for pH values far removed from the optimum pH. Under these circumstances, the forces stabilizing the native protein conformation may be so disturbed that denaturation occurs. In this situation, we cannot expect the normal enzymatic activity to return quickly, if at all, in a practical time scale if the pH is restored to its optimal or near-optimal value.

3.6.2 Enzyme Reaction Rates and Temperature

In any study of chemical kinetics, a recurring theme is the Arrhenius form relating temperature to a reaction-rate constant

$$k = Ae^{-E_a/RT} \tag{3.66}$$

where E_a = activation energy
R = gas-law constant
A = frequency factor
T = absolute temperature

In an Arrhenius plot, log k is graphed against $1/T$ to give a straight line with a slope of $-E_a/R$ [assuming, of course, that Eq. (3.66) holds]. The Arrhenius dependence on temperature is indeed satisfied for the rate constants of many enzyme-catalyzed reactions, as exemplified by the data in Fig. 3.24.

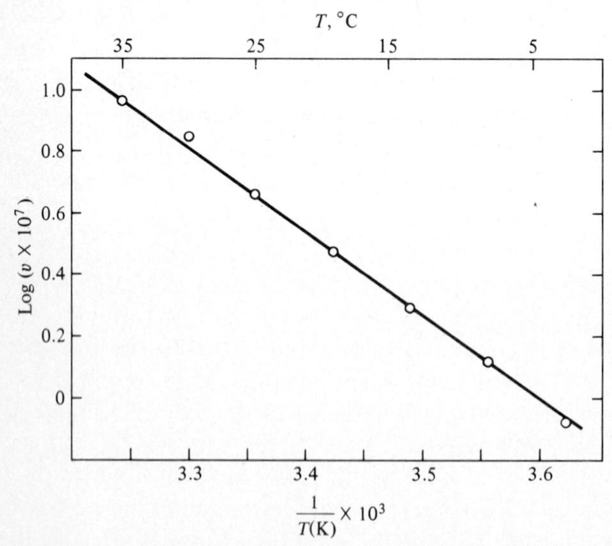

Figure 3.24 Arrhenius plot for an enzyme catalyzed reaction (myosin catalyzed hydrolysis of ATP). *Reprinted from K. J. Laidler, "The Chemical Kinetics of Enzyme Action," p. 197, The Clarendon Press, Oxford, 1958. Data of L. Quellet, K. J. Laidler, M. F. Morales, Molecular Kinetics of Muscle Adenosine Triphosphate, Arch. Biochem. Biophys, vol. 39, p. 37, 1952.)*

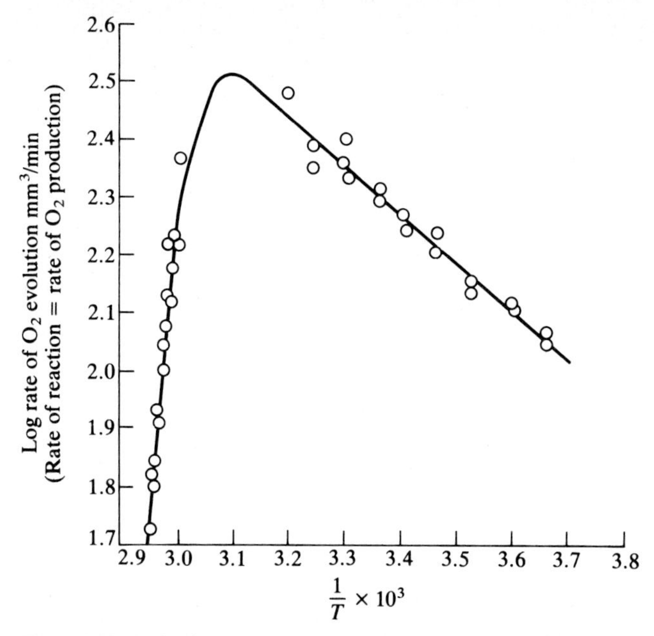

Figure 3.25 Arrhenius rate dependence breaks down at high temperatures, above which enzyme deactivation predominates (H_2O_2 decomposition catalyzed by catalase). The solid curve was calculated from Eq. (3.73) using the parameter values $E = 3.5$ kcal/mol, $\Delta H_d = 55.5$ kcal/mol, $\Delta S_d =$ 168 kcal mol^{-1} K^{-1}, $\beta = 258$ mm^3/min. *(Data from I. W. Sizer, Temperature Activation and Inactivation of the Crystalline Catalase-Hydrogen Peroxide System," J. Biol. Chem., vol. 154, p. 461, 1944.)*

It should be noted, however, that the temperature range of Fig. 3.24 is quite limited. No temperatures significantly larger than the usual biological range were considered. What would happen if we attempted to push the enzyme farther and try for higher rates via higher temperatures? The result would in most cases be disastrous, as the example of Fig. 3.25 illustrates.

For may proteins, denaturation begins to occur at 45 to 50°C and is severe at 55°C. One physical mechanism for this phenomenon is obvious: as the temperature increases, the atoms in the enzyme molecule have greater energies and a greater tendency to move. Eventually, they acquire sufficient energy to overcome the weak interactions holding the globular protein structure together, and deactivation follows.

Thermal deactivation of enzymes may be reversible, irreversible, or a combination of the two. A simple model of *reversible thermal deactivation* often suffices to represent T-activity data for enzyme kinetics over a wide range of temperatures. In this approach we assume that the enzyme exists in inactive (i) and active (a) forms in equilibrium:

$$E_a \rightleftharpoons E_i \tag{3.67}$$

with equilibrium constant

$$\frac{e_i}{e_a} = K_d = \exp\left(\frac{-\Delta G_d}{RT}\right) = \exp\left(\frac{-\Delta H_d}{RT}\right)\exp\left(\frac{\Delta S_d}{R}\right) \tag{3.68}$$

In Eq. (3.68), ΔG_d, ΔH_d, and ΔS_d denote the free energy, enthalpy, and entropy of deactivation, respectively.

Although individual hydrogen bonds are quite weak, typically with bond energies of 3 to 7 kcal/mol, the enthalpy of deactivation of enzymes ΔH_d is quite high: 68 and 73.5 kcal/mol for trypsin and hen egg white lysozyme, respectively. The entropy changes upon deactivation for these enzymes are +213 cal/(mol·K). Due to the large heats of denaturation, the proportion of active enzyme is quite sensitive to small changes in temperature. For such ΔH_d values, the enzyme deactivates almost totally over a range of 30 Celsius degrees.

Since all enzyme present is either in active or inactive forms,

$$e_a + e_i = e_0 \tag{3.69}$$

we may obtain, using Eq. (3.68),

$$e_a = \frac{e_0}{1 + K_d} \tag{3.70}$$

According to transition state theory, the rate for reactions (3.4) at large substrate concentration can be written as

$$v_{\max} = e_a \cdot k(T) \tag{3.71}$$

where

$$k(T) = \alpha\left(\frac{k_B T}{h}\right)e^{\Delta S^*/R}e^{-E/RT} \tag{3.72}$$

k_B and h are Boltzmann's and Planck's constants, respectively, and α is a proportionality constant. Combining Eqs. (3.68) and (3.70) through (3.72), we obtain

$$v_{\max} = \frac{\beta T e^{-E/RT}}{1 + e^{\Delta S_d/R}e^{-\Delta H_d/RT}} \tag{3.73}$$

where the overall proportionality factor β includes α, k_B, h, e_0, and exp $(\Delta S^*/R)$.

Equation (3.73) is the desired relationship for representing behavior like that shown in Fig. 3.25. The solid curve drawn through those data points was in fact calculated from Eq. (3.73) after estimating the parameters β, E, ΔS_d and ΔH_d from the data. The slope at large $1/T$ values is approximately $-E/R$ (the error is equal to T (K), which is usually not significant). The slope of the other straight line obtained at higher temperatures is approximately equal to $(\Delta H_d - E)/R$.

ΔS_d may be estimated after noting that, at the temperature T_{max} where log v_{max} is maximized,

$$K_d(T_{max}) = \frac{E + RT_{max}}{\Delta H_d - E - RT_{max}}. \tag{3.74}$$

(This is obtained simply by setting $d(\log v_{max})/dT$ equal to zero). Since T_{max} is known from the measurements and E and ΔH_d have already been estimated, we can evaluate the right-hand side of Eq. (3.74). Now, knowing $K_d(T_{max})$, we return to Eq. (3.68) and calculate ΔS_d. Finally, the proportionality constant β is chosen to make v_{max} [T_{max} from Eq. (3.73)] the same as the measured value. Some iterative readjustments may be necessary to refine these initial estimates to obtain a good fit of the $T - v_{max}$ data. In particular, results can be quite sensitive to the T_{max} value. Of course other parameters in the reaction rate expression such as the Michaelis constant and an inhibition constant are also functions of temperature. If these parameters are interpreted as equilibrium constants, as is often the case, we expect temperature dependence of the form seen above in Eq. (3.68), giving a straight line ln K (or pK) versus $1/T$ plot from which the standard free energy parameter may be determined.

In some cases data plotted in these coordinates do not give a straight line, or the slope of the line is different in different temperature ranges. Such complications may be due to oversimplification of the interpretation of K_m or K_i, or the catalytic reaction sequence itself. Recall, for example, that K_m for the simplest reaction sequence (3.4) is most properly viewed as a combination of elementary step rate constants as given in Eq. (3.5). If each of these elementary rate constants are in turn written in Arrhenius form [Eq. (3.66)] or in the transition state theory form of Eq. (3.72), a complicated function of temperature results. The existence of more than one intermediate enzyme-substrate or enzyme-product complex is another possible source of complication in temperature dependence of certain rate parameters.

This point can be generalized in a useful way which has been applied productively to test kinetic models and the associated hypothetical reaction sequences employed in kinetic model development. If the kinetic parameters determined at different temperatures do not lie on a straight line in an Arrhenius plot, the model is likely not complete or not correct.

3.7 ENZYME DEACTIVATION

Most of the research literature on enzyme kinetics is devoted to initial rate data and analyses of reversible effects on enzyme activity. In many applications and process settings, however, the rate at which the enzyme activity declines is a critical characteristic. This is especially true when long-term use of the enzyme in a continuous flow reactor is considered (immobilization methods for retaining enzyme molecules in such reactors is a central theme of the next chapter). In such situations, the economic feasibility of the process may hinge on the useful lifetime of the enzyme biocatalyst.

While our focus in this section is on the mechanisms and kinetics of enzyme activity loss, we should recognize that alteration of protein structure and function has many important practical implications. Proteins are often recovered from culture broths by precipitation, an operation that involves changing the proteins' configuration or chemical state by pH or ionic strength adjustment. Later protein purification steps sometimes use antibodies as highly specific sorbents, a process that fails if the antibodies deactivate and lose their ability to recognize and adsorb a particular protein. We shall return to these separation themes in Chap. 11; now we consider the causes, effects, and kinetic representation of enzyme activity decay.

3.7.1 Mechanisms and Manifestations of Protein Denaturation

We know that protein structure is stabilized by weak forces, often giving rise to functionally important molecular flexibility. On the other hand, this weak stabilization implies that proteins are poised near, in an energetic sense, to several alternative and less biologically active configurations. This property is illustrated dramatically by the thermodynamic data for α-lactalbumin in Fig. 3.26. We see there that the free energy difference between the native and completely denatured state is only 9.0 kcal/mol (although several intermediate energy barriers must be passed to effect the transition from one conformation to the other). Consequently, we should recognize that native protein structure is only marginally stable.

It is therefore not surprising that a multitude of physical and chemical parameters can and do cause perturbations in native protein geometrical and chemical structure, with concomitant reductions in activity (Table 3.9). After recognizing the individual parameters which influence protein denaturation, we must recognize that it is not the individual factors listed above but their combinations that determine rates of enzyme deactivation. For example, the sensitivity of a protein to denaturation at elevated temperatures can vary widely with solution pH, and the influence of various temperature-pH combinations may differ tremendously from one protein to another.

We can identify a large set of different but often related protein properties that may be affected by the physicochemical factors listed in Table 3.9. These possible manifestations of denaturing environments include the solubility, tendency for crystallization or gel formation, viscosity, and biological activity (enzyme catalysis, antibody binding) of protein solutions. Just as combinations of influence are important, so too are combinations of effects. For example, it is essential in recovering an enzyme in a precipitation step that the conditions used do not also cause major irreversible loss of enzyme activity.

3.7.2 Deactivation Models and Kinetics

In the simplest model, active enzyme (E_a) molecules undergo an irreversible structural or chemical change to an inactive form (E_i)

$$\mathrm{E}_a \longrightarrow \mathrm{E}_i \tag{3.75}$$

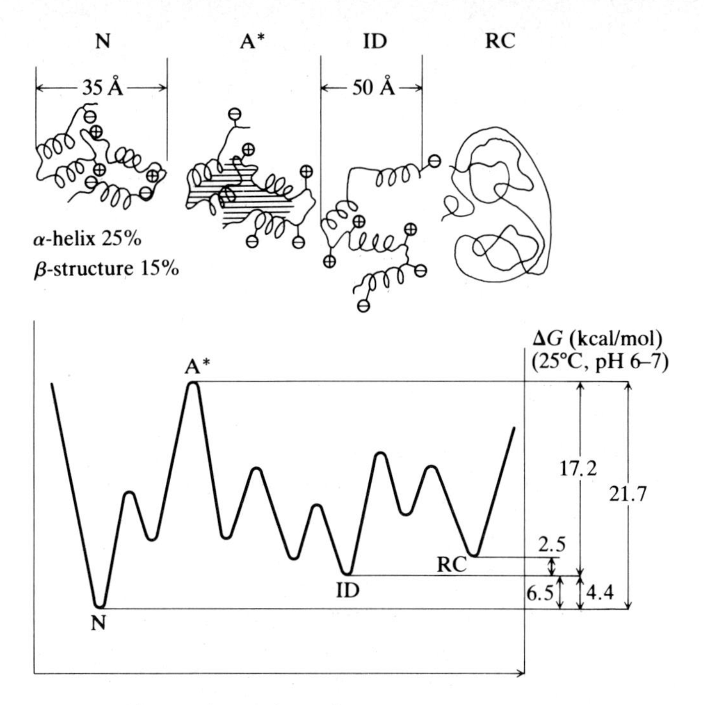

N = native conformation
A* = critically activated state
ID = incompletely disordered conformation
RC = random coil (fully denatured state)

Figure 3.26 Schematic illustration of the stages of unfolding of the protein α-lactalbumin with corresponding thermodynamic data. *(Reprinted by permission from K. Kuwajima, "A Folding Model of α-Lactalbumin Deduced from the Three-state Denaturation Mechanism," J. Mol. Biol., vol. 114, p. 241, 1977.)*

at a rate r_d proportional to the active enzyme concentration

$$r_d = k_d e_a \tag{3.76}$$

Consequently, in a well-mixed closed system (assumed for the moment not to contain any substrate, product, inhibitors, or modulators), the time course of active enzyme concentration is described by

$$\frac{de_a}{dt} = -k_d e_a \tag{3.77}$$

so that

$$\ln\left[\frac{e_a(t)}{e_a(0)}\right] = -k_d t \tag{3.78}$$

Data consistent with this relationship are shown in Fig. 3.27 for the enzyme ATPase under several different conditions. The decay constant k_d depends on temperature in a fashion that can often be described by transition state theory

Table 3.9 Summary of protein denaturants and their effects

HD = highly disordered conformation; ID = incompletely disordered conformation; RC = random coil (fully denatured)

(Reprinted by permission from R. D. Schmid, "Stabilized Soluble Enzymes", Adv. Biochem. Eng., vol. 12, p. 41, 1979)

Denaturant	Target	Driving force	End product
Physical denaturants			
Heat	Hydrogen bonds	Increase of denatured conformations due to increased thermal movement and decreased solvent structure	HD
			Aggregates
		Irreversible covalent modification (e.g., disulfide interchange)	
Cold	Hydrophobic bonds	Altered solvent structure	Aggregates
	solvated groups	Dehydration	Inactive monomers
Mechanical forces	Solvated groups	Changes in solvation and void volume	HD
	void volume	Shearing	Inactive monomers
Radiation	Functional groups (e.g., cySH, peptide bonds)	Decrease of structure-forming interactions after photooxidation or attack by radicals	HD
			Aggregates
Chemical denaturants			
Acids	Buried uncharged groups (e.g., his, peptide bonds)	Decrease of structure-forming ionic interactions	RC
Alkali	Buried uncharged groups [e.g., tyr, cySH, $(cyS)_2$]	Decrease of structure-forming ionic interactions	RC

Organic H-bond-formers	Hydrogen bonds	Decrease of structure-forming H bonds between water and native conformation	RC
Salts	Polar and nonpolar groups	"Salting in"/"salting out" bias of polar and nonpolar groups in solvent of increased DK	HD
Solvents	Nonpolar groups	Solvation of nonpolar groups	Highly ordered Peptide-chains with large helical regions
Surfactants	Hydrophobic domains (all surfactants) and charged groups (ionic surfactants only)	Formation of partially unfolded substructures including micellelike regions	ID Large helical regions
Oxidants	Functional groups (e.g., cySH, met, try, and others)	Decrease of structure-forming and/or functional interactions	Inactivated enzyme Sometimes disordered structure
Heavy metals	Functional groups (e.g., cySH, his, and others)	Masking of groups pertinent to structure or function	Inactivated enzyme
Chelating agents	Cations important for structure or function	Ligand substitution or cation removal	Inactivated enzyme
Biological denaturants			
Proteases	Peptide bonds	Hydrolysis of terminal or other peptide bonds	Oligopeptides, amino acids

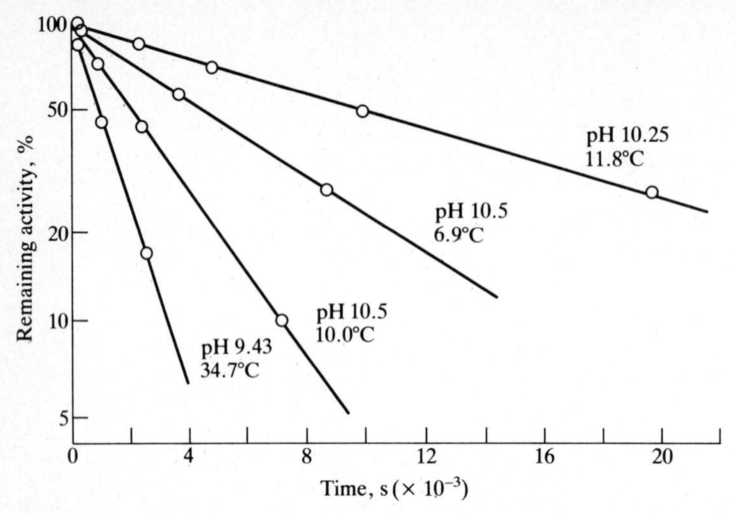

Figure 3.27 Time course of loss of ATPase enzyme activity at several pH and temperature values. *(Reprinted by permission from G. E. Pelletier and L. Ouellet, "Influence of Temperature and pH on Myosin Inactivation," Can. J. Chem., vol. 39, p. 265, 1961.)*

[as in Eq. (3.72) with α equal to unity], which in turn is little different from the Arrhenius form (Eq. 3.66) over the narrow temperature ranges of interest for biological systems. Listed in Table 3.10 are the activation energies E and entropies ΔS^* for k_d for several common enzymes. The magnitude of the activation energy for protein denaturation is typically very large as these data illustrate.

Without enzyme activity, cells cannot function. In some cases, destruction of a very small fraction of the cell's enzymes results in its death. Viewed in this light, the preceding discussion reveals why heat can be used to *sterilize* gases, liquids or solids, i.e., to rid them of microbial life. Very few enzymes (and consequently few microbes) can survive prolonged heating, as we shall explore in greater detail in Chap. 7.

Enzyme deactivation data is usually obtained by exposing the enzyme to denaturing conditions for some time interval *in the absence of substrate*, and then

Table 3.10 Energies and entropies of activation for enzyme denaturation†

Enzyme	pH	Energy of activation, kcal/mol	Entropy of activation ΔS^* (e.u./mol)
Pancreatic lipase	6.0	46.0	68.2
Trypsin	6.5	40.8	44.7
Pepsin	4.83	56–147	unknown
ATPase	7.0	70	150.0

† K. J. Laidler and P. S. Bunting. *The Chemical Kinetics of Enzyme Action*, 2d ed., p. 430, Oxford University Press, London, 1973.

making an initial rate activity assay after returning the enzyme solution to some standard conditions and adding substrate. Enzyme deactivation rates when the enzyme is turning over substrate can be significantly different when, for example, (*a*) free enzyme, enzyme-substrate and/or enzyme-product complexes deactivate at different rates, or (*b*) the substrate and/or product cause deactivation. To illustrate an approach to such situations, we consider a simple case in which substrate binding to enzyme is presumed to stabilize the enzyme (an effect that is qualitatively reasonable in some systems). Thus, only the free enzyme experiences deactivation.

Combining this deactivation model with the simple catalytic reaction sequence used by Michaelis and Menten Eq. (3.4) gives

$$\mathrm{E}_a + \mathrm{S} \underset{k_{-1}}{\overset{k_1}{\rightleftharpoons}} \mathrm{E}_a\mathrm{S} \xrightarrow{k_2} \mathrm{E}_a + \mathrm{P} \tag{3.79a}$$

$$\mathrm{E}_a \xrightarrow{k_d} \mathrm{E}_i. \tag{3.79b}$$

If we apply the reasonable assumption that the deactivation process is much slower than the reactions in Eq. (3.79*a*), invoking the quasi-steady-state approximation for the ($\mathrm{E}_a\mathrm{S}$) complex gives

$$v = \frac{k_2 e_{\mathrm{tot},\,a} s}{K_m + s} \tag{3.80}$$

where $e_{\mathrm{tot},\,a}$ is the total concentration of *active* enzyme both in free and complexed forms. The rate of change of $e_{\mathrm{tot},\,a}$ is given by

$$\frac{de_{\mathrm{tot},\,a}}{dt} = -k_d e_a \tag{3.81}$$

Going back to the quasi-steady-state calculations used in obtaining Eq. (3.80), we can express e_a in terms of $e_{\mathrm{tot},\,a}$ and the catalytic reaction parameters to give, after substitution in Eq. (3.81)

$$\frac{de_{\mathrm{tot},\,a}}{dt} = -\frac{k_d e_{\mathrm{tot},a}}{1 + s/K_m} \tag{3.82}$$

Thus we see that in Eqs. (3.80) and (3.82) that the substrate conversion and enzyme deactivation rates are mutually coupled. In particular, the rate of enzyme deactivation depends on the substrate concentration. If, on the other hand, both E_a and ($\mathrm{E}_a\mathrm{S}$) deactivate at the same rate, then enzyme activity will decline under reaction conditions exactly as observed in a substrate-free deactivation experiment.

Extending these notions to the general case, we can see that, if different forms of the enzyme (free, various complexes, various ionization states, etc.) deactivate at different rates, the overall deactivation rate will in turn depend on any reaction

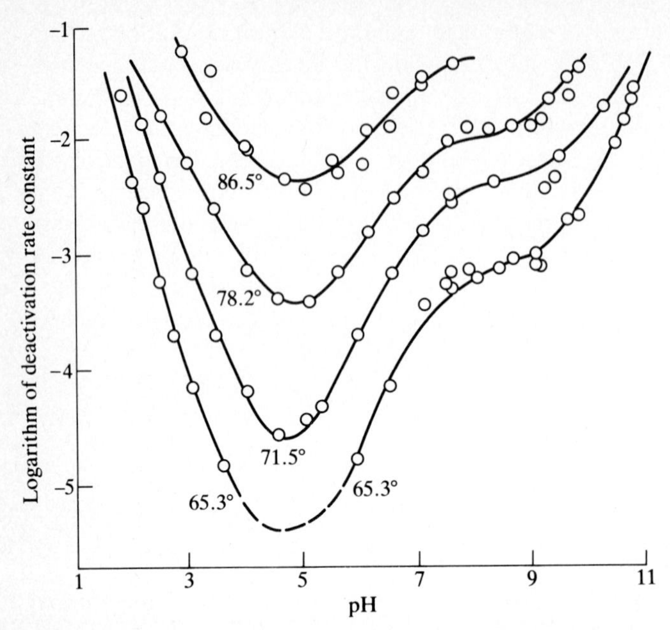

Figure 3.28 Temperature and pH dependence of the first-order rate constant for deactivation of ricin. *(Reprinted by permission from M. Levy and A. E. Benaglia, "The Influence of Temperature and pH upon the rate of Denaturation of Ricin," J. Biol. Chem. vol. 186, p. 829, 1950.)*

parameter (substrate and inhibitor concentrations, pH, etc.) which influences the proportions of the different enzyme forms. For example, the following model

$$\begin{array}{ccccccccc} E_{a1} & \underset{+H^+}{\overset{-H^+}{\rightleftharpoons}} & E_{a2} & \underset{+H^+}{\overset{-H^+}{\rightleftharpoons}} & E_{a3} & \underset{+H^+}{\overset{-H^+}{\rightleftharpoons}} & E_{a4} & \underset{+H^+}{\overset{-H^+}{\rightleftharpoons}} & \cdots \\ k_{d1}\downarrow & K_1 & k_{d2}\downarrow & K_2 & k_{d3}\downarrow & K_3 & k_{d4}\downarrow & K_4 & \\ E_i & & E_i & & E_i & & E_i & & \end{array}$$

has been developed to describe the sometimes complex effects of pH on deactivation rates. Figure 3.28 illustrates an example of such complicated deactivation behavior for ricin (not an enzyme; active protein is assayed by measuring the fraction of protein remaining soluble). A minimum in the rate of deactivation with respect to pH has been observed for several proteins.

Decay of enzyme activity with time does not always follow the first-order model of Eq. (3.78). Convex log (activity) versus time relationships have been measured for several proteins; sometimes these trajectories contain two different linear regions (Fig. 3.29*a*). Several different models for such behavior have been proposed. One which includes parallel reversible and irreversible deactivation has

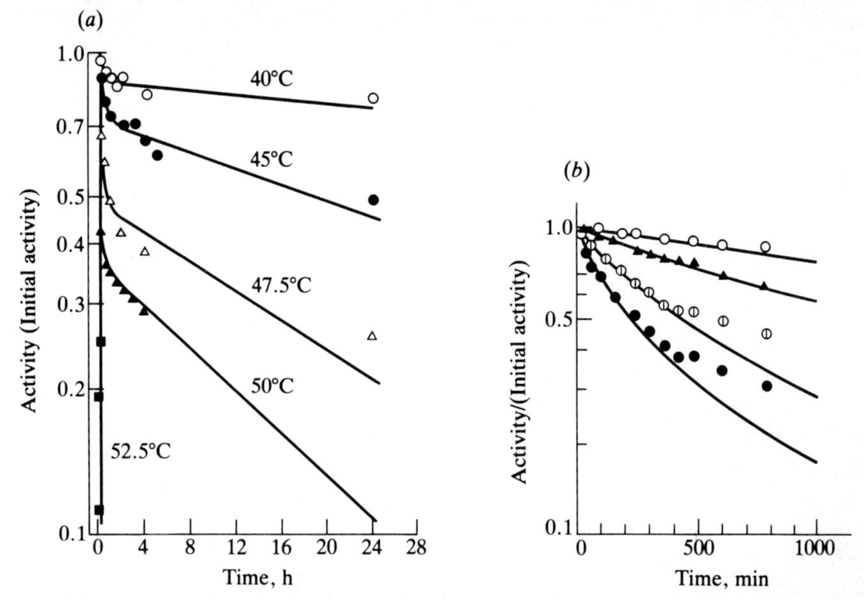

Figure 3.29 Deactivation of (*a*) a luciferase preparation (pH 6.8, temperatures as indicated) and (*b*) α-chymotrypsin in solution (pH 7.8, 40°C, $Ca^{2+} = 10^{-3}$ *M* and initial enzyme concentrations, ○ = 7.31×10^{-7} *M*, △ = 3.65×10^{-6} *M*, ⦶ = 1.46×10^{-5} *M*, ● = 2.92×10^{-5} *M*. *(Reprinted by permission from (a) A. M. Chase, "Studies on Cell Enzyme Systems. IV. The Kinetics of Heat Inactivation of Cypridina Luciferase," J. Gen. Physiol., vol. 33, p. 535, 1950, and (b) Y. Kawamura, K. Nakanishi, R. Matsuno, and T. Kamikubo, "Stability of Immobilized α-Chymotrypsin," Biotech. Bioeng., vol. 23, p. 1219, 1981.)*

been employed for conceptual and quantitative representation of activity decline for several enzymes:

$$\mathrm{E}_a \underset{k_r}{\overset{k_{d1}}{\rightleftarrows}} \mathrm{E}_{i1}, \qquad \mathrm{E}_a \xrightarrow{k_{d2}} \mathrm{E}_{i2} \tag{3.83}$$

Assuming first-order kinetics for each indicated transformation, this model was used to fit the luciferase deactivation data in Fig. 3.29*a*. The solid line near the 45° data, for example, was calculated using the rate parameters $k_{d1} = k_{d2} = 1.02$, and $k_r = 0.02\ \mathrm{h}^{-1}$.

Loss of activity in protease solutions is more complex because these enzymes catalyze their own hydrolysis. This *autodigestion* property should be included in the deactivation model. The following model for deactivation of the protease α-chymotrypsin includes a variation on the theme of Eq. (3.83) plus an autodigestion step:

$$E_a \underset{K}{\rightleftharpoons} E_{i1} \xrightarrow{k_{d1}} E_{i2}$$

$$E_a + E_{i1} \underset{K_m}{\rightleftharpoons} (E_a E_{i1}) \xrightarrow{k_{d2}} E_a + \text{inactive peptide hydrolysis products} \quad (3.84)$$

Certain features of this sequence are supported by direct chemical evidence for this extensively studied enzyme. In particular, notice that only the reversibly inactivated form E_{i1} is susceptible to attack and hydrolysis by the active protease form E_a. Further analysis of this model and its use to calculate the solid curves in Fig. 3.29*b* is suggested in Problem 3.14.

In concluding this summary of enzyme deactivation kinetics, we should mention approaches for irreversible enzyme deactivation by a poison. In the simplest case, we have

$$E_a + \text{poison} \longrightarrow E_d \qquad r_d = k_d e_a \cdot (\text{poison}) \qquad (3.85)$$

However, since poisons often act at the enzyme active site and since poison access may be blocked by complexed substrate, we again must modify the analysis if substrate is present. Then, we should consider the coupled catalysis and deactivation processes and write the model as, for example, reaction (3.79*a*) plus poisoning step (3.85), where E_a now denotes free, uncomplexed enzyme. Similar modifications, while not pursued here, can also be envisioned for the deactivation models given in Eqs. (3.83) and (3.84).

3.7.3 Mechanical Forces Acting on Enzymes

Mechanical forces can disturb the elaborate shape of an enzyme molecule to such a degree that deactivation occurs. Included among such forces are forces created by flowing fluids. Experiments to assess shear effects on enzyme activity have been conducted in flow through a capillary and in a coaxial cylinder viscometer.

Let θ and γ denote the time of exposure to shear and the shear rate (averaged over the cross section in the case of capillary flow), respectively. Experiments on shear denaturation of catalase and urease have shown that activity loss can be correlated as a function of the product $\gamma\theta$ (Fig. 3.30*a*). Thus we see here explicitly that the combination of exposure time and intensity of the denaturation effect determines the extent of deactivation. We should note that experiments on lactate dehydrogenase indicate that it is shear stress, not shear rate, which controls shear deactivation of that enzyme.

The data in Fig. 3.30*b* show clearly that shear deactivation of urease is partially reversible. Curve *A* shows substrate urea concentration in a stagnant urease solution, and curve *B* was measured with the reaction mixture exposed to a shear rate of $1717 \, s^{-1}$ in a coaxial cylinder viscometer. Obviously enzyme activity is reduced by fluid motion. The other data in Fig. 3.30*b* were obtained at

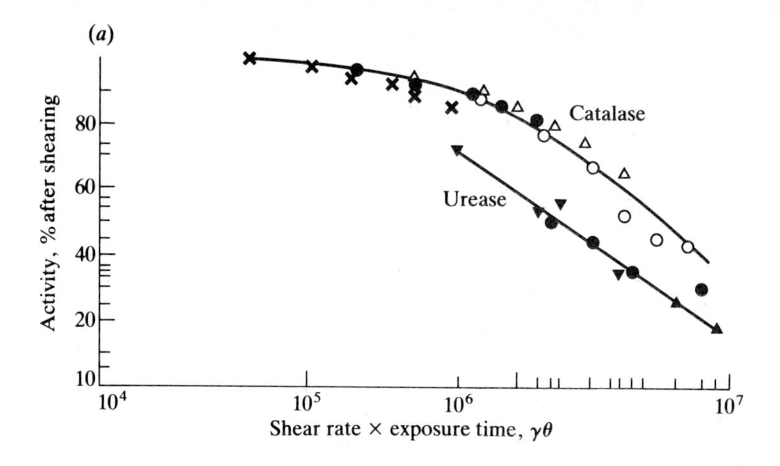

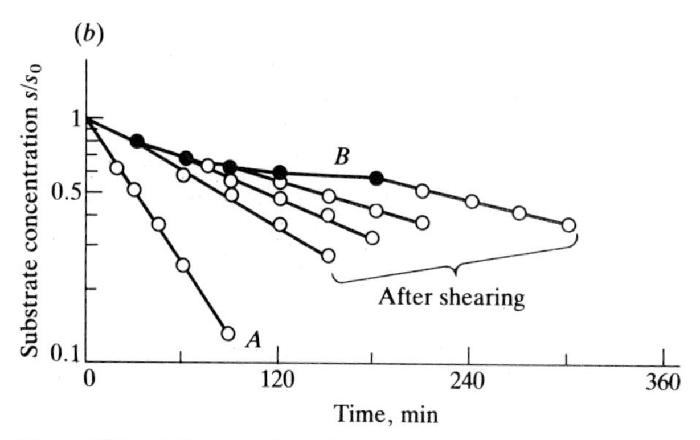

Figure 3.30 (*a*) Deactivation of catalase and urease by exposure to shear. Different symbols denote different shear rate experiments. (*b*) Conversion of urea by urease solution at rest (*A*), under shear (shear rate 1717 s^{-1}) (*B*), and after shearing for different times (0.4 units urease/mL, pH 6.75, 23°C). *(Catalase data reprinted with permission from S. E. Charm and B. L. Wong, "Enzyme Inactivation with Shearing," Biotech. Bioeng., vol. 12, p. 1103, 1970. Urease data reprinted from M. Tirrell and S. Middleman, "Shear Modification of Enzyme Kinetics," Biotech. Bioeng., vol. 17, p. 299, 1975.)*

zero stress after exposure to shear for different time intervals. Here enzyme activity increases immediately after shear is removed, but not to the previous level. The difference has been interpreted as irreversible deactivation.

This characteristic mechanical fragility of enzymes may impose limits on the fluid forces which can be tolerated in enzyme reactors being stirred to increase substrate mass-transfer rates or in an enzyme ultrafiltration system in which increasing membrane throughput causes increased fluid shear and extension just in front of the membrane and passing through it.

Another mechanical force, surface tension, often causes denaturation of proteins and consequent inactivation of enzymes. Since the surface tension of the interface between air and pure water is 80 dyn/cm, foaming or frothing in protein solutions commonly causes denaturation of protein adsorbed at the air–water interface. Liquid–liquid interfaces normally have considerably lower surface tensions. Similarly, the plasma membrane of the cell is believed to have a surface tension of the order of 1 dyn/cm or less. As active proteins are known to exist in such plasma membranes, evidently these very low surface tensions do not deactivate the enzymes. Foam fractionation is a separation technique whereby molecules are concentrated at surfactant–air interfaces without deactivation; here the surfactant lowers the air–liquid surface tension to the order of 1 dyn/cm.

In processing contexts and in some laboratory situations, combinations of different mechanical factors and also chemical processes, such as oxidation, interact to influence the rate of enzyme deactivation. The complicated nature of these interactions is discussed lucidly by Thomas and Dunnill,† in which careful studies of shear effects on the same enzymes as discussed in Fig. 3.30 give much different results (i.e., negligible deactivation due to simple shear alone). In addition to the factors already mentioned, extensional flow, cavitation, local adiabatic heating, metal contamination, and surface denaturation at cavities may influence enzyme deactivation. We must leave these to the references, however. As a general rule of thumb, the following may suffice. If the enzyme is surrounded *in vitro* with essentially the same environment it enjoys *in vivo*, it will be active. If *any* parameter of its environment is altered significantly, loss of activity is likely to occur.

There is a corollary of this idea which finds numerous applications in biochemical technology. If an enzyme is required which is active at extreme temperature or pH values, we should look for an organism which normally lives under these conditions. It will often contain enzymes especially adapted for use in an unusual environment. The development of microbial alkaline-stable enzymes for use in laundry detergents is a good example of this practice, as wash-water pH is typically 9.0 to 9.5.

3.7.4 Strategies for Enzyme Stabilization

Besides trying to identify enzymes that are intrinsically more stable, there are several methods available for improving enzyme stability. These fall into three main categories:

a. Adding stabilizing compounds to the storage and/or reaction medium.
b. Chemically modifying the soluble protein.
c. Immobilizing the protein on or within an insoluble solid or matrix.

In this brief overview, we focus on the first two strategies, leaving discussion of immobilization and its effects on stability for the next chapter.

Chemical additives that can stabilize proteins in some cases include substrates, organic solvents, and salts. Since the active site of an enzyme may be its

† C. R. Thomas and P. Dunnill, "Action of Shear on Enzymes: Studies with Catalase and Urease," *Biotechnol. Bioeng.* **21**: 2279 (1979).

least stable region, presence of the substrate may stabilize the enzyme by "holding" some of the protein in the form of enzyme-substrate complexes. On the other hand, examples of destabilization of enzymes by their substrates are also known. Polyalcohol solvents, which stabilize a number of enzymes, may diminish the tendency for hydrogen bond rupture in the protein.

At low salt concentrations ($<0.1\ M$), salt cations such as Ca^{2+}, Zn^{2+}, Mn^{2+}, Fe^{2+}, Mo^{2+}, and Cu^{2+}, interact specifically with a group of enzymes called *metalloenzymes*. Some of these cations are cofactors, and their presence stabilizes the enzyme. Ca^{2+} is implicated in tertiary structure stabilization in several proteins; by forming ionic bonds with two different amino acid residues, Ca^{2+} can serve as a stabilizing bridge analogous to a disulfide bond.

Specific examples of chemical additive effects on enzyme stability are listed in Table 3.11. Further details are given in Ref. [4]. We should note that the influence of a particular chemical or strategy on one enzyme's stability cannot always be extrapolated to other enzymes and, further, the stabilizing effect depends often on a particular combination of solution composition and temperature. For example, while addition of some salts at low concentrations can stabilize enzymes, higher concentrations usually cause denaturation. A similar comment pertains to organic solvents.

Chemical modification of proteins, an important tool in protein biochemistry, has also been applied with some success to improve enzyme stability. One class of chemical modification strategies involves adding to or modifying the

Table 3.11 Examples of enzyme stabilization

Enzyme	Method	Effect
Glucoamylase	Addition of substrate analogs, glucose, gluconolactone	Enhanced thermal stability
Lactate dehydrogenase	Addition of substrate lactate or effector fructose-diphosphate	Greater thermostability; destabilized by addition of pyruvate substrate
α-Amylase	Addition of 50–70% sorbitol	Better storage and thermal stability
Chymotrypsin	Addition of 50–90% glycerol	Improved resistance to proteolysis
β-Galactosidase	Addition of 5–10% ethanol, 2-propanol	Increased heat stability; similar levels of methanol, *n*-propanol destabilize the enzyme
α-Amylase (*Bacillus caldolyticus*)	Addition of Ca^{2+}	Greatly enhanced thermostability
Trypsin	Addition of polyalanyl (~10 units long) to protein's amino groups	More proteolysis, heat deactivation resistance
Asparaginase	Succinyl substituents added using the acid anhydride	Protease resistance up
Glycogen phosphorylase	Butyl or propyl substituents added using the aldehyde and $NaBH_4$	Enhanced thermal stability
Papain	Cross-linked using glutaraldehyde	Better thermostability

R-groups of certain amino acid residues. In these methods, polyamino acid side chains may be added to amino groups of the native protein. Acylation and reductive alkylation may be used to introduce other substituents.

Bifunctional reagents such as glutaraldehyde are important in the other class of chemical stabilization methods. By cross-linking amino groups, these reagents limit access by proteases and may also serve to lock the protein into an active configuration. Diimides, which cause formation of amino-carboxyl cross-links, may also be used for chemical stabilization. We shall examine these reagents and their reactions in more detail when we consider their use for enzyme immobilization.

3.8 ENZYME REACTIONS IN HETEROGENEOUS SYSTEMS

Our attention to this point has been concentrated on enzymes in solution acting on substrates in solution. This is not a universal situation, as we have already hinted in Fig. 2.27. In that diagram of a procaryote, it was indicated that some of the cell's enzymes are attached to the cell membrane. Similar features are found in eucaryotes. In mitochondria, for example, the enzymes for a very complicated chain of reactions are bound to a convoluted internal-membrane system (see Chap. 5).

Many other combinations of enzyme and substrate physical states arise in nature and technology, as Fig. 3.31 illustrates. The kinetics of enzymes in solution acting on insoluble substrates will be examined next. In the next chapter, we shall analyze reactions of soluble substrates catalyzed by enzymes attached to surfaces.

Sometimes only the soluble form of a substrate which may also exist as a separate phase is suitable for enzyme catalysis. An example is evident in the data shown in Fig. 3.32. Since all molecular species have a finite aqueous solubility, a small amount of substrate will always be in solution to supply the enzyme. This may be so slow a process, however, that the rate is nil for practical puposes. Let us turn next to some examples of the opposite sort.

One of these is the hydrolysis of methyl butyrate by pancreatic lipase, a fat-splitting enzyme which is secreted in the human digestive tract. In contrast to the previous example, here the reaction does not occur until an insoluble form (liquid droplets) of the substrate is available (Fig. 3.33). Apparently the enzyme is active only at the liquid–liquid interface. Recalling the possibility of enzyme denaturation by interfacial tension, it is interesting to note that the bile salts, natural surfactants which are also secreted into the digestive tract, may adsorb on fat droplets and reduce the interfacial forces, just as with surfactant foam fractionation mentioned above.

Other enzymes are active toward both the soluble and insoluble form of the substrate. Trypsin, a protease, is found to digest both free lysozyme and lysozyme adsorbed on the surface of kaolinite. A second enzyme which hydrolyses both

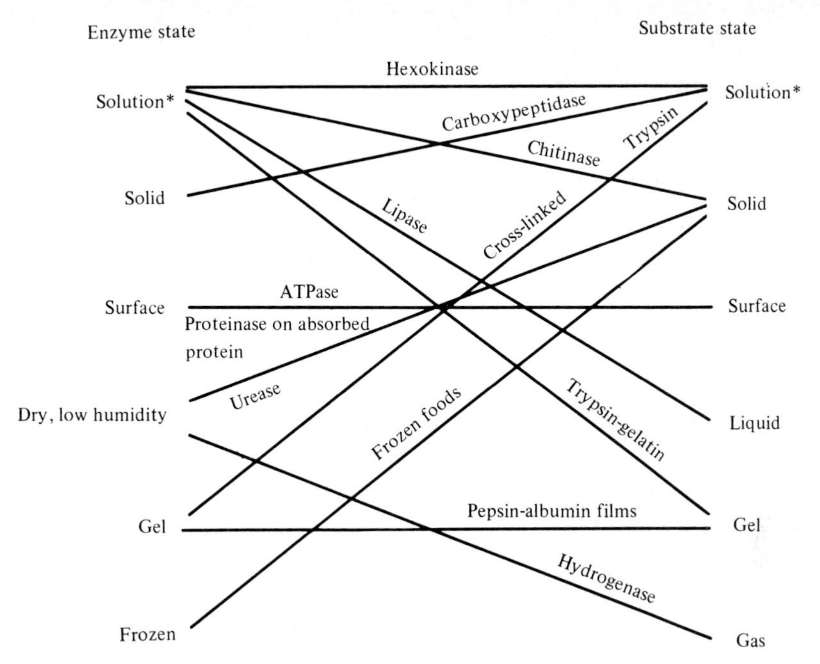

Figure 3.31 Enzymes in several different states catalyze reactions of substrates in various forms. The classical solution case is only one of a broad spectrum of possible enzyme-substrate interactions. *(Reprinted from A. D. McLaren and L. Packer, "Some Aspects of Enzyme Reactions in Heterogeneous Systems," Adv. Enzymol. Rel. Sub. Biochem., vol. 33, p. 245, 1970.)*

soluble and "insoluble" substrates is lysozyme itself. As noted earlier, lysozyme is active in splitting the bacterial cell wall. However, it also catalyzes the breakdown of soluble oligomers derived from the cell-wall polymer (Sec. 3.4.2).

An interesting variation on the kinetic equations derived earlier arises when enzyme in solution acts on an insoluble substrate by absorbing onto the surface of the substrate. In contrast to previous cases, where the reaction rate increases in

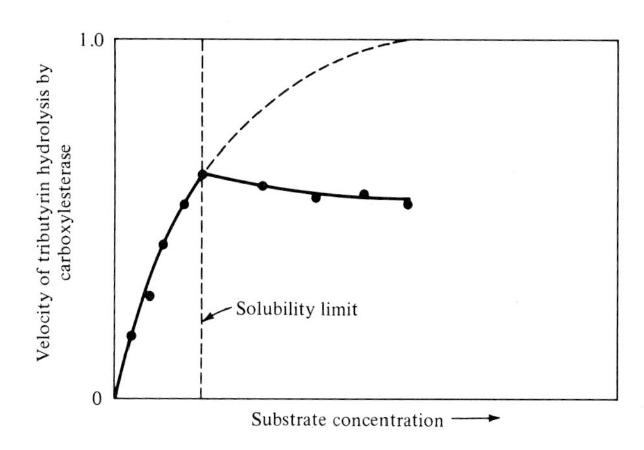

Figure 3.32 Reaction only of soluble-substrate form. *(Reprinted by permission from M. Dixon and E. C. Webb, "Enzymes," 2d ed., p. 90, Academic Press, Inc., New York, 1964.)*

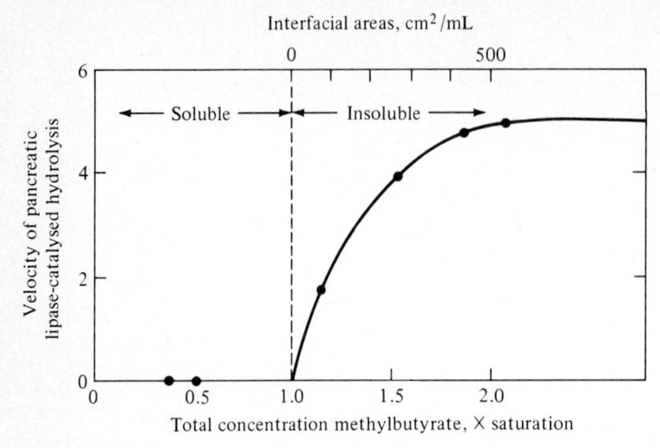

Figure 3.33 Reaction at liquid–liquid interface. *(Reprinted by permission from L. Sarda et P. Desnuelle, "Action de la Lipase Pancréatique sur les Esters en Émulsion," Biochim. Biophys. Acta, vol. 30, p. 513, 1958.)*

direct proportion to total enzyme concentration, a limiting rate is approached as enzyme concentration is increased. This behavior is evident in kinetic data for hydrolysis of a solid cube of protein (specifically thiogel, a cross-linked gelatin) under the action of trypsin (Fig. 3.34).

To develop a reasonable model for heterogeneous kinetics, we begin by turning the tables: the enzyme now adsorbs on the substrate. Letting A denote a vacant site on the substrate surface, we assume the following equilibrium:

$$\mathrm{E} + \mathrm{A} \underset{k_{\mathrm{des}}}{\overset{k_{\mathrm{ads}}}{\rightleftharpoons}} \mathrm{EA} \tag{3.86}$$

If a_0 is the total number of moles of adsorption sites on the substrate surface per unit volume of the reaction mixture, we have

$$a_0 = a + (ea) \tag{3.87}$$

Combining this with the equilibrium relation for enzyme adsorption (3.86) gives

$$(ea) = \frac{a_0 e}{K_A + e} \qquad \text{with} \qquad K_A = \frac{k_{\mathrm{des}}}{k_{\mathrm{ads}}} \tag{3.88}$$

The development is completed by assuming that the hydrolysis is accomplished by irreversible decomposition of the EA complex. Consequently

$$v = k_3(ea) = \frac{k_3 a_0 e}{K_A + e} \tag{3.89}$$

Recalling our notation conventions, e is the concentration of free enzyme and is related to the total concentration e_0 at the start of the experiment by

$$e_0 = e + (ea) \tag{3.90}$$

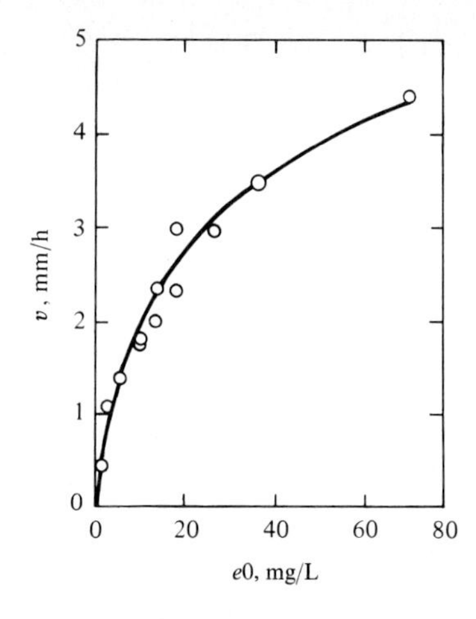

Figure 3.34 Dependence of the rate of disappearance of solid substrate (thiogel) on the concentration e_0 of enzyme in solution. *(Data of A. G. Tsuk and G. Oster, "Determination of Enzyme Activity by a Linear Measurement," Nature (London), vol. 190, p. 721, 1961.)*

If the initial concentration of enzyme is much larger than that of substrate ($e_0 \gg a_0$), we may assume to an excellent degree of approximation that

$$e_0 \approx e \tag{3.91}$$

so that

$$v = \frac{k_3 a_0 e_0}{K_A + e_0} \tag{3.92}$$

The situation $e_0 \gg a_0$ is not at all uncommon for reactions involving solid substrates. For example, the data in Fig. 3.34 were obtained in an experiment for which e_0/a_0 was approximately 4000. This case contrasts sharply with reactions in solution, where s_0 is typically much larger than e_0.

Rate equation (3.92) reveals that a Lineweaver-Burk double-reciprocal plot ($1/v$ vs. $1/e_0$) should be linear. This indeed occurs, as Fig. 3.35 demonstrates. In Fig. 3.35*a*, the data shown in Fig. 3.34 have been replotted. The other part of the figure illustrates a similar result for another soluble-enzyme solid-substrate system.

Many potential microbial nutrients begin as solid particles (in waste streams, lakes, compost piles, etc.). Hydrolysis of these particles by extracellular enzymes is clearly necessary before the cell can absorb or pump the then solubilized nutrient through its own membrane. Also, cellulose hydrolysis by cellulase enzymes requires initial breakdown of insoluble particulates. Hence we can expect the enzyme kinetics in this section to be of importance in such particle-substrate reactors.

Before leaving this topic, we should emphasize two further points: (1) The development above is not necessarily restricted to solid substrates. It has also been applied to a dispersed phase of immiscible liquid substrate, e.g., the system shown in Fig. 3.33; (2) we have ignored possible differences in concentrations between bulk fluid phases and interfaces. These will be examined explicitly in the next chapter.

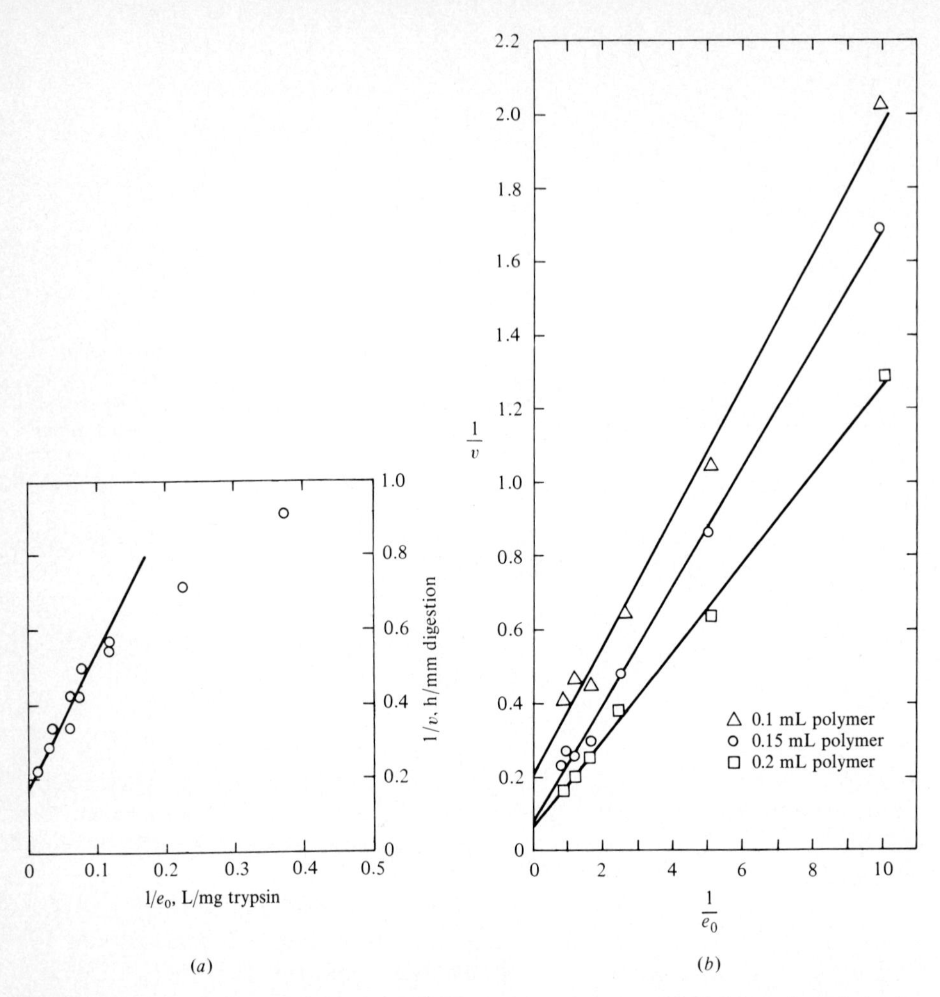

Figure 3.35 (*a*) Double-reciprocal plot of the data in Fig. 3.34. (*b*) Another double-reciprocal plot for digestion of an insoluble substrate (poly-β-hydroxybutyrate particles) by an enzyme (depolymerase of *P. lemoignei*) in solution. *(Reprinted from A. D. McLaren and L. Packer, "Some Aspects of Enzyme Reactions in Heterogeneous Systems," Adv. Enzymol. Rel. Sub. Biochem., vol. 33, p. 245, 1970.)*

PROBLEMS

3.1 Determination of K_m and v_{max} Initial rates of an enzyme-catalyzed reaction for various substrate concentrations are listed in Table 3P1.1.

(*a*) Evaluate v_{max} and K_m by a Lineweaver-Burk plot.
(*b*) Using an Eadie-Hofstee plot, evaluate v_{max} and K_m.
(*c*) Calculate the standard deviation of the slope and intercept for each method.

3.2 Batch enzymic reaction An enzyme with a K_m of 1×10^{-3} M was assayed using an initial substrate concentration of 3×10^{-5} M. After 2 min, 5 percent of the substrate was converted. How much substrate will be converted after 10, 30, and 60 min?

Table 3P1.1

s, mol/L	v, mol/(L·min) × 10^6
4.1×10^{-3}	177
9.5×10^{-4}	173
5.2×10^{-4}	125
1.03×10^{-4}	106
4.9×10^{-5}	80
1.06×10^{-5}	67
5.1×10^{-6}	43

3.3 Multisite enzyme kinetics Suppose that an enzyme has two active sites so that substrate is converted to product via the reaction sequence

$$\mathrm{E} + \mathrm{S} \underset{k_{-1}}{\overset{k_1}{\rightleftharpoons}} (\mathrm{ES}) \qquad (\mathrm{ES}) + \mathrm{S} \underset{k_{-2}}{\overset{k_2}{\rightleftharpoons}} (\mathrm{ESS})$$

$$(\mathrm{ESS}) \xrightarrow[k_3]{} (\mathrm{ES}) + \mathrm{P} \qquad (\mathrm{ES}) \xrightarrow[k_4]{} \mathrm{E} + \mathrm{P}$$

Derive a rate expression for P formation by assuming quasi-steady state for (ES) and for (ESS).

3.4 Multiple enzyme-substrate complexes Multiple complexes are sometimes involved in some enzyme-catalyzed reactions. Assuming the reaction sequence

$$\mathrm{S} + \mathrm{E} \underset{k_2}{\overset{k_1}{\rightleftharpoons}} (\mathrm{ES})_1 \underset{k_4}{\overset{k_3}{\rightleftharpoons}} (\mathrm{ES})_2 \xrightarrow{k_3} \mathrm{P} + \mathrm{E}$$

develop suitable rate expressions using (*a*) the Michaelis equilibrium approach and (*b*) the quasi-steady-state approximations for the complexes.

3.5 Relaxation kinetics with sinusoidal perturbations As Fig. 3P5.1 indicates, when a reaction system at equilibrium (or steady state) is perturbed slightly in a sinusoidal fashion, the concentrations of the

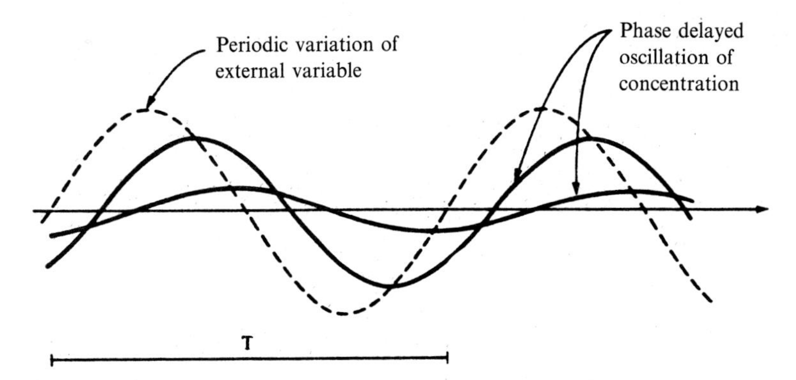

Figure 3P5.1 Response to small-amplitude periodic environmental perturbations.

reacting species also become sinusoidal. Develop the equations necessary to relate the observed concentration fluctuations to the kinetic parameters for the reaction

$$\mathrm{A} \underset{k_2}{\overset{k_1}{\rightleftharpoons}} \mathrm{B} \qquad r = k_1 a - k_2 b$$

when the medium temperature is oscillated according to

$$T = T_0(1 + \alpha \sin wt) \qquad \text{where } \alpha \ll 1$$

Indicate how you would determine k_1 and k_2 from the response of a and b to the periodic perturbations.

3.6 Reversible reactions For the reversible reaction

$$\mathrm{E + S} \underset{k_{-1}}{\overset{k_1}{\rightleftharpoons}} \mathrm{(ES)} \underset{k_{-2}}{\overset{k_2}{\rightleftharpoons}} \mathrm{E + P}$$

show that:

(*a*) The reaction will proceed far to the right only if $k_1 k_2 \gg k_{-1} k_{-2}$.

(*b*) The parameters v_s, v_P, K_s, and K_p are not independent.

(*c*) Under what conditions will a Lineweaver-Burk plot of the equation in part (*b*) yield useful results.

(*d*) Integrate dp/dt above to obtain $p(t)$ in terms of t and the value of p at equilibrium

$$At = Bp + C \ln\left(1 - \frac{p}{p_{\text{eq}}}\right)$$

(*e*) If $K_s = K_p$, show that $B = 0$.

3.7 Enzyme deactivation An enzyme irreversibly denatures according to Eq. (3.77).

(*a*) Show that only $v_{\max}$ and not K_m is affected by such a change.

(*b*) For enzymes acting on insoluble substrates [Eq. (3.92)], show that the converse of part (*a*) will appear to be correct if the investigator is unaware that the active enzyme concentration is changing.

3.8 pH dependence An enzyme-catalyzed reaction irreversibly generates protons according to the equation

$$\mathrm{H_2O + E + S^+ \longrightarrow E + SOH + H^+}$$

If the active form of the enzyme is e^-, and $e^-/e = 1.0$ at pH $= 6.0 = \mathrm{p}K_1$, $e^{2-}/e^- = 1.0$ at pH $= 10 = \mathrm{p}K_2$:

(*a*) Show that the reaction velocity at pH 7.0 is given approximately by

$$v = v_{\max} \frac{sK_1}{(K_s + s)(K_1 + h)} \qquad \text{where } h = [\mathrm{H^+}]$$

(*b*) Integrate the previous equation to show that the time variation of s obeys

$$\alpha \ln \frac{s}{s_0} + \beta(s_0 - s) + \delta(s_0^2 - s^2) = v_m t$$

where α and β depend on the initial pH and initial substrate concentration.

3.9 Inhibitor kinetics A pesticide inhibits the activity of a particular enzyme A, which can therefore be used to assay for the presence of the pesticide in an unknown sample.

(*a*) In the laboratory, the initial rate data shown in Table 3P9.1 were obtained. Is the pesticide a competitive or noncompetitive inhibitor? Evaluate K_i, $v_{\max}$, and K_m.

(*b*) After 50 mL of the same enzyme solution in part (*a*) is mixed with 50 mL of $8 \times 10^{-4}\ M$ substrate and 25 mL of sample, the initial rate observed is 18 μmol/min. What is the pesticide concentration in the unknown *assuming* no other inhibitors or substrates are present in the sample?

Table 3P9.1

s, mol/L	v, mol/(L·min) × 10^6	
	No inhibitor	10^{-5} M inhibitor
3.3×10^{-4}	56	37
5.0×10^{-4}	71	47
6.7×10^{-4}	88	61
1.65×10^{-3}	129	103
2.21×10^{-3}	149	125

3.10 Kinetics with an ionizing cofactor An enzyme requires the presence of a bound cofactor for the enzyme to be in the catalytically active form. The cofactor binds very tightly to the enzyme. A critical group on the cofactor has a pK of pK_c, and the cofactor is only functional when in the deprotonated form. The first pK values of the enzyme active site on either side of the pH of maximum activity are pK_1 and pK_2. Derive (or simply write down) appropriate velocity expressions for the conversion of a single substrate to products when the following conditions apply:

(*a*) $|pK_c| \ll |pK_1| \ll |pK_2|$
(*b*) $|pK_1| \ll |pK_c| \ll |pK_2|$
(*c*) $pK_c = pK_1$

3.11 Substrate activation Derive the rate equation for activation by substrate:

$$E + S \underset{}{\overset{K_S}{\rightleftharpoons}} ES$$

$$E + S \overset{K_{SA}}{\rightleftharpoons} ES_A \underset{S}{\overset{S \quad K'_S}{\rightleftharpoons}} ES_AS \xrightarrow{k_2} ES_A + P$$

Make the usual assumptions: (i) quasi-steady state for intermediates and (ii) concentration of substrate is much greater than that of the enzyme.

3.12 Heat generation in enzyme conversions The maximum temperature rise in a cylindrical plug-flow reactor can be estimated by a closed heat balance around the conversion of substrate on a single flow through the reactor.

(*a*) Writing heat generated ≥ heat gained by flowing medium, calculate the maximum temperature rise obtainable for a single enzyme-catalyzed reaction in terms of heat of reaction per mole reactant ΔH_r, fractional conversion of reactant δ, reactant inlet concentration s_0, liquid heat capacity C_p, and $\Delta T \equiv T_{\text{outlet}} - T_{\text{inlet}}$.

(*b*) For 80 percent hydrolysis of 20% lactose solution ($\Delta H_r = -7100$ cal/g mol), show that the maximum temperature rise is only a few degrees Celsius.

(*c*) When a cooling jacket is applied to a thin enzyme reactor, the exit centerline temperature is approximately given by

$$T^* = \frac{T_{\text{outlet}} - T_{\text{wall}}}{T_{\text{inlet}} - T_{\text{wall}}} = \text{erf}\,\frac{z}{2\sqrt{\alpha t}}$$

where z = column length, t = residence time of fluid, and $\alpha = k_t/\rho C_p$, where k_t = overall thermal conductivity, ρ = packed bed density, and erf is the error function (tabulated for the argument, $z/2\sqrt{\alpha t}$, in any handbook). Show that for the conditions of part (*b*), T^* approaches unity rapidly (calculate T^* when z = 1, 2, 3, 4 in) [W. H. Pitcher, *Immobilized Enzymes for Industrial Reactors*, Academic Press, N.Y. 1975, p. 151.]

3.13 The Hill equation for cooperative binding The following simple model is often used to describe cooperative ($n > 1$) binding of substrates (S) to oligomeric proteins (E_n):

$$nS + E_n \rightleftharpoons E_nS_n$$

Develop a simple graphical method for evaluation of n from measurements of E_nS_n.

3.14 Deactivation of α-chymotrypsin Consider the thermal denaturation plus autodigestion protease deactivation reactions in Eq. (3.84), where K is the equilibrium constant between E_a and E_{i1} and K_m is the dissociation constant of the (E_aE_{i1}) complex. Assuming $(e_a e_{i1}) \ll e_a + e_{i1}$, determine the concentration of potentially active enzyme $e(=e_a + e_{i1})$ versus time in a batch reactor.

REFERENCES

All the biochemistry textbooks listed at the end of Chap. 1 contain introductory material on enzyme kinetics. Mahler and Cordes (Ref. [4] of Chap. 2) is especially strong on some detailed and modern aspects of this subject. For more detailed information, the following books and articles may be consulted.

1. A. Bernhard, *Structure and Function of Enzymes*, W. A. Benjamin, New York, 1968. A short, very readable introduction to enzyme fundamentals. Available in paperback.
2. M. Dixon and E. C. Webb, *Enzymes*, 3d ed., Academic Press Inc., New York, 1979. The most complete single-volume reference on enzymes. In addition to a 200-page chapter on enzyme kinetics, this handbook of the enzymologist considers many other aspects and includes an extensive table of enzymes, the reactions they catalyze, and appropriate references.
3. K. J. Laidler and P. S. Bunting, *The Chemical Kinetics of Enzyme Action*, 2d ed., Oxford University Press, London, 1973. A more detailed treatment of enzyme kinetics. Extensive presentations of experimental evidence for various rate equations.
4. R. D. Schmid, "Stabilized Soluble Enzymes," p. 41 in *Advances in Biochemical Engineering, Vol. 12: Immobilized Enzymes II* (T. K. Ghose, A. Fiechter, and N. Blakebrough, eds.), Springer-Verlag, Berlin, 1979. A superb review article with an excellent summary of protein denaturation causes and effects and a thorough overview of methods for enhancing enzyme stability.
5. F. H. Johnson, H. Eyring, and M. J. Polissar, *The Kinetic Basis of Molecular Biology*, John Wiley & Sons, Inc., New York, 1954. A gem loaded with data and insights on temperature, pH, pressure, and inhibitor effects on enzyme and other biological activities. Application of thermodynamics and absolute rate theory is emphasized.
6. M. Joly, *A Physico-Chemical Approach to the Denaturation of Proteins*, Academic Press, Inc., New York, 1965. While much recent data is obviously not included, this monograph summarizes central concepts and phenomena concerning the causes, manifestations, characterization, and analysis of protein denaturation. A view from a deeper chemical level.
7. A. D. McLaren and L. Packer, "Some Aspects of Enzyme Reactions in Heterogeneous Systems," *Advan. Enzymol.* **33**: 245 (1970). A fascinating review paper covering the entire gamut of heterogeneous enzyme reactions in nature. This article surveys an important but relatively untraveled domain including bound enzymes in the cell and on soil and insoluble substrates.

CHAPTER

FOUR

APPLIED ENZYME CATALYSIS

In this chapter we shall survey some of the applications of enzymes and examine immobilized enzyme catalyst formulations which allow sustained, continuous use of the enzyme. Since the kinetic properties of these biocatalysts depend upon coupled mass transfer and chemical reaction processes, it is also important here to learn how this coupling influences catalyst properties.

All enzymes used in applications are derived from living sources (Table 4.1). Although all living cells produce enzymes, one of the three sources—plant, animal, or microbial—may be favored for a given enzyme or utilization. For example, some enzymes may be available only from animal sources. Enzymes obtained from animals, however, may be relatively expensive, e.g., rennin from calf's stomach, and may depend on other markets, e.g., demand for lamb or beef, for their availability. While some plant enzymes are relatively easy to obtain (papain from papaya), their supply is also governed by food demands. Microbial enzymes are produced by methods which can be scaled up easily. As we shall explore further in Chap. 6, recombinant DNA technology now provides the means to produce many different enzymes, including those not normally synthesized by microorganisms or permanent cell lines, in bacteria, yeast and cultured cells. Moreover, due to the rapid doubling time of microbes compared with plants or animals, microbial processes may be attuned more easily to the current market demands for enzymes. On the other hand, for use in food or drug processes, only those microorganisms certified as safe may be exploited for enzyme production.

While all enzymes used today are derived from living organisms, in this chapter we consider only enzymes which are utilized in the absence of life. Such

Table 4.1 Some enzymes of industrial importance†

Name	Source	Application	Notes	Commercial importance
		Starch-liquifying amylases		
Diastase	Malt	Digestive aid; supplement to bread; syrup	α-Amylase activity, β-amylase activity	+ + +
Takadiastase	*Aspergillus oryzae*	Digestive aid; supplement to bread; syrup	Contains many other enzymes, protease, RNase	+ + +
Amylase	*Bacillus subtilis*	Desizing textiles; syrup; alcohol fermentation industry; glucose production	Crude preparation contains protease	+ + +
Acid-resistant amylase	*Aspergillus niger*	Digestive aid	Optimum pH 4–5	+
		Starch-saccharifying amylases		
Amyloglucosidase	*Rhizopus niveus, A. niger, Endomycopsis fibuliger*	Glucose production		+ + +
		Animal and vegetable proteases		
Trypsin	Animal pancreas	Medical uses; meat tenderizers; beer haze removal		+ + +
Pepsin	Animal stomach	Digestive aid; meat tenderizer		+ + +
α-Chymotrypsin	Animal stomach	Medical uses		
Rennet	Calf stomach	Cheese manufacture		+ + +
Pancreas protease	Animal pancreas	Digestive aid; cleaning; leatherbating; dehairing; feed improvement		+ +
Papain	Papaya	Digestive aid; medical uses; beer haze removal; meat tenderizer		+ + +
Bromelain, ficin	Pineapple, fig	Digestive aid; medical uses; beer haze removal; meat tenderizer		+ +

Name	Source	Application	Notes	Commercial importance
		Microbial proteases		
Protease	*A. oryzae*	Flavoring of sake; haze removal in sake		+
Protease	*A. niger*	Feed, digestive aid	Acid resistant protease; optimum pH 2–3	+ +
Protease	*B. subtilis*	Detergents; removal of gelatin from film (recovery of silver); fish solubles; meat tenderizer	Optimum pH 7.0	+ +
Protease	*Streptomyces griseus*	Detergents; removal of gelatin from film (recovery of silver); fish solubles; meat tenderizer	Optimum pH 8.0	+ +
Varidase	*Streptococcus* sp.	Medical use	Lederle	+ +
Streptokinase	*Streptococcus* sp.		Profibrinolysin	+ +
		Some other commercial enzymes		
Glucose isomerase	*Lactobacillus brevis, Bacillus coagulans, Arthrobacter simplex, Actinoplanes missourensis*	Glucose → fructose	Novo, ICI, Gist Brocades.	+ + +
Penicillinase	*B. subtilis, Bacillus cereus*	Removal of penicillin	Takamine, Schenley	+
Glucose oxidase	*Aspergillus niger*, Dee O, Dee G	For removal of oxygen or glucose from various foods; dried-egg manufacture	Takamine	+
	Penicillium chrysogenum	For glucose determination	Nagase Co.	
Hyaluronidase	Animal, bacteria	Medical use		+
Lipase	Pancreas, mold (*Rhizopus*)	Digestive aid; flavoring of milk products		+

(continued)

Table 4.1 Some enzymes of industrial importance[†] *(continued)*

Name	Source	Application	Notes	Commercial importance
		Some other commercial enzymes		
Cytochrome *c*	Yeast (*Candida*)	Medical use	Sankyo Co.	+
Catalase		Sterilization of milk		
Keratinase	*Streptomyces fradiae*	Removal of hair from hides	Merck Co.	+
5′-Phosphodiesterase	*Penicillium citrinum, S. griseus, B. subtilis*	Inosinic acid and guanylic acid manufacture. (5′-nucleotides)	Yamasa Co., Takeda Co.	+ + +
Adenylic acid deaminase	*A. oryzae*	AMP → IMP	In Takadiastase	+
Microbial rennet	*Mucor* sp.	Cheese manufacture	Meito Sangyo Co	+ +
Naringinase.	*Aspergillus niger*	Removal of bitter taste from citrus juice	Rohm and Haas	+
Laccase	*Coriolus versicolor*	Drying of lacquer		
Cellulase	*Trichoderma koningi*	Digestive aid	Optimum pH 4.6	
	Trichoderma viride	Cellulose hydrolysis	An enzyme mixture	
Invertase	*Saccharomyces cerevisiae*	Confectionaries, to prevent crystallization of sugar; chocolate; high-test molasses		
Pectinase	*Sclerotina libertina*	Increase yield and for clarifying juice	Scrase (Sankyo); Pectinol (Rohm and Hass); Takamine	+ + +
	Coniothyrium diplodiella, Aspergillus oryzae, Aspergillus niger, Aspergillus flavus	Removal of pectin; coffee concentration	Hass); Takamine Pectinase Clarase (Takamine); Filtragol (I. G. Farben)	

[†] Adapted from K. Arima, Microbial Enzyme Production, in M. P. Starr (ed.), *Global Impacts of Applied Microbiology*, pp. 278–279, John Wiley & Sons, Inc., New York, 1964.

biological catalysts include *extracellular enzymes*, secreted by cells in order to degrade polymeric nutrients into molecules small enough to permeate cell walls. Grinding, mashing, lysing, or otherwise killing and splitting whole cells open frees *intracellular enzymes*, which are normally confined within individual cells.

Certain applications of enzymes demand the use of a relatively pure extract. For example, glucose oxidase for desugaring of eggs must be free of any protein-splitting enzymes, and proteolytic enzymes injected into animals for meat tenderization just before slaughter must not contain any compounds which would cause a serious physiological reaction. Other examples requiring relative purity include enzymes in clinical diagnosis and some enzymes in food processing.

Studies of enzyme kinetics have been generally carried out with the purest possible enzyme preparations. As indicated in Chap. 3, such research also involves the fewest possible number of substrates (one if achievable) and a controlled solution with known levels of activators (Ca^{2+}, Mg^{2+}, etc.), cofactors, and inhibitors. The results of these studies provide the clearest picture of enzyme kinetics.

Many useful industrial enzyme preparations are not highly purified. They contain a number of enzymes with different catalytic functions and, under most conditions, are not used with anything approaching either a pure substrate or a completely defined synthetic medium in the sense discussed in Chap. 3. Also, the simultaneous use of several different enzymes may be more efficient than sequential catalysis by a separated series of the enzymes. In spite of this added complexity, such enzyme preparations are kinetically more simple than the integrated living organisms from which they are produced and are thus logically considered before the chapters dealing with cellular metabolism and industrial biotechnological routes of product synthesis.

4.1 APPLICATIONS OF HYDROLYTIC ENZYMES

The action of hydrolytic enzymes is important not only in obvious macroscopic degradations such as food spoilage, starch thinning, and waste treatment, but also in the chemistry of ripening picked green fruit, self-lysis of dead whole cells (autolysis), desirable aging of meat, curing cheeses, preventing beer haze, texturizing candies, treating wounds, and desizing textiles. These and other uses, the enzymes involved and their sources, will occupy our attention in this section.

A general classification of the major hydrolytic enzymes is given in Table 4.2. The three groups of enzymes are those involved in the hydrolysis of ester, glycosidic, and various nitrogen bonds, respectively. Finer classification of all such enzyme groupings are available and in common use. For the present, we shall simply point out that some enzymes catalyze the hydrolysis of a large variety of glucose linkages, for example, whereas other enzymes may catalyze the hydrolysis of only one glucose oligomer. Thus, the name of the enzyme by itself is not necessarily indicative of the precise substrate specificity.

Table 4.2 Hydrolytic enzymes†

Enzyme	Substrate	Hydrolysis product
Esterases:		
Lipases	Glycerides (fats)	Glycerol + fatty acids
Phosphatases:		
Lecithinase	Lecithin	Choline + H_3PO_4 + fat
Pectin esterase	Pectin methyl ester	Methanol + polygalacturonic acid
Carbohydrases:		
Fructosidases	Sucrose	Fructose + glucose
α-Glucosidases (maltase)	Maltose	Glucose
β-Glucosidases (cellobiase)	Cellobiose	Glucose
β-Galactosidases (lactase)	Lactose	Galactose + glucose
Amylases	Starch	Maltose or glucose + maltooligosaccharides
Cellulase	Cellulose	Cellobiose
Cytase	....	Simple sugars
Polygalacturonase	Polygalacturonic acid	Galacturonic acid
Nitrogen-carrying compounds		
Proteinases	Proteins	Polypeptides
Polypeptidases	Proteins	Amino acids
Desaminases:		
Urease	Urea	$CO_2 + NH_3$
Asparaginase	Asparagine	Aspartic acid + NH_3
Deaminases	Amino acids	NH_3 + organic acids

† H. H. Weiser, *Practical Food Microbiology and Technology*, p. 37, Avi Publishing Co., Westport, Conn., 1962.

We should again note here that most enzymes have been named according to the chemical reactions they are observed to catalyze, rather than according to their structure.† Since a one-enzyme–one-reaction uniqueness does not generally exist, enzymes from different plant or animal sources which catalyze a given reaction will not always have the same molecular structure or necessarily the same kinetics. Consequently, maximum reaction rate, Michaelis constant, pH of optimum stability or activity, and other properties will depend on the particular enzyme source used.

Many hydrolases are directed to specific compartments separated from the cytoplasm by membranes. This serves the obvious purpose of protecting essential cytoplasmic bipolymers from degradation. Gram-positive bacteria secrete a variety of hydrolases into their environment. With their double membrane outer envelope, gram-negative bacteria have available the periplasmic space which safely stores a variety of hydrolases. In eucaryotes, hydrolases may be stored

† The international classification scheme of the Enzyme Commision was outlined in Table 3.1. For further information on classification, see M. Dixon and E. C. Webb, *Enzymes*, 3d ed., chap. 5. Academic Press, Inc., New York, 1979.

inside the cell in membrane-enclosed lysosome organelles, reside in the periplasm in microbes like yeast, or be secreted into the environment. Most hydrolytic enzymes used commercially are extracellular microbial products.

A few hydrolases are found in the cytoplasm, however. These serve important recycle functions in the cell's use of chemical resources. Intracellular hydrolases have significant implications in genetic engineering technology (Chap. 6).

Since water is an omnipresent substrate at roughly 55 *M* concentration, and since hydrolytic enzymes are normally associated with degradative reactions, e.g., conversion of starch to sugar, proteins to polypeptides and amino acids, and lipids to their constituent glycerols, fatty acids and phosphate bases, the following discussion is organized around the various possible polymer substrates. It should be noted, however, that several recent experimental processes for polypeptide synthesis have employed these naturally hydrolytic enzymes in media with low water activity. Under these circumstances, the reverse, synthetic reaction can be catalyzed in some cases.

4.1.1 Hydrolysis of Starch and Cellulose

Amylases are extensively applied enzymes which can hydrolyze the glycosidic bonds in starch and related glucose-containing compounds. To appreciate the distinction between the two major types of amylases, we should recall that starch contains straight-chain glucose polymers called amylose and a branched component known as amylopectin. The branched structure is relative more soluble than the linear amylose and is also effective in rapidly raising the viscosity of starch solution. The action of α-amylase reduces the solution viscosity by acting *randomly* along the glucose chain at α-1,4 glycosidic bonds: α-amylase is often called the *starch-liquefying enzyme* for this reason.[†] β-Amylase can attack starch α-1,4 bonds only on the nonreducing ends of the polymer and always produces maltose when a linear chain is hydrolyzed. Because of the characteristic production of the sugar maltose, β-amylase is also called a *saccharifying enzyme.* A soluble mixture of starch and β-amylase yields maltose and a remainder of dextrins, starch remnants with 1,6 linkages on the end. β-Amylase cannot hydrolyze these bonds.

Another saccharifying enzyme, *amyloglucosidase* (also called glucoamylase, among other names) attacks primarily the nonreducing α-1,4 linkages at the ends of starch, glycogen, dextrins, and maltose. (α-1,6 linkages are cleaved by amyloglucosidase at much lower rates.) Sequential treatment with α-amylase and glucoamylase or enzyme mixtures are utilized where pure glucose rather than maltose is desired: in distilleries (as opposed to breweries) and in the manufacture of glucose syrups (corn syrup) and crystalline glucose. It is estimated that 1.35 billion pounds of glucose were produced by this method in the United States

[†] α-Amylase action is apparently random with long-chain substrates but shows subsite structure specificity for shorter oligomers.

Table 4.3 Common applications of amylase preparations†

Industry	Use
Glucose and syrup	Total or partial hydrolysis of corn starch by amyloglucosidase or α-amylase to give a large quantity of sweeteners
Brewing	Conversion of crushed grain starch to maltose (a suitable dissacharide substrate for yeast fermentation)
Breadmaking	*Leavening:* Conversion of sufficient starch to fermentable saccharides needed for carbon dioxide generation
Fruit juice	Hydrolysis of starch causing turbidity due to insolubility
Papermaking	α-Amylase action to liquefy starch coatings to a desired viscosity for application to fibers (variable-weight papers)
Textiles	*Sizing:* α-Amylase activity to liquefy starch; resulting solution used to strengthen warp threads before weaving
	Desizing: α,β-Amylase action to remove size from woven material so that all threads will dye uniformly and fabric will have desired texture
Candy	Production of candy of desired softness

† Adapted from H. H. Weiser, *Practical Food Microbiology and Technology*, p. 37, Avi Publishing Co., Westport, Conn., 1962.

in 1971. This and other applications given in Table 4.3 make these enzymes one of the most important groups in commercial use. The relative proportions of α- and β-amylase selected in various applications depend on the result desired.

The sources of amylases are very numerous. This is not surprising since starch is a common form of carbon fuel for many life forms. Amylases are produced by a number of bacteria and molds; an important example is the amylase produced by *Clostridium acetobutylicum* which is clearly involved in the microbial conversion of polysaccharides to butanol and acetone. Commercial amylase preparations used in human foods are normally obtained from grains, notably barley, wheat, rye, oats, maize, sorghum, and rice. The ratio of saccharifying to liquefying enzyme activity depends not only on the particular grain but also upon whether the grain is germinated. In the production of malt (softened, germinated barley) for brewing, the ungerminated seeds are exposed to a favorable temperature and humidity so that rapid germination occurs, with resulting large increase in α-amylase. The germinated barley is then kiln-dried slowly; this halts all enzyme activity without irreversible inactivation. The dried malt preparation is then ground, and its enormous liquefying and saccharifying power (to convert starches to fermentable sugars) is utilized in the subsequent yeast fermentation.

The other carbohydrases listed in Table 4.1 also cleave glycosidic bonds. Increased yields of glucose from starch have been reported by application of *pullulanase* in conjunction with amylase treatment. Pullulanase hydrolyzes 1-6 glycosidic branching bonds selectively. Splitting lactose into the sweeter components glucose and galactose by *lactase* is common in the manufacture of ice cream products. A related enzyme, *invertase* hydrolyzes sucrose and polysaccharides containing a β-D-fructofuranosyl linkage. The enzyme name derives from the early observation that the hydrolyzed sucrose solution containing fructose and glucose rotates a polarized light beam in the direction opposite that of the original solution. The partially or completely hydrolyzed solution allows two properties desirable in syrup and candy manufacturing: a slightly sweeter taste than sucrose and a much higher sugar concentration before hardening.

Substantial research and development efforts worldwide have focused on enzymatic hydrolysis of cellulose. To begin our overview on this process, we should emphasize that *cellulase*, the name usually used to describe the enzyme material active in depolymerizing cellulose, is a complex mixture of several different enzymes. Furthermore, the different enzymes present and their relative quantities depend upon the microorganism used for cellulase production and, in some situations, on the enzyme production process. As mentioned in the lignocellulosics discussion in Chap. 2, biomass and waste materials from different sources have different physical properties such as crystallinity and surface area and different chemical compositions. Through several different pretreatment processes, these substrates may be modified to enhance their susceptibility to hydrolysis. Thus, the hydrolysis rates and yields achieved in a particular process depend upon the interactions of substrate properties, pretreatment effects, and multiple enzyme activities and modes of attack.

Cellulase systems produced by species of *Trichoderma* fungi are the most thoroughly developed and characterized at present. These systems possess three major classes of enzymes with different substrates and products as indicated in Fig. 4.1. Feedback inhibition of enzyme activity occurs as indicated; kinetic models for this reaction network have typically employed Michaelis-Menten kinetics for each step with strictly competitive or noncompetitive inhibition. Many other microorganisms including the molds *Fusarium solani, Aspergillus niger, Penicillium funicolsum, Sporotrichum pulverulentum, Cellulomonas* species, *Clostridium thermocellum*, and *Clostridium thermosaccharolyticum* bacteria produce cellulases with distinctive activities and properties.

Table 4.4 provides further information of the *T. viride* cellulase system. Notice the use of different "standard" cellulosic substrates for determination of characteristic activities. Also important are the enzyme dimensions, which are comparable to the microfibril size in native cellulose. Thus, enzyme hydrolysis rates depend critically on the crystalline structure of cellulose. Kinetic forms which incorporate the influence of crystallinity will be examined in Example 4.1 below, after a brief overview of chemical and physical pretreatment methods for lignocellulosics.

Insights on the influence of lignin on cellulose digestion are provided by studies of chemical pulping of woods. In papermaking, wood chips are treated with acid sulfite or alkaline sulfate solutions (Kraft process) at elevated temperature and pressure to solubilize lignins and pectins. By varying these conditions for four wood species, pulps with different lignin contents were obtained which were then treated with sulfuric acid to hydrolyze the accessible cellulosics. The

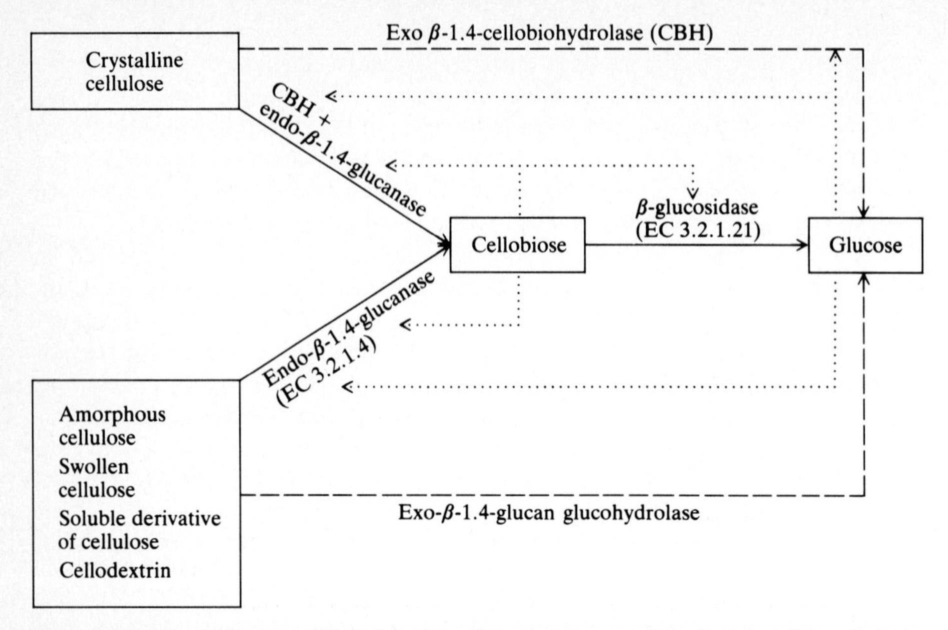

Figure 4.1 Schematic diagram of the substrates, cellulase enzymes, and products in cellulose hydrolysis (→ major reaction; --> side pathways; ··> inhibition effects.) *(Adapted from Y. H. Lee and L. T. Fan, "Properties and Mode of Action of Cellulase," Advances in Biochemical Engineering (A. Fiechter, ed.), vol. 17, p. 101, 1980.)*

Table 4.4 Properties of enzyme classes in the *T. viride* cellulase system

	Enzyme class		
	Endo-β-1,4-glucanase	Exo-β-1,4-cellobiohydrolase	β-Glucosidase (cellobiase)
Standard substrate	Carboxymethylcellulose (CMC)	Avicel†	Cellobiose
Inhibitors	Cellobiose	Glucose (cellobiose)	?
Approximate molecular weight	12,500–52,000	46,000	76,000
Equivalent sphere diameter, Å	34–64	62	76

† Avicel is a commercially available crystallized acid-treated cellulose.

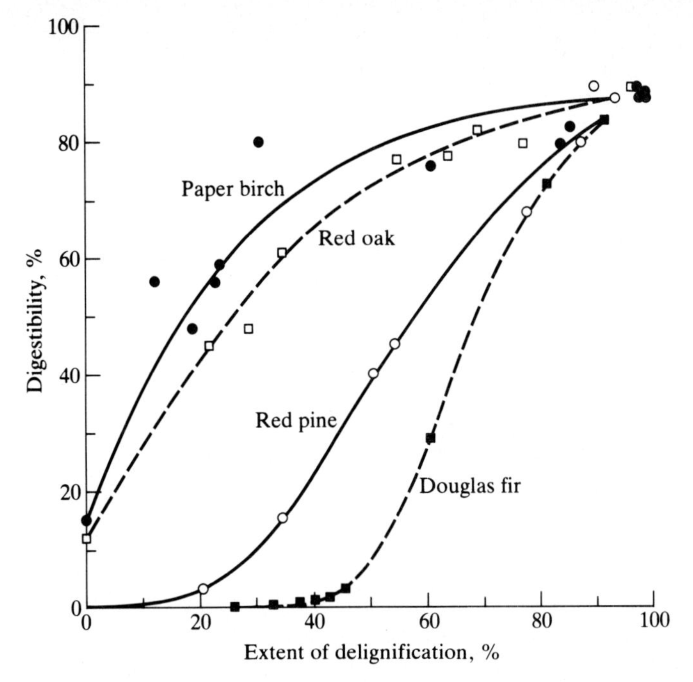

Figure 4.2 Experimental data showing the relationship between delignification and cellulose digestibility for four different wood species. *(Reprinted by permission from A. J. Baker, "Effect of Lignin on the in vitro Digestibility of Wood Pulp," J. Anim. Sci., vol. 37, p. 768, 1973.)*

results of this experiment (Fig. 4.2) clearly reveal the importance of lignin content in restricting biomass digestibility. Other approaches for removing lignins include gaseous SO_2, mineral acid and lignin-degrading enzyme treatment. Ligninase enzymes are produced by several fungi including *Sporotrichum pulverulentum* and *Pleurotus ostreatus*.

Cellulose structure—crystallinity, specific surface area and degree of polymerization—can be altered by a variety of pretreatments such as ball or compression milling, γ-irradiation, pyrolysis, and acidic or caustic chemicals. We shall focus here on pretreatment-crystallinity-hydrolysis kinetics interactions as an example, suggesting further reading in the References for discussions of other aspects. Pretreatment effects on crystallinity are best characterized by x-ray diffraction measurements as exemplified in Fig. 4.3*a*. From such data a *crystallinity index* (CrI) may be calculated using an empirical formula which, for the type of cellulose considered in Fig. 4.3*a*, is given by

$$\mathrm{CrI}(\%) = \left[1 - \frac{\mathrm{I(am)}}{\mathrm{I(002)}}\right] \times 100\% \tag{4.1}$$

where I(am) (am = amorphous) and I(002) are the diffraction intensities at $2\theta =$ 18.50 and 22.5°, respectively. While not a precise measure of the crystalline fraction, the CrI values are indicative of the relative crystallinity content and average

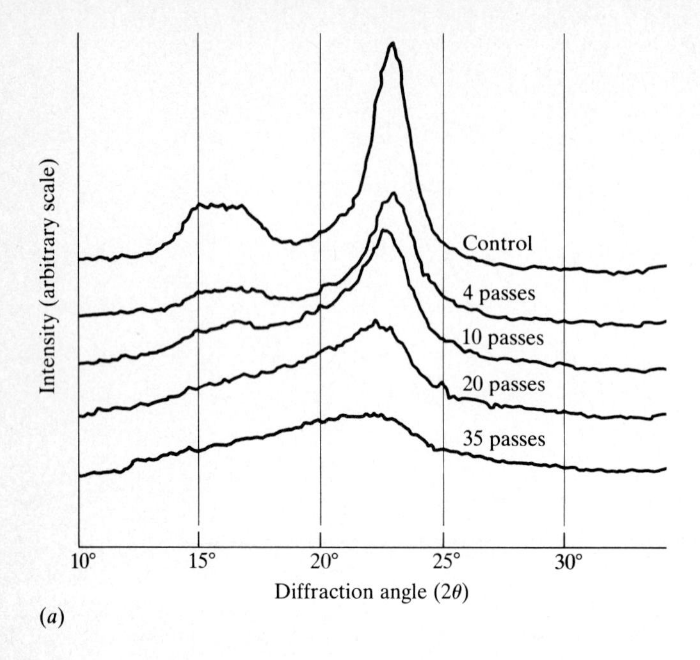

(*a*)

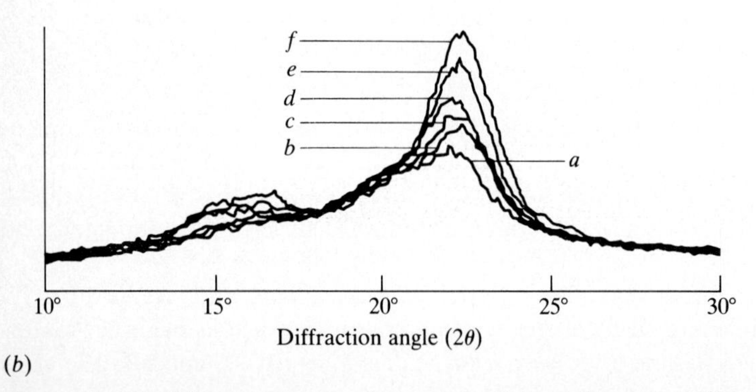

(*b*)

Figure 4.3 X-ray diffraction patterns of (a) Avicel cellulose subjected to different numbers of passes through a two-roll compression mill and (b) after different times in *T. reesei* cellulase (*a* through *f* correspond to the reaction time sequence 0, 3, 6, 12, 24, and 36h). Intensities at the two indicated diffraction angles $2\theta = 18.5$ and $22.5°$ are used to evaluate the crystallinity index CrI. *(Reprinted by permission from (a) D. D. Y. Ryu, S. B. Lee, T. Tassinari, and C. Macy, "Effect of Compression Milling on Cellulose Structure and on Enzymatic Hydrolysis Kinetics," Biotech. Bioeng., vol. 24, p. 1047, 1982; and (b) S. B. Lee, I. H. Kim, D. D. Y. Ryu, and H. Taguchi, "Structural Properties of Cellulose and Cellulase Reaction Mechanism," Biotech. Bioeng., vol. 25, p. 33, 1983.)*

Table 4.5 Effects of repeated compression milling on crystalline structure of Avicel cellulose and on overall kinetic properties of hydrolysis by *T. reesei* cellulase

(Data from D. D. Ryu, S. B. Lee, T. Tassinari, and C. Macy, "Effect of Compression Milling on Cellulose Structure and on Enzymatic Hydrolysis Kinetics", Biotech. Bioeng., vol. 24, p. 1047, 1982.)

Number of passes	CrI, %	Crystallite Size, Å	v_{max}^{app}, μg/mL/min	K_m^{app}, mg/mL	v_{max}^{app}/K_m^{app}, h^{-1}
0	81	38			
4	71	31	19.6	31.3	0.038
10	61	21	22.2	20.0	0.067
20	37	10	23.1	14.8	0.110
35	17	7	19.3	7.7	0.150

crystallite size of different cellulose materials. The experimental results in Fig. 4.3*a* show clearly the effect of successive compression milling treatments in reducing cellulose crystallinity.

Our earlier discussions of cellulose structure and the activities of the cellulase enzyme system suggest that amorphous or paracrystalline regions of the cellulose should be hydrolyzed more easily than the crystalline portions, implying an increase in enzyme hydrolysis rates with decreasing crystallinity. Studies of the cellulose materials considered in Fig. 4.3*a*, treated with crude cellulase from *Trichoderma reesei* MCG-77, confirm this trend (Table 4.5). The parameters in Table 4.5 are apparent overall values in a conventional Michaelis-Menten rate equation, assuming that no cellobiose is present during the initial rate measurement.

Calculation of the time-course of enzymatic hydrolysis of these materials is complicated by the change with time of the relative amorphous and crystalline content of the cellulosic substrate (see Fig. 4.3*b*), The following example illustrates application and extension of the approaches of Chap. 3 to address this problem.

Example 4.1: Influence of crystallinity on enzymatic hydrolysis of cellulose† Many different kinetic models have been formulated for cellulose hydrolysis by cellulase enzyme systems. One which focuses on the effect of cellulose crystallinity presumes that cellulase enzymes (here lumped into a single form E) adsorbs on the cellulose in state E*

$$E \underset{k_d}{\overset{k_{ad}}{\rightleftharpoons}} E^* \tag{4E1.1}$$

Assuming that this adsorption process is at equilibrium and that the enzyme is present at low concentration in solution (under practical conditions $e_0 < 0.5$ mg protein per milliliter) gives

$$e^* = K_D e \tag{4E1.2}$$

† See D. D. Y. Ryu, S. B. Lee, T. Tassinari and C. Macy, "Effect of Compression Milling on Cellulose Structure and on Enzymatic Hydrolysis Kinetics," *Biotech. Bioeng.* vol. 24, p. 1047, 1982.

The model framework is completed by assuming that adsorbed enzyme catalyzes amorphous (S_a) and crystalline (S_c) cellulose hydrolysis according to the following parallel processes:

$$E^* + S_a \underset{k_{-1a}}{\overset{k_{1a}}{\rightleftharpoons}} E^*S_a \xrightarrow{k_{2a}} E^* + P \tag{4E1.3}$$

$$E^* + S_c \underset{k_{-1c}}{\overset{k_{1c}}{\rightleftharpoons}} E^*S_c \xrightarrow{k_{2c}} E^* + P \tag{4E1.4}$$

and that the adsorbed and free cellulase binds to inert materials in the cellulose substrate and to product, respectively,

$$E^* + S_x \underset{k_{-1x}}{\overset{k_{1x}}{\rightleftharpoons}} E^*S_x \tag{4E1.5}$$

$$E + P \underset{k_{-1i}}{\overset{k_{1i}}{\rightleftharpoons}} EP \tag{4E1.6}$$

Applying the quasi-steady-state approximation to all enzyme-substrate complexes and using a mass balance over all enzyme forms gives the following expression for the initial hydrolysis rate of pure cellulosics ($s_x = 0$) at low enzyme concentrations with no product present ($p = 0$):

$$v\bigg|_{t=0} = -\frac{ds}{dt}\bigg|_{s=s_0,\,p=0,\,s_x=0} = \frac{v_{\max}^{\text{app}} s_0}{K_m^{\text{app}} + s_0} \tag{4E1.7}$$

Here, s is the total cellulose concentration ($= s_a + s_c$). The apparent maximum velocity and Michaelis constants are functions of the initial crystallinity CrI_0 [$= s_{c0}/(s_{a0} + s_{c0})$] and the elementary step rate constants as follows

$$v_{\max} \equiv \frac{v_{\max,a}(1 - \text{CrI}_0) + v_{\max,c}\text{CrI}_0(K_a/K_c)}{1 - \text{CrI}_0 + (K_a/K_c)\text{CrI}_0} \tag{4E1.8}$$

$$K_m \equiv \frac{K_a}{1 - \text{CrI}_0 + (K_a/K_c)\text{CrI}_0} \tag{4E1.9}$$

where

$$v_{\max,a} = k_{2a}e_0 \qquad v_{\max,c} = k_{2c}e_0$$

$$K_a = (1 + K_D)(k_{-1a} + k_{2a})/K_D k_{1a} \tag{4E1.10}$$

$$K_a/K_c = \frac{k_{1c}(k_{-1a} + k_{2a})}{k_{1a}(k_{-1c} + k_{2c})}$$

The initial rate expression above clearly displays dependence on initial substrate crystallinity CrI_0. Suitability of this representation of crystallinity effects can be tested simply after noting that the pseudo-first-order rate constant $v_{\max}^{\text{app}}/K_m^{\text{app}}$ and the reciprocal apparent Michaelis constant $1/K_m^{\text{app}}$ are, according to Eqs. (4E1.8, 9), linear functions of CrI_0. Experimental measurements of apparent initial rate parameters and initial substrate crystallinity for several cellulosic substrates are plotted in this form in Fig. 4E1.1. Most of these data are very consistent with the model above, with the greatest deviation observed for the pseudo-first-order rate constant data for Solka Floc.

This result is instructive because it shows how some reasonable hypotheses concerning the physical events taking place can lead to a useful functional form for interpreting kinetic data. While the initial rate expression given above is somewhat complicated, it is clear on physical grounds that the initial hydrolysis rate should depend on each of the parameters which appear in this equation. In addition, the required rate expression to calculate the time course of cellulose hydrolysis in a batch reactor is necessarily more complicated since product inhibition [see Eq. (4E1.6)] must be considered.

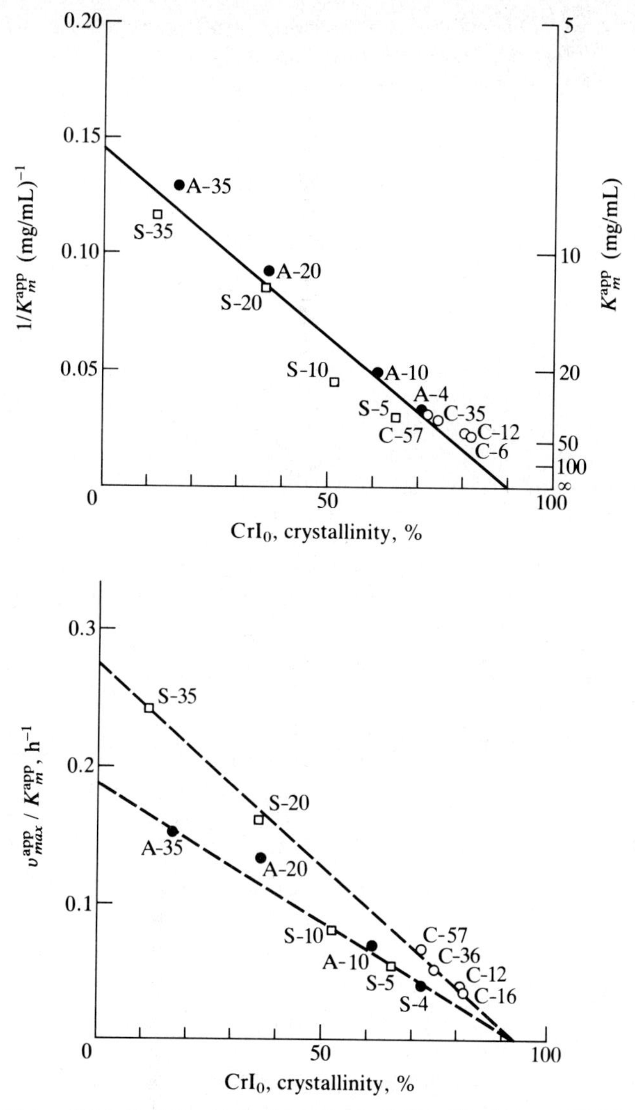

Figure 4E1.1 Experimentally determined relationships between initial substrate crystallinity CrI_0 and the apparent kinetic parameters v_{max}^{app} and K_m^{app} for Avicel (A, ●), cotton (C, ○), and Solka Floc (S, □). Numbers indicate the number of compression milling passes in substrate pretreatment. X-ray diffraction patterns for the Avicel substrates are given in Fig. 4.3*a (Reprinted by permission from D. D. Y. Ryu, S. B. Lee, T. Tassinari, and C. Macy, "Effect of Compression Milling on Cellulose Structure and on Enzyme Hydrolysis Kinetics," Biotech. Bioeng., vol. 24, p. 1047, 1982).*

Furthermore, when calculating the change in substrate concentration with time, it must be recognized that the substrate is a combination of crystalline and amorphous forms which hydrolyze at different rates. Thus, one must calculate the reaction progress not with a single substrate but based upon parallel hydrolysis of two substrates. These hydrolysis processes interact because the crystallinity parameter changes with time as the relative quantities of amorphous and crystalline cellulose change.

This example introduces some of the necessary modeling complexities when dealing with biological substrates and catalysts. A general modeling principle for these systems, to be encountered at many points in our future discussions, is the selection of a biochemical or mechanistic view of the process and a corresponding mathematical representation which focuses on the key variables of interest *in that particular analysis*. Thus, in the example above, the substrate is viewed conceptually as different, idealized forms so that the effect of crystallinity can be represented, associated kinetic parameters can be evaluated, and the change with time of substrate crystallinity in a batch reaction can be calculated. All of the experimental data was obtained for the same cellulase enzyme complex, and, consequently, this model did not consider different enzyme forms and their relative quantities. Clearly, in a study of cellulases from different sources, it would be necessary to consider the different enzymes present and their activities. Ideally we would eventually be able to combine information from several careful experimental and modeling studies, such as the one summarized above, on different facets of the overall bioconversion system to develop a master model which would represent a more complete spectrum of catalyst and substrate properties.

To conclude the overview of cellulose utilization, we should mention that numerous physicochemical alternatives exist for cellulose decomposition. Improved acid hydrolysis processes have been proposed recently in which high temperature/short time contacting is employed to achieve the desired breakdown while minimizing side product formation. Pyrolysis is another approach with significant potential. Other strategies under study, in development or in use for lignocellulose utilization include screw extrusion, treatment with heterogeneous catalysts, fluidized-bed gasification, and the ancient method: combustion.

4.1.2 Proteolytic Enzymes

The varieties and uses of enzymes which attack nitrogen-carrying compounds selectively, especially proteins, are quite large. As with the amylases, the mode of attack on polyamino acids is either on terminal groups (exopeptidases) or internal linkages (endopeptidases).

Since enzymes, the essential catalysts of living organisms, are themselves protein, it is not surprising that protein-splitting enzymes are often initially synthesized in an inactive form. The enzyme is synthesized in an inactive form suitable either for storage or for transport from the site of synthesis to the desired site of activity, as is the case for pepsin, trypsin, chymotrypsin, and carboxypeptidase. Activation of these proteolytic enzymes is then accomplished in one of two ways (Fig. 4.4). It is interesting to note that the activation of pepsin and trypsin is *autocatalytic*: the inactive enzyme precursor is a substrate for the active form of the enzyme, the reaction product being more of the activated enzyme. Equally interesting is the fact that further proteolytic attack is not observed after the initial conversion of inactive trypsin or pepsin to the active form. A second group of enzymes, typically exopeptidases, require one or perhaps several specific metal ions for activation. Dialysis of the enzyme-containing solution with resultant loss

(*a*) From a precursor

$$\underset{MW = 42{,}000}{\text{Pepsinogen}} \xrightarrow{H^+,\ \text{pepsin}} \underset{MW = 38{,}000}{\text{Pepsin}} + \underset{MW \cong 4000}{X}$$

$$\underset{MW \simeq 34{,}000}{\text{Trypsinogen}} \xrightarrow[\text{enterokinase}]{\text{trypsin}} \underset{MW \simeq 34{,}000}{\text{Trypsin}}$$

$$\text{Chymotrypsinogen} \xrightarrow{\text{trypsin}} \text{Chymotrypsin}$$

$$\text{Procarboxypeptidase} \xrightarrow{\text{trypsin}} \text{Carboxypeptidase}$$

(*b*) By presence of a metal ion

Glycylglycine dipeptidase (inactive) + Co^{++} + Mg^{++}

$\longrightarrow$ Glycylglycine peptidase (active)

Leucine aminopeptidase (inactive) + (Mn^{++} or Mg^{++})

$\longrightarrow$ Leucine aminopeptidase (active)

Figure 4.4 Activation of proteolytic enzymes. (*a*) From a precursor, and (*b*) by presence of a metal ion.

of metal ions is a standard form of determining specific metal ion requirements of enzymes and cell life in general.

The commercial sources of proteases include animals (pancreas) and large plants (sap, juices) as well as yeasts, molds, and bacteria. Some of them are listed in Table 4.1, along with the corresponding applications.

The major uses of free proteases occur in dry cleaning, detergents, meat processing (tenderization), cheesemaking (rennin only), tanning, silver recovery from photographic film (pepsin), production of digestive aids, and certain medical treatments of inflammations and virulent wounds. Enzymes were used in laundry aids as early as 1913. During the late 1960s an explosive increase in protease utilization in detergents occurred. The enzymes used facilitate spot removal; they are a mixture of bacterial neutral and alkaline proteases which are active over the pH range of 6.5 to 10 and temperatures from 30 to 60°C. A peak in this enzyme application occurred in 1969 when 30 to 75 percent of all European detergents and about 40 percent of detergents in the United States contained enzymes. However, subsequent warnings from the U.S. Federal Trade Commission caused concern about health hazards from these preparations, and this enzyme market plummeted in 1970 and 1971. Following retraction of the Trade Commission warning and use of modified manufacturing procedures to minimize enzyme dust formation (by "wax coating"), a partial recovery followed, with the 1980 United States demand for detergent bacterial protease estimated at 6 million dollars.

The tenderization of individual meat pieces by commercial tenderizer products depends on proteolytic action of the relatively inexpensive and nonheat labile plant proteases papain and bromelain. Aging of whole meat carcasses prior to cutting and packaging is normally accomplished by controlled partial self-digestion (autolysis) of the bled meat at about 15°C in the presence of ultraviolet light, which acts as a germicidal agent preventing the concurrent surface growth of undesirable microorganisms.

Ground pancrease preparations from different animal sources contain all the digestive proteases, including trypsin as well as lipases and amylases. These obviously digestive mixtures are useful in dehairing animal hides and for simultaneous removal of other noncollagen protein from the hides. Since pepsin itself attacks collagen, the fibrous skin protein which is converted into leather, this proteolytic enzyme is useless in tanning.

In the dairy industries, rennin (or rennet) is the single most important enzyme. It acts by removing a glycopeptide from soluble calcium casein to yield a relatively insoluble calcium paracaseinate, which precipitates to form the desired curd. Other proteases are also effective in converting calcium caseinate to calcium paracaseinate. However, these enzymes normally continue the proteolysis much further, and thus curd formation is prevented since the later degradation products are more soluble. Shortages of animal rennin have stimulated development of suitable microbial rennin enzymes, and these are now used commercially. Genetic engineering methods have also been applied to produce calf rennet in microorganisms.

Clinical and medicinal applications of proteolytic enzymes include both digestive aids and cleansers of serious wounds. Since enzymes are proteins, digestive enzyme aids are suitably coated to protect the enzyme during passage through the stomach, where the acid environment could cause protein denaturation. The sources of enzymes shown in Table 4.1 are all nonhuman. Injection of some foreign proteases (pig trypsin differs from human trypsin) into human beings has been used to reduce tissue inflammation: the highly purified crystallizable form of the enzyme minimizes immune system response. The natural defenses of live cells against protease attack are usually inactivated in dead cells. This convenient difference allows application of solutions of proteases to virulent or oozing wounds: selective liquefaction of the dead tissue and cells is achieved, which facilitates wound drainage and thereby decreases the time needed for healing.

Proteolytic enzymes, especially trypsin, apparently reduce inflammation and swelling associated with internal injuries and infections by dissolving blood clots and extracellular-protein precipitates, by locally activating other body defenses which do the same thing, or both. Some severe lung infections resulting in accumulation of viscous lung deposits have been reduced successfully or eliminated by proteolytic enzyme administration.

4.1.3 Esterase Applications

This group of enzymes carries out the cleavage or synthesis of various ester bonds to yield an acid and an alcohol:

$$R_1COOR_2 + H_2O \longleftrightarrow R_1COOH + R_2OH$$

The most important subgroups of these enzymes are the lipases, which hydrolyze fats into glycerol and fatty acids:

$$\underset{\text{Lipid}}{\begin{matrix} H_2C-OR_1 \\ | \\ HC-OR_2 \\ | \\ H_2C-OR_3 \end{matrix}} \xrightarrow{\text{lipase}} \underset{\text{Glycerol}}{\begin{matrix} H_2COH \\ | \\ HCOH \\ | \\ H_2COH \end{matrix}} + \underset{\text{Fatty acids}}{\begin{matrix} R_1OOH \\ + \\ R_2OOH \\ + \\ R_3OOH \end{matrix}}$$

Pancreatic lipase, secreted into the digestive tract following neutralization of stomach-imparted acidity, splits ingested fats into fatty acids as well as the intermediate products of mono- and diglycerides. As mentioned in Chap. 3, the insolubility of higher-molecular weight fats apparently requires the enzyme to act at the water-fat interface in order to yield an appreciable rate of hydrolysis (see Fig. 3.33).

The specificity of lipases and a second group of esterases known as aliesterases is not extreme. Lipases are active toward hydrolysis of high-molecular weight fats and inactive toward the hydrolysis of fats formed from short-chain fatty acids: the reverse applies to aliesterases. The activity of the former is increased both by surfactants, e.g., bile salts, which tend to stabilize high-surface area emulsions, and by calcium ions, which apparently precipitate the fatty acid hydrolysis products or complex with them. This removes the free fatty acids, which otherwise tend to act on the lipase in an inhibitory manner.

In aerobic waste-digestion processes like those occurring in natural water and in activated-sludge treatments (Chap. 14), mass transfer of sufficient oxygen is necessary to maintain the desired life forms. Since oxygen transport is relatively slow through fats, thin layers of such fats must often be continuously removed from the surface of aerated oxidation tanks in treatment plants; the skimmed fat-rich liquid is then digested by cells which are able to manufacture extracellular lipases. Formation of a continuous oil or fat layer over natural water will often quickly cut off the supply of dissolved oxygen with the resulting death of macroscopic and microscopic oxygen-requiring life which cannot pierce the layer.

In the meat processing industry, the production of relatively fat-free meats can be carried out by partial fat hydrolysis of the meat cuts using a lipase preparation. Drains in food processing or in domestic or industrial waste treatment may be clogged by a mixture of biological materials containing proteins, carbohydrates, and fats. Synthetic mixtures of proteases, various carbohydrases, and lipases have been used successfully to control or remedy such problems.

The second class of esterase enzymes is known collectively as phosphatases; they are intricately involved in the biosynthetic metabolic pathways internal to cells which utilize the chemical free energy of fuel oxidation; practical applications outside living systems are not known.

4.1.4 Enzyme Mixtures, Pectic Enzymes, and Additional Applications

Mixtures of enzymes, either of the same general type, for example, α- and β-amylase, or trypsin and chymotrypsin, or of different types such as found in pancreas extracts, e.g., trypsin, lipase, and amylase, are often used more conveniently or more successfully than single enzyme preparations. Thus, blends of different amylases achieve large yields of saccharified starch suitable for yeast fermentations yielding alcohol; a combination with less β-amylase achieves the desired thinning of starches (as in textile sizing) without too great a saccharification. Similarly, pancreas extracts from different animals contain a mixture of digestive enzymes which together carry out many of the same functions as enzymes in human pancreatic fluids. At least one of the enzyme preparations marketed in detergents in this country contained a combination of both bacterial amylase and proteases, claimed to be more effective in removing certain stains than is protease alone.

Oxidation of galactose to galacturonic acid·followed by dehydration and resulting polymerization yields polygalacturonic acid molecules. Naturally occurring plant molecules containing such polygalacturonic acid species as a major fraction are collectively termed *pectins*. Often the acid forms of the molecules are found esterified with methanol. The two major pectin-hydrolyzing enzymes are pectin methylesterase (pectin esterase, hydrolyzing pectase) and polygalacturonase (pectinase, pectolase), which hydrolyze the methyl ester and the glycosidic linkages, respectively. Major sources are fruits for the former and fungi for the latter.

There are two main applications for *pectic enzymes*. Crushing fruits and vegetables yields juices which have high viscosities, desirable in the production of tomato and citrus juices but not so often for apple cider and other fruit juices. A controlled partial pectin hydrolysis of these juices yields a free-flowing product which retains enough viscosity to prevent undesirable settling of particulate matter. A greater hydrolysis is effected with apple juice: the hydrolyzed product is much more easily filtered to yield a clear juice. If the juices are to be used in jelly manufacture, only the pectin esterase is added. The resulting polyacid hydrolysis product is then gelled by precipitation with calcium ions.

The second important application of pectic enzymes is wine production. Addition of the pectic enzyme mixture to the crushed grapes tends to increase the weight yield of juice, allows extraction of greater color from the grape skin, and permits faster filtering and pressing. Later addition to the fermented product again gives a faster subsequent separation of the wine from the yeast and grape sediment and yields a clear wine with an increased stability (largely resulting

Table 4.6 Hydrolytic enzyme applications which are expanding or in development

Enzyme	Process
Penicillin acylase	Production of semisynthetic penicillin core from natural penicillin G
Lactase	Removal of lactose from whey, milk
Ribonuclease	Production of 5′-nucleotides from RNA
Dextranase	Removal of tooth plaque
Isoamylase	Production of maltose from starch
Keratinase	Modification of wool, hair, leather
Tannase	Removal of tannic acid from foods

from reduction of the suspended protein concentration in the final wine). In both these cases, a major use of pectic enzymes is thus the development of a process stream with a desirable viscosity and filterability. The application includes benefits to process economics and the appearance of the product.

Future utilization of pectic enzymes to treat softwoods is a possibility. When trees of this type, e.g., Norway spruce, are felled, they resist penetration by chemical preservatives. Pretreatment with pectic enzymes has been shown to improve the efficiency of preservative treatment by rendering the wood more permeable.

Several additional hydrolytic enzymes enjoy small-scale applications, and promising new processes using hydrolases are being developed (Table 4.6).

4.2 OTHER APPLICATIONS OF ENZYMES IN SOLUTION

While the hydrolytic-enzyme applications considered above dominate past and present enzyme technology, other enzyme processes currently serve important functions in the food, pharmaceutical, and biochemical industries. Moreover, many new applications are emerging, as we shall see in the remainder of this chapter.

4.2.1 Medical Applications of Enzymes

A rapidly growing area of medicine now involves the use of free or extracellular enzymes. A brief listing of some of the more common or promising enzymes in diagnosis, therapy, and treatment is given in Table 4.7.

An enzyme which is present in protective body fluids such as nasal mucus and tears is *lysozyme*, which hydrolyzes the mucopolysaccharides of a number of (gram-positive) bacterial cell walls. The enzyme is used as an antibacterial agent (it apparently has other catalytic functions as well).

Initial stages of certain diseases and the presence of internal injuries give rise to elevated or depressed levels of enzyme concentrations in the easily sampled body fluids of lymph, blood, and urine. The presence of a given enzyme can

Table 4.7 Some enzymes of importance in medicinal applications

Enzyme	Typical application
Trypsin	Anti-inflammation agent, wound cleanser
Glucose oxidase	Glucose test in blood or urine
Lysozyme (though protein, not affected by trypsin, chymotrypsin, or papain)	Recommended in treatment of certain ulcers, measles, multiple sclerosis, some skin diseases, and postoperative infections (antibacterial agent)
Hyaluronidase (from beef testicles)	Hydrolyses polyhyaluronic acid, a relatively impermeable polymer found between human cells; administered to increase diffusion of coinjected compounds, e.g., antibiotics, adrenaline, heparin, and local anaesthetic in surgery and dentistry
Digestive enzyme (mixtures of amylase, lipase, protease, and cellulases)	Digestive aids
Streptokinase	Anti-inflammatory agent
Streptodornase	Anti-inflammatory agent; also digests DNA, reducing viscosity of wound exudates
Penicillinase	Removal of allergenic form of penicillin from allergic individuals
Urokinase	Prevention and removal of blood clots
Tissue plasminogen activator (TPA)	Dissolution of blood clots
Asparaginase	Anticancer agent

usually be detected quite simply by using an appropriate substrate test. Consequently, assay of enzyme activity in body fluids can be employed as a diagnostic tool. Serum enzyme levels provide useful input in diagnosing a number of cardiac, pancreatic, muscular, bone, and malignant disorders.

The enzyme L-*asparaginase* catalyzes the hydrolysis of L-asparagine

$$\text{L-Asparagine} + H_2 \longrightarrow \text{L-aspartate} + NH_3$$

Since some cancer cells require L-asparagine, their growth can be inhibited by using L-asparaginase to remove this essential nutrient. *E. coli* is known to produce two different asparaginases, only one of which exhibits antilymphoma activity.

As noted earlier, the cell membranes and the intermembrane materials of macroscopic animals are relatively impermeable to many substances; in the latter example this property is due to the presence of a very viscous polyhyaluronic acid. Effective administration of certain drugs and local anaesthetics, e.g., in dental work, is enhanced by simultaneous injection of *hyaluronidase*, which partially hydrolyzes this cellular diffusion barrier.

A number of single-cell species can manufacture *penicillinase* (or β-lactamase) and thus protect themselves from otherwise lethal levels of the antibiotic penicillin. Human beings do not produce penicillinase; the appearance of allergic symptoms in a patient to whom penicillin has been adminstered can be treated with injection of penicillinase solutions which, ideally, convert the drug into a nonallergenic form.

Glucose oxidase catalyzes the oxidation of glucose to gluconic acid

$$\text{Glucose} + O_2 + H_2O \longrightarrow \text{gluconic acid} + H_2O_2$$

Since the resulting hydrogen peroxide is noted easily with an appropriate reducible indicator, glucose oxidase provides a sensitive specific test for the presence of glucose in blood and urine (as in diabetes). Additional medical applications of enzymes will be mentioned later during our discussion of immobilized enzymes.

4.2.2 Nonhydrolytic Enzymes in Current and Developing Industrial Technology

Glucose oxidase, the very useful enzyme just mentioned, finds applications whenever glucose or oxygen removal is desirable. Dried egg powders undergo an undesirable darkening due to a reaction between glucose and protein. This reaction, commonly called a *Browning reaction* or a *Maillard reaction*, can be prevented by addition of glucose oxidase. In the production and storage of orange soft drinks, canned beverages, dried food powders, mayonnaise, salad dressing, and cheese slices, the presence of oxygen (which would otherwise eventually form other oxygenated products associated with undesirable flavors) is usually avoided by addition of glucose oxidase and, in the case of cheese wrapping, glucose itself. Since enzyme activity is normally maintained for a long time at storage temperatures, such enzyme additions also increase the shelf life of food products by continually removing the oxygen which diffuses through the food packaging.

The hydrogen peroxide produced in the glucose oxidase-catalyzed reaction has an antibacterial action; if the presence of hydrogen peroxide is undesirable in the product, *catalase* is added, which catalyzes the reaction

$$2H_2O_2 \longrightarrow 2H_2O + O_2$$

With catalase present this reaction proceeds very quickly [$v_{max}/e_0 = 1.2 \times 10^7\ (s \cdot M)^{-1}$, $K_m = 10^{-7}\ M$]. Because of its rapid action to decompose peroxide, catalase is used in the rinse for some hair dyes.

Besides these established applications, those listed in Table 4.8 are now being studied, and some are being implemented commercially. One of the commercially important cases is use of glucose isomerase to produce fructose from glucose. Since this process and some others in this list are often accomplished with enzymes in an immobilized form, we consider them in greater detail in the next section.

Table 4.8 Some recent applications or promising future uses for nonhydrolytic eazymes[†]

Production of L-malate from fumarate by fumarase
Production of L-aspartate from fumarate by aspartase
Production of ATP from adenine by microbial enzymes
Production of NAD from adenine and nicotinamide by microbial enzymes
Production of fructose from glucose by glucose isomerase
Production of fructose from sorbitol by sorbitol dehydrogenase
Resolution of DL-amino acids by immobilized aminoacylase
Production of L-tyrosine and L-dopa by tyrosine phenol-lyase
Production of L-dopa from the corresponding α-keto acid by transaminase
Enzymatic production of L-tryptophan from indole, pyruvate, and ammonia
Enzymatic production of L-lysine from α-amino-ε-caprolactam
Transformation of saturated fatty acids to polyunsaturated fatty acids by fatty acid desaturase
Production of galactonolactose from whey using glucose oxidase and galactose oxidase

[†] From E. K. Pye and L. B. Wingard (eds.), *Enzyme Engineering*, p. 365, Plenum Press, New York, 1973.

4.3 IMMOBILIZED-ENZYME TECHNOLOGY

Immobilization of an enzyme means that it has been confined or localized so that it can be reused continuously. There are several reasons why immobilization may be desirable: for processing with isolated enzymes, an immobilized form can be retained in the reactor. With enzymes in solution, on the other hand, some enzymes will leave the reactor with the final product. Not only must new enzymes be introduced to replace the lost ones, but enzymes in the product may be undesirable impurities which must be removed. Also, as we saw in Chap. 3, immobilized enzymes may retain their activity longer than those in solution (Fig. 3.6). Finally, an immobilized enzyme may be fixed in position near other enzymes participating in a catalytic sequence, thereby increasing the catalyst efficiency for the multistep conversion.

These characteristics make immobilized enzymes attractive if a very large throughput of substrate is required and/or the enzymes involved are expensive. Moreover, the ability to confine an enzyme in a well-defined, predetermined space provides opportunities for applications unique to immobilized enzymes. In the next section we examine methods of enzyme immobilization. This is followed by a summary of some present and potential applications of immobilized enzymes.

4.3.1 Enzyme Immobilization

Many methods are available for enzyme immobilization. As we shall see below, the immobilization method used greatly influences the properties of the resulting biocatalyst. Thus, the selection of an immobilization strategy derives from process specifications for the catalyst including such parameters as overall catalytic activity, effectiveness of catalyst utilization, deactivation and regeneration characteristics and, of course, cost. Also, toxicity of immobilization reagents should be considered in connection with immobilization process waste disposal and intended application of the immobilized enzyme catalyst.

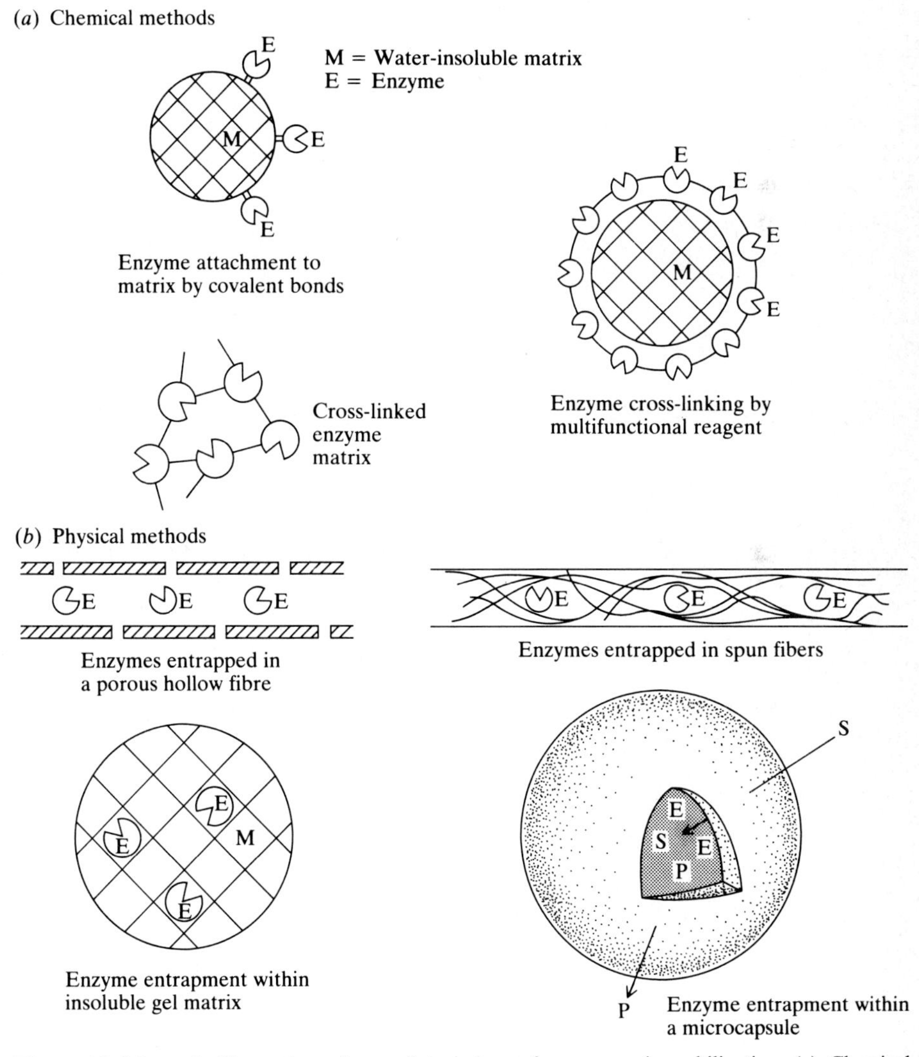

Figure 4.5 Schematic illustration of several techniques for enzyme immobilization. (*a*) Chemical methods, and (*b*) physical methods.

Table 4.9 Interactions and carriers used for enzyme immobilization by adsorption

Interaction	Adsorbents
Physical adsorption	Activated carbon, silica gel, alumina, starch, clay, glass
	Modified materials
	tannin-aminohexyl cellulose, Concanavalin A-Sepharose
Ionic binding	Cation exchangers
	CM-cellulose,† Aberlite, CG-50, Dowex 50
	Anion exchangers
	DEAE-cellulose,† DEAE-Sephadex, polyaminopolystyrene, Amberlite IR-45

† CM = carboxymethyl; DEAE = diethylaminoethyl.

The various methods devised for enzyme immobilization may be subdivided into two general classes: chemical methods, where covalent bonds are formed with the enzyme, and physical methods, where weaker interactions or containment of the enzymes are involved (Fig. 4.5). Enzymes may be adsorbed on a variety of carriers (Table 4.9), offering in some cases the practical convenience of simple regeneration by removal of deactivated enzyme and reloading with fresh, active catalyst. If a support or entrapping material is used, its properties combined with those of the enzyme and the immobilization procedure dictate overall catalyst properties.

Selecting a support for chemical or adsorption immobilization of enzyme depends first upon its surface properties: Will the enzyme adsorb on the surface? Does the material possess functional groups which can be used for bonding to the enzyme? If the native surface is not ideal, can it be chemically modified or coated to facilitate enzyme attachment? Table 4.10 lists several materials which have been employed for covalent enzyme immobilization and some of their interesting surface functional groups. Other materials which have been used as immobilized enzyme supports include ceramics, glass and other metal oxides.

Protocols for covalent enzyme immobilization often begin with a surface modification or activation step. Silanization, coating the surface with organic functional groups using an organofunctional silane reagent (for example, $(CH_3CH_2O)_3Si(CH_2)_3R$ where R is frequently $—NH_2$), is a widely used strategy for initial surface modification of inorganic supports. Such coatings or

Table 4.10 Insoluble materials and some of their surface functional groups useful for covalent enzyme attachment

Natural supports	Synthetic supports
Cellulose (—OH)	Polyacrylamide derivatives
CM-cellulose (—COOH)	(Bio-Gel, Enzacryl) (-aromatic amino)
Agarose (Sepharose) (—OH)	Polyaminopolystyrene ($—NH_2$)
Dextran (Sephadex) (—OH)	Maleic anhydride copolymers

native surface amino groups can be derivatized to aldehyde groups using glutaraldehyde, to arylamine groups using *p*-nitrobenzoylchloride, or to carboxyl groups using succinic anhydride. Another commonly studied surface modification is attachment of flexible spacer arm moieties (for example, *n*-propyl amine) to the support, to which enzyme is subsequently linked. This may render the surface more "flexible" so it can conform to the enzyme's structure, thereby retaining to a greater extent and potentially stabilizing native catalytic activity. Such modifications may also be applied to alter the hydrophobicity or hydrophilicity of the support surface.

Typical examples of immobilization chemistries which utilize these surface functionalities and those appearing in Table 4.10 are summarized in Fig. 4.6. We should note that each step requires suitably adjusted reaction conditions including pH, ionic strength, and reagent concentration. This information is available in the chapter references along with extensive tabulations of specific enzymes which have been immobilized by different methods on various supports. Which functional groups on the protein are involved in covalent linkages to the support surface obviously depends on the immobilization chemistry applied. Taking advantage of the available repertoire of surface modification techniques, typically there are several options for immobilizing a particular enzyme to a particular support. Clearly, attachment to residues near or in the enzyme active site is to be avoided.

When considering selection of a support and adjustment of its surface properties, one must also consider interaction between the support surface and the reaction mixture. This interaction may cause the fluid environment adjacent to the surface in contact with the enzyme to differ substantially from the bulk fluid environment surrounding the immobilized enzyme catalyst. For example, a charged support will increase local concentrations of oppositely charged ions. As a result, the relationship between bulk solution pH and observed catalytic activity can be shifted substantially relative to the pH-activity function observed in solution (Fig. 4.7). Similarly, the hydrophobicity or hydrophilicity of the support will influence the local concentrations of solutes and solvents according to their hydrophobicity/hydrophilicity.

Another important role of the support surface is in defining, directly and indirectly, the molecular environment of the enzyme. That is, to some degree the enzyme will contact, at the molecular level, the support surface. The remainder of the immobilized enzyme molecule will be in contact with the local fluid environment which is influenced by the support as just outlined. The enzyme molecule is structured so as to assume the proper configuration and corresponding activity in its native biological environment. Clearly, it is possible by selection of the support for an immobilized enzyme to attempt to mimic or to alter substantially the local enzyme environment and thereby also to influence the catalytic activity of the enzyme and the effect of any modulating factors on activity (and selectivity and stability). Efforts to understand enzyme-environment interactions at the molecular level and to apply them for improved understanding of enzymes in nature and for optimizing process biocatalysts are active topics of current research and development.

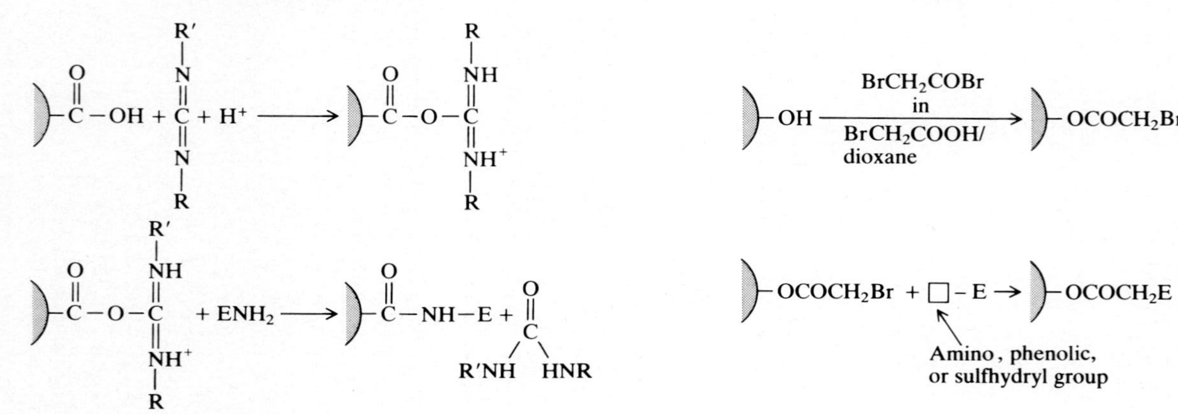

Figure 4.6 Summaries of some of the available procedures for covalent enzyme attachment to surfaces (E denotes enzyme).

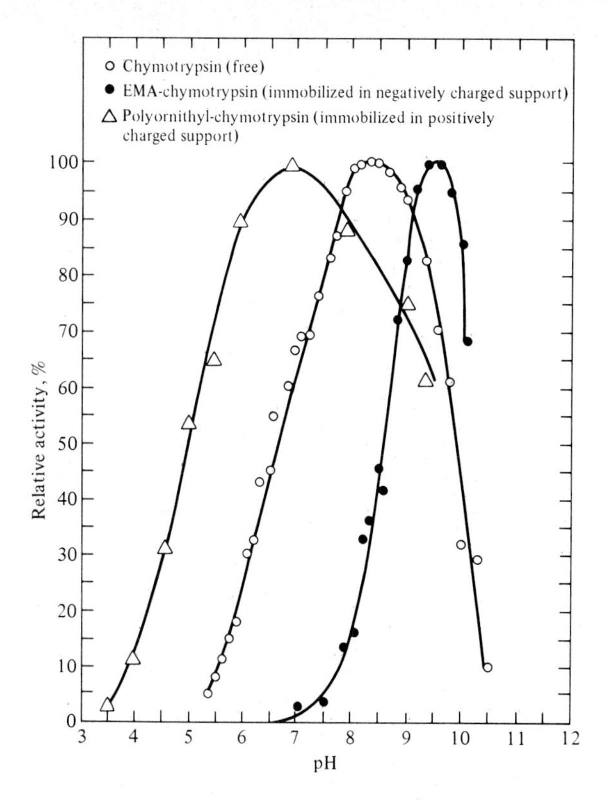

Figure 4.7 Use of a charged support gives an immobilized enzyme different apparent pH dependence than the free enzyme. Substrate is acetyl-L-tyrosine ethyl ester. *(Reprinted with permission from L. Goldstein et al., "A Water-Insoluble Polyanionic Derivative of Trypsin. II. Effect of the Polyelectrolyte Carrier on the Kinetic Behavior of Bound Trypsin," Biochemistry, vol. 3, p. 1913, 1964. Copyright by the American Chemical Society.)*

The physical and mechanical properties of the support are also important. The pore size distribution and porosity of a solid material determines the quantity of enzyme which can be immobilized in the support and the accessibility of substrate to the enzyme attached to the internal surfaces. The mechanical strength of the support influences significantly catalyst suitability for different reactor configurations. For example, solid materials highly susceptible to mechanical attrition are undesirable for reactor environments involving agitated slurries of catalyst particles, while extremely compressible materials are not suitable for large-scale packed column applications. The swelling properties of the support are also important. Additional information on physical properties of many potential enzyme support materials and means for support characterization in these respects are presented in the chapter references.

Enzymes may be cross-linked with several bi- or multifunctional reagents. The most widely used method employs glutaraldehyde which establishes intermolecular (and possible intramolecular) cross-links at amino groups as follows:

$$\text{E—NH}_2 + \text{OHC(CH}_2)_3\text{CHO} + \text{H}_2\text{N—E}'$$

$$\Big\downarrow -2\text{H}_2\text{O}$$

$$\text{E—N=HC(CH}_2)_3\text{CH=N—E}'$$

Other cross-linking reagents include bisbiazobenzidine, cyanuric chloride, and hexamethylene-diisocyanate. Particles of cross-linked enzyme alone are gelatinous and lack the mechanical properties required in many applications. As indicated in Fig. 4.5, improved mechanical and substrate accessibility characteristics may be obtained by absorbing the enzyme onto a carrier and then cross-linking the adsorbed protein.

Enzyme immobilization by physical entrapment has the benefit of applicability to many enzymes and may provide relatively small perturbation of enzyme native structure and function. Entrapment strategies may be divided into two classes: where the enzyme is embedded in a polymer network and where the enzyme in solution is retained by a membrane permeable to substrates, reaction products, or both.

The most widely used system for enzyme entrapment in a polymer lattice is polyacrylamide gel made by cross-linking acrylamide in the presence of enzyme (E) (Fig. 4.8). As indicated, potassium persulfate ($K_2S_2O_8$) may be used as a polymerization initiator and β-dimethylaminopropionitrile (DMAPN) as an accelerator. Reported pore sizes in immobilized enzyme gels prepared by these reagents are in the range 100–400 nm, sufficiently small to retain many enzymes, which have molecular diameters in the range of 300–2000 nm.

In the remaining immobilization procedures, the enzyme stays in solution, but this solution is physically confined so that the enzyme cannot escape. There is a biological analog of these techniques: the lysosome within the cell confines hydrolytic enzymes, which would kill the cell immediately if they were released. In the artificial immobilization method known as *microencapsulation*, enzymes are entrapped in small capsules with diameters ranging up to 300 μm. The capsules are surrounded by spherical membranes which have pores permitting small substrates and product molecules to enter and leave the capsule. The pores are too small, however, for enzymes and other large molecules to penetrate.

Two different types of semipermeable microcapsules can be made. The first has a permanent polymeric membrane. To form this membrane, a copolymerization reaction can be conducted at the interface between an organic phase and a dispersed aqueous phase containing the enzyme. By choosing a water-insoluble monomer as one of the reactants and a monomer slightly soluble in both phases for the other, the copolymerization will occur only in the vicinity of the interface. Another method for making permanent microcapsules involves coacervation, a phase separation in a polymer solution which concentrates polymer at the interface of a microdroplet. Nonpermanent microcapsules can be made by emulsifying an aqueous enzyme solution with a surfactant. This forms capsules surrounded by liquid-surfactant membranes which can then be added to an aqueous substrate solution.

Both microencapsulation methods have the potential to offer a very large surface area (for example, 2500 cm^2 per milliliter of enzyme solution) and the possibility of added specificity: the membrane can be made in some cases to admit some substrates selectively and exclude others. In principle these methods should be applicable to a large variety of enzymes. However, the membrane is a significant mass-transfer barrier, so that the "effectiveness factor" for the enzymes

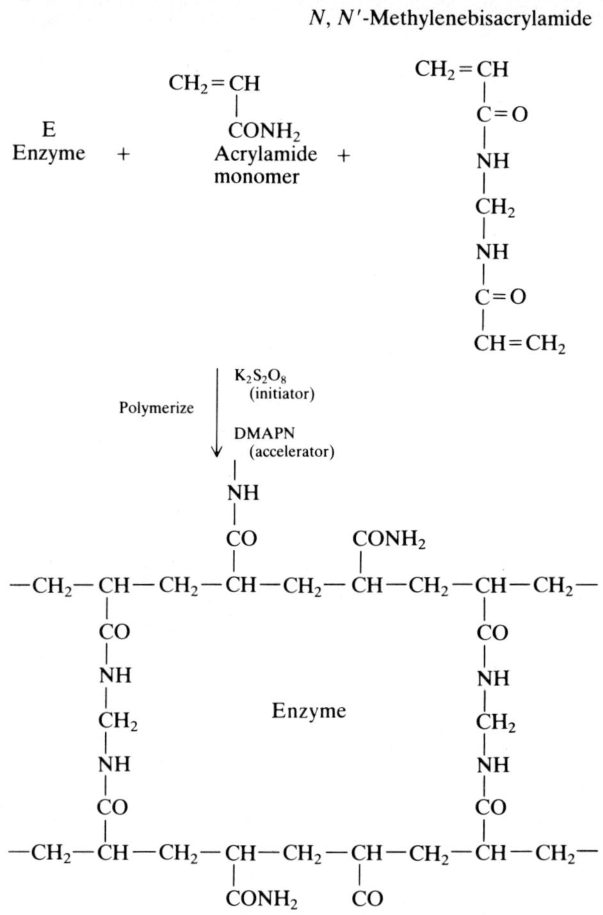

N, N'-Methylenebisacrylamide

E Enzyme + $CH_2{=}CH{-}CONH_2$ Acrylamide monomer + $CH_2{=}CH{-}C({=}O){-}NH{-}CH_2{-}NH{-}C({=}O){-}CH{=}CH_2$

Polymerize ↓ $K_2S_2O_8$ (initiator), DMAPN (accelerator)

$-CH_2-CH(CONHCH_2NHCO-)-CH_2-CH(CONH-)-CH_2-CH(CONH_2)-CH_2-CH(CONHCH_2NHCO-)-CH_2-$

Enzyme

$-CH_2-CH(-)-CH_2-CH(CONH_2)-CH_2-CH(CO-)-CH_2-CH(-)-CH_2-$

Figure 4.8 Summary of one procedure for enzyme immobilization by polymer entrapment. Here the polymer matrix is comprised of cross-linked acrylamide.

may be quite small. Also, these techniques are not applicable when the size of the substrate molecule approaches that of the enzyme.

Semipermeable-membrane filtration devices can be used to contain enzyme while allowing interchange of smaller molecules with a neighboring solution. A continuous flow ultrafiltration concept, illustrated schematically in Fig. 4.9, physically retains the enzyme as in microencapsulation. The surface areas separating the two solutions are substantially lower when ultrafiltration is used rather than microencapsulation, and flow-induced denaturation may occur. The mass-transfer rates available through membranes are small and may limit the overall rate. However, almost any enzyme or combination of enzymes can be immobilized in this way. It is advantageous when the substrate has very high molecular weight or is insoluble, situations where polymer-bound enzymes may typically have a small efficiency. Also, the semipermeable membrane will not permit relatively large product molecules from polymer-hydrolysis reactions to escape from

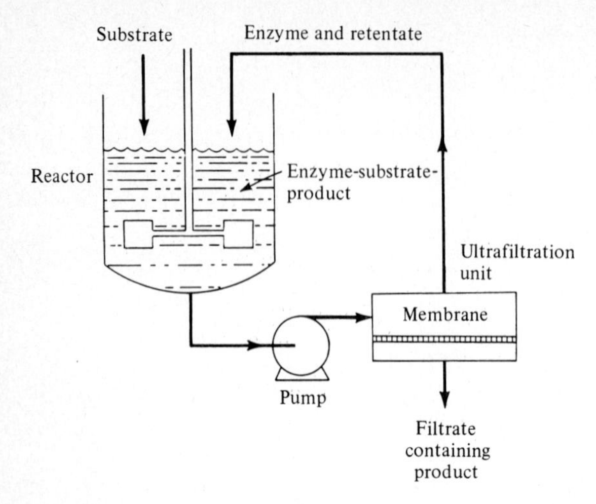

Figure 4.9 Enzyme entrapment on a macroscale. An ultrafiltration membrane is used to retain enzyme and other large molecules in the reactor.

the enzyme solution. This provides an interesting method of controlling the molecular-weight distribution of the products which ultimately leave the reactor.

Several alternate realizations of the basic concept illustrated in Fig. 4.9 are possible and have been studied. The well-mixed fluid reservoir may be replaced by tubing or pipe. Also, the single membrane separation unit may be replaced with a hollow-fiber device in which one fluid stream occupies the internal lumen passages in many parallel hollow fibers of ultrafiltration membrane and the other stream passes through the shell surrounding the fibers. Further examination and quantitative treatment of ultafiltration is provided in Chap. 11.

Important characteristics of different immobilization techniques are summarized in Table 4.11. Here, "enzyme activity" denotes the intrinsic catalytic activity

Table 4.11 Comparisons of the characteristics of different methods of enzyme immobilization

(Reprinted by permission from "Immobilized Enzymes, Research and Development", I. Chibata, ed., p. 72, Kodansha Ltd., Tokyo, 1978).

	Carrier binding method				
Characteristic	Physical adsorption	Ionic binding	Covalent binding	Cross-linking method	Entrapping method
Preparation	Easy	Easy	Difficult	Difficult	Difficult
Enzyme activity	Low	High	High	Moderate	High
Substrate specificity	Unchangeable	Unchangeable	Changeable	Changeable	Unchangeable
Binding force	Weak	Moderate	Strong	Strong	Strong
Regeneration	Possible	Possible	Impossible	Impossible	Impossible
General applicability	Low	Moderate	Moderate	Low	High
Cost of immobilization	Low	Low	High	Moderate	Low

of the immobilized enzyme molecule relative to that enzyme's activity in solution. In general, chemical immobilization methods tend to reduce activity in this sense, since the covalent bonds formed as a result of immobilization may perturb the enzyme's native tertiary structure. On the other hand, such covalent linkages provide strong, stable enzyme attachment and may in some case reduce enzyme deactivation rates and usefully alter enzyme specificity. Entrapment and adsorption immobilization methods typically perturb the enzyme much less and consequently offer retention of enzyme properties resembling those in solution. However, we must recognize that the overall, observed properties for an immobilized enzyme catalyst may, because of mass transfer effects, differ significantly from the intrinsic properties just discussed. This very important point will be developed further and analyzed in Sec. 4.4. Diffusion limitations are generally greater for cross-linked and entrapped enzyme catalysts than for carrier-immobilized enzymes.

4.3.2 Industrial Processes

Several large-scale industrial processes already in operation employ immobilized-enzyme catalysts at some point. Two notable examples are production of high-fructose syrups from corn starch and manufacture of L-amino acids by resolution of racemic amino acid mixtures (containing both D and L optical isomers). Also, immobilized penicillin acylase has been used commerically in manufacture of semisynthetic penicillins. The first application mentioned is the most important economically at present.

D-glucose cannot be substituted directly for sugar (sucrose) because glucose is less sweet. Also, glucose crystallization in concentrated solutions can make subsequent handling and processing difficult. These problems are alleviated considerably by isomerizing some of the glucose to fructose, using the enzyme glucose isomerase:

```
  H     O
   \   //
     C                          CH2OH
     |                            |
 H—C—OH                          C=O
     |                            |
HO—C—H       glucose        HO—C—O
     |       isomerase            |
 H—C—OH     <=======>        H—C—OH
     |                            |
 H—C—OH                      H—C—OH
     |                            |
   CH2OH                        CH2OH
 D-Glucose                    D-Fructose
```

The equilibrium constant of this reaction at 50°C is approximately unity, and this values does not change greatly with temperature since estimates of the heat of the isomerization reaction are of the order of 1 kcal/mol. Consequently, the equilibrium product contains roughly a 1:1 ratio of glucose to fructose. Such a mixture has greater sweetness than glucose alone and is well suited to substitute for sugar in many applications including soft drinks, processed foods, and baking. Sweeter formulations may be prepared by chromatographic separation (Chap. 11) of the isomerization product giving an enrichment of fructose.

An intracellular enzyme, glucose isomerase, is produced by a number of microorganisms, *Arthrobacter* and *Streptomyces* sp. being among the preferred sources. The need to disrupt cells without destroying the enzyme makes glucose isomerase substantially more costly than, say, the extracellular hydrolases. Also, glucose isomerase is very sensitive to several inhibitors. Both these factors suggest that retention of the enzyme in immobilized form under well-controlled reaction conditions is a desirable strategy. Several forms of immobilized glucose isomerase have been prepared, including enzyme fixed in whole cells, which in turn are held in collagen or other flocculating and binding agents. This illustrates another strategy of general utility for immobilizing intracellular enzymes: immobilize entire cells (usually permeabilized to reduce cell envelope mass transfer resistance) or cell lysates containing the enzyme. Immobilized cell systems are considered further in Chap. 9.

A typical process flowsheet for high-fructose corn syrup production using immobilized glucose isomerase is shown in Fig. 4.10. Many of the separation and treatment needs between the saccharification and isomerization units are dictated by basic enzymology: in order to maximize heat stability of the α-amylase used in the liquefaction step (which typically operates at about 105°C), calcium ions are

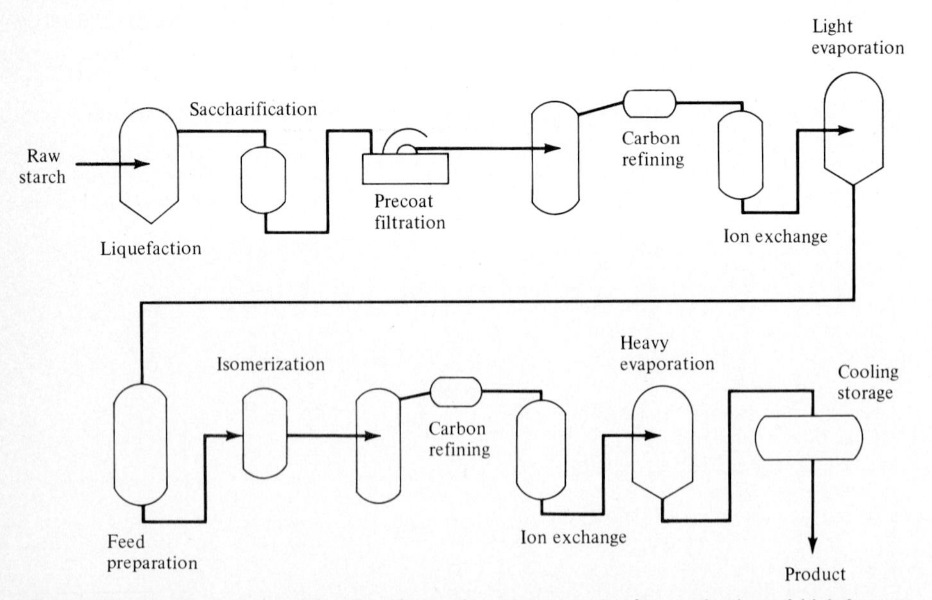

Figure 4.10 Flowsheet summarizing the glucose isomerase process for production of high-fructose corn syrups. *(Redrawn from J. D. Harden, "On-Line Control Optimizes Processing," Food Eng., vol. 44, no. 12, p. 59, 1972).*

required. These ions, on the other hand, inhibit glucose isomerase, and they are removed by ion exchange before the dextrose liquor is fed to the isomerization reactor.

Example 4.2: Immobilized glucose isomerase reactor process parameters† Selection of a catalyst formulation, reactor configuration, and reactor operating conditions involves careful consideration of a large and complicated array of interacting parameters. Some of the properties and criteria which have been considered in previous designs of immobilized glucose isomerase reactors for production of high-fructose corn syrups are listed in Table 4E2.1. The properties of the catalyst selected after

Table 4E2.1 Parameters considered in immobilized glucose reactor design study

(Reprinted by permission from K. Venkatasubramanian and H. S. Harrow, "Design and Operation of a Commercial Immobilized Glucose Isomerase Reactor System," Annals N.Y. Acad. Sci. vol. 326, p. 141, 1979.)

Biochemical characteristics

1. Activity
2. Operational stability (half-life) and activity decay profile
3. Productivity in usage lifetime
4. Optimal substrate concentration
5. Effect of oligosaccharides concentration
6. Effect of dissolved oxygen
7. Minimum and maximum residence times
8. By-product formation
9. pH and temperature sensitivity
10. Storage stability
11. Protein-enzyme elution
12. Microbial growth, if any
13. Reactor effluent quality (composition, color, odor, protein content, pH, etc.)

Mechanical characteristics

1. Particle size, shape and size distribution
2. Density (dry bulk density and wet bulk density)
3. Swelling behavior
4. Compressibility
5. Cohesion
6. Particle attrition

Hydraulic characteristics

1. Pressure drop
2. Mode of flow (upflow versus downflow)
3. Bed compaction
4. Axial dispersion and channeling
5. Residence time distribution
6. Stratification
7. Length-to-diameter ratio
8. Minimum velocity for onset of fluidization

† See K. Venkatasubramanian and H. S. Harrow, "Design and Operation of a Commercial Immobilized Glucose Isomerase Reactor System," *Annals of N.Y. Acad. Sci. vol. 326, p. 141 (1979).*

Table 4E2.2 Physical and catalytic properties of Immobilase (product of ICI) glucose isomerase catalyst

(Adapted from K. Venkatasubramanian and H. S. Harrow, "Design and Operation of a Commercial Immobilized Glucose Isomerase Reactor System," Annals N.Y. Acad. Sci., vol. 326, p. 141, 1979; reprinted by permission.)

Product form	Dry pellets
Appearance	Tan color
Mesh size	Nominally 12 × 20 mesh (cylindrical pellets ~1 mm diameter, 2.5 mm length)
Dry bulk density	40–45 lb/ft^3
Wet bulk density	14 lb/ft^3 ±10%
Characteristic pore size	0.2 μm
Activity assay	40 $m\mu\ g^{-1}$, minimum†
Productivity	Normally 2000 lb 42% d.b. HFCS/lb enzyme in 1000 h
Bed void volume	Estimated at 45%

† One mμ is defined as the amount of catalyst that would convert 10^{-9} moles of substrate per minute at 60°C, pH 8.0.

consideration of these process parameters in a design study at the H. J. Heinz Company are listed in Table 4E2.2. Selection of catalyst particle size is a compromise between the desirability of small particles to reduce diffusion limitations on reaction rates and a preference for large particles to minimize pressure drop through the packed-column immobilized enzyme reactor.

The reaction conditions selected are summarized in Table 4E2.3. Notice the requirement for a high concentration of glucose in the reactor feed. If the feed contains more than 10 percent oligosaccharides, the enzyme activity is reduced substantially. A significant range of residence times is indicated because of catalyst deactivation. As the process is operated, enzyme activity decays and exit conversion would decrease if flow rate were maintained constant. However, in order to maintain constant product quality, the flow rate is reduced to increase residence time as the catalyst deactivates. Quantitative treatment of activity-residence time-conversion relationships for different types of enzyme reactors is considered in Chap. 9.

The hydraulic properties of the packed column, which is operated with downward flow of feed solution, determine column sizing. The packed bed of immobilized enzyme catalyst particles is compressible. Figure 4E2.1 illustrates the relationship between column reactor bed height and pressure

Table 4E2.3 Reaction conditions for glucose syrup isomerization in a packed-column immobilized enzyme reactor

(Reprinted by permission from K. Venkatasubramanian and H. S. Harrow, "Design and Operation of a Commercial Immobilized Glucose Isomerase Reactor System," Annals N.Y. Acad. Sci., vol. 326, p. 141, 1979.)

Dry substance content	40–45%
Dextrose percentage	93–96%
Feed purity	Filtered, carbon-treated, and ion-exchanged
pH	8.2 to 8.5
pH drop	0.2–0.4 units
Temperature	60°C
Activator	0.0004 M Mg^{2+}
Residence time	0.5–4 h

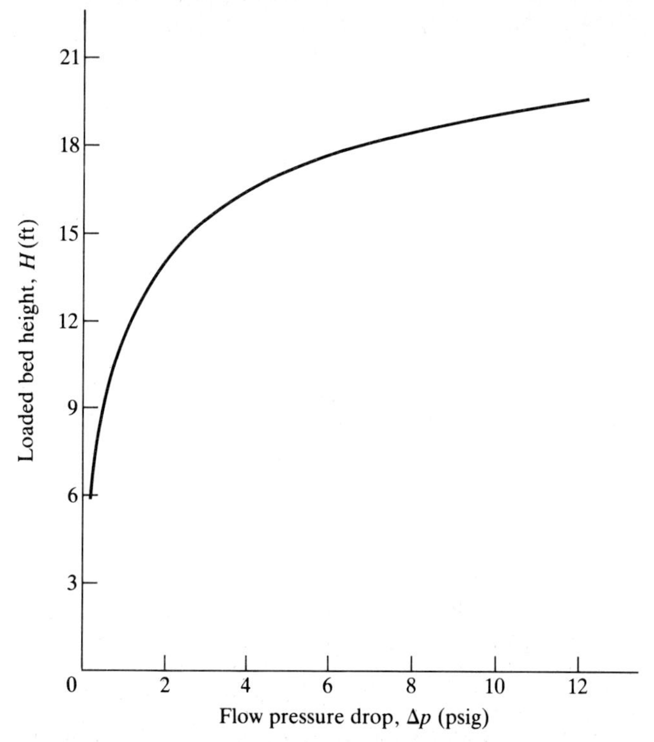

Figure 4E2.1 Relationship between catalyst bed height and overall column pressure drop in a down-flow packed column of immobilized glucose isomerase pellets. *(Reprinted by permission from K. Venkatasubramanian and H. S. Harrow, "Design and Operation of a Commercial Immobilized Glucose Isomerase Reactor System," Annals N.Y. Acad. Sci., vol. 326, p. 1141, 1979.)*

drop through the column at the largest flow rate which is used for fresh catalyst. Above 3 lb/in^2 pressure drop in the column, bed permeability is reduced significantly. Consequently, applying a design criterion of 3 lb/in^2 pressure drop, the maximum bed height which should be used is approximately 15 feet. Combining this criterion with desired reactor residence time and plant capacity implies reactor sizing and the number of reactor columns which should be used. Because of the deactivation processes mentioned earlier, it is preferable to utilize multiple columns and to rotate the service cycle of each to maximize catalyst utilization and plant productivity.

Demand for L-amino acids for food and medical applications has been growing rapidly. While microbial processes for L-amino acids have been developed, considerable efforts have also been devoted to a production scheme based on chemical synthesis. Chemically synthesized amino acids suffer the disadvantage of being racemic mixtures. The D isomer in this mixture is generally of no nutritive value: it is desirable to obtain a product consisting strictly of the physiologically active L isomer.

A process achieving this goal is that of Tanabe Seiyaku Co., Ltd., of Osaka, Japan. This was the first publicly announced commercial process using immobilized enzymes. Essential to the optical resolution scheme used by Tanabe Seiyaku is the following reaction, which is catalyzed by the enzyme aminoacylase:

$$H_2O + \underset{\text{DL-Acylamino acid}}{\text{DL-R}-\underset{\displaystyle NHCOR'}{\underset{|}{CH}}-COOH} \xrightarrow{\text{aminoacylase}} \underset{\text{L-Amino acid}}{\text{L-R}-\underset{\displaystyle NH_2}{\underset{|}{CH}}-COOH} + \underset{\text{D-Acylamino acid}}{\text{D-R}-\underset{\displaystyle NHCOR'}{\underset{|}{CH}}-COOH}$$

Following this step, which is carried out in a column reactor containing immobilized aminoacylase (Fig. 4.11), the desired L-amino acid is separated from the unhydrolyzed acyl D-amino acid based on solubility differences. The D-acylamino acid is then racemized to the DL-acylamino acid, which is recycled to the aminoacylase column. A flow sheet for the process is shown in Fig. 4.12.

I. Chibata, T. Tosa, and coworkers at Tanabe Seiyaku spent considerable effort in an attempt to develop an optimal immobilized-enzyme formulation for use in this process, and some of their results are presented in their paper [16]. Table 4.12 gives a summary of their findings for several different forms of immobilized enzymes. The categories in this table remind us that a variety of factors besides initial activity must be considered when selecting a catalyst for industrial use. In the present case, aminoacylase ionically bound to DEAE-Sephadex was chosen because of its high activity, ease of preparation, regeneration capability, and stability. In 1978 Tanabe Seiyaku reported that this formulation had been used for more than five years with no physical decomposition or reduction in binding activity.

The first commerical process using enzymes immobilized physically by ultrafiltration membranes, conversion of N-acetyl-D,L-methionine to L-methionine using acylase enzyme, was announced in 1983. Numerous other applications of immobilized enzymes to chemical processes have been explored (see Table 4.13). Some of these are already commercialized and others are currently in development.

4.3.3 Medical and Analytical Applications of Immobilized Enzymes

The number of known inborn human metabolic disorders now exceeds 120, and many of them are related to the absence of activity of one particular enzyme normally found in the body. For example, phenylketonuria, a disease leading to mental retardation, is thought to be caused by a deficiency in the enzyme which converts phenylalanine into tyrosine. A possible alternative to the current therapy (a phenylalanine-free diet) is to replace the missing enzyme. However, an enzyme with the same function from a nonhuman source cannot be introduced

Figure 4.11 Immobilized enzyme columns used in the Tanabe Seiyaku Co., Ltd. Process for L-amino acid production. *(Photograph courtesy of Dr. I. Chibata, Tanabe Seiyaku Co., Ltd.)*

directly into the body because it would trigger an adverse immunological response. A possible approach to this problem involves isolation of the enzyme within a microcapsule, fiber, or gel. In this manner, the enzyme may not cause an adverse response but the small substrate can reach it through the gel, hollow fiber, or microcapsule membrane. While membrane-contained enzymes are not susceptible to antibody attack, protein buildup on the *in vivo* membrane surfaces adds mass-transfer resistance and causes decreased efficiency of substrate utilization.

A variant of this idea has been proposed for construction of a compact artificial kidney. In this device, urease and an adsorbent resin or charcoal are

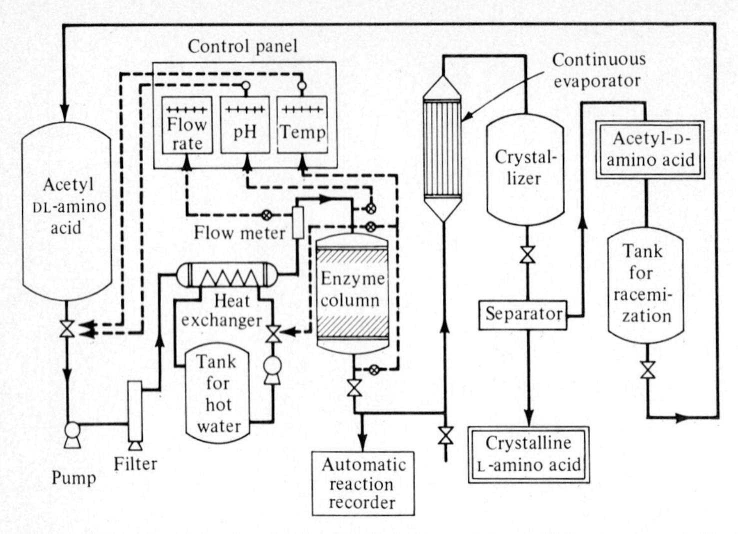

Figure 4.12 Flowsheet for the Tanabe Seiyaku process, which uses immobilized aminoacylase. *(Reprinted by permission from I. Chibata et al., "Preparation and Industrial Application of Immobilized Aminoacylases," in G. Terui (ed.), "Fermentation Technology Today," p. 387, Society of Fermentation Technology, Japan, Osaka, Japan, 1972.)*

Table 4.12 Properties of different forms of aminoacylase (acetyl-DL-methionine as substrate)†

		Immobilized aminoacylases		
Property	Native amino-acylase	Ionic binding (DEAE-Sephadex)	Covalent binding (Iodoacetyl-cellulose)	Entrapped by polyacrylamide
Optimum pH	7.5–8.0	7.0	7.5–8.5	7.0
Optimum temperature, °C	60	72	55	65
Activation energy, kcal/mol	6.9	7.0	3.9	5.3
Optimum Co^{2+}, mM	0.5	0.5	0.5	0.5
K_m, mM	5.7	8.7	6.7	5.0
v_{max}, μmol/h	1.52	3.33	4.65	2.33
Preparation		Easy	Difficult	Difficult
Binding force		Weak	Strong	Strong
Regeneration		Possible	Impossible	Impossible

† From T. Mori, T. Sato, T. Tosa, and I. Chibata, "Studies on Immobilized Enzymes X: Preparation and Properties of Aminoacylase Entrapped into Acrylamide Gel-Lattice," *Enzymologia*, **43**: 213 (1972).

Table 4.13 Summary of current and some potential chemical process applications of immobilized enzymes

(Reprinted by permission from "Immobilized Enzymes, Research and Development," I. Chibata (ed.), p. 164, Kodansha, Ltd., Tokyo, 1978.)

Catalytic reaction	Immobilized enzyme	Chemical process
Oxidation-reduction	L-Amino acid oxidase	Production of D-amino acids
	β-Tyrosinase	Production of L-DOPA
		Production of L-tyrosine
	Δ′-Hydrogenase	Production of prednisolone
	Flavoprotein oxidase	N-oxidation of drugs containing amino or hydrazine groups
Group transfer	Dextransucrase	Production of dextran
	Phosphorylase	Polymerization of glucose
	Polynucleotide phosphorylase	Production of polynucleotides
	Carbamate kinase	Regeneration of ATP
Hydrolysis	Ribonuclease	Synthesis of trinucleotides
	α-Amylase	Production of glucose
	Glucoamylase	Production of glucose
	Cellulase	Production of glucose
	Invertase	Production of invert sugar
	Leucine aminopeptidase	Optical resolution of DL-amino acids
	Carboxypeptidase	Optical resolution of DL-amino acids
	Papain	Hydrolysis of casein
	Penicillin amidase	Production of 6-aminopenicillanic acid
		Synthesis of penicillins and cephalosporins
	Aminoacylase	Optical resolution of DL-amino acids
	AMP deaminase	Production of 5′-inosinic acid
Lyase reaction (asymmetric synthesis)	Aspartase	Production of L-aspartic acid
	Tryptophanase	Production of L-tryptophan
	D-Oxynitrilase	Production of D-mandelonitrile
Isomerization	Glucose isomerase	Production of fructose

encapsulated together, so that ammonia produced by urea decomposition is adsorbed within the microcapsule:

$$\text{Urea} \xrightarrow[\text{into microcapsule}]{\text{diffusion}} \text{urea} \xrightarrow{\text{urease}} HCO_3^- + NH_4^+ \xrightarrow[\text{on resin or charcoal}]{\text{adsorption}}$$

Among the small-scale trials of immobilized enzymes in reaction processes are steroid (see Sec. 2.12) conversions. Cortisol, a useful drug in arthritis treatment, can be made from the cheap precursor 11-deoxycortisol in a column of immobilized 11-β-hydroxylase, following which cortisol can be converted into a superior therapeutic component, prednisolone, using immobilized Δ^1 dehydrogenase in a subsequent packed-bed reactor. Note the extreme specificity of these enzymes. Most commercial steroid transformations are currently achieved by microbial processes.

11-Deoxycortisol $\xrightarrow{\text{Immobilized 11-}\beta\text{-hydroxylase}}$ Cortisol

Cortisol $\xrightarrow{\text{Immobilized } \Delta^1 \text{ dehydrogenase}}$ Prednisolome

Immobilized enzymes have already made important contributions to analytical biochemistry, and others are certain to follow. One example is the immobilized-enzyme electrode, which permits continuous monitoring of small concentrations of a specific biochemical. In the urea electrode (Fig. 4.13), immobilized urease decomposes urea into ions which can be detected using standard electrochemical techniques. Also, using immobilized-enzyme electrodes,

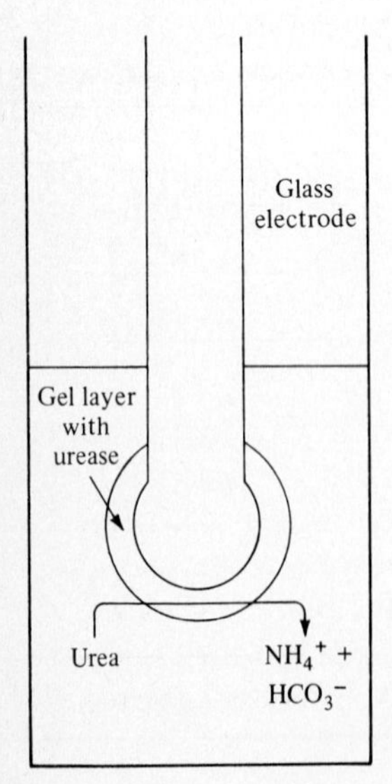

Figure 4.13 A urea electrode constructed by immobilizing urease in a gel attached to a glass electrode.

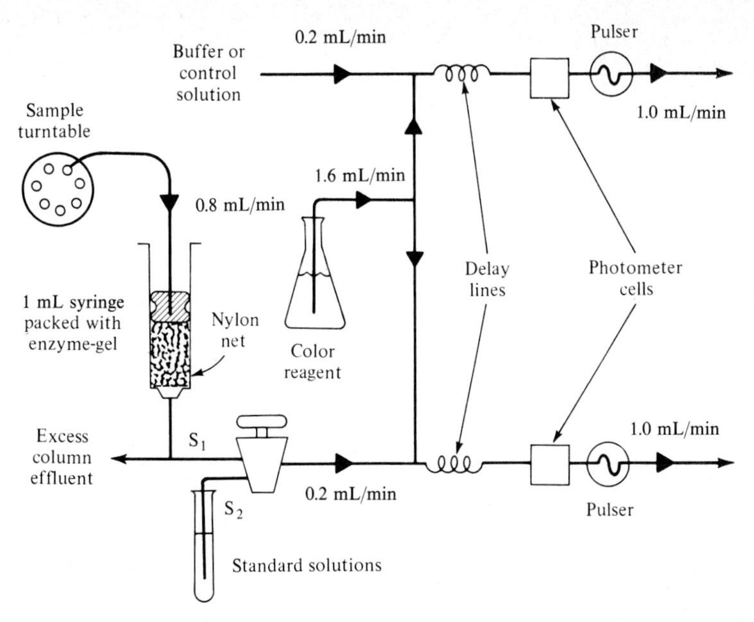

Figure 4.14 Schematic diagram of a system for automated analysis of glucose (using immobilized glucose oxidase) and lactate (using immobilized lactate dehydrogenase). *(Reprinted with permission from G. T. Hicks and S. J. Updike, "The Preparation and Characterization of Lyophilized Polyacrylamide Enzyme Gels for Chemical Analysis," Anal. Chem., vol. 38, p. 726, 1966. Copyright by the American Chemical Society.)*

standard biochemical tests can be automated, as in the schematic diagram in Fig. 4.14. Such a system can be used, for example, to determine glucose or lactate levels by employing immobilized glucose oxidase or lactate dehydrogenase, respectively.

Based on similar strategies, immobilized-enzyme electrodes have now been constructed for many biologically important compounds (Table 4.14). Also, a solid-surface fluorometric method has been developed recently for observation of enzyme reactions in membranes. With this approach, direct assay of the concentration of many enzymes, substrates, and cofactors is possible. A wealth of additional material on this subject will be found in the references.

Immobilization of biochemicals for affinity chromatography will be discussed in Chap. 11. This principle—immobilization of one species which has an extremely high affinity for the material to be removed from solution—permits purification or analysis of enzyme inhibitors, cofactors, antigens, antibodies, and other substances.

4.3.4 Utilization and Regeneration of Cofactors

In addition to affinity separations applications, cofactors are also of interest in the capacity of functioning coenzymes. Only two of the six general classes of enzymes are catalytically active without cofactors. If the other four groups of

Table 4.14 Selected compounds which can be monitored using immobilized-enzyme electrodes

Acetaldehyde	D-Galactose
Acetylcholine	D-Glucose
D-Alanine	D-Glutamate
L-Alanine	L-Gulono-λ-lactone
Aliphatic nitro compounds	Hypoxanthine
Alkaline phosphatase	D-Lactose
L-Arginine	Lactate dehydrogenase
D-Aspartate	L-Lactose
Benzaldehyde	NADH
Cholinesterase	Penicillin
Creatine	Some pesticides
Creatine phosphokinase	L-Phenylalanine
L-Cysteine	Phosphate
Dehydrogenases	Sulfate
Diamines	L-Tryptophan
2-Deoxy-D-glucose	L-Tyrosine
Ethanol	Urea
Formaldehyde	Uric acid
L-Galactonolactone	

enzymes are to enjoy industrial use, efficient large-scale methods must be developed for production, separation, and isolation of organic cofactors. Moreover, special reactor and catalyst designs are required for effective coenzyme utilization and regeneration. Many of these problems are not satisfactorily resolved at present. However, the potential commerical and scientific rewards of a well-developed cofactor technology are sufficiently large to justify considerable current research in this area. Here we shall touch upon some of the concepts for enzyme-coenzyme reactors: additional information on this and other aspects of cofactor applications is available in the references.

By retaining both enzyme and coenzyme in hollow fibers which substrate and product can permeate, the desired reaction can be carried out without loss of enzyme or coenzyme. Figure 4.15 shows schematically a laboratory reactor design based on this concept. Its extension to larger scale is clearly feasible provided suitable activity and stability of the catalytic system can be obtained. One reaction system which has been studied in this apparatus is oxidation of ethanol to acetate in a two-step sequence of reactions which require NAD:

$$\text{Ethanol} + \text{NAD}^+ \xrightarrow[\text{dehydrogenase}]{\text{alcohol}} \text{acetaldehyde} + \text{NADH}$$

$$\text{Acetaldehyde} + \text{NAD}^+ \xrightarrow[\text{dehydrogenase}]{\text{aldehyde}} \text{acetate} + \text{NADH}$$

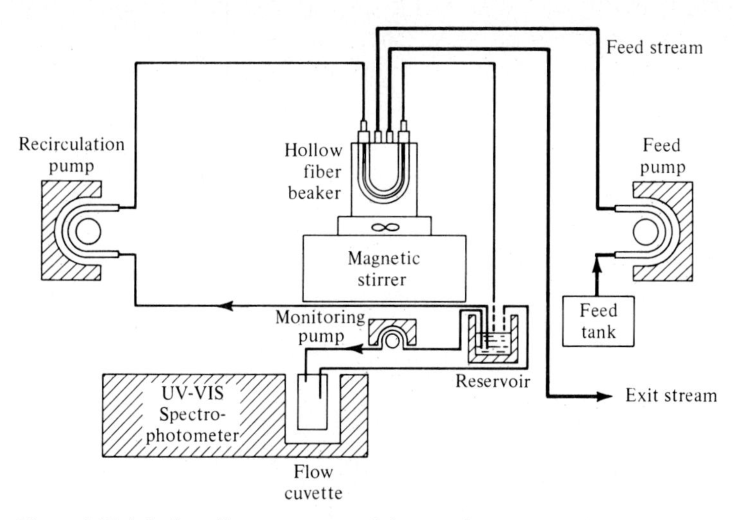

Figure 4.15 A hollow fiber system permitting continuous operation of an enzyme-cofactor reactor.

Clearly, these reactions will take place continuously only if NADH is oxidized continuously, a process which has been accomplished by adding the enzyme pig-heart diaphorase to the catalytic mixture inside the hollow fibers. With this enzyme, O_2 serves as the NADH electron acceptor, and H_2O_2 is formed along with NAD^+ as the coenzyme regeneration reaction occurs. Since hydrogen peroxide causes inactivation of many enzymes, catalase is also included in the enzyme-coenzyme solution. Extension of this scheme to other reactions seems straightforward, although it is not immediately applicable in cases where the substrate molecular weight approaches those of the coenzyme(s) and enzymes involved (Why?). One strategy for retaining the relatively small coenzyme molecules by ultrafiltration is to enlarge artificially the cofactor by covalently bonding it to a soluble polymer such as polyethylene glycol (PEG), dextran or polylysine.

Let us next consider possible enzyme-coenzyme reaction schemes where enzymes and/or their respective coenzymes are bound to insoluble supports. Any such formulation must allow reversible physical contact of enzyme and coenzyme and must provide opportunities for cofactor regeneration. At least three different approaches to this problem can be identified:

1. Immobilize the coenzyme and put enzyme and all necessary substances for regeneration in solution. All participants in the catalytic system except coenzyme will then be contained in the reactor effluent.
2. A complementary idea to 1: the enzyme is immobilized with cofactor in solution and thus in the effluent. Enzymes and additional substrates required for coenzyme regeneration may be passed through this reactor in solution or contained in a separate regeneration reactor, which may itself employ immobilized enzymes.

3. Link the enzyme and coenzyme to each other using a long, flexible molecule as a connection. Such a coenzyme-on-a-string structure is then immobilized. Concepts 1 and 2 have been demonstrated on a laboratory scale, while the feasibility of approach 3 is still under study.

Choice between schemes 1 and 2 for a possible large-scale process depends on a variety of technological and economic factors. Critical among these is the ease, efficiency, and expense with which enzyme, coenzyme and/or regeneration systems can be removed from the product stream and recycled into the reactor. If recovery of none of these species is feasible, for example, concept 1 will have an advantage since most coenzymes are much more expensive than any other part of the catalytic system.

The efficiency with which whole cells utilize and regenerate cofactors suggests another strategy for cofactor-requiring enzyme-catalyzed reaction processes: use an entire immobilized cell preparation and take advantage of the natural cofactor supply and cofactor regeneration system. While this concept is intuitively attractive, its active and stable realization has not been, so far, extremely successful. Problems remain to be solved concerning stability of the immobilized cell system. What if the cofactor-requiring enzyme is not found in a readily immobilized, active cell? In principle, this difficulty can be overcome through application of genetic engineering by inserting the enzyme of interest into a suitable host cell which then serves primarily as a cofactor reservoir and regeneration system. Another fundamental obstacle in implementing this technology is appropriate regulation of transport through the cell envelope. Just as indicated above for the ultrafiltration immobilization of a cofactor-enzyme catalyst system, the entire cell used for this purpose must admit substrates, release products, but not lose essential enzymes and cofactors and other components required for activity and maintenance of the intracellular catalytic systems.

4.4 IMMOBILIZED ENZYME KINETICS

To enable reactor design for immobilized enzyme catalytic systems, it is important to examine in some detail the kinetic properties of immobilized enzymes. The observed catalytic properties of single catalyst particle containing immobilized enzyme or of a membrane reactor in which enzymes are retained by ultrafiltration depend upon interaction of substrate transport and the catalytic activity of the enzyme. In this section we shall focus on mass transport-catalytic reaction coupling and its effect on the performance of a single particle of immobilized enzyme.

To begin our examination of the ways in which mass transport and reaction processes interact to determine the overall activity, deactivation, and parametric dependences of an immobilized enzyme catalyst, consider the schematic diagram in Fig. 4.16. In this diagram, we look at a cross section of a sheet of immobilized

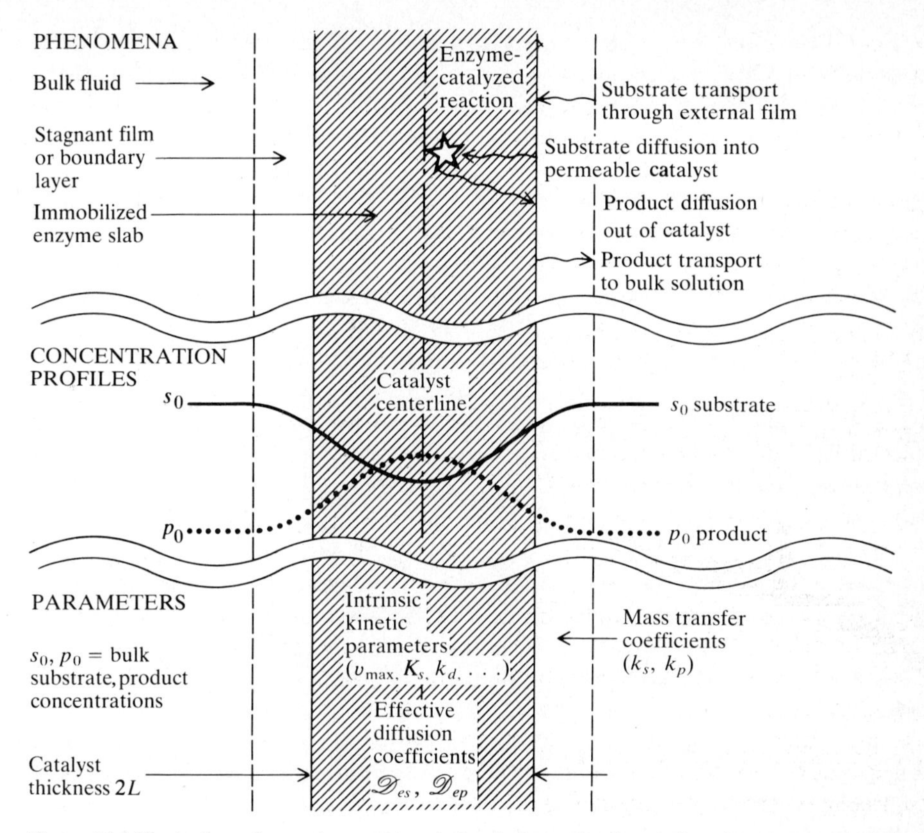

Figure 4.16 Illustration of mass transport and chemical reaction interactions in a symmetric slab of immobilized enzyme. The phenomena summarized in the top section cause the substrate and product concentration profiles sketched in the middle section which can be described quantitatively in terms of the parameters indicated in the bottom section.

enzyme in contact with substrate solution which is flowing over both sides of the sheet. Far from the catalyst, the substrate concentration and other process variables such as pH have values characteristic of the bulk reaction mixture. These are the compositions that we would measure by conventional analytical chemical methods.

Because substrate is consumed within the immobilized enzyme catalyst and product is formed there, concentration gradients arise between the bulk solution far from the catalyst and reaction events which are occurring at the active sites of the immobilized enzyme molecules. Substrate, for example, must be transported from the bulk solution to the outer surface of the catalyst. If the reaction mixture is stagnant, this transport occurs by molecular diffusion. Ordinarily, mixing or flow of substrate solution adds convective transport contributions to the movement of substrate from bulk solution to the external surface. If the immobilized enzyme formulation does not contain enzyme in its internal volume or if substrate cannot penetrate into the internal volume, this external mass transport process is the only one which we need to consider. Often, however, the enzyme is entrapped or impregnated within a permeable matrix. In such a circumstance, most of the catalytic activity is located in and distributed through the interior of the catalyst formulation. Consequently, for such a system, the substrate must diffuse from the outer surface of the catalyst to some internal location where reaction occurs. All of this scenario is inverted for products. In this type of situation, one must consider the intraparticle diffusion processes as well as external mass transport.

As indicated in Fig. 4.16, reaction occurs within the immobilized enzyme sheet at rates which are determined by the concentrations within the sheet. Because of the concentration gradients just described, local reaction rates vary as a function of internal position. The total rate of substrate consumption is the sum of all the substrate consumption rates inside the permeable catalyst. At steady state, this overall rate is also equal to the rate of substrate transport to the catalyst. Clearly, in such a situation, overall rates depend on the interaction between transport processes and catalytic reaction.

Development of an analytical framework for description of these interacting processes and of criteria for assessing their importance is one of the central achievements of chemical reaction engineering. Because of the importance of these interactions in immobilized enzyme and in immobilized cell catalytic processes and because the influences of these interactions have been very often overlooked in research and development activities on immobilized enzymes and cells, we shall consider here the central concepts, analytical appoaches, and results. The topic is treated in further detail in chemical reaction engineering textbooks and several monographs.

4.4.1 Effects of External Mass-Transfer Resistance

In order to introduce central concepts and terminology, we begin with the simplest possible case. Suppose that enzyme is immobilized only on the external surfaces of a slab of support. Then we must consider only mass transport from the bulk solution to the support surface and reaction at that position.

One traditional model, called the *Nernst diffusion layer* in the biochemistry literature and a *stagnant film* in engineering, leads to the following equation for the flux N_s (moles per unit time per unit area) of substrate from the bulk fluid (sometimes called the *pool* by biochemists) to the interface:

$$N_s = k_s(s_0 - s) \tag{4.2}$$

Here, s and s_0 are the substrate concentrations at the interface and in the bulk fluid, respectively, and k_s is the mass-transfer coefficient. A function of physical properties as well as hydrodynamic conditions near the interface, k_s can usually be evaluated from available correlations (discussed in Chap. 8). In particular, k_s increases with increasing liquid flow rate through a packed-column immobilized enzyme reactor.

In steady state, substrate cannot accumulate at the catalyst interface. Consequently, the rate of substrate supply by mass transfer must be exactly counterbalanced by the rate of substrate consumption in the reaction at the interface. Assuming that Michaelis-Menten kinetics applies at the surface, and letting $\bar{v}$ denote the surface reaction rate (moles per unit time per unit area), we have

$$k_s(s_0 - s) = \bar{v} = \frac{v_{\max} s}{K_m + s} \tag{4.3}$$

The number of parameters necessary to specify the system can be reduced from four ($k_s, s_0, v_{\max}$, and K_m) to two (Da and κ) by introducing the dimensionless variables

$$x = \frac{s}{s_0} \qquad \mathrm{Da} = \frac{v_{\max}}{k_s s_0} \qquad \kappa = \frac{K_m}{s_0} \tag{4.4}$$

In terms of these quantities, the substrate mass balance (4.3) becomes

$$\frac{1 - x}{\mathrm{Da}} = \frac{x}{\kappa + x} \tag{4.5}$$

where $0 \leq x \leq 1.0$.

The physical significance of Da, the Damköhler number, should be emphasized:

$$\mathrm{Da} = \frac{v_{\max}}{k_s s_0} = \frac{\text{maximum reaction rate}}{\text{maximum mass-transfer rate}} \tag{4.6}$$

Thus, if Da is much less than unity, the maximum mass-transfer rate is much larger than the maximum rate of reaction (low mass-transfer resistance). When the mass-transfer resistance is large, mass transfer is the limiting process and Da is much greater than 1. These cases are known as the *reaction-limited regime* and the *diffusion-limited regime*, respectively.

Algebraic manipulation of Eq. (4.5) gives a quadratic equation for x, so that an analytical solution is available:

$$x = \frac{\beta}{2}\left(\pm \sqrt{1 + \frac{4\kappa}{\beta^2}} - 1 \right) \qquad 4.7)$$

where

$$\beta \equiv \mathrm{Da} + \kappa - 1 \tag{4.8}$$

The + and − signs are taken for $\beta > 0$, $\beta < 0$, respectively. When $\beta = 0$, $x = \sqrt{\kappa}$. Using this value for s/s_0, either the right- or left-hand side of Eq. (4.5) can be used to evaluate the dimensionless *observed* reaction rate $\bar{v}/v_{\max}$. It should be clear that $\bar{v}$ will not in general exhibit Michaelis-Menten dependence on s_0. Also, it is no longer correct to equate K_m to the substrate concentration $s_{1/2}$ where $\bar{v}$ is equal to one-half the observed maximal reaction rate $v_{\max}^{\text{app}}$. In general, $s_{1/2}$ varies with Da.

In spite of this fact, the parameter $s_{1/2}$ is cited frequently as the apparent $K_m(K_m^{\text{app}})$, and this value is employed as a measure of mass-transfer influence. While this practice may be convenient for the experimentalist who has only to measure $s_{1/2}$, it runs serious risk of misinterpretation. One may be tempted to use the incorrect equation

$$\bar{v}_{\text{obs}} = \frac{v_{\max}^{\text{app}} s_0}{K_m^{\text{app}} + s_0} \tag{4.9}$$

for the observed rate. Moreover, even if this form provides an adequate empirical representation of the observed kinetics in a given situation, it is not generally correct since it ignores the dependence of K_m^{app} on fluid properties and hydrodynamics near the interface. The experimental results shown in Fig. 4.17 makes this dependence evident.

By tradition in chemical engineering, the influence of mass transfer on the overall reaction process is represented using the effectiveness factor η, which is defined physically by

$$\eta = \frac{\text{observed reaction rate}}{\text{rate which would be obtained with no mass-transfer resistance, i.e., surface concentration } s = \text{bulk concentration } s_0} \tag{4.10}$$

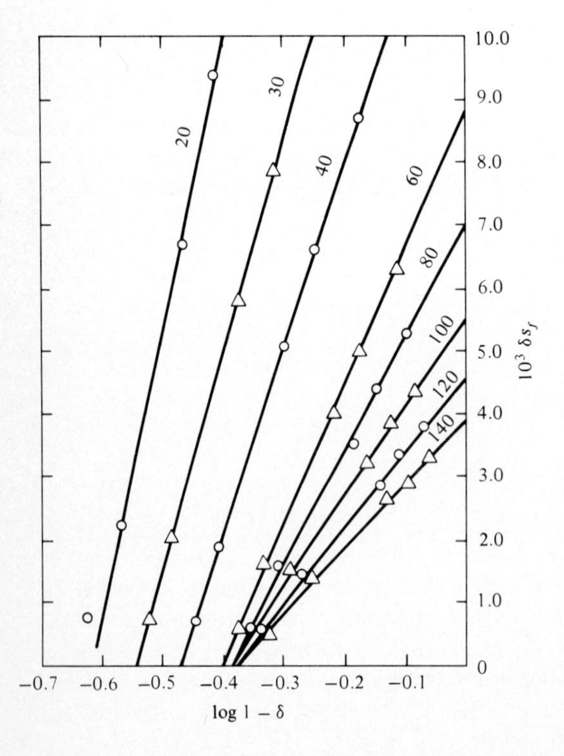

Figure 4.17 Experimental relationships between substrate conversion $\delta\,[= (s_f - s_e)/s_f$, where s_f and s_e are feed and exit substrate concentrations, respectively] and flow rate (numbers on curves, in milliliters per hour) in a packed bed immobilized enzyme reactor (hydrolysis of benzoyl L-arginine ethyl ester by CM-cellulose-ficin). If plug flow and Michaelis-Menten kinetics are assumed, the slope of all the lines should be constant and equal to K_m. The change in slope with flow rate shows a significant mass-transfer effect on overall kinetics. *(Reprinted by permission from M. D. Lilly, W. E. Hornby, and E. M. Crook, "The Kinetics of Carboxymethylcellulose-Ficin in Packed Beds," Biochem, J., vol. 100, p. 718, 1966.)*

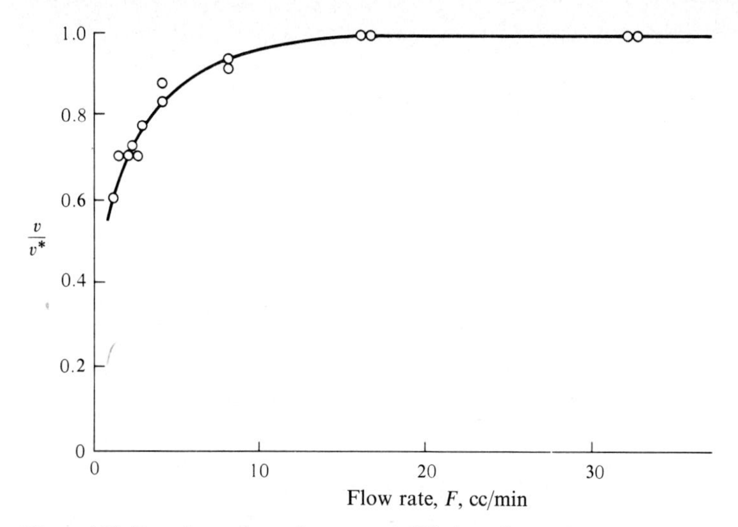

Figure 4.18 Experimental test for external diffusion effects: v^* is the rate obtained at high flow rates through a packed column (hydrolysis of 4×10^{-4} M D, L-N-benzoyl DL-arginine-p-nitroanilide at pH 8, $T = 25°C$ by immobilized trypsin). *(Reprinted by permission from J. R. Ford et al., Recirculation Reactor System for Kinetic Studies of Immobilized Enzymes, in L. B. Wingard, Jr. (ed.), "Enzyme Engineering," Wiley-Interscience, New York, 1972.)*

Consequently, for the problem in question

$$\eta = \frac{x/(\kappa + x)}{1/(\kappa + 1)} \tag{4.11}$$

so that $\eta \leq 1$ and, in general, the effect of increasing mass-transfer resistance is reduction of the observed activity of the catalyst. An experimental demonstration of this phenomenon is related in Fig. 4.18.

For Da approaching zero (very slow reaction relative to maximum mass-transfer rate), Eq. (4.5) shows that x must approach unity, and so for the reaction-limited regime (Da $\rightarrow$ 0)

$$\eta = 1 \qquad \bar{v} = \frac{v_{\max} s_0}{K_m + s_0} \tag{4.12}$$

Here the observed reaction kinetics is the same as the true, intrinsic kinetics at the fluid-solid inteface. If a new immobilized-enzyme surface has been prepared, it is important to determine whether the immobilization procedure has changed the catalytic behavior of the enzyme. The evaluation of $v_{\max}$ and K_m needed for such a determination must occur in experiments for which Da $\ll 1$ in order to avoid disguise by significant mass-transfer resistance.

Intrinsic kinetics is also required for engineering design of immobilized-enzyme reactors, since only with this information can fluid properties, shape of the enzyme support, and mixing characteristics be adequately taken into account.

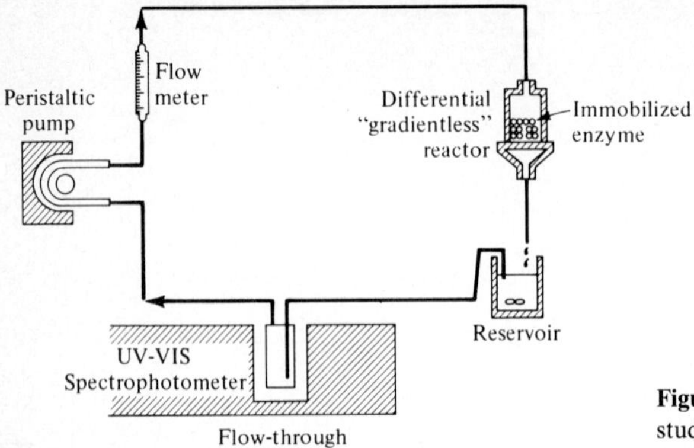

Figure 4.19 Recirculation reactor for study of external mass transfer effects on immobilized enzyme kinetics.

Many experimental reactors designed to operate in the reaction-limited regime have been developed for ordinary heterogeneous catalysts. Some have been adapted for immobilized-enzyme kinetics studies (Fig. 4.19). They all share the common concept of high fluid flow rates near the catalyst to minimize mass-transfer resistance (larger k_s, smaller Da). There are several possible drawbacks to this approach which are peculiar to immobilized enzymes. Fluid forces might cause partial or complete denaturation of the attached enzymes (see earlier discussion in Sec. 3.7.3). Also, relative motion of catalyst paticles may give enzyme loss by particle-particle abrasion.

The diffusion-limited regime of systems with coupling between chemical reaction and mass transfer arises when v_{max} is much larger than $k_s s_0$, so that $\mathrm{Da} \gg 1$. By expanding the square root in Eq. (4.7) and subsequent manipulation, we can derive for the diffusion-limited regime ($\mathrm{Da} \to \infty$, κ finite)

$$\eta = \frac{1 + \kappa}{\mathrm{Da}} \qquad \bar{v} = k_s s_0 \tag{4.13}$$

Thus, so long as Da is very large, the observed rate of reaction $\bar{v}$ is first order in bulk substrate concentration and totally independent of the intrinsic rate parameters v_{max} and K_m. This behavior disguises the true kinetic parameters of the immobilized enzymes. In the diffusion-limited regime, for example, the observed activity for given s_0 is constant even though the enzymes at the interface may be losing activity because of adverse temperature, pH or other conditions. Studies aimed at determining activity retention or denaturation rates of immobilized enzymes should therefore also be conducted as close as possible to the reaction-limited regime.

4.4.2 Analysis of Intraparticle Diffusion and Reaction

As discussed earlier, enzymes are typically immobilized on the internal surfaces of porous supports or entrapped in matrices through which substrate can diffuse. In such systems, calculation of the observed rate of substrate disappearance requires evaluation of the concentration profile of substrate within the profile. To accomplish this we begin by writing a steady-state material balance for a thin section of the permeable catalyst. Assuming enzyme is immobilized within a pellet that has approximately spherical shape, a useful thin section is contained between two concentric spheres of radius r and $r + dr$, respectively (see Fig. 4.20). A shell of differential thickness is considered so that all conditions within the shell may be considered independent of position.

The symbols $\mathscr{D}_{es}$ and v denote substrate effective diffusion coefficient and local rate of substrate utilization, respectively. Both of these quantities differ conceptually from their counterparts in solution, and important quantitative differences may also arise. Considering first the effective diffusion coefficient, we should be aware that diffusion rates of all species through the support are subject to the following influences:

1. Some of the particle cross section is occupied by solid and hence not available for diffusive transport (parameter: particle porosity ε_p).
2. The pore network is complex and entangled so diffusion occurs only in allowed, frequently changing directions (parameter: tortuosity factor τ).
3. Pores may have very small diameters, similar to substrate molecular dimensions (restricted diffusion situation; parameter: K_p/K_r).

The effective diffusion coefficient for substrate $\mathscr{D}_{es}$ may now be written as

$$\mathscr{D}_{es} = \mathscr{D}_{s0} \cdot \frac{\varepsilon_p}{\tau} \cdot \frac{K_p}{K_r} \tag{4.14}$$

where $\mathscr{D}_{s0}$ is the substrate diffusivity in the bulk liquid. The porosity parameter ε_p must be determined for the support in question. Tortuosity factor values are usually assumed to be in the range of 1.4 to 7. If restricted diffusion occurs, a crude estimate of K_p/K_r may be obtained from

$$\frac{K_p}{K_r} \cong \left(1 - \frac{r_{\text{substrate}}}{r_{\text{pore}}}\right)^4 \tag{4.15}$$

where $r_{\text{substrate}}$ and r_{pore} are the substrate molecular (equivalent) radius and characteristic pore radius, respectively. Uncertainties in all of the factors on the right-hand side of Eq. (4.14) make accurate a priori estimation of $\mathscr{D}_{es}$ quite difficult. Consequently, it is often preferable to estimate this quantity from overall kinetic data. We will examine this method after analyzing the coupled diffusion-reaction processes within the immobilized enzyme particle.

We should note also that mass-transfer rates within the particle may depend on the internal surface chemistry and charge. Further, if the reaction involves consumption or production of ionic species, gradients in electrical potential may exist within the particle. Then, these gradients will influence charged species transport rates. Consideration of this situation may be found in the chapter references.

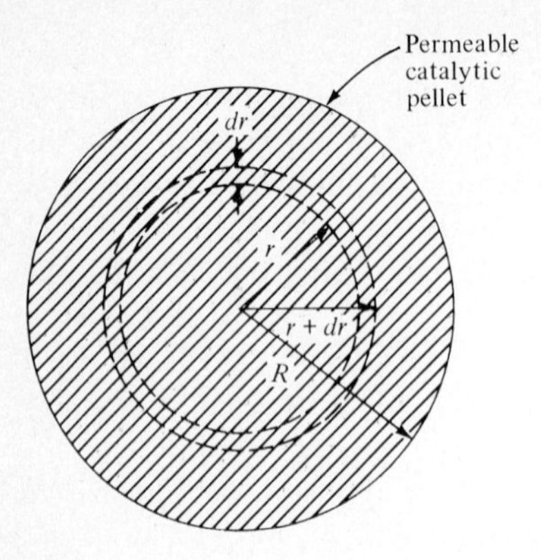

Steady-state mass balance on shell between r and $r + dr$:

$$\left(-\mathscr{D}_{es}\frac{ds}{dr}\cdot 4\pi r^2\right)\Bigg|_r - \left(-\mathscr{D}_{es}\frac{ds}{dr}\cdot 4\pi r^2\right)\Bigg|_{r+dr} = v \cdot 4\pi r^2\,dr$$

Divide by $4\pi\,dr$ and rearrange. Assume effective substrate diffusivity within the permeable pellet $\mathscr{D}_{es}$ is constant.

$$\frac{\mathscr{D}_{es}\left(r^2\left.\dfrac{ds}{dr}\right|_{r+dr} - r^2\left.\dfrac{ds}{dr}\right|_r\right)}{dr} = r^2 v$$

Take limit as $dr \to 0$,

$$\mathscr{D}_{es}\frac{d}{dr}\left(r^2\frac{ds}{dr}\right) = r^2 v$$

or

$$\mathscr{D}_{es}\left(\frac{d^2s}{dr^2} + \frac{2}{r}\frac{ds}{dr}\right) = v$$

Figure 4.20 Development of a steady-state material balance on a spherical, permeable immobilized enzyme pellet using the thin-shell method.

Presuming that the intrinsic, local reaction kinetics of the immobilized-enzyme catalyzed reaction are of Michaelis-Menten form

$$v = \frac{v_{\max} s}{K_m + s} \tag{4.16}$$

we may write the maximum velocity parameter as the product of the enzyme loading e_{imm} (μmol enzyme/g support), the immobilized enzyme specific activity $q_{E,\text{imm}}$ [μmol substrate converted s^{-1} (μmol enzyme)$^{-1}$], and the particle density ρ_p (g support/unit volume of support):

$$v_{\max} = e_{\text{imm}} \cdot \rho_p \cdot q_{E,\text{imm}} \tag{4.17}$$

Recall that immobilization may alter enzyme structure and/or molecular environment, causing $q_{E,\text{imm}}$ and K_m to differ from the corresponding values for enzyme in solution. As discussed in Sec. 4.4.1, it is desirable to determine intrinsic kinetics in situations where mass-transfer effects on overall kinetics are negligible. A criterion for such reaction-limited conditions will emerge from the following analysis.

The steady-state material balance on substrate developed in Fig. 4.20 is the following second-order ordinary differential equation:

$$\mathscr{D}_{es}\left(\frac{d^2s}{dr^2} + \frac{2}{r}\frac{ds}{dr}\right) = v(s) \tag{4.18}$$

The model is incomplete without boundary conditions. Since the concentration profile through the pellet will almost always be symmetrical about the center of the sphere

$$\left.\frac{ds}{dr}\right|_{r=0} = 0 \tag{4.19}$$

We shall assume that the substrate concentration at the external surface of the pellet is equal to the substrate concentration s_0 of the liquid medium bathing the particle. (The case of simultaneous film- and intraparticle mass transfer resistance is treated in the following section.) Consequently,

$$\left. s \right|_{r=R} = s_0 \tag{4.20}$$

The observed overall rate of substrate utilization v_0 by the particle is equal to the substrate diffusive flux into the pellet. Expressed in moles per pellet volume per unit time, v_0 is

$$v_0 = \frac{A_p}{V_p}\left(\mathscr{D}_{es}\left.\frac{ds}{dr}\right|_{r=R}\right) \tag{4.21}$$

where V_p and A_p denote the particle volume and external surface area, respectively. As in the previous section, we present such rates in a dimensionless form which reveals the extent of diffusional effects. The effectiveness factor η is defined by

$$\eta = \frac{v_0}{v(s_0)} = \frac{\text{observed rate}}{\text{rate which would obtain with no concentration gradients in pellet}} \tag{4.22}$$

Unfortunately, the effectiveness factor cannot be easily evaluated analytically when v takes the nonlinear form indicated in Eq. (4.16). In this situation, it is necessary to solve numerically the boundary-value problem posed by Eqs. (4.18–4.20) and then evaluate v_0 using Eq. (4.21). Since such numerical calculations are difficult and time-consuming, we seek to present results in the most compact yet general form. This is achieved here by transforming the above equations into an equivalent dimensionless formulation.

Let $\bar{s} = s/s_0$ and $\bar{r} = r/R$; then Eq. (4.18) can be written in dimensionless form

$$\frac{d^2\bar{s}}{d\bar{r}^2} + \frac{2}{\bar{r}}\frac{d\bar{s}}{d\bar{r}} = \frac{vR^2}{\mathscr{D}_{es}s_0} = 9\phi^2 \frac{\bar{s}}{1 + \beta\bar{s}} \tag{4.23}$$

where the dimensionless parameters ϕ and β are defined by

$$\phi = \frac{R}{3}\sqrt{\frac{v_{\max}/K_m}{\mathscr{D}_{es}}} \qquad \beta = \frac{s_0}{K_m} \tag{4.24}$$

The dimensionless boundary conditions associated with Eq. (4.23) are

$$\bar{s}\Big|_{\bar{r}=1} = 1 \qquad \frac{d\bar{s}}{d\bar{r}}\Big|_{\bar{r}=0} = 0 \tag{4.25}$$

The square of the *Thiele modulus* ϕ has the physical interpretation of a first-order reaction rate $R^3(v_{\max}/K_m)s_0$ divided by a diffusion rate $R\mathscr{D}_{es}s_0$. The magnitude of the saturation parameter β provides a measure of local rate deviations from first-order kinetics, with very large values indicating an approach to zero-order kinetics.

In terms of these dimensionless variables, the effectiveness factor is

$$\eta = \frac{(d\bar{s}/d\bar{r})_{\bar{r}=1}}{3\phi^2[1/(1 + \beta)]} \tag{4.26}$$

Equations (4.23) and (4.25) reveal that $\bar{s}$ is a function of $\bar{r}$, ϕ, and β only, so that $(d\bar{s}/d\bar{r})_{\bar{r}=1}$ depends only β and ϕ. This result in conjunction with Eq. (4.26) implies that η depends only upon the Thiele modulus and saturation parameters

$$\eta = f(\phi, \beta) \tag{4.27}$$

A practical problem arises in the use of effectiveness-factor correlations of the form given in Eq. (4.27): the intrinsic-rate parameters $v_{\max}$ and K_m are frequently unknown. A few simple manipulations readily reveal, however, that uncertainty about the former number need not be a problem. Using Eqs. (4.26) and (4.21), we obtain

$$\phi = \left[\frac{R^2 v_0}{9\eta\mathscr{D}_{es}s_0}(1 + \beta)\right]^{1/2}$$

Substitution of this expression into the right-hand side of Eq. (4.27) gives an implicit relationship between η, β and a new dimensionless *observable modulus* Φ, defined by

$$\Phi = \frac{v_0}{\mathscr{D}_{es}s_0}\left(\frac{V_p}{A_p}\right)^2 \tag{4.28}$$

(Notice that Φ depends on the measurable overall rate v_0; Φ is independent of intrinsic kinetic parameters.) Solution of this implicit equation yields η as a function of β and the observable modulus Φ

$$\eta = g(\Phi, \beta) \tag{4.29}$$

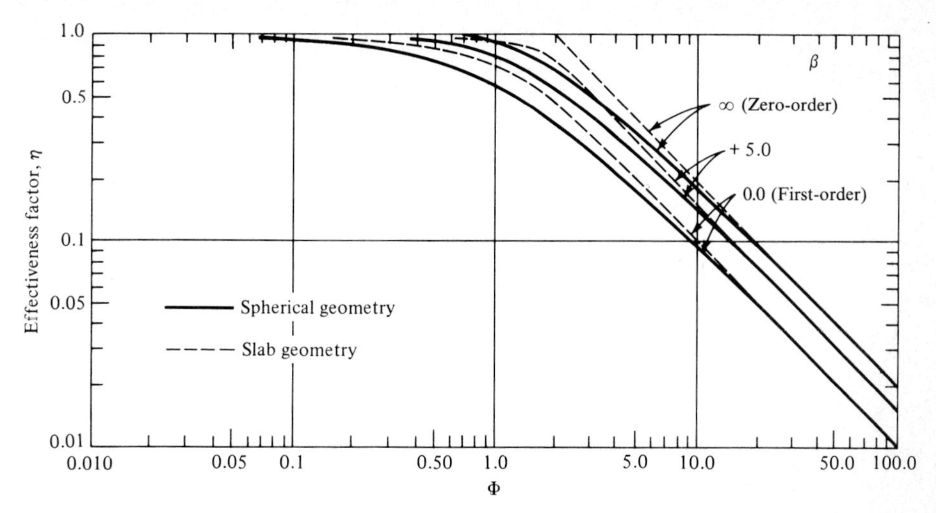

Figure 4.21 Effectiveness factors for immobilized enzyme catalysts with Michaelis-Menten intrinsic kinetics ($\beta = s_0/K_m$). Φ is the observable modulus defined in Eq. (4.28).

Plots of the η-Φ relationship prescribed by Eq. (4.29) for $\beta \to 0$ and $\beta \to \infty$ are given in Fig. 4.21. Since the effectiveness factors for intermediate K_m/s_0 values lie between the curves for these limiting cases, we see that η is relatively insensitive to the remaining intrinsic parameter K_m/s_0.

Before examining analytical solutions based on approximate kinetics, however, we first seek a formula for η which applies for large Φ (or ϕ) and Michaelis-Menten kinetics. When Φ is sufficiently large ($\Phi \geq 3$, see Table 4.15), diffusion of substrate is slow relative to consumption. In such a situation with diffusion-limited rate, it may be assumed that all substrate is utilized in a thin region within the particle adjacent to its exterior surface, so that the effect of curvature can be neglected in Eq. (4.18). This leaves

$$\mathscr{D}_{es}\frac{d^2s}{dr^2} = v \tag{4.30}$$

If the right-hand side of the identity

$$\frac{d^2s}{dr^2} = \frac{1}{2}\frac{d}{ds}\left(\frac{ds}{dr}\right)^2 \tag{4.31}$$

Table 4.15 Criteria for assessing the magnitude of mass-transfer effects on overall kinetics

Criterion	η value	Limiting rate process	Extent of mass-transfer limitation
$\Phi < 0.3$	~ 1	Chemical reaction	Negligible
$\Phi > 3$	$\propto \Phi^{-1}$	Diffusion	Large

is substituted for d^2s/dr^2 in Eq. (4.30), integrating the resulting equation with respect to s yields

$$\left.\frac{ds}{dr}\right|_{r=R} = \left[\frac{2}{\mathscr{D}_{es}}\int_{s_c}^{s_0} v(s)\,ds\right]^{1/2} \tag{4.32}$$

where s_c and s_0 are the substrate concentrations at the particle center ($r = 0$) and external surface ($r = R$), respectively.

In the diffusion-limited case, $s_c \approx 0$, so that the integral in Eq. (4.32) can be evaluated. Combining Eqs. (4.21), (4.22), and (4.32) gives the desired effectiveness-factor expression valid for sufficiently large Φ or ϕ:

$$\eta\Big|_{\text{large }\Phi,\,\phi\text{ (diffusion-limited overall rate)}} = \frac{3\mathscr{D}_{es}}{R}\frac{\left[\dfrac{2}{\mathscr{D}_{es}}\displaystyle\int_0^{s_0} v(s)\,ds\right]^{1/2}}{v(s_0)} \tag{4.33}$$

Evaluation of this formula for the case of Michaelis-Menten kinetics [Eq. (4.16)] yields

$$\eta\big|_{\phi \gg 1} = \frac{1}{\phi}\frac{1+\beta}{\beta}\sqrt{2}\,[\beta - \ln(1+\beta)]^{1/2} \tag{4.34}$$

Continuing now our investigation of Michaelis-Menten kinetics, Fig. 4.21 shows that by assuming first-order kinetics (valid if $s \ll K_m$)

$$v = ks \tag{4.35}$$

($k = v_{\max}/K_m$), a conservative (low) value of η is obtained which is not too inaccurate. This is convenient because the diffusion-reaction model can be solved analytically for the linear rate law of (4.35) to give

$$\eta = \frac{1}{\phi}\left(\frac{1}{\tanh 3\phi} - \frac{1}{\phi}\right) \tag{4.36}$$

where the Thiele modulus ϕ is defined by

$$\phi = \frac{V_p}{A_p}\sqrt{k/\mathscr{D}_{es}} \tag{4.37}$$

Equation (4.37) can be used to obtain the first-order curve shown in Fig. 4.21 by using the relationship valid for first-order kinetics

$$\Phi = \eta\phi^2 \tag{4.38}$$

At the other end of the spectrum of Michaelis-Menten kinetics is the zero-order approximation ($s \gg K_m$), which takes the form

$$v = \begin{cases} k_0 = \text{const} & \text{for } s > 0 \\ 0 & \text{for } s = 0 \end{cases} \tag{4.39}$$

($k_0 = v_{max}$). Solving the boundary-value problem of Eqs. (4.18) to (4.20) with this substrate-utilization function reveals that substrate concentration s depends on radial position r according to

$$s = s_0 - \frac{k_0}{6\mathscr{D}_{es}}(R^2 - r^2) \tag{4.40}$$

This relationship holds as long as s is nonnegative, which will be true from $r = R$ inward to a critical radius $r = R_c$, determined by setting $s = 0$ in Eq. (4.40). This yields

$$\left(\frac{R_c}{R}\right)^2 = 1 - \frac{6\mathscr{D}_{es}s_0}{k_0 R^2} \tag{4.41}$$

Equation (4.41) yields a real root for R_c only if

$$R\sqrt{k_0/6\mathscr{D}_{es}s_0} > 1 \tag{4.42}$$

Consequently, if this inequality is satisfied, there is an interior portion of the particle (from $r = 0$ to $r = R_c$) in which $s = 0$ and $v = 0$. When inequality (4.42) applies, reaction occurs only in an outer shell of the particle ($R_c < r \leq R$), so that

$$\eta = \frac{\frac{4}{3}\pi(R^3 - R_c^3)k_0}{\frac{4}{3}\pi R^3 k_0} = 1 - \left(\frac{R_c}{R}\right)^3 = 1 - \left(1 - \frac{6\mathscr{D}_{es}s_0}{k_0 R^2}\right)^{3/2} \tag{4.43}$$

When condition (4.42) is not fulfilled, a uniform rate of substrate utilization prevails within the entire pellet, given $\eta = 1$.

Another important feature shown in Fig. 4.21 is the insensitivity of the η-ϕ relationship to particle geometry. For example, the effectiveness factor for a first-order reaction in a slab is (ϕ defined by Eq. (4.37); V_p/A_p is half the total slab thickness)

$$\eta = \frac{\tanh \phi}{\phi} \tag{4.44}$$

This function coincides with the one given in Eq. (4.36) to within about 10 percent over the entire ϕ range. Differences are greatest for ϕ near unity and diminish rapidly for larger and smaller values. Because of this geometrical insensitivity, we can make frequent use of the following empirical correlation for the effectiveness factor of an immobilized enzyme slab with intrinsic Michaelis-Menten kinetics

$$\eta = \begin{cases} 1 - \dfrac{\tanh \phi}{\phi}\left(\dfrac{\eta_d}{\tanh \eta_d} - 1\right) & \text{for } \eta_d \leq 1 \\[2ex] \eta_d - \dfrac{\tanh \phi}{\phi}\left(\dfrac{1}{\tanh \eta_d} - 1\right) & \text{for } \eta_d \geq 1 \end{cases} \tag{4.45}$$

where η_d is the asymptotic value of the effectiveness factor calculated from Eq. (4.34). In applying this correlation (developed originally for slab geometry) to other symmetric shapes, it is important always to use the particle volume divided by the external particle surface area as the characteristic length in the Thiele modulus parameter.

The insensitivity of the η function with respect to both reaction order and particle geometry suggests useful general criteria for reaction- and diffusion-limitation, which are summarized in Table 4.15 (see Ref. 21 for further details). These criteria are based entirely on the value of the Φ parameter which, we recall, contains only observable quantities.

Very commonly in nonbiological catalysis η is viewed as a measure of the efficiency of catalyst utilization. Unless η is near unity, greater overall rates can be obtained by subdividing the same amount of catalytic material into smaller pieces, diminishing Φ and thus increasing η. This interpretation is also valuable in the context of biological catalysis, for it indicates how immobilized-enzyme particle size should be controlled to achieve maximum efficiency.

Also, effectiveness factor values near unity are needed for direct observation of intrinsic kinetics. With sufficiently small particles (see Prob. 4.5), we can determine the functional form of the intrinsic immobilized enzyme kinetic expression and identify values of the rate parameters. With this information, we can then utilize the above framework to calculate the observed, overall kinetics for larger particles. Since substrate effective diffusion coefficients and/or bulk concentrations are typically small in aqueous solution, particle sizes sufficiently large to avoid pressure drop problems and flow limitations in packed-bed reactors or particle retention difficulties in slurry and fluidized bed reactors generally imply diffusion-limited operation. Sometimes it is even difficult in the laboratory to operate under reaction-limited conditions. In such cases, the above analytical framework must be used to extract intrinsic kinetic information from diffusion-disguised data. Also, the effective diffusion coefficient for substrate can be obtained from kinetic experiments with the immobilized enzyme as illustrated in the example below.

Example 4.3: Estimation of diffusion and intrinsic kinetic parameters for an immobilized enzyme catalyst We assume at the outset that the intrinsic kinetics have Michaelis-Menten form. Estimation of the intrinsic kinetic parameters v_{max} and K_m and the substrate effective diffusion coefficient $\mathscr{D}_{es}$ can be accomplished based on two sets of kinetic experiments, one employing large particles (giving large ϕ and hence diffusion-limited conditions) and the other using small particles (sufficiently small so that substrate conversion is reaction-limited). For each particle size, we vary s_0, measure the corresponding v_0 values and plot both sets of data using Eadie-Hofstee coordinates (v_0/s_0 vs. v_0; recall Sec. 3.2.2). For sufficiently large values of s_0, the reaction is zero order throughout the catalyst, implying an effectiveness factor of unity and coincidence between the data for the large and small particles [this presumes that Eq. (4.42) is not satisfied for the large particles and that substrate solubility limitations do not preclude experiments in the range $s_0 \gg K_m$]. Thus, for both particle sizes, the intercept on the abscissa is v_{max}.

On the other hand, when external substrate concentration and overall reaction velocities are small, the local, intrinsic kinetics become approximately first-order. Then, v_0/s_0 is equal to v_{max}/K_m

for the smaller, reaction-limited particles, while for the larger, diffusion-limited immobilized enzyme particles

$$\left.\frac{v_0}{s_0}\right|_{\substack{s_0 \ll K_m \\ \phi \gg 1}} = \eta \bigg|_{\substack{\text{first-order} \\ \phi \gg 1}} \frac{v_{\max}}{K_m} = \frac{3}{R}\sqrt{\frac{v_{\max}\mathscr{D}_{es}}{K_m}} \tag{4E3.1}$$

Accordingly, the intercept on the ordinate of the Eadie-Hofstee plot for the small particles is $v_{\max}/K_m$ and the large particle intercept is given by Eq. (4E3.1). Knowing the size of the large particles, evaluation of $v_{\max}$, K_m, and $\mathscr{D}_{es}$ follows from determination of the three different Eadie-Hofstee intercepts.

Application of this procedure to α-chymotrypsin immobilized on CNBr-activated Sepharose 4B is illustrated in Fig. 4E3.1. The immobilized enzyme specific activity $q_E[= v_{\max}/$(moles active enzyme immobilized/unit volume of catalyst)] and K_m values so determined are 213 μmol substrate (ATEE: *N*-acetyl-L-tyrosine ethyl ester) per micromole active enzyme per second and 2.6 m*M*, respectively. These values differ substantially from those measured for the same enzyme in solution: specific

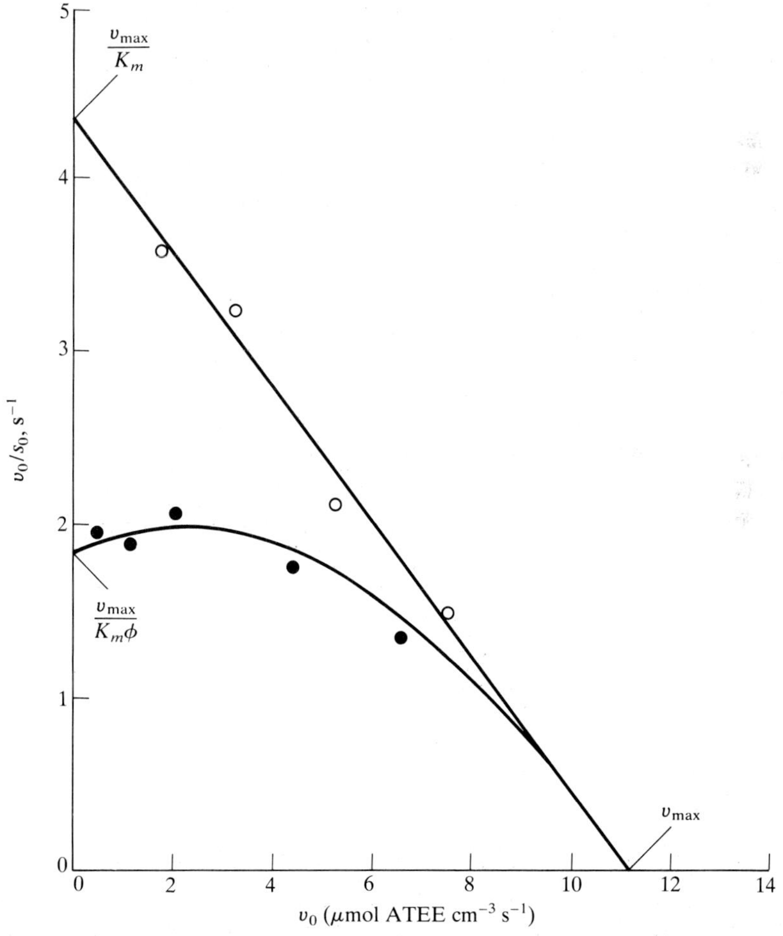

Figure 4.E3.1 Eadie-Hofstee plots of experimental data for α-chymotrypsin immobilized on large ($R = 60$μm; ●) and small ($R = 10$ μm; ○) particles of CNBr-activated Sepharose 4B. *(Plotted points calculated from data in D. S. Clark and J. E. Bailey, Structure-Function Relationships in Immobilized Chymotrypsin Catalysis, Biotech. Bioeng., vol. 25, p. 1027, 1983.)*

activity = 311 μmol ATEE/(μmol active enzyme·s) and $K_m = 0.73$ mM-indicating significant alteration in the enzyme intrinsic catalytic properties caused by immobilization. The estimated effective diffusion coefficient for ATEE in this catalyst is 3.8×10^{-6} cm^2/s. This small value combined with reasonably active immobilized enzyme implies significant diffusion limitation even in particles with 60 μm radius. The estimated observable modulus Φ for this preparation is 2.6.

A common strategy employed in the formulation of catalysts with extremely valuable and highly active supported metal is localization of the supported catalytic material in a thin outer shell on the support. Clearly, the same strategy can be employed in formulating immobilized enzyme catalyst particles sufficiently large to have necessary process properties but without extremely wasteful use of the enzyme. A number of strategies can be envisioned for preparing such catalysts. It has been discovered recently that internal distribution of supported enzyme can be highly nonuniform in catalysts made by impregnating a porous support material with enzyme solution. As explored further in Prob. 4.10 below, such nonuniform loading affects the apparent activity and deactivation properties of the immobilized enzyme catalyst.

Before turning to the problem of combined external and internal mass transfer, we should note that, just as is observed in solution, immobilized enzyme kinetics may differ from the single substrate, irreversible Michaelis-Menten form. While complete discussion of possibilities for other types of intrinsic kinetics must be left to the references, a few qualitative remarks are needed here. First, notice that for a reversible reaction, the minimum substrate concentration in the particle is the equilibrium concentration. Thus, when calculating an asymptotic effectiveness factor as in Eq. (4.33), the lower limit on the integral should be the equilibrium value. Interestingly, stoichiometry within a catalyst pellet can be altered from that in the bulk by the relative diffusion coefficients of the chemical species involved. This phenomenon is illustrated in Prob. 4.8.

As a final anecdote on the intriguing possibilities in such coupled diffusion-reaction systems, let us consider a situation in which the intrinsic kinetics are of substrate-inhibited form. If the external substrate concentration s_0 is greater than the substrate concentration corresponding to maximum rate (see Sec. 3.4.1), as the reaction occurs in the pellet, the decreased substrate concentration will result in a larger local reaction rate than that at the exterior surface of the catalyst particle. In such situations, it is possible under certain conditions to observe effectiveness factors larger than unity. This counterintuitive situation, reasonable after some contemplation, is a possibility for any reaction rate expression exhibiting autocatalytic features—that is, giving larger reaction rates as the extent of reaction increases.

4.4.3 Simultaneous Film and Intraparticle Mass-Transfer Resistances

The previous two sections have examined special cases of the mass transport-chemical reaction interactions sketched in Fig. 4.16. In general, substrate must first traverse the external film or boundary layer and subsequently diffuse into the catalyst where reaction occurs. One goal of our investigation of this general situation will be derivation of guidelines for use of the simpler models considered above. That is, when does internal resistance dominate, when does external resistance dominate, and when must both be considered?

These central points can be addressed adequately by considering a relatively simple problem with slab geometry for the immobilized enzyme catalyst and first-order intrinsic kinetics. In this case, the steady-state intraparticle mass balance on substrate is

$$\mathscr{D}_{es}\frac{d^2s}{dx^2} - ks = 0 \tag{4.46}$$

with centerline symmetry condition

$$\left.\frac{ds}{dx}\right|_{x=0} = 0 \tag{4.47}$$

At the exterior surface of the slab ($x = L$), no substrate accumulates. Transport into this surface by intraparticle diffusion equals transfer out by film transport, giving the final boundary condition:

$$-\mathscr{D}_{es}\left.\frac{ds}{dx}\right|_{x=L} = k_s[s(L) - s_0] \tag{4.48}$$

Solving the problem posed by Eqs. (4.46)–(4.48) gives the effectiveness factor η_s for this situation which can be written in the form

$$\eta_s = \frac{\tanh \phi}{\phi[1 + (\phi \tanh \phi)/\mathrm{Bi}]} \tag{4.49}$$

where the Thiele modulus ϕ is defined as before (see Eq. 4.37; for slab geometry $V_p/A_p = L$). The important new parameter appearing here is the *Biot number* Bi, defined as

$$\mathrm{Bi} = \frac{k_s L}{\mathscr{D}_{es}} = \frac{\text{characteristic film transport rate}}{\text{characteristic intraparticle diffusion rate}} \tag{4.50}$$

It is useful to rearrange Eq. (4.49) into the following form

$$\frac{1}{\eta_s} = \frac{1}{\eta} + \frac{\phi^2}{\mathrm{Bi}} \tag{4.51}$$

where here η ($= \tanh \phi/\phi$) denotes the effectiveness factor for the catalyst without film transport resistance. The reciprocal of the effectiveness factor may be viewed as a measure of the resistance to substrate reaction imposed by substrate transport requirements. Consequently, Eq. (4.51) echoes the familiar rule that the overall resistance equals the sum of resistances in series.

This equation provides the needed basis for deciding when one of the resistances can be neglected. If

$$\frac{\eta\phi^2}{\text{Bi}} = \frac{\Phi}{\text{Bi}} = \frac{kL}{k_s} \ll 1 \tag{4.52}$$

then the influence of external film resistance is negligible. If instead kL/k_s is much larger than unity, internal resistance can be ignored. If neither condition is satisfied, both resistances must be considered. Notice in this case that as ϕ increases, η_s becomes inversely proportional to ϕ^2. Plotting η_s vs. Φ for various Bi values reveals an alternative criterion which does not require knowledge of the intrinsic rate constant: if Bi is of the order 100 or greater, the effects of external resistance are not significant.

The general concepts just summarized apply also to other geometries and intrinsic kinetics. While details must be left to the references, we can see that evaluation of the mass transport steps which significantly influence the overall rate depends on examination of the relative values of characteristic reaction, intraparticle diffusion and external mass transfer rates.

4.4.4 Effects of Inhibitors, Temperature, and pH on Immobilized Enzyme Catalytic Activity and Deactivation

Quantitative representation of enzyme deactivation and of the effects of reaction parameters—concentrations of inhibitors and activators, temperature, pH, ionic strength, and others—on enzyme activity were considered for enzymes in solution in the previous chapter. Generally, we can expect the same types of functional forms to describe the parametric dependence of immobilized enzyme intrinsic activity. However, as noted several times above, it is possible that immobilization may alter the intrinsic properties of the enzyme. Different rate constants may be required for immobilized enzymes, and immobilization may even alter the form of the equation needed to describe the effects of reaction parameters on the immobilized enzyme intrinsic activity. Determination of the response of immobilized enzyme intrinsic catalytic activity to changes in reaction parameters requires careful kinetic studies using the methods described above to be certain to examine intrinsic kinetic behavior and not a combination of mass transfer and catalytic rates.

One major purpose of this section is to point out the influence which mass transport processes can have on the relationship between apparent overall catalytic activity and the environment of the immobilized enzyme catalyst. Examination of first order kinetics suffices to make the necessary observations. First, remember that for an immobilized enzyme catalyst operating in the reaction-limited regime, the observed overall kinetics are the same as the actual, local intrinsic kinetics. Thus, the observed overall kinetics will be first order with rate constant equal to the intrinsic rate constant. On the other hand, for an immobilized enzyme catalyst functioning in the diffusion-limited regime, the apparent overall kinetics will still be first order, but the apparent rate constant will be the square root of the intrinsic rate constant. (In general, if the intrinsic kinetics are nth order, the apparent kinetics under diffusion-limited conditions will be of order $(n + 1)/2$.) Thus, the apparent activation energy will be one-half of the true activation energy. Similarly, the effect of all reaction parameter changes on the observed overall kinetics will be smaller than the actual effects of those parameter changes on the local, intrinsic kinetics. For example, deactivation processes, addition of an inhibitor, or a change in pH which reduces the intrinsic kinetic rate constant by a factor of 4, will only reduce the observed overall rate by a factor of 2 under diffusion-limited conditions.

Therefore, one must be very careful in studying the effects of reaction parameters on the kinetics of immobilized enzyme-catalyzed reactions. Without care to operate under reaction-limited conditions or to extract intrinsic kinetic information from measurements under diffusion-influenced conditions, conclusions will be reached about the immobilized enzyme activity response to reaction conditions which are valid only for that particular catalyst formulation under the particular contacting conditions studied and which do not characterize the intrinsic behavior of the immobilized enzyme. Then, any change in flow rate of substrate solution past the particles, change in particle size, change in particle pore structure or change in the amount of enzyme in the particle or its internal distribution will alter the parametric response of the catalyst from that determined previously. This illustrates the importance and indeed the necessity of identifying properly those attributes of the immobilized enzyme system which are due only to mass-transfer effects and those which reflect the intrinsic properties of the immobilized enzyme.

Unfortunately, such concern has been absent from many published studies of the parametric responses of immobilized enzyme systems. Remembering that most immobilized enzyme catalysts operate under diffusion-influenced if not highly diffusion-limited conditions, this means that many of the observations which have been reported in the literature about deactivation rates and pH and temperature effects on immobilized enzyme kinetics do not describe solely intrinsic immobilized enzyme properties.

Ample evidence exists, however, to conclude that enzyme immobilization can alter the intrinsic kinetics of enzyme deactivation. Several different mechanisms and explanations for this phenomena have been proposed, postulated, and some have been tested experimentally, and we shall briefly survey these next. First, by holding the enzymes in relatively fixed spatial position, immobilization reduces interaction between enzyme molecules which contributes to deactivation by aggregation and to autolysis by proteolytic enzymes. Second, by attachment of the support to several points on the molecule, unfolding of the protein is retarded. Dramatic stability enhancements have been reported based on this strategy in which gel entrapment was applied to attempt to form a local support microstructure complementary to the enzyme surface. Similarly, deactivation caused by dissociation of oligomeric proteins may be reduced by immobilization which stabilizes the active, multiunit structure.

Enzyme stabilization by immobilization may also be caused by the existence of a local environment for the immobilized enzyme which is less damaging than bulk solution conditions. Examples include support surfaces which buffer pH changes, reduce local oxygen solubility, adsorb poisons, or catalyze conversions of denaturants (for example, H_2O_2) to innocuous components. Future extensions of these concepts can be expected as immobilized enzyme structure and function are increasingly understood and approached on a fundamental scientific basis.

CONCLUDING REMARKS

We have now examined some of the current avenues of commercial application of enzyme catalysis. In addition, we have learned how enzymes can be immobilized for continuous use, and we have analyzed the kinetic features of immobilized enzyme catalysts. Next we turn our attention to natural multienzyme systems—living cells in which elaborate looping and branching networks of enzyme-catalyzed reactions occur.

In the next chapter, we examine the stoichiometric and thermodynamic features of cellular reaction networks. Then, the control systems for these reaction pathways are considered in Chap. 6, and the kinetics of cell populations are subsequently explored. Before attempting analysis of biological reactors in Chap. 9, central concepts of energy, mass, and momentum transport are reviewed in Chap. 8. Those interested primarily in enzyme reactors may wish to read next the analysis and design of biological reactors presentation in Chap. 9. Sections 9.1.2 and 9.1.4 specifically address enzyme reactor analysis, and several topics in Sec. 9.6 on multiphase reactors apply to immobilized enzyme bioreactors.

PROBLEMS

4.1 Particulate substrates A suspension of uniform-sized gelatin (polyglycine) particles is mixed with a proteolytic enzyme powder to give a final mixture of α volume fraction gelatin, and e moles of enzyme per liter of suspension.

(*a*) If the particle size is initially d_0, derive expressions for the time-varying production rate of the (assumed) sole product glycine and for the time rate of change of particle size. Assume that the rate of reaction is given by the Michaelis-Menten form for $s_0 \ll e_0$, that is,

$$\frac{\text{Rate}}{\text{Volume}} = \frac{k e_0 s_a}{e_0 + K_E}$$

where s_a is the surface area of substrate per unit volume of reactor.

(*b*) If silver particles in developed photographic film are to be recovered by enzyme digestion of the gelatinous binder, what is the shortest length of contact time between film and enzyme solution needed for total binder hydrolysis?

4.2 Autocatalysis The batch-autocatalytic activation of pepsin [Fig. 4.4] may be presumed to follow Michaelis-Menten kinetics.

(*a*) Show that the maximum rate is reached when

$$s = \frac{K_s(s_0 + e_0 - s)}{K_s + s}$$

(*b*) Show that the course of the reaction in time is

$$-t = \frac{1}{k_3}\left[\frac{K}{s_0 + e_0} \ln\left(\frac{e_0}{s_0}\frac{s}{s_0 + e_0 - s}\right) - \ln\left(1 + \frac{s_0 - s}{e_0}\right)\right]$$

(*c*) Sketch the Lineweaver-Burk or $1/v$ vs. $1/s$ behavior of such an autocatalytic system. What other form of plotting data would be more useful?

4.3 Multiple substrates In industrial enzyme reactors, more than one substrate is often present in the reaction mixture.

(*a*) Show that for a two-substrate solution, the rate of conversion of S_1 and S_2 is given by

$$v_j = \frac{v_{\max,j}}{1 + K_j/s_j(1 + s_i/K_i)} \qquad (j, i) = (1, 2), (2, 1)$$

and thus each substrate acts as a competitive inhibitor for the other.

(*b*) Derive a general form for the total reaction velocity when *m* substrates are present simultaneously.

4.4 Cellulose-hydrolysis kinetics The enzymes which degrade cellulose include a solubilizing enzyme, an enzyme producing the dimer cellobiose, and the cellobiose hydrolyzer β-glucosidase (Fig 4.1). As the earlier two enzymes are inhibited by cellobiose (taken to represent all inhibitory oligomers), a simplified reaction network can be written

$$[G_2] + E_1 \xrightarrow{k_1} E_1[G_2]$$

$$E_1 + G_2 \xrightarrow{k_3} E_1G_2$$

$$E_1[G_2] \longrightarrow E_1 + G_2$$

where $[G_2]$, G_2 are insoluble cellulose and soluble cellobiose and E_1 is indicative of the enzyme involved in the slowest step leading to cellobiose.

(*a*) In a β-glucosidase-deficient system, glucose production can be ignored. Derive appropriate forms giving the explicit dependence of g_2 upon time for competitive inhibition. Repeat for noncompetitive inhibition.

(*b*) Show graphically that a plot of g_2/t vs. $(1/t)\ln(s_0/g_2)$ should provide a means of distinguishing between uninhibited, competitive, and noncompetitive inhibited systems. Why is a plot of g_2 rather than remaining substrate useful here? (Experimentally, the kinetics is noncompetitive: J. A. Howell and J. D. Stuck, "Kinetics of Solka floc Hydrolysis by *Trichoderma viride* Cellulase," *Biotechnol. Bioeng.* **17**: 873, 1975).

4.5 Biocatalyst design for reaction limitation Acetyl-L-tyrosine-L-ethyl ester (ATEE) is hydrolyzed by immobilized α-chymotrypsin at a characteristic overall volumetric rate of 18.4 μmol ATEE$\cdot$cm$^{-3}\cdot$s^{-1}. The effective diffusion coefficient of ATEE has been estimated to be approximately 3.8×10^{-6} cm^2/s. What range of biocatalyst sphere radii is appropriate for studying the immobilized enzyme kinetic properties under reaction-limited conditions?

4.6 Mass-transfer disguise of immobilized enzyme deactivation Consider an immobilized enzyme which loses activity irreversibly as described by Eq. (3.78). You may assume that substrate conversion kinetics are approximately first order.

(*a*) Suppose that the enzyme is immobilized on the outer surface of a solid support impermeable to substrate, and that the Damköhler number at $t = 0$ is large. Plot the variation with time in the effectiveness factor and the reaction rate. If due allowance was not made for mass-transfer effects, what mistaken conclusions might by reached about the effects of immobilization on enzyme stability?

(*b*) Consider the same question for a different situation in which the enzyme is immobilized within a permeable sheet. Assume that external mass-transport resistance is relatively negligible.

4.7 Hysteresis: pH influence on immobilized papain A papain-coated pH electrode surface is observed to generate a pH at the electrode surface (internal pH) which follows one of two different curves vs. external bulk solution pH depending on the direction of pH change. (Reaction: benzoyl-arginine ethyl ester hydrolysis which produces acid). (See Fig. 1 in A. Naparstek, J. L. Romette, J. P. Kernevez, and D. Thomas, *Nature* **249:** 490, 1974.)

(*a*) Assuming $s_0 \gg K_m$ and that the enzyme layer is very thin, show that external mass transfer could account for this behavior by noting that the optimum pH for papain is 6.0.

(*b*) The kinetic constants of the soluble enzyme for this reaction at pH 6.0 and 20°C are $k_2 =$ 19.I.U (μmol BAEE/min/mg E) and $K_m = 5 \cdot 10^{-3}$ M. If the electrode enzyme loading is α mg E/cm^2 electrode surface and this enzyme is only 6 percent as active as the solution enzyme, estimate the values of the mass transfer coefficient for hydrogen ion and the pK_2 of the enzyme which gives the best fit of the data (assume $s \gg K_m$ everywhere).

(*c*) From the estimated value of pK_2 and the given parameters of part (*b*), calculate and plot the value of the effectiveness factor(s) for this situation versus external pH. These results should show some values greater than unity and some multiple solutions at a given pH. (See also J. E. Bailey and M. T. C. Chow, *Biotechnol. Bioeng.* **16:** 1345, 1974.)

4.8 Intrapellet stoichiometry The enzyme-catalyzed reaction

$$\alpha S_1 + \beta S_2 \longrightarrow \gamma P_1 + \delta P_2$$

(α, β, γ, and δ are stoichiometric coefficients) occurs inside a porous immobilized enzyme catalyst pellet.

(*a*) Using the material balances for reactants and products, express all concentrations inside the pellet in terms of the concentration of S_1 (at the same position in the particle), the concentrations at the external particle surface, and the effective diffusion coefficients of the substrates and products.

(*b*) Show that a sparingly soluble reactant will tend to limit reaction rates within the immobilized enzyme particle.

4.9 Reversible glucose isomerization in a packed column The commercially important isomerization of glucose to fructose is conducted in a packed-bed reactor column using immobilized glucose isomerase which exhibits reversible Michaelis-Menten kinetics [Eq. (3.18)].

(*a*) Show that the substitution $\bar{s} = s - s_e$ ($s_e =$ equilibrium concentration) gives a simple Michaelis-Menten equation in $\bar{s}$. What are the expressions for the apparent Michaelis and maximum velocity constants, K_{app} and v_{app}, of this equation?

(*b*) With porosity ε and average flow velocity u_z in a one-dimensional packed bed, the change in the variable $\bar{s}$ with position can be described by

$$\varepsilon u_z \left(\frac{d\bar{s}}{dz}\right) = -(1 - \varepsilon)v$$

with the single boundary condition $\bar{s} = s_0 - s_e$ at $z = 0$. Integrate this equation to obtain a reactor performance equation for the space-time τ defined as $[L(1 - \varepsilon)/\varepsilon u_z] =$ [reactor (catalyst) volume/volumetric flow rate].

(*c*) Consider a slowly deactivating enzyme in the packed-bed reactor of part (*b*) above. If deactivation proceeds such that $e(\text{active}) = e(t = 0)\exp(-k_d t)$, derive from part (*b*) above, appropriately modified, a function $f[(s_0, v_{m,\text{app}}, s(L)]$ which, plotted vs. time t, would allow evaluation of both τ and k_d.

4.10 Nonuniformly loaded immobilized enzyme catalyst Assuming first-order kinetics, compare the overall rates and the effectiveness factors when the same amount of enzyme activity is immobilized in a porous slab, or

(*a*) Distributed uniformly throughout the slab, or

(*b*) Concentrated in an outer layer near the external catalyst surface. No enzyme is present in the inner core, occupying one-half of the slab thickness. Enzyme activity is uniformly distributed in the outer layer.

4.11 Simultaneous kinetic- and mass-transfer parameter evaluations An enzyme is immobilized on the surface of a nonporous solid. Assuming that the external mass-transfer resistance for substrate is not negligible and that the Michaelis-Menten equation describes the intrinsic kinetics:

(*a*) Derive an expression which indicates the explicit form of the coefficients in a Lineweaver-Burk plot. From this result, what is the *apparent* maximal velocity v_{max}^{app} and Michaelis constant K_m^{app} in terms of the real variables v_{max}, K_m, and k_s (mass-transfer coefficient)?

(*b*) If a sufficient range of substrate concentrations is examined, show graphically (sketch) how the parameters v_{max}, K_m, and k_s can be evaluated.

4.12 Immobilized enzyme kinetics with ion partitioning The concentration dependence of a charged species (substrate, inhibitor) in a charged matrix, e.g., that of an enzyme membrane, may be expressed as

$$[C^+]_{\text{membrane}} = [C^+]_{\text{bulk soln}}\, e^{ze\psi/RT}$$

where z = ion charge of species

e = charge of one electron

ψ = electrostatic potential of membrane ($\psi_{\text{soln}} = \text{o}$)

Show that K_s and K_I have apparent values given by

$$K^{\text{app}} = K_{(\text{true})} e^{-ze\psi/RT}$$

if bulk solution concentrations are used. Making allowance for the existence of e, e^-, and e^{2-} in the membrane, show that the maximum initial rate (ignoring mass-transfer effects) is obtained by choosing an enzyme support matrix with a ψ such that

$$1 + \frac{h_0}{K_1\alpha} + \frac{K_2\alpha}{h_0} = \left(\alpha + \frac{s_0}{K_s}\right)\left(\frac{K_2}{h_0} - \frac{h_0}{K_2\alpha^2}\right)$$

where $\alpha = e^{ze\psi/RT}$ and $h_0 = [H^+]$ in bulk solution (e^- is the active enzyme form).

4.13 Microencapsulation: β-galactosidase The kinetics of lactose conversion by β-galactosidase includes a product-inhibition term:

$$v = \frac{v_m s}{K_m(1 + p/K_i) + s}$$

To allow easy recovery of β-galactosidase from milk, the enzyme solution was encapsulated in thin microcapsular cellulose nitrate membranes of diameter $d \approx 30\ \mu\text{m}$ (microcapsules).

(*a*) *If* $v_m = k_p e_0$, $k_p = 0.57\ \mu\text{mol/(mg enzyme}\cdot\text{min)}$, $K_m = 0.54\ \text{m}M$, $K_i = 1.5\ \text{m}M$, construct Lineweaver-Burk plots for p = 0, 0.5, 1.5, 5, and 10 mM ($e_0 = 75$ mg per 100 mL).

(*b*) Without evaluating the formula above, use the graph in part (*a*) to sketch the trajectory of a batch reaction ($1/v$ vs. $1/s$ for this plot) when the initial values are

s_0, mM	0.5	5.0	10.0	5.0
p_0, mM	0	0	0	5.0

(*c*) Show that even under the maximum possible rate of reaction in the microcapsule, the system in part (*a*) is reaction-rate-limited, i.e., diffusion of substrate and product is rapid compared to reaction (see D. T. Wadiak and R. G. Carbonell, Kinetic Behavior of Microencapsulated β-Galactosidase, *Biotechnol. Bioeng.* **17**: 1157, 1975).

4.14 Membrane bioreactor Enzyme is retained in a continuous-flow bioreactor using an ultrafiltration membrane. Active enzyme in the reactor deactivates following first-order kinetics with rate constant k_d. In order to maintain uniform effluent product quality, active enzyme is continuously added to the reactor in order to maintain a constant concentration e_a of active enzyme in the reactor.

A gel layer of enzyme accumulates on the membrane, adding resistance to solution flow. To compensate for this, the pressure drop across the membrane is continuously adjusted in order to maintain a constant flux J of the reaction mixture through the membrane. However, the desired flux J cannot be achieved if the total concentration of enzyme e_{tot} (active + inactive) exceeds a maximum value given by

$$e_{max} = e_s \exp(-J/k_e)$$

where e_s is the saturation concentration of enzyme and k_e is an enzyme mass transfer coefficient.

(*a*) Determine the total enzyme concentration in the reactor as a function of time.

(*b*) The reactor must be shut down and cleaned when the flux falls below J. Determine the maximum reactor operating time t_{max} as a function of J.

(*c*) Determine the value of J which maximizes the total amount of product obtained during an operating period $(0 \le t \le t_{max})$ for $e_0 = 5\ \mathrm{kg\ m^{-3}}$, $e_s = 100\ \mathrm{kg\ m^{-3}}$, $k_e = 0.017\ \mathrm{mm\ s^{-1}}$, $k_d = 0.018\ d^{-1}$.

(*d*) Redo the calculation of part 3 for $e_o = 50\ \mathrm{kg\ m^{-3}}$, all else being equal.

(*e*) How is the calculation in part (*c*) above influenced by the deactivation rate constant k_d? What other economically important parameters are affected by k_d?

REFERENCES

Most of the references given for Chap. 3 have information pertinent to this chapter. Several general sources covering many aspects of enzyme technology are available, including:

1. H. Tauber, *The Chemistry and Technology of Enzymes*, John Wiley & Sons, Inc., New York, 1949.
2. L. A. Underkofler, "Manufacture and Use of Industrial Microbial Enzymes," *Bioeng. Food Process. CEP Symp. Ser.* no. 69, **62**: 11 (1966).
3. W. T. Faith, C. E. Neubeck, and E. T. Reese, "Production and Applications of Enzymes," *Advan. Biochem. Eng.* **1**: 77 (1971).
4. K. Arima, "Microbial Enzyme Production," in M. P. Starr (ed.), *Global Impacts of Applied Microbiology*, John Wiley & Sons, Inc., New York, 1964.
5. Bernard Wolnak and Associates, "The Present and Future Technological Status of Enzymes," *U.S. Dept. Commerce Doc. PB-219636, Natl. Tech. Inf. Serv.*, December 1972.
6. K. Aunstrup, "Enzymes of Industrial Interest: Traditional Products," in D. Perlman (ed.), *Annual Reports on Fermentation Processes*, vol. 2, Academic Press, New York, 1978.
7. T. Godfrey and J. Reichelt, *Industrial Enzymology*, Macmillan Publishers Ltd., London, 1983.

The following references contain information on preparation, characterization and utilization of immobilized enzymes.

8. G. G. Guilbault, *Enzymatic Methods of Analysis*, Pergamon Press, New York, 1970.
9. O. R. Zaborsky, *Immobilized Enzymes*, CRC Press, Cleveland, Ohio, 1973.
10. M. Salmona, C. Saronio and S. Garatlin (eds.), *Insolubilized Enzymes*, Raven Press, New York, 1974.
11. A. C. Olson and C. L. Cooney, *Immobilized Enzymes in Food and Microbial Processes*, Plenum Press, New York, 1974.
12. R. A. Messing (ed.), *Immobilized Enzymes for Industrial Reactors*, Academic Press, Inc., New York, 1975.
13. K. Mosbach (ed.), "Immobilized Enzymes," *Methods in Enzymology*, vol. XLIV (S. P. Colwick and N. O. Kaplan, eds.-in-chief), Academic Press, New York, 1976.
14. I. Chibata (ed.), *Immobilized Enzymes, Research and Development*, Halsted Press, New York, 1978.

15. K. Buchholz (ed.), "Characterization of Immobilized Biocatalysts," DECHEMA Monographs No. 1724–1731, vol. 31, Verlag-Chemie, Weinheim, 1979.
16. I. Chibata, T. Tosa, T. Sato, T. Mori, and Y. Matsuo: "Preparation and Industrial Application of Immobilized Aminoacylases," p. 383 in G. Terui (ed.), "Fermentation Technology Today," *Society of Fermentation Technology*, Japan, 1972.
17. S. S. Wang and C.-K. King, "The Use of Coenzymes in Biochemical Reactors," p. 119 in *Advances in Biochemical Engineering*, vol. 12 (T. K. Ghose, A. Feichter, and N. Blakebrough, eds.), Springer-Verlag, New York, 1979.
18. A. M. Klibanov, "Stabilization of Enzymes by Immobilization," *Analyt. Biochem.* **93**: 1 (1979).

Two useful review articles on cellulose and "cellulase":

19. Y. H. Lee and L. T. Fan: "Properties and Mode of Action of Cellulase," p. 101; Y. H. Lee, L. T. Fan and L-S. Fan: "Kinetics of Hydrolyses of Insoluble Cellulose by Cellulase," p. 131 in *Advances in Biochemical Engineering*, vol. 17, A. Fiechter, (ed.), Springer-Verlag, New York, 1980.

The following sources contain extensive treatment of chemical-reaction–mass-transfer coupling. The first concentrates on biological systems, while synthetic catalysts are emphasized in the others. As indicated in the text, however, much of the latter material can be applied directly to biological reactors following minor changes in terminology and notation.

20. B. Atkinson, *Biochemical Reactors*, Pion Ltd., London, 1974.
21. P. W. Weisz, "Diffusion and Chemical Transformation: An Interdisciplinary Excursion." *Science*, **179**: 433, 1973.
22. C. N. Satterfield, *Mass Transfer in Heterogeneous Catalysis*, MIT Press, Cambridge, Mass., 1970.
23. E. E. Petersen, *Chemical Reaction Analysis*, Prentice-Hall, Inc., Englewood Cliffs, N.J., 1965.
24. R. Aris: *The Mathematical Theory of Diffusion and Reaction in Permeable Catalysts* vol. 1, *The Theory of the Steady State*, vol. 2, *Questions of Uniqueness, Stability and Transient Behavior*, Clarendon Press, Oxford, 1975.

A useful series of books starting with "Enzyme Engineering" and followed by other volumes with the same title plus a number contain papers from the biannual conferences on enzyme engineering sponsored by the Engineering Foundation. The first volume was published by Wiley-Interscience, New York, 1972, and successive volumes have been published by Plenum (New York) and in *Annals of the New York Academy of Sciences.*

CHAPTER

FIVE

METABOLIC STOICHIOMETRY AND ENERGETICS

A living cell is a complex chemical reactor in which more than 1000 independent enzyme-catalyzed reactions occur. Still, the material and energy balance restrictions and thermodynamic principles familiar from analysis of chemical process systems apply equally well to biological systems. In this chapter, we shall examine the cell as a chemical reactor, focusing on the stoichiometric rules and energy flows which characterize living organisms.

The total of all chemical reaction activities which occur in the cell is called *metabolism*. A simplified diagram of some of the major elements of metabolism in the bacterium *Escherichia coli* is shown in Fig. 5.1. We see there that metabolic reactions tend to be organized into sequences called metabolic pathways, and that there is some connection between the pathways by virtue of circular, closed pathways feeding back on themselves and because of pathway branches which connect one reaction sequence with another. All of the arrows in Fig. 5.1 denote one or more enzyme-catalyzed reactions which convert cell constituents (*metabolites*) to different compounds.

A cell produces order (itself and its offspring) from its disorderly surroundings. We now know in some detail how energy from the environment is used to drive this process. Finally, and perhaps of greatest importance in bioprocess engineering, the study of energy exchanges helps explain the major distinction between cell function in the presence and absence of oxygen. As we have already

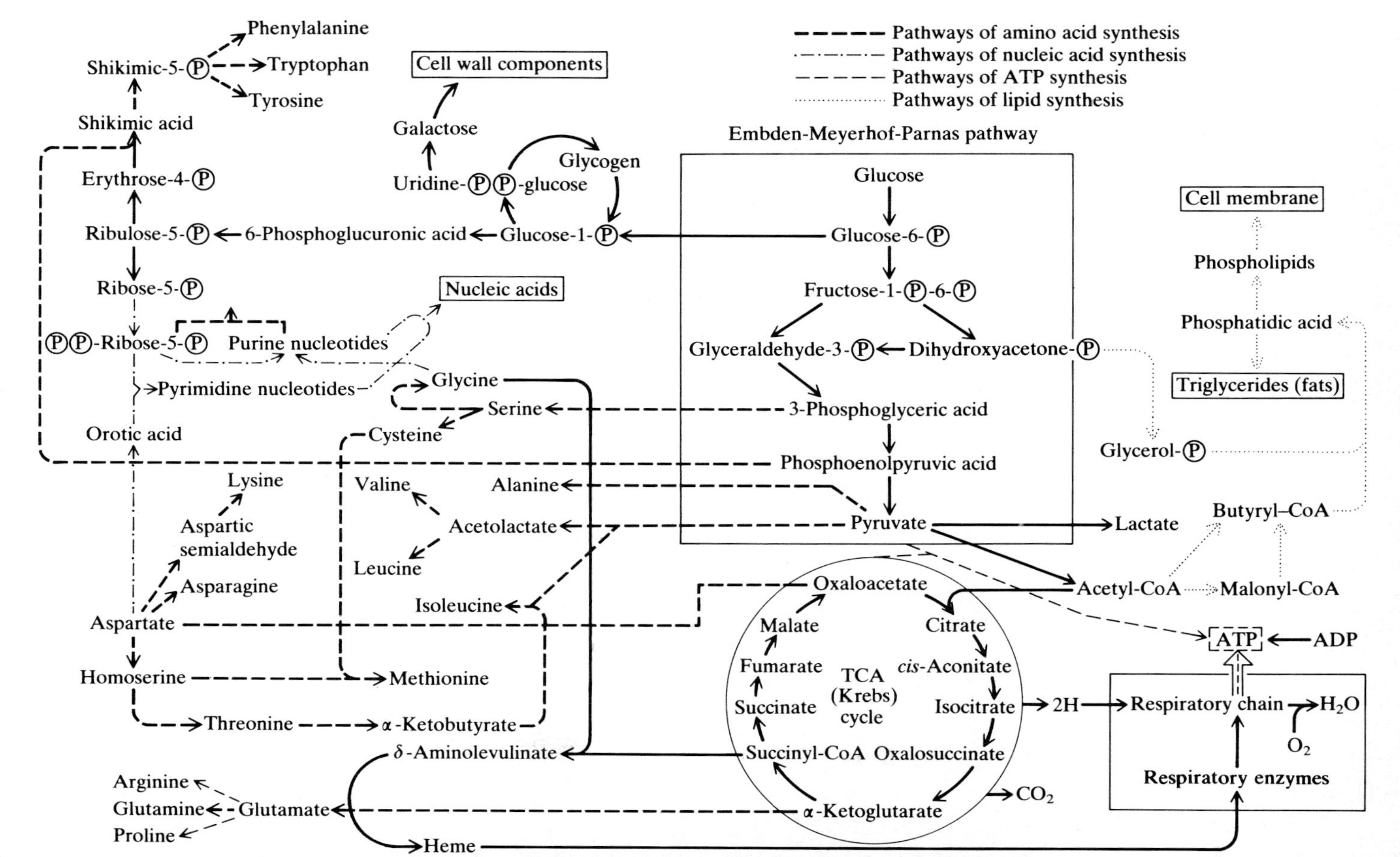

Figure 5.1 A summary of a few of the major metabolic pathways in the bacterium *E. coli*. *(Reprinted by permission from J. D. Watson, "Molecular Biology of the Gene," 2d ed., pp. 96–97, W. A. Benjamin, Inc., New York, 1970.)*

commented, these two conditions are called aerobic and anaerobic, respectively. While some cells (*obligate* or *strict anaerobes*) do not use free oxygen and other cells require it (*obligate aerobes*), a third class of cells can grow in either environment. Yeast is a familiar example of this metabolically versatile third group, known collectively as *facultative anaerobes.*

Two different kinds of energy, light and chemical, are tapped by inhabitants of the microbial world. Organisms which rely on light energy are called *phototrophs*, while *chemotrophs* extract energy by breaking down certain nutrients. Further subdivision of the chemotrophs is possible according to the nature of the energy-yielding nutrient. In particular, *lithotrophs* oxidize inorganic material, and *organotrophs* employ organic nutrients for energy production. Some examples will be mentioned shortly after we have introduced cellular nutrition.

The energy obtained from the environment is typically stored and shuttled in convenient high-energy intermediates such as ATP (Sec. 2.3.1). The cell uses this energy to perform three types of work: chemical synthesis of large or complex molecules (growth), transport of ionic and neutral substances into or out of the cell or its internal organelles, and mechanical work required for cell division and motion. All these processes are, by themselves, nonspontaneous and result in an increase of free energy of the cell. Consequently, they occur when simultaneously coupled to another process which has a negative free-energy change of greater magnitude. We shall return to this theme in Sec. 5.1.

Biosynthesis work is performed with relatively high efficiency of free-energy utilization, typically greater than about 20 percent. The transport work also involves ATP consumption in a process unique to living systems; small molecules and ions can be moved through membranes against a concentration gradient to achieve a ratio of concentrations on the two sides of the membrane as great as 10^5. Mechanical work is evident during cell division and bacterial and protozoal movement. Animal muscle activity and sperm swimming also imply ATP participation; the resulting direct conversion of chemical free energy into mechanical work without such intermediates as electricity or heated gases is also unique to life. Losses during chemical energy conversions in cells result in heat generation which must be considered when engineering processes for cell growth.

In order to grow and reproduce, cells must ingest the raw materials necessary to manufacture membranes, proteins, walls, chromosomes, and other components. From our review of the atoms and compounds of life, four major requirements are evident: carbon, nitrogen, sulfur, and phosphorus; hydrogen and oxygen may be obtained from medium components or in some instances from water. Typical sources of these elements are listed in Table 5.1, and Fig. 5.2 shows a schematic view of their utilization in an imaginary bacterium. This organism would be called a *heterotroph* because its carbon comes from an organic compound. Other microorganisms can use simpler nutrients: *autotrophs* require only CO_2 to supply their carbon needs. By combining these classifications by carbon source with those based on energy source, we can construct appropriate adjectives which are descriptive of the cell's metabolism. Table 5.2 gives this nomenclature and example organisms of each type. The fact that different cells

Table 5.1 Typical sources of elements

Element	Source
Carbon	CO_2, sugars, proteins, fats
Nitrogen	Proteins, NH_3, NO_3^-
Sulfur	Proteins, SO_4^{2-}
Phosphorus	PO_4^{3-}

employ different carbon and energy sources shows clearly that all cells do not possess the same internal chemical machinery. While the differences are important and will be discussed further, we shall concentrate in this chapter on aspects of metabolism which are common to many varieties of living cells.

In Fig. 5.2 the reactions within the cell have been subdivided into three classes: degradation of nutrients, biosynthesis of small molecules, and biosynthesis of large macromolecules. The number of different chemical reactions necessary for sustenance of cell life is of the order of 1000 or more. As we have already commented, each reaction is catalyzed by an enzyme. The enzymes serve the essential function of determining which reactions occur and their relative rates.

In order to appreciate the importance of regulating relative reaction rates, a different view of the cell's chemistry is useful. Many reacting substances within the cell (*metabolites*) can be attacked simultaneously by several different enzymes —perhaps to oxidize them, reduce them, or couple them to other molecules. Thus, the sequences of reactions occurring in the cell intersect and overlap in complex ways, as Fig. 5.1 illustrates. Notice, for example, how pyruvate, the final product of the Embden-Meyerhof (also called Embden-Meyerhof-Parnas or EMP) pathway, can be used in five different pathways. Although the reaction patterns shown in Fig. 5.1 may appear very complex, we should recognize that it

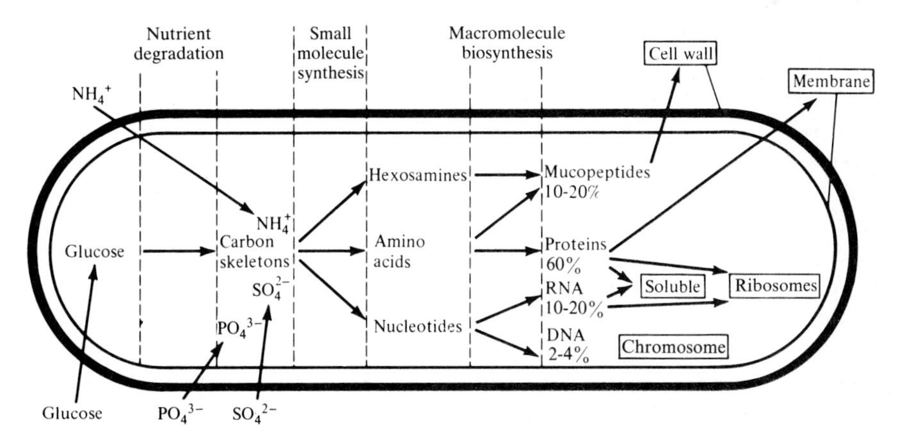

Figure 5.2 Schematic diagram showing synthesis of biological macromolecules from simple nutrients in a bacterium. *(Reprinted by permission from J. Mandelstam and K. McQuillen (eds.), "Biochemistry of Bacterial Growth," 2d ed., p. 4, Blackwell Scientific Publications, Oxford, 1973.)*

Table 5.2 Classification of organisms by carbon and energy source

	Energy source	
Carbon source	Chemical	Light
Organic compounds	Chemoheterotrophs (higher animals, protozoa, fungi, and most bacteria)	Photoheterotrophs (some bacteria, some eucaryotic algae)
CO_2	Chemoautotrophs (some bacteria)	Photoautotrophs (higher plants, eucaryotic algae, blue-green algae, and some bacteria)

is an extremely abbreviated version of the entire story: nowhere near 1000 steps are shown. Also, Fig. 5.1 does not indicate that portions of these metabolic pathways are reversible, so that they help accomplish either synthesis or degradation of a biomolecule. Later in this chapter we shall examine some of these pathways and the enzymatic machinery for controlling them in greater detail.

There is still another omission from Figs. 5.1 and 5.2. We have not shown the release of metabolic end products from the cell. Many of these compounds are substances unnecessary or useless for the cell's function; others, e.g., antibiotics and extracellular enzymes, serve a purpose. From the perspective of humankind, these end products (alcohols, organic and amino acids, antibiotics, and many others) are often valuable, making it profitable to produce them by growing cells. For technological processes, we usually seek a cellular species which is inefficient from a biological point of view in that it produces much more of some biochemical than it can use. Another objective in this chapter, therefore, is to trace the interconversions of various chemical compounds within the cell so that we shall have some idea of what we may be able to manufacture using living organisms. While we shall only scratch the surface of these topics here and in later chapters, we should be able to gain an appreciation of the tremendous complexity and variety of reaction processes which abound in the biological world.

Since the elemental composition of a particular species of microorganism or animal cell is not extremely variable, certain stoichiometric and thermodynamic constraints apply to cellular metabolic activities. The synthesis of a certain amount of cellular material implies the utilization of an amount of carbon, nitrogen, and oxygen from the cell environment simply to account for the new cell mass. Furthermore, if we know the chemical form in which these different elements are supplied in the medium, we can establish additional constraints relating the amount of substrate consumed, the amount of cell material produced, and the amounts of certain products formed. Also, since there is an energy requirement associated with synthesizing all of the necessary components required to produce additional cells, a particular amount of energy generation activity is implied. This in turn, given a particular reaction pathway for obtaining energy in the form of ATP, implies corresponding requirements with respect to the amount of chemical or light energy utilized by the cell.

In fact, material balance constraints and thermodynamic requirements give useful working relationships within a very simplified view of cellular activity, one which is entirely macroscopic and which does not utilize any information about the internal chemical workings of the cell. This macroscopic view, considered further in Sec. 5.10, examines the utilization of substrates by cells to produce more cells and certain products. Based on this perspective, we can develop equations for inferring certain biological reactor conditions from available measurements and for testing the consistency of experimental data.

In this chapter, we will build from fundamental detail to overall macroscopic views following the sequence of the preceding outline. We begin by looking in some detail at the role of ATP, the energy shuttle in the cell, and the types of reactions in which energy is transferred to and taken from this carrier. Also, the importance of oxidation-reduction reactions in the cell is considered along with the electron carrying reagents NADH and NADPH. Then, we turn to certain catabolic reaction pathways—reaction pathways involving the breakdown of certain nutrients, here typically glucose, to determine their energetic and chemical stoichiometry. Photosynthetic mechanisms are reviewed briefly. Subsequently, some of the synthetic pathways of the cell are considered from the same viewpoint. Next, we examine coupling between energy and electron generating pathways and those pathways which utilize these cellular commodities, and, finally, we examine the overall stoichiometric constraints which apply to cell growth. To conclude the chapter, the stoichiometry of product formation and its relationship to cell growth stoichiometry is considered.

5.1 THERMODYNAMIC PRINCIPLES

To get an idea of whether a certain chemical reaction in the cell will run forward or backward, we will use a number of approximations since full analysis of the entire metabolic network is not practical. First, we note that the free-energy change $\Delta G'$ of the single chemical reaction

$$\alpha A + \beta B \rightarrow \gamma C + \delta D \tag{5.1}$$

can be written in the form

$$\Delta G' = \Delta G^{\circ\prime} + RT \ln \left(\frac{c^{\gamma} d^{\delta}}{a^{\alpha} b^{\beta}} \right) \tag{5.2}$$

As before, lowercase letters a, b, ... signify molar concentrations of compounds A, B, ... In writing Eq. (5.2), we have substituted concentrations for activities since biological solutions are typically dilute. Here, primes denote evaluation in aqueous solution at pH 7. Accordingly, under these conditions, the concentrations of water and H^+ are not included in the last term of Eq. (5.2) even when H_2O and H^+ appear in reaction (5.1). The standard free energy change $\Delta G^{\circ\prime}$ thus denotes the free energy change of reaction (5.1) in neutral aqueous solution with all other reactants and products present at 1 M concentration (see Ref. 1).

In a closed system, the reaction will proceed from left to right if and only if $\Delta G'$ is negative. Accordingly, $\Delta G'$ is zero at equilibrium giving the well-known relationship

$$\Delta G^{\circ\prime} = -RT \ln K'_{\text{eq}} \tag{5.3}$$

where

$$K'_{\text{eq}} = \frac{c^{\gamma}_{\text{eq}} d^{\delta}_{\text{eq}}}{a^{\alpha}_{\text{eq}} b^{\beta}_{\text{eq}}} \tag{5.4}$$

Remember that if water or H^+ are involved in the reaction, their concentrations do not enter into the calculation of the right-hand side of Eq. (5.4); the value of K'_{eq} already includes the water and H^+ concentrations (for pH 7).

Occasionally in the following discussion, we will associate negative values of $\Delta G^{\circ\prime}$ with reactions which occur from left to right. This is clearly an approximation since a cell is not a closed system and since reactants are generally not present at 1 *M* concentration. The utility of this approximation in gaining a general understanding of bioenergetic principles is presented lucidly in Lehninger's book *Bioenergetics* [1], which is recommended additional reading.

However, we should recognize possible pitfalls of such a limited approach. In coupled-reaction networks of the kind indicated in Fig. 5.1, the direction of a major reaction path in cell metabolism is often not indicated properly by examining an isolated reaction. For example, consider the reaction between two isomers in the Embden-Meyerhof pathway for glucose breakdown

$$\begin{array}{c} \text{CHO} \\ | \\ \text{CHOH} \\ | \\ \text{CH}_2\text{O—P} \\ \text{Glyceraldehyde} \\ \text{3-phosphate} \end{array} \longrightarrow \begin{array}{c} \text{CH}_2\text{OH} \\ | \\ \text{C=O} \\ | \\ \text{CH}_2\text{O—P} \\ \text{Dihydroxyacetone-P} \\ \end{array} \qquad \Delta G^{\circ\prime} = -1830 \text{ cal/mol} \tag{5.5}$$

where P denotes phosphate. Because of the negative free-energy change, equilibrium favors the dihydroxyacetone by a 22:1 ratio. However, as Fig. 5.1 indicates, when this reaction occurs within the EMP pathway, glyceraldehyde phosphate is continually removed by reactions leading ultimately to pyruvate. As the glyceraldehyde phosphate is tapped off, reaction (5.4) is forced to proceed from right to left in an attempt to maintain equilibrium.

Many biological reactions and energy conversion processes involve oxidation-reduction reactions such as

$$A_{\text{ox}} + B_{\text{red}} \rightleftharpoons A_{\text{red}} + B_{\text{ox}} \tag{5.6}$$

The tendency for this type of reaction to occur is described frequently using the standard potential change $\Delta E^{\circ\prime}$, which in turn may be written

$$\Delta E^{\circ\prime} = E^{\circ\prime}_{(\text{A}_{\text{ox}}/\text{A}_{\text{red}})} - E^{\circ\prime}_{(\text{B}_{\text{ox}}/\text{B}_{\text{red}})} \tag{5.7}$$

where $E^{\circ\prime}_{(A_{ox}/A_{red})}$ is the standard half-cell potential for the half-reaction

$$A_{ox} + 2e^- \longrightarrow A_{red} \tag{5.8}$$

As a reference point for half-cell potential values, the hydrogen half-cell (at pH = 0) is assigned a value of zero:

$$2H^+ + 2e^- \longrightarrow H_2 \qquad E_0 = 0.00 \text{ V (pH} = 0) \tag{5.9}$$

Free energy changes and corresponding potential changes are related by

$$\Delta G' = -n\mathscr{F}\Delta E' \tag{5.10}$$

where n is the number of electrons transferred and Faraday's constant $\mathscr{F}$ is equal to 23.062 kcal/V mol. Equation (5.10) indicates that only redox reactions with positive $\Delta E'$ values proceed from left to right. If we require the value of $\Delta E'$ for nonstandard conditions, Eqs. (5.2) and (5.10) may be used.

5.2 METABOLIC REACTION COUPLING: ATP AND NAD

In Chaps 3 and 4 the kinetics of isolated enzyme systems and the industrial utilization of free enzymes and enzyme preparations have been considered. The majority of the kinetic studies mentioned in Chap. 3 and most of the current important usages cited in Chap. 4 involve hydrolytic enzymes which, by definition, split or degrade larger molecules into smaller components by consuming water as a second substrate. Such degradations proceed with a decrease in free energy of the system and are therefore spontaneous in a closed system. The present section introduces the mechanisms by which an open system, the living cell, is able to couple energy-yielding (*exergonic*) reactions with chemical reactions and other functions which do not occur appreciably unless energy is supplied (*endergonic* processes).

5.2.1 ATP and Other Phosphate Compounds

The structure and some properties of adenosine triphosphate (ATP) have been mentioned already (see Sec. 2.3.1 and Fig. 2.8). As noted there, the enzymatic hydrolysis of ATP to yield ADP and inorganic phosphate has a large negative free-energy change

$$\text{ATP} + H_2O \longrightarrow \text{ADP} + P_i \qquad \Delta G^{\circ\prime} = -7.3 \text{ kcal/mol} \tag{5.11}$$

where P_i denotes inorganic phosphate. Thus, a substantial amount of free energy may be released by the hydrolysis and, by reversing the reaction and adding phosphate to ADP, free energy can be stored for later use. Let us next examine how the latter situation can occur using coupled chemical reactions.

Another example from the Embden-Meyerhof-Parnas pathway serves to illustrate the concept of a *common chemical intermediate*. Conversion of an aldehyde in an aqueous medium to a carboxylic acid results in a free-energy decrease of about 7000 cal/mol. As summarized in Fig. 5.3, this chemical free energy would be completely dissipated in an isolated solution. This does not occur in the living cell. In biochemical glucose oxidation, when 3-phosphoglyceraldehyde is converted into a carboxylic acid (3-phosphoglycerate), ATP is simultaneously regenerated from ADP (see reaction 2 of Fig. 5.3). The free-energy decrease resulting from aldehyde oxidation is coupled by the cell enzymes to the simultaneous regeneration of ATP. Since reaction 2 results in approximately no free-energy change, the free-energy released in oxidation of 3-phosphoglyceraldehyde has been transformed into a so-called *high-energy phosphate bond* in adenosine triphosphate.

The elementary reaction sequence by which the conversions actually occur is shown at the bottom of Fig. 5.3. The central important features of these last two reactions are: (1) the appearance of a common intermediate; the same compound

1 Isolated oxidation of aldehyde to carboxylic acid (aqueous solution),

$$RCHO + H_2O \longrightarrow 2H + RCOO^- + H^+, \qquad \Delta G_1^{\circ\prime} \simeq -7 \text{ kcal/mol}$$

2 Same reactions, coupled to ATP generation (glucose oxidation),

$$RCHO + HPO_4^{2-} + ADP^{3-} \longrightarrow 2H + RCOO^- + ATP^{4-},$$

$$\Delta G_2^{\circ\prime} \simeq 0 \text{ kcal/mol}$$

3 Evidently, *2* − *1* yields,

$$ADP^{3-} + HPO_4^{2-} + H^+ \longrightarrow ATP^{4-} + H_2O, \qquad \Delta G_3^{\circ\prime} \simeq +7 \text{ kcal/mol}$$

The elementary reactions occurring in *2* are,

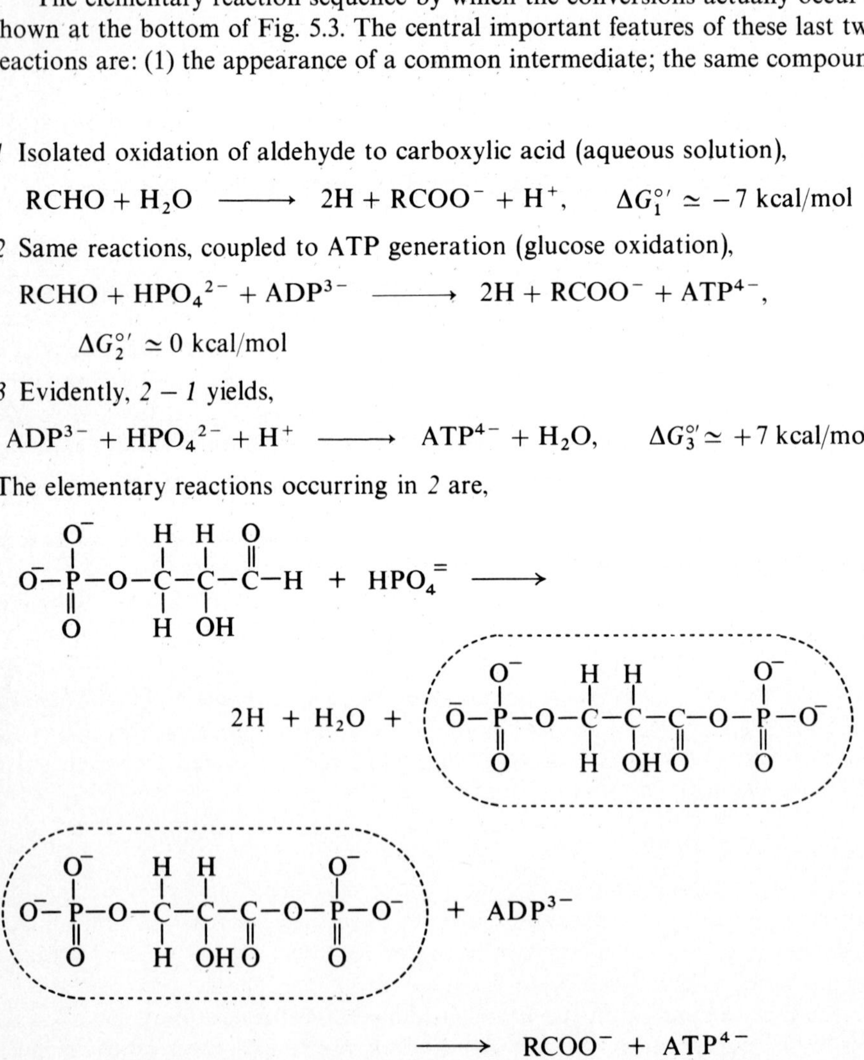

```
 O⁻     H  H  O
 |      |  |  ‖
⁻O—P—O—C—C—C—H  +  HPO₄⁼  ⟶
    ‖      |  |
    O      H  OH

                          O⁻      H  H          O⁻
                          |       |  |          |
        2H + H₂O  +     ⁻O—P—O—C—C—C—O—P—O⁻
                             ‖       |  |  ‖       ‖
                             O       H  OH O       O

 O⁻     H  H          O⁻
 |      |  |          |
⁻O—P—O—C—C—C—O—P—O⁻   +  ADP³⁻
    ‖      |  |  ‖      ‖
    O      H  OH O      O

                          ⟶  RCOO⁻  +  ATP⁴⁻
```

Figure 5.3 Example of reactions coupled via a common intermediate.

is a product of the first reaction and a reactant in the second and (2) the phosphorylated intermediate formed and consumed has a larger free energy of hydrolysis (with phosphate removal) than ATP. The equilibrium of the final reaction lies to the right, or product, side: thus this part of glucose metabolism is one of several points at which the cell regenerates the ATP needed for endergonic processes. This regeneration is accomplished by the conversion of a partially metabolized nutrient into a high-energy phosphorylated intermediate, which then donates a phosphate to ADP via an enzyme-catalyzed reaction.

The phosphorylation of various compounds, including ADP, serves several functions. It provides a useful means of storing considerable fractions of the free energy of fuel oxidation. Free energies of hydrolysis of several compounds called phosphate donors are greater than $\Delta G^{\circ\prime}$ for ATP hydrolysis (for example, phosphoenolpyruvate: $\Delta G^{\circ\prime} = -14.8 \text{ kcal mol}^{-1}$; 1,3-diphosphoglycerate: $\Delta G^{\circ\prime} = -11.8 \text{ kcal mol}^{-1}$). Consequently, hydrolysis of these compounds can be used to drive ADP phosphorylation. Similarly, ATP hydrolysis serves to phosphorylate "low-energy" phosphate compounds. The latter group has $\Delta G^{\circ\prime}$ values below that for ATP hydrolysis (for example, glucose-6-phosphate: $\Delta G^{\circ\prime} = -3.3 \text{ kcal mol}^{-1}$; glycerol-1-phosphate: $\Delta G^{\circ\prime} = -2.20 \text{ kcal mol}^{-1}$).

Yet another function is served by phosphorylation. Highly ionized organic substances are virtually unable to permeate the cell's plasma membranes. The charged phosphorylated compounds which serve as metabolic intermediates may therefore be contained within the cell. In this manner, the maximum amounts of energy and chemical raw materials can be extracted from a nutrient. Typically, the last reaction in production of a metabolic end product is a dephosphorylation. The uncharged waste material can then escape from the cell's interior.

5.2.2 Oxidation and Reduction: Coupling via NAD

We have just reviewed the role of ATP as a shuttle for phosphate groups bound with rather high energies. In this section we shall consider briefly how oxidation and reduction reactions are conducted biologically and introduce the connection between these mechanisms and ATP metabolism.

To begin, we should recall that oxidation of a compound means that it loses electrons and that addition of electrons is reduction of a compound. When an organic compound is oxidized biochemically, it usually loses electrons in the form of hydrogen atoms: consequently, oxidation is synonymous with *dehydrogenation*. Similarly, *hydrogenation* is the usual way of adding electrons, or reducing a compound, e.g., the reduction of pyruvic acid and oxidation of lactic acid

$$\underset{\text{Pyruvic acid}}{\begin{matrix} CH_3 \\ | \\ C{=}O \\ | \\ COOH \end{matrix}} \quad \underset{\text{of lactic acid)}}{\overset{\text{+2H (reduction of pyruvic acid)}}{\underset{-2\text{H (oxidation}}{\rightleftharpoons}}} \quad \underset{\text{Lactic acid}}{\begin{matrix} CH_3 \\ | \\ CHOH \\ | \\ COOH \end{matrix}}$$

(Pyruvic acid is the same compound as pyruvate in Fig. 5.1. Pyruvate refers to the ionized form CH_3COCOO^-, which predominates at biological pH.)

Pairs of hydrogen atoms freed during oxidations or required in reductions are carried by nucleotide derivatives, especially nicotinamide adenine dinucleotide (NAD) (see Fig. 2.9) and its phosphorylated form NADP. These compounds were classified as coenzymes earlier since they usually must be present when an oxidation or reduction is conducted. When hydrogen atoms are needed, for example, the nicotinamide group of reduced NAD can contribute them by undergoing the oxidation

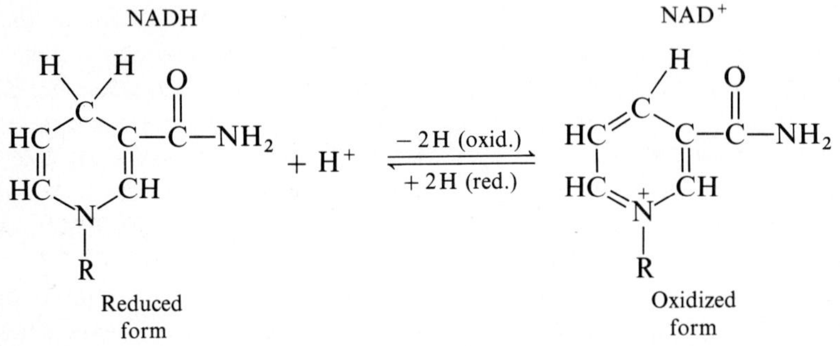

NADH (Reduced form): ring with $\mathrm{C(H)(H)}$, $\mathrm{HC{=}HC}$, $\mathrm{N{-}R}$, CH, $\mathrm{C{-}C({=}O){-}NH_2}$

$$\text{NADH} + \mathrm{H^+} \underset{+2\mathrm{H\ (red.)}}{\overset{-2\mathrm{H\ (oxid.)}}{\rightleftharpoons}} \text{NAD}^+$$

NAD$^+$ (Oxidized form): ring with $\mathrm{C{-}H}$, HC, HC, $\mathrm{\overset{+}{N}{-}R}$, CH, $\mathrm{C{-}C({=}O){-}NH_2}$

This oxidation is readily reversible, so that NAD can also accept electrons (H atoms) when they are made available by oxidation of other compounds. As indicated above, we shall denote the oxidized and reduced forms of NAD as NAD$^+$ and NADH, respectively.

In its role as electron shuttle, NAD serves two major functions. The first is analogous to one of ATP's jobs: reducing power (=electrons $\approx$ H atoms) made available during breakdown of nutrients is carried to biosynthetic reactions. Such a transfer of reducing power is often necessary because the oxidation state of the nutrients to be used for construction of cell components is different from the oxidation state of biosynthesis products. As observed already, the oxidation state of carbon within the cell is approximately the same as carbohydrate (CH_2O). Thus, autotrophic organisms, which employ CO_2 as their carbon source, use considerable reducing power when assimilating carbon

$$CO_2 + 4H \longrightarrow (CH_2O) + H_2O$$

While carbon dominates reducing-power requirements, nitrogen and sulfur assimilation often also demands some of the cell's carrier-bound hydrogen. To estimate these needs, we may assume that cell material contains nitrogen at the oxidation level of ammonia (NH_3), while sulfur's oxidation state is approximately that of sulfide (S^{2-}). For example, use of sulfate as a sulfur source requires considerable reducing power, as suggested by

$$SO_4^{2-} + 8H \longrightarrow S^{2-} + 4H_2O$$

NAD and related pyridine nucleotide compounds carrying hydrogen also participate in ATP formation in aerobic metabolism. As we shall investigate in greater detail in Sec. 5.3, the hydrogen atoms in NADH are combined with oxygen in a cascade of reactions known as the *respiratory chain*. The energy released in this oxidation is sufficient to form three molecules of ATP from ADP.

It is an intriguing fact that all biological systems known, whether anaerobic, aerobic, or photosynthetic, utilize ATP as a central means of accumulating oxidative or radiant energy in a form convenient for driving the endergonic processes of the cell. The remainder of the chapter will follow ATP utilization and electron transfer through progressively more complex pathways of anaerobic, aerobic, and photosynthetic systems. Considerations of oxidation-reduction balances will allow us to derive stoichiometric constraints based on nutrient and cell composition, metabolic pathways, and end products. Since the free energy of fuel oxidation is eventually stored in ATP, the efficiency of ATP utilization in different cell processes gives a useful reflection of relative energy demands by the various cell functions. Further, study of ATP logically allows accounting for both the progress of carbon skeletons and the utilization of the free energy released during their formation. Finally, comparison of the energetics of these different biological systems gives insights into the origin of the relatively high thermodynamic efficiency of these chemical engines.

5.3 CARBON CATABOLISM

Breakdown of nutrients to obtain energy is called *catabolism*. Carbohydrates are by far the most important class of carbonaceous nutrients for fermentations, although some microbial species can also utilize amino acids, hydrocarbons and other compounds. As an illustration of the diversity of the microbial world, we note that almost any carbohydrate or related compound is fermented by some microorganism. Most microorganisms which can employ carbohydrate in a fermentation are capable of fermenting the simple sugar glucose. In the following section we shall examine the major processes for glucose breakdown. There are at least seven different glucose fermentation pathways, and the particular ones used and the end products produced depend on the microorganism involved. Our main concerns here will be observation of examples of energetic and electron shuttling discussed above, overall stoichiometry of the pathway, and identification of intermediates important for biosynthetic pathways.

5.3.1 Embden-Meyerhof-Parnas Pathway

The Embden-Meyerhof-Parnas (EMP) pathway already seen in Fig. 5.1 involves ten enzyme-catalyzed steps which start with glucose and end with pyruvate (Fig. 5.4). In this section we examine this best-known carbon catabolism pathway with emphasis on energetics and reducing power transfer.

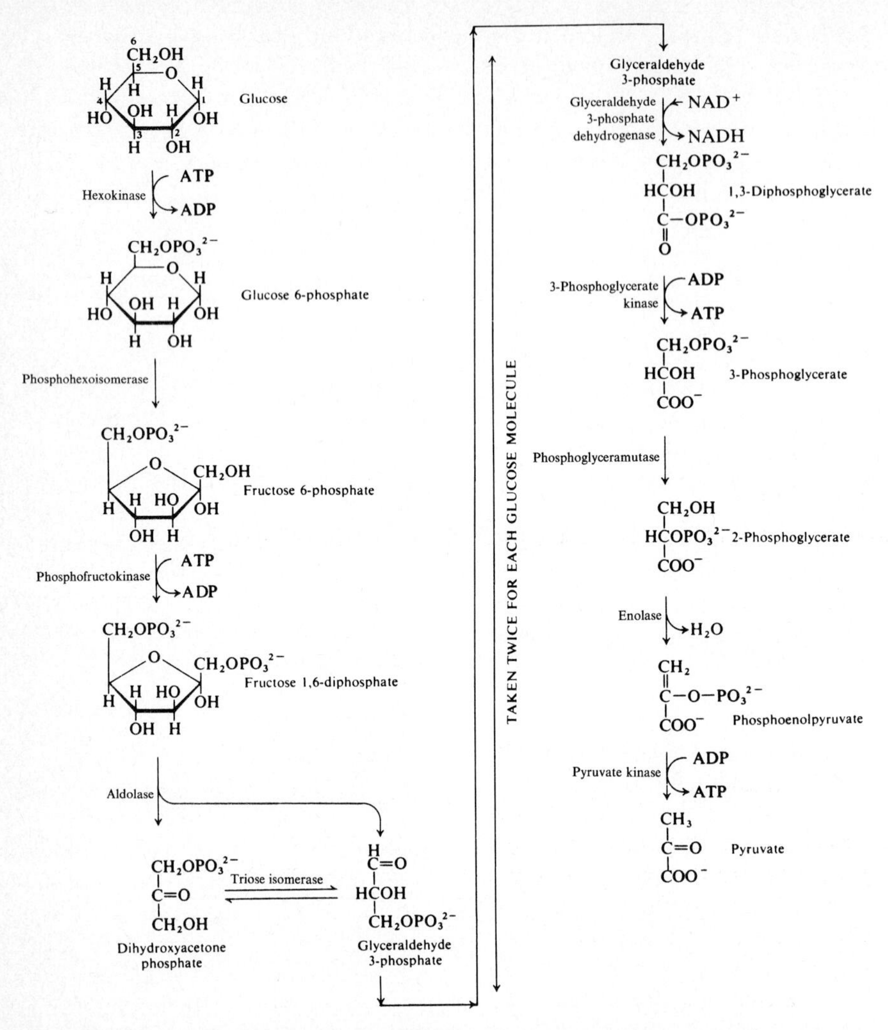

Figure 5.4 The Embden-Meyerhof-Parnas (EMP) pathway. Notice that each six-carbon glucose substrate molecule yields two three-carbon intermediates, each of which passes through the reaction sequence on the right-hand side.

Examination of the EMP pathway reaction sequence in Fig. 5.4 shows that each step is quite simple and involves isomerization, ring splitting, or transfer of a small group such as hydrogen or phosphate. Two moles of pyruvate are produced per mole of glucose passing through the pathway. ATP hydrolysis is coupled with two reactions which would not occur spontaneously otherwise, and two reactions involve sufficiently negative free energies to drive ADP phosphorylation. Because the latter two reactions occur twice for each mole of glucose processed, the overall effect is phosphorylation of ATP. As indicated in Fig. 5.4, dehydrogenation of glyceraldehyde 3-phosphate is coupled with reduction of

NAD^+, and this reaction occurs twice per mole of glucose. Thus, the overall stoichiometry of the EMP pathway is

$$C_6H_{12}O_6 + 2P_i + 2ADP + 2NAD^+ \longrightarrow 2C_3H_4O_3 + 2ATP + 2(NADH + H^+) \qquad (5.12)$$

Stored chemical energy and reducing power result from the overall pathway. Energy storage accomplished by this or other substrate rearrangement pathways is called *substrate-level phosphorylation.*

Another central function of the EMP pathway is provision of carbon skeletons for starting materials in cellular biosynthetic reactions. Not only is pyruvate available, but also intermediates in the pathway are used as biosynthesis substrates (see Fig. 5.1). Consequently, the EMP pathway is one of several which are called *amphibolic* because they serve as carbon skeleton as well as energy sources.

In muscle cells and lactic acid bacteria, among others, the reactions of the EMP pathway are followed by the single step

$$C_3H_4O_3 + NADH + H^+ \longrightarrow C_3H_6O_3 + NAD^+$$

This overall reaction sequence from glucose to lactic acid is called *glycolysis.* It is interesting to compare the free energy change for glycolysis

$$\text{Glucose} + 2P_i + 2ADP \longrightarrow 2 \text{ lactate} + 2ATP + 2H_2O \qquad \Delta G^{\circ\prime} = -32{,}400 \text{ cal/mol} \qquad (5.13)$$

with the corresponding quantity for glucose breakdown alone

$$\text{Glucose} \longrightarrow 2 \text{ lactic acid} \qquad \Delta G^{\circ\prime} = -47{,}000 \text{ cal/mol} \qquad (5.14)$$

A free-energy total of 14.6 kcal, or 7.3 kcal for each mole of ATP generated, has been conserved by the pathway as high-energy phosphate compounds. The apparent efficiency of free-energy transfer is thus about $14/47 \times 100 \approx 31$ percent. Correction of standard-free-energy data for the concentrations and pH prevailing *in vivo* (see Table 5.3) suggests that this estimate is quite low and that the true efficiency is about 53 percent. The reason for this high efficiency is clear after correcting the $\Delta G^{\circ\prime}$ values in Table 5.3 for existing concentrations (see Prob. 5.4): the results indicate that $\Delta G^{\circ\prime}$ is near zero for all of the reaction steps except three (glucose to G6P, F6P to FDP, and PEP to Pyr)(abbreviations defined in Table 5.3). Thus, most reactions in the sequence are near equilibrium and hence readily reversible. The significance of the non-equilibrium steps emerges from examination of enzymatic regulation of the EMP pathway (Sec. 5.7.2).

5.3.2 Other Carbohydrate Catabolic Pathways

We begin with two additional reaction sequences for glucose catabolism. The *pentose phosphate* cycle or pathway (also called the hexose monophosphate pathway or shunt) begins by oxidizing glucose phosphate.

$$\text{Glucose 6-phosphate} + NADP^+ \longrightarrow \text{6-phosphogluconate} + NADPH + H^+$$

Table 5.3 Concentrations of glycolytic intermediates in the human erythrocyte and corresponding standard free energy changes. Note (*a*) concentrations differ from the 1*M* value used in standard free energy calculations and (*b*) $\Delta G^{\circ\prime}$ values are for the glycolytic reaction which consumes the indicated intermediate.[†]

Intermediate	Concentration, μM	Standard free energy change $\Delta G^{\circ\prime}$ (kcal/mol)
Glucose	5000	−4.0
Glucose 6-phosphate(G6P)	83	+0.4
Fructose 6-phosphate(F6P)	14	−3.40
Fructose 1,6-diphosphate(FDP)	31	+5.73
Dihydroxyacetone phosphate(DHP)	138	+1.83
Glyceraldehyde 3-phosphate(GAP)	18.5	−3.0
3-Phosphoglycerate(3PG)	118	+1.06
2-Phosphoglycerate(2PG)	29.5	+0.44
Phosphoenolpyruvate(PEP)	23	−7.5
Pyruvate(Pyr)	51	−6.0
Lactate(Lact)	2900	
ATP	1850	
ADP	138	
Phosphate	1000	

[†] Adapted from A. L. Lehninger, *Biochemistry*. 2d ed., table 16.1, Worth Publishers, Inc., New York, 1975.

A major function of the pentose phosphate pathway is supplying the cell with NADPH which in turn carries electrons to biosynthetic reactions. The total reaction scheme is quite complicated (see p. 456 of Ref. 1), but its overall result is indicated by the following summary of pentose phosphate pathway stoichiometry:

$$\text{Glucose} + 12\text{NADP}^+ + 7\text{H}_2\text{O} + \text{ATP} \longrightarrow 6\text{CO}_2 + \text{P}_i + 12(\text{NADPH} + \text{H}^+) + \text{ADP} \quad (5.15)$$

The net effect is complete oxidation of one mole of glucose 6-phosphate liberating CO_2 and transferring all of the electrons (H) to NADP.

This overall reaction consumes ATP. In order to provide ATP to the cell, an incomplete pentose phosphate cycle may occur with overall stoichiometry

$$3\text{ glucose} + 6\text{NADP}^+ + \text{ATP} \longrightarrow 2\text{ fructose 6-phosphate} + \text{glyceraldehyde 3-phosphate} + 3\text{CO}_2 + 6(\text{NADPH} + \text{H}^+) + \text{ADP} + \text{P}_i \quad (5.16)$$

If next fructose 6-phosphate and glyceralde 3-phosphate follow the balance of the EMP reaction sequence to pyruvate, ADP phosphorylation occurs six times. The net energy yield is therefore 5/3 ATPs for each mole of glucose, still less than the energy yield of the EMP pathway.

However, the pentose phosphate pathway is advantageous in the sense of including ribose 5-phosphate and erythrose 4-phosphate, important precursors for purine and pyrimidine biosynthesis, among its intermediates. These components are absent from the EMP pathway. As a consequence, microorganisms such as lactobacilli which possess only the EMP pathway require a complex medium including pentoses, purines and pyrimidines when grown in anaerobic environments. *E. coli*, on the other hand, is able to grow anaerobically in simpler media by using the EMP and pentose phosphate pathways simultaneously (for example, in the ratio of 75% EMP: 25% pentose phosphate under certain culture conditions). This illustrates the important general concept that cells may possess multiple catabolic pathways for a given nutrient and may employ more than one pathway simultaneously, presumably to optimize growth considering both energy and precursor requirements.

The final glucose catabolic pathway which we will examine is the *Entner-Doudoroff* (ED) pathway. The overall stoichiometry of this reaction sequence is

$$\text{Glucose} + \text{ATP} + \text{NADP}^+ \longrightarrow \text{glyceralde 3-phosphate} + \text{pyruvic acid} + \text{ADP} + \text{NADPH} + \text{H}^+ \quad (5.17)$$

Noting that two moles of ADP can be phosphorylated by reacting one mole of glyceraldehyde 3-phosphate to pyruvate through the same reactions as used for this conversion in the EMP pathway, the energy yield of the ED pathway is relatively poor: one mole ATP per mole of glucose processed. Before turning to catabolic pathways for other nutrients, we should note that the above discussion of catabolism, while covering the major pathways, is far from a complete treatment of all possibilities for bacteria.

We mentioned earlier that microorganisms can utilize many different compounds as carbon source. While details must be left to the references [3], a few examples will serve to illustrate some of the central catabolic strategies.

Utilization of sugars other than glucose often involves conversion in a few reactions either into glucose or into one of the intermediates in a glucose catabolic pathway. This approach is illustrated in Fig. 5.5 which shows the sequence of processing reactions used by *E. coli* to begin assimilation of the disaccharide lactose. The galactose utilization pathway is a subset of this reaction scheme. Notice the cyclic, and hence catalytic, role of uridine diphosphate (UDP). As a second example, we note that in several bacteria fructose is converted by two enzyme-catalyzed reactions into fructose 1,6-diphosphate which may then enter the EMP reaction sequence. Bacterial utilization of pentoses typically involves a series of reactions giving the intermediate xylulose 5-phosphate which then can enter the pentose phosphate pathway or be broken down in the phosphoketolase pathway (Fig. 5.6).

Microbial utilization and conversion of hydrocarbons and other organic compounds is a fascinating and diverse topic. Here we simply note that enzyme systems exist in the microbial world for oxygenating hydrocarbons with or without changes in the carbon skeleton and for converting hydrocarbon substrates into amino acids, lipids, vitamins and other microbial components and products.

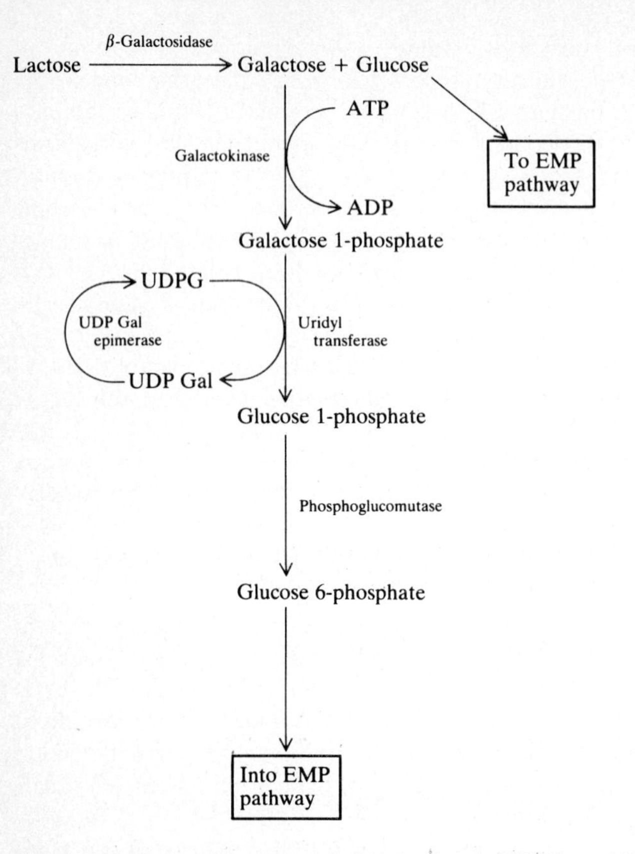

Figure 5.5 Pathway of lactose catabolism in *E. coli* (UDPG = uridine diphosphate glucose; UDP Gal = uridine diphosphate galactose).

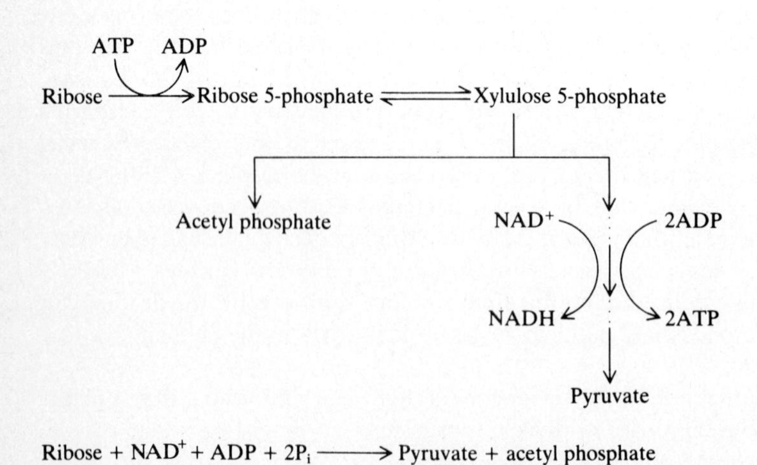

$$\text{Ribose} + \text{NAD}^+ + \text{ADP} + 2\text{P}_i \longrightarrow \text{Pyruvate} + \text{acetyl phosphate} + \text{ATP} + \text{NADH} + \text{H}^+$$

Figure 5.6 Simplified diagram of ribose catabolism via the phosphoketolase pathway.

Furthermore, certain microbes can degrade toxic compounds such as phenols and polychlorinated biphenyls. Obviously, such organisms and the enzymes they synthesize have exciting process potential for specific chemical catalysis and for environmental protection. An introduction to the enzymes and pathways encountered in hydrocarbon metabolism is provided in Doelle's excellent book [3].

5.4 RESPIRATION

Respiration is an energy-producing process in which organic or reduced inorganic compounds are oxidized by inorganic compounds. As Table 5.4 indicates, various bacteria conduct respiration using several different reductants and oxidants. When an oxidant other than oxygen is involved, the process is called *anaerobic respiration*, the term *aerobic respiration* being reserved for the situation typical of eucaryotes and many bacteria where O_2 is the oxidant. Recall that lithotrophs are organisms employing inorganic reductants. Several lithotrophs are evident in Table 5.4. So far as is known today, all lithotrophs are also autotrophs; they obtain carbon from CO_2.

In the most common forms of respiration, an organic compound is oxidized using oxygen. We consider only this case in the remainder of this section, and the term respiration will be used to describe this process. It is convenient to decompose the overall process of respiration into two phases. In the first, organic compounds are oxidized to CO_2, and pairs of hydrogen atoms (electrons) are transferred to NAD. Next, the hydrogen atoms are passed through a sequence of reactions, during which ATP is regenerated from ADP. At the end of their journey, the hydrogen atoms are combined with oxygen to give water. These two phases of biological oxidation will now be examined in greater detail.

Table 5.4 Reductants and oxidants in bacterial respirations†

Reductant	Oxidant	Products	Organism
H_2	O_2	H_2O	Hydrogen bacteria
H_2	SO_4^{2-}	$H_2O + S^{2-}$	*Desulfovibrio*
Organic compounds	O_2	$CO_2 + H_2O$	Many bacteria, all plants and animals
NH_3	O_2	$NO_2^- + H_2O$	Nitrifying bacteria
NO_2^-	O_2	$NO_3^- + H_2O$	Nitrifying bacteria
Organic compounds	NO_3^-	$N_2 + CO_2$	Denitrifying bacteria
Fe^{2+}	O_2	Fe^{3+}	*Ferrobacillus* (iron bacteria)
S^{2-}	O_2	$SO_4^{2-} + H_2O$	*Thiobacillus* (sulfur bacteria)

† From W. R. Sistrom, *Microbial Life*, 2d ed., table 4-2, p. 53, Holt, Rinehart, and Winston, Inc., New York, 1969.

5.4.1 The TCA Cycle

For the moment let us continue the story of carbohydrate metabolism started in Sec. 5.3. All the pathways to pyruvate described there can also operate during respiration. The reactions peculiar to respiration start with pyruvate. In respiration metabolism, pyruvate is not reduced to some end product using the hydrogen atoms obtained during glucose breakdown. Instead, this reducing power is saved for other uses, to be discussed shortly. Moreover, additional reducing power is generated from pyruvate by converting it to an acetic acid derivative (acetyl CoA)

$$CH_3COCOOH + NAD^+ + CoA{-}SH \xrightarrow{\text{pyruvic dehydrogenase complex}} CH_3CO{-}S{-}CoA + CO_2 + NADH + H^+ \quad (5.18)$$

Acetyl CoA† is also a key intermediate in the catabolism of amino acids and fatty acids. Consequently, three classes of biomolecules can be oxidized through acetyl CoA.

The first phase of this oxidation is carried out in a cyclic reaction sequence called the *tricarboxylic acid* or *TCA cycle* (also, Krebs cycle and citric acid cycle; Fig. 5.7). Notice that the two remaining carbon atoms from pyruvate (one was lost as CO_2 in the reaction to acetyl CoA) enter the cycle to create a six-carbon acid from one with four carbons. In one pass around the TCA cycle, however, two-carbon atoms are expelled as CO_2. Thus, this first phase of respiration consumes all the carbon atoms from the original pyruvate substrate.

The overall stoichiometry of the TCA cycle plus reaction (5.12) is

$$C_3H_4O_3 + ADP + P_i + 2H_2O + FAD + 4NAD^+ \longrightarrow 3CO_2 + ATP + FADH_2 + 4(NADH + H^+) \quad (5.19)$$

The TCA cycle as shown in Fig. 5.7 appears to serve a strictly catalytic function: no input of carbon other than the substrate is indicated. As Fig. 5.1 shows, however, the TCA cycle serves a very important function as a pool of precursors for biosynthetic reactions. Consequently, some intermediates in the cycle are constantly being tapped off and must be replaced. This is accomplished by synthesis of oxaloacetic acid from pyruvate or another three-carbon acid. Alternatively, in some microorganisms TCA cycle intermediates are replenished by the *glyoxylate cycle* which has the net effect of producing one molecule of succinate by condensation of two acetate molecules.

5.4.2 The Respiratory Chain

Leaving carbon metabolism, let us next follow the reactions in which hydrogen atoms are oxidized to water, the process from which aerobic cells derive most of

† For coenzyme A see Fig. 2.9.

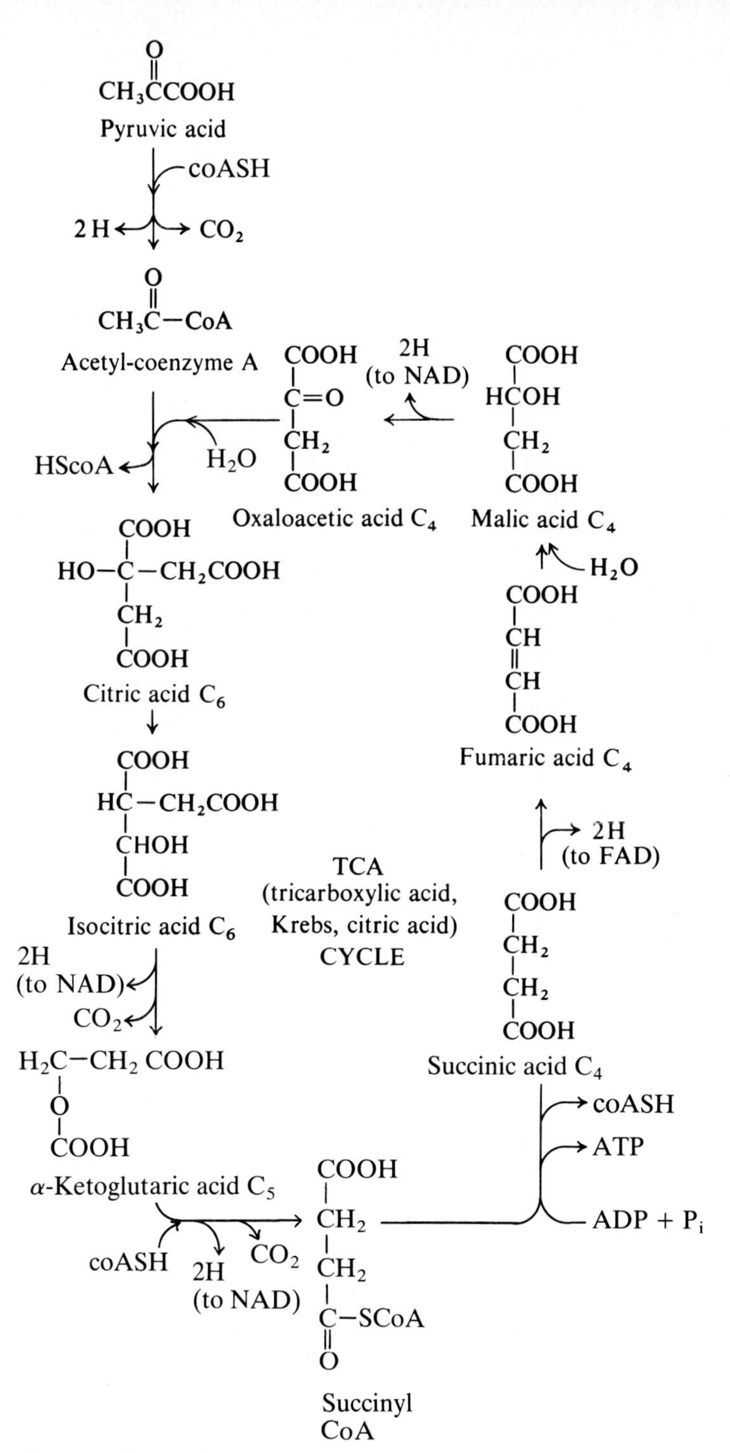

Figure 5.7 The tricarboxylic acid cycle.

their energy. In each pass around the TCA cycle, four pairs of hydrogen atoms are liberated. Three pairs are transferred to NAD and, as outlined in Fig. 5.7 the hydrogen atoms resulting from succinic acid dehydrogenation are transferred to flavin adenine dinucleotide (FAD) (see Fig. 2.9). Some of the reducing power derived from the TCA cycle may be needed in biosynthesis reactions; the rest is used to generate ATP.

In the following discussion of ATP generation during respiration, we shall concentrate on the case in which all the hydrogen atoms obtained during glucose breakdown are available for reaction in the *respiratory chain.* This is the situation which provides the most stored energy for the cell in the form of ATP. Figure 5.8 shows an abbreviated diagram of this reaction sequence and how it ties with the ultimate breakdown of pyruvate via acetyl CoA and the TCA (Krebs) cycle. In this diagram, FP_1 and FP_2 denote two different flavoproteins, which are enzymes containing FAD for transport of electrons. Electrons from NADH $(=FP_1)$ are passed to coenzyme Q (designated as Q in the figure), and in this process one molecule of ADP is regenerated to ATP for each pair of electrons passed. The electrons obtained from succinate dehydrogenation in the TCA cycle are carried by FAD in FP_2 directly to coenzyme Q.

From there, all electrons enjoy the common fate of passing through a sequence of *cytochromes*, proteins containing heme groups which are designated b, c, a and a_3 in Fig. 5.8. Along the way, ATP is generated twice for each electron pair. The process of ATP regeneration in the respiratory chain is called *oxidative phosphorylation.* Ultimately, the hydrogen atoms are combined with dissolved oxygen to yield water as the second final product of the oxidation. If we examine the free-energy changes along the respiratory chain, we find that ATP is regenerated at each point where there is a sufficiently large decrease in electron free energy to more than offset the 7.3 kcal/mol needed to phosphorylate ADP.

The net reactions, then, which describe the respiratory chain are

$$\text{NADH} + \text{H}^+ + \tfrac{1}{2}\text{O}_2 + 3\,\text{ADP} + 3\,\text{P}_i \longrightarrow \text{NAD}^+ + 4\,\text{H}_2\text{O} + 3\,\text{ATP} \tag{5.20}$$

and

$$\text{FADH}_2 + \tfrac{1}{2}\text{O}_2 + 2\,\text{ADP} + 2\,\text{P}_i \longrightarrow \text{FAD} + 3\,\text{H}_2\text{O} + 2\,\text{ATP} \tag{5.21}$$

Respiration potentially makes available much more energy for use by the cell than glycolysis since $\Delta G^{\circ\prime}$ is -686 kcal/mol for the reaction

$$\text{Glucose} + 6\,\text{O}_2 \longrightarrow 6\,\text{CO}_2 + 6\,\text{H}_2\text{O} \tag{5.22}$$

Let us examine how efficiently respiring living systems tap this large source of energy. For comparison with the above reaction, we presume here that glucose is completely oxidized to CO_2 and water via the EMP, TCA, and respiratory chain pathways. We should remember, however, that the EMP pathway and TCA cycle are both amphibolic pathways and that some intermediates are constantly being drawn off in biosynthetic side streams so that glucose substrate is not always

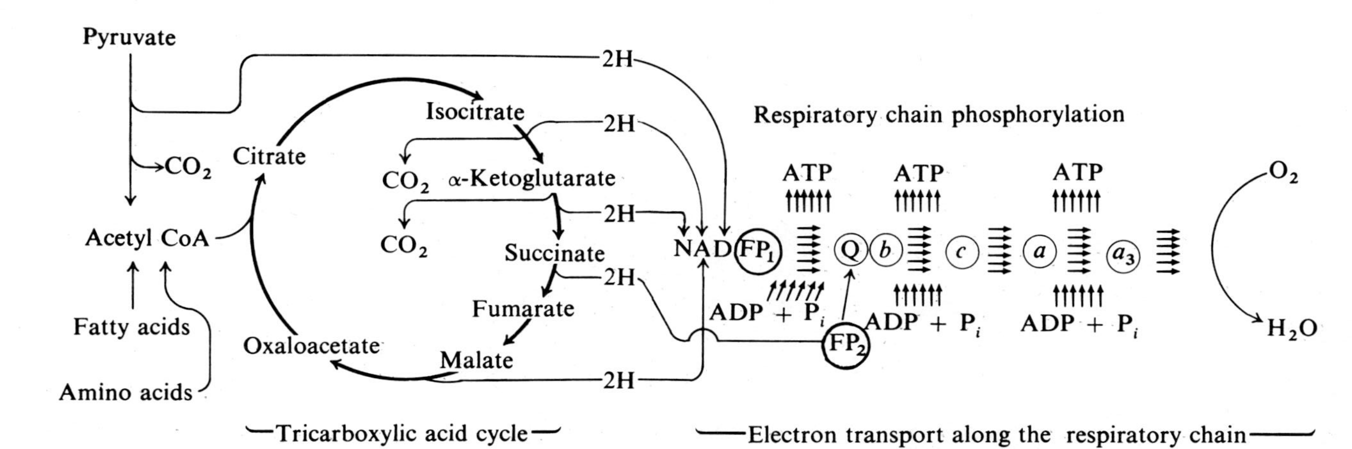

Figure 5.8 High-energy electrons released in oxidation of carbohydrates, fatty acids, and amino acids drive ADP phosphorylation as they move through the respiratory chain: FP_1(NADH), FP_2 (succinate dehydrogenase), Q (coenzyme Q), *b*, *c*, *a* and a_3 (the cytochromes). *(Adapted from A. Lehninger, "Bioenergetics," 2d ed., p. 74, W. A. Benjamin, Inc., Palo Alto, CA, 1974.)*

processed as we now assume. The following calculation gives an upper bound on the ATP yield from glucose in a respiring cell.

Adding Eq. (5.12) (EMP reactions) with two times Eq. (5.19) (TCA cycle) with ten times Eq. (5.20) [oxidation of all $NADH + H^+$ generated in the EMP (two) and TCA reactions (four)] and twice Eq. (5.21) [$FADH_2$ oxidation] gives

$$C_6H_{12}O_6 + 38\,ADP + 38\,P_i \longrightarrow 6CO_2 + 38\,ATP + 44\,H_2O \quad (5.23)$$

Since ATP hydrolysis has a standard free-energy change of -7.3 kcal/mol, the free-energy change of reaction (5.23) is approximately

$$\Delta G^{\circ\prime} \approx (38 \text{ mol ATP/mol glucose})(7.3 \text{ kcal/mol ATP}) = -277 \text{ kcal/mol glucose} \quad (5.24)$$

This is 19 times the energy which the cell captured during glycolysis. As in glycolysis, energy retention is very efficient:

$$\text{Energy capture efficiency} = \frac{277}{686} \approx 40\% \quad (5.25)$$

If this figure is corrected for the nonstandard concentrations within the cell, a rather astounding efficiency estimate of greater than 70 percent results. Most of the remaining energy is dissipated as heat, which must be removed in some fashion to keep the temperature in the physiologically suitable range.

Since *chemical* engines are not commonly studied in engineering thermodynamics, it may be helpful to draw an analogy with the classical example of compressing a gas in a cylinder with a piston. If the gas is compressed rapidly, much energy is lost as heat and therefore cannot be recovered in a subsequent expansion. An analogous process would be burning glucose in air, which is quite inefficient. On the other hand, the reversible compression ideal can be approached by pushing the piston very slowly, so that heat generation is minimized. Similarly, by carrying out glucose oxidation in many steps, where each has a relatively small free-energy change, the living cell is able to approach reversibility and to maximize efficiency for extraction of energy.

The above description of respiratory chain operation applies to a typical eucaryotic cell. In procaryotes, electron transport to oxygen may result in fewer than three phosphorylations. Furthermore, the cytochrome pathway differs in detail from one species to another, and phosphorylation efficiency may vary with growth conditions. The *P/O ratio*, the number of ADP phosphorylations per atom of O_2 consumed, is the parameter used to describe the energy storage activity of the respiratory chain.

Later in this chapter we shall examine how the oxidation state of the substrate and the cell's biosynthetic requirements interact to provide stoichiometric constraints on the activities of amphibolic pathways. Next, however, we should conclude our overview of energy-yielding metabolism with a brief examination of photosynthesis, the energy transduction process on which all life on earth depends.

5.5 PHOTOSYNTHESIS: TAPPING THE ULTIMATE SOURCE

In most respirations, hydrogen atoms are transferred continuously from the fuel to oxygen, with simultaneous release of energy. Photosynthesis is largely the reverse of respiration: energy in the form of light is captured and used for conversion of carbon dioxide to glucose and its polymers. Photosynthesis is the prime supplier of energy for the biosphere. It extracts energy from the only significant source external to our planet, the sun. Also, photosynthesis plays a vital role in closing the carbon and oxygen cycles by reducing the carbon oxidized by respiration:

$$\text{Photosynthesis: } 6CO_2 + 6H_2O + \text{light} \longrightarrow C_6H_{12}O_6 + 6O_2 \qquad (5.26)$$

5.5.1 Light-Harvesting

In procaryotes (cyanobacteria, also known as blue-green algae; green sulfur bacteria; purple sulfur bacteria), photosynthesis occurs in stacked membranes while the organelle called the *chloroplast* conducts photosynthesis for eucaryotes (algae, plants) (Fig. 1.5). Both systems contain *chlorophylls*, complex molecules which strongly absorb visible light (Fig. 5.9). The energy content E_p of 1 mole of photons depends on the frequency (or wavelength), according to

$$E_p = h\nu = h\frac{c}{\lambda} \qquad (5.27)$$

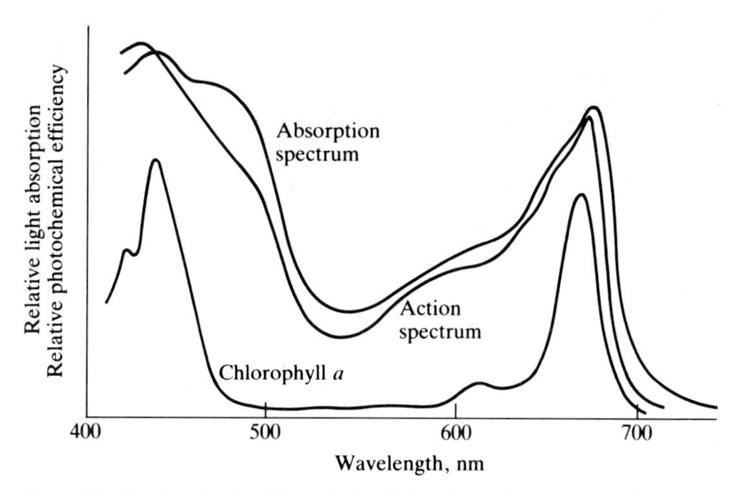

Figure 5.9 Maxima in the chlorophyll *a* light absorption spectrum (*lower curve*) are reflected in the absorption spectrum for a green leaf (*upper curve*) and the photosynthetic activity (action) spectrum of the leaf (*middle curve*). Accessory pigments in the leaf improve light utilization in the 500–600 nm range relative to chlorophyll alone. *(Reprinted by permission from A. Lehninger, "Principles of Biochemistry," p. 655, Worth Publishers, Inc., New York, 1982).*

where h = Planck's constant
ν = frequency of photon
λ = wavelength of photon
c = speed of light in the medium

The two major absorption peaks of Fig. 5.9 correspond to 43.5 kcal/mol (650 nm) and about 67 kcal/mol (430 nm) of photons, respectively. Both these absorption bands evidently correspond to free-energy promotions in excess of that needed to drive a single phosphorylation of ADP.

Two different light-harvesting and reaction systems, called *Photosystem I* and *Photosystem II*, are known. The second is activated by light with wavelengths below 680 nm, while longer wavelengths activate Photosystem I. Both systems are found in all photosynthetic organisms which release oxygen. The photosynthetic bacteria which do not release oxygen contain only Photosystem I.

5.5.2 Electron Transport and Photophosphorylation

In Photosystems I and II, absorption of two light quanta by chlorophyll and other pigments results in excitation of an electron from a reaction center (for example, P700 for Photosystem I; see Fig. 5.10). The high-energy electron is

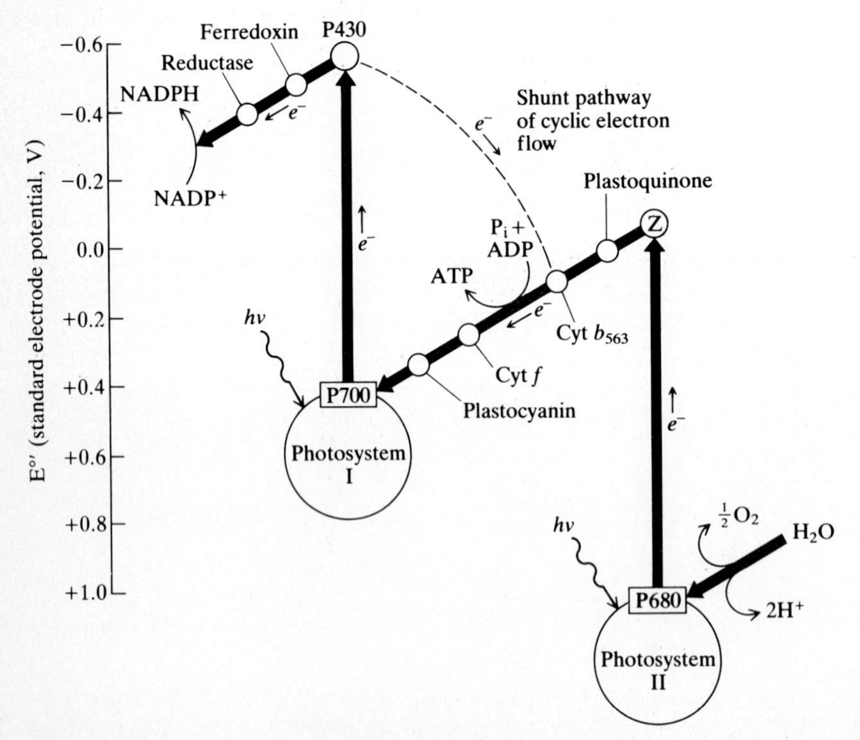

Figure 5.10 Illustration of photosynthetic systems for electron excitation (*vertical arrows*) using absorbed light and for electron transport. *(Reprinted by permission from A. L. Lehninger, "Principles of Biochemistry," p. 657, Worth Publishers, Inc., New York, 1982.)*

passed to the start of an electron-transport chain. The electron then "falls" to lower energy levels (indicated in Fig. 5.10 by increasing value of the standard half-cell potential $E^{\circ\prime}$). In the case of Photosystem I, the electron is ultimately used to reduce $NADP^+$. Excited electrons from Photosystem II are passed to Photosystem I with ADP phosphorylation accomplished once for every excited electron pair transported. Thus, the overall stoichiometry for electron flow in photosynthetic eucaryotes is

$$H_2O + 4h\nu + NADP^+ + ADP + P_i \longrightarrow NADPH + H^+ + \tfrac{1}{2}O_2 + (ATP + H_2O) \quad (5.28)$$

The coupled actions of both systems are necessary to accept electrons from H_2O ($E^{\circ\prime} = +0.82$ V) and donate them to $NADP^+$ ($E^{\circ\prime} = -0.32$ V) using the available light excitation energies. We can also see from Eq. (5.10) why bacteria which use H_2S, for example, as an electron donor can operate with only Photosystem I, since the half-cell standard potential for sulfur reduction is $E^{\circ\prime} = -0.23$.

Additional ATPs may be generated from absorbed light energy via *cyclic photophosphorylation* as shown by the dashed pathway in Fig. 5.10. Here, electrons excited from P700 to P430 flow to *cyt* b_{563} and back to P700, causing ADP phosphorylation in the process. The following simple stoichiometry therefore applies:

$$3h\nu + ADP + P_i \longrightarrow ATP + H_2O \quad (5.29)$$

Use of the energy and reducing power stored by the above mechanisms for synthesis of glucose from CO_2 and H_2O will be described in the next section. Here we note that the free energy change for this photosynthesis reaction

$$6CO_2 + 6H_2O \longrightarrow C_6H_{12}O_6 + 6O_2 \quad (5.30)$$

is +686 kcal/mol. Since two light quanta are required to excite a single electron through both photosystems and since four electrons must be transferred to generate one O_2 molecule, formation of $6O_2$ implies absorption of 48 light quanta. As indicated in Eq. (5.27), the energy associated with 48 moles of photons depends upon their wavelength. Assuming $\nu = 700$ nm, the minimum light energy input required to drive reaction (5.30) using the plant photosynthetic system is 48 mol electrons × (41 kcal/mol) = 1968 kcal. Thus, a rough estimate of efficiency based on light *absorbed* is 686/1968 = 35 percent. Later, in our discussion of overall metabolic stoichiometry, we will examine observed photosynthetic efficiency in plants and microorganisms.

5.6 BIOSYNTHESIS

In the opening pages of this text, we discussed three phenomena which characterize most microbial processes: substrate or nutrient utilization, cell growth and product release. Biosynthesis influences all three. Nutrient requirements are dictated by the cell's need for precursor molecules, stored chemical energy, and reducing power. Some biosynthesis products are released into the environment of the cell. Finally, the rate of cell growth is determined by the net rate of biosynthesis, the rate at which new cellular materials are formed. Cell growth rates vary widely. While the *E. coli* bacterium can double in 20 min, rat-liver cells reproduce over a two- to three-month cycle, and nerve cells do not multiply in adult human beings. Even in the last instance, however, some biosynthesis is necessary for maintenance and repair.

The chemical resources acquired by the cell from catabolism are invested to accomplish biosynthesis. Generally, the reactions are thermodynamically unfavorable and require ATP hydrolysis to ADP or AMP. The pyrophosphate ($P \sim P$) produced in the latter case, upon hydrolysis, provides additional free energy ($\Delta G^{\circ\prime} \cong -7$ kcal/mol) to "drive" the synthetic step. Because nutrient elements are often more oxidized than are their cellular forms, reducing power is also required.

5.6.1 Synthesis of Small Molecules

In these reactions, the monomeric building blocks are constructed. Approximately 70 different compounds are required for this purpose: 4 ribonucleotides, 4 deoxyribonucleotides, 20 amino acids, about 15 monosaccharides and about 20 fatty acids and lipid precursors. Also, ATP, NAD, other carriers, and coenzymes must be manufactured in class II reactions. The products of these reactions are known collectively as the *central intermediary metabolites*. In this section we will survey some of the major biosynthetic routes to the intermediary metabolites and their organization.

The amino acids are conveniently grouped into four families which are distinguished by their chemical structure and by their shared precursors. Also, as we will see in the discussion below of metabolic regulation, these families are often synthesized in a branched, systematically controlled reaction sequence. Synthesis of all the amino acids begins with carbon metabolism intermediates (see the schematic illustration in Fig. 5.11).

Living cells assimilate nitrogen by incorporating it into the amino acids glutamic acid and glutamine. First, glutamic acid is formed by reaction between ammonia and α-ketoglutaric acid, one of the TCA cycle intermediates:

$$\mathrm{HOOC(CH_2)_2COCOOH + NH_4^+ + NADH \xrightarrow[\text{dehydrogenase}]{\text{glutamate}} HOOC(CH_2)_2CHNH_2COOH + NAD^+ + H_2O} \tag{5.31}$$

Additional ammonia can be accepted by adding it to glutamic acid to give glutamine:

$$\mathrm{HOOC(CH_2)_2CHNH_2COOH + NH_4^+ + ATP \xrightarrow[\text{synthase}]{\text{glutamine}} HOOCNH_2CH(CH_2)_2CONH_2 + ADP + P_i + H^+} \qquad \Delta G^{\circ\prime} = -3.9 \text{ kcal/mol} \tag{5.32}$$

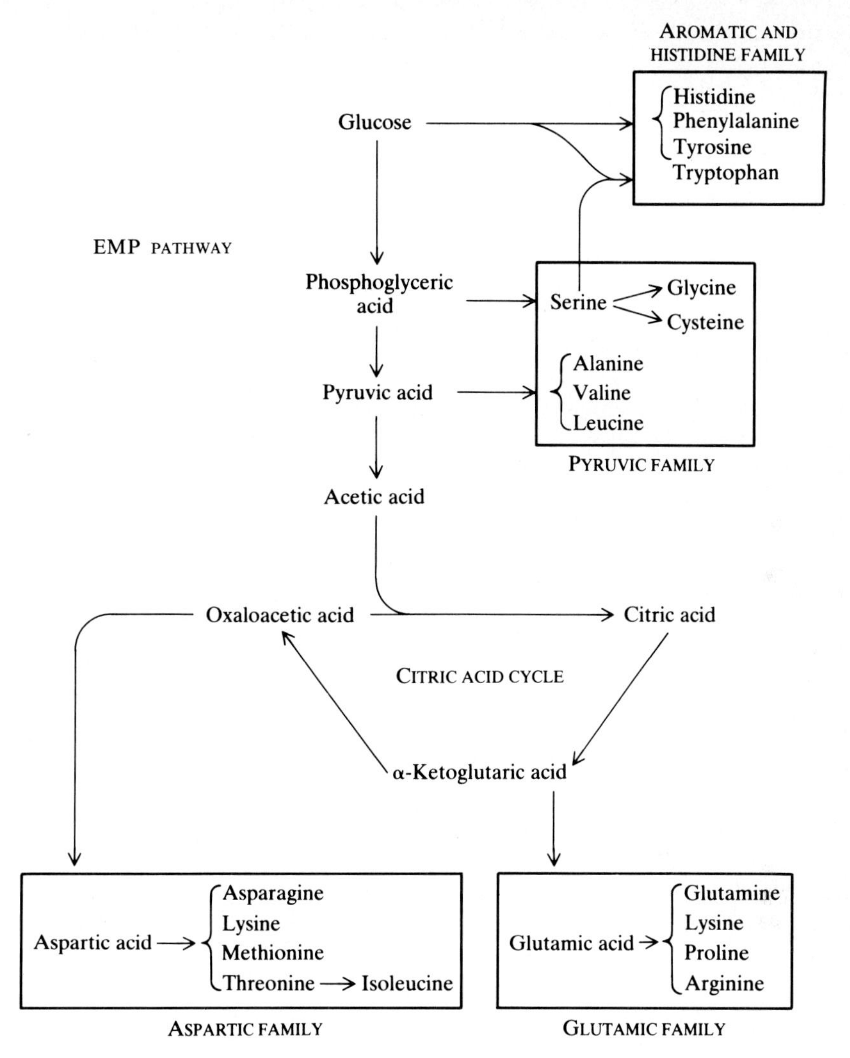

Figure 5.11 The amino fatty acid families and their carbohydrate precursors. *(Reprinted by permission from G. S. Stent and R. Calendar, "Molecular Genetics: An Introductory Narrative," 2d ed., p. 76, W. H. Freeman and Co., San Francisco, 1978.)*

The second reaction, requiring investment of metabolic energy, is used in ammonia deficient environments. In some bacteria, direct amination of pyruvate to alanine occurs with consumption of NADH. Also, certain bacteria can directly aminate fumarate to aspartate. Some microorganisms take in nitrogen in the form of nitrate NO_3^- and free nitrogen N_2, but it is now known that these nutrients are first transformed into ammonia before being assimilated by the above reaction(s).

All of the remaining amino acids are formed by conversion of glutamate or by transfer of its amino group to other carbon skeletons. For example, glutamate is converted to proline in a sequence of two enzyme-catalyzed reactions plus a spontaneous hydrolysis step with overall stoichiometry

$$C_5NH_8O_4^- + ATP + 2(NADPH + H^+) \longrightarrow C_5NH_8O_2^- + ADP + P_i + 2NADP^+ + H_2O \quad (5.33)$$

In most organisms, transaminations from glutamate give alanine and aspartate:

$$\text{Glutamate + oxaloacetate} \longrightarrow \alpha\text{-ketoglutarate + aspartate}$$

$$\text{Glutamate + pyruvate} \longrightarrow \alpha\text{-ketoglutarate + alanine}$$

The coenzyme *pyridoxal phosphate*, derived from vitamin B_6 (pyridoxal), is required for all transaminations.

As a final example of amino acid biosynthesis and of the frequent involvement of ATP and NADPH, some individual steps and the overall structure of the aspartic acid family synthesis pathway are illustrated in Fig. 5.12. Dashed lines in this figure denote regulatory interactions which will be discussed later.

Descriptions of nucleotide biosynthesis reactions in any detail is beyond the scope of this text. Origins of the components of these molecules are indicated in the diagram in Fig. 5.13. Further information may be found in the references.

Concerning fatty acid and other lipid precursor biosynthesis, we note that acetyl CoA and glycerol serve as the starting materials. To illustrate again the requirement for ATP and NADPH, we shall examine the pathway for synthesis of palmitate $[CH_3(CH_2)_{14}COOH]$, the most common of the fatty acids. First, acetyl CoA is carboxylated to yield malonyl CoA.

$$\text{Acetyl CoA} + ATP + CO_2 \xrightarrow[\text{carboxylase}]{\text{acetylCoA}} \underset{(HOOCCH_2COSCoA)}{\text{malonyl CoA}} + ADP + P_i \quad (5.34)$$

Subsequently, seven molecules of malonyl CoA and one of acetyl CoA react to give palmitate. The reaction pathway is a repeated cycle which gives the overall result

$$\text{Acetyl CoA} + 7\text{ malonyl CoA} + 14(NADPH + H^+) \longrightarrow \text{palmitate} + 7CO_2 + 8HSCoA + 14NADP^+ + 6H_2O \quad (5.35)$$

Note that, according to reaction (5.34), one high energy phosphate bond is invested for each malonyl CoA used.

To conclude this brief overview of biosynthesis, we consider routes for synthesis of glucose and related compounds. We have seen above the central role played by glucose catabolism products as precursors for amino acid, nucleotide, and fatty acid synthesis. Hence, organisms which grow on other carbon sources such as CO_2 must convert these substrates into glucose or one of its nearby metabolic products. Also, under conditions of excess carbon supply relative to other nutrients, many cells convert glucose to carbohydrate storage materials for later use.

Gluconeogenesis is the synthesis of glucose in chemotrophs. Interestingly, in the synthesis of glucose from pyruvate, all the intermediates found in the EMP

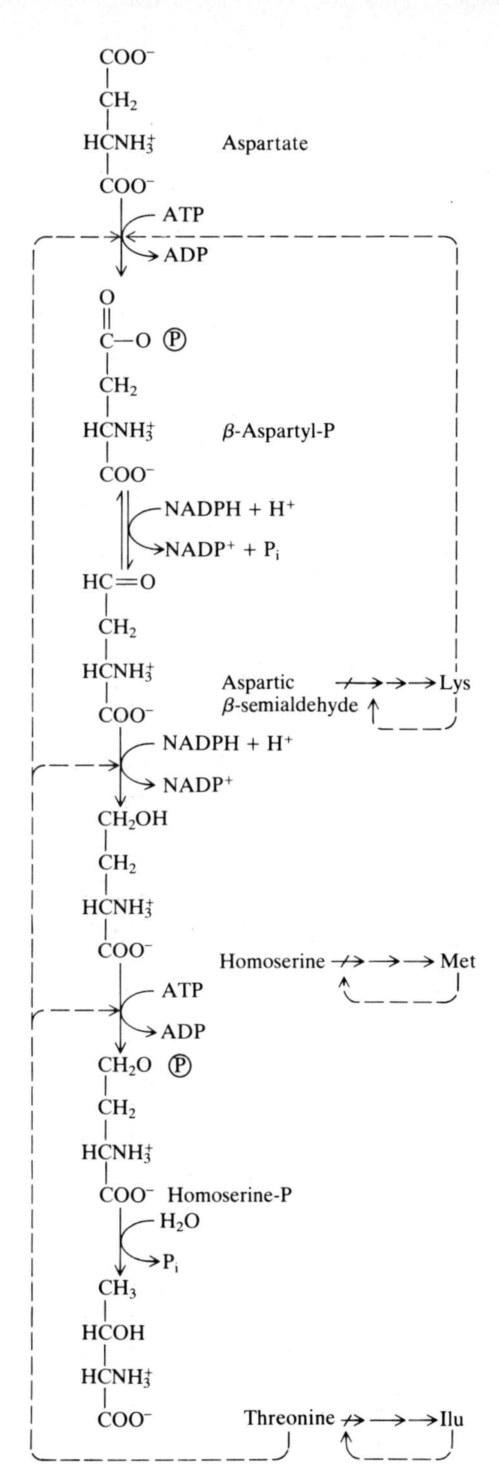

Figure 5.12 Some of the reactions which convert aspartate to lysine, homoserine, methionine, threonine, and isoleucine. *(Reprinted by permission from W. B. Wood, J. H. Wilson, R. M. Benbow, and L. E. Hood, "Biochemistry, A Problems Approach," 2d ed., p. 294, Benjamin/Cummings Publishing Co., Inc., Menlo Park, Ca., 1981.)*

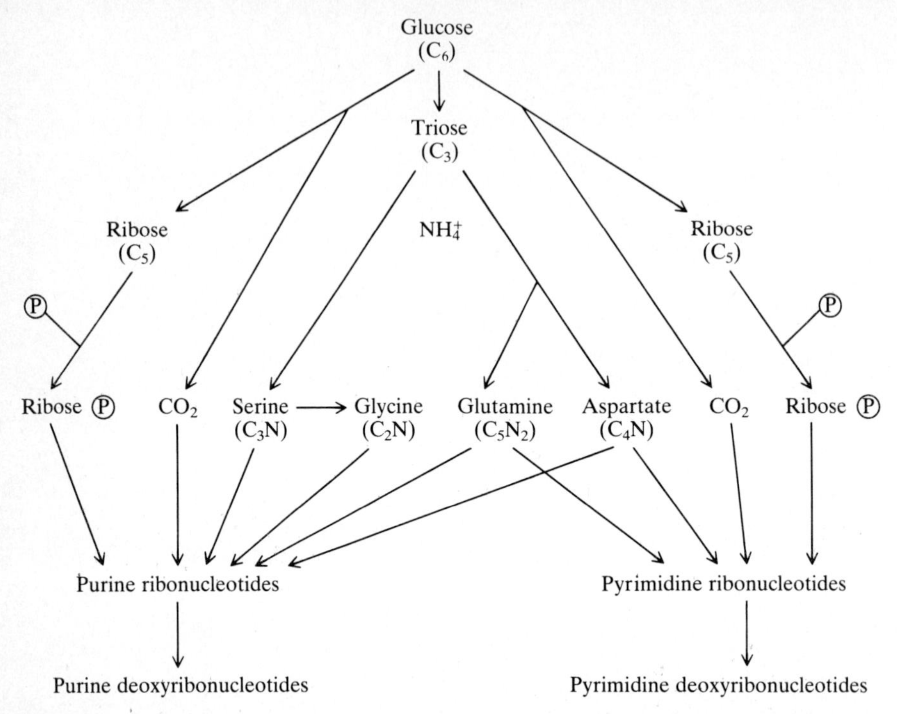

Figure 5.13 Flowchart of precursors into nucleotides and deoxyribonucleotides. *(Reprinted by permission from J. Mandelstam and K. McQuillan (eds.), "Biochemstry of Bacterial Growth," 2d ed., p. 33, Blackwell Scientific Publications, Oxford, 1973.)*

catabolic pathway participate. More important, all enzymes catalyzing reactions near equilibrium in the EMP pathway also catalyze reactions near equilibrium in glucose biosynthesis. In biosynthesis, however, different enzymes catalyze the phosphorylation and dephosphorylation reactions which are somewhat different from their reverse counterparts in glycolysis. These biosynthetic reactions (going upward in Fig. 5.14) proceed spontaneously with a decrease of free energy. The overall stoichiometry of gluconeogenesis shows that the reaction is energetically expensive and that the reaction (below) is not the reverse of the EMP pathway (GTP is guanosine triphosphate).

$$2\text{ pyruvate} + 4\text{ATP} + 2\text{GTP} + 2\text{NADH} + 2\text{H}^+ + 4\text{H}_2\text{O} \longrightarrow \text{glucose} + 2\text{NAD}^+ + 2\text{GDP} + 4\text{ADP} + 6\text{P}_i \quad (5.36)$$

The most important biosynthetic reaction is the synthesis of glucose from CO_2 in plants. The living world as we know it would not exist without this process. Furthermore, the extent and efficiency of these reactions determine the magnitude of the renewable carbonaceous resources available as fuels and bioprocess and chemical feedstocks.

Previously, the *light reactions* of photosynthesis—the absorption of light energy by chlorophyll and other pigments and the flow of excited electrons to

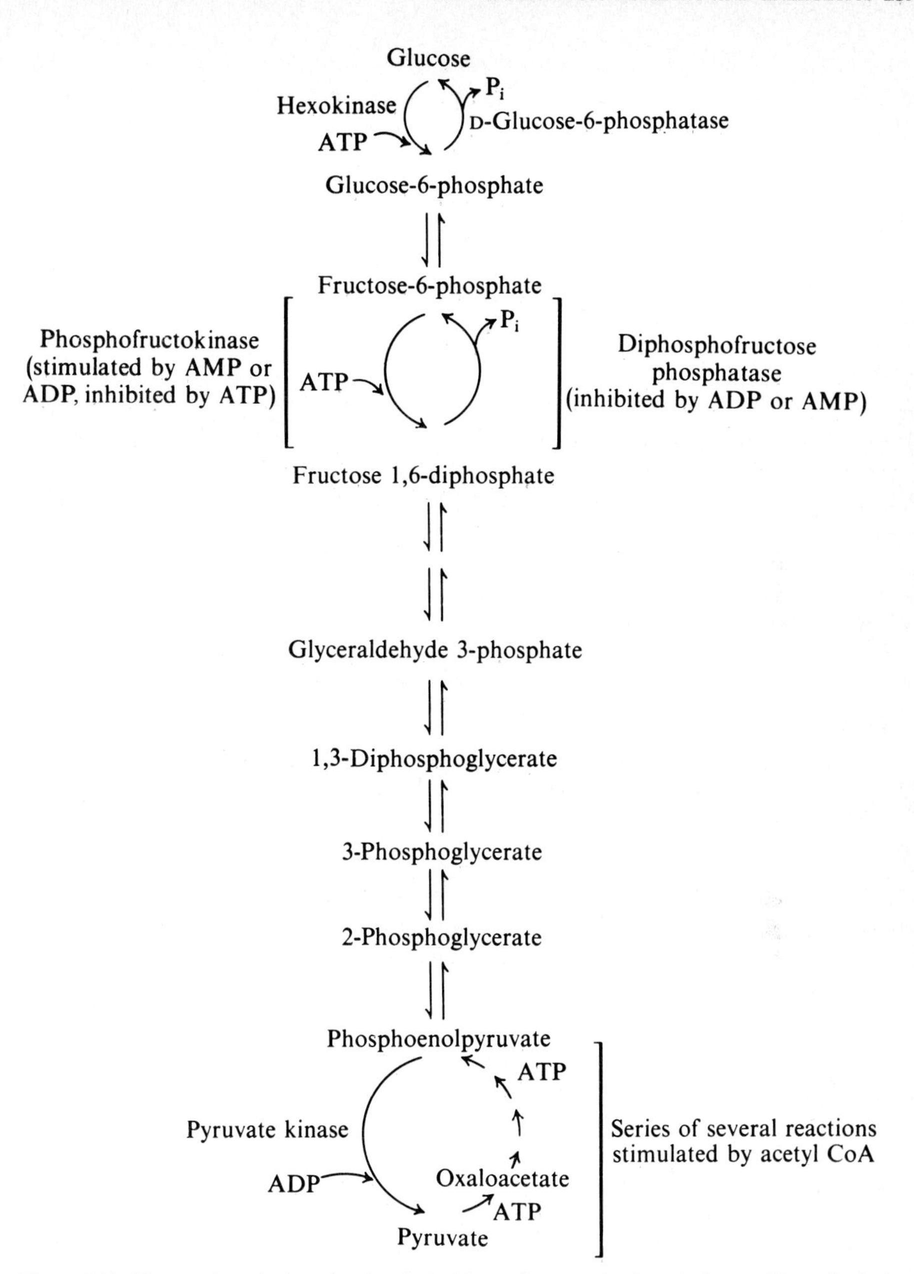

Figure 5.14 Glucose degradation via glycolysis (*descending reactions*) and glucose biosynthesis (*ascending reactions*) share many common reactions, namely those near equilibrium in both pathways. Reactions far from equilibrium are catalyzed by different enzymes (most allosteric) in the two pathways, thereby allowing independent control of glucose synthesis and degradation. *(Adapted from A. Lehninger, "Bioenergetics," 2d ed., p. 129, W. A. Benjamin, Inc., Palo Alto, CA, 1974.)*

generate ATP and NADPH—were described. Here, we consider the photosynthetic *dark reactions*—so called because these reactions will continue in the absence of light so long as a sufficient supply of ATP and NADPH is available.

The CO_2 is initially incorporated by reaction with *ribulose* 1,5-*diphosphate*:

$$\text{Ribulose 1,5-diphosphate} + CO_2 \xrightarrow[\text{carboxylase}]{\text{ribulose diphosphate}} 2\,(\text{glyceraldehyde 3-phosphate}) \quad (5.37)$$

The glyceraldehyde 3-phosphate formed in this step can undergo several reactions, including a sequence which closely resembles gluconeogenesis and which produces glucose. The other reactions of the *Calvin cycle* (Fig. 5.15) regenerate ribulose 1,5-diphosphate, allowing further carbon assimilation via reaction (5.37). Unlike the TCA cycle which is a net source of ATP and NADH, the Calvin cycle requires ATP and NADPH investment. Since one pass around the Calvin cycle serves to incorporate one CO_2, six passes are required for glucose synthesis, giving an overall stoichiometry

$$6CO_2 + 12NADPH + 12H^+ + 18ATP + 12H_2O \longrightarrow C_6H_{12}O_6 + 12NADP^+ + 18ADP + 18P_i \quad (5.38)$$

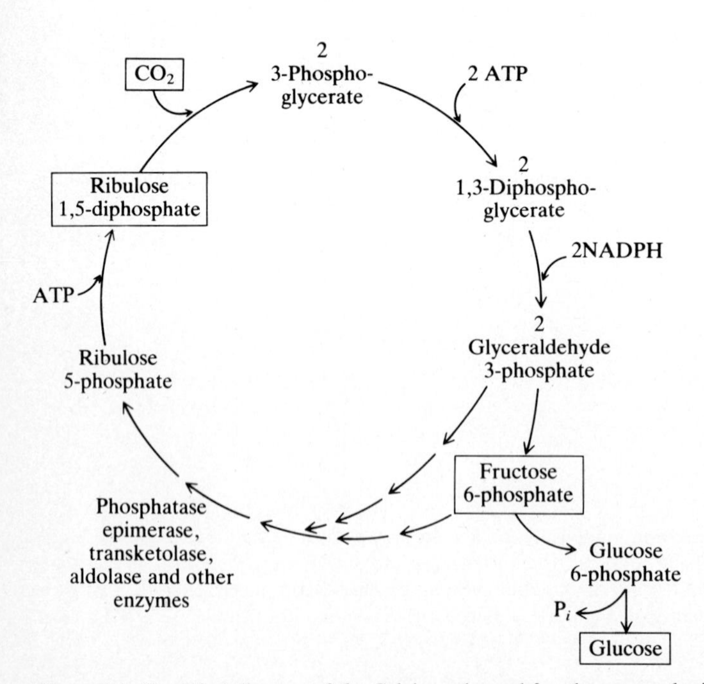

Figure 5.15 Simplified diagram of the Calvin cycle used for glucose synthesis from CO_2.

5.6.2 Macromolecule Synthesis

According to the metabolic roadmap given in this chapter's introduction, the monomeric precursors must next be assembled into the cell's polymeric components. Metabolic energy is again required in large quantities. The free energy so used serves to drive otherwise impossible condensation reactions. Also, extraordinarily large energy inputs enhance the driving force for desired reactions by shifting overall equilibrium far to the product side. This feature is especially important where accuracy is needed, as in the synthesis of particular nucleotide and amino acid sequences in nucleic acids and proteins.

Figure 5.16 shows how energy stored in the phosphate bonds of ATP is mobilized in the form of other nucleoside triphosphates for constructing the four classes of biopolymers. Typical of macromolecular biosynthesis is the hydrolysis of *two* high-energy phosphate bonds in conjunction with a driven reaction. The nucleoside triphosphate is converted to the nucleoside monophosphate plus pyrophosphate which also undergoes hydrolysis. Thus, approximately double the free-energy driving force ($\sim$14 kcal/mol) of a split to the nucleoside diphosphate is available. Some version of this mechanism operates in the synthesis of lipids, RNA, DNA, and glycogen. The common-intermediate principle is again involved, although in a more complicated form. For example, the addition of one glucose unit to glycogen proceeds in six steps linked by overlapping intermediates (Table 5.5).

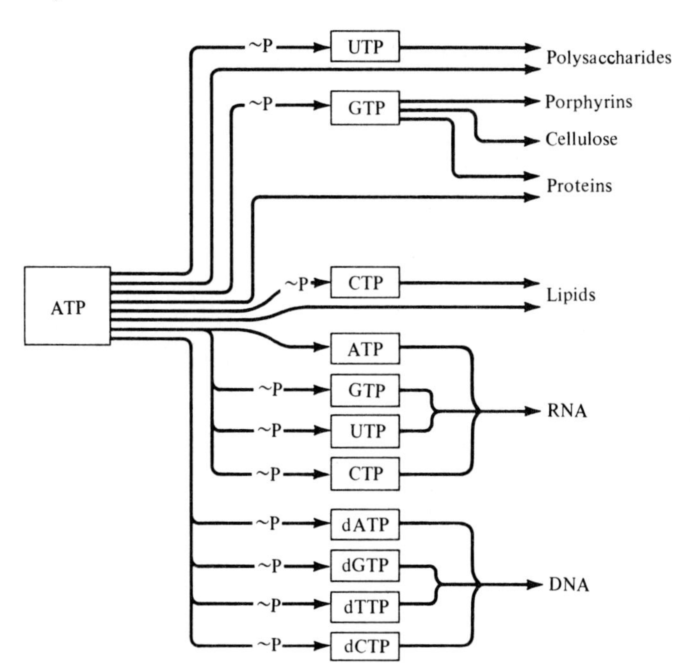

Figure 5.16 The high energy triphosphate forms of the nucleosides participate in various biosynthetic routes. *(Reprinted by permission from A. Lehninger, "Bioenergetics," 2d ed., p. 136, W. A. Benjamin, Inc., Palo Alto, 1975.)*

Table 5.5 Reaction sequence for glycogen biosynthesis in mammalian cells

(ATP·glucose is used in bacteria and plants)

1. Glucose + ATP $\xrightarrow{\text{hexokinase}}$ glucose 6-phosphate + ADP
2. Glucose 6-phosphate $\xrightarrow{\text{phosphoglucomutase}}$ glucose 1-phosphate
3. Glucose 1-phosphate + UTP $\xrightarrow[\text{pyrophosphorylase}]{\text{UDP-glucose}}$ UDP·glucose + pyrophosphate
4. UDP·glucose + glycogen$_n$ $\xrightarrow[\text{synthetase}]{\text{glycogen}}$ glycogen$_{n+1}$ + UDP
5. ATP + UDP $\xrightarrow[\text{diphosphokinase}]{\text{nucleomide}}$ ADP + UTP
6. Pyrophosphate + H_2O $\xrightarrow{\text{pyrophosphatase}}$ 2 phosphate

Sum: glucose + 2ATP + glycogen$_n$ + H_2O $\longrightarrow$ 2ADP$_i$ + 2P. + glycogen$_{n+1}$

† A. L. Lehninger, *Bioenergetics*, 2d ed., p. 140, W. A. Benjamin, Inc., Palo Alto, Ca., 1971.

Naturally, synthesis of informational polymers (RNA, DNA, and proteins) is a considerably more complex process. In both cases, however, the monomer is activated to permit its addition to the polymer chain. For RNA and DNA synthesis, the nucleotides enter the scheme as nucleoside triphosphates. These are incorporated in their respective polymers with a single phosphate, freeing pyrophosphate. Thus, about 14 kcal/mol is invested to push the monomer addition reaction to completion. Activation of amino acids for protein construction can be represented by

$$\text{Amino acid} + \text{ATP} \longrightarrow \underbrace{\text{amino acyl}}_{\text{active amino acid}}\cdot\text{AMP} + \text{P} \sim \text{P} \tag{5.39}$$

GTP is hydrolyzed also when the amino acid adenylate is added to the peptide chain, giving a total of three high-energy phosphate bonds per monomer addition. In Chap. 6, we will examine the mechanisms by which particular monomer sequences in DNAs, RNAs and proteins are constructed.

5.7 TRANSPORT ACROSS CELL MEMBRANES

Controlled transport of ions and molecules between the cell and its environment is essential to normal cell function. Cell membranes provide selective permeability to various medium and cellular constituents, regulating transport processes which serve several important purposes. First, these transport processes maintain intracellular composition and pH in a narrow range consistent with necessary enzyme activities. Membrane transport regulates cell volume, admits and concentrates nutrients, and secretes toxic compounds.

At least three different means of transport across cell membranes are known: passive diffusion, facilitated diffusion, and active transport. Regardless of the mechanism, the transport characteristics of a given membrane with a given substrate are often expressed in terms of the membrane permeability K, which is computed from

$$K = \frac{V}{At} \ln \frac{c_e - c_{i0}}{c_e - c_i(t)} \tag{5.40}$$

where V = cell volume
A = cell external surface area
c_e = external concentration
c_{i0} = interior substrate concentration initially
$c_i(t)$ = interior substrate concentration after elapsed time t

As Eq. (5.40) indicates, the units of permeability are those of a velocity (centimeters per second). The assumptions underlying this equation are rather extreme, however, as we shall explore in the problems.

5.7.1 Passive and Facilitated Diffusion

In *passive diffusion*, material moves across the membrane from regions of high concentration to low concentration (Fig. 5.17). The diffusion rate is proportional to the overall *driving force*, the concentration difference across the membrane. Thermodynamic considerations show that passive diffusion is spontaneous: the free-energy change ΔG° accompanying the transport of material from a region of concentration c_2 to another locale where the concentration is c_1 is given by

$$\Delta G^\circ = RT \ln \frac{c_1}{c_2} \tag{5.41}$$

Since c_1 is smaller than c_2 in passive diffusion, ΔG° is negative. If the component being transferred is charged, Eq. (5.41) should be modified to read

$$\Delta G^\circ = RT \ln \frac{c_1}{c_2} + Z_1 \mathscr{F} \Delta\psi \tag{5.42}$$

where Z_1 = number of charges on transported molecules
$\mathscr{F}$ = Faraday = 23.062 kcal V^{-1} mol^{-1}
$\Delta\psi$ = potential difference across membrane, V

Because of the structure of the plasma membrane which surrounds all cells and the organelles of eucaryotes, not all chemical species penetrate the membrane with equal ease. The diffusion rate of most large molecules is strongly correlated with the molecule's solubility in lipids (Fig. 5.18). This should not surprise us since the central core of the plasma membrane is believed to be mostly lipid

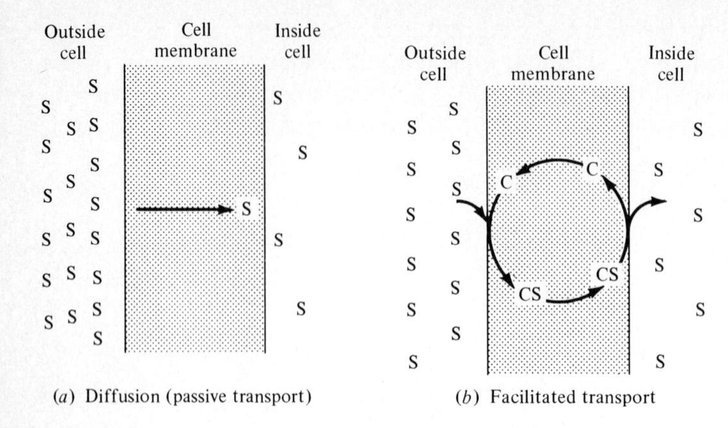

(*a*) Diffusion (passive transport)   (*b*) Facilitated transport

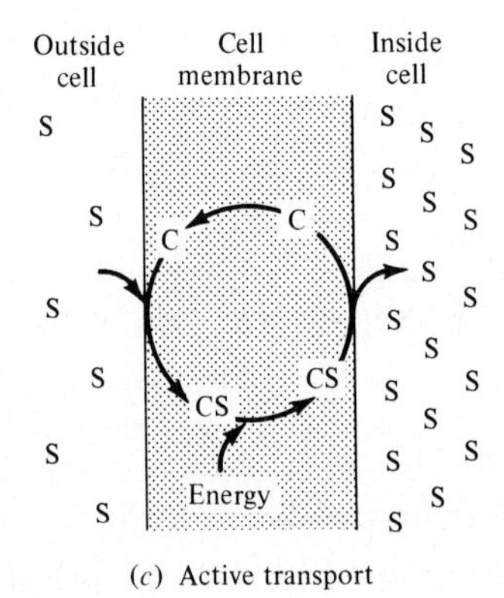

(*c*) Active transport

Figure 5.17 Various modes of membrane transport. (*a*) Diffusion, (*b*) facilitated transport, and (*c*) active transport.

bilayer (recall Fig. 2.2). A variation in the unit-membrane model has been developed to account for the abnormally rapid diffusion of water and other small molecules through the membrane: very small pores through the membrane are envisioned.

Charged molecules and other polar substances have a very small lipid solubility. Thus, these species have little tendency to pass through the membrane by ordinary diffusion. The importance of this property has already been emphasized: most metabolic intermediates are charged and consequently stay within the cell. Still, there are some polar components which move across the membrane barrier quite rapidly.

Facilitated diffusion provides one mechanism for this anomaly. As illustrated schematically in Fig. 5.17, substrate on the outside combines with a *carrier molecule*, which "diffuses" to the other side, where the complex splits, discharging the carried molecule inside the cell. This mode of biological transport has several distinguishing characteristics. Instead of increasing linearly with the exterior

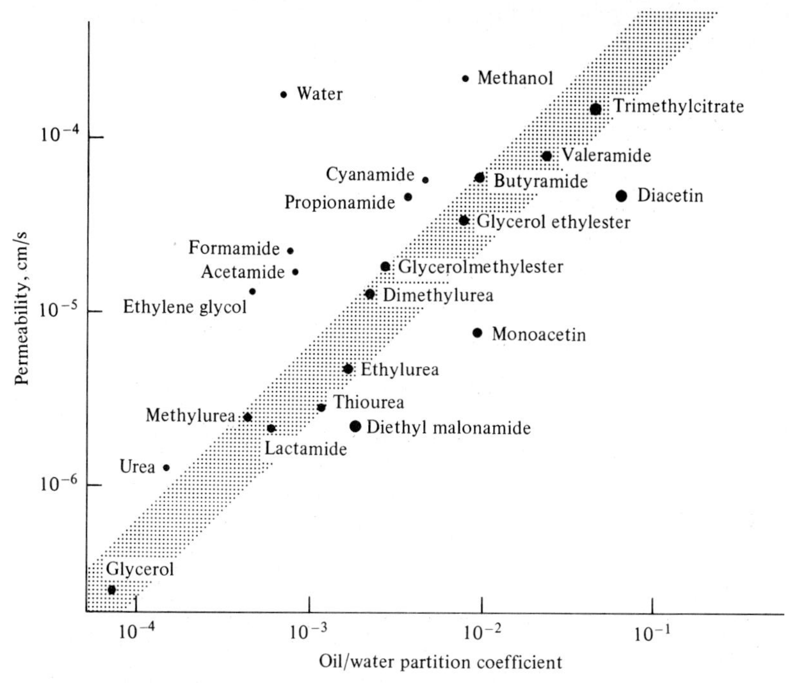

Figure 5.18 Permeability of many substances through the membrane of the alga *Chara* correlates well with lipid solubility (partition coefficient). The sizes of the dots are roughly proportional to the respective molecular sizes. *(After Collander, 1947).*

solute concentration, the transport rate for facilitated transport reaches a saturation, maximum level. Further increases in the overall mass-transfer driving force have no influence on the rate of transport. From our exposure to Michaelis-Menten and other models of enzyme kinetics, it should be obvious why the scheme shown in Fig. 5.17*b* has been postulated as a facilitated-diffusion mechanism.

The other essential properties of facilitated diffusion are also reminiscent of enzyme kinetics: only specific compounds are transported, and specific inhibitors slow the process. Because of this specificity and kinetic behavior, carrier molecules are believed to be proteins. These carriers, many of which have been isolated and characterized, are called *permeases.* Perhaps the best known example of facilitated diffusion is glucose transport in the human erythrocyte (red blood cell). Details on this system and additional information on facilitated transport are available in the references.

5.7.2 Active Transport

As already shown in Fig. 5.17, active transport has two distinguishing characteristics: (1) it moves a component *against* its chemical (or electrochemical) gradient, from regions of low to high concentration; (2) this process requires metabolic energy, as Eq. (5.41) indicates. For example, if an uncharged compound is transported from the cell's environment, where its concentration is 0.001 M, to the interior of the cell, where this compound's concentration is 0.1 M, then

$$\Delta G^\circ = [1.98 \text{ cal/(mol}\cdot\text{K)}](298 \text{ K})(\ln 100) = 2.72 \text{ kcal/mol}$$

Consequently, at least this amount of free energy must be expended to drive the process. Again, specific permease carriers are implicated.

Important in nerve action, active transport enjoys wider application: almost all cells have active-transport systems to maintain a proper balance between K^+, Na^+, and water within the cell. In particular, Na^+ is pumped out of the cell while K^+ is pumped in. The pumping action allows the cell to offset the simultaneous passive diffusion of these ions, which occurs continuously. These ion-transport processes are coupled and driven by ATP (Fig. 5.19). The sodium-potassium pump in red blood cells is an oligomeric protein called Na^+-K^+ ATPase. As indicated in Fig. 5.19, this protein is embedded within and traverses the plasma membrane. Proteins so situated are called *transmembrane proteins.*

The second common class of active-transport systems pumps molecular nutrients such as glucose and amino acids into the cell at far greater rates than can be achieved by passive diffusion. In the cells of higher animals, glucose active transport is dependent on *cotransport* of Na^+ into the cell. Na^+ is pumped back out with simultaneous ATP hydrolysis using the Na^+-K^+ ATPase. Figure 5.20 shows the overall process. A different process called *group translocation* is responsible for glucose active transport in bacteria. In this scheme, glucose is released into the cell's interior in the energized and relatively nonpermeating form, glucose 6-phosphate (Fig. 5.21). The rate of this process is believed to be the limiting step which determines the growth rate of some cells.

Many transport systems in bacteria are driven by proton flows across the plasma membrane. The central role of proton flows in bacterial energetics was

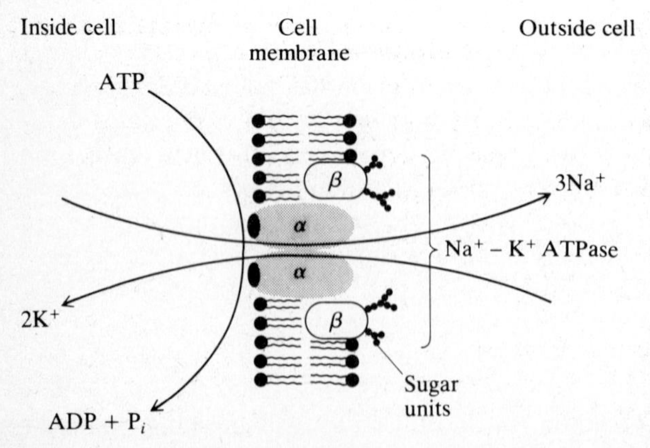

Figure 5.19 The Na^+-K^+ pump (Na^+-K^+ ATPase) transports sodium ions out of the cell and potassium ions in with driving force provided by ATP hydrolysis.

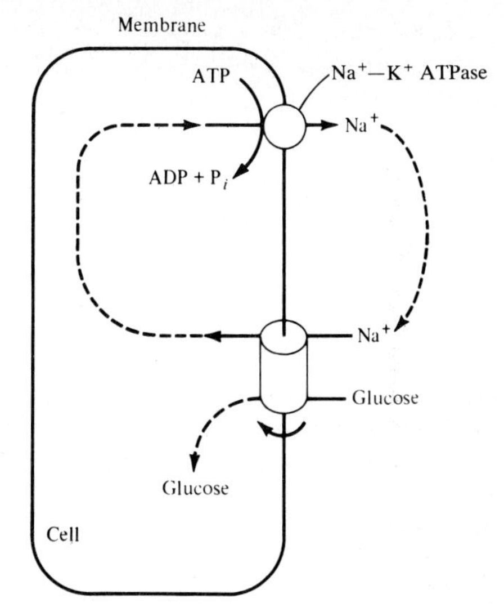

Figure 5.20 Na^+ external to the cell moves through the cell membrane in a passive carrier that also transports glucose. A high external concentration of Na^+ is maintained by active transport. *(Reprinted by permission from A. Lehninger, "Bioenergetics," 2d ed., p. 205, W. A. Benjamin, Inc., Palo Alto, CA, 1974.)*

described in the *chemiosmotic theory* proposed by Peter Mitchell in 1961. Figure 5.22 shows schematically the coupling between proton and lactose transport in respiring *E. coli.* Protons and lactose are cotransported into the cell by *lactose permease* (the product of the *y* gene of the *lac* operon discussed earlier). Simultaneously, protons are transported out of the cell in connection with electron flow through the respiratory chain. Overall, the cell interior is maintained at higher pH than the cell environment.

Eucaryotic cells must possess a hierarchy of transport systems, since the concentration of some components within organelles is maintained at a level other than that in the cell's cytoplasm. This feature serves as a reminder that the inside of a cell, especially a eucaryote, is not a uniform well-mixed pool. On the contrary, living cells are highly organized systems even down to the molecular level.

Many mathematical theories of active transport have been developed, although we shall not pursue them now. Further details are available in the references and the problems.

Before we turn to examination of flows and controls in metabolic reaction sequences, we should show a more complete and detailed illustration of the structure and organization of a cell membrane. In the diagram of a bacterial cell

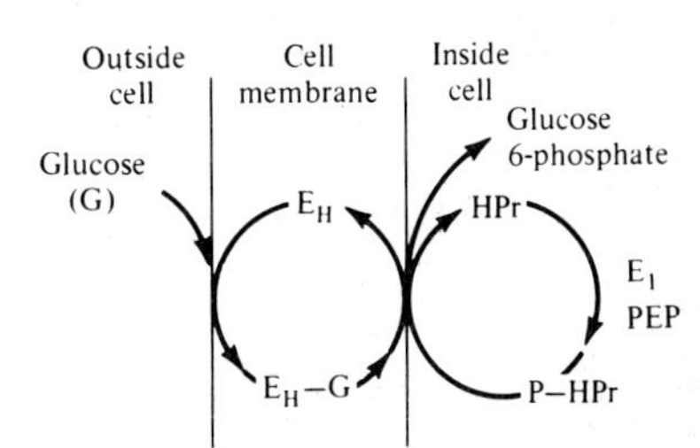

Figure 5.21 Phosphorylation of glucose inside the cell membrane helps maintain a large driving force for glucose membrane transport and also serves to keep the transported glucose moiety in the cell's interior. *(Reprinted by permission from A. L. Lehninger, "Biochemistry", 2d ed., p. 799, Worth Publishers, New York, 1975.)*

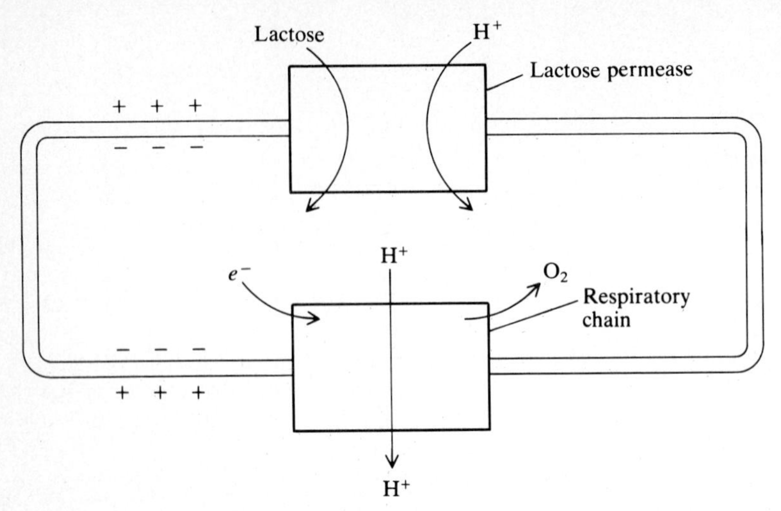

Figure 5.22 Active transport of lactose is coupled to flow of protons down a gradient generated by electron flow through the respiratory chain.

membrane in Fig. 5.23, we see a considerably more elaborate structure than the simple lipid bilayer-protein coat membrane model described in Chap. 2. Here, the membrane is seen to contain numerous proteins embedded within the membrane, some exposed only to one membrane face and others traversing the membrane. These permease and receptor proteins recognize specific compounds in the cell environment and connect with cellular control systems for response to the environment. In bacteria these membrane components permit rapid adjustment

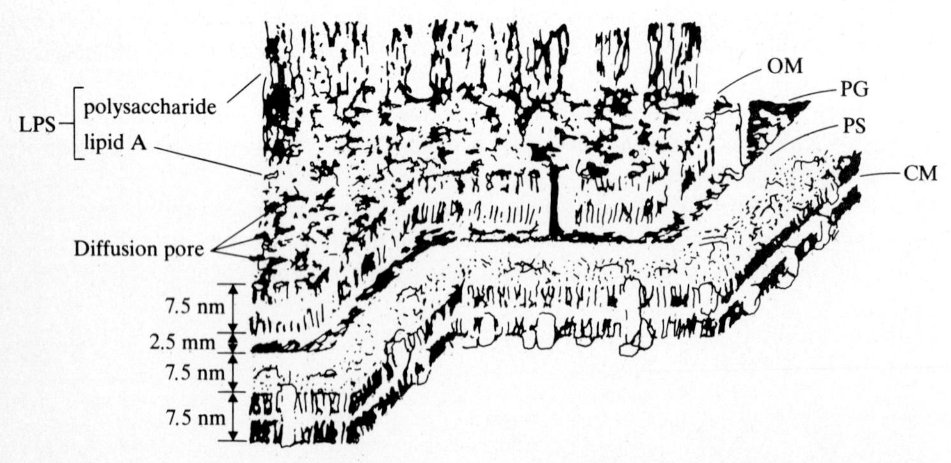

Figure 5.23 The molecular architecture of the *E. coli* cell envelope is shown here in detail. White globular objects on and embedded within the cell membranes denote proteins. Abbreviations: OM, outer membrane; PG, peptidoglycan; PS, periplasmic space; CM, cytoplasmic membrane. *(Reprinted by permission from M. Inouye, Bacterial Outer Membranes, John Wiley, New York, 1979.)*

to changing environmental conditions. Cell–cell recognition important in both tissue organization and response to disease is mediated by surface receptors in higher eucaryotic cells.

Membranes are generally asymmetric, exhibiting different functions and components on the interior and exterior sides. Diffusion of lipids and many membrane proteins in the plane of the membrane is rapid; the lipid diffusivity in several membranes has been estimated as approximately 10^{-8} cm^2/s.

5.8 METABOLIC ORGANIZATION AND REGULATION

The cell, like any other chemical plant, must have controls on its complex array of chemical reactions so that supply and demand for materials, energy, and electrons are balanced and so that resources are used efficiently. Also, the reaction systems must be organized or structured in a fashion consistent with effective control. In this section we examine the ways in which modulation of the activities of key enzymes controls flows through the cell's many branched and looped metabolic reaction pathways. Also, we consider key structural features of the metabolic network. We can gain an appreciation for the importance and efficiency of these controls and structures by examining metabolic activities in an *E. coli* bacterium. In a rich medium, the cells divide every 20 minutes, and they conduct an amazing array of chemical functions with great precision, productivity, and balance (see Table 5.6).

Table 5.6 Biosynthetic activity during a 20-min cell-division cycle of *E. coli*†

Chemical component	Percent of dry weight	Approximate mol wt	Number of molecules per cell
DNA	5	2,000,000,000	1
RNA	10	1,000,000	15,000
Protein	70	60,000	1,700,000
Lipids	10	1,000	15,000,000
Polysaccharides	5	200,000	39,000

Chemical component	Number of molecules synthesized per second	Number of molecules of ATP required to synthesize per second	Percent of total biosynthetic energy required
DNA	0.00083	60,000	2.5
RNA	12.5	75,000	3.1
Protein	1,400	2,120,000	88.0
Lipids	12,500	87,500	3.7
Polysaccharides	32.5	65,000	2.7

† A. L. Lehninger, *Bioenergetics*, 2d ed., p. 123, W. A. Benjamin, Inc., Palo Alto, Ca., 1965.

Considering protein synthesis alone in this case, we see that, on the average, 1400 protein molecules per second are manufactured within the cell. Since proteins are large biopolymers with an average of over 300 covalent bonds, peptide bonds are formed at the rate of 420,000 bonds per second per cell. Further, since proteins are informational biopolymers, their monomeric units must be connected in a definite, predetermined sequence.

The living cell is able to achieve such prodigious rates of protein synthesis at the expense of a very substantial portion of its metabolically derived energy. Bacteria devote almost all their energy to biosynthesis and, of this, roughly 88 percent is channeled into protein synthesis. Table 5.9 shows that approximately $2\frac{1}{2}$ million ATP molecules per second are invested in biosynthesis. Since the total ATP concent of *E. coli* is approximately five million molecules, the total cell inventory is sufficient for only two seconds of work. This gives some indication of the large rates of ATP regeneration which must be maintained in order to sustain the cell, and of the necessity of regulating ATP generation to match ATP requirements. Similar comments apply to primary metabolite precursors, NADH and NADPH.

5.8.1 Key Crossroads and Branch Points in Metabolism

A careful look at metabolic pathways as a network of directed steps reveals key features of its organization. A straight sequence of metabolic reactions is encountered in committed conversions along a particular pathway. The first and last steps of the straight pathway are either starting nutrients, final cellular or excreted products, or branch points from which several alternative reactions emanate. Typically, the first irreversible (significantly negative $\Delta G'$ under intracellular conditions) reaction leading from a branch is an allosterically controlled step. Then, by regulating entry into the subsequent reaction sequence, the cell determines the allocation of metabolites which have several alternative uses by the cell. Viewed in structural terms, these metabolites are the starting points or *nodes* of the metabolic branch. From this perspective, we see that the branch points and modulated irreversible reaction steps are critical elements in determining chemical activities of a cell. This theme will recur several times in the remainder of the text, and these concepts will be very important in formulating simplified engineering descriptions of the cell as a chemical reactor.

We shall first examine the importance of branching reaction structures in a global perspective: all metabolic activities in the cell depend on the flow of reactions through three primary crossroads: glucose-6-phosphate, pyruvate, and acetyl CoA. Figure 5.24 illustrates alternative routes for entering and leaving these crossroads. The relative magnitudes of these flows are regulated by the cell in response to metabolic requirements and the cell's current composition.

On a finer scale, let us review the diverging pathways in amino acid synthesis shown in Fig. 5.12. Here again, the existence of branches followed by straight sequences and subsequent branches is clear. Control locations in this part of the metabolic network will be discussed in the next section.

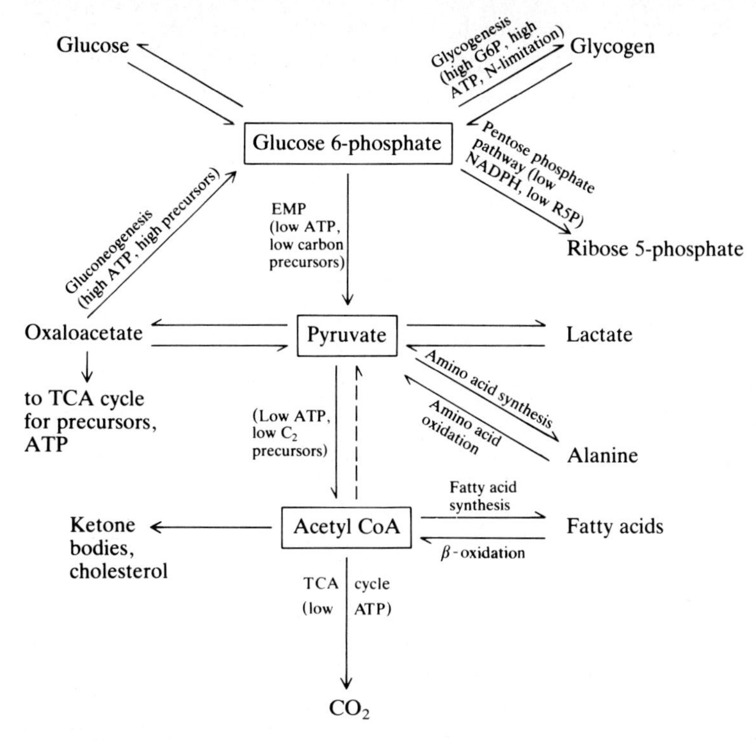

Figure 5.24 Summary of the key crossroads of carbon metabolism. The conversion of acetyl CoA into pyruvate is not possible in mammals.

5.8.2 Enzyme Level Regulation of Metabolism

Allosteric enzymes control flows through metabolic pathways. The concentrations of ATP, NADH, NADPH, and certain key precursors and intermediates are the regulating signals which govern the activities of these controlling enzymes. The influences of some of these signals on flow division at branch points are indicated in general terms in Fig. 5.24. Here we consider representative examples of metabolic regulation at the molecular level. Also, we shall see here that different pathways are sometimes controlled in a coordinated fashion.

Clearly, maintaining a suitable ATP level in the cell is a necessity. A useful indicator of the energetic state of a cell is the adenylate *energy charge* [4], defined as the ratio of ATP concentration to the sum of AMP, ADP, and ATP concentrations:

$$\text{Energy charge} = \frac{[\text{ATP}]}{[\text{AMP}] + [\text{ADP}] + [\text{ATP}]} \tag{5.43}$$

The energy charge of most cells is in the range 0.87 to 0.94.

To see how the ATP level is maintained so precisely, we shall begin with the EMP pathway (Fig. 5.25). The primary regulatory enzyme is phosphofructokinase, which is activated by ADP and inhibited by ATP. Thus, this enzyme slows the flow through the EMP pathway if the energy charge is high and vice versa. Pyruvate kinase, activated by AMP, acts similarly. We recall that glucose phosphorylation and the reactions of F6P and PEP were the three irreversible reactions of the EMP pathway. Now we can see why: the first step gives a charged derivative which is retained effectively by the cell envelope and the other two irreversible steps are control points.

The energy charge also affects the activity of the TCA cycle. Isocitrate dehydrogenase is activated by ADP and inhibited by ATP. Since one of the major

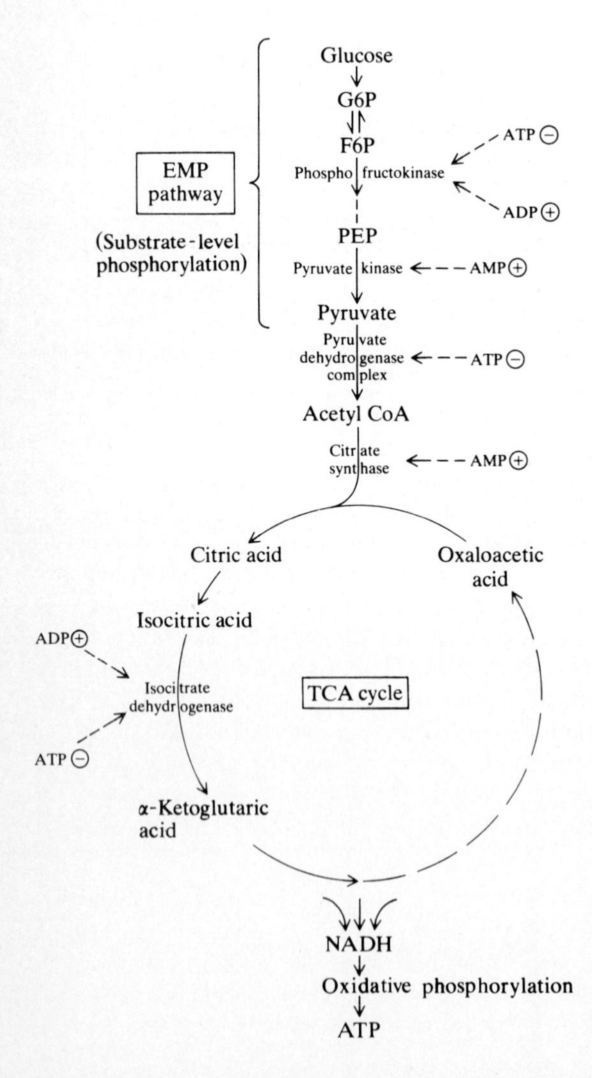

Figure 5.25 Regulation of the EMP pathway and TCA cycle by adenosine phosphate compounds. Dots denote allosteric effects on the indicated enzymes. ⊕ = activation, ⊖ = inhibition.

functions of the EMP and TCA pathways is ADP phosphorylation, which increases the cell's energy charge, we see that the regulatory patterns in Fig. 5.25 are concerted and coordinated forms of feedback inhibition.

Obviously, other controls are required to govern precursor supply and biosynthetic activities. For example, citrate is an allosteric modulator of several enzymes. Citrate increases the inhibitory effect of ATP on phosphofructokinase. This reflects the amphibolic nature of the EMP and TCA reactions: if both energy charge and precursor supply is high, six-carbon molecules can be used elsewhere. Citrate is also a strong activator of the first reaction in fatty acid synthesis [Eq. (5.34)].

Biosynthetic pathways are generally regulated at the first step emanating from a branch point. Additional examples of this strategy are shown in the aspartate family conversion reactions in Fig. 5.12. Dotted lines there denote feedback inhibition. The logic of the reaction network plus controllers is clear: in a small number of steps, several products are formed; the pathways to each amino acid are regulated separately. However, there seems to be a flaw: high levels of lysine or threonine appear to shut down synthesis of homoserine and methionine. This does not occur because there are three different enzymes catalyzing the aspartate to β-aspartyl-phosphate reaction. One of these enzymes is unregulated, one is inhibited by lysine, and the last is inhibited by threonine. Different enzymes catalyzing the same reaction step are called *isozymes*. Similary, there are two isozymes, one unmodulated and the other inhibited by threonine, which convert aspartic semialdehyde to homoserine. With these elaborations, we can now see the successful and efficient control of this reaction network given any combination of surplus products.

In biochemical processes, we may wish to produce large quantities of particular metabolites. Then, the *process* goal of maximum production of these compounds is different from the *survival* goal of the native organism. To make the cell produce well (from our perspective, but very inefficiently from the cell's viewpoint), we often try to modify these controlling enzymes of metabolism so that natural controls fail and metabolite overproduction occurs. We shall pursue this theme further in our study of applied genetics in the next chapter. Our remaining task here is to consider stoichiometric and energetic coupling between pathways and the implied overall stoichiometry of substrate utilization, cell growth, product formation and heat generation. As a prelude to this discussion, we should examine some of the end products of metabolism which are released into the cell environment.

5.9 END PRODUCTS OF METABOLISM

Cells release a variety of chemicals into their environment. In many cases, these products are the result of energy-yielding metabolism. Alcohols and organic acids are frequent examples of catabolic metabolism end products. Other metabolic products, such as antibiotics, toxins, alkaloids, and growth factors, are more complex molecules which serve special functions for the cell. Such products, which are not required for growth in pure culture, are called *secondary metabolites.* In this section we survey briefly some of the pathways which lead from metabolic intermediates to end products and the types of products which different cell types yield under different environmental conditions.

5.9.1 Anaerobic Metabolism (Fermentation) Products

We shall consider first different end products obtained by anaerobic utilization of glucose. Many organisms proceed from glucose via the EMP, HMP and/or ED pathways to pyruvate. The metabolic route from pyruvate to final products can vary significantly. The economically most important fermentation pathway from pyruvate forms ethanol (the *alcohol fermentation*) from pyruvate by the following reaction sequence:

$$\text{Pyruvate} \xrightarrow[\text{decarboxylase}]{\text{pyruvate}} \text{acetaldehyde} + CO_2$$

$$\text{Acetaldehyde} + \text{NADH} + H^+ \xrightarrow[\text{dehydrogenase}]{\text{alcohol}} \text{ethanol} + \text{NAD}^+ \tag{5.44}$$

In these reactions, and those of the other carbohydrate fermentations considered below, NADH formed in the pathways to pyruvate is oxidized. Since anaerobic metabolism does not combine electrons with external electron acceptors as occurs in respiration, a strict oxidation-reduction balance applies to substrate utilization and product formation in fermentation. Another illustration of utilization of NADH in fermentation product formation is the *lactic acid* or *homolactic fermentation*, which proceeds from pyruvate to lactate in a single step:

$$\text{Pyruvate} + \text{NADH} + H^+ \xrightarrow[\text{dehydrogenase}]{\text{lactate}} \text{lactate} + \text{NAD}^+ \tag{5.45}$$

Other carbohydrate fermentations which begin with conversion to pyruvate via the EMP pathway are listed in Table 5.7. An overview of the corresponding reaction pathways is provided in Fig. 5.26. We should also emphasize that the products obtained depend on cultivation conditions as well as organism. For example, in conducting alcohol fermentation with the yeast *Saccharomyces cerevisiae*, changing the pH from 3.0 to 7.0 in a 5% glucose medium decreases ethanol production from 171.5 to 149.5 mmol ethanol/(100 mmol glucose consumed) and increases glycerol yield from 6.16 to 22.2 mmol glycerol/(100 mmol glucose consumed). Addition of sulphite to the medium can further increase glycerol production: this process was practiced on a large scale during World War I.

In closing this brief review of fermentative product formation, we should note that some organisms convert amino acids in the medium into organic end products. These metabolic processes have important implications in production of alcoholic beverages, since higher aliphatic alcohols formed by fermentation of nitrogeneous compounds contribute to end product flavor and aroma.

Table 5.7 Bacterial sugar fermentations which proceed through the Embden-Meyerhof-Parnas pathway[†]

Fermentation class	Principal products from pyruvic acid	Bacterial groups(s) where found
1. Homolactic	Lactic acid	Lactic acid bacteria of genera *Streptococcus, Pediococcus, Lactobacillus* (some species)
2. Mixed acid	Lactic acid, acetic acid, succinic acid, formic acid (or CO_2 and H_2), ethanol	Many enteric bacteria, e.g., *Escherichia, Salmonella, Shigella, Proteus, Yersinia*
a. Butanediol	As in 2 but also 2,3-butanediol	*Aerobacter, Serratia, Aeromonas, Bacillus polymyxa*
3. Butyric acid	Butyric acid, acetic acid, CO_2 and H_2	Many anaerobic sporeformers (*Clostridium*); some nonsporeforming anaerobes (*Butyribacterium*)
a. Butanol-acetone	As in 3 but also butanol, ethanol, acetone, and isopropanol	Certain anaerobic sporeformers (*Clostridium* spp.)
4. Propionic acid	Propionic acid, acetic acid, succinic acid, CO_2	*Propionibacterium, Veillonella*

[†] R. Y. Stanier, M. Doudoroff, and E. A. Adelberg, *The Microbial World*, 3d ed., p. 183, Prentice-Hall, Inc., Englewood Cliffs, N.J., 1970.

5.9.2 Partial Oxidation and Its End Products

Normally, water and carbon dioxide are the metabolic end products of respiration for most aerobic microorganisms. Under abnormal conditions or with a few aerobic microbes, however, the oxidation of organic nutrient is not carried to completion, and end products then accumulate. Since some partial oxidations are of economic importance, we shall briefly review some of them here.

Some partial oxidations are determined by the microorganism's environment. One of these, production of citric acid by the mold *A. niger*, was cited in Sec. 1.3.3. By keeping sugar concentration very high and the concentration of iron (a cofactor for an enzyme which uses citric acid as a substrate) low, the yield of citric acid can be increased greatly (Table 5.8). By growing other fungi and aerobic bacteria under abnormal conditions, other intermediates of the TCA cycle such as α-ketoglutaric acid and fumaric acid are produced.

The acetic acid bacteria are well known for their tendency for incomplete oxidation. Both the *Acetobacter* and *Gluconobacter* genera oxidize ethanol to acetic acid, but, in the absence of ethanol, *Acetobacter* bacteria can oxidize acetic acid to CO_2. The *Gluconobacter*, on the other hand, cannot metabolize acetic acid; some species are known to lack the enzymes necessary for conducting the TCA cycle. The acetic acid end product secreted by Acetobacter growing in ethanol is defined to be *vinegar*. In Table 5.9, other end products available from the acetic acid bacteria are summarized.

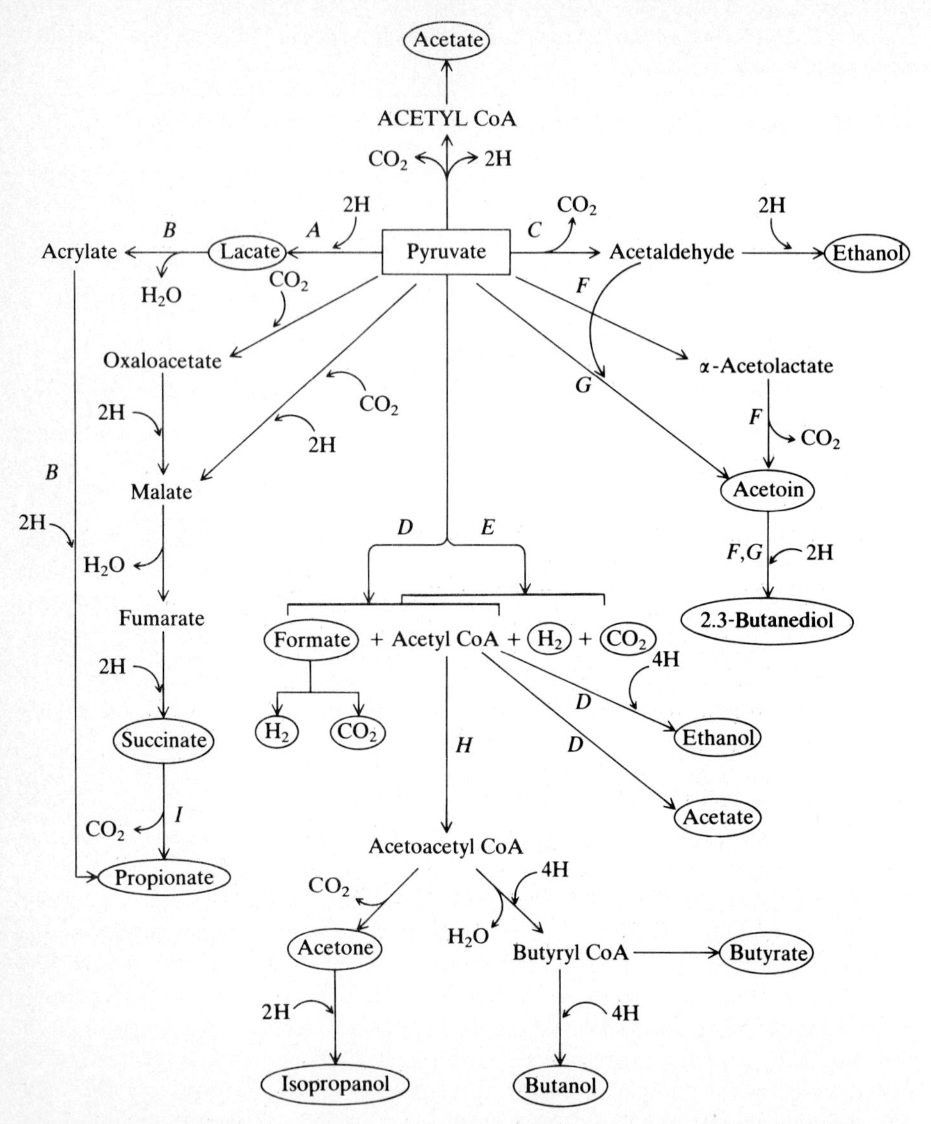

Figure 5.26 End products (*circled*) of microbial fermentations from pyruvate. Letters indicate organisms which conduct these reactions as follows: *A*, Lactic acid bacteria (*Streptococcus, Lactobacillus*); *B*, *Clostridium propionicum*; *C*, Yeast, Acetobacter, Zymomonas, *Sarcina ventriculi*, *Erwinia amylovora*; *D*, Enterobacteriaceae (coli-aerogenes); *E*, Clostridia; *F*, Aerobacter; *G*, Yeast; *H*, Clostridia (butyric, butylic organisms); *I*, Propionic acid bacteria. *(Reprinted by permission from J. Mandelstam and K. McQuillen (eds.), "Biochemistry of Bacterial Growth," 2d ed., p. 166, Blackwell Scientific Publications, 1973.)*

Table 5.8 Iron concentration vs. citric acid yield.†‡ *Aspergillus niger* utilizing sucrose; submerged culture

Iron§ ($FeCl_3$), mg/L	Weight yield; %¶	Iron§ ($FeCl_3$), mg/L	Weight yield; %¶
0.0	67.0	1.00	76.0
0.05	73.0	2.00	71.0
0.50	88.0	5.00	57.0
0.75	79.0	10.00	39.0

† H. J. Peppler (ed.), "Microbial Technology," table 8-1, Reinhold Publishing Corporation, New York, 1967.

‡ Medium composition: sucrose solution purified by ion exchange 3.6 MΩ resistance at 40% concentration, diluted to 14.2% sugar content, KH_2PO_4, 0.014%; $MgSO_4 \cdot 7H_2O$, 0.1%; $(NH_4)_2CO_3$, 0.2%; HCl to pH 2.6. (From U.S. patent 2,970,084.)

§ Supplied as $FeCl_3$.

¶ (g citric acid produced)/(g hexose moiety supplied) × 100.

5.9.3 Secondary Metabolite Synthesis

Microorganisms and other cells synthesize secondary metabolites when the cells and their environment are at appropriate conditions. Generally, these products are not synthesized in substantial quantity during rapid growth. Secondary metabolites are known which vary tremendously in their chemical structure and biological activity. For example, on the order of 600 different antibiotics have been identified. Both *Bacillus subtilis* and *Streptomyces griseus* produce more than 50 different antibiotics!

Many different secondary metabolites and their general routes of synthesis are summarized in Fig. 5.27. As shown, primary metabolites are the raw materials for secondary metabolite synthesis. Also, primary metabolites and the cell's energy charge are important modulators of the reaction pathways leading to secondary metabolite products. We shall consider this topic further in our examination of catabolite repression in Chaps. 6 and 7. The medium phosphate level is another important parameter in regulating secondary metabolite synthesis.

5.10 STOICHIOMETRY OF CELL GROWTH AND PRODUCT FORMATION

Cell growth, although an exquisite and complex process, obeys the laws of conservation of matter. Atoms of carbon, nitrogen, hydrogen, oxygen, and the other elements of life discussed in Chap. 2 are rearranged in the metabolic processes of the cell, but the total amounts of each of these elements incorporated into cell material is equal to the amounts removed from the environment. Further, the amount of some metabolic product formed or the amount of heat released by cell growth is often proportional to the amount of consumption of some substrate or the amount of formation of another product such as CO_2. In this section we

Table 5.9 Useful end products obtained by partial oxidations conducted by the acetic acid bacteria[†]

$$\underset{\text{Ethanol}}{CH_3CH_2OH} + O_2 \longrightarrow \underset{\text{Acetic acid}}{CH_3COOH} + H_2O$$

$$\underset{\text{Propanol}}{CH_3CH_2CH_2OH} + O_2 \longrightarrow \underset{\text{Propionic acid}}{CH_3CH_2COOH} + H_2O$$

$$\underset{\text{Isopropanol}}{(H_3C)_2CHOH} + \tfrac{1}{2}O_2 \longrightarrow \underset{\text{Acetone}}{(H_3C)_2C{=}O} + H_2O$$

$$\underset{\text{Glycerol}}{\begin{array}{l} CH_2OH \\ | \\ CHOH \\ | \\ CH_2OH \end{array}} + \tfrac{1}{2}O_2 \longrightarrow \underset{\text{Dihydroxyacetone}}{\begin{array}{l} CH_2OH \\ | \\ C{=}O \\ | \\ CH_2OH \end{array}} + H_2O$$

$$\underset{\text{2 3 Butanediol}}{\begin{array}{l} CH_3 \\ | \\ (CHOH)_2 \\ | \\ CH_3 \end{array}} + \tfrac{1}{2}O_2 \longrightarrow \underset{\text{Acetoin}}{\begin{array}{l} CH_3 \\ | \\ CHOH \\ | \\ C{=}O \\ | \\ CH_3 \end{array}} + H_2O$$

$$\underset{\text{Mannitol}}{\begin{array}{cccccccccc} & & H & & H & & OH & & OH & \\ & & | & & | & & | & & | & \\ HOCH_2 & - & C & - & C & - & C & - & C & -CH_2OH \\ & & | & & | & & | & & | & \\ & & OH & & OH & & H & & H & \end{array}} + \tfrac{1}{2}O_2 \longrightarrow \underset{\text{Fructose}}{\begin{array}{cccccccccc} & & & & H & & OH & & OH & \\ & & & & | & & | & & | & \\ HOCH_2 & - & C & - & C & - & C & - & C & -CH_2OH \\ & & \| & & | & & | & & | & \\ & & O & & OH & & H & & H & \end{array}} + H_2O$$

$$\underset{\text{Glucose}}{\begin{array}{cccccccccc} & & H & & H & & OH & & H & \\ & & | & & | & & | & & | & \\ HOCH_2 & - & C & - & C & - & C & - & C & -CHO \\ & & | & & | & & | & & | & \\ & & OH & & OH & & H & & OH & \end{array}} + \tfrac{1}{2}O_2 \longrightarrow \underset{\text{Gluconic acid}}{\begin{array}{cccccccccc} & & H & & H & & OH & & H & \\ & & | & & | & & | & & | & \\ HOCH_2 & - & C & - & C & - & C & - & C & -COOH \\ & & | & & | & & | & & | & \\ & & OH & & OH & & H & & OH & \end{array}} + H_2O$$

$$\underset{\text{Gluconic acid}}{\begin{array}{cccccccccc} & & H & & H & & OH & & H & \\ & & | & & | & & | & & | & \\ HOCH_2 & - & C & - & C & - & C & - & C & -COOH \\ & & | & & | & & | & & | & \\ & & OH & & OH & & H & & OH & \end{array}} + \tfrac{1}{2}O_2 \longrightarrow \underset{\text{5-Ketogluconic acid}}{\begin{array}{cccccccccc} & & & & H & & OH & & H & \\ & & & & | & & | & & | & \\ HOCH_2 & - & C & - & C & - & C & - & C & -COOH \\ & & \| & & | & & | & & | & \\ & & O & & OH & & H & & OH & \end{array}} + H_2O$$

[†] R. Y. Stanier, M. Doudoroff, and E. A. Adelberg, *The Microbial World*, 3d ed., p. 211, Prentice-Hall, Inc., Englewood Cliffs. NJ., 1970.

shall examine rigorous material balance constraints as well as approximate, empirical stoichiometric relationships which have proven useful. As we shall see, these stoichiometric considerations have broad implications in biochemical technology ranging from growth medium formulation to computer control and cooling requirements in bioreactors. Also, as in reactor analysis in general, knowledge of stoichiometric relationships is critical in formulating bioreactor material balances and making most effective and systematic use of reaction kinetics (Chap. 7).

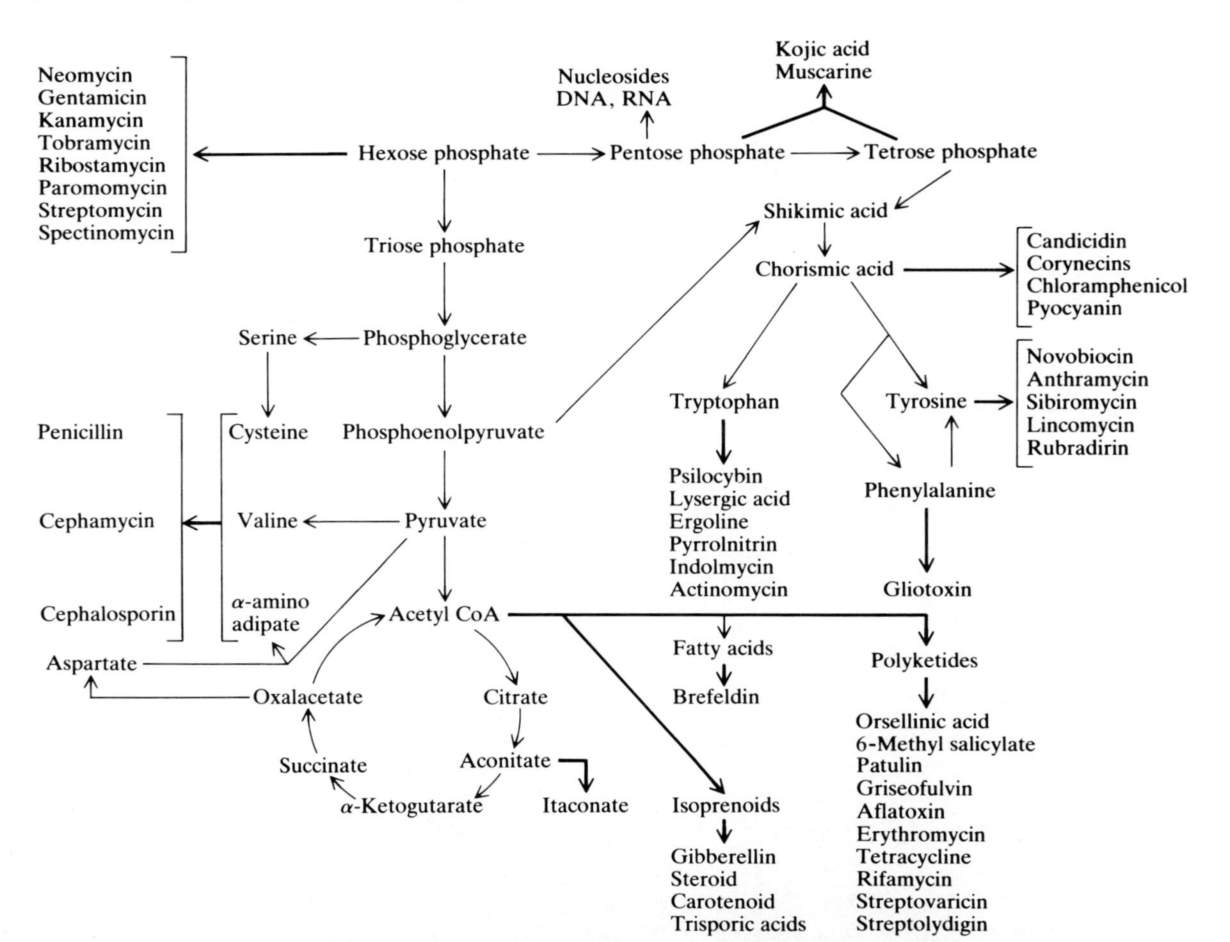

Figure 5.27 Summary of different metabolic routes to secondary metabolites. The heavy arrows denote enzyme-catalyzed reaction sequences and the products at the termini of these arrows are secondary metabolites. *(Reprinted by permission from V. S. Malik, "Microbial Secondary Metabolism," Trends Biochem. Sci., vol. 5, p. 68, 1980.)*

5.10.1 Overall Growth Stoichiometry: Medium Formulation and Yield Factors

Cell growth involves consumption of substrates which provide energy and raw materials required for the synthesis of additional cell mass. Viewed in strictly macroscopic terms, this process requires that the cell environment contains elements needed to form additional cell mass and that the free energy of substrates consumed should exceed the free energy of cells and metabolic products† formed (see Fig. 5.28). Clearly, this implies that the free energy of products formed must be less than the free energy of substrate consumed.

Also, the compounds supplied as nutrients must be compatible with the available enzymatic machinery of the cell for catabolism and biosynthesis. However, in examining cell growth from a strictly macroscopic viewpoint, which is the outlook we wish to develop in much of this section, it can be confusing to interject considerations of metabolic mechanisms. The whole point of the macroscopic perspective indicated in Fig. 5.28 is that certain stoichiometric constraints apply to this growth process *independent of* the particular mechanism or reaction pathways which the "system" of cells employs to effect the overall growth reaction. Consequently, we shall for the moment resolve the question of chemical suitability of certain nutrients by simply restricting our attention to those nutrients which are known to be acceptable substrates for growth of the cell strain of interest.

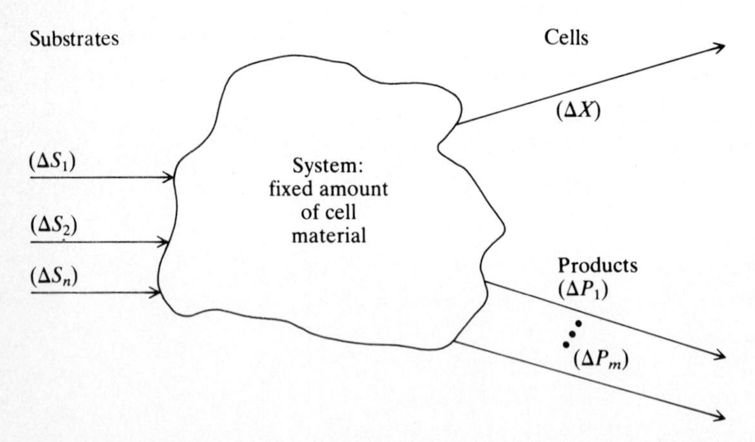

Figure 5.28 Simplest macroscopic view of cell growth. The "system" is defined as some fixed amount of cell material. In the process of growth, with this "system" cell material as a catalyst, substrates are converted into more cells and metabolic products.

† Here we shall adopt the following nomenclature convention; *metabolic product* or *product* refers to a substance, typically an organic compound, different from cell material that is released into the medium, and *nutrients* or *substrates* denote compounds which are depleted from the medium as a result of cell growth or product formation.

The simplest form of stoichiometric constraints on medium formulation are very briefly stated: if we wish to grow a total mass of cells (X) (uppercase letters in parentheses will be used to denote total mass of the substance indicated inside the parentheses) and if the cells contain w_i weight fraction of element i, then, at a minimum, substrates must be provided such that the total amount of element i in all nutrients is $w_i(X)$. Generally, this calculation is most conveniently done on a dry cell mass basis. To accomplish the calculation for many different elements, the cellular content of all of those elements must be known. Table 2.1 is an example of a relatively detailed breakdown for the bacterium *E. coli.* Cell content of the major elements C, N, O and H for several different microorganisms is indicated in Table 5.10. While some variations are evident from one species to another and for the same strain growing in different environments, the composition of microbial cell mass viewed on this elemental basis is quite consistent. The cellular empirical formulas indicated in this table will be explained and applied in the next section.

Medium formulation is actually more complicated for the following reasons. (1) Some substrate elements are released in products, not assimilated into cell material. (2) Rate limitations as well as stoichiometric limitations must be considered. (3) Specific nutrients may be limiting or specific products may be inhibitory due to the metabolic properties of a particular cell strain. The first point will be considered later in this section. The second point is discussed briefly before consideration of yield factors, the simplest quantitative formulation of cell growth stoichiometry.

While we wish in this section to focus only on stoichiometry and not on matters of rates or kinetics, it is necessary to consider both when formulating a strategy for simplifying the stoichiometric representation of cell growth and product formation. The number of nutrient components in the medium is typically very large, and we cannot in practice include all of them in our considerations either of process stoichiometry or of process kinetics. Therefore, we seek a rational simplification of the description of the system based upon the identification of certain key, limiting compounds or elements. We can consider identification of the *stoichiometrically limiting* compound on the basis of growth stoichiometry. That is, if cell growth were to continue according to the overall "reaction" indicated in Fig. 5.28, which substrate would be completely exhausted first? Another type of limiting component can be identified based upon medium composition effects on the cells' growth rate. The *growth-rate limiting* medium component may be identified for a particular strain and environment (temperature, pH, and background of certain other nutrients) in the following operational way. Suppose that cells are growing in a given medium under the prescribed environmental conditions. If the concentration of a particular medium component is increased suddenly, does the growth rate of the cells increase? Often, growth media are designed so that a single component is growth-rate limiting; changing the concentrations of all other medium components by small amounts relative to their base values in the medium does not alter the growth rate of the cells. However, it is possible in general for several substrates to be simultaneously rate limiting.

Table 5.10 Data on elemental composition of several microorganisms. μ denotes specific growth rate = mass of cells formed per unit time/mass of cells (see Chap. 7)

(Reprinted by permission from B. Atkinson and F. Mavituna, "Biochemical Engineering and Biotechnology Handbook," p. 120, Macmillan Publishers Ltd., Surrey, England, 1983).

Microorganism	Limiting nutrient	μ (h^{-1})	Composition (% by wt) C	H	N	O	P	S	Ash	Empirical chemical formula	Formula "molecular" weight
Bacteria			53.0	7.3	12.0	19.0			8	$CH_{1.666}N_{0.20}O_{0.27}$	20.7
Bacteria			47.1	7.8	13.7	31.3				$CH_2N_{0.25}O_{0.5}$	25.5
Aerobacter aerogenes			48.7	7.3	13.9	21.1			8.9	$CH_{1.78}N_{0.24}O_{0.33}$	22.5
Klebsiella aerogenes	Glycerol	0.1	50.6	7.3	13.0	29.0				$CH_{1.74}N_{0.22}O_{0.43}$	23.7
K aerogenes	Glycerol	0.85	50.1	7.3	14.0	28.7				$CH_{1.73}N_{0.24}O_{0.43}$	24.0
Yeast			47.0	6.5	7.5	31.0			8	$CH_{1.66}N_{0.13}O_{0.40}$	23.5
Yeast			50.3	7.4	8.8	33.5				$CH_{1.75}N_{0.15}O_{0.5}$	23.9
Yeast			44.7	6.2	8.5	31.2	1.08	0.6		$CH_{1.64}N_{0.16}O_{0.52}P_{0.01}S_{0.005}$	26.9
Candida utilis	Glucose	0.08	50.0	7.6	11.1	31.3				$CH_{1.82}N_{0.19}O_{0.47}$	24.0
C. utilis	Glucose	0.45	46.9	7.2	10.9	35.0				$CH_{1.84}N_{0.2}O_{0.56}$	25.6
C. utilis	Ethanol	0.06	50.3	7.7	11.0	30.8				$CH_{1.82}N_{0.19}O_{0.46}$	23.9
C. utilis	Ethanol	0.43	47.2	7.3	11.0	34.6				$CH_{1.84}N_{0.2}O_{0.55}$	25.5

The limiting component identified on the basis of growth rate may not be the same as the limiting component from a stoichiometric viewpoint. That is, one compound may limit the growth rate of the cells under certain conditions but it may be exhaustion of a different compound that causes the growth of the cell mass to stop in a batch cultivation. The possible distinction between such a growth-rate limiting component and a stoichiometrically limiting component is often ignored in formulating quantitative treatments of microbial growth. While this practice can sometimes cause difficulties in reactor analysis and confusion in interpreting performance of the bioreactor, we shall adopt at this point a perspective which focuses on a single, limiting nutrient, keeping in mind the possible complications just mentioned. For a particular system and a particular medium, we can always conduct experiments to explore the validity of the assumption that a single component is both growth-rate and stoichiometrically limiting.

It has been observed frequently that the total amount of cell mass formed by cell growth is proportional to the mass of substrate (typically carbon source, energy source or oxygen) utilized (see Fig. 5.29). The yield factor $Y_{X/S}$ is the corresponding ratio defined as

$$Y_{X/S} = \frac{(\Delta X)}{(\Delta S)} \tag{5.46}$$

We should note that the yield factor has dimensional units implied by the units used for cell amount and substrate amount (for example, mass or moles of either quantity). The units appropriate to any reported Y value should be checked

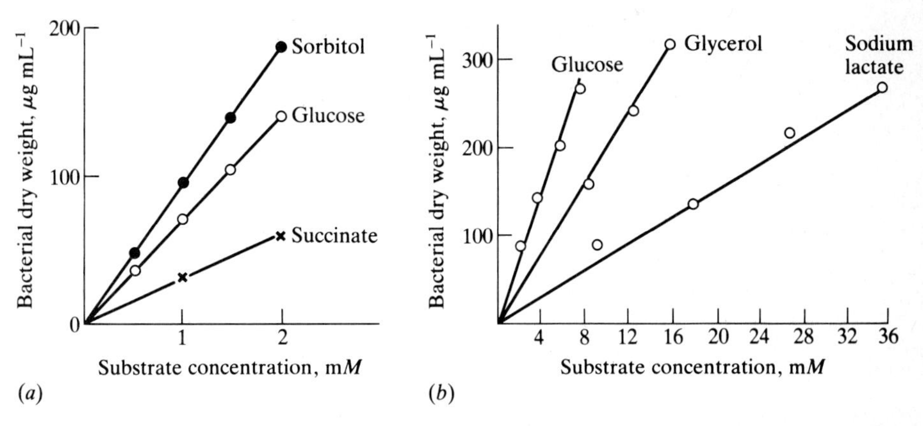

Figure 5.29 Growth yields (total cell mass formed vs. substrate consumed) for bacterial growth on several carbon sources. (*a*) Aerobic growth of *Aerobacter aerogenes*, (*b*) anaerobic growth of *Propionibacterium pentosaceum*. *((a) reprinted by permission from L. P. Hadjipetrou, J. P. Gerrits, F. A. G. Tenlings, and A. H. Stouthamer, "Relation between Energy Production and Growth of Aerobacter aerogenes," J. Gen. Microbiol., vol. 36, p. 139, 1964. (b) reprinted by permission from T. Bauchop and S. R. Elsden, "Growth of Microorganisms in Relation to Their Energy Supply," J. Gen. Microbiol., vol. 23, p. 457, 1960.)*

carefully to ensure consistent application of the yield factor parameter. Considering different substrates S_i in the medium, different corresponding cell growth yield factors Y_{X/S_i} may be defined. Table 5.11 lists yield factors for several different microorganisms grown in media with different carbon sources. Different values listed for each case are based upon substrate change in units of mass, moles, or grams of carbon per mole of substrate.

If the yield factor is approximately constant for a particular cell cultivation system, the relationship it provides is extremely useful, for knowledge either of the substrate or the cell mass concentration change suffices to determine the other quantity based on stoichiometry alone. This means that one of these variables can be expressed in terms of the other, eliminating one variable in reactor

Table 5.11 Summary of yield factors for aerobic growth of different microorganisms on various carbon sources. Y_{X/O_2} is the yield factor relating grams of cells formed per gram of O_2 consumed

[Reprinted by permission from S. Nagai, "Mass and Energy Balances for Microbial Growth Kinetics" in Advances in Biochemical Engineering vol. 11 (T. K. Ghose, A. Fiechter, and N. Blakebrough (eds.), Springer-Verlag, New York, p. 53, 1979.]

		$Y_{X/S}$			Y_{X/O_2}
Organism	Substrate	$\frac{g}{g}$	$\frac{g}{mole}$	$\frac{g}{g\text{—}C}$	$\frac{g}{g}$
Aerobacter aerogenes	Maltose	0.46	149.2	1.03	1.50
	Mannitol	0.52	95.5	1.32	1.18
	Fructose	0.42	76.1	1.05	1.46
	Glucose	0.40	72.7	1.01	1.11
Candida utilis	Glucose	0.51	91.8	1.28	1.32
Penicillium chrysogenum	Glucose	0.43	77.4	1.08	1.35
Pseudomonas fluorescens	Glucose	0.38	68.4	0.95	0.85
Rhodopseudomonas spheroides	Glucose	0.45	81.0	1.12	1.46
Saccharomyces cerevisiae	Glucose	0.50	90.0	1.25	0.97
Aerobacter aerogenes	Ribose	0.35	53.2	0.88	0.98
	Succinate	0.25	29.7	0.62	0.62
	Glycerol	0.45	41.8	1.16	0.97
	Lactate	0.18	16.6	0.46	0.37
	Pyruvate	0.20	17.9	0.49	0.48
	Acetate	0.18	10.5	0.43	0.31
Candida utilis	Acetate	0.36	21.0	0.90	0.70
Pseudomonas fluorescens	Acetate	0.28	16.8	0.70	0.46
Candida utilis	Ethanol	0.68	31.2	1.30	0.61
Pseudomonas fluorescens	Ethanol	0.49	22.5	0.93	0.42
Klebsiella sp.	Methanol	0.38	12.2	1.01	0.56
Methylomonas sp.	Methanol	0.48	15.4	1.28	0.53
Pseudomonas sp.	Methanol	0.41	13.1	1.09	0.44
Methylococcus sp.	Methane	1.01	16.2	1.34	0.29
Pseudomonas sp.	Methane	0.80	12.8	1.06	0.20
Pseudomonas sp.	Methane	0.60	9.6	0.80	0.19
Pseudomonas methanica	Methane	0.56	9.0	0.75	0.17

design calculations. The stoichiometric relationship is also potentially useful for bioreactor monitoring: measurement of substrate concentration implies a corresponding estimate of the cell concentration if initial values of both quantities are known. This is quite significant since substrate levels are often much easier to determine than is biomass concentration. Actually, some product concentration levels, such as CO_2 leaving an aerobic bioreactor, are the most convenient to measure in practice. Here, too, stoichiometry provides a useful structure for relating different bioreactor process variables. Before turning to these points, however, we need to mention yield factor variability and the concept of maintenance metabolism.

There is no guarantee that a yield factor, an empirically defined, apparent stoichiometric ratio, is a constant for a given organism in a given medium. Variations in yield factors, for example with respect to growth rate, are commonly observed. To understand these variations it is sometimes useful to break down substrate consumption into three parts: assimilation into cell mass, provision of energy for cell synthesis, and provision of energy for maintenance. *Maintenance* here refers to the collection of cell energetic requirements for survival, or for preservation of a certain cell state, which are not directly related to or coupled with the synthesis of more cells. Such activities include active transport of ions and other species across cell membranes and replacement synthesis of decayed cell constituents.

For chemoheterotrophs, a single substrate serves as both carbon and energy source, so that total substrate utilization may be written as

$$(\Delta S) = (\Delta S)_{\text{assimilation}} + (\Delta S)_{\text{growth energy}} + (\Delta S)_{\text{maintenance energy}} \tag{5.47}$$

Dividing the equation by (ΔX) gives

$$\frac{1}{Y_{X/S}} = \frac{(\Delta S)_{\text{assimilation}}}{(\Delta X)} + \frac{(\Delta S)_{\text{growth energy}}}{(\Delta X)} + \frac{(\Delta S)_{\text{maintenance energy}}}{(\Delta X)} \tag{5.48}$$

While the true growth yield factor $\Delta X/(\Delta S)_{\text{assimilation}}$ is a relatively constant, stoichiometrically well-defined quantity, the overall yield factor $Y_{X/S}$ is not. The allocation of substrate used into the three components indicated in Eq. (5.48) is variable. A rapidly growing cell population will use more substrate for assimilation and growth energy, while a cell population in a resting or stationary state often consumes substrate for maintenance without any growth (here $Y_{X/S} = 0$).

In the next section we will develop an alternative and more rigorous formulation of cell growth stoichiometry based upon material balances for C, H, N, and O. Also, we shall examine a more detailed view of cell growth that, although still basically in macroscopic form, considers the central role of NADH and ATP in growth metabolism.

5.10.2 Elemental Material Balances for Growth

In order to describe cell growth and related metabolic activities in the same way as used for simple chemical reactions, it is first necessary to establish a chemical formula for dry cell material. If the elemental composition of a particular strain growing under particular conditions is known (see, for example, Table 5.10), the ratios of subscripts in the empirical cell formula $C_\theta H_\alpha O_\beta N_\delta$ are determined. In order to establish a unique cell formula and corresponding formula weight, it has proved convenient to employ a formula which contains one gram-atom of carbon. That is, we choose $\theta = 1$ and then fix α, β, and δ so that the formula is consistent with known relative elemental weight content of the cells. The cell formulas in Table 5.10 have been written using this convention. One *C-mole of cells* is by definition the quantity of cells containing one gram-atom (12.011 grams) carbon, and corresponds to the cell formula weight with the carbon subscript θ taken as unity.

As the simplest illustration of the elemental balance approach to stoichiometry, we consider first aerobic growth without product formation, that is, where the only products of the growth reaction are cells, CO_2 and H_2O. Then, writing the carbon source and nitrogen source chemical formulas as CH_xO_y and $H_lO_mN_n$, respectively, the growth reaction equation is

$$a'CH_xO_y + b'O_2 + c'H_lO_mN_n \longrightarrow CH_\alpha O_\beta N_\delta + d'H_2O + e'CO_2 \quad (5.49)$$

Without loss of generality (since stoichiometric coefficients may always be multiplied by the same constant), the coefficient of cells is taken as unity.

Balances on the four elements in Eq. (5.49) provide four relationships among the five unknown stoichiometric coefficients a', b', c', d', and e':

$$\begin{aligned} &\text{C:}\ a' = 1 + e' \\ &\text{H:}\ a'x + c'l = \alpha + 2d' \\ &\text{O:}\ a'y + 2b' + c'm = \beta + d' + 2e' \\ &\text{N:}\ c'n = \delta \end{aligned} \quad (5.50)$$

An additional relationship may be provided by experimental determination of the *respiratory quotient*, or RQ, for the growth reaction. The respiratory quotient is defined as the molar ratio of CO_2 formed to O_2 consumed:

$$\text{Respiratory quotient} = \text{RQ} = \frac{\text{moles } CO_2 \text{ formed}}{\text{moles } O_2 \text{ consumed}} \quad (5.51)$$

For the growth reaction in Eq. (5.49) above,

$$\text{RQ} = e'/b' \quad (5.52)$$

Thus, if the RQ is known, Eq. (5.52) plus Eqs. (5.50) provide five equations for the five unknown stoichiometric coefficients. Possible difficulties with extreme sensitivity of the calculated stoichiometry to small errors in the RQ measurement are considered in Prob. 5.16.

The chemical formula for cell composition and the stoichiometric coefficients in Eq. (5.49) in general change as a function of cell environment. We must remember that this is a phenomenological relationship, used here to describe the combined effects of a large number of independent chemical reactions. Insights into the influences on overall stoichiometry and bridges to our earlier discussions of metabolism can be obtained by decomposing the overall growth reaction into several reactions steps which correspond to different metabolic functions (but not necessarily to specific metabolic pathways).

We shall next consider a more detailed stoichiometric representation for aerobic growth of a chemoheterotrophic organism. To each reaction we associate a reaction involving ATP generation or utilization. These have been written separately here, and not simply added to the associated reaction involving C, H, O, and N rearrangement, because of practical difficulties with ATP stoichiometry which will be discussed shortly.

Energy source dissimilation

$$a\mathrm{CH}_x\mathrm{O}_y + \text{coefficient}\cdot\mathrm{H_2O} + \frac{\gamma_s}{2}a\mathrm{NAD}^+ \longrightarrow a\mathrm{CO}_2 + \frac{\gamma_s}{2}a(\mathrm{NADH} + \mathrm{H}^+) \quad (5.53a)$$

$$\varepsilon_s a(\mathrm{ADP} + \mathrm{P}_i) \longrightarrow \varepsilon_s a(\mathrm{ATP} + \mathrm{H_2O}) \quad (5.53b)$$

Oxidative phosphorylation

$$2b(\mathrm{NADH} + \mathrm{H}^+) + b\mathrm{O}_2 \longrightarrow 2b\mathrm{NAD}^+ + 2b\mathrm{H_2O} \quad (5.54a)$$

$$2b(\mathrm{P/O})(\mathrm{ADP} + \mathrm{P}_i) \longrightarrow 2b(\mathrm{P/O})(\mathrm{ATP} + \mathrm{H_2O}) \quad (5.54b)$$

Biosynthesis

$$(1+\sigma)\mathrm{CH}_x\mathrm{O}_y + \frac{\delta}{n}\mathrm{H}_l\mathrm{O}_m\mathrm{N}_n + \tfrac{1}{2}\left(\gamma_B - \gamma_S(1+\sigma) - \frac{\delta}{n}\gamma_N\right)(\mathrm{NADH} + \mathrm{H}^+) \longrightarrow \mathrm{CH}_\alpha\mathrm{O}_\beta\mathrm{N}_\delta + \sigma\mathrm{CO}_2 + (\text{coefficient})\cdot\mathrm{H_2O} \quad (5.55a)$$

$$\frac{MW_B}{Y_{\mathrm{ATP}}^{\max}}(\mathrm{ATP} + \mathrm{H_2O}) \longrightarrow \frac{MW_B}{Y_{\mathrm{ATP}}^{\max}}(\mathrm{ADP} + \mathrm{P}_i) \quad (5.55b)$$

Maintenance and dissipation

$$c(\mathrm{ATP} + \mathrm{H_2O}) \longrightarrow c(\mathrm{ADP} + \mathrm{P}_i) \quad (5.56)$$

Interconnections among these reactions are provided by the constraints that, under the conditions of interest here, it may be assumed that no ATP and no NADH accumulates. Formation must be balanced by utilization.

The cost of a more complex but more revealing representation of chemical changes accompanying cell growth is a dramatic increase in the number of stoichiometric parameters. First, notice that the stoichiometric coefficients are written taking into account atom balances for each reaction. Next, notice that the stoichiometric coefficient of biomass ($\mathrm{CH}_\alpha\mathrm{O}_\beta\mathrm{N}_\gamma$) has been taken to be unity. The coefficients a, b, and c which appear throughout other reactions are therefore used to denote the extent to which these reactions occur *relative to* the growth reaction. Rather than writing out the value in detail, the stoichiometric coefficients of water have been written simply as "coefficient" in two cases above since water consumption or production is usually not significant relative to the amount present. Here ε_s denotes the number of substrate-level phosphorylations per carbon mole passing through dissimilation metabolism.

The parameters γ_B, γ_S, and γ_N denote the *degrees of reductance* of biomass, carbon source, and nitrogen source, respectively. The reductance degree is the number of equivalents of available electrons per g atom carbon, based on carbon = +4, hydrogen = +1, oxygen = −2 and nitrogen = −3. Thus, the degree of reductance of a compound with formula $CH_rO_sN_v$ is $\gamma = 4 + r - 2s - 3v$. Based on this convention, the reductance degrees of CO_2, H_2O, and NH_3 are all zero.

The yield coefficient Y_{ATP}^{max} is the mass of cells formed per mole of ATP consumed for biosynthesis. Efforts have been made to evaluate this quantity based upon known cell composition and known biosynthetic pathways (see Ref. 13). Such calculations indicate, for example, that Y_{ATP}^{max} is 28.8 g cells/(gmol ATP) for growth of *E. coli* on glucose and inorganic salts. The molecular weight (B denotes biomass) in Eq. (5.55*b*) is needed for conversion from mass to molar units.

As noted earlier, P/O is the number of ADP phosphorylations per atom of O_2 consumed; this parameter characterizes the efficiency of oxidative phosphorylation and in general varies depending on cultivation conditions. Calculations based upon an average microbial composition estimate maximum theoretical P/O values of 2.25, 2.50, and 3.00 for growth on acetate, malate, and glucose, respectively.

Direct experimental measurement of these quantities is complicated by the difficulty of isolating experimentally a particular ATP generation or utilization pathway and the dependence of the relative extents of ATP formation and use on culture environment. The major problems here arise because of ATP utilization to drive membrane transport processes, to synthesize molecules which are degraded by intracellular hydrolases, and other not fully characterized processes sometimes called futile cycles. All of these ATP demands, grouped together here under the title "maintenance and dissipation" and indicated above by the single "reaction" (5.56), influence other ATP generation and utilization processes.

Since ATP generation must match ATP utilization, alteration in the relative extent or weight of maintenance ATP consumption, indicated above by the coefficient c, forces shifts in other reaction extents. Experimental evidence shows that maintenance requirements are not dictated by stoichiometric considerations but depend instead on the rates of metabolic processes. Accordingly, we must defer quantitative treatment of maintenance until we explore cell kinetics in Chap. 7.

It may offer useful perspective at this point to mention the following observations: the overall cell mass yield based on ATP utilization is quite similar for many substrates and organisms grown anaerobically. The average value is about

$Y_{ATP} = 10.7$ g cells/(mol ATP). In chemoheterotraphic anaerobes, typically much more substrate is consumed for energy production than for assimilation into biomass. In one study of anaerobic growth of bakers' yeast, *Saccharomyces cerevisiae*, in rich medium with glucose carbon source, 98 percent of the carbon consumed is used for energy production and only 2 percent for cell mass. It has proved quite difficult to determine accurately ATP generation in aerobic growth. Available data suggests P/O ratios in the range 0.5 to 1.8—much lower than theoretical estimates. Studies of the bacterium *Aerobacter cloacae* in aerobic minimal medium showed that 55 percent of the glucose consumed was assimilated, and 45 percent of the glucose utilized was used for energy production. As these figures suggest, the cell mass yield per amount of substrate consumed is typically greater under aerobic conditions. For example, for *Streptococcus faecalis* grown in glucose medium, the yield factor $Y_{X/S}$ [g cells/mol glucose)] is 21.5 under anaerobic conditions and 58.2 for aerobic growth. We will examine the implications of these allocations of substrate for product formation stoichiometry in the next section.

5.10.3 Product Formation Stoichiometry

A variety of metabolic end products is released into the growth medium or accumulated intracellularly. The pertinent stoichiometries for product formation may be classified usefully into the following four classes; the first three correspond to the fermentations classification formulated by Elmer L. Gaden, Jr. in 1955.

1. The main product appears as a result of primary energy metabolism. Example: ethanol production during anaerobic growth of yeast.
2. The main product arises indirectly from energy metabolism. Example: citric acid formation during aerobic mold cultivation.
3. The product is a *secondary metabolite*. Example: penicillin production in aerated mold culture.
4. Biotransformation. The product is obtained from substrate through one or more reactions catalyzed by enzymes in the cells. Example: steroid hydroxylation.

Class 1 processes have a relatively simple stoichiometric description. Product appears in relatively constant proportions as cell mass accumulates and substrate is consumed. Here, the processes of substrate utilization, cell mass synthesis, and product formation are linked as in a simple, single chemical reaction. This may be obtained by adding product, denoted CH_vO_w, to the right-hand side of Eq. (5.49), giving

$$a'CH_xO_y + b'O_2 + c'H_lO_mN_n \longrightarrow CH_\alpha O_\beta N_\delta + d'H_2O + e'CO_2 + f'CH_vO_w \quad (5.57)$$

(Extension of the formalism to include nitrogen-containing products is straightforward.)

A corresponding yield factor may be written, taking into account the use of molar units in the yield factor parameter. Thus, if the substrate is glucose which contains six carbon atoms per molecule, the number of glucose molecules used when reaction (5.57) occurs is $a'/6$. Similarly, if the product contains n_p carbon atoms per molecule, f'/n_p product molecules are formed. Thus, the molar yield factor $Y_{P/S}$ (moles product formed/moles substrate consumed) may be written

$$Y_{P/S} = \frac{f'}{a'}\frac{n_s}{n_p} \tag{5.58}$$

Here, n_s denotes the number of carbon atoms in a substrate molecule.

For class 2 situations the simple stoichiometry of Eq. (5.57) does not apply. Product formation is not necessarily proportional to substrate utilization or cell mass increase. Representation of the stoichiometry in such a case requires an independent reaction equation for product formation. It is instructive to consider the addition of a product formation step to the growth reactions given in Eqs. (5.53) through (5.56) above:

Product formation

$$dCH_xO_y + (\text{coefficient})H_2O + \tfrac{1}{2}(\gamma_s - z\gamma_p)NAD^+ \longrightarrow$$

$$dCH_yO_w + d(1 - z)CO_2 + d\tfrac{1}{2}(\gamma_s - z\gamma_p)(NADH + H^+) \tag{5.59a}$$

$$\varepsilon_p d(ADP + P_i) \longrightarrow \varepsilon_p d(ATP + H_2O) \tag{5.59b}$$

Here, z denotes the carbon fraction of substrate used for product formation which is found in the product. The number of ATPs generated by product formation is denoted ε_p; this coefficient may be negative, indicating that ATP is utilized rather than generated in connection with product formation metabolism. Similarly, if the coefficient $\gamma_s - z\gamma_p$ is negative, reducing equivalents stored as NADH (or NADPH—at this low level of metabolic detail there is no reason to distinguish between these two carriers) are utilized in concert with product formation.

The stoichiometric descriptions appropriate for classes 3 and 4 cases depend on the particular substrates and products involved. Product formation in these cases is typically completely uncoupled from cell growth. Secondary metabolite accumulation is dictated by kinetic regulation and activity of the cells.

Examination of product formation stoichiometry offers several useful insights for engineering analysis. One application is estimation of upper bounds for product yields. These numbers can be very useful in preliminary economic analysis (see Chap. 12). Consider the anaerobic fermentation of glucose to ethanol as an initial example. (The reactions given above for aerobic metabolism may be adapted for anaerobic situations by deleting O_2 everywhere and dropping oxidative phosphorylation reactions.) The best possible case is utilization of all substrate for product formation alone. Assuming that cell growth is negligible, then, Eq. (5.57) becomes simply

$$CH_2O \longrightarrow \tfrac{1}{3}CO_2 + \tfrac{2}{3}CH_3O_{0.5} \qquad (5.60)$$

showing that at most 2/3 of the substrate carbon used can appear in the product. Recalling Eq. (5.58), the corresponding upper bound on the molar yield factor is $(Y_{P/S})_{max} = (2/3)(6/2) = 2$.

The same approach can be applied in all other types of product formation situations, including those in which multiple products are formed. We can determine the upper bound for product yield by assuming that no cells or other products accumulate. For example, writing the reaction for penicillin synthesis from precursors,

$$1.5\text{ glucose} + H_2SO_4 + 2NH_3 + \text{phenylacetate} \longrightarrow \text{penicillin G} + CO_2 + 8H_2O \qquad (5.61)$$

we can determine that the maximum possible yield of penicillin is 1.2 g penicillin/(g glucose). Actual yields are much lower (ca. 0.1 g/g), indicating useful remaining potential for organism and process improvement. Yields much lower than the upper bound value indicate that significant substrate utilization supports growth, maintenance or synthesis of other products.

Examining the interconnection between energy metabolism and product formation as represented in Eqs. (5.53) through (5.56) and (5.59) can provide useful insights into the impact of maintenance requirements on product yields. Suppose that ethanol is produced from glucose anaerobically in a medium containing ammonia as a nitrogen source. Substrate-level phosphorylation [Eq. (5.53*b*)] is the only source of ATP. Thus, a relatively high maintenance requirement means a relatively large amount of glucose will be used in dissimilation reactions. This in turn generates relatively more NADH, which is not used to any great degree in biosynthesis, Eq. (5.55*a*), since the reductance degrees of substrate and biomass are quite similar. The stored electrons are used in product formation, Eq. (5.59*a*), resulting in relatively large accumulation in the medium of a product (ethanol), more reduced than the substrate. Conversely, in aerobic systems where classes 3 or 4 product formation is often the goal, increased maintenance requirements decrease product yields by increasing substrate utilization for energy production.

Clearly, there are many different ways in which metabolic subreactions such as Eqs. (5.53) through (5.56) and (5.59) could be formulated. Modifications in the stoichiometric formulation are needed for example, if multiple carbon and/or nitrogen sources are used or if the product contains nitrogen. Also, stoichiometric statements could equally well be written based on particular metabolic pathways. Indeed, approaches of this type, combined with measurements of the amounts of substrates consumed and products formed, have been applied usefully to ascertain the relative activities of parallel and branching metabolic pathways. Since the basic principles and methods used in such cases correspond closely to that described above, we shall leave to the problems further consideration of alternative stoichiometric formulations. Next, we extend our discussion of stoichiometry to consider metabolic heat generation.

5.10.4 Metabolic Energy Stoichiometry: Heat Generation and Yield Factor Estimates

Cells use chemical energy quite efficiently but, like any real process, some of the energy in the substrates is released as heat. Metabolic heat generation dictates cooling requirements for bioreactors which harbor cells. Cellular heat production is primarily the result of energy and growth metabolism. Consequently, it is reasonable to expect an approximately proportional relationship between heat generated and energy substrate utilized. Accordingly, we define a yield factor Y_Δ (grams of cell mass per kilocalorie of heat evolved) analogous to the earlier yield coefficients. If Y_s is grams of cell mass produced per gram substrate consumed and ΔH_s and ΔH_c are the heats of combustion (kilocalories per gram) of substrate and of cell mass material, respectively we can write

$$Y_\Delta(\text{g cell/kcal}) = \frac{Y_s(\text{g cell/g substrate})}{(\Delta H_s - Y_s \Delta H_c)(\text{kcal/g substrate})} \tag{5.62}$$

This arises from approximate energy balances over two pathways, as shown in Fig. 5.30, for aerobic growth. Provided that the *predominant* oxidant is oxygen, the heat generation ΔH_s per gram of substrate completely oxidized minus $Y_s \Delta H_c$, the heat obtained by combustion of cells (and extracellular-fluid residue) grown from the same amount of substrate, will reasonably approximate the heat generation per gram of substrate consumed in the fermentation which produces cells, H_2O and CO_2.

The accuracy of the assumptions underlying Fig. 5.30 and Eq. (5.62) is exemplified by data for yeast growing on *n*-paraffins. By experiment, ΔH_s and ΔH_c were found to be 11.4 and 4.7 kcal/g, respectively, giving

$$\frac{1}{Y_\Delta} = \frac{11{,}400}{Y_s} - 4700$$

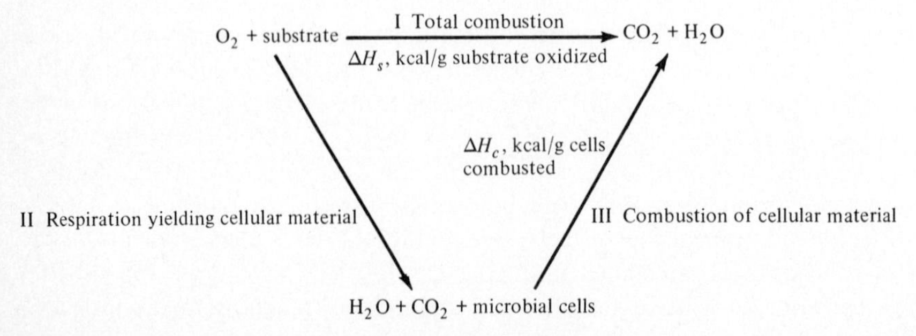

Figure 5.30 Approximate heat balance in substrate consumption.

which is Eq. (5.62) inverted. Direct measurement of Y_s and Y_Δ produced the following results [16]:

	Y_Δ	
Y_s, experiment	From Eq. (5.62)	Measured
1.25	4400	4640
1.03	6400	6060
1.09	5800	5830

In the absence of experimental data, the heat of combustion of cells (and other compounds, for that matter) can be estimated based upon the following empirical observation: the energy obtained by transferring electrons from a compound which has reductance degree γ_s to a compound such as CO_2 or CH_4, which has zero reductance degree, is given by $K \cdot \gamma_s$, where the value of K is in the range 26 to 31 kcal/(electron equivalent). Since molecular oxygen O_2 accepts four electrons during respiration, this corresponds to 104–124 kcal/mol O_2 consumed. Data for two bacteria, a yeast, and a mold grown on different media are summarized in Fig. 5.31. While the scatter is nontrivial, the general trend is clear.

As an example, we shall estimate the heat of combustion of *Pseudomonas fluorescens* growing in glucose medium. First, a reaction for cell combustion is

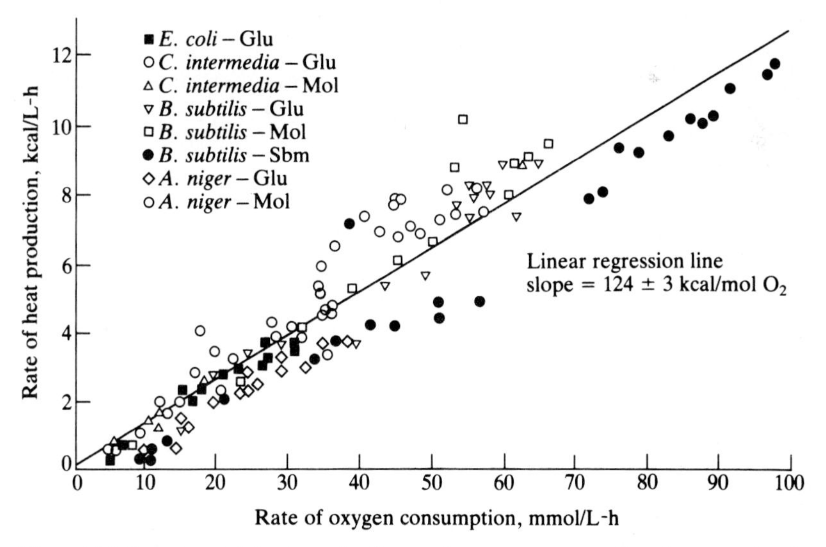

Figure 5.31 Experimental heat production and oxygen consumption for several microorganisms grown in different media. (Glu, glucose medium; Mol, molasses medium; Sbm, soy bean medium). *(Reprinted by permission from C. L. Cooney, D. I. C. Wang, and R. I. Mateles, "Measurement of Heat Evolution and Correlation with Oxygen Consumption during Microbial Growth," Biotech. Bioeng., vol. 11, p. 269, 1968.)*

written using the measured cell composition, assuming the combustion products are CO_2, H_2O, and N_2:

$$\underset{\text{Cells}}{CH_{1.66}N_{0.20}O_{0.27}} + 1.28\,O_2 \longrightarrow CO_2 + 0.10\,N_2 + 0.83\,H_2O \qquad (5.63)$$

Assuming a heat of combustion of 104 kcal per mole O_2, the heat released by combustion of bacteria as in Eq. (5.63) is

$$\frac{(1.28 \text{ mol } O_2)(104 \text{ kcal/mol } O_2)}{[12 + (1.66)(1) + (0.20)(14) + (0.27)(16)]\text{g}} = 6.41 \frac{\text{kcal}}{\text{g}} \qquad (5.64)$$

However, cell dry weight measured experimentally includes ash. If the cell's dry matter contains 10 percent ash, then the heat of cell combustion is

$$\Delta H_c \cong (0.90)\left(\frac{6.41 \text{ kcal}}{\text{g}}\right) = 5.8 \frac{\text{kcal}}{\text{g dry cell mass}} \qquad (5.65)$$

The value of Y_Δ will depend both upon the particular microbial species (through ΔH_c) and upon the substrate consumed (through ΔH_s). A sampling of calculated values appears in Table 5.12. Note, in general, that hydrocarbons produce more heat than partially oxygenated species [$Y_\Delta(CH_4) < Y_\Delta(CH_3OH)$, $Y_\Delta(n\text{-alkanes}) < (Y_\Delta(\text{glucose})$]. Thus, more reduced substrates imply greater heat removal demands on the bioreactor. The economic implications of this energetic stoichiometry are considered in Chap. 12.

In fermentations or waste-treatment processes, the substrates are frequently a mixture of many compounds, each having a different heat of combustion. Writing the standard free energy of combustion as $\Delta G°$, Servizi and Bogan [18] noted that the stoichiometric ratio (m_0 = moles O_2 needed per complete oxidation of

Table 5.12 Comparison of yield coefficients for bacteria grown on various carbon sources

Substrate	Y_s, g cell/g substrate	Y_{O_2}, g cell/g O_2 consumed	Y_Δ, g cell/kcal
Malate	0.34	1.02	0.30
Acetate	0.36	0.70	0.21
Glucose equivalents (molasses, starch, cellulose)	0.51	1.47	0.42
Methanol	0.40	0.44	0.12
Ethanol	0.68	0.61	0.18
Isopropanol	0.43	0.23	0.074
n-Paraffins	1.03	0.50	0.16
Methane	0.62	0.20	0.061

† From B. J. Abbott and A. Clamen, "The Relationship of Substrate, Growth Rate, and Maintenance Coefficient to Single Cell Protein Production," *Biotech. Bioeng.*, **15**: 117, 1973.

1 mol of substrate) is proportional to ΔG° for a number of compounds. In particular

$$\Delta G^\circ = -116 m_0 \qquad \text{kcal/mol substrate} \tag{5.66}$$

for carbohydrates, TCA cycle intermediates and some products of glycolysis, and

$$\Delta G^\circ = -104 m_0 \qquad \text{kcal/mol substrate} \tag{5.67}$$

for aromatics, alcohols, and aliphatic acids. (Hattori [Ref. 4 of Chap. 13] notes that if the chemical oxygen demand (COD) in grams of oxygen per mole is already known, $m_0 = \text{COD}/32$ which may be substituted directly into these equations.) The average free energy in complete oxidation of multiple substrates is simply

$$\overline{\Delta G^\circ} = \frac{\sum_i (M_i \Delta G_i^\circ)}{\sum_i M_i} \qquad \text{kcal/mole} \tag{5.68}$$

where i = species

M_i = moles of ith substrate

ΔG_i° = free energy for complete oxidation of substrate i

Further, it appears that the yield coefficient Y for a reactor containing a number of unicellular species can be defined in the same manner as for the single species (bacteria or yeast) of the preceding discussion. Servizi and Bogan [18] determined the average yield coefficient $\bar{Y}_s$ (grams of cell per mole of substrate) for the multiple-species/multiple-substrate system *activated sludge* (Chap. 14) and found (see Fig. 5.32)

$$\bar{Y}_s = 0.108\overline{\Delta G^\circ} \qquad \text{g cells/mol substrate} \tag{5.69}$$

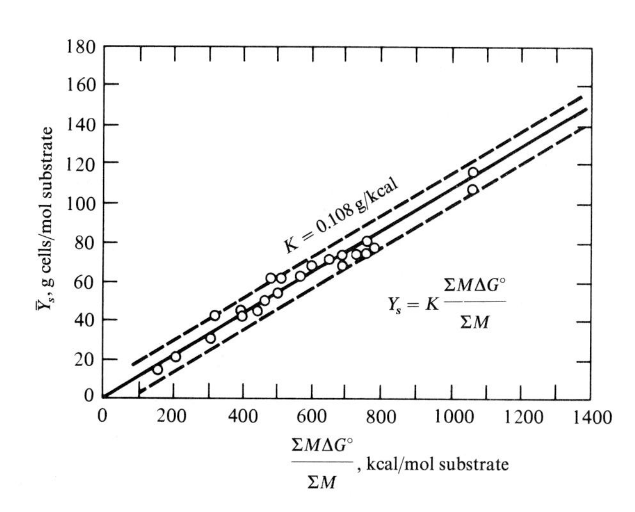

Figure 5.32 Average yield coefficient for activated sludge growth as a function of the average substrate free energy of oxidation. *(Reprinted from J. A. Servizi and R. H. Bogan, "Thermodynamic Aspects of Biological Oxidation and Synthesis," J. Water Pollut. Control Fed., vol. 36, p. 607, 1961.)*

Thus, given $\bar{Y}_s$, the other quantities in Eq. (5.62) can be determined easily experimentally, and hence the microbial heat generation yield factor Y_Δ can again be calculated.

It is reasonable to expect the coefficient in Eq. (5.69) to depend upon the microbial species involved, so that again direct measurement on the appropriate species is desirable. However, the original argument of Servizi and Bogan that $\bar{Y}_s \propto \overline{\Delta G^\circ}$ is based on an assumed linearity between the yield coefficient $\bar{Y}_s$ and the number $\bar{N}_{\text{ATP}}$ of molecules of ATP synthesized per mole of substrate consumed. Since $\bar{Y}_s$ = grams of cell per gram of substrate, the coefficient of proportionality between $\bar{Y}_s$ and $\bar{N}_{\text{ATP}}$ is simply the grams of cells produced per mole of ATP utilized. This latter value is known to be approximately constant for a wide range of anaerobic species but is more variable for aerobes (Sec. 5.9.2). Equation (5.69) may provide a useful estimate in general for mixed-substrate/mixed-population systems in the absence of better data. It is probably accurate only to the order of ± 20 to 30 percent, as can be seen best from the data for the correlation displayed in Fig. 5.32.

The relationship mentioned above between available electrons and energy content of a substance may be used to obtain an estimate of the upper bound of organic product yield from organic substrate. Defining ζ_p as the fraction of available electrons in the substrate included in the product, we can evaluate ζ_p as

$$\zeta_p = \frac{\text{(available electrons/g product)}}{\text{(available electrons/g substrate)}} \cdot \left(\frac{\text{g product}}{\text{g substrate}} \right) \tag{5.70}$$

The last term on the right-hand side is simply the mass yield factor $Y_{\text{P/S}}$. The number of available electrons per gram of an organic substance may be written as $\sigma\gamma/12$ where σ is the mass fraction of carbon in the compound and γ is the degree of reductance of this carbon. Substituting this relationship into Eq (5.70) gives

$$\zeta_p = \frac{\sigma_p \gamma_p}{\sigma_s \gamma_s} Y_{\text{P/S}} \tag{5.71}$$

Since the substrate and product energies are approximately proportional to their available electron content, ζ_p has the alternative interpretation as the energy yield of product from substrate. As such, it follows that ζ_p must be no greater than unity. Accordingly, an upper bound on product mass yield, $Y_{\text{P/S}}^{\max}$, may be obtained by taking ζ_p equal to unity in Eq. (5.71):

$$Y_{\text{P/S}}^{\max} = \frac{\sigma_s \gamma_s}{\sigma_p \gamma_p} \tag{5.72}$$

Values of $Y_{\text{P/S}}^{\max}$ for several different substrates and products are listed in Table 5.13.

Table 5.13 Upper bounds for product mass yields $Y_{P/S}^{max}$ (g product/g substrate) estimated from Eq. (5.72)

(Reprinted by permission from V. K. Eroshin and I. G. Minkevich, "On the Upper Limit of Mass Yield of an Organic Product from an Organic Substrate," Biotech. Bioeng., vol. 24, p. 2263, 1982)

	Substrate		
Product	*n*-Alkanes	Ethanol	Glucose
Citric acid	4.55	2.77	1.42
Polysaccharides	3.21	1.96	1.00
Acetic acid	3.21	1.96	1.00
Lysine	2.23	1.36	0.70
Streptomycin	....	....	0.66
Ethanol	1.64	....	0.51
Triglycerides	1.18	0.72	0.37

5.10.5 Photosynthesis Stoichiometry

Yields for photosynthetic reduction of CO_2 to carbohydrate are important first for food production and secondly for determining rates at which biomass process feedstocks can be renewed. Based upon known pathways of electron flow and photosynthetic reactions, a theoretical estimate for photosynthetic efficiency (free energy stored as glucose ÷ free energy of absorbed (700 nm) radiation × 100 percent) of 35 percent was obtained. This figure is an idealized upper bound, however, because it does not take into account the fraction of incident solar radiation that is useful for photosynthesis (~0.43), light reflection and leaf shading reductions (by a factor of 0.80 or less), and dark respiration (effects vary but a representative factor is 0.67). Multiplying these factors by the theoretical photosynthetic efficiency estimate gives a (high) practical efficiency estimate of 8 percent.

The corresponding amount of biomass (as starch) produced per unit area (m^2) per unit time (say, one day), is given by

$$\left(\frac{\text{biomass produced (g starch)}}{\text{m}^2\ \text{day}}\right)$$

$$= (\text{solar flux: kcal m}^{-2}\ \text{day}^{-1}) \times (\text{photosynthetic efficiency} \times 0.01) \times \left(\frac{1\ \text{mol glucose}}{686\ \text{kcal}}\right) \times \frac{(180 - 18)\text{g starch}}{\text{mol glucose}} \tag{5.73}$$

The solar flux varies with global location and season. For example, the overall U.S. average solar flux value is 3930 kcal m^{-2} day^{-1}. However, the daily average

Table 5.14 Measured annual average and maximal photosynthetic productivity and efficiency for several different plants

(Data from J. A. Bassham, "Increasing Crop Production Through More Controlled Photosynthesis," Science, vol. 197, p. 630, 1977. Reprinted by permission.)

	Annual average		Measured maximum	
Plant	Yield, g m^{-2} day^{-1}	Efficiency, %	Yield. g m^{-2} day^{-1}	Efficiency, %
C_3 plants				
Alfalfa	8	0.7	23	1.4
Sugar beet	9	0.8	31	1.9
Chlorella			28	1.7
C_4 plants				
Sugarcane	31	2.8	38	2.4
Sorghum	10	0.9	51	3.2
Corn (*Zea mays*)			52	3.2

value in the U.S. southwest in the summer is 6775 kcal m^{-2} day^{-1}. The last term in Eq. (5.73) includes the loss of one H_2O for each glucose incorporated into starch.

Equation (5.73) may be used in connection with measured biomass productivities and solar fluxes to calculate the actual photosynthetic efficiencies for various plants and algae (Table 5.14). These actual values are in the range of 0.8 to 3.2 percent, substantially below the theoretically based estimate. Notice also that a group of photosynthetic plants called "C_4" have significantly greater efficiencies than those grouped as "C_3." We shall next briefly examine the biochemical basis for this difference.

The phenomenon of photorespiration, use of oxygen with CO_2 production in the presence of light, is an important consideration in the overall growth efficiencies of photosynthetic organisms. In this process, carbon fixation reactions in the plant give the intermediate riboluse 1,5-diphosphate (see Sec. 5.6.1) which is degraded in photorespiration as follows:

$$\text{Ribulose 1,5-diphosphate} + O_2 \xrightarrow[\text{carboxylase}]{\text{ribulose diphosphate}} \text{phosphoglycolate} + \text{3-phosphoglycerate} + 2H^+ \quad (5.74)$$

The phosphoglycolate is next hydrolyzed, yielding glycolate which is converted to glycine and other products in microbodies in the plant cell.

This oxidation process produces neither ATP nor stored electrons and therefore wastes photosynthetic energy. However, the oxidation reaction (5.74) is inhibited by high concentrations of CO_2. The C_4 plants maintain relatively high CO_2 concentrations in cells in which the Calvin cycle (Fig. 5.15) occurs by an interesting device. In C_4 plants, mesophyll cells in contact with the air conduct a number of reactions with the following overall result:

Mesophyll cells

$$\text{Pyruvate} + CO_2 + \text{ATP} + 2H_2O \longrightarrow \text{malate} + \text{AMP} + 2P_i \quad (5.75)$$

The malate so formed enters an adjacent bundle-sheath in which the Calvin cycle reactions take place. There the CO_2 incorporated into malate above is released by the reaction

$$\text{Malate} \longrightarrow \text{pyruvate} + CO_2 \quad (5.76)$$

The net result of this cycle is use of two high-energy phosphate bonds to maintain high CO_2 concentrations in the environment of the ribulose diphosphate carboxylase enzyme, thereby minimizing photorespiration. This corresponds to use of 12 additional ATP per mole of glucose formed by photosynthesis. This process clearly involves a trade-off between energy consumption to drive the C_4-plant CO_2 transport reaction cycle and energy loss due to enhanced photorespiration without this device. Interestingly, the photorespiration reaction (5.74) has a higher activation energy than the carboxylation reaction (5.37) also catalyzed by ribulose diphosphate carboxylase. This likely explains why C_4 plants are more abundant in the tropics while C_3 plants predominate in temperate climates.

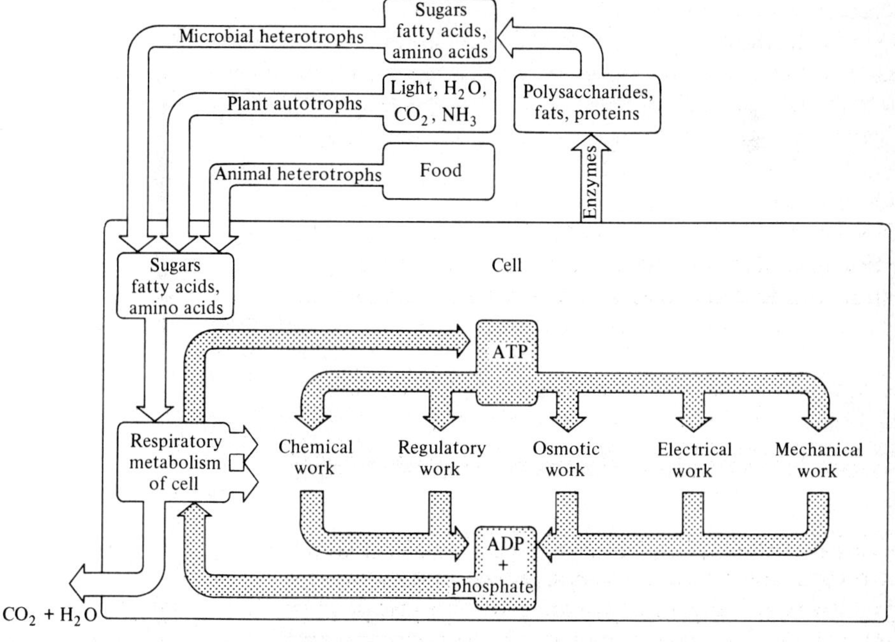

Figure 5.33 Overall schematic diagram of energy flows in the cell and its environment. (From "Cell Structure and Function," 2d ed., p. 26, by Ariel G. Loewy and Philip Siekevitz, Copyright © 1963, 1969 by Holt, Rinehart and Winston, Inc. Reprinted by permission of Holt, Rinehart and Winston.)

5.11 CONCLUDING REMARKS

We now know that extremely complicated reaction processes abound in the living cell. In spite of the complexities, a few central features occur quite often and provide a helpful unifying link between all aspects of cell metabolism. Perhaps the most important common aspects of cell biochemistry are ATP's role as an energy carrier (Fig. 5.33) and the reducing power shuttle, NAD. We have seen how stoichiometric relationships at many levels can be used to interrelate the chemical changes which result from cell metabolism.

In the next chapter, we finish our introductory study of the cell's operation. After that, we shall tackle the difficult problem of analyzing, interpreting, and designing processes for exploitation of biological reactions.

PROBLEMS

5.1 Equilibrium The standard-free-energy change of the reaction phosphoenolpyruvate + ADP → pyruvate + ATP is −7.50 kcal. If the reaction is initiated with 6.0 m*M* ADP, 6.0 m*M* phosphoenolpyruvate, 6.0 m*M* ATP, and no pyruvate, what will the final concentrations be?

5.2 Mineral function For each element or species in Table 5P2.1 identify the important function(s) within the cell and the predominant chemical form for each function. Is it necessary in such an experiment for every element in the resulting ash to have a biochemical function? (See C. N. Frey, *Ind. Eng. Chem.*, **22**; 1154, 1930.)

Table 5P2.1 Approximate ash content from complete oxidation of yeasts

Species	Approximate content, %	Species	Approximate content, %
Phosphorus pentoxide	50	Silicon oxide	1
Potassium oxide	35	Sodium oxide	1
Magnesium oxide	5	Sulfur trioxide	0.5
Calcium oxide	1	Chlorine, iron	Trace

5.3 Standard-free-energy changes The free-energy change of a reaction ΔG is related to the half-cell potentials $E_1 - E_2$ of the individual half reactions

$$\Delta G = -nF\,\Delta E = -n\mathscr{F}(E_1 - E_2)$$

where n is the number of electrons transferred per mole reactant converted, $\mathscr{F}$ is Faraday's constant ≈ 23 kcal/(V·mol), and E_i ($i = 1, 2$) is given by the Nernst equation

$$E_i = E_{i0} - \frac{RT}{n\mathscr{F}} \ln \frac{s_{i,\text{red}}}{s_{i,\text{ox}}}$$

where s_i = concentration of a reactant in oxidized or reduced form, and E_{i0} = half-cell potential when $s_{i,\text{red}}/s_{i,\text{ox}} = 1.0$ compared to the half cell for H:

$$S_{i,\text{red}} \longrightarrow S_{i,\text{ox}} + ne^- \qquad E_0 \neq 0$$

$$H_2 \longrightarrow 2H^+ + 2e^- \qquad E_0 \equiv 0 \text{ by definition}$$

Some half-cell potentials of interest are given in Table 5P3.1.

Table 5P3.1

Half cell†	n	E_0, V
Acetate + CO_2 + $2H^+$ $\longrightarrow$ pyruvate + $2H_2O$	2	−0.70
Acetoacetate + $2H^+$ $\longrightarrow$ β-hydroxybutyrate	2	−0.27
Pyruvate + $2H^+$ $\longrightarrow$ lactate	2	−0.19
S + $2H^+$ $\longrightarrow$ H_2S	2	−0.23
$\frac{1}{2}O_2$ + $2H^+$ $\longrightarrow$ H_2O	2	0.82
$NAD(P)^+$ + $2H^+$ $\longrightarrow$ NAD(P)H + H^+	2	−0.32
FAD + $2H^+$ $\longrightarrow$ $FADH_2$ (free coenzyme)	2	−0.18
Chlorophyll$^+$ $\longrightarrow$ chlorophyll* + e^-	1	−0.2‡
Chlorophyll$^+$ $\longrightarrow$ chlorophyll	1	0.9‡
NO_3^- + $2H^+$ $\longrightarrow$ NO_2^- + H_2O	2	0.42
SO_4^{2-} + $2H^+$ $\longrightarrow$ SO_3^{2-} + H_2O	2	0.48
Acetate + $2H^+$ $\longrightarrow$ acetaldehyde + H_2O	2	−0.60
Acetaldehyde + $2H^+$ $\longrightarrow$ ethanol	2	−0.20

† Selected values from W. B. Wood, J. H. Wilson, R. M. Benbow, and L. E. Hood, *Biochemistry: A Problems Approach*, pp. 190–191, W. A. Benjamin, Inc., Palo Alto, Calif., 1974.

‡ Eucaryotic photosynthesis, noncyclic portion.

From the standard half-cell potentials given, at standard conditions show that the following overall reactions are (not) possible:

(*a*) Electron transfer from excited chlorophyll

(*b*) Oxidation of β-hydroxybutyrate with oxygen

(*c*) Oxidation of acetaldehyde by acetaldehyde (dismutation).

5.4 $\Delta G'$ of cellular conditions (*a*) From Table 5.3, calculate the free-energy change for each step in glycolysis in the human erythrocyte.

(*b*) From these data and the free energy of ATP hydrolysis to ADP under the same conditions, establish the need for ATP to drive the particular reaction steps in which it participates.

(*c*) From your calculations, determine the overall free energy change $\Delta G'$ for glycolysis at cellular conditions.

5.5 Network kinetics A cell carries out many reactions simultaneously, yet often the apparent behavior of a sequence of reactions is that of only a few.

(*a*) Show that, at steady state, the mathematical expression for the rate of P formation for the postulated Michaelis-Menten sequence

$$E + S \rightleftharpoons ES \longrightarrow E + P$$

is indistinguishable from that for the following equilibrated sequence:

$$S + E_1 \rightleftharpoons S_2 + E_1$$

$$S_2 + E_2 \rightleftharpoons S_3 + E_2$$

$$S_3 + E_3 \rightleftharpoons E_3S_3 \longrightarrow E_3 + P$$

(*b*) The 10 reactions in series in Fig. 5.5 can be represented by only three kinetic equations describing

$$A \xrightarrow{k_1} B \xrightarrow{k_2} C \xrightarrow{k_3} D \quad \text{(lactate)}.$$

Identify A, B, and C.

Table 5P5.1

	$\Delta G^{\circ\prime}$, kcal/mol
Phosphoenolpyruvate	−12.8
1,3-Diphosphoglycerate	−11.8
Phosphocreatinine	−10.5
Acetyl phosphate	−10.1
ATP (terminal bond)	−7.3
Glucose 1-phosphate	−5.0
Fructose 6-phosphate	−3.8
Glucose 6-phosphate	−3.3
3-Phosphoglycerate	−3.1
Glycerol 1-phosphate	−2.3

(*c*) The standard free energies of hydrolysis of phosphate compounds are given in Table 5P5.1. If the ATP/ADP ratio in the cell diminishes by a factor of 10 (with sum ATP + ADP = const), what happens to the free-energy differences calculated in Prob. 5.4 for the three steps (*b*) above? What (un)useful effect results for the cell?

5.6 ATP regeneration The cell drives many chemical reactions with unfavorable equilibrium constants by the coupled hydrolysis of ATP to ADP or AMP. The use of isolated enzymes for similar potential syntheses of interest would require (eventually) the transfer of a phosphate group from a system with a higher free energy of hydrolysis (in order to regenerate ATP). For the phosphate compounds listed in Prob. 5.5:

(*a*) What is the cost per pound of each compound with a higher free energy of hydrolysis? (Consult any biochemical catalogue.)

(*b*) To produce ATP from ADP at the levels cited in Table 5.3, what concentration of the following phosphorylated species is needed: glucose 1-phosphate, fructose 6-phosphate, or glycerol 1-phosphate?

5.7 NAD regeneration in enzymatic steroid transformation A technique for steroid conversion with 20 β-hydroxysteroid dehydrogenase (20 β-HSDH) (P. Cremonesi et al., "Enzymatic preparation of 20 β-Hydroxysteroids in a two phase system," Biotech. Bioeng., **17**: 1101, 1975) consisted of three steps:

Steroid hydrogenation:

$$\text{Cortisone} + \text{NADH} + \text{H}^+ \xrightleftharpoons{20\ \beta\text{-HSDH}} \text{NAD}^+ + \text{H}_2 \cdot \text{cortisone} \tag{1}$$

Ethanol oxidation:

$$\text{C}_2\text{H}_5\text{OH} + \text{NAD}^+ \xrightleftharpoons[\text{alcohol dehydrogenase}]{} \text{CH}_3\text{CHO} + \text{NADH} + \text{H}^+ \tag{2}$$

Acetaldehyde removal:

$$\text{CH}_3\text{CHO} + \underset{\text{Semicarbazide}}{\text{NH}_2\text{NHCONH}_2} \longrightarrow \text{CH}_3\text{CHNNHCONH}_3 \tag{3}$$

Reactions (1) and (2) together have an equilibrium constant near unity.

(*a*) Taking the equilibrium for (1) and (2) to be unity, evaluate the thermodynamically possible conversion of cortisone assuming that reaction (3) (which must also be reversible at equilibrium) has an equilibrium constant of 10, 10^3, 10^5, 10^{10} in units of reciprocal concentration. Plot the result as cortisone conversion vs. log K.

(*b*) Since steroids are relatively insoluble (saturation at 10^{-4} to 10^{-5} mol/l in water), the overall amount of cortisone in the system was increased by the addition of organic solvents, which formed a *two-phase* system, thus allowing the organic phase to hold relatively high concentrations of steroid (cortisone solubility = 0.160 g/100 mL in butyl acetate) which acted as a reservoir for the aqueous phase. Taking ε = volume fraction of butyl acetate in water–butyl acetate emulsion, and 10^{-4} mol/L = cortisone solubility in water, what is the maximum conversion possible if H_2·cortisone is insoluble in the solvent (assume same solubility in water as cortisone) and K for reaction (3) is 10, 10^3, 10^5? Repeat for H_2·cortisone solubility = 0.160 g/100 mL butyl acetate.

(*c*) Enumerate the thermodynamic devices used by these authors to drive the desired reaction (1).

5.8 Permeability Derive Eq. (5.40). State all assumptions.

5.9 Network mass balance You have an absentminded friend who unfortunately is not a very good experimenter. He was asked to run a fermentation in a laboratory course but neglected to weigh the glucose he added and also forgot to analyze for ethanol. He has found that the fermentation produced the compounds shown in Table 5P9.1. He knows that all glucose ($C_6H_{12}O_6$) is oxidized to pyruvic acid ($CH_3COCOOH$) by the EMP pathway and that no other products except ethanol are formed. He comes to you for help in salvaging something out of this mess. How many moles of ethanol should you tell him have been formed? (State your assumptions.)

Table 5P9.1

	Moles
Lactic acid ($CH_3CHOHCOOH$)	10
Acetic acid (CH_3COOH)	5
Carbon dioxide (CO_2)	15
Hydrogen (H_2)	10

5.10 Definition of life Comment on the (lack of) necessity for each word in the following definition (J. Perrett, *New Biol.*, **12**: 68, 1952): "Life is a potentially self-perpetuating open system of linked organic reactions, catalysed stepwise and almost isothermally by complex and specific organic catalysts which are themselves produced by the system." For a more recent discussion, see J. D. Bernal, *Theoretical and Mathematical Biology*, p. 96ff. Blaisdell Publ. Co., N.Y., 1965.

5.11 Free energy: electron-transfer basis The fact that intracellular precursors are typically ions and that electron transfer is intimately associated with free-energy transfer in glycolysis, TCA cycle, mitochondrial, and other typical metabolic processes suggests that energetics might most fundamentally be based on units of electron transfer. McCarty ["Energetics of Organic Matter Degradation," in *Water Pollution Microbiology*, R. Mitchell (ed.), p. 91, Wiley-Interscience, New York, 1972] suggests the use of a yield coefficient Y_e (grams of cell synthesized per electron equivalent) defined by

$$Y_e \equiv \frac{C}{hA}$$

where C = grams of cells used to generate an electron equivalent

h = 1.0 (usually) = number of electron moles actually transferred from a donor molecule divided by electron equivalents per mole (in proper half-cell reaction)

A = electron equivalents of electron donor converted for energy per electron equivalent of cells synthesized

From the presumed cell formula $C_5H_7O_2N$, the value of C is calculated to be 5.65, assuming the half reaction for oxidation to be

$$\tfrac{1}{20}C_5H_7O_2N + \tfrac{9}{20}H_2O = \tfrac{1}{5}CO_2 + \tfrac{1}{20}HCO_3^- + \tfrac{1}{20}NH_4^+ + H^+ + e^-$$

(*a*) Verify the value of C above.

Table 5P11.1[†] **Cell-yield coefficients estimated from energetics of substrate oxidation**
$K = 0.6$, ammonia = nitrogen source

Electron donor	Electron acceptor	ΔG_r,[‡] kcal	ΔG_p, kcal	A, calc	Y_e, calc, g/electron equiv
Acetate	O_2	−25.28	1.94	0.71	7.96
	NO_3^-[§]	−23.74	1.94	0.76	7.43
	SO_4^{2-}	−1.52	1.94	11.8	0.48
	CO_2	−0.85	1.94	21.1	0.27
Glucose	O_2	−28.70	−1.48	0.38	14.90
	CO_2	−4.26	−1.48	2.58	2.19
Ethanol	O_2	−26.27	0.95	0.58	9.76
	CO_2	−1.83	0.95	8.3	0.67

[†] P. L. McCarty, in R. Mitchell (ed.), *Water Pollution Microbiology*, p. 107, Wiley-Interscience, New York, 1972.
[‡] Products and reactants at unit thermodynamic activity except pH 7.
[§] Reduction to N_2.

(*b*) Verify the values of A and Y_e in Table 5P11.1 from additional data in the table.

Keeping the same value of $K = 0.60$, calculations of A were typically within 50 percent of measured values for 25 systems, about half anaerobic and half aerobic.

The value of A needed in the definition of Y_e above is estimated from

$$A = \frac{-\Delta G_p/K + \Delta G_c + \Delta G_n/K}{K\,\Delta G_r}$$

where ΔG_p = free energy required to convert carbon source used for cell synthesis to intermediate level

ΔG_n = free energy required to convert inorganic nitrogen source into ammonia, the oxidation state of nitrogen in cellular material

ΔG_c = free energy required to convert both intermediate-level carbon and ammonia into cellular material ≈ 7.5 kcal/electron equiv of cells (including inefficiencies)

K = average efficiency of free-energy transfer (range = 0.4 to 0.8 in heterotropic or autotrophic bacteria)

ΔG_r = free-energy change per electron equivalent of substrate oxidized

Depending on the nitrogen source, the value of ΔG_n is 0 (ammonia), 3.25 (nitrite), 4.17 (nitrate) or 3.78 (N_2) kcal/mol (using assumed cell stoichiometry above).

5.12 Carried-mediated transport Suppose component A is insoluble in a membrane of thickness L, and that its concentrations on either side of the membrane are a_1 and a_2. Carrier B, which exists only in the membrane, forms a complex AB with A at the membrane surfaces. Assuming that complex formation is always at equilibrium *at each surface* (not within membrane) and that the equilibrium constants are K_1 and K_2, what is the flux of A from side 1 to side 2 if AB has a diffusivity $\mathcal{D}$. Under what conditions (if any) can active transport occur?

5.13 Growth stoichiometry and yield coefficients (*a*) Assuming that cells can convert two-thirds (wt/wt) of the substrate carbon (alkane or glucose) to biomass, calculate the "stoichiometric" coefficients for hexadecane or glucose utilization:

Hexadecane:

$$C_{16}H_{34} + \alpha_1 O_2 + \alpha_2 NH_3 \longrightarrow \beta_1(C_{4.4}H_{7.3}N_{0.86}O_{1.2}) + \beta_2 CO_2 + \beta_3 H_2O$$

Glucose:

$$C_6H_{12}O_6 + \alpha_1' O_2 + \alpha_2' NH_3 \longrightarrow \beta_1'(C_{4.4}H_{7.3}N_{0.86}O_{1.2}) + \beta_2' CO_2 + \beta_3' H_2O$$

(*b*) Calculate the three yield coefficients Y_s, Y_0, Y_Δ. Note that Y_0 and Y_Δ for hydrocarbon are below those for glucose even when identical carbon-to-biomass conversion efficiencies are assumed.

5.14 In vitro metabolic reactions A yeast extract is incubated with 200 mmol of D-glucose, 20 mmol of ATP, 2 mmol of NAD^+ and 20 mmol of phosphate (the extract has all the enzymes needed for glucose conversion to ethanol). What is the glucose and ethanol content of this mixture at equilibrium? How can the fermentation be made to proceed to completion?

5.15 Energetic efficiency of glucose storage in glycogen Compare the energy required to store one glucose molecule in glycogen with the energy obtained by subsequent degradation of that stored glucose by (*a*) fermentation or (*b*) respiration. From an energetic viewpoint, which conditions favor glycogen accumulation?

5.16 Singular metabolic stoichiometry Measurement of the respiratory quotient [RQ; Eq. (5.51)] is frequently used or proposed for estimation of metabolic stoichiometric coefficients. (*a*) Using a quasi-steady-state reductance balance (this means $d[NADH + H^+]/dt = 0$) and the definition of RQ, develop a system of two linear algebraic equations for the unknowns a and b which appear in Eqs. (5.53) and (5.54). (*b*) Evaluate and discuss in physical terms conditions under which the coefficient matrix in the equations for a and b becomes singular (i.e., has zero determinant). (*c*) For yeast ($\gamma = 4.14$) growth on glucose ($\gamma_s = 4.0$) and ammonia ($\gamma_N = 0$), CO_2 evolution associated with biosynthesis has been reported to be small ($\sigma = 0.095$). Calculate a and b for RQ values of 1.05 and 1.06. Comment on the practical utility of RQ measurements for determining metabolic stoichiometry for this system [12].

REFERENCES

Bioenergetics

References 4, 5, and 6 in Chap. 2 are recommended as are the following sources:

1. Albert L. Lehninger, *Bioenergetics*, 2d ed., W. A. Benjamin, Inc., New York, 1971. A superbly written paperback covering the central topic of the present chapter.
2. L. Peusner, *Concepts in Bioenergetics*, Prentice-Hall, Inc., Engelwood Cliffs, N.J., 1974. Another view of bioenergetics, this time combined with a more complete and self-contained treatment of thermodynamics.

Metabolism

See Chap. 2 references plus the following:

3. H. W. Doelle, *Bacterial Metabolism*, 2d ed., Academic Press, New York, 1975. An excellent summary of important metabolic processes in various bacteria.
4. Daniel E. Atkinson, *Cellular Energy Metabolism and Its Regulation*, Academic Press, New York, 1977. A thought-provoking monograph which emphasizes the role of energy charge in metabolic control.

Growth and product formation stoichiometry

5. Bernhard Atkinson and Ferda Mavituna, *Biochemical Engineering and Biotechnology Handbook*, Macmillan Publishers Ltd., Surrey, England, 1983. A superb compilation of well-organized data and information on this and many other topics of interest.
6. S. John Pirt, *Principles of Microbe and Cell Cultivation*, Blackwell Scientific Publications, Oxford, 1975. Lucid explanation of yield factors and many other aspects of cell growth.
7. D. Herbert, "Stoichiometric Aspects of Microbial Growth," p. 1, in *Continuous Culture 6: Applications and New Fields*, A. C. R. Dean, D. C. Ellwood, C. G. T. Evans, and J. Melling (eds.),

Ellis Horwood Ltd., Chichester, England, 1976. Clear explanation of the C-mole concept and its significance.

8. J. A. Roels, "Simple Model for the Energetics of Growth on Substrates with Different Degrees of Reduction," *Biotech. Bioeng.* **22**: 33, 1980.
9. J. A. Roels, "Bioengineering Report: Application of Macroscopic Principles to Microbial Metabolism," *Biotech. Bioeng.* **22**: 1437, 1980.
10. L. E. Erickson, I. G. Minkevich, and V. K. Eroshin, "Application of Mass and Energy Balance Regularities in Fermentation," *Biotech. Bioeng.* **20**: 1595, 1978.
11. S. Nagai, "Mass and Energy Balances for Microbial Growth Kinetics," p. 49, *Advances in Biochemical Engineering* **11** [T. K. Ghose, A. Fiechter and N. Blakebrough, (eds.)], Springer-Verlag, Berlin, 1979.
12. G. Stephanopoulos, K.-Y. San, and R. Grosz, "Studies on On-Line Bioreactor Identification. I-IV," *Biotech. Bioeng.* **26**: 1176, 1189, 1198, 1209; 1984.
13. A. H. Stouthamer, "A Theoretical Study on the Amount of ATP Required for the Synthesis of Microbial Cell Material," *Anton. van Leeuwenhoek* **39**: 545, 1973.
14. I. G. Minkevich, "Mass-Energy Balance for Microbial Product Synthesis—Biochemical and Cultural Aspects," *Biotech. Bioeng.* **25**: 1267, 1983.
15. C. L. Cooney and F. Acevedo, "Theoretical Conversion Yields for Penicillin Synthesis," *Biotech. Bioeng.* **19**: 1449, 1977.

Metabolic heat generation

See above references 5, 6, 8, 9, 11, 14, Ref. 4 in Chap. 14, and the following:

16. B. J. Abbott and A. Clamen, "The Relationship of Substrate, Growth Rate, and Maintenance Coefficient to Single Cell Protein Production," *Biotech. Bioeng.* **15**: 117, 1973.
17. M. Kanazawa, "The Production of Yeast from *n*-Paraffins," p. 438 in S. Tannenbaum and D. I. C. Wang (eds.), *Single Cell Protein II*, MIT Press, Cambridge, MA, 1975.
18. J. A. Servizi and R. H. Bogan, "Thermodynamic Aspects of Biological Oxidation and Synthesis," *J. Water Poll. Control Fed.* **36**: 607, 1961.

CHAPTER

SIX

MOLECULAR GENETICS AND CONTROL SYSTEMS

The previous chapters have discussed enzyme kinetics and the overall features of some common cellular reaction networks. In a simpler chemical process plant, only a few products are synthesized, and the *steady-state* operation usually sought requires relatively minimal controls. The complex chemical plant which is the living cell is continually in a *transient* state; the cell alters the proportions of substrates, enzyme catalysts, RNA, cofactors, and other constituents as it proceeds through the stages of its *life cycle*. Synthesis and regulation of the systems which control cellular dynamics are the subjects of this chapter. Important applications of these topics occur in viral infections of cultures, natural and induced mutations, genetic manipulations, and the versatility with which some cultures utilize multiple substrate feeds.

6.1 MOLECULAR GENETICS

To discuss cell genetics and expression of genetic information at the molecular level, we must return to the composition and structure of DNA, the three RNA varieties, and proteins presented in Chap. 2. In order to appreciate the significance of the mechanisms discussed below, we should remember that proteins, particularly in their role as enzymes, directly determine how the cell operates by determining which chemical reactions occur. Consequently, in a sense the enzyme makeup of a cell defines the cell's metabolic processes: specifying the concentration and kinds of enzymes present determines the cell's reaction-network characteristics. Also, other proteins in the cell serve important functions (recall Table 2.6). Thus the inheritance which a dividing cell leaves to daughter cells must include the information for directing the production of the same protein constituents as found in the parent cell. The study of transmission of such information is called *genetics*. We shall next investigate how the nucleotide sequence in deoxyribonucleic acid (DNA), a nonrepetitive and therefore informational biopolymer, determines the course of protein synthesis in the cell.

6.1.1 The Processes of Gene Expression

We already know that the DNA molecule is a very long double helix consisting of two chains. Both chains are polymers constructed from four nucleotides, each characterized by a particular nitrogeneous base: adenine (A), guanine (G), cytosine (C), or thymine (T). Moreover, the two chains in the DNA double helix have complementary sequences: A in one strand is always paired with T in the other, and G and C are likewise paired.

A *gene* is a DNA segment that codes for a functional product. *Expression* of a gene to form the amino acid sequence of the corresponding protein (the *gene product*) proceeds by a two-step process in which a corresponding *messenger ribonucleic acid* [*message RNA* (mRNA)] serves as an intermediate. In the first step of gene expression, which is called *transcription*, a complex oligomeric enzyme called *RNA polymerase* catalyzes synthesis of an mRNA molecule using the gene as a template. The second step of gene expression uses the information contained in the nucleotide sequence of an mRNA molecule to direct the synthesis of a peptide chain with a corresponding amino acid sequence. This process, which is known as *translation*, involves many different components including *ribosomes*, the sites and organizers of peptide synthesis, and several different *transfer RNA* (tRNA) molecules, which bring chemically activated amino acids into the peptide synthesis process in the sequence prescribed by the mRNA. These processes as well as the *genetic code* which connects nucleotide sequence to amino acid sequence are next explained. This discussion focuses on major points and omits details which may be important in some situations; further reading in the References is highly recommended.

Transcription begins by binding of the enzyme RNA polymerase to a signal sequence called a *promoter* on one of the complementary double strands of DNA. The strand possessing this enzyme binding sequence becomes the *DNA template strand* for synthesis of an mRNA molecule. After some unwinding of the DNA double helix in the vicinity of the RNA polymerase-DNA promoter complex, the polymerase moves along the DNA template strand in the 3′ to 5′ direction. Using DNA-RNA base pairing rules (recall Sec. 2.3.2 or see Fig. 6.1), the nucleotide sequence in the DNA template strand is employed to construct a complementary nucleotide sequence in a (single-stranded) mRNA molecule. The mRNA being synthesized runs antiparallel to the DNA template strand; that is,

DNA 5′ { —∫∫— GCCTTTATGGAAATTATACGTTAAAGC —∫∫— 3′
3′ —∫∫— CGGAAATACCTTTAATATGCAATTTCG —∫∫— 5′

Transcription ↓ (RNA polymerase)

mRNA 5′ — GCC·UUU·AUG·GAA·AUU·AUA·CGU·UAA·AGC·—3′

Pairing rules:	A T C G	DNA
	U A G C	mRNA

Figure 6.1 Schematic diagram illustrating transcription by RNA polymerase enzyme of the DNA template strand (lower strand in this illustration) to synthesize a complementary nucleotide sequence in the corresponding mRNA molecule.

the mRNA is synthesized from the 5′ to the 3′ end. In the example illustrated schematically in Fig. 6.1, the bottom strand of DNA serves as the template. Transcription ceases, that is, mRNA synthesis stops, when the RNA polymerase enzyme reaches a nucleotide sequence in the template strand that signals for *termination.* Before turning to the next steps, we should note that RNA polymerase also binds to DNA in a nonspecific fashion at sites which are not promoters. Such nonspecifically bound polymerase does not initiate or conduct transcription.

The lengths of the mRNA molecules vary from about 300 nucleotide units to more than 3000. Often, the mRNA corresponds to one gene on the DNA chain. In some cases the mRNA molecule carries the genetic message from a group of related genes called an *operon* (see Sec. 6.1.4). As we noted when discussing the energetics of biosynthesis in Chap. 5, an RNA chain is built up by splitting pyrophosphate from the triphosphate form of the nucelotides. Consequently, the equivalent of two high-energy phosphate bonds is invested in every RNA phosphodiester bond. This ensures that the synthesis reaction proceeds to "completion" and provides us with some indication of the importance of biopolymer synthesis to normal cell function.

Before reviewing the translation process, we must look more carefully at tRNA. It is known that each specific tRNA molecule can carry only one corresponding amino acid. Fig. 6.2 shows the secondary structure for alanine tRNA from yeast. Attachment of an alanyl residue to its tRNA first requires activation to alanyl acyl-AMP, as depicted in Eq. (5.39). The activated residue is then bound to this particular tRNA by an enzyme specific for both the residue and tRNA. This ensures that only alanine is linked to alanine tRNA.

Now we can turn to the specificity of RNA. At the base of the lower loop of the tRNA molecule is a sequence of three nucleotides called an *anticodon.* Such a structure is found in all tRNAs so far studied. This anticodon base sequence is complementary to a three-nucleotide segment of mRNA known as a *codon.* Biochemical research has revealed that each codon is a "word" in the genetic

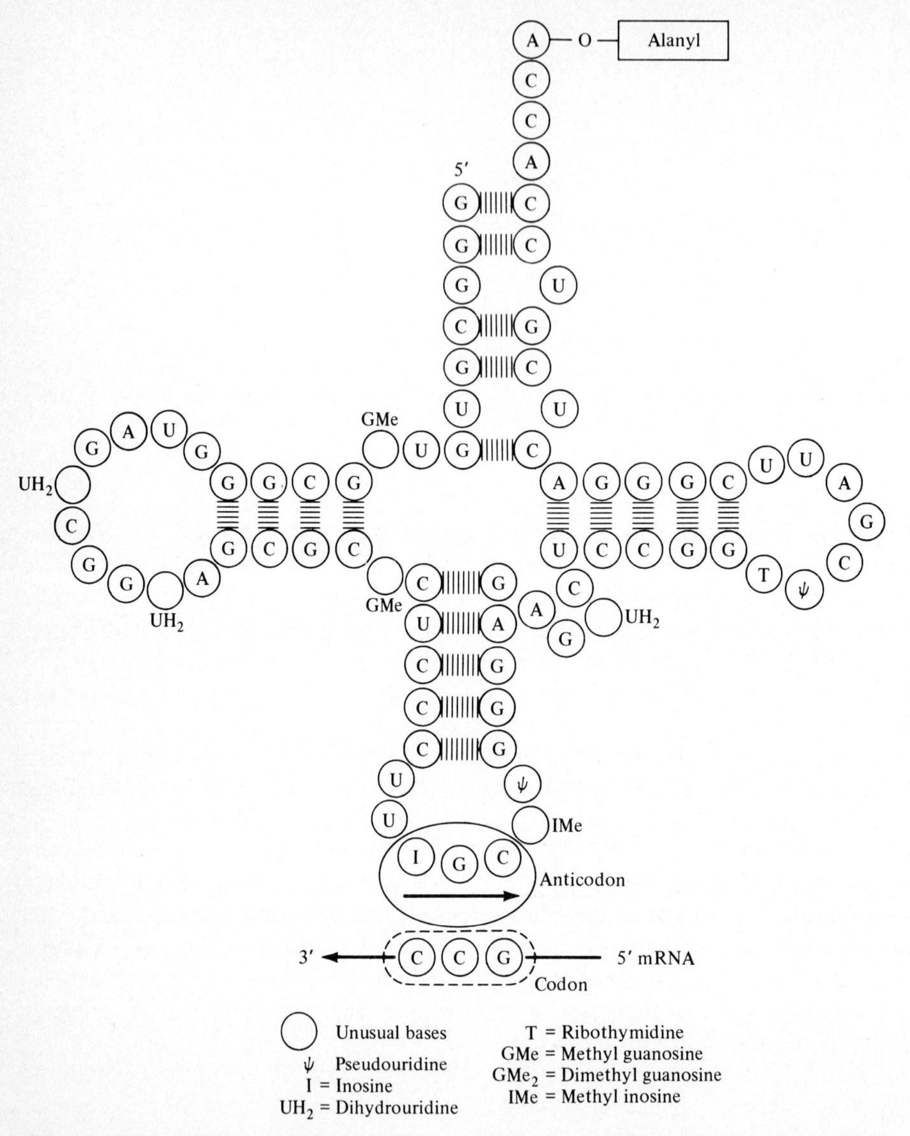

Figure 6.2 The anticodon of alanine tRNA recognizes the complementary three-base codon in mRNA. *(Reprinted by permission from J. D. Watson, "Molecular Biology of the Gene," 2d ed., p. 361, W. A. Benjamin, Inc., New York, 1970.)*

message; each codon specifies one amino acid. Since there are only four letters in the chemical alphabet of RNA (the four bases A, C, G, U) and there must be at least 20 words (one for each amino acid), a language or code is required for transmission of genetic information. Amazingly enough, the *genetic code* has been completely deciphered; further, it appears to be essentially universal in all living organisms—microbe, plant, and animal (Fig. 6.3).

first	second: U	C	A	G	third
U	Phe	Ser	Tyr	Cys	U
	Phe	Ser	Tyr	Cys	C
	Leu	Ser	Stop	Stop	A
	Leu	Ser	Stop	Trp	G
C	Leu	Pro	His	Arg	U
	Leu	Pro	His	Arg	C
	Leu	Pro	Gln	Arg	A
	Leu	Pro	Gln	Arg	G
A	Ile	Thr	Asn	Ser	U
	Ile	Thr	Asn	Ser	C
	Ile	Thr	Lys	Arg	A
	Met	Thr	Lys	Arg	G
G	Val	Ala	Asp	Gly	U
	Val	Ala	Asp	Gly	C
	Val	Ala	Glu	Gly	A
	Val	Ala	Glu	Gly	G

Figure 6.3 The genetic code. Three codons (Stop) mean stop synthesis of the peptide chain. Examination of this chart reveals that change in one base (especially the third) in a codon often gives a similar amino acid, a "fail-safe" feature.

To begin the translation process, the smaller of the two ribosome subunits binds to the mRNA molecule at a specific sequence called the *ribosome binding site*. This site is generally located a characteristic distance, which differs between procaryotes and eucaryotes, "upstream" from the AUG codon which almost always begins the polypeptide coding sequence. Here the term "upstream" means that the site is in the 5′ direction relative to the AUG codon. Starting from the AUG codon and proceeding in the 5′ to the 3′ direction is the sequence of codons which specifies the amino acid sequence of the polypeptide.

After the mRNA has formed a complex with the complete ribosome, tRNAs carrying the amino acids corresponding to the first two mRNA codons are attached to the complex. With the amino acid sequence arranged, a peptide bond is formed between the carboxyl group of the first amino acid and the amino group of the second. The ribosome then shifts in the 3′-direction by one codon, and tRNA for the first amino acid is released. The third amino acid, carried by its tRNA, then binds to the complex with subsequent attachment to the elongating peptide chain. This polypeptide elongation process (Fig. 6.4) continues until a stop codon is encountered (see Fig. 6.3), at which point the polypeptide is released and the ribosome and mRNA separate. The first amino acid residue is on the amino-terminal end of the polypeptide, and the last amino acid residue added is on the polypeptide's carboxyl terminus.

A different view of translation which focuses on the transmission of information from the mRNA nucleotide sequence to the corresponding polypeptide

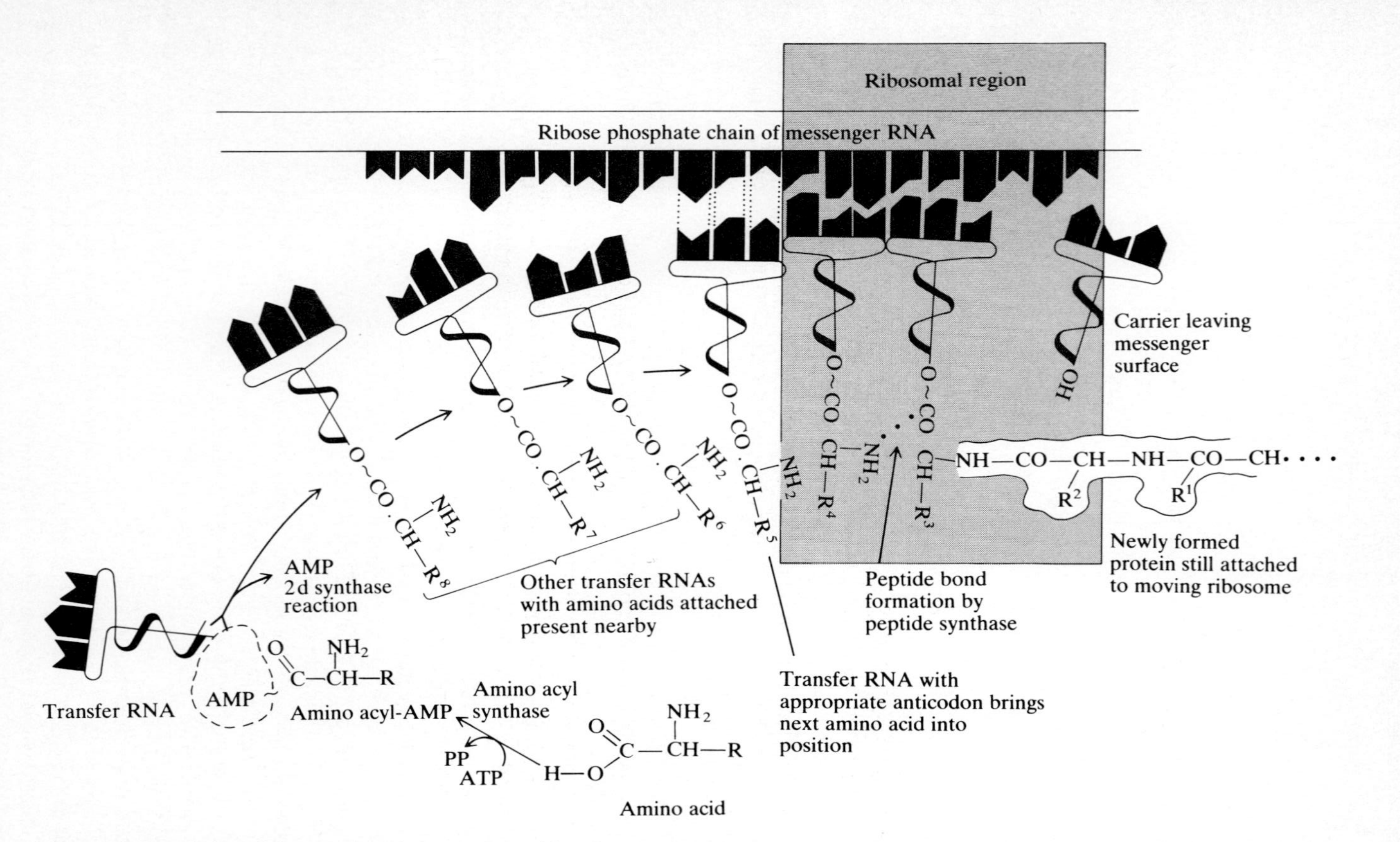

Figure 6.4 Translation of the genetic code into polypeptide primary structure occurs via specific interactions between mRNA and tRNAs at the ribosome. *(Adapted from N. A. Edwards and K. A. Hassall, "Cellular Biochemistry & Physiology," p. 342, McGraw-Hill Publishing Company Ltd., London, 1971.)*

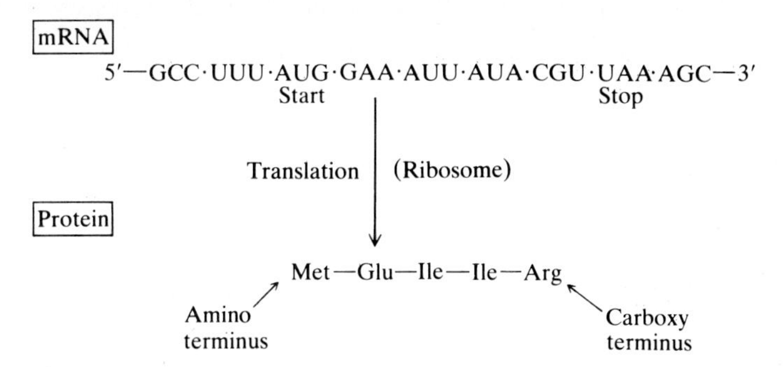

Figure 6.5 Schematic illustration of the translation at the ribosome of nucleotide sequence information in mRNA into the amino acid sequence of the corresponding polypeptide. This is the same mRNA as shown in Fig. 6.1.

amino acid sequence is given in Fig. 6.5. Here, the hypothetical mRNA obtained by the transcription process in Fig. 6.1 is translated to yield the corresponding polypeptide. Notice the role of AUG both as an indicator of where the *coding sequence* for the protein starts and as an instruction to install methionine (met) at the amino terminus of the polypeptide. Also notice that the mRNA is not all protein-coding information. The ribosome binding site which precedes the coding sequence has already been mentioned; other sequences, including coding sequences with another set of "start" and "stop" codons, may lie downstream (in the 3′ direction) from this coding region.

A second summary of the expression process, reviewing and emphasizing the control elements, is worthwhile since proper orchestration of these controls is essential for success in recombinant DNA technology. An overall schematic of expression which tracks the transmission of genetic information and the functions of control sequences is shown in Fig. 6.6. At the outer flanks of the gene are the promoter and terminator sequences which control initiation and termination of transcription. The DNA must also contain the translation control signals (ribosome binding site, start and stop codons) so that these can be transcribed into the mRNA. Finally, during translation, the portion of mRNA complementary to the *structural gene*, which contains the code for a particular amino acid sequence, is employed to guide polypeptide synthesis.

Often more than one ribosome is attached to a single mRNA chain at a time. Polyribosomes are clusters containing several individual ribosomes. This arrangement, which can be highly organized, permits simultaneous translation of information carried by different portions of mRNA and thus lends speed and efficiency to protein synthesis.

The molecules of tRNA and *ribosomal ribonucleic acid* (rRNA), like mRNA, are synthesized using a portion of the DNA chain as a template. Thus, the information carried by DNA includes the structure of the translation devices for protein synthesis as well as the coded information for protein primary structure.

We should recall at this point that the interactions between the amino acids in the resulting sequence determine (Sec. 2.4) secondary, tertiary, and even the

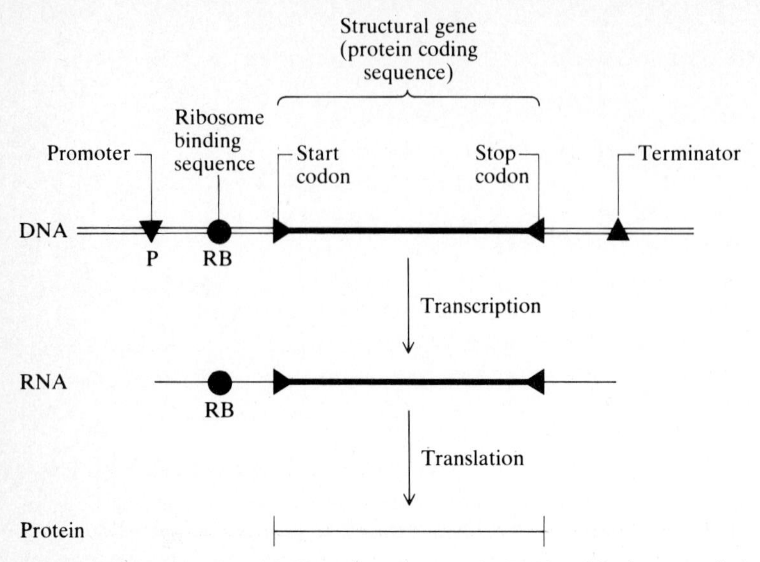

Figure 6.6 The control and information sequences in DNA guide the transcription and translation processes which result in gene expression.

quaternary structure of proteins. The importance of weak interactions between atoms at all stages of genetic information storage, transmittal, and action should now be apparent.

Before discussing how genetic information is passed from generation to generation and how this process can be disturbed, we should recognize that the protein-synthesis mechanisms of eucaryotes and procaryotes are different. Synthesis of mRNA and translation of mRNA to proteins on the ribosomes occur at almost the same point within the nuclear region of the bacterial cell. In eucaryotes transcription occurs in the nucleus. The mRNA molecules, after modifications discussed next, diffuse through pores in the nuclear membrane into the cyctoplasm where, at the ribosomes, translation is accomplished. In eucaryotes and to a lesser degree in procaryotes, the translation product is frequently modified to give the final, functional form of the molecule. Posttranslational modification of proteins is considered in Sec. 6.1.3. Finally, we should note that timing of the various events involved in protein synthesis is different in eucaryotes and procaryotes. More on this will follow in Sec. 6.5.

6.1.2 Split Genes and mRNA Modification in Eucaryotes

Eucaryotic messenger RNA is modified at the 5′ end to add a structure called a *cap* which helps stabilize the mRNA against attack by phosphatases and nucleases. Attached at the 3′ end of most mRNAs in eucaryotes are *poly A* tails which contain 150 to 200 residues of deoxyadenylate. This property can be used to advantage to isolate eucaryotic mRNAs using a column packed with particles to which poly T oligonucleotides are covalently attached.

It is now conclusively established that many eucaryotic genes are mosaics of nucleotide sequences which code for protein structure and of sequences which do not. Sequences of the former class are sometimes called *exons*. The interjected noncoding sequences are designated *introns* or *intervening sequences* (IVS). As the name implies, intervening sequences are found between exons, separating protein synthesis coding sequences with what appears superficially to be useless information. Clearly, eucaryotic cells must have mechanisms to ensure that exons and introns are identified and distinguished during gene expression so that only the exon sequences are ultimately employed to direct translation.

The basic features of eucaryotic gene structure and expression are illustrated in Fig. 6.7. At the top of the diagram we see a hypothetical segment of a eucaryotic structural gene with exon and intron domains. During transcription, the DNA template strand guides synthesis of an mRNA (the primary transcript) which is complementary to the complete template. That is, both the exon and intron sequences are transcribed into this mRNA.

Next follows a number of *splicing* steps, apparently unique to eucaryotes, in which the regions in the primary transcript derived from the introns are clipped

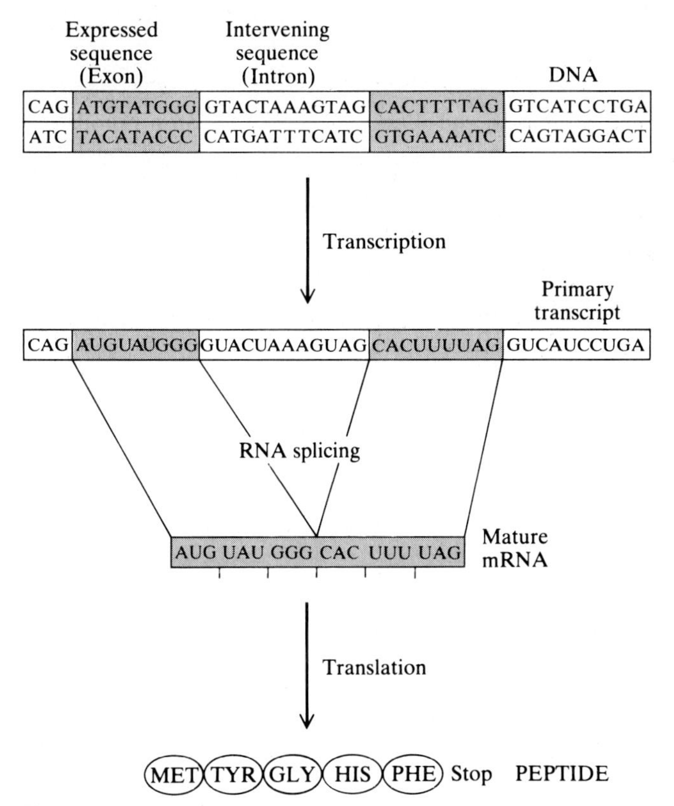

Figure 6.7 In eucaryotic gene expression, the primary transcript mRNA is spliced to obtain the mature mRNA which is translated. An abbreviated hypothetical example is shown here.

out. The remaining regions, all complementary to exon sequences, are spliced together in the same order as found in the primary transcript. The *mature* mRNA which results may then be translated as described above to give the protein gene product.

According to currently available data, the assumption that almost all eucaryotic genes contain introns is a useful working hypothesis. Some exceptions are known and include the human α-interferon genes. Introns can be very large, and the total amount of intron DNA in eucaryotes is probably more than one-half of the cell's total DNA.

The biological function of intervening sequences remains somewhat obscure as of this writing. These regions may participate in regulation of eucaryotic gene expression. On a much longer time frame, introns may provide a mechanism for relatively large-scale changes in genes by adding or deleting an entire exon or intron sequence. It has been proposed that exons may code for protein subdomains and that shuffling exons thereby might provide a useful mechanism for cells to make and thereby evaluate the utility of new proteins. For technological purposes, introns pose problems in expressing eucaryotic genes in procaryotic hosts, since procaryotes do not have an RNA splicing apparatus. We shall see how this difficulty can be circumvented in our overview of recombinant DNA technology in Sec. 6.3.

6.1.3 Posttranslational Modifications of Proteins

The polypeptide produced by mRNA translation is often not the final, biologically active form of the molecule. The N-terminal methionine is sometimes removed. Two cysteine residues may be oxidized to give a disulfide bond which can be a significant factor in protein tertiary structure as discussed in Chap. 2. Attachment of carbohydrate moieties to the side chains of asparagine, serine, and threonine yields glycoproteins. Other types of polypeptide chemical modifications include hydroxylation, phosphorylation, and acetylation.

Some polypeptides have at their N terminus a short (15–30 residues) sequence of hydrophobic residues. These *signal sequences* are important components in protein transport across cell membranes. Secreted proteins generally contain such signal sequences. We attach the prefix "pre-" to the name of a protein to indicate the form which includes the signal sequence (for example, prelysozyme). Before the preprotein has crossed the membrane, its signal sequence is removed to give the mature protein (lysozyme for this example).

Proteins may contain other amino acid sequences which are cleaved away to yield the active form of the molecule. This device and the protein secretion process just outlined allow the cell to compartmentalize protein synthesis and activation of different specific proteins at different locations within or outside the cell. Preproinsulin, for example, has a 24-residue signal sequence at its amino terminus followed by 21 residues called the B chain, then by 21 residues called the C chain, and finally 30 residues of the A chain at the carboxy terminus. The signal sequence is cleaved during secretion leaving proinsulin (A-C-B chains) which is further cleaved enzymatically to obtain insulin which consists of the A and B chains connected by two intrachain disulfide linkages. The A and B chains in insulin are not connected by any peptide bonds (Fig. 6.8).

6.1.4 Induction and Repression: Control of Protein Synthesis

That living cells possess intricate control systems to ensure efficient use of material and energy resources is already known to us. We have previously explored how activation and inhibition of enzyme activity by metabolites helps channel these intermediates through the complex network of cellular reactions. In this section, we explore another level of control, which must be carefully distinguished from those discussed earlier. *Activation and inhibition influence the catalytic activity of enzymes* already present in the cell, but they do not alter the amount of enzyme present. Thus, these control modes operate at the enzyme level. *Induction and repression*, the control devices of interest now, *cause change in the rates of protein sythesis* (and consequently in the amount of enzyme present) and act at the level of the gene. Perhaps the greatest similarity between these two kinds of controls is their sensitivity to the concentration of relatively small molecules.

We shall concentrate here on genetic-level controls of protein synthesis in bacteria because the molecular mechanisms are best understood for these organisms. Analogous controls are found in higher cells which also possess additional, less well understood regulatory systems which govern cell differentiation in the development of the organism.

Before delving into the details of the mechanisms, it may be helpful to place protein synthesis regulation in microbiological perspective. We should recall from Chap. 1 that bacteria are free-living, isolated cells which consequently have almost no influence on their external environment. Therefore, they must be very adaptable: it may be essential for their survival that they function efficiently in variable surroundings.

Many bacteria indeed possess such versatility. They can synthesize enzyme systems for effective utilization of many different types of nutrients. An extreme example is the bacterium *Pseudomonas multivorans* which can utilize as its sole source of carbon more than 90 different compounds including carbohydrates and carbohydrate derivatives, fatty acids, dicarboxylic and other organic acids, primary alcohols, amino acids and other nitrogenous compounds, and ring compounds such as phenol. Different enzymes are generally required for each nutrient. Consequently, the bacterium must carry genetic information for them all. The sum total of all information carried in the chromosomes is called the cell's *genotype*.

However, such a bacterium will not require the same mix of enzymes for all these nutrients, and synthesis of any extra protein represents a waste of valuable energy and intermediary metabolites. Thus, for maximum efficiency in a particular environment, only a portion of the total genetic information is expressed (actually synthesized) as protein. The term *phenotype* denotes the observable features of an organism. In view of the previous comments, the phenotype arises

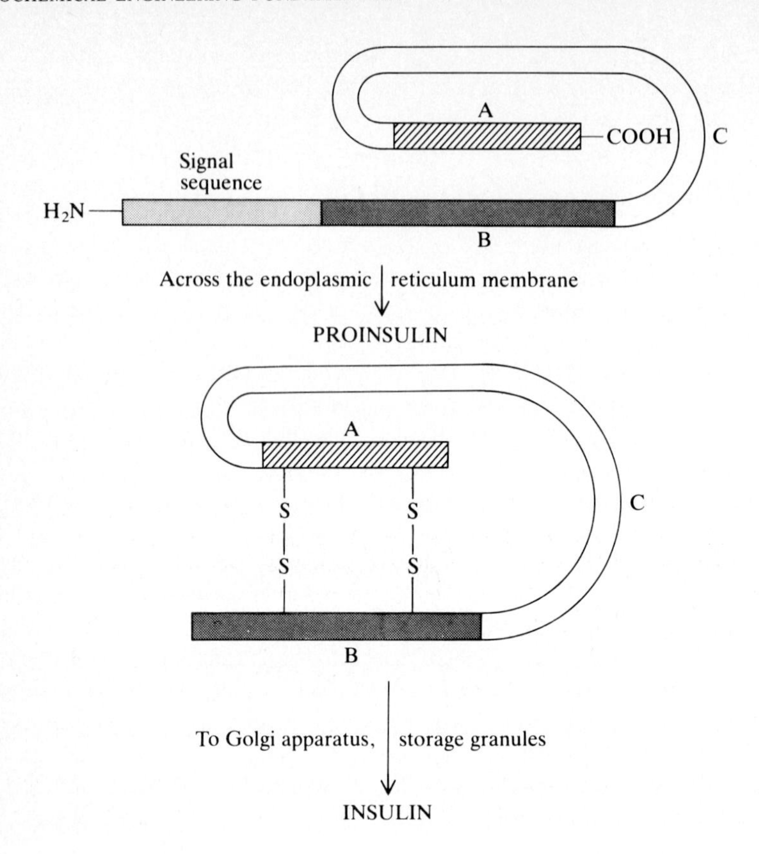

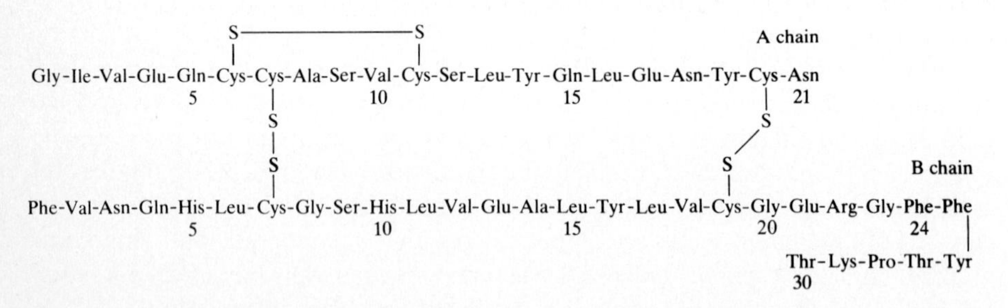

Figure 6.8 The active form of insulin is obtained by several posttranslational steps involving signal sequence removal, disulfide bond formation, and cleavage of the *C* chain peptide. The amino acid sequence for the human insulin *A* and *B* chains are shown.

from a combination of two factors, the genotype and the organism's environment. *Constitutive enzymes* are synthesized independent of the cell's surroundings.

The biosynthesis rates of *inducible enzymes*, however, are sensitive to environmental influence. A common example is β-galactosidase synthesis. This enzyme catalyzes the hydrolysis of the disaccharide lactose to its component monosaccharides, glucose and galactose. The reaction is essential if the cell is to employ lactose as a nutrient since only the hydrolysis products can be used in subsequent reactions. The *E. coli* bacterial cell regulates synthesis of β-galactosidase in response to the need for this enzyme. As the lactose concentration in the medium increases from zero, larger amounts of the enzyme are produced up to a maximum, fully induced level (Fig. 6.9). Thus, the substrate *induces* the formation of the enzyme.

An analogous situation arises with the *repressible enzymes*. For example, although *E. coli* can make all the enzymes necessary to synthesize all 20 amino acids from simpler precursors, these particular enzymes are not found in significant quantity when the amino acids are supplied to the cell as nutrients. In this instance the end product of a biosynthetic pathway *represses* synthesis of the enzymes for that pathway. This phenomenon is demonstrated in Fig. 6.9 using histidine synthesis as an example.

Useful models for induction and repression, developed by Monod and his coworkers in France, are given in Fig. 6.10, which reveal the basic similarities of the two mechanisms. Both regulate gene expression at the level of transcription. In both, a regulator gene produces a protein which interacts with a specific *operator* DNA sequence. If the modulator is bound to the operator sequence, RNA polymerase does not bind to the promoter. Transcription is blocked, and the structural gene is not expressed.

The regulator gene in the induction model (Fig. 6.10*a*) produces a repressor molecule which can prevent enzyme production. When the inducer is present, it

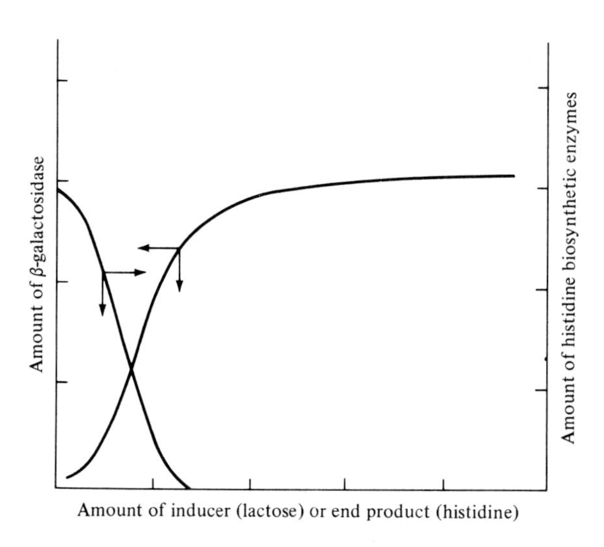

Figure 6.9 Cell content of β-galactosidase increases with increasing concentration of inducer (lactose) in the growth medium. Increased medium concentration of a repressor (histidine) decreases cellular content of enzymes catalyzing repressor biosynthesis. *(Reprinted by permission from J. D. Watson, "Molecular Biology of the Gene," 2d ed., p. 439, W. A. Benjamin, Inc., New York, 1970.)*

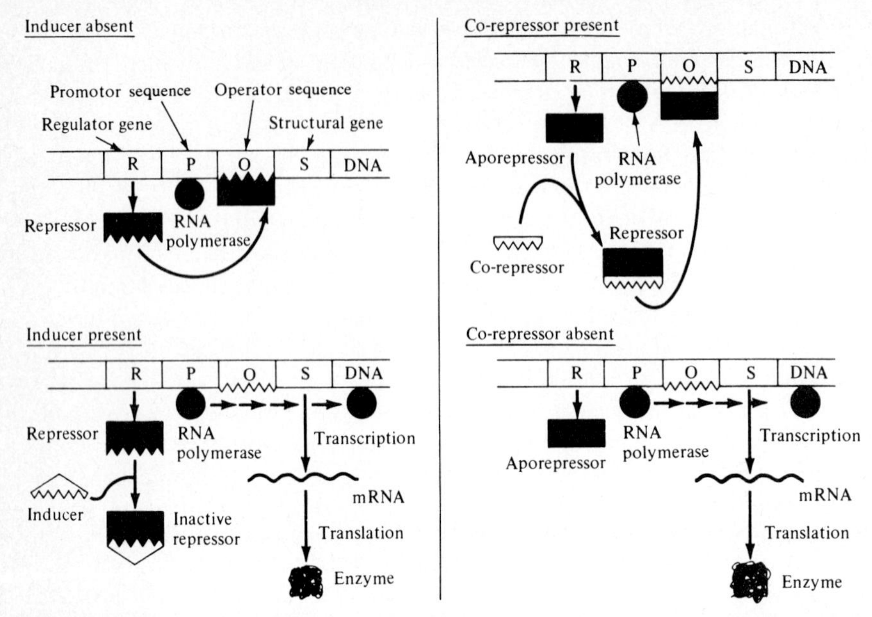

Figure 6.10 (*Left*) An inducer (for example galactose) acts by inactivating repressor so that the repressor does not bind to the operator sequence and block transcription of the structural gene for β-galactosidase, for example. (*Right*) A (co-) repressor (e.g., histidine) combines with an aporepressor to form an active repressor, which can block expression of the structural gene by binding at the operator. In this manner synthesis of repressible enzymes is controlled. *[Adapted from A. L. Demain, Theoretical and Applied Aspects of Enzyme Regulation and Biosynthesis in Microbial Cells, in L. B. Wingard, Jr. (ed.), "Enzyme Engineering," Interscience, New York, 1972.]*

binds with the repressor to form an inactive complex, which does not interfere with subsequent DNA transcription. In the repression model of Fig. 6.10*b*, the regulator gene must complex with another molecule to yield a repressor. Without the corepressor (histidine in the example above), enzyme synthesis proceeds.

Such regulated promoters may simultaneously control synthesis of several enzymes. For example, when production of β-galactosidase is induced, so are two other proteins. One of these is a galactoside permease, which participates in active transport of β-galactosides. Obviously, there is a logical connection between the functions of these two proteins, and coordinating their biosynthesis therefore makes good sense. Such a collection of several genes which are regulated together is called an *operon*. The system just mentioned is the *lac operon*. Quantitative mathematical analysis of the regulatory properties of the *lac* promoter-operator is considered later in Sec. 7.5.3.

Although this discussion has emphasized regulation of bacterial enzyme synthesis, similar mechanisms apply for control of synthesis of other proteins in bacteria and in higher cells. We should be aware in a general sense that the set of proteins synthesized by cells and the associated biological and catalytic activities can and often do change in response to changes in the cell's environment. This adaptation ability poses problems in formulation of kinetics for cell reactions and in bioreactor design which are unparalleled in reaction engineering for synthetic catalysis.

There are several variations of these themes of genetic level control. While we cannot pursue them all, *catabolite repression* should be mentioned. If *E. coli* bacteria are cultivated in a medium containing glucose and a less easily metabolized carbon source such as lactose, the glucose is preferentially consumed. Under these circumstances lactose does *not* induce β-galactosidase. In the cell growing rapidly on glucose, formation of cyclic AMP is inhibited. Cyclic AMP (Fig. 2.8) levels influence promoter activity in the lac operon, such that induction is blocked or moderated when intracellular cyclic AMP levels are low. Since the amount of cyclic AMP is reduced by catabolic products of glucose, this control action is called catabolite repression.

It should be emphasized that catabolite repression can occur in the absence of glucose. Typically, a bacterium growing in a mixture of carbon sources will selectively utilize the best (the one that gives the fastest growth rate) and catabolically repress utilization of the less useful nutrients. What would happen when the first carbon source is exhausted? (See Chap. 7, fermentation kinetics).

Catabolite repression also plays a central role in product synthesis. Often, if cells are growing rapidly, the enzymes required for production of secondary metabolites are not synthesized. This poses interesting challenges and important opportunities for the biochemical engineer: what are the medium design and process operating strategies which can minimize catabolite repression and maximize secondary metabolite accumulation? Answers to this question will be discussed in Chaps. 7, 9, and 10.

6.1.5 DNA Replication and Mutation

Since DNA contains the information required for proper development and function of the living cell, it is essential that there be an almost foolproof mechanism for copying DNA. When a cell reproduces, each of the offspring must receive a complete set of genetic data in the form of DNA. We have already noted that the double-helical model of DNA structure provides a ready hypothesis for faithful reproduction of the molecule (Fig. 6.11). After the original strands unwind, new complementary chains are assembled, so that ultimately two intact DNA molecules result. As a result of this scheme, each offspring molecule contains one nucleotide chain from the parent molecule. This presumably helps minimize errors in the replication process.

Actually, DNA replication occurs in a slightly more complex fashion than indicated in Fig. 6.11. It is not built up in one continuous sequential pass; instead an enzyme called *DNA polymerase* conducts synthesis of daughter strands in the $5' \rightarrow 3'$ direction. This proceeds relatively smoothly on the parental strand running $3' \rightarrow 5'$. On the other parental strand, DNA polymerase adds fragments of the daughter strand in the $5' \rightarrow 3'$ direction which are then covalently coupled by the enzyme *DNA ligase*. This device achieves overall daughter strand synthesis in the $3' \rightarrow 5'$ direction.

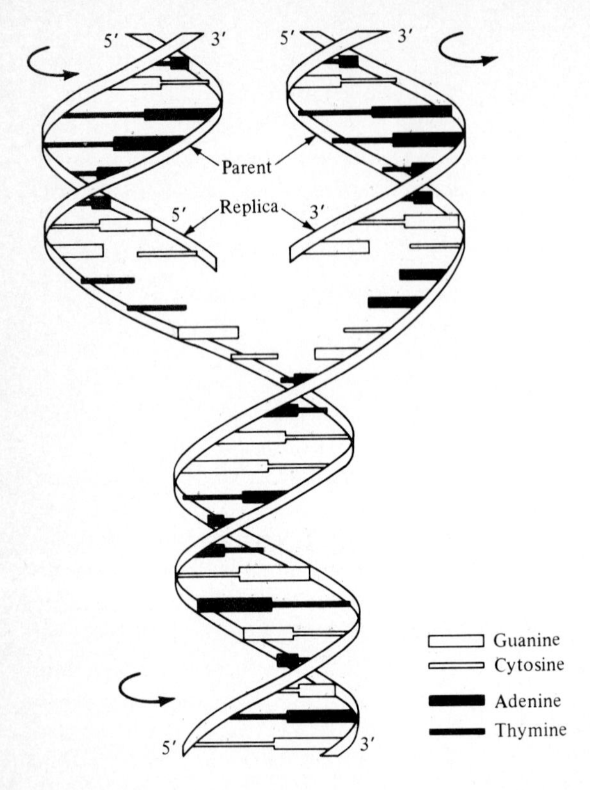

Figure 6.11 A simplified diagram of DNA replication. As the parent strands separate, complementary strands are added to each parent, resulting in two daughter molecules identical to the parent. Notice that each daughter molecule contains one strand from the parent. *(From "Cell Structure and Function," 2d ed., p. 145, by Ariel G. Loewy and Philip Siekevitz, Copyright © 1963, 1969 by Holt, Rinehart and Winston, Inc., Reprinted by permission of Holt, Rinehart and Winston.)*

We shall next briefly outline some differences in DNA storage and duplication between procaryotes and eucaryotes. In the ill-defined nuclear region of procaryotes, there is a single *chromosome*, or carrier of genetic information, which consists of a circular double strand of DNA. This huge molecule is 1.2 mm long, about 20 angstroms thick, and has molecular weight on the order of 2.8 billion. This provides enough storage capacity to code for about 2000 different proteins. The circular nature of *E. coli* DNA was first demonstrated by genetic studies aimed at finding the relative position of the genes. A chromosome in eucaryotes consists of a DNA molecule associated with proteins and possibly some RNA. As Table 6.1 reveals, eucaryotic cells typically contain several chromosomes. Some eucaryotes, such as yeasts, may be either *haploid* (each chromosome present once) or *diploid* (each cell contains two of each chromosome except possibly sex).

A *mutation* is a change in DNA which is passed to succeeding generations. In molecular terms, mutation involves an alteration in the nucleotide sequence of DNA. Several possibilities are given in Fig. 6.12. To some extent, mutation is a spontaneous process which is constantly occurring. The rate of spontaneous mutation is rather low, however. A typical value is 1 error for every 10^6 gene duplications.

The importance of a gene-copying error depends on its nature. In *missense mutation*, a codon for one amino acid is altered so that a different amino acid is

Table 6.1 Normal chromosome number in various species

Species	Number
Procaryotes (haploid)	
Bacteria	1
Eucaryotes (diploid)	
Red clover	14
Honeybee	16
Baker's yeast (*Saccharomyces cerevisiae*)	17
Frog	26
Hydra	30
Cat	38
Rat	42
Man	46
Chicken	78

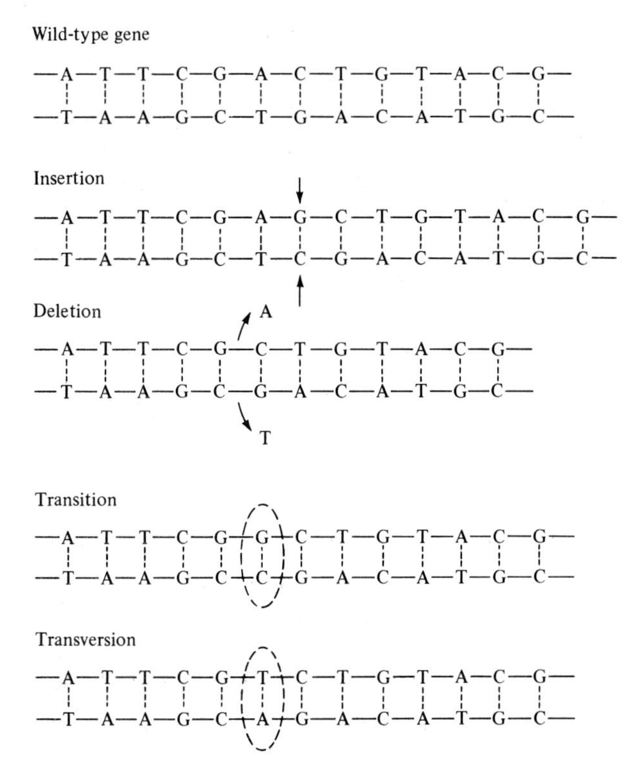

Figure 6.12 Different types of mutation in the base pair sequence of a DNA molecule.

inserted into the protein. This type of mutation leads, for example, to the abnormal hemoglobin characteristic of sickle-cell anemia in human beings. In this instance the abnormal protein contains Val instead of Glu. Referring to the genetic code in Fig. 6.3, we see the codons GAA and GAG code for Glu while valine is called for by the codons GUA and GUG. Apparently then, the switch of one base pair in a human chromosome can lead to serious genetic disease. Other alterations in codons can give rise to stop codons and cause premature termination of peptide synthesis; they are known as *nonsense mutations.*

There are several postulated mechanisms for spontaneous mutation. First, the nucleotide bases of DNA have several different structural forms, known as *tautomers.* Although the configurations given in Fig. 2.7 are believed to predominate, shifts to other tautomeric forms could cause errors in base pairing. Another possible cause of spontaneous mutation is interference with the enzymes necessary for DNA synthesis and repair. Finally, some intermediates of normal cellular metabolism, e.g., peroxides, nitrous acid, and formaldehyde, are *mutagens,* i.e., chemicals which interfere with DNA function.

The action of chemical mutagens on DNA has been widely studied by growing cells in environments rich in such agents. Among the several classes of known mutagens are base analogs, which are compounds with structures similar to the bases normally found in DNA. Consequently, the analogs rather than the proper bases may be incorporated during synthesis of a DNA chain. Other types of mutagens and their mode of interference are listed in Table 6.2.

Another common cause of mutation is radiation. In particular, ultraviolet light is strongly absorbed by DNA, to such an extent that exposure to ultraviolet light rapidly kills most cells. The surviving cells exhibit a high rate of mutation. All cells possess enzymatic machinery to repair DNA occasionally damaged by ultraviolet light. These repair enzymes, in a rather elaborate process, replace the damaged segment, which contains covalently linked pyrimidine residues.

The phenomenon of mutation occupies several important niches in biochemical engineering practice. Returning to Fig. 1.10, we recall how mutations in microorganisms can improve them for our use. Thus, mutagens and exposure to ultraviolet light have been employed in attempts to obtain mutated, more productive protists. Essential in any strain-development activities are effective means for identifying and isolating mutants with specific characteristics. Table 6.3 summarizes some of the basic techniques for this purpose. Other methods for genetic manipulation of production organisms are discussed in Sec. 6.2 and 6.3.

Table 6.2 Chemical mutagens and their mode of action[†]

Chemical agent	Mutagenic effect
Base analogs	Incorporation into DNA in place of natural bases
Nitrous acid	Deamination of purine, pyrimidine bases of DNA
Proflavin, acridine orange	Intercalation between stacked bases of DNA
Alkylating agents	Depurination of DNA

[†] Adapted from R. Y. Stanier, M. Doudoroff, and E. A. Adelberg, *The Microbial World*, p. 418, Prentice-Hall, Inc., Englewood Cliffs, N.J., 1970.

Table 6.3 Methods for detection and selection of different bacterial mutants†

Type of mutant	Selection methods	Detection
Able to use as carbon source a compound not utilizable by the wild type	Plate on agar containing the compound in question as the only available carbon source	Plating method is absolutely selective; only the desired type will form colonies
Resistant to inhibitory chemical agents, such as penicillin, streptomycin, sulfonamides, dyes	Plate on agar containing the inhibitor	Plating method is absolutely selective; only the desired type will form colonies
Resistant to bacteriophage	Plate on agar previously spread with a suspension of phage	Plating method is absolutely selective; only the desired type will form colonies
Auxotrophic (requirement for one or more growth factors not required by wild-type cells)	Incubate in growth medium lacking the growth factor in question but containing penicillin; wild-type cells multiply and most are killed; auxotrophic mutants are unable to multiply without the growth factor and survive (penicillin kills only actively growing cells)	Plate on agar lacking the growth factor; mark the few wild-type colonies that appear, then add layer of agar containing growth factor; auxotrophs form colonies only after addition of growth factor (delayed-enrichment method)
Unable to ferment a given sugar	Apply penicillin technique as above, using sugar in question as only fermentable carbon source	Plate on agar containing sugar in question, plus chemical indicator that changes color in the presence of fermenting cells, e.g., acid-base indicator; mutant colonies appear a contrasting color
Temperature-sensitive DNA synthesis	Incubate cell suspension at 42°C for 15 min, add 5-bromouracil (5-BU), and continue incubation an additional 60 min; irradiate with light of 310 nm wavelength (cells that have incorporated 5-BU into their DNA are killed; cells that have failed to replicate their DNA at 42°C are spared)	Test colonies for ability to grow at 42°C, e.g., by replica plating
Sensitive to ultraviolet irradiation (unable to repair ultraviolet-light damage)	Infect culture with ultraviolet-light-inactivated bacteriophage; normal cells repair the ultraviolet-light lesions of the phage and are subsequently killed by the phage; repair-deficient cells are spared	Make suspensions from isolated colonies, measure survival at several ultraviolet light doses

† R. Y. Stanier, M. Doudoroff, and E. A. Adelberg, *The Microbial World*, 3d. ed., p. 450, Prentice-Hall, Inc., Englewood Cliffs, N.J., 1970.

On the other hand, mutations can create processing difficulties. Successful operation of a microbial reactor often requires the availability of a pure strain of microorganisms with known characteristics. Consequently, *stock cultures* of the needed microbe must be maintained. Degradation of these cultures by mutation is always a possibility, so that regular testing of the stock is necessary to preserve its integrity. Other practical problems concerned with mutations and other types of genetic instabilities will be explored in the following chapters

6.1.6 Overview of Information Flow in the Cell

A familiar cliché asserts that one picture is worth a thousand words. Figure 6.13 proves the point: here in one compact schematic many of the cell's control and information-carrying channels are summarized. Their interrelationships are

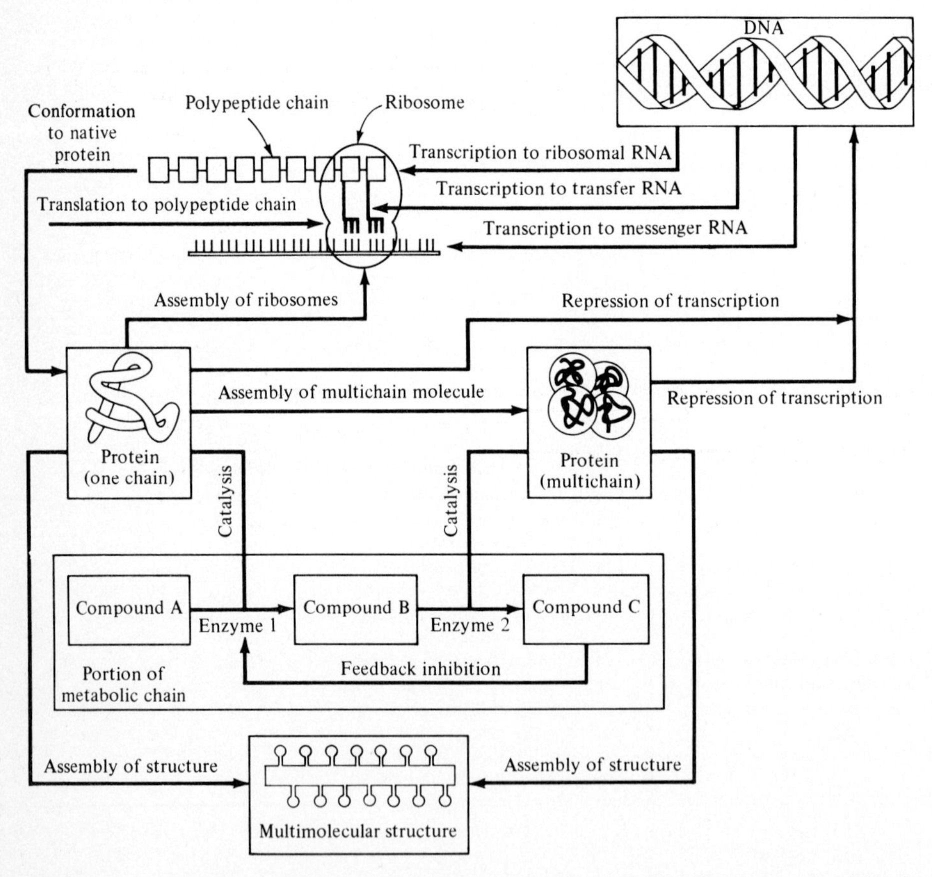

Figure 6.13 A hierarchy of control and information transmittal and processing systems is evident in this schematic illustration of information flow in the cell. *(From "Cell Structure and Function," 2d ed., p. 31, by Ariel G. Loewy and Philip Siekevitz. Copyright © 1963, 1969 by Holt, Rinehart and Winston, Inc., Reprinted by permission of Holt, Rinehart and Winston.)*

clearly revealed. Notice that a type of cascade control is embodied here: activation and inhibition (at the enzyme level) allow rapid adjustment to short-term changes in the cell's chemical balance, whereas induction and repression (at the gene level) readjust the entire metabolic pattern when a long-term disturbance in the cell's environment appears.

6.2 ALTERATION OF CELLULAR DNA

In addition to mutation, a variety of processes change the genetic material within a cell. These have both positive and negative implications from the standpoint of industrial biotechnology. From the former perspective, any method for altering a cell's DNA content provides an opportunity for development of more productive strains. On the other hand, an unplanned modification of a desired species' DNA can cause expensive failures of industrial bioprocesses.

Much of what follows pertains only to bacteria, *E. coli* serving as the subject of most research on these topics. Cell fusion methods for genetic modification are summarized below, and Sec. 6.4 is devoted to recombinant DNA techniques and their application. We shall not discuss here the elaborate and complex processes of sexual recombination which operate in many eucaryotic organisms. A brief introduction to this topic within the context of commercial microbes is provided in Elander [4].

6.2.1 Virus and Phages: Lysogeny and Transduction

We know that viruses are agents of human disease such as common colds, influenza, smallpox, polio, yellow fever, mumps, measles, and many other ailments. Also, several commercial biological processes employing bacteria (including cheese-making and antibiotic production) can be severely disrupted by viral infection of the bacteria. The subgroup of viruses which infect bacteria are called *bacteriophages* or just *phages*.

Viruses are constructed of protein and nucleic acid and also may include lipoproteins. The function of the protein is to house and protect the nucleic acid viral component and sometimes to attach the virus to a living cell. The nucleic acid component, which may be either DNA or RNA, is ultimately responsible for the infection and its aftermath. In Table 6.4 the diverse properties of several viruses are summarized; the viral nucleic acid may be single-stranded or double-stranded and is often circular.

Outside a living cell of the proper type, the virus is an inert particle which cannot reproduce by itself. Multiplication of the virus occurs after it has infected a host cell. Thus, viruses are parasites. A brief summary of the basic differences between viruses and living cells is provided in Table 6.5.

Next we shall consider the typical "life cycle" of a virus using as an example the virus called *phage* λ, which infects the *E. coli* bacterium. After attaching to the

Table 6.4 Characteristics of several kinds of viruses†

Type of virus	Size, nm	Shape, composition, and comment
Animal:		
Cubic symmetry:		
Poliomyelitis	30	Consists of 1 molecule RNA (mol wt 2×10^6) in a spiral, surrounded by protein macromolecules 6 nm diam. arranged as an icosahedron with no retaining membrane; particle mol wt 10×10^6
Helical symmetry:		
Influenza	100	Consists of 1 molecule RNA (mol wt 2×10^6) as a nucleoprotein macromolecule arranged in a helix, the whole coiled and enclosed in a lipoprotein sheath; particle mol wt 100×10^6
Plant:		
Rods:		
Tobacco mosaic	300×15	Whole virus particle rod-shaped, consisting of 1 molecule RNA (mol wt 2×10^6) associated with protein macromolecules arranged in a helix; particle mol wt 39×10^6
Sphere:		
Tomato bushy stunt	30	Icosahedron, consisting of 16% RNA (mol wt 1.6×10^6) and protein; particle mol wt 9×10^6
Insect:		
Silkworm	280×40	Actual virus rod-shaped; DNA constitutes about 8% of dry weight, but in vivo the virus rods are embedded in polyhedral crystalline aggregates of protein 0.5–15 μm diameter
Bacteriophages:		
Double-stranded:		
DNA		
T-even of *E. coli*		Tadpole-shaped phage with DNA (mol wt 130×10^6) confined to head; the protein (some contractile), long tail fibrils involved in attachment to host cell; particle mol wt 250×10^6
Head:	90×60	
Tail:	100×25	
Tail fibrils:	130×2.5	
Single-stranded:		
DNA		
ϕX174 of *E. coli*	22	Dodecahedron with 12 subunits; DNA (mol wt 1.6×10^6) 25% dry weight; particle mol wt 6.2×10^6
RNA f2 of *E. coli*	20	Polyhedron containing RNA (3×10^{-12} μg/virus) and protein; nucleic acid content probably similar to that of ϕX174

† S. Aiba, A. E. Humphrey, and N. F. Millis, *Biochemical Engineering*, 2d ed., p. 19, Academic Press, Inc., New York, 1973.

Table 6.5 Viral vs. cellular characteristics

Virus	Cell
Only one kind of nucleic acid	Contains DNA and RNAs
Contains only a few enzymes	Contains thousands of enzymes
Reproduced by assembly of nucleic acid and protein components synthesized by host cell	Reproduces itself in orderly, controlled fashion

cell wall, the phage injects its DNA into the cell's interior. At this point two alternative courses are possible. One option is a state of *lysogeny*, where the phage DNA, now called the *prophage*, is integrated into the bacterial chromosome. If this occurs, the cell lives and reproduces normally, at the same time copying the prophage and creating more lysogenic cells. Phages, like λ, which can enter into a lysogenic relationship with their hosts are called *temperate phages.*

The other possible outcome of temperate-phage infection, called the *lytic cycle*, invariably kills the host cell. The lytic cycle always results from infection by *lytic phages* such as phage T_2. During the lytic cycle the phage is said to be in a vegetative state: the phage DNA literally takes over control of the cell. It first directs the ribosomes to synthesize enzymes which destroy the host cell's DNA (this is not a universal feature of viral infections, however) and which will multiply the phage DNA. Then the protein components necessary to create an intact phage particle are synthesized. These proteins spontaneously join with phage DNA to form complete, highly organized bacteriophage particles. After this self-assembly of many new phages, the enzyme lysozyme is synthesized. As we have already learned, this enzyme attacks the murein cell wall of bacteria. Subsequently, lysis occurs; the cell breaks apart, freeing many phage particles.

From this brief survey it should be clear why phage infestation of commercial cultures can be a serious problem. The phage particles are very small, they can multiply rapidly in the right environment, and they can hide in the relatively dormant state of lysogeny.

Occasionally, during reproduction of the phage within its host, some phage particles are formed which contain a small portion (≈ 1 to 2 percent) of the host cell's chromosome. When one of these *transducing phages* injects its DNA into another bacterium, the DNA derived from the first host may cross over with a fragment of the new host's chromosome. In this process, called *transduction*, the genetic characteristic of the new host may be altered. Crossing-over generally occurs only between DNA segments which contain homologous nucleotide sequences and which govern similar characteristics. For example, a transducing phage could carry a small DNA fragment permitting lactose utilization from a lactose-metabolizing *E. coli* strain to an *E. coli* strain unable to use lactose. Figure 6.14 provides a simplified schematic illustration of transduction as well as crossing-over of DNA segments.

Transduction is known to occur with many actinomycetes. In one study involving *Streptomyces griseus*, phage transferred the ability to synthesize the

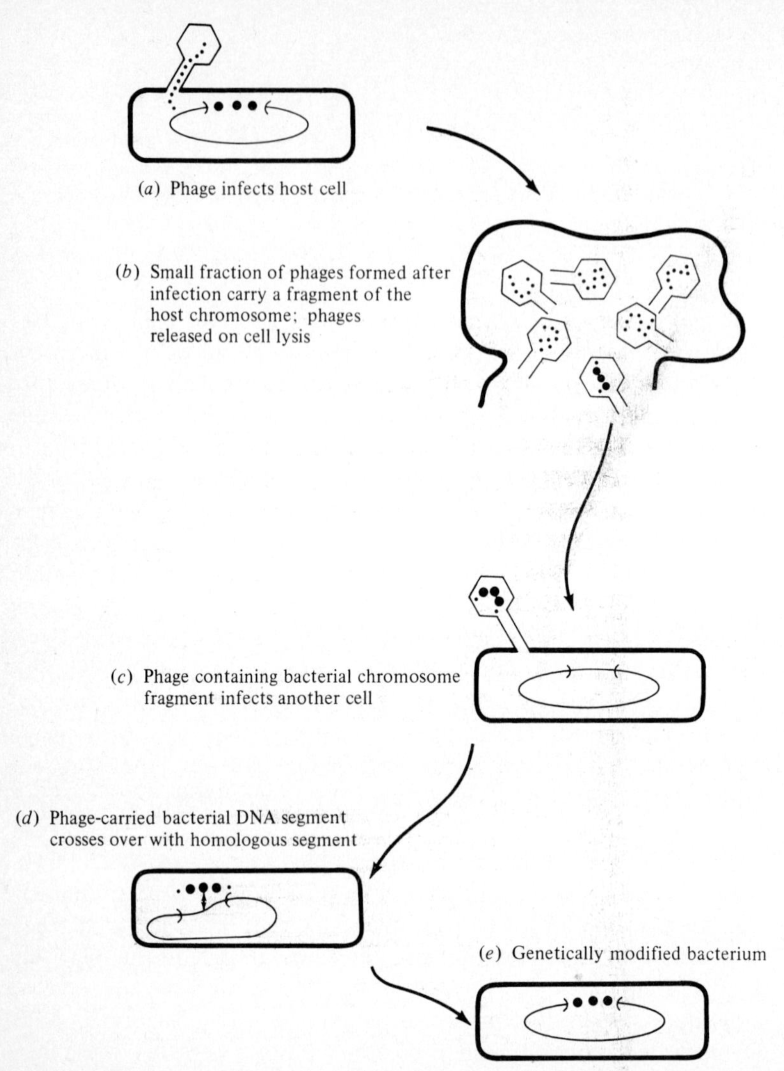

Figure 6.14 Schematic diagram of bacterial chromosome modification via transduction, a process mediated by bacteriophage.

antibiotic streptomycin from a producing to a nonproducing strain. Some of the transduced variants produced more of the drug than the original synthesizing strains.

6.2.2 Bacterial Transformation and Conjugation

Transformation is a process of genetic transfer by free DNA. Here a double-stranded DNA fragment enters recipient cells which, because of their physiological state, are competent to take up external DNA. If the transformed fragment is similar to a fragment of the recipient's DNA, alteration of the recipient chromosome by crossing-over occurs rapidly.

Besides chromosome fragments, *plasmids* may be introduced into living bacteria by transformation. We recall from Sec. 2.3 that a plasmid is a DNA molecule that is separate from the bacterial chromosome and duplicates independently. Plasmids are typically relatively small circular molecules with molecular weights of the order of 10^6 to 10^8 (see Fig. 6.15). Transformation by plasmids is an integral part of the technique for DNA manipulation discussed in Sec. 6.3.

While usually nonessential for normal cell function, plasmids can confer useful properties upon the cell they inhabit. For example, plasmids called *R factors* have been identified as the substances responsible for bacterial resistance to antibiotics.

Closely related to plasmids are *episomes*, DNA molecules which may exist either integrated into the cell chromosome or separate from it. One of the better-known episomes is the F, or fertility, factor which characterizes the partners in

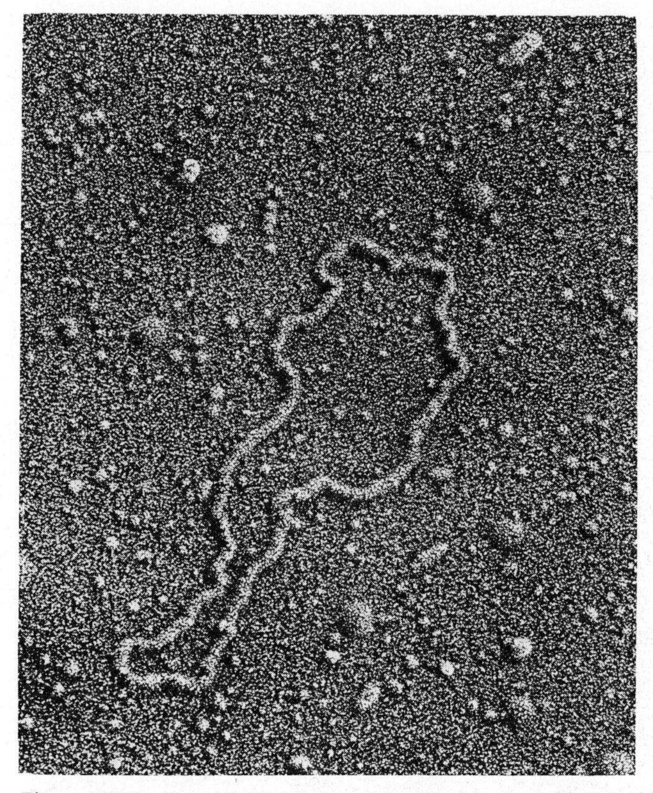

Figure 6.15 A plasmid (pSC101, electron micrograph × 230,000). This circular DNA molecule, which exists and replicates independently from the bacterial chromosome, was an important component in early recombinant DNA experiments. *(Electron micrograph by Dr. S. N. Cohen.)*

conjugating *E. coli* cells. During *conjugation* of *E. coli*, cells containing the F factor (F^+ cells) transmit it to F^- (lacking F) cells, and occasionally some of the chromosome of the F^+ partner is transferred to the F^- cell. R (resistance) factors may be transmitted in a similar fashion. Certain F^+ mutants, denoted Hfr, act differently, however; they do not infect the F^- cell with the F factor, but the Hfr partner does donate much of its chromosome. Transmitted segments may then cross over with the F^- chromosome to yield genetically altered strains.

Numerous experiments have revealed that the F^+ cell chromosome enters the F^- cell in linear order in the form of a thread. If the transfer is interrupted at different times, chromosomal threads of different lengths are found in the F^- cells. Comparison of the characteristics of recipient populations exposed to different amounts of injected chromosome then reveals the relative locations of genetic information storage on the (F^+ cell) chromosome. In this fashion, detailed genetic maps like that in Fig. 6.16 have been constructed. Notations around this map indicate the relative loci of various mutant characteristics. For example, *thr* refers to a mutant which requires threonine for growth. Many of the other symbols are defined in Watson [7, Chap. 2].

6.2.3 Cell Fusion

Multiple genetic modifications, including crossing of genetic material from different species, can be achieved by fusing different cells together. As the flowchart for microbial cell fusion in Fig. 6.17 indicates, the first step is preparation of *protoplasts*, cells that have been stripped of their outer cell wall and which are contained only in the cell membrane. Various cell wall hydrolase enzymes are applied to remove the cell walls, and hypotonic medium is employed at this stage to minimize breakage of the osmotically sensitive protoplasts. Incubation of the protoplasts results in cell wall regeneration and reversion to normal cell morphology. This is a critical property since, after fusing protoplasts from different strains, we wish to recover an organism which can be cultured further by normal methods, potentially on a large scale. Protoplast fusion is induced and accomplished in solutions of polyethylene glycol (PEG), and selection methods similar to those used in traditional microbial genetics (Table 6.3) are used to isolate protoplast fusion products.

The microbes that have been successfully fused by this type of protocol include different Penicillia, Aspergilli, *Streptomyces* species, *Cephalosporium* species, yeasts, and bacilli. Interspecies crosses have been obtained between Pencillia, between Aspergilli, between *Saccharomyces cerevisiae* and *Saccharomyces diastaticus*, between *Condida tropicalis* and *Saccharomycopsis fibuligera*, and others. While crosses between species which are significantly different are often relatively unstable, reverting to one of the parental forms after a few generations, protoplast fusion is an exciting method for accomplishing multiple genetic alterations in a single step. One objective of current research on protoplast fusion is a combination of fast-growing *Streptomyces* with highly mutated production organisms which are high producing but slow growing.

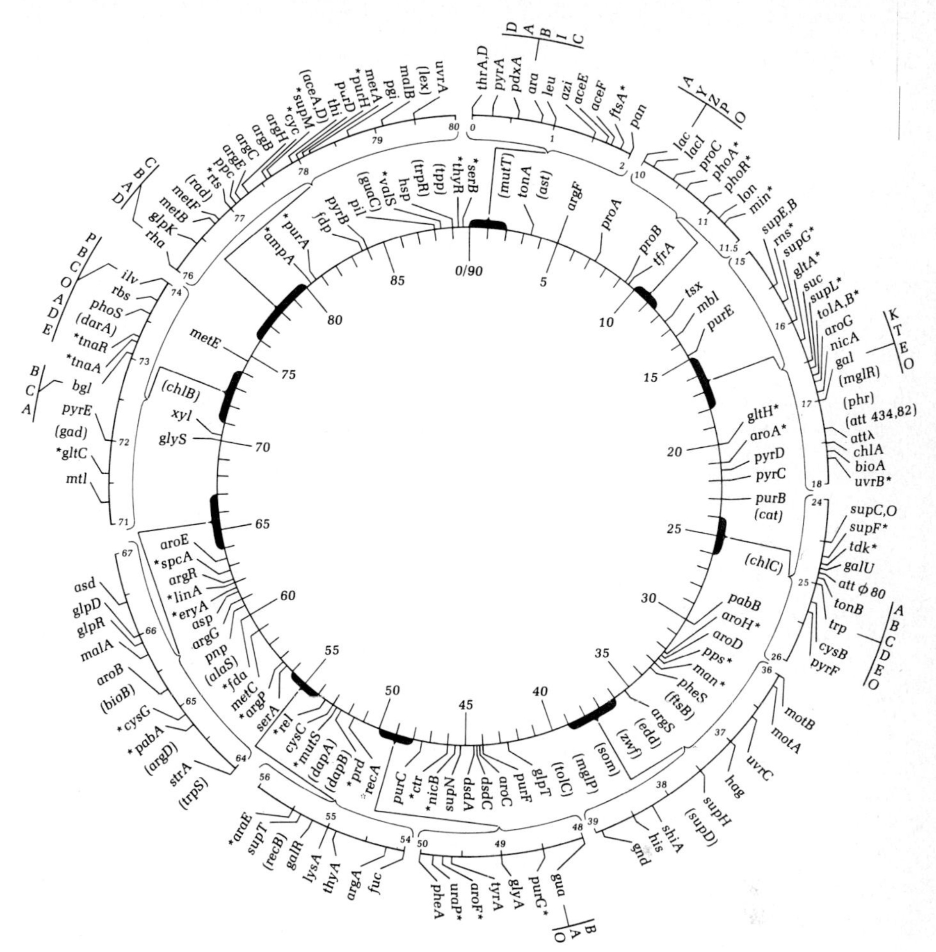

Figure 6.16 A genetic map of *E. coli* K12 showing the relative location of various genes in the circular chromosome. *(Reprinted from A. L. Taylor and C. D. Trotter, "Revised Linkage Map of* E. coli," *Bacteriol. Rev., vol. 31, p. 332, 1967).*

Cell fusion may also be used to obtain genetically crossed animal or plant cells. An example of this type of manipulation, which has numerous current and potential practical implications, is the formation of *hybridoma* cells by fusion of an antibody-producing white cell with a myeloma (skin cancer) cell of a mouse or other animal. Each hybridoma cell synthesizes a single molecular species of antibody. Cells cultured from a single such hybridoma produce a *monoclonal antibody*, in pure form. In contrast, when a protein or other antigen is injected in a rabbit to make an antibody, many different antibodies, specific for different regions on the antigen molecule, are synthesized. Monoclonal antibodies, already important research reagents, are the bases of many diagnostic methods in current practice or under development. In immobilized form, monoclonal antibodies provide highly specific affinity adsorbents for laboratory and large-scale purification

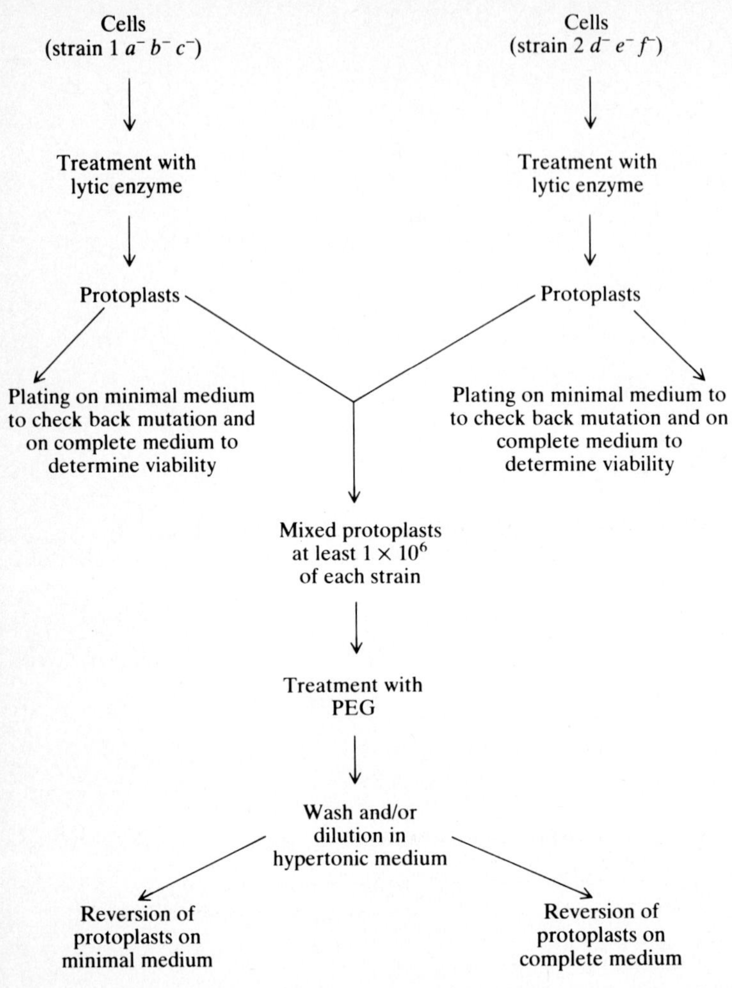

Figure 6.17 Summary of the steps for making and isolating new strains resulting from fusion of protoplasts derived from two different parental strains. *(Reprinted from J. F. Peberdy, "Protoplast Fusion—A Tool for Genetic Manipulation and Breeding in Industrial Microorganisms," Enzyme Microb. Technol., vol. 2, p. 25, 1980.)*

of antigen. Also, monoclonal antibodies may be widely used in the future for tumor imaging and for drug delivery to specific cell types in the body.

Small quantities of monoclonal antibodies are conveniently prepared by inoculating the peritoneal cavity of compatible mice with hybridoma cells. The resulting tumor secretes monoclonal antibodies into the surrounding fluid at levels from 1 to 20 mg per milliliter. Hybridoma cells can also be grown in culture. Searches for new reactor designs to increase cell density and antibody productivity and for ways to minimize medium costs to make large-scale monoclonal antibody production more economical are currently high priority research and development activities.

6.3 COMMERCIAL APPLICATIONS OF MICROBIAL GENETICS AND MUTANT POPULATIONS

Basic knowledge of molecular biology and cellular control systems has several important practical implications. Some were considered in the previous section; if the location of a desirable or unwanted gene can be determined, a variety of techniques can be applied to attempt modification of that particular gene. Another tactic is induced mutation to a more productive species. The fruits of this approach have already been mentioned in Chap. 1, where we discussed increased penicillin yields based on improved strains obtained by a sequence of mutations. The final portions of this section will outline several other microbial processes where development of special mutant species has played a major role.

Often, improved productivity of a mutant microorganism has a straightforward interpretation in terms of basic cellular control systems. Understanding of the mechanisms controlling biosynthesis of the desired product has another important application: deciding how to formulate and control the growth medium so that productivity is maximized. We consider such environmental manipulations next.

6.3.1 Cellular Control Systems: Implications for Medium Formulation

There are two different approaches to altering cellular productivity via choice of medium composition: (1) add inducers and (2) decrease amount of repressor present. Both approaches are rather obvious on the surface, but in practice some sophistication is required to achieve optimal results.

Beyond the enzyme's substrate itself, nonmetabolizable substrate analogs can be extremely effective inducers for enzyme production. For example, β-galactosidase specific activity in *E. coli* can be increased more than 1000-fold by galactosides. In addition to the isopropyl β-D-thiogalactoside sometimes used for β-galactosidase induction, the following substrate analogs have proved effective (enzyme induced follows in parenthesis): *N*-acetylacetamide (amidase), methicillin (penicillin-β-lactamase), and malonic acid (maleate isomerase).

Catabolite repression is known to decrease product biosynthesis in many important processes. Availability of a rapidly consumed substrate such as glucose at high concentrations fosters rapid growth but limits production of the antibiotics penicillin, mitomycin, bacitracin, and streptomycin. Great improvements in product yields can be obtained for penicillin by intentionally fostering diauxic growth (see Fig. 7.14) by using glucose plus a slowly metabolized sugar such as lactose in the growth medium. The desired biomass is grown on the glucose fraction, and a phase of product synthesis follows as the lactose is consumed. An alternative strategy for achieving the same effect is slow addition of glucose to the medium.

Other examples of repressors whose concentrations can be directly controlled by medium formulation include inorganic phosphate (represses phosphatase synthesis in *E. coli* and nuclease production in *Aspergillus quernicus*) and ammonia (represses urease biosynthesis by *Proteus retgeri*). For example, the alkaline phosphatase content of *E. coli* can be increased from about zero up to roughly 5 percent of the cell protein by limiting the phosphate content of the growth medium. In a like manner, avoiding high concentrations of amino acids or sulfate greatly increases protease synthesis by bacteria or *Aspergillus niger*, respectively.

Maintenance of low repressor concentrations is more difficult when the repressor is synthesized by the microorganism. One useful approach in this case is to alter the cell's membrane so that the repressing substance quickly diffuses from the cell's interior to the medium. This strategy is one component of a very productive process for glutamate manufacture using *Corynebacterium glutamicum*. Here, limiting the biotin concentration in the medium permits excretion of glutamate into the medium. Intracellular accumulations of other amino acids are greatest when biotin is present in excess. While a detailed explanation of the effect is elusive, there is some evidence relating biotin effects to phospholipid components of the cell membrane.

In the next section we consider another means of ameliorating the negative effects of repressors.

6.3.2 Utilization of Auxotrophic Mutants

Auxotrophic mutants (see Table 6.3 for selection and detection methods) lack enzyme activity for one or more steps in a biosynthetic pathway. As a result, one or several end products of the pathway are not synthesized. In order for such a mutant to survive, it must be fed the unsynthesized metabolites. For example, a tryptophan auxotroph must be fed tryptophan in order to grow. A strain which synthesizes its own tryptophan is said to be *prototrophic* for tryptophan.

Because the missing metabolites are supplied in the growth medium and are not synthesized by the auxotroph, the engineer rather than the microorganism has control of the metabolites' concentrations. Clearly this can be a desirable situation from a practical viewpoint when the unsynthesized end product is a repressor. As Fig. 6.18 shows, such auxotropic mutants lacking repressor synthesis can be made to overproduce† an intermediate metabolite. By keeping the concentration of repressor E low in the medium, feedback inhibition and repression of pathway enzymes are minimized. The intermediate C (normally the substrate for enzyme c, which is absent from the mutant) will then achieve much higher concentrations than in the native organism. As indicated at the close of the previous section, additional manipulations may then be necessary to allow intermediate C to diffuse through the cell membrane into the medium.

† "Overproduce," as used here, is relative to the optimal allocation of biosynthetic intermediates for cell growth and maintenance. Usually such an optimum from the cell's perspective does not lead to high yields of commercially important end products, hence the motivation for industrial use of mutant species with intentionally imperfect internal controls.

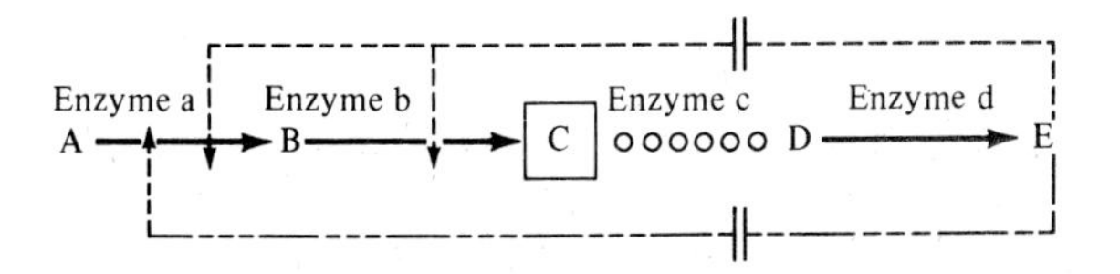

Figure 6.18 Auxotrophic mutants can be used to enhance production of intermediates in a metabolic pathway. In this hypothetical example, the mutant lacks enzyme c and therefore does not synthesize the repressor E. *(Reprinted from A. L. Demain, "Overproduction of Microbial Metabolites and Enzymes due to Alteration of Regulation," in T. K. Ghose and A. Fiechter (eds.), "Advances in Biochemical Engineering 1," p. 120, Springer-Verlag, New York 1971.)*

The role of auxotrophy in commercial L-lysine manufacture using *C. glutamicum* is illustrated in Fig. 6.19. The productive mutant lacks homoserine dehydrogenase, so that the inhibitory effect of threonine on lysine synthesis (via aspartokinase) is eliminated. Since the auxotropic mutant does not synthesize threonine or methionine, both these amino acids must be supplied in the growth medium.

Comparison of the aspartate pathway controls for *C. glutamicum* in Fig. 6.19 with the corresponding pathway in *E. coli* (Fig. 5.12) makes an important point: organisms with similar biosynthetic pathways do not necessarily utilize identical control systems. Specifically, notice in Fig. 5.12 that lysine inhibits its own production via the reaction leading to dihydrodipicolinic acid. The corresponding inhibitition step is absent from the (unmutated) parent strain of *C. glutamicum.* Also, the aspartokinase system found in the *Corynebacterium* apparently differs

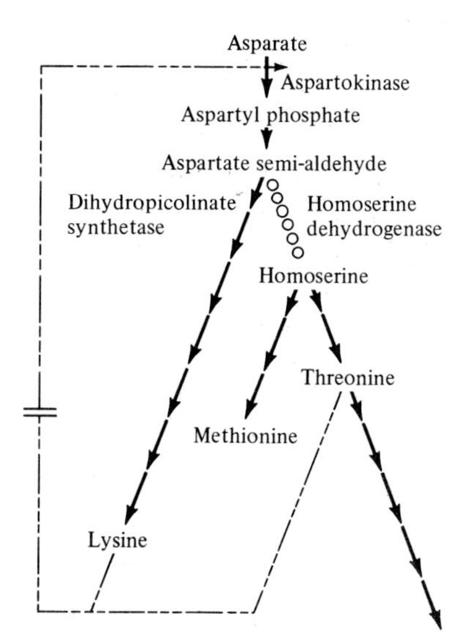

Figure 6.19 A mutant of *C. glutamicum* lacking the enzyme homoserine dehydrogenase permits enhanced production of L-lysine. *(Reprinted from A. L. Demain, "Overproduction of Microbial Metabolies and Enzymes due to Alteration of Regulation," in T. K. Ghose and A. Fiechter (eds.), "Advances in Biochemical Engineering 1," p. 122, Springer-Verlag, New York, 1971.)*

from the *E. coli* aspartokinases in that the former is not repressed unless *both* lysine and threonine are present (*concerted* or *multivalent* feedback inhibition). whereas one of the aspartokinase isozymes in the *E. coli* system is repressed by lysine alone. Both these differences confer commercial advantage to the *Corynebacterium.*

Similar strategies have been successfully employed to obtain productive microbial processes for manufacture of the flavor-enhancing purine nucleotides guanosine monophosphate (GMP), inosine monophosphate, (IMP), and xanthine monophosphate (XMP). As Fig. 6.20 illustrates, minimization of AMP and

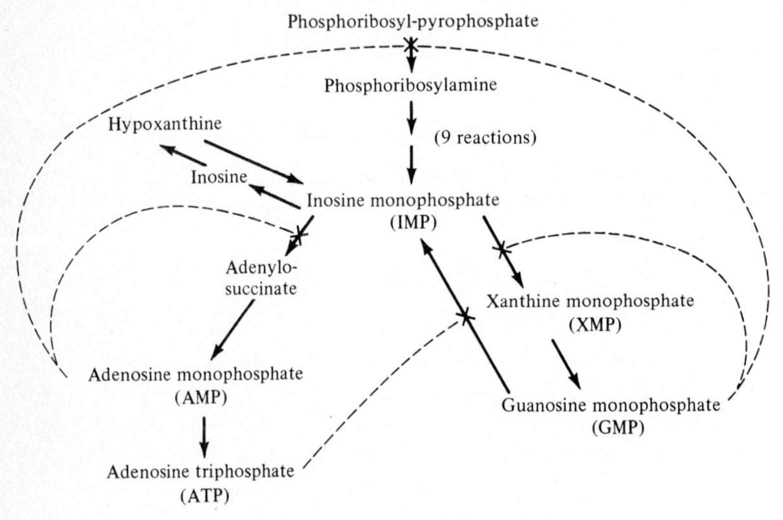

(*a*) Control of nucleotide biosynthesis

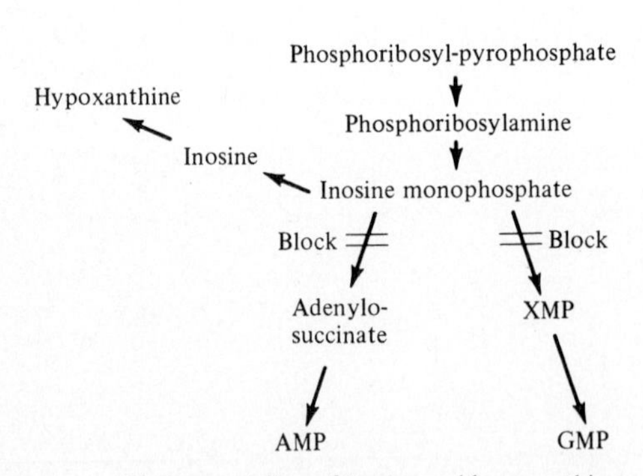

(*b*) Accumulation of inosine and hypoxanthine

Figure 6.20 (*a*) In a normal cell, AMP and GMP exert feedback inhibition at several points in the nucleotide biosynthesis reaction sequence. (*b*) In a mutant which does not synthesize AMP and GMP, the concentration of these components may be kept low by proper medium formulation. This permits enhanced yields of inosine and hypoxanthine. *(Reprinted from S. Aiba, A. E. Humphrey, and N. Millis, "Biochemical Engineering," 2d ed., p. 84, University of Tokyo Press, Tokyo, 1973.)*

GMP concentrations by use of an auxotrophic mutant of *Brevibacterium ammoniagenes* permits accumulation of inosine and hypoxanthine. Appearance of these products in the medium is enhanced by adding a small amount of manganese (Mn^{2+} concentration $\approx 10\ \mu g/L$) to the medium. Evidence to date indicates that the manganese changes the cell-membrane permeability.

6.3.3 Mutants with Altered Regulatory Systems

Several other types of genetic manipulation and selection have provided commercially superior strains by altering controls at the enzyme and/or at the gene level. Mutants of this type can be used to increase yields of metabolites and enzymes.

To obtain overproduction of a metabolite which acts as an inhibitor and/or repressor of its biosynthesis, we seek mutant organisms whose relevant allosteric enzymes and operons are insensitive to the metabolite's presence. Such mutants are often isolated using antimetabolites, which are toxic analogs of the metabolite in question. Normal cells will not usually grow in a medium containing antimetabolite since the antimetabolite represses or inhibits biosynthesis of the necessary metabolite without serving as a substitute for the unsynthesized metabolite in subsequent pathways. On the other hand, strains with deficient feedback controls will not alter their biosynthesis patterns or rates in response to the antimetabolite and will therefore survive in its presence. Table 6.6 lists some of the microbial products whose yields can be enhanced by this technique.

Although resistance to an antimetabolite may involve a variety of control system alterations, one possibility is elimination or reduction of repression of

Table 6.6 Mutants providing enhanced yields of these end products may be selected using the corresponding antimetabolites†

End product	Antimetabolite used	End product	Antimetabolite used
Arginine	Canavanine	Threonine	α-Amino, β-hydroxyvalerate
Phenylalanine	*p*-Fluorophenylalanine		
	Thienylalanine	Methionine	Ethionine
Tyrosine	*p*-Fluorophenylalanine		Norleucine
	Thienylalanine		α-Methylmethionine
	D-Tyrosine		L-Methionine-DL-sulfoximine
Tryptophan	5-Methyltryptophan		
	6-Methyltryptophan	Histidine	2-Thiazolealanine
Valine	α-Aminobutyrate		1,2,3-Triazole-3-alanine
Isoleucine	Valine	Proline	3,4-Dehydroproline
Leucine	Trifluoroleucine	Adenine	2,6-Diaminopurine
	4-Azaleucine	Uracil	5-Fluorouracil

† A. L. Demain, "Overproduction of Microbial Metabolites due to Alteration in Regulation," p. 113 in T. K. Ghose and A. Fiechter (eds.), *Adv. in Biochemical Engineering* 1, Springer-Verlag, New York, 1971.

enzymes in the metabolite's biosynthesis pathway. In this context, antimetabolites can be used to discover mutants with unusually large concentrations of biosynthetic enzymes. Such a strategy would be appropriate, for example, in manufacturing enzymes for subsequent *in vitro* biosynthetic processes.

Other methods can be employed to identify and isolate *constitutive mutants*. In these species, normally induced or repressed enzymes are produced whether or not inducer or repressor is present. Among the enzymes whose yields can be improved with constitutive mutants are β-galactosidase, catalase, phosphatases, proteases, homoserine dehydrogenase, invertase, histidase, penicillinase, and amidase.

Throughout the previous discussion we have concentrated on various strategies for increasing yield of a desired compound. Additional avenues of application of such molecular biological principles include biosynthesis of derivatives of the original products and generation of completely new end products with pharmaceutical and other uses. Brief summaries of these topics are given in Refs. 4–6. An important example of such an application is biosynthesis of the tetracycline derivative 6-demethyl tetracycline by a mutant of *Streptomyces auriofaciens*. More stable under acidic conditions than the usual methylated form, this derivative is one of the dominant commercial forms of tetracycline antibiotic.

6.4 RECOMBINANT DNA TECHNOLOGY

In the late 1970s a set of biochemical methodologies emerged which made *genetic engineering* possible. The tools of recombinant DNA, which have revolutionized the potential of biotechnology for mankind's benefit, permit precise construction of new genetic instructions which can be inserted into and utilized by living cells. Through recombinant DNA technology, our knowledge of DNA function, gene organization, gene expression regulation, and protein primary structure is advancing at a dizzying pace, promising major breakthroughs in understanding and combating disease. The opportunity to introduce totally new DNA into cells also has created unique industrial microorganisms able to synthesize valuable proteins. In this section our emphasis will be application of genetic engineering in manufacturing and on use of the bacterium *Escherichia coli* as the host organism.

Before outlining some of the key reagents and methods of genetic engineering, it is useful to examine an overview of the process and to introduce some new terminology. A brief schematic of a method for *cloning* a segment of DNA is shown in Fig. 6.21. Here cloning means obtaining a colony of genetically identical cells which contain the DNA segment of interest. This segment, called "foreign DNA" in Fig. 6.21, is obtained by cutting a larger piece of DNA into fragments using specific endonuclease enzymes or by enzymatic or chemical synthesis. The foreign DNA is joined in the test tube (*in vitro*) to a *vector* or vehicle which will carry the foreign DNA as a passenger into a bacterial cell. In this example the vector is a bacterial plasmid which has been manipulated to facili-

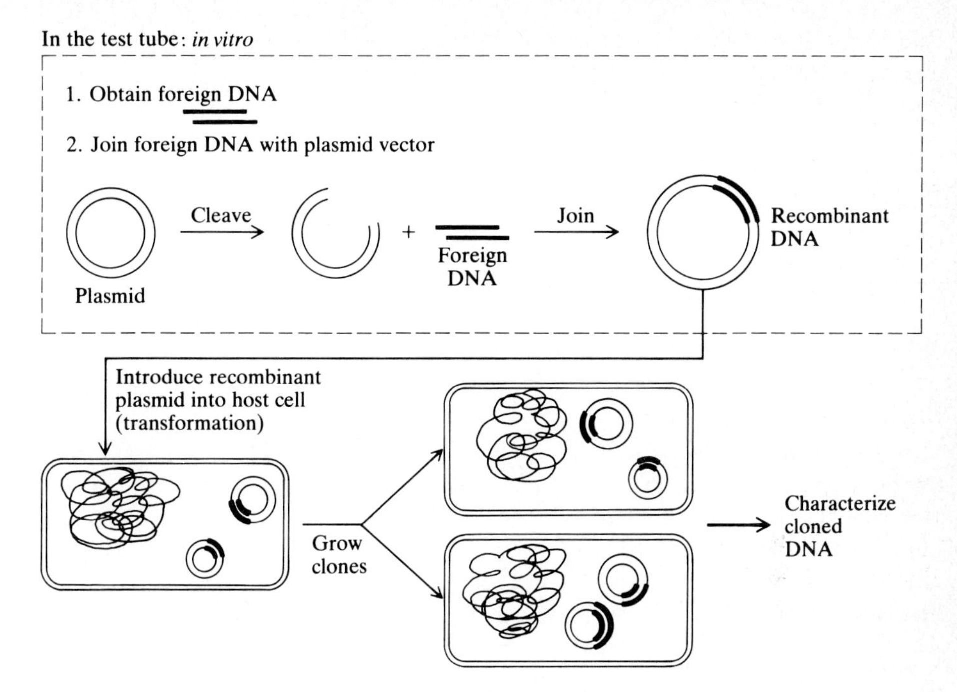

Figure 6.21 Schematic illustration of the major steps in cloning a foreign DNA segment.

tate the cloning process. The *recombinant DNA molecule* formed by the foreign DNA-plasmid construct is then introduced into a bacterial cell by transformation. Using methods described below, the clone containing the plasmid which includes the foreign DNA segment is identified.

Cloning a DNA fragment provides enough of that DNA sequence for detailed analysis and for use as a reagent in subsequent biochemical and genetic manipulations. As an example of use of cloned DNA for analysis, consider the status of research on fibroblast interferon (IFN-β) in the late 1970s. Supplies of this protein were extremely limited, so that, even with the most sensitive research instruments for amino acid sequencing, it was only possible to determine the first few amino acids at the amino terminus of the protein. However, once a fibroblast interferon gene was cloned and its nucleotide sequence determined, the complete amino acid sequence of the corresponding protein followed directly by use of the genetic code! Examples showing use of cloned DNAs for construction of strains which express foreign proteins at high levels will be considered in Sec. 6.4.4. Next we examine several enzymes which are crucial to recombinant DNA technology.

6.4.1 Enzymes for Manipulating DNA

One key to recent breakthroughs in genetic engineering is the discovery and the subsequent commercial supply of several enzymes which can be used to cut, alter, and join DNA molecules in the test tube. Prominent among these crucial enzymes are the *restriction endonucleases* which recognize and cut specific nucleotide sequences within the DNA molecule. For example the restriction enzyme called *Eco*RI recognizes the double-stranded six-nucleotide sequence

$$\begin{array}{c} \cdots\!-\!G\overset{\downarrow}{-}A\!-\!A\!-\!T\!-\!T\!-\!C\!-\!\cdots \\ \cdots\!-\!C\!-\!T\!-\!T\!-\!A\!-\!A\underset{\uparrow}{-}G\!-\!\cdots \end{array}$$

and cleaves each strand between the G and A residues as marked above by arrows. The two ends may then separate

$$\begin{array}{lcr} \cdots\!-\!G & \qquad\qquad & A\!-\!A\!-\!T\!-\!T\!-\!C\cdots \\ \cdots\!-\!C\!-\!T\!-\!T\!-\!A\!-\!A & \qquad\qquad & G\cdots \end{array}$$

Notice that in this case each end contains a short single-stranded segment. These are called "sticky ends" or "cohesive termini" since the two ends are complementary single-stranded nucleotide sequences which will naturally tend to base pair and to join, by hydrogen bonding, to each other. Whether or not these hydrogen bonded sequences tend to stay together or separate can be adjusted by manipulating the temperature or the pH of the solution.

More than 100 different restriction enzymes have now been identified. Some of the more commonly used ones, their *recognition sites*, and the points of DNA backbone hydrolysis are listed in Table 6.7. We should observe here that not all restriction enzymes leave sticky ends. For example, *Hpa*I cuts "straight across" a six-nucleotide recognition site, leaving double-stranded *blunt ends.*

Restriction enzyme nomenclature is based on the organism in which the restriction enzyme was discovered. The first, and capitalized, letter is the first letter of the genus of the source organism, and the next two or three letters are the first two or three letters of the species of this organism. The Roman numeral denotes the order of discovery. For example, *Bgl*II is the second restriction enzyme discovered in the bacterium *Bacillus globigii.* Sometimes additional letters are added to define more precisely the strain of origin. *Eco*RI, for example, was first isolated from *E. coli* RY13.

The utility of restriction enzymes in recombinant DNA technology derives from their specificity in cleaving DNA at particular sites. Also, because recognition sites of 6 or 4 nucleotides in length are not found too frequently in a DNA molecule, the DNA fragments obtained by restriction enzyme hydrolysis, typically several hundred nucleotides in length, are sufficiently long to contain useful genetic information and sufficiently short to allow convenient physical and biochemical *in vitro* manipulation.

Our treatment of other important enzymes in recombinant DNA technology is necessarily brief. First, we should mention *DNA ligase.* Suppose the sticky ends obtained by earlier *Eco*RI hydrolysis of different DNA molecules are allowed to *anneal*—to come together and base pair. This configuration is not *covalently closed*: the phosphodiester bonds (indicated by dashes) are missing between the

Table 6.7 Recognition sites and cleavage points (arrows) for some restriction endonuclease enzymes

Enzyme	Target site
*Bam*HI	G↓G A T C C C C T A G↑G
*Bgl*II	A↓G A T C T T C T A G↑A
*Eco*RI	G↓A A T T C C T T A A↑G
*Hae*III	G G↓C C C C↑G G
*Hin*dIII	A↓A G C T T T T C G A↑A
*Hpa*I	G T T↓A A C C A A↑T T G
*Pst*I	C T G C A↓G G↑A C G T C

G and A residues in both strands. DNA ligase catalyzes the condensation of a 3′-hydroxyl group with a 5′-phosphate group to add these missing links.

Annealed fragments

···—G A—A—T—T—C—···

···—C—T—T—A—A G—···

are covalently closed by DNA ligase

···—G—A—A—T—T—C—···

···—C—T—T—A—A—G—···

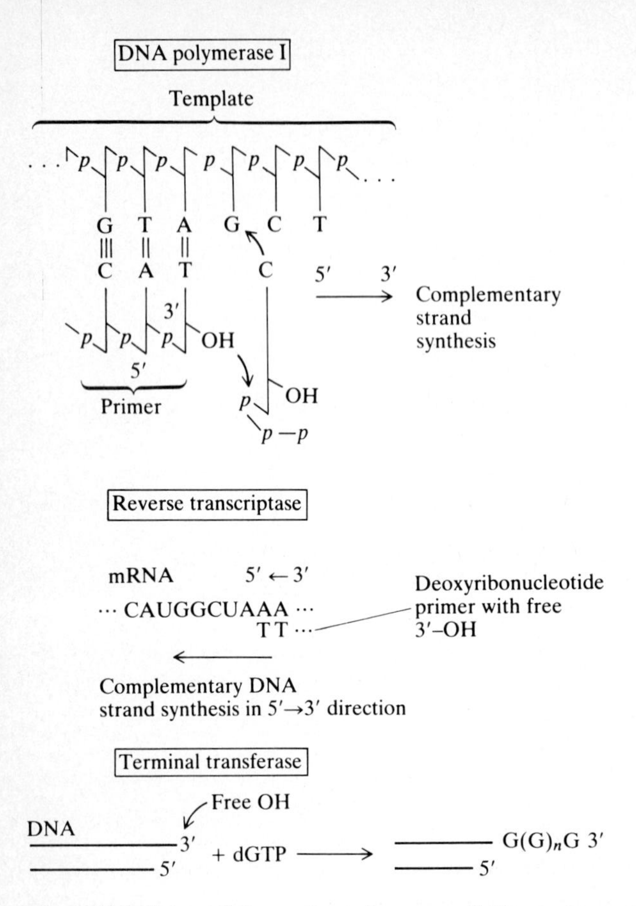

Figure 6.22 Characteristic reactions of some synthetic enzymes useful in recombinant DNA technology.

The final enzymes we shall consider all catalyze synthesis of oligonucleotides (see Fig. 6.22). *DNA polymerase I* (DNA *pol*I) uses a template strand of DNA to direct the synthesis of a complementary strand in the $5' \rightarrow 3'$ direction. Triphosphates of all four of the deoxyribonucleotides are required reagents, as is a *primer* which terminates with a 3′-hydroxyl. *Reverse transcriptase* accomplishes the relatively rare feat of reversing the normal flow of genetic information: this enzyme synthesizes a single DNA strand complementary to an mRNA template. Joining different DNA fragments is often facilitated by adding to them complementary *homopolymeric tails.* Given a single nucleotide triphosphate, the enzyme *terminal transferase* will repeatedly add this nucleotide to the 3′-OH termini of DNA molecules.

A few additional enzymes important in genetic engineering methods will be mentioned below. Next we turn our attention to the required characteristics of vectors for recombinant DNA.

6.4.2 Vectors for *Escherichia coli*

Vectors are DNA molecules which provide propagation of a DNA fragment in a growing cell population. A useful vector for cloning should have the following properties:

1 Ability to replicate in the host cell
2 Ability to accommodate foreign DNA of various sizes without damage to replication functions
3 Easy insertion in host cell after *in vitro* DNA manipulations
4 Contains a *selection marker* to facilitate rapid, positive selection of cells which contain the vector
5 Contains only one target site for one or more different restriction endonucleases

Two different classes of vectors—plasmids and bacteriophages—have been developed for cloning DNA in *E. coli*. We will concentrate here on plasmid vectors.

As mentioned earlier, plasmids are introduced into *E. coli* cells by transformation. Since a transformed *E. coli* cell may receive only a single plasmid and since we require a clone of many identical plasmid-containing cells to isolate, characterize, and utilize DNA sequences of interest, it is absolutely essential that the plasmid be able to replicate in the growing bacterial cell. This requires inclusion in the plasmid of an *origin of replication*—a nucleotide sequence, about 600 bp long for *E. coli* plasmids, which directs and regulates replication so that each cell contains a reasonably consistent number (typically around 30) of plasmid copies

Selection markers are also important. By including, for example, a gene for antibiotic resistance in the plasmid, we can rapidly and positively select for cells containing plasmids. This is done by growing the cells on medium containing a level of the antibiotic which kills plasmid-free cells but which allows growth of cells containing plasmids and hence the antibiotic resistance gene. Another common strategy for selection is to use a mutant host lacking an enzyme required for growth in a particular medium and to include on the recombinant plasmid the normal gene for that enzyme. Here the enzyme expressed from the plasmid gene *complements* the genetic lesion in the host.

Selection markers in plasmids can serve two quite distinct functions in recombinant DNA technology. First, laboratory investigations are greatly accelerated by positive selection procedures which rapidly identify vector-containing colonies. Second, in subsequent growth of cells containing the plasmid, imposing selection pressure—by adding antibiotic to the growth medium, for example—minimizes competition from any plasmid-free cells that may be born during population growth.

Plasmid vectors for cloning in *E. coli* have been highly developed in recent years. Among the more popular plasmids is pBR322 (Fig. 6.23) which includes a number of unique restriction enzyme sites, genes for tetracycline and for ampicillin resistance, and an origin of replication which allows amplification of the

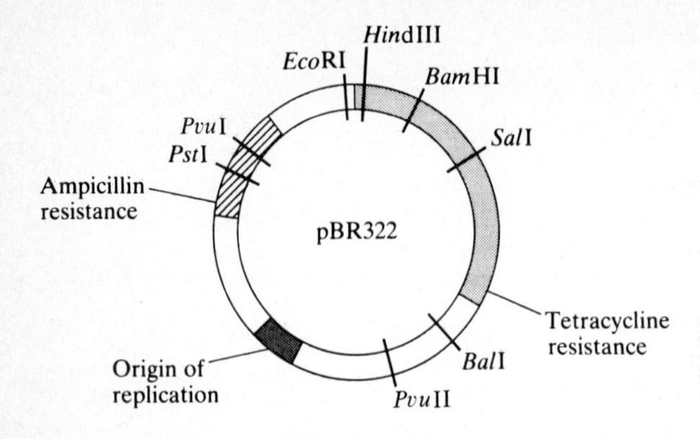

Figure 6.23 Map of the plasmid pBR322 showing genetic loci and some of the unique restriction enzyme sites. The complete nucleotide sequence of the 4362 base-pair plasmid is known.

plasmid. Here "amplification" means a large increase in the ratio of plasmid to chromosomal DNA. Amplification is possible using pBR322 and related plasmids since the plasmid can replicate in the absence of protein synthesis in the host cell but the chromosome cannot. Consequently, by adding an inhibitor of protein synthesis such as chloramphenicol to the culture medium, plasmid (and hence cloned DNA) content of the cell may be increased from around 30 to over 1000 copies. This provides large quantities of cloned DNA for characterization and for use as a reagent in later DNA constructions. Inserts of variable length up to an upper limit of 15 kb pairs may be cloned in pBR322 and related plasmid vectors.

Bacteriophage lambda has been extensively developed as a cloning vector. About 25 percent of the wild-type λ genome can be deleted without imparing the replicative and lytic functions of the phage, permitting cloning of around 12 kb pairs of foreign DNA. Much larger DNA segments—up to 50 kb—may be cloned through use of hybrid plasmid-phage λ vectors called *cosmids* which may be packaged *in vitro* into λ phage heads and tails. For this reason, bacteriophage vectors are preferred when cloning *libraries* consisting of large fragments of some or all of a eucaryotic genome. Phage with a single-stranded genome such as M13 have also been manipulated to obtain useful vectors for cloning. The single-stranded clones so produced are especially convenient for DNA nucleotide sequencing using the Sanger method. DNA sequencing and other techniques for characterizing cloned DNAs are summarized in the next section.

6.4.3 Characterization of Cloned DNAs

Suppose now that recombinant plasmids have been transformed into a population of *E. coli* cells, and those cells containing plasmids have been identified by growth on agar plates made with selective medium. In this section we briefly review the methods by which the clones so obtained may be screened and the DNA inserts from interesting clones analyzed in detail.

First, we can distinguish between clones which have plasmids with inserts and those which contain only reclosed plasmids by applying *insertional inactivation* during the cloning procedure. Suppose, for example, that DNA fragments were inserted at the *Pst*I restriction enzyme site in plasmid pBR322. Since inserts at this position interrupt the gene for ampicillin resistance, cells containing recombinant plasmids will be ampicillin sensitive and tetracycline resistant. On the other hand, cells with insert-free plasmids will be resistant to both ampicillin and tetracycline.

Consequently, clones containing recombinant plasmids may be identified by first growing on tetracycline plates (to eliminate the large background of cells with no plasmids). Then, the *replica plating* method is applied in which a felt pad or nitrocellulose paper is pressed on the original, master plate and then pressed onto a second plate, inoculating the replica plate with an exact copy of the colony distribution on the master. In this example, the medium used for the replica plates would contain ampicillin. Colonies present on the master plate which do not grow in the replica plate are those which contain plasmids with inserts in the ampicillin resistance gene.

Screening for specific nucleotide sequences in the clone may be accomplished by several different procedures which are all based on hybridization of denatured (i.e., single-stranded) DNA from the clone with a nucleic acid *probe*, typically labelled with ^{32}P to make it radioactive. The nucleotide sequence of the probe, whether mRNA or DNA, is complementary to a portion of the sequence of the DNA segment of interest. mRNA probes are obtained by isolating mRNA pools from cells enriched in the message of interest, while DNA probes are obtained by an earlier cloning step or perhaps by direct chemical synthesis.

In the in situ colony hybridization method, a replica plate of the clones is grown on nitrocellulose paper laid over the Petri plate with nutrient medium. Cells are then lysed by treatment with lysozyme, and NaOH is used to denature the DNA. The filter paper is then removed, contacted with the probe, washed, and exposed to x-ray film. Spots appear on the film where probe hybridization to complementary DNA has occurred, identifying interesting clones for further study.

Plasmid DNA may be isolated by ultracentrifugation in a cesium chloride density gradient, then treated with a restriction enzyme. The resulting DNA fragments provide a kind of fingerprint of the plasmid DNA. These fragments may be sized by electrophoresis in 0.5 to 1.5 percent agarose gels, where the fluorescent dye ethidium bromide, specific for double-stranded nucleic acids, is frequently used to visualize the bands corresponding to DNA fragments of different sizes. By analysis of the fragment sizes obtained with other restriction enzymes and by multiple restriction enzymes together, a *restriction map* showing the relative positions of various restriction sites can be generated. Restriction mapping is extremely useful to verify DNA insertions at particular target points in the plasmid and also provides a basis for further manipulation of the recombinant plasmid.

The radiolabelled probes mentioned earlier may be used to identify particular restriction fragments of interest. In the *Southern blotting* method, the DNA in the gel is denatured with alkali, and the gel is covered with cellulose nitrate filter paper. Buffer solution drawn from moist filter paper below this sandwich to dry filter paper above it carries DNA from the gel to the cellulose nitrate paper where it binds. The bound, denatured DNA is then exposed to probe and then to x-ray film as before to identify gel bands containing nucleotide sequences complementary to that of the probe.

Current methods of DNA biochemistry are sufficiently sensitive that a few nanograms of DNA obtained by cutting a band out of a gel may be used for cloning or for further analysis. Clearly the most detailed characterization of a DNA fragment is determination of its nucleotide sequence. Two different methods for rapid and efficient sequencing of DNA, the Maxam-Gilbert and Sanger techniques, were invented in the 1970s and have had tremendous impact in the biological sciences and biotechnology.

In the Maxam-Gilbert method, the double-stranded DNA is first labelled radioactively at one end of each strand. The DNA is then denatured, and a preparation of one of the two types of strands is divided into four aliquots, each of which is treated with different chemicals. Each of the four chemical treatments destroys selectively one or two bases and cleaves the DNA at those points. It is critical to the method that this destruction be incomplete—ideally we would like the chemical treatment to cleave each strand only once. The fragments resulting from these four parallel treatments are then run on a long polyacrylamide gel. Gel band positions are visualized by exposure to x-ray film. Reading down the four parallel lanes of the gel, the nucleotide sequence follows directly (Fig. 6.24). Sequences of the order of 200 nucleotides may be routinely determined in a single experiment using this method. The Sanger DNA sequencing method is summarized in Ref. 21; this approach is usually used when sequencing extensive domains of DNA.

Often we wish to screen a diverse set of clones for expression of a particular gene product, that is, a particular protein. The basis for most such methods is an antibody for the particular protein. After making the antibody radioactive (by labelling it with I^{125}, for example) antibody solution may be contacted with colonies on a plate which are then washed and visualized on x-ray film to identify colonies which contain the protein of interest. Alternatively, an enzyme may be linked to the antibody so that the enzyme is localized wherever antigen is encountered. Then, using a chromogenic substrate (one that changes or generates color after conversion by the enzyme), colonies containing the target protein are readily identified. More sensitive immunochemical methods which are able to detect as few as 1–5 molecules of a protein per *E. coli* cell are described in the references.

While it is impossible to convey a detailed picture of these methods for identifying and characterizing desired clones, hopefully the previous comments suffice to show the time-consuming and painstaking work required to screen clones and to monitor each step during construction of new recombinant molecules. Although cloned gene expression, the subject of the next section, is a topic of more immediate interest to biochemical engineers, we should recognize that cloning the gene is often a major bottleneck in a program to manufacture a foreign protein in *E. coli* or another host. Advances in laboratory methods and

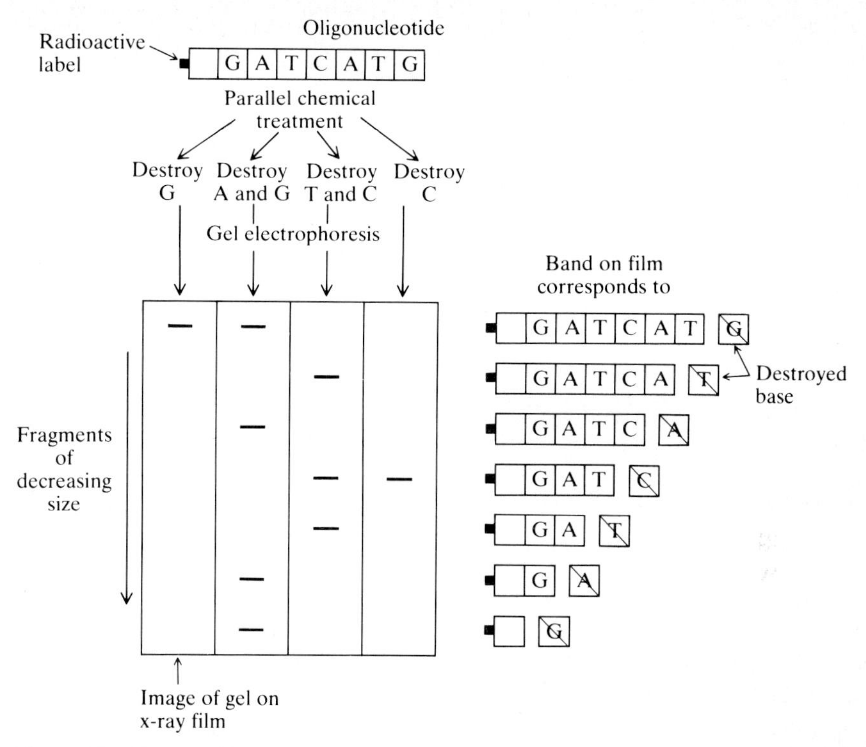

Figure 6.24 Maxam-Gilbert DNA sequencing: application of four different reagents which attack different bases gives a mixture of fragments with relative lengths corresponding to the destroyed bases. The pattern of bands on the four lane gel permit direct reading of the corresponding nucleotide sequence running, from top to bottom, toward the labelled end of the starting oligonucleotide.

automation will continue to accelerate the "unit operations" of molecular genetics, however, promising more rapid access to an expanding pool of useful and interesting genes and regulatory sequences.

6.4.4 Expression of Eucaryotic Proteins in *E. coli*

The first generation of applications of genetic engineering in biotechnology involves synthesis of useful eucaryotic proteins in *E. coli* cells. Here we shall first consider how to obtain a gene suitable for expressing a eucaryotic protein. Next, the desired properties of an expression vector will be discussed. This section concludes with a few comments on screening for gene expression and on genetic and product protein stability.

Recalling Fig. 6.7 and the associated discussion, we can anticipate a fundamental difficulty in direct expression of eucaryotic genes in *E. coli*. Procaryotes such as *E. coli* do not possess the biochemical machinery to do RNA splicing. Accordingly, a eucaryotic gene placed directly in *E. coli* (with appropriate bacterial expression controls to be discussed next) will not be properly expressed.

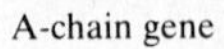

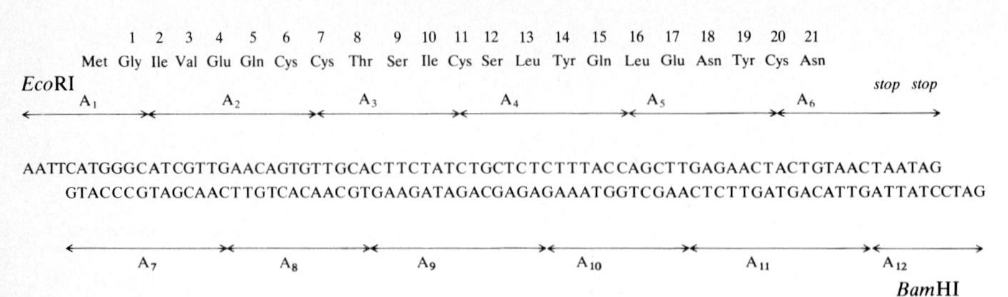

Figure 6.25 This chemically synthesized gene for the A-chain of human insulin (amino acid sequence shown on top) was assembled from twelve oligodeoxyribonucleotides designated A_1 through A_{12}. *(Reprinted from R. Crea, A. Kraszewski, T. Hirose, and K. Itakura, "Chemical Synthesis of Genes for Human Insulin," PNAS (US), vol. 75, p. 5765, 1978.)*

Sequences in the mRNA derived from intervening sequences in the eucaryotic gene will be translated in *E. coli*, producing a polypeptide generally much different from the desired eucaryotic product.

Accordingly, we must make somehow a bacterial gene for the eucaryotic protein—a gene that has only protein sequence coding information without any introns. There are two quite different strategies for accomplishing this. The first is chemical synthesis of the desired nucleotide sequence. While details of chemical synthesis of genes are best left for the references, we should observe that DNA synthesis chemistry as of this writing does not permit direct synthesis of the entire gene. Instead, overlapping oligonucleotides are designed which will tend to self-assemble by base pairing into a complete double stranded coding sequence. As an example, Fig. 6.25 shows the gene for the 21 amino acid A-chain of human insulin that was assembled from 12 different oligodeoxyribonucleotides ranging from 10 to 15 nucleotides in length. A number of properties related to cloning and gene expression including codon usage, avoidance of termination signals, and inclusion of convenient restriction enzyme target ends should also be considered when designing the synthetic gene.

The capability for chemical synthesis of genes or gene fragments raises the fascinating possibility of *protein engineering*. Now we can in principle alter at will any amino acids in a protein with the possibility of increasing the biological activity and the process stability of that protein. Furthermore, we can direct cells to synthesize novel (so far as we know) amino acid sequences and possibly create new catalysts, drugs, and food ingredients. Unfortunately, we know so little about the connection between protein amino acid sequence and protein function that this potential for improving function for applications remains largely untapped at present.

The other way to obtain a bacterial gene for a eucaryotic protein is to work backward from the mature, spliced eucaryotic messenger RNA. As shown schematically in Fig. 6.26, the mRNA for the desired protein provides a template from which the enzyme reverse transcriptase can synthesize a complementary

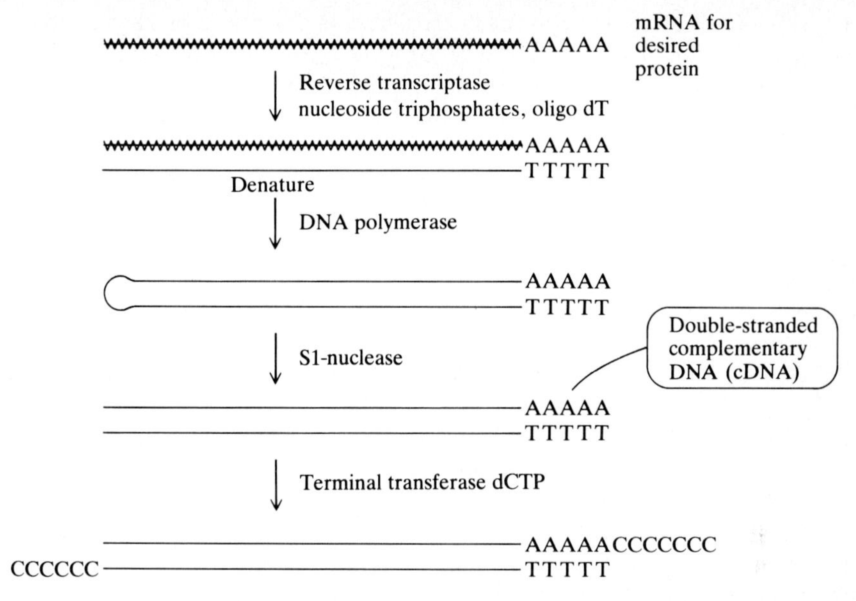

Figure 6.26 Summary of the procedure for synthesizing a double-stranded DNA molecule with nucleotide sequence complementary to the nucleotide sequence of an mRNA molecule (wavy line). The final step prepares the complementary DNA (cDNA) molecule for cloning.

DNA strand. After separating the mRNA and single DNA strand, the complementary DNA strand is added using DNA polymerase, and the enzyme S1 nuclease then serves to clip off a small single-stranded loop at one end. The result is a double-stranded complementary DNA (cDNA) molecule. In the last step in Fig. 6.26, terminal transferase has been used to add homopolymeric tails of deoxycytidine (dC) to the cDNA in anticipation of cloning (pBR322 cleaved by *Pst*I with deoxyguanylate (dG) homopolymeric tails is a favorite vector for this purpose).

For effective expression, the structural gene must be bracketed by nucleotide sequences that provide, before the gene, transcription initiation (promoter sequence) and translation start (ribosome binding site and start codon) and, after the gene, translation stop (stop codon) and transcription termination (recall Fig. 6.6). Expression vectors include these control sequences as well as an origin of replication and at least one selection marker. Cloning and DNA synthesis methods have been used to make a large set of expression control sequences available for application in expression vectors. The *lac*, *trp* and λp_L promoters are some of the more frequently used.

Generally it is desirable to use a controlled promoter that can be turned on by some environmental change such as addition of inducer, depletion of repressor, or temperature shift. In a well-designed expression system, the cloned gene product may constitute up to 70 percent of the total cellular protein (10–25 percent is a more typical figure). Since this product serves no useful purpose for the host bacterium, diversion of so much biosynthetic activity to product synthesis seriously limits the host cell's capacity for growth and even survival. Accordingly, it is common practice to operate a reactor first so that cloned gene expression is off, permitting more rapid growth to a high cell density, and then to alter cultivation conditions to maximize cloned gene expression.

When foreign proteins are expressed at high levels in *E. coli*, the product tends to accumulate in the cells as dense refractile bodies (Fig 6.27). These refractile bodies are primarily product protein but some bacterial proteins are included as well. The product protein in this form is typically denatured and cross-linked to other protein molecules by disulfide bonds. Thus, while the concentration of product into these particles facilitates some aspects of product recovery, additional processing steps to stabilize the product and refold it to active form are required.

Several types of instability may limit the net productivity of a recombinant culture. *E. coli* contains intracellular proteases which somehow recognize and degrade small abnormal proteins. Serving in normal cells to recycle amino acid building blocks from *E. coli* peptides containing synthesis errors, in recombinant

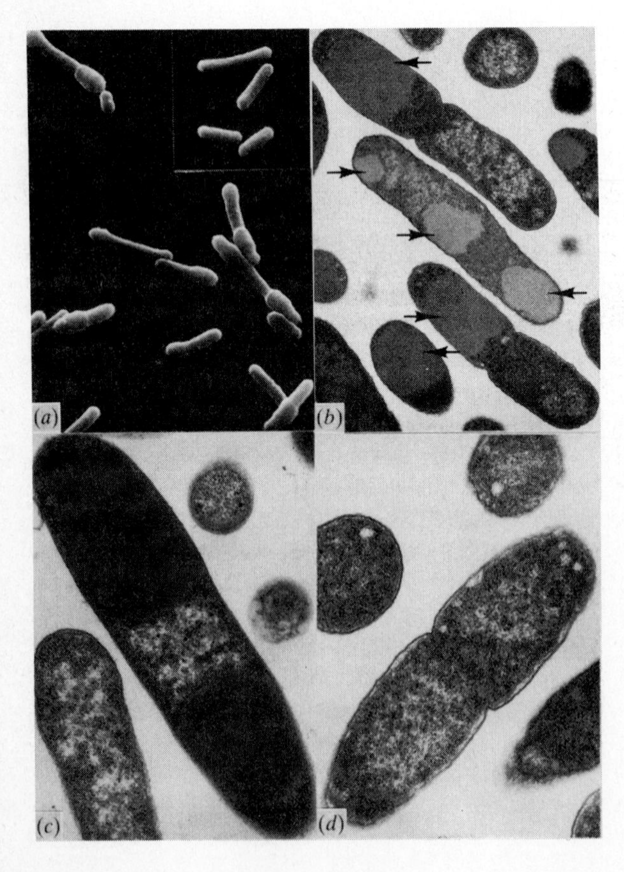

Figure 6.27 (*a*) Scanning electron micrograph of recombinant *E. coli* cells producing trp polypeptide-proinsulin chimeric protein. Elongated morphology with bulges is apparent by comparison with plasmid-free *E. coli* (inset, ×5300). Intracellular inclusion bodies in transmission electron micrographs of (*b*) recombinant *E. coli* cells producing β-galactosidase—insulin A chain chimeric protein (×17,500) and (*c*) recombinant *E. coli* producing trp polypeptide-proinsulin protein (×30,000). Plasmid-free cells (*d*), (×30,000) contain no inclusion bodies. *(Reprinted by permission from D. C. Williams, R. M. Van Frank, W. L. Muth, and J. P. Burnett, "Cytoplasmic Inclusion Bodies in Escherichia coli Producing Biosynthetic Human Insulin Protein," Science, vol. 215, p. 687, 1982.)*

systems these proteases may rapidly degrade the product protein. A common strategy that has been employed to minimize product instability is genetic engineering to obtain synthesis of a *fusion protein* consisting primarily of an *E. coli* protein to which, usually at the carboxy terminus, the foreign protein is appended. This method was used to stabilize several relatively small polypeptides such as somatostatin (14 amino acids) and the human insulin A (21 amino acids) and B (30 amino acids) chains. In each of these cases, a methionine residue was used as a connector between the *E. coli* protein fragment and the human peptide so that, by subsequent treating with cyanogen bromide which cleaves proteins at methionine, the active human peptide was obtained. (Fortunately, none of the polypeptides mentioned contain internal methionine residues!)

Two different forms of genetic instability can damage recombinant populations. First, a certain fraction of newborn cells may be born without plasmids. In this instance of *segregational instability*, plasmid-free cells may then outgrow the plasmid-containing, productive cells unless selection pressure is applied. Even with selection pressure, segregational instability reduces the overall growth rate and productivity of the recombinant population. The other type of instability, called *structural instability*, results in inability of cells to synthesize active cloned gene product. This can be caused by mutation either in the structural cloned gene or in the associated sequences which control and make possible cloned gene expression. Unfortunately, using selection pressure does not control this form of instability.

6.4.5 Genetic Engineering Using Other Host Organisms

E. coli is the workhorse of cloning and genetic engineering because it is the organism we know best at the molecular level. However, *E. coli* is not a familiar industrial microorganism; it is gram-negative and so has toxic lipopolysaccharides on its outer surface (*E. coli* infections in the bloodstream have a 50 percent fatality rate), and it does not ordinarily secrete proteins into the surrounding medium. Also, as a procaryote, *E. coli* cannot conduct RNA splicing or post-translation protein modification characteristic of eucaryotic cells. For these and other reasons, there is great interest in developing cloning and expression systems for organisms other than *E. coli*. In all cases it is important to keep in mind the following potential barriers to expression of a foreign gene:

1. Host cell nuclease attack on foreign DNA or RNA
2. Vector replication malfunction
3. Poor activity of promoter or transcription terminator
4. Incomplete mRNA splicing
5. Inefficient translation
6. Protease attack.

Also, we need an efficient procedure for introducing the vector into host cells with high yields.

Methods exist for molecular coloning and gene expression in several bacteria. Perhaps the most extensively studied is *Bacillus subtilis*, a gram positive, nonpathogenic, nonparasitic, industrial microorganism used before the advent of genetic engineering to manufacture several enzymes and polypeptide antibiotics. *B. subtilis* secretes some of its proteins into the surrounding medium. Product secretion is a desirable feature for some genetic engineering applications since secreted proteins in the medium will not be contaminated by large quantities of closely related intracellular proteins. Also, if the product protein is secreted, it may be possible to obtain higher product densities than is possible with an intracellular protein (what is an upper bound on the amount of product per unit volume of culture of an intracellular protein?)

Several plasmids and bacteriophages are available for cloning in *B. subtilis*. Transformation, transduction, and protoplast fusion methods can be used to introduce foreign DNA. Several mammalian proteins including insulin and interferons have been successfully expressed in *B. subtilis*, but, as of this writing, export of foreign proteins from *B. subtilis* remains problematic.

Cloning technology is also available for several *Pseudomonas* and *Streptomyces* strains. A combination of genetic engineering and mutation/selection methods has generated *Pseudomonas* species with new metabolic capabilities—for example, the ability to grow on ordinarily toxic chlorinated hydrocarbons as their sole carbon source. The tremendous current commercial importance of *Streptomyces* organisms motivates intensive research on genetic engineering methods useful in these microbes. Potential benefits include strains which give higher yields of industrial enzymes or which synthesize new semisynthetic or hybrid antibiotics.

The last examples introduce a new theme—*metabolic engineering*—which deserves further comment. With genetic engineering we can not only produce proteins useful in their own right, but we can also add to a living cell more of some particular enzymes, regulatory proteins, permeases, or any other protein. Further, we may be able to give to the cell completely new enzymatic, regulatory, or transport activity which we would never expect to find in nature or through use of random mutagenesis. Accordingly, we now have the capability to redesign and rebuild in a carefully controlled and directed fashion some parts of the metabolic network of the cell. On a long-term basis, this will be a major application of genetic engineering. More immediate progress in this area is limited by our insufficient knowledge of certain important enzymatic pathways and of their rate-limiting steps. Also, our understanding of control of the overall metabolic network is lacking so that the effect of a perturbation in one pathway on others cannot be predicted. Finally, new genetic engineering methods are needed so that large numbers of new genes can be propagated and expressed in recombinant cells.

Great progress has been achieved in recent years in genetic engineering in eucaryotic cells. The yeast *Saccharomyces cerevisiae* has been extensively studied because it is capable of expressing directly some eucaryotic and some procaryotic genes, because a powerful genetic system exists for this organism, and because yeast can be transformed efficiently with naked DNA. Yeast also conducts at least some posttranslational modification typical of eucaryotes and secretes some of its proteins into the medium. For example, hepatitis B Australia antigen particles, apparently glycosylated and aggregated as in humans with the disease, have been produced in recombinant *S. cerevisiae.* Human immune interferon (INF-γ) was successfully expressed, processed, and secreted by genetically engineered yeast.

Cloning and gene expression in mammalian cells has so far been based on vectors derived from simian virus 40 (SV40). The covalently closed circular DNA genome of this virus can replicate independently in mammalian cells or may be integrated into the host cell chromosomes. In addition to providing an origin of replication that functions in mammalian cells, promoters from SV40 may be used to obtain cloned gene expression.

Because of all the available methodology and its rapid growth, *E. coli* remains the organism of choice for cloning and characterizing DNAs. Thus, when the final objective is to propagate the vector and express its genes in another organism, it is convenient to construct and to utilize a *shuttle vector* which can be propagated both in *E. coli* and in the other organism of choice. Clearly, a shuttle vector must contain two origins of replication—one for each host organism. Shown in Fig. 6.28 is the shuttle vector developed to express immune interferon (INF-γ) in monkey cells. Here a 342-base pair fragment containing the origin of replication and the "late" promoter from SV40 is ligated to a cDNA fragment containing the pre-INF-γ coding sequence and to a portion of pBR322 which includes the ampicillin resistance gene and the *E. coli* origin of replication. Transfection (infection of cells with naked viral DNA) of the transformed monkey cell line COS-7 gave 50–100 INF-γ activity units per milliliter of culture fluid after three to four days. This example illustrates a major potential advantage of genetically engineered mammalian cell hosts for synthesis of mammalian proteins—all

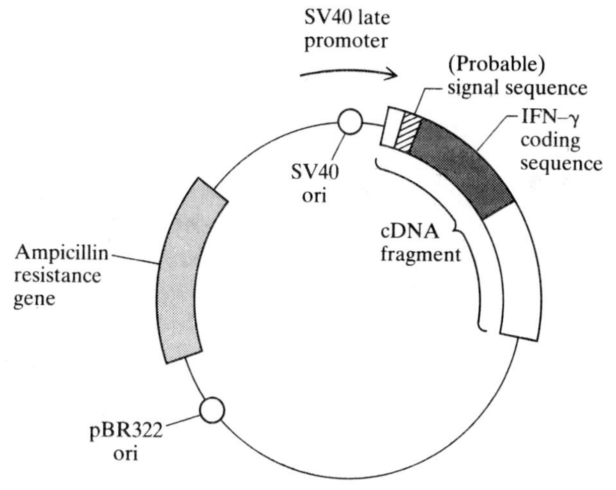

Figure 6.28 Diagram of the shuttle vector for *E. coli* and monkey cells which gives expression and secretion of immune interferon in monkey cells (see Ref. 23).

processing and secretion steps should mimic more closely, if not exactly, those steps which occur in normal in vivo synthesis and transport of the protein.

Efforts to introduce and express foreign genes in whole plants and animals are achieving increasing success. The technology involved is beyond the scope of this text. Clearly the potential for benefit to mankind is great. Along with promise, the technical developments have been accompanied by an emerging literature on the potential impact of biotechnology on society (Ref. 24).

6.4.6 Concluding Remarks

The checklist in Table 6.8 serves as an excellent review for the topics of this section. We should remember that cloning the desired gene and expressing it effectively are done in separate, sequential operations. The importance of monitoring DNA manipulations by repeated restriction mapping and DNA sequencing cannot be overemphasized.

The overview presented in this section, because of the scope of the overall text, is greatly abbreviated. Given the potential significance of recombinant DNA technology in future activities in biotechnology, further study of the underlying molecular biology and methods is essential to the biochemical engineer. Knowledge in these areas is a prerequisite to being able to read the contemporary biological sciences literature and to communicate with collaborating molecular biologists to develop an optimized organism and process. Gene expression levels, intracellular protease activities, genetic stability, and other factors which determine the productivity of recombinant populations all depend on the nucleotide sequence of the vector, on the genetic properties of the host cell, on medium composition, and on bioreactor configuration and operating conditions. Combined efforts of biochemical engineers and molecular biologists are essential

Table 6.8 Summary of the sequence of steps required to clone a gene and express its protein product using recombinant DNA technology

1. Obtain DNA
2. Attach to plasmid vector
3. Transform
4. Identify desired clone
5. Grow culture and isolate plasmid
6. Determine DNA sequence of cloned DNA fragment
7. Design and construct expression plasmid
8. Transform
9. Identify desired clone
10. Grow culture and isolate plasmid
11. Check DNA sequence
12. Transform
13. Grow culture for protein production

to understanding causes and effects and interactions in genetically engineered organisms.

6.5 GROWTH AND REPRODUCTION OF A SINGLE CELL

We may define *growth* of an organism as an orderly increase in all its chemical constituents. When most single-celled organisms grow, they eventually divide. Consequently, growth of a population usually implies an increase in the *number* of cells as well as the *mass* of all cellular material: either parameter may be used to investigate cell growth quantitatively. Actually, several different measures of mass are in use. Macromolecular dry mass includes only the cell's macromolecular components. This material plus the pool of low molecular weight materials in the cell is included in a total dry mass measurement. Cell volume or total cell mass is total dry mass plus water in the cell.

In the following chapter we shall attempt to understand growth when many cells coexist in a nutrient solution. A simpler objective faces us in this section. Here we shall review the events of the cell cycle, which is the time interval between the formation of one cell by division of its mother cell and the subsequent division of that one cell. Study of the cell cycle is a necessary preliminary to Chap. 7 for two reasons: (1) it will help us appreciate the difficulty of analyzing what happens when many coexisting cells are simultaneously growing and dividing, and (2) on the other hand, a rudimentary knowledge of the cell cycle will aid us in developing the necessary models in Chap. 7.

The life cycle of a single cell may be represented schematically by either a circular or a linear map (see Fig. 6.29). In either case, cell division is the key

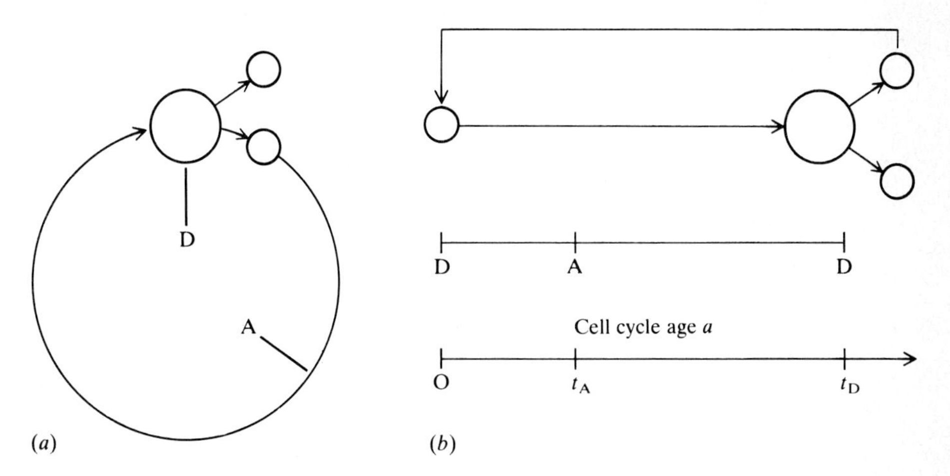

Figure 6.29 The events that occur between birth of a newborn cell and cell division may be represented in (*a*) circular or (*b*) linear diagrams of the cell cycle. "D" denotes cell division. "A" denotes a point of initiation of DNA synthesis for this hypothetical cell.

benchmark. This point is conventionally the right-hand end of a linear diagram of the cell cycle, with the left-hand end denoting the smallest or youngest progeny cell. To describe the relative timing of other benchmarks of the cell cycle such as initiation and cessation of DNA synthesis, it is often convenient to introduce a parameter a called the *age* or the *cell cycle age* of the cell. This coordinate typically varies during the cell cycle from zero for newborn cells to a value of t_D, the generation or doubling time of the organism, for dividing cells. In this introductory discussion, we have implicitly assumed that cells divide in half to give identical progeny. While true in binary fission, other situations can arise in budding, exemplified by the cell cycle of baker's yeast, *S. cerevisiae.*

6.5.1 Experimental Methods: Flow Cytometry and Synchronous Cultures

While the interested reader is referred to Mitchison [2] for details on experimental techniques, we should at least outline the principal methods employed for experimental elucidation of the cell cycle. The most obvious approach, that of observation of a single cell as it passes through its cycle, is of limited value. It is time-consuming and the information it offers is restricted because measurements are difficult. Only in special cases is the method of much value.

It would obviously be simpler to perform biochemical analyses on a large number of cells. Unfortunately, if we have many cells in suspension, it is highly probable that at any instant different individual cells are at different stages of the cell cycle; i.e., some of the suspended cells are very young, some are mature and on the brink of dividing, and others are in between. In such a heterogeneous population of individuals, with most conventional methods one measures averages and not characteristics of individuals.

Methods of flow cytometry can be used to determine rapidly certain characteristics of individual cells within heterogeneous cell populations. A sample of cells is diluted and sometimes treated to label particular biochemical components. Then, this sample flows through a chamber on which measurements are made on individual cells as they pass through. Possible measurements include total single-cell volume by light scattering or by changes in flow stream resistivity. Light scattering may also be used to characterize single-cell morphology and internal structure. A broad repertoire of fluorescence labeling methods have been developed for flow cytometry. Different fluorescent markers can be applied to measure total cellular DNA, double-stranded RNA, protein, particular enzymes, and cell surface components. The instrumentation involved is discussed further in Sec. 10.3.2.

One single-cell measurement by these flow methods requires about 1 ms. Consequently, analysis of 10^5 cells requires only a few minutes. Data generated in this fashion are typically displayed as a distribution of some single-cell parameter (relative number of cells or frequency in the sample vs. single-cell light scattering intensity or fluorescence). For example, Fig. 6.30 shows experimental data on the distribution of single-cell DNA content in a *Saccharomyces cerevisiae* population grown in a continuous-flow reactor (see Chap. 7). Such data is intrinsically useful since it characterizes heterogeneity in the cell population. In addition, we can frequently relate the shape of the measured frequency function to important parameters or charactertistics of single-cell growth and division.

In order to apply conventional biochemical assay methods to study of the cell cycle, the ingenious technique of *synchronous culture* has been developed. In the modern practice of the *selection-synchrony* version of this method, the first step is collection of cells which are at the same stage of the cell cycle. This can be achieved by attaching cells to a solid support and washing off newborn daughter cells as they are formed. The cells leaving the column at any instant are all of the same age, hence synchronous. Another practice involves density-gradient centrifugation of a cell suspension: cells at different stages of the cell cycle generally separate because of differences in size and density. Then, the *homogeneous population* from a particular density zone is placed in a nutrient solution and allowed to grow.

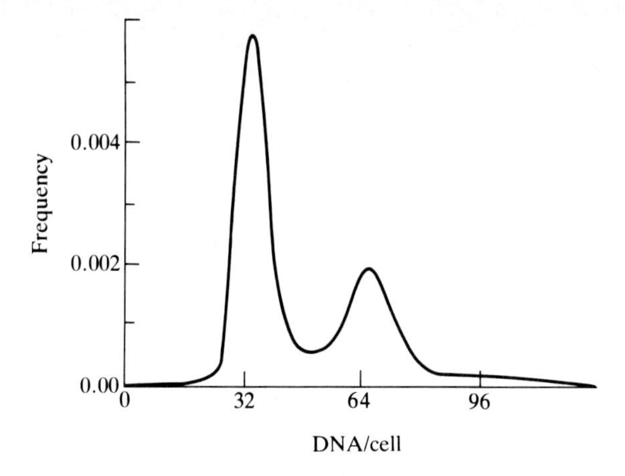

Figure 6.30 Relative cellular DNA content vs. relative number of cells in a population of baker's yeast (*Saccharomyces cerevisiae*) cells. Relative fluorescence of DNA of individual cells stained with propidium iodide was measured for 10^5 cells using a flow cytometer.

If we then monitor the number concentration of cells as a function of time, a stepwise increase in cell number is expected. This should occur because once every cell cycle, the number of cells should double. The experimental results in Fig. 6.31 illustrate this behavior. For at least the first two cycles, the population of cells appears to remain homogeneous and synchronized: all the cells are at approximately the same point in their cycle at any instant of time. As already observed in Chap. 1, a homogeneous population is convenient to study experimentally since it provides a useful amplification (millions of cells instead of one) of individual characteristics. In this manner many features of the cell cycle can be studied in detail.

Another common procedure in synchronous cultures is *induction synchronization.* Here, some environmental conditions are varied periodically in order to force a population into a synchronized stepwise growth pattern. Since the normal metabolic processes of the cells may be disturbed by such treatment, selection synchrony is a preferable practice for probing cellular physiology.

We should note carefully that in Fig. 6.31 synchrony seems to be fading as growth proceeds. The initially stepwise increases in cell numbers gradually shifts to a more continuous pattern of growth. This occurs because the cell population is losing its homogeneity. Behind this tendency toward heterogeneity is the stochastic nature of cell growth and multiplication. The length of the cell cycle is not a definite, fixed quantity for any single cell. Thus, two initially identical daughter cells may divide at different times. However, the length of the *average* life cycle of a population of billions of cells is a well-defined number which is very useful in reactor design (Chap. 7).

In the next two sections we will examine patterns of growth and macromolecular synthesis in an example procaryote, *E. coli*, and in the eucaryotic microorganism *Saccharomyces cerevisiae.* A central consideration in each of these discussions will be coordination between DNA replication and cell division. Obviously, before one cell can divide to give two progeny, a complete set of replicated

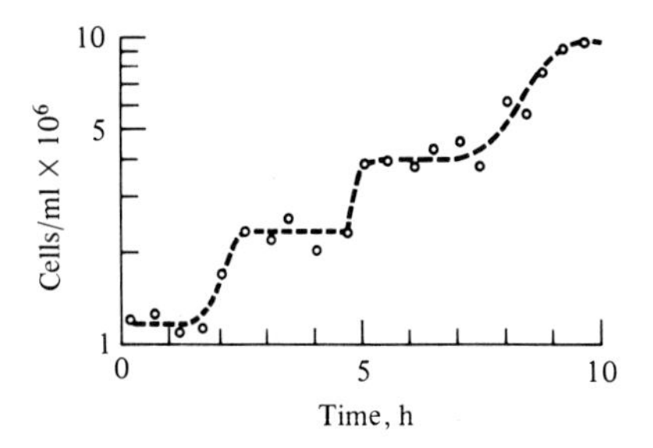

Figure 6.31 Stepwise increases in cell numbers occur during synchronous culture of the yeast *Schizosaccharomyces pombe. (Reprinted from J. M. Mitchison and W. S. Vincent, "A Method of Making Synchronous Cell Cultures by Density Gradient Centrifugation," in I. L. Cameron and G. M. Padilla (eds.), "Cell Synchrony," p. 328, Academic Press, New York, 1966.)*

genetic instructions—the cell's genome—must be synthesized. Thus, we can expect that DNA replication and cell division must be closely coupled in order for the species to survive.

6.5.2 The Cell Cycle of *E. coli*

The simple procaryote *E. coli* divides by binary fission and can, in rich medium, attain doubling times of 20 minutes or less. *E. coli* is a rod-shaped bacterium whose length provides a reasonable measure of cell volume and cell mass. Microscopic observations of elongation of individual *E. coli* cells indicate that length increase occurs continuously over the cell cycle and appears to be well approximated by an increasing exponential function of time. Total protein increases through the cell cycle in a similar fashion.

E. coli exhibits patterns of DNA synthesis during the cell cycle which can be quite complex and which differ substantially from replication timing in eucaryotic cells. Coordination of DNA synthesis and cell division in *E. coli* depends on a combination of regulatory and kinetic properties which are summarized next.

Under many growth conditions giving different doubling times, the time for replication of the *E. coli* chromosome is approximately constant at a value around 40 minutes. During DNA synthesis[†] the rate of advance of a replication fork (a point on the DNA where strand separation and DNA synthesis is occurring) is also constant. Hence, the rate of DNA synthesis is directly proportional to the number of active replication forks. DNA replication always begins at a particular initiation point, the *origin of replication*, on the chromosome. Two forks move in opposite directions from the origin of replication around the circular *E. coli* chromosome and converge at the terminus, at which point replication is complete.

Cell division control is coupled with DNA synthesis: the bacterial cell divides approximately twenty minutes after a replication fork reaches the terminus of the chromosome. As summarized in Fig. 6.32, the situation is relatively simple when *E. coli* grows in a medium which supports one cell division per hour. DNA synthesis takes place at a constant rate for the first forty minutes of the cell cycle, and the cell synthesizes no DNA during the final twenty minutes of growth before division. However, growth of an *E. coli* cell with a doubling time of fifty minutes (say at a slightly higher temperature) implies a more complicated pattern of DNA synthesis in which a new set of replication forks begin synthesizing DNA 10 minutes before the cell divides. Consequently, DNA is synthesized during the final 10 minutes of cell growth at double the rate observed at any time in the slower growth case. We see that the kinetics of DNA synthesis in *E. coli*, and presumably in certain other bacteria, is complicated and exhibits discontinuities depending on the cell age and the generation time.

Relatively few data exist at present on RNA synthesis during the cell cycle. The available evidence suggests that it is continuous. However, the composition of an RNA fraction believed to be mRNA has been observed to fluctuate. Consequently, the synthesis of individual proteins may also be expected to oscillate over several cycles.

Such behavior has indeed been observed in a number of experimental studies dealing mostly with inducible and repressible enzymes. Synthesis of a large number of inducible enzymes can be induced in bacteria at any point in the cell cycle. Usually the rate of induced enzyme synthesis shows a stepwise increase in synchronous culture. This is presumably related to doubling of the number of genes for the enzyme as DNA replication occurs. Several possibilities arise in the case of repressed enzymes. If bacteria are grown under conditions where end-product repression is either minimal or maximal, the rate of enzyme synthesis is constant for an interval equal to the cell-cycle time and then doubles. On the other hand, there is regular periodic pattern of enzyme synthesis for intermediate levels of repression. Several experimental findings suggest that this periodic behavior results from a stable oscillation in the cell's feedback control system.

[†] Throughout this discussion, "DNA synthesis" refers only to synthesis of chromosomal DNA. Plasmid replication will be considered later.

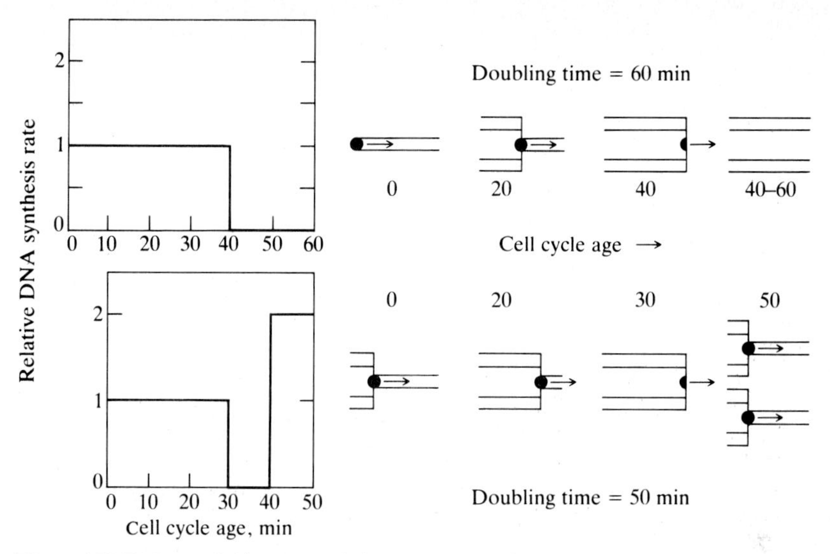

Figure 6.32 DNA synthesis rates and chromosome configurations in *E. coli* as a function of cell cycle age and of cell doubling time. The diagrams on the right represent one-half of the *E. coli* chromosome from the origin of replication to the replication terminus. Black dots denote replication forks.

6.5.3 The Eucaryotic Cell Cycle

At the outset we should recall that there is no such thing as a typical eucaryote. This imaginary cell, however, does serve as a useful device when discussing general features common to many eucaryotes, and we shall employ it in that sense here.

To appreciate some of the possible differences, consider the increase in mass during the cell cycle of mouse cells and of cells of an amoeba (Fig. 6.33). While the mouse cells show a linear or exponential increase of dry mass with time, the growth rate of the amoeba declines as the cell grows and is essentially zero during cell division. The second pattern is more typical of eucaryotes in general. Notice also the time scale for one generation of these eucaryotic cells—of the order of one day. Some eucaryotic microorganisms like yeast, however, can divide about once per hour under optimum conditions.

In the eucaryotic cell cycle, there is more differentiation between various parts of the cycle than in procaryotes. The standard pattern of the eucaryotic cycle is given in Fig. 6.34. It is subdivided into four phases, G_1, S, G_2, and M, whose relative durations are indicated in the figure. In the G_1 phase, protein and RNA are actively synthesized, but DNA is not. Following the G_1 phase, the chromosomes are replicated in the S phase. The significance of the G_2 phase is not fully understood at present. After its completion, cell division starts (M phase). Division in eucaryotes proceeds by a well-orchestrated process called *mitosis*. During mitosis, the two sets of chromosomes are separated and partitioned into each of the progeny cells. In nongrowing or very slowly growing cultures, sometimes cells enter a resting state which is designated G_0.

In contrast to the procaryotic linear pattern, enzyme synthesis is almost always periodic in eucaryotes. A particular enzyme increases sharply in amount at a definite point of the cell cycle. The available evidence supports sequential transcription of DNA as the mechanism underlying this behavior.

The organization of the eucaryotic cell cycle and an example of asymmetric cell division is illustrated by brief examination of the life cycle of the budding yeast *S. cerevisiae*. This is an important industrial microorganism in brewing, baking, food manufacture, and genetic engineering. As illustrated in Fig. 6.35, a mother cell grows a daughter cell as a bud and, upon division, the mother

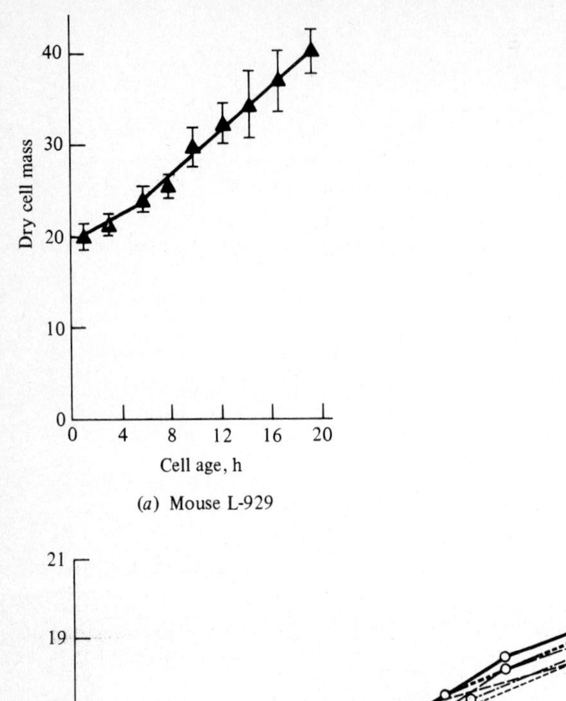

(*a*) Mouse L-929

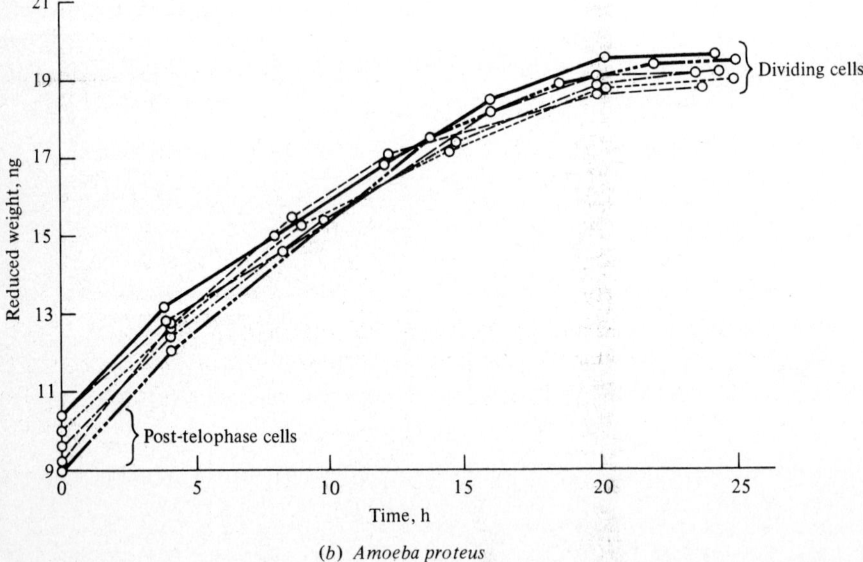

(*b*) *Amoeba proteus*

Figure 6.33 Increase in mass of (*a*) mouse tissue cells and (*b*) *Amoeba proteus* cells during their respective cell cycles. The different curves in (*b*) are for six different individual cells. *((a) Reprinted from D. Killander and A. Zetterberg, "Quantitative Cytochemical Studies on Interphase Growth. I. Determination of DNA, RNA, and Mass Content of Age Determined Mouse Fibroblasts in vitro and of Intercellular Variation in Generation Time," Exp. Cell Res., vol. 38, p. 272, 1965; (b) Reprinted from D. M. Prescott, "Relations Between Cell Growth and Cell Division. I. Reduced Weight, Cell Volume, Protein Content, and Nuclear Volume of* Amoeba proteus *from Division to Division," Exp. Cell Res., vol. 9, p. 328, 1955.)*

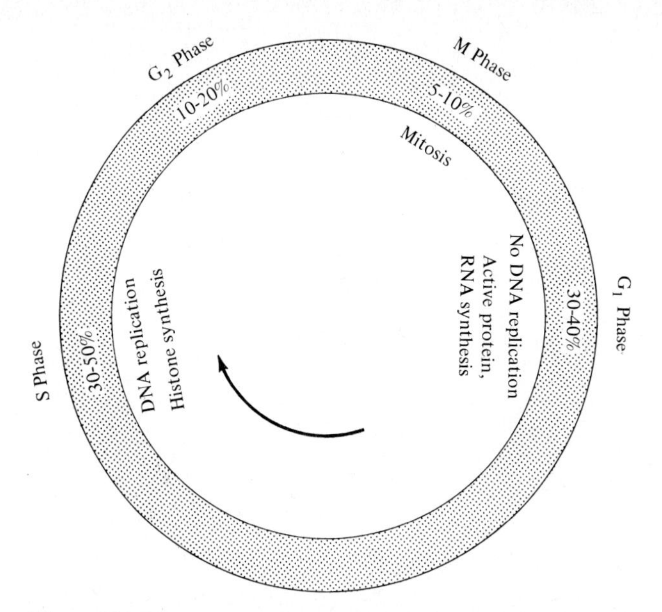

Figure 6.34 The sequence of events during a typical eucaryotic cell cycle.

cell recycles to reenter the budding cycle quickly while the daughter, usually smaller than the mother, must grow for some time to achieve the condition called "start" which marks the beginning of the budding cycle.

Initiation of the S phase coincides approximately with bud emergence. All unbudded cells are in the G_1 phase. Recalling now the distribution of single-cell DNA contents for *S. cerevisiae* shown in Fig. 6.30, we can associate the first mode of the distribution with cells containing one genome, that is, cells in the G_1 phase. The second mode corresponds to cells with two genomes which are in either G_2 or M phase. By computer analysis of distributions such as this, we can rapidly determine the relative durations of G_1, S, and $G_2 + M$ intervals in *S. cerevisiae* and so study the influences of genetic changes and different environments on cell cycle regulation.

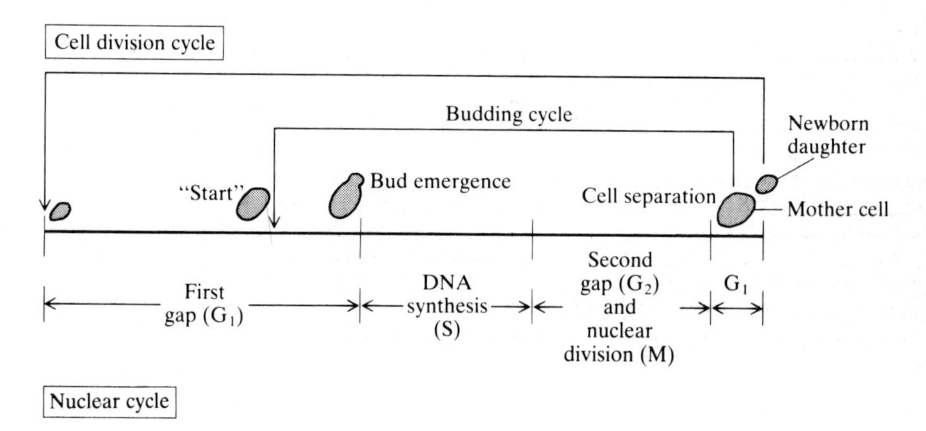

Figure 6.35 Summary of the cell cycle of the budding yeast *S. cerevisiae*.

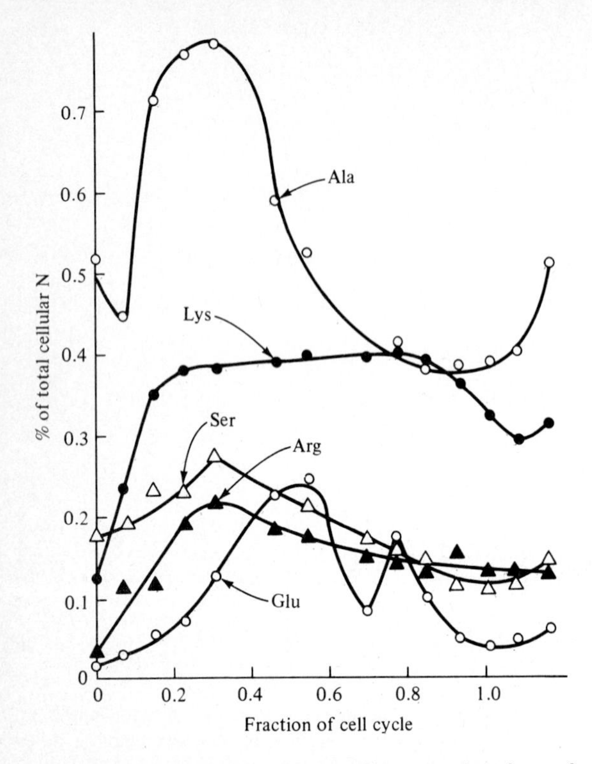

Figure 6.36 Complex variations occur in the proportions of free amino acids during the cell cycle of *Chlorella pyrenoidsa. (Reprinted by permission from T. A. Hare and R. R. Schmidt, "Nitrogen Metabolism During Synchronous Growth of Chlorella: II. Free-, Peptide-, and Protein-Amino Acid Distribution," J. Cell. Physiol., vol. 75, p. 73, 1970).*

In this section we have discovered a few of the intricacies of the cell cycle. Actually, disappointingly little is known about it. From our overview of cell metabolism in Chap. 5., we realize that an extremely complicated network of chemical reactions must be carefully coordinated in order to utilize nutrient efficiently. While much of the biochemistry of these reactions has been discovered, we have very little idea about how they are executed *in vivo.* Figure 6.36 reveals some intriguing data suggesting that there are indeed variations in the *timing* of many interacting reaction sequences within the cell. Thus, cellular reactions are often localized not only in space, as in immobilization of an enzyme at a membrane surface, but also in time. In view of these complexities, we must be both careful and humble as we face the task of mathematically representing the kinetic processes of living cells.

This completes our foundation in biology. With the tools of the past chapters in hand, we are prepared to undertake engineering of process systems employing living cells. Analysis, modeling, and design of these biological processes will occupy most of our attention for the remainder of the text. In the next chapter, we consider the growth kinetics of cells.

PROBLEMS

6.1 Mutation frequency Mutation processes lead to a new species, the *mutant*. As the DNA material is simply a chemical polymer, it presumably undergoes chemical and physical changes just like any other molecule.

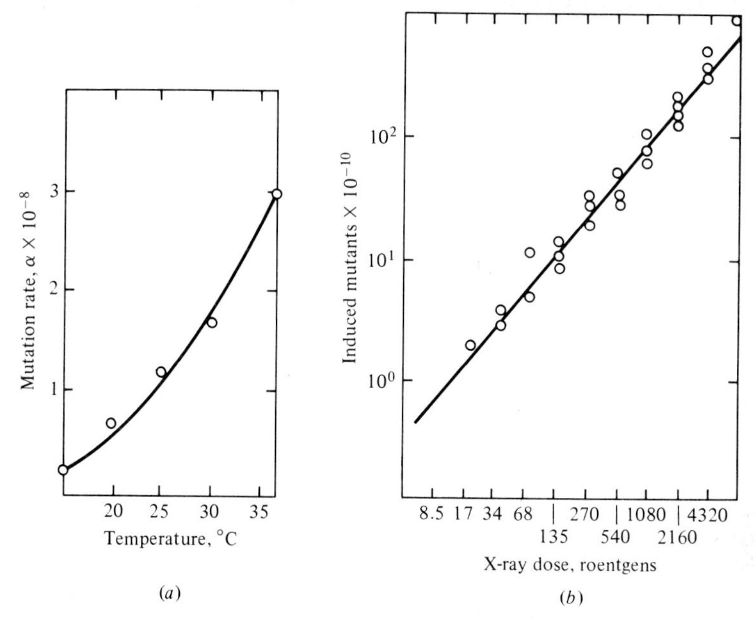

Figure 6P1.1 Mutation rate in *E. coli* increases with temperature and ionizing radiation. (*a*) The effect of temperature on the incidence of mutation *his*$^-$ (inability to synthesize histidine) to *his*$^+$ (ability to synthesize it). (*b*) The number of *met*-2$^+$ mutants (able to synthesize methionine) induced by exposing *met*-2$^-$ colonies (nonsynthesizers) to increasingly large doses of x-rays. *(R. Sager and F. J. Ryon, "Cell Heredity: An Analysis of the Mechanisms of Hereditary at the Cellular Level," John Wiley & Sons, Inc., New York, 1961.)*

(*a*) From the data in Fig. 6P.1.1*a* calculate the activation energy of the mutation rate E_m and the preexponential factor k_m°, where $k_m = k_m^\circ \exp(-E_m/RT)$. How do these values compare with those noted in your freshman chemistry text for simpler chemical species undergoing chemical reactions?

(*b*) Radiation absorption follows Beer's law: every increment of target absorbs the same fraction of radiation incident upon that increment; therefore $I(x) = I_0 e^{-\alpha x}$. Figure 6P1.1*b* shows a linear response of mutation appearance vs. initial intensity I_0. Under what condition has all the sample in Fig. 6P1.1*b* received an equal radiation dose? Would you expect to observe a linearity in the mutant-occurrence-vs.-dose curve when not all cells had received an equal dose?

(*c*) The mutation rate in Fig. 6P1.1 refers to *E. coli* populations mutated from *his*$^-$ (cannot synthesize histidine) to *his*$^+$ (Fig. 6P1.1*a*), and *met*-2$^-$ (methionine) to *met*-2$^+$. If all DNA alterations occur with the same probability as those deduced here, what is the total mutation rate for DNA of *E. coli*? State your assumptions clearly. Are these mutations of genotype or phenotype?

6.2 Mutation repair and thermodynamics Provided that a large number of individuals are involved, the kinetics of mutation can be usefully described by conventional mass-action kinetics.

(*a*) In a bacterial population of 1000 L of 3×10^7 cells per milliliter, the mutation of one gene g_1 to a second type g_2 occurs with a frequency of 10^{-8} per cell division, and the mutation from g_2 to g_1 occurs at a frequency 10^{-6} per cell division. Calculate the "equilibrium" concentrations of each species.

(*b*) In cells able to repair damaged DNA enzymatically, it is reasonable to suppose that, since the enzyme exists, there is an equilibrium between normal and damaged DNA:

$$\text{Normal} \underset{\substack{k_{-1} \\ E_{\text{repair}}}}{\overset{k_1}{\rightleftharpoons}} \text{damaged}$$

Suppose ultraviolet light opens a second forward path parallel to the first with rate constant k^1. Show that the equilibrium population fraction of damaged DNA rises from $k_1/(k_1 + k_{-1})$ to $(k_1 + k^1)/(k_1 + k^1 + k_{-1})$.

(*c*) Show that when ultraviolet light is cut off, assuming that the K_m of the repair enzyme is much larger than the total DNA level, the average concentration of damaged DNA returns to its original value at a rate proportional to DNA(damaged)-DNA(damaged)$_{\text{equil, no UV}}$.

6.3 A very simple repressor kinetics model J. Maynard Smith sketches the kinetic control network shown in Fig. 6P3.1.

(*a*) Assuming steady concentrations of P and gene, develop a kinetic model which could (if solved) predict the time behavior of RNA, M, and Z. Include a loss rate for RNA proportional to its concentration. Proceed as far as possible in obtaining a full transient solution.

(*b*) From the discussion of this and preceding chapters, indicate whether diffusion should also be considered and what weakenesses this model may have.

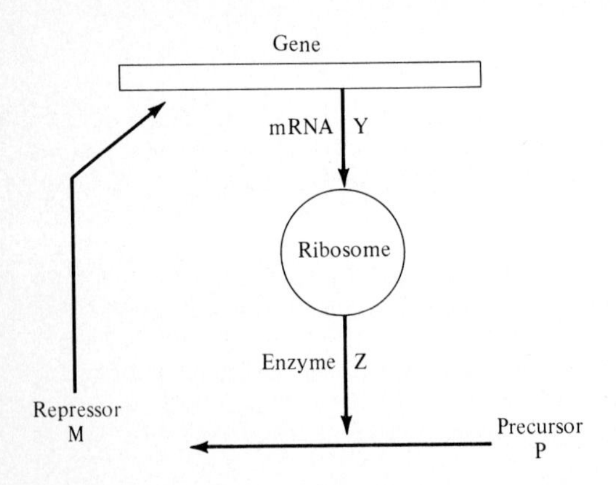

Figure 6P3.1 A simple repressor model. *(From J. Maynard Smith, "Mathematical Ideas in Biology," fig. 30, Cambridge University Press, London, 1971.)*

6.4 Mutation and the genetic code Amino acid changes as diagrammed below have been observed in mutant proteins. In each case, describe the codons involved and identify the type of mutation.

(*a*) His → Tyr; His → Arg

(*b*) Glu → Lys; Glu → Val

(*c*) Arg → Thr; Arg → Met; Arg → Trp

6.5 Cell organization: multienzyme systems Many cellular reactions involve the enzyme-catalyzed formation of an intermediate S_2 between S_1 and S_3, where the equilibrium concentration of S_2 is small relative to S_1 and S_3:

$$S_1 \underset{k_{-1}}{\overset{k_1}{\rightleftharpoons}} S_2 \underset{k_{-2}}{\overset{k_2}{\rightleftharpoons}} S_3$$

If S_3, etc., are subsequently consumed by the next enzyme-catalyzed reaction, we may neglect the reverse of the second reaction.

(*a*) Assuming $s_i \ll K_i$ for each enzyme, show that the steady-state rate of reaction is given by

$$\text{Rate} = \frac{k_2 s_1}{k_{-1}/k_1 + k_2/k_1} \equiv k s_1$$

if concentrations of S_1 and S_2, E_1 and E_2 are spatially uniform.

(*b*) If E_1 exists on one face of a permeable slab and E_2 on the other, show that the steady-state reaction rate is now

$$\text{Rate} = \frac{k s_1}{1 + kLk_{-1}/k_1 \mathscr{D}}$$

and that this is less than $s_{2,\text{equil}} \mathscr{D}/L$, where L is the distance of separation between planes and $s_{2,\text{equil}}$ is the concentration of S_2 equilibrated with the value of s_1.

(*c*) A typical oxygen consumption rate is 10^{-8} mol/(s · cm^3). Assuming internal metabolic rates to be of the same magnitude at every step, calculate the *maximum* separation between E_1 and E_2 which would allow this rate for values for $s_{2,\text{equil}} = 10^{-4}$ to 10^{-12} mol/cm^3 (P. B. Weisz, "Enzymatic Reaction Sequences and Cytological Dimensions," *Nature*, **195**: 772, 1962).

6.6 Cell organization: ATP production and utilization The intact cell contains localized factories (organelles) which must evidently supply the entire cell with various products. An interesting case which may be usefully examined in one dimension is that of the bull sperm: only the midpiece regenerates ATP, and the attached tail section consumes much of it during motion.

(*a*) Assuming a zero-order reaction for ATP consumption at each point in the tail and a uniform concentration c_0 in the midpiece, show that the ATP profile in the tail is $\bar{c} = 1 - z(1 + \phi^2/2) + \phi^2 z^2/2$, where ϕ is the Thiele modulus of the tail, z the dimensionless distance from the midpiece, and $\bar{c}$ the dimensionless concentration.

(*b*) Each moving sperm consumes oxygen at 3.7 to 5.0×10^{-18} mol/s for motility. With the aid of the following additional data, calculate the minimum ATP content of sperm needed to maintain diffusive transport to the tail.

ATP diffusivity (corrected for tail water content and tortuosity) $= 3.6 \times 10^{-6}$ cm^2/s
Midpiece volume $= 1.3 \times 10^{-12}$ cm^3 tail length $= 5 \times 10^{-3}$ cm
Tail cross-sectional area (corrected for sheath and fibrous matter) $= 3 \times 10^{-10}$ cm^2
ATP yield per O_2 consumed $= 6$

(*c*) The average observed ATP content is 200×10^{-18} mol per sperm. It has been claimed that one-third to one-half of ATP consumption is used for cell mitochondrial maintenance. Does your calculation in part (*b*) indicate the need for consideration of other than passive diffusional transport for proper internal distribution of ATP from the mitochondria localized in the midpiece (A. C. Nevo and R. Rikmenspoel, "Diffusion of ATP in Sperm Flagella," *J. Theoret. Biol.*, **26**: 11, 1970).

6.7 Target theory In an attempt to connect known molecular biology with viral, spore, or cell deactivation, the concept has been developed that "in each cell or in each organism, there are a number of 'targets,' and that changes occur as a consequence of random 'hits' on these targets" (J. M. Smith, *Mathematical Ideas in Biology*, Cambridge University Press, London 1971, p. 87). Consider the following:

Suppose that a cell containing N targets is exposed to a dose of K particles. Let the probability that a particular particle will hit a particular target be p; clearly p is very small.

The probability that a particular target is not hit by a particular particle is $1 - p$.

Hence the probability that a particular target is not hit by any one of the K particles is

$$(1 - p)^K \approx e^{-Kp}$$

(*a*) Noting that if K is large but p is small, $Kp \approx (K - 1)p \approx (K - 2)p$, establish the validity of the above approximation for this typical case (large K, small p).

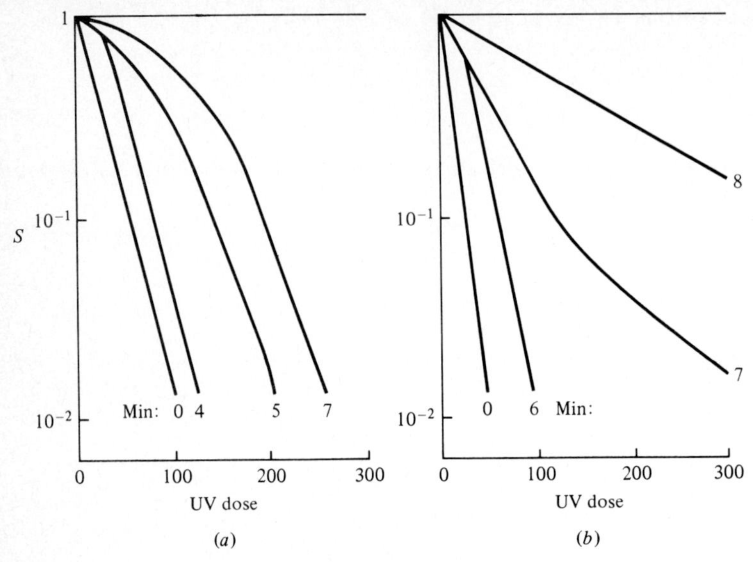

Figure 6P7.1 Observed values of infectivity S against dose for bacteria irradiated for different lengths of time after infection: (a) bacteriophage T7; (b) bacteriophage T2. *(After S. Benzer, Resistance to Ultraviolet Light as an Index to the Reproduction of Bacteriophage, J. Bacteriol., vol. 63, p. 59, 1952.)*

(b) Show that when the proportion of damaged cells is small (damaged means sustaining one or more hits), "the proportion of cells damaged is NKp, i.e., it is proportional to the dose." If there are, in contrast, few undamaged cells, show that, for this formulation, the survivor fraction s varies as $\ln s = -NKp$.

(c) When more than one hit is required to damage a cell, the same arguments as used above can be applied. "A bacterium is infected by r similar bacteriophages, each containing N essential genes. Provided that at each of the N loci one gene is undamaged, a functional bacteriophage can be produced by recombination" (above reference). Show that if "the probability that any particular target escapes is small," $\ln s = N \ln r - NpK$ (here the r targets are the number of virus copies within the cell at the time of irradiation).

(d) Show that your result in part (c) fits the data of Fig. 6P7.1a and evaluate each parameter possible. What phenomena might account for the behavior of Fig. 6P7.1b?

6.8 Fate of mutant gene in breeding population Mutations occur frequently in populations. The persistence of a mutant gene is surprisingly high, yet the permanence low. For mating diploid populations of N members, at the instant a mutant gene (allele) appears, it has a gene frequency $1/2N$. The total population N is greater than the effective number involved in breeding N_e.

(a) For a mutation which is *selectively neutral* (mutant has no advantage or disadvantage vs. original species), the mean number of generations until loss (extinction) of the mutant gene is $t_l = 2(N_e/N) \ln 2N$, and the mean number of generations until fixation (defined as the achievement of gene frequency $= 1.0$) is $t_f = 4N_e$. Here t_l is averaged over all *loss* results only and t_f over only all fixations. Taking $N_e/N = 0.5$ for estimation purposes and a doubling time of 1 h, calculate t_l and t_f for 1 L of mating-cell population with density 10^7 cells per milliliter.

(b) The distribution of values for t_f and t_l are relatively small and large with respect to the appropriate mean, as you can show by calculation from

$$\text{var } t_l \approx 4.58 N_e^2 \qquad \text{var } t_f \approx \frac{16 N_e^2}{N} - \left(2 \frac{N_e}{N} \ln 2N\right)^2$$

(*c*) The probabilty that the mutant becomes fixed in the population is

$$u = \frac{1 - \exp\left[-2(N_e/N)s_1\right]}{1 - \exp\left(-4N_e s_1\right)}$$

where s_1 is the selective advantage of the mutant vs. original gene. Show for the typical conditions in microbial populations that $u = 2s_1 N_e/N$. Thus, even when the selective advantage is large ($s_1 = 1$ percent $= 10^{-2}$), the likelihood that it will eventually dominate the culture is small (M. Kimura and T. Ohta, *Theoretical Aspects of Population Genetics*, Princeton University Press, Princeton, N.J., 1971.)

6.9 Natural selection The relative fitness w of a particular genotype x may be conveniently defined as the ratio of the survival rate for genotype x divided by the survival rate for the mean genotype $\bar{x}$ of the population. (The survival rate is the ratio of occurrence frequency at a point in time divided by the same term evaluated one generation earlier.) Suppose that the fitness of the genotype x varies linearly with the average phenotype expression: $w(x) = 1 - \alpha(\bar{x} - x)$, where α is simply a measure of the *intensity* with which a difference $\bar{x} - x$ affects $w(x)$, $\alpha > 0$. By definition, $g(x)$ is the normalized original genotype distribution $[\int g(x)\,dx = 1.0]$. One generation later, the distribution has become $g'(x) \equiv w(x)g(x)$.

(*a*) Evaluate $\bar{x}'$ in terms of the mean $\bar{x}$ and variance σ^2 of the original population [recall that mean $\equiv \int xg(x)\,dx$, variance $\equiv \int (x - \bar{x})^2 g(x)\,dx$].

(*b*) Show that the change in $\bar{x}$ in one generation is proportional to σ^2, that is, "the rate of evolution is proportional to the genetic variance of the population" (fundamental theorem of natural selection) (E. O. Wilson and W. H. Bossert, *Primer of Population Biology*, pp. 79–83, Sinauer Associates, Stamford, Conn., 1971).

(*c*) When w varies other than linearly with x, Wilson and Bossert claim that the value $\bar{x}' - \bar{x}$ is still "in some way proportional to the genotypic variance." Pick several reasonable forms for $w = f(x)$ and prove or disprove this statement.

6.10 Multiple operator sites: regulation of the λP_R promoter Repressor protein for the phage λP_R promoter can bind to three different operator sites designated O_R1, O_R2, and O_R3. As indicated in Fig. 6P10.1, binding of repressor to any site except O_R3 blocks the P_R promoter. Assume that repressor-operator binding is at equilibrium and that the equilibrium constants for repressor binding to the three sites are K_1, K_2, and K_3, respectively.

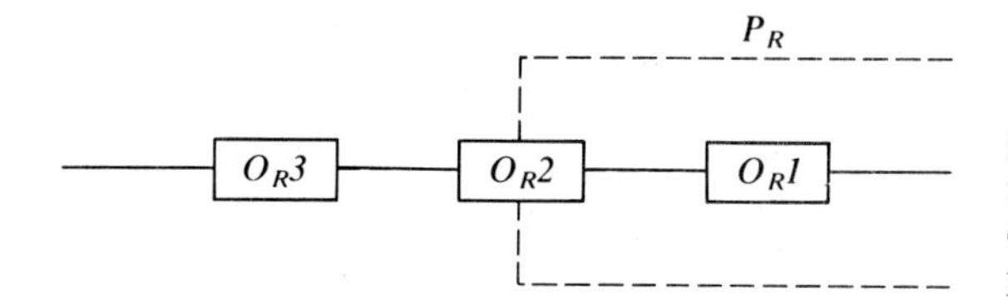

Figure 6P10.1 Relative positions of the P_R promoter and the three operators O_R1, O_R2, and O_R3 in phage λ.

(*a*) Develop a general equation for the fraction of P_R promoters accessible for RNA polymerase binding in terms of the equilibrium constants and the total amount of repressor protein and $P_R O_R$ sequences in the cell.

(*b*) Evaluate the fraction of accessible promoters for a cell containing 25 $P_R O_R$ sequences and 70 repressor molecules. K_1, K_2, and K_3 are $125 \times 10^8\ M^{-1}$, $1.25 \times 10^8\ M^{-1}$, and $1 \times 10^9\ M^{-1}$, respectively. What is the physical significance of this result?

(*c*) The previous analysis was based on classical thermodynamics. When calculating interactions of small numbers of different molecules in cells in populations, a different approach based on statistical thermodynamics is more rigorous. Reconsider the previous questions in light of the theory presented in O. G. Berg and C. Blomberg: "Mass Action Relations in vivo with Application to the *Lac* Operon," *J. Theoret. Biol.*, **67**: 523 (1977).

6.11 Procaryote and eucaryote DNA replication rates (*a*) Unwinding of double-stranded DNA is required during replication. Supposing there are two swivel points at which twist in the molecule can be relieved, what is the rotation rate in rpm of the DNA at these swivel points for *E. coli* in which the entire chromosome is synthesized in 41 min. (*b*) The S-phase in a mammalian cell line, with genome containing 1.1 m of double-stranded DNA, lasts 4.8 h. If DNA chain growth rates are the same as in *E. coli*, how many replication forks are active simultaneously?

6.12 Recombinantion of DNA fragments Two double-stranded DNA molecules, 1 and 2, are cleaved by the restriction endonuclease *Eco*RI. (One strand of each of these DNAs is shown below.) The resulting fragments are mixed, allowed to recombine, and covalently closed with DNA ligase. List all of the possible reaction products.

DNA1:

(5′) CGATAGAATTCAGTCAA (3′)

DNA 2:

(5′)GAATTCGGTTCGAATTCG (3′)

6.13 Restriction mapping The relative positions of restriction sites in a DNA fragment are determined by complete or partial digestion of the DNA by the restriction enzymes and subsequent size analysis by gel electrophoresis. A *Hind*III fragment of size 3.0 kb is treated by *Eco*RI, by *Bgl*II, and by a mixture of these enzymes, respectively. Gel electrophoresis shows the products consist of two fragments of sizes 1.4 and 1.6 kb, three fragments of sizes 0.4, 0.9, and 1.7 kb, and four fragments of sizes 0.4, 0.5, 0.9, and 1.2 kb, respectively. Determine the restriction map for this fragment.

6.14 Primer for interferon cDNA To maximize the probability of cloning cDNA which codes for fibroblast interferon (IFN-β), it is useful to use an oligodeoxyribonucleotide primer complementary to the 5′-end of the mRNA coding sequence. The first four amino acids at the amino-terminus of the protein are

Met-Ser-Tyr-Asn

List all of the possible corresponding mRNA nucleotide sequences and the set of primers which should be considered. Be careful to identify 3′- and 5′-termini in each case.

6.15 Intracellular protein production by recombinant cells Order of magnitude estimates of the bioreactor capacity needed to produce a specified amount of a cloned gene product which accumulates in the cell can be made quite simply. Suppose that a recombinant *E. coli* strain accumulates a foreign protein expressed from a plasmid gene at a level of 30 percent of total cell protein. If the cells can be grown to a density of 10^9 cells/mL, how much protein product can be made per liter of bioreactor capacity?

REFERENCES

All of the references given in Chaps. 1 and 2 have worthwhile material on various aspects of this chapter. Also recommended are:

1. G. S. Stent and R. Calendar, *Molecular Genetics: An Introductory Narrative*, 2d ed., W. H. Freeman & Company, San Francisco, 1978. Excellent presentation of central concepts in molecular genetics in historical context with emphasis on experimental bases for current principles.
2. B. Lewin, *Genes*, 2d ed. John Wiley & Sons, New York, 1985. Extremely broad, detailed, and clearly presented summary of genes and their replication and function.
3. B. Alberts, D. Bray, J. Lewis, M. Raff, K. Roberts, and J. D. Watson, *Molecular Biology of the Cell*, Garland Publishing Inc., New York, 1983. Approached now from a molecular perspective, cell biology has reemerged as one of the most exciting areas of biological science research.

An introduction to mutation and applied genetics in microbial processing is provided in the following references:

4. R. P. Elander, "Applications of Microbial Genetics to Industrial Fermentation," p. 89 in *Fermentation Advances*, (D. Perlman, ed.), Academic Press, Inc., New York, 1969.
5. A. L. Demain, "Overproduction of Microbial Metabolites and Enzymes Due to Alteration of Regulation," *Adv. Biochem. Eng.*, **1**: 113, 1971.
6. W. T. Dobrazanski, "Microbial Genetics in Pharmacy," *Chem. Br.*, **10**: 386, 1974.
7. A. L. Demain and N. A. Solomon (eds.), *Biology of Industrial Microorganisms*, Benjamin Cummings & Co., Menlo Park, CA, 1984. This recent volume summarizes metabolism, control systems and genetics of microorganisms of contemporary industrial importance. It is one of few currently available sources which provide detailed coverage of microbiology of molds and *Streptomyces*.

Cell fusion methods and hybridomas for synthesis of monoclonal antibodies are reviewed in

8. J. F. Peberdy, "Protoplast Fusion—A Tool for Genetic Manipulation and Breeding in Industrial Microorganisms," *Enzyme & Microbial Tech.*, **2**: 25, 1980.
9. C. Milstein, "Monoclonal Antibodies," *Scientific American*, **243**: 66, 1980.
10. J. E. Boyd, K. James, and D. B. L. McClelland, "Human Monoclonal Antibodies—Production and Potential," *Trends in Biotech.*, **2**: 70, 1984.

Specific examples of genetics application, all in D. Perlman (ed.), *Fermentation Advances*, Academic Press, Inc., New York, 1969:

11. A. B. Pardee, "Enzyme Production by Bacteria," p. 3.
12. K. Veda, "Some Fundamental Problems of Continuous L-Glutamic Acid Fermentations," p. 43.
13. A. Furuya, M. Misawa, T. Nara, S. Abe, and S. Kinoshita, "Metabolic Controls of Accumulations of Amino Acids and Nucelotides," p. 177.

The literature at all levels on recombinant DNA technology is exploding in parallel with scientific and technological advances. The following list is graded from introductory, relatively nontechnical reviews to summaries of methods and research results.

14. S. N. Cohen, "The Manipulation of Genes," *Scientific American*, **233**: 24, 1975.
15. W. Gilbert and L. Villa-Komaroff: "Useful Proteins from Recombinant Bacteria," *Scientific American*, **242**: 74, 1980.
16. R. Wetzel: "Applications of Recombinant DNA Technology," *American Scientist*, **68**: 664, 1980.
17. J. D. Watson, J. Tooze, and D. T. Kurz, *Recombinant DNA: A Short Course*, W. H. Freeman and Company, New York, 1983.
18. D. M. Glover, *Genetic Engineering: Cloning DNA*, Chapman & Hall, New York, 1980.
19. R. W. Old and S. B. Primrose, *Principles of Gene Manipulation*, 2d ed., University of California Press, Berkeley, 1981.
20. R. L. Rodriguez and R. C. Tait, *Recombinant DNA Techniques: An Introduction*, Addison-Wesley Publishing Co., Reading Massachusetts, 1983.
21. R. Wu, L. Grossman, and K. Moldane (eds.), "*Recombinant DNA*," *Methods in Enzymology*, vols. 68 (1979), 100, and 101 (1983), Academic Press, New York.
22. F. Sanger, S. Nicklen, and A. R. Coulson, "DNA Sequencing with Chain-Terminating Inhibitors," *Proc. Nat. Acad. Sci. USA*, **74**: 5463, 1977.
23. P. W. Gray et al., "Expression of Human Immune Interferon cDNA in *E. coli* and Monkey Cells," *Nature*, **295**: 503, 1982.

The early scientific and social debates that accompanied the emergence of recombinant DNA methods are well documented and summarized in:

24. J. D. Watson and J. Tooze, *The DNA Story*, W. M. Freeman & Co., San Francisco, 1981.

Surveys of cell cycle studies and the cell cycles of *E. coli* and *S. cerevisiae* in particular:

25. J. M. Mitchison, *The Biology of the Cell Cycle*, Cambridge University Press, London, 1971.
26. J. L. Ingraham, O. Maaløe, and F. C. Neidhardt, *Growth of the Bacterial Cell*, Sinauer Associates, Inc., Sunderland, Massachusetts, 1983.
27. L. H. Hartwell, "*Saccharomyces cerevisiae* Cell Cycle," *Bact. Rev.*, **38**: 164, 1974.
28. J. C. Lievense and H. C. Lim, "The Growth and Dynamics of *Saccharomyces cerevisiae*," p. 211 in *Annual Reports on Fermentation Processes*, vol. 5, G. T. Tsao (ed.), Academic Press, New York, 1982.

CHAPTER

SEVEN

KINETICS OF SUBSTRATE UTILIZATION, PRODUCT FORMATION, AND BIOMASS PRODUCTION IN CELL CULTURES

When a small quantity of living cells is added to a liquid solution of essential nutrients at a suitable temperature and pH, the cells will grow. The growth processes of interest to us have two different manifestations according to the morphology of the cells involved. For unicellular organisms which divide as they grow, increases in *biomass* (mass of living matter) are accompanied by increases in the number of cells present. This case, which confronts us with a problem in *population* growth, will occupy most of our attention in this chapter. When considering growth of molds, however, the situation is quite different. Here the length and number of mycelia increase as the organism grows. The growing mold thus increases in size and density but not necessarily in numbers.

Associated with cell growth are two other processes: uptake of some material from the cell's environment and release of metabolic end products into the surroundings. As we shall see below, the rates of these processes vary widely as growth occurs. While a general predictive capability will elude us, past experience has shown that among the many cell cultivation processes known, a few general patterns of substrate utilization and product formation occur frequently. By reviewing them, we shall be better prepared to cope with new problems in application of growing cells.

The complexity of the kinetic description which is required and appropriate depends in turn on the complexity of the physical situation and the intended application of the kinetics. Before considering specific cases and their kinetic representation, we should have some awareness of the spectrum of possibilities for reactions in cell populations. In this discussion we shall focus on cell growth, but similar concepts, comments, and situations apply to other aspects of chemical reactions in cellular processes.

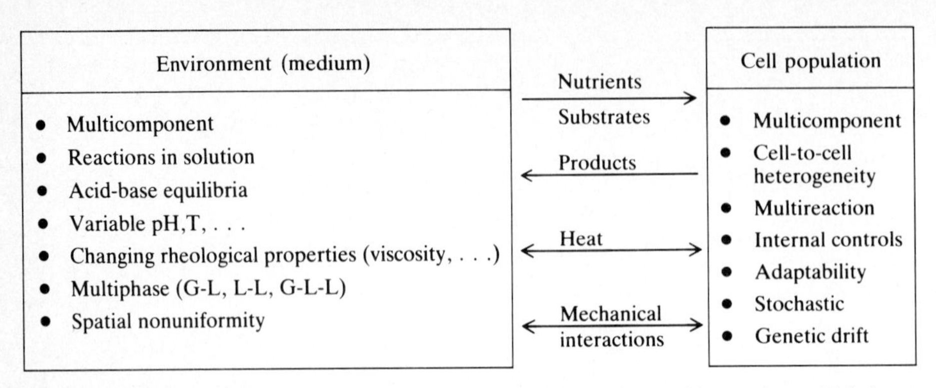

Figure 7.1 Summary of some of the important paramaters, phenomena, and interactions which determine cell population kinetics.

Let's begin by thinking about a growing cell population in the most general and hence most complicated terms. Then, we shall consider simplifications that can usefully be introduced and the situations in which they apply. Summarized in Fig. 7.1 are some of the parameters, phenomena, and interactions which influence the kinetic behavior of cell populations. First, we must recognize that two interacting systems are involved—the biological phase consisting of a cell population and the environmental phase or growth medium. Cells consume nutrients and convert substrates from the environment into products. The cells generate heat and, in turn, the medium temperature sets the temperature of the cells. Mechanical interactions occur through hydrostatic pressure and flow effects from the medium to the cells and from changes in medium viscosity due to accumulation of cells and of cellular metabolic products.

Turning now to the different factors important in the medium, we must note that it is a multicomponent system which must contain all of the required nutrients for cell growth and which will accumulate, as the cells grow, various end products of cellular metabolism. Reactions can occur in the medium solution modifying the form of cell products. Hydrolysis of penicillin is one example. Often, the cells consume or produce components which influence the acidity of the environment, and the interplay of cellular consumption with acid-base equilibria determines medium pH which in turn influences cellular activities and transport processes. During the course of cellular reactions, the broth temperature, pH, ionic strength, and rheological properties may change with time.

The cellular environment is often a multiphase system consisting of liquid with dispersed gas bubbles, of a liquid–liquid system in which immiscible liquid phases are present, or sometimes a three-phase system with two liquid and one gas phase. Important recent developments in bioreactor research and development involve intentional addition of substances to the environment in order to create multiple liquid or solid phases to accomplish desired separations simultaneously with cellular reactions. Because of the scale of the tanks in which biological reactions are conducted and the high viscosity and non-Newtonian nature of the broth in some situations, it is not uncommon for conditions within the biological reactor to differ from point to point in space. All of these environmental variables and parameters can influence cell kinetics significantly.

Turning next to the important features of the cell population, we recall from our previous discussions that each individual cell is a complicated multicomponent system which is frequently not spatially homogeneous even at the single cell level. Many independent chemical reactions occur simultaneously in each cell, subject to a complex set of internal controls. These internal controls endow the cells with the capability and propensity to adapt the activity and even the types of chemical reaction steps which occur in the cell as a function of cellular environment. In long term cultivation of a cell population, spontaneous mutations may accumulate or the reactor system may impose a selection pressure which results in a slow drift in the genetic makeup of the strain. Furthermore, in a growing cell population, one encounters significant cell-to-cell heterogeneity. That is, on the single-cell level, different cells in the population at a given point in time and a given small region in space vary with respect to age (some are newborn, relatively young cells while some are older and some are dividing) and with respect to chemical activity. As we saw in our discussion of the cell cycle in the previous chapter, cells of different ages are often characterized by different types of metabolic functions and activities.

Clearly, it is not practical or possible to try to formulate a kinetic model which includes all of the features and details mentioned in Fig. 7.1. We shall next examine some of the approximations which allow simplification of this picture and useful representations of cell population kinetics. First, with respect to the environment, it is common practice to formulate the growth medium so that all components but one are present at sufficiently high concentrations that changes in their concentrations do not significantly affect overall rates. Thus, a single component becomes the rate-limiting nutrient, and we need consider only the concentration of this one component when analyzing the effects of medium composition on cell growth kinetics. Occasionally, it is necessary to include other medium components, such as an inhibitory product which accumulates in the medium, in order to obtain a suitable description of cell kinetics. Concerning the other environmental parameters, often it is reasonable to assume that changes in these parameters do not significantly affect microbial kinetics over the time scale or within the range of variation encountered in a typical experiment or typical process situation. Also, external controls on the bioreactor may regulate and maintain constant some environmental parameters such as pH, temperature, and dissolved oxygen concentration. On the other hand, it may be necessary in some cases to include a multicomponent and multivariable description of the environment in the model in order to represent adequately the desired range of kinetic behavior.

A conceptual view of different approximations and representations which are useful for the cellular phase of the system is summarized in Fig. 7.2. This perspective, first presented by Arnold Fredrickson and Henry Tsuchiya, classifies approaches to microbial systems according to the number of components used in

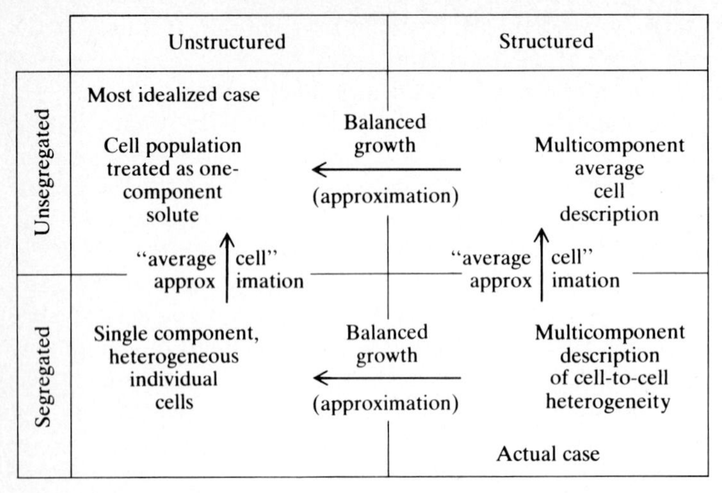

Figure 7.2 Different perspectives for cell population kinetic representations.

the cellular representation and whether or not the cells are viewed as a heterogeneous collection of discrete entities, as they really are, or instead as some kind of average cell which becomes almost the same conceptually as a component in solution. Cellular representations which are multicomponent are called *structured*, and single component representations are designated *unstructured*. Consideration of discrete, heterogeneous cells constitutes a *segregated* viewpoint, while an *unsegregated* perspective considers average cellular properties. As indicated in Fig. 7.2 the actual situation is a structured, segregated one. If cell-to-cell heterogeneity does not substantially influence kinetic processes of interest, one may introduce the "average cell approximation" and simplify the segregated viewpoint to an unsegregated perspective. In a growth state called balanced growth, all cellular synthesis activities are coordinated in such a way that the average cellular composition is not affected by proliferation of the population. In this situation, models which ignore the multicomponent nature of cells may be adequate. As we shall see below, it is common to assume the most idealized situation, namely an unsegregated, unstructured model, when analyzing and describing growth of cellular populations.

However, we shall also see a number of contexts in which there are advantages to a more complicated view of the biological phase of the system. First, in unsteady-state growth as obtains, for example, in conventional batch fermentation, balanced growth conditions are approximated only during a portion of the process, and it is well documented that cellular composition and the types of reactions conducted by the population can change substantially over the course of the batch process. In such a case, more detailed models can be useful. Furthermore, by employing a structured model, one can utilize directly in the kinetic representation known features of the biochemical reaction network in the cell. Similarly, known features of the cell cycle may be readily incorporated and used to enhance the validity and range of application of a kinetic model for a cell population by use of a segregated approach. In segregated models, single-cell kinetics and regulatory features are essential and central features.

Living cells are extremely small systems, and they do not contain large numbers of molecules of any chemical component. Remembering that we usually deal with numbers of molecules of the order of 10^{23}; the numbers listed in Table 5.6 are quite small. As an extreme case, since there is only one DNA molecule in a slowly growing bacterium, how can we justify speaking of a "DNA concentration in the cell"? Similar comments could be made about trace ions, the contents of organelles, and many other components within the cell. Although some models described below will employ continuum representations of intracellular events, we must view these as engineering approximations of a typical cell *in* a population of cells as is shown below. Because of the small numbers of molecules involved, reactions and mass transfer processes within individual cells should be treated as random events. Although stochastic population models can be constructed to attempt to treat such phenomena, these models have not shown significant advantages relative to simpler deterministic ones.

The precision of a deterministic description of a population as small as a typical inoculum (10^5 cells) conveniently illustrates the predictability of cell-population behavior. For example, let the distribution of generation times t resulting from sampling *individual cells* be the normal distribution

$$P(t) = \frac{1}{\sqrt{2\pi}\,\sigma} \exp\left[\frac{-(t-\bar{t})^2}{2\sigma^2}\right]$$

and take $\bar{t} = 1$ and the standard deviation σ to be 0.5. Then, with 95 percent confidence, we know that the generation time of a single cell is

$$t_{\text{one cell}} = \bar{t} \pm 1.96\sigma = 1.0 \pm 0.98$$

Now consider a cell suspension containing many cells, say m. Assuming that each of the m cells grows independently of the others, the cell suspension contains m independent samples. Then the 95 percent confidence limit on generation time for this *population* is $2\bar{\sigma}_{\text{pop}}$, where

$$\bar{\sigma}_{\text{pop}} = \frac{0.98\sigma}{\sqrt{m}}$$

Consequently, the uncertainty in the population generation time diminishes rapidly as m becomes large. Taking $\sigma = 0.5$ and $\bar{t} = 1.0$ as before, for example, the 95 percent confidence limits for the population $\bar{t}$ vary with m as given below:

m	95% confidence limits on $\bar{t}$
1	$t = 1 \pm 0.98$
10^4	$t = 1 \pm 0.0098$
10^8	$t = 1 \pm 0.000098$

Such an exercise simply reminds us that, as with other stochastic processes involving relatively large numbers of events (such as fluid-transport phenomena or chemical reactions), the course of a change in population characteristics can be predicted quite precisely even when the standard deviations for individual characteristics are large. Following the progress of an inoculum growing from 10^4 to 10^8 cells per milliliter evidently involves sufficient averaging at all stages to give a well-defined value to the population generation time. In this same sense, we may speak sensibly of the rate of DNA synthesis in a typical cell *in a population* even though in each individual cell only one or two molecules are being assembled at an instantaneous rate which may be quite different from the mean.

It is impossible to discuss kinetics in the absence of some reactor configuration for measuring or evaluating those kinetics. In the following section, we shall briefly summarize the pertinent material balances for two ideal types of bioreactors. Then, we shall begin our consideration of kinetics in the upper left-hand corner of Fig. 7.2, dealing with the simplest types of growth, substrate utilization, and product formation kinetics models. In subsequent sections of the chapter, we will explore other domains of the matrix in Fig. 7.2 and try to obtain some appreciation of the value of different types of perspectives, conceptual and mathematical, on cell population kinetics. In general, formulating a useful kinetic model of a cell population is an art which requires consideration of the intended end use of the kinetic model, careful thought and experiment to identify the key variables and parameters which influence the processes of primary interest, and some conceptual and mathematical agility in translating this qualitative view of the system into a workable mathematical representation.

7.1 IDEAL REACTORS FOR KINETICS MEASUREMENTS

It is difficult to obtain useful kinetic information on microbial populations from reactors that have spatially nonuniform conditions. Hence it is desirable to study kinetics in reactors that are well mixed. We examine here well-mixed batch reactors and well-mixed continuous flow reactors. Other configurations, including fed-batch reactors and reactors with nonideal mixing, are considered in Chap. 9.

7.1.1 The Ideal Batch Reactor

Many biochemical processes involve batch growth of cell populations. After *seeding* a liquid *medium* with an *inoculum* of living cells, nothing (except possibly some gas) is added to the *culture* or removed from it as growth proceeds. Typically in such a reactor, the concentrations of nutrients, cells, and products vary with time as growth proceeds.

A material balance on moles of component i shows that the rate of accumulation of component i, given by the time derivative of the total amount of component i in the reactor, must be equal to the net rate of formation of component i due to chemical reactions in the vessel. Thus,

$$\frac{d}{dt}\left[\begin{pmatrix}\text{culture}\\ \text{volume}\end{pmatrix}\begin{pmatrix}\text{molar concentration}\\ \text{of component } i\end{pmatrix}\right]$$

$$=\begin{pmatrix}\text{culture}\\ \text{volume}\end{pmatrix}\cdot\left(\frac{\text{moles } i \text{ formed by reaction}}{\text{unit culture volume}\cdot\text{unit time}}\right) \tag{7.1}$$

or

$$\frac{d}{dt}\left(V_R\cdot c_i\right)=V_R\cdot r_{f_i} \tag{7.2}$$

where

$$V_R = \text{culture volume}$$

$$c_i = \frac{\text{moles } i}{\text{unit culture volume}}$$

$$r_{f_i} = \frac{\text{moles } i \text{ formed by reaction}}{\text{unit culture volume}\cdot\text{unit time}}$$

If no liquid is added to or removed from the reactor and if gas stripping of culture liquid is negligible, then V_R is constant, and Eq. (7.2) reduces to

$$\frac{dc_i}{dt}=r_{f_i} \tag{7.3}$$

We should note here that a similar balance may be formulated in terms of mass or number density of a component. Also, if component i is contained in a gas stream entering or leaving the reactor, then corresponding terms giving the net rate of component i addition to the reactor by gas flow must by added to the above material balances.

We can see from Eq. (7.3) that measurement of the time rate of change of component i concentration allows direct determination of the overall rate of i formation due to reactions (including those in the cell) which take place in the batch reactor. In general, the rate of formation r_{f_i} depends upon the state of the cell population (composition, morphology and age distribution, etc.) and all environmental parameters which influence the rates of reactions in the cells and in the medium. However, as we discussed in this chapter's introduction, simplifications which include only the most influential parameters are usually introduced into the kinetic representation. In subsequent sections, we shall see how reaction stoichiometry and reaction rates are related to the rate of formation.

7.1.2 The Ideal Continuous-Flow Stirred-Tank Reactor (CSTR)

Two diagrams of this process are shown in Fig. 7.3*a*, which illustrates a laboratory implementation, and Fig. 7.3*b*, which is a schematic diagram of a completely mixed continuous stirred-tank reactor (CSTR). Such configurations for cultivation of cells are frequently called *chemostats.* As this figure suggests, mixing is supplied by means of an impeller, rising gas bubbles, or both. We assume that this mixing is so vigorous that each phase of the vessel contents is of uniform composition; i.e., the concentrations in any phase do not vary with position inside the reactor. The schematic indicates an important implication of this complete-mixing assumption: the liquid effluent has the same composition as the reactor contents.

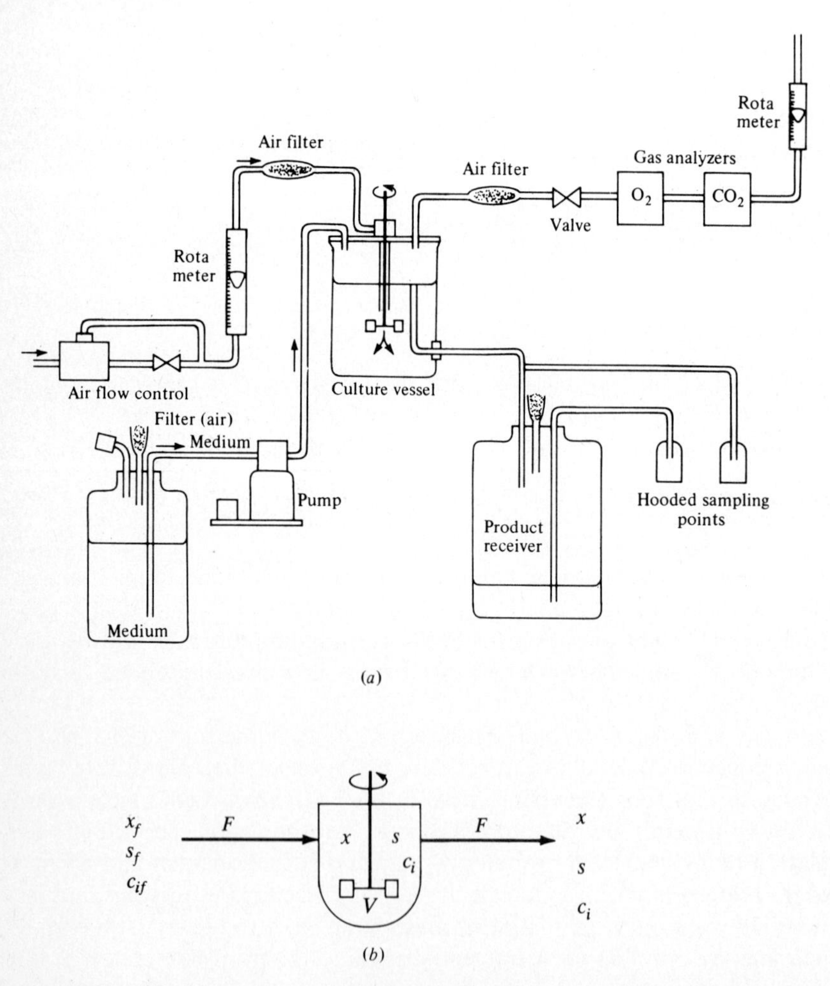

Figure 7.3 CSTRs for continuous cultivation of cell populations: (*a*) major components of a laboratory CSTR for growing cells; (*b*) the notation usually used in modeling and analysis of these reactors.

Because of complete mixing, the dissolved-oxygen concentration is the same throughout the bulk liquid phase. This is of crucial importance in considering aerated CSTRs, for it means that we can often decouple the aerator or agitator design from consideration of the reaction processes. So long as the aeration system maintains dissolved oxygen in the CSTR above the limiting concentration, we can view analysis of the cell-kinetics aspects of the system as essentially a separate problem. Similar logic is also usually applied to the heat-transfer problems which can accompany microbial growth. So long as the vessel is well stirred, has adequate heat removal capacity, and is equipped with a satisfactory temperature controller, we can assume that it is isothermal at the desired temperature and proceed with investigation of microbial reaction processes. These approaches are implicitly adopted in much of the discussion which follows.

In the steady state, where all concentrations within the vessel are independent of time, we can apply the following material balance to any component of the system:

$$\begin{bmatrix}\text{Rate of addition}\\ \text{to reactor}\end{bmatrix} - \begin{bmatrix}\text{rate of removal}\\ \text{from reactor}\end{bmatrix} + \begin{bmatrix}\text{rate of formation}\\ \text{within reactor}\end{bmatrix} = 0 \quad (7.4)$$

Letting V_R denote the total volume of culture within the reactor as before, the steady-state CSTR material balance may be written

$$F(c_{if} - c_i) + V_R r_{f_i} = 0 \quad (7.5)$$

where F = volumetric flow rate of feed and effluent liquid streams
c_{if} = component i molar concentration in the feed stream
c_i = component i concentration in the reaction mixture and in the effluent stream.

Rearranging Eq. (7.5),

$$r_{f_i} = \frac{F}{V_R}(c_i - c_{if}) = D(c_i - c_{if}) \quad (7.6)$$

we see that the rate of formation may be easily evaluated based upon measurements of the (steady) inlet and exit concentrations. As noted in Eq. (7.6), the parameter D, called the *dilution rate* and defined by

$$D = \frac{F}{V_R} \quad (7.7)$$

characterizes the holding time or processing rate of the CSTR. The dilution rate is equal to the number of tank liquid volumes which pass through the vessel per unit time. D is the reciprocal of the mean holding time or mean residence time more familiar in chemical reaction engineering. Because D is almost universally employed in the biochemical engineering literature, however, we shall consistently employ the dilution rate concept and notation.

Comparing Eqs. (7.3) and (7.6), we can see that kinetics determinations are more straightforward for a CSTR since we do not need to measure the time dependence of a concentration and then differentiate data as batch reactor analysis requires. Studying cell population kinetics in a CSTR has another important advantage in that the cell population can adjust to a steady environment and achieve or closely approximate a state of balanced growth. Thus, there is the benefit of obtaining a relatively well-defined, reproducible state for the cells. This is more difficult to achieve routinely in batch cultivation. On the other hand, batch experiments can be done in a flask, several at a time, in an incubator shaker. The equipment required for CSTR experiments is more expensive and more complicated. Achieving steady-state conditions in a biological CSTR may take hours or days, magnifying the possibility of contamination which ruins the experiment. Finally, since the large-scale process will likely involve batch operation with all of the attendant complications of transient, unbalanced growth and different metabolic responses and activities at different times in the batch, kinetic models based on steady-state CSTR data may not be appropriate. Again, the guiding principle in experimental study and mathematical representation of microbial and higher cell population kinetics is careful definition of the intended application and scope of use of the kinetics. Based upon these requirements, the experimental and mathematical modeling programs can be formulated.

We next turn to the simplest class of cell population growth models. Because the central principles of these kinetic representations were discovered and have been most fully characterized by chemostat experiments, we focus our attention first on growth in a CSTR.

7.2 KINETICS OF BALANCED GROWTH

In our beginning attempts to model and understand cell population kinetics, we shall use unstructured models in which only cell mass or number concentrations will be employed to characterize the biophase. The net rate of cell mass growth, r_x, is often written as μx where x is the cell mass per unit culture volume and μ, which has the units of reciprocal time, is the *specific growth rate* of the cells. Using this representation in a steady-state CSTR material balance for cell mass gives

$$Dx_f = (D - \mu)x \tag{7.8}$$

Often the liquid feed stream to a continuous culture consists only of sterile nutrient, so that $x_f = 0$. In this case Eq. (7.8) reveals that a nonzero cell population can be maintained only when

$$D = \mu \tag{7.9}$$

i.e., when the culture has adjusted so that its specific growth rate is equal to the dilution rate. When Eq. (7.9) is satisfied, it appears that Eq. (7.8) does not determine x when the feed is sterile. Experiments with continuous culture of *Bacillus linens* have confirmed the indeterminate nature of the population level. After a steady continuous operation was achieved at the 6-h point (Fig. 7.4), two subsequent interruptions of the culture were imposed. In each case, a portion of the

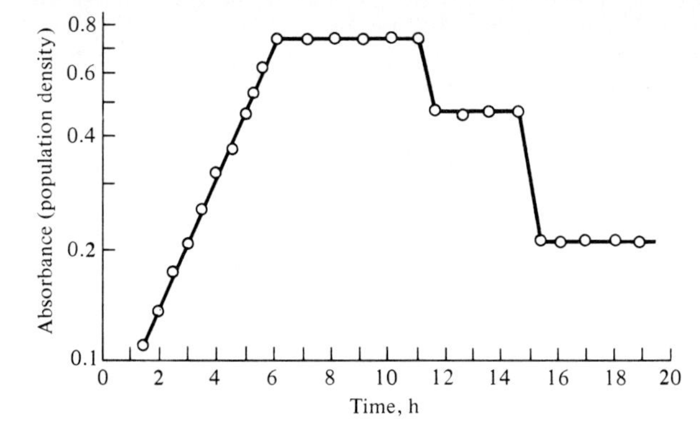

Figure 7.4 As Eq. (7.9) indicates, the population concentration in continuous culture is indeterminate so long as μ is constant (*B. linens* at 26°C with $D = 0.417\ h^{-1}$). *(Reprinted from R. K. Finn and R. E. Wilson, "Population Dynamics of a Continuous Propagator for Microorganisms," J. Agri. Food Chem., vol. 2, p. 66, 1954. Copyright by the American Chemical Society.)*

reactor contents consisting of cells plus medium was removed and replaced by medium alone. Following each interruption, the system achieved a new steady population of different size.

7.2.1 Monod Growth Kinetics

The behavior displayed in Fig. 7.4 can occur only when the specific growth rate is independent of x and s. (Here s denotes the mass concentration of limiting substrate). The indeterminate nature of the system disappears if a particular nutrient is growth-rate limiting.

Before discussing details of the dependence of growth rates on nutrient supply, we should review the general ideas and practices for construction of cell-culture media. We distinguish two types of media according to their makeup. A *synthetic medium* is one in which chemical composition is well defined. As Table 7.1 shows, such media can be constructed by supplementing a mineral base with the necessary carbon, nitrogen, and energy sources as well as any necessary vitamins. In addition to providing necessary ions for proper cell function, the mineral base of the medium also contains buffering compounds to reduce large pH fluctuations during growth. *Complex media* contain materials of undefined composition. For example, in Table 7.1, medium 4 is complex because the exact chemical makeup of the yeast extract is unknown. Other common complex media include beef broth, blood-infusion broth, corn-steep liquor, and sewage.

The general goal in making a medium is to support good growth and/or high rates of product synthesis. Contrary to intuitive expectation, this does *not* necessarily mean that all nutrients should be supplied in great excess. For one thing, excessive concentration of a nutrient can inhibit or even poison cell growth. Moreover, if the cells grow too extensively, their accumulated metabolic end products will often disrupt the normal biochemical processes of the cells. Consequently, it is common practice to limit total growth by limiting the amount of one nutrient in the medium.

If the concentration of one essential medium constituent is varied while the concentrations of all other medium components are kept constant, the growth

Table 7.1 Some examples of synthetic and complex media[†]

Common ingredients (Mineral base)	Additional ingredients			
	Medium 1[‡]	Medium 2[‡]	Medium 3[‡]	Medium 4[§]
Water, 1 L K_2HPO_4, 1 g $MgSO_4 \cdot 7H_2O$, 200 mg $FeSO_4 \cdot 7H_2O$, 10 mg $CaCl_2$, 10 mg Trace elements (Mn, Mo, Cu, Co, Zn) as inorganic salts, 0.02–0.5 mg of each	NH_4Cl, 1 g	Glucose,[¶] 5 g NH_4Cl, 1 g	Glucose, 5 g NH_4Cl, 1 g Nicotinic acid, 0.1 mg	Glucose, 5 g Yeast extract, 5 g

[†] R. Y. Stanier, M. Doudoroff, and E. A. Adelberg, *The Microbial World*, 3d ed., p. 79, Prentice-Hall, Inc., Englewood Cliffs, N.J., 1970.
[‡] Synthetic.
[§] Complex.
[¶] If the media are sterilized by autoclaving, the glucose should be sterilized separately and added aseptically. When sugars are heated in the presence of other ingredients, especially phosphates, they are partially decomposed to substances that are very toxic to some microorganisms.

rate typically changes in a hyperbolic fashion, as Fig. 7.5 shows. A functional relationship between the specific growth rate μ and an essential compound's concentration was proposed by Monod in 1942. Of the same form as the Langmuir adsorption isotherm (1918) and the standard rate equation for enzyme-catalyzed reactions with a single substrate (Henri in 1902 and Michaelis and Menten in 1913), the Monod equation states that

$$\mu = \frac{\mu_{\max} s}{K_s + s} \tag{7.10}$$

Here $\mu_{\max}$ is the maximum growth rate achievable when $s \gg K_s$ and the concentrations of all other essential nutrients are unchanged. K_s is that value of the limiting nutrient concentration at which the specific growth rate is half its maximum value; roughly speaking, it is the division between the lower concentration range, where μ is strongly (linearly) dependent on s, and the higher range, where μ becomes independent of s. As shown in Fig. 7.5, K_s values for *E. coli* strains growing in glucose- and tryptophan-limited media are 0.22×10^{-4} *M* and 1.1 ng/mL, respectively.

From our earlier examination of the cell's biochemistry, it is apparent that the Monod equation is probably a great oversimplification. As in other areas of engineering, however, this is a case where a relatively simple equation reasonably expresses interrelationships even though the physical meaning of the model parameters is unknown or perhaps does not exist. In some special instances, however, we may attach physical significance to the Monod equation. One of the

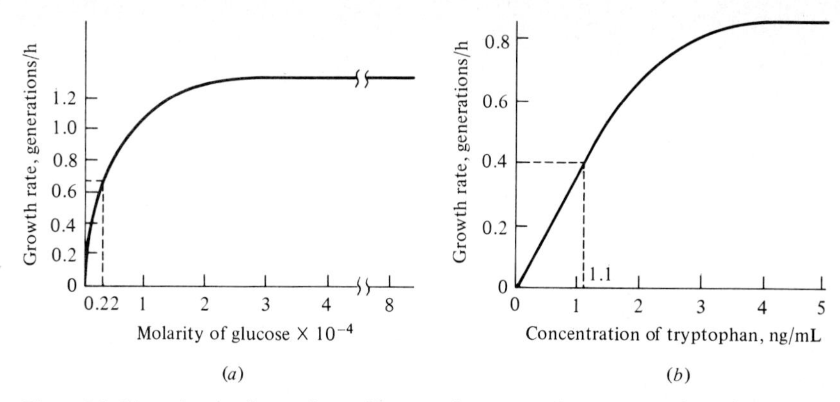

Figure 7.5 Dependence of *E. coli* specific growth rate on the concentration of the growth-limiting nutrient: (*a*) glucose medium and (*b*) tryptophan medium (for a tryptophan-requiring mutant). *(From R. Y. Stanier, M. Doudoroff, and E. A. Adelberg, "The Microbial World" 3d ed., p. 315, 1970. Reprinted by permission of Prentice-Hall, Inc., Englewood Cliffs, New Jersey.)*

clearest of these involves growth rate limitation by permease-mediated membrane transport rate (recall Sec. 5.7).

The particular form of Eq. (7.10) is appealing; its simplicity urges several warnings upon the user. First, the value of K_s is often rather small. Thus s can be $\gg K_s$ rather easily, and the term $s/(K_s + s)$ may be regarded simply as an adequate decription for calculating the deviation of μ from $\mu_{\max}$ as the concentration s becomes smaller. The relation also suggests that the specific growth rate is finite ($\mu \neq 0$) for any finite concentration of the rate-limiting component. Generally, this implied behavior is not well tested for $s \ll K_s$.

When population growth rate is related to limiting nutrient equation by a relationship such as the one proposed by Monod, definite connections emerge among reactor operating conditions and microbial kinetic and stoichiometric parameters. To show this, we write a mass balance on limiting substrate which couples to the cell mass balance since μ depends on s. In the substrate balance we make use of the yield factor $Y_{X/S}$, introduced in Sec. 5.10.1,

$$Y_{X/S} = \frac{\text{mass of cells formed}}{\text{mass of substrate consumed}} \tag{7.11}$$

The steady-state mass balance on substrate is then

$$D(s_f - s) - \frac{1}{Y_{X/S}} \mu x = 0$$

or, taking μ from Eq. (7.10),

$$D(s_f - s) - \frac{\mu_{\max} s x}{Y_{X/S}(s + K_s)} = 0 \tag{7.12}$$

The corresponding cell mass balance is

$$\left(\frac{\mu_{max} s}{K_s + s} - D\right) x + D x_f = 0 \tag{7.13}$$

Equations (7.12) and (7.13) are often called the *Monod chemostat model.*

For the common case of sterile feed ($x_f = 0$), these equations can readily be solved for x and s to yield

$$x_{\text{sterile feed}} = Y_{X/S}\left(s_f - \frac{DK_s}{\mu_{max} - D}\right) \tag{7.14}$$

and

$$s_{\text{sterile feed}} = \frac{DK_s}{\mu_{max} - D} \tag{7.15}$$

Equations (7.14) and (7.15) contain the explicit steady-state dependence of x and s on flow rate ($D = F/V$). For very slow flows at a given volume, $D \to 0$; thus s tends to zero. Since nearly all feed substrate is consumed by the cells, the resulting effluent cell-mass concentration is $x = s_f Y_{X/S}$.

As D increases continuously, s increases first linearly with D and then still more rapidly as $D \to \mu_{max}$. The cell-mass concentration x declines with the same functional behavior: first linearly in D, then diminishing more rapidly as $D \to \mu_{max}$. At some point as D approaches μ_{max}, x becomes zero. The dilution rate D has just surpassed the maximum possible growth rate, and the only steady-state solution is $x = 0$. This condition of loss of all cells at steady state, termed *wash-out*, occurs for D greater than D_{max}, where, if $x = 0$ in Eq. (7.14)

$$D_{max} = \frac{\mu_{max} s_f}{K_s + s_f} \tag{7.16}$$

The character of these solutions for substrate and cell-mass concentration is shown in Fig. 7.6 for the following parameter values:

$$\mu_{max} = 1.0\ \text{h}^{-1} \qquad Y_{X/S} = 0.5 \qquad K_s = 0.2\ \text{g/L} \qquad s_f = 10\ g/\text{L}$$

Notice that near washout the reactor is very sensitive to variations in D; a small change in D gives a relatively large shift in x and/or s.

This sensitivity must be kept in mind if production of cell mass is the objective of the continuous cultivation. The rate of cell production per unit reactor volume is Dx. This quantity is illustrated in Fig. 7.6, and there is a sharp maximum. We can compute the maximal cell output rate by solving

$$\frac{d(Dx)}{dD} = 0 \tag{7.17}$$

where Eq. (7.14) is used to write x as a function of D. Solution of Eq. (7.17) gives

$$D_{\text{max output}} = \mu_{max}\left(1 - \sqrt{\frac{K_s}{K_s + s_f}}\right) \tag{7.18}$$

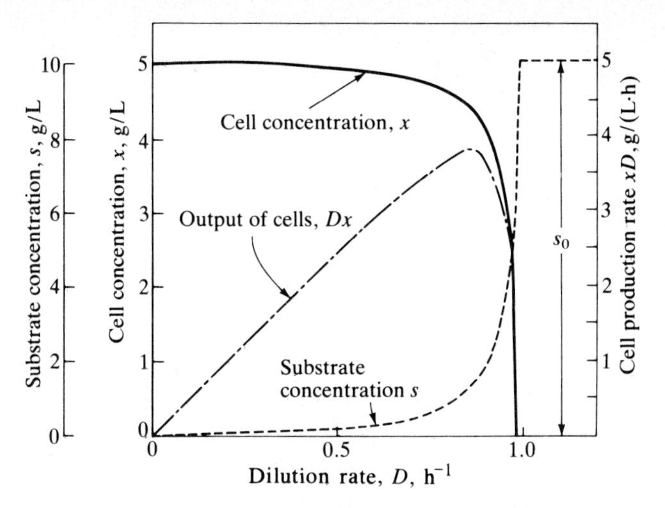

Figure 7.6 Dependence of effluent substrate concentration s, cell concentration x, and cell production rate xD on continuous culture dilution rate D as computed from the Monod chemostat model ($\mu_{max} = 1\ h^{-1}$, $K_s = 0.2$ g/L, $Y_{X/S} = 0.5$, $s_f = 10$ g/L).

If $s_f \gg K_s$, as often is the case, the value of $D_{max\ output}$ approaches μ_{max} and consequently is near washout. This situation, evident in Fig. 7.6, may require us to forgo maximal biomass production in order to avoid the region of large sensitivity. Inclusion of the practical aspects of sensitivity, controllability, and reliability into optimization problems such as this must not be forgotten in the pursuit of more easily quantified objectives.

When analyzing the production of end products from a continuous fermentation, we introduce another yield coefficient $Y_{P/X}$, defined by

$$Y_{P/X} = \frac{\text{mass of product formed}}{\text{increase in cell mass}} \tag{7.19}$$

From our earlier discussions of product formation stoichiometry in Chap. 5, we are already aware that $Y_{P/X}$ is constant for a type I fermentation but will vary for other cases. Use of the product yield coefficient allows us to write the steady-state product mass balance in the form

$$D(p_f - p) + Y_{P/X}\mu x = 0 \tag{7.20}$$

so that

$$p = p_f + \frac{Y_{P/X}\mu x}{D} \tag{7.21}$$

Combined with earlier equations relating μ and x to process parameters, Eq. (7.21) allows calculation of the concentration of product in the effluent. The rate of product output is then given by pD, which is maximixed for constant $Y_{P/X}$ when D has the value specified in Eq. (7.18). Thus, the goal of maximum product

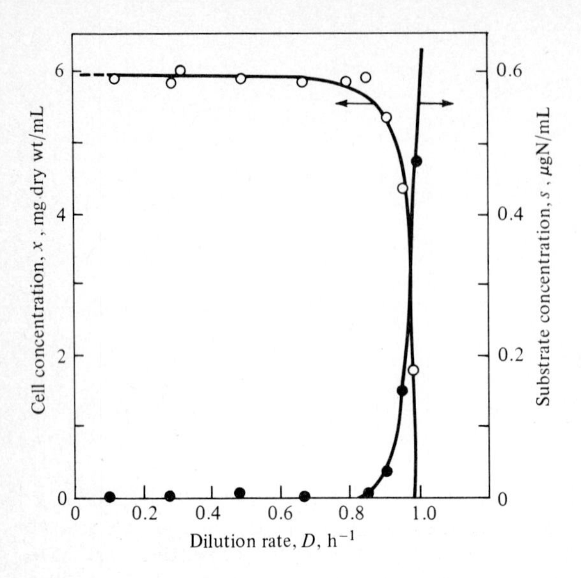

Figure 7.7 Experimental continuous culture data qualitatively consistent with the Monod model (*A. aerogenes*). *(Replotted from D. Herbert, "A Theoretical Analysis of Continuous Culture Systems," in "Society of Chemical Industries Monograph 12," p. 247, London, 1961.)*

output also must be reconciled with sensitivity considerations. (If $Y_{P/X} = f(D)$, what value of D maximizes pD?)

An experimental example of continuous-culture behavior is shown in Fig. 7.7. In the culture of the bacterium *Aerobacter aerogenes*, the agreement between experiment and the simple model just outlined is qualitatively correct. However, both the observed cell-mass production and the observed substrate concentration remain approximately constant over a wider range of conditions than suggested in Fig. 7.6. Other data reveal breakdowns in the above model at very high and very low dilution rates. We shall next examine these two extreme cases in turn and attempt to understand why the Monod model might fail for them.

7.2.2 Kinetic Implications of Endogeneous and Maintenance Metabolism

The data presented in Fig. 7.8 for *A. aerogenes* show a marked decline in cell concentration as dilution rate becomes small. Similar behavior has also been observed for the food yeast *Torula utilis*. This trend, which is contrary to the Monod chemostat model, can be explained by including the possibility of *endogeneous metabolism* in the model. By endogeneous metabolism we mean that there are reactions in cells which consume cell substance. Thus, we might write

$$r_x = \frac{\mu_{max} s x}{s + K_s} - k_e x \tag{7.22}$$

to account for this effect. Notice that the additional term in Eq. (7.22) can also be interpreted formally as a cell death rate.

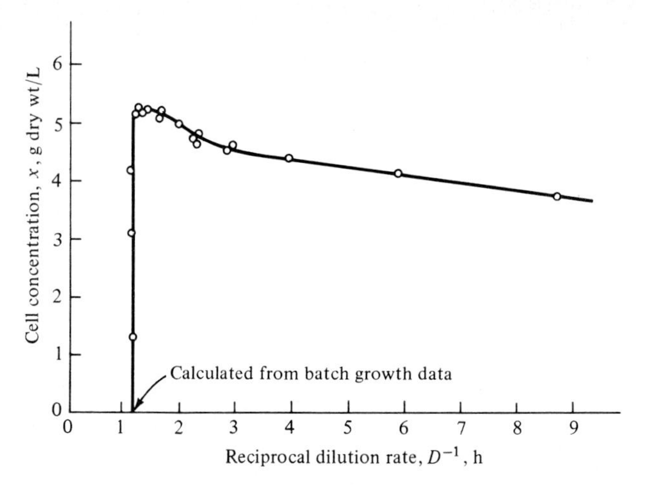

Figure 7.8 Cell concentration decreases as the dilution rate is decreased, a trend contrary to the Monod chemostat model (continuous cultivation of *A. aerogenes* in glycerol medium). *(Replotted from D. Herbert, "Continuous Culture of Microorganisms: Some Theoretical Aspects" in "Continuous Culture of Microorganisms: A Symposium," p. 48, Publishing House of the Czechoslovakia Academy of Sciences, Prague, 1958.)*

Such a modification of the Monod model is also consistent with other available data. For example, if the rate of respiration of an aerobic culture is proportional to the rate of substrate utilization

$$r_{\text{resp}} = \frac{\beta \mu_{\max} s x}{s + K_s} \tag{7.23}$$

then from Eqs. (7.8), (7.22), and (7.23) it follows that the specific respiration rate is

$$\frac{r_{\text{resp}}}{x} = \beta(D + k_e) \tag{7.24}$$

Figure 7.9 displays experimental data which agrees nicely with Eq. (7.24).

Observed variations in the yield coefficient Y with D also support a growth rate of the form given in Eq. (7.22). If the rate of substrate disappearance is

$$-r_s = \frac{1}{Y'_{X/S}} \frac{\mu_{\max} s x}{s + K_s} \tag{7.25}$$

where $Y'_{X/S}$ is the "true" coefficient, using Eqs. (7.22) and (7.25) and the definition of $Y_{X/S}$ (remember $Y_{X/S}$ is an overall stoichiometric coefficient equal to the total mass of cells produced divided by the total mass of substrate consumed, so that with sterile feed $-r_s = Dx/Y_{X/S}$) gives

$$Y_{X/S} = \frac{D Y'_{X/S}}{D + k_e} \tag{7.26}$$

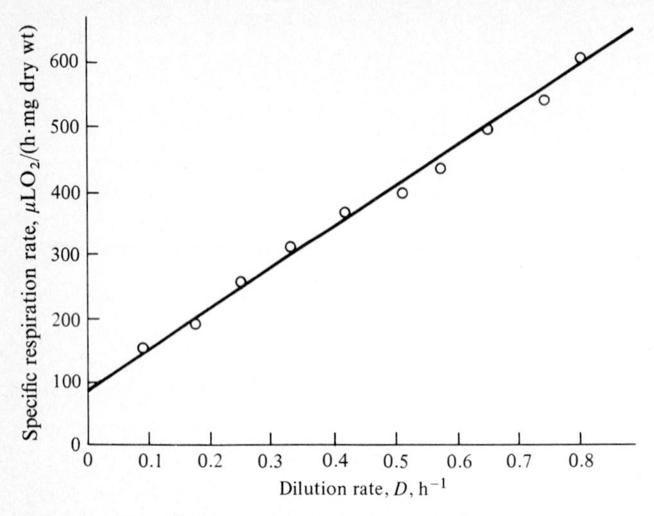

Figure 7.9 The specific respiration rate of the bacterium *A. aerogenes* in continuous culture depends linearly on dilution rate as Eq. (7.24) indicates. *(Replotted from D. Herbert, "Continuous Culture of Microorganisms: Some Theoretical Aspects" in "Continuous Culture of Microorganisms: A Symposium," P. 49, Publishing House of the Czechoslovakia Academy of Sciences, Prague, 1958.)*

Such dependence of $Y_{X/S}$ on D has been verified experimentally for several microorganisms.

Another possible scenario, introduced earlier in the discussion of cellular stoichiometry in Section 5.10.1, is parallel consumption of substrate for growth and for other energetic, or *maintenance*, requirements. In this case, the rate of substrate utilization is

$$-r_s = \frac{1}{Y'_{X/S}} \mu x + mx \tag{7.27}$$

where m is the specific maintenance rate. Presuming now that r_x is equal to μx, it follows that

$$Y_{X/S} = \frac{Y'_{X/S} D}{D + mY'_{X/S}} \tag{7.28}$$

which is exactly the same functional relationship between $Y_{X/S}$ and D as was found for the endogeneous metabolism model [Eq. (7.26)].

At large dilution rates, continuous-culture behavior can deviate significantly from the Monod chemostat model, as Fig. 7.10 shows. Not only is the predicted cell concentration in error near washout, but the population may be maintained at substantially larger dilution rates than theory indicates. Moreover, the yield coefficient decreases as D approaches the critical maximum value. Among the probable explanations for these difficulties is the relatively high substrate concentration which prevails at large dilution rates. Under these circumstances, the substrate may not limit growth, and the members of the cell population may shift

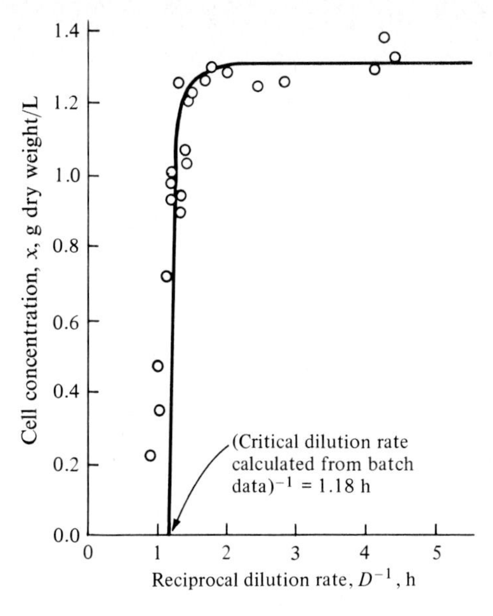

Figure 7.10 This experimental data on continuous growth of *Aerobacter cloacae* shows nonzero cell concentrations at dilution rates exceeding the calculated critical dilution rate. *(Replotted from D. Herbert et al., "The Continuous Culture of Bacteria: A Theoretical and Experimental Study," J. Gen. Microbiol., Vol. 14, p. 601, 1956.)*

their metabolic patterns in recognition of some other limiting factor in their environment. A second possibility is imperfect mixing, a topic explored in Chap. 9.

7.2.3 Other Forms of Growth Kinetics

Other related forms of specific growth rate dependence have been proposed which in particular instances give better fits to experimental data. For example, Teissier, Moser, and Contois suggest the following models:

Tessier

$$\mu = \mu_{\max}(1 - e^{-s/K_s}) \tag{7.29}$$

Moser

$$\mu = \mu_{\max}(1 + K_s s^{-\lambda})^{-1} \tag{7.30}$$

Contois

$$\mu = \mu_{\max} \frac{s}{Bx + s} \tag{7.31}$$

The first two examples render algebraic solution of the growth equations much more difficult than the Monod form. The equation of Contois contains an apparent Michaelis constant which is proportional to biomass concentration x. This last term will therefore diminish the maximum growth rate as the population density increases, eventually leading to $\mu \propto x^{-1}$.

The specific growth rate may be inhibited by medium constituents such as substrate or product. An example due to Andrews [11] proposes that *substrate inhibition* be treated by the form

$$\mu = \mu_{\max} \frac{s}{K_i + s + s^2/K_p} \tag{7.32}$$

Alcohol fermentation provides an example of *product inhibition*; the anerobic glucose fermentation by yeast has been treated by Aiba, Shoda, and Nagatani [9, 10] with specific-growth function of the type

$$\mu = \mu_{\max} \frac{s}{K_i + s} \frac{K_p}{K_p + p} \tag{7.33}$$

It is possible that two (or more) substrates may simultaneously be growth-limiting. While few data are available, a Monod dependence on each limiting nutrient may be proposed, so that

$$\mu = \mu_{\max} \frac{s_1}{K_1 + s_1} \frac{s_2}{K_2 + s_2} \cdots \tag{7.34}$$

In the absence of convincing data for this form, we may regard it simply as a useful indicator that growth depends on several limiting nutrients.

7.2.4 Other Environmental Effects on Growth Kinetics

During balanced growth, only a single parameter μ [or the population doubling time $\bar{t}_d (= \ln 2/\mu)$] is required to characterize population growth kinetics. For this reason, the magnitude of the specific growth rate μ is widely used to describe the influence of the cells' environment on the cells' performance. Consider first the influence of temperature: the range of temperature capable of supporting life as we know it lies between roughly -5 and 95°C. Procaryotes can be classified according to the temperature interval in which they grow. As seen in Table 7.2,

Table 7.2 Classification of microorganisms in terms of growth-rate dependence on temperature†

Group	Temperature, °C		
	Minimum	Optimum	Maximum
Thermophiles	40 to 45	55 to 75	60 to 80
Mesophiles	10 to 15	30 to 45	35 to 47
Psychrophiles:			
Obligate	−5 to 5	15 to 18	19 to 22
Facultative	−5 to 5	25 to 30	30 to 35

† R. Y. Stanier, M. Doudoroff, and E. A. Adelberg. *The Microbial World.* 3d ed., p. 316, Prentice-Hall, Inc., Englewood Cliffs, N.J., 1970.

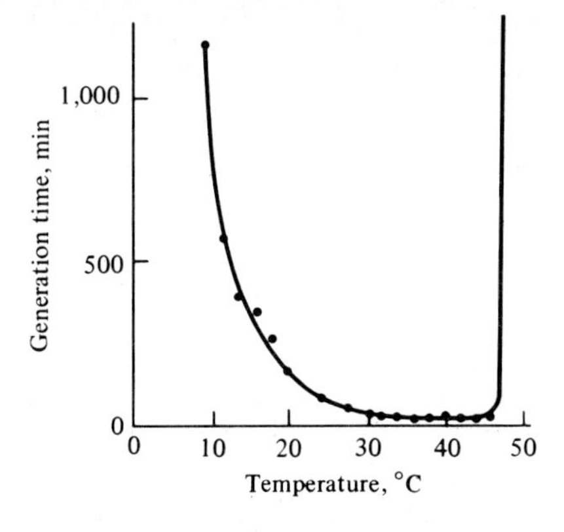

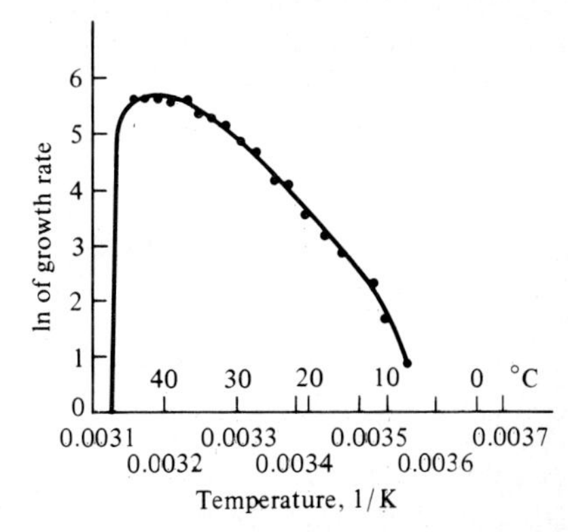

Figure 7.11 (*a*) Up to a point, the growth rate of *E. coli* increases with increasing temperature, but the cells die if the temperature is too high. (*b*) An Arrhenius plot of the data in (*a*). *(From R. Y. Stanier, M. Doudoroff, and E. A. Adelberg, "The Microbial World," 3d ed., pp. 316, 317, 1970. Reprinted by permission of Prentice-Hall, Inc., Englewood Cliffs, New Jersey.)*

each class has an optimum temperature where growth is maximal and upper and lower temperature bounds beyond which the population cannot grow.

The data in Fig. 7.11 for growth of *E. coli* dramatically illustrate the strong influence of temperature. Notice in the Arrhenius plot that classical Arrhenius behavior appears at low temperature whereas there is a rapid decrease in growth rate as the temperature approaches the upper limit for survival of the bacterium. The similarity of the temperature dependence for growth in Fig. 7.11 with the enzyme-activity-temperature relationship depicted in Fig. 3.25 is unescapable. Accordingly, the temperature dependence of specific growth rates may frequently be expressed in equation form by Eq. (3.73). Apparently at low temperatures the metabolic activity of the cell increases with increasing temperature as the activities of its enzymes rise. When the most thermally sensitive essential protein denatures, however, the cell dies. This hypothesis has been confirmed in several instances by genetic studies in which mutation of a single gene has caused a large change in the maximum tolerable temperature for a microorganism.

Since protein configuration and activity are pH dependent, we expect cellular transport processes, reactions, and hence growth rates to depend on pH. Bacterial growth rates are usually maximum in the range of pH from 6.5 to 7.5. This is exemplified by the data for *E. coli* shown in Fig. 7.12. There are exceptions, however, including acidophilic bacteria which grow at pH 2.0. Typically, a change in 1.5 to 2.0 pH units above or below the pH for maximum growth rate results in negligible growth (Fig. 7.12). Yeasts grow best in the pH range 4 to 5, while pH optima for mold growth are usually between 5 and 7. Yeasts and molds will grow over a wide range of pH from about 3.0 to 8.5. As the example in Fig. 7.12 illustrates, the pH for maximum growth rate typically increases with increasing temperature.

Specific growth rates of microorganisms are also influenced by the thermodynamic activity of water (influenced in turn by solutes in the medium) and by

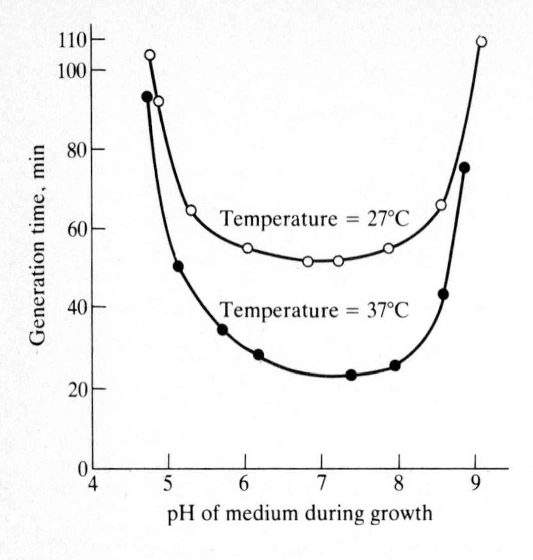

Figure 7.12 Effect of pH and temperature on the generation time of *E. coli*. *[Reprinted by permission from J. R. Norris and D. W. Ribbons (eds.), "Methods in Microbiology," vol. 2. Academic Press, New York, 1970.]*

hydrostatic pressure. Obviously, dissolved oxygen concentration is an important determinant of growth rates of aerobic organisms. We shall examine cellular oxygen requirements and kinetics influences in our discussion of reactor aeration in Chap. 8.

7.3 TRANSIENT GROWTH KINETICS

During certain intervals in batch cultivation or during startup or disturbances to continuous flow reactors, cell populations grow in a transient state in which more complicated kinetic behavior may be observed. Different types of kinetic models may be required to describe different transient growth situations. Here we examine some examples of transient growth kinetics, with emphasis on the common case of batch growth. Further discussion of transient cell kinetics is included in our consideration of bioreactor dynamics in Sec. 9.2.

7.3.1 Growth-Cycle Phases for Batch Cultivation

In a typical batch process the number of living cells varies with time, as shown in Fig. 7.13. After a *lag phase*, where no increase in cell numbers is evident, a period of rapid growth ensues, during which the cell numbers increase exponentially with time. Although this stage of batch culture is often called the *logarithmic phase*, we prefer the more descriptive term *exponential growth*, which we shall use in the following discussion. Naturally in a closed vessel the cells cannot multiply indefinitely, and a *stationary phase* follows the period of exponential growth. At this point the population achieves its maximum size. Eventually a decline in cell numbers occurs during the *death phase*. Here an exponential decrease in the number of living individuals is often observed.

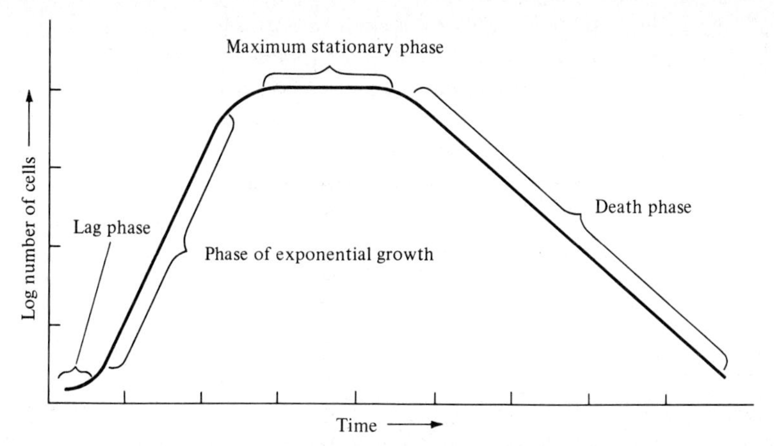

Figure 7.13 Typical growth curve for batch cell cultivation.

Each phase is of potential importance in microbiological processes. For example, the general objective of a good design may be to minimize the length of the lag phase and to maximize the rate and length of the exponential phase, the last objective being achieved by slowing the onset of the transition to stationary growth. Reaching the largest possible cell density at the end of the process is frequently very important. To achieve such goals, we should understand the variables which influence the phases of batch growth. Each phase will be discussed individually next, and, where appropriate, mathematical models of the important phenomena will be considered. This discussion will be augmented by Sec. 7.3.2, where we consider mathematical models of the entire batch-growth process.

The length of lag observed when fresh medium is inoculated depends on both the changes in nutrient composition (if any) experienced by the cells and the age and size of the inoculum. The shock of rapid switch to a new environment has several effects on the living cell. First, recall that the modes of control and regulation of enzyme activity include an adaptive characteristic: when presented with a new nutrient, its assimilation is achieved by cell production of new enzymes. Thus transfer of a glucose-bred culture in its exponential phase to a lactose medium will necessarily result in a time interval of insignificant cell-division rate while the enzymes and cofactors for the lactose metabolic pathway are synthesized in the cell. (What would happen if a lactose-bred culture were transferred to a glucose-lactose medium?) Similarly, variation in the concentration of nutrients may cause a lag phase. If the new nutrient medium is richer in a limiting nutrient, some time and nutrient will be expended in nonmultiplicative growth while larger concentrations of metabolizing enzymes are created. A decreased nutrient level may result in no lag at all; the exponential rate may resume immediately but at a slower pace.

Many of the intracellular enzymes require activation by small molecules (vitamins, cofactors) or ions (activators) which may have appreciable permeability through the cell membrane. Transfer of a small culture volume or inoculum to a large volume of medium will cause outward diffusion of these requisites for catalysis into the bulk medium if the new medium is lacking in these species or differs appreciably in ionic strength. The rate of growth will fall, corresponding to the lower concentrations of such species inside the cell, and again a lag will appear while new machinery to generate such activators is assembled. If essential activators are diluted (vitamins and ions which the cell cannot produce internally), the total level of cell activity must diminish irrevocably.

The growth stage of the inoculum, which itself comes from a smaller scale batch cultivation, exerts a strong influence on the length of the lag phase. The size of the transferred inoculum is also an important variable, as already mentioned for the loss upon transfer of such diffusible species as vitamins and activators. Thus while a young cell population shows no lag upon transfer into a medium rich in metabolic intermediates such as amino acids, the same inoculum transferred into ammonium sulfate medium loses these vital intermediates into solution. With cultures in exponential growth at the time of transfer, their original medium may already contain a reasonable bulk concentration of intermediates, and the dilution on transfer will have a lesser effect. Transfer of an old culture (approaching or in stationary phase) into ammonium sulfate medium results in a longer lag.

Multiple lag phases may sometimes be observed when the medium contains multiple carbon sources (Fig. 7.14). This phenomenon, known as *diauxic growth*, is caused by a shift in metabolic patterns in the midst of growth. After one carbon substrate is exhausted, the cell must divert its energies from growth to "retool" for the new carbon supply. A possible explanation for this serial utilization phenomenon is catabolite repression, disccussed in Sec. 6.1.4.

Design to minimize culture and process times normally includes minimization of the lag times associated with each new batch culture. From the previous

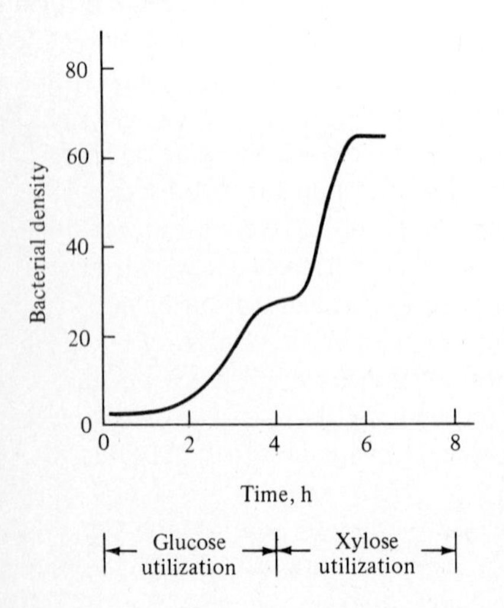

Figure 7.14 In a medium containing initially equal amounts of glucose and xylose, diauxic growth of *E. coli* is observed in batch culture. *(From R. Y. Stanier, M. Doudoroff, and E. A. Adelberg, "The Microbial World," 3d ed., p. 308, 1970. Reprinted by permission of Prentice-Hall, Inc., Englewood Cliffs, New Jersey.)*

general discussion and other relevant data (see references), the following generalizations can be drawn:

1. The inoculating culture should be as active as possible and the inoculation carried out in the exponential-growth phase.
2. The culture medium used to grow the inoculum should be as close as possible to the final full-scale fermentation composition.
3. Use of reasonably large inoculum (order of 5–10 percent of the new medium volume) is recommended to avoid undue loss by diffusion of required intermediate or activators.

At the end of the lag phase the population of microorganisms is well adjusted to its new environment. The cells can then multiply rapidly, and cell mass, or the number of living cells, doubles regularly with time. The equations

$$\frac{dx}{dt} = \mu x \qquad \text{or} \qquad \frac{1}{x}\frac{dx}{dt} = \mu \tag{7.35a}$$

with

$$x = x_0 \qquad \text{at} \qquad t = t_{\text{lag}} \tag{7.35b}$$

describe the increase in cell numbers during this period. Thus the rate of increase in x is proportional to x. From the integrated form of Eqs. (7.35)

$$\ln\frac{x}{x_0} = \mu(t - t_{\text{lag}}) \qquad \text{or} \qquad x = x_0 e^{\mu(t - t_{\text{lag}})} \qquad t > t_{\text{lag}} \tag{7.36}$$

we can readily deduce that time interval $\bar{t}_d$ required to double the population is given by

$$\bar{t}_d = \frac{\ln 2}{\mu} \tag{7.37}$$

As is the case for steady state growth in a CSTR, only a single parameter μ (or $\bar{t}_d$) is required to characterize the population during exponential batch growth. To a reasonable approximation, growth is balanced during this stage of batch cultivation. Consequently, useful balanced growth kinetic data may be obtained from batch experiments provided attention is confined to the exponential growth phase.

Deviations from exponential growth eventually arise when some significant variable such as nutrient level or toxin concentration achieves a value which can no longer support the maximum growth rate. Exhaustion of a particular critical nutrient may appear rather sharply at a given time since the cells are rapidly increasing the *total* rate of nutrient consumption in the exponential-growth phase. To formulate a rough analysis of this event, we suppose that the rate of nutrient A consumption is proportional to the mass concentration of living cells until the stationary phase is reached:

$$\frac{da}{dt} = -k_a x \tag{7.38}$$

Next we assume that exponential growth continues unabated until the stationary phase is reached, and we take the time when exponential growth begins as time zero. Then

$$x = x_0 e^{\mu t} \tag{7.39}$$

where x_0 is the mass concentration of living cells when exponential growth starts.

If the concentration of A at time zero is a_0, we can determine from Eqs. (7.38) and (7.39) that A is completely consumed when

$$a_0 = \frac{k_a}{\mu}(x_s - x_0) \tag{7.40}$$

where x_s is the mass of the population when A is exhausted and the population enters the stationary phase. Consequently, x_s is the maximum population size achieved during the batch culture (review Fig. 7.13). Rearranging Eq. (7.40), we find the maximum population in the case of nutrient depletion to be given by

$$x_s = x_0 + \frac{\mu}{k_a} a_0 \tag{7.41}$$

Linear dependence of x_s on initial nutrient level has been observed experimentally in many cases (often x_0 is so small that it is imperceptible). Figure 7.15 illustrates this behavior. In the plot of n (the number density of cells at stationary phase) for the bacterium *A aerogenes* vs. lactose concentration, however, distinct variations from Eq. (7.41) are obvious. Apparently other factors can influence the onset of the stationary phase and the size the maximum population.

If a toxin accumulates which slows the rate of growth from exponential, an equation of the form

$$\frac{dx}{dt} = kx[1 - f(\text{toxin concentration})]$$

is found useful. In the particular case where toxin linearly decreases the growth rate, we have a specific growth rate

$$\frac{1}{x}\frac{dx}{dt} = k(1 - bc_t) \tag{7.42}$$

where c_t is toxin concentration and b is a constant. A plausible assumption is that the rate of toxin production depends only on x and is proportional to it

$$\frac{dc_t}{dt} = qx \tag{7.43}$$

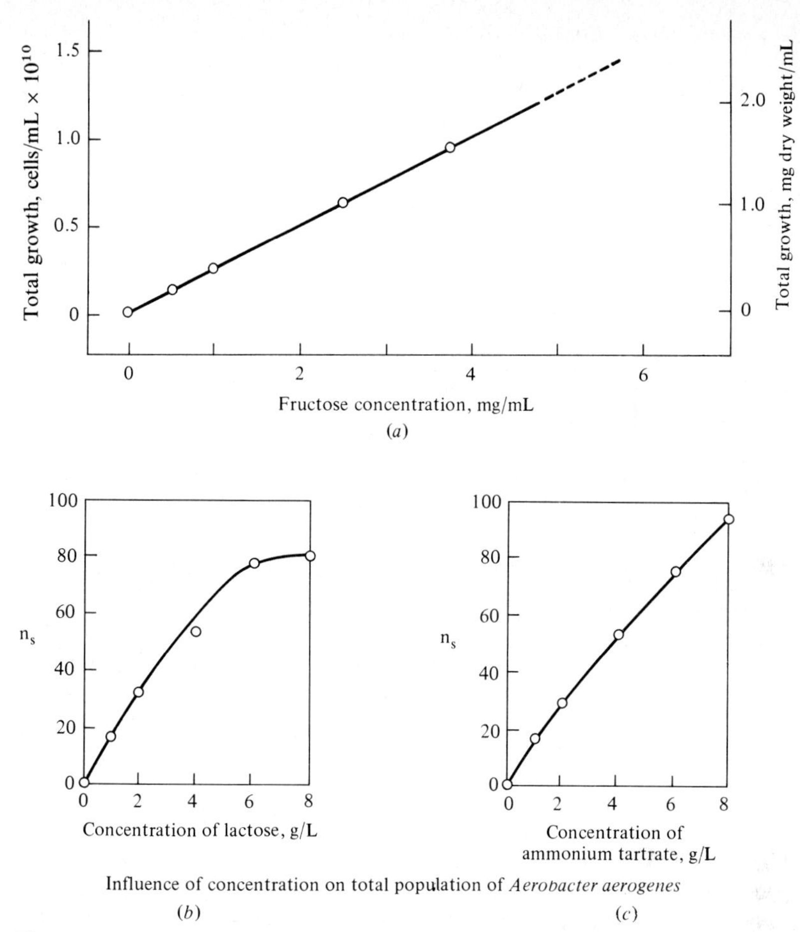

Figure 7.15 Dependence of maximum batch population size on the initial concentration of growth-limiting nutrient: (*a*) *Pseudomomas* sp. in fructose medium, and *A. Aerogenes* in (*b*) lactose and (*c*) ammonium tartrate media. *[(a) From R. Y. Stanier, M. Doudoroff, and E. A. Adelberg, "The Microbial World" 3d ed., p. 313, 1970. Reprinted by permission of Prentice-Hall, Inc., Englewood Cliffs, New Jersey; (b, c) reprinted by permission from A. C. R. Dean and C. Hinshelwood, "Growth, Function and Regulation in Bacterial Cells," p. 72, Oxford University Press, London, 1956.]*

so that $c_t = q \int_0^t x\,dt$ (assuming $c_t = 0$ at $t = 0$) and the growth equation becomes

$$\frac{dx}{dt} = kx\left(1 - bq \int_0^t x\,dt\right) \tag{7.44}$$

The instantaneous value of the effective specific-growth rate μ_{eff}

$$\mu_{\text{eff}} = \frac{1}{x}\frac{dx}{dt} = k\left(1 - bq \int_0^t x\,dt\right) \tag{7.45}$$

diminishes more and more rapidly with time, i.e.,

$$\frac{d\mu_{\text{eff}}}{dt} = -kbqx < 0 \qquad \frac{d^2\mu_{\text{eff}}}{dt} = -kbq\frac{dx}{dt} < 0 \tag{7.46}$$

Growth halts when

$$\frac{1}{bq} = \int_0^t x\,dt \tag{7.47}$$

Equation (7.42) indicates that growth ceases only when c_t reaches a particular level $c_t = 1/b$. Dilution of a given toxified medium or addition of a nonnutritive substance which complexes with a given toxin should allow further growth and a consequent increase in the maximal stationary phase biomass concentration x_s. If growth halts due to nutrient exhaustion, dilution with a nonnutritive volume produces no change in x_s. These criteria may be roughly utilized to determine the cause of growth decline and eventual halt. Exact criteria are more difficult to ascertain: growth of nutrient-limited populations does slow somewhat before total exhaustion, as will be seen later, and the growth rate of a poisoned population may become imperceptibly slow long before dx/dt is exactly zero.

The expected dependence of maximum population on the initial level of a given nutrient is sketched in Fig. 7.16. Diminution of nutrient concentration eventually brings the culture to a maximum size which is linear in the initial critical nutrient concentration. Here nutrient depletion apparently causes the cessation of exponential growth. Conversely, a rise in initial nutrient supply may eventually yield an x_s value apparently independent of the nutrient level. This suggests accumulation of toxic products or the existence of some other limiting nutrient as the determining factor.

We should not lose sight of the fate of individual cells when examining the population. In general, the population is *not* homogeneous, and the batch growth curve is a gross overview of a very complex system. For example, during the exponential-growth phase, some cells are dividing and giving birth to very young cells at the same time others are growing and maturing. Since cells of different ages generally have distinct sizes and chemical compositions, we could view a cell

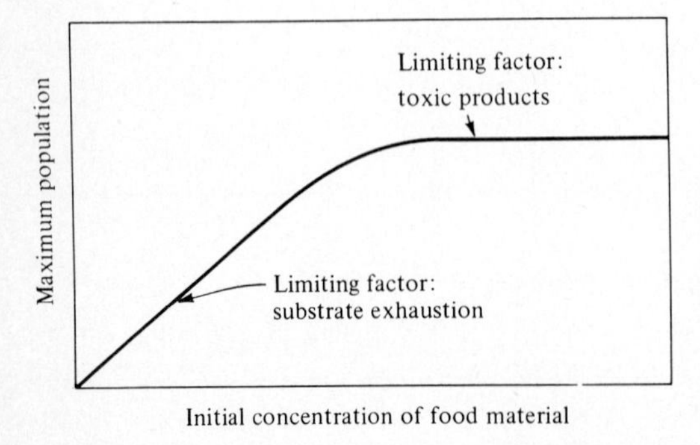

Figure 7.16 Typical relationship between initial nutrient concentration and the maximum cell population in batch culture. When the nutrient concentration is large and has no influence on the maximum population, accumulation of toxic products may be the factor which limits population size.

of a given age as a distinct "species." From this perspective, then, culture of one type of microorganism leads to a population containing a great variety of "species."

The diversity among individual cells becomes increasingly apparent during the stationary and death phases. Some cells are dividing during the stationary phase while others die. Often the dead cells lyse (break open), and the carbohydrates, amino acids and other components freed from the lysed cells are then used as nutrients by the remaining living members of the population. Such cannibalistic events help maintain the population size during the stationary phase. Eventually, due to nutrient depletion and toxic-product buildup, however, the population cannot sustain itself, and the death phase begins.

Relatively few studies have been made on the death phase of cell cultures, perhaps because many industrial batch processes are terminated before the death phase begins. Usually death of the population is assumed to follow an exponential decay

$$x = x_s e^{-k_d t} \tag{7.48}$$

where now t denotes the time elapsed since the onset of the death phase. This relationship implies that the number of cells which die at any time is a constant fraction of those living.

One physical interpretation of exponential population decay states that there are random lethal events which occur in the culture. When one of these happens to a cell, it dies. An obvious objection to this interpretation is that the past history of the population is neglected. Dean and Hinshelwood [3] suggest that not only do living cells prey on dead ones as the population stabilizes and declines but that competing portions of the cell's interacting metabolic machinery also prey on each other as they compete for scarce intermediates. An extension of their argument leads to a rationalization of the exponential-decay law. Other models of population decline will be discussed in Sec. 7.7.

To increase our understanding of the batch growth cycle and to provide background useful for the forthcoming mathematical representations of cell-population growth, it is worthwhile to review some of the available data on changes in the size and chemical makeup of a population undergoing batch growth. Again we must emphasize that this information does not apply for an individual cell but rather to the diverse collection of cells which constitute the population at any time.

Dean, Hinshelwood, and other investigators made numerous observations on composition variations in cultures of *A. aerogenes* bacteria. Taking the inoculum from a stationary population of a glucose-exhausted culture minimizes the lag phase in a glucose-containing medium. The resulting data are displayed in Fig. 7.17. Notice that early in the growth cycle the average cell size and the RNA/DNA ratio pass through high maxima. Although the amount of DNA per unit cell mass and the ratio of the protein to RNA remain relatively constant, when they are taken with the other information, we can see an early picture of the population gearing up its metabolic capacity to exploit the increased nutrient

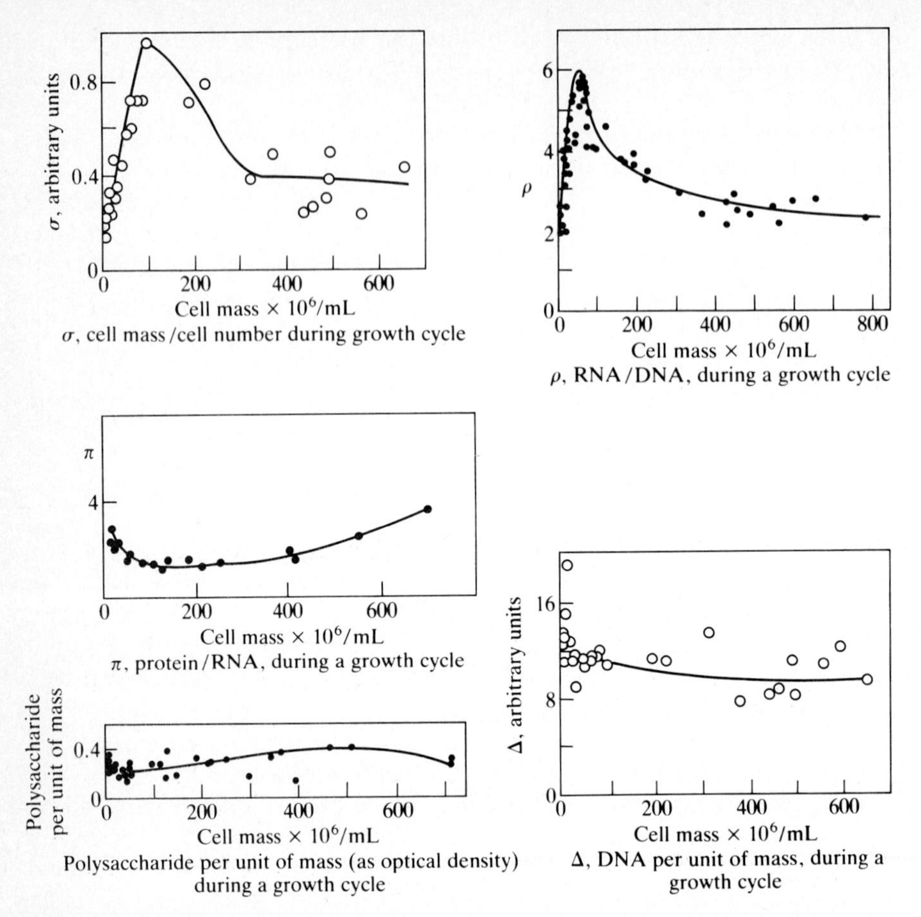

Figure 7.17 Changes in cell characteristics during batch growth of *A. aerogenes*. *(Reprinted by permission from A. C. R. Dean and C. Hinshelwood, "Growth, Function and Regulation in Bacterial Cells," pp. 87–89, Oxford University Press, London, 1966.)*

available in its new environment and to generate a larger pool of diffusible intermediates.

Another interesting phenomena which will be important when we examine end-product rates of formation is evidenced in Fig. 7.18. The average RNA concentration within a cell increases directly with population growth rate. It should be emphasized that this behavior is observed only for growth-rate variations caused by differences in nutrient medium. If, for example, the growth rate is altered by changing the temperature, RNA concentration during the exponential phase does not seem to be altered significantly.

Finally let us focus closer on the population's characteristics and consider how enzyme activities vary during the growth cycle. While two hydrogenase enzymes in *A. aerogenes* vary little during the exponential growth phase, significant changes in asparagine deaminase specific activity are observed. The initial

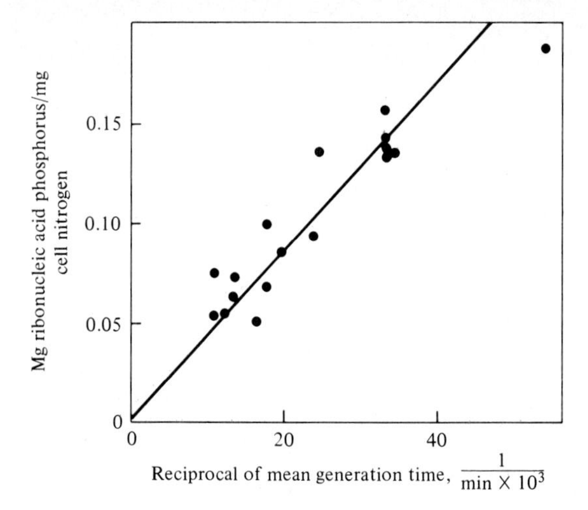

Figure 7.18 These data, which were obtained using a variety of carbon-sources, show a direct correspondence between growth rate and the cells' ribonucleic acid content (*A. aerogenes*). *(Adapted from A. C. R. Dean and C. Hinshelwood, "Growth, Function and Regulation in Bacterial Cells," p. 92, Oxford University Press, London, 1966.)*

drop is attributed to dilution of the inoculum while the rise near the end of the growth cycle is believed due to the decrease in medium pH at that stage.

This observation prompts us to emphasize once more that in a batch culture we must never lose sight of the overall *system* comprising a cell population in a fluid medium of changing composition. In a sense we may view batch growth as a function of the starting conditions of both cell and medium: whatever we observe once the batch process started depends on *both* phases. Thus, it is really somewhat of a misnomer to speak of properties of cell populations in batch growth because these properties are intimately dependent on the interaction between the medium and the population.

7.3.2 Unstructured Batch Growth Models

In the simplest approach to modeling batch culture, we suppose that the rate of increase in cell mass is a function of the cell mass only. Thus

$$\frac{dx}{dt} = f(x) \tag{7.49}$$

As we shall soon see, such a form does not require us to neglect changes occurring in the medium during growth.

One of the simpler models belonging to the general form given in Eq. (7.49) is *Malthus' law*, which uses

$$f(x) = \mu x \tag{7.50}$$

with μ as a constant. We immediately recognize this as the growth rate characteristic of exponential growth, which we have already treated. Malthus' prediction of doom resulting from unbridled population growth has not (yet?) been realized, and transition to a stationary population is generally observed for microbial populations.

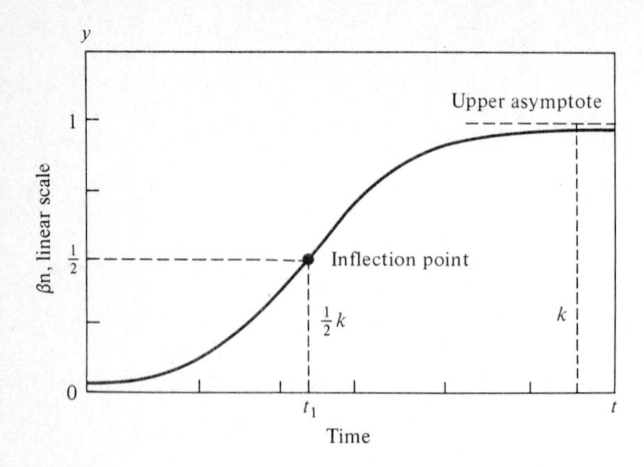

Figure 7.19 The logistic curve ($k > 0$, $\beta > 0$).

Verlhurst in 1844 and Pearl and Reed in 1920 contributed to a theory which included an inhibiting factor to population growth. Assuming that inhibition is proportional to x^2, they used

$$\frac{dx}{dt} = kx(1 - \beta x) \qquad x(0) = x_0 \tag{7.51}$$

This is a Riccati equation which can be easily integrated to give the *logistic curve*

$$x = \frac{x_0 e^{kt}}{1 - \beta x_0(1 - e^{kt})} \tag{7.52}$$

As illustrated schematically in Fig. 7.19, the logistic curve is sigmoidal and leads to a stationary population of size $x_s = 1/\beta$.

One possible interpretation of the logistic curve can be formulated by assuming that the production rate of a toxin is proportional to the population growth rate

$$\frac{dc_t}{dt} = \alpha \frac{dx}{dt} \tag{7.53}$$

so that if

$$c_t(0) = 0 \tag{7.54}$$

$$c_t = \alpha(x - x_0) \tag{7.55}$$

In the usual case where x_0 is negligible relative to x, substitution of Eq. (7.55) into (7.43) gives an equation of the same form as (7.51).

Another class of unstructured models which approach a stationary population level can be formulated based on limiting nutrient exhaustion. Assuming a constant overall yield factor $Y_{X/S}$ and $\mu = \mu(s)$, such as Monod kinetics, nutrient and cell material balances can be combined into a single equation of the form of Eq. (7.49) (see Prob. 7.14).

A drawback of the logistic equation is its failure to predict a phase of decline after the stationary population has exhausted all available resources. This feature is found in one model developed by Volterra early in this century. In this model, an integral term of the form

$$\int_0^t K(t, r)x(r)\,dr \tag{7.56}$$

is added to Eq. (7.50). Physically we may interpret such a term as a crude recognition that the history $K(t, r)$ of a population influences its growth rate. Through the integral in (7.56), all past values of the population density can influence the growth rate at time t. If K is independent of t, a term such as (7.56) may be viewed as representative of the influence of a component of the culture whose concentration follows

$$\frac{dc}{dt} = |K(t)x(t)| \qquad c(0) = 0 \tag{7.57}$$

Specifically, let us suppose that K is a constant equal to K_0 and that the history or memory term (7.57) is added to Eq. (7.51). Then the population size is described by

$$\frac{dx}{dt} = kx(1 - \beta x) + K_0 \left| \int_0^t x(r)\,dr \right| \qquad x(0) = x_0 \tag{7.58}$$

As Eq. (7.53) indicates, the sign of K_0 is taken as negative for an inhibitor and positive for a compound which promotes growth. This problem can be solved numerically with results as given in Fig. 7.20. So long as K_0 is negative, the population size declines after passing through the maximum.

The unstructured growth models we have just examined have several weaknesses. They show no lag phase and give us no insight into the variables which influence growth. Also, they make no attempt to utilize or recognize knowledge about cellular metabolism and regulation. In later sections we shall seek some understanding of how the complex biochemistry of the cell is connected with observed batch-growth phenomena.

7.3.3 Growth of Filamentous Organisms

In the previous portions of this chapter we have been concerned with analysis of microbial populations in which increase in biomass is accompanied by an increase in the number of cells. A different situation prevails in molds and other filamentous organisms, where the mass and morphology of a mold pellet or pulp varies as growth proceeds. Several experimental studies of batch submerged culture have indicated that biomass increases at a slower than exponential rate with proportionality to $(\text{time})^3$ providing a reasonable representation of the data. Such a growth pattern can be rationalized from observations of one- and two-dimensional mold cultures. In the first instance, the rate of increase in the colony

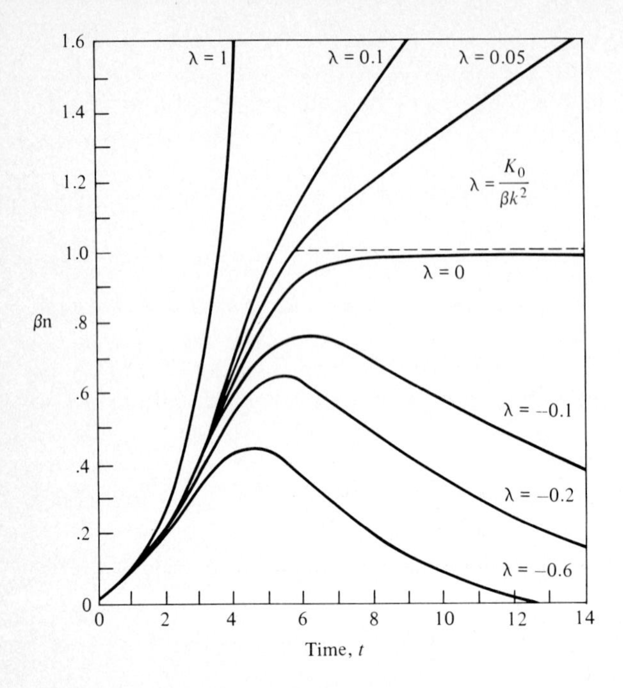

Figure 7.20 Response of the Volterra model, which includes the population history.

length is constant, while the radius of the mold colony increases at a constant rate in surface culture (two-dimensional).

Extrapolating to a spherical pellet growing in submerged culture, let us assume that

$$\frac{dR}{dt} = k_g = \text{const} \tag{7.59}$$

where R now denotes the pellet radius. Since the biomass M is

$$M = \rho\tfrac{4}{3}\pi R^3 \tag{7.60}$$

we have from Eqs. (7.59) and (7.60)

$$\frac{dM}{dt} = \rho 4\pi R^2 \frac{dR}{dt} = k_g 4\pi R^2 \rho \tag{7.61}$$

Eliminating R from (7.61) using (7.60) leaves

$$\frac{dM}{dt} = \gamma M^{2/3} \tag{7.62}$$

where

$$\gamma = k_g[36\pi\rho]^{1/3} \tag{7.63}$$

Integrating Eq. (7.62) with an initial biomass of M_0 yields

$$M = \left(M_0^{1/3} + \frac{\gamma t}{3}\right)^3 \tag{7.64}$$

Since M_0 is usually quite small relative to M, Eq. (7..64) gives the cubic dependence of M on t mentioned above.

A complete analysis of filamentous organism growth should also consider the kinetics of pellet formation. This occurs due to agglomeration of spores and subsequent growth or by outgrowth from an individual spore. Research to date has shown that many properties of the organism and its growth environment interact to influence pellet formation (Fig. 7.21). Because the mechanisms are

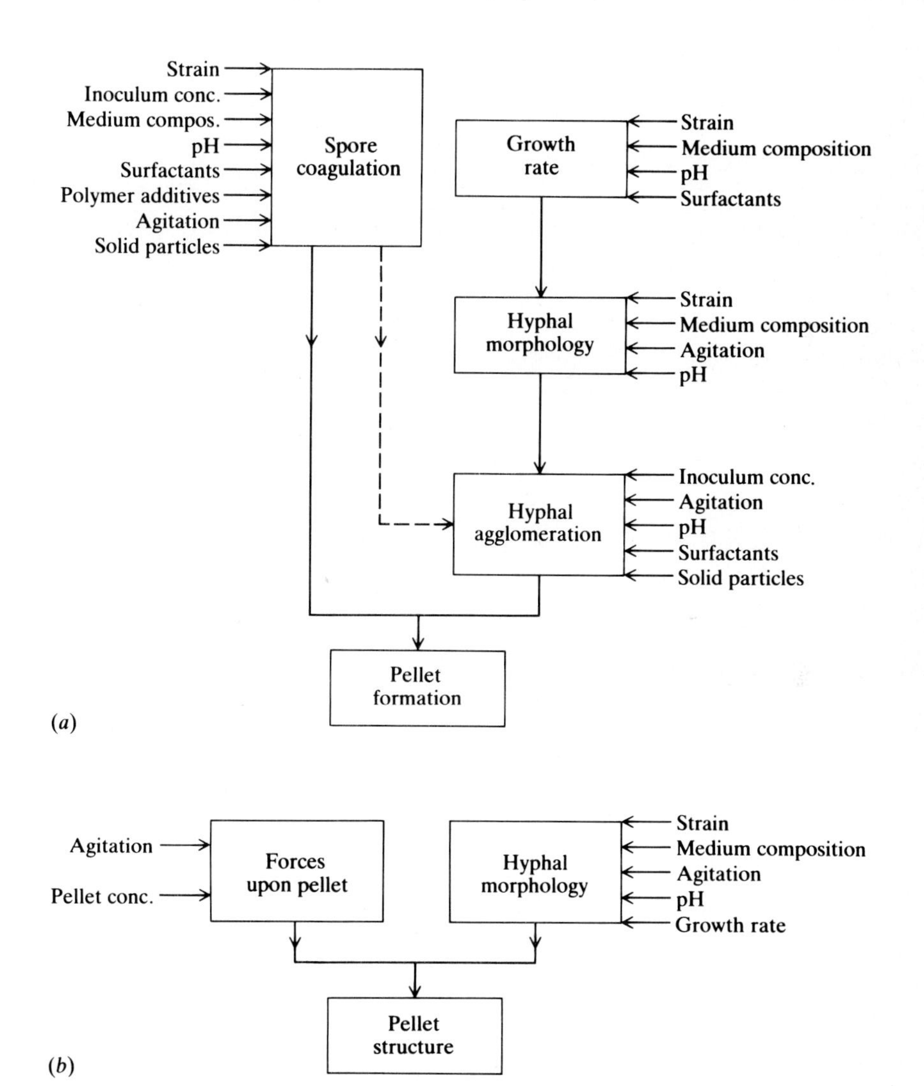

Figure 7.21 Summary of the factors which interact to determine (*a*) pellet formation and (*b*) pellet structure during cultivation of mycelial organisms *(Reprinted by permission from B. Metz and N. W. F. Kossen, "Biotechnology Review: The Growth of Molds in the Form of Pellets—A Literature Review," Biotechnol. Bioeng., vol. 192, p. 781, 1977.)*

complicated and not sufficently well understood or documented, general kinetic models for pellet formation have not yet been developed.

When considering growth of existing pellets, the model outlined above must be viewed as a crude approximation of reality. Pellet size, morphology, and internal structure, all of which are expected to influence pellet kinetics, are determined by interactions among agitation intensity, pellet concentration, organism properties, and medium composition [12]. Overall kinetics of pellets often depend upon diffusion-chemical reaction interactions which were analyzed previously in Sec. 4.4.

7.4 STRUCTURED KINETIC MODELS

Unstructured models of the types considered above describe only the quantity of the biological phase. Such models do not recognize nor represent the composition or what we might call the quality of the biophase. In situations in which the cell population composition changes significantly and in which these composition changes influence kinetics, structured models should be used. Since it is not practical to write material balances on every cell component, we must select skillfully the key variables and processes of major interest in a particular application when formulating a (necessarily approximate) structured kinetic model. As will be illustrated in the examples presented in the following sections, there are several different conceptual bases on which structured models may be built.

The biophase variables employed in structured models are typically the mass (x_j) or molar (c_j) concentrations *per unit volume of biophase* [15]. For a well-mixed reactor, writing a material balance on component j gives

$$\frac{d}{dt}\left[\frac{1}{\rho_c} V_R x c_j\right] = \frac{1}{\rho_c} V_R x r_{f_j} + \frac{1}{\rho_c} \Phi_x c_j \tag{7.65}$$

where ρ_c = mass density of cells (cell mass/unit volume cells)
r_{f_j} = molar rate of formation of component j (moles $j \cdot$ time$^{-1} \cdot$ (unit volume cells)$^{-1}$)
Φ_x = mass of cells added to the reactor by flow $\cdot$ time^{-1}
V_R = culture volume
x = cell mass concentration (cell mass/unit volume culture)
c_j = moles j/unit volume cells.

In writing the last term in Eq. (7.65), it is assumed that any cells added (for example, by recycle) or removed from the reactor have the same state as the cell population in the reactor. If this condition is not met (for example, in the second or later tank in a CSTR cascade), the balance equation must be modified.

Assuming that the density of cells ρ_c and volume of culture V_R are time-invariant, Eq. (7.65) may be rewritten in the following form after carrying out the indicated differentiation:

$$\frac{dc_j}{dt} = r_{f_j} - c_j\left(\frac{1}{x}\frac{dx}{dt}\right) + c_j\frac{\Phi_x}{xV_R} \tag{7.66}$$

For a batch reactor, Φ_x is zero, and the quantity in parenthesis on the right-hand side of Eq. (7.66) is simply the specific growth rate μ, giving

$$\frac{dc_j}{dt} = r_{f_j} - \mu c_j \tag{7.67}$$

The physical interpretation of the r_{f_j} term is clear; the $-\mu c_j$ term represents reduction in concentration caused by dilution from growth of the cell population. Evaluation of the terms on the right-hand side of Eq. (7.66) for a CSTR with sterile feed gives, interestingly, Eq. (7.67) again.

7.4.1 Compartmental Models

In the simplest structured models, the biomass is compartmentalized into a small number of components. Sometimes these components have an approximate biochemical definition, as in a synthetic component (RNA and precursors) and a structural component (DNA and proteins). Alternatively, the compartments may be defined as an assimilatory component and a synthetic component. Another interpretation which has been used to guide formulation of simple compartmental models is that of a bottleneck in metabolism.

Structured models containing a small number of variables may also be justified on the basis of some concepts from systems dynamics. Each class of reaction and transport events which occur in a cell population—molecular collisions, chemical reactions, diffusion, turnover of RNA, protein synthesis, increase in cell number, completion of a batch process, spontaneous mutation—has a characteristic *relaxation time*, or time to reach steady state after a perturbation (recall Sec. 3.3.1). These relaxation times range from fractions of a second to hours. What is important in modeling, from a systems dynamics viewpoint, is the relationship between the time scale of changes in cells' environment (τ_E) and the spectrum of relaxation times of cellular processes. Those cellular processes that respond very fast to environmental changes (small relaxation time/τ_E ratio) may be assumed to be in quasi-steady state (Sec. 3.2.1). For the other extreme, where the relaxation time for some cellular process is very large compared to τ_E, that class of cellular processes may be assumed to be "frozen" at the initial state. For example, significant genetic changes are not usually expected during a single batch cultivation.

It has been found for many complicated systems that there are typically only two or three system relaxation times in the same range as the time scale for environmental changes, justifying approximating the dynamics of a large system using a low order model with only two or three system variables. However, it is often difficult or impossible to relate some of these variables to measurable, physical quantities characteristic of the system. We consider next some examples of simple, compartmental structured kinetic models of cell population growth.

Williams [16] proposed a two-compartment model which reproduces several aspects of batch growth dynamics surprisingly well. The major postulates of this model are the following:

1. The synthetic (1) portion is produced by uptake of external nutrient S with (component 1 mass/substrate mass) yield coefficient Y. The rate of cell synthetic component formation is first order in total cell density x (cell mass/culture volume) and nutrient mass concentrations (substrate mass/culture volume).
2. The structural-genetic (2) cell component is produced from component 1 at a rate proportional to $\rho_1 \cdot \rho_2 (\rho_i =$ mass i/unit volume of cells).
3. Doubling of component 2 is necessary and sufficient for cell division. Accordingly, the cell number density is proportional to the density of component 2 in the culture.
4. Biomass is comprised solely of components 1 and 2.

Restated in equation form for a well-mixed batch reactor, the model is

$$\frac{ds}{dt} = -\frac{1}{Y} k_1 s x \tag{7.68}$$

$$\frac{dx}{dt} = k_1 s x \tag{7.69}$$

$$\frac{d\rho_1}{dt} = k_1 s(\rho_1 + \rho_2) - k_2 \rho_1 \rho_2 - \mu \rho_1 \tag{7.70}$$

$$\frac{d\rho_2}{dt} = k_2 \rho_1 \rho_2 - \mu \rho_2 \tag{7.71}$$

Noting that

$$\mu = k_1 s \tag{7.72}$$

and

$$\rho_1 + \rho_2 = \rho_c = \text{const} \tag{7.73}$$

Eq. (7.71) may be rewritten as

$$\frac{df_2}{dt} = -k_1 s f_2 + (k_2 \rho_c) f_2 (1 - f_2) \tag{7.74}$$

where f_2 is ρ_2/ρ_c, the fraction of the cell mass comprised of component 2. The first two equations in this model can be used to write s in terms of x. When this result is substituted in Eq. (7.69), an equation identical in form to Eq. (7.51) is obtained. Consequently, x will follow the logistic equation given above in Eq. (7.52).

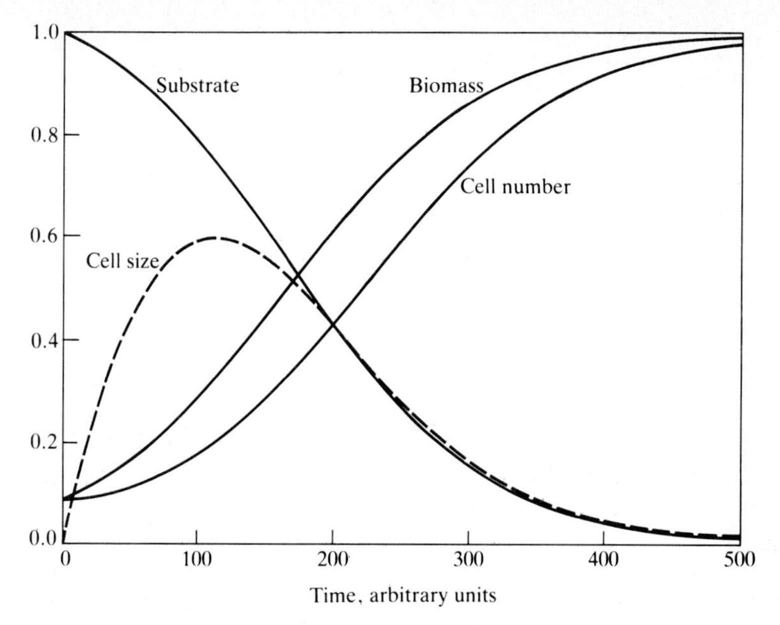

Figure 7.22 Batch growth simulation with stationary phase inoculum based on Williams' model. Ordinate shows dimensionless substrate (s/s_0), biomass (x/Ys_0), and cell number (f_2x/Ys_0) concentrations. Dimensionless cell size increase relative to stationary cells ($= -1 + 1/f_2$) is also shown. (Parameters: $k_1 = 0.0125$, $k_2\rho_c = 0.025$, $Y = 0.5$, $x_0 = 0.05$, $s_0 = 1, f_2(0) = 1$.)

The model has been used to simulate batch growth using a stationary, nutrient-exhausted inoculum, which corresponds to $\rho_1 = 0$. The results, given in Fig. 7.22, mimic several frequently observed features of batch microbial cultures including

1. Existence of a lag phase in which cell size increases.
2. An exponential-growth phase where cell size is maximum.
3. A change in composition of the cells during the growth cycle. Since these changes are evident even during the exponential phase, apparently growth is never exactly balanced.
4. A stationary phase with relatively small cells.

Notice in particular that the changes with time in cell mass density and in cell number density are *not* parallel nor proportional; that is, single-cell mass is *not* constant but depends on growth rate and even on growth-rate history. Detailed studies of microbial population growth based on the *E. coli* and *S. cerevisiae* cell cycles have shown similar results [30, 31]. Based on these cases, it appears that the dynamics of cell mass changes may often be accurately approximated by relatively simple models, while calculating the dynamics of cell numbers is much more complicated.

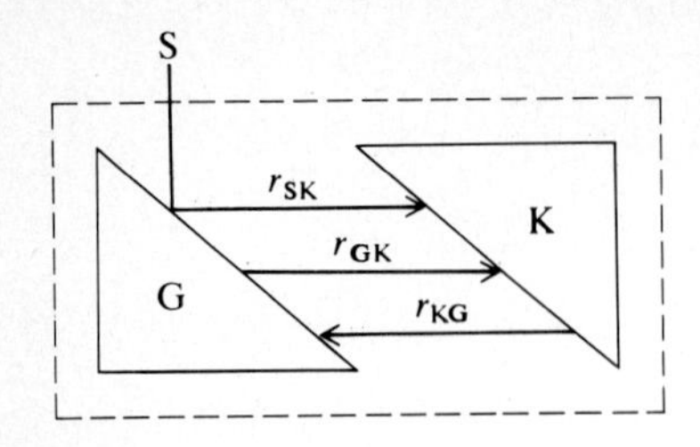

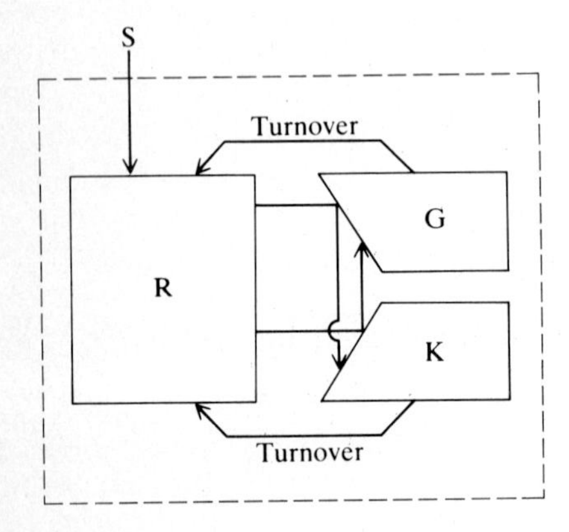

Figure 7.23 Schematic diagrams of the two-and three-compartment models described by Harder and Roels [17].

Some elaborations on the above model are summarized in the sketches in Fig. 7.23. In the two-compartment model of Harder and Roels [17], the G component corresponds to enzymes which catalyze conversion of substrate to biopolymer building blocks, and the balance of the cells' constituents are in component K which is formed from nutrient S at a rate which depends on the quantity of G. Interconversion of G and K by polymerization and depolymerization reactions introduces maintenance metabolism into this simple two-component cell. The three-component cell in Fig. 7.23 has constituents with more well-defined biochemical identity: K denotes RNA, G protein, and R the remaining biomass [17]. Turnover of the K and G compartments adds features of maintenance to the model.

Compared with some of the more elaborate structured models examined below, these small compartmental models are relatively uncomplicated mathematically and contain relatively few kinetic parameters. These models have been used with some success in describing different unbalanced growth situations. On the other hand, the absence in some cases of a clear biochemical definition of the compartment components makes model validation and parameter estimation more difficult. With such models, we cannot make use of known features of the cell's metabolic pathways, molecular controls and cell cycle operation. To use this knowledge in kinetic modeling, cell representations with more structure—that is, more components and intracellular reactions and interactions—are required.

7.4.2 Metabolic Models

In this section we shall summarize and examine the attributes of two different structured kinetic models which incorporate some aspects of cell metabolism. These illustrate some of the benefits and drawbacks of more detailed kinetic models. These examples also reveal that, as we incorporate more biological detail in a model, the model becomes more specific to a particular organism or process. The key metabolic features of the particular system being modeled, sometimes available from the biological sciences or biotechnology literature, must be known and must be considered when formulating the kinetic model. In a certain sense, the more detailed the model becomes, the more we must know a priori about the organism. Otherwise, there is too much freedom in proposing kinetic forms and parameter values, and these many unknown model features cannot be determined well from (limited) experimental data. Although some of these and other highly structured models outlined later are based on a single-cell viewpoint, the deterministic framework usually adopted implies treatment of the average cell behavior in a large population of cells.

The first metabolic model we shall examine is the result of studies of Bijkerk and Hall [18] and Pamment, Hall, and Barford [19] on aerobic growth of the budding yeast *S. cerevisiae.* The model is based on the following assumptions, most of which derive from known cell cycle control, metabolic, and enzyme expression regulation features of the organism.

Model assumptions

1. Segregated, single cell units are not considered.
2. The limiting substrate S is both carbon and energy source; E denotes ethanol.
3. Biomass consists of two parts, A and B.
4. A mass conducts substrate uptake and energy production; B mass does cell synthesis and division.
5. Accumulation of energy and metabolites occurs during the G_1 phase of the cell cycle, is carried out by A mass, and is described by conversion of A mass to B mass [see Eqs. (7.75) and (7.76) below.]
6. DNA replication, mitosis, and cell division (S, G_2, and M phases) occur over a fixed time interval, are carried out by B mass, and described by conversion of B mass to A mass [Eq. (7.77) below].
7. All of the enzymes in the fermentative pathway are grouped together and represented as E_f. The respiratory pathway enzymes, denoted E_r, are similarly treated.
8. Two different control modes influence production of each enzyme system. First, enzymes are produced at rates proportional to the current metabolic flux through that enzyme system. Second, there is an additional adaptive production of each enzyme system which is proportional to the difference between the existing enzyme concentration and a "target" value (e_F and e_R

for fermentation and respiration, respectively). The target values are proportional to the flux which would occur if enzyme concentrations were not limiting.

9. The rates of glycolysis and respiration are linearly dependent on e_f/e_F and e_r/e_R, respectively.
10. Glycolytic enzymes are produced as a consequence of flux through the respiratory pathway (to provide enzymes for gluconeogenesis, biosynthesis and utilization of intracellular carbohydrate reserves).

This viewpoint of growth can be restated by the following stoichiometry (subscripts W here denote variables which are treated here and later in mass units; otherwise molar units, or activity units in the case of enzymes, are assumed):

Fermentation:

$$A_W + a_1 S_W + \mathrm{E}_f \xrightarrow{r_A} 2B_W + a_2 \mathrm{E}_W + \mathrm{E}_f + \mathrm{CO}_2 \tag{7.75}$$

Respiration:

$$A_W + a_3 \mathrm{E}_W + \mathrm{O}_2 + \mathrm{E}_r + \mathrm{E}_f \xrightarrow{r_B} 2B_W + \mathrm{E}_r + \mathrm{E}_f + \mathrm{CO}_2 \tag{7.76}$$

Division:

$$B_W \xrightarrow{r_C} A_W \tag{7.77}$$

The rates r_A, r_B, r_C are given by

$$r_A = \frac{k_1 as}{K_s + s}\left(\frac{e_f}{e_F}\right) \tag{7.78}$$

$$r_B = \frac{k_2 ae}{K_E + e}\left(\frac{e_r}{e_R}\right) \tag{7.79}$$

$$r_C = Kb \tag{7.80}$$

and the target enzyme concentration are

$$e_R = \frac{k_h k_2 ae}{K_E + e} \tag{7.81}$$

$$e_F = \frac{k_f k_1 as}{K_s + s} + \frac{k_7 k_2 ae}{K_E + e} \tag{7.82}$$

The rates of fermentative and respiratory enzyme production are taken to be

$$(r_f)_{\text{fermentative enzymes}} = r_A + k_5 r_c + k_4(e_F - e_f) \tag{7.83}$$

$$(r_f)_{\text{respiratory enzymes}} = r_B + k_6(e_R - e_r) \tag{7.84}$$

Table 7.3 Model parameters for aerobic growth of *S. cerevisiae* (Ref. 19)

a_1 (g/g)	5.95	K_S (g/L)	0.50
a_2 (g/g)	2.50	K_E (g/L)	0.02
a_3 (g/g)	1.50	k_4 (h^{-1})	0.00
k_1 (h^{-1})	5.00	k_5 (h^{-1})	0.075
k_2 (h^{-1})	0.26	k_6 (h^{-1})	0.225
K (h^{-1})	0.50	k_7 (h^{-1})	1.33
k_f (h)	2.37†	s_f (g/L)	9.20
k_h (h)	0.244†		

† The values of k_f and k_h have been calculated from other model parameters using equations from Ref. 19.

respectively. In these equations, e_r and e_f denote enzyme activity per unit volume of culture. Consequently, this structured model is *not* based on "intrinsic" or intracellular concentrations as discussed above.

Based upon comparison of model simulations and data from batch cultivation experiments, the parameter values listed in Table 7.3 were identified. Only the parameter a_3, the yield coefficient for respiratory growth, was modified slightly (reduced from 1.79 in batch culture to 1.50 for continuous cultivation) in application of this kinetic model to steady-state continuous cultivation experiments. This is based in part on known physiology of this organism, since batch cultures grown on ethanol require the gluconeogenic pathway which is not required in a CSTR fed with glucose.

Figure 7.24*a* shows the calculated cell mass density $x(= a + b)$ and specific activities of fermentation (e_f/x) and respiratory (e_r/x) enzymes as functions of dilution rate for steady-state continuous culture of *S. cerevisiae* using the parameters in Table 7.3. The solid squares are experimental measurements of x which closely match the model results. Notice that the experimental and calculated D-x relationship is much different here from the simpler form obtained from Monod's unstructured model (Figure 7.6). Figure 7.24*b* gives experimental data on the change in specific activity of two fermentative pathway-associated enzymes, $NADP^+$-associated glutamate dehydrogenase and alcohol dehydrogenase, with CSTR dilution rate. The trends are qualitatively identical to those in the model calculation of e_f/x. Dilution rate effects on specific activities of two respiration-associated enzymes, malate dehydrogenase and isocitrate lyase, obtained from experiment (Fig. 7.24*c*) exhibit a maximum with respect to D as the model simulations indicate for e_r/x.

These results and others in Ref. 18 clearly demonstrate some of the benefits in applying more structured descriptions of cell population growth. The above model has successfully simulated fed-batch cultures (see Sec. 9.1.1) of *S. cerevisiae* and also describes well lag times in batch growth experiments.

As a final example of this type, and one which represents the most comprehensive structured model so far proposed for a microbial system, we consider the

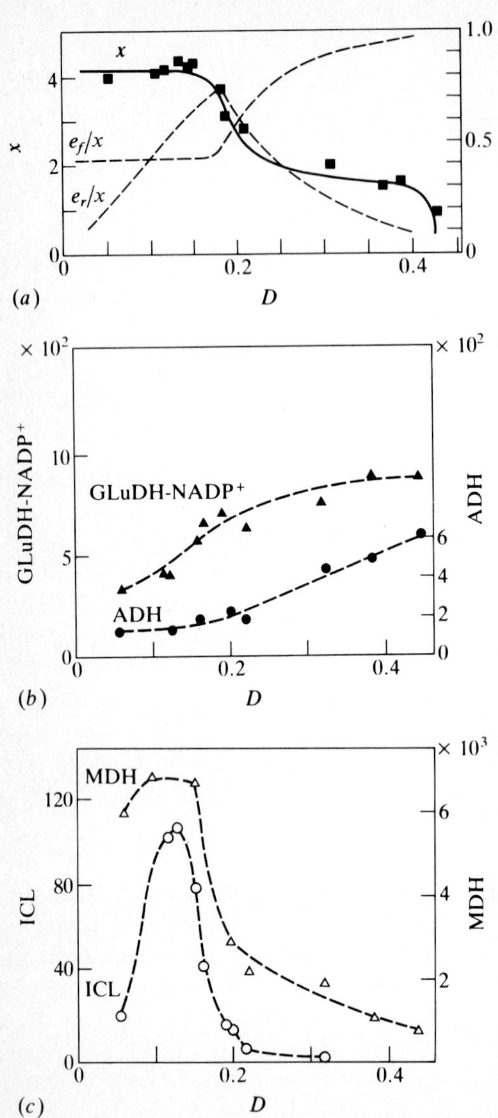

Figure 7.24 (a) Model calculations of *S. cerevisiae* cell mass density x (solid line) and specific enzyme activities e_f/x and e_r/x (dashed lines) versus dilution rate $D(\mathrm{h}^{-1})$ in a steady-state CSTR. Points show experimentally measured values of x. (b) Experimental data on specific activities of the typical fermentative enzymes, $NADP^+$-associated glutamate dehydrogenase (GluDH—$NADP^+$), and alcohol dehydrogenase (ADH). (c) Experimentally measured specific activities of respiration-associated enzymes malate dehydrogenase (MDH) and isocitrate lyase. *(Reprinted by permission from N. B. Pamment, R. J. Hall, and J. P. Barford, "Mathematical Modeling of Lag Phases in Microbial Growth," Biotechnol. Bioeng., vol. 20, p. 345, 1978.)*

single-cell model for *E. coli* formulated by Shuler and colleagues [24]. Figure 7.25 shows schematically the pooled metabolites and biopolymers which are included in this model and the reactions (solid lines) which are presumed to occur in *E. coli* B/r A. Dashed lines denote information flow by which reaction kinetics are modulated. The model includes cross-wall formation reactions so that the simulated cell generation time is a natural *output* of this model rather than an input as it is for many other structured models. Since this model has been formulated to describe growth under either carbon or nitrogen source limitation, substantial structure relating to transport and assimilation of these nutrients is

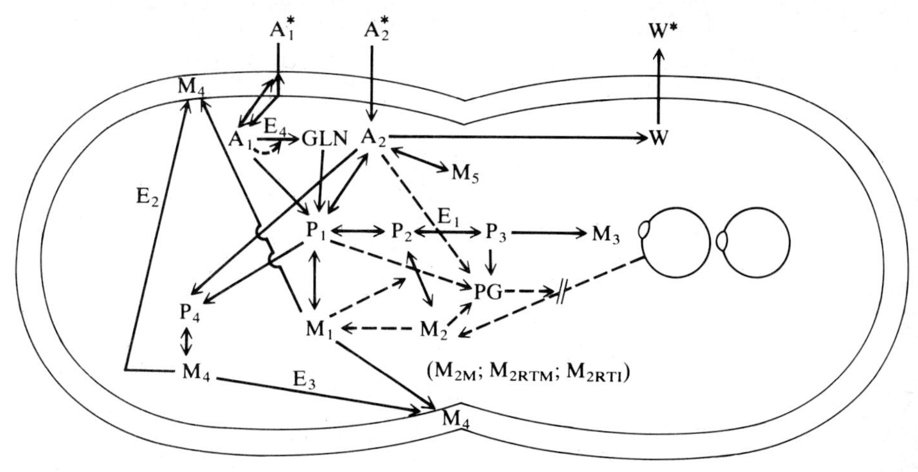

A_1 = ammonium ion
A_2 = glucose (and associated compounds in the cell)
W = waste products (CO_2, H_2O, and acetate) formed from energy metabolism during aerobic growth
P_1 = amino acids
P_2 = ribonucleotides
P_3 = deoxyribonucleotides
P_4 = cell envelope precursors
M_1 = protein (both cytoplasmic and envelope)
M_{2RTI} = immature "stable" RNA
M_{2RTM} = mature "stable" RNA (rRNA and tRNA—assume 85% rRNA throughout)
M_{2M} = messenger RNA
M_3 = DNA
M_4 = nonprotein part of cell envelope (assume 16.7% peptidoglycan, 47.6% lipid, and 35.7% polysaccharide)
M_5 = glycogen
PG = ppGpp
E_1 = enzymes in the conversion of P_2 to P_3
E_2, E_3 = molecules involved in directing cross-wall formation and cell envelope synthesis
GLN = glutamine
E_4 = glutamine synthetase
*—the material is present in the external environment

Figure 7.25 Schematic diagram of a highly structured single-cell model for growth of *E. coli* B/r A on glucose-ammonium salts medium. *(Reprinted by permission from M. L. Shuler and M. M. Domach, "Mathematical Models of the Growth of Individual Cells," in H. W. Blanch, E. T. Papoutsakis and G. Stephanopoulos (eds.), "Foundations of Biochemical Engineering," p. 101, American Chemical Society, Washington, 1983.)*

included. ATP generation and utilization is considered in writing the reactions in Figure 7.25.

Not represented in Figure 7.25 are important model details concerning regulation of initiation of DNA synthesis. A repressor protein RP is synthesized in a short period ("burst synthesis") and is neutralized by an antirepressor protein ARP which is synthesized at a rate proportional to cell envelope synthesis. When the free RP level is sufficiently low, transcription forming O-RNA, required for replication initiation, begins.

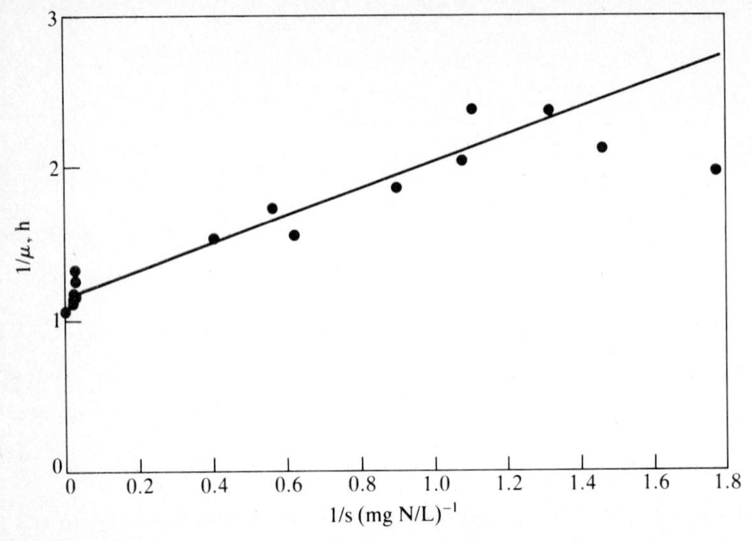

Figure 7.26 Double reciprocal plot of specific growth rate versus ammonium ion concentration for nitrogen-limited growth of *E. coli* B/r A: (———, model calculations; ●, experimental data). *(Reprinted by permission from M. L. Shuler and M. M. Domach, "Mathematical Models of the Growth of Individual Cells," in H. W. Blanch, E. T. Papoutsakis and G. Stephanopoulos (eds.), "Foundations of Biochemical Engineering," p. 101, American Chemical Society, Washington, 1983.)*

Unfortunately, it is not possible to describe this model in full in a text of this scope. Reference 20 should be consulted for further details. The model contains of the order of 100 stoichiometric and kinetic parameters, almost all of which can be determined from previous literature on the biochemistry of *E. coli* growth. While the required information for formulating a model like this is great, so too are the model's capabilities.

The model simulates reasonably well the time of chromosome replication initiation and other key cell cycle properties over a broad range of growth rates. Shown in Fig. 7.26 is a double reciprocal plot of limiting nutrient, here ammonium ion, concentration vs. single cell specific growth rate for the model (line) and from experiment (points). Notice that both model and experiment indicate multiple mechanisms for ammonium utilization, manifested here by the change in slope at very large nutrient concentrations (small $1/s$). The growth rate, cell glycogen content, and cell size calculated from the model (Fig. 7.27) agree very well with experimental results. Williams' model described earlier predicts larger cells at larger growth rates but the present model, with much more biochemical and metabolic structure, is able to show that cell size is also a function of which nutrient is growth-rate limiting.

7.4.3 Modeling Cell Growth as an Optimum Process

Studies of microbial growth on mixtures of different substrates have shown that the organisms typically consume preferentially the substrate providing most

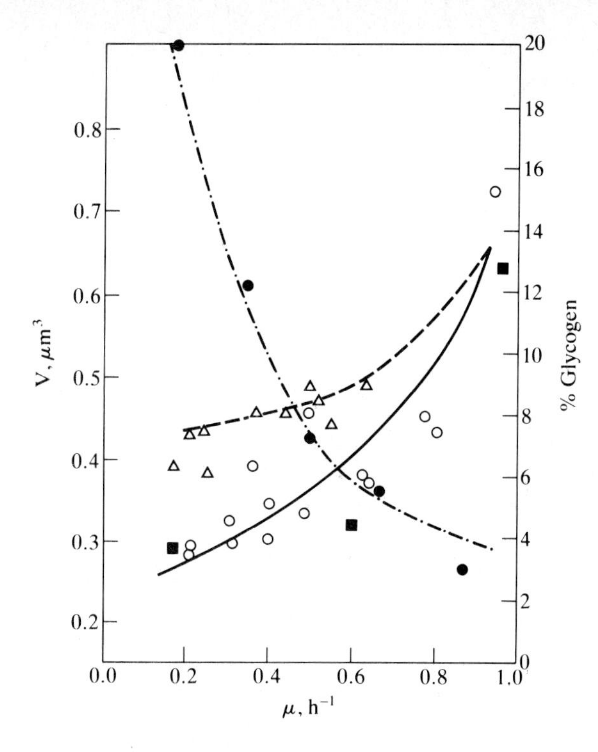

Figure 7.27 Comparison of single-cell model results with experimental data: ———; cell volume from model, glucose-limited growth (○, ■ = expt.); - - - ; cell volume from model, ammonium-limited growth (△ = expt.); -·-·-, % glycogen from model, ammonium-limited growth (● = expt.) *[Reprinted by permission from M. L. Shuler and M. M. Domach, "Mathematical Models of the Growth of Individual Cells," in H. W. Blanch, E. T. Papoutsakis, and G. Stephanopoulos (eds.), "Foundations of Biochemical Engineering," p. 101, American Chemical Society, Washington, 1983.)*

rapid growth. The molecular bases for this kinetic behavior are the induction, repression, inhibition and activation processes discussed in Chaps. 3 and 6. Although kinetic models may be formulated based on these mechanisms (Sec. 7.5.3), many parameters and detailed knowledge of the mechanism are required. An alternative approach, developed by Ramkrishna and colleagues, describes the effects of cellular regulatory processes as the outcome of an optimization strategy [21, 22]. The central idea here is that, through natural selection and evolution, biological systems have acquired the ability to exploit their environment in an optimum fashion. From this *cybernetic model* perspective, the actions of metabolic control systems are deducible as the optimum solution to a problem of resource allocation for achieving maximal growth.

An optimization algorithm based on the matching law for resource allocation has proved quite useful in modeling cell growth on mixed substrates. If the return from utilization of resource Z_i is $\Theta_i(Z_i)$ and the total return (= sum of Θ_i) is maximized subject to the constraint that the sum of Z_i's is equal to a fixed value, the following matching law must be satisfied:

$$\frac{d\Theta_1}{dZ_1} = \frac{d\Theta_2}{dZ_2} = \frac{d\Theta_S}{dZ_S} = \cdots \tag{7.85}$$

Suppose that the population growth rate depends upon the intracellular concentrations $e_1, e_2, \ldots$ of a set of enzymes which are required for assimilation of substrates $S_1, S_2, \ldots$, respectively. In particular, we shall write the contribution of substrate S_i to biomass formation as

$$r_{x_i} = r_i e_i v_i \tag{7.86}$$

where

$$r_i = \frac{\mu_i s_i x}{K_i + s_i} \tag{7.87}$$

The weighting factor v_i is determined by matching law consideration of the energy yield of consumption of different substrates which suggests

$$v_i = \frac{\nu_i r_i e_i}{\sum_j \nu_j r_j e_j} \tag{7.88}$$

where ν_j is the energy released per unit mass of substrate j consumed. Applying the matching law to allocation of cellular resources for enzyme synthesis, the rate of synthesis of enzyme E_i is taken to be

$$r_{f,E_i} = \frac{\alpha_i s_i x}{K_{E_i}} + s_i \mu_i \tag{7.89}$$

where

$$\mu_i = \frac{r_i e_i}{\sum_j r_j e_j} \tag{7.90}$$

A significant attribute of this model is the ability to determine all kinetic parameters from experiments using single substrates. All aspects of population

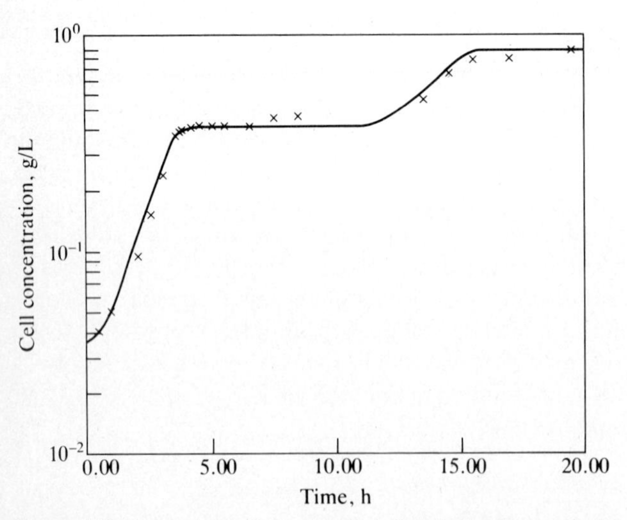

Figure 7.28 Comparison of experimental diauxic batch growth data (×) for *Klebsiella oxyloca* in arabinose-lactose medium with cybernetic model calculations (—). *(Reprinted by permission from D. S. Kompala, D. Ramkrishna, and G. T. Tsao, "Cybernetic Modeling of Microbial Growth on Multiple Substrates," Biotechnol. Bioeng,. vol. 26, p. 1272, 1984.)*

growth kinetics in mixed substrate environments are implied by the optimization strategy. Fig. 7.28 shows model results and experimental data for growth of the bacterium *Klebsiella oxyloca* in a medium containing arabinose and lactose. Corresponding model parameters, identified in separate experiments, are listed in Ref. 22. Here the model successfully simulates a long lag time between growth on lactose and slower growth on arabinose. Other investigations of models of this class have shown excellent qualitative and quantitative simulations of triauxic growth on three substrates in batch culture and of steady-state and transient behavior of microbial CSTRs. Extensions of the cybernetic approach considering other optimization strategies and applications to product formation should prove interesting. This modeling methodology poses an intriguing alternative to models which represent metabolism as a simplified chemical reaction network.

7.5 PRODUCT FORMATION KINETICS

The types of kinetic descriptions that may be employed for product formation by cell populations parallel those used to describe cell population growth. Structured and unstructured approaches are available. As we shall see in Sec. 7.5.3, it is possible to formulate useful models of protein synthesis kinetics at the molecular level, taking advantage of contemporary understanding of molecular controls.

7.5.1 Unstructured Models

The simplest types of product formation kinetics arise when there is a simple stoichiometric connection between product formation and substrate utilization or cell growth. Then, the product formation rate r_{f_p} may be written

$$r_{f_p} = -Y_{\mathrm{P/S}} r_{f_s} \tag{7.91}$$

or

$$r_{f_p} = Y_{\mathrm{P/X}} r_{f_x} \tag{7.92}$$

respectively. Such cases arise in type I fermentations discussed in Sec. 5.9.3. The alcohol fermentation, shown in batch culture in Figure 7.29, is an example of this class. Such product formation kinetics are sometimes called *growth associated.*

In many fermentations, especially those involving secondary metabolites, significant product formation does not occur until relatively late in a batch cultivation, perhaps approaching or into stationary phase. The penicillin fermentation data in Fig. 7.30 exemplify such behavior. Occasionally a simple *nongrowth associated* model suffices for product formation kinetics in such cases. In such a model, production rate is proportional to cell concentration rather than growth rate.

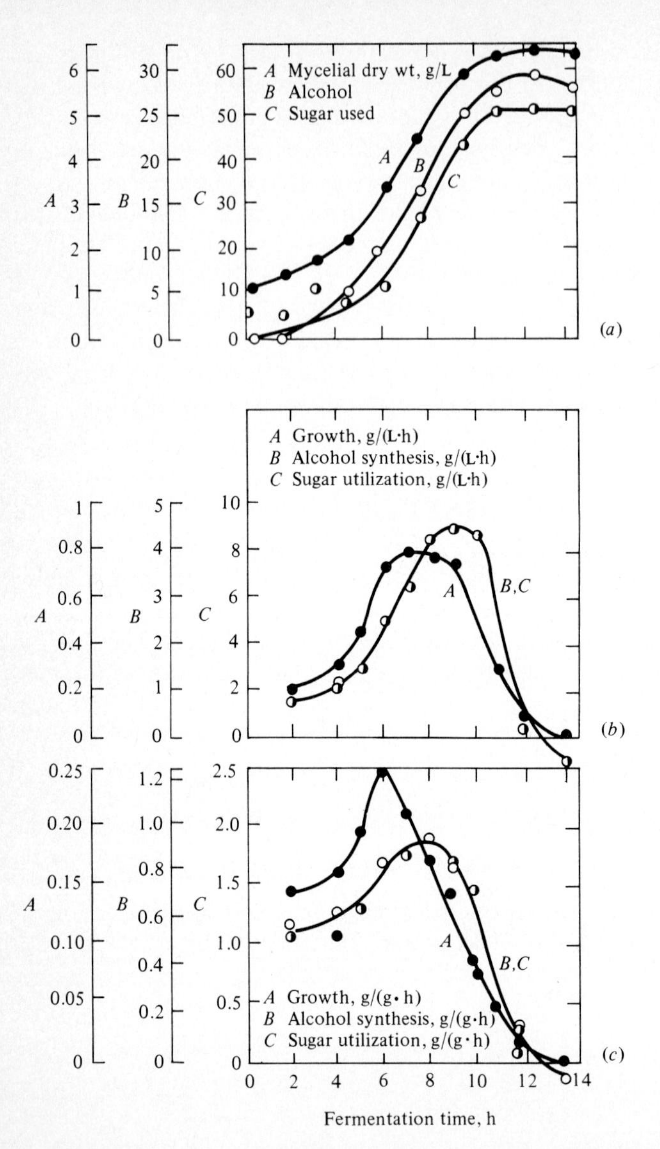

Figure 7.29 The alcohol fermentation exhibits simple growth-associated product formation kinetics (*a*) time course, (*b*) volumetric rates, and (*c*) specific rates. *[Reprinted from R. Leudeking, "Fermentation Process Kinetics," in N. Blakebrough (ed.), "Biochemical and Biological Engineering," vol. 1. p. 203, Academic Press, Inc. (London) Ltd., London, 1967.]*

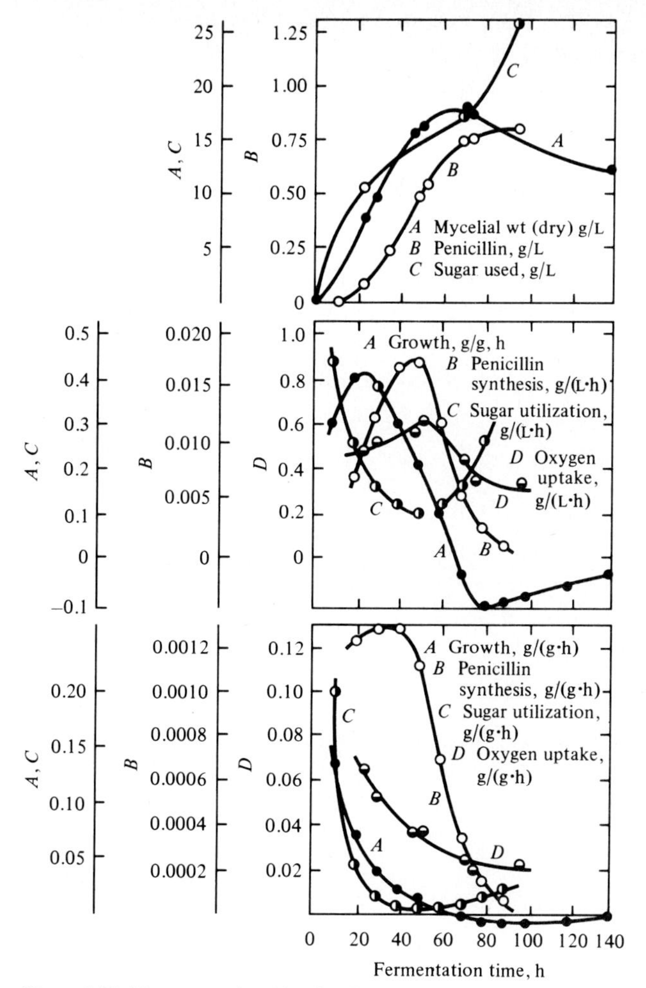

Figure 7.30 More complex kinetics for production of penicillin, a secondary metabolite: (*a*) time course, (*b*) volumetric rates, and (*c*) specific rates. *[Reprinted from R. Leudeking, "Fermentation Process Kinetics," in N. Blakebrough (ed.), "Biochemical and Biological Engineering," vol. 1, p. 205, Academic Press, Inc. (London) Ltd., London, 1967.]*

The now classic study of Leudeking and Piret† on the lactic acid fermentation by *Lactobacillus delbruekii* indicated product formation kinetics which combined growth-associated and nongrowth-associated contributions:

$$r_{f_p} = \alpha r_{f_x} + \beta x \tag{7.93}$$

This two-parameter kinetic expression, often termed *Leudeking-Piret kinetics*, has proved extremely useful and versatile in fitting product formation data from

†R. Leudeking and E. L. Piret, A Kinetic Study of the Lactic Acid Fermentation, *J. Biochem. Microbiol. Technol. Eng.*, **1**: 393, 1959.

many different fermentations. This is an expected kinetic form when the product is the result of energy-yielding metabolism, as in several anaerobic fermentations. In this case the first and second terms in Eq. (7.93) may be identified with energy used for growth and for maintenance, respectively (see, for example, Ref. 5).

Example 7.1: Sequential Parameter Estimation for a Simple Batch Fermentation We consider here a batch fermentation model in which growth is described by the logistic Eq. (7.51) and product formation is described by Leudeking-Piret kinetics [Eq. 7.93)]

$$\frac{dp}{dt} = \alpha \frac{dx}{dt} + \beta x \tag{7E1.1}$$

Substrate utilization kinetics are given by the following equation which considers substrate conversion to cell mass, to product, and substrate consumption for maintenance:

$$\frac{ds}{dt} = -\frac{1}{Y'_{X/S}}\frac{dx}{dt} - \frac{1}{Y'_{P/S}}\frac{dp}{dt} - k_e x \tag{7E1.2}$$

Substituting Eq. (7.91) into Eq. (7E1.2) allows the substrate material balance to be rewritten as

$$\frac{ds}{dt} = -\gamma \frac{dx}{dt} - \eta x \tag{7E1.3}$$

where

$$\gamma = \frac{1}{Y'_{X/S}} + \frac{\alpha}{Y'_{P/S}} \qquad \eta = \frac{\beta}{Y'_{P/S}} + k_e \tag{7E1.4}$$

This and similar models can describe many fermentations of practical interest including those in which multiple products are formed [25]. Also, a convenient set of graphical methods for sequential estimation of model parameters has been presented which is summarized next.

Rearranging Eq. (7.52) gives

$$\ln\left[\frac{x(t)/x_s}{1 - x(t)/x_s}\right] = kt - \ln\left[\frac{x_s}{x_0} - 1\right] \tag{7E1.5}$$

showing that, after x_s is determined by inspection, a plot of

$$\ln\left(\frac{x(t)/x_s}{1 - x(t)/x_s}\right)$$

vs. t has a slope of k and an intercept which can be used to evaluate x_0. Turning next to product formation kinetic parameters, Leudeking-Piret kinetics in the stationary phase of batch culture imply

$$\beta = \frac{(dp/dt)_{\text{stationary phase}}}{x_s} \tag{7E1.6}$$

Integrating Eq. (7E1.1) with x given by Eq. (7.51) gives

$$p(t) - p_0 - \beta\left(\frac{x_s}{k}\right)\left[1 - \frac{x_0}{x_s}(1 - e^{kt})\right] = \alpha[x(t) - x_0] \tag{7E1.7}$$

allowing determination of α by a plot of the left-hand side of Eq. (7E1.7) vs. $[x(t) - x_0]$. The parameters γ and η in the substrate utilization equation are evaluated in the same manner as presented for α and β.

Table 7E1.1 Kinetic model parameter values for several different fermentations producing extracellular polysaccharides

(Reprinted by permission from D. F. Ollis, "A Simple Batch Fermentation Model: Theme and Variations," Annals N.Y. Acad. Sci., vol. 413, p. 144, 1983).

	Product	
Fermentation	β	α
Xanthan gum	0.155 gP/gX-h	1.83 gP/gX
Pullulan		
(pH 4.5)	0	89 wt%/(g-day/100mL)
(pH 5.5)	0	135 wt%/(g-day/100mL)
(pH 6.5)	0	110 wt%/(g-day/100mL)
Alginic acid	0	1.60 gP/gX
Pseudomonas sp	10^{-3} gP/ODU-h	0

	Substrate		Biomass
Fermentation	γ	η	k
Xanthan gum	2.0 gS/gX	0.284 gS/gX-h	0.15 h^{-1}
Pullulan			
(pH 4.5)	....	....	1.12 h^{-1}
(pH 5.5)	....	....	0.89 h^{-1}
(pH 6.5)	....	....	1.12 h^{-1}
Alginic acid	6.6 gS/gX	0.015 gS/gX-h	0.12 h^{-1}
Pseudomonas sp	0.165%(w/v)/ODU	2.8×10^{-2}%(w/v)/ODU	0.31 h^{-1}

Table 7E1.1 lists the kinetic parameter values determined by this procedure for four different fermentations producing extracellular polysaccharides. These results show that extracellular biopolymer production kinetics may be growth-associated (pullulan and polyalginate), nongrowth-associated (biopolymer from *Pseudomonas* sp.) or mixed (xanthan gum).

The time course of product concentration in the medium during a batch fermentation can be quite complicated, and more than one product may be formed and interconverted. F. H. Deindoerfer introduced a classification of many of these possibilities which serves to illustrate some possible cases (Table 7.4). In some situations, complicated product formation kinetics may result from changes in cell metabolic operation during the reaction. A good example of this (with substantial practical and historical importance) is the production of acetone and butanol by the bacterium *Clostridium acetobutylicum* (Fig. 7.31). In the initial period of a batch fermentation, glucose is converted to acetic and butyric acids which subsequently are converted, along with glucose, to acetone and butanol. Product formation by cells may also occur simultaneously with chemical reactions of the product in the medium, such as spontaneous hydrolysis of penicillin. Kinetic description of such complications may require consideration of additional reactions in the model for product formation.

Table 7.4 Deindoerfer's classification of fermentation patterns†

Type	Description
Simple	Nutrients converted to products in a fixed stoichiometry without accumulation of intermediates
Simultaneous	Nutrients converted to products in variable stoichiometric proportion without accumulation of intermediates
Consecutive	Nutrients converted to product with accumulation of an intermediate
Stepwise	Nutrients completely converted to intermediate before conversion to product, or selectively converted to product in preferential order

† F. H. Deindoerfer, *Adv. Appl. Microbiol.*, **2**: 321, 1960.

7.5.2 Chemically Structured Product Formation Kinetics Models

Compared to kinetic studies of single-cell or cell population growth, relatively little development of structured kinetic models for product formation has so far occurred. We can expect, however, many such models to emerge in the future as natural extensions of current structured growth models. In this section we consider an interesting secondary metabolite production process which serves as a useful introduction to several important aspects of many antibiotic fermentations as well as to new considerations in kinetic modeling.

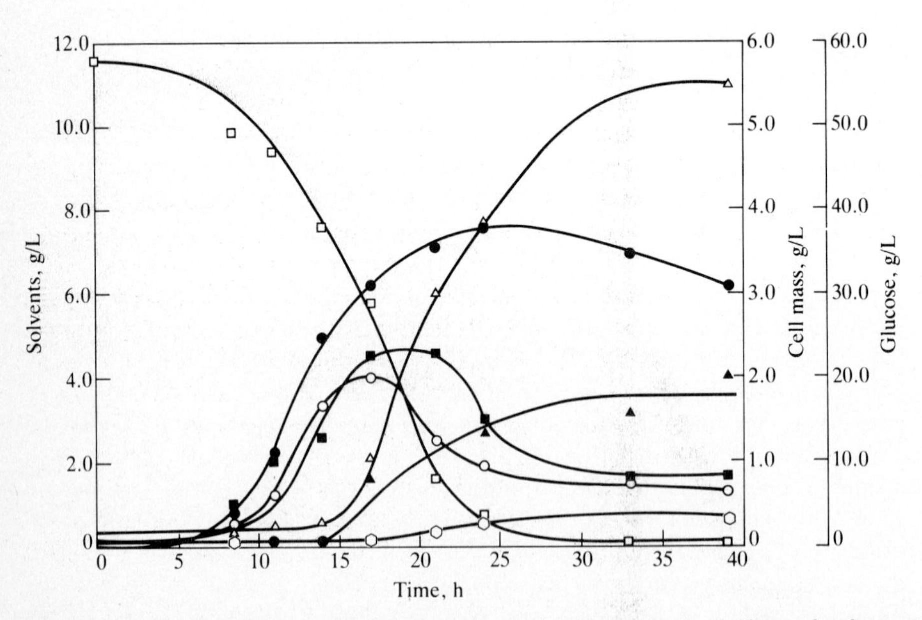

Figure 7.31 Experimental data for product formation during batch growth of *Cl. acetobutylicum* at pH 5.0 (□ glucose, ● cell mass, ■ acetic acid, ○ butyric acid, ▲ acetone, △ butanol, ⬡ ethanol). *(Reprinted by permission from J. M. Costa and A. R. Moreira, "Growth Inhibition Kinetics for the Acetone-Butanol Fermentation," ACS Symposium Series vol. 207, p. 501, 1983.)*

Synthesis of several microbial antibiotics and other secondary metabolites is inhibited by high concentrations of phosphate. Since phosphate is required for growth, an optimum phosphate level presumably exists. Such an optimum with respect to initial phosphate medium concentration was observed, for example, in production of alkaloids by *Claviceps purpurea* (Fig. 7.32).

The experimental data in Fig. 7.32 is well described by the following kinetic model for batch growth and product formation [26]:

Growth

$$\frac{dx}{dt} = k_1[1 - \exp(-K_1 p_i)]x - k_2 x^2 \tag{7.94}$$

Extracellular phosphate

$$\frac{dp}{dt} = -k_3 \frac{p}{p + K_2} x + (Y_{P/X} + p_i)k_2 x^2 \tag{7.95}$$

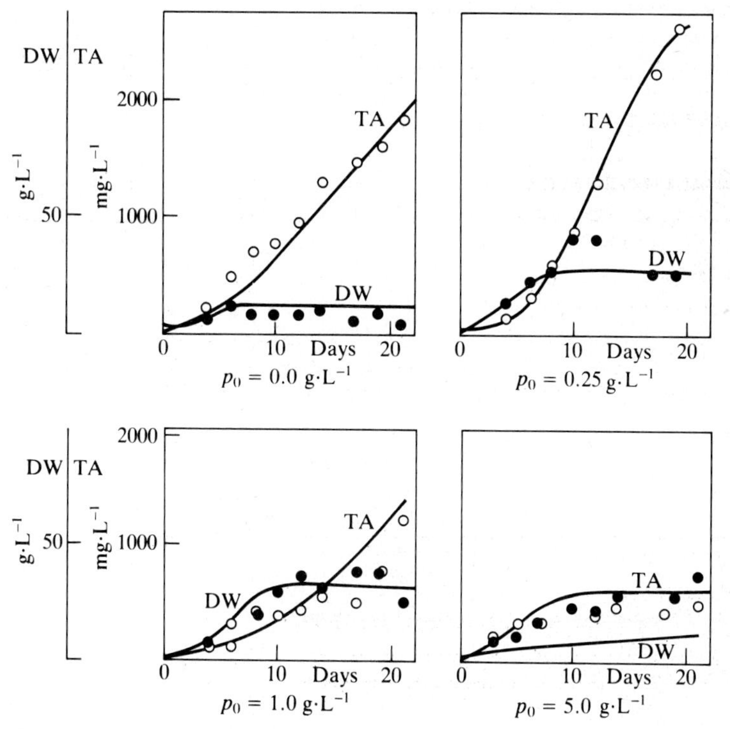

Figure 7.32 Batch cultivation data showing *C. purpurea* mass density (g dry weight/L, ●) and total alkaloids production (mg/L, ○) for different initial medium phosphate concentrations. Solid curves computed from mathematical model described in the text. *(Reprinted by permission from S. Pažoutova, J. Votruba, and Z. Řeháček," A Mathematical Model of Growth and Alkaloid Production in the Submerged Culture of* Claviceps purpurea," *Biotechnol. Bioeng., vol. 23, p. 2837, 1981.)*

Intracellular phosphate

$$\frac{dp_i}{dt} = k_3 \frac{p}{p + K_2} - k_1(Y_{P/X} + p_i)[1 - \exp(-K_1 p_i)] \tag{7.96}$$

Product formation

$$\frac{da}{dt} = k_4 \frac{K_3 x}{K_3 + p_i^2} \tag{7.97}$$

In this model the specific growth rate is described by the Teissier equation based on *intracellular* phosphate concentration p_i (g KH_2PO_4/g biomass; an intrinsic cell composition variable. All other composition variables in this model have units of grams per liter of culture). The term k_2x^2 is used to describe the rate of cell lysis.

This lysis releases phosphate into the medium in quantity proportional to the cell mass phosphate content $Y_{P/X}$ and the intracellular phosphate concentration p_i as indicated by the second term on the right-hand side of Eq. (7.95). The first term describes active transport of phosphate into the cells with saturation kinetics. The intracellular phosphate material balance (7.96) is based on Eq. (7.67), noting that the rate of formation of intracellular phosphate is the difference between the rate of phosphate transport into the cell and the rate of phosphate incorporation into cell material. The product formation kinetic model includes inhibition by phosphate in a form suggested by earlier modeling studies of repression of phosphatase enzyme.

The solid curves in Fig. 7.32 were calculated using the above model and the parameter values listed in Table 7.5. While the growth model is not very accurate at high medium phosphate levels, the time course and maximum accumulation of

Table 7.5 Parameter values for the structured kinetic model of alkaloids production by *C. Purpurea*

(Reprinted by permission from S. Pažoutova, J. Vortruba, and Z. Řeháček, "A Mathematical Model of Growth and Alkaloid Production in the Submerged Culture of Claviceps purpurea," *Biotechnol. Bioeng., vol. 23, p. 2837, 1981.)*

Parameter	Dimension	Value	Individual confidence limit
k_1	day^{-1}	0.5	0.058
k_2	L/g day	0.016	0.0024
k_3	day^{-1}	0.0575	0.017
k_4	mg/g day	6.028	1.69
$Y_{P/X}$	....	0.0025	0.0024
K_1	....	1.87×10^{-4}	1.73×10^{-4}
K_2	g/L	4.29×10^{-4}	6.67×10^{-4}
K_3	....	4.65×10^{-4}	9.84×10^{-5}

total alkaloids are well represented by this structured model. Based on this model, it was found that total alkaloids production is maximized when the initial medium phosphate concentration is 0.17 g/L. Model simulations of various phosphate feeding strategies, in which phosphate is added at several different times in the batch, did not indicate significant increases in alkaloids production for this fermentation relative to conventional batch fermentation with optimized initial medium phosphate concentration. In other situations, considered in later chapters, fed batch operation can provide substantial benefits.

7.5.3 Product Formation Kinetics Based on Molecular Mechanisms: Genetically Structured Models

The more closely a kinetic description represents the actual chemical events which occur, the more robust is the kinetic model. Here "robust" means that the model is more likely to give good results when applied to conditions different from those used to evaluate the model and determine its parameters. In the case of protein synthesis, important in practice in manufacture of enzymes, hormones, and other commercial polypeptides, our knowledge of the pertinent mechanisms permits formulation of kinetic models at the level of molecular events and interactions. Such models should be extremely robust and should prove valuable for environmental and genetic optimization.

A material balance on mRNA obtained by transcription of a particular gene G may be written as follows (Ref. 27) (brackets are used here to indicate molar concentrations within the cell; mols/unit volume cell):

$$\frac{d[\mathrm{mRNA}]}{dt} = k_p \eta [\mathrm{G}] - k_d[\mathrm{mRNA}] - \mu[\mathrm{mRNA}] \tag{7.98}$$

where k_p is the overall transcription rate constant and k_d is the rate constant for deactivation and/or destruction of the mRNA, both processes presumed first-order. The promoter utilization efficiency η is introduced in Eq. (7.98) to account for modulation of transcription by an operator system. Indeed, in many cases of practical interest, evaluation of η in terms of effector protein level and inducer or repressor concentration is the critical part of the modeling problem.

In a similar fashion, a material balance can be written for the intracellular concentration of the product of translation, namely the protein of interest.

$$\frac{d[\mathrm{P}]}{dt} = k_q \xi [\mathrm{mRNA}] - k_e[\mathrm{P}] - \mu[\mathrm{P}] \tag{7.99}$$

Here the synthesis rate is proportional to the concentration of mRNA coding for the protein and the efficiency ξ of utilization of this mRNA at the ribosomes (for example, changing the nucleotide sequence of the ribosome binding site in the message would be expected to change the value of ξ). Deactivation of active protein is again presumed first-order.

Table 7.6 Parameter values for gene expression kinetics in *E. coli*.

These are average values which may require adjustment for a particular gene and corresponding protein. The specific growth rate μ is presumed to have units of h^{-1}. *(Adapted from S. B. Lee and J. E. Bailey, "Analysis of Growth Rate Effects on Productivity of Recombinant* Escherichia coli *Populations," Biotech. Bioeng. vol. 26, p. 66, 1984.)*

Parameter	Value	Units
k_p	$2400/(233\,\mu^{-2} + 78)$	$\min^{-1}$
k_q	$3600\alpha/(82.5\,\mu^{-1} + 145)$	$\min^{-1}$
	$\alpha = 1$ for $\mu > \ln 2$	
	$\alpha = \mu/\ln 2$ for $\mu < \ln 2$	
k_d	0.46	$\min^{-1}$
k_e	0.07	$\min^{-1}$

For balanced growth, the time derivatives of [mRNA] and [P] are both zero, and the two resulting algebraic equations may be used to obtain

$$[\text{P}] = \frac{k_p k_q [G] \xi \eta}{(k_d + \mu)(k_e + \mu)} \tag{7.100}$$

This equation expresses the intracellular product concentration directly in terms of parameters which characterize gene expression and mRNA and product stability in the cell. Characteristic, average values of the model parameters for protein synthesis in *E. coli* are listed in Table 7.6. Included in this tabulation are the estimated growth rate dependencies of the transcription and translation rate parameters.

Currently available data suggests that, in many cases, gene expression rates in *E. coli* are limited at the level of transcription. Consequently, the rate of expression of an operator-regulated gene depends on the efficiency of transcription of that gene which in turn is determined by interactions of modulating species at operator sites and RNA polymerase binding. As an example analysis of such a system, we shall consider the *E. coli lac* promoter-operator, and we shall focus on the influence of *lac* repressor, inducer, and DNA interactions on gene expression, and thus β-galactosidase activity [28].

The amount of enzyme produced is proportional to the promoter utilization efficiency η which in turn is proportional to the probability that the operator site O is not bound to repressor protein R. Thus

$$[\text{P}] = K \cdot \frac{[O]}{[O]_0} \tag{7.101}$$

where $[O]_0$ is the total concentration of operator sites in the *E. coli* cell. The fraction appearing on the right-hand side of Eq. (7.101) can be evaluated by considering the following interactions among R, O, other nonspecific binding sites on DNA designated D, and inducer I, all of which are presumed to be at equilibrium:

$$R + O \rightleftharpoons RO \qquad K_A \tag{7.102a}$$

$$R + D \rightleftharpoons RD \qquad K_B \tag{7.102b}$$

$$R + I \rightleftharpoons RI \qquad K_C \tag{7.102c}$$

$$RI + O \rightleftharpoons RIO \qquad K_D \tag{7.102d}$$

$$RI + D \rightleftharpoons RID \qquad K_E \tag{7.102e}$$

Species O can exist in three different forms so that an overall balance on O gives

$$[O]_0 = [O] + [RO] + [RIO] \tag{7.103}$$

Similarly, an overall balance on R may be written

$$[R]_0 = [R] + [RI] + [RO] + [RD] + [RIO] + [RID] \tag{7.104}$$

Since the concentration of nonspecific binding sites D on the DNA is much greater than the total concentration of R, we may assume that

$$[D] \cong [D]_0 \tag{7.105}$$

Using this approximation and the equilibrium relationships for the interactions in Eq. (7.102) in Eqs. (7.104) and (7.105) and neglecting terms multiplied by $[O]_0/[R]_0$, we find

$$[\mathrm{P}] = K \frac{1 + K_B[D]_0 + K_C[I](1 + K_E[D]_0)}{1 + K_B[D]_0 + K_C[I](1 + K_E[D]_0) + (K_A + K_C K_D[I])[R]_0} \tag{7.106}$$

Experimental information from the literature provides values, at least in an order of magnitude sense, for the parameters in this equation [28]

$$\begin{aligned} [D]_0 &= 4 \times 10^{-2} M & K_A &= 2 \times 10^{12} M^{-1} \\ K_B &= 1 \times 10^{3} M^{-1} & K_C &= 1 \times 10^{7} M^{-1} \\ K_D &= 2 \times 10^{9} M^{-1} & K_E &= 1.5 \times 10^{4} M^{-1} \end{aligned} \tag{7.107}$$

While a large quantity of experimental data is required to establish such a model, consideration of molecular interactions in formulating a kinetic model has several benefits. The parameters in the model have precise physical interpretations and can often be measured in separate, in vitro experiments. Also, models at this level possess what has been called *genetic structure*. That is, a specific genetic change, say a change in the nucleotide sequence of the *lac* operator, has an unambiguous effect on specific model parameters—in particular, the parameters K_A and K_D which characterize interactions with the *lac* operator nucleotide sequence. As our knowledge of DNA-protein interactions improves, it may be possible to calculate the influence of certain nucleotide sequence changes on

Table 7.7 Calculated and experimentally measured values of β-galactosidase activity under control of the *lac* promoter-operator for wild-type *E. coli* and for three mutants (i^-, i^q, i^{sq}) producing different intracellular levels of *lac* repressor R($[R]_0 = 2 \times 10^{-8}$ M)

(Reprinted by permission from J. E. Bailey, M. Hjortso, S. B. Lee, and F. Srienc, "Kinetics of Product Formation and Plasmid Segregation in Recombinant Microbial Populations," Annals NY Acad. Sci., vol. 413, p. 71, 1983.)

Genotype	Repressor concentration	Inducer	β-Galactosidase activity[a] Calculated	Experimental	Reference
i^-	0	−	113	100–140	33
i^-	0	+	113	100–130	33
i^+	$[R]_0$	−	0.1	<0.1	33
i^+	$[R]_0$	+	100	100	33
i^q	$10[R]_0$	−	0.011	0.01–0.014	34
i^q	$10[R]_0$	+	62	65	35
i^{sq}	$50[R]_0$	−	0.002	0.003–0.004	34
i^{sq}	$50[R]_0$	+	23	25	35

[a] Normalized to 100% for induced wild-type (i^+), inducer concentration 10^{-3} *M*.

corresponding DNA-protein binding constants. Then, a kinetic model formulated at the molecular level will allow mapping of nucleotide sequence into cell population productivity. The potential of such capability for rational, quantitative optimization of the genetic makeup of the organism (or at least cloned DNA inserts) is clear.

As an example, we shall examine the use of Eq. (7.106) to calculate the effect of changed repressor levels on induced and uninduced transcription rates of the chromosomal *lac* promoter-operator in *E. coli.* Corresponding data is available for *E. coli.* mutant strains i^-, i^q, and i^{sq}. Since the intracellular repressor concentrations have been measured for these mutants, these measured values may be used in the model directly, and none of the remaining model parameters are changed. As shown in Table 7.7. the model successfully represents the effects of changes in repressor content of the cells under inducing and noninducing cultivation environments. Additional examples of molecular level regulation models are available in the chapter references.

7.5.4 Product Formation Kinetics by Filamentous Organisms

The kinetics of product formation and substrate utilization by molds and other filamentous microbes is usually quite complex. A typical example is the batch fermentation of penicillin. As Fig. 7.30 shows, there are three different regimes of substrate uptake. For the first 20 h, rapid sugar utilization accompanies active growth of mold. As the growth curve passes through the stationary phase, substrate utilization is slow while product-formation rates are at their peak. Follow-

Table 7.8 Stages in batch production of oxytetracycline by *Streptomyces remosus* in submerged culture[†]

Stage	Comment
1 Lag	Lasts about 90 min when the inoculum is small; metabolic activity not measurable
2 Growth of primary mycelium	Depending on inoculum, lasts from 10 to 25 h; respiration, nucleic acid synthesis, and other metabolic activities at very high level; pyruvic acid concentration at peak; no antibiotic formed
3 Fragmentation of primary mycelium	Lasts about 10 h, during which growth of the mycelium ceases, respiration and nucleic acid synthesis decrease, and pyruvic acid concentration drops to very low value
4 Growth of secondary mycelium	During the 25 h of this phase the secondary mycelium grows to level 2 to 4 times higher than in stage 2; filaments now thin, as contrasted to earlier thick ones; antibiotic production increases rapidly now, nucleic acid synthesis is renewed, but respiration continues to decrease; sugars and ammonia nitrogen rapidly depleted; pyruvic acid concentration may increase slightly
5 Stationary phase	Growth ceases and metabolic acitivity is low as the cycle enters this stage; antibiotic production continues for a time, but on a specific basis its rate is comparatively low

[†] R. Luedeking, p. 208 in N. Blakebrough (ed.), *Biochemical and Biological Engineering*, vol. 1, Academic Press, Inc., New York, 1967; see also J. Doskacil, B. Sikyta, J. Kasparova, D. Poskocilova, and J. Zajicek, *J. Gen. Microbiol.*, **18**: 302, 1958.

ing this portion of the fermentation, sugar is again consumed rapidly until it is exhausted.

Several other antibiotic fermentations exhibit similar patterns, although there is no general rule applicable to all such fermentations. In streptomycin production, for example, the maximum product synthesis rate is found later in the cycle than for penicillin, and no final rapid sugar-uptake phase occurs. Another antibiotic fermentation involving filamentous organisms, oxytetraycline produced by *Streptomyces rimosus*, is of interest because it introduces the importance of *morphology* for satisfactory performance of the fermentation. Table 7.8 summarizes the major features of this fermentation.

Listed below are some of the experimental findings on filamentous organisms and product release which illustrate some of the peculiar features of these systems.

1. There is an optimal initial substrate concentration for maximum product yield in a fixed batch time. If there is too much substrate, a "fast fermentation" results, in which little product is formed and substrate is used primarily for biomass production. If there is too little substrate, not enough biomass is formed to manufacture product in the production phase.

2. Product formation is maximized by minimizing the branching in actively growing hyphae in the inoculum. However, the smaller the degree of branching, the longer the lag phase and hence the longer the required batch time. Consequently, there is also an optimal condition for the inoculum.
3. Since fermentations conducted by molds and other mycelial microorganisms are aerobic, it would seem that mixing of a submerged culture should be as intense as possible in order to promote vigorous oxygen transfer to the mold. This is not the case for the penicillin fermentation, where an intermediate degree of agitation has been found to maximize penicillin yields. There are several possible explanations. The morphology of the mold is affected by the applied forces; more vigorous mixing promotes branching of the hyphae, which is believed to decrease product formation. Alternatively, there may well be an optimal rate of oxygen supply, but this is extremely difficult to determine because of the complexity of the system. (Mixing effects are reconsidered in Chap. 8.)

Example 7.2: A Morphologically Structured Kinetic Model for Cephalosporin C Production [14] Experimental studies of production of the antibiotic cephalosporin C(CPC) by submerged culture of the mold *Cephalosporium acremonium* have shown three morphological forms: hyphae (h), swollen hyphal fragments (s), and arthrospores (a). These three forms are interconverted according to medium composition as indicated schematically in Fig. 7E2.1. As indicated in the figure, CPC is produced primarily by the swollen hyphal fragments. CPC synthesis rate is directly related to the activity of CPC-synthesis enzymes which are induced by intracellular methionine and which are repressed by glucose.

Based on this information, the following structured model for batch cultivation has been formulated:

Mold growth and differentiation

$$\frac{dx_h}{dt} = \frac{\mu_m g}{K_g + g} x_h - \beta(g, m)x_h - k_D x_h \tag{7E2.1}$$

$$\frac{dx_s}{dt} = \beta(g, m)x_h - \gamma(g)x_s - k_D x_s \tag{7E2.2}$$

$$\frac{dx_a}{dt} = \gamma(g)x_s - k_D x_a \tag{7E2.3}$$

where

$$\beta(g, m) = \left(k_{11} + \frac{k_{12}m}{K_m + m}\right)\left(\frac{g}{K_G + g}\right) \tag{7E2.4}$$

$$\gamma(g) = k_{21} + k_{22}\left(\frac{g}{K_g + g}\right) \tag{7E2.5}$$

Note that the total biomass is partitioned here into the three different morphological types observed experimentally. Medium glucose and methionine mass concentrations are given by g and m, respectively.

The substrate utilization model presumes glucose and methionine uptake by the h and s morphological states only.

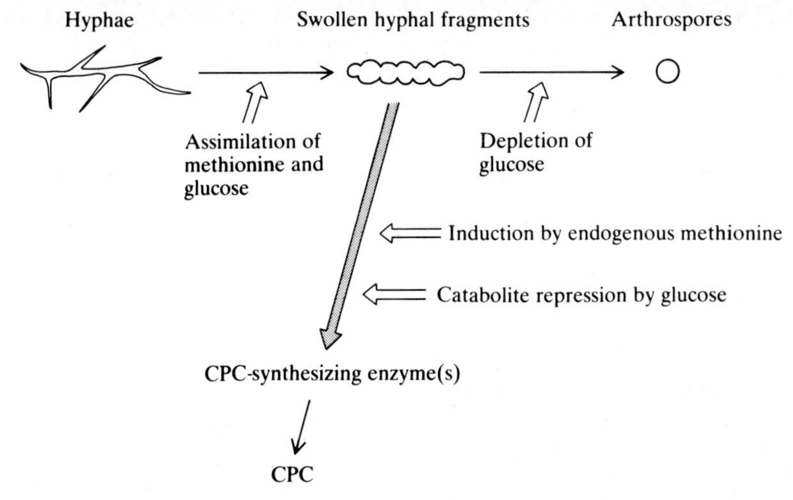

Figure 7E2.1 Schematic diagram of the different mold morphologies, their transitions, and cephalosporin C (CPC) production by *C. acremonium. (Reprinted by permission from M. Matsumura, T. Imanaka, T. Yoshida, and H. Taguchi, "Modeling of Cephalosporin C Production and Its Application to Fed-Batch Culture", J. Ferment. Tech. (Japan), vol. 59, p. 115, 1981.)*

Glucose and methionine utilization

$$\frac{dg}{dt} = -\frac{\mu_m g x_h}{Y_{H/G}(g + K_g)} - \frac{\nu_m g x_s}{g + K_g} \tag{7E2.6}$$

$$\frac{dm}{dt} = -U_h x_h - U_s x_s \tag{7E2.7}$$

where

$$U_h = \frac{U_{mh} m}{m + K_m} \qquad U_s = \frac{U_{ms} m}{m + K_m} \tag{7E2.8}$$

Eqs. (7E2.1) through (7E2.8) have been used to calculate batch growth and substrate utilization, and model parameters have been adjusted by a simplex optimization method to maximize model correspondence with experimental data. In this way, the group 1 parameter values listed in Table 7E2.1 were determined.

Because of the significance of intracellular methionine in regulating expression of CPC-synthetic enzymes, the kinetic model has also been chemically structured to evaluate the concentrations m_{ih}, m_{is}, and m_{ie} (all in units μmol/g cell) of methionine inside the h, s, and a morphological forms, respectively. The material balances used are

Intracellular methionine

$$\frac{dm_{ih}}{dt} = U_h + V_{h,\mathrm{syn}}^{\max} - k_{3h} m_{ih} - \frac{V_{h,\mathrm{util}}^{\max} g}{K_g + g} m_{ih} - \frac{\mu_m g}{K_g + g} m_{ih} \tag{7E2.9}$$

$$\frac{dm_{is}}{dt} = U_s + V_{s,\mathrm{syn}}^{\max} - k_{3s} m_{is} - \frac{V_{s,\mathrm{util}}^{\max} g}{K_s + g} m_{is} + \frac{\beta(g, m) x_h}{x_s} (m_{ih} - m_{is}) \tag{7E2.10}$$

$$\frac{dm_{ia}}{dt} = -k_{3a} m_{ia} + \frac{\gamma(g) x_s}{x_a} (m_{is} - m_{ia}) \tag{7E2.11}$$

Table 7E2.1 Estimated parameter values for cephalosporin C product formation model

(Reprinted by permission from M. Matsumura, T. Imanaka, T. Yoshida, and H. Taguchi, "Modeling of Cephalosporin C Production and its Application to Fed-Batch Culture," J. Ferment. Tech. (Japan), vol. 59, p. 115, 1981.)

Group 1		Group 2		Group 3	
μ_m	0.069 (h^{-1})	$V^{\max}_{h,\text{syn}}$	8.7 (μmol-met/g-cell/h)	V_{mE}	1.2 (mg-CPC/g-cell/h^3)
Y_G	0.45 (g-cell/g-glu)	k_{3h}	7.1 (h^{-1})	K_g	12.0 (μmol-met/g-cell)
K_G	0.05 (g-glu/L)	$V^{\max}_{h,\text{util}}$	48.6 (h^{-1})	t_I	13.0 (h)
k_{11}	0.015 (h^{-1})	$V^{\max}_{s,\text{syn}}$	6.6 (μmol-met/g-cell/h)	κ	0.29 (−)
k_{12}	0.024 (h^{-1})	k_{3s}	1.4 (h^{-1})	α	0.15 (−)
k_{21}	0.004 (h^{-1})	$V^{\max}_{s,\text{util}}$	2.2 (h^{-1})	η	1.6 (−)
k_{22}	0.028 (h^{-1})	k_{3a}	7.8 (h^{-1})	n	5.1 (−)
K_M	0.001 (g-met/L)			k_{ED}	0.38 (h^{-1})
v_m	0.023 (g-glu/g-cell/h)			k_{PD}	0.012 (h^{-1})
U_{mh}	0.0012 (g-met/g-cell/h)				
U_{ms}	0.0098 (g-met/g-cell/h)				
k_D	0.014 (h^{-1})				

Group 1: parameters related to growth and substrate consumption
Group 2: parameters related to accumulation of endogenous methionine
Group 3: parameters related to formation of CPC-synthesizing enzymes and CPC production

In formulating these equations, it has been assumed that the methionine biosynthesis rate is inversely proportional to the intracellular methionine concentration and that presence of glucose in the medium increases the rate of methionine utilization for protein synthesis. The average intracellular methionine concentration is

$$\bar{m}_i = \frac{x_h m_{ih} + x_s m_{is} + x_a m_{ia}}{x} \tag{7E2.12}$$

where

$$x = x_h + x_s + x_a \tag{7E2.13}$$

Now, the model is complete with respect to growth, substrate utilization, and intracellular methionine level, and new parameters introduced in Eqs. (7E2.9) through (7E2.11) may now be estimated by comparison of measured and calculated $\bar{m}_i$ values. These estimates appear in the group 2 column in Table 7E2.1.

Product synthesis rate is presumed to depend on activity of CPC-synthetic enzymes designated e(mg CPC·h^{-1}·g cell^{-1}) which is described by the following equation:

Enzyme synthesis

$$\frac{de}{dt} = \frac{1}{x_s}\left(\frac{V_{mE} m_{is} x_s}{m_{is} + K_E}\right)_{t-t_I} \cdot Q - \left(k_{ED} + \beta(g, m)\frac{x_h}{x_s}\right) \cdot e \tag{7E2.14}$$

where

$$Q = \frac{1 + (\kappa/\alpha^n)g}{1 + (\kappa/\alpha^n)(1 + \eta)g^n} \tag{7E2.15}$$

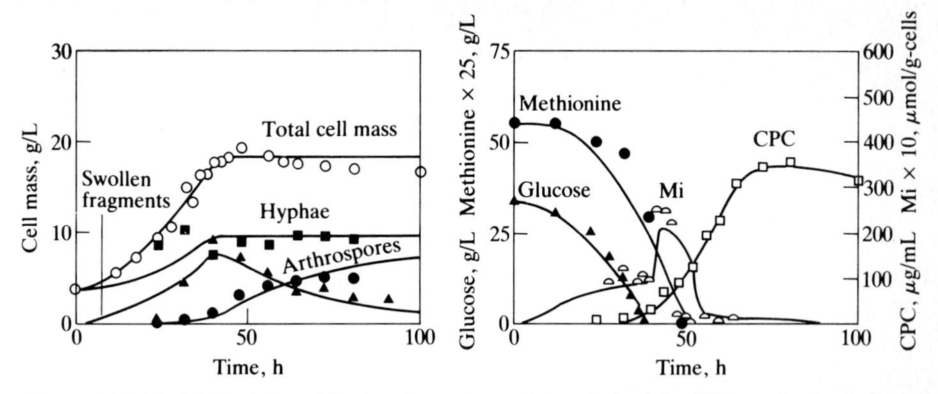

Figure 7E2.2 Model simulations (lines) and experimental data (points) for CPC production in batch culture of *C. acremonium.* M_i = intracellular methionine. *(Reprinted by permission from M. Matsumura, T. Imamaka, T. Yoshida, and H. Taguchi, "Modeling of Cephalosporin C Production and Its Application to Fed-Batch Culture," J. Ferment. Tech. (Japan), vol. 59, p. 115, 1981.)*

Q represents catabolite repression by glucose. The subscript $(t - t_I)$ denotes evaluation at time $t - t_I$; this time shift is introduced here as a way of representing the required lag time between induction and gene expression. Finally, the rate of product CPC formation is described in this model by

Product formation

$$\frac{dp}{dt} = ex_s - k_{PD}p \tag{7E2.16}$$

Parameters appearing in Eqs. (7E2.15) and (7E2.16) estimated from product formation data are listed as Group 3 in Table 7E2.1.

Figure 7E2.2 shows experimental (points) and model simulations (lines) of a batch cultivation. Note the increase and decline with time in the swollen fragment(s) morphological form and the buildup of arthrospores late in the fermentation. The model simulates very well the unusual trajectory of intracellular methionine concentration. However, after formulating a model of this complexity and devoting large efforts to parameter estimation, the model should be useful for simulations of a broad

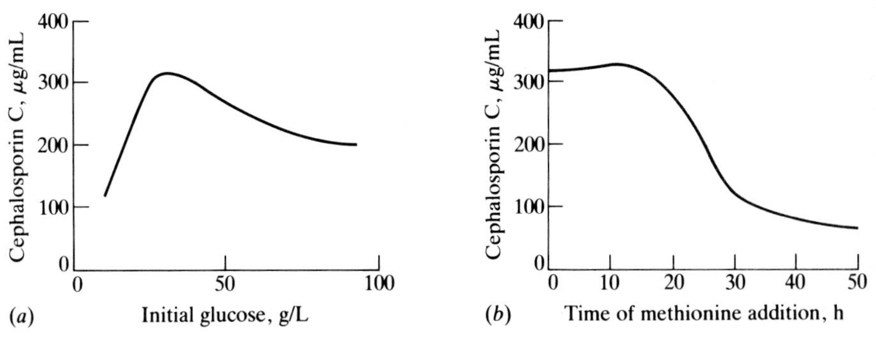

Figure 7E2.3 Model simulation of the effect of (*a*) initial glucose concentration and (*b*) the time of methionine addition on CPC production. *(Reprinted by permission from M. Matsumura, T. Imanaka, T. Yoshida, and H. Taguchi, "Modeling of Cephalosporin C Production and its Application to Fed-Batch Culture," J. Ferment. Tech. (Japan), vol. 59, p. 115, 1981.)*

range of different cultivation conditions. The present model fulfills this requirement quite well. Plotted in Fig. 7E2.3 is the dependency of total CPC production on initial glucose concentration (*a*) and the time of addition of methionine to the culture (*b*) as calculated from the model. The former has a clear maximum in agreement with experimental observations, and the latter indicates the advantage of early methionine addition as was shown also in experiment. The model was also utilized to predict that feeding both glucose and methionine during batch cultivation significantly improved CPC production while feeding of either component individually had relatively small effects. Subsequent experiments confirmed these findings.

7.6 SEGREGATED KINETIC MODELS OF GROWTH AND PRODUCT FORMATION

Cell populations are typically heterogeneous with respect to individual cell size, age, growth rates, and other properties. Because of the closer correspondence with this physical reality, it is appealing to include consideration of the distribution of different single-cell types and corresponding biochemical activities in formulating kinetic models for cell populations. Another benefit of this approach is the opportunity of connecting cell cycle features and other single-cell properties with overall characteristics of the population. The major disadvantage of segregated kinetic models is their mathematical complexity. Because of this, we focus here on the age distribution and correlated distributions for a population of cells that divide by binary fission. Derivations of the population balance equations used in segregated models and solution of these models for several microbial systems are provided in Refs. 29 to 31.

The frequency function of cell age, $W(a)$, has the following significance: $W(a)\,da$ is the fraction of cells with ages between a and $a + da$. In this discussion, a will denote cell cycle age with age zero assigned to newborn cells. Then, dividing cells have age t_D, the cell cycle time or doubling time for a *single cell.* For balanced exponential growth of the *population,* the population balance based upon cell age has the following form

$$\frac{dW(a)}{da} = -\mu W(a) \tag{7.108}$$

where μ is the *overall* specific growth rate *of the population.* Since all cells in the population have ages between zero and t_D, W must also satisfy

$$\int_0^{t_D} W(a)\,da = 1 \tag{7.109}$$

In several instances cell division does not always produce two viable progeny. This may occur because of some aging or death process. Such a case also arises in growth in selective medium of a population containing unstable plasmids. Letting Θ denote the number (usually fractional) of nonviable progeny produced by binary fission of a viable cell, a cell balance relating dividing and newborn viable cells gives

$$W(O) = (2 - \Theta)W(t_D) \tag{7.110}$$

The solution of Eq. (7.108) subject to conditions (7.109) and (7.110) is

$$W(a) = \left(\frac{2-\Theta}{1-\Theta}\right)\mu e^{-\mu a} \tag{7.111}$$

Substituting Eq. (7.111) into Eq. (7.110) yields the following relationship between overall growth rate μ and cell generation time

$$\mu = \frac{\ln(2-\Theta)}{t_D} \tag{7.112}$$

Since the overall doubling time of the population $\bar{t}_D$ is given by

$$\bar{t}_D = \ln 2/\mu \tag{7.113}$$

we see that

$$\frac{\text{Single cell generation time}}{\text{Population doubling time}} = \frac{t_D}{\bar{t}_D} = \frac{\ln(2-\Theta)}{\ln 2} \tag{7.114}$$

Thus, if Θ is not zero, t_D is less than $\bar{t}_D$. This relatively simple example introduces a very important concept: overall kinetic properties of the population do *not* necessarily correspond to kinetic properties at the single cell level.

In many cases the age density function, which cannot be directly observed, may be used to calculate other frequency functions and average population properties with direct physical significance. Suppose, for example, that some cell components or products are synthesized at a constant rate by cells with cell cycle ages between a_1 and a_2. For example, in eucaryotes DNA is synthesized only during the S-phase interval. It has also been reported that several enzymes are synthesized in certain organisms only in certain short subintervals of the cell cycle. Then the product formation rate is proportional to

$$F_{1,2} = \int_{a_1}^{a_2} W(a)\,da \tag{7.115}$$

which for the case analyzed above is given by

$$F_{1,2} = \left(\frac{2-\Theta}{1-\Theta}\right)[e^{-\mu a_1} - e^{-\mu a_2}] \tag{7.116}$$

For any cellular quantity x (say cell mass, protein content, volume, perhaps product content) which has a one-to-one correspondence with cell cycle age a (that is, a can be written or implicitly defined as function of x), the frequency function for a may be transformed to the frequency function for x according to

$$W_x(x) = W(a(x))\left|\frac{da(x)}{dx}\right| \tag{7.117}$$

Suppose, for example, that the mass of an individual cell increase linearly with time during the cell cycle

$$m(a) = m_0 + ak \tag{7.118}$$

so that

$$a = \frac{m - m_0}{k} \qquad \left| \frac{da}{dm} \right| = \frac{1}{k} \tag{7.119}$$

Taking W from Eq. (7.111) and using Eq. (7.119) in Eq. (7.117) gives

$$W_m(m) = \left(\frac{2 - \Theta}{1 - \Theta} \right) \frac{\mu}{k} \exp \left[-\frac{\mu}{k} (m - m_0) \right] \tag{7.120}$$

Again the possible distinction between single-cell and overall population kinetics should be mentioned. Single cells may increase in mass according to Eq. (7.118) (this has been reported for several organisms) while the *total* mass of the cell population increases exponentially as in, for example, Eq. (7.37) describing balanced batch growth. One of the rewards of segregated kinetic model formulation and solution is the opportunity of establishing explicit relationships among single-cell kinetic and regulation parameters and the parameters which characterize overall population kinetics.

Presuming that the density of cell material is constant, Eq. (7.120) may also be interpreted as a volume distribution which could be measured experimentally using a Coulter counter or light-scattering flow cytometry measurements (see Chap. 10). More commonly population-average quantities are measured. These may be calculated from the corresponding frequency functions using

$$\bar{x} = \int_0^{\infty (\text{or } x_{\max})} x W_x(x)\, dx \tag{7.121}$$

In other contexts we may usefully view heterogeneity in a cell population from different perspectives. Several models of product formation by microbial population have been proposed which consider long-term aging effects. The product formation model of Ping Shu [32] expresses the specific product formation rate by cells that have spent a *total* time θ in the reactor by $\sum_{i=1}^{q} A_i e^{-k_i \theta}$. The product concentration at time t in a batch fermentation is then calculated from

$$p(t) = \int_0^t M(\theta, t) \left(\int_0^\theta \sum_{i=1}^{q} A_i e^{-k_i \theta'}\, d\theta' \right) d\theta \tag{7.122}$$

where $M(\theta, t)$ is the frequency function at time t of the concentration of cell mass which has been in the reactor for duration θ. Since cell material with age θ at time t is formed at time $t - \theta$, it follows that

$$M(\theta, t) = r_x(t - \theta) \tag{7.123}$$

This model is quite flexible and can successfully fit many different patterns of product accumulation observed during batch fermentation.

7.7 THERMAL-DEATH KINETICS OF CELLS AND SPORES

The activity of cells, spores, and viruses in air or liquids can be diminished by destruction (heat, radiation, or chemical or mechanical means), removal (filtration or centrifugation), or inhibition (refrigeration, dessication, water depletion, chemicals). The predominant modes for liquid sterilization are heating (industrial) and chlorination (municipal use of water), with particulate filtration the major tool for air sterilization. The latter two processes depend on mass-transfer phenomena discussed in the following chapter; the kinetics of cell and spore inactivation is the topic of this section.

Figure 7.33 illustrates experimental data for vegetative-cell and spore death over a range of temperatures. In general, cells have a greater susceptibility for destruction than spores, consistent with earlier qualitative remarks concerning endospore formation as a survival mechanism of some cells. Similarly, viruses such as the bacteriophages are also inactivated by thermal treatment, so that thermal sterilization in the fermentation industry decreases both the viable microbial population of the liquid-nutrient stream and its virus count, or *titer*.

Before we present some useful equations for estimation of the population reduction rates in sterilization, a few cautionary remarks are in order. The death of a particular cell is probably due to the thermal denaturation of one or more kinds of essential proteins such as enzymes. The kinetics of such cooperative transitions for these molecules may be rather complex in time. Also, the rate of the molecular processes ultimately leading to cell death will depend on environment composition, including the solvent concentration itself. For example, the temperature needed to achieve coagulation (denaturation followed by massive irreversible cross-linking of the denatured protein) of the egg protein albumin increases with decreasing water content as shown in Table 7.9.

These observations and others suggest that it may be more accurate to consider the deactivation of the water-deficient structures such as viruses and spores as requiring some kind of initial hydration, following which the molecular transitions such as denaturation may occur; a similar effect of relative humidity on the survival rate of active bacteriophage is found [34]. A combination of deactivating influences may produce a nonadditive response; e.g., simultaneous desiccation and heat treatment may be less effective than supposed from the individual efficiencies of the processes.

A cell population will have a range of cell ages, as discussed previously. Further, the nature of the cell wall and the relative importance of various metabolic pathways to the cell change with cell and culture age. Thus, the resistance of a species to thermal (or other) destruction will depend on its *history*, a vague term which is often difficult to make more quantitative. For example, an exponential-phase population has relatively permeable cell walls, which allow rapid exchange of solutes with the external medium. If these solutes affect protein stability, a consequent difference of exponential-phase response may be expected.

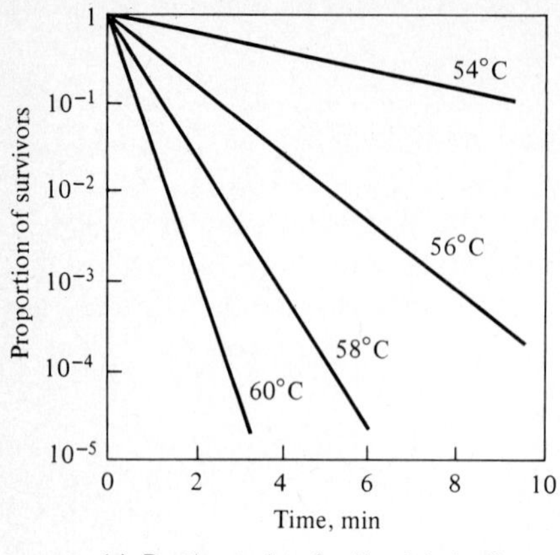

(a) Death rate data for *E. coli* in buffer

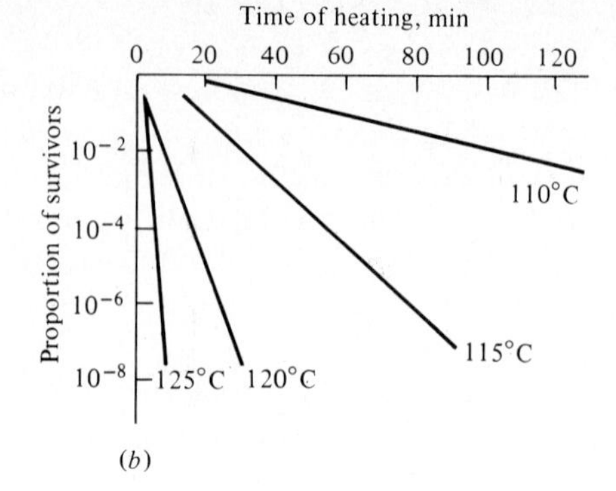

(b) Inactivation of *Bacillus subtilis* spores

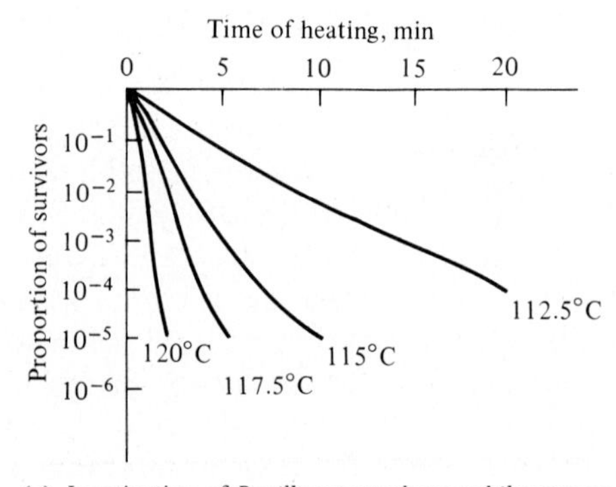

(c) Inactivation of *Bacillus stearothermophilus* spores

Figure 7.33 Experimental data for thermal death of *E. coli* (*a*) and inactivation of vegetative spores of *Bacillus subtilis* (*b*) and *B. sterothermophilus* (*c*). *[(a) Reprinted from S. Aiba, A. E. Humphrey, and N. F. Millis, "Biochemical Engineering," 2d ed., p. 241, University of Tokyo Press, Tokyo, 1973; (b, c) Reprinted from H. Burton and D. Jayne-Williams, "Sterilized Milk," in J. Hawthorn and J. M. Leitch (eds), "Recent Advances in Food Science, vol. 2: Processing," p. 107, Butterworths & Co., Publishers Ltd., London, 1962.]*

The most reliable data come from measurements on systems approximating those of the application as closely as possible.

To begin analysis of death-rate kinetics, note that the straight lines through the data in the semilog plots in Fig. 7.33 are consistent with a first-order decay for the *viable* population level n

$$\frac{dn}{dt} = -k_d n \tag{7.124}$$

which, for constant k_d, yields

$$n(t) = n_0 e^{-k_d t} \tag{7.125}$$

Table 7.9 Time required for coagulation of albumin as a function of water content†

Water content, %	Approximate coagulation T, °C
50	56
25	76
15	96
5	149
0	165

† Data from M. Frobisher, *Fundamentals of Microbiology*, p. 259, W. B. Saunders Company, Philadelphia, 1968.

where n_0 is the concentration of spores or vegetative cells at $t = 0$. The slope of the semilog plots is $-k_d$; a plot of ln k_d vs. reciprocal temperature provides another straight line, so that the temperature dependence of k_d may be taken as that of the familiar Arrhenius form

$$k_d = k_{d0} e^{-E_d/RT} \tag{7.126}$$

The range of E_d values for many spores and vegetative cells lies between 50 and 100 kcal/mol. The historical term known as the *decimal reduction factor* $D_r = 2.303/k_d$ is simply the time needed to reduce the viable population by a factor of 10.

These kinetic descriptions work reasonably well provided the concentration of the spores or cells is a large number in the statistical sense of the word (Sec. 7.1). Deviations from such a prediction become more probable as the numbers involved diminish since the standard deviation of, for example, the normal distribution increases as $1/n$. In a deterministic model, the fraction of the population remaining viable (assuming no current growth under the applied lethal conditions) is seen from the previous equation to be

$$\text{Viable fraction} = \frac{n}{n_0} = e^{-k_d t} \tag{7.127}$$

If each organism's fate is independent of the others in that no reproduction occurs, and if the lethal condition is uniformly applied, a stochastic approach to sterilization [33] shows that the probability that at any time t the remaining population contains N viable organisms is given by

$$P_N(t) = \frac{N_0!}{(N_0 - N)!N!} (e^{-k_d t})^N (1 - e^{-k_d t})^{(N_0 - N)} \tag{7.128}$$

with N_0 the initial *viable number* of organisms in the fluid being sterilized.

The k_d which appears here is the same as the rate constant in Eq. (7.124): in the stochastic model k_d may be interpreted as the reciprocal of the mean life span

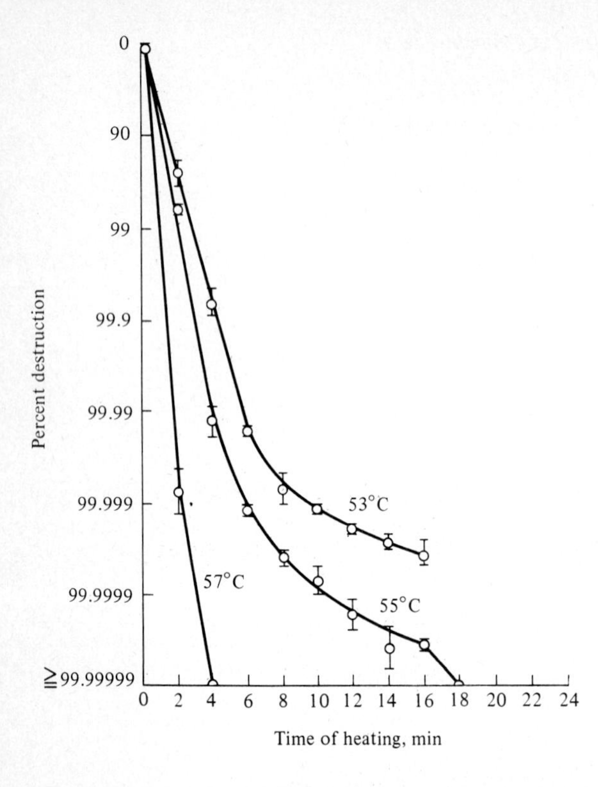

Figure 7.34 Thermal destruction of *S. aureus*. *(Reprinted by permission from G. C. Walker and L. G. Harmon, Thermal Resistance of* Staphylococcus aureus *in Milk, Whey, and Phosphate Buffer," Appl. Microbiol., vol. 14, p. 584, 1966.)*

of the organism. As the number of organisms falls to a low value, the assumptions of a uniform population characterizable by a single k_d value begins to fail. For example, the data in Fig. 7.34 for *Staphlococcus aureus* in neutral phosphate buffer show consistent positive deviations from the random-distribution result of the above equation, suggesting that a minor fraction of the population is more resistant to thermal death than the vast majority.

While considerable numbers of surviving organisms are acceptable in some applications (viable bacterial concentrations in Grade A pasteurized milk must be less than 30,000 cells per milliliter), more stringent requirements apply, for example, to many pure culture fermentations. In such cases it is important to examine the extinction probability, i.e., the probability that all organisms are inactivated. Setting $N = 0$ in Eq. (7.128) gives

$$\text{Extinction probability} = P_0(t) = 1 - (1 - e^{-k_d t})^{N_0} \tag{7.129}$$

from which it follows that the probability of at least one organism's survival is

$$1 - P_0(t) = (1 - e^{-k_d t})^{N_0} \tag{7.130}$$

For the usual situation of $N_0 \gg 1$, this becomes

$$1 - P_0(t) = 1 - e^{-N_t} \tag{7.131}$$

where

$$N_t = N_0 e^{-k_d t} \tag{7.132}$$

We can interpret $1 - P_0$ physically as the fraction of sterilizations which are expected to fail to produce a contaminant-free product.

As we know from our studies of metabolism, living cells often have alternate pathways for tapping an energy source or conducting a biosynthesis. The presence of redundancy alters the chances for organism survival. A statistical analysis shows that the expected value of the population size is

$$E[N(t)] = N_0(1 - \{1 - \exp[\bar{k}_d(t)t]\}^r)^\omega \tag{7.133}$$

where ω = number of kinds of essential subcellular structures
r = number of units of each structure in each organism
$\bar{k}_d(t)$ = mean specific death rate for each kind of structure

Note that this expression expands into a series of decaying exponentials and is not thus expected to be linear over the major fraction of a semilog plot.

In a similar manner, a cell solution containing m different varieties of cells (or spores or viruses) may most simply behave as independent deterministic species, the resulting total viable population being the sum of the individual results following Eq. (7.125) with the appropriate individual values of k_{di}:

$$n(t) = \sum_{i=1}^{m} n_i(t) = \sum_{i=1}^{m} n_{i0} e^{-k_{di} t} \tag{7.134}$$

Again a straight-line semilog plot is not expected. Further, the individual k_{di} values are not routinely available; some methods for estimating the upper and lower bounds on the values of $n(t)/n_0$ from initial data are discussed by Hutchinson and Luss.† The validity of these estimates depends on the accuracy of the initial death data and requires evaluation of the second derivative d^2n/dt^2 at $t = 0$.

7.8 CONCLUDING REMARKS

As discussed at the outset of this chapter, the ultimate objective of cell kinetics is quantitative description of the combined effects of genetic makeup and environment on rate processes in the population. At present we have a collection of partial models, some considering fine-scale genetic variations, others representing

† P. Hutchinson and D. Luss, "Lumping of Mixtures with Many First-Order Reactions," *Chem. Eng. J.*, **1**: 129, 1970.

subtle medium-composition effects, and some describing cell-to-cell heterogeneity in the culture. Given limitations in current kinetic frameworks, the biochemical engineer must carefully identify the intended applications of the kinetic description, and fashion accordingly the form and depth of the kinetic model. Tremendous advances in biological science research and in computational capability will combine in the future to bring to fruition more reliable, more comprehensive, and more mechanistic models for cell kinetics.

In this chapter we have studied the interrelationships between biomass increase, substrate uptake, and product formation. From experimental data, the general features of growth for cell populations and filamentous organisms were discerned. Mathematical models of growth helpful for understanding and design of processes utilizing such organisms have been summarized and reviewed. In Chap. 9, we shall consider design and analysis of a variety of continuous and batch biological reactors. Throughout that discussion we shall rely heavily on the rate expressions presented in this chapter.

Before proceeding to examination of biological reactors, additional kinetic phenomena must be studied: transport processes in macroscopic systems. This subject deals with the transport of oxygen, chlorine, and other gases into liquids for such purposes as aeration and chlorination. Such transport is fundamentally coupled to the forms of natural or forced fluid convection which predominate in the bioreactor. These in turn lead to considerations of power consumption and reactor scale-up under various conditions. Heat-transfer processes are critical in sterilization and bioreactor operation. These matters and others are examined in the next chapter.

PROBLEMS

7.1 Fermentation batch kinetics Grape juice (natural) is a nearly complete nutrient for yeast. Perform the following batch fermentation.

(*a*) Purchase a simple beer- or wine-making kit or make your own. For the wine you need grape juice (free of settleable solids for this experiment), sugar, a yeast source, a 1-gal vessel, a small beaker, and hose for an airlock.

(*b*) Start-up: suspend active yeast in 100 mL of lukewarm (previously boiled) water and stir well to disperse cells. Add to fermentor (sterilized with boiling water or sodium sulfite solutions) 1 qt grape concentration, 3 lb sugar, and sterilized water (slightly above room temperature) to three-quarters full. Be sure in advance to leave sufficient head space for a small hydrometer (allow for hydrometer introduction through fermentor top). Add yeast suspension, cap tightly, and shake thoroughly for 30 s. Set hydrometer in fermentation liquid. Insert one-hole cork with exit to transfer tube and airlock; set water level in latter ~1 cm above hose outlet.

(*c*) *Record* fermentation progress in time (1 week) by monitoring (1) CO_2 bubble rate (constant hose depth below airlock) (devise a means of measuring volume per CO_2 bubble) and (2) hydrometer reading (do not lose airseal).

(*d*) Plot and discuss data in terms of apparent growth phases and relation between ethanol and CO_2 production (state your assumptions clearly). From knowledge of total sugar added initially (grape + crystallized), and of the dry-cell mass weight at the end of fermentation and your data of part (*c*), demonstrate a carbon mass balance, stating assumptions.

(*e*) It has been said that taste and smell are very sensitive measuring devices. Try your product!

7.2 Temperature variations of growth Johnson, Eyring, and Polissar (*The Kinetic Basis of Molecular Biology*, John Wiley & Sons, N.Y., 1954) represent growth of *E. coli* between 18 and 46°C by the following equation for the specific growth rate μ:

$$\mu = \frac{\alpha T \exp\left(-\frac{\Delta H_1}{RT}\right)}{1 + \exp\left(\frac{\Delta S}{R} - \frac{\Delta H_2}{RT}\right)}$$

where $\alpha = 0.3612e^{24.04}$
$\Delta H_1 = 15$ kcal/mol
$\Delta H_2 = 150$ kcal/mol
$\Delta S = 476.46$ cal/(g mol·K)

(*a*) Plot this function as ln μ vs. $1/T$.

(*b*) Show that this equation can be represented as the product of two functions whose form is suggested by the plot in part (*a*). What explanation(s) rationalizes these two individual functions and the value of the above parameters?

(*c*) In this interpretation of $\mu = f(T)$, what implicit assumptions are made with regard to irreversible deactivations?

7.3 Single- and multiple-substrate kinetics A culture is grown in a simple medium including 0.3% wt/vol of glucose; at time $t = 0$, it is inoculated into a larger sterile volume of the identical medium. The optical density (OD) at 420 nm vs. time following inoculation is given below in column 1 of Table 7P3.1. The same species cultured in a complex medium is inoculated into a mixture of 0.15% wt/vol glucose and 0.15% wt/vol lactose; the subsequent OD-vs.-time data are given in column 2. If OD (420 nm) is linear in cell density with 0.175 equal to 0.1 mg dry weight of cells per milliliter, evaluate the maximum specific growth rate μ_{max}, the lag time t_{lag}, and the overall yield factors Y (grams of cell per grams of substrate), assuming substrate exhaustion in each case. Explain the shape of the growth curves in each case.

Table 7P3.1†

	OD			OD	
Time, h	(1)	(2)	Time, h	(1)	(2)
0	0.06	0.06	4.5	0.44‡	0.43
0.5	0.08	0.06	5.0	0.52‡	0.48
1.0	0.11	0.06	5.5	0.52‡	0.50
1.5	0.14	0.07	6.0		0.52
2.0	0.20	0.10	6.5		0.30‡
2.5	0.26	0.13	7.0		0.42‡
3.0	0.37	0.18	7.5		0.50‡
3.5	0.49	0.26	8.0		0.50‡
4.0	0.35‡	0.32			

† Table adapted from D. Kerridge and K. Tipton (eds.), *Biochemical Reasoning*, prob. 39, W. B. Benjamin, Inc., Menlo Park, Calif., 1972.

‡ Sample diluted twofold before OD measurement.

7.4 Death or deactivation kinetics: chlorination A number of death or deactivation rate equations have appeared in the literature; an early result was an exponential decay for anthrax spore deactivation by 5% phenol (H. Chick, "Investigation of the Laws of Disinfection," *J. Hyg.*, **8**: 698, 1908). (The

Table 7P4.1 *E. coli* survival, %[†]
pH 8.5, 2 to 5°C

Cl, mg/L	Time of contact, min				
	0.5	2	5	10	20
0.14	52	11	0.7		
0.07	80	56	30	0.5	0
0.05	95	85	65	21	0.31

[†] Adapted from G. M. Fair, J. C. Geyer, and D. A. Okan, *Water and Wastewater Engineering*, pp. 31–39, John Wiley & Sons, Inc., New York, 1968.

equation for deactivation: $dN/dt = -kN$ is often referred to as *Chick's law*.) Three other forms which have been used are

Logistic:
$$-\frac{dN}{dt} = kN + k'N(N_0 - N)$$

t^2:
$$-\frac{dN}{dt} = ktN$$

Retardant:
$$-\frac{dN}{dt} = \frac{k}{1 + \alpha t} N$$

Thus, for these three, the apparent first-order rate constant changes continually with time.

(*a*) Integrate each form and sketch the shape of the curve which would be obtained on an appropriate plot in each case, indicating how the parameters in each model would be evaluated from such a graph.

(*b*) Which law, if any, do the data in Table 7P4.1 for *E. coli* follow?

(*c*) The time to fixed percentage kill or inactivation has been correlated by the form (concentration)$^\alpha$ (time) = const for a given disinfectant. For chlorine (as HOCl), $\alpha = 0.86$ (*E. coli*), 3 (adenovirus), 1 (poliomyelitis virus), 2 (Coxsackie virus A). Why would you (not) expect such variation of α?

7.5 Cell and product kinetics A generalized form of the logistic equation is proposed by Konak ("An Equation for Batch Bacterial Growth," *Biotech. Bioeng.*, **17**: 271, 1975)

$$\frac{1}{N_\infty^{a+b}} \frac{dN}{dt} = k\left(\frac{N}{N_\infty}\right)^a \left(1 - \frac{N}{N_\infty}\right)^b$$

where N_∞ = stationary-phase population
$a, b,$ = const
N = cell mass

(*a*) Show that the maximum growth rate occurs at $N/N_\infty = a/(a + b)$ and that its value is given by

$$\left.\frac{dN}{dt}\right|_{\max} = \frac{kN_\infty^{a+b} a^a b^b}{(a + b)^{a+b}}$$

(*b*) Indicate by sketches and the previous equation(s) how you would evaluate each parameter of the model.

(*c*) Show that by replacing N by P on the left-hand side only and choosing a and b appropriately, a product-generating equation results which yields growth-associated, nongrowth-associated, or growth- and nongrowth-associated behavior analogous to the model of Luedeking and Piret in this chapter.

7.6 Cell networks and product-formation kinetics (*a*) Summarize why quantification of product formation kinetics is often more difficult than substrate-consumption or biomass-production kinetics.

(*b*) For an antibiotic, vitamin, or amino acid of your choice, research and write a short paper outlining the microbial synthesis steps, identifying, where possible, slow vs. equilibrated transformations and the presence of enzyme inhibition-activation and induction-repression. As a final portion, propose or discuss (if known) the kinetics of product formation based on the synthesis sequence.

7.7 Thiele modulus of the cell The Thiele moduli of many cells and their inner organelles (membrances, mitochondria, ribosomes, etc.) appear to be at or just below unity (P. B. Weisz, "Diffusion and Chemical Transformation" *Science*, **179**: 433, 1973). Assuming that the cell may produce enzyme levels well above or below this value, comment quantitatively on why this might (not) reasonably be expected. For simplicity, assume that only one kind of enzyme exists. Include consideration of the rate per volume, the total growth rate, and the cost of enzyme synthesis. Note that the first statement implies that cell kinetics measured in absence of external mass-transfer influences (Chap. 4) will provide true kinetic data, as has been assumed throughout Chap. 7.

7.8 Double-inhibition kinetics The growth of *Candida utilis* on sodium acetate appears to be inhibited by the substrate acetate and also pH (J. V. Jackson and V. H. Edwards "Kinetics of Substrate Inhibition of Exponential Yeast Growth," *Biotech. Bioeng.*, **17**: 943, 1975).

(*a*) With the following assumptions, write down the form for the growth rate in terms of appropriate constants and the hydrogen-ion and total-substrate concentration.

1. The substrate-inhibition function in terms of total substrate is analogous to that for enzyme-substrate noncompetitive inhibition, i.e., that of Andrews, Eq. (7.32).
2. The maximum growth rate μ_{max} is a similar function of hydrogen ion concentration.
3. The inhibitor in the substrate term is the protonated form; i.e., there is an equilibrium between HS (active inhibitor) $\rightleftharpoons S^-$ (inactive) $+ H^+$.

(*b*) Show on a Lineweaver-Burk plot for $1/\mu$ vs. $1/s$ how the curve(s) would change with pH. How would you evaluate the parameters of this model?

A number of processes involved in water treatment are similarly affected by pH. An example of anaerobic digester kinetics appears in Sec. 14.4.6.

7.9 Monod kinetics in CSTRs Consider an organism which follows the Monod equation where $\mu_{max} = 0.5\ h^{-1}$ and $K_s = 2$ g/L.

(*a*) In a continuous perfectly mixed vessel at steady state with no cell death, if $s_0 = 50$ g/L and $Y = 1$(g cells/g substrate), what dilution rate D will give the maximum total rate of cell production?

(*b*) For the same value of D using tanks of the same size in series, how many vessels will be required to reduce the substrate concentration to 1 g/L?

7.10 CSTR design with inhibition (*a*) Construct a plot of x and s vs. D.

$$r_x = \frac{\mu_{max} s x}{K_s + s + iK_s/K_i}$$

where
$$\begin{aligned} s_0 &= 10 \text{ g/L} \\ x_0 &= 0 \\ K_s &= 1 \text{ g/L} \\ K_i &= 0.01 \text{ g/L} \\ i &= 0.05 \text{ g/L} \\ \mu_{max} &= 0.5\ h^{-1} \\ Y_{X/S} &= 0.1 \text{ g cells/g substrate} \end{aligned}$$

Plot on the same chart the values of x and s when $i = 0$.

(*b*) Suppose a CSTR was designed to operate for inhibitor-free conditions. Develop equations for the ratios (s/s_i) and $xD/(xD)_i$, that is, for the ratio of predicted to observed substrate concentrations and biomass rates. How does the presence of the inhibitor alter reactor behavior with regard to $(xD)_{max}$ and washout?

(*c*) Repeat parts (*a*) and (*b*) for the case of i increasing with x, taking $i = x/10$.

7.11 Yield coefficient When a negative term is included in cellular kinetics in order to model endogenous metabolism or maintenance energy, the resulting equation (assuming excess substrate) can be written

$$-\frac{1}{Y}\mu x = -\frac{1}{Y'}\mu x - k_e x$$

where Y' = growth yield, grams of cell produced per gram of substrate consumed for growth
k_e = grams of substrate consumed for maintenance energy per gram of cell
Y = apparent yield, grams of substrate consumed per gram of cell

Refer to Fig. 7P.11.1 for information.

(*a*) Show that in a CSTR, x and s are related by

$$x = \frac{DY'}{k_e Y' + D}(s_0 - s)$$

(*b*) What relation must hold between k_e, Y', and D in this circumstance?

(*c*) Does the Fig. 7P.11.1 for *A. aerogenes* growing on glycerol satisfy this model? Be quantitative.

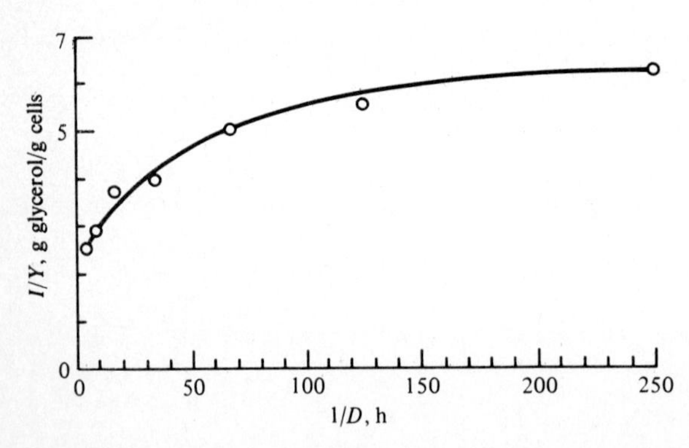

Figure 7P11.1 Apparent yields vs. reciprocal dilution rate. *[From Kirsh and Sykes, Prog. Ind. Microbiol., vol 9, p. 155, 1971, from data of D. W. Tempest, D. Herbert, and P. J. Phipps, in E. O. Powell et al. (eds.), "Microbial Physiology and Continuous Culture," HMSO, London, 1967.]*

7.12 Metabolic stoichiometry and rates A simplified representation of metabolic pathways involved in citric acid production by *Candida lipolytica* is shown in Fig. 7P12.1. Here, substantial stoichiometric detail is retained for key reactions in the TCA and glyoxylate cycles. The quantities v_i denote the specific rates of carbon flow between metabolite pools; ψ_{car}, ψ_{pro}, and α_i values are stoichiometric constants; μ is specific growth rate; v is the specific rate of glucose uptake; and Q_i values denote specific product formation rates (S. Aiba and M. Matsuoka, "Identification of Metabolic Model: Citrate Production from Glucose by *Candida lipolytica*," *Biotech. Bioeng.*, **21**: 1373, 1979).

(*a*) Assuming quasi-steady state for all metabolites, establish constraints relating internal rates with observable overall rates of glucose consumption, CO_2 formation, and production of citrate and isocitrate.

(*b*) Write an overall carbon balance. Is this equation independent of those developed in part (*a*)?

(*c*) In order to evaluate the most important pathways in citrate production, the following models have been proposed: (I) glyoxylate cycle absent ($v_7 = 0$); (II) TCA cycle blocked ($v_8 = 0$). Using the equations developed in parts (*a*) and (*b*) and the experimental data in Table I of Aiba and Matsuoka, calculate the intracellular metabolic fluxes for these two models at $D = 0.122$ and 0.0769 h^{-1}. Based on these calculations, which of the two proposed models appears more reasonable?

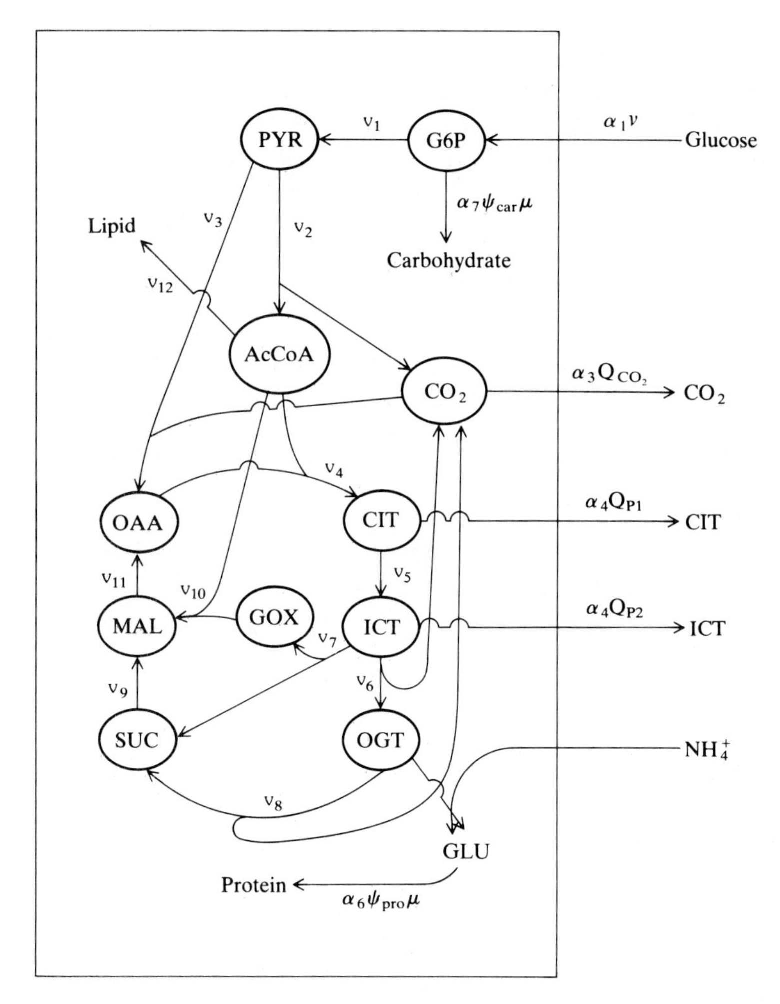

Figure 7P12.1 Stoichiometric representation for citrate production (AcCoA = acetyl-CoA; CIT = citrate; G6P = glucose 6-phosphate; GLU = glutamate; ICT = isocitrate; MAL = malate; OAA = oxaloacetate; OGT = α-ketoglutarate; PYR = pyruvate; SUC = succinate. *(Reprinted by permission from S. Aiba and M. Matsuoka, Identification of Metabolic Model: Citrate Production from Glucose by Candida lipolytica, Biotech. Bioeng., vol. 21, p. 1373, 1979.)*

7.13 Dilution rate for maximum productivity and for washout (*a*) From Eqs. (7.16) and (7.18), show that *D*(max output)/*D*(washout) depends only on (s_f/K_s), and is given by

$$(D_{m0}/D_{w0}) = [1 + (1 + s_f/K_s)^{-1/2}]^{-1}$$

(*b*) Derive the following limiting forms of the relation in (*a*) above:

$$\text{if } (s_f/K_s) \ll 1,\ D_{m0}/D_{w0} \simeq \frac{1}{2}\left[1 + \frac{1}{4}\frac{s_f}{K_s}\right]$$

$$\text{if } (s_f K_s) \gg 1,\ D_{m0}/D_{w0} \simeq 1 - \sqrt{\frac{K_s}{s_f}}\left[1 - \frac{1}{2}\left(\frac{K_s}{s_f}\right)\right]$$

(*c*) Comment on the ease of stable reactor operation and control for the two limiting cases of part (*b*).

(*d*) If $Y_{P/X} = f(D)$, derive an expression for $f(D)$ and its derivative, df/dD, which must be true at maximum productivity, $(pD)_{max}$.

7.14 Inhibition kinetics If a microbial species is inhibited by a volatile product (such as ethanol), the growth rate can be increased by the removal of the inhibiting product via the continuous evacuation of the vapor space above the fermentor.

(*a*) For a batch fermentation following Eq. (7.33), show that the time course of substrate level for the overall reaction: S → 0.3P + cell mass is

$$\frac{ds}{dt} = \frac{-[x_0 + Y_{X/S}(s_0 - s)]}{Y_{X/S}} \mu_{max} \frac{s}{K_s + s} \frac{K_I}{K_I + 0.3(s_0 - s)}$$

where x is biomass (grams per liter) and a constant yield factor $Y_{X/S}$ is assumed.

(*b*) Integrate this result (by the method of partial fractions, for example) and develop a form for the ratio of biomass densities $x(t$, no inhibitor$)/x(t$, inhibitor$)$.

(*c*) Evaluate the biomass-ratio parameter of part (*b*) vs. time when $x_0 = 10^{-6}$ g/mL, $Y_{X/S} = 0.1$ g cell/g substrate, $K_s = 0.22$ g/L, $\mu_{max} = 0.408\ h^{-1}$, $K_I = 16$ g/L for two substrate levels: $s_0 = 5.0$ g/L and $s_0 = 70$ g/L, [molecular weight S = 3 (molecular weight P)].

7.15 Lag phase in batch cultivation (*a*) Data for batch growth of *Aerobacter aerogenes* in ammonium sulfate medium shows that the time spent in lag phase (t_{lag}) is approximately proportional to inoculum volume and inversely proportional to inoculum cell density [3]. Dean and Hinshelwood [3] have proposed a model for these observations which is based on the hypothesis that the lag period ends when some critical substance *in the cell* reaches a threshold concentration c'. They further suggest that, during the lag phase, the concentration c of this critical substance varies with time as

$$c = aV + a'n_0 t + a''t \tag{1}$$

where V = volume of old medium transferred

a = concentration of critical substance per unit volume of old medium x (old volume/new volume)

n_0 = number of cells per unit new volume (assumed constant since growth rate is negligible in lag phase)

a' = average increase in cell critical substance (due to production by other cells) per time per cell

a'' = increase in critical substance due to internal cell production

Evaluate t_{lag} using this equation and comment on the relationship between the result and the experimentally observed trends. Comment on the assumptions required to obtain Eq. (1).

(*b*) The state of the inoculum population is known to influence the duration of the lag phase. Suppose that the inoculum contains n_0 viable cells and n_d dead cells and that the viable cells grow with constant specific growth rate μ immediately following inoculation. Evaluate μ_{app}, the measured, apparent specific growth rate based on the total cell density, as a function of time. Using this result, comment on the effects of inoculum viability and inoculum size on apparent lag time.

7.16 Microbial population growth A new microorganism has been discovered which at each cell division yields three daughters. From the growth rate data given below calculate the mean time between successive cell divisions.

t, h	dry wt, g/L
0.	0.10
0.5	0.15
1.0	0.23
1.5	0.34
2.0	0.51

7.17 Biotransformation kinetics A monolayer of cells adhering to a solid surface is catalyzing transformation of substrate S to product P while growing to a negligible extent. Suppose that the rate of substrate utilization per unit surface is described by

$$\bar{v} = \frac{\bar{v}_{max} \cdot s}{K_s + s}$$

where, because P inhibits the reaction,

$$K_s = K_{s0}(1 + p/K_i).$$

In the above equations, s and p denote concentrations of S and P, respectively, *at the surface.*

S and P concentrations in the bulk fluid far from the surface are s_0 and p_0, respectively. Mass transfer coefficients for S and P are h_s and h_p, respectively.

If every mole of S converted produces $Y_{P/S}$ moles of product P, evaluate the rate of P formation per unit surface in terms of the observable bulk concentrations s_0 and p_0.

7.18 Product formation and maintenance models It has been argued extensively that the Luedeking-Piret model relating growth and product formation [Eq. (7.93)]

$$r_{fp} = \alpha\mu x + \beta x$$

and the maintenance energy model set forth by Pirt (*Proc. Royal Soc.*, Series B, **163**: 224–231, 1965)

$$\frac{1}{Y} = \frac{1}{Y'} + \frac{m}{\mu}$$

where Y = the observed cell yield based on substrate
Y' = the theoretical, or maintenance-free, growth yield coefficient
m = maintenance coefficient (gm substrate/gm cell·h)

are equivalent. Do you agree? Substantiate your answer.

7.19 Williams' structured model of growth Let us follow Williams' idea and divide the biomass into two parts: Part 1 is composed of intermediates, enzymes, and other entities involved in the formation of materials used for synthesis of structural and genetic material; it is Williams' synthetic part of the biomass. Part 2 is the structural and genetic part of the biomass. Suppose that biomass of part 1 is a fraction f_1 of the total biomass and biomass of part 2 is a fraction f_2 of the total biomass; these two fractions sum to unity.

Let the model equations written in intrinsic form be

$$\frac{df_1}{dt} = \alpha\frac{k_1 s f_2}{K_1 + s} - \frac{k_2 f_1 f_2}{K_2 + f_1} - \mu f_1$$

$$\frac{df_2}{dt} = \beta\frac{k_2 f_1 f_2}{K_2 + f_1} - \mu f_2$$

$$\text{where } \mu = \alpha \frac{k_1 s f_2}{K_1 + s} + (\beta - 1)\frac{k_2 f_1 f_2}{K_2 + f_1}$$

$\alpha \equiv$ amount of synthetic biomass formed upon consumption of one unit of mass of the rate-limiting substrate

$\beta \equiv$ amount of structural/genetic biomass formed upon consumption of one unit mass of synthetic biomass.

(*a*) Consider *balanced growth* for which $\dot{f}_1 = \dot{f}_2 = 0$; this will be achieved if s can be maintained constant for a sufficient length of time. How does μ depend on s in balanced growth?

(*b*) If $K_2 \gg 1$ and $k_2 \gg \alpha k_1 K_2$, show that Monod's model will be approximately valid for balanced growth. What are the approximate values of f_1 and f_2 in this case?

(*c*) Let $x_1 \equiv cf_1$ and $x_2 \equiv cf_2$, where $c = x_1 + x_2$. Write the differential equations for x_1, x_2, and s for a chemostat situation. Also write the differential equation for n, the population density; assume that γ, the amount of structural/genetic material per cell, is a constant.

(*d*) For a batch growth situation, construct plots of

$$\log c \text{ vs. } t$$

$$\log n \text{ vs. } t$$

and

$$\bar{m} \equiv c/n \text{ (mean cell size) vs. } t$$

for two different initial values of f_1: $f_1 = 0.40$ and $f_1 = 0.05$. In both cases, take

$$c_0 = 2 \times 10^{-3} \text{ g/L and}$$

$$s_0 = 5.0 \text{ g/L}$$

Use the following set of constants:

$$\alpha = \beta = 0.50$$

$$k_1 = 6.0 \text{ h}^{-1} \qquad k_2 = 3.0 \text{ h}^{-1}$$

$$K_1 = 0.2 \text{ g/L} \qquad K_2 = 0.25$$

$$\text{mass per cell} = 1.1 \times 10^{-13} \text{ g/cell}$$

(*e*) Use the model parameters in (*d*) to construct a graph showing the variation of μ with s for a balanced growth situation. Plot on the same graph the variation of μ with s predicted by Monod's model. Choose the Monod model constants μ_m and K_s such that

(i) the asymptotic value of μ at large values of s and

(ii) the derivative of μ with respect to s at s = 0 are the same for the Monod model as for the general model.

(*f*) For a steady chemostat with $s_f = 5.0$ g/L, construct a graph showing the variation with dilution rate of c, s, and $\bar{m}$.

(*g*) How would c_1, s, and $\bar{m}$ respond to a chemostat "shift-up" from, $D = 0.1 \text{ h}^{-1}$ to $D = 0.3 \text{ h}^{-1}$? How would they respond to a "shift-down" from $D = 0.3^{-1}$ to $D = 0.1 \text{ h}^{-1}$? Assume steady state, balanced growth just before the shift.

(*h*) Discuss the results in terms of comparisons with trends actually observed with bacteria.

REFERENCES

The following journals report recent research on cell population kinetics and many other aspects of biochemical engineering and applied biology: *Biotechnology and Bioengineering, Enzyme and Microbial Technology, Biotechnology Letters, Journal of Fermentation Technology, Journal of Chemical Technology and Biotechnology, Journal of Applied Chemistry and Biotechnology, Process Biochemistry, Trends in Biotechnology, Agricultural and Biological Chemistry, Applied Microbiology, Applied Biochemistry and Microbiology*, and others. The series *Advances in Biochemical Engineering and Biotech-*

nology (originally *Advances in Biochemical Engineering*) offers excellent review papers which often anticipate future technology. Other rich sources of review and research articles are the *Annual Reports on Fermentation Processes*, the volumes on *Biochemical Engineering* published in the *Annals of the New York Academy of Science*, and the annuals *Progress in Industrial Microbiology* and *Society for General Microbiology Symposium*.

1. N. Blakebrough (ed.), *Biochemical and Biological Engineering Science*, vol. 1, Academic Press, Inc. New York, 1967. A useful reference for several aspects of biological reactor design. Chapter 6, Fermentation Process Kinetics, by R. Luedeking, is recommended.
2. S. Aiba, A. E. Humphrey, and N. F. Mills, *Biochemical Engineering*, 2d ed., Academic Press, Inc. New York, 1973. This comprehensive text on fermentation treats batch kinetics in Chap. 4 and offers an extended discussion of continuous fermentation in Chap. 5. Chapter 9 reviews liquid sterilization.
3. A. C. R. Dean and C. N. Hinshelwood, *Growth, Function and Regulation in Bacterial Cells*, Oxford University Press, London, 1966. In addition to discussing the mathematical analysis of reaction networks, this useful monograph presents a large body of experimental information on other aspects of growth and regulation. Highly recommended reading; the only major weakness is undue skepticism about the significance of molecular biology.
4. P. S. S. Dawson (ed.), *Microbial Growth*, Dowden, Hutchinson, & Ross, Inc., Stroudsburg, Pa., 1974. A fascinating collection of many of the classic papers dealing with microbial growth, including Monod's original paper.
5. J. A. Roels, *Energetics and Kinetics in Biotechnology*, Elsevier, Amsterdam, 1983. An excellent integrated treatment of macroscopic thermodynamics and kinetic models for microbial growth and product formation.
6. H. W. Blanch, E. T. Papoutsakis, and G. Stephanopoulos (eds.), "Foundations of Biochemical Engineering: Kinetics and Thermodynamics in Biological Systems," *ACS Symposium Series 207*, American Chemical Society, Washington, D.C., 1983. A comprehensive collection of review papers on kinetics from enzymes to cell populations.

Review papers on mathematical modeling of cell growth which are especially useful:

7. H. M. Tsuchiya, A. G. Fredrickson, and R. Aris, "Dynamics of Microbial Cell Populations," *Adv. Chem. Eng.*, **6**: 125, 1966.
8. A. G. Fredrickson, R. D. Megee, III, and H. M. Tsuchiya, "Mathematical Models for Fermentation Processes," *Adv. Appl. Microbiol.*, **23**: 419, 1970.

Papers considering product and substrate inhibition of microbial growth:

9. S. Aiba, M. Shoda, and M. Nagatani, "Kinetics of Product Inhibition in Alcohol Fermentation," *Biotechnol. Bioeng.*, **10**: 845, 1968.
10. S. Aiba and M. Shoda,"Reassessment of the Product Inhibition in Alcohol Fermentation," *J. Ferment. Technol. Jpn.*, **47**: 790, 1969.
11. J. F. Andrews, "A Mathematical Model for the Continuous Culture of Microorganisms Utilizing Inhibitory Substrates," *Biotechnol. Bioeng.*, **10**: 707, 1968.

Growth and product formation for filamentous organisms:

12. B. Metz and N. W. F. Kossen, "The Growth of Molds in the Form of Pellets—A Literature Review," *Biotechnol. Bioeng.*, **19**: 781, 1977.
13. R. D. Megee, S. Kinoshita, A. G. Fredrickson, and H. M. Tsuchiya, "Differentiation and Product Formation in Molds," *Biotechnol. Bioeng.*, **12**: 771, 1970.
14. M. Matsumura, T. Imanaka, T. Yoshida, and H. Taguchi, "Modeling of Cephalosporin C Production and Its Application to Fed-batch Culture," *J. Ferm. Tech. Jpn.*, **59**: 115, 1981.

Structured growth models. See Refs. 5–8 above and

15. A. G. Fredrickson, "Formulation of Structured Growth Models," *Biotechnol. Bioeng.*, **18**: 1481, 1976.
16. F. M. Williams, "A Model of Cell Growth Dynamics," *J. Theoret. Biol.*, **15**: 190, 1967.
17. A. Harder and J. A. Roels, "Application of Simple Structured Models in Bioengineering," p. 55. In *Advances in Biochemical Engineering*, vol. 21, A. Fiechter (ed.), Springer-Verlag, Berlin, 1982.
18. A. H. E. Bijkerk and R. J. Hall, "A Mechanistic Model of the Aerobic Growth of *Saccharomyces cerevisiae*," *Biotechnol. Bioeng.*, **19**: 267, 1977.
19. N. B. Pamment, R. J. Hall, and J. P. Barford, "Mathematical Modeling of Lag Phases in Microbial Growth," *Biotechnol. Bioeng.*, **30**: 349, 1978.
20. M. L. Shuler and M. M. Domach, "Mathematical Models of the Growth of Individual Cells," p. 101 in *Foundations of Biochemical Engineering*, H. W. Blanch, E. T. Papoutsakis, and G. Stephanopoulos (eds.), American Chemical Society, Washington, D. C., 1983.
21. D. Ramkrishna, "A Cybernetic Perspective of Microbial Growth," p. 161 in *Foundations of Biochemical Engineering*, H. W. Blanch, E. T. Papoutsakis, and G. Stephanopoulos (eds.), American Chemical Society, Washington, D.C., 1983.
22. D. S. Kompala, D. Ramkrishna, and G. T. Tsao, "Cybernetic Modeling of Microbial Growth on Multiple Substrates," *Biotechnol. Bioeng.*, **26**: 1272, 1984.

Product formation kinetics:

23. E. L. Gaden, *Chem. Ind. Rev.* **1955**: 154; *J. Biochem. Microbiol. Technol. Eng.*, **1**: 413, 1959.
24. F. H. Deindoerfer, "Fermentation Kinetics and Model Processes," *Adv. Appl. Microbiol.*, **2**: 321, 1960.
25. D. F. Ollis, "A Simple Batch Fermentation Model: Theme and Variations," *Ann. N.Y. Acad. Sci.*, **413**: 144, 1983.
26. S. Pažoutova, J. Votruba, and Z. Řeháček, "A Mathematical Model of Growth and Alkaloid Production in the Submerged Culture of *Claviceps purpurea*," *Biotechnol. Bioeng.*, **23**: 2837, 1981.
27. S. B. Lee and J. E. Bailey, "Analysis of Growth Rate Effects on Productivity of Recombinant *Escherichia coli* Populations," *Biotechnol. Bioeng.*, **26**: 66, 1984.
28. S. B. Lee and J. E. Bailey, "Genetically Structured Models for *lac* Promoter-Operator Function in the *Escherichia coli* Chromosome and in Multicopy Plasmids: *lac* Operator Function," *Biotechnol, Bioeng.*, **26**: 1372, 1984,

Segregated kinetic models:

29. D. Ramkrishna, "Statistical Models of Cell Populations," *Adv. in Biochem. Eng.*, **11**: 1, 1979.
30. Y. Nishimura and J. E. Bailey, "On the Dynamics of Cooper-Helmstetter-Donachie Procaryote Populations," *Math. Biosci.*, **51**: 505, 1980.
31. M. A. Hjortso and J. E. Bailey, "Steady-State Growth of Budding Yeast Populations in Well-Mixed Continuous Flow Microbial Reactors," *Math. Biosci.*, **60**: 235, 1982.
32. P. Shu, "Mathematical Models for the Product Accumulation in Microbial Processes," *J. Biochem. Microbiol, Technol. Eng.*, **3**: 95. 1961.

An excellent introduction to the literature on sterilization:

33. N. Blakebrough, "Preservation of Biological Materials Especially by Heat Treatment," in N. Blakebrough (ed.), *Biochemical and Biological Engineering Science*, vol. 2, Academic Press, Inc., New York, 1968.

The factors affecting organisms' susceptibility to sterilization are reviewed in Chaps. 20 and 21 of Frobisher (Ref. 2 of Chap. 1); consideration of phage destruction:

34. Hango et al, "Phage Contamination and Control," in *Microbial Production of Amino Acids*, Kodansha Ltd., Tokyo, and John Wiley & Sons, Inc., New York, 1973.

CHAPTER

EIGHT

TRANSPORT PHENOMENA IN BIOPROCESS SYSTEMS

The previous chapters have considered progressively larger scales of distance: from molecular through cellular to fluid volumes containing millions or billions of cells per milliliter. As the sources and sinks of entities such as nutrients, cells, and metabolic products become further separated in space, the probability increases that some physical-transport phenomena, rather than a chemical rate, will influence or even dominate the overall rate of solute processing in the reaction volume under consideration. Indeed, according to the argument of Weisz [1], cells and their component catalytic assemblies operate at Thiele moduli near unity; they are operating at the maximum possible rate without any serious diffusional limitation. If, in bioprocess circumstances, a richer supply of carbon nutrients is created, evidently the aerobic cell will be able to utilize them fully only if oxygen can also be maintained at a higher concentration in the direct vicinity of the cell. This situation may call for increased *gas-liquid mass transfer* of oxygen, which has sparingly small solubility in aqueous solutions, to the culture.

Evidently, the boundary demarcating aerobic from anaerobic activity depends upon the local bulk-oxygen concentration, the O_2 diffusion coefficient, and the local respiration rates in the aerobic region. This line divides the viable from dying cells in strict aerobes such as mold in mycelial pellets or tissue cells in cancer tumors; it determines the depth of aerobic activity near lake surfaces; and it divides the cohabitating aerobes from anaerobic microbial communities in soil particles. Thus, while the modern roots of biological-process oxygen mass transfer began with World War II penicillin production in the 1940s, its implications are now established to include many natural processes such as food spoilage via undesired oxidation and lake eutrophication due either to inadequate system aeration by natural oxygen supplies or to an excessive concentration of material such as phosphate or nitrate.

Other sparingly soluble gases are also of fermentation interest. Methane and other light hydrocarbons have been explored as gaseous substrates for single cell protein production; in this demanding conversion, *both* oxygen and methane must be dissolving continuously at rates sufficient to meet the biological demand. Methane transfer out of solution is important in anaerobic waste treatment, at the metabolic end of which light carboxylic acids (primarily acetic acid) are decarboxylated to give the corresponding alkanes.

Carbon dioxide is generated in nearly all microbial activity. In spite of its large solubility, the interconversion between gaseous and the various forms of dissolved carbon dioxide (CO_2, H_2CO_3, HCO_3^-, CO_3^{2-}) couples its mass transfer rate to the pH variation; this topic figures importantly in controlling the pH of acid-sensitive anaerobic digestors (Chapter 14) where CO_2 and CH_4 removal occur simultaneously.

Liquid–liquid mass transfer is important in SCP production from liquid hydrocarbon feedstocks, as well as in fermentation recovery operations; e.g., filtered or whole broth extraction of pharmaceuticals (Chapter 11) employing organic solvents.

Renewable resource bioconversions, such as the use of cellulosic, hemicellulosic, and lignin fractions of agricultural and forest wastes as fermentation feedstocks, typically involve rate processes (biomass solubilization, liquefaction, hydrolysis) limited by available particulate substrate surface areas and solute diffusion rates. Other topics also involving *liquid–solid mass transfer* include various sorption and chromatographic methods for product recovery and purification, and liquid phase oxygen transfer to mold pellets or beads and biofilms containing immobilized cells.

Operation at high cell densities may often result in mass-transfer limited conditions, as observed in reactors as diverse as laboratory shake flasks or large scale fermentors for penicillin or extracellular biopolymers (xanthan gum) or activated sludge waste plants. The process engineer must, accordingly, know when transport phenomena or biological kinetics are rate-limiting in order properly to design bioreactors.

Strong coupling often occurs between solute diffusion and momentum transport or chemical reactions or (even more complex) both. The case of diffusion and reaction interaction has been considered in Chap. 4. In such circumstances, the Thiele modulus and a saturation parameter K_s/s_0 provide the unifying parameters needed to completely describe cell and enzyme performance; i.e., effectiveness factor, for such systems. Unfortunately, the variety of circumstances under which mass transfer couples with momentum transfer, i.e., fluid mechanics, is enormous; indeed, it is the substance of a major fraction of the chemical engineering literature. For this text, we content ourselves with fundamental concepts and tabulated formulas for calculation or estimation of the appropriate mass-transfer coefficients for solutes.

A final brief section of this chapter concerns instances where heat transfer may provide an important transport effect which strongly influences the bioprocess system's behavior through spatial temperature inhomogeneity. Examples here include relatively exothermic fermentation processes, such as trickling-filter operation for wine-vinegar production or wastewater treatment, and that gardener's delight, the compost heap (municipal dump, etc.) and other solid-state fermentations.

8.1 GAS-LIQUID MASS TRANSFER IN CELLULAR SYSTEMS

The general nature of the mass-transfer problem of primary concern in this chapter is shown schematically in Fig. 8.1. A sparingly soluble gas, usually oxygen, is transferred from a source, say a rising air bubble, into a liquid phase containing cells. (Any other sparingly soluble substrate, e.g., the liquid hydrocarbons used in hydrocarbon fermentations, will give the same general picture.) The oxygen must pass through a series of transport resistances, the relative magnitudes of which depend on bubble (droplet) hydrodynamics, temperature, cellular activity and density, solution composition, interfacial phenomena, and other factors.

These arise from different combinations of the following resistances (Fig. 8.1):

1. Diffusion from bulk gas to the gas-liquid interface
2. Movement through the gas-liquid interface
3. Diffusion of the solute through the relatively unmixed liquid region adjacent to the bubble into the well-mixed bulk liquid
4. Transport of the solute through the bulk liquid to a second relatively unmixed liquid region surrounding the cells
5. Transport through the second unmixed liquid region associated with the cells
6. Diffusive transport into the cellular floc, mycelia, or soil particle
7. Transport across cell envelope and to intracellular reaction site

All of these resistances appear in Fig. 8.1. When the organisms take the form of individual cells, the sixth resistance disappears. Microbial cells themselves have some tendency to adsorb at interfaces. Thus, cells may preferentially gather at the vicinity of the gas-bubble-liquid interface. Then, the diffusing solute oxygen passes through only one unmixed liquid region and no bulk liquid before reaching the cell. In this situation, the bulk dissolved O_2 concentration does not represent the oxygen supply for the respiring microbes.

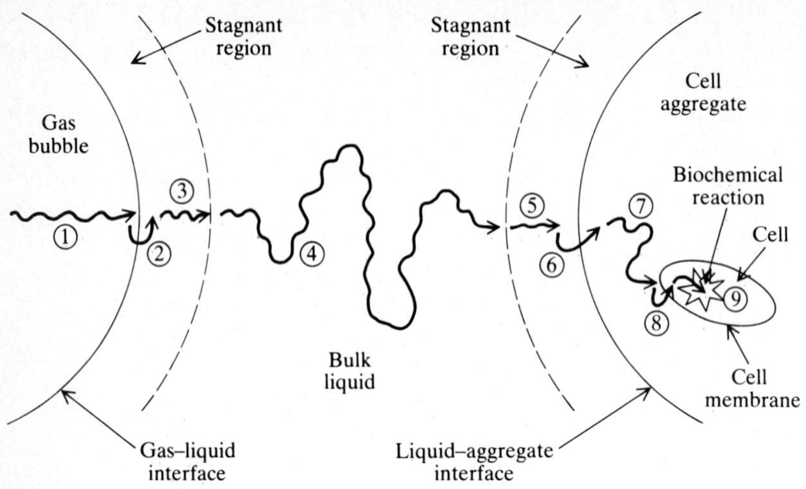

Figure 8.1 Schematic diagram of steps involved in transport of oxygen from a gas bubble to inside a cell.

Similarly, in the microbial utilization of other sparingly soluble substrates such as hydrocarbon droplets, cell adsorption on or near the hydrocarbon-emulsion interface has been frequently observed. A reactor model for this situation is considered in Chap. 9.

The variety of macroscopic physical configurations by which gas–liquid contacting can be effected is indicated in Fig. 8.2. In general, we can distinguish fluid motions induced by freely rising or falling bubbles or particles from fluid motions which occur as the result of applied forces other than the external gravity field (forced convection). The distinction is not clear-cut; gas–liquid mixing in a slowly stirred semibatch system may have equal contributions from naturally convected bubbles and from mechanical stirring. The central importance of hydrodynamics requires us to examine the interplay between fluid motions and mass transfer. Before beginning this survey, some comments and definitions regarding mass transfer are in order.

8.1.1 Basic Mass-Transfer Concepts

The solubility of oxygen in aqueous solutions under 1 atm of air and near ambient temperature is of the order of 10 parts per million (ppm) (Table 8.1). An actively respiring yeast population may have an oxygen consumption rate of the order 0.3 g of oxygen per hour per gram of dry cell mass. The peak oxygen consumption for a population density of 10^9 cells per milliliter is estimated by assuming the cells to have volumes of 10^{-10} mL, of which 80 percent is water. The absolute oxygen demand becomes

$$\frac{0.3 \text{ g O}_2}{\text{g dry mass} \cdot \text{h}}\left(10^9 \frac{\text{cells}}{\text{mL}}\right)(10^{-10} \text{ mL})\left(1 \frac{\text{g cell mass}}{\text{cm}^3}\right)\left(0.2 \frac{\text{g dry cell mass}}{\text{g cell mass}}\right)$$

$$= 6 \times 10^{-3} \text{ g/(mL} \cdot \text{h)} = 6 \text{ g O}_2\text{/(L} \cdot \text{h)}$$

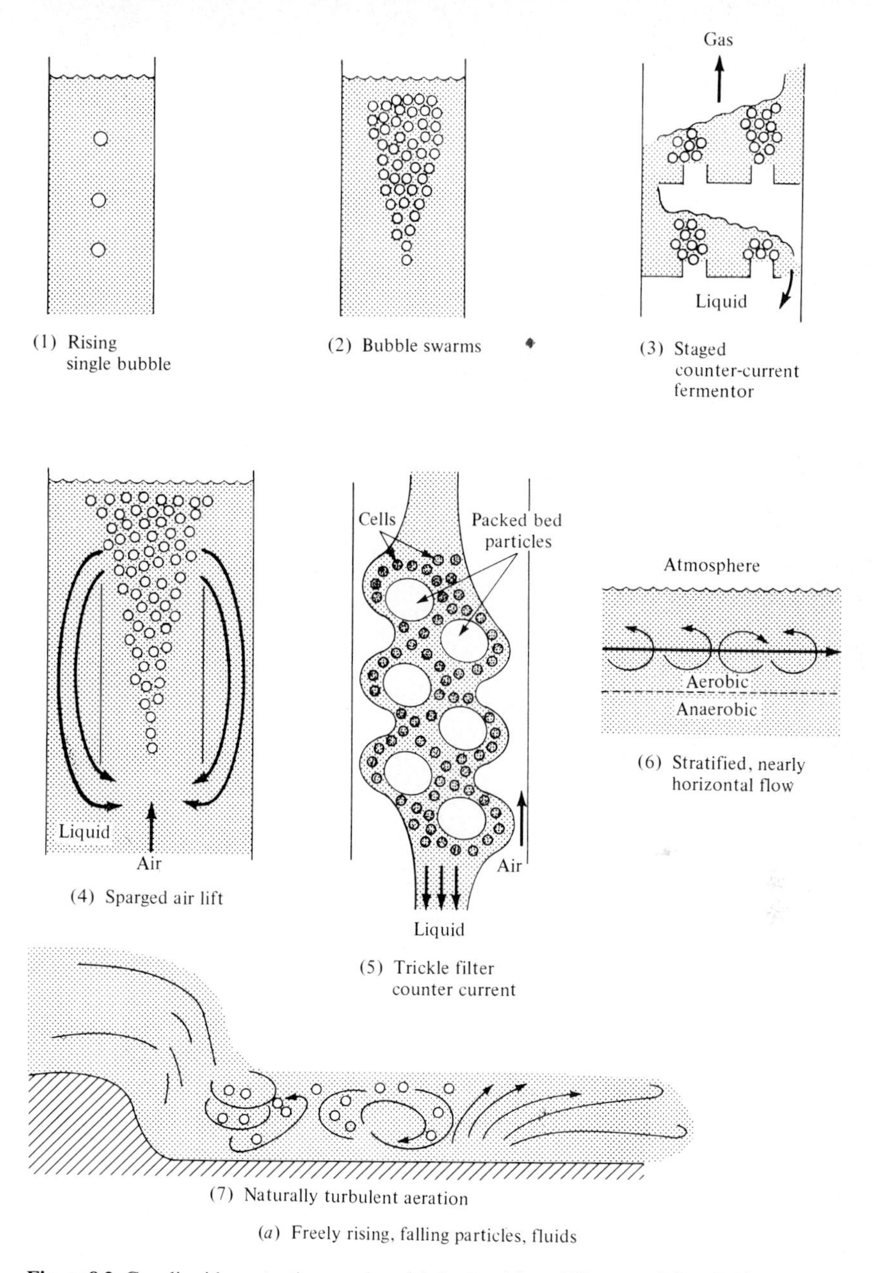

Figure 8.2 Gas–liquid contacting modes: (*a*) freely rising, falling particles, fluids.

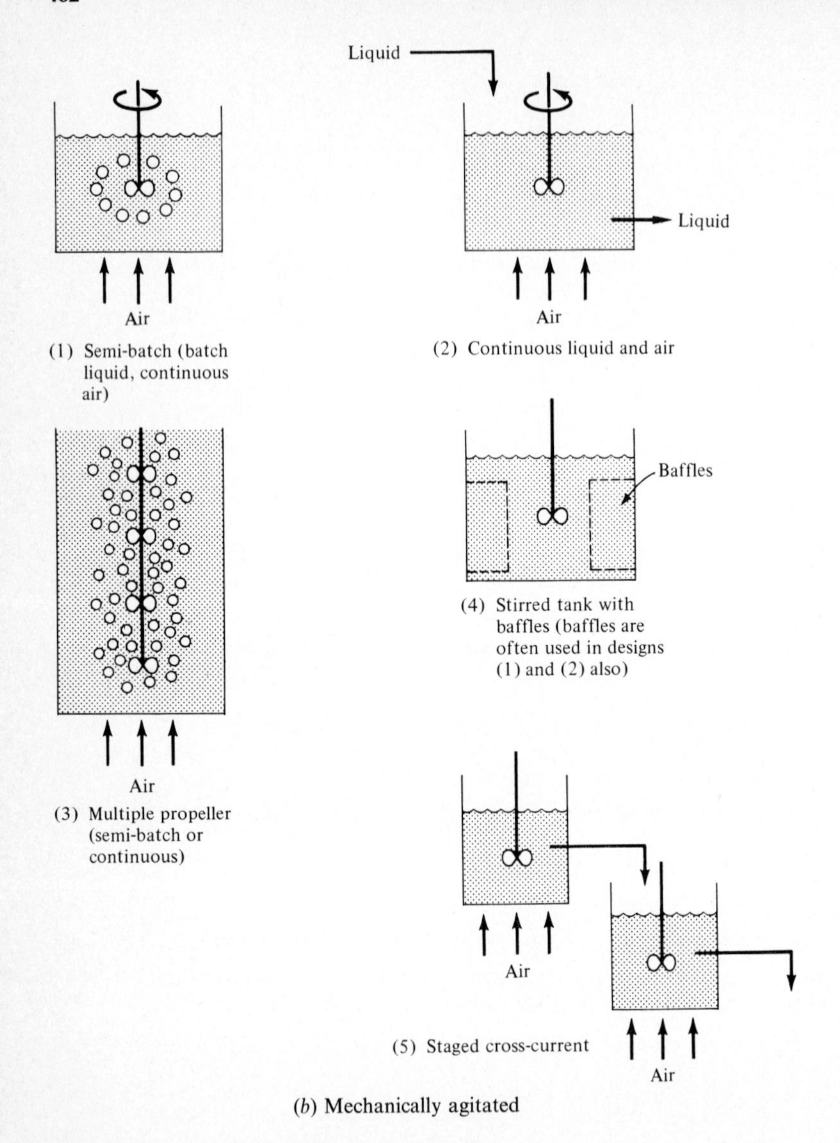

Figure 8.2 *(continued)* (*b*) mechanically agitated.

Thus, the actively respiring population consumes oxygen at a rate which is of the order of 750 times the O_2 saturation value per hour. Since the inventory of dissolved gas is relatively small, it must be continuously added to the liquid in order to maintain a viable cell population. This is not a trivial task since the low oxygen solubility guarantees that the concentration difference which drives the transfer of oxygen from one zone to another is always very small.

For sparingly soluble species such as oxygen or hydrocarbons in water, the two equilibrated interfacial concentrations c_{gi} and c_{li} on the gas and liquid sides,

Table 8.1 Solubility of O_2 at 1 atm in water at various temperatures and solutions of salt or acid at 25°C†

Temp, °C	Water, O_2 mmol/L	Temp. °C	Water, O_2 mmol/L
0	2.18	25	1.26
10	1.70	30	1.16
15	1.54	35	1.09
20	1.38	40	1.03

Aqueous solutions at 25°C

Electrolyte conc, M	O_2, mmol/L		
	HCl	H_2SO_4	NaCl
0.0	1.26	1.26	1.26
0.5	1.21	1.21	1.07
1.0	1.16	1.12	0.89
2.0	1.12	1.02	0.71

† Data from *International Critical Tables*, vol. III, p. 271, McGraw-Hill Book Company, New York, 1928, and F. Todt, *Electrochemische Sauerstoffmessungen*, W. de Guy and Co., Berlin, 1958.

respectively, may typically be related through a linear partition-law relationship such as Henry's law

$$Mc_{li} = c_{gi} \tag{8.1}$$

provided that the solute exchange rate across the interface is much larger than the net transfer rate, as is typically the case: at 1 atm of air and 25°C, the O_2 collision rate at the surface is of the order of 10^{24} molecules per square centimeter per second, a value greatly in excess of the net flux for typical microbial consumption requirements cited above.

At steady state, the oxygen transfer rate to the gas-liquid interface equals its transfer rate through the liquid-side film (Fig. 8.1). Taking c_g and c_l to be the oxygen concentrations in the bulk gas and bulk liquid respectively, we can write the two equal transfer rates

$$\begin{aligned} \text{Oxygen flux} &= \text{mol } O_2/(\text{cm}^2 \cdot \text{s}) \\ &= k_g(c_g - c_{gi}) \qquad \text{gas side} \\ &= k_l(c_{li} - c_l) \qquad \text{liquid side} \end{aligned} \tag{8.2}$$

Since the interfacial concentrations are usually not accessible in mass-transfer measurements, resort is made to mass-transfer expressions in terms of the overall mass-transfer coefficient K_l and the overall concentration driving force $c_l^* - c_l$, where c_l^* is the liquid-phase concentration which is in equilibrium with the bulk gas phase

$$Mc_l^* \equiv c_g \tag{8.3}$$

In terms of these overall quantities, the solute flux is given by

$$\text{Flux} = K_l(c_l^* - c_l) \tag{8.4}$$

Utilization of Eqs. (8.1), (8.2), (8.3), and (8.4) results in the following well-known relationship between the overall mass-transfer coefficient K_l and the physical parameters of the two-film transport problem, k_g, k_l, and M:

$$\frac{1}{K_l} = \frac{1}{k_l} + \frac{1}{Mk_g} \tag{8.5}$$

For sparingly soluble species, M is much larger than unity. Further, k_g is typically considerably larger than k_l. Under these circumstances we see from Eq. (8.5) that K_l is approximately equal to k_l. Thus, essentially all the resistance to mass transfer lies on the liquid-film side.

The oxygen-transfer rate per unit of reactor volume Q_{O_2} is given by

$$Q_{O_2} = \text{oxygen absorption rate} = \frac{\text{(flux)(interfacial area)}}{\text{reactor liquid volume}}$$

$$= k_l(c_l^* - c_l)\frac{A}{V} \tag{8.6}$$

$$= k_l a'(c_l^* - c_l)$$

where $a' = A/V$ is the gas–liquid interfacial area per unit liquid volume and the approximation $K_l \approx k_l$ just discussed has been invoked. Since our major emphasis in this chapter is aeration, we shall concentrate on oxygen transfer and henceforth use k_l in place of K_l as the appropriate mass-transfer coefficient. The symbol a, which appears in several correlations, is the gas–liquid interfacial area per unit volume of bioreactor (gas + liquid) contents. Head space gas is not included.

It is important to recognize that Q_{O_2} is defined "at a point." It is a local volumetric rate of O_2 consumption; the average volumetric rate of oxygen utilization (moles per time per volume) $\bar{Q}_{O_2}$ in an entire liquid volume V is given by

$$\bar{Q}_{O_2} = \frac{1}{V}\int_0^V Q_{O_2}\,dV \tag{8.7}$$

In general, $\bar{Q}_{O_2}$ is equal to Q_{O_2} only if hydrodynamic conditions, interfacial area/volume, and oxygen concentrations are uniform throughout the vessel.

For example, a complete description of the phenomena underlying the observed average transfer rate in a bioreactor depends on power input per unit volume, fluid and dispersion rheology, sparger characterization, and gross flow patterns in the vessel. Figure 8.3 indicates the relationship between observed average transfer rate and the causative phenomena. Since we generally lack

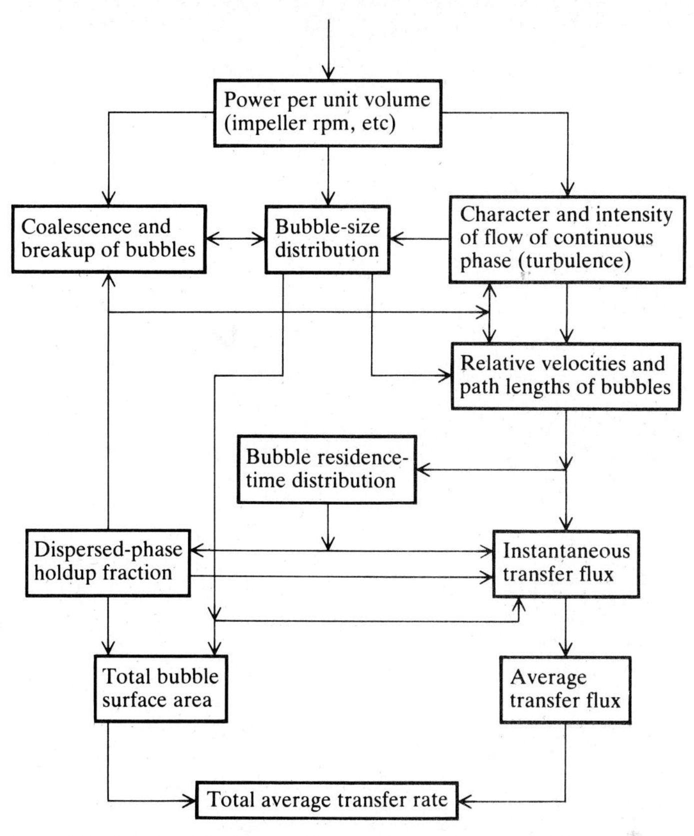

Figure 8.3 Relationships between input agitation intensity and resultant gas transfer rate. *(After W. Resnick and B. Gal-Or, Adv. Chem. Eng., vol. 7, p. 295 (1968). Reprinted by permission of Academic Press.)*

crucial fundamental information on coalescence and redispersion rates, bubble size and residence time distribution, we are typically forced to develop correlations based on appropriate averages of bubble size, holdup (gas volume fraction), gas bubble and liquid residence times, etc.

In Table 8.1 we saw that c_l^* is determined by the temperature and composition of the medium. Composition influences become more complicated when the dissolved gas can undergo liquid-phase reaction. This is the case for carbon dioxide, which may exist in the liquid phase in any of four forms: CO_2, H_2CO_3, HCO_3^-, and CO_3^{2-}. The equilibrium relations

$$K_1 = \frac{[H^+][HCO_3^-]}{[CO_2] + [H_2CO_3]} = 10^{-6.3}\ M \tag{8.8}$$

$$K_2 = \frac{[H^+][CO_3^{2-}]}{[HCO_3^-]} = 10^{-10.25}\ M \tag{8.9}$$

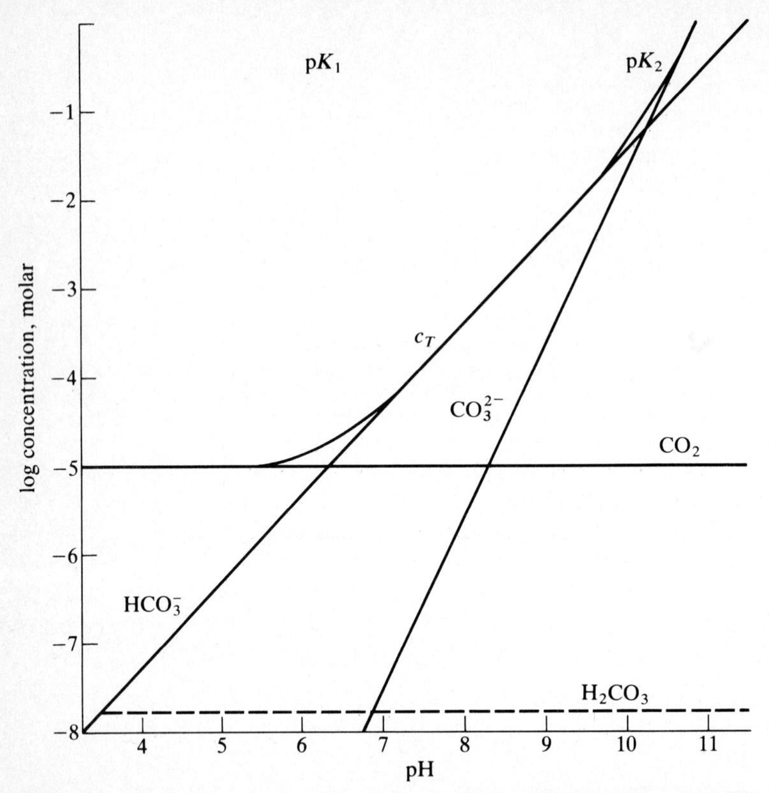

Figure 8.4 Equilibrium concentrations of dissolved CO_2, HCO_3^-, $CO_3^=$ and H_2CO_3. c_T is total concentration of all four forms of CO_2. [$p_{CO_2} = 10^{-3.5}$ atm (ambient concentration); pH adjusted with strong acid or strong base.] *(After W. Stumm and J. J. Morgan, "Aquatic Chemistry," John Wiley and Sons, N.Y. p. 127, 1970.)*

(values at 25°C) indicate that the total dissolved carbon concentration, c_T, as carbon dioxide is quite pH sensitive:

$$c_T = [CO_2] + [H_2CO_3] + [HCO_3^-] + [CO_3^{2-}]$$

$$= c_0\left[1 + \frac{K_1}{[H^+]} + \frac{K_1K_2}{[H^+]^2}\right] \tag{8.10}$$

This relation appears in Fig. 8.4, showing that below pH 5, nearly all carbon is dissolved molecular CO_2, while bicarbonate dominates when $7 < pH < 9$ and carbonate for $pH > 11$. Only the dissolved CO_2 molecule is transported across the gas–liquid interface, and we may again write Eq. (8.2) for the interfacial transfer rate.

The coupling of reaction and mass transfer may occur under neutral to basic conditions. While the reversible reaction (8.11) is rapid,

$$H_2CO_3 \rightleftharpoons HCO_3^- + H^+ \tag{8.11}$$

$$K_{eq}(T = 28°C) = \frac{[H^+][HCO_3^-]}{[H_2CO_3]} = 2.5 \times 10^{-4} \text{ mol/L} \tag{8.12}$$

the important reaction

$$H_2CO_3 \underset{k_{-1}}{\overset{k_1}{\rightleftharpoons}} CO_2 + H_2O \tag{8.13}$$

is far slower, with $k_1 = 20\ s^{-1}$ and $k_{-1} = 0.03\ s^{-1}$ (25°C). Thus, depending on circumstances, the slow step in CO_2 removal to the gas phase could be chemical [Eq. (8.13)] or physical [CO_2 (dissolved) $\rightarrow$ CO_2 (gas)].

8.1.2 Rates of Metabolic Oxygen Utilization

In design of aerobic biological reactors we frequently use correlations of data more or less approximating the situation of interest to establish whether the slowest process step is the oxygen transfer rate or the rate of cellular utilization of oxygen (or other limiting substrate). The maximum possible mass-transfer rate is simply that found by setting $c_l = 0$: all oxygen entering the bulk solution is assumed to be rapidly consumed. The maximum possible oxygen utilization rate is seen from Chap. 7 to be $x\mu_{max}/Y_{O_2}$, where x is cell density and Y_{O_2} is the ratio of moles of cell carbon formed per mole of oxygen consumed.

Evidently, if $k_l a' c_l^*$ is much larger than $x\mu_{max}/Y_{O_2}$, the main resistance to increased oxygen consumption is microbial metabolism and the reaction appears to be biochemically limited. Conversely, the reverse inequality apparently leads to c_l near zero, and the reactor seems to be in the mass-transfer-limited mode.

The situation is actually slightly more complicated. At steady state, the oxygen absorption and consumption rates must balance:

$$Q_{O_2} = \text{absorption} = \text{consumption}$$

$$k_l a'(c_l^* - c_l) = \frac{x\mu}{Y_{O_2}} \tag{8.14}$$

Assuming that the dependence of μ on c_l is known, we can use Eq. (8.14) to evaluate c_l and hence the rate of oxygen utilization.

In general, above some critical bulk oxygen concentration, the cell metabolic machinery is saturated with oxygen. In this case, sufficient oxygen is available to accept immediately all electron pairs which pass through the respiratory chain, so that some other biochemical process within the cell is rate-limiting (Chap. 5). For example, if the oxygen dependence of the specific growth rate μ follows the Monod form, then

$$Y_{O_2} k_l a'(c_l^* - c_l) = x\mu_{max} \frac{c_l}{K_{O_2} + c_l} \tag{8.15}$$

A general solution to an equation of this form was given in Sec. 4.4.1., but here for the sake of illustration we assume that the value of c_l is considerably less than c_l^*. This is not an uncommon situation in biological reactors. Subject to the assumption that $c_l \ll c_l^*$, c_l is easily seen to be

$$c_l = c_l^* \left[\frac{Y_{O_2} K_{O_2} k_l a'/x\mu_{max}}{1 - Y_{O_2} c_l^* k_l a'/x\mu_{max}} \right] \tag{8.16}$$

Table 8.2 Typical values of $c_{O_2,cr}$ in the presence of substrate[†]

Organism	Temp, °C	$c_{O_2,cr}$, mmol/L
Azotobacter vinelandii	30	0.018–0.049
E. coli	37.8	0.0082
	15	0.0031
Serratia marcescens	31	~0.015
Pseudomonas denitrificans	30	~0.009
Yeast	34.8	0.0046
	20	0.0037
Penicillium chrysogenum	24	~0.022
	30	~0.009
Aspergillus oryzae	30	~0.020

[†] Summarized by R. K. Finn, p. 81 in N. Blakebrough (ed.), *Biochemical and Biological Engineering Science*, vol. 1, Academic Press, Inc., New York, 1967.

If the resulting value of c_l is greater than the critical oxygen value c_{cr} (about $3K_{O_2}$), the rate of microbial oxygen utilization is limited by some other factor, e.g., low concentration of another substrate, even though the bulk solution has a dissolved oxygen level considerably below the saturation value. The critical oxygen values for organisms lie in the range of 0.003 to 0.05 mmol/L (Table 8.2) or of the order of 0.1 to 10 percent of the solubility values in Table 8.1, that is, 0.5 to 50 percent of the air saturation values. For the higher critical oxygen values such as obtained for *Penicillium* molds, oxygen mass transfer is evidently extremely important.

Many factors can influence the total microbial oxygen demand $x\mu/Y_{O_2}$, which in turn sets the minimum values of $k_l a'$ needed for process design through Eq. (8.14). The more important of these are cell species, culture growth phase, carbon nutrients, pH, and the nature of the desired microbial process, i.e., substrate utilization, biomass production, or product yield (Chap. 7).

In the batch-system results of Fig. 8.5, a maximum in *specific* O_2 demand occurs in the early exponential phase although x is larger at a later time. A peak in the product $x\mu$, and thus the total oxygen demand, occurs near the end of the exponential phase and the approach to the stationary phase; this is later than the time of achievement of the largest specific growth rate.

The carbon nutrient affects oxygen demand in a major way. For example, glucose is generally metabolized more rapidly than other carbohydrate substances. Peak oxygen demands of 4.9, 6.7, and 13.4 mol/(L · h) have been observed for *Penicillium* mold utilizing lactose, sucrose, and glucose, respectively [2].

The component parts of oxygen utilization by the cell include cell maintenance, respiratory oxidation for further growth (more biosynthesis), and oxida-

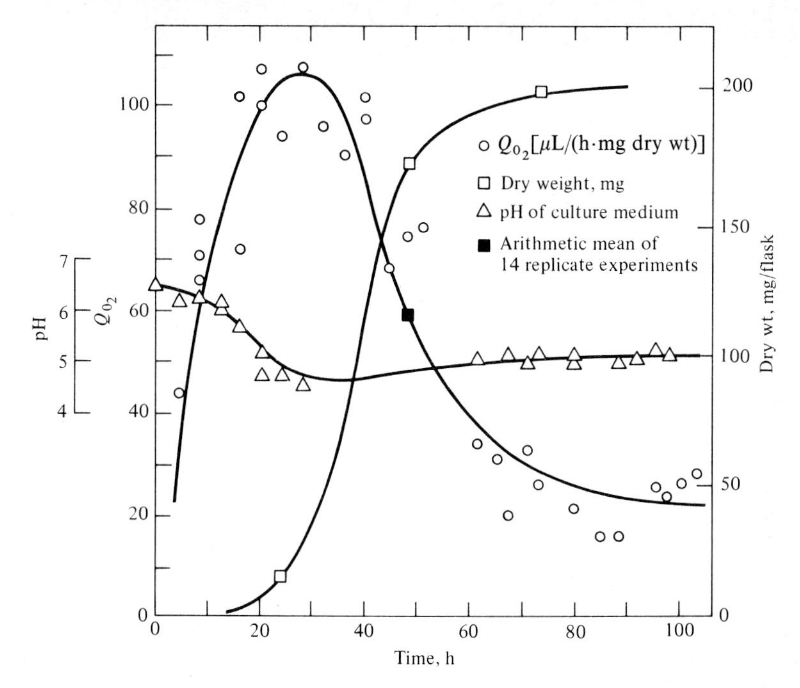

Figure 8.5 Oxygen utilization rate in batch culture of *Myrothecium verrucaria [Reprinted from R. T. Darby and D. R. Goddard, Am. J. Bot., vol. 37, p. 379 (1950).]*

tion of substrates into related metabolic end products. In examining metabolic stoichiometry in Chap. 5, we have seen that oxygen utilization for growth is typically coupled directly to the amount of carbon-source substrate consumed. Furthermore, more reduced substrates such as paraffins and methane require greater oxygen uptake by the cell than substrates such as glucose which have approximately the same carbon oxidation state as the cell. For example, the yield factors $Y_{O_2/C}$ giving moles oxygen used per mole of carbon source metabolized are 1.34, 1.0, and 0.4 for typical microorganisms growing on methane, paraffins, and carbohydrates, respectively.

Oxygen may also be consumed as a reactant in a *biotransformation.* For example, 5-ketogluconic acid production from glucose by batch cultivation of *Acetobacter* begins with a growth phase in which some medium glucose is converted to gluconic acid. Here O_2 use for both growth and product formation occurs. After glucose exhaustion, growth ceases, and gluconic acid is converted to 5-ketogluconic acid with stoichiometry

$$C_6H_{12}O_7 + \tfrac{1}{2}O_2 \rightarrow C_6H_{10}O_7 + H_2O \tag{8.17}$$

In the final phase of the process which involves only this biotransformation, oxygen demand is directly coupled to product formation through the stoichiometry of Eq. (8.17).

8.2 DETERMINATION OF OXYGEN TRANSFER RATES

Ideally, oxygen transfer rates should be measured in biological reactors which include the nutrient broth and cell population(s) of interest. As this requires all the accoutrements for inoculum and medium preparation, prevention of contamination, and environmental control for the cell culture, it is an inconvenient and troublesome way to conduct mass-transfer experiments. Consequently, a common strategy for study of oxygen transfer rates is to use synthetic systems which approximate bioreaction conditions without the complications of a living culture. In such approaches, the major objective is to elucidate the dependence of $k_l a'$ on hydrodynamics.

In order for such synthetic media to represent the cellular broth of interest reliably, the following properties of the synthetic media and actual broth should be identical:

1. Solution viscosity and other rheological characteristics (see Sec. 8.8).
2. Gas–liquid interfacial resistance.
3. Bubble coalescense tendencies
4. Oxygen solubility and diffusivity.

In general, the usefulness of a synthetic (cell-free) situation for approximating a bioreaction situation depends on the degree to which these conditions are met. Experiments with oxygen absorption into pure water, for example, satisfy few of these criteria. We shall examine the quantitative influence of fluid viscosity, surface-active agents, and nature of mixing shortly.

8.2.1 Measurement of $k_l a'$ Using Gas–Liquid Reactions

Considering now the transport paths in Fig. 8.1, we see that if oxygen is consumed by chemical reaction in the bulk liquid at a sufficiently large rate we will find $c_l \approx 0$. Then the bulk-phase chemical-reaction rate is equal to $k_l a' c_l^*$, from which the $k_l a'$ value readily follows. A common bulk-phase oxygen sink in many previous mass-transfer studies is the oxidation of sodium sulfite to sulfate in the presence of catalytic metal ions such as Co^{2+}:

$$SO_3^{2-} + \tfrac{1}{2}O_2 \xrightarrow{\text{catalyst}} SO_4^{2-} \tag{8.18}$$

The kinetics of the rate of oxidation of sulfite solutions to sulfate is complex. The reaction orders for oxygen and sulfite depend on the catalyst used and its concentration, apparently implying a nontrivial series of elementary steps leading to the overall result above. Regardless of the reaction order, the condition sufficient to ensure that the chemical reaction occurs to a negligible extent in the liquid film adhering to each bubble (and thus represents the situations in Fig. 8.1) is a negligible total reaction rate in the film compared with the mass-transfer rate

$k_l(c^* - c)$. If ζ denotes the mass-transfer film thickness, this criterion can be restated mathematically as

$$\zeta \times \text{rate}\Big|_{\text{film}} < k_l(c^* - c) \tag{8.19}$$

The rate in the film will be less than that corresponding to bulk sulfite and saturation oxygen levels, i.e., rate (c^*, sulfite$_{\text{bulk}}$), and so in terms of these measurable or calculable quantities (c^*, sulfite$_{\text{bulk}}$) a conservative criterion for negligible film reaction is

$$\zeta \times \text{rate}(c^*, \text{sulfite}_{\text{bulk}}) < k_l(c^* - c) \tag{8.20}$$

The "thickness" of the mass-transfer film is given by

$$\zeta = \frac{\mathscr{D}_{O_2}}{k_l} \tag{8.21}$$

Assuming the reaction to be of order α_1 in oxygen and α_2 in sulfite leads to the inequality

$$\frac{\mathscr{D}_{O_2}[k_r(c^*)^{\alpha_1}\text{sulfite}^{\alpha_2}]}{k_l} < k_l(c^* - c) \tag{8.22}$$

Thus

$$k_l > \left[\frac{\mathscr{D}_{O_2}[k_r(c^*)^{\alpha_1}\text{sulfite}^{\alpha_2}]}{c^* - c}\right]^{1/2} \tag{8.23}$$

An illustrative example of Danckwerts' considers an experiment using 10^{-5} M cobalt catalyst (known to give $\alpha_1 = 2$) and sufficient sulfite (say 0.5 M) for α_2 to be 0. For $c \ll c^*$, the above inequality becomes

$$k_l > [\mathscr{D}_{O_2} k_r c^*]^{1/2} \tag{8.24}$$

Taking $\mathscr{D}_{O_2} = 1.6 \times 10^{-5}$ cm^2/s, k_r for cobalt catalyst $= 0.85 \times 10^8$ cm^3/(g mol $\cdot$ s), $c^* = 1.35 \times 10^{-7}$ g mol/cm^3 gives

$$k_l \gg 0.01 \text{ cm/s}$$

A less effective catalyst (smaller k_r) would reduce the right-hand side.

Reference to the correlations to be presented later in this section indicates that for large bubbles in water, $k_l \sim 0.04$ cm/s, and for small bubbles $k_l \sim$ 0.01 cm/s. Thus this inequality places a minimum size on the bubbles which may be used for such an interpretation. Smaller bubbles rising more slowly will have an appreciable reaction rate in the adhering fluid film under these specific conditions. Similarly, larger bubbles in media more viscous than water will exhibit reduced mass-transfer coefficients. However, an enhancement factor E to account for film reaction can be calculated provided the reaction-rate constant and reaction order are known [3–5].

A closing series of caveats for sulfite oxidation is illuminating: the rate constant k_r depends on (1) the catalyst and its concentration, (2) the ionic strength of the solution, (3) the presence of catalytic impurities, and (4) the solution pH; for example, in 10^{-5} M cobalt, k_r increases by a factor of 10 between pH 7.50 and 8.50 at 20°C. (The overall reaction generates H^+, so that base must be added to maintain pH constant).

In spite of these difficulties, the literature contains examples of reactors where $k_l a'$ determined from sulfite measurements for a given sparger, stirring rate, etc., correlate closely with the $k_l a'$ values observed in an actual fermentation (counterexamples are also evident). Assuming that the configuration in Fig. 8.1 represents the bioreaction of interest, there is essentially no O_2 consumption in the bubble liquid-side film. Thus, any chemical measure of O_2 absorption attempting to simulate such cell broths must, inter alia, satisfy the fundamental inequality (8.19) above. On the other hand, if growing cells are concentrated in the bubble liquid-side film, a different model chemical reaction situation may be required.

We can measure oxygen transfer rates in several other ways. If the experimental system is strictly a batch operation, with no addition or removal of liquid or gas, $\bar{Q}_{O_2}$ is revealed by monitoring the gas volume or pressure changes with time. Also, as discussed in the next section, measurement of c_l aids in $k_l a'$ estimation.

When gas is continuously added to and removed from the liquid, we use the following O_2 mass balance on the gas phase to determine $\bar{Q}_{O_2}$:

$$\bar{Q}_{O_2} = [F_{g_{\text{inlet}}}(p_{O_2})_{\text{inlet}} - F_{g_{\text{exit}}}(p_{O_2})_{\text{exit}}]/VRT \tag{8.25}$$

Here F_g is the volumetric gas flow rate and p_{O_2} is the partial pressure of O_2. [What assumptions are necessary to justify Eq. (8.25)? Are they generally valid for bioreaction processes?]

We would like to use these $\bar{Q}_{O_2}$ values to determine $k_l a'$, but, as Eqs. (8.6) and (8.7) and the associated discussion reveal, this requires uniformity of conditions within the vessel, so that the local and average oxygen utilization rates are identical. Consequently, stirred vessels [Fig. 8.2*b*(2)] are frequently employed in laboratory mass-transfer studies for biological-reactor design.

When the problem of spatial uniformity has been resolved, $k_l a'$ can be extracted from Eq. (8.6) if c_l^* and c_l are known. The first of these is available from solubility data such as Table 8.1, and direct measurement of c_l is now feasible (even in pure microbial cultures) with the polarographic sterilizable oxygen electrode. The operating principles of this electrode, which produces a current proportional to local dissolved oxygen partial pressure, are described in Chap. 10. An additional method for $k_l a'$ estimation based on dynamic measurements with an oxygen electrode is also described in Chap. 10.

In many reactor configurations or processes of natural origin, the local oxygen transfer rate varies with position. If such variations occur in the vessel in which mass-transfer rates are measured, the observed vessel-averaged value of $k_l a'$ cannot properly be used effectively in scale-up, the process of transferring laboratory scale results into large capacity units. Design methods for such scale-up are considered later in this chapter. Variations of dissolved O_2 within a "homogeneous" phase (bulk fluid, mold pellet, microbial films, etc.) have been examined with miniaturized oxygen probes, the sensing head being of the order of 10 micrometers in diameter. Applications of this instrument include study of local oxygen profiles in mold pellets and the determination of diffusion coefficients in microbial aggregates.

8.3 MASS TRANSFER FOR FREELY RISING OR FALLING BODIES

The rate of material exchange between different regions is governed by the equations of change which describe conservation of mass, conservation of species (such as oxygen), and the momentum balance. When the equations of change are rendered dimensionless in distance, velocity, and concentrations for situations where the density difference between the two contacting phases provides the major driving force for fluid motion, three dimensionless parameters appear in the final expressions. These are the Grashof, Sherwood, and Schmidt numbers, which, for mass transfer, are defined by

$$\text{Grashof number} = \text{Gr} = \frac{D^3 \rho_l (\rho_l - \rho_g)}{\mu_l^2} \tag{8.26a}$$

$$\text{Sherwood number} = \frac{k_l D}{\mathscr{D}_{O_2}} \tag{8.26b}$$

$$\text{Schmidt number} = \text{Sc} = \frac{\mu_l}{\rho_l \mathscr{D}_{O_2}} \tag{8.26c}$$

where D is a characteristic dimension and μ_l is the viscosity of the continuous phase. Consequently, we expect mass-transfer-coefficient correlations for such convective motion to involve only these three groups.

8.3.1 Mass-Transfer Coefficients for Bubbles and Bubble Swarms

The mass-transfer coefficient for a bubble, for example, is the proportionality constant between the total bubble flux and the overall driving force, $c_l^* - c_l$. The local flux at the gas-liquid surface is $-\mathscr{D}_{O_2}(\partial c/\partial z)_{z=0}$ (valid for low mass-transfer rates), where z is the coordinate measured into the liquid phase. Thus,

$$k_l = \frac{-1}{c_l^* - c_l} \mathscr{D}_{O_2} \left. \frac{\partial c}{\partial z} \right|_{z=0} \tag{8.27}$$

or, nondimensionally,

$$\text{Sh} = \frac{k_l D}{\mathscr{D}_{O_2}} = \frac{-1}{1 - \bar{c}_l} \left(\frac{\partial \bar{c}}{\partial \bar{z}} \right)_{\bar{z}=0} \tag{8.28}$$

Near the gas–liquid interface, the dimensionless concentration $\bar{c}$ has a solution from the transport equations of the form

$$\bar{c} = f(\bar{z}, \text{Sh}, \text{Sc}, \text{Gr}) \tag{8.29}$$

Using this expression to evaluate the derivative $(\partial\bar{c}/\partial\bar{z})_{\bar{z}=0}$ in Eq. (8.28), which is possible in principle, leaves the desired mass-transfer coefficient k_l in the form of the Sherwood number

$$\text{Sh} = \frac{k_l D}{\mathcal{D}_{O_2}} = g(\text{Sc}, \text{Gr}) \tag{8.30}$$

Thus, the dimensionless mass-transfer coefficient Sh is a function of only the two parameters Sc and Gr. Here D denotes characteristic bubble diameter.

Correlations for mass-transfer coefficients for falling or rising bubbles, droplets, or solids have appeared in the literature using other dimensionless groups such as the Reynolds number ($\text{Re} = \rho_l Du/\mu_l$) or the Peclet number ($\text{Pe} = uD/\mathcal{D}_{O_2}$). The velocity u applied here is the velocity of the gas bubble relative to the liquid velocity. In both instances, when an expression for the characteristic velocity u in terms of the density difference $\Delta\rho = (\rho_l - \rho_g)$ is substituted, the final result depends only on Gr and Sc according to Eq. (8.30).

Mass transfer from an isolated sphere with a rigid interface, a reasonable approximation to small bubbles in a fermentation broth containing surface-active agents, may be determined theoretically for the case $\text{Re} = \rho_l Du/\mu_l \ll 1$ and $\text{Pe} \equiv uD/\mathcal{D}_{O_2} \gg 1$. (Thus $uD/\mathcal{D}_{O_2} \gg 1 \gg \rho_l Du/\mu_l$, which implies that $\mu_l/\rho_l\mathcal{D}_{O_2} = \text{Sc} \gg 1$. Is the converse true?) In aqueous liquids, since the kinematic viscosity $\nu = \mu_l/\rho_l$ is about 10^{-2} cm^2/s and $\mathcal{D}_{O_2}$ is of the order of 10^{-5} cm^2/s, the Schmidt number is typically of the order of 10^3. Consequently, for Re of the order of 10^{-1}–10^{-2}, the theoretical result of Eq. (8.31) applies.

$$\text{Sh} = 1.01\ \text{Pe}^{1/3} = 1.01(uD/\mathcal{D}_{O_2})^{1/3} \tag{8.31}$$

For small Reynolds numbers for which this prediction applies, the terminal velocity u_t of a sphere is given by

$$u_t = \frac{D\ \Delta\rho^2\ g}{18\mu_l} \tag{8.32}$$

Replacing u in Eq. (8.31) with u_t from Eq. (8.32) gives

$$\text{Sh} = 1.01\left(\frac{D^3\ \Delta\rho\ g}{18\mu_l\mathcal{D}_{O_2}}\right)^{1/3} = 1.01\left(\frac{D^3\rho_l\ \Delta\rho\ g}{18\mu_l^2}\right)^{1/3}\left(\frac{\mu_l}{\rho\mathcal{D}_{O_2}}\right)^{1/3} = 0.39\ \text{Gr}^{1/3}\text{Sc}^{1/3} \tag{8.33}$$

[The grouping $(D^3\ \Delta\rho\ g)/\mu_l\mathcal{D}_{O_2}$ is also known as the Rayleigh number Ra.] Notice here that $\text{Sh} = f(\text{Gr}, \text{Sc})$, as expected.

For a larger Reynolds number, the single-bubble result for a noncirculating sphere in laminar flow is

$$\text{Sh} = 2.0 + 0.60\ \text{Re}^{1/2}\text{Sc}^{1/3} \qquad \text{Re} \gg 1 \tag{8.34}$$

Note that Sh varies as the square root rather than the cube root of the velocity, indicating that a different hydrodynamic regime is present. Again, replacement of u by an appropriate terminal velocity expression will yield Sh = f(Gr, Sc).

In many industrial air-sparged reactors (Fig. 8.2 configurations), air bubbles are produced in swarms or clusters of sufficient intimacy that single isolated-bubble hydrodynamics and mass-transfer results fail to describe fluid motion and mass transport accurately in the vicinity of the gas–liquid interface. Calderbank and Moo-Young [10] report that two correlations are sufficient to describe their data for absorption of sparingly soluble gases into liquids which consume the gas chemically. Two distinct regimes of bubble-swarm mass-transfer are evident, the division between them being indicated by a critical bubble diameter D_c. In the absence of surfactants $D_c \approx 2.5$ mm. Bubbles larger than this are typically encountered with pure water in agitated tanks and in sieve-plate columns. Smaller bubbles are frequently found in sintered-plate columns and in agitated vessels containing hydrophilic solutes in aqueous solution.

For $D < D_c = 2.5$ mm

$$\text{Sh} = \frac{k_l D}{\mathscr{D}_{O_2}} = 0.31\ \text{Gr}^{1/3}\text{Sc}^{1/3} = 0.31\ \text{Ra}^{1/3} \tag{8.35}$$

For $D > D_c = 2.5$ mm

$$\text{Sh} = \frac{k_l D}{\mathscr{D}_{O_2}} = 0.42\ \text{Gr}^{1/3}\text{Sc}^{1/2} \tag{8.36}$$

Thus Eqs. (8.33) and (8.35) indicate that in bubble swarms, the mass-transfer coefficient for the same Schmidt and Grashof numbers is reduced about 20 percent compared with the isolated single-bubble case with an immobile surface. Equation (8.36) has also been verified for air-lift operation, Fig. 8.2a, using a coefficient of 0.50 rather than 0.42.

The change of Schmidt number exponent in Eq. (8.36) indicates a changed hydrodynamic regime from Eq. (8.35). For Newtonian fluids, i.e., viscosity = constant, independent of shear rate due to stirring speed, bubble velocity, etc., the transition from the $D < D_c$ region to the $D > D_c$ regime is accompanied by a change of bubble shape from nearly spherical (small bubbles) to hemispheric and caplike shapes. For further discussion of bubble hydrodynamics in these swarms, see Ref. [10]. The transition value of D varies with surfactant; values as high as 7.0 mm have been reported. In some non-Newtonian fluids, which will be discussed further later, transition with D is much more gradual than the abrupt change observed for Newtonian fluids.

Mass-transfer results for small particles show that as the density difference $\Delta\rho$ diminishes, the Sherwood number approaches 2.0 as a lower limit. For individual cells, clumps, flocs, etc., as well as for gas oil or other hydrocarbon dispersions, a more accurate form of the Sherwood number is

$$\text{Sh} = \frac{k_l D}{\mathscr{D}_{O_2}} = 2.0 + 0.31\ \text{Ra}^{1/3} \tag{8.37}$$

or

$$\frac{k_l}{\mathscr{D}_{O_2}} = \frac{2.0}{D} + 0.31\left[\frac{\Delta\rho\, g}{\mu_l \mathscr{D}_{O_2}}\right]^{1/3} \tag{8.38}$$

Thus the relative importance of the pure-diffusion result ($\Delta\rho \equiv 0$, $k_l = 2.0\mathscr{D}_{O_2}/D$) vs. the buoyancy term diminishes as particle size increases. For an isolated cell, $2\mathscr{D}_{O_2}/D$ is of the order of 10^{-1} cm/s compared with 10^{-2} cm/s for the Raleigh number term; the mass transfer near its surface therefore resembles that for a sphere in a more or less stagnant medium. Larger diameters due to flocs, films, etc., lead to greater relative contributions from the second term.

8.3.2 Estimation of Dispersed Phase Interfacial Area and Holdup

Having evaluated k_l from the appropriate previous formulas, we still must determine the interfacial area a' per unit volume. The value of a' can be estimated from sparger orifice diameter, overall reactor information, or photographic data, among other means. If bubble residence time in the reactor is t_b, volumetric flow rate per orifice is F_0, and total number of (equal) orifices is n, then the interfacial area per unit volume a' (neglecting coalescence and change of D with hydrostatic head or absorption) is given by

$$a' = \frac{1}{\text{volume}} nF_0 t_b \frac{\pi D^2}{\pi D^3/6} = \frac{nF_0 t_b}{V}\frac{6}{D} \tag{8.39}$$

In the following discussion, then, we consider the factors which appear on the right-hand side of Eq. (8.39). In particular, we will examine in detail the physical processes which determine bubble size. Based on these, we will explore the feasibility of predicting bubble size as a function of operating conditions, contactor design, and fluid properties. While introduced here in the context of rising bubbles and bubble swarms, many of the concepts described are also central in determining transport properties in vessels with mechanical agitation.

There are three main factors which interact to determine the size of bubbles in bioreactors (similar comments apply also to dispersion of a sparingly soluble second liquid phase). These are bubble formation, bubble breakup, and bubble coalescence. Bubble formation is dictated by instabilities in the gas stream entering the liquid phase which result in this stream breaking into discrete bubbles rather than flowing through the vessel as a continuous stream. Bubble breakup depends on the competition between surface tension, which stabilizes the bubble, and local fluid forces, which tend to tear the bubble apart.

The probability of bubble coalescence depends on the properties of the gas–liquid interface. In the predominantly aqueous mixtures commonly encountered in bioprocessing, coalescence properties are determined primarily by liquid phase solutes such as fatty acids, polyalcohols, electrolytes, and ketones. Addition of these components suppresses coalescence. In subsequent discussions, it will be useful to consider two limiting cases: coalescing dispersions (e.g., air–pure water) and noncoalescing dispersions (e.g., air–water with electrolyte).

Let us first consider the process of bubble formation as gas flows at volumetric flow rate F_0 through a single orifice of diameter d. The diameter of the initially formed bubble will be denoted D_0. Two regimes can be identified. At low gas flow rates, bubbles form one at a time at the orifice. The simplest analysis of this situation is based on a force balance for a bubble leaving an orifice. Bubble departure occurs when the buoyant force $(\pi D_0^3 \, \Delta\rho \, g)/6$ equals the restraining force $\pi\sigma d$:

$$\frac{g \, \Delta\rho \, D_0^3}{\sigma d} = 6 \tag{8.40}$$

More elaborated theories are summarized in Ref. [21].

At some critical gas flow rate F_0^*, there is a transition from departure of single gas bubbles from the orifice to appearance of a gas jet at the orifice. Precise predictions are not presently possible, but this critical gas flow rate falls in the range indicated by

$$\sqrt{\frac{2\pi\sigma d}{16\rho_g}} \le F_0^* \le 20.4 \frac{\sigma}{\rho_l} \sqrt{\frac{d}{g}} \tag{8.41}$$

At gas flow rates greater than F_0^*, initial bubble formation occurs by breakup of the gas jet due to instability of the interface. Based on stability theory developed first by Rayleigh, the expected bubble diameter for gas jet breakup in laminar liquid flow is approximately

$$D_0 = d(12\pi/0.485)^{1/3} = 4.27d \tag{8.42}$$

provided [21]

$$\frac{\sigma \rho_l d}{\mu_l} > 36 \tag{8.43}$$

This inequality is usually satisfied under bioprocess conditions.

For sparging into viscous broths, liquid viscosity rather than bubble surface tension provides the predominant resistance to new bubble formation. Where the bubbles are formed, the ratio of bubble to sparger orifice diameter, (D/d), is given by

$$\left(\frac{D}{d}\right) = 3.23 \, \mathrm{Re}_0^{-0.1} \, \mathrm{Fr}_0^{0.21} \tag{8.44}$$

where the orifice Reynolds and Froude numbers are given by Eqs. (8.45) and (8.46):

$$\mathrm{Re}_0 = \frac{4\rho_c F_0^*}{\pi \mu_l d} \tag{8.45}$$

$$\mathrm{Fr}_0 = \frac{(F_0^*)^2}{d^5 \cdot g} \tag{8.46}$$

In some circumstances, the bubbles found in the gas–liquid dispersion are smaller than those formed at the gas distributor. This occurs because of instability of bubbles under the forces applied on the bubbles by the moving continuous phase. For a dispersed liquid or gas phase in a continuous liquid phase, the maximum size of dispersed-phase diameters for either freely rising (falling) or agitated configurations is due to a balance of opposing forces:

1. The dynamic pressure τ (the sum of shearing and normal stress differences) tends to draw out the droplets into shapes which eventually disintegrate into smaller pieces; this subdivision process is resisted by the following two forces.
2. The surface-tension forces σ/D of the particle tend to restore the droplet to spherical shape (minimum surface-energy configuration).
3. The viscous resistance of the dispersed phase to deformation is proportional to the term $\mu_d D^{-1}\sqrt{\tau/\rho_d}$, where subscript d indicates a dispersed-phase property.

In gas–liquid systems, term 3 will be negligible compared with term 2. In liquid–liquid contactors the last term should be relatively larger, but a later result of Example 9.2 indicates again that the forces in term 2 appear to predominate even for these all-liquid systems.

The last two restoring forces diminish as $D^{-\beta}$, where β is a positive number. Thus, at some critical diameter D_c, the dynamic pressure will override the two countering resistances and rearrange the bubble or droplet into smaller portions. At the critical diameter, evidently the following equality holds:

$$m_1 \frac{\sigma}{D} + m_2 \cdot \mu_d D^{-1}\left[\frac{\tau}{\rho_d}\right]^{1/2} = \tau \tag{8.47}$$

where m_1 and m_2 are constants.

If the surface-tension forces are much more significant than the viscous forces, as argued, then at the critical bubble size

$$m_1 \frac{\sigma}{D_c} = \tau \qquad \text{or} \qquad m_1 = \tau \frac{D_c}{\sigma} \qquad \text{or} \qquad D_c = \frac{m_1 \sigma}{\tau} \tag{8.48}$$

Equation (8.48) states that the maximum stable bubble size is a dimensionless constant (m_1) times surface tension divided by dynamic pressure. Thus, with greater dynamic pressure, increasingly smaller bubbles will be broken up.

Theoretical relations or correlations describing the relationship between the maximum stable bubble (or drop) size and fluid and flow properties typically employ a dimensionless group based on the form of Eq. (8.48). The Weber number We is defined by

$$\text{We} = \tau \frac{D}{\sigma} \tag{8.49}$$

The *critical* Weber number We_c is the value of We for $D = D_c$ which, according to Eq. (8.48), is a characteristic constant. Experiments for clean air–water systems and theoretical calculations indicate that We_c is approximately unity (actually 1.05).

In order to calculate the maximum stable bubble size using these concepts, we need to determine a suitable value of the dynamic pressure τ for different flow situations. For freely rising bubbles, τ is given by

$$\tau = \frac{\rho_l u_t^2}{2} \tag{8.50}$$

where u_t is the bubble terminal velocity, given by Eq. (8.32) for spherical bubbles. In a complicated turbulent flow, estimation of the dynamic pressure is difficult. Turbulence is expected in bubble columns, for example, near aeration nozzles. Faced with this problem, we now consider some general concepts in turbulence which will be useful in several contexts which follow.

In the statistical theory of turbulence formulated by Kolmogorov and others, the turbulent flow field is regarded as a collection of superposed eddies or velocity fluctuations characterized by their fluctuation frequency (or length scale) and magnitude. The largest vortex elements or primary eddies have the scale of the main flow. These largest eddies are unstable and disintegrate into smaller eddies which are unstable and disintegrate into still smaller eddies and so on. Kinetic energy flows through this cascade from largest eddies to the smallest eddies until ultimately this energy is dissipated as heat. As the energy is transferred through this cascade, the directional character of the primary eddies, which depends on the geometry of the vessel, entering jets, mixers, and the like, decays. Kolmogorov's theory asserts that the smaller eddies are statistically independent of the primary eddies and are locally isotropic (spatially uniform). The smallest vortices which dissipate the turbulence energy have length scale λ_0 given by

$$\lambda_0 = \frac{\mu_l^{3/4}}{\rho_l^{1/2}}\left(\frac{P}{V_l}\right) \tag{8.51}$$

where P/V_l is the power input per unit volume.

In examining the effects of turbulence, we commonly use time-averaged quantities. The rms velocity $u_{\text{rms}} \equiv \langle u^2(t)\rangle^{1/2}$ ($\langle\ \rangle$ denotes time averaging over the instantaneous velocity fluctuations) reflects the typical average magnitude of the local velocity variations. For length scales l much smaller than the scale of the primary eddies and much greater than λ_0, the rms velocity of vortices with characteristic size l is given by

$$u_{\text{rms}}\Big|_{\text{eddies of scale } l} = \alpha\left(\frac{P}{V_l}\right)^{1/3}\left(\frac{l}{\rho_l}\right)^{1/3} \tag{8.52}$$

where α is a constant.

Returning now to a suitable choice of dynamic pressure for use in the Weber number in turbulent flows, we can use

$$\tau = \rho_l \left(u_{\text{rms}} \Big|_{\substack{\text{eddies of} \\ \text{scale } D_c}} \right)^2 \tag{8.53}$$

since this gives a measure of the turbulent shear stress which acts on a bubble of size D_c. Combining Eqs. (8.49), (8.52), and (8.53), we obtain

$$D_c = \alpha' \frac{\sigma^{0.6}}{(P/V_l)^{0.4} \rho_l^{0.2}} \tag{8.54}$$

where α' is a constant. Thus, the maximum stable bubble size is reduced if the power dissipation per unit volume is increased.

It is significant that, according to the theory of isotropic turbulence, the power input per unit volume is a key parameter in determining the scales of eddies obtained and the intensity of turbulent velocity fluctuations of length scales comparable to bubble and drop sizes. Local isotropic turbulence is an idealized situation not always obtained in practice; however, it is important to remember this physical view of the mechanism by which energy input to a process in the form of gas compression or mechanical agitation is ultimately transmitted to bubbles, drops, flocs, and mycelial pellets. Furthermore, Eqs. (8.52) and (8.54) above prepare us well to expect important effects of P/V_l on mass transfer coefficients in sparged towers and agitated tanks. Another important aspect of this theory deserves special emphasis: what matters is the local energy dissipated per unit volume, regardless of the means by which that energy is delivered to the mixture (for example injection of compressed gas versus mechanical mixing, one impeller or two, etc.). Again, this is an idealization, but it is one that is consistent with experimental observations in some cases.

Having considered coalescence, bubble formation, and bubble breakup separately, let us now examine the different possible outcomes of interaction of these processes in a sparged column. First we shall suppose that bubble (or droplet) coalescence is slow in the two-phase dispersion considered. If the initial bubble diameter D_0 is less than the maximum stable bubble diameter D_c evaluated under conditions of greatest dynamic pressure (typically in the region of bubble formation in a sparged column), the characteristic bubble diameter is D_0 (Table 8.3). If D_0 exceeds this maximum stable diameter, bubble breakup tending toward a characteristic diameter equal to D_c is expected.

On the other hand, if coalescence occurs rapidly, bubbles initially formed will coalesce and grow in size until they exceed the maximum stable bubble size after which breakup occurs. In this case, except in the region of the sparger, the initial bubble size D_0 has little influence on bubble size in the vessel. Remembering that turbulent velocity fluctuations generally vary from point to point in the vessel so that D_c does also, the coalescing system is characterized by a tendency at each point toward local coalescence-breakup equilibrium with characteristic bubble size given by D_c.

Table 8.3 Characteristic bubble diameter D depends on bubble coalescence properties and on the relationship between bubble diameter at formation (D_0) and the maximum stable bubble diameter (D_c)

Condition	Noncoalescing	Coalescing
$D_0 < D_c$	$D \sim D_0$	Transition toward dispersion equilibrium
$D_0 > D_c$	$D \sim D_c \vert_{\text{sparger (stirrer)}}$	$D \sim D_c \vert_{\text{local}}$

These observations have important implications for equipment design. In a noncoalescing system, energy dissipated for gas dispersion is most efficiently expended at the point of initial bubble formation and dispersion. Uniform dissipation of energy for dispersion is best when coalescence is important. These facts have motivated invention, characterization, and application of many alternative contacting and mixing configurations for bioreactors as we shall see in Chap. 9. These points and other related qualitative ones may be at present the most practically useful results from the preceding discussion of mechanisms. In fact, spatial inhomogeneities in flow patterns, turbulence properties, and gas/liquid volume fractions are so complicated in most situations that quantitative prediction is difficult, requiring recourse to correlations to obtain useful numbers. Armed with the physical insight just gained, however, we now should be alert to check the basis for various correlations before applying them. A correlation based on data from a clean air–water, and hence coalescing system, will likely have little relevance for a process containing a relatively noncoalescent two-phase mixture.

Returning now to the factors in Eq. (8.39) which determine interfacial area per unit volume a', we consider the bubble residence time t_b. This time may be estimated from the bubble rise velocity integrated over the reactor height h_r

$$t_b = \int_0^{h_r} \frac{dz}{u_b(z)} \approx \frac{h_r}{u_t} \tag{8.55}$$

where in the approximate expression on the right-hand side the bubble rise velocity has been taken to be the bubble terminal velocity. For isolated small bubbles at small Reynolds numbers the terminal velocity given in Eq. (8.32) can be used. In the case of large, spherical-cap shaped bubbles (diameter D) in Newtonian fluids, the terminal rise velocity to be used in these calculations is

$$u_t = 0.711(gD)^{1/2} = 22.26\sqrt{D} \qquad \text{cm/s} \tag{8.56}$$

For bubble clouds or swarms, calculation of the characteristic bubble rise velocity is more difficult, since neighboring bubbles influence each other's motion and since bubble coalescence and breakup may occur. As a crude approximation, comparison of Eqs. (8.33) and (8.35) suggests that for identical Sc and Sh,

$$\left[\frac{u_t \text{ (bubble cloud)}}{u_t \text{ (single bubble)}}\right]^{1/3} = \frac{0.31}{0.39} \tag{8.57}$$

so that

$$u_t \text{ (bubble cloud)} \approx 0.50 u_t \text{ (single bubble)} \tag{8.58}$$

Any real dispersion will generally contain a distribution of bubble sizes. This raises the question of suitable definition of a characteristic or mean size. The value of D for Eqs. (8.35) to (8.38) is the surface-averaged, or Sauter mean bubble diameter D_{sm}:

$$D_{sm} = \frac{\sum m_j D_j^3}{\sum m_j D_j^2} \tag{8.59}$$

where m_j is the number of bubbles of diameter D_j.

The quantity $nF_0 t_b$ in Eq. (8.39) is the total bubble volume in the reactor. The bubble volume per reactor volume is known as the holdup H (volume gas per volume reactor). If the holdup value is available from other laboratory, plant, or literature correlations (Example 8.1), it is used directly in

$$a = H \frac{6}{D} \tag{8.60}$$

When writing mass balances on the liquid phase volume only, it was convenient to define an interfacial area per *liquid volume*, a', as in Eqs. (8.14) and (8.15). A second common quantity is a, the interfacial area per (liquid + gas) volume. These two interfacial measures are related through the holdup H:

$$a'(1 - H) = a \tag{8.61}$$

Thus, in using correlations for mass transfer, care must be taken to note whether the original reference calculated $k_l a$ or $k_l a'$.

Example 8.1 Holdup correlations

Bubble column†:

$$H/(1 - H)^4 = 0.20(\text{Bo})^{1/8}(\text{Ga})^{1/12}\text{Fr} \tag{8E1.1}$$

where Bo = Bond no. = $g d_t^2 \rho_c / \sigma$
Ga = Galileo no. = $g d_t^3 / u_c^2$
Fr = Froude no. = $u_G / \sqrt{g d_t}$
u_G = gas superficial velocity
d_t = tower (tank) diameter

† M. Chakravarty, S. Begum, H. D. Singh, J. N. Baruah, and M. S. Iyengar., *Biotech. Bioeng. Symp.*, **4**: 363, 1973.

Laboratory-scale gas-lift column†: Holdup interior to draft tube (sparged):

$$H_1 = \left[(\mu - \mu_{H_2O})^{2.75} + 1.61\,\frac{73.3 - \sigma}{74.1 - \sigma}\right]10^{-4}u^{0.88} \tag{8E1.2}$$

Holdup in annulus:

$$H_2 = 1.23 \times 10^{-2}\left[\frac{74.2 - \sigma}{79.3 - \sigma}\right]\mu_c^{0.45}\left[\frac{A_{\text{int}}}{A_{\text{ann}}}\right]^{1.08}\mu^{1.38} \tag{8E1.3}$$

Holdup above baffle:

$$H_3 = 7.5 \times 10^{-3}u^{0.88} \tag{8E1.4}$$

Total column holdup:

$$H = 0.003u^{0.88}$$

where μ_l = liquid viscosity at column temperature (cP)
μ_{H_2O} = water viscosity at column temperature (cP)
σ = gas-liquid surface tension (dyne/cm)
u = superficial gas velocity (cm/s)
A_{int} = cross-sectional area of draft tube (cm^2)
A_{ann} = cross-sectional area of annulus (cm^2)

Laboratory-scale gas-lift column‡ in draft tube:

$$H_d = \frac{u_d}{1.065u_d + u_\gamma} = \text{void fraction} \tag{8E1.5}$$

where u_d is the superficial gas velocity in draft tube and, using volumetric flows,

$$\gamma = \frac{\text{gas flow rate}}{\text{gas + liquid flow rate}}$$

$$u_\gamma = \begin{cases} 32 \text{ cm/s} & \gamma < 0.43 \\ 257(\gamma - 0.43) + 32 & \gamma > 0.43 \end{cases}$$

Agitated tank§: diameter ≈ height for $\text{Re}_i^{0.7}\,(N_i D_i/u)^{0.3} < 2 \times 10^4$

$$H = \sqrt{\frac{u}{u_t}}\sqrt{H} + 0.015a_0 \qquad \text{and} \qquad a_0 = 1.44\,\frac{P^{0.4}\rho^{0.2}}{\sigma^{0.6}}\left(\frac{u}{u_t}\right)^{1/2}$$

For $\text{Re}_i^{0.7}\,(N_i D_i/u)^{0.3} > 2 \times 10^4$

$$H = \frac{a_1}{a_0}\sqrt{\frac{u}{u_t}}\sqrt{H} + 0.015a_1$$

and

$$\log\frac{2.3a_1}{a_0} = 1.95 \times 10^{-5}\,\text{Re}_i^{0.7}\left(\frac{N_i D_i}{u}\right)^{0.3}$$

† M. Chakravarty, S. Begum, H. D. Singh, J. N. Barrah, and M. S. Iyengar, *Biotech. Bioeng. Symp.* **4**: 373, 1973.

‡ R. T. Hatch, Ph.D. Thesis in Food Science and Nutrition, p. 150, Massachusetts Institute of Technology, Cambridge, Mass., 1973.

§ P. H. Calderbank, *Trans. Inst. Chem. Eng.*, **36**: 443 1958.

where a_0, a_1 = interfacial area per unit volume of broth
Re_i = impeller Reynolds number = $\rho N_i D_i^2/\mu_c$
u = superficial gas velocity (empty-tank basis)
u_t = bubble rise velocity

and N_i, D_i, P, ρ, σ are as in the text (see Sec. 8.4).

Agitated vessels†: For air in water Richards' data can be represented by

$$\left(\frac{P}{V}\right)^{0.4} u^{1/2} = 7.63H + 2.37$$

where P = horsepower (hp)
V = ungassed liquid volume, m^3
u = superficial velocity, m/h
H = volume void fraction (valid for $0.02 < H < 0.2$)

8.4 FORCED CONVECTIVE MASS TRANSFER

Vigorous mechanical mixing of air–liquid dispersions is often necessary to obtain economic rates of biomass increase, substrate consumption, or product formation. The concerns of this section are again relationships between appropriate variables allowing estimation of mass-transfer coefficients k_l and/or a or a', the interfacial area per appropriate volume.

8.4.1 General Concepts and Key Dimensionless Groups

The functions served by mechanical agitation augment (and in some cases dominate) the influences of convection driven by freely rising or falling dispersed phases:

1. The high dynamic pressure near the impeller tip or other mixer devices produce small bubbles, thereby increasing a' locally. Provided that the rate of bubble coalescence is not correspondingly increased elsewhere in the vessel, the result is an increased value of the volumetric average value of a'.
2. The fermentation fluid may contain a suspension of solid or other liquid phases which may tend to rise or fall in the vessel. Mechanical mixing provides a more uniform volumetric dispersion of these phases in the bulk liquid. For hydrocarbon dispersions, k_l contains a term proportional to the cube root of the phase-density difference $(\rho_{H_2O} - \rho_{HC})^{1/3}$ [recall Eq. (8.38)]; the resulting small mass-transfer coefficient for hydrocarbon-substrate-limited fermentations is increased by agitation.
3. For gas bubbles of given size in vigorously agitated vessels, k_l does not vary significantly with power input since the relative gas or fluid velocity is dominated by density differences. (Why is this true?) The agitator turbulence,

† J. W. Richards, *Prog. Ind. Microbiol.*, **3**: 143, 1961.

however, will decrease D and thus increase a' for a given holdup; note that this result will change the size of the bubbles and thus k_l through the influence of D.

4. The maximum size of loosely aggregated mycelia, microbial slimes, mold pellets, etc., may be diminished by agitation, thus maintaining a smaller microbial Thiele modulus (Sec. 4.4) and again rendering the vessel more uniformly mixed with respect to the liquid phase. Examples of decreased yields of desired products have been reported at relatively high agitation rates; these may be due to cellular or extracellular enzyme damage, mixing interference with morphological development and differentiation, etc.
5. The liquid-cell suspension may be so viscous that only mechanical agitation provides any degree of bulk-liquid mixing (considered further in Sec. 8.8).

In forced convection, the action of the applied mechanical work produces some characteristic velocity against which other motions can be scaled. For impeller agitation, two scales exist: the rms fluid velocity fluctuation u_{rms}, and the impeller tip velocity u_i, which is proportional to $N_i D_i$, where N_i is the impeller rotation rate in revolutions per unit time, and D_i is the impeller diameter.

Reduction of the forced-convection balances for total mass, species, and momentum produces the following dimensionless groups using u_{rms} as the characteristic velocity:

$$\text{Sherwood number} = \text{Sh} = \frac{k_l D}{\mathscr{D}_{O_2}} \tag{8.62a}$$

$$\text{Schmidt number} = \text{Sc} = \mu_l / \rho_l \mathscr{D}_{O_2} \tag{8.62b}$$

$$\text{Reynolds number} = \text{Re} = \rho_l D u_{\text{rms}} / \mu_l \tag{8.62c}$$

$$\text{Froude number} = \text{Fr} = u_{\text{rms}}^2 / gD \tag{8.62d}$$

Alternately, in stirred systems, the characteristic dimension may be taken as the impeller diameter D_i, and the reference velocity is $N_i D_i$. The subscripts i remind us that the scaling is to the impeller rather than the gas, liquid, or solid particles present in the dispersion. In this case, the appropriate Reynolds and Froude numbers are given by

$$\text{Re}_i = \frac{\rho_l D_i}{\mu_l} N_i D_i = \frac{\rho_l N_i D_i^2}{\mu_l} \qquad \text{and} \qquad \text{Fr}_i = \frac{N_i^2 D_i}{g} \tag{8.63}$$

The Froude number has received other definitions. For mass transfer into a suspension of "neutrally buoyant" particles, the following relationship has been suggested:

$$\text{Fr}_L = \frac{N_i^2 D_i^2}{gL} \tag{8.64}$$

where L is the reactor height. As the Froude number represents the contribution of free-surface dynamics vs. mechanical mixing, the distance of the surface from the tank bottom could logically enter in its description. When the stirred volume refers to the mixing of two phases (continuous and dispersed) of different densities, e.g., hydrocarbon droplets in aqueous phases, the following definition of a modified Froude number may be useful:

$$\mathrm{Fr}_{\text{two phase}} = \frac{\rho N_i^2 D_i^2}{\Delta\rho\, gL} \tag{8.65}$$

It is clear that close attention should be paid to both the form of correlations and the definitions of the groups involved in literature reports; a correlation should never be used or cited without careful group definitions.

8.4.2 Correlations for Mass-Transfer Coefficients and Interfacial Area

The mass-transfer coefficient of gases depends largely on the hydrodynamics of the liquid film near the bubble; this in turn is dominated by the natural convection buoyancy forces and the turbulent Reynolds number during most of the bubble residence time; thus correlations for freely rising bubbles are most usefully associated with the Reynolds number of Eq. (8.62*c*).

In sufficiently large reactors fitted with baffles to maximize mixing rates within the continuous phase (Chap. 9), the influence of free-surface effects (Froude number) becomes unimportant. Free-surface gas exchange can be significant in bench-scale bioreactors, but this contribution fades to insignificance as reactor scale increases. In the absence of such surface influences, the dimensionless solutions for the velocity and concentration fields yield

$$\bar{c} = f(\bar{z}, \mathrm{Re}, \mathrm{Sc}, \mathrm{Sh}) \tag{8.66}$$

so that the dependence of the Sherwood number is

$$\mathrm{Sh} = \frac{k_l D}{\mathscr{D}_{O_2}} = g(\mathrm{Re}, \mathrm{Sc})$$

Data of Calderbank's [11] give the correlation

$$\mathrm{Sh}\ (\text{turbulent aeration}) = 0.13\ \mathrm{Sc}^{1/3}\mathrm{Re}^{3/4} \tag{8.67}$$

In terms of power input per unit reactor volume, using relation (8.52) we can show that the variation of $k_l(\mathrm{Sh})$ with P/V is

$$\mathrm{Sh} \propto \left(\frac{P}{V}\right)^{1/4} \tag{8.68}$$

Thus

$$k_l = 0.13\left(\frac{\alpha^3 \mu_l (P/V)}{\rho_l D}\right)^{1/4} \mathrm{Sc}^{-2/3} \tag{8.69}$$

Once turbulence has been achieved, so that the previous equation applies, the specific increase in k_l with P diminishes rapidly, as is seen by evaluation of

$$\frac{1}{k_l}\frac{dk_l}{dP} = \frac{1}{4P} \tag{8.70}$$

From the previous discussion of bubble breakup, we expect to be able to correlate the maximum stable bubble diameter D_c as a function of σ and the variables dominating dynamic stress. Since this value is closely related to the characteristic actual bubble diameter D_{sm} in many cases, it is not surprising that correlations for D_{sm} have a similar form. The inclusion of additional terms depending on the gas holdup H and dispersed phase viscosity μ_d in these correlations indicate important but not dominant contributions from other processes. The earlier caveat about comparing coalescence properties in the correlation basis experiment and the system of interest stands also in connection with the correlations in Example 8.2.

Example 8.2 Correlations for maximum (D_c) or Sauter mean (D_{sm}) bubble or droplet diameters

For freely rising bubbles, experimental values[†] are

$$D_c = \left(1.452 \times 10^{-2} \frac{\sigma}{\Delta\rho}\right)^{1/2} \quad \text{cm} \tag{8E2.1}$$

For agitated vessels, on a power-per-unit-volume basis we list the results of several experiments:

Experiment 1 liquid–liquid:[‡]

$$D_{sm} = 0.224 \frac{\sigma^{0.6}}{\rho_l^{0.2}(P/V)^{0.4}} H^{0.5}\left(\frac{\mu_d}{\mu_c}\right)^{0.25} \tag{8E2.2}$$

Experiment 2 gas–liquid electrolyte:[‡]

$$D_{sm} = 2.25 \frac{\sigma^{0.6}}{\rho_l^{0.2}(P/V)^{0.4}} H^{0.4}\left(\frac{\mu_g}{\mu_l}\right)^{0.25} \tag{8E2.3}$$

Experiment 3 gas in alcohol solutions:[¶]

$$D_{sm} = 1.90 \frac{\sigma^{0.6}}{\rho_l^{0.2}(P/V)^{0.4}} H^{0.65}\left(\frac{\mu_g}{\mu_l}\right)^{0.25} \tag{8E2.4}$$

For gases in viscous liquids,[§]

$$D_{sm} = 0.7 \frac{\sigma^{0.6}}{(P/V)^{0.4}\rho_l^{0.2}}\left(\frac{\mu_l}{\mu_g}\right)^{0.1} \tag{8E2.5}$$

[†] S. Hu and R. C. Kintner, "The Fall of Single Liquid Drops Through Water," *AIChE J.*, **1**: 42, 1955.

[‡] P. H. Calderbank, *Trans. Inst. Chem. Eng.*, **36**: 443, 1958.

[¶] J. A. McDonough, W. J. Tomme, and C. D. Holland, "Formulation of Interfacial Areas in Immiscible Liquids by Orifice Mixers," *AIChE J.*, **6**: 615, 1960.

[§] S. M. Bhavaraju, T. W. F. Russell, and H. W. Blanch, "Design of Gas Sparged Devices for Viscous Liquid Systems," *AIChE J.*, **24**: 454 (1978).

For agitated vessels, impeller variables [26]

$$\frac{P}{V} \approx \text{const} \frac{\rho_l N_i^3 D_i^5}{D_i^3} \qquad \text{for} \qquad \text{Re}_i \geq 10^4 \tag{8E2.6}$$

$$\frac{D_c}{D_i} = (\text{const})\left(\frac{\sigma}{N_i^2 D_i^3 \rho_l}\right)^{0.6} \tag{8E2.7}$$

For turbulent pipe flow [11]

$$\frac{D}{D_{\text{pipe}}} = (\text{const})(\text{We}_{\text{pipe}}^{-0.6}\text{Re}_{\text{pipe}}^{-0.1}) = (\text{const})\left(\frac{\sigma}{\rho_l \langle u^2 \rangle D_{\text{pipe}}}\right)^{0.6} \cdot \left(\frac{\mu_l}{D_{\text{pipe}} u_{\text{rms}} \rho_l}\right)^{0.1} \tag{8E2.8}$$

where

$$\text{We}_{\text{pipe}} \equiv \left(\frac{\sigma}{\rho_l \langle u^2 \rangle D_{\text{pipe}}}\right)^{-1} \qquad \text{and} \qquad \text{Re}_{\text{pipe}} = \frac{D_{\text{pipe}} u_{\text{rms}} \rho_l}{\mu_l} \tag{8E2.9}$$

For flow through an orifice in pipe flow (measured one foot downstream)†

$$\frac{D_{\text{pipe}}}{D} = 21.6\left(\frac{D_{\text{orifice}}}{D_{\text{pipe}}}\right)^{3.73} \text{H}^{0.121}\text{We}_{\text{pipe}}^{-0.722}\text{Re}_{\text{pipe}}^{-0.065} \tag{8E2.10}$$

where We_{pipe} and Re_{pipe} are as in Eq. (8E2.9).

Given D_c or D_{sm} and the holdup H, the value of a is calculated from Eq. (8.60). For complex situations like those applying to most of the macroscopic contactor situations in Fig. 8.2, H must be measured directly or obtained from correlations for similar configurations. A representative sampling of such correlations was given in Example 8.1.

8.5 OVERALL $k_l a'$ ESTIMATES AND POWER REQUIREMENTS FOR SPARGED AND AGITATED VESSELS

In this section we summarize different experimental findings on the volumetric mass transfer coefficient $k_l a'$. Experimental results are often reported in this form because of lack of knowledge of a' directly. Also, in some cases this combined parameter is used to account for other effects such as long residence times of small bubbles in highly viscous fermentations. These bubbles become depleted of oxygen and therefore contribute little to oxygen transfer. Thus, an optically determined a' value may not represent the interfacial area per unit volume of oxygen-containing bubbles.

We have already noted that increasing power input can reduce bubble size and thereby increase interfacial area. Here, we cite methods for calculating power input in terms of the gas sparging and agitation parameters. Also, we consider the common case of simultaneous gas sparging and mechanical agitation.

† J. A. McDonough, W. J. Tomme, and C. D. Holland, "Formulation of Interfacial Areas in Immiscible Liquids by Orifice Mixers," *AIChE J.*, **6**: 615, 1960.

Many studies of gas-liquid mass transfer in low viscosity fluids in agitated vessels have been reviewed by Van't Riet [9]. Results of many experiments in many different vessels with different mixer configurations are all fit within 20 to 40 percent by the relationships

Stirred vessel, water, coalescing:

$$k_l a = 2.6 \times 10^{-2}\left(\frac{P}{V}\right)^{0.4} (u_{gs})^{0.5}\ (\mathrm{s}^{-1}) \tag{8.71}$$

$$(V \leq 2600L;\ 500 < P/V < 10{,}000\ \mathrm{W/m^2})$$

Stirred vessel, water, noncoalescing:

$$k_l a = 2.0 \times 10^{-3}\left(\frac{P}{V}\right)^{0.7} (u_{gs})^{0.2}(\mathrm{s}^{-1}) \tag{8.72}$$

$$(2 < V < 4400L;\ 500 < P/V < 10{,}000\ \mathrm{W/m^2})$$

Here u_{gs} is the superficial gas velocity which is equal to the gas feed volumetric flow rate divided by vessel cross section area times the gas holdup. The ranges of vessel volumes and volumetric power input considered in obtaining these correlations is indicated with each.

It is significant to note that, consistent with the concepts of turbulence discussed earlier, these correlations have been applied (within the indicated 20 to 40 percent) regardless of the type of stirrer (turbines, paddles, propellers, rods, self-inducing agitators) and the number of stirrers. Stirrer position also seems to be immaterial unless the stirrer is close to the bottom of the vessel (less than the stirrer diameter), which decreases dissipated power, or close to the surface, which results in air entrainment and lower power consumption.

Similar examination of mass transfer data from bubble columns shows:

Bubble column, water, coalescing:

$$k_l a = 0.32(u_{gs})^{0.7} \tag{8.73}$$

If noncoalescing conditions exist in a bubble column, no general correlation can be presented since sparger construction influences $k_l a'$.

As indications of the other types of correlations which have been proposed, and because *a priori* determination of power consumption or gas superficial velocity may not be simple, we summarize next several additional correlations. For gas transfer in a bubble column, Akita and Yoshida [12] reported

$$\frac{k_l(ad_t^2)}{D} = 0.6(\mathrm{Sc})^{1/2}\mathrm{Bo}^{0.62}\mathrm{Ga}^{0.31}H^{1.1} \tag{8.74}$$

where the Bond number ($\mathrm{Bo} \equiv gd_t^2\rho_c/\sigma$) and Galileo number ($\mathrm{Ga} = gd_t^3/u_l^2$) are referred to the tower diameter d_t. This equation is found accurate for $d_t \leq 60$ cm, and also useful if, for $d_t > 60$ cm (0.6 m), the value of 0.6 m for d_t is used in Eq. (8.74).

In the same spirit, the volumetric mass transfer coefficient, $k_l a$, for an airlift column is given by Bello et al. [13] as Eq. (8.75)

$$k_l a = \frac{0.0005 \cdot (P/V)^{0.8}}{(1 + A_d/A_r)} \tag{8.75}$$

where A_d/A_r = ratio of areas of downcomer and riser sections and P/V = aeration power input/volume. This correlation is useful, but hides the fact that mass transfer, holdup, etc., are not the same in the riser headspace and downcomer portions of the air lift equipment.

All correlations have a limited range of applicability. For example, the strong dependence of $k_l a'$ on P/V indicated in Eq. (8.75) vanishes at sufficiently small P/V values ($P/V < 1$) since bubble fluid dynamics are here dominated by natural convection driven by bouyancy.

A motionless mixer may be used to subdivide and remix repeatedly the liquid phase, as well as to maintain the upward bubble flow and reduce or eliminate large bubbles or air "slugs." Wang and Fan [14] suggest the volumetric mass-transfer correlation

$$k_l a = 0.12 u_l^{0.624} \left(\frac{u_g}{1.99 u_g + 47.1} \right) (\mathrm{s}^{-1}) \tag{8.76}$$

where u_g, u_l are gas and liquid superficial velocities (cm/s).

A potential shortcoming of using any correlation is the frequent implicit assumption of uniformity of power dissipation and/or $k_l a'$ in the contactor in which data underlying the correlation was taken or in the vessel to be designed or analyzed. As an indication of potential difficulties in this connection, consider the variations in mean flow velocities measured in a standard agitated, baffled tank. The lines in Fig. 8.6 indicate circulation patterns and the numbers give local average velocities as a fraction of the impeller tip velocity.

Several studies of sparged column contactors vividly illustrate important spatial variations in $k_l a'$. Figure 8.7 shows experimental measurements of dissolved oxygen axial profiles in a bubble column and in a three phase fluidized bed containing particles of diameter 0.1 cm. In both cases, there is an entry zone near the sparger in which oxygen transfer rates are relatively large and a later zone of much smaller transfer rates. This may be due to a transition from bubble sizes dominated by sparger conditions to bubble sizes dictated by coalescence-breakup equilibrium. Other indicated points in Fig. 8.7 are calculated from a mathematical model which assumes plug flow of gas through the column through two zones with different volumetric mass-transfer coefficients for each zone. The interface between the two zones was estimated to be around 33 cm above the sparger. Decreasing $k_l a'$ values over the first 27.6 cm above the sparger, then constant $k_l a'$ were successfully employed in another model for a different bubble column contactor.

A clear pitfall exists here for scale-up. The entry region identified above will contribute significantly in small laboratory systems but will constitute only a small fraction of the vessel volume in a tall, large-scale column.

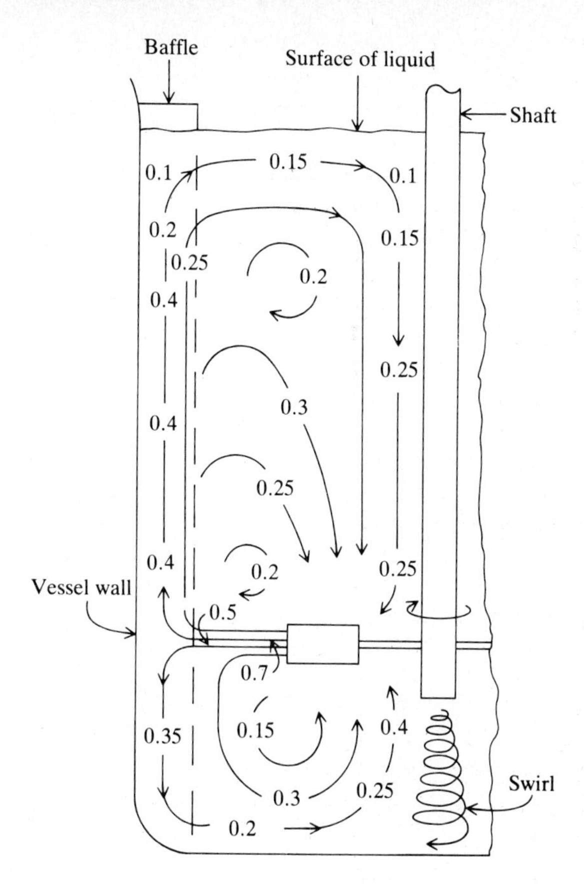

Figure 8.6 Average flow velocities (as fraction of impeller tip speed = 5.2 ft/s) and circulation patterns in water in a 12" high tank stirred at 200 rpm. *(W. L. McCabe and J. C. Smith, Unit Operations in Chem. Eng., 3d ed., 1976, p. 234, McGraw-Hill, New York.)*

Next we consider calculation of power requirements to achieve desired gas sparging and mechanical agitation rates. Our emphasis here is on important overall concepts and trends; more detailed treatments considering energy losses in process equipment components are available in the references. For gas sparging into a column, the power used in compression to sparge a gas volumetric flow rate F_0, at pressure p_1, is

$$P_g = \rho_g F_0 \left[\frac{RT}{(MW)} \ln \frac{p_1}{p_2} + \alpha \frac{u_0^2}{2} \right] \tag{8.77}$$

where p_2 is the pressure at the top of the vessel and u_0 is the gas velocity at sparger orifice. The fraction of gas kinetic energy transferred to the liquid, α, is typically about 0.06.

The power consumption for stirring nonaerated fluids depends upon fluid properties ρ_l and μ_l, the stirrer rotation rate N_i and diameter D_i, and the drag coefficient of the impeller C_{D_i}. The latter is expected to vary with impeller Reynolds number in a different manner for each flow regime: laminar, transition,

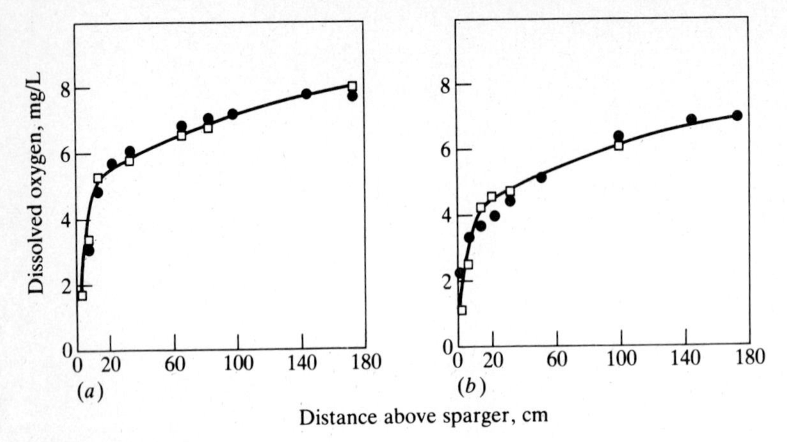

Figure 8.7 Experimental data (●) showing variation of dissolved oxygen content as a function of distance above the sparger for (*a*) a bubble column ($u_l = 7.5$ cm s^{-1}, $u_g = 28$ cm s^{-1}) and (*b*) a three-phase fluidized bed containing 0.1 cm diameter particles ($u_l = 7.5$ cm s^{-1}, $u_g = 20$ s^{-1}). The points (□) were calculated from a mathematical model. Lines drawn to show trends. *(Reprinted by permission from M. Alvarez-Cuenca and M. A. Nerenberg, Adv. in Biotechnol. vol. 1, p. 477 (1980), M. Moo-Young (ed.), Pergamon Press, p. 477.)*

or turbulent. A well-known study by Rushton, Costich, and Everett [26] is summarized in Fig. 8.8*a* for three impeller geometries. The data are plotted as a dimensionless power input, the power number P_{no}, vs. impeller Reynolds number Re_i:

$$P_{\text{no}} = \frac{Pg}{\rho_l N_i^3 D_i^5} \tag{8.78}$$

In the turbulent regime, the power input is independent of Re_i,

$$P \propto N_i^3 D_i^5 \qquad P_{\text{no}} = \text{const}$$

whereas in laminar flow, the relation is more nearly given by

$$P \propto N_i^2 D_i^3 \qquad \text{or} \qquad P_{\text{no}} \propto \frac{1}{\text{Re}_i}$$

The proportionality constant in each case depends on the impeller geometry. It is interesting to note the strong similarity between this figure and the plot of the friction factor in tube flow. In the latter case, the friction factor varies as 1/Re in laminar flow (as does the power number vs. Re_i), and, in turbulent flow, f tends to a nearly constant value which has a larger magnitude for pipes with rough walls. Similarly, as the agitator geometry becomes less "smooth," Fig. 8.8*a*, the power number reaches a higher turbulent plateau value. The latter case is more complicated since the presence of the tank walls and baffles will also exert an effect on the measured power input P.

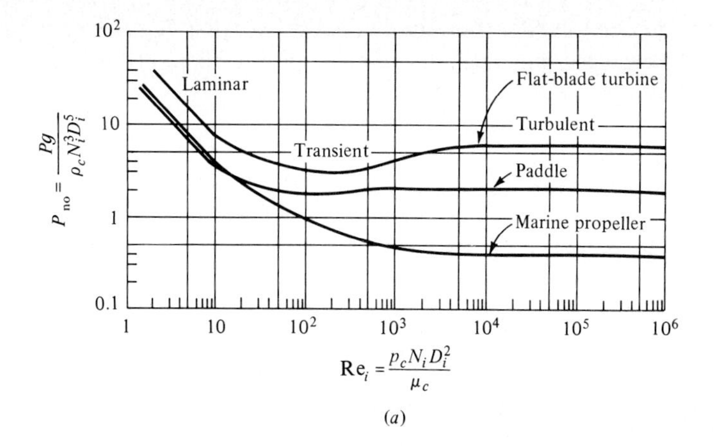

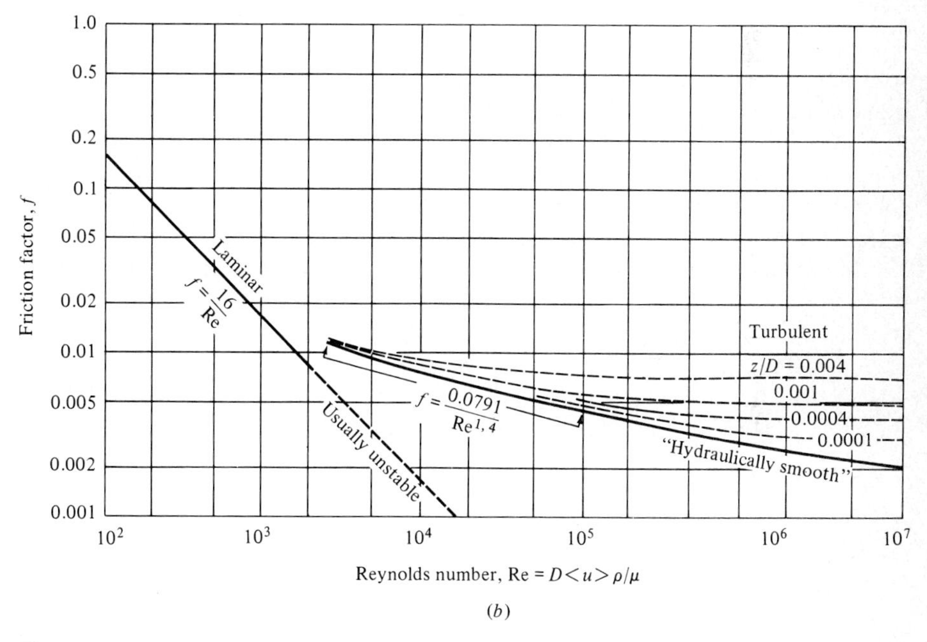

Figure 8.8 (*a*) Power number vs. Reynolds number (impeller) for various impeller geometries. (*b*) Pipe friction factor f vs. Reynolds number, Re. z equals height of surface roughness peaks. *[(a) Reprinted from S. Aiba, A. E. Humphrey and N. F. Millis, "Biochemical Engineering" 2d ed., p. 174, Univ. Tokyo Press, Tokyo, 1973; modified from J. H. Rushton, E. W. Costich, and H. J. Everett, "Power Characteristics of Mixing Impellers, part 2," Chem. Eng. Prog., vol. 46, p. 467, 1950. (b) Reprinted by permission from W. L. McCabe and J. C. Smith "Unit Operations of Chemical Engineering" McGraw-Hill, New York, 1954 (original curves from L. F. Moody, Trans. ASME, vol. 66, p. 671, 1944).]*

When the agitated vessel is simultaneously aerated, the power requirements for agitation decrease. The ratio of power requirements in aerated vs. nonaerated vessels, P_a/P vs. a dimensionless aeration rate N_a

$$N_a = \frac{F_g}{N_i D_i^3} \tag{8.79}$$

(where F_g is volumetric gas rate) has been correlated, as shown in Fig. 8.9.

Except for the most rapidly changing part of the curve, these forms can be fitted to

$$\frac{P_a(N_a) - P_a(N_a = \infty)}{P - P_a(N_a = \infty)} = e^{-mN_a} \tag{8.80}$$

where m = const. An alternative form which is also useful in turbulent aeration of non-Newtonian fluids is due to Michel and Miller [28]:

$$P_a = m'\left(\frac{P^2 N_i D_i^3}{F_g^{0.56}}\right)^{0.45} = m'\left(\frac{P^2 (N_i D_i^3)^{0.44}}{N_a^{0.56}}\right)^{0.45} \tag{8.81}$$

where m' = const. In both the above correlations, P is the nonaerated power input of the earlier chapter formulas.

From the relations of the previous paragraph, at constant N_i and D_i, the power input diminishes with increased N_a, that is, increased air flow F_g. This effect appears partially due to the decrease in average density of the fluid being agitated. Uniformity of bulk mixing diminishes with increasing N_a.

In several bioreactor designs, mixing is provided by injection of a liquid jet into the vessel. In this case the power dissipation may be estimated from (see Ref. 29)

$$P = \frac{8\rho_l F_l^3}{\pi^2 D_j^{0.4}} \tag{8.82}$$

where D_j is the jet diameter.

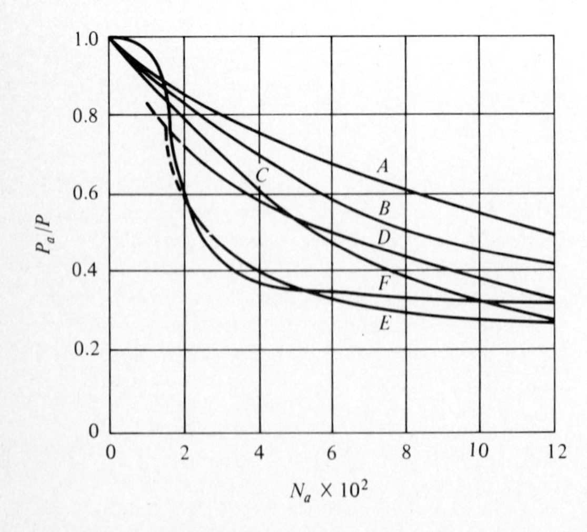

Figure 8.9 Ratio of power requirement for aerated vs. nonaerated systems as a function of N_a (see text): *A*, flat blade turbine (8 blades); *B*, vaned disc (8 vanes); *C*, vaned disc (6 vanes); *D*, vaned disc (16 vanes); *E*, vaned disc (4 vanes); *F*, paddle. *(Reprinted by permission from Y. Ohyama and K. Endoh, "Power Characteristics of Gas-Liquid Contacting Mixers," Chem. Eng., Japan, vol. 19, p. 2, 1955.)*

8.6 MASS TRANSFER ACROSS FREE SURFACES

Gas transfer through gas-liquid free surfaces (Fig. 8.10) plays a major role in oxygen supply and CO_2 removal from animal cell cultures. Surface mass transfer is also important in shake flask and small-scale stirred bioreactors for microbial cultivations. Transport across free liquid surfaces is essential for stream reaeration and respiration of aerobic life near the sea surface and in lake communities. Free-surface mass transfer is also important in many industrial microbial processes employing trickle-bed reactors, e.g., wine-vinegar manufacture and wastewater treatment. In the former cases, the depth of oxygen transfer depends on the scale of eddy motions near the liquid surface. Mass transfer into or out of falling-liquid films has been studied frequently, though not often under conditions appropriate to microbial processes. This circumstance is considered first.

The area-integrated absorption rate for a falling laminar liquid film of thickness h, length L, and width W and with zero initial concentration of dissolved gas is given by

$$\text{Integrated absorption rate (moles/unit time)} = WLc_l^*\left(\frac{4\mathscr{D}_{O_2}u_{\max}}{\pi L}\right)^{1/2} \tag{8.83}$$

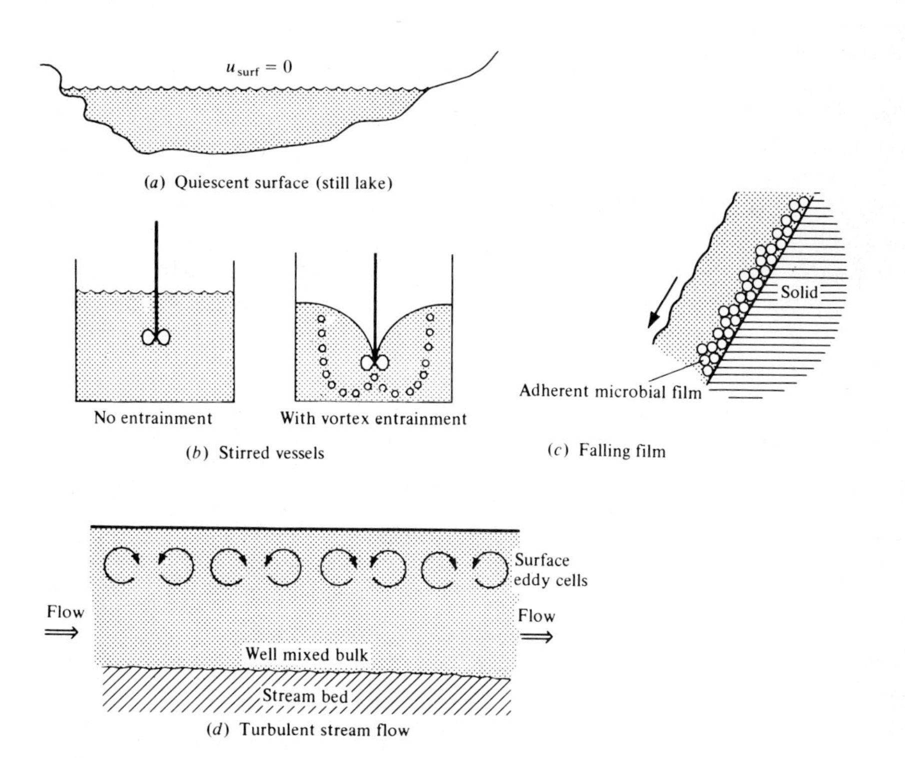

Figure 8.10 Free surface operation configurations: (*a*) quiescent surface, (*b*) stirred vessels, (*c*) falling film, (*d*) turbulent stream flow.

where u_{max} is the free-surface velocity. In the derivation, it is assumed that the solute concentration near the solid boundary never departs from zero; i.e., the diffusing solute does not "penetrate" the entire film thickness during the falling-time interval [5].

We define the Reynolds number for the situation as

$$\text{Re} = \frac{u_{max} R_h \rho}{\mu_l}$$

where the hydraulic radius R_h is used as the length scale

$$R_h = \frac{Wh}{2W + 2h} \approx \frac{h}{2} \qquad \text{if } h \ll W$$

From the definition for the mass-transfer coefficient

$$\text{Integrated absorption rate} = k_l(c_l^* - c_l)WL \tag{8.84}$$

For c_l approximately zero relative to c_l^*, Eqs. (8.83) and (8.84) can be rewritten in the form $\text{Sh} = f(\text{Sc}, \text{Re})$:

$$\text{Sh} = \frac{k_l h}{\mathscr{D}_{O_2}} = 2b(\text{Sc Re})^{1/2} \tag{8.85}$$

where h is the length scale for the Sherwood number and $b = (L/h)^{1/2}$. Thus, Sh varies as $\text{Re}^{1/2}$.

Livansky et al. [18] studied CO_2 absorption into aqueous films moving down a slope of known area using water, algal suspensions, and nutrient medium as absorbing fluids. The value of k_l at $\text{Re} = 7$ to 8×10^3 was the same for all three fluids; only the algal suspensions were studied at different Reynolds numbers. (These films may have been turbulent.) Their results can be described by

$$k_l = 4 \times 10^{-5}\ \text{Re}^{2/3} \qquad \text{for } 2000 \leq \text{Re} \leq 8000$$

In turbulent flowing streams, the scale of circulation is important since this scale determines the depth to which fluid carries fresh, nearly saturated liquid from the surface into the bulk liquid. If we imagine a circulating eddy of length and depth Λ, as shown in Fig. 8.11, the average Sherwood number for mass transfer under turbulent conditions can be defined analogously to Eq. (8.28) as

$$\overline{\text{Sh}} = \frac{\bar{k}_l \Lambda}{\mathscr{D}_{O_2}} = \int_0^{\Lambda} \left(\frac{\partial c}{\partial \bar{z}}\right)_{\bar{z}=0} d\bar{w}$$

where $\bar{z}$ and $\bar{w}$ are dimensionless coordinates scaled by Λ, and $\bar{k}_l$ is the average value over the eddy length.

The rate of mass transfer is ultimately dependent on $\mathscr{D}_{O_2}$ locally and on the rate at which fluid near the surface is renewed by the circulation pattern. An

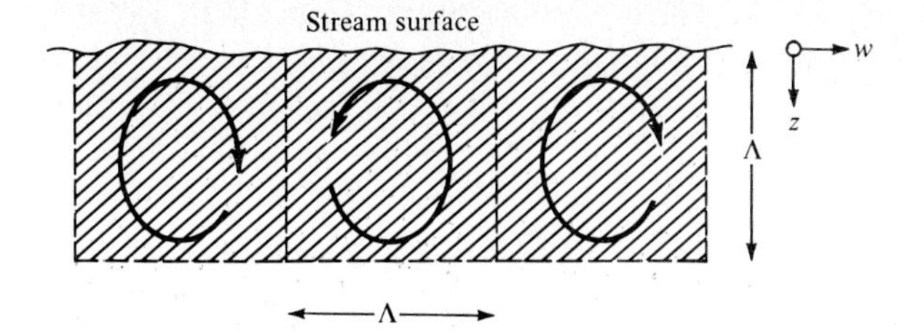

Figure 8.11 Sketch of circulating eddies near a free liquid surface.

early derivation due to Higbie† argued that for fluid elements with identical residence times τ at the gas–liquid surface the mass-transfer coefficient k_l should be

$$\bar{k}_l = \left[\frac{\mathscr{D}_{O_2}}{\pi\tau}\right]^{1/2} \tag{8.86}$$

This form has been extended by Danckwerts for distributions of surface residence times, the result still giving $\bar{k}_l \propto (\mathscr{D}_{O_2})^{1/2}$. A flowing turbulent stream of average flow velocity $\langle u_w \rangle$ has been suggested to have a renewal time τ equal to the ratio of stream depth h to average velocity $\langle u_w \rangle$:

$$\tau_{\text{stream}} = \frac{h}{\langle u_w \rangle}$$

thus predicting

$$k_l = \left(\frac{\mathscr{D}_{O_2}}{\pi}\frac{\langle u_w \rangle}{h}\right)^{1/2}$$

If the stream has width W, the interfacial area a per stream volume is

$$\frac{W(1)}{W(1)(h)} = \frac{1}{h}$$

Thus

$$k_l a = \left(\frac{\mathscr{D}_{O_2}}{\pi}\frac{\langle u_w \rangle}{h^3}\right)^{1/2} \qquad \text{(O'Connor-Dubbins)} \tag{8.87}$$

a form which has had reasonable success for describing reaeration of oxygen-deficient lakes and streams. Another treatment accounting for the variation of k_l with position w in the eddy provides

$$\bar{k}_l = 1.46\left(\frac{\mathscr{D}_{O_2} u_{\text{rms}}}{\Lambda}\right)^{1/2} \tag{8.88}$$

† R. Higbie, *Trans. AIChE*, **35**: 365 (1935).

where u_{rms} is the rms velocity in the eddy circulation. In general, it appears reasonable to take Λ and u_{rms} proportional to mean stream depth and mean stream velocity, respectively:

$$\gamma \times \text{depth} = \Lambda \qquad \gamma \times \text{mean stream velocity} = u_{rms}$$

with the same constant of proportionality γ.

The motion of waves at air-water interfaces is known to influence gas transfer strongly. Here, in contrast to the preceding treatments, the gas-flow patterns, e.g., average and turbulent velocity components of wind which drives ocean waves, are of major importance. However, the general topic is too complex for this text (see the references).

8.7 OTHER FACTORS AFFECTING $k_l a'$

From the definitions of k_l and a' and consideration of the factors responsible for the thickness of the mass-transfer resistance zone near bubble and droplet surfaces, k_l and a' will be influenced by alteration of the values of liquid-phase solute diffusivity $\mathscr{D}_{O_2}$, continuous-phase viscosity μ_c, and the gas-liquid interfacial resistance. The liquid "viscosity" may vary with shear rate; this non-Newtonian behavior is of sufficient importance to be discussed separately (Sec. 8.8). The remaining influences are summarized in this section.

8.7.1 Estimation of Diffusivities

The Wilke-Chang correlation is a useful means of estimating (usually to better than 10 to 15 percent) the diffusion coefficient of small molecules in low-molecular-weight solvents:

$$\mathscr{D} = 7.4 \times 10^{-8} \frac{T(x_a M)^{1/2}}{\mu_l V_m^{0.6}} \qquad \text{cm}^2/\text{s} \tag{8.89}$$

where M = solute molecular weight

V_m = molecular volume of solute at boiling point, $\text{cm}^3/\text{g mol}$

μ_l = liquid viscosity

The parameter x_a represents the association factor for the solvent of interest; some values for x_a are 2.6 (H_2O), 1.9 (methanol), 1.5 (ethanol), and 1.0 (benzene, ether, and heptane).

The diffusion coefficient will vary with ionic strength (as does solubility, Table 8.1) and with concentration of solutes which change the solution viscosity. Provided that the solute-solvent interactions are not altered in the latter case, the relation

$$\mathscr{D}_1 \mu_{c1} = \mathscr{D}_{\text{ref}} \mu_{c,\,\text{ref}}$$

provides a useful scale to correct for changes in solution viscosity from a reference point, say that of pure water, with temperature held constant.

Table 8.4 Diffusion coefficients in microbial film

Biomass	Reactor	$\mathscr{D}_{\text{meas}}$, cm^2/s $\times 10^5$	% of H_2O value
Bacterial slimes	Rotating tube	1.5	70
	Submerged slide	0.04	2
Zoogloea ramigera	Fluidized reactor	0.21	8

The diffusion coefficient in microbial aggregates is usually less than that in pure water (2.25×10^{-5} cm^2/s), as summarized in Table 8.4 for oxygen. A recent study examining other variables concluded that the O_2 diffusion coefficient in waste-treatment microbial aggregates decreases from the pure H_2O value in the range 20:1 to 5:1 with increased aggregate lifetime in the reactor and with increased C/N ratio of the entering wasted substrates.

8.7.2 Ionic Strength

The precise resolution of ionic-strength influences into all pertinent factors appears difficult. An examination of Newtonian fluids gives the physical-absorption result

$$K_l a = \lambda \left(\frac{P_a}{V_l}\right)^n \left(\frac{F_g}{A}\right)^m \left(\frac{\rho_l^{0.533} \mathscr{D}_{O_2}^{2/3}}{\sigma^{0.6} \mu_l^{1/3}}\right) \tag{8.90}$$

where V_l = liquid volume
P_a = power input during aeration
A = reactor cross section perpendicular to flow rate F_g

An empirical fit for λ, n, and m vs. ionic strength $I(=\frac{1}{2}\sum Z_i^2 c_i$, i = species, Z_i = species charge) is possible:

$$\lambda = 18.9 - \frac{28.7I'}{0.276 + I'} \qquad I' = \begin{cases} I & \text{if} \quad 0 \le I \le 0.40 \text{ g ion/L} \\ 0.40 & \text{if} \quad I > 0.40 \text{ g ion/L} \end{cases}$$

$$n = \begin{cases} 0.40 + \dfrac{0.862I'}{0.274 + I'} & I' = I \le 0.4 \\ 0.90 & I > 0.4 \end{cases}$$

$$m = \begin{cases} \text{monotonic increasing} & \\ \text{from } 0.35 & I = 0 \\ \text{to } 0.39 & I \ge 0.4 \end{cases}$$

for $K_l a$ in s^{-1}, P_a in ft-lb_f/min, V_l in ft^3, F_g in ft^3/s, A in ft^2, and physical properties in cgs units.

8.7.3 Surface-Active Agents

As discussed in Chap. 2, many biochemicals are amphipathic, i.e., contain strongly hydrophobic and hydrophilic moieties which tend to concentrate at gas–liquid and liquid–liquid interfaces. In various phases of fermentations, cells secrete species such as polypeptides which may behave like surfactants, at times leading to foaming tendencies in aerated vessels. Addition of chemical antifoams also affects interfacial resistances to mass transfer, though typically in a manner opposing that of surfactants.

Adsorption of surfactants at the phase interface is a spontaneous process; the interfacial free energy and thus the surface tension σ is reduced relative to the original value. From the correlations in Example 8.2 the values of D_{Sauter} and D_c are expected to decrease, leading to higher values of the interfacial area per volume a'.

This tendency for a' to increase is countered by the effect of surfactant films on the mass-transfer coefficient k_l. The adsorption of a macromolecular film results in a stagnant, rigid interface. The decreases in k_l discussed below are thought to be due to either or both of two mechanisms: (1) the ease of liquid movement near the interface is reduced due to the decreased mobility of the interface; thus, the variety of mass-transfer theories based on estimating exchange rates of small fluid elements between the surface and the bulk will predict a decreased mass-transfer coefficient (see Refs. 3 and 4 for further discussion); (2) like the cell membrane itself, the molecular film is expected to contribute a resistance of its own, which may cause a departure from the presumed gas–liquid equilibration in the plane of the interface.

Addition of 10 ppm sodium lauryl sulfate (SLS) reduced k_l for oxygen transfer by 56 percent versus pure water. A constant or plateau value of k_l was observed at all higher surfactant concentrations. The surface area per volume a' increased slowly throughout the range of SLS concentrations from 0 to 75 ppm, with a minimum of $k_l a'$ at about 10 ppm surfactant.

The product $k_l a'$ has been observed to increase continuously with surfactant addition in a turbine aerator. Inspection of the data for the reported ratio of a' (surfactant)/a' (no surfactant) and the corresponding ratio for the product $k_l a'$ shows that while a' increased 400 percent for addition of 4.0 ppm sodium dodecyl sulfate, $k_l a'$ increased only about 15 percent, implying a decrease of about 71 percent in the value of k_l.

This reduction observed in both sets of data is in agreement with results of others. For a variety of sparingly soluble gases, the average plateau values of k_l upon surfactant addition correspond to reductions of k_l by a factor of 60 percent. The sodium dodecyl sulfate data of Benedek and Heideger [33] suggest that turbine-agitated aeration corresponds to the transition region between the two correlations presented earlier in Eqs. (8.35) and (8.36); a similar result appears to be the case for aeration with sieve trays (Danckwerts). This serves to warn the reader that such correlations are useful estimates but should be replaced by experimental values from more pertinent equipment when possible.

8.8 NON-NEWTONIAN FLUIDS

For fluids, the ratio of shear stress τ_s to the velocity gradient du/dy is defined as the viscosity η_v (the subscript v distinguishes the viscosity from the earlier effectiveness factor η). Thus,

$$\tau_s = -\eta_v \frac{du}{dy} \tag{8.91}$$

A plot of τ vs. du/dy for a Newtonian fluid is linear and passes through the origin. A variety of non-Newtonian behaviors has been observed in steady flows for liquid solutions of polymers and/or suspensions of dispersed solids or liquids; the features of the more common of these are summarized in Fig. 8.12, where shear rate $\dot{\gamma}$ is used rather than du/dy.

8.8.1 Models and Parameters for Non-Newtonian Fluids

The dilatant, Newtonian, and pseudoplastic behaviors are examples of the general Ostwald-de Waele or power-law formulations for fluids

$$\tau_s = -\eta_0|\dot{\gamma}|^{n-1}\dot{\gamma} = -(\text{apparent viscosity})(\dot{\gamma}) = -\eta_v\dot{\gamma} \tag{8.92}$$

$$\text{where } n \begin{cases} > 1 \rightarrow \text{dilatant} \\ = 1 \rightarrow \text{Newtonian} \\ < 1 \rightarrow \text{pseudoplastic} \end{cases}$$

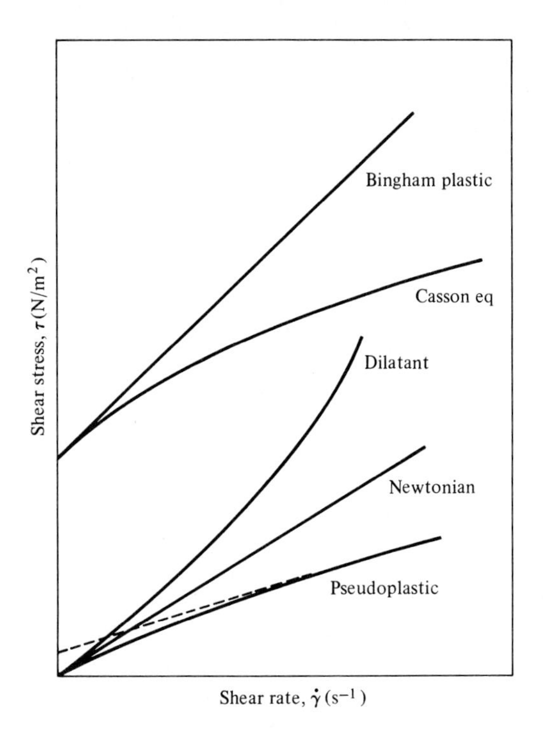

Figure 8.12 Stress vs. shear-rate behavior of Newtonian and common non-Newtonian fluid models. *(Reprinted from J. A. Roels, J. van den Berg, and R. M. Voncken, "The Rheology of Mycelial Broths," Biotech. Bioeng., vol. 16, p. 181, 1974.)*

Some systems appear not to produce a motion until some finite yield stress τ_0 has been applied. For Bingham plastic fluids, the form

$$\tau_s = \tau_0 - \eta_0 |\dot{\gamma}|^{n-1}\dot{\gamma} \qquad \text{where} \qquad n = 1, \tau_0 \neq 0, \tau_s > \tau_0 \tag{8.93}$$

is useful.

The last form with finite τ_0 and n less than unity would give a curve similar to the Casson equation, which is given by

$$\tau_s^{1/2} = \tau_0^{1/2} - K(\dot{\gamma}^{1/2}) \tag{8.94}$$

In the subsequent discussion of mass-transfer coefficients, power input, mixing, etc., we shall refer to various fermentation or other fluid systems of interest as being (apparently) pseudoplastic, Newtonian, etc., and indicate correlations between the fluid-model parameters and the former quantities of interest. The fluid descriptions in Fig. 8.12 refer to behavior in steady shear flows. Under unsteady-state conditions, such as follow a step change in the applied shear rate, time-dependent responses in apparent viscosity η_v are often observed, thus demanding a more structured fluid model (in the same sense as the Chap. 7 cell kinetics models) to describe transient situations. Such transient states may more accurately apply to turbine agitation and turbulent mixing in non-Newtonian systems. These more structured models are a relatively difficult and undeveloped area of mechanics; we simply insert the caveat here that our understanding of the factors responsible for non-Newtonian behavior and their description is weaker than the theories for the previous sections of this chapter.

Non-Newtonian behavior may arise in at least two distinct cases: (1) suspensions of small particles and (2) solutions of macromolecules. It is apparent that the two cases become similar as molecular diameters increase above 50 to 100 Å or as particle diameters fall from the order of micron sizes.

8.8.2 Suspensions

Various theories predict that a dilute suspension of spheres should remain Newtonian, the effective viscosity $\eta_{v,\text{eff}}$ of the suspensions being given by

$$\frac{\eta_{v,\text{eff}}}{\eta_{\text{solvent}}} = 1 + \tfrac{5}{2}\phi + b\phi^2 + \cdots \tag{8.95}$$

where ϕ is the volume fraction solids and b is of the order of 6 to 8. For a bacterial density as high as 10^9 cells per milliliter and a cell diameter of 3×10^{-4} cm, the value of ϕ is about 5×10^{-3}: the effect of the cells is apparently negligible. Higher volume fractions occur in filtration operations such as dewatering (in product recovery from cell broths) and in fermentations producing the mold pellets or mycelia of previous discussions.

At higher volume fractions, solid suspensions may exhibit a yield stress; e.g., aqueous slurries of nuclear fuel particles with diameters in the micrometer range appear usefully modeled by the Bingham plastic model. Mycelial fermentations of *Streptomyces griseus* appear to follow a Bingham form except at very low shear rates; this may be important in aeration. From Eq. (8.50) note that the maximum stress for a rising bubble is associated with the terminal rise velocity. Thus, in Bingham fluids, a sufficiently small bubble will not exert the yield stress on the surrounding fluid and it will remain fixed in the same fluid element for long times; i.e., it would be expected to circulate in the vessel with the fluid rather than following the usually rising path of larger bubbles.

The Casson equation has been fruitfully applied to descriptions of blood flow. Red blood cells form aggregates, the size of which diminishes with increasing shear forces. The apparent viscosity η_v diminishes with increased shear, as seen from Fig. 8.12 for the Casson equation. In floc-forming fermentations, we may expect stirrer shear forces to control the average size of the microbial flocs; hence the Casson equation may be useful in some such fermentations.

A study of rheology of penicillin broths vs. time over a large range of shear rates used a turbine impeller for viscometry studies rather than a rotating cylinder, the two cited advantages being (*a*) the stirring prevents phase separation, which otherwise would affect the reliability of the measurement: without stirring a thin "cell-free" layer of liquid is formed at the wall of the cylinder, and (*b*) the shear rate of the impeller is a simple function of the impeller speed and is independent of the rheology of the liquid as has been shown previously [38, pp. 188–189]. This research revealed that the Bingham and power-law models were inadequate to represent the results over the full range of shear rates investigated. A reasonable modification of the Casson equation was derived:

$$(M_\tau)^{1/2} = (M_{\tau 0})^{1/2}\left(1 + \frac{0.69(M_{\tau 0})^{1/2} - 1.1}{(M_{\tau 0})^{1/2}}(N_i)^{1/2}\right) \tag{8.96}$$

where M_τ is proportional to stress: $M_{\tau 0} = 64D_i^3/2\pi K$, N_i is the impeller rotation rate (revolutions per second) and K is the ratio of shear rate to impeller speed (constant). The data agreed well with the above equation though a slightly better fit resulted if the second term on the right-hand side was modified to

$$\frac{0.193(M_{\tau 0})^{0.75} + 0.1}{(M_{\tau 0})^{1/2}}(N_i)^{1/2}$$

The importance of the result is threefold: (1) The rheological data were taken over a sufficient range of shear rates to *discriminate* between models which are similar over restricted shear-rate ranges. (2) The results showed that the power and Bingham laws were most inadequate at low shear rates; this range is likely to be important in determining tank-mixing uniformity; i.e., extrapolation of the Bingham and power laws to this situation would lead to large errors (mixing problems in vessels are discussed in Chap. 9 in connection with fluid circulation-time and residence-time distributions). (3) The modified Casson equation was shown to predict a factor dependent on the mold *morphology* (shape) through the ratio of length to diameter of the filaments. This result (and other theories for nonspherical particles) provide an important potential connection between reactor-design calculations (analysis) and morphology (observation) since the latter subject has been examined for a range of species and conditions in the literature.

The power-law model appears to fit data for mold filaments over the higher range of shear rates. It also describes suspensions of paper pulp; 4 percent paper pulp in water suspensions have power-law parameters $n = 0.575$ and $\eta_0 = 0.418\ \text{lb}_f \cdot \text{s}^n/\text{ft}^2$. Investigations of oxygen transfer with pulp suspensions have been used to simulate some fermentation conditions: if the overall oxygen uptake is determined by mass-transfer limitations in high shear regions, this may be useful. The previous paragraph casts more doubt on such simulation studies where bulk mixing influences O_2 uptake by the microorganism(s).

In continuous-flow systems with several reactors in series (Chap. 9), the tank number replaces time as an indication of population growth phase, and changes in rheology with tank number may be encountered. For example, two-stage cultivation of *Candida utilis* [41] at large flow rates led to second-tank conditions such that the specific biomass growth rate became so much larger than specific rate of cell multiplication that an increase in average cell size occurred. A change of cell and aggregate morphology was also noted, indicating again a possible connection between morphology and rheology.

8.8.3 Macromolecular Solutions

A number of microbes secrete extracellular polysaccharides and related biopolymeric derivatives. Examples include xanthan gum, polyalginic acid, and pullulan. The first is widely used in modifying viscosity of processed foods, in stabilizing suspensions, and as a major candidate for oil field recovery operations. These secreted biopolymers render the fermentation fluid non-Newtonian; the fluids are characteristically described by a simple power law behavior [Eq. (8.92)] where $n < 1$ (pseudoplastic).

The time-varying biopolymer concentration in a batch fermentation gives rise to time-varying fluid and hence transport properties (mass and heat transfer). As correlations for transport coefficients in non-Newtonian fluids are typically expressed in terms of the power law parameters, n and η_0, we seek a relation between biopolymer concentration and rheological parameters.

For xanthan gum, the polymer concentration, $[P]$, and the power law parameter, η_0, are related by a correlation of the form

$$\eta_0 = A \cdot [P]^B \tag{8.97}$$

where $B \sim 2.5$.

For a number of microbial exobiopolymers, the time-varying power law exponent n and consistency η_0 do not vary independently in a fermentation but are related by a correlation of the form

$$\ln \eta_0(t) = C + D \cdot n(t) \tag{8.98}$$

With $[P(t)]$ available from a biological kinetic model, Eqs. (8.97) and (8.98) provide a simple description of the time evolution of $n(t)$ and $\eta_0[P(t)]$. These values can then be used to predict heat and mass transfer coefficients, as well as mixing and power consumption characteristics [39, 40].

8.8.4 Power Consumption and Mass Transfer in Non-Newtonian Fermentations

An instantaneous impeller Reynolds number Re_i' may be defined according to Calderbank [11]:

$$\mathrm{Re}_i' = \frac{D_i^2 N_i^{2-n} \rho}{0.1 K_c} \left(\frac{n}{6n+2} \right)^n \tag{8.99}$$

where K_c, the consistency index, equals the shear stress τ at a shear rate of $1\ \mathrm{s}^{-1}$.

For the nonaerated non-Newtonian broths of *Endomyces*, the power number [Eq. (8.78)] was correlated with the Reynolds number Re_i' as shown in Fig. 8.13. The curves plotted there can be represented by the relationship

$$P_{\mathrm{no}} = k(\mathrm{Re}_i')^x \left(\frac{D_i}{D_T} \right)^y \left(\frac{W}{D_T} \right)^z \tag{8.100}$$

where D_i = impeller diameter
D_T = tank diameter
W = impeller width

and k, y, and z depend on the range of Re_i':

	Re_i'		
	<10	10–50	>50
k	32	11	9
x	−0.9	−0.4	−0.05
y	−1.7	−1.7	−1.2
z	0.4	0.5	0.9

The similarity of Fig. 8.13 to Rushton's data and the pipe friction factor in Fig. 8.8*a*, *b* is clear.

In aeration studies by the same group, the correlation of Michel and Miller (used earlier for Newtonian fluids) fitted the results in the turbulent regime ($\mathrm{Re}_i' > 50$):

$$P_g \propto \left(\frac{P^2 N_i D_i^3}{F_g^{0.56}} \right)^{0.45} \tag{8.101}$$

The laminar and second regimes ($\mathrm{Re}_i' < 50$) appear reasonably approximated by use of a smaller exponent:

$$P_g \propto \left(\frac{P^2 N_i D_i^3}{F_g^{0.56}} \right)^{0.27} \tag{8.102}$$

The two curves meet at a value of $P^2 N_i D_i^3 / F_g^{0.56} = 2 \times 10^{-2}$, and slightly different proportionality coefficients result for the different impellers used.

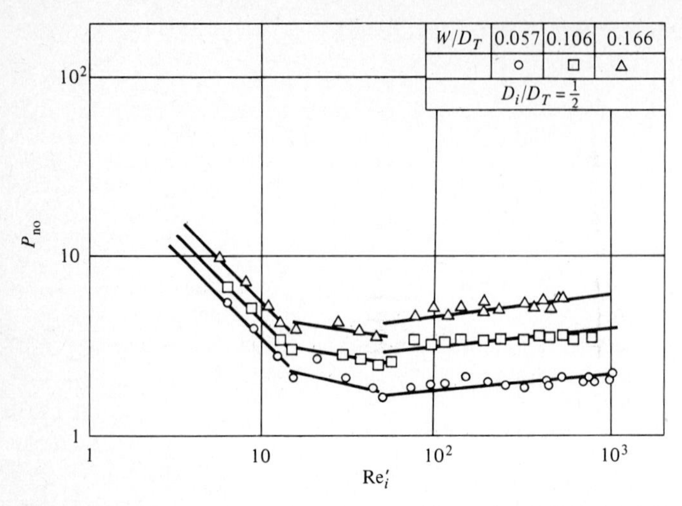

Figure 8.13 Power number vs. impeller Reynolds number, Re'_i (see text) for three different impellers in nonaerated systems. *(Reprinted from H. Taguchi and S. Miyamoto, "Power Requirements in Non-Newtonian Fermentation Broth," Biotech. Bioeng., vol. 8, p. 43, 1966.)*

The dilution of mycelial cultures by small volume percents of water (10 to 25 percent) produces marked changes in k_l and power-number parameters. The implications for power-input variations with dilution rate in continuous culture may be significant.

The influence of the microbial population depends on the physical situation which applies (floating single cells, cell aggregates, cells localized near particle or bubble surface, etc.) and the influence of the microbial particles on the fluid properties. The value of $k_l a$ has been observed to diminish 90 percent as the concentration of mycelial *Aspergillus niger* increased from 0.02 to 2.5 percent. Enhancement of O_2 mass transfer is observed in the presence of suspensions of *Candida intermedia, Pseudomonas ovalis,* or 0.3-μm alumina particles. Not only do cells and alumina give similar results, but through use of oxidative phosphorylation (Chap. 5) inhibitors, it is established that the enhancements are typically 40 percent vs. water and are independent of cell viability. It has been recently shown that the effect of the particles is to alter the hydrodynamics near the gas-liquid interface in such a way as to decrease the mass transfer resistance of the adhering fluid film near the bubble [44].

At the high shear rates needed to mix phases and promote mass transfer, diminutions in microbial activity have frequently been observed. For example, the viability of a relatively large cell, the protozoon *Tetrahymena*, began to be seriously altered by disruption at shear rates $>1200 \text{ s}^{-1}$. In these experiments the maximum shear rate characterized by the impeller tip speed appeared to be a more important variable than the turbulent Reynolds number or power input per unit volume.

The impeller can have other effects: the size of the extant microbial aggregates may be reduced, as mentioned earlier. Taguchi and Yoshida† divided the experimentally observed reduction of myce-

† H. Taguchi and T. Yoshida, *J. Ferment. Technol.*, **46**:814, 1968.

lial pellet size into two phenomena: (1) chipping small pellicules off the larger pellet and (2) directly rupturing the spherical shape of the pellet. The time evolution of particle diameter D due to the first process appeared to be governed by

$$\frac{dD}{dt} = -k_i(N_i D_i)^{5.5} D^{5.7} \tag{8.103}$$

while the second process could be described as a first-order decay

$$\frac{dN_p}{dt} = -k_r N_p \tag{8.104}$$

where N_p is the number of nondisrupted pellets remaining. The rate coefficient k_r was correlated with D, N_i, and D_i to give

$$k_r = +(\text{const})(D^{3.2} N_i^{6.65} D_i^{8.72}) \tag{8.105}$$

Assuming that the turbulent Reynolds stress rather than the viscous stress is responsible for pellet rupture and that the pellet resistance depends directly on the measurable tensile strengths of the pellets, Taguchi et al. argue that the last equation may have some theoretical basis.

Regarding the oxygen transfer rate in paper-pulp suspensions, impeller and tank geometry were found to be important, in agreement with the earlier results of Taguchi and Miyamoto [42] for aerated and nonaerated *Endomyces* suspensions. In 1.6 percent pulp suspensions, the product $k_l a$ was described by

$$\frac{k_l a}{N_i} = 0.113\left(\frac{D_T^2 h}{WL(D_i - W)}\right)^{1.437} (N_i t)^{-1.087}\left(\frac{D_i}{D_T}\right)^{1.021} \tag{8.106}$$

where t = characteristic mixing time of vessel
h = liquid height
W = impeller width

and D_i, D_T, and L are as before.

In summary, a number of factors including bubble and cell dimensions, fluid properties and rheology, agitator and tank geometry, and power input determine mass-transfer coefficients and surface area per unit volume. The combination of these estimates with cell kinetics of the previous chapter and notions of mixing and macroscopic reactor configurations of Chap. 9 is important in assembling a complete reactor design. This state of knowledge is obviously wishful; the previous relations in this section are but the beginning of work needed to design confidently such reactors from first principles. As a closing example of some of the complexities yet to be unraveled, we note the time changes of the relations between bubbles, particulate substrate, and cells that accompany the fermentation of *Candida petrophilum* on *n*-hexadecane, as shown in Fig. 8.14. The description of this gas-liquid-liquid-cell system by the authors† is illuminating:

> During the first period of fermentation, oil droplets are relatively large and cells attach to oil droplets rather than to air bubbles. Air bubbles are unstable and easily renewed. The $k_l a$ value can be kept at the maximum level associated with the fermentor. During the second phase, oil

† A. Mimura, I. Takeda, and R. Wakasa, "Some Characteristic Phenomena of Oxygen Transfer in Hydrocarbon Fermentation," *Biotechnol. Bioeng. Symp. 4*, pt. 1, p. 467, 1973.

droplets become smaller and cells are adsorbed onto the surface of the oil droplets, forming dense flocs. The flocs tend to attach to the surface of the air bubbles, but they can be easily separated with agitation.

The $k_l a$ value is decreased continuously as fermentation proceeds. The third phase is the last half period of the logarithmic growth phase, in which yeast is growing rapidly, although we cannot observe any oil droplets microscopically in the culture liquid. At this point the $k_l a$ value reaches its minimum throughout the fermentation.... [Air] bubbles are covered with yeast cells, and the bubbles come together with cells as an intermedium. They are very stable and float on the surface of the culture liquid. *n*-Paraffin is completely exhausted in the fourth phase. The cells are dispersed uniformly throughout the culture liquid. In this period the nature of the culture liquid may be similar to that in a carbohydrate fermentation, and $k_l a$ recovers its initial levels.

An analytical model including a few of the features of this fermentation type is considered in Chap. 9 (Example 9.2).

8.9 SCALING OF MASS-TRANSFER EQUIPMENT

As discussed by Oldshue,† the various quantities which may influence the product $k_l a$ in an agitated industrial reactor do not scale in the same way with reactor size or impeller rate.

1. The turbulent Reynolds number Re_t determines u_{rms} and thus bubble mass-transfer coefficients k_l

$$\mathrm{Re}_t = \frac{\rho u_{\mathrm{rms}} D}{\mu} \propto \frac{\rho D}{\mu}\left(\frac{D}{\rho}\frac{P}{V}\right)^{1/3} \tag{8.107}$$

2. The impeller tip velocity $\pi N_i D_i$ determines the maximum shear rate $\dot{\gamma}$, which in turn influences both maximum stable bubble or microbial floc size (Sec. 8.3) and damage to viable cells (Sec. 8.8.4).
3. The power input per unit volume P/V through Re_t determines mass-transfer coefficients and particulate sizes. In laminar and transition regimes of aerators

$$P \propto N_i^2 D_i^3 \qquad \text{from Fig. 8.8}b$$

For turbulent regimes, the power number is constant; thus

$$P \propto N_i^3 D_i^5$$

Then taking V_{reactor} to scale with D_i^3 gives

$$\frac{P}{V} \propto \begin{cases} N_i^2 & \text{laminar, transition aeration} \\ N_i^3 D_i^2 & \text{turbulent aeration} \end{cases}$$

† J. Oldshue, *Biotech. Bioeng.*, **8**:3, 1966.

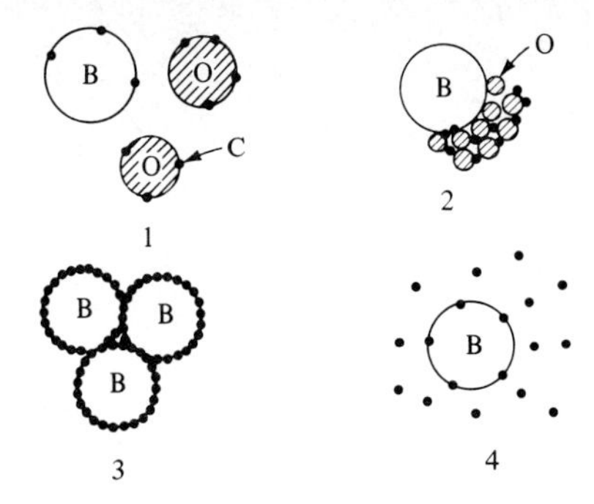

Figure 8.14 Different relationships among air bubbles (B), *n*-hexadecane droplets (O), and yeast cells (C) at different stages of batch culture of the yeast *Candida petrophilum* (1 = log phase, 2 = first half of exponential phase, 3 = second half of exponential phase, 4 = after n-hexadecane exhaustion). *(Reprinted from A. Mimuro, I. Takeda, and R. Wakasa, "Some Characteristic Phenomena of Oxygen Transfer in Hydrocarbon Fermentation," in B. Sikyta, A. Prokop, and M. Novak (eds.). Adv. Microbial. Eng., part 7, p. 467, Wiley-Interscience, New York, 1973.)*

4. The power input *during* aeration is

$$P_a = m'\left[\frac{P^2(N_i D_i^3)^{0.44}}{N_a^{0.56}}\right]^{0.45} \tag{8.100}$$

Thus,

$$\frac{P_a}{V} \approx m'\left[\frac{P^2}{V^{2.22}}\frac{(N_i D_i^3)^{0.44}}{N_a^{0.56}}\right]^{0.45} \tag{8.109}$$

which will determine the motor size needed during fermentation.

5. If the vessel liquid is well mixed internally, a characteristic circulation time exists. The liquid recirculation flow rate F_l through the impeller region varies as a cross-sectional area πD_i^2 and tank average impeller velocity varies as $N_i D_i$. Thus

$$\frac{F_l}{V} \propto \frac{N_i D_i^3}{D_i^3} = N_i \qquad \text{time}^{-1}$$

a quantity of importance since it is inversely proportional to the time that fluid may spend away from the homogenizing influence of the impeller.

Given all these different quantities which can influence the process and which have different dependencies on agitation parameters, which one(s) should be used as the basis for scale-up? Here by "basis for scale-up" we mean the quantity which, by choice of operating conditions in the larger unit, will be maintained at the same value as in the smaller scale unit. For example, if we scale-up on the basis of constant power per unit volume, for mechanical agitators in the turbulent regime this means

$$N_{i1}^3 D_{i1}^2 = N_{i2}^3 D_{i2}^2 \tag{8.110}$$

where 1 and 2 denote the values in the small and large scale vessel, respectively. In scale-up of the early penicillin bioreactors, maintaining constant power input per unit volume (around 1 hp per 100 gal) plus vessel geometric similarity were used. As indicated in Fig. 8.15*a* this gives very similar yields of penicillin for vessels from 5 liters to 200 gallons. However, from the curves in Fig. 8.15*a*, we see

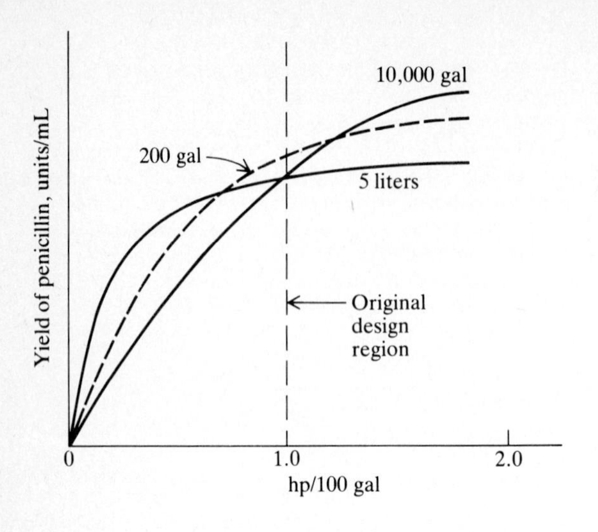

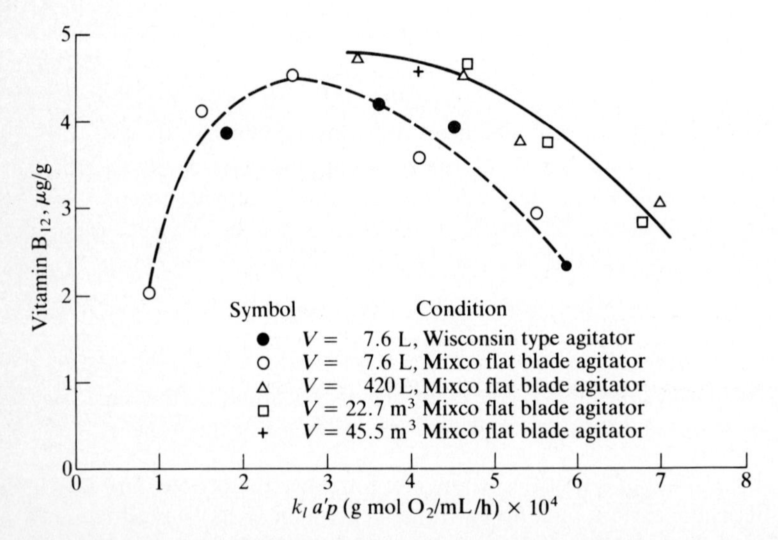

Figure 8.15 (*a*) Penicillin yields vs. power input in various sized vessels *(After E. Gaden, Sci. Rep. Ist Super. Sanita, vol. 1, p. 61, 1961)*. (*b*) Vitamin B_{12} yield (μg/g) vs. mass-transfer group ($k_l ap$) (see text) *(After W. H. Bartholomew, Adv. Appl. Microbiology, vol. 2, p. 289, 1960)*.

that, at different P/V values, there are significant differences in yield at different scales.

Another frequently applied basis for scale-up is constant volumetric transfer coefficient $k_l a$. Fig. 8.15*b* shows vitamin B_{12} yields from bacterial fermentations at different scales versus corresponding values of ($k_l ap$). The total pressure is included here to correct for the greater driving force for oxygen transfer at higher pressures which are encountered in large-scale bioreactors. The correlation between yield and $k_l ap$ is good, although data for the benchtop unit lie generally below those for fermentation in larger vessels. Notice also that, in contrast to earlier discussions emphasizing oxygen transport limits on growth and the need for large $k_l a$ values, these data show a distinct maximum with respect to $k_l a p$. This probably occurs because at the same $k_l a$ value, other mixing and flow characteristics such as maximum shear rate or circulation time are generally *not* the same in vessels of different scale.

Therefore, if we scale up on one basis, we must be aware that other mixing and flow properties are different. This point is dramatically illustrated in an example of Oldshue† which considers scale-up from an 80 L to a 10,000 L agitated bioreactor. Here, D_i increases fivefold and V increases by a factor of 125. The 1.0 in each column of Table 8.5 indicates the property which is kept the same in the large tank as in the small tank. Values for each property have been normalized by the values for the 80-L tank, so the "small-scale" column shows all 1.0s. For example, scale-up based on constant P/V will increase the maximum shear rate by 70 percent ($N_i D_i = 1.7$) and increase the circulation time about threefold. On the other hand, scale-up based on circulation time (that is, F_l/V) requires 3125 times more power in the large tank!

The different dependencies of important transport properties on agitator design makes scale-up of agitated vessels something of an art. We must try to select as a scale-up basis the transport property most critical to the performance of the bioprocess. This is not easy given the potentially sensitive and diverse responses of cells to each of the transport phenomena influenced by the mixer size and rotation speed.

The total oxygen consumption rate of the vessel is found by combining $k_l a$ with an appropriate macroscopic description of the vessel. If the bulk liquid composition is uniform and the bubbles are uniformly dispersed throughout the vessel, the mass-transfer rate is simply

$$Vk_l a(c^* - c_{\text{liq}})_e = \text{oxygen consumption rate, mol } O_2/\text{s} \tag{8.111}$$

Table 8.5 Relationship between properties for scale-up†

Property	Small scale, 80 L	Large scale, 10^4 L			
P	1.0	125	3125	25	0.2
P/V	**1.0**	**1.0**	25	0.2	0.0016
N_i	1.0	0.34	1.0	0.2	0.04
D_i	1.0	5.0	5.0	5.0	5.0
F_l	1.0	42.5	125	25	5.0
F_l/V	**1.0**	0.34	**1.0**	0.2	0.04
$N_i D_i$	**1.0**	1.7	5.0	**1.0**	0.2
Re_i	**1.0**	8.5	25	5.0	**1.0**

† S. Y. Oldshue: Fermentation Mixing Scale-up Techniques, *Biotech. Bioeng.*, **8**: 3, 1966.

where subscript e refers to gas exit compositions (a consequence of the perfect mixing assumption is that exit-stream compositions equal reactor compositions, Chap. 9).

When the bubbles rise in plug flow through the vessel but the impeller still maintains perfect mixing in the liquid phase, c^* varies with position. Over a differential reactor height dz, the instantaneous loss of oxygen from the bubble is

$$HA\,dz\frac{dp_{O_2}}{dt}\frac{1}{RT} = \frac{\text{gas vol}}{\text{reactor vol}}\left(\begin{matrix}\text{differential}\\ \text{reactor volume}\end{matrix}\right)\left(\begin{matrix}\text{conc. rate of}\\ \text{change in bubble}\end{matrix}\right) \tag{8.112}$$

which equals

$$-k_l a(c_l^* - c_b)A\,dz \tag{8.113}$$

the mass transfer rate into the liquid. Since $p_{O_2} = Mc_l^*$ (M is Henry's law constant), we have

$$\frac{H}{RT}\frac{dp_{O_2}}{dt} = \frac{HM}{RT}\frac{dc_l^*}{dt} = -k_l a(c_l^* - c_b) \tag{8.114}$$

For constant bubble-rise velocity u_b, $dt = dz/u_b$, and the z variation of c^* is seen to be

$$\ln\frac{(c_l^* - c_b)z}{(c_l^* - c_b)_{\text{inlet}}} = \frac{-k_l aRT}{HM}\frac{z}{u_b} \tag{8.115}$$

or

$$(c_l^* - c_b)_z = (c_l^* - c_b)_{\text{inlet}}\exp\left(-\frac{k_l aRTz}{HMu_b}\right) \tag{8.116}$$

The overall mass-transfer rate in the volume Ah is therefore

$$\int_0^h k_l a(c_l^* - c_b)(z)A\,dz = \frac{HMu_b}{RT}(c_l^* - c_b)_{\text{inlet}}A\left[1 - \exp\left(\frac{-k_l aRTh}{HMu_b}\right)\right] \tag{8.117}$$

The interaction of mixing, fermentation kinetics, and mass transfer is considered in the remaining text chapters.

8.10 HEAT TRANSFER

In biological reactors, heat may be added or removed from a microbial fluid for the following reasons:

1. It is desired to sterilize a liquid reactor feed by heating in a batch or continuous-flow vessel. Thus, the temperature desired must be high enough to kill essentially all organisms in the total holding time (Sec. 9.9.4).
2. If the heat generated in substrate conversion is inadequate to maintain the desired temperature level, heat must be added. For example, the reactor is an

anaerobic sewage-sludge digestor which operates best between, say, 55 and 60°C (Sec. 14.4.7).

3. The conversion of substrate generates excess heat with respect to optimal reactor conditions for, e.g., maintenance of viable cells, so heat must be removed, as in most microbial fermentation processes.
4. The water content of a cell sludge is to be reduced by drying.

The first three relate to cell viability and metabolism and will therefore be of concern here. The last example is a unit operation, drying, which is covered in most texts on engineering unit operations.

The heat is transferred between the bioprocess fluid to or from a second fluid in several ways, i.e., with externally jacketed vessels, coils inserted in a larger vessel, flow through a heat exchanger, or by evaporation or condensation of water and other volatile components of the cell-containing fluid. Examples of such configurations are shown in Fig. 8.16. Temperature fluctuations between atmosphere and thermally stratified lakes and land also clearly involve heat transfer, the resulting temperatures determining the habitable niches for species. The present section focuses on heat transfer in process reactors.

Assuming that transfer rates and changes in other forms of energy are negligible, the fundamental steady-state equation in heat transfer relates the total

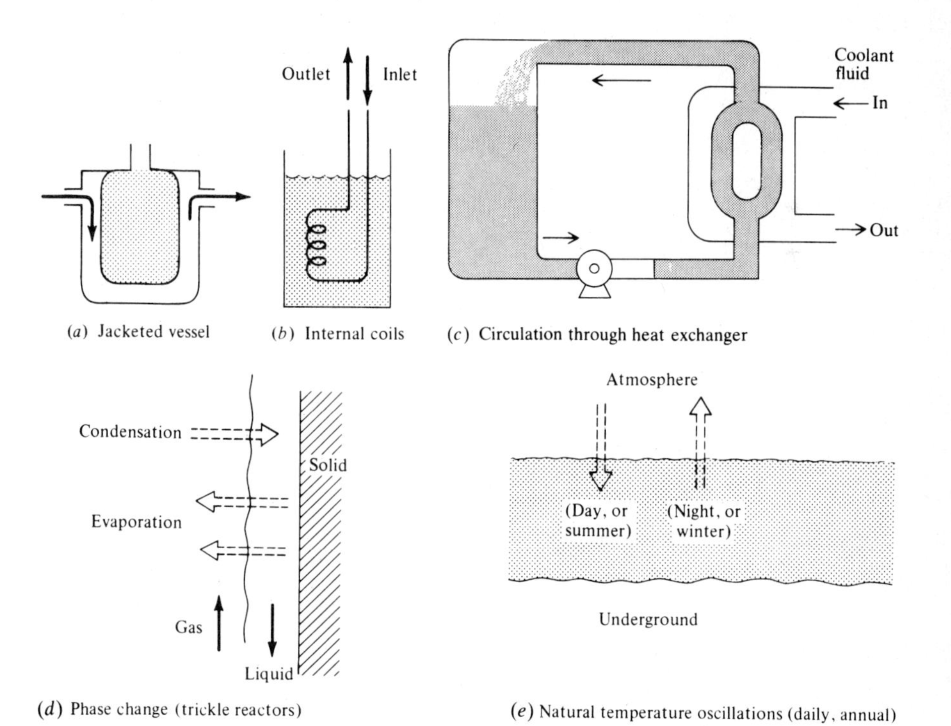

Figure 8.16 Examples of heat-transfer configurations: (*a*) jacketed vessel, (*b*) internal coils, (*c*) heat exchanger, (*d*) phase change of mass, and (*e*) natural temperature oscillations.

rate at which heat is generated to its rate or removal through some heat-transfer surface; thus

$$\text{Net generation rate} = \text{removal rate} = \bar{h}A\,\Delta T \tag{8.118}$$

where ΔT = characteristic temperature difference between bioprocess and cooling or heating fluid
A = heat-transfer surface area
$\bar{h}$ = overall heat-transfer coefficient

As with mass transfer, most of the resistance to heat transfer resides in a relatively quiescent thin fluid near the solid heating-cooling boundary, the bulk fluid being frequently well mixed and thus approximately isothermal. Our main concerns in this chapter are development of an overall energy balance and review of useful predictive formulas for $\bar{h}$ for various heater-cooler-sterilizer systems of interest. Methods for estimation of the heat load accompanying microbial growth have already been considered in Chap. 5.

The heat-transfer coefficients (Table 8.6) for boiling water and condensing vapors (steam, typically) make such fluids convenient in sterilization "reactors" (Sec. 9.9.4). Where lower temperatures are needed, as in heated anaerobic sludge digestors, a nonboiling water stream is useful. Viscous liquids exhibit greater heat-transfer resistances than water; as with mass transfer, this is due to lesser degree of bulk-fluid interchange with wall fluid and also to reduced thermal conductivity (analogous to $\mathscr{D}_{O_2}$).

Inspection of Fig. 8.16 reveals several design problems associated with heat transfer in biochemical reactors. The externally jacketed system has a heat-transfer area A which varies as the tank diameter (say impeller diameter) D_i^2. The volumetric heating or cooling demand of a reactor scales as D_i^3 if the overall

Table 8.6 General magnitude of heat-transfer coefficient h[†]

	h, kcal/(m$^2 \cdot$ h $\cdot$ °C)[‡]
Free convection:	
Gases	3–20
Liquids	100–600
Boiling water	1000–20,000
Forced convection:	
Gases	10–100
Viscous liquids	50–500
Water	500–10,000
Condensing vapors:	1000–100,000

[†] Data from H. Gröber, S. Erk, and U. Grigull, *Wärmeübertragung*, 3d ed., p. 158, Springer-Verlag, Berlin, 1955.

[‡] Multiplication by 0.204 gives h in units of Btu/(ft$^2 \cdot$ h $\cdot$ °F).

microbial reaction rates and power input per unit volume are unchanged. Thus, jacketed vessels, sufficient in rendering laboratory reactors isothermal, must frequently be replaced by reactors with internal or external coils in heating, e.g., anaerobic sludge digestion, or cooling (e.g., hydrocarbon fermentation for single-cell protein production) of large-scale reactors.

The presence of such internal piping clearly alters mixing patterns, fluid velocities, and perhaps bubble-coalescence rates. The complexity of such a situation reminds us of the need to perform measurements in reactors approximating the desired configurations, as we may expect *a priori* design of such reactors to be more uncertain than for simpler systems. (Problems of vessel mixing patterns are discussed in the text and problems of Chap. 9). For example, the correlations in Sec. 8.10.1 indicate that the heat-transfer coefficient $h[\text{Btu}/(\text{ft}^2\cdot\text{h})$ or $\text{kcal}/(\text{m}^2\cdot\text{h})]$ changes as we shift configurations from a single cooling coil perpendicular to fluid flow to a set of staggered rows of coils in a tube coil perpendicular to the fluid-flow direction. Thus, the presence of the first row of coils alters the flow pattern past subsequent tube rows.

In very large scale systems with large heat loads, such as bacterial growth on methanol in a 1500 m^3 reactor, internal coils become inadequate for cooling. Then, circulation through an external heat exchanger, or through an exchanger integral to a loop vessel configuration, is necessary. Here is an example where cooling loads, in concert with other considerations such as required power input for aeration and mixing, dictate a need for bioreactor designs substantially different from traditional agitated tank configurations. Several alternative vessel, contacting, and mixing configurations are summarized in Chap. 9.

Heat generation and removal rates are known with sufficient accuracy for some detailed heat-balance considerations to be possible, provided we clearly understand the basis on which such calculations are made and take appropriate precautions in terms of overdesign to allow for some uncertainty. The following section discusses the estimation of the heat-transfer demand (analogous to an oxygen demand); the subsequent sections discuss the conductance h.

Determination of the process heat transfer requirements begins with consideration of an overall energy balance. In a constant pressure system with negligible changes in potential and kinetic energies, the energy balance can be cast in terms of enthalpy changes, i.e., the heats of chemical transformation or phase transformation (e.g., evaporation, condensation), the sensible-heat flow in mass streams, and the heat transfer to or from second fluids acting as heating or cooling devices. Let

Q_{met} = heat-generation rate from cell growth and maintenance
Q_{ag} = heat-generation rate due to reactor mechanical agitation
Q_{gas} = heat-generation rate from aeration power input
Q_{acc} = heat-accumulation rate
Q_{exch} = heat-transfer rate to surroundings or exchanger
Q_{evap} = rate of heat loss by evaporation
Q_{sen} = rate of sensible-ethalpy gain of streams (exit − entrance)

Then

$$Q_{met} + Q_{ag} + Q_{gas} = Q_{acc} + Q_{exch} + Q_{evap} + Q_{sen} \tag{8.119}$$

Cooney, Wang, and Mateles [52] have utilized such a balance to calculate Q_{met} from measurement of Q_{acc} through monitoring the initial transient temperature rise of a nearly isolated fermentor. Under such an experiment, Q_{evap} and Q_{sen} are quite small, and Q_{exch} represents an important term compared with the difference of the larger individual rates $Q_{acc} - Q_{ag}$. As just mentioned, Q_{acc} was monitored calorimetrically, while Q_{ag} was calculated at each gas flow and impeller rate from the correlation of Michel and Miller (Eq. (8.81)].

In a fermentor design, $Q_{acc} = 0$ for a steady-state system [although temperature programming a batch reactor for optimal product yields (Sec. 10.7) may provide an additional complication]. Q_{ag} is estimated for ungassed or gassed systems using the appropriate power correlation presented earlier.

Neglecting Q_{evap} (which may be an important mechanism in trickle reactors) and Q_{sen} for the moment, the important remaining quantity is Q_{met}. Methods for estimating or measuring Q_{met} were presented in Chap. 5. In design of larger-scale reactors, the choice of an operating temperature and flow conditions will determine Q_{evap} and Q_{sen}, as the choice of agitator speed and diameter will determine Q_{ag} (corrected for the chosen aeration rate). Sparger design and gas flow rate will determine Q_{gas}. The remaining terms are Q_{acc} and Q_{exch}. Whether or not the reactor is operated isothermally, at each instant

$$\underbrace{Q_{exch}}_{\substack{\text{Heater}\\ \text{or cooler}\\ \text{duty}}} = \underbrace{Q_{met} + Q_{ag} + Q_{gas}}_{\text{Generation}} - \underbrace{Q_{acc}}_{\text{Accumulation}} - \underbrace{(Q_{sen} + Q_{evap})}_{\substack{\text{Removal by other}\\ \text{than a solid heat}\\ \text{exchange surface}}} \tag{8.120}$$

This last equation sets the heat-transfer magnitude needed to maintain the desired temperature and rate of heat accumulation, if any.

We can use Eq. (7.35*a*) to represent the instantaneous mass-generation rate per unit volume in a batch reactor

$$\frac{dx}{dt} = \mu x \tag{7.35a}$$

and the corresponding instantaneous microbial heat-generation rate Q_{met} (heat/time) is evidently given by

$$Q_{met} = V_{reactor}\,\mu x \frac{1}{Y_\Delta} \tag{8.121}$$

where Y_Δ is the heat generation coefficient (g cell/kcal) considered in Sec. 5.10.4. Methods for estimating Y_Δ are presented there as are illustrative data showing greater metabolic heat generation for utilization of more reduced substrates (Table 5.12).

Table 8.7 Effect of substrate and yield coefficients on operating costs of fermentation

	Cost, cents per pound of cells			
Substrate	Substrate	O_2 transfer	Heat removal	Total
Maleate (as waste)	0	0.46	0.75	1.2
Glucose equivalents (molasses)	3.9	0.23	0.54	4.7
Paraffins	4.0	0.97	1.4	6.4
Methanol	5.0	1.2	1.9	8.1
Methane	1.6	3.3	3.7	8.6
Ethanol	8.8	0.75	1.3	11.0
Isopropanol	11.6	2.7	3.1	17.4
Acetate	16.7	0.62	1.1	18.4

† B. J. Abbott and A. Clamen, The Relationship of Substrate, Growth Rate, and Maintenance Coefficient to Single Cell Protein Production, *Biotech. Bioeng.*, **15**: 117, 1973.

The corresponding equation for a continuous-flow isothermal reaction at steady state is

$$Y_\Delta Q_{\text{met}} = V_{\text{reactor}} \mu x = (s_0 - s) Y_{X/S} F - x k_e \tag{8.122}$$

$$\frac{Q_{\text{met}}}{V_{\text{reactor}}} = \left[(s_0 - s) Y_{X/S} D - \frac{x k_e}{V_{\text{reactor}}} \right] \Big/ Y_\Delta \tag{8.123}$$

Recall that, as before, $Y_{X/S}$ may depend on the culture age in a batch reactor and upon dilution rate D in a continuous-flow system. The dependence of $Y_{X/S}$ (and thus Y_Δ) on a cell maintenance appears in Eq. (7.26).

Some economic estimates from Abbott and Clamen [43] as of 1973 indicate that both heat-transfer and mass-transfer (oxygen) operating costs of bacterial cell production from the substrates in Table 8.7 are appreciable fractions of total costs.

8.10.1 Heat-Transfer Correlations

From Eq. (8.118) and the overall heat balance [Eq. (8.119)] for heating, cooling, or sterilizing, the general working equation for heat-transfer design is

$$Q_{\text{exch}} = \bar{h} A \, \Delta T \tag{8.118a}$$

An expression for the overall heat-transfer coefficient $\bar{h}$, analogous to the earlier overall coefficient K_l for gas-liquid mass transfer, is required. For steady-state heat transfer through a flat wall of thickness L_w separating the fermentation fluid at $T_{\text{bulk},1}$ from heating or cooling fluid at $T_{\text{bulk},2}$ continuity of heat flux demands

$$h_{w1}(T_{\text{bulk},1} - T_{\text{wall},1}) = k_s \left(\frac{T_{\text{wall},1} - T_{\text{wall},2}}{L_w} \right)$$

$$= h_{w2}(T_{\text{wall},2} - T_{\text{bulk},2}) \qquad \text{kcal/(cm}^2 \cdot \text{h)} \tag{8.124}$$

where k_s is the thermal conductivity of the wall in kilocalories per centimeter per second per Celsius degree. In terms of an overall heat-transfer coefficient $\bar{h}$, defined by

$$\text{heat flux} = \bar{h}(T_{\text{bulk},1} - T_{\text{bulk},2}) \tag{8.125}$$

rearrangement of the previous equations yields:

$$\frac{1}{\bar{h}} = \frac{1}{h_{w1}} + \frac{1}{k_s} + \frac{1}{h_{w2}} \qquad \text{(planar wall)} \tag{8.126}$$

In clear analogy with our mass-transfer discussions, the overall resistance $1/\bar{h}$ is the sum of three resistances in series. For heat transfer across a cylindrical-tube wall in heating or cooling coils, the cross-sectional area for heat transfer changes continuously through the wall. In this instance the appropriate equation for $\bar{h}$ is

$$\frac{1}{\bar{h}_o d_o} = \frac{1}{h_o d_o} + \frac{\ln(d_o/d_i)}{2k_s} + \frac{1}{h_i d_i} \qquad \text{(tube wall)} \tag{8.127}$$

where d_i and d_o are the tube inside and outside diameters, respectively. Note the use of subscript o for $\bar{h}_o$ since it reminds us to use the outside tube surface as the basis for a heat-transfer area. The thermal conductivity k_s of the solid depends on the material; e.g., at 100°C, $k_s = 0.908$ cal/(s·cm·K) (copper) and 0.107 cal/(s·cm·K) (steel): k_s increases slowly with diminishing temperature. Appropriate values for different heat exchanger materials are found in standard engineering handbooks.

The analysis of momentum and heat transfer at either fluid-solid interface gives the individual-side heat-exchange coefficients (h_{w1}, h_{w2}) or (h_o, h_i) in Eq. (8.126) and (8.127). Where such individual coefficients vary along the heat-transfer surface, an overall local heat-transfer coefficient is defined by equations such as (8.126) and (8.127), and a detailed integration over the heat-transfer area is needed to calculate the total heat transferred.

For fluid-wall heat transfer, the important dimensionless groups are the following:

$$\text{Nusselt number} = \text{Nu} = \frac{hd}{k_f} \tag{8.128a}$$

$$\text{Prandtl number} = \text{Pr} = \frac{C_p \mu}{k_f} \tag{8.128b}$$

$$\text{Brinkman number} = \text{Br} = \frac{\mu u^2}{k_f(T_{\text{bulk}} - T_{\text{wall}})} \tag{8.128c}$$

$$\text{Froude number} = \text{Fr} = \frac{u^2}{gd} \tag{8.128d}$$

and

$$\text{Reynolds number} = \text{Re} = \frac{\rho u d}{\mu} \tag{8.128e}$$

where k_f = thermal conductivity of fluid, cal/(s·cm·°C)
C_p = heat capacity, cal/(g·°C)
d = distance (tube diameter or spacing, cm)
μ = viscosity (poise)
u = velocity (cm/s)
g = gravitational constant (g·cm/s^2)

As discussed elsewhere [5], the Brinkman number represents heat production by viscous dissipation divided by heat transport by conduction and may usually be neglected at the heat-exchanger surface for our purposes. (In the impeller-tip vicinity, this number becomes important.) Similarly, in a baffled vessel or one with an off-center stirrer, the Froude number is usually negligible.

The heat-transfer coefficient h, rendered dimensionless as Nu, is a function of Pr and Re:

$$\mathrm{Nu} = f(\mathrm{Pr}, \mathrm{Re}) \tag{8.129a}$$

As hydrodynamics may vary with the aspect of the exchange surface, i.e., the length L to diameter d ratio, L/d, correlations are also available in the form

$$\mathrm{Nu} = f'\left(\mathrm{Pr}, \mathrm{Re}, \frac{L}{d}\right) \tag{8.129b}$$

Temperature variations induce variations of fluid properties at different points near the heat-transfer surface. As liquid viscosity is the most important of these, the ratio

$$\mu_b/\mu_0 = \text{viscosity}\ (T_{\text{bulk fluid}})/\text{viscosity}(T_{\text{wall}})$$

is a useful correlating variable:

$$\mathrm{Nu} = f''\left(\mathrm{Pr}, \mathrm{Re}, \frac{L}{d}, \frac{\mu_b}{\mu_0}\right) \tag{8.129c}$$

As h (and therefore Nu) may be defined as a local transfer coefficient or as one which has been averaged over the surface in several possible ways, care must be taken to use the appropriate ΔT_{loc} or ΔT_{av} with Nu_{loc} or Nu_{av} from literature correlations.

For fluids of viscosity near that of water, a useful correlation in turbulent flow (heating or cooling) is†

$$\mathrm{Nu} = \frac{hd}{k} = 0.023\ \mathrm{Re}^{0.8}\mathrm{Pr}^{0.4} \tag{8.130}$$

† W. H. McAdams, *Heat Transmission*, 3d ed., p. 152. McGraw-Hill Book Company, New York, 1954.

which appears valid when

$10^4 \le \text{Re} \le 1.2 \times 10^5$ turbulent

$0.7 \le \text{Pr} \le 120$ valid for all liquids except molten metals

$\frac{L}{d} \ge 60$ long tubes

A modification due to Seider and Tate† incorporates an allowance for larger temperature differences; it appears useful for estimating heat transfer with viscous fluids (such as oils):

$$\text{Nu} = \frac{hd}{k} = 0.027\,\text{Re}^{0.8}\text{Pr}^{0.33}\left(\frac{\mu_b}{\mu_0}\right)^{0.14} \tag{8.131}$$

When natural convection is also important due to the presence of nonuniform fluid density, the Grashof number

$$\text{Gr} = \frac{d^3 g\,\Delta\rho\,\rho_{\text{av}}}{\mu^2} \tag{8.132}$$

appears in the correlation‡ for the liquid flowing in horizontal tubes:

$$\text{Nu} = 1.75\left[\frac{d}{L}\cdot\text{Pr Re} + 0.04\left(\frac{d}{L}\,\text{Gr Pr}\right)^{0.75}\right]^{1/3}\left(\frac{\mu_b}{\mu_0}\right)^{0.14} \tag{8.133}$$

(For vertical tubes, the viscosity ratio is replaced by 1.0 and the constant 0.04 by 0.0722.)

When the fluid is known to be non-Newtonian, the forms change. Two equations§ which have been used for pseudoplastic fluids (Sec. 8.8) are

$$\text{Nu} = \frac{hd}{k} = 2.0\left(\frac{d}{L}\,\text{Re Pr}\right)^{1/3}\left[\frac{\eta_v(\text{bulk})}{\eta_v(\text{wall})}\left(\frac{3 + 1/n}{4}\right)\frac{1}{2}\right]^{0.14} \tag{8.134a}$$

or

$$\text{Nu} = \frac{hd}{k} = 1.75\left(\frac{d}{L}\,\text{Re Pr}\right)^{1/3}\left(\frac{3n + 1}{4n}\right)^{1/3} \tag{8.134b}$$

where η_v is the apparent viscosity (Eq. 8.92) evaluated at the bulk fluid or wall temperature. Note that for a Newtonian fluid ($n = 1$) without large bulk-wall temperature differences, Eq. (8.134*b*) reduces to Eq. (8.133). The viscosity variations with temperature, however, are explicitly represented in Eq. (8.134*a*) in the same form as Eq. (8.133).

A wide variety of reactor heat-transfer surface and flow configurations are possible. Some correlations for several of these are given in Example 8.3; others

† F. E. N. Seider and G. E. Tate, "Heat Transfer and Pressure Drop of Liquids in Tubes," *Ind. Eng. Chem.*, **28**:1429, 1936.

‡ R. C. Martinelli and L. M. Boelter, *AIChE Mtg.*, 1942 (cited in McAdams, op. cit.).

§ S. E. Charm and E. W. Merrill, "Heat Transfer Coefficients in Straight Tubes for Pseudoplastic Food Materials in Streamline Flow," *Food Res.*, **24**:319, 1959.

can be found in standard heat-transfer texts and in Chap. 3 of Charm (ref. 30 of Chap. 9). Some example calculations for heat-transfer coefficients and heat-exchanger duties are considered in the problem section. Microbial fluids occasionally deposit a residue on the heater surface leading to fouling (time-varying wall heat-transfer coefficient) and a decrease in h for the fluid-side.

Example 8.3 Heat transfer correlations

Natural convection from vertical plane or cylinder:†

$$\mathrm{Nu} \equiv \frac{hL}{k} = c(\mathrm{Gr\ Pr})^a \tag{8E3.1}$$

where L is the plate length or cylinder diameter, all parameters are evaluated at $(T_{\text{bulk}} + T_{\text{wall}})/2$, and

$$3.5 \times 10^7 \leq \mathrm{Gr\ Pr} \leq 10^{12} \qquad c = 0.13,\ a = \tfrac{1}{3}\ \text{(turbulent)}$$

$$10^4 \leq \mathrm{Gr\ Pr} \leq 3.5 \times 10^7 \qquad c = 0.55,\ a = \tfrac{1}{4}\ \text{(laminar)}$$

Heat transfer in concentric annuli:‡

Streamline flow:

$$\mathrm{Nu} \equiv \frac{hd}{k} = \left(\frac{d_o}{d_i}\right)^{0.8} (\mathrm{Re\ Pr})^{0.45} \left(\frac{d}{L}\right)^{0.45} \mathrm{Gr}^{0.5} \tag{8E3.2}$$

where d_o, d_i = outside and inside diameters

Turbulent flow:

$$\frac{h}{c_p G} = \begin{cases} \dfrac{0.23\ \mathrm{Pr}^{-2/3}(\mu_b/\mu_0)^{0.14}}{[(d_o - d_i)/\mu_b]^{0.2}} & \text{outer wall} \\[2ex] \dfrac{(0.023)(0.87)(d_o/d_i)^{0.53}}{[(d_o - d_i)G/\mu_b]^{0.2}} & \text{inner wall} \end{cases} \tag{8E3.3}$$

Gravity flow over horizontal tube surfaces:

$$h = 65\left(\frac{w}{\mu_l L d_o}\right)^{1/3} \quad \mathrm{Btu/(h \cdot ft \cdot {}^\circ F)} \qquad \text{if } \frac{w}{2\mu_c L} < 525 \tag{8E3.4}$$

where w = liquid flow rate
L = tube length
d_o = outside diameter

Turbulent flow in tubes:§

$$\mathrm{Nu} = \frac{hd}{k} = 0.023\ \mathrm{Re}^{0.8}\ \mathrm{Pr}^b \tag{8E3.5}$$

$$b = \begin{cases} 0.4 & \text{for heating} \\ 0.3 & \text{for cooling} \end{cases}$$

† W. J. King, "Free Convection," *Mech. Eng.*, **54**:347, 1932.

‡ C. C. Monrad and J. F. Pelton, in W. H. McAdams, "Heat Transmission," McGraw-Hill Book Company, New York, 1954.

§ F. W. Dittus and C. M. K. Boelter, *Univ. Calif. Publ. Eng.*, **2**:443, 1930 (see McAdams, *Heat Transmission*).

Flow perpendicular to isolated cylinder:†

$$\mathrm{Nu} = \frac{hd}{k} = (\mathrm{Pr})(0.35 + 0.56\ \mathrm{Re}^{0.52}) \tag{8.E3.6}$$

Flow perpendicular to one row of tubes centered 2d apart:†

$$\mathrm{Nu} = \frac{hd}{k} = 0.21\ \mathrm{Re}_m^{0.6}\ \mathrm{Pr}^{1/3} \tag{8E3.7}$$

where Re_m is Re evaluated at $u_{\max}$ and $2d$ is defined as

direction of flow → ⊕ ⊕ ⊕ ($2d$ = spacing between tube centers)

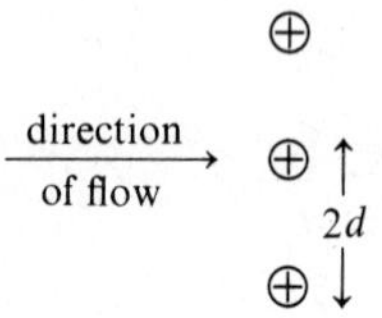

Staggered successive rows of the above type:†

Same as Eq. (8E3.7), but coefficient 0.21 replaced by 0.27 for 3 rows, 0.30 for 5 rows, 0.33 for 10 tube rows or more.

The subject of transport of heat and mass is, we reiterate, an enormously developed area. The present chapter has provided some conceptual guidelines for estimating the quantities of interest. The literature contains a vast number of references for heat- and mass-transfer correlations under a variety of experimental conditions, as indicated in some of the general references of this chapter. Where possible, use of correlations from experimental configurations most apropos to the situation of interest should be practiced, always taking note of the margin of (un)certainty of the correlation.

8.11 STERILIZATION OF GASES AND LIQUIDS BY FILTRATION

Previous sections discussed use of elevated temperature to effect the desired level of sterilization of a liquid. High temperatures can damage medium components, and heat sterilization of gases is not economical. A common alternative approach applicable equally to gases and liquids is the use of appropriate filters to remove undesirable viable cells and, where possible, viruses from the appropriate process stream.

Filters may be made from sintered porcelain, asbestos fiber mats, or synthetic microporous polymer membrane. While the first two categories are important historically, nearly all filtration today associated with sterilization relies on the

† S. E. Charm, *Fundamentals of Food Engineering*, 2d ed., Chap. 4. Avi Publishing, Westport, Conn., 1971.

use of polymeric microporous membranes. These filters may now be routinely used to render sterile a gas flow or dilute liquid suspension entering a bioreactor.

The value of these membrane filters derives from several characteristics:

1. The porous membranes formed, typically from stable gels, have extremely uniform porosity, thereby providing absolute retention of all particles above a certain size (controllable by pore size modification).
2. The extreme porosity (often 70–80%) and thinness (ca. 100 microns) provide low flow resistance, thus allowing a high solvent (water) flux. For example, one liter can be passed through a filter having 0.2 micron pores and 10 cm^2 of filter area in 2–3 min.
3. The membrane filter materials (including cellulose nitrate, cellulose acetate, vinyl polymers, polyamides, and fluorocarbons) are all steam sterilizable and stable against most aqueous suspensions and many organic materials.
4. The quality of the manufactured membrane is easily tested by challenge with a suspension of nearly uniformly sized viable microorganisms. For example, 0.22 micron pore filters are tested with a suspension of *Pseudomonas aeruginosa* bacteria, and *Serratia marcescens* effectively probe a 0.45 micron pore membrane. Viral strains may be used to test filters with very small pore diameters, but uncertainties in virus culture techniques weaken this test for quality control.

The filters of this section are used to remove trace particulates from air or liquid streams, typically to render them sterile (pharmaceuticals) or at least free from pathogens (beverages). Filtration for removal of particles from concentrated suspensions such as fermentation broths is discussed in Chapter 11.

PROBLEMS

8.1 Oxygen diffusivities in protein solutions Stroeve [53] noted that an 1881 derivation of James Clerk Maxwell's (*Treatise on Electricity and Magnetism*, vol. 1, 3d ed.) for diffusion through a fluid containing spherical obstructions simplified to the form below when the obstructions were impermeable:

$$\frac{\mathcal{D}}{\mathcal{D}_0} = \frac{2(1-f)}{2+f}$$

where $\mathcal{D}$ = apparent diffusivity in suspensions
$\mathcal{D}_0$ = apparent diffusivity in pure fluid
f = volume fraction of obstructions

He found that the form gave reasonable fit to experimental data provided that f was defined as $f = f_p + f_b$, where f_p is the volume fraction of protein and f_b the volume fraction of water physically immobilized on the protein surface. Taking the dimensions of the protein to be those of hydrated hemoglobin (spheroid 65 by 55 by 55 Å), calculate and plot $\mathcal{D}/\mathcal{D}_0$ vs. f_p (not f) assuming no, one, or two monolayers of immobilized water around the protein (range of f_p is 0.1 to 0.5). Compare your results with the following measured values for methemoglobin and comment:

$\mathscr{D}/\mathscr{D}_0$	0.69	0.43	0.17
f_p	0.1	0.2	0.4

8.2 Mass-transfer coefficient Determine k_l for the following conditions:

Liquid volume = 10 L

Turbine impeller diameter = 10 cm

Vessel diameter = 50 cm

Speed (rev/min) = 200

Air – medium binary diffusion coefficient = 0.5×10^{-5} cm^2/s

Airflow rate = 2L/min

Medium density = 1.2g/cm^3

Medium viscosity = 0.01 g/(cm · s)

8.3 Oxygen transfer, nonagitated Consider a 0.5L unstirred aerated chemostat with 10 orifices mounted in the bottom. If each is 1 mm in diameter and has an aiflow rate of 5 mL/min, what specific cell growth rate will be maintained if oxygen is limiting? Neglect breakup and coalescence and assume the medium is sufficiently dilute for it to behave like pure water.

$$\mu_{max} = 0.5 \text{ h}^{-1} \qquad K_s = 0.1 \text{ m}M \qquad \sigma = 72 \text{ g/s}^2$$

$$g = 980 \text{ cm/s}^2 \qquad \mu_{gas} = 2 \times 10^{-4} \text{ g/(cm·s)} \qquad \mathscr{D} = 0.5 \times 10^{-5} \text{ cm}^2\text{/s}$$

$$\mu_{liq} = 10^{-2} \text{ g/(cm·s)} \qquad \rho_{gas} = 1.4 \text{ g/L} \qquad H_L = 10 \text{ cm}$$

$$Y_{O/X} = 1 \text{ g } O_2\text{/g cell} \qquad x = 1.0 \text{ g cells/L}$$

8.4 "Variation" of $k_l a$ with temperature Surface renewal theory provides that the mass transfer coefficient k_l varies as $(\mathscr{D})^{1/2}$. For diffusion in liquids, the Stokes-Einstein relation gives $\mathscr{D}\mu/T$ = constant. Thus the variation of k_l with temperature is predicted to follow that of $(T/\mu)^{1/2}$.

(*a*) Using any reference text for the viscosity of water vs. temperature, calculate and plot the expected variation of $k_l(T)/k_l(T = 15°\text{C})$ in the range 15 to 60°C.

(*b*) The equilibrium dissolved oxygen levels (Table 8.1) vary with temperature. Assuming that the interfacial area/volume a is not temperature-dependent, calculate and plot the ratio $\gamma \equiv [k_l ac^*(T)/k_l ac^*(T = 15°\text{C})]$ as in part (*a*).

Comment on the utility of the nearly constant value of γ predicted in part (*b*). Experimentally, this constancy has been confirmed. [54].

8.5 Batch reactor: growth vs. mass transfer limitation A batch fermentation is conducted at 35°C. Experiments with sodium sulfite oxidation indicate that $k_l a' c_l^* = 0.1$ mol/(L-h). The culture has a doubling time, in exponential growth, of 30 min, and an oxygen yield coefficient of 0.6 g cells/gO_2.

(*a*) Calculate the exponential specific growth rate, μ.

(*b*) Use Eq. (8.14) to calculate the dissolved oxygen level, c_l, as the cells increase from $x_0 = 10^{-6}$ g/mL. Plot c_l vs. x. At what biomass level is c_l predicted to become zero?

(*c*) In reality, c_l does not become zero. Rather, μ becomes a function of c_l at low dissolved oxygen levels as in, for example, Eq. (8.15). Use Eq. (8.15) and the same parameter values as above to calculate c_l vs. x. Take $K_{O_2} = 0.05$ m mol/L. (Easiest to assume c_l and calculate x.)

(*d*) For $c_l > 0.9c_l^*$, we have a *growth-limited* condition, while $c_l < 0.1c_l^*$ gives a *mass-transfer* limited condition. Use Eq. (8.15) and your graph in part (*c*) to deduce the ranges of x values corresponding to *growth* and *mass-transfer* limited operation.

(*e*) The right hand side of Eq. (8.15) is simply (dx/dt). Plot (dx/dt) vs. x on the same graph as part (*c*). What expression gives x at $(dx/dt)_{max}$?

8.6 Effect of pressure Oxygen transfer has been increased by augmenting the oxygen partial pressure, p_{O_2}, as may be easily done by use of pure oxygen in place of air feeds. Operation at higher pressures has also been proposed. Unfortunately, cultures may exhibit oxygen inhibition due, for example, to the formation of excess active forms of oxygen within the cell which may damage functions which require a local reducing atmosphere.

(*a*) Assume that growth-dependence on oxygen can be represented as

$$\mu = \frac{\mu_{max} \cdot c_l}{K_{O_2} + c_l + c_l^2/K_i}$$

or

$$\frac{\mu}{\mu_{max}} = \frac{\bar{c}}{1 + \bar{c} + \gamma\bar{c}^2}$$

where $\gamma \equiv K_{O_2}/K_i$ and $\bar{c} \equiv c_l/K_{O_2}$. Show that the maximum specific growth rate occurs at $\bar{c} = \gamma^{-1/2}$.

(*b*) Using Eq. (8.15), modified to include the oxygen inhibition term in part (*a*) above, derive an equation that will predict what oxygen pressure, p_{O_2}, should be used at any x value to maintain maximum cell growth. Since $\mu(\max)$ = constant, give also the required time dependence for $p_{O_2}(t)$.

8.7 Metabolic product export (*a*) At times 1, 2, 3, and 4 h in an L-aspartate producing fermentation, the extracellular product levels are measured as 1, 2, 3, and 4 g/L, respectively. If the surfactant cetylpyridinium chloride is added at $t = 0$, the corresponding measured product levels in the medium are 12, 22, 30, and 35 g/L. Can you say what fundamental process(es) kinetically govern the appearance of product in the original fermentation? in the surfactant-modified medium? (data from Ref. 55).

(*b*) Amino acid export in some *E. coli* appears to be a balance between simultaneous passive transport out of the cell and active transport into the cell. Write a rate equation for amino acid export which is governed by these two phenomena. Using radiolabeled carbon in the amino acid, outline some initial rate measurement experiments by which you could determine all of the parameters in your proposed rate equation.

Mutation programs to eliminate catabolite repression and active transport uptake can lead to transport-limited product export (See Ref. 56).

8.8 Microbead immobilized mycelia The adsorption of spores of a *Penicillium chrysogenum* strain to 300–500 μm porous particles was used to create an immobilized mycelial catalyst, the characteristics of which are compared below with a mycelial suspension culture. Both were propagated in the same bubble column bioreactor.

(*a*) Assuming (roughly) that k_l varies as $(\mu)^{-1/2}$, what was the ratio of viscosities in the cultures: μ (suspension)/μ(immobilized)? What, in your view, allows a broth with a higher biomass level to exhibit a lower viscosity?

(b) Discuss the trade-offs in costs of suspension culture vs. immobilized culture which would lead to choice of the most economic process [57].

	Suspension	Immobilized
x_{max}(g/L)	17.0	29.0
p_{max}(g/L)	2.0	5.5
$k_l a \Delta c$ (m moles O_2/L-h-atm)	50–100	100–350
Basis: Power input (kW/m^3)	2.3	2.3
Oxygen transfer economy (kg O_2/kWh)	0.21	0.48
Specific energy consumption (kWh/g pen G)	0.12	0.07

8.9 Scale-up Methods Table 8.5 provides a scale-up basis under various circumstances (e.g., constant P/V).

(*a*) Discuss the (un)desirable effects on other operating variables resulting from scaling-up at constant reference of (P/V), N_i, F_l/V, N_iD_i, or Re_i.

(*b*) It has been suggested that scale-up is best done by keeping k_la and fluid shear (tip velocity) constant, in particular $D_i/T \sim 0.25 - 0.4$ and $N_iD_i = 0.5$ m/s. Use the correlations of this chapter to indicate how (P/V), N_i, F_l/V and Re_i would vary under scale-up with these mass transfer and shear guidelines [58].

8.10 Scaling parameters in aeration (*a*) In agitated aeration, $\mathrm{Sh} = \alpha \mathrm{Re}_i^{m_1}\mathrm{Sc}^{m_2}$ according to many correlations. Assuming constant bubble size, show that achievement of identical values of k_l in two different vessels, e.g., small (I) and large (II) requires that the impeller speed in revolutions per minute N_i scale as follows:

$$\frac{N_i(\mathrm{II})}{N_i(\mathrm{I})} = \left[\frac{D_i(\mathrm{I})}{D_i(\mathrm{II})}\right]^{2-1/m_1}$$

(*b*) Consequently, establish that constant k_l implies

$$\frac{(P/V)_{\mathrm{II}}}{(P/V)_{\mathrm{I}}} = \left[\frac{D_i(\mathrm{I})}{D_i(\mathrm{II})}\right]^{4-3/m_1}$$

For the turbulent correlation in the text, what fortuitous result arises in the previous equation?

(*c*) For conditions where the bubble size itself is determined by impeller conditions, what relations hold for $N_i(\mathrm{II})/N_i(\mathrm{I})$?, $(P/V)_{\mathrm{II}}/(P/V)_{\mathrm{I}}$?

8.11 Bubble-column performance (*a*) Estimate a, H, and k_l for a bubble column under the following conditions:

Gas flow = 20 std ft^3/min

Liquid flow = 25 gal/min (water)

Column ID = 16 in

Average bubble diameter $D = 0.25$ in

(*b*) An alternate correlation [59] for bubble swarm (liquid or liquid-liquid mass transfer) is

$$\mathrm{Sh} = 2.0 + 0.0187\left[\mathrm{Re}^{0.484}\mathrm{Sc}^{0.339}\left(\frac{Dg^{1/3}}{\mathscr{D}^{2/3}}\right)^{0.072}\right]^{1.61}$$

Compare the explicit dependence of each physical parameter with that of the text formula. Evaluate k_l again and the percentage difference between the two estimates.

8.12 Stream reaeration (rapids and ponds) A moving stream might be approximated by alternating deep and shallow segments of the same width. If the "deep" and "shallow" segments have depths h_D, h_S and lengths l_D, l_S:

(*a*) What is the ratio of aeration mass-transfer coefficients for the shallow to deep segments?

(*b*) Lumping biological activity into a single species, develop an analytic description for substrate utilization by aerobic species of this ponds-rapids configuration, assuming oxygen transfer to be limiting. State your assumptions clearly.

(*c*) Develop expressions (making simplifying assumptions if needed) for the fraction of total microbial growth occurring in the pond and the fraction of total oxygen transfer occurring in the rapids.

8.13 Simplified stream reaeration: (Streeter-Phelps equation) Stream reaeration can be described as a simple plug-flow phenomenon under conditions where organic sedimentation, sediment reactions, and loss of organic volatiles is unimportant. A balance on organic matter S in a stream of velocity u gives

$$\frac{\partial s}{\partial t} = -u\frac{\partial s}{\partial z} - \mu_{\max}S$$

and the oxygen balance is

$$\frac{\partial c_O}{\partial t} = \underset{\text{Gradient}}{-u\frac{\partial c_O}{\partial z}} + \underset{\text{Gain by mass transfer}}{\bar{k}_{O_2}(c^*_{O_2} - c_{O_2})} - \underset{\text{Loss by microbial oxidation}}{\mu_{max} s Y_{O/S}}$$

$$+ \text{ photosynthesis rate} - \text{algal respiration rate} - \text{sedimentation rate}$$

(*a*) If oxygen is always in excess for the *microbes* oxidizing the nutrients, show that

$$s(z) = s_{z=0} \exp\left(-\mu_{max}\frac{z}{u}\right)$$

(*b*) At steady state, neglecting photosynthesis, algal respiration, and sedimentation, show that the oxygen-concentration profile satisfies the Streeter-Phelps equation

$$c^*_{O_2} - c_{O_2}(z) = \frac{Y_{O/S}\mu_{max}s_{z=0}}{\bar{k}_{O_2} - Y_{O/S}\mu_{max}}\left[\exp\left(-Y_{O/S}\mu_{max}\frac{z}{u}\right) - \exp\left(-\frac{\bar{k}_{O_2}z}{u}\right)\right]$$

$$+ (c^*_{O_2} - c_{O_2,z=0})\exp\left(-\frac{\bar{k}_{O_2}z}{u}\right)$$

where $\bar{k}'_{O_2} = \bar{k}_{O_2}/l$
$\bar{k}_{O_2}$ = eddy-averaged oxygen-transfer coefficient [from (8.88)]
l = stream depth

(*c*) Establish analytically that for the Streeter-Phelps treatment, $c^*_{O_2} - c_{O_2}(z)$, known as the *oxygen deficit*, has a single minimum at

$$c^*_{O_2} - c_{O_2}(z) = \frac{Y_{O/S}\mu_{max}s_{z=0}}{\bar{k}'_{O_2}}\exp\left(-Y_{O/S}\mu_{max}\frac{z}{u}\right)$$

What is the downstream distance z corresponding to this point of maximum oxygen deficit? Repeat this derivation including constant photosynthesis, algal respiration, and sedimentation rates.

8.14 Heat transfer, bubble size For the stoichiometry given in Prob. 5.13, (*a*) suppose that a batch aerobic fermentor is run in a cylindrical tank of 6-ft diameter with 130 ft of 1-in-diameter cooling coil arranged on 2-in spacing. Assume that all heat transfer is through the coils and that the bubble sparger always maintains a sufficient $K_l a$ value so growth is not oxygen-limited. If the average fluid velocity perpendicular to the coiled tubes is 10 percent of the impeller-tip velocity, what is the minimum speed in revolutions per minute needed for heat transfer if the average cooling liquid temperature is 18°C and the fermentor should operate no higher than 28°C? Repeat this calculation for cell densities of 10^6, 10^7, 10^8, and 10^9 cells per milliliter (impeller diameter = 4.5 ft, thickness = $\frac{1}{2}$ in, height = 6 in, single paddle).

(*b*) At 10^9 cells per milliliter, the aeration rate is such that 10 percent of the entering oxygen is consumed by the cells. For a poorly designed sparger, bubble size is too large. What stirrer speed (revolutions per minute) is needed to give adequate bubble size? Can this reasonably be achieved with one large paddle? Would a better design include a second much shorter, high-speed paddle just above the sparger with, for example, $D_2 = 0.2D_1$, $N_2 = 10N_1$?

8.15 Dimensionless groups: Buckingham π theorem In heat transfer by forced flow of fluid over a tube surface, the parameters which are physically important in determining the fluid-side heat-transfer coefficient (a conductance h) are the characteristic diameter of the pipe D, the fluid velocity u, and viscosity μ (in poise), density ρ, specific heat at constant pressure C_p (in calories per mole per degree), and the thermal conductivity of the fluid k_f (in calories per second per centimeter per degree). The Buckingham π theorem states that "the functional relationship between q quantities whose units can be given in terms of p fundamental units can be written as a function of $q - p$ dimensionless groups. The fundamental units in heat transfer are mass m, length l, time t, and temperature T.

(*a*) Express h, D, u, μ, ρ, C_p, and k in terms of such units, i.e., variable $= m^a l^b t^c T^d$.

(*b*) Four dimensionless groups are commonly formed from these variables: Nu ($\equiv hD/k_f$), Re ($\equiv \rho Du/\mu$), Pr ($\equiv C_p\mu/k_f$) and Stanton number St ($\equiv h/uC_p\rho$). Comment. By inspection, what does the Stanton number represent?

8.16 Power-law fluids: starch hydrolysis Pastes resulting from cooking 1% wt/vol amylopectin in water are pseudoplastic, thus following the power law, $\tau = \eta\gamma^n$, with $n < 1.0$. For batch α-amylase hydrolysis of this paste, the following changes with increasing time were observed. [60]:

η, dyn·s^n/cm^2	0.32	0.26	0.20	0.14	0.10	0.03
n	0.73	0.75	0.78	0.83	0.85	0.98

(*a*) Show that these data can be described by

$$\tau_0 = \eta\gamma_0^n \qquad \text{where} \quad \tau_0, \gamma_0 = \text{const}$$

(*b*) Establish with an appropriate graph that all τ-vs-γ curves at each degree of hydrolysis pass through a common point.

(*c*) How would the power input at fixed rotation speed vary with time? Could this parameter be used for on-line batch-process control?

8.17 Power input vs. impeller speed in non-Newtonian fluids For non-Newtonian fluids with power-law indices n less than unity, the shear rate γ may be taken to be proportional to the impeller rotation rate N_i [61]. Show that:

(*a*) For a fluid between concentric cylindrical surfaces, shear stress on the outer cylinder (as the inner-cylinder rotation speed varies) changes according to $(d\tau/dN_i) \propto N_i^{n-1}$.

(*b*) The Reynolds number, $\text{Re} = D_i^2 N_i \rho/\eta_v$, where η_v is the apparent viscosity (shear-rate-dependent), varies as $\text{Re} \propto N_i^{2-n}$.

(*c*) For Reynolds numbers defined above, the data for power number ($\equiv Pg_c/D_i^5 N_i^3 \rho$) vs. Reynolds number fall on or just below that correlation for the Newtonian-fluid values. Thus if P_0 varies as Re^α, establish that the power input P varies as $N_i^{3+\alpha(2-n)}$.

(*d*) From the previous information, how would you evaluate the proportionality constant between γ and N_i for a non-Newtonian fluid?

8.18 Hydrocarbon-fermentation phases (*a*) For the quotation of Mimura et al. (and Fig. 8.14) describing the time course of a particular hydrocarbon fermentation, write a mathematical description of growth and substrate(s) utilization in each phase of the batch fermentation. For each "phase," indicate quantitatively where the controlling resistance(s) to growth may lie, i.e., hydrocarbon solubilization, oxygen transfer, cell metabolism, etc.

(*b*) Postulate various reasons why the cell-hydrocarbon-droplet-air-bubble-solvent system adopts each configuration mentioned by the authors. What obvious experiments are suggested by this direct observation? How would you discriminate between or prove the hypotheses advanced?

8.19 Double-substrate design (both gases) Hamer et al. ["SCP Production from Methane," p. 362 in *Single Cell Protein II*, S. R. Tannenbaum and D. I. C. Wang (eds.), MIT Press, Cambridge, Mass., 1975] report that the utilization of methane in the presence of oxygen in a continuous-flow continuously sparged fermentor can be described by the double Michaelis-Menten form of cell growth:

$$r_x = \mu_{\max} x \frac{c_1}{K_1 + c_1} \frac{c_2}{K_2 + c_2}$$

where 1 is oxygen, and 2 methane.

(*a*) Assuming constant yield coefficients, Y_1 and Y_2 (grams of cell per gram of substrate i, $i = 1, 2$), write down the steady-state balances for a sterile-feed system for cells and substrates 1 and 2.

Take the overall mass-transfer conductance to be $K_{li}a$ and assume both liquid and gas phases to be completely mixed.

(*b*) As the dilution rate increases, wash-out will again occur. Show by graphical or analytical evaluation that washout occurs at about $D \approx 0.72\ h^{-1}$ for the following parameter values: $K_1 = K_2 = 5 \times 10^{-4}$ g/L, $Y_i = 1.25$ g cell/g O_2, $Y_2 = 2.0$ g cell/g substrate, $K_{l1}a = K_{l2}a = 100\ h^{-1}$, $c_1^* = 0.015$ g/L, $c_2^* = 0.007$ g/L, $\mu_{max} = 0.8\ h^{-1}$.

REFERENCES

Diffusion:

1. P. Weisz, "Diffusion and Chemical Transformation: An Interdisciplinary Excursion," *Science*, **179**: 433 (1973).

Oxygen demand of cultures and oxygen solubility:

2. R. K. Finn, "Agitation and Aeration," p. 69 in N. Blakebrough (ed.), *Biological Engineering Science*, vol. 1, Academic Press, Inc., New York, 1967.

Interactions between diffusion through liquid films and chemical reactions:

3. P. V. Danckwerts, *Gas-Liquid Reactions*, McGraw Hill Book Company, New York, 1970.
4. G. Astarita, *Mass Transfer with Chemical Reaction*, Elsevier Publishing Company, Amsterdam, 1967.
5. R. B. Bird, W. E. Stewart, and E. N. Lightfoot, *Transport Phenomena*, John Wiley and Son, Inc., New York, 1960.

Fermentation fluids:

6. H. Taguchi, "The Nature of Fermentation Fluids," *Adv. Biochem. Eng.* **1**: (1971).

Problems of scaling microbial reactors:

7. E. L. Gaden, Jr. (ed.), *Biotech. Bioeng.*, **8**: 1966 (Entire volume).

Mass-transfer examples in microbial reactors:

8. A. Moser, in *Proc. Int. Symp. Adv. Microb. Eng.*, **1**: 295–580, 1973.
9. K. Van't Riet, "Mass Transfer in Fermentation," *Trends in Biotechnology*, **1**(4): 113, 1983.

Mass-transfer correlations for oxygen transfer:

10. P. H. Calderbank and M. Moo-Young, "The Continuous Phase Heat and Mass Transfer Properties of Dispersions." *Chem. Eng. Sci.*, **16:** 39, 1961.
11. P. H. Calderbank, "Mass Transfer in Fermentation Equipment," p. 102 in N. Blakebrough (ed.), *Biochemical and Biological Engineering Science*, vol. 1, Academic Press Inc., New York, 1967.
12. K. Akita and F. Yoshida, "Bubble Size, Interfacial Area, and Liquid-Phase Mass Transfer Coefficient in Bubble Columns," *I & EC Process Des. Develop*, **13**: 84, 1974.
13. R. A. Bello, C. W. Robinson, and M. Moo-Young, "Mass Transfer and Liquid Mixing in External Circulating Loop Contactors," *Adv. Biotech.*, **1**: 547, 1981.
14. K. B. Wang and L. T. Fan, "Mass Transfer in Bubble Columns Packed with Motionless Mixers," *Chem Eng. Sci.* **33**: 945, 1978.

Batch oxygen demand:

15. R. T. Darby and D. R. Goddard, "Studies of the Respiration of the Mycelium of the Fungus *Myrothecoum verracaria*," *A. J. Bot.*, **37**: 379, 1950.

$k_l a$ by oxygen-electrode transient response:

16. W. C. Wernan and C. R. Wilke, "New Method for Evaluation of Dissolved Oxygen Response for $K_L a$ Determination," *Biotech. Bioeng.*, **15**: 571, 1973.

Absorption into aqueous films:
17. J. Briffaud and M. Engasser, "Growth and Excretion Kinetics in a Trickle-Flow Fermentor," *Biotech. Bioeng*, **21**, 2093 (1979).
18. K. Livansky, B. Prokes, F. Kihrt, and V. Benes, "Some Problems of CO_2 Absorption by Algae Suspensions," *Biotech. Bioeng. Symp. 4*, p. 513, 1973.

Methane-utilization stoichiometry:
19. D. L. Klass, J. J. Iandolo, and S. J. Knabel, "Key Process Factors in the Microbial Conversion of Methane to Protein," *CEP Symp. Ser.*, [93], **65**: 72, 1969.

Bubble-orifice-diameter correlations:
20. D. W. van Krevelen and P. J. Hoftijzer, "Studies of Gas-Bubble Formation: Calculation of Interfacial Area in Bubble Contactors," *Chem. Eng. Prog.*, **46**: 29, 1950.
21. K. Schügerl and J. Lücke, "Bubble Column Bioreactors," p. 1 in *Advances in Biochemical Engineering, Vol. 7*, T. K. Ghose, A. Fiechter, and N. Blakebrough, ed., Springer-Verlag, Berlin, 1977.

Photographic determination of a, H, D_{Sauter}:
22. P. H. Calderbank and J. Rennie, *Int. Symp. Distill.* (*Inst. Chem. Eng.*), 1960.

Shape-velocity dependence of large bubbles in Eq. (8.36):
23. R. M. Davies and G. I. Taylor, *Proc. Roy. Soc.*, **A200**: 375, 1956.

Free-surface mass transfer to falling film: See Ref 5, p. 540.

Free-surface mass-transfer to turbulent-stream surfaces:
24. G. E. Fortescue and J. R. A. Pearson, "On Gas Absorption into a Turbulent Liquid," *Chem. Eng. Sci.*, **22**, 1163, 1967.
25. D. J. O'Connor and W. Dobbins, "The Mechanism of Reaeration in Natural Streams," *J. Sanit. Eng. Div., Proc. ASCE*, **82**: SA6, 1966.

Turbine power number vs. Re (impeller):
26. J. H. Rushton, E. W. Costich, and H. J. Everett, "Power Characteristics of Mixing Impellers," pt. 2, *Chem. Eng. Prog.*, **46**: 467, 1950.

Aerated vs. nonaerated power requirements:
27. Y. Ohyama and K. Endoh, "Power Characteristics of Gas-Liquid Contacting Mixers, *Chem. Eng. Jpn.*, **19**: 2, 1955.
28. B. J. Michel and S. A. Miller, "Power Requirements of Gas-Liquid Agitated Systems," *AIChE J.*, **8**: 262, 1962.

Mixing by liquid jet injection:
29. H. Blenke, "Loop Reactors," p. 121 in *Advances in Biochemical Engineering, Vol 13*, T. K. Ghose, A. Fiechter, and N. Blakebrough, eds., Springer-Verlag, Berlin, 1979.

Weber number correlations, references: See Ref. 11 above.

Oxygen diffusivities in microbial films:
30. J. V. Matson and W. G. Characklis, "Oxygen Diffusion through Microbial Aggregates," *77th AIChE Meet.*, Pittsburgh, June 1973.

Ionic Strength influence on $k_l a$:
31. C. W. Robinson and C. R. Wilke, "Oxygen Absorption in Stirred tanks: A Correlation for Ionic Strength," *Biotech. Bioeng.*, **15**: 755, 1973.

Surfactants and mass transfer:

32. W. W. Eckenfelder, Jr., and E. L. Barnhart, "The Effect of Organic Substances on the Transfer of Oxygen from Air Bubbles into Water," *AIChE J.*, **7**: 631, 1961.
33. A. Benedek and W. J. Heideger, "Effect of Additives on Mass Transfer in Turbine Aeration," *Biotech. Bioeng.*, **13**: 663, 1971.
34. D. N. Bull and L. L. Kempe, "Influence of Surface Active Agents on Oxygen Absorption to the Free Interface in a Stirred Fermentor," *Biotech. Bioeng.*, **13**: 529, 1971.
35. S. Aida and K. Toda, "The Effect of Surface Active Agents on Oxygen Absorption in Bubble Aeration I," *J. Gen. Appl. Microbiol.*, **7**: 100, 1963.
36. K. H. Mancy and D. A. Okun, "Effect of Surface Active Agents on the Rate of Oxygen Transfer," *Adv. Biol. Waste Treat.*, p. 111 (1963); see also papers by McKeown and Okun, Timson and Dunn, and Carber in this same reference.

Rheology of microbial broths:

37. A. Leduy, A. A. Marson, and B. Corpal, "A Study of the Rheological Properties of a Non-Newtonian Fermentation Broth," *Biotech. Bioeng.*, **16**: 61, 1974 (*A. pullulans* example).
38. J. A. Roels, J. van den Berg, and R. M. Voncken, "The Rheology of Mycelial Broths," *Biotech. Bioeng.*, **16**: 181, 1974 (Penicillin broth example).
39 N. Thompson and D. F. Ollis, "Evolution of Power Law Parameters for Xanthan and Pullulan Batch Fermentations," *Biotech. Bioeng.*, **22**: 875, 1980.
40. H. T. Chang and D. F. Ollis, "Generalized Power Law for Polysaccharide Solutions," *Biotech. Bioeng.*, **24**: 2309, 1982.

Morphology change with tanks in series:

41. D. Vrana: "Some Morphological and Physiological Properties of *Candida utilis* Growing 'Hypertrophically' in Excess of Substrate in a Two-Stage Continuous Cultivation," *Biotech. Bioeng. Symp.* **4**: 161, 1973.

Power-number evolution in non-Newtonian fermentation:

42. H. Taguchi and S. Miyamoto, "Power Requirement in Non-Newtonian Fermentation Broth," *Biotech. Bioeng.*, **8**: 43, 1966.

Cell density and mass transfer:

43. M. R. Brierley and R. Steel, "Agitation-Aeration in Submerged Fermentation, pt. 2: Effect of Solid Dispersed Phase on Oxygen Absorption in a Fermentor," *Appl. Microbiol.*, **7**: 57, 1959 (*A. Niger*).
44. G. F. Andrews, J. P. Fonta, E. Marrota, and P. Stroeve, "The Effects of Cells on Oxygen Transfer Coefficients," *Chem. Eng. J.*, **29**: B39, B47, 1984.

Agitation and cell damage: See Ref. 7 above and

45. M. Midler and R. K. Finn, "A Model System for Evaluating Shear in the Design of Stirred Fermentors," *Biotech. Bioeng.*, **8**: 71, 1966.

Paper-pulp suspensions:

46. N. Blakebrough and K. Sambamurthy, "Mass Transfer and Mixing Rates in Fermentation Vessels," *Biotech. Bioeng.*, **8**: 25, 1966.

Fibrous Filtration:

47. S. K. Friedlander, "Aerosol Filtration by Fibrous Filters," p. 49 in *Biochemical and Biochemical Engineering Science*, Vol. 1. (N. Blakebrough, ed.), Academic Press, New York, 1967.
48. C. N. Davies (ed.), *Air Filtration*, Academic Press, New York, 1973.

Membrane Filtration:
49. J. L. Dwyer, "Filtration in the Food, Beverage, and Pharmaceutical Industries," p. 121 in *Filtration: Principles and Practices, Part II*, Marcel Dekker, New York, 1979.

Heat transfer: Dimensional analysis: pp. 396ff in Ref. 5 and
50. A. I. Brown and S. M. Macro, *Introduction to Heat Transfer*, pp. 85–95, McGraw-Hill Book Company, New York, 1958.

Fermentation heat transfer cost:
51. B. J. Abbott and A. Clamen, "The Relationships of Substrate, Growth Rate, and Maintenance Coefficient to Single Cell Protein Production," *Biotech. Bioeng.*, **15**: 117, 1973.

Fermentation enthalpy balance:
52. C. L. Cooney, D. I. C. Wang, and R. I. Mateles, "Measurement of Heat Evolution and Correlation with Oxygen Consumption during Microbial Growth," *Biotech. Bioeng.*, **11**: 269, 1968.

Problems

53. P. Stroeve, "On the Diffusion of Gases in Protein Solutions," *Ind. Eng. Chem. Fundam.*, **14**: 140, 1975.
54. S. Aiba, J. Koizumi, J. S. Ru and S. N. Mukhopadhyay, "The Effect of Temperature on $k_l a$ in Thermophilic Cultivation of *Bacillus stearothermophilus*," *Biotech. Bioeng.*, **26**: 1136, 1984.
55. S. Fukui and Ishida, *Microbial Production of Amino Acids*, Kodansha Ltd., Tokyo and John Wiley, New York, 1972.
56. D. E. Rancourt, J. T. Stephenson, G. A. Vickell, and J. M. Wood, "Proline Excretion by *Escherichia coli* K12," *Biotech. Bioeng.*, **26**: 74, 1984.
57. K. Gbewonyo and D. I. C. Wang, "Enhancing Gas-Liquid Mass Transfer Rates in Non-Newtonian Fermentations by Confining Mycelial Growth to Microbeads in a Bubble Column," *Biotech. Bioeng.*, **25**: 2873, 1983.
58. J. Oldshue, "Fermentation. Mixing Scale-Up Techniques" *Biotech. Bioeng.*, **8**: 3, 1966.
59. G. A. Hughmark, "Holdup and Mass Transfer in Bubble Columns," *Ind. Eng. Chem. Process Des. Dev.*, **6**: 218, 1967.
60. Angel Cruz, "Kinetics and Shear Viscosity of Enzyme Hydrolyzed Starch Pastes," Ph.D. thesis, Princeton University, Princeton, N.J., 1976.
61. A. B. Metzner, R. H. Feehs, H. L. Ramos, R. E. Otto, and J. D. Tuthill, "Agitation of Viscous Newtonian and Non-Newtonian Fluids," *AICHE J.*, **7**: 3, 1961.

CHAPTER

NINE

DESIGN AND ANALYSIS OF BIOLOGICAL REACTORS

Knowledge of biological reaction kinetics and mass transfer, our primary concerns in Chaps. 7 and 8, is essential for understanding how biological reactors work. In order to assemble a complete portrait of biological reactor operation, however, it is necessary to integrate these two fundamental phenomena with the gas and liquid mixing and contacting patterns in the unit. Different design and scale-up procedures are required for reactors with different flow and mixing characteristics. Consequently, our major task in this chapter is to blend these various ingredients to obtain a coherent overall strategy and analysis of biological reactors.

In our considerations of cell kinetics in Chap. 7, the complex multiphase, interactive nature of cellular bioreactors was indicated. In that context, we examined different types of approximations which could be introduced in order to simplify the kinetic description of the cell population to a practical, workable level while at the same time trying to minimize errors introduced by the approximations. Similar problems and needs face us in biological reactor design and analysis. Now, we examine the interaction of the complex cellular kinetic features discussed earlier with an also complicated fluid flow, mixing, and heat transfer situation. We must now consider the effect of scale or size of the reactor on the mixing, flow, and heat and mass transfer patterns inside the reactor and how different flow and transport fields will influence and interact with biocatalyst kinetics. In this chapter we shall focus on different descriptions of contacting in the reactor and the interaction of contacting patterns with biochemical reactions. In a fashion analogous to that considered earlier for cellular kinetics, we will need to apply approximations judiciously in order to obtain a workable reactor description.

In addressing questions which shall arise in this chapter of approximation strategies for describing bioreactors, it will be extremely useful to consider *relative time scales* and *relative length scales.* The spectrum of time and length scales which we encounter in bioreactor design and analysis is extremely large as suggested by the characteristic times and lengths in Figs. 9.1 and 9.2. A key to analysis of bioreactors is identification of the time and length scales for the phenomena of central interest in a particular reactor design context. Then, it is often possible to analyze phenomena with time or length scales much smaller or much larger than those characteristic of the process of main interest using relatively simple approximations. We have already encountered these ideas in our discussion of the quasi-steady state approximation in Chap. 3 and of structured models in Sec. 7.4.1. Here, comparison of length and time scales shall be used repeatedly in formulating a clear conceptual picture of the bioreactor processes which are most important in the reactor description for a particular design or analysis objective.

Also involved in decisions on appropriate reactor descriptions is the availability of experimental methods for characterizing the transport and reaction processes of interest, and the ability to solve the mathematical models based on

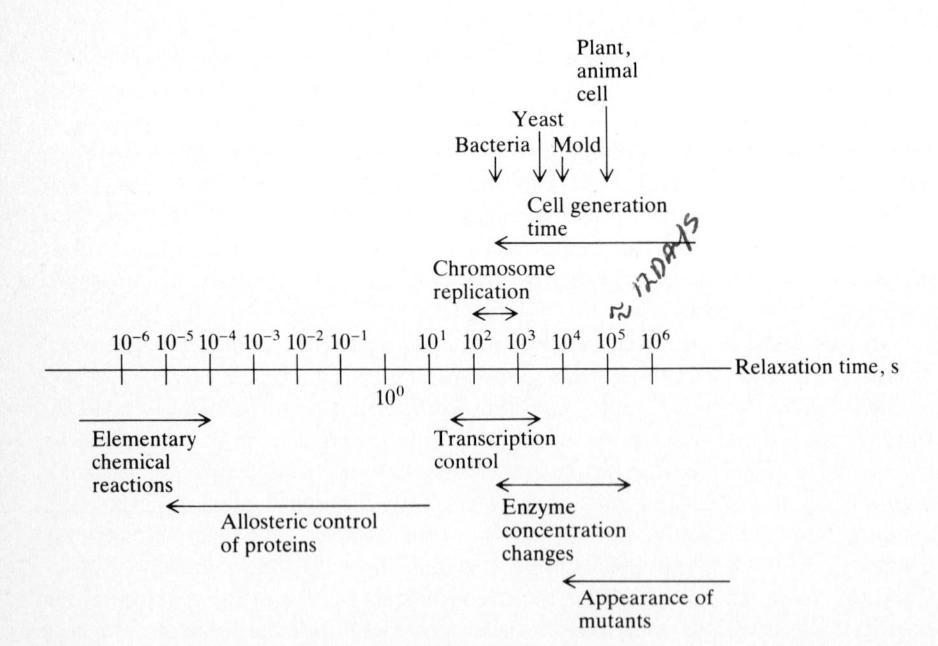

Figure 9.1 Characteristic times for biological responses important in bioreactor engineering *(After Roels [9]).*

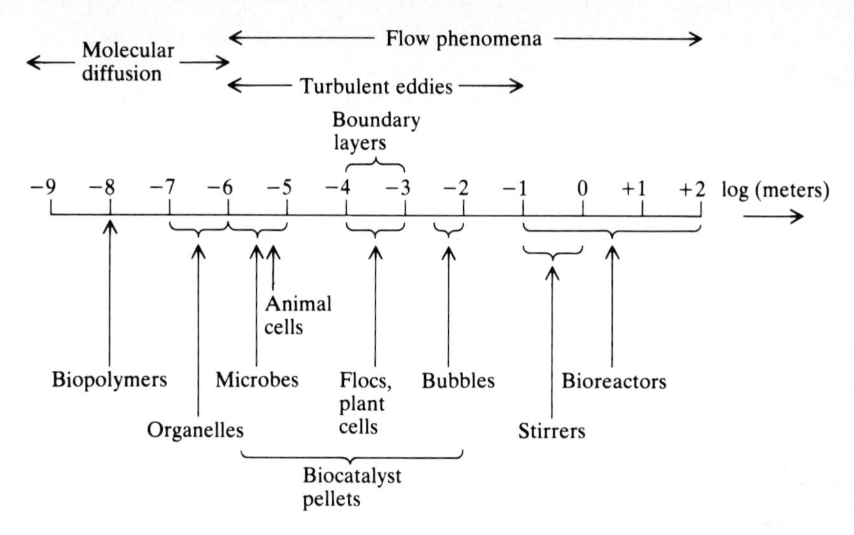

Figure 9.2 Characteristic length scales encountered in bioreactor design and analysis *(After Kossen [8])*.

certain reactor descriptions within reasonable time by currently available numerical methods. We shall see that our ability to arrive at predictive bioreactor models is best for the relatively simple situation of one or a few reactions catalyzed on one or a few enzyme catalysts and that, in the case of cellular reactors with complex multiphase mixing, our lack of knowledge of the structure of the cellular kinetics and of the structure of the physical flow situation hinders predictive analysis at this time. This clearly indicates a major need for additional fundamental research on bioreaction and interacting multiphase flow phenomena. Without more fundamental understanding of these processes, we shall always be faced with a considerable amount of empiricism and need for many scale-up experiments in order to arrive at what may be processes far from optimal.

To begin our consideration of bioreactor design and analysis, we consider elaborations on the ideal batch and continuous-flow stirred-tank (CSTR) reactors introduced in connection with our kinetics discussions in Chap. 7.

9.1 IDEAL BIOREACTORS

In Sec. 7.1, we introduced ideal well-mixed bioreactors. In these systems, mixing is presumed to be sufficiently intense and uniform such that reaction conditions and biocatalyst levels are effectively homogeneous throughout the reactor. This approximation will be valid if any gradients which do exist are sufficiently small so that the reaction rate locally for a given cell or biocatalyst particle is not changed significantly as that catalyst particle moves from one domain of the reactor to another. Alternatively, if the catalyst particles circulate through different regions of the reactor very rapidly with respect to the characteristic response time of the catalyzed reaction to changing conditions, then calculating reactor performance based on the assumption of average, uniform conditions throughout will usually be satisfactory. Conditions like the ones described may be met in laboratory reactors and even pilot scale reactors, depending upon the process involved. In growth of dense cultures of filamentous organisms or organisms producing extracellular polymer, however, highly non-Newtonian conditions are encountered in which, even in small benchtop reactors, ideal mixing is not approximated.

Examination of the theory of ideal completely mixed bioreactors is important for several reasons. First, such reactors provide well-defined conditions for kinetic studies in the laboratory. Second, such models may frequently be used with reasonable success even when the conditions required for validity of these models are not well satisfied. Finally, such ideal mixing configurations provide a starting point for examination and characterization of nonideal mixing and reactors with significant spatial nonuniformities in reaction conditions. As we shall see, we can sometimes calculate and simulate in the laboratory nonidealities in large-scale reactors using model systems comprised of interconnected ideal reactors. In this section, we first elaborate on the ideal batch and ideal CSTR reactors discussed in Sec. 7.1. Finally we consider an ideal plug-flow reactor in which negligible backmixing occurs.

9.1.1 Fed-Batch Reactors

It is often desirable to add liquid streams to a batch bioreactor as the reaction process occurs. This can be done to add precursors for desired products, to add regulating compounds such as inducers at a desired point in the batch operation, to maintain low nutrient levels to minimize catabolite repression, or to extend the stationary phase by nutrient addition to obtain additional product. When a liquid feed stream enters the reactor, the culture volume is also altered, and this must be taken into account in the equations used to describe the reactor. Letting $F(t)$ denote the volumetric flow rate of the entering feed stream at time t and $c_{if}(t)$ denote the concentration of component i in this entering stream, the material balance on component i takes the following form:

$$\frac{d}{dt}[V_R \cdot c_i] = V_R \cdot r_{f_i} + F(t) \cdot c_{if} \tag{9.1}$$

Assuming that the densities of the entering liquid stream and of the culture fluid are both equal to ρ, a total mass balance on the reactor contents takes the following form:

$$\frac{d}{dt}[\rho \cdot V_R] = \rho \cdot F(t) \tag{9.2}$$

(How would this equation be modified to take into account different feed and reactor content densities resulting from, say, aeration of the reactor contents?) Assuming that the density ρ does not change substantially with time during batch operation, Eq. (9.2) becomes simply

$$\frac{dV_R}{dt} = F(t) \tag{9.3}$$

Carrying out the differentiation indicated on the left-hand side of Eq. (9.1) (remembering that now V_R is a function of time), substituting for dV_R/dt using Eq. (9.3), and rearranging the result gives a useful working form of the component i material balance

$$\frac{dc_i}{dt} = \frac{F(t)}{V_R}[c_{if} - c_i] + r_{f_i} \tag{9.4}$$

Eqs. (9.3) and (9.4) are the mass and component balance equations which describe this system. Assuming that suitable kinetic expressions r_{f_i} are available, these equations can be used to simulate the effect of different batch feeding strategies $F(t)$ on reactor performance.

9.1.2 Enzyme-Catalyzed Reactions in CSTRs

CSTRs used for enzyme-catalyzed reactions assume a variety of configurations (Fig. 9.3), depending on the method employed to provide the necessary enzyme activity. In the simplest design (a), enzymes are continuously added to and removed from the reactor via the feed and effluent lines. Obviously this approach is practical only when the enzymes are so inexpensive that they are expendable.

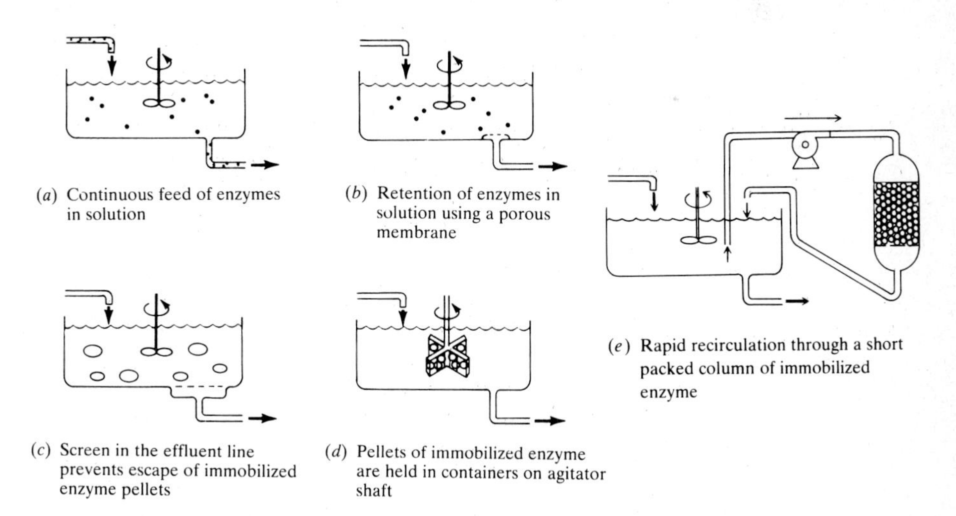

Figure 9.3 Schematic diagrams of CSTR designs for enzyme-catalyzed reactions.

Use of more costly enzymes requires that they be retained in the reactor or recycled. Recalling our discussions of enzyme immobilization in Chap. 4 suggests several possibilities, the first of these (Fig. 9.3*b*) employs an ultrafiltration membrane in the effluent stream with pores sufficiently small to prevent escape of the relatively large enzyme molecules in solution. A screen in the effluent line suffices if the enzyme is immobilized on insoluble particles which are suspended in the reaction mixture as a slurry (Fig. 9.3*c*). Another approach for physical retention of immobilized enzymes within the vessel is shown in Fig. 9.3*d*, where the enzyme is held in screen baskets attached to the agitator shaft. This configuration, which has also been widely used for study of gas-phase reactions on supported-metal catalysts, is intended to minimize mass-transfer resistance between the liquid phase and the immobilized-enzyme pellets. A more conveniently implemented arrangement for achieving the same objective is circulation of reaction mixture from a well-mixed reservoir through a short packed column of immobilized enzyme (Fig. 9.3*e*). So long as the recirculation rate is sufficiently large so that only very small conversion (ca. $< 1\%$) occurs in a single fluid pass through the column, this overall system is equivalent to a CSTR reactor [4]. Thus, this design is especially convenient for laboratory kinetics studies.

Enzyme recycle is feasible only when the enzymes can be readily recovered from the product stream leaving the reactor. Two promising approaches to this problem are containment of enzyme inside liquid-surfactant or phospholipid membranes and immobilization of the enzymes on magnetic supports.

Regardless of which strategy is employed, the common objective is maintenance of the desired enzyme concentration within the CSTR. Assuming that this has been accomplished, we can concentrate our attention on computation of the effluent substrate and product concentrations. The basic principles and general material balances discussed above for microbial growth are applicable, as are additional constraints implied by the relatively simple stoichiometry of these reactions.

For example, for the single reaction

$$\mathrm{S} \longrightarrow \mathrm{P}$$

1 mol of P is formed for each mole of S which reacts, so that the feed (s_0, p_0) and effluent concentrations (s, p) are related by

$$s_0 - s = p - p_0 \tag{9.5}$$

With Eq. (9.5), reaction-rate expressions $v(s, p)$ which are functions of both s and p can be written in terms of s only, simplifying the necessary algebra. The substrate mass balance in this case takes the form

$$F(s_0 - s) - V_R v(s, p_0 + s_0 - s) = 0 \tag{9.6}$$

Table 9.1 gives solutions to this equation for a variety of kinetic forms. These formulas are easiest to use in an indirect fashion: insert the desired substrate conversion into the right-hand side and compute the required residence time

Table 9.1 Relationships among substrate conversions $\delta = (s_0 - s)/s_0$, mean residence time, and catalyst concentration for enzyme-catalyzed reactions in a CSTR†

Reaction-rate expression for v	CSTR design expression for $Vv_{\max}/F$
Michaelis-Menten:	
$\dfrac{v_{\max}s}{K_m + s}$	$\delta\left(\dfrac{K_m}{1-\delta} + s_0\right)$
Reversible Michaelis-Menten:	
$\dfrac{v_{\max}(s - p/K)}{K_m + s + K_m p/K_p}$	$\dfrac{\delta\left[K_m + s_0 - \delta s_0 + \dfrac{K_m(p_0 + s_0\delta)}{K_p}\right]}{1 - \delta(1 + 1/K)}$
Competitive product inhibition:	
$\dfrac{v_{\max}s}{a + K_m(1 + pK_i)}$	$\dfrac{\delta\left[K_m + s_0 - \delta s_0 + \dfrac{K_m(p_0 + s_0\delta)}{K_i}\right]}{1 - \delta}$
Substrate inhibition:	
$\dfrac{v_{\max}}{1 + K_m/s + s/K_i}$	$\delta s_0\left[1 + \dfrac{K_m}{s_0(1-\delta)} + \dfrac{(1-\delta)s_0}{K_i}\right]$

† R. A. Messing, *Immobilized Enzymes for Industrial Reactors*, p. 158, table 1, Academic Press, New York, 1975.

and/or enzyme concentration. Design equations for more complicated reactions and kinetics are obtained by similar methods.

9.1.3 CSTR Cell Reactors with Recycle and Wall Growth

Addition of a cell separator (see Chap. 11) and a recycle stream containing concentrated cells to a CSTR can be used to increase biomass and product yield per unit reactor volume per unit time. Adopting the notation shown in Fig. 9.4, we take F_0 and F_r as the feed and recycle volumetric flow rates and x_1, x_0, and x as the reactor, recycle-stream, and product-stream biomass concentrations, respectively. These concentrations often differ due to a separator, such as a settling

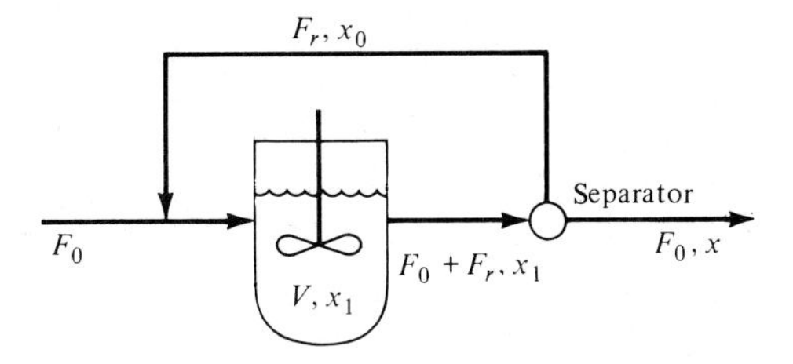

Figure 9.4 Schematic diagram of CSTR with recycle.

basin, at the point where the reactor effluent stream is split. With $a = F_r/F_0$ and $b = x_0/x_1$, the steady-state biomass-conservation equation for the recycle system is

$$F_r x_0 + \mu x_1 V_R - (F_0 + F_r)x_1 = 0 \tag{9.7}$$

so that the overall or external dilution rate D, which is F_0/V_R, is

$$D = \frac{\mu}{1 - a(b-1)} \tag{9.8}$$

Since the microorganisms in the recycle stream are usually more concentrated than in the reactor effluent, $b > 1$. Then Eq. (9.8) reveals that, with recycle, the dilution rate is larger than the organism's specific growth rate. Thus, with organisms growing at the same rate, use of recycle permits processing of more feed material per unit time and reactor volume than in the nonrecycle situation. This feature of recycle is used to great advantage in biological waste-treatment processes, considered in further detail in Chap. 14. (What is the effect of recycle if $b = 1$? What physical interpretation can you provide for your answer?)

Additional important benefits of recycle are revealed by a few manipulations of the system mass balances. Assuming a constant yield factor, the substrate balance is

$$D(s_0 - s) - \frac{\mu x_1}{Y} = 0 \tag{9.9}$$

Combining this equation with Eq. (9.8) we find that μx_1, the biomass production rate per unit reactor volume, is

$$\mu x_1 = \frac{\mu Y(s_0 - s)}{1 - a(b-1)} \tag{9.10}$$

This is greater than the nonrecycle production rate by a factor of $[1 - a(b-1)]^{-1}$. If we assume that μ follows Monod kinetics, we can also show that recycle increases the washout dilution rate by this same factor.

Experiments with CSTRs propagating cell populations sometimes allow higher dilution rates without washout than the theory of the ideal CSTR indicates (recall Sec. 7.1.2). This phenomenon can occur because of wall growth. There may be several solid films of organisms at different points in the vessel. Such colonies can arise, for example, above the liquid level, where splashed droplets have hit the vessel walls, or in crevices and crannies in relatively stagnant zones of the reactor. If we assume that cells on the film at the vessel wall have concentration x_f which is constant with time, reproduction in the film implies addition of cells from the wall into the stirred liquid. In such a situation the steady-state continuous-reactor mass balances take the general form

$$Dx = \mu x + \mu_f x_f \tag{9.11}$$

$$D(s_0 - s) = \frac{1}{Y}\mu x + \frac{1}{Y_f}\mu_f x_f \tag{9.12}$$

where μ_f and Y_f are the specific growth rate and yield factor in the film, respectively. These may differ from the corresponding bulk-liquid parameters μ and Y for a variety of reasons, including diffusion-reaction interactions.

The important thing to notice here is that the $\mu_f x_f$ term in Eq. (9.11) is a source term which is not seriously dependent on D, so that wall growth functions as a second, nonsterile feed which prevents washout. We should note in this connection that laboratory reactors have much larger surface-to-volume ratios than their commercial-sized counterparts, so that in systems involving wall growth, extra care is necessary in scaling up from laboratory data on microbial kinetics.

9.1.4 The Ideal Plug-Flow Tubular Reactor

When fluid moves through a large pipe or channel with sufficiently large Reynolds number (e.g., >2100 in a pipe), it approximates *plug flow*, which means that there is no variation of axial velocity over the cross section. If we assume that plug flow approximately describes fluid movement through the reactor, we can formulate the mass balance on the plug-flow tubular reactor (PFTR) easily using the *differential-section* approach. As Fig. 9.5 suggests, the steady-state conservation equation is applied to a thin slice of the tubular reactor taken perpendicular to the reactor axis. Considering an arbitrary component C, the mass balance on the thin section is

$$Auc\Big|_z - Auc\Big|_{z+\Delta z} + A\,\Delta z\, r_{fc}\Big|_z = 0 \tag{9.13}$$

where r_{fc} is the rate of formation of species C in terms of amount per unit volume per unit time. Rearranging and dividing by $A\,\Delta z$ yields

$$\frac{uc|_{z+\Delta z} - uc|_z}{\Delta z} = r_{fc} \tag{9.14}$$

Taking the limit of this equation and recalling the definition of the derivative gives the final form

$$\frac{d}{dz}(uc) = r_{fc} \tag{9.15}$$

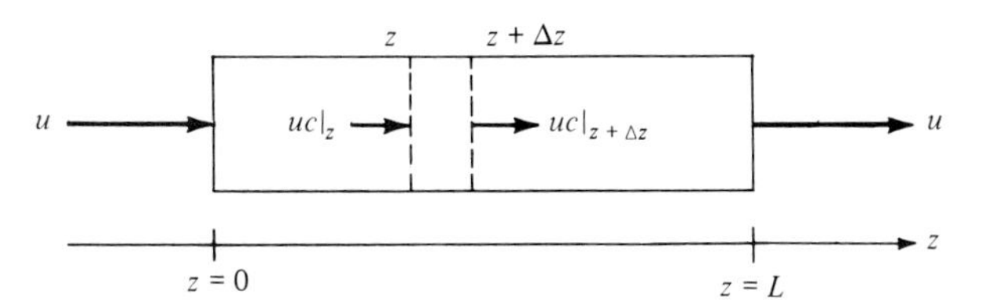

Figure 9.5 Plug-flow reactor.

So long as the reaction does not cause a change in fluid density (would this assumption be valid for a microbial process?), the axial velocity is constant and Eq. (9.15) becomes

$$u \frac{dc}{dz} = r_{fc} \tag{9.16}$$

The quantity z/u is equal to the time required for a small slice of fluid to move from the reactor entrance to axial position z. If we use this transit time t

$$t = \frac{z}{u} \tag{9.17}$$

as a new independent variable, the mass-balance equation (9.16) can be rewritten as

$$\frac{dc}{dt} = r_{fc} \tag{9.18}$$

which is *exactly the same as the batch-reactor mass balance.* This mathematical demonstration can be supplemented by a physical argument: in plug flow with constant velocity, each thin slice of fluid moves through the vessel with absolutely no interaction with neighboring slices. The system is totally segregated, with each thin slice behaving the same as a batch reactor. Consequently, if the initial charge in a batch reactor has the same composition as the feed to the plug-flow reactor, and if the mean residence time L/u in the tube is the same as the batch reaction time, the tube effluent is identical to the batch-reactor product. The boundary condition appropriate for this model is

$$c\Big|_{z=0} = c_0 \tag{9.19}$$

where $z = 0$ denotes the reactor inlet and c_0 is the C concentration in the feed.

As an example, we shall assume that the kinetics used in the Monod chemostat are applicable in the PFTR (or the equivalent batch reactor). The mass balances on cells and substrate in the form of Eq. (9.18) are

$$\frac{dx}{dt} = \frac{\mu_{\max} x s}{s + K_s} \tag{9.20}$$

$$\frac{ds}{dt} = -\frac{1}{Y} \frac{\mu_{\max} x s}{s + K_s} \tag{9.21}$$

with initial conditions

$$x(0) = x_0 \qquad s(0) = s_0 \tag{9.22}$$

On physical grounds or by manipulations with Eq. (9.20) and (9.21), we can see that the s and x concentrations are bound by the stoichiometric relationship

$$x + Ys = x_0 + Ys_0 \tag{9.23}$$

Using Eq. (9.23) to express x in terms of s and substituting this into Eq. (9.20) gives the single ordinary differential equation

$$\frac{ds}{dt} = \frac{\mu_{\max}}{Y} \frac{[x_0 + Y(s_0 - s)]s}{s + K_s} \tag{9.24}$$

Integration of this equation subject to condition (9.22) can be achieved analytically, with the result

$$x_0 + Y(s_0 + K_s) \ln \frac{x_0 + Y(s - s_0)}{x_0} - K_s Y \ln \frac{s}{s_0} = \mu_{\max} t(x_0 + Ys_0) \tag{9.25}$$

The effluent substrate concentration is the s value corresponding to $t = L/u$; then x is found with Eq. (9.23). If viewed as the result of a batch reaction, the kinetics of Eq. (9.20) here shows no lag or death phases but does reach a stationary phase.

In contrast to a CSTR, sterile feed to a PFTR automatically implies zero biomass concentration in the effluent: plug flow prevents a slice of fluid moving through the vessel from ever being inoculated. One way to circumvent this problem is by recycle, so that the incoming stream is inoculated before entering the vessel.

The plug-flow material balances may be readily integrated to relate exit conversion to total reactor residence time L/u for several different common forms of *enzyme* kinetics. Results of such calculations are summarized in Table 9.2.

The relative performance characteristics of ideal CSTRs and PFTRs depend upon the reaction network involved and the corresponding kinetics. For a single reaction with ordinary kinetics (decreasing rate with increasing substrate conversion, such as Michaelis-Menten kinetics), the PFTR provides greater substrate conversion and higher product concentration than the CSTR of equal volume. The opposite is true if the kinetics are autocatalytic (higher rates with decreasing substrate concentration). For microbial processes, the PFTR typically maximizes effluent product concentration. However, the requirement of continuous inoculation and practical difficulties with gas exchange for PFTRs often results in use of their analog, the batch reactor, when high final-product concentration is important. For exponential microbial growth, the CSTR is more efficient than a PFTR or batch reactor. Investigation of the performance of ideal PFTR and CSTR reactors for various simple reaction networks is a major theme of the reaction engineering texts listed in the chapter references. Additional comparisons are explored in the problems.

Table 9.2 Relationships among substrate conversion $\delta = (s_0 - s)/s_0$ and reactor design parameters for enzyme-catalyzed reactions in a PFTR[†]

Enzymes in solution[‡] or in immobilized pellets with negligible mass-transfer limitations

Reaction rate expression for v	PFTR design expression for $\frac{1-\varepsilon}{\varepsilon}\frac{Lv_{max}}{u}$
Michaelis-Menten:	
$\frac{v_{max}s}{K_m + s}$	$s_0\delta - K_m \ln(1-\delta)$
Reversible Michaelis-Menten:	
$\frac{v_{max}(s - p/K)}{K_m + s + K_m p/K_p}$	$s_0\left(1 - \frac{K_m}{K_p}\right)\left[\frac{\delta}{b} + \frac{1}{b^2}\ln(1-b\delta)\right] - \left(K_m + s_0 + \frac{K_m}{K_p}p_0\right)\frac{1}{b}\ln(1-b\delta)$ where $b = \frac{K+1}{K}$
Competitive product inhibition:	
$\frac{v_{max}s}{s + K_m(1 + p/K_i)}$	$s_0\left(1 - \frac{K_m}{K_i}\right)[\delta + \ln(1-\delta)] - \left(K_m + s_0 + \frac{K_m}{K_i}p_0\right)\ln(1-\delta)$
Substrate inhibition:	
$\frac{v_{max}}{1 + K_m/s + s/K_i}$	$s_0\delta - K_m\ln(1-\delta) + \frac{s_0^2}{K_i}\left(\delta - \frac{\delta^2}{2}\right)$

[†] Adapted from R. A. Messing, *Immobilized Enzymes for Industrial Reactors*, p. 158, Academic Press, Inc., New York, 1975.

[‡] The $(1-\varepsilon)/\varepsilon$ factor in these equations should be set equal to unity for enzymes in solution.

9.2 REACTOR DYNAMICS

In this section we consider dynamic characteristics of bioreactors. Although dynamics of CSTRs are the primary focus here, many of the concepts and principles introduced can be applied to other reactor configurations. We first develop the equations needed to describe unsteady-state reactor performance and then examine use of those equations to characterize transient behavior of the bioreactor. We shall see that successful application of the approaches presented here is often limited by lack of kinetic models which are accurate under transient operating conditions.

9.2.1 Dynamic Models

For dynamic studies of CSTRs the conservation equation (Eq. 7.4) must be modified to give the corresponding unsteady-state mass balance:

$$\frac{d}{dt}(\text{total amount in the reactor}) = \text{rate of addition to reactor}$$

$$- \text{ rate of removal from reactor} + \text{rate of formation within reactor}$$

Thus, for a well-stirred vessel we have for component i

$$\frac{d}{dt}(V_R \cdot c_i) = F(c_{IF} - c_i) + V_R r_{fi} \tag{9.26}$$

Assuming that the feed stream and reactor contents have equal density, equality of inlet and outlet volumetric flow rates means the volume of the reactor contents V_R is constant, allowing rearrangement of Eq. (9.26) into the form

$$\frac{dc_i}{dt} = D(c_{if} - c_i) + r_{fi} \tag{9.27}$$

This unsteady-state material balance is the starting point for characterization of reactor dynamics. Before introducing some general mathematical tools useful in dynamics analysis, we should comment on the new considerations required to justify the use of a CSTR model to calculate dynamics of a bioreactor. At this point we shall focus entirely on mixing phenomena, saving for later consideration of the required biological kinetics model for transient analysis. As mentioned in Chap. 8, one characteristic parameter of mixing in a vessel is the mixing or circulation time. This is an order of magnitude indication of the time required for an element of fluid to return to a similar region of the reactor after circulating around the reactor according to the existing flow patterns. In order to apply a CSTR model, it is important that this circulation time be short relative to other characteristic times concerning the CSTR.

In particular, a new characteristic time is introduced when we examine dynamic behavior of a CSTR. Now, there is the possibility of time-varying feed rate or feed concentration. In order for the ideal mixing approximation to apply, it is necessary that the circulation time be much less than the characteristic time scale for fluctuations in the feed stream or, for that matter, in any other entering streams such as base additions for pH control. Exactly the same consideration applies to the use of the perfect mixing assumption for the case of fed-batch reactors discussed earlier.

Eq. (9.27) applies to each component considered in the bioreactor model. In all but the simplest case, then, the dynamic reactor model consists of a set of equations of the form of Eq. (9.27) which are usually coupled through the rate of formation terms r_{f_i}. That is, in general the rate of formation of component i may depend on the concentrations of all of the other components in the reactor. It will be very convenient to introduce vector-matrix notation at this point to simplify writing large sets of equations.

We shall adopt the notation convention that lowercase boldface letters like **c** denote vectors and that uppercase boldface letters like **A** denote matrices. Then, the set of equations indicated by Eq. (9.27) may be written in the form

$$\frac{d\mathbf{c}(t)}{dt} = \mathbf{f}(\mathbf{c}(t), \mathbf{p}) \tag{9.28}$$

where **c** is a vector of concentrations with dimension m (with m elements or components; an m-vector) equal to the number of components considered in the model. Here **p** denotes a q-vector of model parameters, such as feed concentrations, dilution rate, and kinetic parameters. The ith component of the vector-valued function **f** is equal to the right-hand side of Eq. (9.27).

Since the system described by Eq. (9.28) is in general nonlinear, we usually cannot go too far in our analysis without resorting to some approximations. Often we are interested in dynamic properties of the system near a particular steady state $\mathbf{c}_s$. In the notation of Eq. (9.28), the steady state concentration vector $\mathbf{c}_s$ must satisfy

$$\mathbf{f}(\mathbf{c}_s, \mathbf{p}) = \mathbf{0} \tag{9.29}$$

We can attempt to determine behavior near $\mathbf{c}_s$ by expanding the right-hand side of (9.28) in a Taylor's series about $\mathbf{c}_s$ and neglecting all terms of second order and higher in the deviations $c_i(t) - c_{is}$, since they are presumed small. Then we obtain the following linear approximation for our system:

$$\frac{d\mathbf{x}(t)}{dt} = \mathbf{A}\mathbf{x}(t) \tag{9.30}$$

where the vector $\mathbf{x}(t)$ denotes the vector of deviations from the steady state $\mathbf{c}_s$:

$$\mathbf{x}(t) = \mathbf{c}(t) - \mathbf{c}_s \tag{9.31}$$

The element a_{ij} in the ith column of the matrix **A** is defined by

$$a_{ij} = \frac{\partial f_i(\mathbf{c}_s, \mathbf{p})}{\partial c_j} \tag{9.32}$$

We should emphasize that **A** corresponds to some particular steady state. Some systems have more than one steady state for a given **p** and this usually implies that a different **A** matrix corresponds to each steady state.

The dynamic properties of the linearized system are relatively easy to determine since all solutions of Eq. (9.31) usually take the form

$$\mathbf{x}(t) = \sum_{i=1}^{m} \alpha_i \boldsymbol{\beta}_i e^{\lambda_i t} \tag{9.33}$$

The quantities $\boldsymbol{\beta}_i$ and λ_i are the corresponding pairs of eigenvectors and eigenvalues of **A**. Thus $\lambda = \lambda_i$ satisfies the characteristic equation

$$\det(\mathbf{A} - \lambda\mathbf{I}) = 0 \tag{9.34}$$

(**I** is the identity matrix), and the $\boldsymbol{\beta}_i$ satisfy

$$(\mathbf{A} - \lambda_i\mathbf{I})\boldsymbol{\beta}_i = 0 \qquad i = 1, \ldots, m \tag{9.35}$$

The α_i are constants to be chosen to fulfill the specified initial conditions; consequently they satisfy the linear algebraic equations

$$\sum_{i=1}^{m} \alpha_i \boldsymbol{\beta}_i = \mathbf{x}(0) \tag{9.36}$$

where $\mathbf{x}(0)$ is a specified vector of initial deviations.

This linearized dynamic model of the reactor provides a systematic framework for identification of characteristic response times of the system. Within the framework of local dynamics which are considered here, we can see from Eq. (9.33) that the time scales for decay of disturbances from the reference steady state $\mathbf{c}_s$ are characterized by the eigenvalues λ_i of the matrix $\mathbf{A}$. Thus, in terms of local behavior, the system exhibits a spectrum of characteristic times indicated approximately by

$$t_{c_i} = |\lambda_i|^{-1} \qquad i = 1, \ldots, m \tag{9.37}$$

These values can be used to examine the relative magnitudes of reactor time scales with time scales for input variations, for example.

While the approach leading to the time-scale estimates given in Eq. (9.37) is systematic and locally rigorous, the time scale estimates so obtained are difficult to assign a particular physical significance. The eigenvalues λ_i are in general functions of all entries of the matrix $\mathbf{A}$ and, as such, depend upon the entire vector of steady-state operating conditions $\mathbf{c}_s$ and the entire parameter vector $\mathbf{p}$. This situation does not allow convenient comparison of time scales for mixing, for reaction, and for other interactions in the system. In the case of a CSTR, the eigenvalues may be shown to have the form $-D +$ (a value characteristic of the reaction network).† Although providing some guidance, this relationship does not unravel the complex relationships between other parameters and the λ_i. Accordingly, some judgment, experience, and some art is needed in developing a reasonable yet not altogether mathematically rigorous approach for identification of different characteristic lengths and time ratios. Of course one systematic method for achieving this which is well known in chemical engineering is transformation of all system equations to dimensionless form, followed by rearrangements to identify dimensionless groups which characterize the system's behavior. Frequently, such dimensionless groups take the form of ratios of characteristic length or time scales.

9.2.2 Stability

Next we shall examine how the dynamic characteristics of the reactor depend upon the function $\mathbf{f}$ and the selected parameter values $\mathbf{p}$. For our purposes, the *local stability* of a particular steady state $\mathbf{c}_s$ will be of greatest concern. If a steady state is locally asymptotically stable, the system concentrations will return to that steady state after a small disturbance has moved those concentrations slightly away from the reference steady state of interest. For an *unstable* steady state, the concentrations considered will "run away" from the steady-state values following certain small disturbances. To be sure to avoid ambiguity, we shall restate these definitions in more formal mathematical language.

We shall say that the steady state $\mathbf{c}_s$ is *locally asymptotically stable* if $\lim_{t\to\infty} \mathbf{c}(t) \to \mathbf{c}_s$ provided that the initial state $\mathbf{c}_0$ is sufficiently close to $\mathbf{c}_s$. [Our mathematical measure of closeness for vectors is the Euclidean norm, defined by

$$|\mathbf{c}| = \left(\sum_{i=1}^{m} c_i^2 \right)^{1/2}$$

Then "$\mathbf{c}_0$ sufficiently close to $\mathbf{c}_s$" means that $|\mathbf{c}_0 - \mathbf{c}_s|$ is a sufficiently small real number.] The steady state $\mathbf{c}_s$ is *globally asymptotically stable* if $\lim_{t\to\infty} \mathbf{c}(t) = \mathbf{c}_s$ for any choice of $\mathbf{c}_0$ (except ridiculous ones like negative concentrations). If $\mathbf{c}_s$ is an *unstable* steady state, some initial states $\mathbf{c}_0$ arbitrarily close to $\mathbf{c}_s$ lead to trajectories $\mathbf{c}(t)$ which do not approach or stay arbitrarily close to $\mathbf{c}_s$. Thus, in the case of instability, the magnitude of the deviation $\mathbf{x}(\mathbf{t})$ tends to increase from its initial value for some initial deviations.

Local stability is determined in most cases by the eigenvalues λ_i of the matrix $\mathbf{A}$ defined in Eq. (9.32). The steady state $\mathbf{c}_s$ is locally asymptotically stable if all eigenvalues of $\mathbf{A}$ have negative real parts:

$$\operatorname{Re}(\lambda_i) < 0 \qquad i = 1, \ldots, m \tag{9.38}$$

† M. Fjeld, O. A. Asbjørnsen, and K. J. Åstrom, "Reaction Invariants and their Importance in the Analysis of Eigenvectors, State Observability, and Controllability of the Continuous Stirred Tank Reactor," *Chem Eng. Sci.*, **29**:1917, 1974.

On the other hand, $\mathbf{c}_s$ is unstable if any eigenvalue has positive real part:

$$\operatorname{Re}(\lambda_j) > 0 \qquad \text{any } j \tag{9.39}$$

These results are eminently reasonable in view of Eq. (9.33). If the largest real part of the system eigenvalues is equal to zero, a *critical case* arises and further analysis is needed to determine local system dynamic characteristics (see, for example, Ref. 10). Fortunately, we need not compute all the eigenvalues to check the inequalities listed above. First, suppose that the determinant in Eq. (9.34) has been expanded to provide an mth-order algebraic equation.

$$\lambda^m + B_1\lambda^{m-1} + \cdots + B_{m-1}\lambda + B_m = 0 \tag{9.40}$$

Now we can apply the Hurwitz criterion,[†] which asserts that all roots of (9.40) have negative real parts if and only if the following conditions are met:

$$\begin{gathered}
B_1 > 0 \\
\det\begin{bmatrix} B_1 & B_3 \\ 1 & B_2 \end{bmatrix} > 0 \\
\det\begin{bmatrix} B_1 & B_3 & B_5 \\ 1 & B_2 & B_4 \\ 0 & B_1 & B_3 \end{bmatrix} > 0 \\
\vdots \\
\det\begin{bmatrix} B_1 & B_3 & B_5 & \cdots & 0 \\ 1 & B_2 & B_4 & \cdots & 0 \\ 0 & B_1 & B_3 & \cdots & 0 \\ 0 & 1 & B_2 & \cdots & 0 \\ \hdashline \cdots & \cdots & \cdots & \cdots & B_m \end{bmatrix} > 0
\end{gathered} \tag{9.41}$$

As an example, we shall investigate the situation where a single substrate limits growth and examine the dynamic version of the Monod chemostat model. Application of the general unsteady-state mass balance (9.27) to both biophase and substrate and use of Monod's expression (7.10) for the specific growth rate yields

$$\frac{dx}{dt} = D(x_0 - x) + \frac{\mu_{\max} s x}{s + K_s} \tag{9.42a}$$

and

$$\frac{ds}{dt} = D(s_0 - s) - \frac{1}{Y_{X/S}} \frac{\mu_{\max} s x}{x + K_s} \tag{9.42b}$$

For the case of sterile feed ($x_0 = 0$), there are *two* possible steady states, the non-trivial one given earlier in Eqs. (7.14) and (7.15) and the "washout" solution

[†] C. F. Walter, "Kinetic and Biological and Biochemical Control Mechanisms," p. 335 in E. Kun and S. Grisola (eds.), *Biochemical Regulatory Mechanisms in Eucaryotic Cells*, John Wiley & Sons, Inc., New York, 1972.

$x = 0$, $s = s_0$. We can determine which of these steady states will be observed in a continuous culture by determining their stability. Local stability can be studied using the linearized form of Eqs. (9.42). The results of such a local-stability study of the Monod chemostat are summarized below.

	$D > \dfrac{\mu_{\max} s_0}{K_s + s_0}$	$D < \dfrac{\mu_{\max} s_0}{K_s + s_0}$
Nontrivial steady-state [Eqs. (7.14, 7.15)]	Unstable	Stable
Washout steady-state	Stable	Unstable

Another prediction of this analysis is that concentrations cannot approach their steady-state values in a damped oscillatory fashion. As such oscillatory phenomena have been observed experimentally, this substrate-and-cell model is insufficent to predict all dynamic features of some reactors.

Other weaknesses in the dynamic model of the Monod chemostat are known. It predicts instantaneous response of the specific growth rate to a change in substrate concentration: experimentally, a lag is present (see Prob. 10.6). Moreover, growth-rate hysteresis and variations in the yield factor have been established. Steady oscillations have been found in several experimental studies (Fig. 9.6). Consequently, while the Monod chemostat model is quite successful for steady-state purposes in many cases, it has numerous drawbacks as a dynamic representation.

By introducing additional variables into the model, i.e., by giving it more "structure," some of the phenomena unexplained by the Monod model can be accounted for. The need for structured models in such cases rests on conceptual

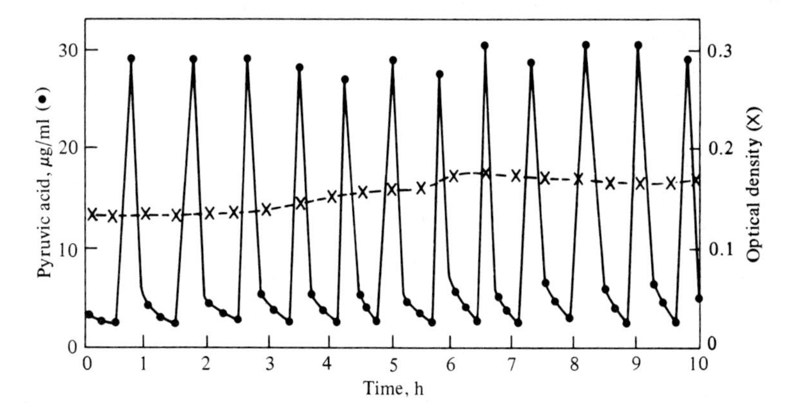

Figure 9.6 These sustained oscillations of pyruvate concentration (●) were observed in continuous culture of *E. coli*. Notice that the cell concentration (crosses) remains approximately constant. *(Reprinted from B. Sikyta, "Continuous Cultivation of Microorganisms," Suom. Kemistil., vol. 38, p. 180, 1965.)*

points mentioned earlier in Chap. 7 and in this chapter's introduction. In a transient situation, the balanced-growth approximation does not apply if the time scale of the environmental changes are comparable to the time scale for biological response (e.g., by induction or repression of enzyme synthesis). In such situations, the biological kinetics model should be expanded to include more components (or pseudocomponents; Sec. 7.4) of the cell phase.

We have already examined the two-component structured model of Williams in Sec. 7.4.1. Applied to continuous culture dynamics, this model reproduces several experimental features not anticipated by the Monod model. Extending the analogy between the Monod growth-rate equation (7.10) and enzyme kinetics, Jeffreson and Smith [11] include in their dynamic chemostat model an intermediate species which is an analog of the enzyme-substrate complex. Ramkrishna, Fredrickson, and Tsuchiya [12] consider an inhibitor of cell growth in their dynamic model. In a sense, this approach can be viewed as adding more structure to the nutrient phase.

A completely different viewpoint has been taken by Lee, Jackman, and Schroeder [13], who consider the influence of flocculation on the overall growth process. We have already observed that the individual cells in many microbial systems form aggregates called flocs. Metabolic processes within such flocs could presumably be different from those in individual dispersed cells; nutrients, for example, would have to diffuse into the floc to reach cells in its interior. Thus, the biophase is viewed as having two components (flocs and individuals) with different kinetics but also with the possibility of interchange of individual cells between the two different morphological forms of behavior. The resulting model exhibits overall yield-factor fluctuations, growth-rate hysteresis, and slower responses than the Monod model—all more compatible with experimental findings than Monod's model.

Yet another conceptual attack is apparent in the model of Young, Bruley, and Bungay [14]. They propose that because of resistances in the mass transport processes which bring nutrient into the cell, the substrate concentration within the cell is not equal at every instant to the external nutrient concentration, and it is the former quantity which directly influences the cell's growth rate. The model based on this view point exhibits lags in response to environmental changes, as has often been observed experimentally.

In closing our review of chemostat dynamics, we should note another potentially important phenomenon not embodied in the Monod model. In situations where excessive nutrient inhibits growth, the specific growth-rate expression given in Eq. (7.33)

$$\mu = \frac{\mu_{\max} s}{K_s + s + s^2/K_p}$$

should be used. A chemostat with this specific growth rate can behave significantly differently from the classical Monod chemostat: now there can be three steady states for some operating conditions. Dynamic behavior for such a system can be complex, and nonlinear effects not considered in a local stability analysis can be quite important. It has been suggested that this model with its unusual characteristics may help explain the operating difficulties which are common in anaerobic digestion processes. Some aspects of substrate-inhibition effects in CSTRs will be explored further in Chap. 14.

Mixed culture systems involving multiple cellular species can exhibit complicated dynamic behavior. We shall investigate these types of bioreaction systems in detail in Chap. 13. There, we shall also introduce additional general mathematical methods and results useful for analyzing and describing reactor dynamics.

9.3 REACTORS WITH NONIDEAL MIXING

Now we depart from the ideal cases of completely mixed tanks or plug-flow tubular reactors, situations which can be approximated under small-scale laboratory conditions, and consider more realistic conditions encountered in larger scale process reactors. We shall be concerned in this section with methods to characterize mixing and flow patterns in reaction vessels, with application of this knowledge for reactor design, and with examination of some of the interactions which arise between biological or biocatalyzed reactions and the mixing and flow patterns in the vessel. First, we consider mixing times in agitated tanks to introduce important time scales, to show the existence of large-scale circulation patterns in reactor vessels, and to get some feeling for orders of magnitudes of the circulation times encountered in different bioreactor situations.

9.3.1 Mixing Times in Agitated Tanks

The mixing time denotes the time required for the tank composition to achieve a specified level of homogeneity following addition of a tracer pulse at a single point in the vessel. The tracer might be a salt solution, an acid or base, or a heated or cooled pulse of fluid. The circulation characteristics of the vessel and mixing time can be measured by continuously monitoring the tracer concentration at one or several points in the vessel. As shown schematically in Fig. 9.7, different types of reactor internals and agitators give rise to different circulation and mixing time characteristics. In the sketched responses in Fig. 9.7, periodic patterns in the tracer concentration are evident, indicating a characteristic number of bulk circulations of fluid required before achieving composition homogeneity. The circulation time is also important because it indicates approximately the characteristic time interval during which a cell or biocatalyst suspended in the agitated fluid will circulate through different regions of the reactor, possibly encountering different reaction conditions along the way. Then, as mentioned before, one must consider whether or not the fluctuations encountered are of sufficient magnitude and on an appropriate time scale to influence local kinetic behavior significantly. We shall return in the conclusion of this section on mixing to examination of some experimental studies of mixing effects on biocatalyst performance.

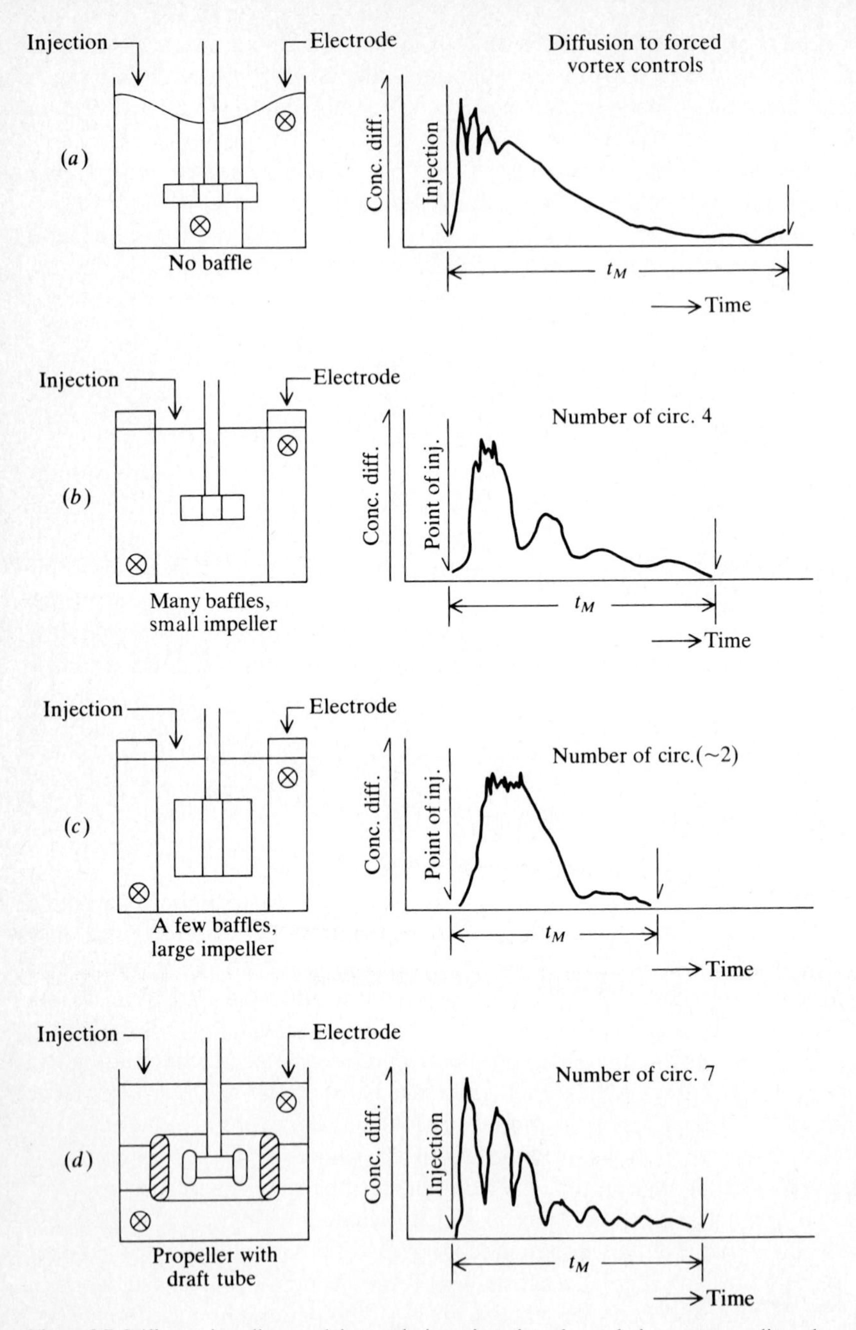

Figure 9.7 Different impellers and internals in agitated tanks and the corresponding characteristic response patterns observed following pulse injection of tracer. The latter are plotted as deviations from the tracer concentration in the final, completely mixed state *(After Nagata [15])*.

Mixing time correlations developed for Newtonian fluids and mycelial cultures and experimental data on mixing times in microbial polysaccharide solutions may be found in the review of Charles [18]. Mixing times of 29 to 104 s were measured in fermentation tanks of size 2.5 to 160 m^3. In some cases mixing times of several minutes have been reported. Studies of a 25 m^3 deep jet aeration system revealed a mixing time of 80 s in water. Charles reported mixing times of around 6 min in 1% xanthan solution at 300 rpm with no air flow, decreasing to around one minute at 500 rpm and with 0.25% air flow. On the other hand, mixing times in the range of 2 to 3 s have been mentioned in small reactors. These figures give a sense of the order of magnitude range which might be expected at different reactor scales in different types of bioreactor fluids. Here our concern is with large-scale fluid circulation and possible composition and temperature nonuniformities. Finer scale considerations having to do with turbulence and its interaction with mass transfer and cells is provided in Chap. 8.

Clearly consideration of a single circulation time in, say, an agitated tank is a conceptual approximation. If we monitor the sojourn of various parcels of fluid from the impeller region, different paths through the vessel will be followed by different fluid parcels, giving rise to correspondingly different circulation times. Bryant [19] has described how the *circulation time distribution* $f_c(t)$ can be experimentally determined by use of a small, neutrally bouyant radio transmitter and a monitoring antenna placed in the vessel. By definition, $f_c(t)\,dt$ is the fraction of circulations which have circulation time between t and $t + dt$. Bryant indicates that the circulation time distribution for agitated tanks can usually be well approximated by the functional form of a log-normal distribution

$$f_c(t) = \frac{1}{\sigma_l\sqrt{2\pi}} \exp\left[-\frac{(\ln t - t_l)^2}{2\sigma_l^2}\right] \tag{9.43}$$

The two parameters in this representation, the log-mean circulation time t_l and log-mean circulation time standard deviation σ_l, are related to the mean circulation time $\bar{t}$ and standard deviation σ of f_c by

$$\bar{t} = e^{(t_l + \sigma_l^2/2)} \tag{9.44a}$$

$$\sigma = \bar{t}^2(e^{\sigma_l^2} - 1) \tag{9.44b}$$

Experimental measurements give $\bar{t}$ and σ which can then be used in Eq. (9.44) to determine parameter values for f_c in Eq. (9.43). Later in this chapter we shall apply the circulation time distribution concept to calculate effects of fluid circulation on overall bioreactor performance.

9.3.2 Residence Time Distributions

Let us now try to imagine what happens to a small parcel of fluid after it has entered a continuous-flow bioreactor. Because of mixing in the vessel, this fluid will be broken into smaller parts, which separate and disperse throughout the vessel. Thus, some fraction of this fluid element will rapidly find its way to the

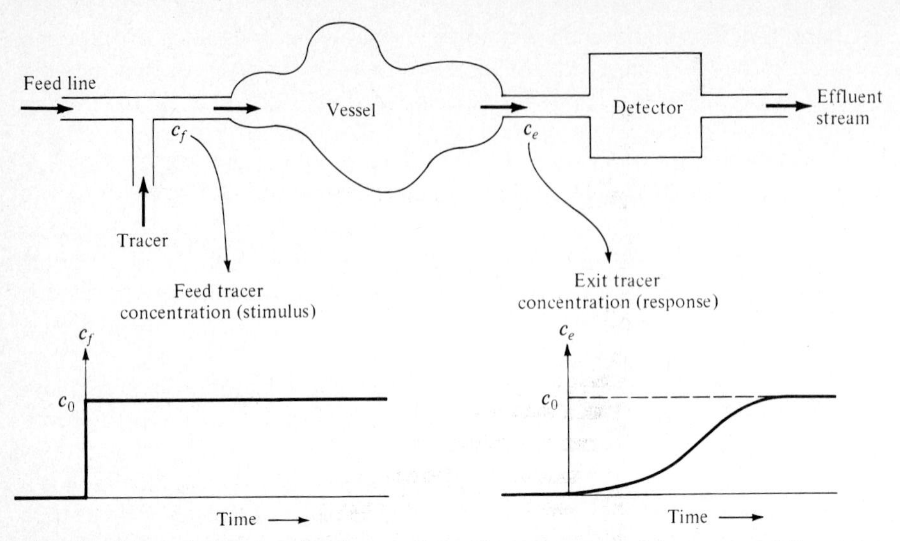

Figure 9.8 Schematic diagram showing experimental measurement of the response ($\mathscr{F}$ curve) to a step tracer input.

effluent stream, while other portions of it will wander about the vessel for varying times before entering the exit pipe. Viewed differently, this scenario indicates that the effluent stream is a mixture of fluid elements which have resided in the reactor for different lengths of time. Determination of the distribution of these residence times in the exit stream is a valuable indicator of the mixing and flow patterns within the vessel. Methods for determining the residence time distribution are reviewed next.

We shall consider first an arbitrary vessel with one feed and one effluent line, and it will be assumed for the moment that there is no back diffusion of vessel fluid into the feed line or of effluent fluid into the vessel. In order to probe the mixing characteristics of the vessel, we conduct a *stimulus-response* experiment using an inert tracer: at some datum time designated $t = 0$, we introduce tracer at concentration c^* into the feed line and maintain this tracer feed for $t > 0$. Then we monitor the system response (in this case the exit tracer concentration) to this specific stimulus. Figure 9.8 shows schematically the general features of this experiment, as well as the shape of a typical exit concentration response $c(t)$.

Under these conditions, the response to a unit-step tracer input $c_0(t) = H(t)$ where

$$H(t) = \begin{cases} 0 & t < 0 \\ 1 & t \geq 0 \end{cases} \tag{9.45}$$

is obtained by dividing the $c(t)$ function obtained in the above experiment by the tracer feed concentration c^* used in that experiment. The result, the unit-step response of the mixing vessel, is called the $\mathscr{F}$ function (Fig. 9.8):

$$\mathscr{F}(t) = \frac{c(t)}{c^*} = \text{response to unit-step input of tracer} \tag{9.46}$$

What is needed in many cases is not the $\mathscr{F}$ function but the residence-time-distribution (RTD) function† $\mathscr{E}(t)$, which is defined by

$$\mathscr{E}(t)\,dt = \text{fraction of fluid in exit stream which has been in vessel for time between } t \text{ and } t + dt \tag{9.47}$$

Thus, for example, the fraction of the exit stream which has resided in the vessel for times smaller than t is

$$\int_0^t \mathscr{E}(x)\,dx$$

It follows from definition (9.47) that

$$\int_0^\infty \mathscr{E}(x)\,dx = 1 \tag{9.48}$$

A simple thought experiment will now serve to clarify the relationship between the $\mathscr{E}$ and $\mathscr{F}$ functions. Returning to the stimulus-response experiment of Fig. 9.8, let us imagine the vessel contents to consist of two different types of fluid. Fluid I contains tracer at concentration c^*, and fluid II is devoid of tracer. Consequently, all elements of fluid I must have entered the vessel at some time greater than zero. Then, any fluid I in the effluent at time t has been in the system for a time less than t. On the other hand, fluid II had to be in the vessel at $t = 0$ since only fluid I has entered since then. All fluid II elements in the effluent at time t consequently have residence times greater than t. Assuming that we know the $\mathscr{E}$ function, we can write the exit tracer concentration $c(t)$ as the sum of the fluid I and fluid II contributions:

$$c(t) = c^* \cdot \int_0^t \mathscr{E}(x)\,dx + 0 \cdot \int_t^\infty \mathscr{E}(x)\,dx \tag{9.49}$$

Combining Eqs. (9.49) and (9.46) produces the desired relationship

$$\mathscr{F}(t) = \int_0^t \mathscr{E}(x)\,dx \tag{9.50}$$

which can be differentiated with respect to t to provide the alternative form

$$\frac{d\mathscr{F}(t)}{dt} = \mathscr{E}(t) \tag{9.51}$$

We note first from Eq. (9.51) that $\mathscr{E}(t)$ can be obtained by differentiating an experimentally determined $\mathscr{F}$ curve. Also, the theory of linear systems states that

† Standard terminology from statistics would indicate that $\mathscr{E}$ is a density function with $\mathscr{F}$ the corresponding distribution. The language above is so firmly embedded in the reaction engineering literature, however, that it would cause confusion to alter it here.

the time derivative of the unit-step response is the unit-impulse response, which reveals that $\mathscr{E}(t)$ can be interpreted as the reponse of the vessel to the input of a unit tracer impulse at time zero. While an impulse is a mathematical idealization, we can approximate it experimentally by introducing a given amount of tracer into the vessel in a short pulse of high concentration.

In addition to the experimental methods just described for determining the RTD function, we can sometimes evaluate it if a mathematical or conceptual model of the mixing process is available. Considering the ideal CSTR as an example, the unsteady-state mass balance on (nonreactive) tracer is

$$\frac{dc}{dt} = \frac{F}{V_R}(c_0 - c) \tag{9.52}$$

To determine the result of an $\mathscr{F}$ experiment for this system, we take

$$c(0) = 0 \tag{9.53}$$

$$c_0(t) = c^* \qquad t \geq 0 \tag{9.54}$$

The solution to Eq. (9.52) under conditions (9.53) and (9.54) reveals

$$\mathscr{F}(t) = \frac{c(t)}{c^*} = 1 - e^{-(F/V_R)t} \tag{9.55}$$

Applying formula (9.51) to the result in Eq. (9.55) reveals that the RTD for a CSTR is

$$\mathscr{E}(t) = \frac{F}{V_R} e^{-Ft/V_R} \tag{9.56}$$

The physical perspective of the PFR introduced above readily reveals its RTD. If a tracer pulse is introduced in the feed, it flows through the vessel without mixing with adjacent fluid and emerges after a time L/u. Thus, the tracer pulse in the exit has exactly the same form as the pulse fed into the PFR, except that it is shifted in time by one vessel holding time. Deviation from such behavior is evidence of breakdown in the plug-flow assumption.

Often when dealing with distribution functions such as $\mathscr{E}(t)$, it is helpful to consider the moments of the distribution. The kth moment of $\mathscr{E}(t)$ is defined by

$$m_k = \int_0^\infty t^k \mathscr{E}(t)\,dt \qquad k = 0, 1, 2, \ldots \tag{9.57}$$

Since a unit amount of tracer is introduced into a vessel to observe its $\mathscr{E}$ curve, and since all tracer eventually must leave the vessel, we know that

$$m_0 = 1 \tag{9.58}$$

The first moment m_1 is the mean of the RTD, or the *mean residence time* $\bar{t}$. Under the conditions of zero back diffusion stated at the start of this section, it can be proved [2] for a single phase fluid in an *arbitrary* vessel that

$$\bar{t} = m_1 = \frac{V_R}{F} \tag{9.59}$$

which says that the mean residence time is identical to the nominal holding time of the vessel. This relationship does not apply for a single phase in a multiphase mixture nor to a single phase in a vessel with heterogeneous catalyst or adsorbent. The second moment m_2 is most often employed in terms of the distribution's variance $\sigma^2 = m_2 - m_1^2$, which is the average of the squares of deviations from the mean residence time.

Related functions useful in mixing and reactor analysis include the *internal age distribution* function $I(t)$, where $I(t)\,dt$ denotes the fraction of fluid *within* the vessel which has been in the vessel for a time between t and $t + dt$. A mass balance can be applied to obtain

$$I(t) = \bar{t}^{-1}[1 - \mathscr{F}(t)]. \tag{9.60}$$

The intensity function $\Lambda(t)$, defined such that $\Lambda(t)\,dt$ is the probability that a fluid element which has been in the reactor for time t leaves in the next short time interval dt, is especially useful in diagnosing deviations from ideal mixing regimes. $\Lambda(t)$ is related to the functions introduced earlier by [16]

$$\Lambda(t) = \frac{\mathscr{E}(t)}{1 - \mathscr{F}(t)} = -\frac{d \ln [1 - \mathscr{F}(t)]}{dt} \tag{9.61}$$

For an ideal CSTR, $\Lambda(t)$ is a constant. Figure 9.9 shows how $\Lambda(t)$ behaves for several types of nonidealities in stirred vessels. In general, whenever $\Lambda(t)$ has a maximum and subsequent decrease, the mixing vessel has stagnant regions or regions of fast and slow flows from inlet to effluent.

Although we cannot discuss all the details here, it is now well established that the RTD does not characterize *all* aspects of mixing (further discussion of this point from a variety of perspectives will be found in the references). The

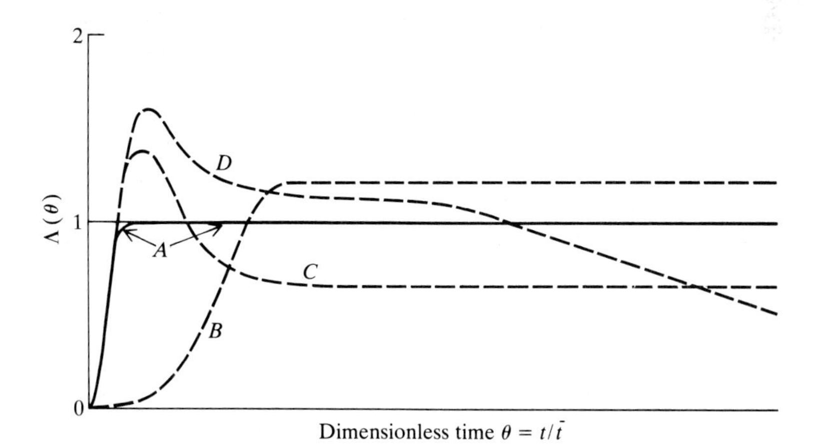

Figure 9.9 Intensity functions indicative of different types of imperfect mixing in stirred tank reactors (Ref. 16): *A*, Short delay between inlet and outlet (normal case); *B*, Delay between inlet and outlet due to insufficient stirring; *C*, Bypass between inlet and outlet; *D*, Bypass between inlet and outlet and stagnant regions due to insufficient stirring.

RTD indicates how long various "pieces" of effluent fluid have been in the reactor, but it does *not* tell us when fluid elements of different ages are intermixed in the vessel.

This point can perhaps be clarified by considering two limiting cases. In the first instance, suppose that fluid elements of all ages are constantly being mixed together. In other words, the incoming feed material immediately comes into intimate contact with other fluid elements of all ages. Such a situation, usually termed a state of *maximum mixedness*, prevails in the ideal CSTR. At the other extreme, fluid elements of different ages do not intermix at all while in the vessel and come together only when they are withdrawn in the effluent stream. In this case, which is called *complete segregation*, reaction proceeds independently in each fluid element: the reaction processes in one segregated clump of fluid are unaffected by the reaction conditions and rates prevailing in nearby fluid elements. Between these two limiting situations falls a continuum of small-scale mixing, or *micromixing*.

The RTD of a reactor is completely independent of its micromixing characteristics. Often, this is not a limitation because micromixing has a small effect on reactor performance. On the other hand, micromixing can influence reactor performance significantly in special situations. It has been suggested [16] that the sensitivity of reactor performance to micromixing can be usefully assessed by calculating reactor performance under the special cases of maximum mixedness and complete segregation. If substantial difference is obtained, the reactor is sensitive to micromixing and will be difficult to scale up. In such situations, predictability of scale-up will be enhanced by using a PFTR or something approximating it (see next section) because PFTR performance is micromixing insensitive regardless of the reaction network or kinetics.

In a reactor with complete segregation, each independent fluid element behaves like a small batch reactor. The effluent fluid is a blend of the products of these batch reactors, which have stayed in the system for different lengths of time. Restating this in mathematical terms, let $c_{ib}(t)$ be the concentration of component i in a batch reactor after an elapsed time t, where the initial reaction mixture in the batch system has the same composition as the flow-reactor feed steam. So long as significant heating or volume change is not caused by the reaction(s), it makes no difference whether one or many different reactions are occurring. A fraction $\mathscr{E}(t)\,dt$ of the reactor effluent contains fluid elements with residence times near t and hence concentrations near $c_{ib}(t)$. Summing over all these fractions gives the exit concentration c_i for the completely segregated reactor:

$$c_i = \int_0^\infty c_{ib}(t)\mathscr{E}(t)\,dt \tag{9.62}$$

In deriving Eq. (9.62), we assumed that reaction conditions (T, pH, dissolved oxygen, etc.) were effectively uniform in the "small-batch reactor" *and* in the mixing vessel characterized by the RTD. Here "effectively uniform" means that any differences in conditions which do exist have negligible or acceptably small effects on the bioreaction processes of interest. This assumption may well break down in large-scale reactors in which substantial inhomogeneities in reaction conditions occur. The RTD for the vessel as a whole does not define how fluid within the vessel circulates through domains of different reaction conditions. Consequently, Eq. (9.62) can be used only to provide an estimate under conditions where appreciable nonuniformities exist in the bioreactor.

Certainly the individual cells or flocs of cells found in some biochemical reactors are segregrated, so that application of Eq. (9.62) seems appealing. However, a rather subtle pitfall exists here. We must remember that living cells contain sophisticated control systems, with which they adapt and respond to their environment. Consequently, the changes which occur during batch culture reflect the combined and interactive influences of the medium and biological phases. Compositions in both of these phases change during the batch in a directly coupled fashion. Consequently, if a cell or floc in a flow system is to behave like the same small batch reactor observed in a batch experiment, it is necessary in general that the cell's environment (the surrounding fluid) also remain segregated in the flow system. If changes in medium composition do not play a critical role in the batch biological reactions, this requirement can be loosened and use of Eq. (9.62) can be better rationalized. Examples of such instances shall appear later in this chapter.

Returning now to the influence of micromixing, consider a single half-order irreversible reaction.

$$\mathrm{S} \longrightarrow \mathrm{P} \qquad r = ks^{1/2} \tag{9.63}$$

occurring in a stirred vessel which is completely segregated but has the same RTD as a CSTR. We might view the half-order reaction as an approximation to the Michaelis-Menten form over a rather narrow range of substrate concentrations. By computing $s(t)$ for reaction (9.63) in a batch reactor and using this with Eq. (9.56) in Eq. (9.62) we find

$$s = s_0\left\{1 - \frac{k\bar{t}}{s_0}\left[1 - \exp\left(-\frac{2s_0}{k\bar{t}}\right)\right]\right\} \tag{9.64}$$

On the other hand, if the same reaction takes place in an ideal CSTR, which has by definition micromixing at the maximum mixedness limit, the effluent substrate concentration is

$$s = s_0\left\{1 - \frac{(k\bar{t})^2}{2s_0}\left[-1 + \sqrt{1 + \frac{4s_0}{(k\bar{t})^2}}\right]\right\} \tag{9.65}$$

A general method for calculating conversion under maximum mixedness conditions in vessels with arbitrary residence times is described in Ref. 20.

From a practical design viewpoint, it is fortunate that reactor performance is often not too sensitive to micromixing. For example, for an irreversible second-order reaction in a CSTR, the maximum difference between the conversions at complete segregation and maximum mixedness is less than 10 percent. Thus, even in cases where it is known to be inexact, Eq. (9.62) may provide an adequate approximation of reaction performance. [By series expansion of Eqs. (9.64) and (9.65), develop an expression for the relative difference, $|1 - s(9.64)/s(9.65)|$.]

For all single reactions with order less than unity, maximum substrate conversion is achieved at maximum mixedness. Complete segregation provides greatest conversions for reaction orders greater than one. As may be justified using the superposition principle for linear systems, the degree of micromixing does not influence first-order reactions. Thus, for one or more reactions with first-order kinetics, Eq. (9.62) may be used independent of micromixing state.

Another exploitation of RTD data is the evaluation of parameters in various nonideal-flow reactor models. A variety of such models is considered in the next section.

9.3.3 Models for Nonideal Reactors

Obviously the RTD contains useful information about flow and mixing within the vessel. One way the $\mathscr{E}$ and $\mathscr{F}$ functions can be used is to assess the extent of deviation from an idealized reactor. For example, we calculated $\mathscr{E}$ and $\mathscr{F}$ above for an ideal CSTR. By comparing these curves with those for a real vessel, we can get an idea of how well the actual system approximates complete mixing.

Different kinds of nonidealities often have distinctive manifestations in the observed response functions. We can gain some appreciation of them by constructing models representing various sorts of deviations from the idealized mixing system. Examples are shown in Fig. 9.10. Case (*a*) is the ideal CSTR, case (*b*)

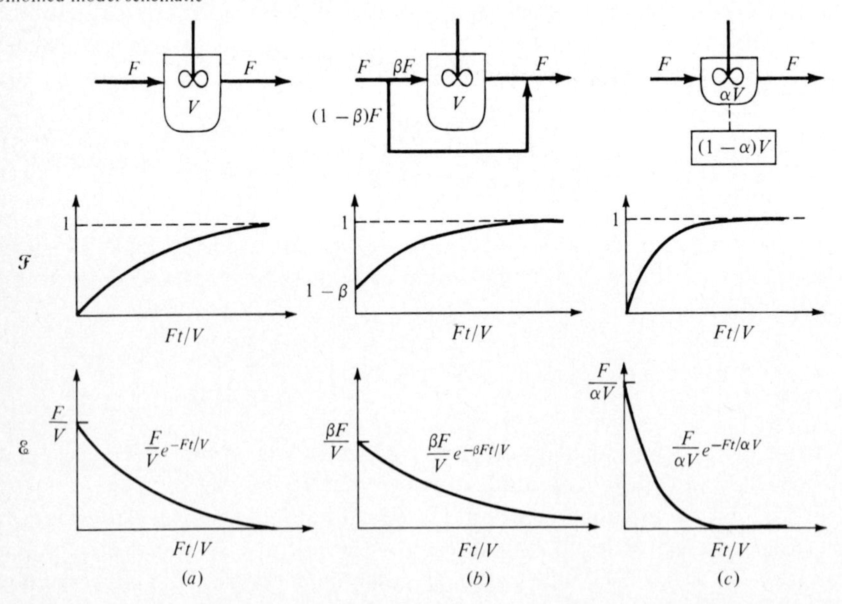

Figure 9.10 $\mathscr{F}$ and $\mathscr{E}$ functions for (*a*) an ideal CSTR, (*b*) a CSTR with bypassing, and (*c*) a CSTR with a dead zone.

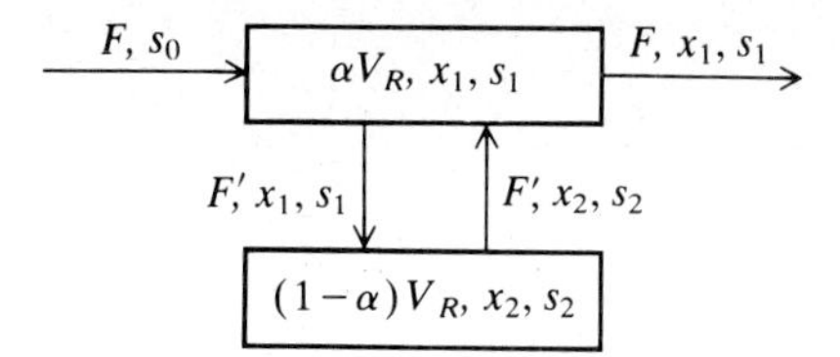

Figure 9.11 Model for an incompletely mixed CSTR with a stagnant region.

involves bypassing of the feed stream, and there is a dead volume $(1-\alpha)V_R$ in case (c). When there is bypassing, tracer appears immediately in the $\mathscr{F}$ function, while a dead region results in faster decay of the $\mathscr{E}$ curve than in the ideal case.

Another useful and commonly applied model for a nonideal continuous-flow stirred reactor uses two interconnected ideal CSTRs (Fig. 9.11). Here the reactor contents have been divided into two smaller, completely mixed regions. The feed and effluent streams pass through region 1, whose volume is a fraction α of the total reactor volume. In turn, region 1 exchanges material with stagnant region 2 at volumetric flow rate F'. If we assume Monod growth kinetics with constant yield factor, the following mass balances describe steady-state conditions in this system:

$$x_1 = Y_{X/S}(s_f - s_1) \qquad \text{region 1 substrate} \tag{9.66a}$$

$$x_2 - x_1 = Y_{X/S}(s_1 - s_2) \qquad \text{region 2 substrate} \tag{9.66b}$$

$$x_2 + \alpha\gamma\mu_{\max}\frac{s_1}{K_s + s_1}x_1 = (1 + \gamma D)x_1 \qquad \text{region 1 cells} \tag{9.66c}$$

$$x_1 + (1-\alpha)\gamma\mu_{\max}\frac{s_2}{K_s + s_2}x_2 = x_2 \qquad \text{region 2 cells} \tag{9.66d}$$

where

$$\gamma = \frac{V}{F'} \qquad D = \frac{F}{V} = \text{nominal dilution rate} \tag{9.67}$$

The dilution rate at washout for this model is obtained by setting $s_0 = s_1 = s_2$ in Eqs. (9.66) to obtain

$$D_{\text{washout}} = \frac{\mu_{\max}s_f}{K_s + s_f}\left[1 + \frac{(1-\alpha)^2\mu_{\max}s_0}{K_s + s_0 - (1-\alpha)\gamma\mu_{\max}s_0}\right] \tag{9.68}$$

Since the first expression on the right-hand side is identical to the washout dilution rate for the perfectly mixed system [Eq. (7.16)], the effect of incomplete mixing is to increase the value of D_{washout}. If the bracket on the right-hand side of Eq. (9.68) happens to be negative, it indicates that washout is impossible. (What does this mean physically?)

Figure 9.12 shows the effluent cell concentration x_1 and biomass productivity x_1D as functions of the dilution rate for $\gamma = 0.5$ and a variety of volume fractions α. In this figure, which was computed using the K_s and s_0 values from Fig. 7.10, the change in the x-vs.-D curve as α changes from 1 (perfect mixing)

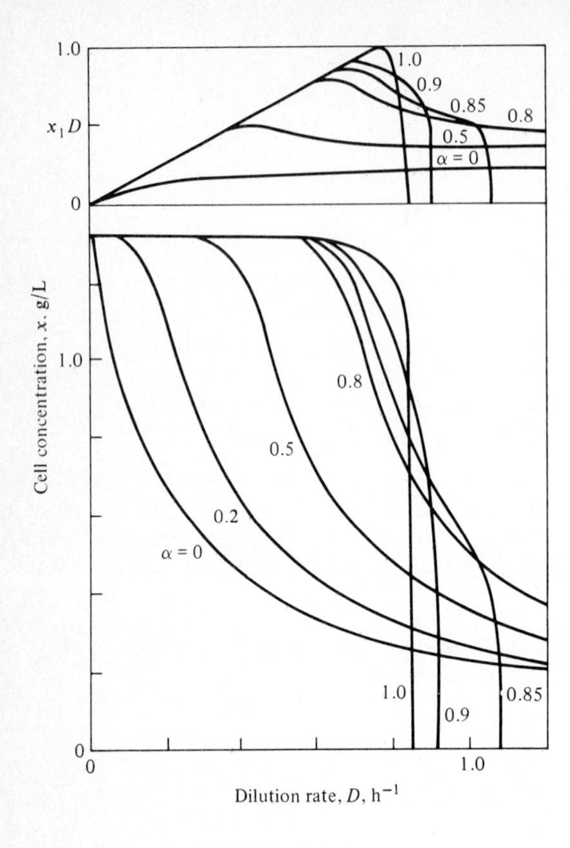

Figure 9.12 Exit cell concentration x_1 as a function of dilution rate D and active zone volume fraction γ for dead zone model of Fig. 9.11. The upper portion of the figure shows biomass production rate $x_1 \cdot D(\gamma = 0.5)$.

to 0.9 or 0.85 (small stagnant zone) is very similar to the difference between the chemostat theoretical and experimental data given in Fig. 7.10. Moreover, the value $\alpha = 0.9$ corresponds roughly to the vessel volume fraction above the impeller in the experimental reactor used to obtain Fig. 7.10.

In examining RTDs so far we have concentrated on reactors with continuous flow of medium and cells, but we should remember that a gassed-batch bioreactor has continuous throughput of gas. Thus, RTD measurements have been used to characterize gas holdup and mixing in batch reactors. These studies indicate major effects of type of contactor employed and reaction mixture rheology. For example, a CSTR in series with a PFR gives RTD behavior closely approximating measured gas RTDs in mechanically agitated tanks for water, water–salt, and *Saccharomyces cerevisiae* suspensions which all have low viscosities. On the other hand, in high viscosity solutions in agitated tanks, more complicated gas RTD behavior is evident. Figure 9.13 shows the measured internal age distribution for sparged gas flow through a 10 g/L suspension of *Penicillium chrysogenum* in an agitated vessel of standard dimensions. Here, a plug-flow element must be added to the model of Fig. 9.11 in order to fit the experimental result.

This type of model, commonly called a *combined model* or a *mixed model*, consists of interconnected idealized reactor types. It is but one of a wide class of

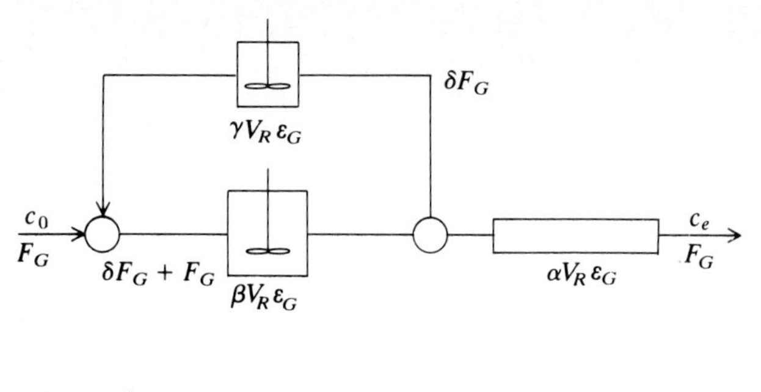

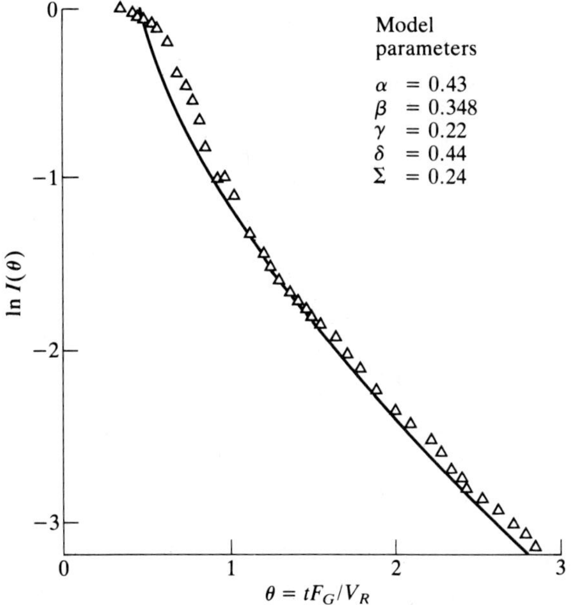

Figure 9.13 Experimentally measured internal age distribution for gas in an agitated vessel (Δ). The solid line is the internal age distribution calculated from the mixed model shown above with the parameter values indicated. *(Reprinted by permission from M. Popović, A. Papalexiou, and M. Reuß, "Gas Residence Time Distribution in Stirred Tank Reactors," VI International Fermentation Symposium, London, Canada, 1980.)*

possible representations of incomplete mixing. Combined models have the advantages of serving at once as a useful conceptual tool for thinking about flow and mixing in the reactor, as a basis for computing the performance of the nonideal vessel as a reactor, and as a guide for constructing laboratory reactors to study effects of nonidealities in large-scale equipment.

Once a combined model such as any of those shown above has been adopted to describe a particular nonideal bioreactor, the combined model can be used to calculate the performance of the vessel as a reactor. That is, the reaction kinetic expressions and material balances on substrates, products, and reaction effectors are written and solved in the context of the combined model network. This procedure has already been illustrated above in Eqs. (9.66) for the combined model in Fig. 9.11 and simple Monod kinetics. However, we must be cautious and recognize the potential limitations in this approach, since a combined model, while adjusted to be consistent with the actual reactor RTD, will not in general possess identical micromixing and circulation characteristics compared to the real system. Accordingly, the combined model may fail to approximate reasonable system performance as a reactor. For example, the combined models shown in Fig. 9.11 will give the same RTD if the plug-flow tube element is put at the start rather than at the end of the network, but these two configurations will not give the same results for nonlinear kinetics. However, they will give the same output for a reaction network consisting entirely of first-order kinetics or for a first-order process such as mass transfer.

Although still subject to the limitations just described, mixed models can be extended in some situations to encompass nonuniform reaction conditions. Often there is a reasonable intuitive physical correspondence between the different ideal regions in a combined model and actual physical domains and environments within a bioreactor. For example, we have already mentioned in connection with the combined model of Fig. 9.11 that one region might represent the impeller zone and the other region the bulk of the tank. If air sparging enters the system near the impeller, it is reasonable to consider the dissolved oxygen level to be relatively high in that part of the combined model and the dissolved oxygen level to be relatively low in the bulk of the tank far from the impeller in a large-scale reactor. Thus, we could superimpose on the different interconnected tanks estimates or measured values of reaction conditions in different zones of a bioreactor and take these into account when calculating reactor performance.

In such situations, our knowledge of the biological kinetics may be inadequate to simulate reaction behavior properly under such fluctuating conditions. We shall return to this topic in the following section. To the extent that a combined model with different reaction conditions in different ideal elements is useful in describing the behavior of real, large-scale bioreactors, we can see now one approach to the problem of *scale-down*. Scale-down refers to a reasonable method for designing small-scale experiments in the laboratory to attempt to simulate and to study operation of nonideal large-scale reactors. Since the ideal systems, especially CSTRs, can be well approximated on a small scale, using a set of small reactors and pumps and interconnections, we can set up on a small scale in the laboratory the physical counterpart of the combined model. We can, in this model, scaled-down system, apply different levels of mixing intensity in different tanks, different aeration levels, different pH values, and other tank-to-tank variations in order to try to study and simulate the effects of such nonuniformities on performance and on biological reaction kinetics in large-scale systems. This approach is applicable to batch reactors as well as to continuous-flow reactors.

Different types of combined models involving a sequence of ideal CSTRs (Fig. 9.14) are usually used to simulate staged and column bioreactors and mixing configurations which more closely approximate plug flow. We can calculate the RTDs for these models by again setting up tracer mass balances on the

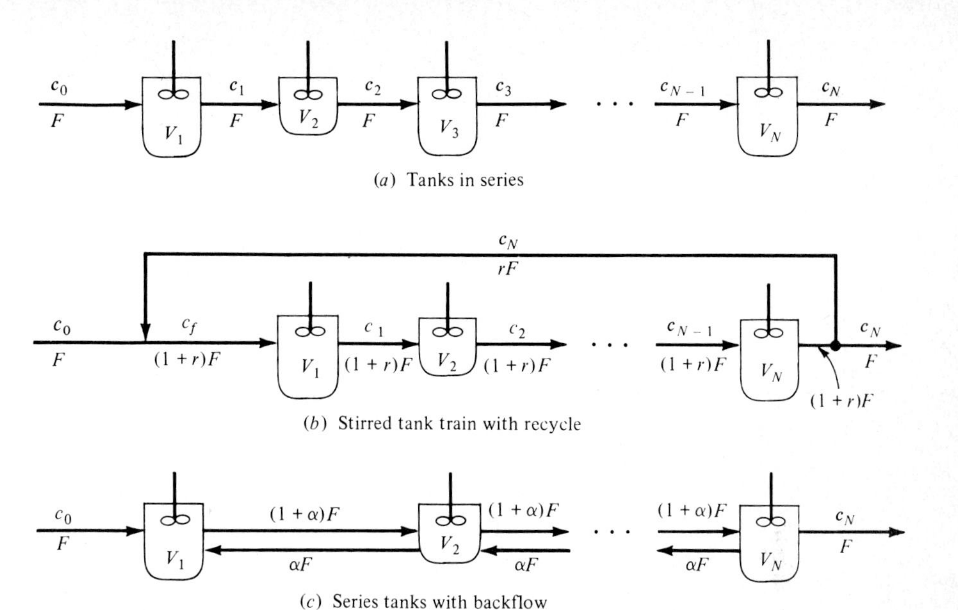

Figure 9.14 Schematic diagrams of different tanks in series designs and models: (*a*) tanks in series; (*b*) stirred tank train with recycle; and (*c*) series tanks with back flow.

stirred-vessel cascade and evaluating the response to a unit-impulse input. Carrying out the necessary algebra for the system of Fig. 9.14*a* subject to the usual assumption that the total reactor volume V_R is divided into N equally sized tanks

$$V_1 = V_2 = \cdots = V_N = \frac{V_R}{N} \tag{9.69}$$

gives

$$\mathscr{E}(t) = \frac{N^N}{(N-1)!}\left(\frac{NF}{V_R}\right)^{N-1} t^{N-1} \exp\left(-\frac{NF}{V_R}t\right) \tag{9.70}$$

Plots of this function for a variety of N values are given in Fig. 9.15. The shift in $\mathscr{E}(t)$ with increasing N from an exponential decay to a pulse at $t = V_R/F$ is apparent.

The variance of the RTD in Eq. (9.70) is

$$\sigma^2 = \frac{1}{N} \tag{9.71}$$

This relationship is useful in developing a series CSTR model for an arbitrary vessel whose RTD has been experimentally determined. Taking the total staged-CSTR system volume V_R and flow rate F as in the actual process makes the mean residence time for the model match that of the real vessel. Taking N equal to the reciprocal of the experimentally measured variance for the real vessel RTD then

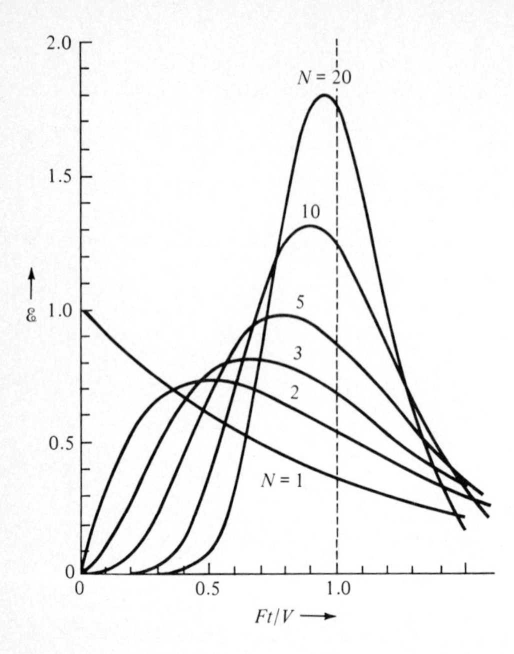

Figure 9.15 RTDs for N ideal CSTRs in series (V_R = total system volume, individual reactor volumes = V_R/N).

ensures that the series-CSTR model RTD has the same second moment as the vessel of interest. If the vessel's RTD has a shape similar to those given in Fig. 9.15, the tanks-in-series model so derived will probably provide an adequate approximation of the real reactor's performance. (Remember that identical RTDs guarantee identical reactor performances only for total segregation or first-order kinetics). Notice as $N \to \infty$, the RTD for this model approaches that for an ideal PFR.

Figure 9.16 shows the results of experimental RTD studies of cocurrent perforated plate towers. In both experiments three plates separated the column into four sections, but in case 1 the plates had 3-mm holes; 2-mm holes were used in case 2, with the total hole area held at 10 percent in both instances. Comparing the experimental RTDs with the theoretical result for $N = 4$ in Fig. 9.16 reveals that the plates with 2-mm holes provide good staging. Evidently there is some backflow through the plate perforations when they exceed the 2-mm size. Thus, in this case RTD data provide a useful design guideline for preserving the desired segregation in the tower.

Sometimes the addition of a growing microbial phase to a reactor dramatically alters its RTD. In the measured RTD of Fig. 9.17*a*, the eight-plate tower with no growing organisms exhibits a clear staging effect. When baker's yeast is grown in the same system, however, the RTD measured by a variety of tracers closely corresponds to an ideal CSTR (solid curve in Fig. 9.17*b*) rather than to an eight-CSTR cascade. This breakdown in staging effect is apparently due to sedimentation of the yeast suspension. By changing the column design so that four plates had only 3 percent hole area, the RTD with yeast growth became nearly that of a four-CSTR cascade.

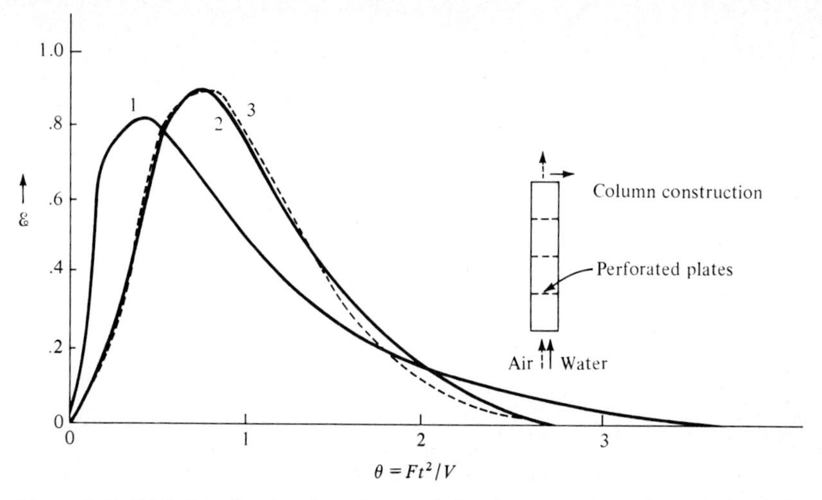

Figure 9.16 RTD data showing flow characteristics of a cocurrent perforated plate column. (1) Hole diameter = 3 mm, plate-area fraction = 0.0981, (2) hole diameter = 2 mm, area fraction = 0.0975, (3) calculated from four CSTRs-in-series model. *(Reprinted from A. Ketai et al., "Performance of a Perforated Plate Column as a Multistage Continuous Fermentor," Biotech. Bioeng., vol. 11, p. 911, 1969.)*

Next we examine the performance of the CSTR cascades in Fig. 9.14 as bioreactors. The mass balance on an arbitrary component c in the jth tank is

$$Fc_{j-1} - Fc_j + V_j r_{fc}\Big|_{\text{tank } j} = 0 \tag{9.72}$$

For example, if we consider microbial growth in the tanks-in-series system with non-sterile feed ($x_f \neq 0$), the biomass balances for tanks 1 through N are

$$F(x_f - x_1) + V_1\mu_1 x_1 = 0$$

$$F(x_{j-1} - x_j) + V_j\mu_j x_j = 0 \qquad j = 2, 3, \cdots, N \tag{9.73}$$

These equations can be solved recursively to yield

$$x_1 = \frac{Fx_f}{F - V_1\mu_1} \tag{9.74}$$

and

$$x_j = \frac{F^{j-1}x_1}{(F - \mu_2 V_2)(F - \mu_3 V_3)\cdots(F - \mu_j V_j)} \qquad j = 2, \ldots, N \tag{9.75}$$

For the case of equal volumes, the simpler form

$$\frac{D_1^{j-1}x_1}{(D_1 - \mu_2)(D_1 - \mu_3)\cdots(D_1 - \mu_j)} \tag{9.76}$$

results from Eq. (9.75). Here D_1 is the dilution rate of an individual tank $(=F/V_1)$.

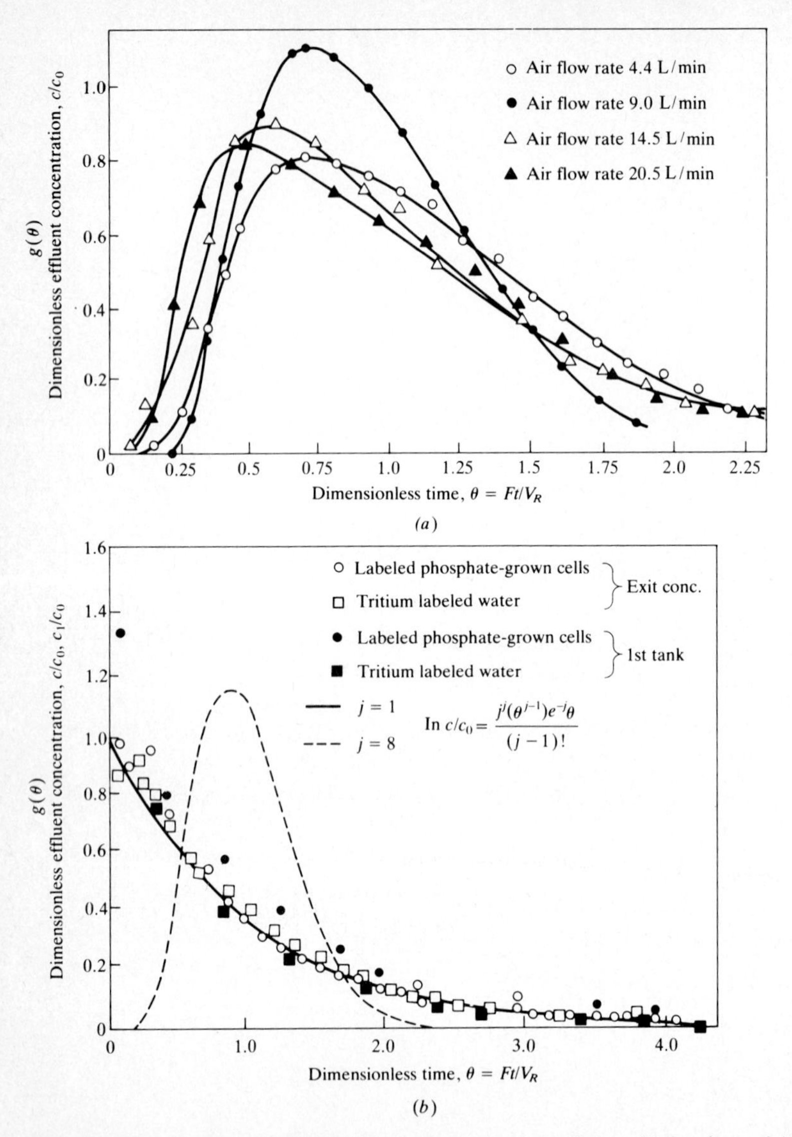

Figure 9.17 Measurement of RTD in an eight-plate tower (hole void fraction = 0.15) by different methods and under different operating conditions. (*a*) Salt tracer data show that the increased aeration flattens and broadens the residence time distribution. (*b*) Yeast cells present in the column; both labeled water and labeled cells used as tracers. *(A. Prokop et al., "Design and Physical Characteristics of a Multistage, Continuous Tower Fermentor," Biotech. Bioeng., vol. 11, p. 945, 1969.)*

We can deduce the effect of staging and recycle on the washout dilution rate D_{max} by examining the system of Fig. 9.14*b*. Assuming Monod growth kinetics with a maintenance term [Eq. (7.22)], sterile feed, equal volumes, and a constant yield factor, it has been shown that D_{max} satisfies

$$\frac{\mu_{max}}{D_{max}} = \frac{N(1 + K_s/s_f)\{1 - [r/(1 + r)]^{1/N}\}}{1 - (k_c/\mu_{max})(1 + K_s/s_f)} \tag{9.77}$$

where as usual D is defined in terms of the overall process

$$D = \frac{F}{V_R} = \frac{F}{NV_1} \equiv \frac{D^*}{N} \tag{9.78}$$

Setting $r = 0$ in Eq. (9.77) gives the critical dilution rate for the equal-volume cascade without recycle. If in addition to $r = 0$, we take $k_c = 0$ so that the growth kinetics is of classical Monod form, Eq. (9.77) reduces to the familiar expression [see Eq. (7.16)].

$$D^*_{max} = \frac{\mu_{max} s_0}{K_s + s_0} \tag{9.79}$$

This result is expected on intuitive grounds: if washout occurs in the first tank in the equal-volume train, it will also prevail in the second, third, ..., and Nth tank. (What happens in the nonequal-volume case if $F > V_j D^*_{max}$ for $j = 1, 2, \ldots, k - 1$ and $F < V_k D^*_{max}$?)

A close conceptual analog of the CSTR cascade with backflow (Fig. 9.14*c*) is the *dispersion model.* A modification of the ideal PFTR, the dispersion model is derived by considering an axial diffusion process which is superimposed on the convective flow through the tube. Returning to the thin-section model described in Fig. 9.5, we add a dispersion flow $A(D_z dc/dz)_z$ into the section and subtract a similar term $A(-D_z dc/dz)_{z+\Delta z}$ for diffusion out of the thin slice on the left-hand side of Eq. (9.13). The same manipulations and limiting processes as followed that equation produce the dispersion-model mass balance

$$\frac{d(uc)}{dz} = \frac{d}{dz}\left(D_z \frac{dc}{dz}\right) + r_{fc} \tag{9.80}$$

Usually u and the *effective axial dispersion coefficient* D_z are approximately constant for liquids, so that Eq. (9.80) becomes

$$u\frac{dc}{dz} = D_z \frac{d^2c}{dz^2} + r_{fc} \tag{9.81}$$

Since Eq. (9.81) is a second-order differential equation, two boundary conditions are required. The generally accepted ones are

$$uc_f = \left(uc - D_z \frac{dc}{dz}\right)_{z=0} \tag{9.82}$$

$$\left.\frac{dc}{dz}\right|_{z=1} = 0 \tag{9.83}$$

When D_z is not too large, which is often the case, the complicated condition (9.82) can be replaced by

$$c\Big|_{z=0} = c_f \tag{9.84}$$

In this case the variance of the dispersion-model RTD is

$$\sigma^2 = \frac{2}{\text{Pe}}\left[1 - \frac{1}{\text{Pe}}(1 - e^{-\text{Pe}})\right] \tag{9.85}$$

where the *axial Peclet number* Pe is defined by

$$\text{Pe} = \frac{uL}{D_z} \tag{9.86}$$

Physically Pe may be regarded as a measure of the importance of convective mass trasport ($\approx uc$) relative to mass transport by dispersion ($\approx D_z c/L$). Considering the limits of Eq. (9.85) as $\text{Pe} \to 0$ and $\text{Pe} \to \infty$, we see that the former case gives $\sigma^2 = 1$, which is identical to the ideal CSTR variance. In the limit of very large Peclet numbers, $\sigma^2 \to 0$, which corresponds to plug flow. As with the tanks-in-series model, Eq. (9.85) can be used to evaluate Pe for the dispersion model from an experimentally determined σ^2. When Monod kinetics with maintenance is used, the dilution rate $D_{\max}$ at washout for the dispersion model is given by

$$\frac{\mu_{\max}}{D_{\max}} = \frac{\frac{1}{4}\,\text{Pe}\,(1 + K_s/s_f)}{1 - (k_e/\mu_{\max})(1 + K_s/s_f)} \tag{9.87}$$

As the relative influence of dispersion becomes vanishingly small ($\text{Pe} \to 0$), $D_{\max}$ decreases to zero. This agrees intuitively with the notion that an ideal PFTR with sterile feed will not support a biological population.

It is important to realize that the axial dispersion coefficient D_z is *not* usually equal to molecular diffusivity. It is a modeling parameter which, when chosen properly, allows the dispersion model to represent some of the mixing effects of several physical phenomena. We shall mention three here: laminar flow in tubes, turbulent flow in pipes, and flow in packed beds.

If axial and radial diffusion characterized by diffusivity $\mathscr{D}$ is superposed on axial convective transport by laminar flow, it can be shown that an effective axial dispersion coefficient is given by

$$D_z = \mathscr{D} + \frac{\bar{u}^2 d_t^2}{192\mathscr{D}} \tag{9.88}$$

Here $\bar{u}$ denotes the average axial velocity ($=\frac{1}{2}$ centerline velocity). The dispersion model with D_z given by Eq. (9.88) provides an excellent approximation to the RTD for laminar flow with diffusion provided the tube is long enough ($L \gg d_t^2/40\mathscr{D}$).

In turbulent flow in pipes (Re > 2100), motion of macroscopic eddies of fluid provides an important mechanism for mass, momentum, and energy transport. The effects of eddy transport closely resemble those of molecular-diffusion processes, but turbulent-diffusion fluxes are usually much greater in magnitude than their molecular counterparts. Consequently, the effective dispersion coefficient in this case depends mostly on the state of fluid flow. While the turbulent-flow Peclet number is typically of the order of 3, it falls with decreasing Reynolds number. This trend in turbulent flow as well as a variety of laminar-flow data are displayed in Fig. 9.18. Perhaps the most important biological reactor involving flow in empty tubes is the continuous liquid sterilizer, which is examined in detail in the following section.

In some reactors with immobilized biocatalysts the tube is packed with particles. These particles are fixed in the bed while the fluid flows around the particles and through the tube. Because fluid is constrained to flow in the interstices between pellets, a fluid element passing axially through the bed undergoes something like a random walk in the radial dimension. The effect of this particle-interrupted sojourn on the RTD of the system can be described by the dispersion

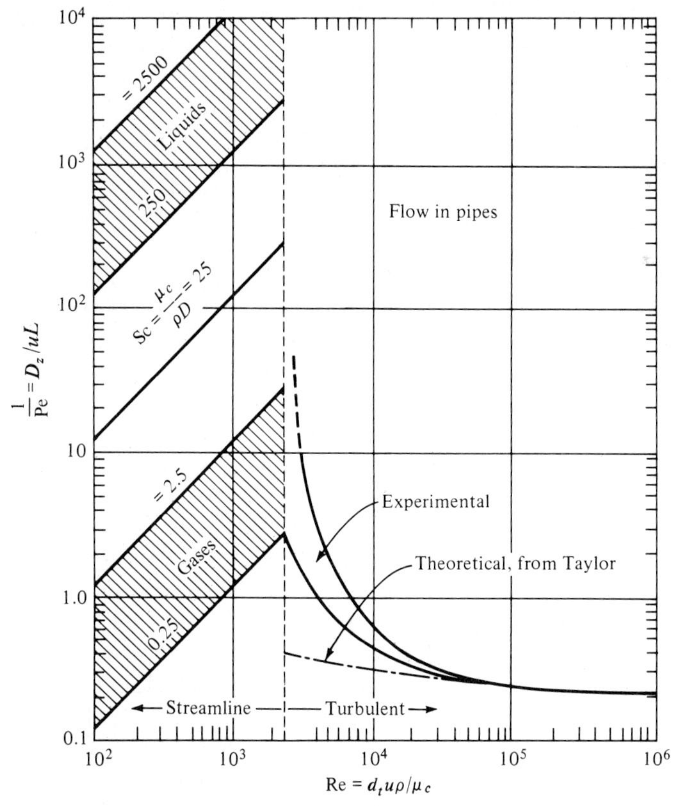

Figure 9.18 Correlations for the axial Peclet number (Pe) in terms of the Reynolds (Re) and Schmidt (Sc) numbers for fluid flow in pipes.

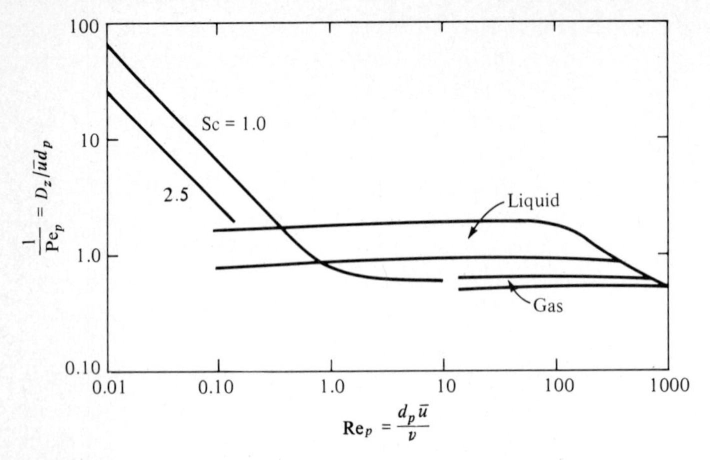

Figure 9.19 Correlations for the axial Peclet number Pe_p for flow in packed pipes. Notice that the dimensionless groups are based upon particle diameter d_p rather than the pipe diameter d_t.

model. Theory suggests that an axial Peclet number based on particle diameter, Pe_p, defined by

$$\text{Pe}_p = \frac{\bar{u}d_p}{D_z} \tag{9.89}$$

is approximately 2 in this instance, where d_p is the pellet diameter. Experimental studies confirm this result under some circumstanes, but dependence of Pe_p on Sc and Re_p is apparent in Fig. 9.19.

In closing this discussion of dispersion-producing processes, we should note that there are many additional possibilities, including pipe bends and gradations in depth contours in rivers and streams. Theory is of little help in identifying D_z for these complicated flow situations, and we must consequently rely on measured RTD data to determine an appropriate D_z value.

The tanks-in-series and dispersion models are both one-parameter (N and D_z, respectively) nonideal mixing models. In different fashions each spans a continuum of mixing and segregation states ranging from ideal CSTR to ideal PFTR. Thus we are faced with a choice of which model to use. In terms of convenience of computation and analysis, the tanks-in-series representation is usually far superior. Also application of the dispersion model is prone to difficulty when backmixing (σ^2) is large. However, Fig. 9.14 shows that a very large number of tanks is necessary to represent situations near plug flow. Consequently, as a general rule of thumb, the dispersion model is usually preferable for small deviations from plug flow (say σ^2 of the order of 0.05 and smaller), while the tanks-in-series formulation is superior when there is substantial backmixing ($\sigma^2 \gtrsim 0.2$). This latter case is typical of fermentors and biological waste-treatment basins. The first is encountered in continuous sterilization of liquids, considered later in this chapter.

9.3.4 Mixing-Bioreaction Interactions

Flow and transport phenomena on different scales influence the kinetic behavior of cells. Effects imposed at a certain length scale (recall Fig. 9.2) can influence the observed kinetics of cell populations in different ways. It is important to recognize this connection so that kinetic measurements and models can be developed under conditions which will resemble in some senses those encountered in the large-scale reactor. In this section we examine some of the interactions of cellular kinetics and mixing in bioreactors.

We shall begin at the largest scale in which bulk circulation patterns carry the cells into different regions of the reactor which, as noted before, typically have different dissolved oxygen and turbulence levels in large-scale equipment. The role of bulk circulation in influencing overall cellular kinetics can be appreciated using a simple example. Suppose that oxygen is supplied locally at some region of the reactor (as occurs in a stirred tank) and that utilization of oxygen occurs in segregated fluid elements circulating in the reactor. This is not an unreasonable view for highly viscous fermentation situations. If one starts with a saturated oxygen concentration in water of 0.3 mol m^{-3} and considers reasonable oxygen uptake rates in the range of 10–100 mol h^{-1} m^{-3}, oxygen will be exhausted in a segregated fluid element after 11 to 50 s. This is of the same order of magnitude as the mixing time in a large-scale reactor.

In order to put this impact of fluid circulation in a more quantitative and general perspective and to take into account the statistical distribution of circulation times, we shall examine the performance characteristics of a batch reactor in which the circulation time distribution is $f_c(t)$. Further, we assume that a zero-order reaction occurs in completely segregated fluid elements. Extension to other types of local reaction kinetics such as Michaelis-Menten kinetics is straightforward and only slightly complicates the calculations involved. The qualitative conclusions, however, remain the same. For such a zero-order reaction, with rate constant k_0 and with initial reactant concentration s_0, the concentration in a fluid element varies with time according to

$$s(t) = \begin{cases} s_0 - k_0 t & t \leq t_e = k_0/s_0 \\ 0 & t > t_e \end{cases} \tag{9.90}$$

Here we define time t_e for nutrient exhaustion which is equal to k_0/s_0. We shall define F here as the fraction of time a fluid element spends in conditions where nutrient has been completely exhausted. This fraction can be computed using

$$F = \int_{t_e}^{\infty} f_c(t)\left(\frac{t - t_e}{\bar{t}}\right) dt \tag{9.91}$$

where $\bar{t}$ denotes the mean circulation time. Assuming that $f_c(t)$ is a log-normal distribution [Eq. (9.43)], Eq. (9.91) has been evaluated for different relative values of the parameters $\bar{t}$, t_e, and the standard deviation σ of the mixing time distribution. Results indicated in Fig. 9.20 illustrate clearly that exposure of cells to starvation conditions increases as the mean circulation time and as the

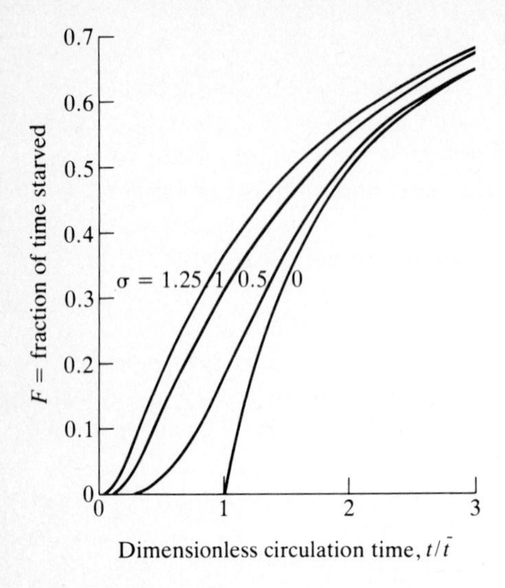

Figure 9.20 The fraction of time cells in circulating fluid elements are starved as a function of the dimensionless circulation time and the coefficient of variation of the circulation time distribution (Ref. 19).

standard deviation of circulation times increase. The quantity $1 - F$ may be interpreted as the yield of the bioreactor relative to the yield that would have been obtained with nutrient provided at all points in the vessel. This calculation indicates that, for a given circulation time, it is preferable to design the reactor so as to minimize the dispersion of mixing times about the mean.

The model just considered, while informative, does not take into account the effects which transient environmental conditions encountered during sojourns around the reactor may have upon cellular kinetics. As noted in Fig. 9.1, the time scales for circulation in large scale bioreactors are comparable to the time scales for certain metabolic processes and adjustments, indicating that kinetic models developed under much different mixing conditions in a small reactor may not apply when greater mixing times and greater reaction environment fluctuations are encountered at larger scale.

Experiments conducted with the on-line fluorometer system described in Sec. 10.2 clearly illustrate the similarity of mixing and biological response times. Mixing times in an aerated (0.5% v/v) 70-L, mechanical agitated fermentor with 40-L working volume have been measured by injecting pulses of quinine in 0.05 M H_2SO_4 solution and measuring transients in culture fluorescence. This measurement characterizes fluid circulation and mixing processes in the reactor. Subsequently, fluorescence responses were measured after pulse addition of glucose to a yeast culture. Here, the fluorescence is produced by intracellular reduced pyridine nucleotides, the levels of which are highly sensitive to the rate of glucose reaction in the cell. Thus, the second measurement provides information on the response time of the complete sequence of transport processes (bulk mixing → diffusion to cells → transport into cell) and metabolic reactions involved in nutrient use.

Table 9.3 shows the mixing time and the overall substrate utilization response time in this reactor at different stirring speeds. Interestingly, the biological

Table 9.3 Comparison of reactor mixing times and yeast glucose response time (all in seconds) as a function of mixing rate in a 70-L (40-L working volume) stirred bioreactor with 0.5 v/v/m aeration.

(Reprinted by permission from A. Einsele, D. L. Ristroph, and A. E. Humphrey, "Mixing Times and Glucose Uptake Measured with a Fluorometer," Biotech. Bioeng., vol. 20, p. 1487, 1978.)

	200 rpm	500 rpm	700 rpm
Intracellular fluorescence response time	8.5	6.8	5.9
Bulk mixing response time	4.2	2.5	1.5
Difference = biological response time	4.3	4.3	4.4

response time for local glucose uptake and utilization is, as might be expected on intuitive grounds, essentially the same in all three cases, while the mixing time is reduced as agitation speed is increased. Another very significant fact indicated by these data is the similarity in order of magnitude for the response time for local glucose uptake and utilization and the mixing time for the reactor. This indicates that, as cells circulate in the reactor and encounter spatially inhomogeneous conditions, the metabolism of the cell may always be in a transient state since it does not respond much more rapidly than the characteristic time scale for environmental fluctuations due to liquid circulation.

Several experimental investigations have explored the influence of transient conditions on metabolism and noted significant effects. In a closed tubular-loop fermentor in which oscillations in dissolved oxygen levels are encountered around the loop, a culture of *Candida tropicalis* showed increased biomass yield based on substrate, reduced oxygen utilization for biomass production, and lower respiratory quotient. Opposite effects were observed in imposed oscillations in dissolved oxygen level in a continuous culture in *Pseudomonas methylotropha* ASI with methanol and carbon and energy source. In this case, decreased biomass to substrate yields and decreased growth rates were observed. In another experimental study, sinusoidal fluctuations of dissolved oxygen level with a period of 2 min. and a mean of 30% of air saturation were imposed on a culture of *Penicillium chrysogenum* P1. In this case, the specific penicillin production rate decreased significantly, resembling the effect expected at a lower mean dissolved oxygen level. This shift indicates a significant transient nonlinear effect on the bioreaction.

Turning now to the smaller scale of turbulent velocity fluctuations, we can expect important interactions between the turbulence intensity at different scales and the morphology (and thereby potentially the metabolic state) of certain organisms. This connection is expected to be most important for those organisms that grow to a size scale comparable to the turbulence scales expected. These scales range from the largest eddies, on the scale of the height of a turbine blade in an agitator, say 0.1 m, to the smallest eddies which are produced by the cascade of transmission of turbulent energies (Sec. 8.3). In agitated bioreactor systems, this smallest eddy size is of the order of 20–100 μm. To determine the biological structures influenced by velocity fluctuations on this scale, we can note in Fig. 9.2 that flocs of microorganisms and mycelial aggregates are intermediate in the size spectrum of turbulence and therefore will be substantially influenced by mixing intensity and the distribution of turbulence fields encountered in the reactor.

We have already discussed the effects which overall cellular aggregate size can have on overall rates due to diffusion limitation. Here we shall concentrate on more subtle effects which involve changes in morphological structure and metabolic state which substantially influence the fermentation. One interesting example of this class is the influence of mixing intensity upon growth, product formation, and nucleotide leakage from various mutants of *Aspergillus niger* which produce high levels of citric acid. As indicated in Fig. 9.21, the total biomass productivity increases with increasing agitation strength, but the dependence of citric acid production on agitation intensity is more complicated, exhibiting strong maxima with respect to agitation intensity for the three mutants considered. Parallel studies of nucleotide release as a function of agitation speed showed patterns which depended upon the strain. For some strains, increasing mixer speed gave higher levels of released nucleotides while for one strain the opposite and unexpected effect of decreasing nucleotide release at higher mixer speeds was noted. The effect of mixer speed on morphology of the mycelium for this unusual mutant is shown in Fig. 9.22. Here in micrographs of the organisms

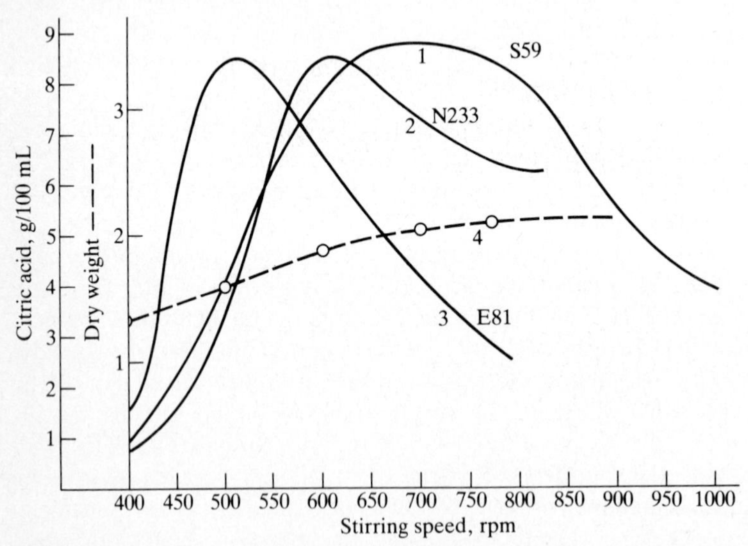

Figure 9.21 Curves 1 through 3 show citric acid accumulation after seven days' fermentation at different turbine agitation rates for *Aspergillus niger* strains S59, N233, and E81, respectively. Curve 4 is the corresponding biomass dry weight for strain S59. *(Reprinted by permission from E. Ujcová, Z. Fencl, M. Musi'lková, and L. Siechert, "Dependence of Release of Nucleotides from Fungi on Fermentor Turbine Speed," Biotech. Bioeng., vol. 22, p. 237, 1980.)*

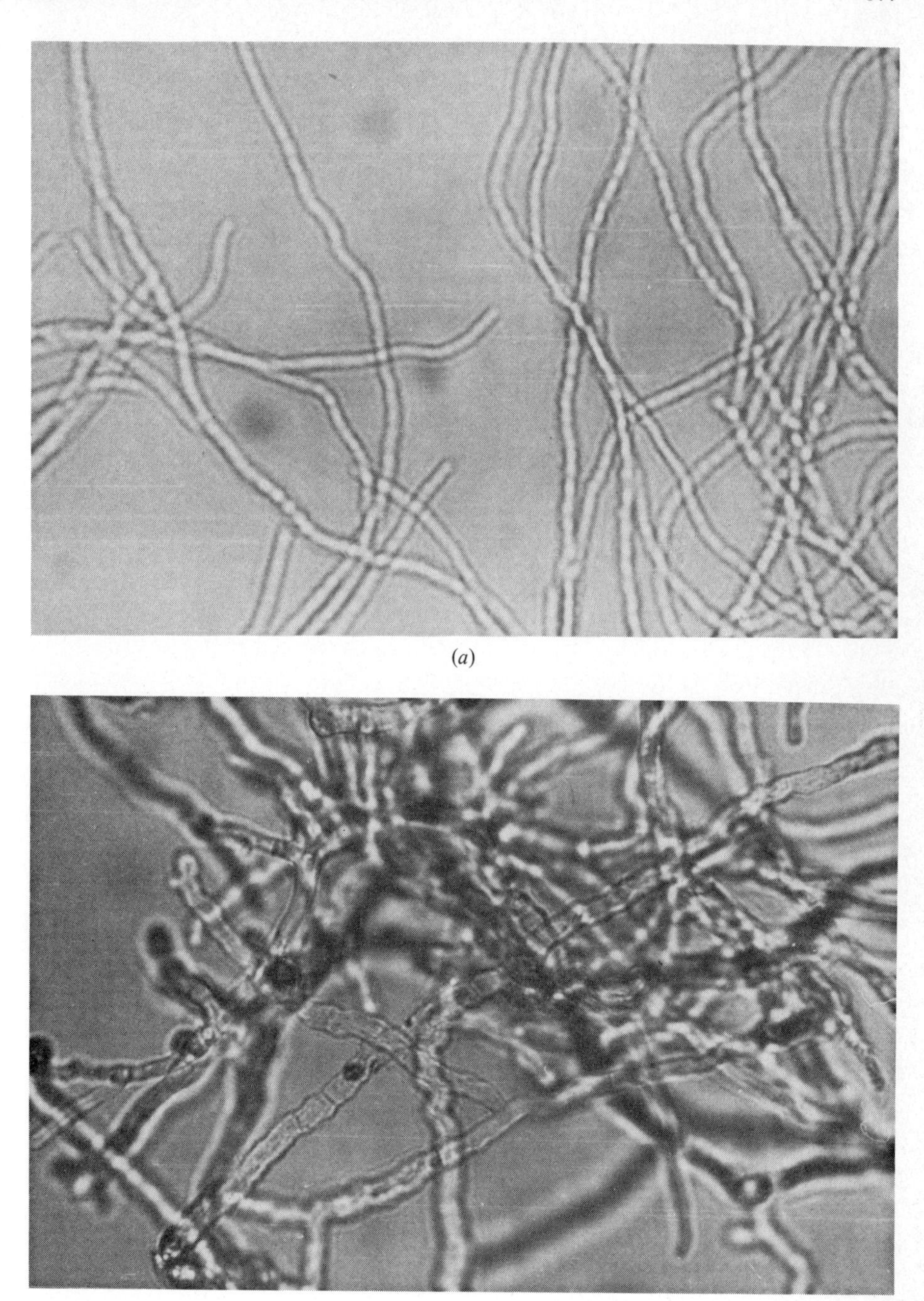

(*a*)

(*b*)

Figure 9.22 Microphotographs (400×) showing different mycelial morphology and septation in *A. Niger* S59 cultivated at (*a*) 400 and (*b*) 1200 rpm. *(Reprinted by permission from E. Ujcová, Z. Fencl, M. Musi'lková, and L. Seichert, "Dependence of Release of Nucleotides from Fungi on Fermentor Turbine Speed," Biotech. Bioeng., vol. 22, p. 237, 1980.)*

we see long, thin, and sporadically branched filaments with relatively few septa at low agitation speeds, while the hyphae are thicker, highly septated, and more densely branched and twisted at higher agitation speeds. This example points out that mixing intensity can have a profound effect on the organisms which influence process kinetics in a substantial way.

In the discussions and examples in this section, we have tried to indicate that, because of the potentially complicated response of growing cells to physical and chemical environment and our inability to develop a generally applicable kinetic model valid for all circumstances, we must be careful to recognize the interaction between chemical and physical environment provided in a certain bioreactor and the corresponding kinetic description which is appropriate and indeed necessary. The difficulty of reliable scale-up in cases of strong sensitivity of kinetic behavior to environmental fluctuations should be obvious. In such cases, we must look to reactor designs which provide reaction environments which are as well defined as possible—not necessarily uniform; this may be extremely difficult on a large scale. However, by use of some of the alternative bioreactor configurations considered in Sec. 9.7, we can reduce the variance of mixing times and the degree of environmental fluctuations in order to achieve a better defined contacting situation for which we can seek correspondingly valid kinetic descriptions. Basically, we cannot expect success in rational, predictable scale-up of bioreactors until we have better understanding of the transport processes at several scales and the bioreaction kinetics at several levels and can then synthesize these to calculate accurately reactor performance.

Example 9.1: Reactor modeling and optimization for production of α-galactosidase by a *Monascus* sp. mold The enzyme α-galactosidase may be useful in the beet-sugar industry because it can decompose raffinose, an inhibitor of sucrose crystallization. In a fascinating series of papers, Imanaka, Kaieda, Sato, and Taguchi† have investigated production of this intracellular enzyme by a mold they isolated from soil. Their original work deserves serious study: here we summarize some of the major results of their investigations.

As a first step in developing highly productive continuous processes for enzyme synthesis, batch and continuous cultures of the mold were cultivated under a variety of conditions. We list next the major findings from the batch experiments:

1. Among 20 different carbon sources including glucose, fructose, mannitol, and starch, only four sugars were effective in inducing high α-galactosidase activity. The strong inducers are galactose, melibiose, raffinose, and stachyose.
2. Ammonium nitrate gave more enzyme production than the alternative nitrogen sources urea, KNO_3, $(NH_4)_2SO_4$, and peptone. The optimal NH_4NO_3 concentration in the medium is between 0.3 and 0.5 percent by weight.
3. When grown in a galactose medium, the cell mass is directly proportional to the α-galactosidase activity. When a mixture of glucose and galactose was used as the carbon source, diauxic growth

† The material in this example is drawn chiefly from T. Imanaka, T. Kaieda, K. Sato, and H. Taguchi, "α-Galactosidase Production in Batch and Continuous Culture and a Kinetic Model for Enzyme Production," *J. Ferment. Technol. (Japan)*, **50**:633, 1972, and T. Imanaka, T. Kaieda and H. Taguchi, "Optimization of α-Galactosidase Production by Mold, II, III," *J. Ferment. Technol. (Japan)*, **51**:423, 431, 1973.

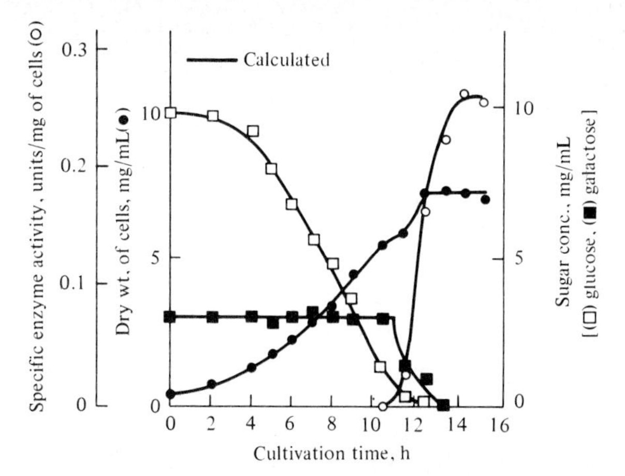

Figure 9E1.1 Results of batch cultivation of a *Monascus* sp. mold in a mixture of glucose and galactose [initial medium composition: glucose 1% (by weight), galactose 0.3%, NH_4NO_3 0.5%, KH_2PO_4 0.5%, $MgSO_4 \cdot 7H_2O$ 0.1%, yeast extract 0.01%]. The inoculum was grown in a glucose medium. The initial conditions used in the calculations were $x = 5 \times 10^{-4}$ g/mL, $s_1 = 1 \times 10^{-2}$ g/mL, $s_2 = 3 \times 10^{-3}$ gm/mL, $s_{2I} = 0$ μg/mg cell, $rs_{2I} = 0.910$ μg/mg cell, $e = 0$ units/mg cell. *[Reprinted from T. Imanaka et al., "Unsteady-state Analysis of a Kinetic Model for Cell Growth and α-Galactosidase Production in Mold," J. Ferment. Tech. (Japan), vol. 51, p. 423, 1973.]*

was observed (see Fig. 9E1.1); almost no galactose is consumed until the glucose is nearly exhausted.

4. Figure 9E1.1 also shows that α-galactosidase production does not start until the glucose is almost gone. Separate experiments revealed that glucose concentrations greater than 0.05 percent by weight repress synthesis of the enzyme.

Two different series of steady-state continuous-culture experiments were conducted in a single CSTR. In the first series, the dilution rate was initially at a very low level, and it was increased slowly in a stepwise fashion (shifted up) with observations of steady-state behavior at each D along the way. The data so observed are plotted in Fig. 9E1.2. Especially interesting is the discontinuous jump evident at $D = 0.142$ h^{-1}. Below this dilution rate, galactose is being consumed and α-galactosidase is synthesized. When D is increased above 0.142 h^{-1}, however, both these activities stop and glucose alone is utilized as the mold's carbon source. Evidently, this jump is a manifestation of the glucose effect, already seen in batch culture of this organism. When cultivated under relatively large specific growth rates (large D's), the mold preferentially feeds on glucose.

Figure 9E1.3 illustrates the results of a similar series of experiments, except that here the CSTR was started up at a high dilution rate. Then, in a sequence of shift-down steps, the dilution rate was gradually decreased. While less sharp than the previous case, another discontinuity occurs, this time around $D = 0.008$ h^{-1}. Below that critical dilution rate, enzyme is produced and galactose is assimilated, while no α-galactosidase activity is evident for $D > 0.008$ h^{-1}. This is in marked contrast to the shift-up experimental results, where enzyme production was apparent up to $D = 0.142$ h^{-1}.

Replotting some of the data from the previous two figures in Fig. 9E1.4 clearly shows that this system exhibits multiple, stable steady states between $D = 0.008$ and $D = 0.142$ h^{-1}. For dilution rates between these limits, whether or not α-galactosidase is produced depends upon how the reactor is started up. Shifting down into this range results in no enzyme synthesis, while shifting up will provide α-galactosidase production.

Based upon these and other experiments, a mathematical model for substrate utilization, cell growth, and product synthesis was developed. In most respects, the individual model equations in

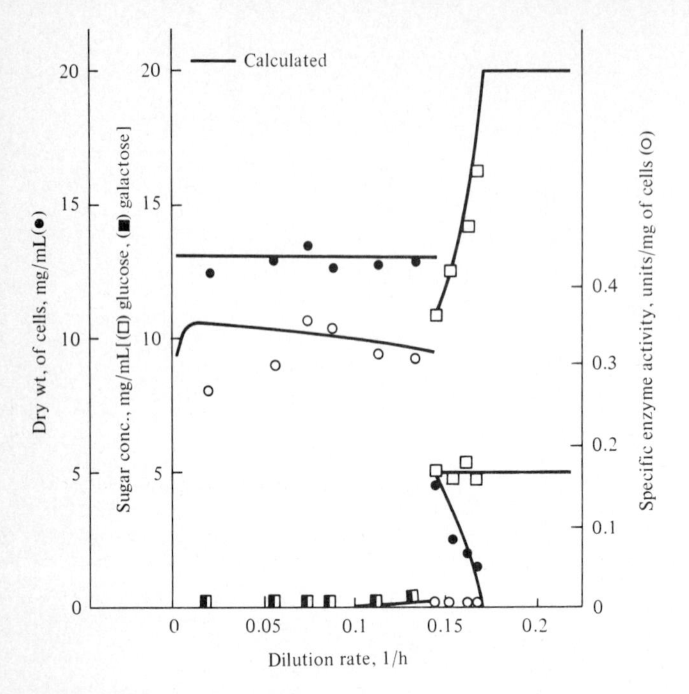

Figure 9E1.2 Steady-state cell and substrate concentrations and specific enzyme activity observed during gradual shift up of dilution rate for continuous culture (30°C). Initially the medium contains 2% glucose and 0.5% galactose. *[Reprinted from T. Imanaka et al., "Optimization of α-Galactosidase Production by Mold," J. Ferment. Tech. (Japan), vol. 50, p. 633, 1972.]*

Table 9E1.1 are familiar from our earlier studies: the specific growth rate μ_2 based on galactose is of Monod form, while the specific growth rate of glucose μ_1 includes competitive inhibition by galactose. All the constants in these growth-rate functions were evaluated for two different media from continuous-culture experiments. Parameters labeled G in Table 9E1.2 correspond to a glucose medium (20 g glucose, 5 g NH_4NO_3, 5 g KH_2PO_4, 1 g $MgSO_4 \cdot 7H_2O$, 0.1 g yeast extract in 1000 mL tap water at pH 4.5) while the p subscripts refer to a galactose medium advantageous for enzyme production (5 g galactose, 5 g NH_4NO_3, 5 g KH_2PO_4, 1 g $MgSO_4 \cdot 7H_2O$ in 1000 mL tap water, pH 4.5).

The model for enzyme production is based upon the operon theory of induction, studied in Chap. 6. The specific rate of α-galactosidase synthesis is proportional to the intracellular concentration of mRNA which codes for that enzyme. This mRNA is assumed to decompose by a first-order reaction and is produced provided the intracellular concentration of repressor R is smaller than a threshold value r_c. Below this threshold value, lower r values cause increased specific rates of mRNA synthesis. The repressor is formed at constant specific rate k_2 and decomposes with first-order specific rate $k_2 r$. Repressor concentration is also reduced by complexing with the inducer, intracellular galactose.

The rate of galactose transport into the cell is given by the term in the intracellular galactose mass balance with coefficient U. To take into account the glucose effect this transport term is set equal to zero whenever the glucose concentration s_1 exceeds a critical value s_{1c}, which is taken to be 2.25×10^{-4} g/mL.

Little information is available for direct evaluation of the rate constants in the operon model. Values for k_3 and k_7 were assigned based on the assumption that the repressor and mRNA half-lives are 40 and 5 min, respectively. The other parameter values listed in Table 9E1.2 were estimated by trial and error to achieve a reasonable fit to the experimental data.

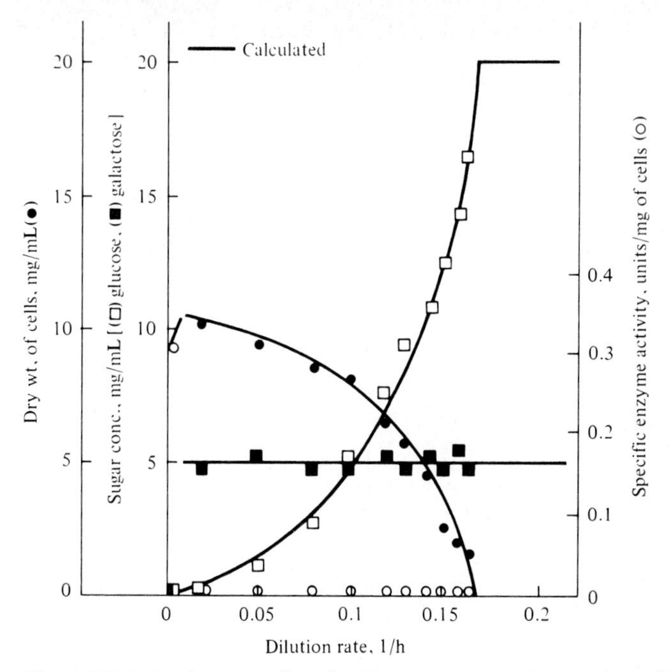

Figure 9E1.3 Steady-state cell and substrate concentrations and specific enzyme activity observed during gradual shift down of dilution rate for continuous culture (30°C). Initially the medium contains 2% glucose and 0.5% galactose. *[Reprinted from T. Imanaka et al., "Optimization of α-Galactosidase Production by Mold," J. Ferment. Tech. (Japan), vol. 50, p. 633, 1972.]*

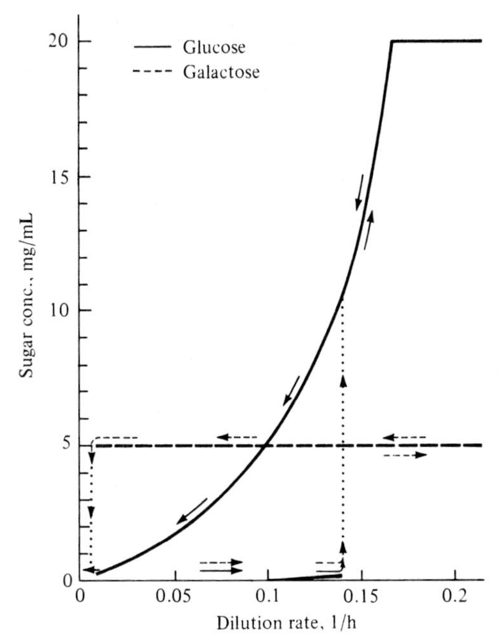

Figure 9E1.4 This replot of some of the curves from Figs. 9E1.2 and 9E1.3 displays steady-state multiplicity and hysteresis in continuous culture. Two stable steady states occur for dilution rates between 0.008 and 0.142 h^{-1}. Which is obtained depends on reactor start-up. *[Reprinted from T. Imanaka et al., "Optimization of α-Galactosidase Production by Mold," J. Ferment. Tech. (Japan), vol. 50, p. 633, 1972.]*

Table 9E1.1 Mathematical Model for α-galactosidase production†‡

Substrate utilization

$$\frac{ds_j}{dt} = -\frac{1}{Y_j}\mu_j x \quad \text{for } j = 1\text{(glucose)}, 2\text{(galactose)} \qquad (9E2.1)$$

Biomass growth

$$\frac{dx}{dt} = (\mu_1 + \mu_2)x \qquad (9E2.2)$$

where

$$\mu_1 = \frac{\mu_{\max,1} s_1}{s_1 + K_1(1 + s_2/K_i)} \qquad (9E2.3)$$

and

$$\mu_2 = \frac{\mu_{\max,2} s_2}{s_2 + K_2} \qquad (9E2.4)$$

Operon model for enzyme production

Intracellular galactose:

$$\frac{d}{dt}(s_{2I}\cdot x) = Ux\left(\frac{G_2 s_2}{K_{m2} + s_2} - s_{2I}\right) - k_1 s_{2I}\cdot x \quad \text{for } s_1 < s_{1c} \qquad (9E2.5a)$$

$$= -k_1 s_{2I}\cdot x \quad \text{for } s_1 \geq s_{1c} \qquad (9E2.5b)$$

Repressor:

$$\frac{d}{dt}(r\cdot x) = k_2 s - k_3 r\cdot x - k_4 r\cdot s_{2I}\cdot x + k_5(rs_{2I})x \qquad (9W2.6)$$

Galactose-repressor complex:

$$\frac{d}{dt}[(rs_{2I})x] = k_4 r\cdot s_{2I}\cdot x - k_5(rs_{2I})x \qquad (9E2.7)$$

mRNA for galactosidase:

$$\frac{d}{dt}(m\cdot x) = \begin{cases} k_6(r_c - r)x - k_7 m\cdot x & \text{for } r_c > r \quad (9E2.8a) \\ -k_7 m\cdot x & \text{for } r \geq r_c \quad (9E2.8b) \end{cases}$$

Enzyme:

$$\frac{d}{dt}(e\cdot x) = k_8 m\cdot x \qquad (9E2.9)$$

† T. Imanaka, T. Kaieda, K. Sato, and H. Taguchi, *J. Ferment. Technol.* (*Japan*), **50**: 633, 1972.

‡ Concentration variables are x = biomass, s_1 = glucose, s_2 = extracellular galactose, s_{2I} = intracellular galactose, r = intracellular repressor, (rs_{2I}) = intracellular inducer-repressor complex, m = intracellular mRNA, and e = intracellular α-galactosidase. The remaining symbols are kinetic, yield, and transport parameters.

This model is certainly attractive because it includes substantial structure which is heavily based on established biological principles. On the other hand, we could object to the large number of adjustable parameters it contains. Several tests can be applied to investigate the suitability of this model. One is based on the following question: Can other models based on different assumptions but containing a similar number of adjustable constants fit the data equally well? If so, we cannot place much confidence in this particular form. Imanaka et al. conducted several such tests, including cases in which (1) intracellular galactose concentration is proportional to galactose concentration in the medium, or (2) repressor formation is proportional to intracellular glucose, or (3) rate of mRNA formation is inversely proportional to concentration of repressor. Any of these modifications in the model resulted in serious discrepancies with the experimental observations.

Table 9E1.2 Parameter values for the mathematical model of α-galactosidase production†

Entries in the right column were evaluated experimentally; the remaining parameters were adjusted to fit the batch and continuous-culture results; G = glucose medium, 30°C, p = galactose medium, 35°C.

$k_1 = 40\ h^{-1}$	$\mu_{max,1_G} = 0.215\ h^{-1}$
$k_2 = 1$ mg/(mg cells · h)	$\mu_{max,2_G} = 0.208\ h^{-1}$
$k_3 = 1\ h^{-1}$	$K_{1_G} = 1.54 \times 10^{-4}$ g/mL
$k_4 = 0.1$ mg cells/(mg · h)	$K_{2_G} = 2.58 \times 10^{-4}$ g/mL
$k_5 = 1 \times 10^{-4}\ h^{-1}$	$\mu_{max,1_p} = 0.190\ h^{-1}$
$k_6 = 1\ h^{-1}$	$\mu_{max,2_p} = 0.162\ h^{-1}$
$k_7 = 8\ h^{-1}$	$K_{1_p} = 1.45 \times 10^{-4}$ g/mL
$k_{8_G} = 3.2787$ units/(mg mRNA · h)	$K_{2_p} = 3.07 \times 10^{-4}$ g/mL
$k_{8_p} = 5.0442$ units/(mg mRNA · h)	$K_i = 1.39 \times 10^{-4}$ g/mL
$U = 100\ h^{-1}$	$Y_{1_G} = 0.530$
$G_2 = 1$ mg/mg cells	$Y_{2_G} = 0.516$
$K_{m2} = 1 \times 10^{-8}$ mg/mg cells	$Y_{1_p} = 0.377$
$s_{1c} = 2.25 \times 10^{-4}$ g/mL	$Y_{2_p} = 0.361$
$r_c = 0.934$ mg/mg cells	

† T. Imanaka, T. Kaieda, K. Sato, and H. Taguchi, *J. Ferment. Technol.*, **50**: 558, 1972.

The other tests of the model which Imanaka et al. considered involved its ability to fit data collected under a wide variety of operating conditions. Indeed, this is the raison d'être for a complex, structured kinetic model, for if the model is sufficiently complete, it can be used to determine optimal operating conditions for the process. Here too the model performed extremely well. The solid curves in Fig. 9E1.1 were computed using this model with initial conditions as indicated in the figure, and the fit is very good.

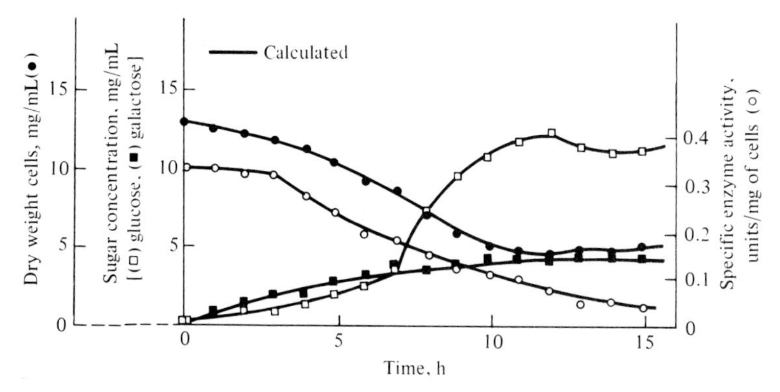

Figure 9E1.5 Transient behavior of CSTR continuous culture following a change in dilution rate from $D = 0.140\ h^{-1}$ to $D = 0.142\ h^{-1}$. The medium contained two carbon sources: glucose (2%) and galactose (0.5%). Initial conditions used in the calculation were $x = 1.30 \times 10^{-2}$ g/mL, $s_1 = 2.23 \times 10^{-4}$ g/mL, $s_2 = 5.01 \times 10^{-5}$ g/mL, $s_{2I} = 2.5$ μg/mg cell, $r = 0.718$ μg/mg cell, $(rs_{2I}) = 1.28$ μg/mg cell, $m = 1.04 \times 10^{-2}$ μg/mg cell, $e = 0.297$ units/mg cell. *[Reprinted from T. Imanaka et al., "Unsteady-state Analysis of a Kinetic Model for Cell Growth and α-Galactosidase Production by Mold," J. Ferment. Tech. (Japan), vol. 51, p. 423, 1973.]*

Table 9E1.3 Optimized operating conditions for six different continuous fermentor configurations[†‡]

System	Type of fermentation process	Optimum operating conditions, D values in h^{-1}	Specific enzyme activity, units/mg of cells	Enzyme productivity, units/(L · h)
1	Glu + Gal 30°C	$D = 0.142$ Shift-up system	0.325	554
2	Glu + Gal 35°C	$D = 0.121$ Shift-up system	0.500	559
3	Glu, Gal 30°C 35°C	$D_1 = 0.200$ $D_2 = 0.250$ $\bar{D} = 0.118$	0.293	415
4	Glu, Gal 30°C 35°C	$D_1 = 0.133$ $D_2 = 0.286$ $\bar{D} = 0.097$	0.500	582
5	Glu, Gal 30°C 30°C 35°C	$D_1 = 0.193$ $D_2 = 2.342$ $D_3 = 0.286$ $\bar{D} = 0.117$	0.500	702
6	Glu, Gal 30°C 35°C	$D_1 = 0.178$ $D_2 = 0.286$ $\bar{D} = 0.117$ $C = 1.34$	0.500	702
		$D_1 = 0.266$ $D_2 = 0.286$ $\bar{D} = 0.145$ $C = 200$	0.500	870

[†] T. Imanaka, T. Kaieda, and H. Taguchi, *J. Ferment. Technol. (Japan)*, **51**: 558, 1973.
[‡] Glucose = 2 percent, galactose = 0.5 percent of the total medium.

The model equations for an unsteady-state CSTR can be obtained simply by adding terms of the form D (inlet concentration–concentration in the reactor) to each of the batch equations in Table 9E1.1. Steady-state CSTR mass balances are then obtained by setting all time-derivative terms equal to zero. With these equations, the solid lines in Figs. 9E1.2 and 9E1.3 were calculated. The hysteresis and jump phenomena observed experimentally are clearly well represented by the model. As a final test before turning to reactor optimization, the model was used to compute the transient behavior of the CSTR following shift-up. The results predicted by the model as well as experimental data are displayed in Fig. 9E1.5: here the agreement between measured and calculated responses, including the overshoot of glucose and undershoot of cell concentrations, is quite dramatic.

With the model of Table 9E1.1 thus well established, it was used to compute the enzyme productivity of a variety of continuous-reactor configurations (Table 9E1.3). For each design, the volumes of the various reactors were adjusted to maximize enzyme production. The overall dilution rate D indicated in the table is defined as the total medium flow rate into the system divided by the total volume of all reactors in the process.

We shall examine a few details of systems 3 and 4. In these, as in systems 5 and 6, the basic idea is to grow a large cell concentration in the first part of the system using relatively cheap glucose only. Then a galactose medium is added at a later stage to induce enzyme production. Because enzyme induction takes some time, it is necessary to take into account adaptation of the intracellular reaction systems in the later stage. For this purpose, the induction stage is treated as a completely segregated system, and its effluent enzyme activity is computed using Eq. (9.62), which can be rewritten

$$e_e = \int_0^\infty e_b(t)\mathscr{E}(t)\,dt \tag{9E1.1}$$

where $e_b(t)$ is the enzyme concentration at time t computed from the batch-reactor model. In the batch calculations the initial conditions used are the concentrations in the feed to the induction stage.

Using this procedure, the enzyme productivity ($=De_ex$) was computed for the two-stage system (3), with results shown in Fig. 9E1.6. The model revealed that maximum productivity would be obtained with $D_1 = 0.20\ \text{h}^{-1}$, and experiments performed under that condition showed nearly exactly the same D_2 dependence as predicted by the model. In system 4, a bioreactor designed to approximate plug-flow conditions was used for the second stage. Again, the model results (line) agreed very well with experimental data (dots) for this process (Fig. 9E1.7).

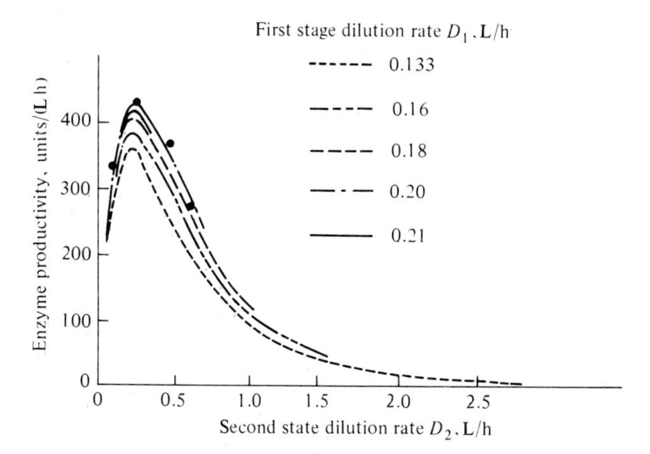

Figure 9E1.6 Calculated and experimental enzyme productivity in a two-CSTR continuous culture (System 3 from Table 9E1.3). *[Reprinted from T. Imanaka et al., "Optimization of α-Galactosidase Production in Multi-stage Continuous Culture of Mold, J. Ferment. Tech. (Japan), vol. 51, p. 431, 1973.]*

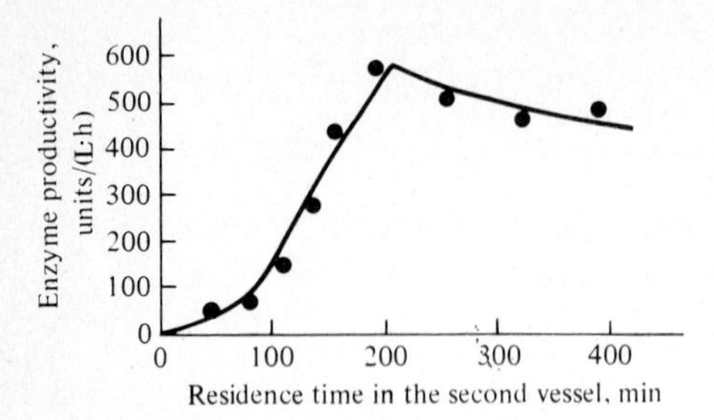

Figure 9E1.7 Productivity of enzyme in a CSTR-tubular fermentor cascade (system 4 from Table 9E1.3). *[Reprinted from T. Imanaka et al., "Optimization of α-Galactosidase Production in Multi-stage Continuous Culture of Mold," J. Ferment. Tech. (Japan), vol. 51, p. 431, 1973.]*

Returning now to Table 9E1.3, we see that well-chosen staged cultures provide substantially greater enzyme production than the best single process. For example, system 6 produces 55 percent more enzyme than the single CSTR. Clearly the availability of a sound kinetic model and the general tools of reactor design have here proved essential ingredients in formulating and optimizing a superior continuous-reaction process.

9.4 STERILIZATION REACTORS

Liquids, usually aqueous, can be sterilized by several means, including radiation (ultraviolet, x-rays), sonication, filtration, heating, and chemical addition. Only the last two are widely used in large-scale processes. However, small amounts of liquids containing sensitive vitamins and other complex molecules are sometimes sterilized by passage through porous membranes. In this section we shall concentrate on design of heat-treatment processes.

Requirements for destruction of viable microbes and viruses vary widely depending upon the material and its intended use. In some instances, e.g., sauerkraut manufacture and biological wastewater treatment, microorganisms naturally present in the process fluid are responsible for desirable reactions. Inhibitors for the growth of unwanted organisms are rapidly evolved in alcohol, vinegar, and silage production, so that here too sterilization requirements are not extreme. Milk pasteurization involves killing most but not all actively growing microbes. More severe treatment of milk is not practiced because degradation of desired components results. Trade-offs between destruction of useful compounds and death of unwanted organisms play a major role in choice and design of sterilization and pasteurization equipment.

Pure-culture fermentations, tissue culture, and some food products require more stringent measures. Essentially all contaminating microbial life must be excluded from the system, although the degree of "perfection" also varies somewhat. Economic considerations might indicate, for example, that a contamination probability [$1 - P_0$ in Eq. 7.131)] of 10^{-2} is acceptable for a batch fermentation process. In this case we would expect 1 batch out of every 100 to be lost due to contamination. We could accept this if the loss were comparable to the cost of additional sterilization capacity.

Much more severe requirements hold in the canning industry. A single surviving spore of *Clostridium botulinum* may cause lethal contamination, so that virtually complete elimination is required. Typically, a design criterion in this situation specifies that the spore survival probability $1 - P_0$ be reduced to less than 10^{-12}. This example illustrates the importance of small deviations from essentially complete conversion of substrate (spores and vegetative cells) into products (inactive spores, dead cells) in sterilization reactors. Thus, careful sterilization-reactor design can clearly be critical. Continuous sterilization processes are examined after analysis of the batch case.

9.4.1 Batch Sterilization

Let us begin by considering a well-mixed closed volume containing a cell or spore suspension. The fluid is to be sterilized by heating, and then cooled to a suitable temperature for subsequent processing. The concentration of surviving organisms resulting from this process can readily be computed starting from Eqs. (7.125) and (7.126):

$$\frac{dn}{dt} = -k_{d0}e^{-E_d/RT(t)}n \tag{9.92}$$

where we have explicitly included time variations in the fluid temperature. Separating the variables in Eq. (9.92) and integrating, we find

$$\ln\frac{n_0}{n_f} = \int_0^{t_t} e^{-E_d/RT(t)}\,dt \tag{9.93}$$

where the f subscript denotes final conditions.

Common batch-sterilization designs include one or more of the following heat sources: steam sparging (bubbling of live steam through the medium), electrical heating, and heating or cooling with a two-fluid heat exchanger. Deindoerfer and Humphrey† have associated with each heating or cooling mode a particular time-temperature profile; these functions are shown in Table 9.4. The integral on the right-hand side of Eq. (9.93) can be evaluated by segmentation into three intervals of heating, holding, and cooling. A total of four integral forms arise: constant temperature and each of the three transient modes (hyperbolic, linear, and exponential).

Constant temperature

$$\ln\frac{n_f}{n_0} = \int_0^{t_f} k_{d0}e^{-E_d/RT}\,dt = k_{d0}t_f e^{-E_d/RT} \tag{9.94}$$

Hyperbolic

$$\ln\frac{n_f}{n_0} = \frac{k_{d0}a(E_d/RT_0)}{(a+b)^2}e^{-(E_d/RT_0)[b/(a+b)]}$$
$$\times\left[E_2\left\{\frac{E_d}{RT_0}\left[\frac{1+bt_f}{1+(a+b)t_f} - \frac{b}{a+b}\right]\right\} - E_2\left(\frac{E_d}{RT_0}\frac{a}{a+b}\right)\right] \tag{9.95}$$

† F. H. Deindoerfer and A. E. Humphrey, "Analytical Method for Calculating Heat Sterilization Times," *Appl. Microbiol.*, 7:256 (1959).

Table 9.4 Temperature-time profiles in batch sterilization[†]

Type of heat transfer	Temperature-time profile	Parameters
Steam sparging	$T = T_0\left(1.0 + \frac{at}{1+bt}\right)$ hyperbolic	$a = \frac{hs}{MT_0\rho C_p}$ $b = \frac{s}{M}$
Electrical heating	$T = T_0(1.0 + at)$ linear	$a = \frac{q}{MT_0\rho C_p}$
Steam (heat exchanger)	$T = T_N(1 + be^{-at})$ exponential	$a = \frac{UA}{Mc}$ $b = \frac{T_0 - T_N}{T_N}$
Coolant (heat exchanger)	$T = T_{e0}(1 + be^{-at})$ exponential	$a = \frac{wc'}{M\rho C_p}(1 - e^{-UA/wc'})$ $b = \frac{T_0 - T_{e0}}{T_{e0}}$

where h = enthalpy differences between steam at sparger temperature and raw medium temperature
s = steam mass flow rate
M = initial medium mass
T_0 = initial medium temperature
q = rate of heat transfer, kcal per unit time
U = overall heat-transfer coefficient, kcal/(m^2·h·°C)
A = heat-transfer area, m^2
T_N = temperature of heat source
w = coolant mass flow rate
c' = coolant specific heat
T_{e0} = coolant inlet temperature
ρ = medium density
C_p = medium heat capacity

† After F. H. Deindoerfer and A. E. Humphrey, *Appl. Microbiol.*, **7**: 256, 1959.

Linear increasing

$$\ln \frac{n_f}{n_0} = \frac{k_{d0} E_d}{RT_0 a}\left(E_2 \frac{E_d/RT_0}{1 + at_f} - E_2 \frac{E_d}{RT_0}\right) \tag{9.96}$$

Exponential

$$\ln \frac{n_f}{n_0} = \frac{k_{d0}}{a}\left[E_1\left(\frac{E_d/RT_H}{1+b}\right) - E_1\left(\frac{E_d/RT_H}{1+be^{-at_f}}\right)\right] - \frac{k_{d0}}{a} e^{-E_d/RT_H}\left[-E_1\left(\frac{E_d}{RT_H}\frac{b}{1+b}\right) + E_1\left(\frac{E_d}{RT_H}\frac{be^{-at_f}}{1+be^{-at_f}}\right)\right] \tag{9.97}$$

where E_n is the exponential integral

$$E_n(z) = \int_z^\infty \frac{e^{-w}\,dw}{w^n} \tag{9.98}$$

a tabulated function available in many handbooks and computer packages. The result for cooling is obtained by substituting T_{c0} for T_H and using the definitions of a and b appropriate for cooling (Table 9.4).

In each case, the final result yields ln (n_f/n_0), the logarithm of the ratio of final to initial concentrations. If, for example, electrical heating is followed by holding at an elevated temperature and subsequent liquid-coolant heat exchange, we can write the ratio of final to initial viable-cell concentrations in the form

$$\ln \frac{n_f}{n_0} = \ln \left[\frac{n_f}{n_0(\text{coolant})} \frac{n_f(\text{holding})}{n_0(\text{holding})} \frac{n_f(\text{electrical})}{n_0(\text{electrical})} \right] \tag{9.99}$$

since $n_0(\text{coolant}) = n_f(\text{holding})$ and $n_0(\text{holding}) = n_f(\text{electrical})$. Rewriting Eq. (9.99) as

$$\ln \frac{n_f}{n_0} = \ln \frac{n_f(c)}{n_0(c)} + \ln \frac{n_f(h)}{n_0(h)} + \ln \frac{n_f(e)}{n_0(e)} \tag{9.100}$$

we see that the overall result ln (n_f/n_0) is obtained by adding the three appropriate individual solutions above, each evaluated for the particular time interval in that mode of operation.

Another situation which may be usefully examined analytically is thermal sterilization of solids or stagnant fluids. Such processes are important from several perspectives. One significant application is destruction of toxic organisms in sealed food containers. Also, it is necessary to minimize all viable microorganisms in the closed container which could decompose or otherwise spoil the product. Solid particles or microbial aggregates are often found suspended in liquids to be sterilized. We should recognize that these forms tend to protect organisms in their interior from thermal destruction and that more extensive heating is therefore required when such solids are present.

Analysis of both of these processes, at least for simple container and particulate geometries, can be reduced to a two-step recipe analogous to the procedure followed above. First, we solve a transient heat-conduction problem to determine the temperature in the solid as a function of position and time. Assuming constant thermal conductivity k, this problem has the general form

$$\rho C_p \frac{\partial T}{\partial t} = k \nabla^2 T \tag{9.101}$$

where ∇^2 is the Laplacian operator and ρC_p is as defined in Table 9.4. In addition to the partial differential equation (9.101), T is specified as a function of position within the solid at time zero, and the temperature at the external solid surface is known as a function of time for $t > 0$.

Considering a spherical solid, for example, the following equations must be solved to determine the interior temperatures for $t > 0$:

$$\frac{\partial T}{\partial t} = \frac{k}{\rho C_p} \frac{1}{r^2} \frac{\partial}{\partial r}\left(r^2 \frac{\partial T}{\partial r}\right) \tag{9.102}$$

$$T(r, 0) = f(r) \qquad 0 \le r \le R \tag{9.103}$$

$$T(R, t) = g(t) \qquad t > 0 \tag{9.104}$$

where r is distance from the sphere's center and f and g are prescribed functions. If the sphere is initially at a uniform temperature T_0, and if the surface temperature $T(t, R)$ is maintained at a constant value T_1 for $t > 0$, this problem can be solved by separation of variables to obtain

$$T(r, t) = T_1 + \frac{2R(T_1 - T_0)}{\pi r} \sum_{n=1}^{\infty} \frac{(-1)^n}{n} \sin \frac{n\pi r}{R} \exp\left(-\frac{k}{\rho C_p} \frac{n^2\pi^2 t}{R^2}\right) \tag{9.105}$$

The temperature at the center of the sphere ($r = 0$) can be deduced from Eq. (9.105) by taking the limit $r \to 0$. This result is

$$T(0, t) = T_1 + 2(T_1 - T_0) \sum_{n=1}^{\infty} (-1)^n \exp\left(-\frac{k}{\rho C_p} \frac{n^2\pi^2 t}{R^2}\right) \tag{9.106}$$

Figure 9.23 shows temperature distributions computed from these formula for a variety of elapsed times. In the context of sterilization, it is critical to note that there is a time lag between the imposition of a high temperature at the

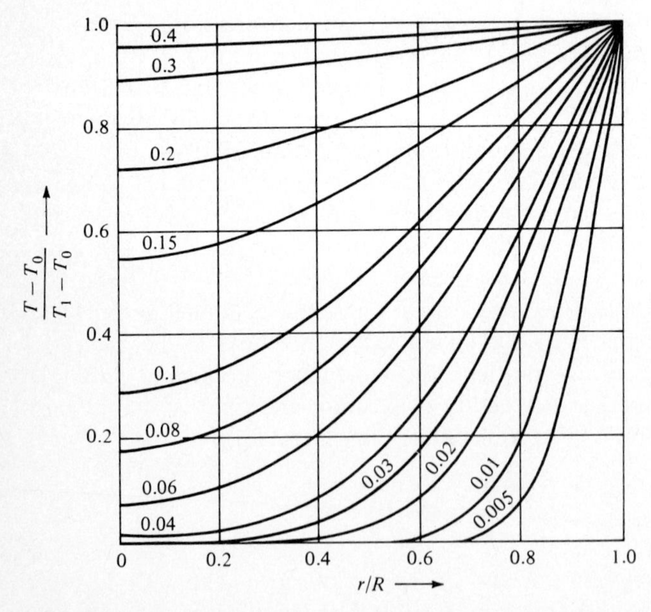

Figure 9.23 Temperature profiles within a sphere of radius R as a function of dimensionless time $kt/\rho C_p R^2$ (given as a parameter on the curves). Initially the sphere temperature is T_0 throughout, and the outer-surface temperature for $t > 0$ *is* T_1.

surface and achievement of similar temperatures everywhere within the solid. Consequently, the kill of organisms at the sphere's center will generally be less complete than near the surface. This observation remains qualitatively correct for many other geometries and initial and boundary conditions. While we cannot delve further here into the details of transient heat conduction under various conditions, a wealth of additional information, theory, and analytical solutions is available in Carslaw and Jaeger [29].

Once the time-temperature-position relationship has been determined, the resulting destruction of microbes and spores can be calculated. Two different design approaches have been used in the food-processing industry. In the first, the organism concentration at the center is considered. Since the center heats most slowly, presumably if the center is adequately treated, the remainder of the solid is sufficiently sterilized. To compute the surviving organisms at the center we need only insert the temperature-time function evaluated at that position into Eq. (9.93) and compute the integral. For this step, numerical or graphical means are often necessary since $T(t)$ is usually a rather complex expression, such as Eq. (9.106).

Another approach is to conduct the previous calculations many times in order to determine survivor concentrations at each point within the solid. Then by integrating these concentrations over the volume, the number of survivors or probability of survivors can be evaluated. Obviously, this calculation is somewhat tedious although it is straightforward in principle. Consequently, several shortcut design procedures based on this whole-container philosophy have been developed for use in the food industry. Since their explanation requires substantial additional vocabulary and definitions, we refer the interested reader to Charm [30] for a thorough discussion.

While batch sterilization enjoys the advantages of being a relatively simple process, it suffers from several drawbacks. One is the time required for heating and cooling. Related to this disadvantage is another: the extent of thermal damage to desirable components. Many vitamins are destroyed by heating, and proteins can be denatured at elevated temperatures. While the destruction of these components often follows the same kinetics as organism death [Eq. (9.92)], it is important to recognize that the activation energies for these undesirable side reactions are typically much smaller than for the sterilization "reaction." The values listed in Table 9.5, for example, are less than the 50 to 100 kcal/g mol magnitudes which usually characterize spore and cell destruction.

Since the desired reaction here has a higher activation energy than the side reaction, increasing the temperature has the beneficial effect of increasing the ratio of desired rate to undesired rate. This means that if unfavorable Browning reactions or damage to susceptible compounds are to be avoided, the sterilization process should operate at the highest feasible temperature and for the shortest time (HTST = high temperature, short time) which provides the necessary organism death. The slow heating and cooling portions of batch sterilization do not achieve these objectives. Continuous sterilization, considered next, is much better suited for achieving HTST conditions.

Table 9.5 Approximate activation energies for some undesirable side reactions resulting from heat treatment

Reaction	Activation energy E, kcal/g mol
Browning (or Maillard) reaction between proteins and carbohydrates	31.2
Destruction of vitamin B_1	21.0
Destruction of riboflavin (B_2)	23.6
Denaturation of peroxidase	23.6

9.4.2 Continuous Sterilization

Two basic types of continuous-sterilizer designs are shown schematically in Fig. 9.24. In Fig. 9.24*a*, direct heating is provided by steam injection, with the heated fluid then passing through a holding section before cooling by expansion. The system in Fig. 9.24*b* features indirect heating using a plate heat exchanger. A number of variations in each type are possible with most featuring high ratios of heat-exchange surface to process volume. Consequently, continuous sterilizers provide relatively rapid heating and cooling so that HTST conditions can be realized.

Two different design approaches are available if we assume that the temperature in the continuous sterilizer is approximately uniform. This may be valid at least in the holding sections. In the first method we apply the dispersion model

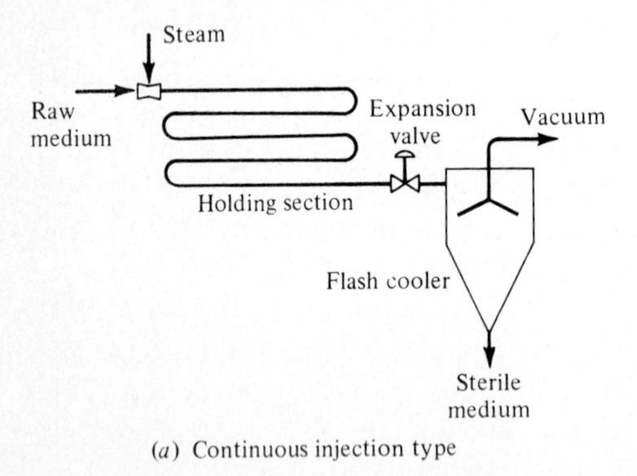

(*a*) Continuous injection type

(*b*) Continuous plate exchanger type

Figure 9.24 Two different continuous sterilizer designs: (*a*) direct steam injection (*b*) plate heat exchanger. *(Reprinted from S. Aiba, A. E. Humphrey and N. F. Millis, "Biochemical Engineering," 2d ed., p. 257, University of Tokyo Press, Tokyo, 1973.)*

described earlier. After c in Eqs. (9.81) to (9.83) has been identified with n, we take $r_{fn} = -k_d n$; then the organism concentration throughout from the tubular reactor with dispersion can be determined analytically. The resulting value of n at the sterilizer effluent is given by

$$\frac{n(L)}{n_0} = \frac{4y \exp[\text{Pe}/2]}{(1+y)^2 \exp[(\text{Pe})(y)/2] - (1-y)^2 \exp[-(\text{Pe})(y)/2]} \tag{9.107}$$

with

$$y = \left(1 + \frac{4\,\text{Da}}{\text{Pe}}\right)^{1/2} \tag{9.108}$$

where here the Damköhler number Da is defined by

$$\text{Da} = \frac{k_d L}{u} \tag{9.109}$$

For small deviations from plug flow (Pe^{-1} small), Eq. (9.107) reduces to the simpler form

$$\frac{n(L)}{n_0} = \exp\left(-\text{Da} + \frac{\text{Da}^2}{\text{Pe}}\right) \tag{9.110}$$

These solutions are conveniently displayed as a plot of remaining viable fraction $n(L)/n_0$ vs. the dimensionless group Da for various values of the Peclet number (Fig. 9.25). From this plot we can see that, as $\text{Pe} \to \infty$ so that ideal plug flow is approximated, the desired degree of medium sterility can be achieved with the shortest possible sterilizer. Consequently, the flow system should be designed to keep dispersion at a minimum.

If flow through the isothermal continuous sterilizer cannot be well represented with the dispersion model, we can make use of general RTD theory. It is difficult to think of a better example of a completely segregated reactor than a suspension of cells or spores subjected to heat treatment. Consequently, provided back diffusion of organisms into the sterilizer feed line is negligible, we can compute the effluent surviving-organism concentration n using Eq. (9.62). Rewriting this equation in terms of organism concentration, we have

$$n = \int_0^\infty n_b(t)\mathscr{E}(t)\,dt \tag{9.111}$$

where $\mathscr{E}(t)$ is the RTD of the continuous sterilizer. Recalling that $n_b(t)$ is the organism concentration at time t in a batch sterilizer with $n_b(0) = n_o$, we can use Eq. (7.125) in Eq. (9.111) to obtain

$$\frac{n}{n_0} = \int_0^\infty \mathscr{E}(t)e^{-k_d t}\,dt \tag{9.112}$$

In some instances evaluation of the right-hand side of Eq. (9.112) is facilitated by noting that it is formally identical to the Laplace transform of $\mathscr{E}$ with the usual Laplace transform parameter s replaced by k_d.

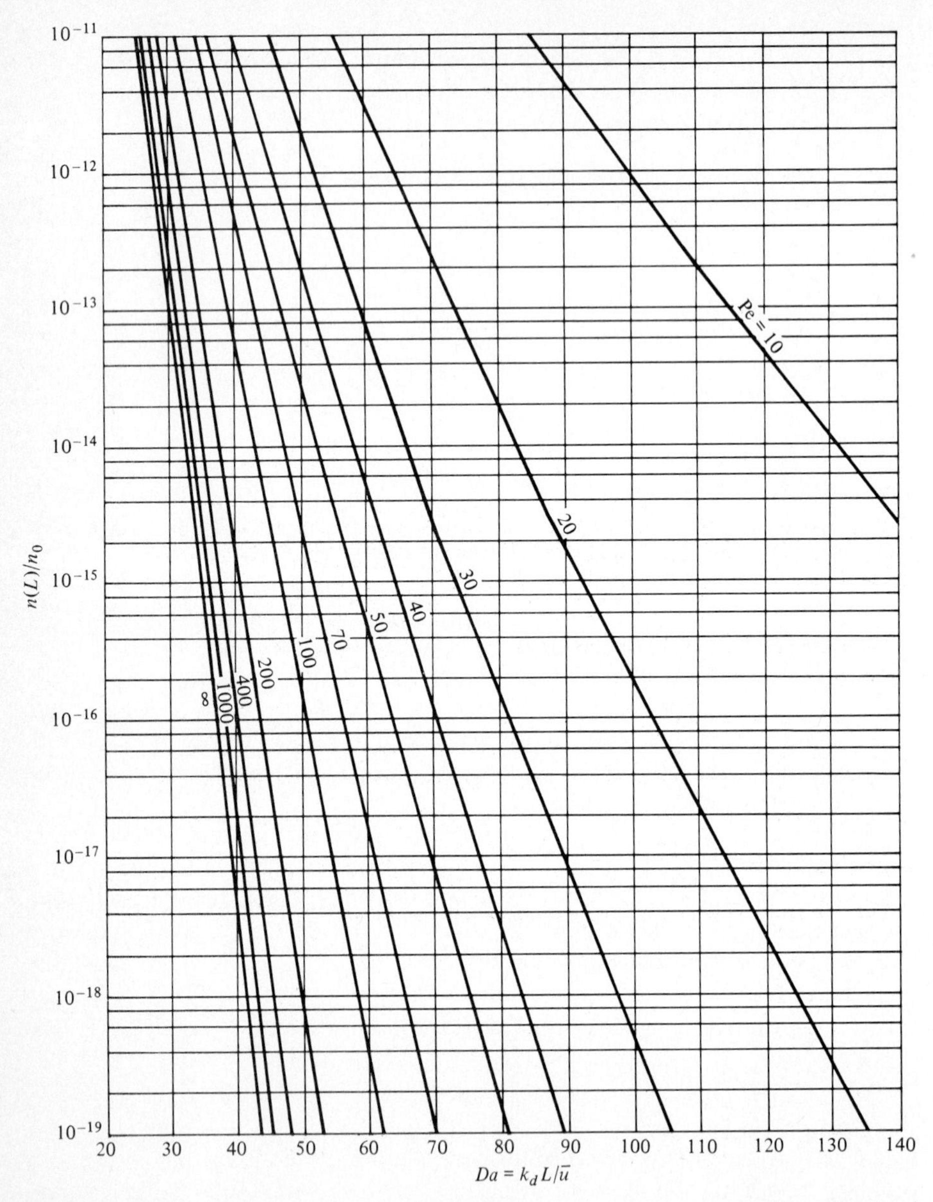

Figure 9.25 Effect of axial dispersion on organism destruction in a continuous sterilizer. *(Reprinted from S. Aiba, A. E. Humphrey, and N. F. Millis, "Biochemical Engineering, 2d ed., p. 263, University of Tokyo Press, Tokyo, 1973.)*

Before leaving the topic of continuous sterilization, we should mention its several advantages in addition to the HTST feature already discussed. First, continuous processing typically requires less labor than batch operations. Also, continuous treatment units provide a more uniform, reproducible effluent than batch sterilization. This is extremely important in the food industry, where small changes in treatment can change the taste of the product.

When practiced in the fermentation industry, batch sterilization is usually done in the fermentor itself (*in situ*). Heating and cooling rates here depend upon the surface-to-volume ratio of the fermentor, and this ratio in turn usually changes during scale-up. The sensitivity of sterilization effects on organisms and medium components to equipment size can be reduced by using continuous sterilization. If continuous sterilization is used, it is not necessary to design the fermentor also to fulfill the requirements of a good batch sterilizer.

Of course there are some drawbacks to continuous sterilization: direct steam heating can add excess water to the medium, and heat exchangers used for indirect heating or cooling can be fouled by suspended solids. Also, continuous sterilization tends to cause foaming of fermentation media. Additional details on the operation trade-offs of sterilizers are available in Ref. 33 of Chap. 7.

9.5 IMMOBILIZED BIOCATALYSTS

Performance of a bioreactor depends directly and critically upon the properties of the biocatalysts employed. Earlier, we discussed different ways in which enzymes can be applied to provide useful catalytic functions. Also, we have already examined genetic modification of cells in order to improve their productivities for desired compounds. Now, before investigating additional reactor types, we must introduce another form of biocatalyst application.

Our main topic in this section is immobilized cell catalysis. The central feature of immobilized cell systems is the use of some confining or binding structure to constrain the cells in a particular region of the reactor. As we have already seen for immobilized enzymes, such immobilization methods for cells may involve entrapment in or attachment to small particles or might be achieved by use of larger-scale barriers to cell transport.

Before considering methods to achieve cell immobilization and some of the experiments conducted to date with immobilized cell catalysts, we will review some of the motivations for cell immobilization. One benefit of cell immobilization is attainment of higher cell densities than in suspended cell systems. There are several other additional reasons for using immobilized cells in batch bioreactors. Some mammalian cell lines grow only if attached to a surface. For these types of cells, immobilization is a rule rather than an option. Second, immobilized cells can be used as the basis for specific electrodes to measure concentrations of nutrients, metabolites, drugs, and toxic chemicals in bioreactors and in other process and clinical contexts. In addition, immobilization can be used to control cell morphology and broth rheology. By confining microbial growth to the interstices and surfaces of support particles, the rheological and mass-transfer properties of the broth are relatively well defined and do not change as much during the batch process as occurs with some organisms during batch growth. Cell immobilization offers continuous processing without organism washout or genetic drift and also allows in some cases continuous separation of product and removal of reaction inhibitors. As with other immobilized catalysts, the costs of immobilization may be appreciable and may more than offset these cited advantages.

In the case of continuous-flow bioreactors, the motivation for using immobilized cells is to extend the time which the catalytic functions of the cells can be employed to accomplish a desired chemical reaction or chemical reaction sequence. In contrast to immobilized enzyme catalysis, immobilized cell catalysts can be used to achieve a broad spectrum of catalytic functions varying substantially in complexity. As sketched in the diagram in Fig. 9.26, immobilized cell catalytic activities can be classified based upon the level of metabolism that remains active and which is applied to achieve process objectives. When cells are used in a suspended state and leave the reactor with the effluent stream, some ongoing growth of cells is required in order to prevent washout. Indeed, the increase in washout dilution rate achieved with cell concentration and recycle discussed earlier in this chapter represents a macroscopic form of partial cell immobilization. By this means, the throughput of the process can be increased in a way that is partially decoupled from the rate of growth of the organisms. If the cells are completely retained within the reactor, however, the complete spectrum of catalytic complexity diagrammed in Fig. 9.26 may possibly be available since there is no requirement for growth. In fact, growth is often undesired in some of

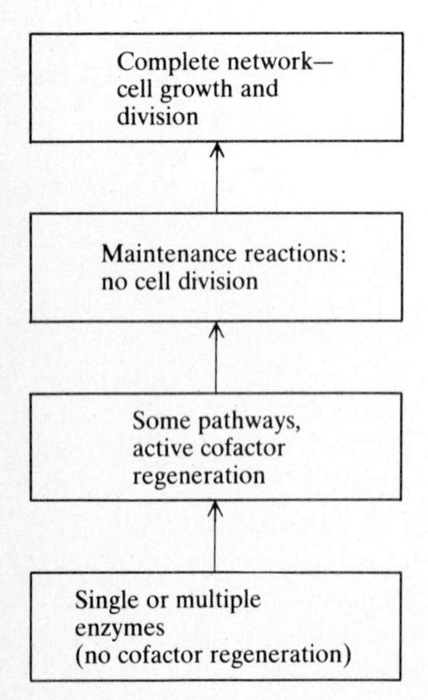

Figure 9.26 Levels of reaction network complexity encountered in immobilized cell catalysis.

these cases since growth can lead to reduced yields and since accumulated cells can cause mechanical and flow disruptions in the reactor.

The simplest level of catalysis provided by immobilized cells is activity of one or several enzymes but without any involvement of cofactors. Here, the immobilized cell essentially serves as an immobilized enzyme catalyst, with the entire cell used in order to minimize treatment and processing costs in formulating the immobilized enzyme catalyst. However, possible difficulties with extra mass-transfer resistance provided by the cell envelope and with degradation of the desired enzyme by intracellular proteases must be considered in such applications. On the other hand, some enzymes may be stabilized by remaining in their native cellular environment.

Next, a nongrowing immobilized cell system can provide catalysis through a multistep pathway involving cofactor utilization and regeneration. In principle, so long as all cofactors, enzymes, substrates, and regenerating chemicals are present, there is no reason why associated synthetic reactions and cell growth need to occur. If the desired cofactor-using pathway already exists in an organism, it may be advantageous to consider using the existing functional pathway rather than attempt to engineer a different type of cofactor utilization and regeneration process.

In the next level of catalytic complexity considered, biosynthesis and maintenance reactions occur, but there is insignificant cell growth and cell division. Finally, the most complicated situation is the one in which the full metabolic network is active and the cells are growing and dividing. This case arises, for example, in the immobilized cell batch applications mentioned earlier. It might also be employed in a continuous reactor context as a means of regenerating the biocatalyst, since some organisms do not respond well to a nongrowth environment and lose activity rapidly under such conditions. As we shall see later in examples, alternating reaction conditions between one of the less complex catalytic levels and full growth conditions is a useful strategy for obtaining extended catalyst service.

There are two different advantages in using a catalyst which carries out the desired functions without a large number of additional reactions. First is the consideration of yield. Clearly, if a substrate can be converted stoichiometrically to a certain product by action of a specific microbial pathway without concomitant growth of cells or production of other end products, yields will be increased. Another potential advantage is higher overall reaction rate. The characteristic times to achieve significant conversion of a substrate increases dramatically as one progresses from the simplest, single enzyme level in Fig. 9.26 to the most complex, full-growth state. In the former situation, rates are characterized by rates of single enzyme-catalyzed steps, with time scales on the order of minutes, typically. On the other hand, cell doubling times, which characterize the complete network reaction time, are in the range of hours to days for all but the most rapidly growing bacteria. Thus, as one moves up the ladder of metabolic complexity displayed in Fig. 9.26, the residence time required in the reactor to achieve substantial substrate conversion will also in general tend to increase significantly. Therefore, it is a substantial potential advantage to be able to use the enzymes and cofactors needed to accomplish a particular conversion and to design the reactor residence time accordingly.

With this background framework in mind, we turn next to formulation and characterization of immobilized cell catalysts, after which some examples illustrating the spectrum of catalytic possibilities with these systems will be examined.

9.5.1 Formulation and Characterization of Immobilized Cell Biocatalysts

In this section, we review methods which have been applied to immobilize cells and the characteristics of the resulting catalysts that are important in different process contexts. Entrapment in polymeric networks is the most commonly applied method for cell immobilization. Table 9.6 lists different mechanisms which have been used to form the entrapping network. By far the most widely used cell entrapment method involves ionic cross-links in a layer or bead of alginate, a natural polysaccharide material. The gentleness of this gellation procedure, in contrast to chemical polymerizations, results in much higher initial viability of the immobilized cells. Fig. 9.27 illustrates schematically one protocol for preparation of spherical calcium-alginate beads containing immobilized cells.

When selecting a network entrapping method and the particular conditions applied to formulate the network, the resulting properties of the immobilized cell catalyst must be considered. Some of the major parameters of process interest are listed in Table 9.7. Both the chemical and mechanical characteristics of the network can influence its permeability to substrates, inhibitors, products, and other medium components. The mechanical properties of the network may constrain the type of catalyst particle geometries that are possible. Compressibility and other attributes of mechanical strength are much more important in large-scale practice than in laboratory study. Many of the polysaccharide beads commonly applied in laboratory research because of their biocompatibility are not well suited for large-scale application because of their high compressibility which limits the utility of these materials to shallow columns. Similarly, particles which are impact sensitive are not suitable for use in contacting schemes such as agitated slurry reactors in which the catalyst beads must withstand exposure to collisions without substantial attrition or breakage.

Table 9.6 Polymeric networks used for cell immobilization

Network formation	Cross-links	Examples
Precipitation	Nonspecific	Collagen, polystyrene, carrageenan
Ion-exchange gelation	Ionic	Al-alginates
Polycondensation	Covalent-heteropolar	Epoxy resins
Polymerization	Covalent-homeopolar	PAAm, PMAAm

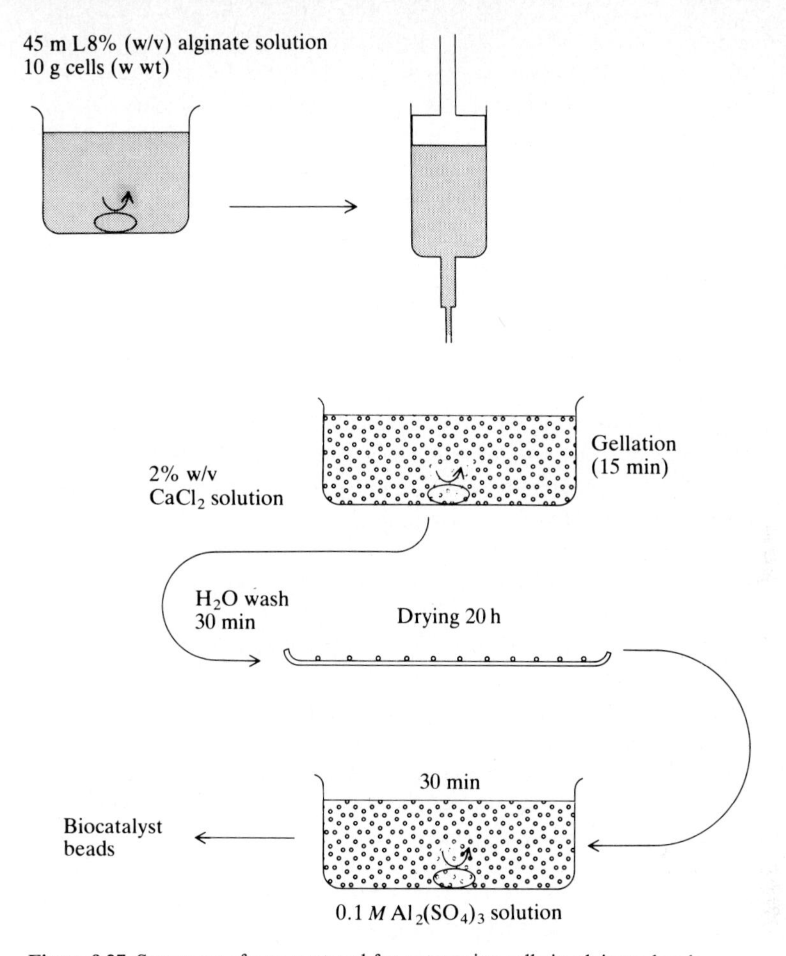

Figure 9.27 Summary of one protocol for entrapping cells in alginate beads.

Table 9.7 Important parameters of polymer networks used for cell entrapment

Mechanical	Chemical
Permeability	
Geometry	Toxicity
Compressibility, strength	Hydrophobic/philic
Shear sensitivity	Ionic composition

The chemical properties listed may influence catalyst activity directly through action on the cell's metabolic controls and indirectly by influence on the local microenvironment of the cells. For example, if the polymeric network contains toxic chemicals, degradations in cellular activities can be expected. The charge, ionic composition, and hydrophobicity of the network may all influence the partition coefficients of reaction mixture components and alter the concentrations in contact with the cells relative to the concentrations delivered in the bulk reaction mixture solution at the exterior of the catalyst particles.

Another method of cell entrapment is entanglement in convoluted or porous structures. Compressed stainless steel screens have been used to support the growth of microorganisms. Also, mold spores inoculated into autoclaved beads of porous filter aid material have been used as an inoculum for batch fermentations. Cells have also been entrapped within the macroporous regions in asymmetric hollow fibers. In addition to these local entrapment methods, we should remember that complete or partial cell entrapment can be achieved on a macroscopic scale by use of cell separation and recycle to the reactor. Alternatively, relatively small substrates and products may be added to or removed from a cell suspension through a dialysis or other type of ultrafiltration membrane placed in the cell suspension.

Many different types of cells have been immobilized by adsorption or covalent attachment to the surfaces of various solid support materials. Examples of immobilization of cells ranging from bacteria to mammalian tissue cells on natural and synthetic materials are summarized in Table 9.8. Further information on these and other immobilization methods may be found in the general immobilized cell references listed at the end of this chapter and in the specific references associated with the examples presented next.

Table 9.8 Examples of cells immobilized by attachment to surfaces

Cells	Support
Adsorption adhesion	
E. coli, Cl. acetobutylicum	Ion exchange resin
Asp. oryzae	Modified cellulose
Streptomyces	Modified Sephadex
Lactobacilli, yeast	Gelatin
S. carlsbergensis	PVC
Pseudomonas sp.	Anthracite
Tissue cells	Polysaccharide
Covalent bonding	
B. subtilis	Agarose and carbodiimide
E. coli	Ti(IV) oxide

9.5.2 Applications of Immobilized Cell Biocatalysts

In this summary of examples illustrating the different levels of immobilized cell catalysis, we shall proceed from the simplest to the most complicated situation. By immobilizing whole cells which have been suitably permeabilized [by treatment with lytic enzymes, by exposure to alcohols or dimethylsulfoxide (DMSO)], individual enzymes in approximately their natural environment may be immobilized for continuous process use. Many such immobilized enzyme/immobilized cell catalysts have been prepared and some are now in commercial use. This strategy for enzyme immobilization has the advantage of being relatively simple compared to isolation of enzyme from the cells and subsequent immobilization, and, perhaps of greater practical significance, this method may enhance immobilized enzyme stability.

Shown in Fig. 9.28 is experimental data on deactivation of aspartase and fumarase enzyme activity in immobilized *E coli* cells and in suspended intact cells. For both enzymes, loss of activity is greatly retarded in the immobilized cell formulations. There is a possibility that some of this apparent stability may be due to mass-transfer limitations as was discussed in Sec. 4.4; that is, if the immobilized enzyme/cell catalyst is operating under severely diffusion-limited kinetics, observed deactivation will be substantially slower than actual deactivation. However, in this particular system, further experiments by Chibata and colleagues point to a true stabilizing effect of enzyme retention in the native cellular environment. Intact *E. coli* cells were sonicated and the soluble and precipitate fractions were subsequently separated and immobilized. The aspartase activity in the immobilized soluble fraction decayed much more rapidly than the aspartase activity in the immobilized precipitate fraction. Furthermore, intact cells treated

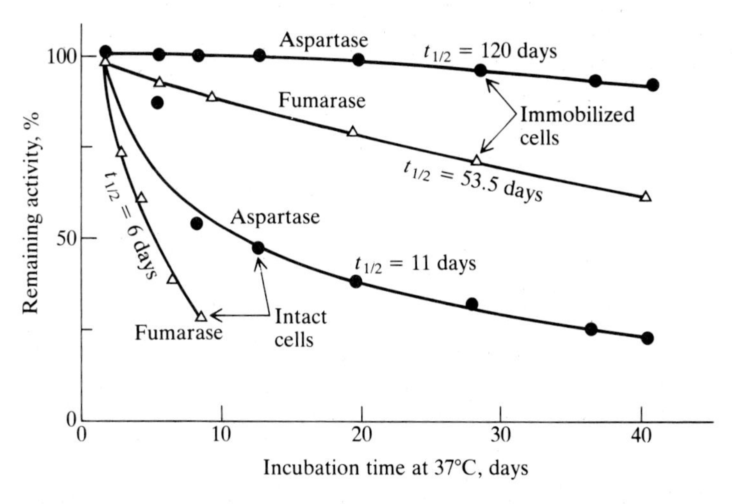

Figure 9.28 Comparison of deactivation of aspartase and fumarase in suspended cells and in immobilized cells *(Reprinted by permission from I. Chibata (ed.), "Immobilized Enzymes," p. 140, Kodansha Ltd., Tokyo, 1978.)*

with solubilizing agents for membrane-bound enzymes (deoxycholate or Triton X-100) yielded precipitate and soluble fractions which both lost activity rapidly after immobilization. These experiments suggest that aspartase is stabilized by binding or association with cellular membranes or granules. However, it should be noted that the opposite effect has been observed in whole-cell enzyme immobilization. Glucose isomerase in permeabilized immobilized cells of some organisms loses activity more rapidly than the same enzyme isolated from the cells. This phenomenon has been attributed to proteolytic attack on the enzyme by intracellular enzymes.

The subject of intracellular hydrolytic activities is very important across the entire spectrum of immobilized cell applications. As discussed in Chap. 6, cells require for efficient function certain intracellular hydrolytic activities. Loss of enzyme activity by intracellular hydrolysis as well as loss of cellular membrane and transport integrity by other degradative processes clearly play key roles in determining the useful lifetime of an immobilized cell catalyst. Greater understanding of these intracellular degradative processes, along with other types of deactivation events due to protein unfolding and membrane damage due to chemical attack from medium components, is required in order to optimize the cells genetically and to formulate the best immobilized cell preparation and corresponding reactor configuration and operating conditions. Indeed, we should note here parenthetically that decades have been invested in developing useful strains within the context of suspended cell cultivation and that long-term efforts may be needed to achieve the genetic modifications required to yield especially useful organisms for immobilized cell applications.

At the next level of catalytic complexity is multistep bioconversion involving cofactor use and regeneration. The most widely studied systems of this type involve transformation of glucose to economically important end-products in anaerobic environments. Glucose conversion to ethanol by immobilized yeast (*Saccharomyces cerevisiae*) or bacteria (primarily *Zymomonas mobilis*) have been studied extensively, and at least one process based upon immobilized cell biocatalysis has been demonstrated at pilot scale [34].

Studies of glucose to ethanol conversion using immobilized cells originated with investigations of continuous brewing, in which highly flocculent yeast strains permit retention of cells which could be effectively immobilized in a fluidized-bed reactor (Sec. 9.6.4). Subsequently, yeast immobilized by adsorption on a gelatin film, and by entrapment in κ-carrageenan, alginate, or polyacrylamide have been reported. Experiments with a column of *Saccharomyces carlsbergensis* cells immobilized in κ-carrageenan showed a startup transient period of around seven days after which the concentration of viable cells in the gel was constant as were the effluent glucose and ethanol concentrations. Subsequent process improvements have lead to increased product concentrations exceeding 100 g ethanol/L.

The Kyowa Hakko Kogyo Company of Japan has announced a pilot-scale process for ethanol production using immobilized yeast cells in fluidized-bed reactors (Sec. 9.6.4). The flow sheet for this process is given in Fig. 9.29. Results

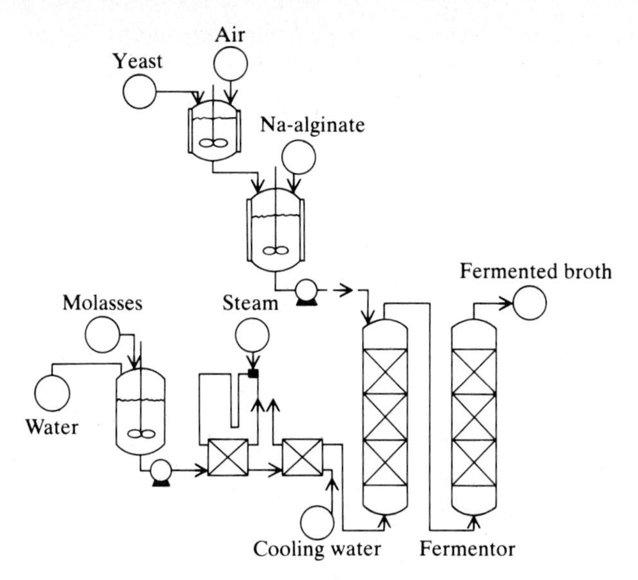

Figure 9.29 Process schematic for ethanol production using immobilized yeast [34].

of three months of initial trials include production of 8.5% v/v of ethanol from diluted cane molasses feed (14% glucose concentration) at 30°C at a space velocity (volume feed · volume $gel^{-1} \cdot h^{-1}$) of 0.4 to 0.5 h^{-1}. Sugar-to-ethanol yield was 95% of the theoretical maximum and the ethanol productivity based on total column volume was ca 20 g L^{-1} h^{-1}.

A more complicated situation arises in conversion of glucose to acetone, butanol and ethanol using the bacterium *Clostridium acetobutylicum* and related strains. This organism possesses a complex regulatory system for its catabolic pathways, details of which remain incompletely understood in spite of long experience with this organism in traditional suspended culture fermentations. Under certain environmental and cell states, organic acids are the primary end-products of glucose metabolism, while in other situations the desired organic solvents are produced. Consequently, this is a more complicated and difficult system than the glucose to ethanol conversion accomplished by yeast, since here we must consider how to manipulate the final product selectivity in the desired fashion.

The feasibility of continuous, long-term bioconversion of glucose to organic solvents has been demonstrated in research by Förberg and coworkers [35]. In this study, *Cl. acetobutylicum* entrapped in calcium alginate gel was fed with a glucose medium which lacked essential growth factors. Intermittently, at intervals starting at two to four hours and subsequently increased to eight hours, the culture was fed a complete growth medium in order to regenerate deactivated catalyst. Using this operating strategy, which combines the multistep bioconversion, nongrowth operation with brief intervals of full cell growth, stable operation of the system for an interval of eight weeks was reported. Overall biomass production was greatly suppressed and the yield of butanol per unit amount of

Table 9.9 Comparison of butanol formation yields and productivities for immobilized† and suspended‡ cells of *Clostridium acetobutylicum*

	$Y_{P/X}$ $\left(\frac{\text{g butanol}}{\text{g cells (dry wt)}}\right)$	$Y_{P/S}$ $\left(\frac{\text{g butanol}}{\text{g glucose}}\right)$	r_{fp} $\left(\frac{\text{g butanol}}{\text{L}\cdot\text{day}}\right)$
Growing immobilized cells (full medium)	1.3–1.9	0.11	
Immobilized cells (intermittent dosing of full nutrient medium)	50	0.20	16.8
Batch culture, suspended cells	3.6	0.19	6.9

† C. Förberg, S.-O. Enfors, and L. Häggström, *Eur. Appl. Microbiol. Biotech.* **17**: 143, 1983.
‡ A. R. Moreira, D. C. Ulmer, and J. C. Linden, *Biotech. Bioeng. Symp.* no. 11: 567, 1981.

glucose consumed was increased significantly relative to a comparison case in which immobilized cells were exposed to full growth medium. Listed in Table 9.9 are representative parameters of the immobilized cell experiments with full growth medium, for immobilized cells with intermittent dosing of growth medium alternating with only conversion medium, and results from a conventional batch fermentation using the suspended bacterium.

While further research and development is needed to improve the immobilized cell system, this example clearly indicates the opportunity for altering kinetic and stoichiometric behavior by appropriate choice of cell formulation and medium manipulation, providing additional options for engineering manipulation and optimization of a biocatalytic process.

These fermentation examples introduce the challenging problem of gas transport in immobilized cell catalysts and associated reactors. Here we shall consider only the local problem of transport in the catalyst. Suppose that the bioconversion can be described stoichiometrically as a single chemical reaction and that the transport of substrate S and product P in the catalyst can be characterized by effective diffusivities D_s and D_p, respectively. It is easy to show using the governing material balances and boundary conditions on the catalyst that the concentration of product c_p and concentration of substrate c_s within the catalyst formulation are related by the following equation, in which α denotes the number of moles of product formed for each mole of substrate consumed

$$c_p = c_{p0} + \frac{\alpha D_s}{D_p}(c_{s0} - c_s) \tag{9.113}$$

Here subscripts 0 denote conditions at the external boundary of the catalyst. Several useful observations can be made simply by setting c_s equal to zero and calculating thereby, based only on stoichiometric and transport considerations, the maximum level of product concentration which can be obtained in such a catalyst.

This simple equation has important ramifications in cases of gas consumption and production. First, supposing that the substrate is a sparingly soluble gas such as oxygen, we see from Eq. (9.113) that only a small increase in product concentration can be expected in reaction in a single catalyst particle. Considering now the case of formation of a moderately soluble gas such as CO_2, it is evident that, using a highly soluble substrate such as glucose, a concentration of CO_2 may be obtained within the catalyst which exceeds the saturation concentration, giving rise to nucleation and growth of gas bubbles within the catalyst. This phenomenon has been reported to cause severe mechanical difficulties in immobilized cell reactors, resulting in bursting of networks containing entrapped CO_2-producing cells and of cartridges containing yeast cells immobilized in hollow fibers.

Accordingly, we have problems with limitations on product formation for sparingly soluble substrates and with removal of sparingly soluble products. These concerns dictate use of small thicknesses of catalyst in order to avoid ineffective use of interior catalyst in the case of a sparingly soluble gas and in order to avoid excessive buildup of sparingly soluble products in that situation. This short catalyst thickness can be achieved either by use of small particles or by formulation of the catalyst so as to contain supported cells or entrapped cells only in a thin outer film on the outer surface of a larger particle or object which facilitates mechanical retention in the process. Addressing requirements for supply and removal of sparingly soluble compounds is a major consideration also at the reactor design level.

Interestingly, experimental studies of ethanol production from glucose using immobilized yeast have indicated different overall conversion rates in immobilized cells compared to suspended cells incubated under similar conditions. In some of these investigations, diffusion limitations and their effects are possibly implicated, but enhancements in specific rates of ethanol production of the order of 30–50 percent have been observed in carefully controlled comparison experiments under identical reaction conditions and with no mass-transfer limitations. Alteration in local water activity by the presence of the supporting matrix or surface has been hypothesized as one cause of such phenomena. Other possibilities include altered chemical microenvironments due to the support as mentioned earlier.

However, it would not be surprising to encounter substantial metabolic adjustments to the unusual environments which cells encounter in the immobilized state. Binding or adherence to a solid surface may be recognized by specific receptors on the cell which alter metabolic function. Similarly, interference with normal morphological development by multipoint binding to a surface, by fibers of the entrapment matrix, or by cell–cell contact may elicit metabolic responses not evident in relatively unconcentrated suspended cell environments. Furthermore, certain extracellular products will be present in such dense immobilized matrices at much higher concentrations than in suspended cell systems. Certainly, we might expect that microbial cells ordinarily functioning in a relatively independent state may exhibit altered catalytic properties when grown in a more dense, tissuelike form in an immobilized cell preparation. Therefore, it is necessary to characterize carefully the intrinsic kinetics of immobilized cell catalysts

Table 9.10 Examples of biosynthesis processes conducted by immobilized cells

Substrates	Immobilized cells	Product	Comments
Glucose, inorganic ammonia, metal ions	*Corynebacterium glutamicum*[1]	Glutamic acid	15 g product/L in 144 h
Pantothenic acid, cysteine, ATP, $MgSO_4$	*Brevibacterium ammoniagenes*[1]	Coenzyme A	500 μg product/mL
Glucose medium	*Penicillium chrysogenum*[1]	Penicillin	1.5 units $mL^{-1}h^{-1}$
1% peptone medium	*Bacillus* sp (KY 4515)†	Bacitracin	16–19 units mL^{-1}
1% meat extract, 0.05% yeast extract medium	*Bacillus subtilis* FERM-P No. 2040[1]	α-Amylase	15,000 units/mL
LB broth	*Escherichia coli* C600 (pBR322)‡	β-Lactamase	1×10^{-11} units/cell/h 8 units/mL/h

† Entrapped in polyacrylamide gel.
‡ Entrapped in hollow-fiber membranes.

and to anticipate possible kinetic alterations due to immobilization. Kinetics based upon suspended cell studies under identical conditions may not be applicable. Even more complicated cellular adjustments in terms of morphology and biochemical activities can be expected as we consider more complicated biocatalytic functions.

In view of the complexities and early stage of development of immobilized cell catalysis for multistep conversions, it is not surprising that relatively little has been done so far on use of immobilized cells to achieve biosynthesis of metabolites and biological polymers. Features of some of the experimental studies of biosynthesis by immobilized cells are indicated in Table 9.10. In some cases, there was little cell growth, while in other cases, intermittent full growth medium feeding was employed to extend useful activity of the cells. Full cell growth was allowed in some of these studies. Of course, if the cells are growing at some finite rate, either washout or lysis of cells must occur at a rate which balances the growth rate in order to maintain a steady-state level of functioning cells. In the case of animal cells cultivated on solid surfaces, regulatory mechanisms built into the cells arrest growth approximately at the stage of monolayer development and switch the metabolism of the cells from growth to product formation. There is much still to be learned to achieve the feasible ideal of immobilized cell systems for synthesis of complex products at high yields and high rates in continuous reactors.

9.6 MULTIPHASE BIOREACTORS

Under many circumstances it is approximately valid to treat bioreactors, which almost always contain multiple phases in the form of cells, low solubility substrates or products, gas bubbles, or catalyst particles, as effectively homogeneous. There are several situations, however, in which it is important to take the multiphase nature of the reactor contents into account when designing the bioreactor or analyzing its performance. Our focus in this section is on such reactors and some of the approaches which may be applied to describe them conceptually and mathematically.

9.6.1 Conversion of Heterogeneous Substrates

Bioreactors containing heterogeneous substrates are encountered in utilization of starch and cellulose particles, in conversion of steroids, and in growth on paraffinic hydrocarbons. In such cases, we must establish or make assumptions about the location of the reaction—with the major possibilities being either at the phase interface or in bulk solution using a very small amount of dissolved substrate. In some cases, both situations arise. For example, digestion of cellulose particles begins by adsorbed cellulase enzymes followed by conversion of glucose in solution by microorganisms. In reactors involving heterogeneous substrates, the interfacial area per unit reactor volume is typically a central parameter which depends on substrate pretreatment in some cases and most directly upon reactor operating conditions in others. The following example of microbial growth on an insoluble carbon source provides a specific illustration of the different types of considerations which are frequently necessary in analysis, design, and operation of such reactors.

Example 9.2: Agitated-CSTR design for a liquid-hydrocarbon fermentation† Some microorganisms, e.g., the yeast *Candida lipolytica*, will grow on dodecane and other paraffinic hydrocarbons which are practically insoluble in water. Two alternative mechanisms have been proposed for microbial growth at the hydrocarbon-aqueous-phase interface; (1) cells (characteristic diameter D_c) much smaller than dispersed hydrocarbon droplets (diameter D_h) cluster around the paraffin drops; on the other hand, (2) if $D_h \ll D_c$ we may presume that droplets adsorb onto the outer surfaces of the relatively large microorganisms.

The closing observations in Sec. 8.4 on aerated microbial hydrocarbon fermentations indicate that the adsorption and flocculation relations between air bubbles, cells, and hydrocarbon droplets change over the course of a batch fermentation. For a continuous process, however, we may expect to operate in one particular growth mode, thus a single bubble-cell-hydrocarbon droplet configuration may be predominant, rendering quantitative description easier.

Moo-Young and coworkers have considered these two situations and have proposed the following modified Monod growth-rate equations which include the effect of surface-area availability:

Case I, $D_c \ll D_h$

$$\mu = \mu_{\max} \frac{s/D_h}{K_s + s/D_h} \tag{9E2.1}$$

† Drawn from M. Moo-Young, "Microbial Reactor Design for Synthetic Protein Production," *Can. J. Chem. Eng.*, **53**: 113, 1975.

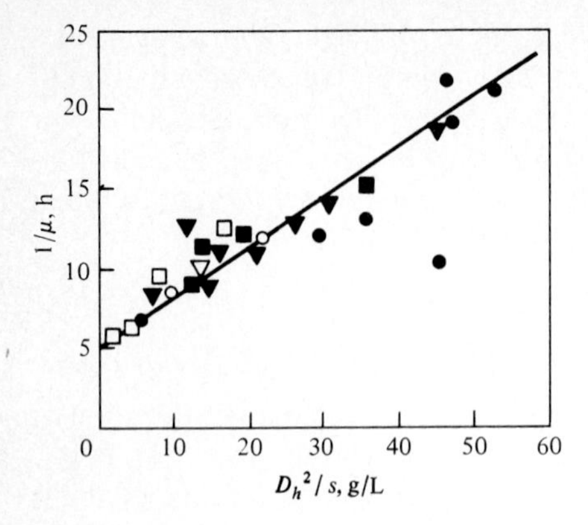

Figure 9E2.1 This double-reciprocal plot of experimental data for growth rates of *C. lipolytica* yeast on dodecane is consistent with the model assuming hydrocarbon droplet attachment to microorganisms. *(Reprinted from M. Moo-Young, "Microbial Reactor Design for Synthetic Protein Production," Can. J. Chem. Eng., vol. 53, p. 113, 1975.)*

Case II, $D_c \gg D_h$

$$\mu = \mu_{\max} \frac{s/(D_h)^2}{K_s'' + s/(D_h)^2} \tag{9E2.2}$$

where K_s' and K_s'' are modified K_s values. As the Lineweaver-Burk plot in Fig. 9E2.1 indicates, experimental data for the *C. lipolytica*-dodecane system support the second hypothesized mechanism.

Equations useful for CSTR design can be developed by relating input agitator power to dispersed hydrocarbon drop size (Chap. 8). If the diameter D_h of the substrate hydrocarbon droplets is determined by shear at the impeller-tip region, the correlations in Example 8.2 predict that

$$D_h = C\left(\frac{P}{V}\right)^{-0.4} \tag{9E2.3}$$

which was found to fit the experimental data of this study when $C = 0.023$. Substitution of this formula into Eq. (9E2.2) yields

$$\mu = \mu_{\max} \frac{s(P/V)^{0.8}}{K_s''' + s(P/V)^{0.8}} \tag{9E2.4}$$

Using this specific growth rate in the ideal CSTR model with constant yield factor gives for sterile feed

$$x = Y\left[s_0 - \frac{DK_s'''}{\mu_{\max} - D}\left(\frac{P}{V}\right)^{-0.8}\right] \tag{9E2.5}$$

and

$$D_{\text{max output}} = \mu \frac{K_s'''}{K_s''' + s_0\left(\frac{P}{V}\right)^{0.8}} \tag{9E2.6}$$

Figure 9E2.2 shows the biomass production rate Dx computed using Eq. (9E2.5). Those plots, as well as Eqs. (9E2.5) and (9E2.6), clearly show the importance of considering interactions between biological reactions and fluid mechanics in the design and analysis of microbial reactors.

In spite of the changing adsorption patterns in batch hydrocarbon fermentations referred to in Sec. 8.4 and Fig. 8.14, a model assuming that the cells of *C. lipolytica* growing on hydrocarbon droplets adsorb continuously on the hydrocarbon-droplet surface until a cell monolayer is formed gives some agreement with batch-fermentation data (See L. E. Erickson, A. E. Humphrey, and A. Prokop, "Growth Models of Cultures with Two Liquid Phases. I. Substrate Dissolved in Dispersed Phase," *Biotech. Bioeng.*, **11**: 449, 1969).

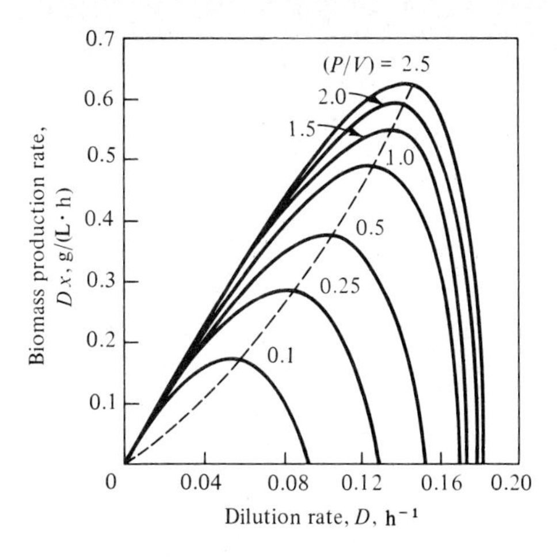

Figure 9E2.2 Biomass productivity for liquid-hydrocarbon fermentation as a function of dilution rate and agitator power input per unit volume P/V (hp/m^3).

9.6.2 Packed-Bed Reactors

Columns packed with immobilized biocatalyst particles currently enjoy several applications and additional uses are expected. In such reactors, which are called packed-bed or fixed-bed reactors, immobilized enzymes are used for glucose isomerization, for selective penicillin hydrolysis, and for selective reactive separation of racemic mixtures of amino acids. Many immobilized cell systems have also been examined in packed-bed configurations.

The simplest and often quite useful description of packed-bed reactor performance uses a plug-flow reactor model modified to account for the influence of the packed catalyst on flow and kinetics features. The superficial flow velocity through the reactor is equal to the volumetric flow of the feed divided by the void cross-sectional area which is the total cross-sectional area times the void fraction ε. The appropriate rate expression for use in the tubular reactor material balance is based upon use of effectiveness factors as described in Chap. 4. For example, considering a single reaction $S \rightarrow P$ with intrinsic rate $v = v(s, p)$ the rate of product formation per unit volume of immobilized biocatalyst pellet at a point in the reactor is:

$$v\Big|_{\text{overall/unit volume of pellet}} = \eta(s_s, p_s)v(s_s, p_s) \tag{9.114}$$

where s_s and p_s are the substrate and product concentrations at the exterior pellet surface at that position inside the reactor. In general, the effectiveness factor η, which accounts for intraparticle diffusion, and the rate expression v depend upon both s_s and p_s, as indicated.

If mass-transfer resistance between the bulk liquid phase and the pellet surface is next examined, a steady-state material balance on substrate over the pellet gives for a spherical catalyst pellet of radius R:

Rate of substrate diffusion out of bulk liquid

= rate of substrate disappearance by reaction within pellet

or

$$4\pi R^2 k_s(s - s_s) = \tfrac{4}{3}\pi R^3 \eta(s_s, p_s)v(s_s, p_s) \tag{9.115}$$

With this equation and reaction stoichiometry, s_s and p_s can be evaluated in terms of the bulk liquid concentration s. Substituting these expressions into Eq. (9.114) then gives the total rate of substrate disappearance per unit volume of catalyst pellets in terms of the bulk fluid substrate concentration.

In writing the material balance on a differential slice of the plug-flow packed-bed reactor, we must remember that bulk fluid exists only in a fraction ε of this volume and catalyst particles occupy a fraction $1 - \varepsilon$ of this volume. The substrate concentration s used in this equation is the concentration per unit fluid volume. Accordingly, the material balance on substrate becomes

$$u\frac{ds}{dz} = -\left(\frac{1-\varepsilon}{\varepsilon}\right)\eta(s_s, p_s)v(s_s, p_s) \tag{9.116}$$

where, as noted above, the quantities on the right-hand side can be evaluated in terms of s, allowing integration of Eq. (9.116) for given values of feed substrate and product concentration.

The situation is greatly simplified if intraparticle and external mass-transfer resistances are negligible, since these conditions imply $\eta \to 1$ and $s_s \to s$, respectively. In such circumstances, the governing mass balances can be integrated analytically, with results as indicated previously in Table 9.2 for plug-flow reactors.

As mentioned earlier in connection with our discussions of the dispersion model for flow reactors, flow around the particles in a packed bed and mixing in the interstitial voids of the reactor create a small amount of backmixing which may cause deviations from ideal plug-flow behavior. In these cases, the dispersion model with dispersion coefficient evaluated as discussed earlier or a tanks-in-series model may be applied. The effect of a small amount of dispersion on reactor performance relative to ideal plug-flow behavior was discussed already in connection with sterilization reactors. Other approaches to mathematical modeling of packed-bed systems with different levels of backmixing and different interactions between the fixed and flowing phases are available in the chapter references.

9.6.3 Bubble-Column Bioreactors

By bubble-column bioreactors we mean reactors with large aspect ratio (height to diameter ratio) which take the form of columns instead of more squat tanks typical of agitated vessels. Also, in such reactors, mixing is supplied entirely by forcing compressed gas into the reactor which then rises through the liquid. Reactors of this type have been used for many years in the chemical industry because of their advantages of relatively low capital cost, their simple mechanical configuration, and reduced operating costs based on lower energy requirements. While relatively unfamiliar in the biological processing industries, tower bioreactors have been used on a large scale for beer production and for vinegar manufacture. Also, related tower designs are essential elements of very large scale processes which have emerged for cultivation of microorganisms (single-cell protein or SCP) for use as animal feed.

In some cases a single column, which may contain internal plates or even agitators in some or all stages, is used. The reactors may be used in a batch mode or operated continuously with cocurrent or countercurrent flow of liquid relative to the rising gas. In several recent designs, called air lift or pressure cycle reactors, an external loop is used to provide fluid circulation. Such loops have the advantages of permitting high efficiency heat exchange, a major need for large-scale microbial cultivation on paraffinic or methanol substrates. Also, the circulation loop enhances definition of the flow and mixing properties in the vessel. Properties of bubble columns with and without external loops and single or multiple stages have been extensively investigated and described in a series of studies by Schügerl and colleagues [36, 37]. Here we shall only introduce some of the concepts which can be applied to formulate equations for designing such reactors and for analyzing experimental data obtained in such systems.

We saw in Chap. 8 that for a sufficient density of rapidly growing aerobic organisms, the overall growth rate is typically limited by the rate of oxygen transfer from the gas bubbles into the liquid phase. Analysis of this limiting rate process requires knowledge of liquid and gas mixing within the tower. Studies on air–water sparged columns without any recycle loops have shown that if (*a*) gas flow rates are large relative to the liquid and (*b*) column height L and diameter d_t are of similar magnitude, both liquid and gas phases are well mixed. Conversely, for the more typical long columns, Eq. (8.115) gives the column height $L(=z$ there) to obtain a desired amount of O_2 transfer.

For the integrated form (8.115) to be valid, it is necessary to maintain the interfacial area factor a nearly constant along the tower. This in turn requires that the gas remain in bubbling flow. Air–water experiments reveal that the gas bubbles rising through the liquid will coalesce into slugs if the gas volume fraction ε exceeds a critical value ε_{max}, which is roughly 0.3. The requirement that the gas volume fraction remain less than ε_{max} can be translated into a design specification for column diameter by noting that any point in the tower

$$F_G = u_G \varepsilon \frac{\pi d_t^2}{4} \tag{9.117}$$

where F_G and u_G are the gas volumetric flow rate and linear velocity, respectively. We may reasonably assume u_G is the terminal velocity u_t of a single gas bubble in a stagnant liquid and that F_G is roughly the same as the feed gas flow rate F_{Gf}. The latter assumption is rationalized on the grounds that O_2 consumed from the bubbles is at least partially replaced by CO_2. Under these conditions Eq. (9.117) reveals that ε is smaller than ε_{max} so long as

$$d_t \geq 2\left(\frac{F_G}{u_G \varepsilon_{max}}\right)^{1/2} \tag{9.118}$$

which can be used to estimate sizing of the tower. Not considered in this elementary analysis is rising bubble growth due to reduction in hydrostatic head. This may increase u_t and, in the extreme, cause a transition to cap-shaped bubbles.

Other means of providing small gas bubbles throughout the tower include insertion of perforated plates and/or impellers or other internal devices. These internals break up any coalesced gas slugs and thereby preserve a large gas–liquid contact area. Further details on alternative mechanical designs for bioreactors are presented in Sec. 9.7.3.

Figure 9.30 illustrates schematically the conceptual framework applied to formulate a 2-phase mathematical model for a concurrent flow tower-loop reactor (airlift bioreactor). In the tower section on the right-hand side, two-phase flow of gas and lqiuid occurs. After gas separation at the top of the column, a liquid stream is recycled through the loop on the left to the bottom of the reactor which is also the point of gas sparging into the system. The system is described by treating the liquid and gas as separate phases which ascend the tower section with different linear velocities u_L and u_G, respectively. The volume fractions occupied by liquid and gas are designated ε_L and ε_G, respectively, and these values are determined from use of gas holdup correlations or measurements. Superimposed on bulk convective axial transport in plug flow is axial dispersion in both the gas and liquid phase with different dispersion coefficients. Substrate and oxygen utilization occurs due to microbial reactions in the fluid phase and oxygen is depleted from the gas phase by virtue of oxygen transfer into the liquid.

Examining now a differential slice of the tower section between position z and position $z + dz$ (see Fig. 9.30*b*), we can derive the following material balances on oxygen concentration in the liquid phase and oxygen mole fraction in the gas phase by extensions of the methods already illustrated for treating steady-state plug-flow reactors. The final forms of these mass balances are given by

Liquid phase

$$\frac{\partial c_l(z,t)}{\partial t} = D_l(t)\frac{\partial^2 c_l(z,t)}{\partial^2 z} - u_L(t)\frac{\partial c_l(z,t)}{\partial z} - r_f(x, s, c_l, z, t) + k_l(z,t)a(z,t)[c_l^*(z,t) - c_l(z,t)] \tag{9.119}$$

Gas phase

$$p(z,t)\frac{\partial x_g(z,t)}{\partial t} = D_g(t)\frac{\partial}{\partial z}\left(p(z,t)\frac{\partial x_g(z,t)}{\partial z}\right) - \frac{\partial}{\partial z}\left(p(z,t)x_g(z,t)u_G(z,t) - \frac{RT\varepsilon_L(t)}{M_{O_2}\varepsilon_G(t)}\right) - k_l(z,t)a(z,t)[c_l^*(z,t) - c_l(z,t)] \tag{9.120}$$

Combined with analogous material balances for substrate consumption and cell growth, and similar balances for the liquid phase in the recycle loop, these equations can be used for description of unsteady-state mass-transfer experiments, batch operation of the bioreactor, or steady-state and dynamic behavior of the

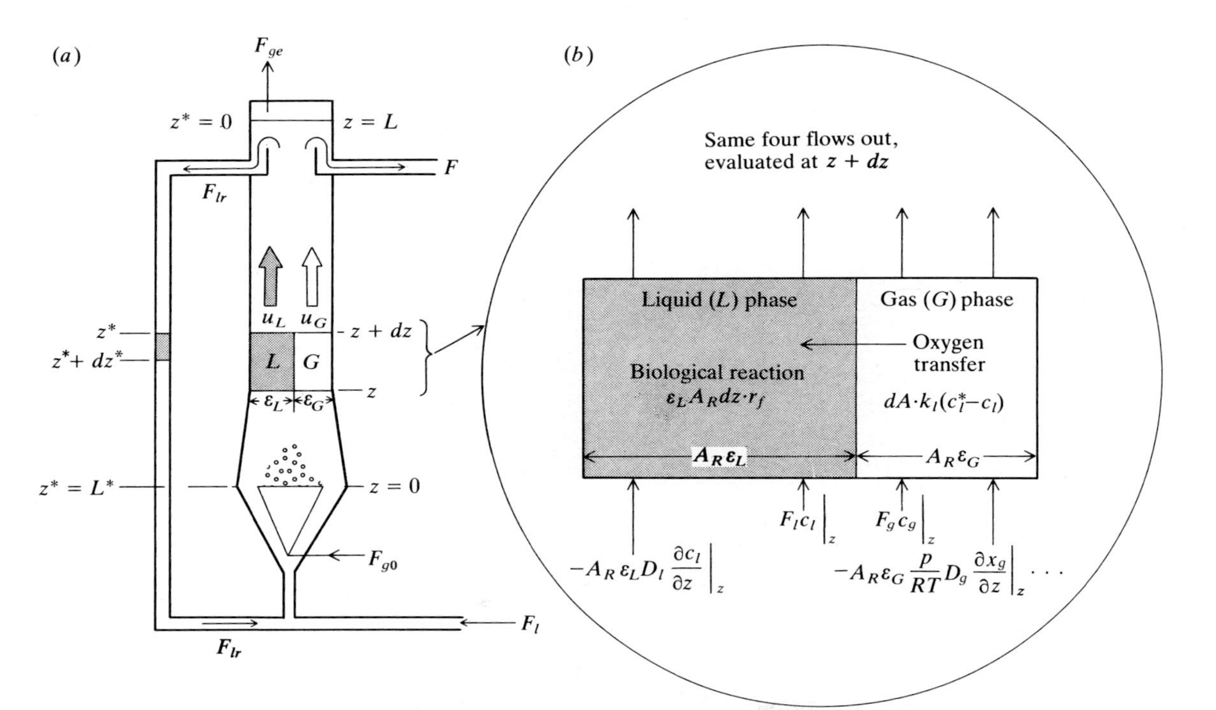

Figure 9.30 (*a*) Schematic diagram showing framework of a mathematical model for a bubble column reactor with recycle and (*b*) detail of a differential section showing terms considered in deriving gas (g, G) and liquid (l, L) phase material balances (Ref. 37).

tower bioreactor with continuous feed and product removal. Extensive experimental studies and parameter evaluation for this model are described in Ref. 37.

9.6.4 Fluidized-Bed Bioreactors

Fluidized-bed reactors generally are column geometries as considered in the last section and, if gas supply or removal is involved, require consideration of gas flow and mass transfer similar to those just discussed. However, in a fluidized-bed reactor an additional catalyst phase appears.

In a fluidized-bed tower reactor like the one shown in Fig. 9.31, liquid flows upward through a long vertical cylinder. Heterogeneous biocatalyst particles (flocculated organisms, pellets of immobilized enzymes or cells) are suspended by drag forces exerted by the rising liquid. Entrained catalyst pellets are realeased at the top of the tower by the reduced liquid drag at the expanding cross section and fed back into the tower. Thus, by a careful balance between operating conditions and organism characteristics, the biocatalyst is retained in the reactor while the medium flows through it continuously.

Fluidized-bed biological reactors are considerably more complex that the CSTR and PFTR varieties so far examined. For example, in tower fermentors used for continuous beer production, there is a gradient of yeast flocs through the unit. Near the bottom, the organism concentration (centrifuged wet weight per weight of broth) may reach 35%, while the yeast concentration drops to 5 or 10% at the top of the tower. Moreover, there is a progressive change in medium characteristics along the reactor. Easily fermented sugars (glucose, fructose, sucrose, some maltoses) are consumed first, near the feed point, thereby lowering the medium density. In the middle and upper portions of the tower, the yeast flocs ferment maltotriose and additional maltose. This scenario of rapid initial fermentation followed by slower reactions involving less desirable substrates is consistent with the experimental data shown in Fig. 9.32.

A rudimentary model for such fluidized reactors can be developed by assuming that (1) the biological catalyst particles (microbial flocs or immobilized-enzyme pellets) are uniform in size, (2) the fluid-phase density is a function of substrate concentration, (3) the liquid phase moves upward through the vessel in plug flow, (4) substrate-utilization rates are first-order in biomass concentration but zero-order in substrate concentration, and (5) the catalyst-particle Reynolds number based on the terminal velocity is small enough to justify Stoke's law (recall Example 1.1). Assumptions (4) and (5) are reasonable for many applications, and (1) to (3) may be adequate approximations.

Under these assumptions, the substrate conservation equation follows the form of Eq. (9.15):

$$\frac{d(su)}{dz} = -kx$$

or

$$u\frac{ds}{dz} + s\frac{du}{dz} = -kx \tag{9.121}$$

Figure 9.31 A tower fermentor used for continuous brewing. *(Courtesy of A. P. V. Co., Ltd.)*

For Stoke's flow, the concentration of the suspended biomass can be related to the liquid flow velocity in a fluidized bed by

$$x = \rho_0 \left[1 - \left(\frac{u}{u_t} \right)^{1/4.65} \right] \tag{9.122}$$

where ρ_0 is the microbial density on a dry weight basis and u_t is the terminal velocity of a sphere in Stoke's flow (Eq. (8.34)). (Should the dry-weight or wet-weight biomass density be used in the terminal-velocity formula?) Note that in the context of the fluidized bed we assume that local biomass concentration is

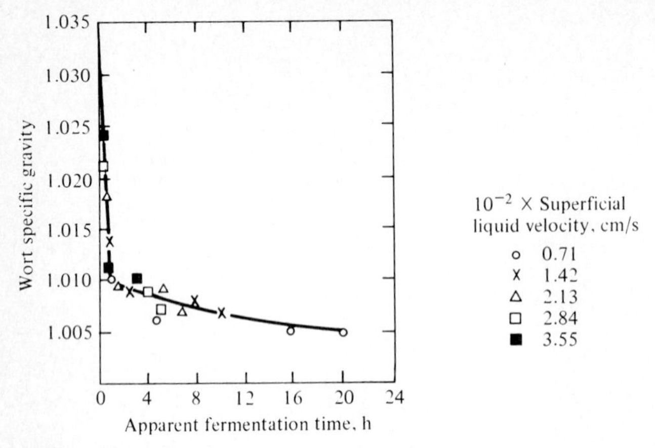

Figure 9.32 Column fermentor data for yeast growth on a sugar mixture show an initially rapid rate of substrate utilization (as revealed by change in wort specific gravity), followed by a period of much slower utilization rates. *(Reprinted from R. N. Greenshield and E. L. Smith, "Tower-Fermentation Systems and Their Applications," The Chemical Engineer, May 1971, 182.)*

dictated entirely by hydrodynamic factors rather than the biochemical-reaction-metabolism features emphasized for other reactor types. Substituting Eq. (9.122) into (9.121) leaves two unknowns, s and u as functions of position z in the tower.

We complete the model by applying Eq. (9.15) to total mass ($r_f = 0$) to reveal

$$\frac{d}{dz}(\rho u) = 0 \tag{9.123}$$

Expanding (9.123) and using $\rho = \rho(s)$ gives

$$\rho(s)\frac{du}{dz} + \left(u\frac{d\rho}{ds}\right)\frac{ds}{dz} = 0 \tag{9.124}$$

To cast the model in standard form suitable for numerical integration, we may now view Eqs. (9.121) and (9.124) as simultaneous algebraic equations in the unknowns ds/dz and du/dz. Solving this algebraic set, which need not be written out in full here, gives ds/dz and du/dz in terms of s and u: a set of two simultaneous differential equations to be integrated with the initial conditions

$$s(0) = s_f \qquad u(0) = u_f = \frac{F_f}{A_f} \tag{9.125}$$

where A_f is the tower cross section at the bottom. The effluent substrate concentration s_e is $s(z = L)$.

All this is much simplified if we assume that whatever fluid density changes occur do not affect u significantly. With u independent of position, Eq. (9.121) integrates directly, with the result

$$s_c = s_f - k\rho_0\left[1 - \left(\frac{u}{u_t}\right)^{1/4.65}\right]\frac{L}{u} \tag{9.126}$$

where x from Eq. (9.122) has been inserted and L is the tower height. Such linear dependence of substrate concentration on mean reaction time L/u is apparent in at least portions of Fig. 9.32 (if we also assume a linear relationship between s and ρ).

Unfortunately, however, the data in Fig. 9.32 suggest that our most serious error in this model is lumping of many substrates: the various sugars consumed during the anaerobic alcohol fermentation have been grouped together into a single hypothetical or average substrate in our model. In doing this, we have no way to include the glucose effect, which plays a very important role in the operation of continuous tower fermentors for brewing.

It is usually desirable to attempt to maintain plug flow of the reaction mixture through the fluidized bed. Instabilities in the flow patterns within the bed can in some situations cause significant backmixing which results in deterioration of reactor performance. Increased backmixing is more likely as the reactor capacity increases by increasing diameter and as the fluid-flow rate through the reactor is decreased. In many instances in biological fluidized-bed reactors, relatively small fluid linear velocities are necessary because small catalyst particles are used and the density difference between the fluid and the catalyst particles is low. Also, lower fluid velocities give greater catalyst concentrations in the reactor. It has been shown that insertion of static mixing elements into a fluidized-bed bioreactor can substantially improve the bed expansion characteristics and reduce undesired fluid backmixing [44].

Since packed-bed reactors provide a closer approximation to plug-flow, the question may arise as to the advantages and motivations for use of fluidized-bed bioreactors. One major consideration in this connection is contacting of the reaction mixture with gases. It is very difficult to provide effective aeration of packed-bed reactors at significant scale, and, in cases where gaseous products such as CO_2 are produced, to prevent excessive buildup of gas in later portions of the reactor. A fluidized-bed reactor provides a less constrained flow environment for gas–liquid–solid contacting and mass transport. Providing good gas and liquid supply to biocatalysts is also a major advantage of trickle-bed reactors, our final topic in this section.

9.6.5 Trickle-Bed Reactors

Trickle-bed reactors are three-phase systems containing a packed bed of heterogeneous catalyst and flowing gas and liquid phases. One (or more) reactant is provided in each feed liquid and gas phase, so that biochemical reaction depends on contacting of liquid, containing the sparingly soluble reactant from the gas phase, with the catalyst surface. Accordingly, the performance of such reactors is substantially influenced by the physical state of gas–liquid flow through the fixed bed and by the associated mass-transfer processes.

The important physical characteristics of such a reactor are the surface area of the packing, the efficiency of wetting of the catalyst by the flowing liquid phase, the gas–liquid flow pattern, mass transfer of sparingly soluble reactants from the gas to the liquid phase, mass transfer of both reactants to the catalyst surface, and, in the case of a porous or permeable catalyst, diffusion of reactants to the intraparticle catalytic sites.

One of the first applications of trickle-bed bioreactors which remains in practice today is the trickling biological filter used for wastewater treatment. In this system, which is described in more detail in Chap. 14, a rotating distributor sprays the liquid waste stream over a circular bed of gravel on which microbial films adhere. The liquid trickles down and through the packed bed in approximately laminar flow, while air rises through the bed by natural convection due to heat generated by the microbial reaction. A very similar operating design has been used for manufacture of vinegar (biological oxidation of ethanol to acetic acid) in a rectangular column packed with wood chips. For such a laminar liquid flow case, assuming a simplified geometry such as a plane sheet, a detailed model of the flow and transport processes can be formulated and solved. This problem is addressed further in the case study presented by Atkinson [5].

In applying biocatalysts in industrial practice, other trickle-bed reactor configurations involving cocurrent upflow or downflow of the two phases can be employed. Such reactors have long been used in the petroleum and petrochemical industry for hydrocracking, hydrotreating, and other multiphase reaction processes. When specifying operating conditions and formulating design models for such reactors, it is important to remember that, depending upon the gas and

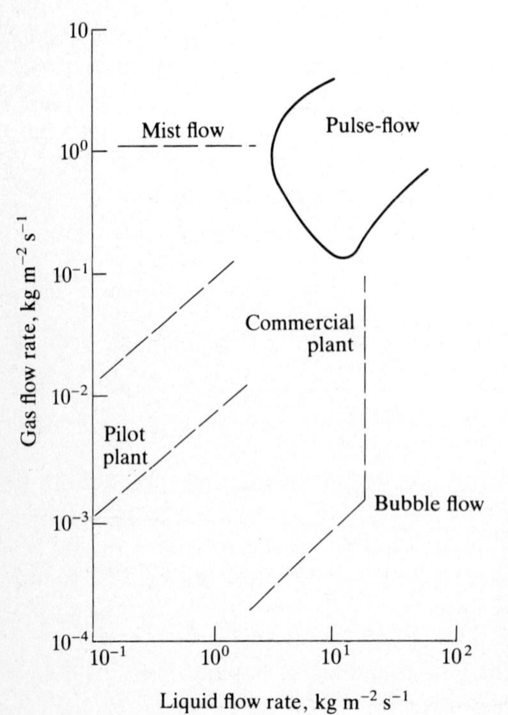

Figure 9.33 Two-phase flow regimes for cocurrent gas–liquid flow in packed beds. *(Reprinted by permission from J. G. van de Vusse and J. A. Wesselingh, "Multiphase Reactors," p. 561 in "Chemical Reaction Engineering: Survey Papers" (4th International/6th European Symposium on Chemical Reaction Engineering), DECHEMA, Frankfurt, 1976.)*

liquid flow rate (and to some degree on other properties of the fluids), a broad spectrum of two-phase flow regimes ranging from continuous liquid phase with dispersed bubbles to continuous gas phase with dispersed droplets or mist can be obtained (Fig. 9.33). Also, an unstable flow regime in which pulses of gas and liquid alternatively pass as slugs through the reactor is evident. The regions in Fig. 9.33 labelled pilot plant and commercial plant are taken from experience in petroleum processing. Some bioreactor operating regimes involve slow air flow rate. In trickling-filter biological waste treatment, for example, the small exothermicity of the reaction drives aeration upflow by natural convection.

The types of reactor models applicable here resemble those considered earlier in this section. Typically, the system is divided conceptually into a solid phase in contact with a covering liquid film in contact with the gas phase. This is a direct extension of the two-phase bubble-column already mentioned. Then, transport between and through the phases is considered along with any diffusion limitations on reaction rates. An excellent case study illustrating engineering treatment of a biological trickle-bed reactor has been reported by Briffaud and Engasser based on their studies of a trickle-flow fixed-film bioreactor for conversion of glucose to citric acid [46].

As is always the case in selecting a certain reactor configuration, many design, construction, and operating properties must be considered. Table 9.11 lists the comparative advantages and disadvantages of trickle-flow, stirred-slurry, and sparged-slurry-column or fluidized-bed reactors for achieving three-phase contacting and reaction.

Table 9.11 Comparison of design and operating characteristics of different three-phase reactor configurations.

(Reprinted by permission from J. F. van de Vusse and J. A. Wesselingh, "Multiphase Reactors," p. 561 in "Chemical Reaction Engineering: Survey Papers" (4th International/6th European Symposium on Chemical Reaction Engineering), DECHEMA, Frankfurt, 1976.)

+ GOOD
– POOR

	Reactor		
Problem areas	Trickle-flow	Stirred slurry	Sparged slurry column
Staging	++	–	+
Pressure drop	–	+	+
Maxium flow rates	–	+	++
Heat removal	(+)	+	+
Catalyst replacement	–	+	+
Catalyst attrition	(+)	–	(+)
Catalyst utilization	–	+	+
Ease of construction	+	–	++
Scaling-up	(+)	+	–

9.7 FERMENTATION TECHNOLOGY

In order to gain some appreciation of different practical aspects of bioreactor design and operation and the process context in which such reaction processes occur, we shall consider in this section some aspects of industrial practice in use of microbial bioreactors (which are traditionally called fermentors). Our primary emphasis will be on standard materials and methods used in batch-fermentation processes. At the close of this section, we will examine some of the alternative reactor designs which have been evaluated in small scale and in some cases have been implemented on extremely large scales.

Figure 9.34 is a schematic illustration of the important components of a typical fermentation process. Selection of a suitable medium has already been mentioned (Sec. 7.1.2), as have means of sterilizing it (Sec. 9.4) and any necessary gases (Chap. 8). Additional comments on medium formulation are provided in the following section. Although some influences of the inoculum on process behavior were discussed in Chap. 7, the microbiological problems encountered in this step require further investigation here. Section 9.7.2 will concentrate on the design of the fermentation vessel itself. Instrumentation and control are examined in Chap. 10; cell and product recovery operations are considered subsequently in Chap. 11.

9.7.1 Medium Formation

A variety of factors must be considered when formulating a fermentation medium. One relates to cellular stoichiometry and the desired amount of biomass to be produced. The basic concept here is simply a material balance: during the

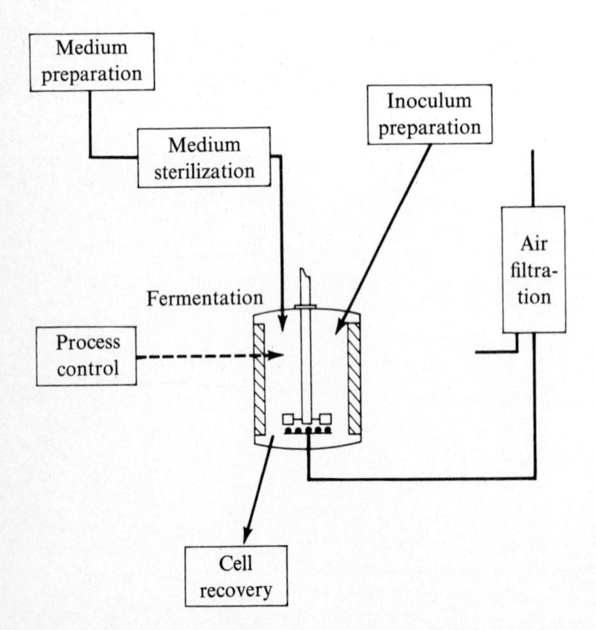

Figure 9.34 The basic operations involved in a typical aerobic fermentation.

course of cellular growth, small organic and inorganic molecules such as glucose and ammonia are converted into biomass. Nutrients (reactants) must be provided in sufficient quantities and proper proportions for a specified amount of biomass (products) to be synthesized. Computation of the necessary amounts of various substrates clearly requires knowledge of the product (biomass) composition. Typical elementary compositions of several different microorganisms were listed in Table 5.10. Also important in medium formulation is provision of necessary minerals (Table 9.12).

Once the elemental requirements have been calculated, choices still remain of the chemical compounds used to supply the necessary elements.Many commercially important microorganisms are chemoheterotrophs whose energy and carbon needs are satisfied by simple sugars. Instead of purified sugars, crude sources such as beet, cane, or corn molasses (50 to 70 percent fermentable sugars) are frequently used as carbon and energy sources in industrial fermentation media. In some instances process wastes like whey and cannery wastes provide cheap yet satisfactory carbon sources for fermentations. For example, one type of food yeast is grown commercially using a byproduct of papermaking, sulfite waste liquor, which contains about 2 percent fermentable hexoses and pentoses.

A variety of possible nitrogen sources are available including ammonia, urea, and nitrate. If the microorganism produces proteolytic enzymes, however, it can obtain necessary nitrogen from a variety of relatively crude proteinaceous sources. Among the possibilities for such crude sources are distiller's solubles, cereal grains, peptones, meat scraps, soybean meal, casein, cereal grains, yeast extracts, cottonseed meal, peanut-oil meal, linseed-oil meal, and corn-steep liquor. Especially important in penicillin fermentation media, corn-steep liquor is a

Table 9.12 Inorganic constituents of different microorganisms

	g/100 g dry weight		
Element	Bacteria	Fungi	Yeast
Phosphorus	2.0–3.0	0.4–4.5	0.8–2.6
Sulphur	0.2–1.0	0.1–0.5	0.01–0.24
Potassium	1.0–4.5	0.2–2.5	1.0–4.0
Magnesium	0.1–0.5	0.1–0.3	0.1–0.5
Sodium	0.5–1.0	0.02–0.5	0.01–0.1
Calcium	0.01–1.1	0.1–1.4	0.1–0.3
Iron	0.02–0.2	0.1–0.2	0.01–0.5
Copper	0.01–0.02	....	0.002–0.01
Manganese	0.001–0.01	....	0.0005–0.007
Molybdenum	....	....	0.0001–0.0002
Total ash	7–21	2–8	5–10

† From S. Aiba, A. E. Humphrey, and N. F. Millis, *Biochemical Engineering*, 2d ed., p. 29, Academic Press, Inc., New York, 1973.

concentrated (50 percent solids) aqueous waste resulting from the steeping of corn to make corn starch, gluten, and other products.

We have mentioned in earlier chapters that some microorganisms require an external source of some amino acids and growth factors. Other microbes which do not have a strict requirement for such medium adjuncts are frequently more productive if nonessential growth factors and nitrogen and carbon sources are provided. In industrial practice growth factors are typically provided by some of the crude-medium components already mentioned, e.g., corn-steep liquor or yeast autolysate. Similarly, these crude preparations often supply many of the minerals necessary for cell function. Other minerals are added to the medium as necessary.

When product formation is the major objective of a fermentation, *precursors* may be added to the medium to improve yield or quality. Generally the precursor molecule or a closely related derivative is incorporated into the fermentation product molecule. Specific examples of precursor applications include benzoic acids for production of novobiocins, phenylacetic acid for manufacture of penicillin G, and 5,6-dimethylbenzimidazole for vitamin B_{12} fermentation. In addition to these well-defined precursors, crude media components like corn-steep liquor may also provide useful precursor compounds.

Additional detail on selection and formulation of fermentation media are given in the References. We now turn our attention to other aspects of fermentation technology.

9.7.2 Design and Operation of a Typical Aseptic, Aerobic Fermentation Process

While common features are evident among most commercial fermentation processes, significantly different process designs as well as operating practices arise, often as a result of varying sensitivity to contamination by undesirable organisms. If it is necessary to avoid any intrusion, the fermentation must be operated on an *aseptic* basis so that a pure culture is maintained. In some situations, e.g., yeast growth at low pH or fermentation of hydrocarbons by carefully selected bacterial strains, aseptic precautions can be relaxed somewhat since operating conditions discourage growth of many potential contaminants.

This section will concentrate on aseptic practices since they impose the greatest demands on the ingenuity and thoroughness of the biochemical engineer. Our discussion follows the general sequence of events in the operation of a batch fermentation, beginning with development of an inoculum from a stock culture.

Preparation of an inoculum requires careful proliferation of relatively few cells to a dense suspension of from 1 to 20 percent of the volume of the production fermentor. This involves a stepwise procedure of increasing scale. The starting point is a *stock culture*, a carefully maintained collection of a particular microbial strain. Since the strain may be the result of extensive screening and mutation searches and may constitute a significant competitive advantage in the industry, it is imperative that the integrity of the production species be preserved. The usual strategy for achieving maximal genetic stability in a stock strain is to minimize its metabolic activities during storage. Microorganisms are usually maintained in the desired dormant state by *lyophilization* (freeze-drying) of a liquid cell suspension or by thoroughly drying dispersions of spores or cells on sterile soil or sand. Highly mutated stock organisms are frequently susceptible to *back mutation* and other types of undesirable genetic instability. Thus, constant checks on stock cultures are essential.

If we take lyophilized culture as an example, the next step in inoculum preparation is suspension of the cells in a sterile liquid. A drop of this suspension is then transferred to the surface of an agar *slope* or *slant*, made by solidifying a sterile nutrient medium in an inclined test tube using agar, a polysaccharide derived from seaweed. After incubation to obtain sufficient growth, the cells are again suspended in liquid and added either to a larger agar surface in a flat-sided Blake or Roux bottle or transferred to a shake flask. These flasks are agitated in machines which shake them in rotary or reciprocal patterns to promote submerged growth with adequate transport of gases to and from the organisms. Several successive steps with increasingly larger flasks usually are required before proceeding to the next step. All the transfers described above must be accomplished under sterile conditions. Rooms especially designed to permit sterilization and maintenance of aseptic conditions and controlled temperature are used in the fermentation industry for these delicate operations.

Further proliferation of the culture is next accomplished in one or more seed vessels, small fermentors with many of the instrumentation and control systems typical of large production units. Conditions in these reactors are chosen to maximize growth of the culture.

Since at this point we have moved from conditions typical of a microbiology laboratory into the plant environment, it is well to pause and consider some of the special design features required to maintain aseptic conditions. First, the system must be arranged to permit independent sterilization of its components. As an example of the extreme care this requires in design and operations, consider the problem of transferring the inoculant from the seed tank to the production fermentor. The schematic in Fig. 9.35 illustrates the required services and

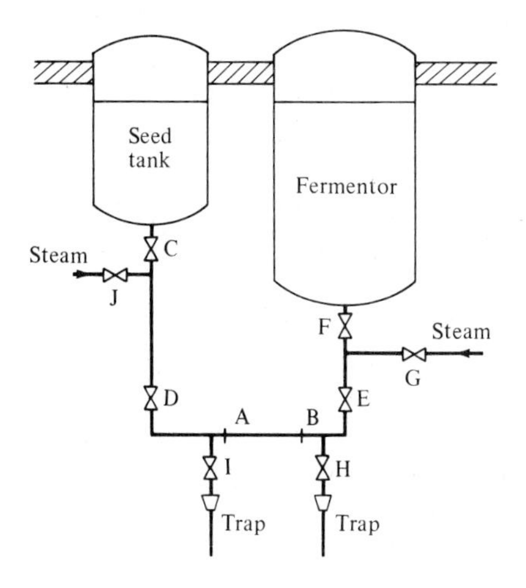

Figure 9.35 Valve and piping configuration for aspectic inoculation of a large-scale fermentor. Operation sequence: (1) install pipe section *AB*; (2) sterilize connection with 15 lb/in^2 gauge steam for 20 min valves *D* to *J* open; valve *C* remains closed; condensate collects in steam traps in branches *H* and *I*; (3) cool fermentor under sterile air pressure with valves *C*, *G*, *H*, *I*, *J* closed and valves *D*, *E*, *F* left open: sterile medium fills connection; (4) increase pressure in seed tank to 10 lb/in^2 gauge; lower fermentor pressure to 2 lb/in^2 gauge; (5) transfer inoculum by opening valve *C*; (6) steam-seal fermentor-seed-tank connections by closing *C* and *F* and opening *G* and *J*; steam and condensate is bled from partially open *D* and *E*.

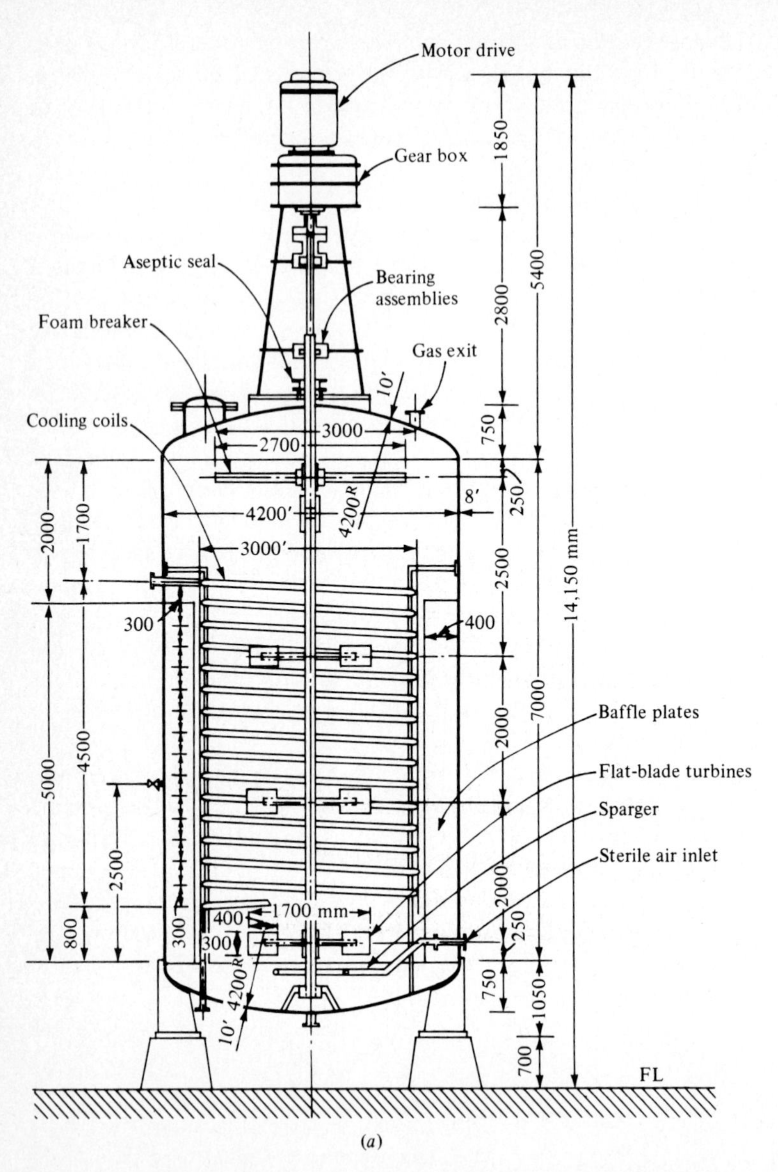

Figure 9.36 (*a*) Cutaway diagram of a 100,000-liter fermentor used for penicillin production. (*b*) Photograph looking down into a large production fermentor. *[(a) Reprinted from S. Aiba, A. E. Humphrey and N. F. Millis, "Biochemical Engineering," 2d ed., p. 304, University of Tokyo Press, Tokyo, 1973. (b) Reprinted from R. Müller and K. Kieslich, "Technology of the Microbiological Preparation to Organic Substances," Angew. Chem. Intl. Edit. Eng., vol. 5, p. 653, 1966.]*

(b)

Figure 9.36 (*Continued*)

sequence of events. All valves in the system must be easy to maintain, clean, and sterilize. For these reasons ball valves are quite popular. Several other general principles of aseptic process design are shown in Fig. 9.35. Specifically, all vessel connections should be steam-sealed: no direct connections between sterile and nonsterile portions of the system should be allowed. Maintenance of a positive pressure on the system ensures that leakage will be outward rather than inward.

The physical characteristics of a typical commercial fermentation vessel are shown in Fig. 9.36. These vessels are usually constructed from stainless steel to minimize corrosion problems and contamination of the fermentation broth by unwanted metallic ions (recall the discussion in Sec. 5.9.2 on the influence of iron ions in the citric acid fermentation). Care must be taken here and elsewhere in the overall process to avoid dead spaces, crevices, and other niches where solids resistant to sterilization can accumulate and where microbial films can grow. All-welded vessel construction with polished welds helps to minimize these problems.

The agitator assembly is designed to meet the mixing and aeration requirements already discussed in Chap. 8. Special attention to the design and maintenance of the aseptic seal is essential to avoid contamination. Although only one impeller is required in laboratory-scale fermentation, several may be necessary in a large commercial vessel.

Typically only 70 to 80 percent of the vessel volume is filled with liquid, with a gas space occupying the top portion of the tank. Often the combined action of aeration and agitation of the liquid promotes the formation of a foam on the liquid surface, especially if the medium contains high concentrations of peptides. Foam impedes gas mass transfer from the broth to the head space, forcing foam out of the vessel and contaminating the system when collapsed foam reenters the fermentor. In Fig. 9.36, a supplementary agitator located in the head space serves to destroy the foam. For especially persistent foams, chemical agents called *antifoams* are added to the broth. These compounds destabilize the foam by reducing surface tension. As noted in Chap. 8, the interfacial characteristics of antifoams can decrease the rate of oxygen transfer.

Several functions can be served by the heat-transfer coils within the vessel. If the medium is to be sterilized batchwise within the fermentor, these coils must have adequate capacity for the necessary heating and cooling. The heat-exchange system must also be able to handle the peak process load, which includes the combined effects of microbial activity and viscous dissipation from mixing (Sec. 8.10). Typical heat-transfer coefficients for uninoculated medium are about the same as for water, while a dense mycelial broth may exhibit a coefficient more typical of a paste.

Fermentation in the production vessel may take from less than 1 day to more than 2 weeks, 4 to 5 days being typical of many antibiotic manufacturing processes. During this interval, operating conditions must be carefully maintained or varied in a predetermined manner. More details on this operating practice will be provided in Chap. 10. Nevertheless, it should be evident from the general review already given that aseptic fermentation is an expensive, time-consuming proposition. Obviously, loss of product from one batch cycle can be extremely costly, so that the previous emphasis accorded sterilization is well taken.

9.7.3 Alternate Bioreactor Configurations

A number of factors, summarized in Table 9.13, have motivated development of new types of bioreactors. Many of these factors were encountered in the development by ICI of an extremely large-scale SCP process. This reactor has a total volume of 2300 m^3 (a column of 7-m diameter and 60-m height within an effective reactor volume of 1560 m^3). Furthermore, in this reactor organisms are grown on methanol, resulting in extremely large heat release. A conventional

Table 9.13 Factors motivating development of new types of reactors [50]

1. Construction of very large reactors
1.1. Design problems
1.2. High power requirement (P/V = constant)
1.3. High costs of energy and cooling water
1.4. Problems of heat removal
2. Reduction of specific capital costs
3. Reduction of specific energy costs
4. Avoidance of cell damage
5. Reduction of substrate losses (due to evaporation and respiration)
6. Increase of substrate conversion

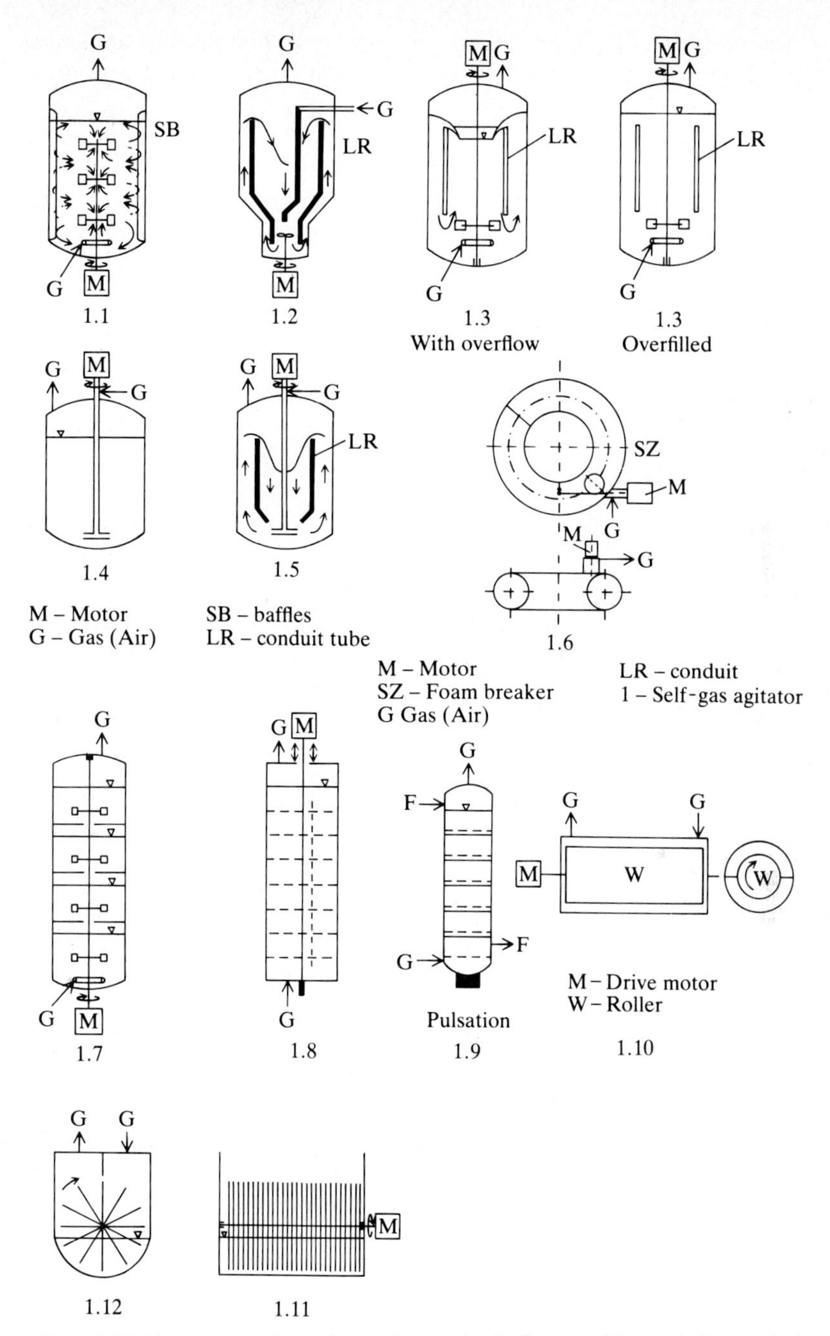

Figure 9.37 Bioreactor configurations using mechanically moved internals for energy input [50].

sparged, mechanically agitated stirred-tank fermentor is completely impractical on this scale. This led to the development of a new airlift design mentioned in the previous section.

A number of other alternative reactor configurations have been proposed and examined in various scales ranging from laboratory to pilot to full scale. An excellent summary of many of these designs and their properties with respect to efficiency of energy use for gas dispersion and mixing have been presented in a

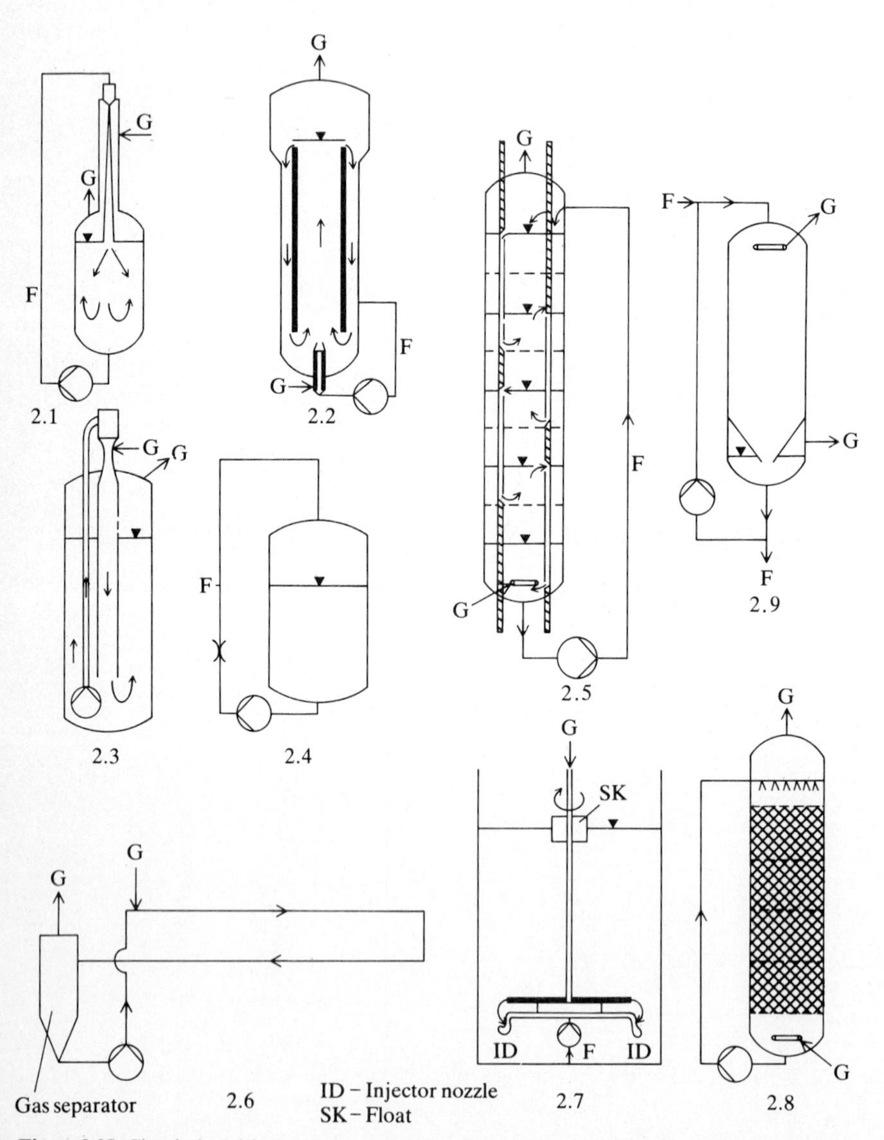

Figure 9.38 Circulation driven by an external pump provides energy input in these types of reactors [50].

review article by Schügerl [50]. In that review, bioreactors have been classified according to three main classes of energy input.

One class of bioreactors employs mechanically moved internals for energy input (Figure 9.37). In several of these designs an internal draft tube provides a defined circulation pattern to achieve loop flow patterns. Design 1.6 is a horizontal loop with foam separator which is completely filled with the gas–liquid mixture. In case 1.9, pulsation of the liquid flow occurs, and in configuration 1.11, viewed here from the side, disks on a rotating shaft are intermittently dipped into the liquid contents at the bottom of the reactor.

Figure 9.38 shows another set of reactor configurations in which energy is provided by liquid circulation using an external pump. In design 2.1 the liquid jet is injected downward and plunges into the liquid reactor contents, while in design 2.7 the injection nozzles for fluid are at the tips of a rotating bar.

The final group of reactor types, where energy input is provided in the form of compressed gas, are sketched in Fig. 9.39. In most of these designs, reactor

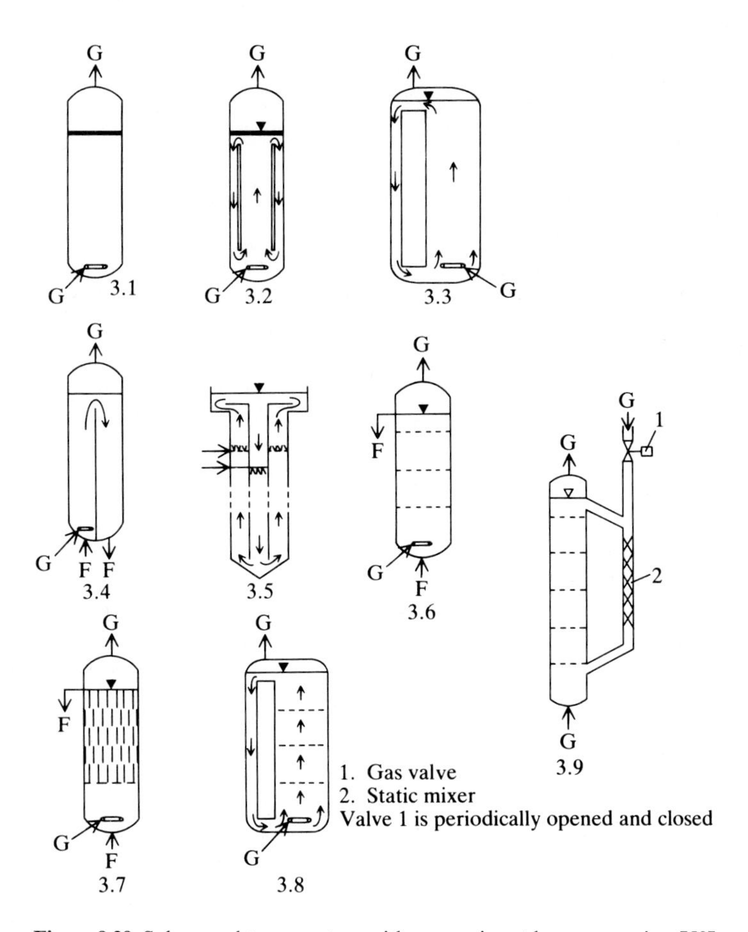

Figure 9.39 Submerged-type reactors with energy input by compression [50].

internals are provided for redispersion of gas and several different loop-flow configurations are evident. Further details on all of these reactor configurations and additional references are available in Schügerl's review [50].

9.8 ANIMAL AND PLANT CELL REACTOR TECHNOLOGY

Growth of animal cells in culture is currently used for manufacture of several products including vaccines, the proteolytic enzyme urokinase, monoclonal antibodies, and interferons. Such processes also have substantial potential for production of other lymphokines (a group of proteins which regulate certain aspects of the immune system), other enzymes, growth factors, clotting factors, and hormones. The advent of recombinant DNA technology introduces competition with animal cell cultivation for some of these products but also presents new possibilities for product manufacture using animal cell culture. On one hand, the opportunity of expressing foreign proteins in microorganisms means that microbial processes can now be used to manufacture these proteins in significant quantity. However, as explained in section 6.4.4, protein naturally synthesized in animal cells are often subjected to several different types of posttranslational modifications which are not accomplished in procaryotes. This means that for molecules requiring such posttranslational modification for activity or stability, eucaryotic hosts are necessary. Furthermore, problems with proper protein folding and proteolytic attack makes expression of some eucaryotic proteins difficult in procaryotic hosts. With improving methods for expression of foreign genes in animal cells, the prospects for industrial recombinant host-vector systems using animal cells are increasing.

Many useful and interesting chemicals can conceivably be synthesized in cultures of plant cells. Also, cultivation of plant cells may facilitate genetic engineering of plants and may ultimately allow the regeneration of whole crop plants from tissue originated in culture.

There are several common features in cultivation of plant and animal cells which complicate reactor design for their cultivation and for manufacture of their products. Many of the cell types of interest exist naturally as dense packings of similar cells. Such cell tissues in their native state are contacted with fluids of the organism of specific composition and containing a variety of regulatory molecules which can strongly influence cellular function. Growth rates of the cells may naturally be very low. The challenges faced in submerged cultivation of these cells is to provide an acceptable environment for growth of the cells—many types of plant and animal cells when placed in culture do not grow at all. Once this obstacle has been surmounted, the goal is to determine cultivation conditions which allow growth of cells to high densities in the shortest possible time, while retaining the metabolic capability of the cells to carry out the desired reactions.

In this section, we will concentrate primarily on approaches to environmental support of animal cells and a brief summary of some experiences with these organisms. Developments in plant cell culture are at a relatively early stage and will be more briefly summarized at the end of this section. Because of the expense of animal cell cultivation, because of slow growth rates which prolong experimental studies, and because of some neglect, quantitative characterization of the kinetic properties of animal cells is practically nonexistent. In many cases, limiting nutrients are not known, and descriptions of growth and product formation kinetics as functions of nutrient and inhibitor concentrations, pH, temperature, and other environmental variables are not generally available. Although the influence of mechanical forces on animal and plant cell cultures is recognized as a critical factor in reactor design, controlled studies designed to characterize in engineering terms the connection between mechanical forces and cell survival, growth, and morphological state are lacking. Instead, hardware and catalyst configurations have been invented which provide enhancements in growth rates, greater cell densities, or more product. Although substantial advances have been accomplished based on this empirical approach, there is room for great improvement based on new and improved designs once the kinetic behavior and engineering properties of these cells are more thoroughly understood.

9.8.1 Environmental Requirements for Animal Cell Cultivation

Culture media for animal cells is relatively complicated and expensive compared with microbial media. Antibiotics are usually included to reduce problems with microbial contamination. Animal serum is included in most media at concentrations from 5 to 20 vol % in order to promote cell replication. Some but not all of the functions provided by serum in animal cell growth medium have been identified. Serum helps to generalize the utility of the given medium for growth of different types of animal cells which may have different nutritional and growth factor requirements. It has been observed that nutritional requirements can depend upon cultivation conditions, and in these cases use of a rich and very abundant medium component helps to compensate for uncertainties in particular nutritional requirements. For reasons not yet understood, cells grown in media with greater serum content tend to be more resistant to mechanical damage.

There are, however, a number of problems associated with serum use in cell cultivation media. First, serum is the major cost in large-scale cell production. For example, serum, which can cost up to $300/L, contributes 80 percent of the material cost when used at a level of 10 percent. Serum also represents a major source of contamination to the culture by viruses, mycoplasma (parasitic or pathogenetic gram-negative procaryotes), and bacteria. Serum can contain inhibitors which interfere with virus replication (for vaccine production) or enzyme production. As in the undefined natural complexes added to microbial growth media, serum represents a variable and somewhat unpredictable medium component. Finally, serum introduces pyrogenic (fever-producing) contaminants into the medium which complicate product recovery. The albumin protein background contributed by serum also interferes with recovery of protein products present in low concentrations such as monoclonal antibodies, motivating the use of serum-free medium in order to facilitate recovery of the desired product.

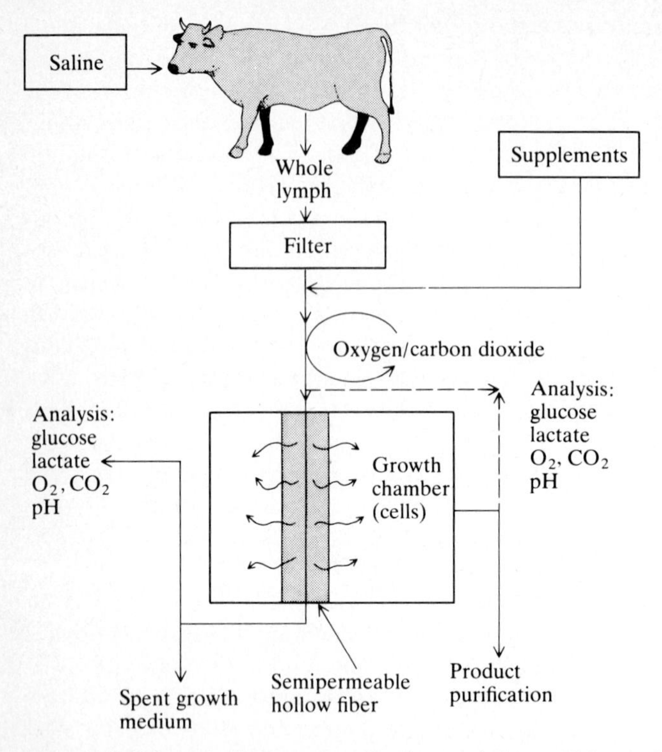

Figure 9.40 Schematic diagram of a cell culture technique which uses filtered cow lymph to provide medium components. *(Reprinted by permission of Bio-Response, Inc.)*

One novel strategy which has been proposed for reducing the cost of complex adjuncts for cell culture media is illustrated in Fig. 9.40. Here, whole lymph from a live cow is filtered and subsequently contacted with cells in the growth chamber using a hollow fiber ultrafilter. Provisions for adding supplements to complete the medium and for gas exchange are indicated.

Substantial advances have been made in formulation of a variety of hormonally defined, serum-free media. Such media offer the potential advantages of optimal tuning of the mixture of the medium for the particular organism and cultivation conditions, elimination of contamination problems, and more reproducibility. However, defined medium has the disadvantages of higher cost and of possible long-term effects on the organisms which are not easily anticipated or discovered during an initial short-term evaluation of the medium.

Animal cells do not possess cells walls to provide mechanical strength and are larger than microbial cells. Consequently, there are constraints on the forces which can be applied to a cell culture reactor in order to mix the cells, or cell-carrying or -containing particles, to maintain uniform environmental conditions and to accelerate transfer of oxygen into the culture. Animal cells require oxygen at a level of approximately 25–40 percent of air saturation, and design of oxygen supply systems for large-scale animal cell cultivation is an ongoing engineering problem. The demands imposed on the reactor engineer are moderated here because of the relatively slow growth and metabolic rates of the cells.

Oxygen requirements for a culture of typical density 10^6 cells mL^{-1} is in the range of 0.05 mmol $O_2\,L^{-1}\,h^{-1}$ to 0.6 mmol $O_2\,L^{-1}\,h^{-1}$. In small vessels characteristic of laboratory cultures, this oxygen requirement can be met simply by oxygen diffusion from the gas in the head space of the vessel into the bulk liquid. This gas overlay exchange is insufficient to meet oxygen needs of the culture in larger scale. Here direct sparging of gas into the fluid has been used successfully in some cases. In other instances, this direct sparging appears to damage the cells and may also cause excessive foaming, especially in media containing large quantities of serum. An alternative approach which has been proposed is immersion in the culture fluid of silicon tubing with oxygen diffusing from the flowing gas in the tubing into the medium without bubble formation. The overall mass-transfer coefficient for oxygen transfer through silicon tubing is around 0.35 mmol $O_2\,atm^{-1}\,cm^{-2}\,h^{-1}$ from which the amount of tubing required can be calculated. As the scale of cultivation increases, the practical feasibility of providing the necessary tubing becomes increasingly problematic.

In relatively advanced technology for animal cell cultivation, the pH of the culture is monitored and controlled, typically at values near 7.0. Today, there is very little experience with highly instrumented cell culture reactors to which the fullest possible set of data analysis and control strategies have been applied. As we shall investigate in the following chapter, there are many available instruments and approaches from microbial bioreactors which can be applied in the future to improve performance of cell culture reactors.

pH control in cell culture systems has been provided by a number of different approaches. Use of buffers in the medium can moderate pH changes which often occur due to lactic acid production by the cells. pH can also be manipulated by adjustment of the CO_2 content of the gas in contact with the culture. Direct addition of base is another possible approach. In reactors with continuous or intermittent exchange of reactor medium with fresh medium, pH variations can also be moderated.

9.8.2 Reactors for Large-Scale Production Using Animal Cells

Different types of animal cells can be divided into two broad classes depending upon their need for attachment to a solid surface for growth. Cells from the blood stream, lymph tissue, tumors, and many transformed cells can be adapted for growth in suspension culture. Other types of animal cells must be anchored to a compatible solid surface in order to grow. Furthermore, in the latter case, a contact inhibition regulatory mechanism usually prohibits growth beyond single monolayer coverage of the surface. On a small scale, required agitation is provided by use of a magnetic stirrer and a small "spinner flask" for suspended cells and by use of a cylindrical "roller bottle" partly filled with medium which rolls on its side horizontally at about one revolution per minute.

Many reactor designs for larger scale cultivation of animal cells in suspension resemble microbial bioreactors. Often the flat-bladed turbine mixers common in microbial processes are replaced by propeller agitators. In another quite effective design at the 100-L scale, "sails" of monofilament fiber are used in place of impeller blades to provide mixing of the culture suspension. Not often encountered in microbial cultivation, the Vibromixer has been used in animal cell bioreactors. This mixer is a circular flat disk containing several holes, mounted horizontally on the bottom of the shaft in the fluid. Rapid vertical oscillation in shaft position vibrates the perforated disk vertically in the fluid, creating a circulation pattern in the vessel. Airlift designs similar to those discussed earlier have also been used for animal cell cultivation.

A different perspective is appropriate when using the term "large-scale" in connection with animal cell cultivation. Here, because of the high value of the product and the high operating expense, a reactor as small as 10 liters may qualify as large scale, and a reactor of 100 liters certainly does.

Another factor favoring the use of relatively small reactors for animal cell cultivation is the relatively slow growth rate of these cells. Shown in Fig. 9.41 is data on growth in batch cultivation of the human liver adenocarcinoma cell line SK-HEP-1. Table 9.14 summarizes observed specific growth rate and oxygen uptake rates for several cell lines. In the first two cases, the cells were grown in suspension in an airlift fermentor. The final two cell types mentioned are anchorage-dependence lines, cultivation of which will be considered shortly.

Two different types of encapsulation methods have been demonstrated for achieving very high local cell densities. Instead of growing suspended cells, the cells are immobilized by entrapment using ultrafiltration membranes or by microencapsulation. Although conceptually similar to cell immobilization methods discussed earlier for microbial cells, the opportunities and the requirements are somewhat different for animal cells which are relatively large. In one quite promising microencapsulation process [60], cells are first entrapped in alginate beads (see Fig. 9.27). A surface coating of polylysine is applied and cross-linked, after which the alginate gel can be dissolved and removed. In this formulation, the cells are suspended within an entrapping network of polylysine, and, after growth, can pack the interior to cell densities approaching that of tissue. These strategies offer promise of higher volumetric productivites in animal cell reactors.

Special requirements arise in scaling up bioreactors for anchorage-dependent cells. The surface-to-volume ratio for a roller bottle is low, making scale-up by use of larger bottles quite impractical. A number of different methods have been demonstrated for increasing surface-to-volume (S/V) ratio in a bioreactor and thereby increasing the volumetric density of anchorage-dependent cells. Several of the possible designs and their S/V ratios are illustrated schematically in Fig. 9.42. Among these options, greatest attention recently has concentrated on application of microcarriers, small beads on which cells grow, which can be suspended in cultivation fluid. A major advantage of the microcarrier approach is the opportunity of using many of the same reactors and contacting designs for micro-

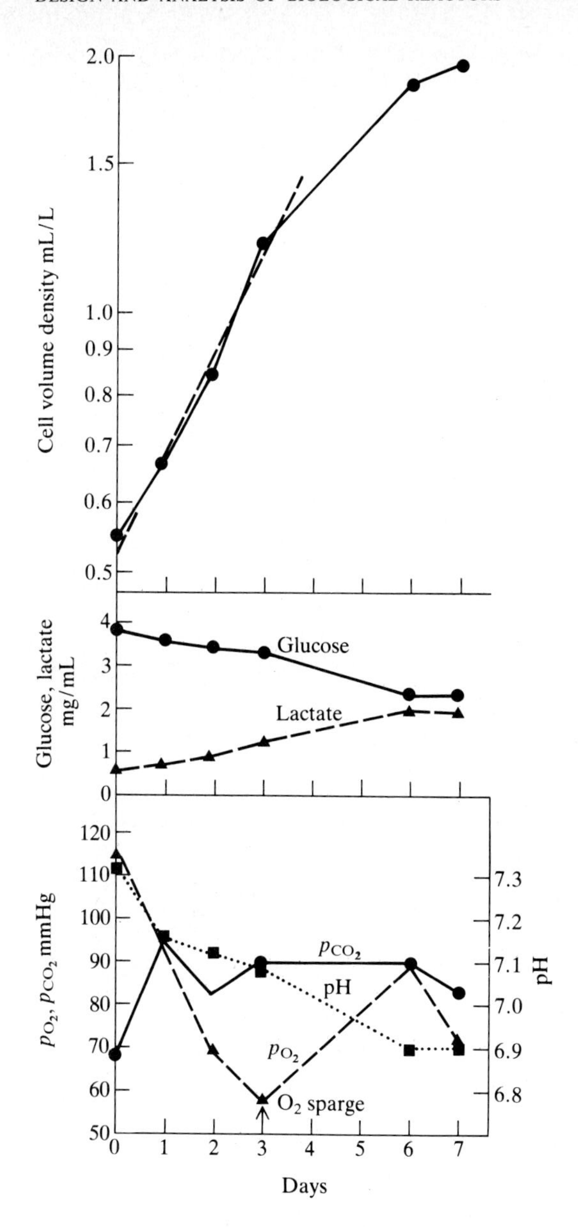

Figure 9.41 Time course of batch cultivation of SK-HEP-1 (human liver adenocarcinoma) cells in 100 L suspension culture. Dashed line in upper frame corresponds to exponential growth with 62 h doubling time. After three days, a sparge of 0.2 mL/min 100% O_2 was initiated. *(Reprinted by permission from W. R. Tolbert, R. A. Schoenfled, C. Lewis, and J. Feder, "Large-Scale Mammalian Cell Culture: Design and Use of an Economical Batch Suspension System," Biotech. Bioeng., vol. 24, p. 1671, 1982.)*

carrier suspensions as are used for cultivation of animal cells which are not anchorage-dependent. The first demonstrated microcarriers were charged dextran beads. Inhibition observed with the early beads was subsequently eliminated by modifying bead composition and decreasing bead surface charge. For example, one commercially available microcarrier consists of a core matrix of cross-linked dextran to which is covalently bound a surface layer of denatured collagen. Reported bead densities in cell culture reactors range from 3 g/L to 25 g/L.

Table 9.14 Summary of specific growth rates and specific oxygen uptake rates for different cell lines. Continuous culture was conducted in a bubble-column reactor.
(Reprinted by permission from H. W. D. Katinger and W. Scheirer, "Status and Developments of Animal Cell Technology using Suspension Culture Techniques," Acta Biotechnologica vol. 2, p. 3, 1982.)

Cell type	Cultivation technique applied	Specific oxygen uptake (Q_{O_2}), (μmol $O_2/10^6$ cells × hour) × 10^{-3}	Specific growth rate, $\mu(h^{-1})$	p_{O_2}, mmHg
BHK 21c 13 Baby hamster kidney	Continuous culture	85	0.018	60
		110	0.033	60
Human lymphoblastoid cells (Namalwa)	Continuous culture	54	0.012	50
		75	0.020	50
Mouse L 929	Microcarrier	20	Stationary phase at monolayer coverage	60–80
		65	Exponential growth phase	ca. 60
Human foreskin FS-4	Microcarrier	50	0.010	ca. 60
		85	0.025	ca. 60

When propagating animal cells from an original stock culture to a high cell density in a large reactor, a number of sequential steps of increasing scale are used. Summarized in Fig. 9.43 is a schematic flow sheet showing the cultivation steps typically utilized for a free suspension culture and for an anchorage-dependent culture. Note that, compared to microbial systems, the inoculum size for the next largest reactor is quite large.

A number of different operating modes have been applied in animal cell cultivation. In addition to the simple batch process, repetitive batch operations have been conducted in which a fraction of the culture at the end of the batch is used as the inoculum for a refilled reactor which is then grown to maturity. Fed batch operation, which adds nutrient at some preprogrammed rate or based upon process measurements as the cultivation takes place, is another possibility. An additional option is a perfusion system in which cells are retained in the reactor while some of the reactor medium is removed and new medium is added. This is an example of the macroscale immobilization methods for cells discussed earlier. Finally, continuous culture with continuous removal of cells as well as medium has been conducted. Medium removal helps reduce cell inhibition caused by metabolic product accumulation. For example, ammonia inhibits cell growth at concentrations above 4 m*M*.

The rate at which products appear can be quite different depending upon the process and pathway by which the product is formed. In some cases, products appear as the cells grow. In other cases, particular precursors are added after

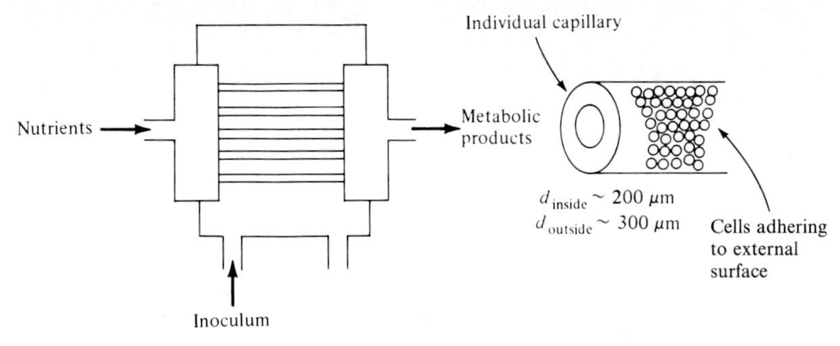

(*a*) Artificial capillary bundles (frequently ultrafiltration devices) ($S/V = 31\ \text{cm}^{-1}$)

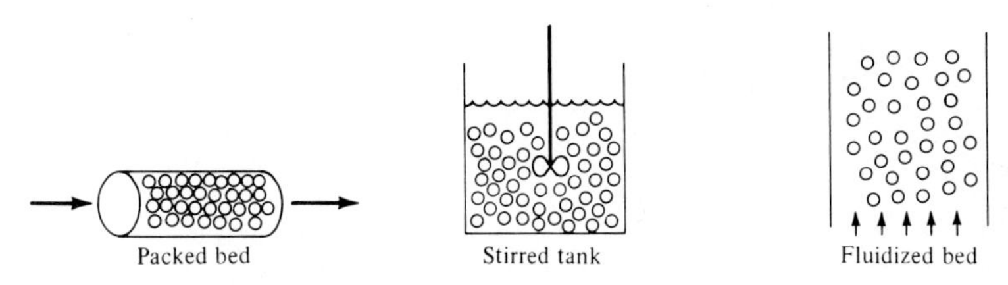

(*b*) Bead supports ($S/V = 120\ \text{cm}^{-1}$ at 20 g beads/L)

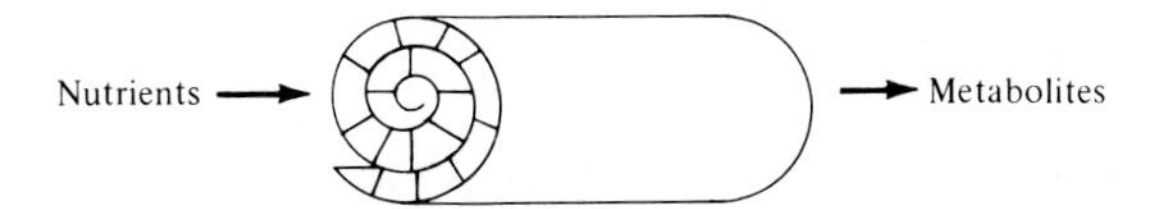

(*c*) Coiled sheet (with spacers to maintain sheet separation) ($S/V = 4\ \text{cm}^{-1}$)

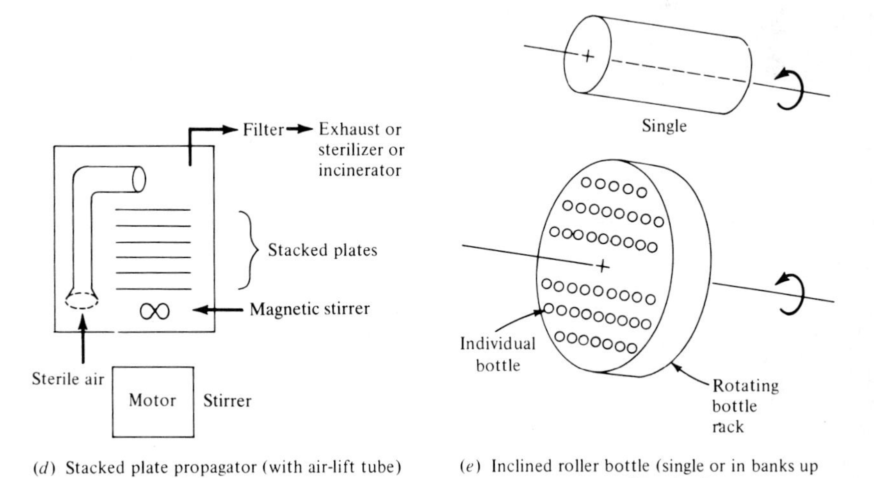

(*d*) Stacked plate propagator (with air-lift tube) ($S/V = 1.7\ \text{cm}^{-1}$)

(*e*) Inclined roller bottle (single or in banks up to several hundred) ($S/V = 0.2–0.7\ \text{cm}^{-1}$)

Figure 9.42 High surface area configurations for tissue culture.

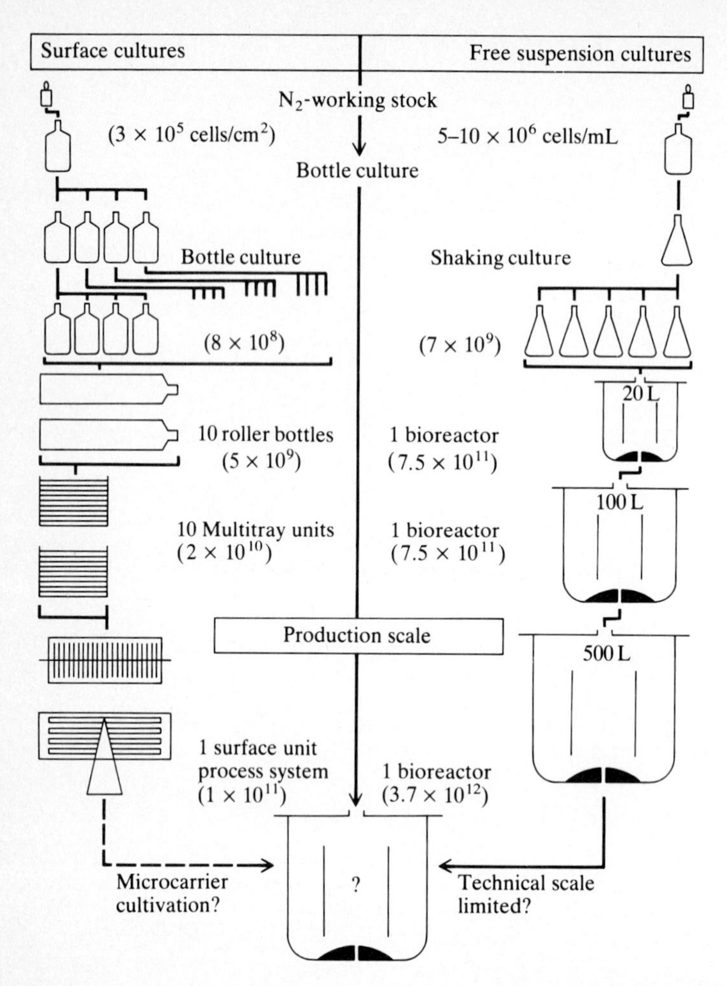

Figure 9.43 Scale-up of animal cell cultivation using BHK (baby hamster kidney) 21 cells as an example. Numbers in parentheses denote total amount of cells produced at the indicated stage. *(Reprinted by permission from H. W. D. Katinger and W. Scheirer, "Status and Developments of Animal Cell Technology using Suspension Culture Techniques," Acta Biotechnologica, vol. 2, p. 3, 1982.)*

some interval of growth giving rise to later product formation. Quite complicated kinetics can be observed in virus production in which the cells are serving essentially as a growth medium for the virus. In fact, it is common in the literature concerned with animal cell cultivation for virus manufacture to call the cell layer the "substrate." The infective virus is often added after the culture has grown to a high density. There then follows a subsequent life cycle of growth of the virus as was discussed in Sec. 6.2.1. This involves a lag for penetration of the virus, synthesis, and assembly of the virus components, followed by virus appearance in the cells or the medium.

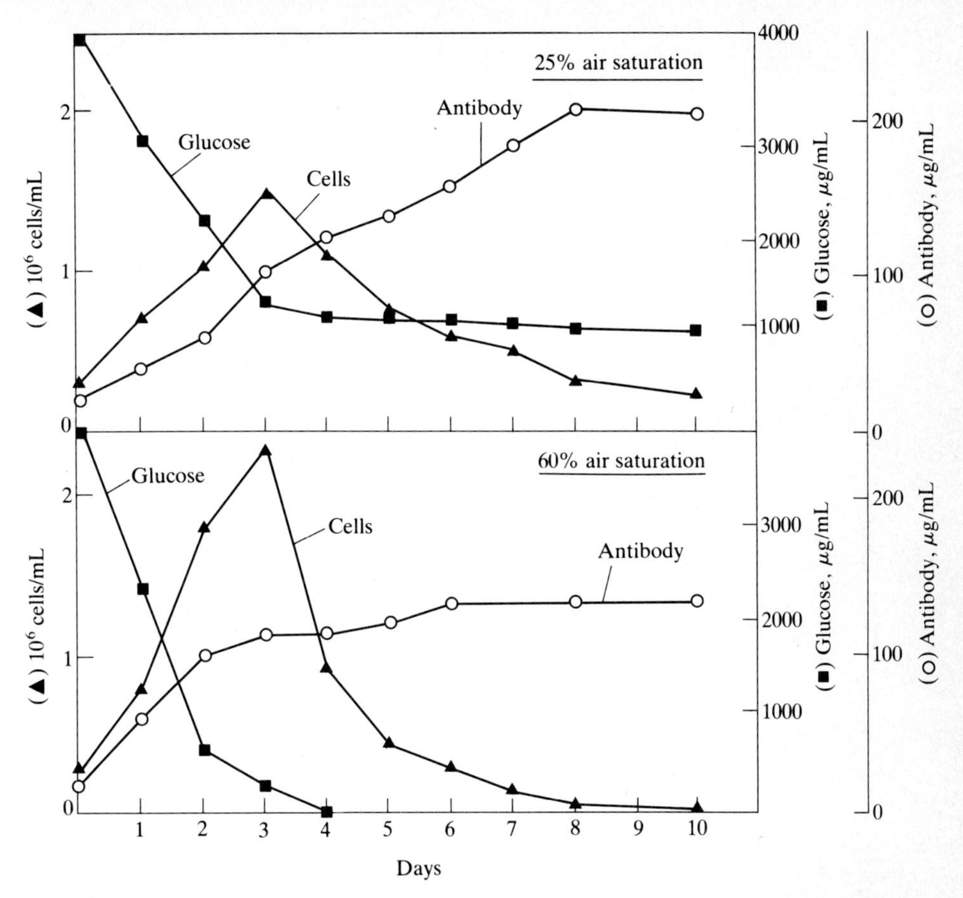

Figure 9.44 Time trajectories of hybridoma growth and monoclonal antibody production at different dissolved oxygen concentrations in a 3-liter fermentor. *(Reprinted by permission from S. Reuveny, D. Valez, F. Riske, J. D. Macmillan, and L. Miller, "Production of Monoclonal Antibodies in Culture," E.S.C.A.T. Mtg., Italy, May, 1984.)*

An example of monoclonal antibody production strongly illustrates the potential for reactor optimization. A mouse-mouse hybridoma line has been propagated in batch, fed-batch and continuous culture to produce a MAb for a surface antigen of *Rhizobium japonica* NR-7 cells.† Medium studies showed that an inexpensive medium could be used, consisting of Dulbecco's Modified Eagle Medium, plus 0.25% Primatone RL, 0.01% Pluronic polyol F-68 and as little as 1% fetal bovine serum, achieving 2×10^6 cells/mL in suspension culture with doubling times of 24 h. While a detailed kinetic model was not suggested, the results (Figs. 9.44 and 9.45) are intriguing. Maintenance of diminished oxygen

† S. Reuveny, D. Velez, F. Riske, J. D. Macmillan and L. Miller, "Production of Monoclonal Antibodies in Culture," E.S.C.A.T. Mtg., Italy, May, 1984.

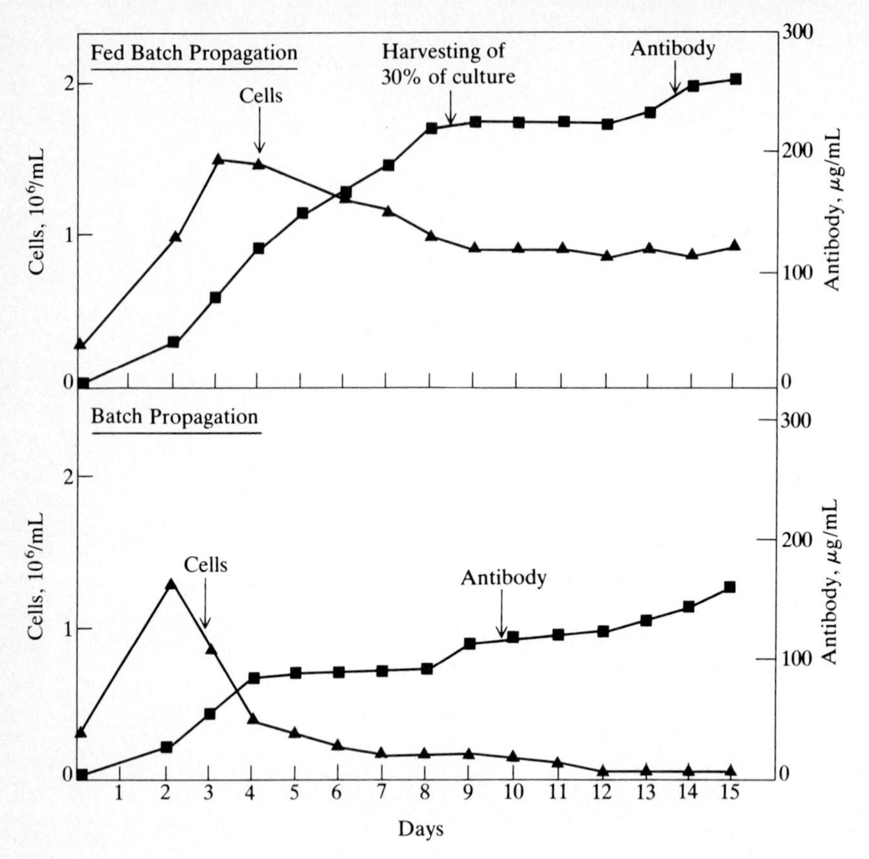

Figure 9.45 Comparison between batch and fed-batch production of monoclonal antibodies. Cells were grown in 100 mL spinner flasks. In fed-batch cultivation, 5–10 percent of growth medium was added for 10^6 cells every 24 h and 30 percent of the culture was harvested periodically. *(Reprinted by permission from S. Reuveny, D. Valez, F. Riske, J. D. Macmillan, and L. Miller, "Production of Monoclonal Antibodies in Culture," E.S.C.A.T. Mtg., Italy, May, 1984.)*

saturation (25 vs. 60 percent air saturation) provided a lower maximum cell density but a longer lived culture (Fig. 9.44) with a resultant higher final MAb titer.

In both conditions, production of a cell growth inhibitor was hypothesized in which case a fed-batch system should improve operation by stepwise dilution of inhibitor. The comparison of batch vs. fed-batch behavior (Fig. 9.45) is consistent with this notion; both cell viability and (consequent) antibody titer reached maximum values in the fed-batch operating mode. While MAb production for small quantities may be done most efficiently in the mouse peritoneal cavity, large-scale production points to bioreactor cultivation of hybridoma cells with emphasis on inexpensive nutrient medium, long cell viability, and optimization based on maximal antibody production. Hollow fiber propagation and perfusion culture are among the promising modes of larger-scale operation for hybridoma growth and product (MAb) formation.

9.8.3 Plant Cell Cultivation

Plant tissue dissected from the interior of plant organs after washing and disinfection can be cultured on agar containing a suitable growth medium. Nutrient media for plant cells are usually comprised of a mixture of inorganic salts with sucrose or glucose as carbon and energy source. (Plant cells are usually propagated chemoheterotrophically rather than photoautotrophically.) This basic medium is typically augmented with particular plant growth regulators, phytohormones, vitamins, amino acids, and the sugar alcohol inositol. Plant cells so propagated require oxygen for division so that aeration is necessary either by diffusion from the top surface of the culture or by sparged gas. The increase of cell number during batch cultivation follows the traditional pattern discussed in Chap. 7, except it is possible to obtain only three to four doublings of the population, corresponding to an increase in mass by a factor of 10 to 15 times the initial inoculum mass.

Plant tissue culture can achieve several potentially useful functions. Proceeding from the chemically most simple to the most complicated, enzymes and metabolic pathways in plants can be used to achieve biotransformation. One example of this class is 12-β hydroxylation of digitoxin to digoxin, an important heart stimulant drug, using cells of foxglove (*Digitalis*). Next, plant cells can be used for biosynthesis of complex compounds from simpler precursors. Plant tissue cultures can accomplish de novo synthesis of complex molecules, typically secondary metabolites, from simple media. Plant secondary metabolites of potential commercial interest include pharmaceuticals (currently plant-derived compounds account for about one-fourth of all U.S. prescriptions), dyes such as shikonin, gums, a group of natural insecticides called pyrethrins, and flavors and fragrances like vanilla and jasmine. Also, plant tissue culture provides a technology for rapid genetic engineering of plants for crop improvement and potentially for regeneration of crop plants. Compared with alternative technologies of growing plants in soil and harvesting their products, plant tissue culture may offer the advantage of more intensive production and more reliable and predictable supply.

When plant cells are grown in suspension culture, many different cell forms are present, including isolated cells and large aggregates. It has been noted that cells in plant cell aggregates are differentiated to some degree, with some cells in the interior of a clump being morphologically distinct from those on the outside. In some cases product formation does not occur from isolated single cells but results entirely from cells associated with the aggregates. This suggests that some degree of differentiation is necessary in certain circumstances to achieve the metabolic conditions necessary for product synthesis. In other situations, preservation of an undifferentiated state of the plant cells is the desired objective. Better understanding of the controls of differentiation, of the effects of differentiation, and

Table 9.15 Doubling times and yield coefficients for cultured plant cells

(Reprinted by permission from G. Wilson, "Continuous Culture of Plant Cells Using the Chemostat Principle," Adv. in Biochem. Eng. (A. Fiechter, ed.) vol. 16, p. 1, 1980.)

Cell culture	Limiting nutrient	Yield coefficient (10^6 cells/μmole)	Doubling time (h)
Galium	Phosphate (PO_4^-)	3.95	40
Galium	Phosphate (PO_4^-)	2.47	25
Acer	Phosphate (PO_4^-)	3.47	182
Acer	Phosphate (PO_4^-)	1.28	36
Acer	Nitrate (NO_3^-)	0.263	109
Acer	Glucose	0.039	109

of ways to regulate differentiation in culture by adjustment of growth conditions are the primary technical unknowns in the field at present.

Relatively little information is available concerning kinetic parameters of plant cell growth in culture. An exception is the body of results developed by Wilson and coworkers based upon chemostat plant cell cultivation. Table 9.15 lists the yield coefficients and doubling times for two different plant cell types grown in media with different limiting components. Notice the extremely long doubling times, again a problem in experimental studies of plant cell culture and in economical exploitation of plant cells as biocatalysts. Estimates of the substrate saturation or Monod constant K_S for *acer* (sycamore) plant cells in culture on the limiting substrates NO_3^-, glucose, and PO_4^- are 0.13, 0.5, and 0.032 mM, respectively [66].

Suspended plant cell culture beyond the shake flask scale has been conducted in mechanically agitated tanks and also in bubble column reactors. In order to obtain greater volumetric density of cells, to allow reuse of cells, to permit containment and retention of cells in desired morphological and differentiation states, and to allow better control of contacting betweeen medium and cells, several research groups have investigated immobilized plant cell reactors. Plant cells entrapped in hollow fiber membranes and in alginate and other types of matrices in beads have been investigated [67, 68]. Many products of plant metabolism accumulate intracellularly. This potential handicap for immobilized cell processes may be resolved by alternating permeabilization of cells and product recovery with exposure to growth and production medium [68].

As was noted in batch cultivation for secondary metabolite production using microorganisms, it is necessary for these products to separate in time or in space a growth phase from a phase of secondary metabolite production. This strategy is evident in the first large-scale, commercial plant cell culture process. In this process, developed by Mitsui Petrochemical Industries, the dye and pharmaceutical shikonin is obtained by culturing *Lithospermum erythrorhizon* over a period of longer than three weeks in three successive operations (Fig. 9.46). After growing

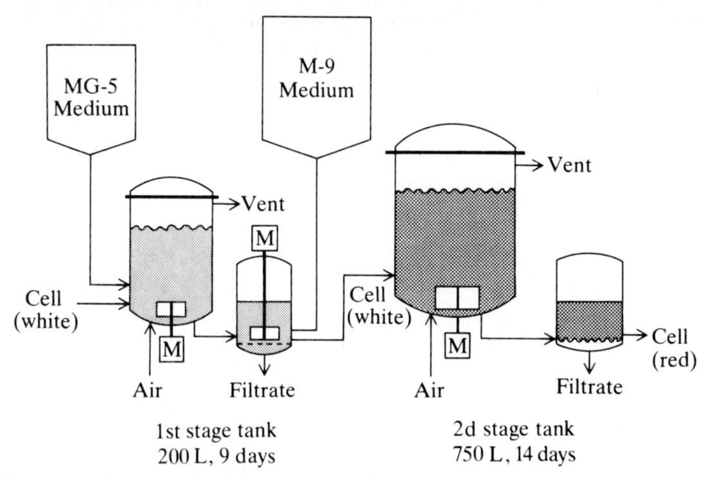

Figure 9.46 Processing steps in manufacture of the secondary metabolite shikonin using plant tissue *(Lithospermum erthrorhizon)* culture. *(Reprinted by permission from Mary Ellen Curtin, "Harvesting Profitable Products from Plant Tissue Culture," Bio/Technology, October, 1983, p. 649.)*

cells in an initial aerated tank, they are transferred to a smaller tank and a medium denoted M-9 is added which stimulates production of shikonin. Subsequent transfer to a third tank followed by reaction for 14 days results in accumulation of the desired product in the cells. Process yields have not been reported by the manufacturer, but independent estimates place the productivity of a single batch at 5 kg product. Obviously, the use of labor, capital, and expensive medium in the quantities involved in this process necessitate a high-priced product to justify commercialization. The selling price of shikonin as of October, 1983 was $4000/kg!

9.9 CONCLUDING REMARKS

In this chapter we have examined the reactor design principles and reactor technologies which enjoy current application and which will likely be used in the future for product manufacture using microorganisms and higher cells. We have tried to indicate how knowledge of the kinetics of the biological reactions and flow and mixing processes in the reactor can be synthesized to achieve an overall mathematical description of the process useful for design and optimization. This is the traditional and very successful strategy for chemical reaction engineering that has emerged from the chemical and petroleum processing industries. Unfortunately, due to our inadequate knowledge of biological kinetics, use of complex growth media with undefined composition and effects, use of reactors with poorly defined flow and mixing characteristics, and the lack of concerted effort toward more systematic and predictive reactor design, one often does not find reactor operation or design specification accomplished using the systematic methodology outlined in this chapter. Instead, process scale up proceeds through a succession of experimental investigations at increasing scales and costs, but with much smaller jumps in scale from one test to the next than can now be predictably accomplished in other types of chemical reactors. In designing a batch reactor for a biological application, for example, one is presently more likely to worry about providing a certain mass transfer capability than with calculating optimal operating strategies based upon knowledge of the kinetics.

Experience in those industries where chemical reaction engineering has been systematically applied shows that reductions in capital costs, lower operating expenses, fewer byproducts, and greater yields can be obtained. Also, it is not uncommon in the petroleum industry now to translate from bench to full scale, jumping in scale by factors of 10,000 based upon solid foundations in chemical reaction engineering. As opportunities for manufacture of many new products using new organisms and new processes increase in the future, we can expect a more central role for bioreactor engineering to increase the efficiency of scale-up, design, and operation of these reactors, to improve their reliability, and to accelerate commercialization.

Our theme in the next chapter is instrumentation and control, focusing mostly on bioreactors. Here too, our lack of knowledge about the fundamental transport and reaction processes in the reactor limit the types of control methods we bring to bear to obtain the best process performance. While the topics of the next chapter are considered primarily in the context of bioreactors, most apply equally well to upstream and downstream processing units. Following our consideration of instrumentation and control, we shall examine the essential separation processes which are used to convert the mixture emerging from the bioreactor to a stream of sufficiently pure useful product.

PROBLEMS

9.1 CSTR analysis (*a*) Verify each CSTR design equation of Table 9.1.

(*b*) For each CSTR case, indicate graphically how you would evaluate all terms in the reaction rate expression from an appropriate plot of CSTR performance.

9.2 PFTR design (*a*) Validate the results of Table 9.2 by direct integration.

(*b*) What variables are most convenient to use for each PFTR design equation? Show by sketches how each kinetic parameter in the reaction-rate expression can be evaluated.

(*c*) In general, do CSTRs lend themselves more easily to parameter evaluation than PFTRs? Why (not)?

9.3 Single-cell protein production *Methylomonos methanolica* grown on methanol at 30°C, pH 6.0 was observed to obey the following kinetic parameter values; $\mu_{max} = 0.53\ h^{-1}$, Y_s (methanol yield coefficient) = 0.48 g/g, $Y_{O_2} = 0.53$ g/g, carbon-conversion ($c_{biomass}/c_{methanol}$) efficiency = 0.57 g/g, oxygen quotient = 0.90 mol O_2 per mole CH_3OH, respiratory quotient (RQ) = 0.52 mol CO_2 per mole O_2, k_e = maintenance coefficient = 0.35 g CH_3OH/g, K_s = 2.0 mg/L [M. Dostolek and N. Molin, "Studies of Biomass Production of Methanol Oxidizing Bacteria" p.385 in *Single Cell Protein II*, S. R. Tannenbaum and D. I. C. Wang (eds.), MIT Press, Cambridge, Mass., 1975]. The specified yield factors correspond to a dilution rate of 0.52 h^{-1}.

(*a*) Write down the equations for CSTR growth which describe the rates of cell-mass production, oxygen consumption, and CO_2 production vs. dilution rate.

(*b*) Plot x, xD, and s vs. D (h^{-1}) and locate the predicted maximum values of x and xD when $s_{inlet} = 7.96$ g/L.

(*c*) Display the variation in the oxygen consumption rate vs. D on the same graph. What stirrer power input per unit volume would be needed to operate the reactor at the maximum productivity $(xD)_{max}$? State your assumptions.

(*d*) On the same graph, plot the predicted heat-generation rate per volume vs. D.

(*e*) The specific growth rate μ is observed to be diminished by substrate in an approximately linear fashion from its maximum value of 0.53 h^{-1} at $s = 3$ g/L to 0 at 13 g/L (extrapolated estimate). Repeat part (*b*) taking this design information into account.

(*f*) Under what exit conditions would several equal tanks be better than a single tank of the same total volume?

9.4 Liquid sterilization Set up the equation to determine the probability of complete sterility ($n < 1.0$) in a continuous sterilizer if plug flow is maintained and:

Specific death rate = 10 min^{-1} sterilizer volume = 10 L

Medium flow rate = 10 L/min temperature of medium = 131°C

Collection time = 5 min original cell concentration $n_0 = 10^3$ L^{-1}

What would you expect to happen to this probability if all the variables given above were held constant but the sterilizer were shortened and widened? Explain your reasoning and show pertinent equations.

9.5 Growth with variable yield coefficient Derive equations for $x(D)$ and $s(D)$ analogous to Eqs. (7.14) and (7.15) for the case where Y is given by Eq. (7.26).

9.6 Homogeneous and film reactor A feed contains a suspension of inert particles as well as substrate for an anaerobic fermentation. The vessel agitation is sufficient to keep the particles suspended and well dispersed. The microbial (single) population partitions itself between the particles and the bulk solution by a linear isotherm:

$$x_s(\text{cells/cm}^2) = Kx_{\text{bulk}}\ (\text{cells/mL})$$

where K has units of mL/cm^2. The adsorption process does alter the maximum specific growth rate μ_{max} but not K_s (assuming the Monod form is valid).

(*a*) Evaluate over all feasible dilution rates when d(inert) = 0.1 mm and volume fraction = 1 percent (1) the ratio of substrate utilization in the presence of inert particles to that occurring in their absence and (2) the influence of the suspended particles on reactor washout.

(*b*) How would you design a cell-recovery scheme at the reactor exit to maximize biomass recovery?

9.7 Rapid K_s measurement Williamson and McCarty ("A Rapid Measurement of Monod Half Velocity Coefficients for Bacterial Kinetics", *Biotech. Bioeng.*, **17**, 915, 1975) developed a relatively rapid means of determining K_s in microbial kinetics. A small, concentrated feed stream enters the microbial reactor, giving rise to a negligible volume increase over several hours. For values of s allowing less than the maximum possible substrate-utilization rate by the population, $\mu_{max} V/Y$, steady state was achieved in less than 1 h.

(*a*) Show that a Lineweaver-Burk type of plot (rate^{-1} vs. $1/s$ at "steady state") allows evaluation of K_s.

(*b*) Over what range of sampling times is the above analysis valid?

9.8 Fed batch reactor (*a*) Assuming exponential growth for cells at $\mu = \mu_{max}$, develop a general form for $x(t)$ starting with Eq. (9.4).

(*b*) Under what conditions will x increase, decrease, or remain constant?

(*c*) If the feed function $F(t)$ is alternated to be $F = F_0$ $(t_0 < t < t_1)$, $= 0$ $(t_1 < t < t_2)$, F_0 $(t_2 < t < t_3)$, etc., derive expressions for x for $t < t_1$, for $t_1 < t < t_2$, and for any large t.

(*d*) Suppose a methanol utilizing strain has a specific growth rate given by Eq. (7.32) (Monod form plus substrate inhibition). Why could operation in fed batch mode be superior to operation as a simple batch reactor with full charge of methanol present at $t = 0$?

9.9 Serial substrate utilization With multiple substrates, substrate consumption patterns in staged tanks or in towers may be complex as mentioned in the text. Consider a very simple model for diauxic growth on two substrates, glucose (G) and a second carbohydrate (S), such that

$$r_x = \mu(g, s)x$$

$$= \left[\frac{\mu_G \cdot g}{K_G + g} + \frac{\mu_S \cdot s}{K_S + s} \cdot \frac{K_R}{K_R + g}\right] x$$

(*a*) For typical conditions involving glucose repression, we expect that $\mu_S \gtrsim \mu$ and $K_S \sim K_G \gg K_R$. Sketch the expected variation of g, s, and $\ln x$ vs. time in a batch fermentation, showing clearly the relation between changes in the concentrations of biomass and the two substrates G and S. [Take $(g_0, s_0) \gg (K_S, K_G)$].

(*b*) For diauxic fermentations aimed at complete utilization of cell substrates, could you usefully use a plug flow reactor?, Tanks (chemostats)-in-series? A single chemostat? Why (not)?

(*c*) Since $K_R \ll K_G$, K_S, the growth equation can be conveniently simplified depending on whether $g > K_R$ or $K_R > g \simeq 0$. Develop a simplified plug flow description and integrate it analytically to give $x(z)$, $g(z)$ and $s(z)$.

(*d*) For $\mu_G = 1\ \text{h}^{-1} = 1.1\ \mu_S$, $K_G = K_S = 10\ \text{m}M$, $K_R = 0.1\ \text{m}M$, $x_0 = 0.1$ g/L, $Y_G = Y_S = 0.5$ g/g, $g_0 = s_0 = 0.1\ \text{m}M$, plot the results of part (*c*) for u_0 (flow velocity)/L (reactor length) $= 1\ \text{h}^{-1}$, $3\ \text{h}^{-1}$, and $5\ \text{h}^{-1}$.

(*e*) Vertical tower fermentors may have microbial populations which agglomerate and slowly settle. How would you modify the simple equation of part (*b*) to account for cells which settle at an average velocity $u_c (> u_0)$ (Net settling velocity $= u_c - u_0$).

9.10 Chemostat design from batch data The mass balances for component C in batch (or plug flow reactor) and chemostat (CSTR) are:

$$\text{Batch (or plug)} \quad \frac{dc}{dt} = f(c) \tag{i}$$

$$\text{steady-state chemostat} \qquad D(c_0 - c) = -f(c) \tag{ii}$$

Since $f(c)$ is available from (i) by differentiation, a plot of $f(c)$ vs. c can be constructed from batch data. Similarly, equation (ii) indicates that a plot of $D(c - c_0)$ vs. c will intersect $f(c)$ vs. c at c^* corresponding to the solution of equation (ii). (This technique, which provides the solution c^* for any given D, is applicable *only* if the fermentation can be described by a single variable; e.g., c.) (This solution technique was used by R. Luedeking and E. L Piret "Transient and Steady States in Continuous Fermentation; Theory and Experiment," *J. Biochem. Microbiol. Technol. Eng.*, **1**, 431, 1959). Its usefulness is seen in the following comparison of bacterial concentration (UOD/mL) from their paper

Measurement	Graphical solution from batch data
5.37	5.35
2.4	2.3
8.0	7.15

(*a*) Consider the logistic equation, $dx/dt = \mu x(1 - x/x_{max})$. With $\mu = 1\ \text{h}^{-1}$, $x_{max} = 10$ g/L and x_0 (in feed) $= 0$. Solve for x^* graphically for the cases $D = 1.5\ \text{h}^{-1}$, $0.75\ \text{h}^{-1}$, and $0.25\ \text{h}^{-1}$ using the method discussed above. Verify your results by direct solution analytically.

(*b*) For fermentors in series, the outlet from one chemostat is the inlet condition to the next. Show graphically how you would evaluate x_3^*, the biomass concentration in the third CSTR in a three-reactor cascade with an overall dilution rate of $D = 0.75\ \text{h}^{-1}$.

(*c*) Inspect the data for diauxic growth (Fig. 7.14) carefully. Differentiate it graphically and plot dx/dt vs. x. How does this plot differ from the shape for the simple logistic form? Sketch, using the graphical design procedure above, a solution for 5 tanks in series which will consume all of the glucose and most of the seond carbohydrate as well.

9.11 Fluidized beds of immobilized enzymes (*a*) Chinloy (PhD thesis, Princeton University, 1976) examined the conversion vs. reciprocal space velocity θ^{-1} (milliliter per gram of catalyst particles per second) obtained in packed and fluidized beds of protease immobilized on nonporous stainless-steel particles. In both cases, the data fell on the same straight line passing through the origin. Assuming $s_0 \gg K_m$, show that the data are not mass-transfer-influenced and that the specific rate constant can be evaluated directly from the data if the enzyme loading per catalyst particle (milligrams E per gram of catalyst) is known.

(*b*) Gelf and Boudrant ("Preliminary Study of a Fluidized Bed Enzyme Reactor" *Biochim. Biophys. Acta*, **334,** 467, 1974), studied hydrolysis of benzoylarginine ethyl ester (BAEE) by papain immobilized in porous 170 to 250 μm particles. The following parameter values were reported:

Soluble papain: $K_m = 5 \times 10^{-3}\ M$

$$v_{\max} = 19 \text{ IU } [\mu\text{mol BAEE}/(\text{min} \cdot \text{mg}) \text{ at pH 6.0 and } 20°\text{C}]$$

Immobilized papain: K_m (apparent) $= 1.2 \times 10^{-2}\ M$

$$v_{\max} = 0.05 \text{ IU } [\mu\text{mol}/(\text{min} \cdot \text{mg of } support)]$$

The porous support was largely iron oxide particles. From the data in Fig. 9.P11.1 determine whether these studies were influenced by external or internal mass transfer. State your assumptions. The catalyst charge in the fluidized bed was 10 g of particles; it was fabricated from a mixture including 100 mg crystalline papain per 30 g oxide.

(*c*) Discuss how you would design a fluidized-bed reactor of immobilized proteases for hydrolysis of (1) BAEE, (2) casein, (3) 1-μm-diameter gelatin particles.

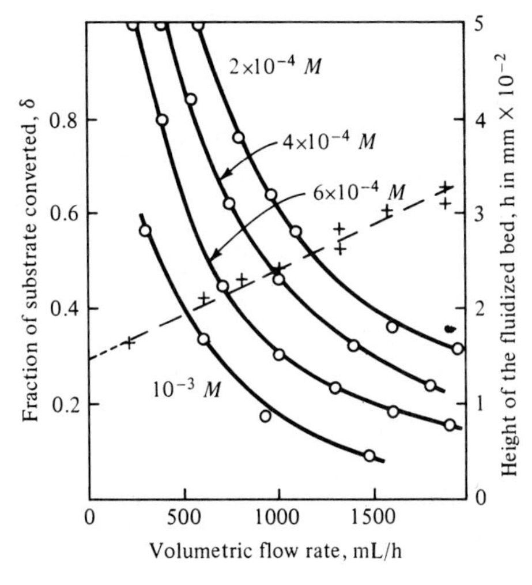

Figure 9P11.1 Fraction of substrate converted (O—O) and height of fluidized bed (+ — +) vs. volumetric flow rate. Molarities of inlet substrate concentrations are indicated on the curves. *[From G. Gelf and J. Boudrant, Enzymes Immobilized on a Magnetic Support, Biochim. Biophys. Acta, vol. 334, p. 468, 1974.]*

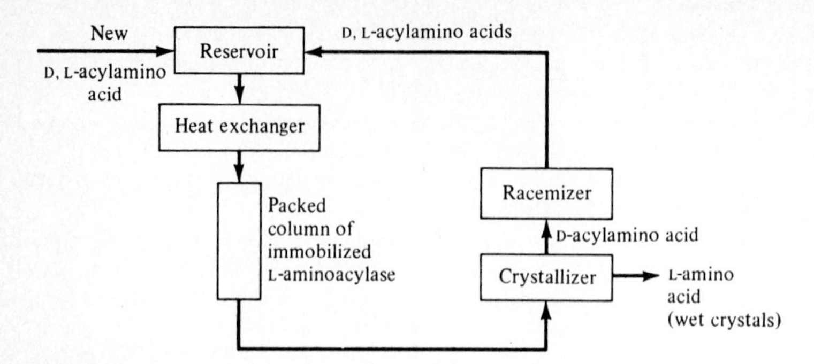

Figure 9P12.1 Catalyzed resolution of amino acids. *[From T. Tosa, T. Mori, N. Fuse, and I. Shibata, Enzymologia, vol. 31, p. 225, 1966.]*

9.12 Optically pure amino acids The process developed by Tosa *et al.* ("Studies on Continuous Enzyme Reactors II. Preparation of DEAE-Cellulose Aminoacylase Columns and Continuous Optical Resolution of Acetyl-d,l,-methionine," *Enzymologia*, **31**, 225, 1966) can be represented schematically by Fig. 9P12.1. Assume for the moment that v_{max} is independent of pH. The initial racemic amino acid solution is acetylated by reaction with acetic anhydride; the L-aminoacylase column reverses the acetylation reaction for the L-amino acid, which is then crystallized in alcohol solutions.

(*a*) Assume that the initial amino acid concentration is $\gg K_m$, develop an expression for the fractional L-amino acid conversion achieved by the enzyme column in plug flow. Repeat including axial dispersion.

(*b*) The racemization reaction may be taken to be first order reversible, so that the rate is proportional to $c_{\text{D acid}}$-$c^*_{\text{D acid}}$, where $c^*_{\text{D acid}}$ is equilibrium D acid level for the solution. If 90 percent of the L acid and 1 percent of the D acid is removed in the wet-crystal stream, along with 10 percent of the entering aqueous phase, what racemization CSTR volume is needed to achieve 95 percent approach to equilibrium?

(*c*) The enzymatic deacetylation step releases acetic acid into the solution. If the p*K*'s leading to enzyme deactivation are p$K_1 = 5$ and p$K_2 = 8$, what entering pH would give maximum conversion for a 10^{-6} *M*, 10^{-4} *M*, or 10^{-1} *M* feed mixture? (Assume plug flow.) State your assumptions clearly.

9.13 Digestion of insoluble substrates As an example of processes involved with digestion of particulate substrates, the following unit-operation sequence for yeast growth on newsprint has been suggested; mechanical grinding, acid hydrolysis, medium neutralization, addition of additional minor nutrients for yeast growth, yeast fermentor (aerobic), vacuum filtration to separate liquid from cell mass.

(*a*) Sketch the flow scheme above, indicating by arrows points of addition and by circles each unit operation. Include solids conveyers and liquid-pump locations where needed.

(*b*) From any human physiology test, sketch the human food-digestion process in a similar manner.

(*c*) Bionics is the study of natural systems with an eye toward development of synthetic analogs. Discuss similarities and differences between processes (*a*) and (*b*). How might you design a solids handling scheme for part (*a*) using the "conveyer" type in part (*b*)?

9.14 Cell maintenance: washout at small *S* For some populations, a minimum level of substrate may be needed to achieve a nontrivial steady state. As an example, consider the system with kinetics of the form:

$$r_x = \frac{\mu_{max} s x}{K_s + s} - k_e x \qquad r_s = \frac{-1}{Y_s} \frac{\mu_{max} s x}{K_s + s}$$

Assuming that the design basis underlying Prob. 9.10 is valid:

(*a*) Show that at substrate level below $k_e K_s/(\mu_{max} - k_e)$ the only steady state in a CSTR system is $x = 0$.

(*b*) If $\mu_{max} = 0.5\,h^{-1}$, $K_s = 0.2$ g/L, $k_e = 0.1\,h^{-1}$, and $Y_s = 0.6$ g cell/g substrate, plot $(dx/dt)_{batch}$ vs. x and prove by direct solution of the above equations that $dx/dt = 0$ for low s and for $D >$ D (washout).

(*c*) Determine the stability of each steady state to small perturbations.

9.15 Whey fermentation The fermentation of whey lactose to lactic acid by *Lactobacillum bulgaricus* at 44°C and pH 5.6 has been observed to fit the model of Luedeking and Piret [Eq (7.93)] provided the following modifications are made:

1. The maximal growth rate is $\mu_{max}^{\circ}\left(1 - \dfrac{p}{p_{max}}\right)$

2. $\mu_{max}^{\circ} = 0.48\,h^{-1}$, $p_{max} = 5\%$ at $p \leq 3.8\%$

   $\mu_{max}^{\circ} = 1.1\,h^{-1}$, $p_{max} = 4.3\%$ at $p > 3.8\%$

3. Parameters for continuous fermentation are: $\alpha = 2.2$
   $\beta = 0.2\,h^{-1}$
   $Y = 0.88$ g product/g substrate
   $K_s = 50$ mg/L

(*a*) Write down the equations for s, x, and p in continuous fermentation.

(*b*) Assuming steady-state behavior, show that at a total retention time of 15 h, two equal stages are better than one, but three produce essentially no further improvement in the reduction of substrate level. Is the same result true for biomass? (Consider the case $s_o = 5.0\%$, $x_0 = p_0 = 0$).

(*c*) Keller and Gerhardt ("Continuous Lactic Acid Fermentation of Whey to Produce a Ruminant Feed Supplement High in Crude Protein" *Biotech. Bioeng.*, **17**, 997, 1975.) note that when s_0 is less than 5 percent, product inhibition is not particularly strong, thus arguing that: "from a practical standpoint,..., cheddar cheese whey (4.9% lactose, 0.2% lactic acid) might be fermented adequately in a single stage fermentor, whereas cottage cheese whey (5.8% lactose, 0.7% lactic acid) benefits from an additional stage." Illustrate the magnitude of this benefit by repeating part (*b*) design using $s_0 = 5.8\%$, $p_0 = 0.7\%$.

(*b*) These authors also point out that addition of sugar would reduce the amount of water which it would be necessary to remove to get a fixed mass of product. How would sugar addition affect a reactor design strategy?

9.16 Staged fermentations: hydrocarbons Let us suppose that you have the batch growth curve for the hydrocarbon fermentation described so vividly by Mimura et al. in Sec. 8.8. In Prob. 8.11, you identified the probable controlling resistances of each of the fermentation phases observed. Your company has decided (in your absence) to scale up this fermentation by n tanks in series, where $n \equiv$ number of tanks $\equiv$ number of distinct cell-bubble-substrate configurational phases reported in Fig. 8.14. For each phase, write down the controlling resistance(s) and discuss quantitatively how you would scale the reactor volume, power inputs, etc., to obtain a scale factor of 5000 from laboratory to process units.

9.17 Penicillin fermentation The results shown in Fig. 9P17.1 were obtained from a *Penicillium chrysogenum* fermentation for penicillin production. The experiment was run in a well-stirred ten-liter fed-batch fermentor, aerated at 0.2 VVM (gas volumes per fermentor volume per minute). The growth and production medium contained initially (in grams per liter):

KH_2PO_4: 3.0; Na_2SO_4: 0.9; MgCl: 0.25; $MgSO_4 \cdot H_2O$: 0.05; Glucose: 10; NH_4Cl:2
2. In addition, glucose was fed continuously and NH_4OH was used for pH control. Benzil penicillin has the following formula; $C_{16}H_{18}N_2O_4S$.

(*a*) Explain the profiles for cell mass, penicillin, glucose and NH_3 in this fed-batch fermentation.

(*b*) Notice that very unexpectedly, the biosynthesis of penicillin came to a rapid halt even though this organism has the capability for synthesizing five times the amount of penicillin accumulated at the point its synthesis ceased. As chief trouble shooter for Antibiotics Unlimited, Ltd., you are requested to solve the mystery of why penicillin synthesis stopped.

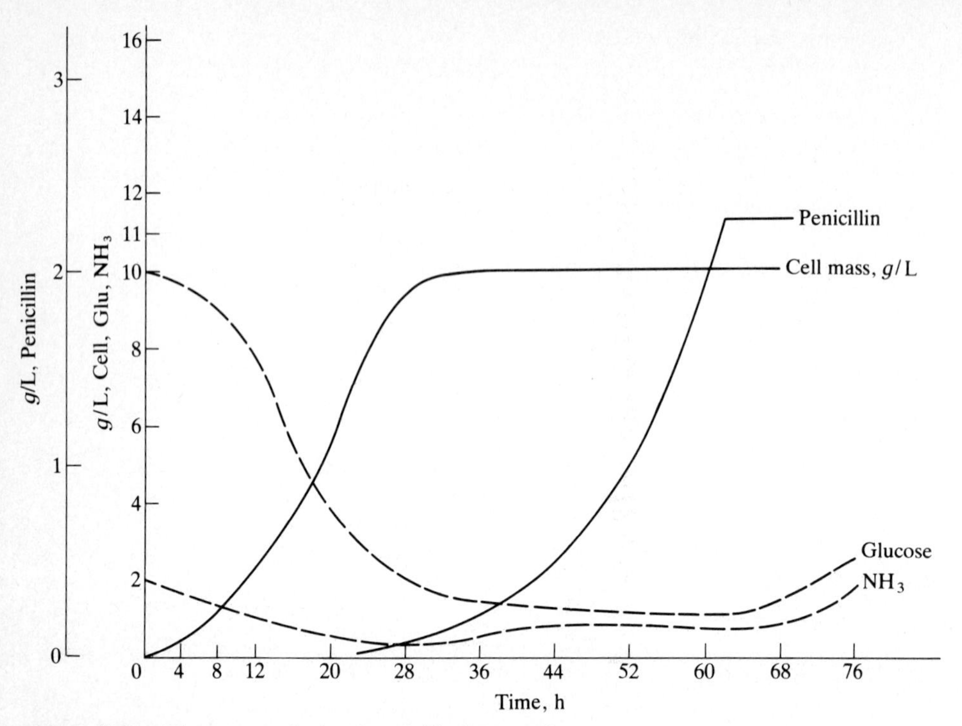

Figure 9P17.1 Time course of a batch penicillin fermentation.

9.18 Cyclic batch operations It is difficult to administer continuously low, controlled liquid feed rates, whereas the comparable periodic stepwise addition of substrate feed removal of fermentor fluid is easier. Suppose microorganism growth is limited by a single substrate in a Monod fashion, and that cell growth is proportional to the time rate of substrate change. At time = 0, the fermentation is begun (negligible lag). At time t, a volume V_1 is removed from the fermentor of liquid volume V, it is replaced by an identical volume of fresh feed at concentrations_f.

(*a*) Show that at time t (before volume removal), the substrate concentration s is given by

$$\left[1+\frac{K_s}{(x_0/Y)+s_0}\right]\ln\left[1+\frac{1-s/s_0}{x_0/(Ys_0)}\right]-\frac{K_s}{(x_0/Y)+s_0}\ln\left[\frac{s}{s_0}\right]=\mu_m t$$

(*b*) At time t, the volume V_1 is removed, now feed is added, and the process repeats. Evaluate s_0, x_0 in terms of s, x_t from this volume operation.

(*c*) Then establish that the substrate utilization efficiency

$$\beta \equiv (s_f - s_1)/s_f = 1 - \frac{\phi}{1-\phi}\left(e^{\mu(Vt/V_1)s_f\phi/K_s}[1-\phi]^{s_f/K_s}-1\right)^{-1}$$

where $\phi = V_1/V$.

(*d*) As $\phi \to 0$, show that this efficiency becomes identical with the continuous culture result

$$\beta = 1 - \frac{K_s}{s_f[\mu_m(V/V_1)t-1]}$$

where Vt/V_1 is the reciprocal dilution rate.

(*e*) Show graphically that at $s_f/K_s = 1.0$, finite values of ϕ (discrete operation) give higher β values than the continuous reactor if $(\mu_m Vt/V_1) > 4$.

9.19 Fluidized bed tissue culture It is proposed to produce tissue culture biomass continuously on nonporous beads in a fluidized bed (Fig. 9P19.1). The proposed fluidizing nutrient liquid moves

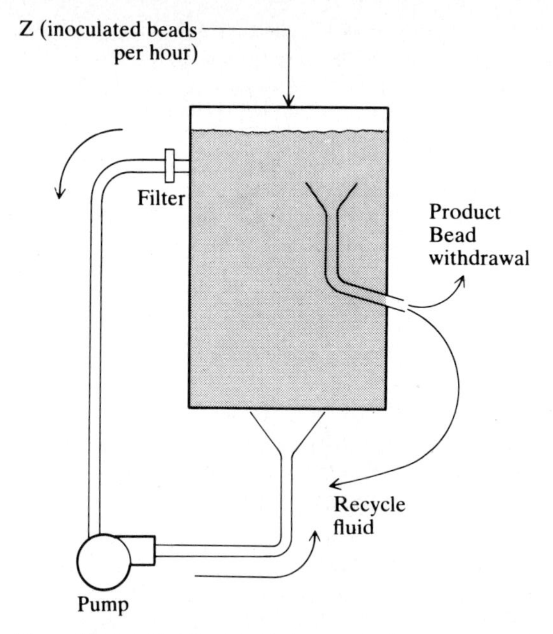

Figure 9P19.1 Fluidized bed microcarrier bioreactor for tissue culture.

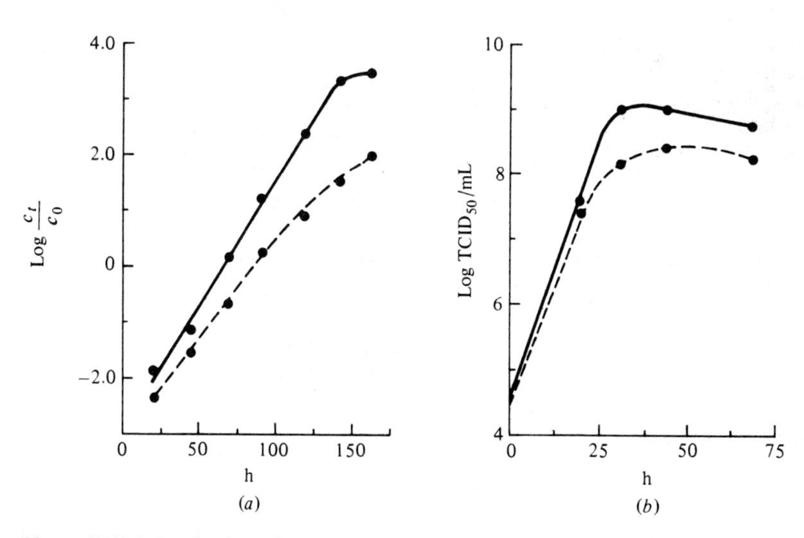

Figure 9P19.2 Production of cell number and viral titer vs. time. (*a*) c_t (cells attached at time t)/c_0 (cells inoculated at time $t = 0$) vs. time (——— 75% O_2 saturation; ------- 5% O_2 saturation). (*b*) TCID$_{50}$ (≡ virus dose yielding 50 percent tissue culture infection dose per milliliter) vs. time (——— 13 × 10^5 cells/milliliter initial cell concentration; ------- 3.5 × 10^5 cells/milliliter initial cell concentration). *[Reprinted from A. L. van Wezel, Microcarrier Cultures of Animal Cells, in "Tissue Culture: Methods and Applications," p. 372, Academic Press, Inc., New York, 1973.]*

upward, passing out of the reactor through a screen to retain beads, and then through a pump. The choice of beads with ($\rho_{bead} - \rho_{H_2O}$) small allows operation at a low fluidizing volume rate and use of a pump which does not damage the macromolecules in the nutrient medium. For liquid fluidized reactors, assume that the liquid moves in plug flow while the beads are perfectly mixed and distributed throughout the reactor. Beads passing over the exit funnel settle into it and are continuously removed; the exit fluid being returned by a subsequent filter.

(*a*) Assuming the data of Fig. 9P19.2 to typify cell growth kinetics on each bead with $\mu = 0$ for $t > 160$ h [due to achievement of a complete monolayer (contact inhibition)], develop a steady state expression for the exit biomass concentration per bead in terms of the fluid bed void volume ε, the feed rate of inoculated beads (Z per hour), the reactor volume V_R and the appropriate growth rate law. The entering beads have c_0 cells/bead.

(*b*) Repeat (*a*) for the situation where the cells per entering bead have a distribution given by

$$\text{Prob of } (y) \text{ cells/bead} = Ae^{-(y-y_m)^2} = Ae$$

where A = normalizing factor

(*c*) How would you design a "continuous," sterile transport system for adding new beads and for recovering the exit liquid volume as assumed?

9.20 Tissue culture support inhibition It has been observed that high concentrations (cm^2/cm^3) of solid supports for tissue culture may be deleterious to growth because of adsorption on these surfaces of growth enhancing factors and serum protein from the support medium, adsorption rendering these factors inaccessible (C. B. Horng and W. McLimens, *Biotech. Bioeng.*, **17**: 713, 1975). Consider a well-stirred microcarrier tissue culture reactor employing nearly neutrally bouyant beads as the cell supports.

(*a*) Assuming that the cells grow logarithmically (all nutrients present in excess) until the available surface area is covered (see Figure 9P19.2), develop an expression for the total biomass in the system versus time.

(*b*) Now assume that the growth rate also depends in a Monod function fashion upon the concentration of a growth-enhancing factor, S_g, which is not consumed. Derive the analogous result for biomass versus time when the growth enhancing factor S_g reversibly adsorbs on the solid in an inactive form following a linear partition law:

$$s_g(l) = Ks_g(\text{adsorbed})$$

(*c*) For a batch system, assuming that growth exhaustion results from lack of further bead surface, show that for a finite culture time (say 6 days) there is an optimal initial loading of beads into the tissue culture inoculum. Develop an expression for this value.

(*d*) Since the beads act in some sense as an inhibitor, it would seem logical to consider several stirred tanks in series rather than one tank. What problems would face the experimenter searching to set up such a system?

9.21 Penicillin-V deacylation: multiple reactions Enzymatic deacylation of penicillin-V to produce the desired 6-aminopenicillanic acid (Chap. 12) for production of semisynthetic penicillins involves a reaction network represented below:

$$A^-(\text{pen.-V}) \xrightarrow{r_1} P^-(\text{6-APA}) + \text{QOH(phenoxyacetic acid)}$$

$$A^- \xrightarrow{r_2} R^{2-}\ (\text{inactive product}) + H^+$$

$$P^- \xrightarrow{r_3} HS^-\ (\text{inactive ring cleavage product})$$

$$QOH\ {}_2 \overset{K_1}{\rightleftarrows} QO^- + H^+$$

$$P^- + H^+ \overset{K_2}{\rightleftarrows} \text{HP (stable penicillin-V)}$$

An optimal pH exists because protonation of P^- gives stable HP, but too acidic conditions inhibit the reaction rate r_1.

(*a*) Assuming r_1 given by Michaelis-Menten kinetics with noncompetitive H^+ inhibition, and r_2 and r_3 given by simple first-order irreversible forms, write down the equations governing these three rates and two (assumed) equilibria.

(*b*) For a steady-state CSTR system with perfect enzyme recycle, derive an expression giving the feed level of A^- which maximizes fractional conversion to P^-.

(*c*) For very high conversion of an expensive starting material, use of several CSTR's in series appears appropriate. Discuss tactically the advantages and disadvantages attending such a series arrangement. Assume r_2, r_3 are about 2-5 percent of the rate r_1 *in the first tank.*

(L. G. Karlsen and J. Villadsen, "Optimization of a Reactor Assembly for the Production of 6-APA from Penicillin-V," *Biotech. Bioeng.*, **26**: 1485, 1984).

9.22 Hollow fiber reactor productivity Different reactor types can yield profoundly different microbial densities and productivities. For example, β-lactamase specific production activity is less in hollow fiber systems than in continuous culture by a factor of five:

Reactor	Productivity (units E/cell-h)
Shaker flask	1×10^{-10}
Hollow fiber	2×10^{-11}

(*a*) If the biomass level in the hollow fiber reactor is (typically) 1000 times greater than in suspension culture, calculate the productivity of each reactor in units/(reactor volume – h), assuming a shake flask biomass level of $x = 10^9$ cells/mL.

(*b*) Cell lysis in the hollow fiber, as well as protein export, accounted for some increases in β-lactamase release. Describe a program by which you would determine cell lysis kinetics, which could then be used to describe β-lactamase production by excretion and lytic release. (See D. S. Inloes et al, "Hollow Fiber Membrane Bioreactors Using Immobilized *E. coli* for Protein Synthesis," *Biotech. Bioeng.*, **25**: 2653, 1983).

9.23 Batch production of non-growth associated product Batch production typically involves biomass production and product formation, with substrate being consumed by each process. Suppose that batch growth is modelled by the logistic equation, where x_{max} is set by the initial value of a second substrate s_2, (not used in product formation). Then for a nongrowth associated product (e.g. L-glutamate from *Micrococcus glutamicus*), we may write;

$$dx/dt = \mu x(1 - x/x_{max})$$
$$dp/dt = nx$$
$$ds/dt = -(dp/dt)/Y_p - (dx/dt)/Y_X$$

(*a*) Integrate each equation to obtain $x(t)$, $p(t)$, $s(t)$.

(*b*) Discuss your strategy in setting x_{max} if your objective is (i) to maximize $p(\theta)$ or (ii) to maximize $p(\theta)/\Delta s(\theta)$, where θ is a fixed, end-of-fermentation time.

(*c*) Under what circumstances would you choose objective (i) or (ii) in part (*b*)?

REFERENCES

Many of the references given for Chap. 7 contain substantial material on reactor design and analysis. Additional general presentations are available in the following texts:

1. O. Levenspiel, *Chemical Reaction Engineering*, 2d ed., John Wiley & Sons, Inc., New York, 1972. While many other aspects of reactor design are included, this is perhaps the best single source for material on mixing and RTDs.
2. J. M. Smith, *Chemical Engineering Kinetics*, 2d ed., McGraw-Hill Book Company, New York, 1970. One of the most popular general texts in the field, made richer by many worked examples illustrating applications to real reactors.

3. J. J. Carberry, *Chemical and Catalytic Reaction Engineering*, McGraw-Hill Book Company, New York, 1976. Rich in information with special emphasis on heterogeneous catalysis and multiphase reactors.
4. C. G. Hill, Jr., *An Introduction to Chemical Engineering Kinetics & Reactor Design*, John Wiley & Sons, Inc., New York, 1977. Lucid presentation of many topics from kinetics to reactor engineering.

These sources provide broad treatment of many aspects of bioreactor design and analysis.

5. B. Atkinson, *Biological Reactors*, Pion Limited, London, 1974.
6. L. E. Erickson and G. Stephanopoulos, "Biological Reactors," chap. 13 in *Chemical Reaction and Reactor Engineering*, J. J. Carberry and A. Varma (eds.), Marcel Dekker, Inc., New York, 1985.

Wall growth effects in biological CSTRs are considered in:

7. J. A. Howell, C. T. Chi, and U. Pawlowsky, "Effect of Wall Growth on Scale-Up Problems and Dynamic Operating Characteristics of the Biological Aerator," Biotech Bioeng., **14**: 253, 1972.

Excellent overviews of time and length scales and their importance in bioreactors are given by:

8. N. W. F. Kossen, "Models in Bioreactor Design," p. 23 in *Computer Applications in Fermentation Technology*, Society of Chemical Industry, London, 1982.
9. J. A. Roels, "Mathematical Models and the Design of Biochemical Reactors," *J. Chem. Tech. Biotechnol.*, **32**: 59, 1982.

Elementary stability theory and reactor dynamics

10. N. R. Amundson and R. Aris, "An Analysis of Chemical Reactor Stability and Control," Parts I–III. *Chem. Eng. Sci.* **7**: 121, 1958.
11. C. P. Jeffreson and J. M. Smith, "Stationary and Nonstationary Models of Bacterial Kinetics in Well-Mixed Flow Reactors," *Chem. Eng. Sci.*, **28**: 629, 1973.
12. D. Ramkrishna, A. G. Fredrickson, and H. M. Tsuchiya, "Dynamics of Microbial Propagation: Models Considering Inhibitors and Variable Cell Composition," *Biotech. Bioeng.*, **9**: 129, 1967.
13. S. S. Lee, A. P. Jackman, and E. D. Schroeder, "A Two-State Microbial Growth Kinetic Model." *Water Res.*, **9**: 491, 1975.
14. T. B. Young, D. F. Bruley, and H. R. Bungay, III, "A Dynamic Mathematical Model of the Chemostat," *Biotech. Bioeng.*, **12**: 747, 1970.

Nonideal mixing in reactors

15. S. Nagata, *Mixing: Principles and Applications*, Wiley, New York, 1975.
16. R. Shinnar, "Residence Time and Contact Time Distributions in Chemical Reactor Design," in *Chemical Reaction and Reactor Engineering*, J. J. Carberry and A. Varma (eds), Marcel Dekker, Inc., New York, 1985.
17. J. Villermaux, "Mixing in Chemical Reactors," p. 135 in *Chemical Reaction Engineering–Plenary Lectures*, J. Wei and C. Georgakis (eds.), American Chemical Society, Washington, D.C., 1983.
18. M. Charles, "Technical Aspects of the Rheological Properties of Microbial Cultures," p. 1 in *Advances in Biochemical Engineering*, vol. 8, T. K. Ghose, A. Fiechter, and N. Blakebrough (eds.), Springer-Verlag, New York, 1978.
19. J. Bryant, "The Characterization of Mixing in Fermenters," p.101 in *Advances in Biochemical Engineering*, vol. 5, T. K. Ghose, A. Fiechter, N. Blakebrough (eds.), Springer-Verlag, New York, 1977.
20. Th. N. Zweitering, "The Degree of Mixing in Continuous Flow Systems," *Chem. Eng. Sci.*, **11**: 1, 1959.
21. M. Popovic, A. Papalexiou and M. Reuss, "Gas Residence Time Distribution in Stirred Tank Bioreactors," *Chem. Eng. Sci.*, **38**: 2015, 1983.

Nonideal reactor models

22. C. G. Sinclair and D. E. Brown, "The Effect of Incomplete Mixing on the Analysis of the Static Behavior of Continuous Cultures," *Biotech. Bioeng.*, **12**: 1001, 1970.

23. K. Gschwend, A. Fiechter, and F. Widmer, "Oxygen Transfer in a Loop Reactor for Viscous Non-Newtonian Biosystems," *J. Ferment. Technol.*, **61**: 491, 1983.

Mixing-bioreaction interactions

24. F. Vardar, "Problems of Mass and Momentum Transfer in Large Fermenters," *Process Biochem.*, **18**: 21, 1983.
25. E. Ujcová, Z. Fencl, M. Musi'lková, and L. Seichert, "Dependence of Release of Nucleotides from Fungi on Fermentor Turbine Speed," *Biotech. Bioeng.*, **22**: 237. 1980.
26. J. C. van Suihdam and B. Metz, "Influence of Engineering Variables upon the Morphology of Filamentous Molds," *Biotech. Bioeng.*, **23**: 111, 1981.
27. D. I. C. Wang and R. C. J. Fewkes: "Effect of Operating and Geometric Parameters on the Behavior of Non-Newtonian, Mycelial, Antibiotic Fermentations," *Devel. Ind. Microbiol.*, **18**: 39, 1977.
28. F. Vardar and M. D. Lilly, "Effect of Cycling Dissolved Oxygen Concentrations on Product Formation in Penicillin Fermentations," *European J. Appl. Microbiol. Biotechnol.*, **14**: 203, 1982.

The standard reference for unsteady-state heat conduction in solids

29. H. S. Carslaw and J. C. Jaeger, *Conduction of Heat in Solids*, 2d ed., Clarendon Press, Oxford, 1959.

Heat-treatment design for food processing

30. S. E. Charm, *The Fundamentals of Food Engineering*, 2d ed., Avi Publishing Co., Inc., Westport, Conn., 1971.

Formulation, characterization, and application of immobilized cell biocatalysts are included in Refs. 14 and 15 of Chap. 4. Other useful sources include the following:

31. K. Venkatsubramanian (ed.), "Immobilized Microbial Cells", ACS Symposium Series 106, American Chemical Society, Washington, D.C., 1979.
32. T. R. Jack and J. E. Zajic, "The Immobilization of Whole Cells," p. 125 in *Advances in Biochemical Engineering*, vol 5, T. K. Ghose, A. Fiechter, and N. Blakebrough (eds.), Springer-Verlag, New York, 1977.
33. K. Gbewonyo and D. I. C. Wang, "Confining Mycelial Growth to Porous Microbeads. A Novel Technique to Alter the Morphology of Non-Newtonian Mycelial Cultures," *Biotech. Bioeng.*, **25**: 967, 1983.
34. M. Nagashima, M. Azuma, and S. Noguchi, "Technology Developments in Biomass Alcohol Production in Japan: Continuous Alcohol Production with Immobilized Microbial Cells," *Ann. N. Y. Acad. Sci.*, **413**: 457, 1983.
35. C. Förberg, S-O. Enfors, and L. Häggström, "Control of Immobilized, Non-Growing Cells for Continuous Production of Metabolites," *Eur. J. Appl. Microbiol. Biotechnol.*, **17**: 143, 1983

Tower and air-lift reactors

36. K. Schügerl, J. Lücke, and U. Oels, "Bubble Column Bioreactors (Tower Bioreactors Without Mechanical Agitation)," p. 1 in *Advances in Biochemical Engineering*, vol. 7, T. K. Ghose, A. Fiechter, and N. Blakebrough (eds.), Springer-Verlag, New York, 1977.
37. K. Schügerl, "Characterization and Performance of Single- and Multistage Tower Reactors with Outer Loop for Cell Mass Production," p. 93 in *Advances in Biochemical Engineering*, **22**, A. Fiechter (ed.), Springer-Verlag, New York, 1982.
38. A. Kitai, H. Tone, and A. Ozaki, "Performance of a Perforated Plate Column as a Multistage Continuous Fermentor," *Biotech. Bioeng.*, **11**: 911, 1969.
39. A. Prokop et al., "Design and Physical Characteristics of a Multistage Continuous Tower Fermentor," *Biotech. Bioeng.*, **11**: 945, 1969.
40. E. A. Falch and E. L. Gaden, Jr., "A Continuous, Multistage Tower Fermentor, I: Design and Performance Tests," *Biotech. Bioeng.*, **11**: 927, 1969: "II: Analysis of Reactor Performance," ibid., **12**: 465, 1970.

41. R. G. Ault et al., "Biological and Biochemical Aspects of Tower Fermentation," *J. Inst. Brewing*, **75**: 260, 1969.
42. R. N. Greenshields and E. L. Smith, "Tower-Fermentation Systems and Their Applications," *Chem. Eng. Lond.*, May 1971, p. 182.

Fluidized-bed and trickle-bed reactors

43. C. D. Scott, "Fluidized-Bed Bioreactors Using a Flocculating Strain of *Zymomonas mobilis* for Ethanol Production," *Annals N.Y. Acad. Sci.*, **413**: 448, 1983.
44. E. Flaschel, P.-F. Fauquex, and A. Renken, "Zum Verhalten van Flüssigkeits-Wirbelschichten mit Einbauten," *Chem.-Ing.-Tech.* **54**: 54, 1982.
45. J. G. van de Vusse and J. A. Wesselingh, "Multiphase Reactors," p. 561 in 4th International/6th European Symposium on Chemical Reaction Engineering, vol. 2, DECHEMA, *Deutsche Gesellschaft für Chemisches Apparatewesen*, e.V., Frankfurt, 1976.
46. J. Briffaud and J. Engasser, "Citric Acid Production from Glucose. II. Growth and Excretion Kinetics in a Trickle-Flow Fermentor," *Biotech. Bioeng.*, **21**: 2093 (1979).

Fermentation technology

47. H. J. Peppler and D. Perlman (eds.), *Microbial Technology*, 2d ed., vols. I and II, Academic Press, Inc., New York, 1979. An unusually rich collection of review articles on processes, technology, products, and markets.
48. L. E. Casida, Jr., *Industrial Microbiology*, John Wiley & Sons, Inc., New York, 1968. A good qualitative review of products, processes, and techniques in the fermentation industry, including patents and economics. Numerous photographic illustrations give a good impression of industrial practice.
49. F. C. Webb, *Biochemical Engineering*, D. Van Nostrand Company, Ltd., London, 1964. Besides discussing some microbial products and equipment-design considerations, this reference is distinguished by inclusion of several other important subjects, e.g., colloids, emulsions, redox potentials, chemical disinfection, dehydration, radiation, and vaccine manufacture
50. K. Schügerl, "New Bioreactors for Aerobic Processes," *Int. Chem. Eng.* **22**: 591, 1982.

Animal cell culture

51. J. Feder and W. R. Tolbert, "The Large-Scale Cultivation of Mammalian Cells," *Scientific American*, **248**: 36, 1983.
52. W. R. Tolbert, R. A. Schoenfeld, C. Lewis, and J. Feder, "Large-Scale Mammalian Cell Culture: Design and Use of an Economical Batch Suspension System," *Biotech. Bioeng.*, **24**: 1671, 1982.
53. A. Fiechter, "Batch and Continuous Culture of Microbial, Plant and Animal Cells," p. 453 in *Biotechnology*, vol. 1, H.-J. Rehm and G. Reed (eds.), Verlag Chemie, Weinheim, 1982.
54. H. W. D. Katinger and W. Scheirer, "Status and Developments of Amimal Cell Technology using Suspension Culture Techniques," *Acta Biotechnologica*, **2**: 3, 1982.
55. M. W. Glacken, R. J. Fleischaker, and A. J. Sinskey, "Large-scale Production of Mammalian Cells and Their Products: Engineering Principles and Barriers to Scale-up." *Annals N.Y. Acad. Sci.*, **413**: 355, 1983.
56. P. F. Kruse, Jr., and M. K. Patterson, Jr., *Tissue Culture: Methods and Applications*, Academic Press, Inc., 1973.
57. I. S. Johnson and G. B. Boder, "Metabolites from Animal and Plant Cell Culture," *Adv. Microbiol.*, **15**: 215, 1973.
58. R. Acton and J. D. Lynn (eds.), *Cell Culture and its Application*, Academic Press, New York, 1977.
59. R. T. Acton and J. D. Lynn, "Description and Operation of a Large-Scale, Mammalian Cell, Suspension Culture Facility," in *Advances in Biochemical Engineering*, vol. 7, T. K. Ghose, A. Fiechter, and N. Blakebrough (eds.), Springer-Verlag, New York, 1977.
60. F. Lim and R. D. Moss, "Microencapsulation of Living Cells and Tissues," *J. Pharm. Sci.*, **70**: 351, 1981.
61. K. Ku, M. J. Kuo, J. Delente, B. S. Wildi, and J. Feder, "Development of a Hollow-Fiber System for Large-Scale Culture of Mammalian Cells," *Biotech. Bioeng.*, **23**: 79, 1981.

62. E. M. Scattergood, A. J. Schlabach, W. J. McAleer, and M. R. Hilleman: "Scale-Up of Chick Cell Growth on Microcarriers in Fermenters for Vaccine Production," *Annals N. Y. Acad. Sci.*, **413**: 332, 1983.
63. R. J. White, F. Klein, J. A. Chen, and R. M. Stroshane, "Large-Scale Production of Human Interferons," in *Annual Reports on Fermentation Processes*, vol. 4, G. T. Tsao (ed.), Academic Press, 1980.

Plant cell culture:
64. H. E. Street (ed.), *Tissue Culture and Plant Science*, Academic Press, London, 1974.
65. A. Fiechter (ed.), *Advances in Biochemical Engineering*, vols. 16, 18, Springer-Verlag, Berlin, 1980.
66. G. Wilson, "Continuous Culture of Plant Cells Using the Chemostat Principle," p. 1 in *Advances in Biochemical Engineering*, vol. 16, A. Fiechter (ed.), Springer-Verlag, New York, 1980.
67. M. L. Shuler, O. P. Sahai, and G. A. Hallsby, "Entrapped Plant Cell Tissue Cultures," *Annals N.Y. Acad. Sci.*, **413**: 373, 1983.
68. P. Brodelius, "Production of Biochemicals with Immobilized Plant Cells: Possibilities and Problems," *Annals N.Y. Acad. Sci.*, **413**: 383, 1983.
69. M. E. Curtin, "Harvesting Profitable Products from Plant Tissue Culture," *Bio/Technology*, **1**: 649, 1983.

CHAPTER

TEN

INSTRUMENTATION AND CONTROL

We have seen repeatedly that the activity and useful lifetime of an enzyme catalyst or cell population depends directly on the catalyst environment. Accordingly, in order to develop and optimize biological reactors and in order to operate them most efficiently, it is critical that the state of the catalyst environment be monitored and controlled and that the response of the catalyst to the environment be determined. Achieving these goals requires three different functions: measurement, analysis of measurement data, and control. In this chapter we will examine currently available reactor instrumentation and its application. After a brief look at the rapidly changing technologies for data acquisition and computing, we shall summarize some of the strategies for data analysis and process control. Although this chapter's presentation emphasizes bioreactor instrumentation and control, the principles described also apply to downstream processing and to feedstock preparation.

10.1 PHYSICAL AND CHEMICAL SENSORS FOR THE MEDIUM AND GASES

Figure 10.1 shows a recently prepared schematic summary of biochemical reactor instrumentation. A striking feature of this illustration is the predominance of measurements of medium and gas chemical and physical properties and the shortage of measurements of cell properties. Consequently, one of the major goals, if not requirements, of bioreactor data analysis is estimation of cell properties based on the available physiochemical measurements of the gas streams and the medium. In this section we will concentrate on instrumentation for on-line physical and chemical monitoring of bioreactors.

10.1.1 Sensors of the Physical Environment

The major physical process parameters that influence cellular function and process economics and which can be monitored continuously are temperature, pressure, agitator shaft power, impeller speed, broth viscosity, gas and liquid flow rates, foaming, and reactor contents volume or mass. In small laboratory reactors, only temperatures and air-feed flow rates are commonly measured. Pressure measurement and regulation is common on larger fermentors.

The most widely used temperature sensor is the thermistor, a semiconductor device which exhibits changing resistance as a function of temperature. Although the temperature-resistance relationship is nonlinear, this is not a serious difficulty over the narrow temperature range of interest for most fermentations (25–45°C). Other possible temperature sensors are the platinum resistance sensor, thermometer bulbs (Hg in stainless steel), and thermocouples.

Pressure monitoring is important during sterilization, and maintaining a positive reactor head pressure (around 1.2 atm absolute) can aid in preserving asepsis. Pressure also influences gas solubility. In fermentation reactors, diaphragm gauges are usually used to monitor pressure. These produce a pneumatic signal which may be transduced if necessary to an electrical signal.

Several different types of measurements can be made to monitor power input in mechanically agitated vessels. A Hall effect wattmeter measures at the drive motor armature the total energy consumed by the agitator. A torsion dynamometer may also be used to measure shaft power input. A disadvantage of both of these measurement methods is inclusion of frictional losses in shaft bearings and seals. For example, in a study of mixing in a 270-liter fermentor with 200 liters working volume, it was found that 30% of the energy used by the motor was lost between the motor and the internal impeller shaft. This loss factor was also observed to be an increasing function of agitator speed. Direct measurement of impeller power input to the reactor fluid may be achieved using balanced strain gauges mounted on the impeller shaft inside the reactor.

On-line devices for measuring broth viscosity and other rheological properties are not well developed. One possible strategy is measurement of power consumption at several different impeller speeds. Also, a dynamic method has been proposed in which shaft power input is monitored during and after a brief (less than 30 s) shutoff in agitator drive power. As sketched in Fig. 10.1, Newtonian and non-Newtonian broths have been observed to respond differently during such a brief agitation transient.

Several different instruments are available for measuring flow rates of gases (air feed, exhaust gas). The simplest, a variable area flowmeter such as a rotameter, provides visual readout or may be fitted with a transducer to give an

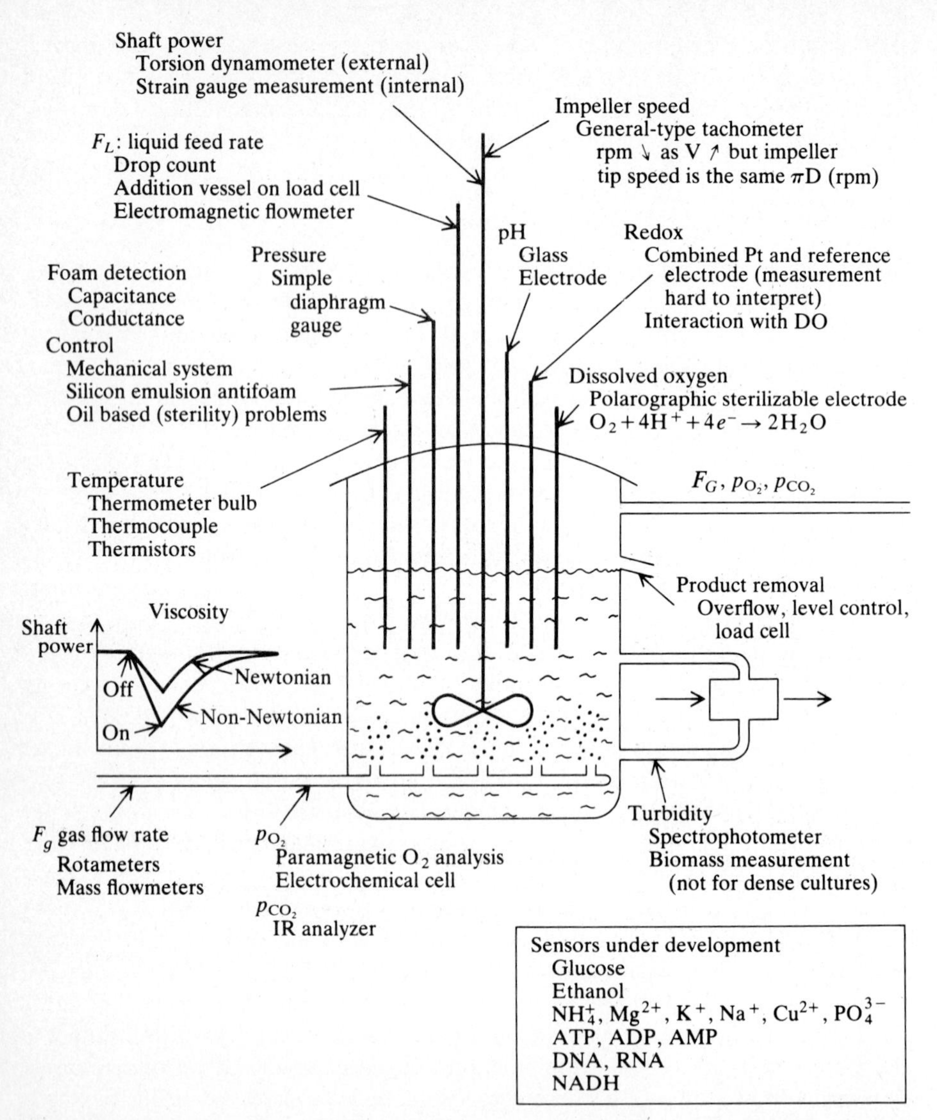

Figure 10.1 Schematic summary of biochemical reactor instrumentation. *[Reprinted by permission from L. E. Erickson and G. Stephanopoulos, "Biological Reactors," Chap. 13 in "Chemical Reaction and Reactor Engineering," J. J. Carberry and A. Varma (eds.), Marcel Dekker, Inc., New York, 1985.]*

electrical output. Thermal mass flowmeters are increasingly popular, especially for lab and pilot-scale reactors. In these devices, gas flows through a heated section of tubing, and the temperature difference across this heated section is directly related to mass flow rate. These instruments have accuracies on the order of 1% of full-scale and are most useful for flow rates less than 500 L/min. Also available are laminar flow measurement devices which determine flow based on differential pressure drop across a matrix device which divides the total flow into multiple parallel capillary flows. Gas flow rate measurements are important since these quantities are used frequently in material balancing calculations (Sec. 10.5).

Liquid flow rates can be monitored with electromagnetic flowmeters, but these are not used widely due to their cost. Occasionally, especially in laboratory scale studies, one relies on a metering or other well-calibrated pump to provide the desired liquid flow rate. Alternatively, liquid can be added to the reactor in discrete doses of well-defined volume or mass. Long-term monitoring of net flow into the vessel may be achieved by continuous weighing of the reactor and its liquid contents using a strain gauge (vessels >250 L) or scale (smaller vessels). Alternatively, a liquid level sensor based on a capacitance probe may be used to monitor reactor liquid content. Such capacitance probes or a conductance probe may also be used to detect buildup of foam on the top surface of the reactor contents. In some situations an external loop of circulating broth is used for measurements (see below), to effect product removal and cell recycle, or for heat and/or gas exchange as discussed in Chaps. 9 and 14. Here the presence of suspended particulates and changing broth rheology severely complicate liquid flow rate measurement.

10.1.2 Medium Chemical Sensors

Electrodes which can be repeatedly steam-sterilized in place are now available for pH, redox potential (E_h) and dissolved oxygen and CO_2 partial pressures. The most widely used and reliable probe among these is the pH electrode, which is generally a single unit glass-reference electrode design. A schematic diagram of a pH electrode designed for autoclave sterilization is shown in Fig. 10.2. Electrodes for *in situ* sterilization must include a housing to provide pressure balance during sterilization or pressurized bioreactor operation. Measurement of medium redox potential is possible using a combined platinum and reference electrode. Combined pH-redox probes are available. While the influence of pH on biochemical kinetics is clearly established and the physical significance of a pH measurement is straightforward, interpretation of redox potential measurements and understanding the relationship between redox potential and cell activity can be difficult. One promising application of redox measurements is in monitoring low contents of dissolved oxygen (<1 ppm) in anaerobic processes ($-450\text{ mV} < E_h < -150\text{ mV}$) where product formation may be quite sensitive to E_h.

The various types of dissolved oxygen probes now available are of galvanic (potentiometric) or polarographic (amperometric or Clark) types. These electrodes measure the partial pressure (or activity) of the dissolved oxygen and not the dissolved oxygen concentration. In both designs, an oxygen-permeable membrane usually separates the electrode internals from the medium fluid (Fig. 10.3). Also, both designs share the common feature of reduction of oxygen at the cathode surface.

Cathode

$$\tfrac{1}{2}O_2 + H_2O + 2e^- \xrightarrow{\text{Pt}} 2OH^- \tag{10.1}$$

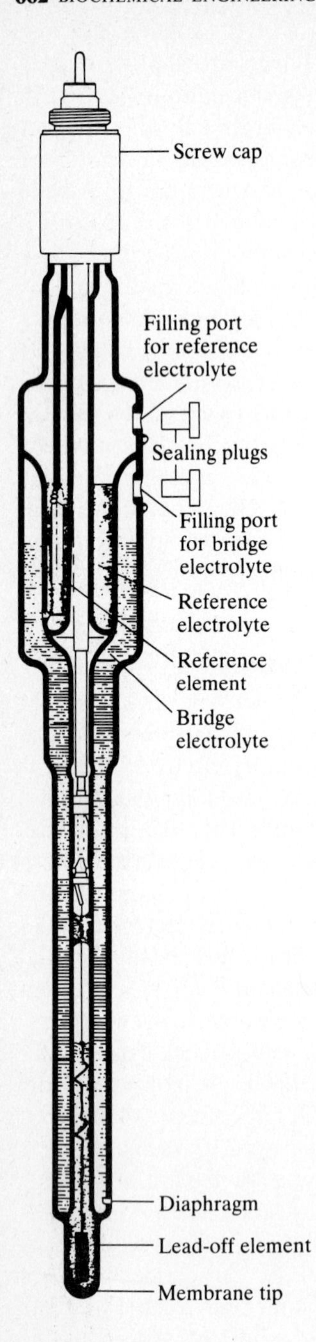

Figure 10.2 Schematic diagram of a combination pH electrode designed for autoclave steam sterilization. *(Ingold type 465. Illustration courtesy of Ingold Electrodes Inc.)*

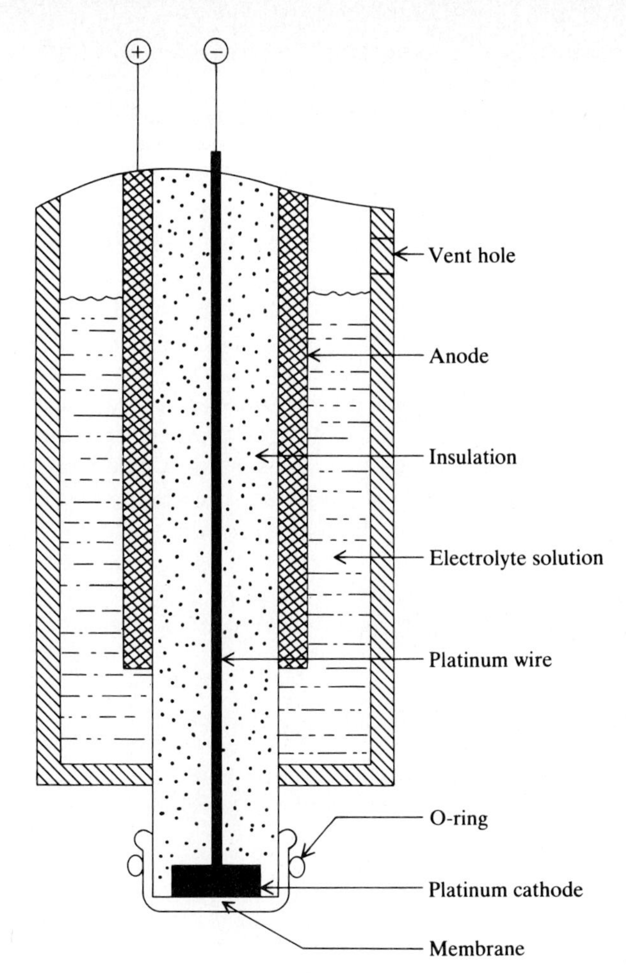

Figure 10.3 Schematic diagram of major components of an electrochemical probe for measurement of dissolved oxygen partial pressure. *(Reprinted by permission from N. S. Wang and G. Stephanopoulos, "Computer Applications to Fermentation Processes," CRC Critical Reviews in Biotechnology, vol. 2, p. 1. © CRC Press, Inc., 1974. Used by permission of CRC Press, Inc.)*

The reaction at the anode in a galvanic electrode

Anode (galvanic)

$$Pb \longrightarrow Pb^{2+} + 2e^- \tag{10.2}$$

completes the cell from which a small amount of current is drawn to provide a voltage measurement which in turn is correlated to the oxygen flux reaching the cathode surface. In a polarographic type of oxygen electrode, a constant voltage is applied across the cathode [Eq. (10.1)] and anode

Anode (polarographic)

$$Ag + Cl^- \longrightarrow AgCl + e^- \tag{10.3}$$

and the resulting current, which depends on the oxygen flux to the cathode, is measured. Drift caused by accumulation of hydroxyl or metal ions or chloride depletion is a common drawback of both electrode types. External fouling of the membrane surface may also contribute to drift.

In steady state, the oxygen flux at the cathode depends upon a series of transport steps in which oxygen moves from the bulk liquid to the outer membrane surface, diffuses through the membrane, and finally diffuses through the electrolyte solution to the cathode surface where reaction occurs effectively instantaneously. To the extent that the first step limits the overall transport rate, and thus the oxygen flux to the cathode, the electrode output will depend on fluid properties (e.g., viscosity) and local hydrodynamic conditions near the electrode. For this reason it has been recommended, for example, that the fluid velocity at the tip of a polarographic electrode should be at least 0.55 m/s. Sensitivity of the electrode output to external boundary layer transport can also be reduced by using a less permeable membrane. (Why?) This approach has the disadvantage of introducing additional time delay in the instantaneous electrode response to transients in dissolved oxygen partial pressures. The characteristic response time of membrane-covered dissolved oxygen sensors is quite long (10–100 s). However, as shown in the following example, transient probe measurements may still be applied to characterize mass-transfer properties of bioreactors provided the influence of diffusion through the probe membrane is included in the analysis.

Example 10.1: Electrochemical determination of $k_l a$ An oxygen electrode is inserted into a steadily aerated batch or continuous-flow reactor. After a steady electrode response has been achieved, the oxygen-carrying flow is suddenly replaced by an equivalent nitrogen flow, which then strips the oxygen from solution in a transient manner. Under these circumstances, the voltage response of the probe is given by [11]

$$E(t) = E_0\left[\frac{\tau^{1/2}\exp(-\beta t)}{\sin \tau^{1/2}} - 2\sum_{n=1}^{\infty}\frac{(-1)^n \exp(-n^2\pi^2\mathscr{D}_{O_2}t/L^2)}{1 - n^2\pi^2/\tau}\right] \tag{10E1.1}$$

where $\tau = \beta L^2/\mathscr{D}_{O_2}$
$E(t)$ = probe voltage at time t
E_0 = probe voltage at time 0
F_g = gas flow rate
V = liquid volume in tank
M = Henry's law constant
L = thickness of probe membrane
$\mathscr{D}_{O_2}$ = oxygen diffusivity in probe membrane = P_m/L
= permeability/membrane thickness, and

$$\frac{1}{\beta} = \text{time constant of system (reactor)} = \frac{1}{k_l a} + \frac{VRT}{F_g M} \tag{10E1.2}$$

If the system has a large time constant (small $k_l a$), the value of β is determined from electrode-voltage values for times much greater than the electrode response itself, in which case the series contribution above is negligible. For larger values of $k_l a$ (50–500 h^{-1}), Wernan and Wilke [11] suggest using the slope of the electrode response at the inflection point (where the second derivative of E with respect to t vanishes). This method was found to give results as accurate as computer-aided curve fitting using the entire voltage-time response. For various electrode relaxation times, the value of the inverse relaxation time is read from a graph of this parameter vs. the slope of E vs. t at the inflection point.

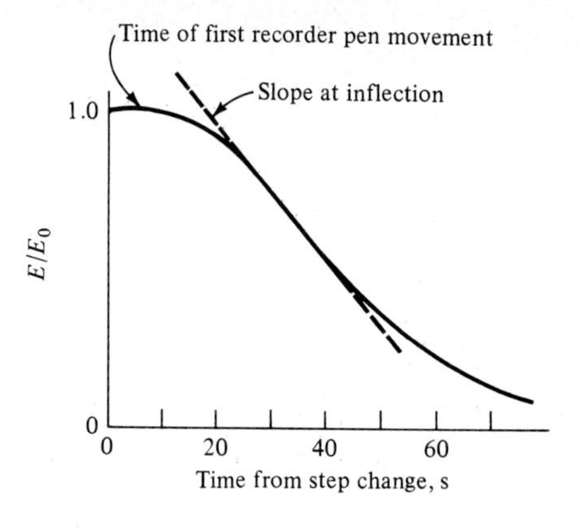

Figure 10E1.1 The inflection-point slope is determined graphically from the electrode response following a switch from sparging of an oxygen-carrying gas to nitrogen sparging. *(Reprinted from W. C. Wernan and C. R. Wilke, "New Method for Evaluation of Dissolved Oxygen Response for k_La Determination," Biotech. Bioeng., vol. 15, p. 571, 1973.)*

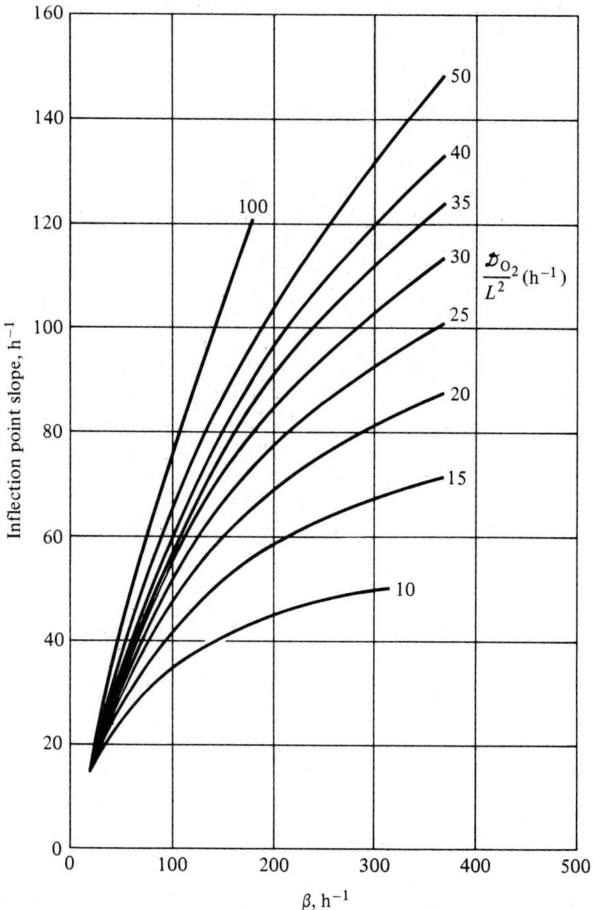

Figure 10E1.2 From this graph, the value β may be determined from the inflection-point slope and membrane-transport parameters. k_la then follows from Eq. (10E1.2). *(Reprinted from W. C. Wernan and C. R. Wilke, "New Method for Evaluation of Dissolved Oxygen Response for k_La Determination," Biotech. Bioeng., vol. 15, p. 571, 1973.)*

Evaluation of the desired value, $k_l a$, then proceeds in three steps:

1. From the E-vs.-t electrode-response curve (Fig. 10E1.1) evaluate the slope at the inflection point (steepest point).
2. From the generalized display of Eq. (10E1.2) in Fig. 10E1.2, obtain β, given the slope and the electrode time constant.
3. Evaluate $k_l a$ from the definition of the system time constant $(1/\beta)$ above.

It is evident that the usefulness of this technique depends on the accuracy with which the electrode response is known, particularly for large $k_l a$ values.

The presence of high solute concentrations may alter oxygen solubility (Table 8.1), the oxygen diffusion coefficient and the electrode itself (thus the electrode response). A complete dissection of these influences clearly requires calibration by independent means.

Steam-sterilizable electrochemical probes for dissolved CO_2 partial pressure have been introduced relatively recently. The CO_2 probe produced by Ingold, for example, determines p_{CO_2} by measuring the pH of a standard bicarbonate solution which is separated from the process fluid by a gas-permeable membrane. Calibration is accomplished by measuring pH after substitution of a reference buffer solution for the bicarbonate solution.

Another class of methods for on-line assay of volatile medium components and dissolved gases is based upon immersion of a length of tubing, permeable to the component(s) of interest, in the fluid to be analyzed. Continuous flow of a carrier gas through the tubing sweeps the compounds which penetrate the tubing to a gas analysis device (see next section), where the measurement is conducted. This approach suffers from substantial measurement delays (2–10 min) and, therefore, is not optimal for monitoring rapid transients in concentrations.

Several *biosensors* have been developed for assay of specific components in the liquid phase. These are based on coupling the action of immobilized enzymes or cells with an analytical device which detects a particular product of the biocatalyzed reaction. Examples of combining enzyme-catalyzed reactions with electrochemical detectors were discussed earlier in Sec. 4.3.3. Also studied extensively are enzyme thermisters, in which the heat released by the enzyme-catalyzed reaction is detected by a nearby calorimeter. Table 10.1 summarizes some of the compounds which have been assayed by this method and the corresponding enzymes employed. Other possibilities for biosensor development using immobilized enzymes include enzyme transitors in which reaction products (for example, hydrogen) cause changes in the electronic properties of solid-state devices (for example, silicon chips with an SiO_2-layer covered with a Pd film).

The spectrum of biosensor designs and configurations can be enlarged by considering a broader class of biocatalyzed reactions, including multistep or coupled reactions, to generate the detected component. Immobilized cells provide a convenient means in many cases for transforming the component to be assayed into a suitable detectable compound. Immobilized whole-cell respiratory activity (assay by oxygen electrode) and production or consumption of electroactive metabolites by whole cells (assayed by fuel cell electrode or by pH or CO_2 electrodes) have been used as the bases for design and application of biosensors

Table 10.1 Examples of analytical applications of enzyme thermistors
(Adapted from B. Danielsson and K. Mosbach, "The Prospects for Enzyme-Coupled Probes in Fermentation," p. 137, in "Computer Applications in Fermentation Technology," Society of Chemical Industry, London, 1982.)

Substance analyzed	Immobilized enzyme	Concentration range or detection limit, mmole/L
Ascorbic acid	Ascorbic acid oxidase	0.05–0.6
Albumin (antigen)	Immobilized antibodies + enzyme-linked antigen	10^{-10}
ATP	Apyrase or hexokinase	1–8
Cellobiose	β-glucosidase + glucose oxidase/catalase	0.05–5
Cephalosporin	Cephalosporinase	0.005–10
Cholesterol	Cholesterol oxidase	0.03–0.15
Cholesterol esters	Choleserol esterase + cholesterol oxidase	0.03–0.15
Creatinine	Creatinine iminohydrolase	0.01–10
Ethanol	Alcohol oxidase	0.01–1
Galactose	Galactose oxidase	0.01–1
Gentamicin (antigen)	Immobilized antibodies + enzyme-linked antigen	0.1 μg/mL
Glucose	Glucose oxidase/catalase	0.002–0.8
Heavy metal ions (e.g., Pb^{2+})	Urease	10^{-6}
Insecticides (e.g., parathion)	Acetylcholinesterase	5×10^{-3}
Insulin (antigen)	Immobilized antigen + enzyme-linked antigen	0.1–1.0 unit/mL
Lactate	Lactate 2-monooxygenase	0.01–1
Lactose	Lactase and glucose oxidase/catalase	0.05–10
Oxalic acid	Oxalate oxidase	0.005–0.5
Penicillin G	Penicillinase	0.01–500
Phenol (substrate)	Tyrosinase	0.01–1
Sucrose	Invertase	0.05–100
Triglycerides	Lipase, lipoprotein	0.1–5
Urea	Urease	0.01–500
Uric acid	Uricase	0.5–4

containing immobilized whole cells. Table 10.2 summarizes the properties of a number of whole-cell sensors which are described in greater detail in Ref. 12.

An interesting and promising alternative strategy for formulating specific "affinity sensors" for individual metabolites has been described and developed by J. S. Schultz and coworkers [13]. The requirements for this method are a specific binding agent for the component to be assayed, availability of a suitably labeled component which competes for the same specific binding agent, and a means of separating binding agent from the solution to be assayed. For example, a fiber optic fluorescence probe has been constructed for glucose analysis by immobilizing conconavalin A (Con-A), a protein from jack bean which selectively binds sugars, on the internal surfaces of a measurement chamber. The chamber, separated from the assayed solution by a dialysis membrane permeable to glucose,

Table 10.2 Summary of microbial sensors for process fluid analysis

Component assayed	Organism employed	Immobilization method	Measurement principle	Detection device
Acetic acid, ethanol	*Trichosporon brassicae*	Attachment to porous acetylcellulose	Simultaneous O_2 consumption	O_2 electrode
Ammonia	*Nitrosomonas* and *Nitrobacter* species	Rentention by gas-permeable membrane	Simultaneous O_2 consumption	O_2 electrode
Cephalosporin	*Citrobacter freundii*	Collagen membrane entrapment	Liberation of H^+	pH electrode
Formic acid	*Clostridium butyricum*	Agar entrapment on acetylcellulose filter	Production of H_2	Fuel cell electrode
Glucose	*Pseudomonas fluorescens*	Collagen membrane entrapment	Simultaneous O_2 consumption	O_2 electrode
Glutamic acid	*Escherichia coli*	Rentention by cellophane membrane	CO_2 production	CO_2 electrode

also contains dextran labeled with the fluorochrome fluoroscein isothiocyanate (FITC). The membrane used is impermeable to the FITC-dextran which competes with glucose for binding to Con-A:

$$\text{Glucose} + \text{Con-A} \rightleftharpoons \text{Con-A-glucose}$$

$$\text{FITC-Dextran} + \text{Con-A} \rightleftharpoons \text{Con-A-FITC-dextran}$$

Thus, the amount of unbound FITC-dextran, and hence the measured fluorescence emission intensity, is a function of the solution glucose concentration. Interference by other solutes which also bind to the specific binding agent (maltose, sucrose, and fructose to Con-A, for example) pose potential problems for this approach, but the concept involved is intriguing, quite general, and should enjoy further development.

A further concern in use of any sensor employing enzymes, cells, or other biochemicals is deactivation of the sensor during reactor sterilization. Mechanical designs which allow aseptic removal and insertion of the sensor in the reactor interior have now been developed to address this potential problem. Also, as in all sensors which depend on transport of the monitored component through a membrane, membrane fouling by cells or medium components and external mass transport resistance can cause drift or shifts in calibration of the sensor.

10.1.3 Gas Analysis

The concentration of CO_2 in the exhaust gas from a cell reactor is indicative of respiratory or fermentative activity of the organisms and hence is one of the most useful and widely applied measurements in monitoring and controlling a cell bioreactor. CO_2 content in bioreactor gas streams is most commonly monitored using an infrared spectrophotometer. The gas sample stream must be dessicated carefully before entering the instrument to avoid damage to the sample cell windows. Gas stream CO_2 concentration may also be measured using thermal conductivity, gas chromatography, or mass spectrometry.

Gas stream oxygen partial pressure is usually measured using a paramagnetic analyzer. Here too, elimination of water vapor in the sample stream is essential to minimize drift, and the sample stream flow rate must be controlled carefully for consistent measurements. Paramagnetic analyzers are also quite sensitive to small changes in total atmospheric pressure, requiring simultaneous monitoring of barometric pressure for compensation in oxygen analysis. Drift in readings which necessitate on-line recalibration is a frequent occurrence with paramagnetic analyzers when applied to fermentations.

Gas chromatography (GC) can be applied to analyze several components of the exhaust gas stream including O_2, CO_2, CH_4 (e.g., in anaerobic methane generation), and H_2 (from *Hydrogenomonas* cultures, for example). Also, by determining the gas phase partial pressure of volatile components such as ethanol, acetaldehyde, and carboxylic acids, GC measurements provide useful information on the status of the fermentation and on the liquid phase concentrations of these compounds. The requirement of intermittent injection of samples (ca. 15 min apart) limits the utility of GC measurements for monitoring process transients.

Mass spectrometry (MS) is enjoying increasing popularity for monitoring gas stream composition. Lower-priced instruments are making MS more accessible for research applications, and reliable, robust process instruments have made mass spectrometry more practical for industrial application. MS instruments offer rapid response times (<1 min), high sensitivity (around 10^{-5} M detection limit), capability to analyze several components essentially simultaneously, linear response over a broad concentration range, and negligible calibration drift. Because of the expense of MS instruments, it is often desirable to interface the analyzer to several bioreactors and use a computer-controlled switching manifold to cycle sample streams from different reactors into the MS (Fig. 10.4). As indicated in this schematic diagram (only three fermentors are shown, but a single mass spec can support up to 30), the same computer may also be used for process control.

Often, standard values (20.91 % O_2, 0.03 % CO_2) are assumed for the feed air composition, but it is sometimes more reliable to measure feed gas composition directly by including a feed gas sample stream in the manifolding arrangement as indicated in Fig. 10.4. Of course, the merits of sharing analyzer instrumentation by use of such multiplexing and manifold arrangements are not limited to cases in which mass spectrometry analyzers are applied.

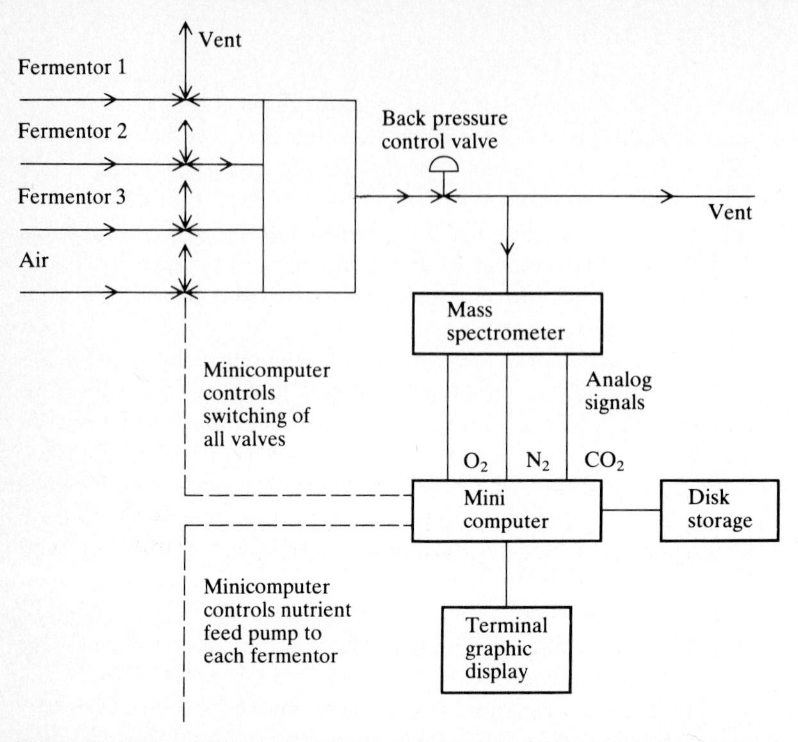

Figure 10.4 Schematic illustration of a computer-controlled sample selection system for time-shared use of a mass spectrometer. *(Reprinted by permission from R. C. Buckland and H. Fastert, "Analysis of Fermentation Exhaust Gas Using a Mass Spectrometer," p. 119, in "Computer Applications in Fermentation Technology," Society of Chemical Industry, London, 1982.)*

10.2 ON-LINE SENSORS FOR CELL PROPERTIES

Unfortunately, there are few instruments for continuous monitoring of cell properties in a bioreactor. The most basic measurement needed is total biomass content or concentration or, better still, active biomass concentration. Although a number of possible methods exist, no approach has yet been invented which provides such data reliably, consistently, and for a broad class of organisms and media.

Optical methods based upon light absorbance (spectrophotometry) or scattering (nephelometry, reflectance measurement) have been investigated widely. A sample stream from the reactor may be circulated through a spectrophotometer. A potential difficulty here is the nonlinearity between optical density and biomass concentration above O.D. = 0.5 or 0.5 g biomass/L. Consequently, sample stream dilution or a shorter light path may be used for measurement of dense cultures. Figure 10.5 shows a flow-through cuvette design developed by Lim and colleagues which has proved very convenient in a number of laboratories. Alternatively, probes which can be inserted into the process fluid for optical cell density measurements have been developed.

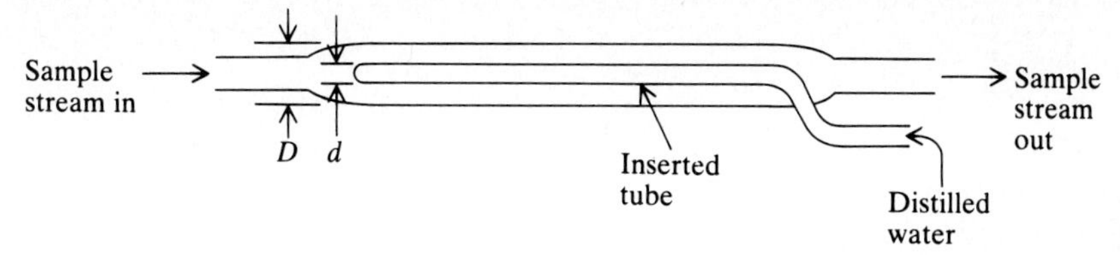

Figure 10.5 Schematic diagram of a flow-through cuvette with internal effective dilution by means of an inserted tube containing distilled water. With $d = 8$ mm, $D = 12$ mm, this device provides a linear OD_{600}-cell density relationship for cell densities exceeding 1.8 g/L. *(Adapted from C. Lee and H. Lim, "New Device for Continuously Monitoring the Optical Density of Concentrated Microbial Cultures," Biotech. Bioeng., vol. 22, p. 636, 1980.)*

The only continuous monitoring strategy so far developed that provides information on the biochemical or metabolic state of the cell population is in situ fluorometry. Ultraviolet light (366 nm wavelength) is directed into the culture. Excited by this incident UV radiation, reduced pyridine nucleotides (NADH and NADPH) fluoresce with a maximum intensity at approximately 460 nm. The fluorescence emitted from the culture is measured with a suitable detector such as a photodiode or photomultiplier. Originally, these measurements were made through quartz windows installed in the walls of laboratory fermentors. The advent of fluorescence probes which can be used in standard electrode ports in fermentation vessels should increase investigations and application of culture fluorescence measurements (Fig. 10.6).

Culture fluorescence intensity depends on cell density, average cell metabolic state and fluorescence emissions, and light absorption by the medium. Experiments in particulate-free media have shown that culture fluorescence measurements provide useful information on biomass concentration, oxygen transfer and

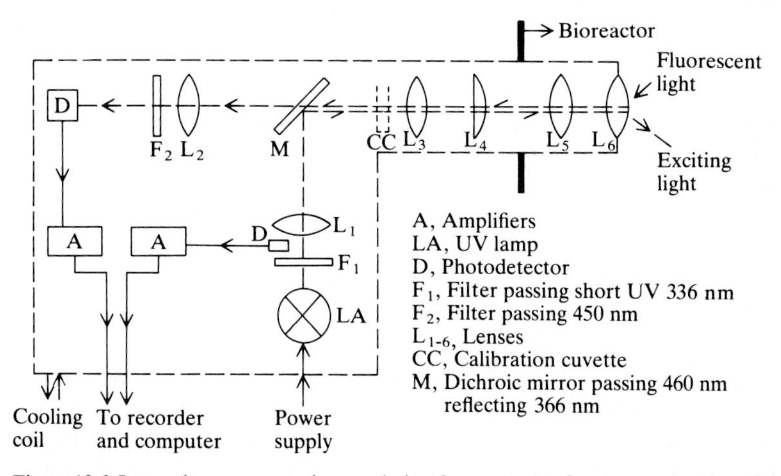

Figure 10.6 Internal components in one design for an *in situ* fluorescence probe. *(Reprinted by permission from W. Beyeler, A. Einsele and A. Fiechter, "On-Line Measurements of Culture Fluorescence: Method and Application," European J. Appl. Microbial. Biotechnol., vol. 13, p. 10, 1981.)*

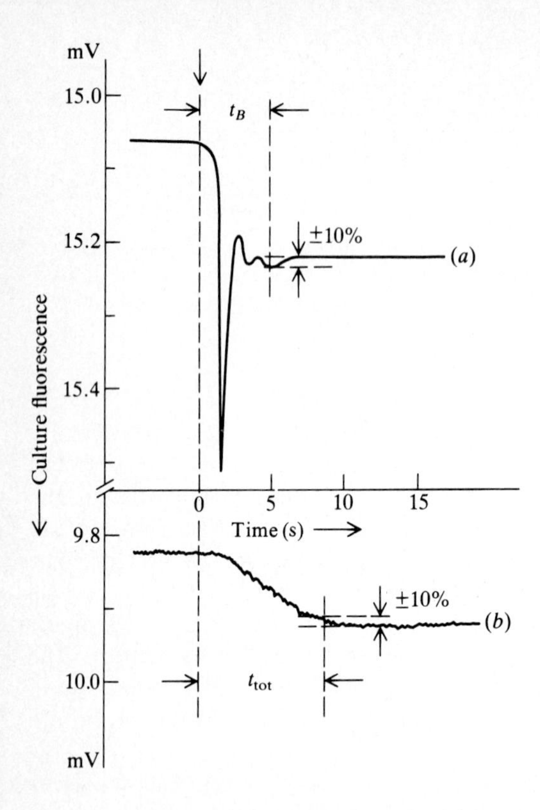

Figure 10.7 Transients in fluorescence measured (*a*) after pulse addition of quinine to 0.5 *M* H_2SO_4 solution, which characterizes bulk liquid mixing, and (*b*) after glucose addition to a yeast culture, which characterizes liquid mixing plus cellular glucose uptake. *(Reprinted by permission from A. Einsele, D. L. Ristroph and A. E. Humphrey, "Mixing Times and Glucose Uptake Measured with a Fluorometer," Biotech. Bioeng., vol. 20, p. 1487, 1978.)*

reactor mixing times, substrate exhaustion, and metabolic transients. For example, Einsele and coworkers compared the dynamics for liquid mixing in a 40-liter working volume fermentor agitated mechanically at 200 rpm with the dynamics of mixing plus glucose uptake by yeast cells. For the first measurement, fluorescence of quinine pulsed into 0.05 *M* H_2SO_4 in the reactor was monitored (this solute has approximately the same fluorescent properties as NADH). The results, shown in part *a* of Fig. 10.7, exhibit oscillations representative of a periodic circulation pattern in the vessel and provide clear evidence of significant dynamic delays in achieving new steady-state conditions in the reactor. Part *b* of Fig. 10.7 is the reduced pyridine nucleotide fluorescence from a yeast culture following a pulse of glucose added to the reactor at time zero. Here the response time is longer, indicating significant dynamic delay in glucose uptake by the cells. The implications of these features in the context of scale-up were mentioned earlier in Section 9.3.4.

Another illustration of an alternative application of culture fluorescence is monitoring of cellular metabolic state to control the fermentation. In an experimental study of *Candida utilis* grown on ethanol in a fed-batch process, culture fluorescence was used to estimate the time of substrate ethanol exhaustion and to control pulse feeding of ethanol into the process. Shown in Fig. 10.8 are some cycles of culture fluorescence, respiratory quotient, and ethanol concentration

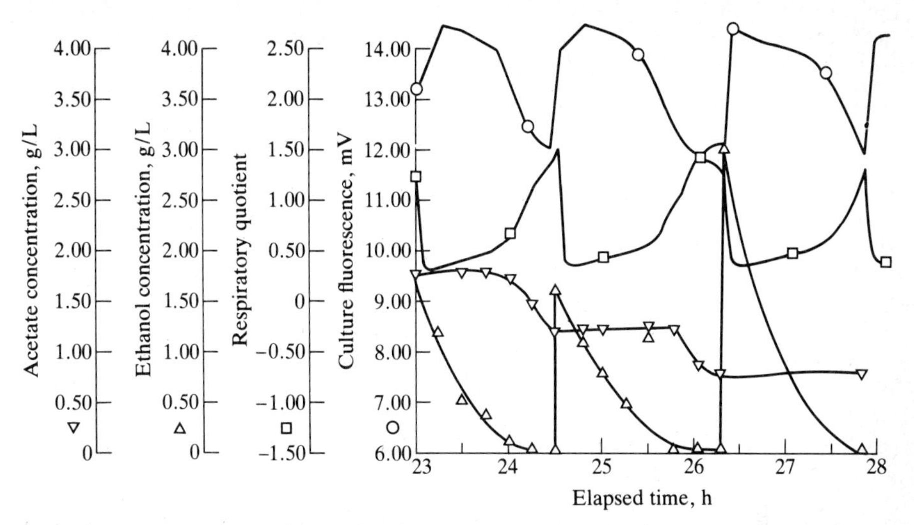

Figure 10.8 Relationship among culture fluorescence (○; mV), ethanol concentration (△; g/L), respiratory quotient (□) and acetate concentration (◇; g/L) in a culture of *C. utilis* fed with ethanol pulses. *(Reprinted by permission from C. M. Watteeuw, W. B. Armiger, D. L. Ristroph, and A. E. Humphrey, "Production of Single Cell Protein from Ethanol by a Fed-Batch Process," Biotech. Bioeng., vol. 21, p. 1221, 1979.)*

during this fed-batch operation. The circles are used to identify the culture fluorescence curve—the measurements are obtained continuously. In the portion of the experiment shown here, ethanol pulses of 1.85 g/L were added at 23 and 24.5 h into the batch, while the pulse at 26.15 h added 3 g ethanol per liter of culture. Reduction in culture fluorescence accompanying ethanol exhaustion is clear.

In all applications of direct optical measurements in cell cultures, a number of potential problems arise which can interfere with interpretation of the measurements. For example, the optical surfaces in contact with the process fluid may become fouled with cells or medium components. Gas bubbles and particulates in the multiphase reaction fluid may interrupt or interfere with the desired measurement and, in the case of fluorescence, certain medium components or products may fluoresce at the same wavelengths useful for monitoring intracellular state, complicating interpretation of the measurements. However, in view of the importance of determining the biomass concentration and cellular metabolic state for monitoring, control, and optimization of the process, these and other optical methods can be expected to enjoy expanding applications in the future.

The following section is devoted to some of the intermittent, off-line analyses of cell and medium properties which are useful in bioprocess technologies. We should recognize that the dividing line between "continuous, on-line" monitoring and "intermittent, off-line" measurements is somewhat diffuse. If the time between successive off-line analyses is less than the characteristic time scale for changes in the measured quantity, then the intermittent, somewhat delayed measurement is practically as useful as a continuous, real-time measurement. Since the time scales for changes in enzyme, cell culture, and reaction fluid properties may range from minutes to hours to days, certain off-line assays provide important information for process monitoring and control.

10.3 OFF-LINE ANALYTICAL METHODS

In this section we consider some of the measurement principles and methods applied to determine the properties of process fluids, biocatalysts, and biosorbants. Possible methods span the entire spectrum of analytical chemistry, spectroscopy, and biochemistry, making anything approaching a complete presentation impossible in this context. Here, we emphasize certain new methods relating to cell property measurements which have potential for process monitoring and control applications and also provide an overview of other types of commonly applied analyses.

10.3.1 Measurements of Medium Properties

After withdrawing a sample from a bioreactor or separation unit, a solid-liquid separation is accomplished by centrifugation or filtration in order to remove cells and any other particulate matter from the fluid phase sample. The analyses conducted subsequently, of course, depend upon the particular application; analytical methods which perform satisfactorily for defined medium may not be accurate or appropriate for analyses in undefined medium which may contain interfering components.

The desired measurements in a bioreactor are the concentrations of substrates and components influencing rates, and the concentrations of reaction products and inhibitors. For fermentation, analyses of the carbon and nitrogen sources are often desirable. Also, it may be necessary or useful to determine the levels of certain ions such as magnesium or phosphorus in the medium. Products of cellular processes vary over a broad range of chemical complexity and properties, from small organic compounds such as ethanol to more complex structures such as penicillin to biological macromolecules such as enzymes and other proteins. Accordingly, the spectrum of appropriate methods for product assay is extremely broad.

Liquid-phase quantitative analysis is based usually upon light refraction (measured with a refractive index detector), absorption of light at a particular wavelength (measured with spectrophotometer) or fluorescence due to excitation at one wavelength and subsequent emission at a longer wavelength (measured with a spectrofluorometer). Sugars, for example, do not absorb light strongly and do not fluoresce but do alter solution refractive index. On the other hand, protein and nucleic acids lend themselves to spectrophotometric and spectrofluorometric detection. Fluorescence measurements are usually more sensitive and allow measurement of lower concentrations. However, spectrofluorometers are more expensive than spectrophotometers, making spectrophotometric measurements very popular. In order to avoid interference from other compounds in solution, separation or concentration of the component to be analyzed from other solutes is often necessary. Sometimes this can be accomplished by chemical treatment to decompose or to precipitate the desired interfering compounds. For example, RNA is extracted from cell lysates with $HClO_4$ (perchloric acid) at 37°C and analyzed by the orcinol method for ribose. Interfering sugars are removed during the extraction.

Finer-scale separation among related compounds by chromatographic methods is also commonly applied in medium chemical analysis. The basic principle of chromatography, which is discussed in greater detail in Sec. 11.4 below, is selective retention or retardation of certain compounds by an immobile phase in a column due to preferential attraction of these components for the immobile phase relative to other solutes. For example, in analyzing mixtures of sugars such as maltose and glucose, the different affinities for these two sugars for the primary amino groups on the surface of the support material in a commercially prepared carbohydrate column is used to separate the sugars in an HPLC (high performance liquid chromatography) apparatus. The different sugars emerge from the column at different times, and they may be then detected and quantified separately using a refractive index detector. Many other separations based on HPLC methods are useful in medium analyses. Also, separations accomplished under atmospheric pressure using ion exchange chromatography or size partitioning chromatography are useful in resolving mixtures of related components before their individual quantification.

As noted previously in Chap. 4, a useful strategy for chemical analysis is selective conversion of the component of interest to a readily measurable product. This is the basis for one of the standard laboratory methods for glucose assay, in which the enzymes glucose oxidase and peroxidase selectively convert glucose to a colored compound which can be assayed spectrophotometrically.

$$\text{Glucose} + 2\text{H}_2\text{O} \xrightarrow[\text{oxidase}]{\text{glucose}} \text{gluconic acid} + 2\text{H}_2\text{O}_2$$

$$\underset{\text{(colorless)}}{\text{H}_2\text{O}_2 + o\text{-dianisidine}} \xrightarrow{\text{peroxidase}} \underset{\text{(brown)}}{\text{oxidized } o\text{-dianisidine}}$$

Ion-specific electrodes have become important tools in assaying certain biologically important ions. Cellular nitrogen content can be determined with an ammonia electrode either directly (NH_3, or after chemical modification by pH adjustment, NH_4^+), reduction (NO_2^-, NO_3^{2-}), digestion (amino acids), or enzymatic modification (urea). Ion specific electrodes are also available for analysis of many other ions which influence biochemical structure and function including potassium, sodium, and calcium.

Occasionally, as an alternative to determining the concentration of a particular compound in solution, the measurement determines the compound's biological activity. Assay of penicillin in fermentation broths by this method has been a standard procedure in the pharmaceutical industry. Here, the size of a zone of dead bacteria around a porous disc soaked with the solution to be assayed provides an indication of penicillin activity in the solution. It is very common to analyze enzyme content by measuring the activity of that enzyme. This functional assay is accomplished by exposing the sample solution to a standard enzyme substrate under standard conditions, then measuring the rate of substrate disappearance, or product appearance, often spectrophotometrically or by fluorescence.

An alternative set of analytical procedures is based upon volatilization of the components of interest and their measurement in the gas phase. This can be done for glucose, for example, by forming its TMS (trimethysilyl) derivative which can be vaporized in the injection chamber of a gas chromatograph and the product detected by a flame ionization detector. Determination of the contents of relatively volatile components such as ethanol, acetone, and butanol in fermentation fluids by this method is quite straightforward.

Analytical laboratories which support pilot- and production-scale fermentation facilities often contain one or more automated wet chemical analyzers. These automatically partition, dilute, and process a sample to carry out several chemical analyses. The response time of such instrumentation is 10 to 30 minutes, sufficient in many cases to be useful for monitoring of bioreactions in progress.

10.3.2 Analysis of Cell Population Composition

Analytical methods for cell populations can be categorized in much the same way as were mathematical models for cell population kinetics in Chap. 7 (recall Fig. 7.2). Most classical measurements of biochemistry provide population-averaged and thus unsegregated data on the cell population. Measurements of this type can be extended to a very large number of cellular constituents, even to the level of particular proteins, RNA molecules, and DNA molecules and sequences. If the experimental measurement is made on a single-cell basis, or can be used to infer single-cell information, so that the distribution of single-cell properties is obtained, the data may be said to be segregated.

We shall first discuss measurements of nonsegregated type. The most coarse of such measurements, after determination of total cell mass or number density, is analysis of the elemental composition of the cells including carbon, hydrogen, and nitrogen. Automatic analyzers such as the apparatus for determination of total nitrogen have been developed for bulk sample assays of this type. Specific ions are also known to play an important role in biological processes, and it is common to see total levels of iron, magnesium, phosphorus, and calcium reported as well. Determination of total protein, total RNA content, total DNA content, and other average macromolecular content of the cells can be accomplished by well-established methods. Shown, for example, in Fig. 10.9 are some of the procedures useful in analysis of the macromolecular composition of yeast. It is beyond the scope of this text to describe each of the individual analytical methods for each class of macromolecules in detail; they can be found in Refs. 14 and 15 as well as in many research publications dealing with composition measurements during cellular growth. Measurements of this type provide the

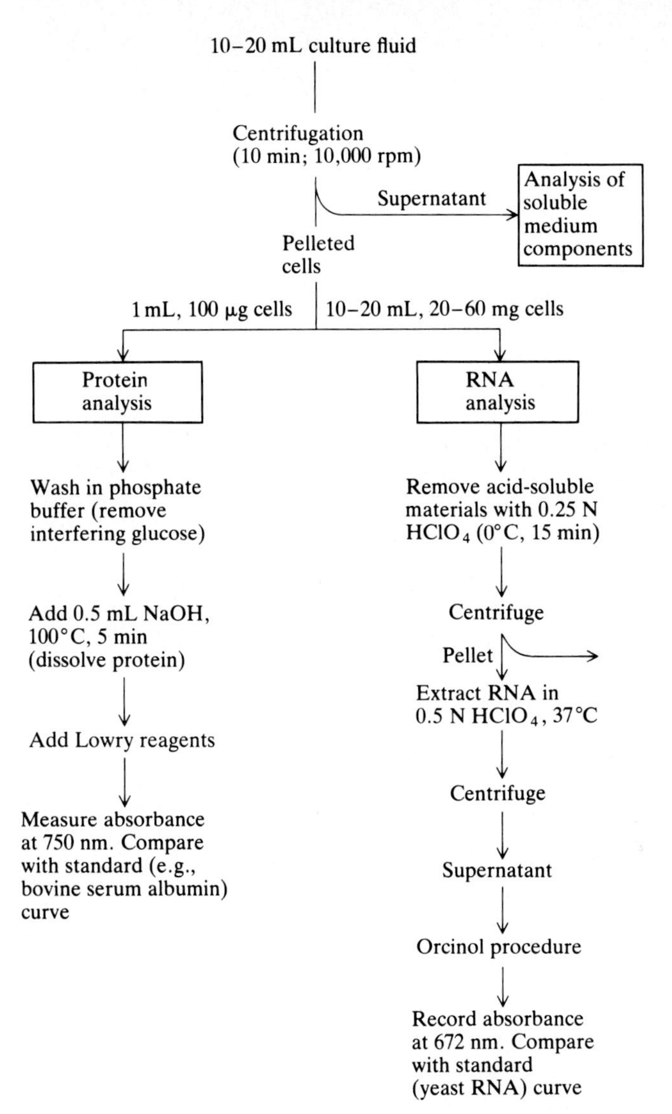

Figure 10.9 Flowchart of analysis procedures for determination of the (population-average) protein and RNA content of yeast cells. Details on the Lowry and orcinol procedures may be found in Refs. 14, 15.

experimental basis for results such as those shown in Fig. 7.17. Although a given method of protein or DNA analysis, for example, will often apply for many different types of organisms, it may be necessary to adapt analytical methods to the particular species under investigation.

Population-average cell content of particular proteins can be determined in several different ways. First, for enzymes, activity assays are used to monitor the

Table 10.3 Characteristic substrates and measurement principles for assay of activity of key enzymes in the catabolic pathways of *Clostridium acetobutylicum*

Enzyme	Substrate	Reaction	Measurement
Acetoacetate decarboxylase	Acetoacetate	Acetoacetate $\longrightarrow$ acetone + CO_2	Consumption of acetoacetate by spectrophotometry ($\Delta OD_{270\,nm}$)
Phosphofructokinase	Fructose-6-phosphate	Reaction coupled with aldolase, triosephosphate isomerase, and α-glycerophosphate	Consumption of NADH by fluorometry (excitation 340 nm, emission 460 nm)
Hydrogenase	Methylviologen	Dehydrogenase methylviologen + $2H^+$ $\longrightarrow$ methylviologen (ox) + H_2	Oxidation of methylviologen which produces H_2 gas is followed manometrically
Pyruvate-ferredoxin oxidoreductase	Acetyl-CoA	$^{14}CO_2$ + acetyl-CoA + ferredoxin(red) $\longleftrightarrow$ ^{14}pyruvate + CoA + ferredoxin(ox)	Production of pyruvate by scintillation counter (radioactivity of ^{14}C)
Butyrokinase	Butyrophosphate	Butyrophosphate + hydroxylamine $\longrightarrow$ butyrohydroxamic acid + $HOPO_3^-$	Production of butyrohydroxamic acid by spectrophotometry (ΔOD at 540 nm)

changes in enzyme levels during process operation. Table 10.3 lists the characteristic reactions and measurement principles involved in assaying levels of key enzymes in the metabolic pathways from glucose to organic solvents in the bacterium *Clostridium acetobutylicum* used for fermentative production of acetone and butanol. Shown in Fig. 10.10 are the time courses of activity of several key enzymes in this organism during batch cultivation. Here it is seen that the activity levels of the enzymes associated with acids production decline late in the fermentation, while there is an increase in enzyme activity associated with solvents production late in the batch. Based upon information of this type, alterations in metabolism may be more directly correlated with strain and bioreactor operating parameters in order to optimize the organism and the process conditions.

Individual proteins can sometimes be analyzed by protein chromatography (see Sec. 11.4) or by examining the relative intensities of bands obtained by electrophoresis of a protein mixture. Increased resolution and extreme sensitivity to many different protein levels can be achieved by two-dimensional gel electrophoresis in which proteins are separated in one direction on the basis of their size

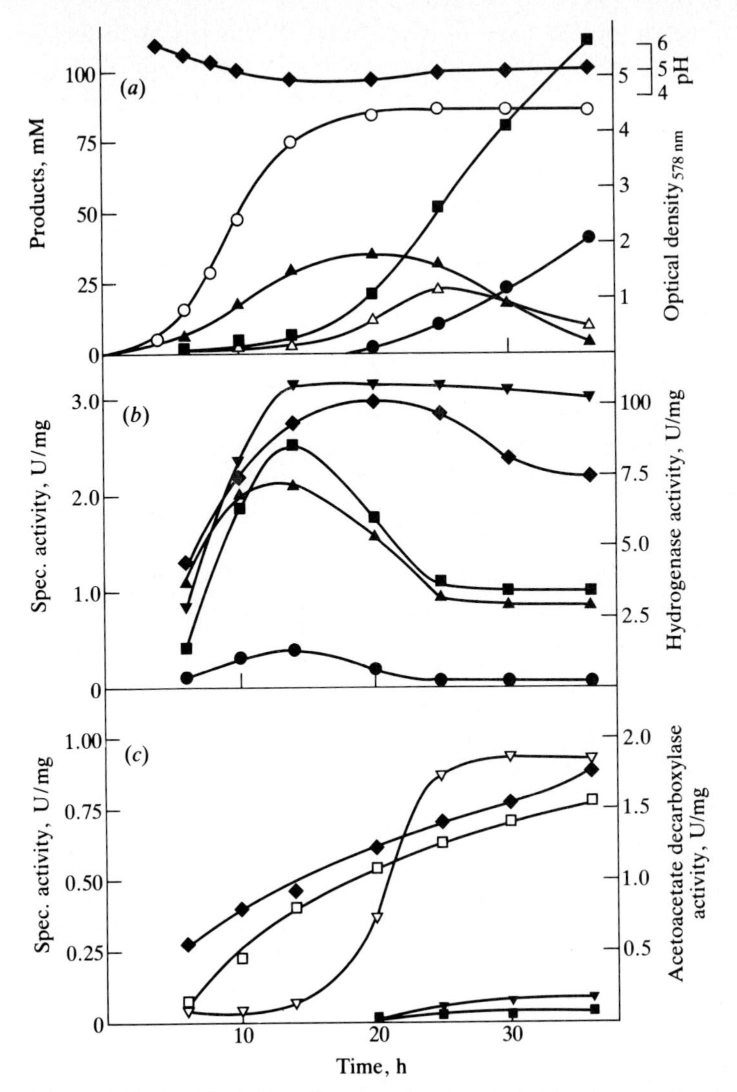

Figure 10.10 Growth of *Clostridium acetobutylicum* in batch culture and level of enzymes involved in product formation. (a) Growth parameter and products formed. Optical density, ○; pH-value, ◆; acetate, △; butyrate, ▲; acetone, ●; and butanol, ■. (b) Level of enzymes involved in the formation of acetate and butyrate and of hydrogenase. Phosphotransacetylase, ●; acetate kinase, ■, phosphotransbutyrylase, ◆; butyrate kinase, ▲; and hydrogenase ▼. (c) Level of enzymes involved in the formation of acetone and butanol. Acetoacetyl-CoA: CoA-transferase with acetate, (◆) or butyrate (□) as acceptor; acetoacetate decarboxylase, ▽; butyraldehyde dehydrogenase, ■; and butanol dehydrogenase, ▼. *(Reprinted by permission from W. Andersch, H. Bahl, and G. Gottschalk, "Level of Enzymes Involved in Acetate, Butyrate, Acetone and Butanol Formation by* Clostridium acetobutylicum," *Eur. J. Appl. Microbiol. Biotechnol., vol. 18, p. 327, 1983.)*

and in a second direction on the basis of their charge. When separated in this way, different proteins tend to move to different spots in the two-dimensional plane, making it possible to identify and quantify a large number of proteins simultaneously. For example, this method was used to study the rates of synthesis of 140 different proteins during growth of the bacterium *E. coli* [17].

Another basis for analysis of individual proteins is binding of antibodies to a particular region on an individual protein molecule [18]. If antibody is available for the protein of interest, analyses based upon precipitation, detection of radioactively labeled antibodies, or the amount of enzyme activity which can be linked to a particular protein by an antibody [the enzyme-linked-immunosorbent assay (ELISA) method] may be used to quantify the amount of the individual protein present. Such methods may also be applied to analyze cellular content of other components or of macromolecular structures against which specific antibodies can be made. Antibody labels are used frequently to determine the existence on a cell surface of particular types of molecules or structures and to quantify in some cases the amount of these components on the cell exterior. Labeling of cell surface compounds and subsequent measurement may be conducted without killing the organisms, a feature which may be useful in screening or selection during strain improvement by mutation. Such methods are also convenient for distinguishing between species in a mixed culture, since organisms usually carry specific surface markers which can be identified separately and quantified with specific antibodies.

The importance of plasmids as the carriers of the genetic instructions for product synthesis in recombinant organisms makes assay of cellular plasmid content a potentially important measurement. The most rigorous method of plasmid quantification in bacteria is done by isolating all DNA from the organism, then separating plasmid DNA from chromosomal DNA in a cesium chloride gradient using an ultracentrifuge. The relative quantities of chromosomal and plasmid DNA can be examined in several ways. For example, if a radioactive preparation of DNA was used, fractions can be collected from the bottom of the tube and analyzed for radioactivity using a scintillation counter. Determination of plasmid DNA content in yeast or animal cells may be accomplished by a hybridization assay using a labeled probe complementary to a nucleotide sequence unique to the plasmid. Alternatively, the gene for a particular enzyme, the activity of which is easy to assay, may be included on the plasmid as a marker, and the activity of this enzyme used to estimate the plasmid content of the organisms. This latter method has been implemented in bacteria, yeast, and animal cell recombinant strains.

All of the measurements discussed above provide information on cellular composition and to some degree on tendencies in metabolism, but they do not directly indicate the current metabolic state or energetic state of the organism. Measurement of cellular ATP content can be carried out with a Biometer that measures luminescence produced by a reaction requiring ATP and catalyzed by the enzyme luciferase. Since ATP levels change rapidly as a function of cellular environment and metabolic activity, it is necessary that samples of the cell population be quenched rapidly in phosphoric acid in order to preserve their ATP content before this or alternative ATP analyses. Since ATP is absent from nonviable cells, measurements of the ATP level can also be interpreted usefully in some cases as a measure of the metabolically active biomass in the population.

High resolution nuclear magnetic resonance (NMR) measurements of ^{31}P have been successfully applied to determine intracellular ATP, ADP, sugar phosphate, polyphosphate, and pH values. Several different microorganisms have been studied in this fashion including the bacteria *E. coli* [19] and *Clostridium thermocellum* [20] and the yeasts *Saccharomyces cerevisiae*, *Candida utilis*, and *Zygosaccharomyces bailii* [21]. In addition, tracer isotopes such as ^{13}C and ^{15}N may be used to observe functioning of intracellular pathways of carbon and nitrogen metabolism via NMR [23, 24].

It has been observed in several fermentations with mycelial microorganisms that the process productivity and kinetics are correlated with the morphological state of the mold or actinomycete. Example 7.2 presented some data and discussion of the connection between morphology and product formation for cephalosporin production using the mold *Cephalosporium acremonium.* Direct observation and quantitative monitoring of mycelial morphology is quite difficult and time-consuming since repeated microscopic observations and human or computerized image analysis are involved. It has been observed that the filtration properties of a suspension of mycelia are influenced by the mycelial morphology, and this principle has been used by Wang and collaborators [25, 26] and refined by Lim and colleagues [27] to formulate an ingenious mycelium morphology and biomass probe based upon a batch filtration measurement. A small sample of culture suspension is filtered, and the filtrate volume and cake thickness are monitored continuously. Based upon previously established correlations between filtration characteristics and the morphological properties and density of the particular mold considered, these data provide a basis for intermittent on-line monitoring of the progress of the fermentation. (see Prob. 11.6)

There are several different methods available for measuring and characterizing the distribution of single-cell characteristics in a population of single-celled organisms. Microscopic observation can give some approximate indications and, coupled with image analysis methods, quantitative information can be obtained, although gathering data on a sufficiently large number of cells to have a good statistical sample is rather difficult. More suitable for rapid measurements of properties of large numbers of individual cells are flow measurement methods of which there are two general types. In instruments utilizing the Coulter principle, the volume of individual cells is detected as cells suspended in a sample stream of an electrolyte solution flowing through a small orifice across which resistivity is measured. For spherical particles, the alteration in resistivity across the orifice may be correlated directly with the volume of the spherical particle, allowing many particles to be sized as they flow rapidly through the orifice. Alterations in particle morphology can cause some difficulties in interpretation of the measurements, but still this is a useful approach for obtaining the size distribution in a cell population.

A richer class of measurements is possible using a flow cytometer. In this instrument, a dilute cell suspension again flows through a measuring section and, in this case, optical measurements are conducted. As indicated in the schematic diagram of Fig. 10.11, the cell sample stream is irradiated by a laser or other light source and the light absorption, scatter and/or fluorescence is measured on a single-cell basis. Light-scattering measurements may be used to obtain information on the cell size distribution. Since right-angle light-scattering intensity is sensitive to intracellular morphology, this measurement has been applied to monitor the accumulation in individual bacterial cells of refractile particles consisting of the storage carbohydrate polyhydroxybutyrate [28]. Individual cell macromolecular composition has been measured for microorganisms and animal cells by applying specific fluorescent dyes which label the macromolecular pool of interest including total cellular protein, double-stranded RNA and cellular DNA [29, 30]. Accumulation of an intracellular fluorescent product produced under the action of a single enzyme can be monitored on the single-cell level in such an instrument, allowing assay of individual enzyme activity in individual cells, study of *in vitro* enzyme kinetics and, by cloning the gene for this enzyme on a plasmid, characterization of single-cell plasmid content [31]. Flow cytometry measurements may also be used to differentiate and quantify multiple species in a mixed culture and to detect the presence of contaminant in a fermentation inoculum [32, 33]. Since flow cytometry provides not only average information but gives a distribution of single-cell characteristics in the population, the data is rich and provides detailed insight into the state of the microbial population.

Although most applications of flow cytometry to study fermentation processes have involved single-parameter measurements, it is possible to effect simultaneous multiple measurements on individual cells, gaining even further detailed information on the cell population. Data of this type, considering two-parameter measurements as an example, take the form of a surface indicating frequency or relative number of cells as a function of the coordinates in an underlying plane representing the measured quantities. Figure 10.12 shows example data of this type in which the measured single-cell properties are light-scattering intensity—related to cell size—and intracellular fluorescence produced by the action of an enzyme encoded on a plasmid gene. Thus, the amount of fluorescent product accumulated may be correlated with the existence and even with the number of plasmids in the yeast cell. The data in Fig. 10.12 show two clear peaks which represent different types of cells in this culture. Those with relatively small fluorescence intensity are cells without plasmids which exhibit a low degree of natural fluorescence under the measurement conditions applied. The population with larger fluorescence, evident as a distinct mode or mountain in this sort of measurement, represents cells containing plasmid and therefore the enzyme required to generate an additional fluorescence response. Based upon this type of measurement, the proportion of cells with and without plasmid in the culture can be very rapidly assayed and further information can be extracted on plasmid replication and segregation in the recombinant strain.

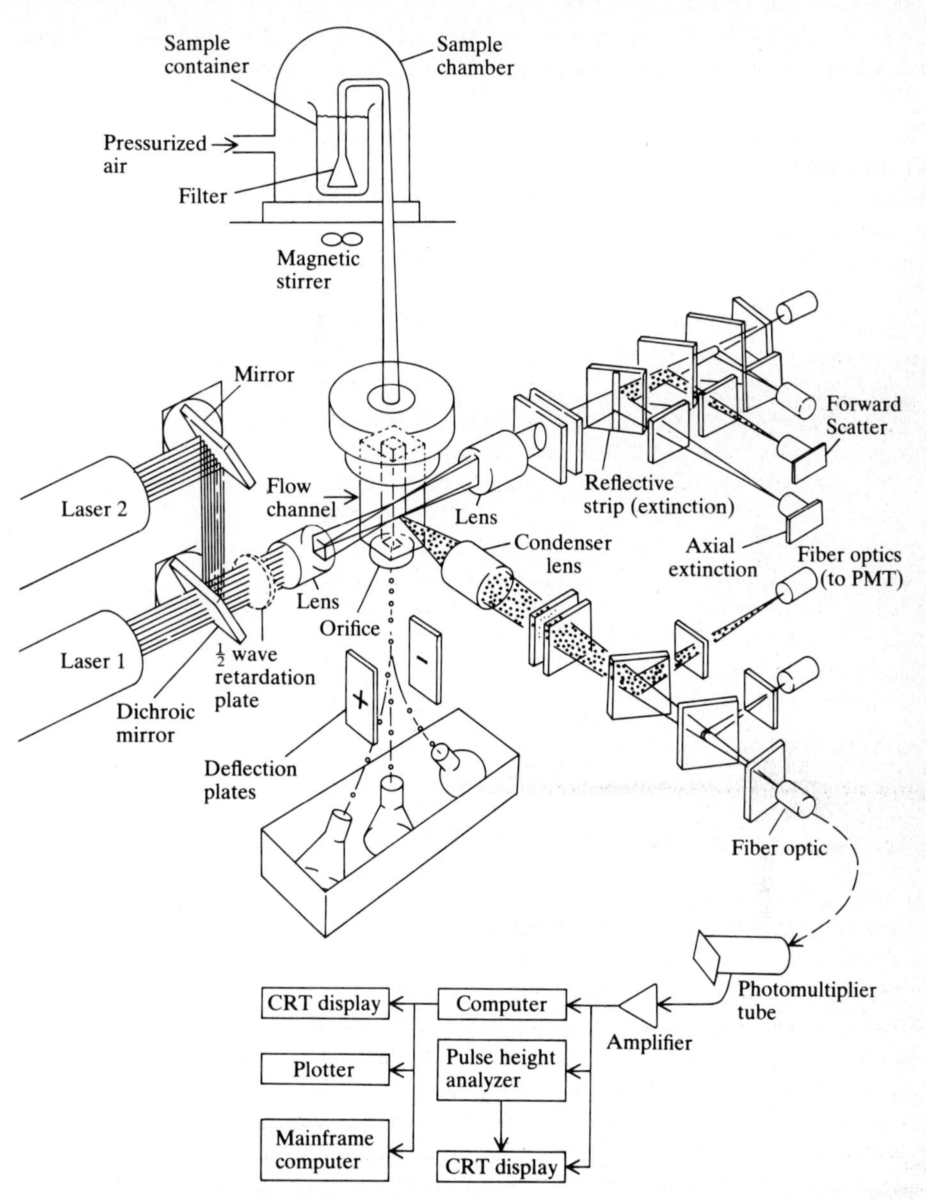

Figure 10.11 Schematic diagram of a flow cytometer showing the paths of the flowing suspended cell sample, the laser beams, optical filters and detectors for measuring single-cell properties, and the electronic signals generated, analyzed, and stored. Diagnonally placed square elements are dichroic filters (reflect certain wavelengths and pass others), and orthogonally placed elements are barrier filters based on wavelength, polarization, etc. Many instruments contain multiple photomultiplier tube detectors and associated electronics, permitting multiple simultaneous measurements on each cell. *(Based on Ortho Instruments Cytofluorograph System 50H.)*

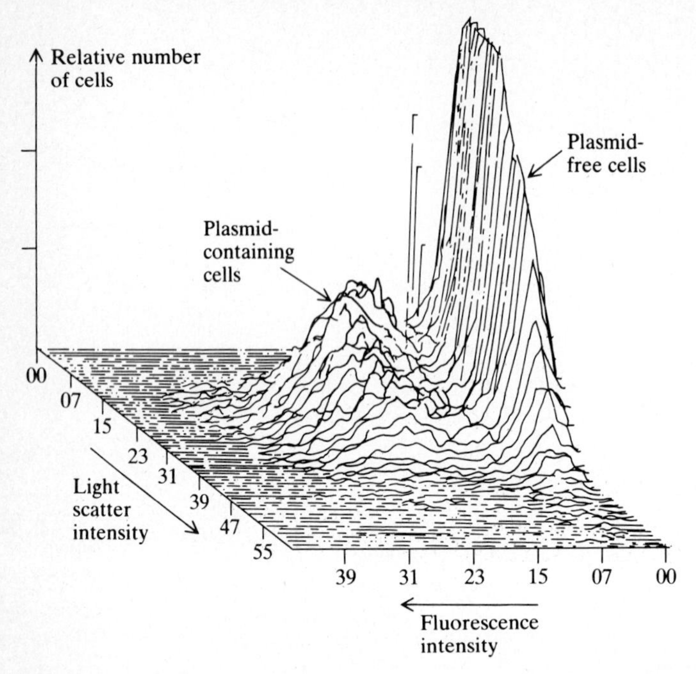

Figure 10.12 Two-parameter characterization of a population of recombinant *Saccharomyces cerevisiae* by flow cytometry. Axes of the basal plane denote single-cell light-scattering intensity (correlated with cell size) and single-cell fluorescence intensity which here is correlated with cellular plasmid content. Here, plasmid-free and plasmid-containing cells are readily distinguished based on fluorescence.

10.4 COMPUTERS AND INTERFACES

There are a number of advantages to be gained by coupling process instruments to digital computers. First, the computer can enhance *data acquisition* functions in several respects. Improved reliability and accuracy can be obtained by using statistical methods and digital filtering. Readings from several parallel sensors can be compared and analyzed to provide on-line recalibration and to identify sensor failure. With a computer, the number and sophistication of analysis systems can be increased. For example, a computer-controlled system may take samples automatically, conduct a chromatographic analysis, and interpret the results, using internally stored calibrations or algorithms to give output directly in convenient units. Although simple signal conditioning and correcting operations such as linearization can be done with particular electronic circuits, these functions are readily accomplished using a computer without the need for additional specific hardware. Another advantage of computers with respect to data acquisition is the ability to store large quantities of measured results in digital form which may be accessed conveniently, analyzed, and displayed later.

Using computers, *data analysis and interpretation* can be enhanced greatly. Results of several measurements may be combined to calculate instantaneously quantities such as oxygen utilization rate and respiratory quotient. Advanced state and parameter estimation methods may also be applied on-line to provide additional useful information on process status from the limited measurements available. More specifics and some examples of computer applications for data analysis are presented in the next section.

Computers expand opportunities tremendously for *improved process control and optimization.* One computer can replace many conventional analog controllers and control many individual valuables such as pH and temperature using standard feedback algorithms. Furthermore, more sophisticated multivariable control methods may be implemented easily with a computer. Controlled variables may include derived quantities such as RQ when a computer is applied. Computer methods may be used to evaluate and improve process mathematical models which may then be employed for determining optimum operating conditions and strategies. Then, the computer provides the memory and computation capability to implement the optimization method, such as variation of nutrient feeding rate or pH during a batch fermentation.

Operation of a batch process requires a carefully controlled and coordinated sequence of valve openings and closings and pump starts and stops. While all of these functions have been done by various timers and relays in earlier technology, they may now be managed efficiently by computer. Use of a computer to manage such switching operations during batch process operation becomes essential if we wish to optimize the scheduling of a number of parallel batch processes (e.g., fermentors) which feed sequentially to downstrean batch processes (e.g., precipitation, chromatography, and so forth). We shall examine elementary process regulation and more ambitious control objectives and strategies in Sec. 10.6 and 10.7, respectively.

Before turning to these interesting domains of computer application, we shall examine briefly some of the principles of digital computers and computer interfaces. Our objective here is to introduce some generic concepts and, by example, to illustrate specific realizations of different types of computer-process configurations. Because improvements and cost reductions in computer hardware and software are proceeding presently at a rapid rate, any specific computer system is probably outdated by the time its description has been published—certainly in book form. Thus, we should view the examples here and in the remainder of the chapter as the kinds of things which can be done, recognizing that, as of this reading, there are probably cheaper, more efficient ways of doing the same thing or something even more effective.

10.4.1 Elements of Digital Computers

The basic components of a digital computer are shown in the block diagram in Fig. 10.13. The central processing unit (CPU) accepts instructions from a stored program through its control unit and performs the indicated arithmetic and

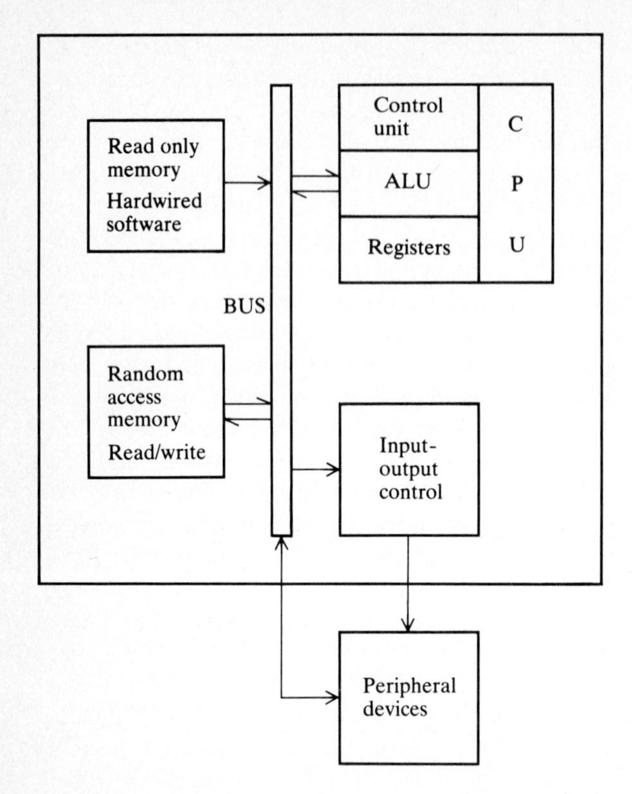

Figure 10.13 Basic functional elements in a computer. CPU denotes central processing unit and ALU stands for arithmetic and logic unit. The different elements communicate through a bus system. *(Adapted from W. A. Hampel, "Application of Microcomputers in the Study of Microbial Processes," p. 1, in "Advances in Biochemical Engineering, Vol. 13," [T. K. Ghose, A. Fiechter, and N. Blakebrough (eds.)], Springer-Verlag, Berlin, 1979.)*

logical operations in the arithmetic and logic unit (ALU), using internal registers for short-term storage. Operations in the CPU are controlled and synchronized by an internal quartz oscillator clock. The cycle time of the CPU, which may range from less than 10 to 10^4 ns (10^{-9} s), combined with a number of bytes (each byte contains eight bits, a binary number with value 0 or 1) processed per cycle (the word size), determines the speed of computation in the CPU. For example, microcomputers available in 1980 employed 8-bit words. By 1982 16-bit microcomputers were available, and 32-bit machines were manufactured by several companies in 1984.

Memory for storage of program instructions and data is provided in several different forms. Read-only memory (ROM) contains fixed instruction sets such as compilers and interpreter programs, while random access memory (RAM) is used for short-term storage of programs, input data, and computational results. The CPU reads frequently from and writes on the RAM during computer operation. As of the mid-1980s, popular microcomputers had RAM capacities from 64,000 bytes (64 kilobytes or just 64 K) to 512 K and beyond. Additional memory is usually available also in one or more peripheral devices. Common external mass memory devices for use with microcomputers are magnetic tape cassettes (storage 80–300 K, access time ~10 s), floppy disks (100–600 K, access time ~0.5 s) and hard disks (500–20,000 K, access time ~3 s: all numbers as of 1984).

The input-control allows the computer to communicate with external peripheral devices. Within the computer, a bus system interconnects the CPU, memory, and input-output control segments.

The functional elements described above are common to all computers, but the speed and memory capacity are determined by the particular hardware configuration. Based on these parameters, computers are often classified into supercomputers, mainframe computers, minicomputers, and microcomputers, with this list ordered from largest, fastest, and most costly to smallest, slowest, and least expensive. Definition of the boundaries between these different classes of computers is constantly shifting; today's microcomputers have the power of mainframe computers of the 1970s. The availability of tremendous computing capacity at low cost is driving a revolution of new computer uses in consumer products, communications, information processing, scientific instrumentation, and in biotechnology. While computer-coupled fermentors were a novelty in the 1960s, we can expect in the not too distant future that almost every bioreactor, analytical instrument, and other bioprocess unit will be monitored and controlled by digital computers.

10.4.2 Computer Interfaces and Peripheral Devices

The storage and arithmetic and logic capabilities of a computer are worthless unless the computer is connected to or interfaced with something else. We can classify computer interfaces as follows according to the object connected with the computer:

1. Computer → computer: communication
2. Operator → computer: instruction
3. Computer → operator: information
4. Sensor → computer: input
5. Computer → actuator: manipulation

We shall next consider each of these types of interface and their significance in computer applications in biochemical engineering. Before turning to particular interface classes, however, we should make the general comment that communications hardware and software controls applied to a particular type of interface are not standardized, so that different computers, peripheral devices, and sensors cannot be "plugged-in" to each other arbitrarily. Care must be exercised to ensure that units to be interconnected have compatible input-output functions.

Computer-computer connections are important because different types of computers and digital devices have different costs and capabilities. Overall efficiency is maximized and cost minimized by using the minimum computing power for the major task at hand, communicating with a higher-level computer when more rapid or complex computational operations or larger storage are needed. A proposed hierarchy of computing levels applicable to fermentation research and development is shown in Fig. 10.14. Here, at the first and lowest level, which

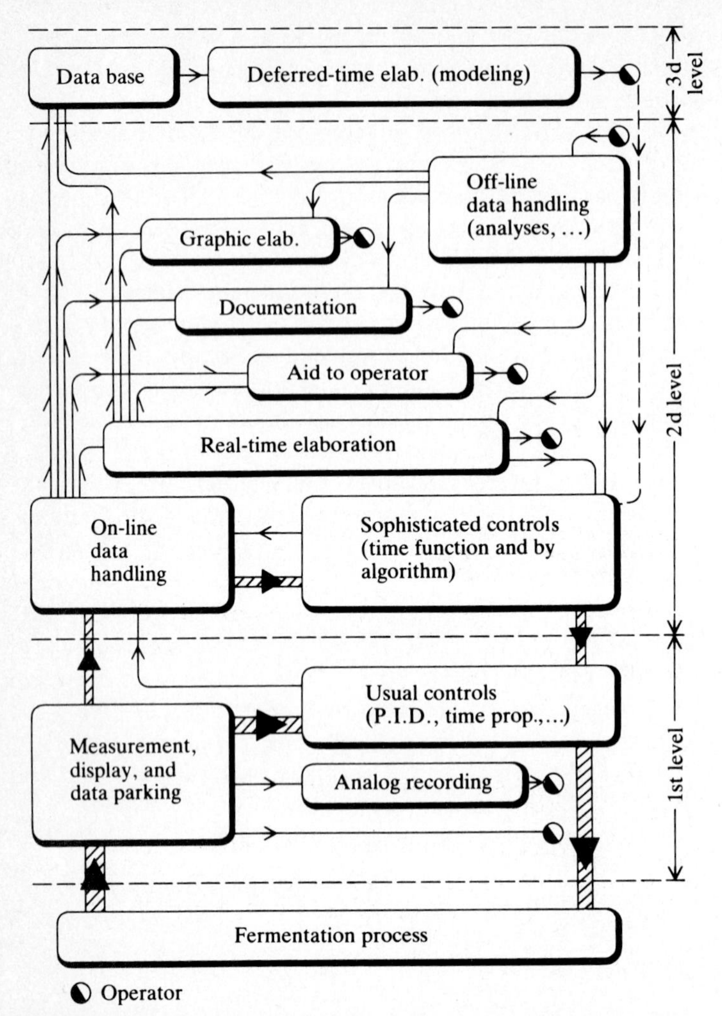

Figure 10.14 A useful breakdown of functions in a hierarchical computer system. Level 1, 2, and 3 functions correspond approximately to microcomputer, minicomputer, and mainframe computer hardware, respectively. *(Reprinted by permission from L. Valentini, F. Andreoni, M. Buvoli, and P. Pennella, "A Computer System for Fermentation Research: Objectives, Fulfillment Criteria and Use," p. 175 in "Computer Applications in Fermentation Technology," Society of Chemical Industry, London, 1982.)*

could be handled by a microcomputer, measurements from the fermentation process are displayed and stored and conventional single-loop regulatory control functions (see Sec. 10.6) are provided. A mini-computer might manage the second level where more detailed data analysis and further operator interaction occurs. Also, the calculations needed for more sophisticated control algorithms could be done at this level.

The first- and many of the second-level functions are conducted in *real time*, which means that there is negligible lag between input of information or instructions to the computer and output or control action based on that information. At the highest or third level, a mainframe computer may be used to solve complicated model equations, estimate model parameters from obtained data, and evaluate advanced control and optimization strategies. Here, information supplied from the first- and second-level computers would be used, but the results would not necessarily be obtained in real time. Instead, the goal may be improvement of process operation at some future time.

Communication at the "person-machine interface" between operator and computer takes several different forms. Computer-to-operator communication may occur using a digital meter, a CRT (cathode-ray tube display), printer, plotter or, recently, computer-generated speech. The form of output displayed on a CRT printer or plotter may be text or graphical format shown in single or several colors. Operator input to the computer is often using a keyboard, directly communicating with the computer in the case of a terminal or indirectly using a keypunch. Other modes for operator input include switches, touch-sensitive screens, and "mouse" devices which, when rolled on a hard surface, translate correspondingly a cursor on a CRT display. Also, speech recognition capabilities, the object of intensive contemporary R & D activities, already permit a small set of instructions to be entered verbally. Input of prerecorded programs or data is done through one of the mass memory peripherals mentioned earlier.

The convenience and flexibility of operator-computer interactions combined with the computer's ability to store and manipulate rapidly large quantities of data have several significant benefits in biotechnology as in other areas of process technology. Process and controller status can be communicated to the process operator in clear, efficient fashion, giving the operator the option of examining particular aspects of the process in greater detail on command. Alarms and cues generated by the computer can alert the operator to existing or emerging problems in the process. Also, the operator possesses great flexibility in altering controller settings or, if needed, even the controller algorithms and configuration.

In addition, safety and regulatory policies usually require gathering and maintaining detailed records on each batch of a drug. Such record-keeping and report generation is facilitated greatly if a computer is linked to the process. Reports on raw material and energy consumption, cycle times, yields and inventories important in R & D, in plant management, and in overall economic optimization for the company may be prepared more rapidly and with much less effort if the process units are computer-coupled. Benefits solely from improved record-keeping and report generation have provided the economic justification for installation of process computers in several pharmaceutical companies.

Sensor-to-computer interfaces depend on the measured quantity and the available forms of sensor output. Some important process operating variables have only "on" or "off" states such as running of a pump or compressor or an open or closed valve. These may be read by the computer directly in digital form. Another type of digital input is a pulse which is generated by a tachometer or mass-flow meter for a given amount of shaft rotation or mass. Rates are then determined by counting pulses per unit time interval.

Many sensors monitor parameters which have a continuous range of values. Traditionally, in addition to an analog meter readout on the measuring unit, an analog current or voltage output is provided for input to a stripchart recorder or perhaps an electronic controller. These analog signals may be input to a computer through an *analog-to-digital* (A/D) *converter*. Usually, the analog signal provided by the analysis instrument must be conditioned suitably—that is, translated, amplified, or attenuated and perhaps converted from current to voltage—before it is input to the A/D converter which is designed to accept analog

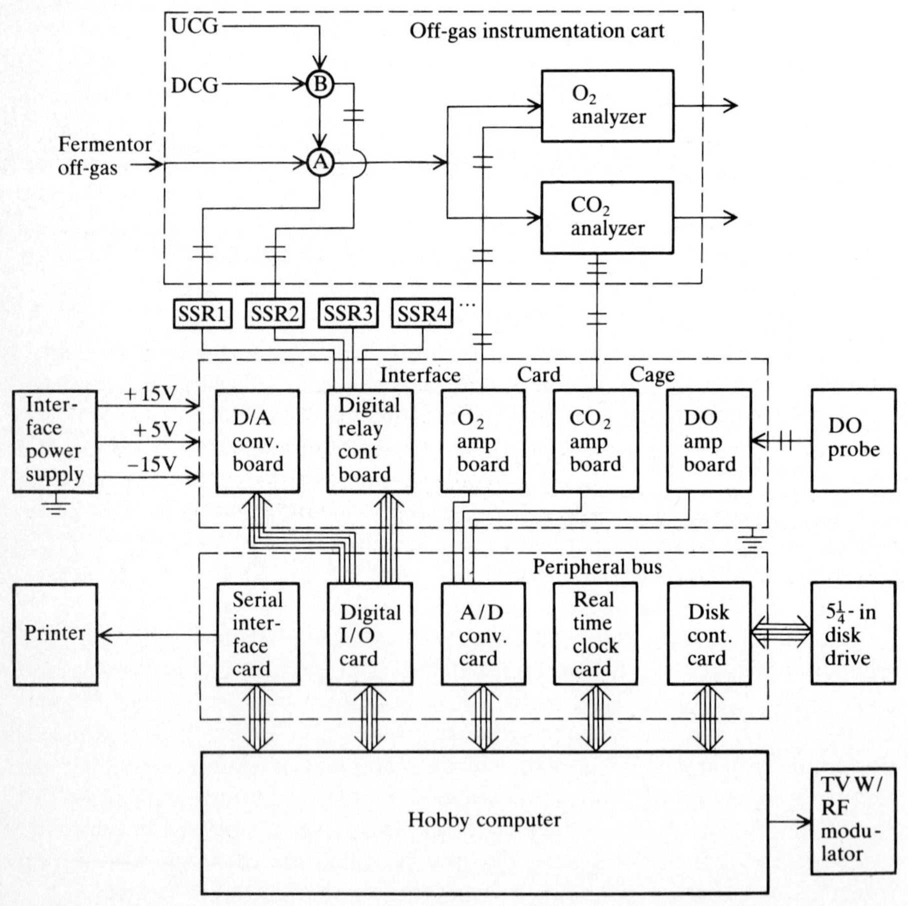

Figure 10.15 Schematic diagram of a gas analysis system based on a microcomputer. UCG and DCG denote upscale and downscale calibration gases which may be introduced intermittently into the gas analyzers in place of fermentor off-gas. This switching is controlled by the computer through solenoid switch relays (SSR1, etc.). *(Reprinted by permission from E. H. Forrest, N. B. Jansen, M. C. Flickinger, and G. T. Tsao, "A Simple Hobby Computer-Based Off-Gas Analysis System," Biotech. Bioeng., vol. 23, p. 455, 1981.)*

signals over a prespecified range. Resolution of the analog signal is directly related to the number of bits employed. If the digital output has n bits, the input analog range will be divided into 2^n discrete subintervals by the converter.

An example of a laboratory data acquisition system based on a microcomputer is illustrated schematically in Fig. 10.15. As mentioned, the O_2 and CO_2 analyzer analog outputs are adapted before going to the A/D converter. Interestingly, notice that an amplifier board has been used to input directly the voltage generated by the dissolved oxygen probe without the use of a separate instrument or readout unit for DO. By replacing some of the functions of analog instruments and controllers, a microcomputer system can provide all of the advantages discussed at lower cost than the traditional set of analog instrumentation.

Two different approaches have been used in connecting several analog inputs to a computer. The first uses a separate A/D converter for each input (there are usually several such converters on a single A/D board). Alternatively, a scanning unit containing several relays (relay multiplexer) may be used to switch among several analog inputs, feeding the selected input to a single A/D converter. The former approach permits more rapid sampling and reading of each analog input at the cost of a larger number of A/D converters. Increasingly, analysis instruments are being equipped with digital outputs, usually in the form of a *binary coded digital* (BCD) *signal*, which may be used for computer input.

Consideration of computer-to-actuator interfaces parallels that just described for physical device-to-computer communication. Digital outputs control electrical on/off switches (relays) and stepping motors. *Digital-to-analog* (D/A) *converters* send to actuators analog signals corresponding to digital value output of the computer. Figure 10.16 shows the interconnections and interfaces used in computer coupling to a pilot plant. While the complexity and scope of the two systems clearly differ, the major components, types of information, and signal flows are very similar to those in the laboratory gas analysis microcomputer module in Fig. 10.15.

10.4.3 Software Systems

The *software*—the set of programmed instructions which govern the operation of the computer, its interfaces and its peripheral devices—is a critical component of a computer-coupled fermentation process. The software dictates how the computer, and perhaps ultimately the process, will perform. Which data is displayed and stored in what format, what operator inputs or interventions are necessary or allowable, perhaps how the process is operated, are determined by the software. It has long been recognized by those with experience in the field of computer-coupled instruments and processes that good software is a central key to a successful system, and that software development can be the major task and a significant expense in installing a process computer. Recent downward trends in hardware prices certainly reinforce this theme. Consequently, it behooves the laboratory researcher and plant manager alike to examine carefully the

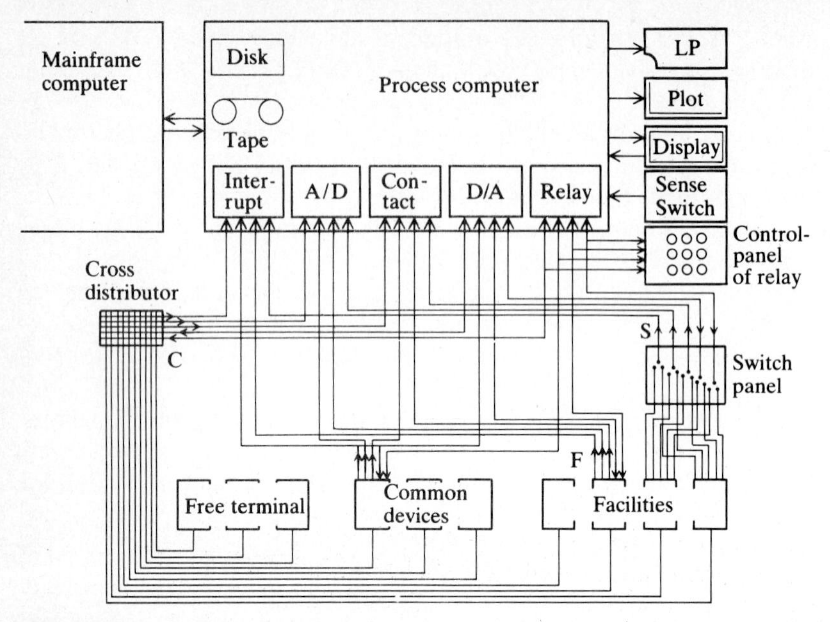

Figure 10.16 Diagram of a computer system for monitoring and control of a fermentation pilot plant. *Abbreviations*: LP, line printer; F, fixed wiring; S, switching system; C, cross distributor. Contact sensors receive information on measurement equipment and input from the fermentor operator. *(Reprinted by permission from M. Meiners and W. Rapmundt, "Some Practical Aspects of Computer Applications in a Fermentor Hall," Biotech. Bioeng., vol. 25, p. 809, 1983.)*

capabilities and flexibility of existing software for alternative hardware and to consider carefully the ultimate costs of not paying what may seem initially like a high price for system software.

The *operating system* of the computer controls program execution, file storage management, inventory and allocation of memory, and coordination of these functions. There are three different types of computer programs: *utility programs* which start up the system and create files, *language programs* to permit use of high level languages (BASIC, FORTRAN, APL and others), and *applications programs* for accomplishing particular computations and other tasks. Some objectives of applications programs such as data acquisition, operator information, and report generation have already been discussed, and others including data handling and process control are considered in the remainder of this chapter. Examples of particular algorithms and application program functions abound in the chapter references.

An important consideration in selecting a hardware-software system is its ability to do time-sharing or multitask operations. With this capability, the computer system essentially can run several programs simultaneously, thereby observing, analyzing, and controlling several different process units at the same time. Also, new programs can be written and old ones debugged or modified while the computer continues to interact with one or more processes. Multitask capability is essential at the second and third levels displayed in Fig. 10.14, but the dedicated microcomputer data acquisition and analysis system in Fig. 10.15 does not require time-sharing features.

Having surveyed the basic concepts important in process computer systems, we now examine some uses of computers and of data which they have acquired to define the operating state of the process and to provide the desired environmental conditions for maximizing bioprocess productivity.

10.5 DATA ANALYSIS

Although the available measurements for a bioreactor are limited, these can be used in concert with defining equations, overall mass and energy balances, and process mathematical models to deduce the values of process variables and parameters. As A. E. Humphrey suggested in 1971, the available measurements taken together provide a "gateway" into other aspects of the process which are not directly observed. We have already discussed how measurements of inlet and exit gas flow rates and composition may be used to calculate the average volumetric mass-transfer coefficient, $\overline{k_l a}$, for a bioreactor (see Sec. 8.2.1). Also, in Sec. 5.10 we saw how macroscopic balances on cellular growth could be used systematically to calculate macroscopic flows based upon known overall cellular reaction stoichiometry. Figure 10.17 shows in flowchart form how measured quantities may be used to calculate related process properties associated with mixing and aeration and with cell growth and metabolism.

On-line estimation of biomass concentration and of specific growth rates during fermentation has been a central objective for data analysis methodologies. In this section, we will use this biomass estimation problem as a prototype for illustration of several different approaches to bioreactor data analysis.

10.5.1 Data Smoothing and Interpolation

Often, measurements obtained from process instruments are noisy. Significant measurement fluctuations make the instrument outputs unsuitable for use as an accurate, instantaneous value. Some sort of signal conditioning or smoothing is required in such cases.

The simplest way to smooth data is to apply a first-order (analogous to an RC) filter which attenuates input fluctuations smoothly and increasingly as the fluctuation frequency increases. Such a filter is characterized by a time constant t_f: inputs with frequencies much smaller than t_f^{-1} are attenuated to negligible magnitude in the output, and fluctuations with frequencies much greater than t_f^{-1} pass essentially unaffected. In discrete-time form, convenient for digital computer implementation, such a filter is given by [2]

$$w_k = a_0 u_k + a_1 u_{k-1} - b_1 w_{k-1} \tag{10.4}$$

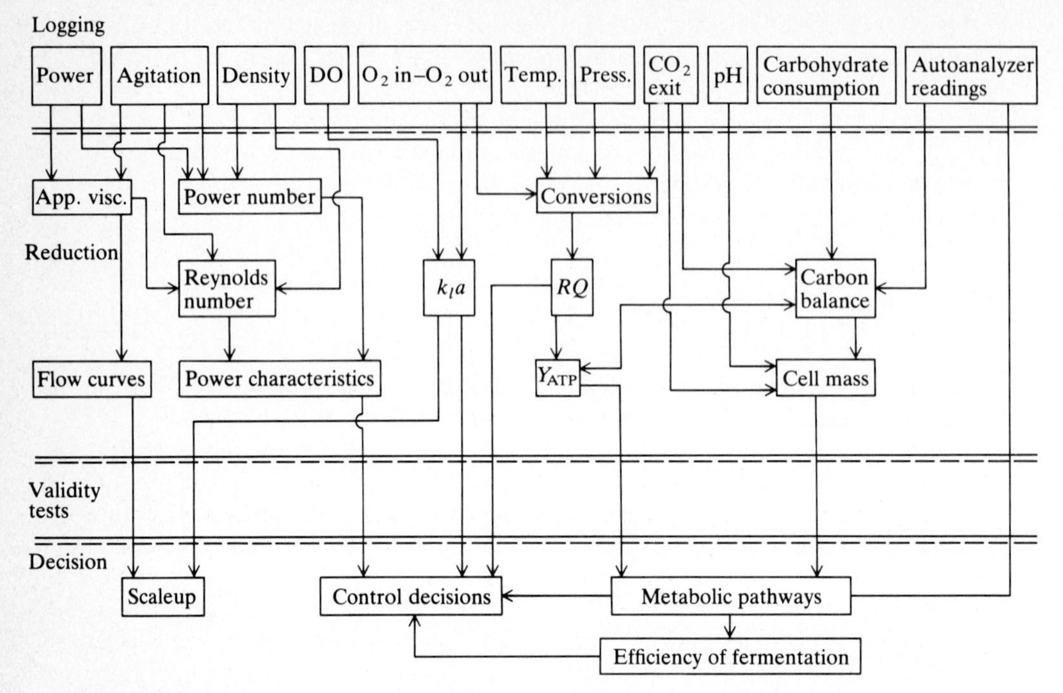

Figure 10.17 Primary measurements, shown on top, may be used to calculate many related process properties and parameters. *(Reprinted by permission from L. P. Tannen and L. K. Nyiri, "Instrumentation of Fermentation Systems," p. 331 in "Microbial Technology," 2d ed., Vol. II [H. J. Peppler and D. Perlman (eds.)], Academic Press, New York, 1979.)*

where u is the filter input (noisy signal), w is the filtered output, and subscripts k and $k - 1$ denote the current and the previous discrete sampling times. a_0, a_1, and b_1 are parameters which determine the filter characteristics. In particular, t_f is related to the *sampling time* T_s (the time interval between samples) and the parameter b_1 by the equation

$$t_f = \frac{T_s(1 - b_1)}{2(1 + b_1)} \tag{10.5}$$

Somewhat simpler to implement, yet giving similar filtering characteristics, is a *moving average* calculation. Here, we take measurements more frequently, and average some set, say 10, of the sequence of measurements to obtain a representative value over the time for all of those measurements. This is quite feasible for bioreactions where significant changes in measured quantities often occur over a much longer time scale than the time required for the measurement.

A more sophisticated filter which also provides an interpolation polynomial has been proposed for fermentation applications by Jefferis and coworkers [35]. We shall consider estimation of biomass density and growth rate from noisy, intermittent turbidity measurements as an example. The cell mass density within some time interval τ backward from the current sampling time is represented as a second-order polynomial in time:

$$x(t) = \alpha_1 + \alpha_1 t + \alpha_3 t^2 \tag{10.6}$$

The objective is to estimate values of the coefficients α_1, α_2, and α_3 using the present measured value of x, $x(t_k)$, previously measured values, and possibly information on measurement noise as well. One approach is to effect a least-squares deviations fit of Eq. (10.6) to the data over the time interval $t_k - \tau$ to t_k. A recursive least-squares filtering method for this purpose is described in Ref. 35. Other techniques for filtering noisy data are discussed in the general references at chapter's end.

10.5.2 State and Parameter Estimation

Neglecting accumulation of oxygen in the reactor, the oxygen material balance on a batch reactor becomes

$$\frac{F_f c_{O_2,f} - F_e c_{O_2,e}}{V} \equiv Q_{O_2} = M_{O_2/X} x + \frac{1}{Y_{X/O_2}} \frac{dx}{dt} \tag{10.7}$$

where f and e denote feed and exit values, respectively. In this equation, oxygen use for both growth and maintenance metabolism is considered. By measurements on the feed and exit gas streams, the *oxygen utilization rate* Q_{O_2} (this quantity is often also indicated as OUR in the literature) may be determined experimentally. With $Q_{O_2}(t)$, a known function of time and assuming the coefficients $M_{O_2/X}$ and Y_{X/O_2} to be constants, Eq. (10.7) may be integrated to obtain

$$x(t) = \exp(-M_{O_2/X} Y_{X/O_2} t) \times \left[x(0) + \int_0^t Y_{X/O_2} \exp(M_{O_2/X} Y_{X/O_2} \tau) Q_{O_2}(\tau)\, d\tau \right] \tag{10.8}$$

This equation may now be used to estimate values of $x(t)$ from $Q_{O_2}(t)$ measurements. Then, the growth rate dx/dt and specific growth rate follow directly from Eq. (10.7).

However, before applying this to a particular fermentation, the oxygen stoichiometric parameters must be determined. A convenient equation for this purpose may be obtained by integrating Eq. (10.7) with respect to time and rearranging to obtain [29]

$$\frac{\int_0^t Q_{O_2}(\tau)\, d\tau}{\int_0^t x(\tau)\, d\tau} = M_{O_2/X} + \frac{1}{Y_{X/O_2}} \left[\frac{x(t) - x(0)}{\int_0^t x(\tau)\, d\tau} \right] \tag{10.9}$$

Using measured values of $Q_{O_2}(t)$ and $x(t)$ from an off-line experiment, the parameters $M_{O_2/X}$ and Y_{X/O_2} may be determined from a linear plot of the left-hand side of Eq. (10.9) vs. the bracketed quantity on the right.

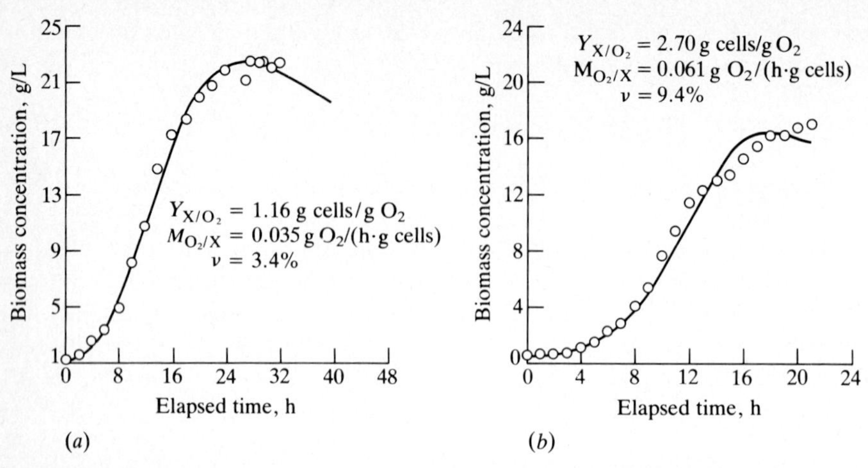

Figure 10.18 Comparison between measured (*dots*) and estimated (*line*) biomass concentrations for batch growth of (*a*) *Streptomyces* sp. and (*b*) *Saccharomyces cerevisiae*. (*Reprinted by permission from D. W. Zabriskie and A. E. Humphrey, "Real-Time Estimation of Aerobic Batch Fermentation Biomass Concentration by Component Balancing," AIChE J., vol. 24, p. 138, 1978.*)

Zabriskie and Humphrey [36] applied this approach to batch cultivation of several microorganisms. Figure 10.18 shows the results for (a) a *Streptomyces* fermentation and (b) for growth of *Saccharomyces cerevisiae*. Yield factor and maintenance coefficient values for each case are shown, as is the coefficient of variation v between experimental data (dots) and biomass concentration estimates (continuous line) based on Eq. (10.8). We see that this estimation procedure performs well for the *Streptomyces* cultivation but that, in the *S. cerevisiae* experiment, the data deviate qualitatively from the estimates. There is evidence of diauxic growth in the measurements which is not indicated by the estimates derived from material balancing. These discrepancies have been attributed to variations in $M_{O_2/X}$ and Y_{X/O_2} which accompany shifts in glucose utilization metabolism during cultivation [36].

This particular example is illustrative of an entire class of data analysis approaches based on macroscopic balances. These methods often perform well as gateways from measurements to derived process variable values. Difficulties may arise, however, because of error accumulation.

General methods for estimating the state of a process, that is the values of the variables which appear in a differential equation mathematical model of the process, have been developed by systems engineers and applied mathematicians. Like the material balancing approaches, these more advanced estimation methods take into account known relationships among process variables. However, these methods also consider noise effects and error propagation—in fact, the mathematical bases for modern estimators rests in the theory of stochastic processes.

Presentation of modern multivariable *estimators* such as the *extended Kalman filter* is outside the scope of this text. We should, however, note several attributes of the extended Kalman filter applied to bioreactors as revealed in the computational and experimental studies of San and Stephanopoulos [37, 38]:

1. The estimator responds rapidly and accurately to changes in process parameters.
2. The estimator is not sensitive to errors in the starting estimates of the state variables.
3. The state estimation method can be extended in a straightforward fashion to provide estimates of time-varying process parameters.
4. Error propagation and growth is not a problem with this type of estimator.

Figure 10.19 shows one computational test of the extended Kalman filter estimation method proposed by San and Stephanopoulos [31]. Here, a mathematical model of a chemostat was subjected to a series of step increases in dilution rate, and the model variables were corrupted by computer-generated noise before entering the estimator. The resulting changes in biomass density and specific growth rate are tracked accurately, even after the dilution rate has exceeded μ_{max} and washout has started to occur.

Process measurements and process variables and parameters deduced from the measurement data provide the information needed to operate the process. In

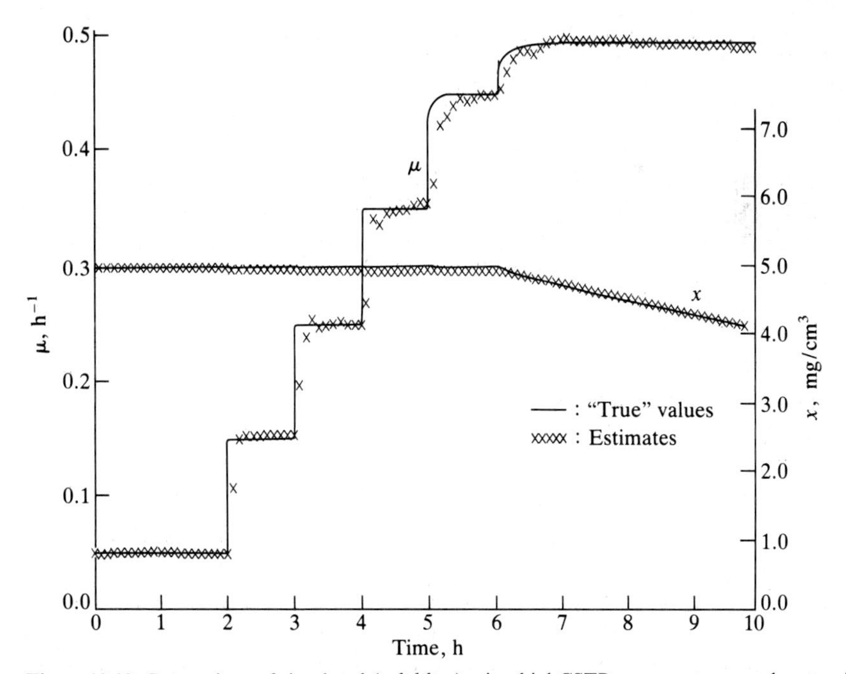

Figure 10.19 Comparison of simulated (*solid line*) microbial CSTR reponse to several successive shifts up in dilution rate compared to estimated values (*crosses*) obtained using an extended Kalman filter. *(Reprinted by permission from K.-Y. San and G. Stephanopoulos, "Studies of On-Line Bioreactor Identification II. Numerical and Experimental Results," Biotech. Bioeng., vol. 26, p. 1189, 1984.)*

the next sections, we examine different strategies for maintaining and manipulating process operating conditions for good performance.

10.6 PROCESS CONTROL

Obtaining satisfactory process performance requires maintenance of operating conditions at design values. Because of unpredictable upsets which invariably enter a process due to fluctuations in pumping rates, flow patterns, mixer speed and other operating conditions, and because of chemical changes within the process, control action is usually required to maintain specified conditions. Sometimes we can improve batch reactor performance by varying process conditions such as pH or temperature in a predetermined fashion as the reaction occurs. Here, too, controls are required in order to carry out the desired batch operation program. In this section we first examine controls to maintain desired values of measured variables. Then, we consider control based on estimated process conditions.

10.6.1 Direct Regulatory Control

Using control of a bioreactor as an example, we frequently wish to control pH, temperature, aeration rate, agitation speed, and perhaps dissolved oxygen partial pressure at specified values. Since all these quantities can be measured on-line, regulation of each of these process state variables can be accomplished using a conventional *feedback controller*, the basic components of which are summarized in Fig. 10.20. Here, the controlled or output variable, say pH, is measured, and the analyzer output is sent to a controller where the measured output value is compared to the desired, or set point, value. Based on the deviation between desired and measured value, control action is determined by some control algorithm.

The controller may be a person who monitors instrument readout and decides what to do. More often, the controller is a pneumatic, electronic, or digital

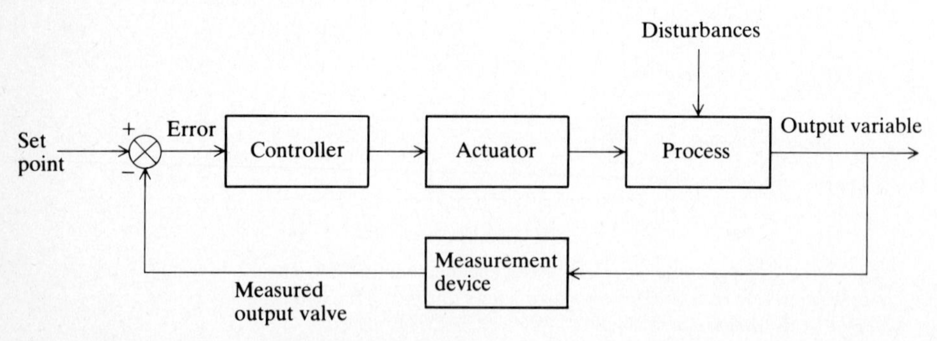

Figure 10.20 Elements in a feedback control system.

computer device. The simplest type of control is on-off control. Here, the actuator is turned on when the error exceeds a specified value and turned off when the error falls below another specified, threshold level, or vice versa. Such controls are used when the activator or control element, which acts physically on the process to cause corrective action, is an on-off device such as a single-speed pump. Thus, in on-off pH control, a pump which feeds base to the fermentor is turned on when the pH falls a certain amount (typically 0.25 pH units) below the set point pH. When the pH reaches a certain level (set point + 0.25 pH units) above the set point, the pump is turned off. (At this point, an acid feed pump may be turned on. However, this is usually not necessary since the effect of most fermentations is to lower medium pH.) Temperature control is often accomplished in a similar way, at least in small-scale reactors.

If the control element provides a continuous range of outputs, such as a variable-speed motor on the fermentor impellor shaft or a continuously adjustable valve on feed air supply, it is common to use *proportional-integral-derivative* (PID) control or some variation of this algorithm. For PID control, the controller output o is given by

$$o(t) = o_s + K_c\left[e(t) + \frac{1}{\tau_I}\int_0^t e(w)\,dw + \tau_D\frac{de(t)}{dt}\right] \tag{10.10}$$

Here, o_s is the nominal controller output corresponding to operation at the undisturbed, design condition, and $e(t)$ denotes the error

$$e(t) = (\text{set point value} - \text{measured value}) \text{ at time } t \tag{10.11}$$

As indicated in Eq. (10.10), the three different terms in the bracket contribute to control action in proportion to the error (P), the integral of the error (I) and the derivative of the error (D). Relative weighting between these three control modes is determined by the parameters τ_I and τ_D which are called the *integral time* and the *derivative time*, respectively. The overall "strength" of control action is determined by the magnitude of the *proportional gain* K_c. If the derivative mode is absent ($\tau_D = 0$), the controller is called simply proportional-integral (PI). Other possibilities are obvious.

A properly adjusted controller of this type often provides excellent regulation of the measured variable. Poorly adjusted, a feedback controller can destablize the system, causing undesirable, accentuated fluctuations. How to set or "tune" a PID controller and its relatives is a central theme of many texts on process control (see, for example, Refs. 1–3).

Common practice is simultaneous use of several controllers of this type, one using temperature measurements to change cooling rate, one regulating pH, and so forth. Because of decreasing costs of digital computer hardware, it is increasingly effective to use a single microcomputer as the controller for several such single-variable feedback control loops. If the computer output (usually after a D/A converter or a relay) is used directly to drive the actuator, the system is said to be in *direct digital control* (DDC).

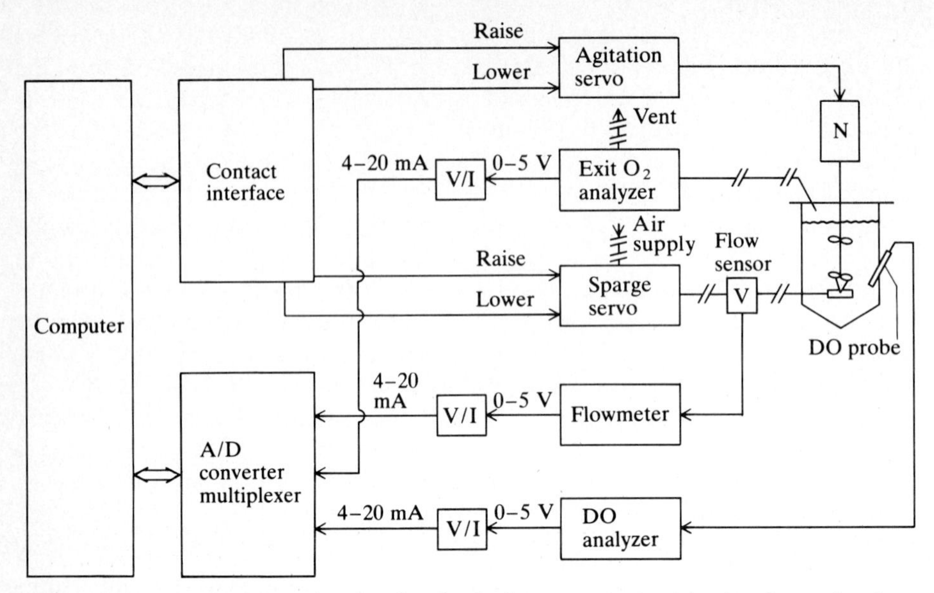

Figure 10.21 A computer control system for dissolved oxygen concentration based on exhaust gas and dissolved oxygen analyses. *(Reprinted by permission from L. P. Tannen and L. K. Nyiri, "Instrumentation of Fermentation Systems," p. 331 in "Microbial Technology," 2d ed., Vol. II [H. J. Peppler and D. Perlman (eds.)], Academic Press, New York, 1979.)*

A distinct advantage of using a computer as a controller is the opportunity to combine the data analysis capabilities of the computer with the flexibility of manipulating more than one process input to achieve control. An example of such a system is shown schematically in Fig. 10.21. Here, measurements of dissolved oxygen (DO) level and exit gas O_2 concentration allow on-line estimation of $k_l a$. This information can then be used in concert with the DO measurement to manipulate agitation rate and/or gas feed rate to control DO at the desired level. This is a DDC system. Notice that the A/D converter accepts current signals here, requiring conversion of instrument output voltages to currents (the V/I converter). It is preferable to transmit electrical analog signals in current form unless the transmission lines are very short (e.g., in a research laboratory) in order to avoid significant line losses. The theme of applying calculated process states and parameters for control is extended in the next section.

10.6.2 Cascade Control of Metabolism

The ultimate objective of any cellular bioreactor control scheme is to provide an environment or an environment history which drives the metabolic controls in the organism to maximize production of the desired compounds. Accordingly, instead of thinking about keeping pH or temperature at some particular value, it may be more useful to consider controlling the bioreactor so that the culture

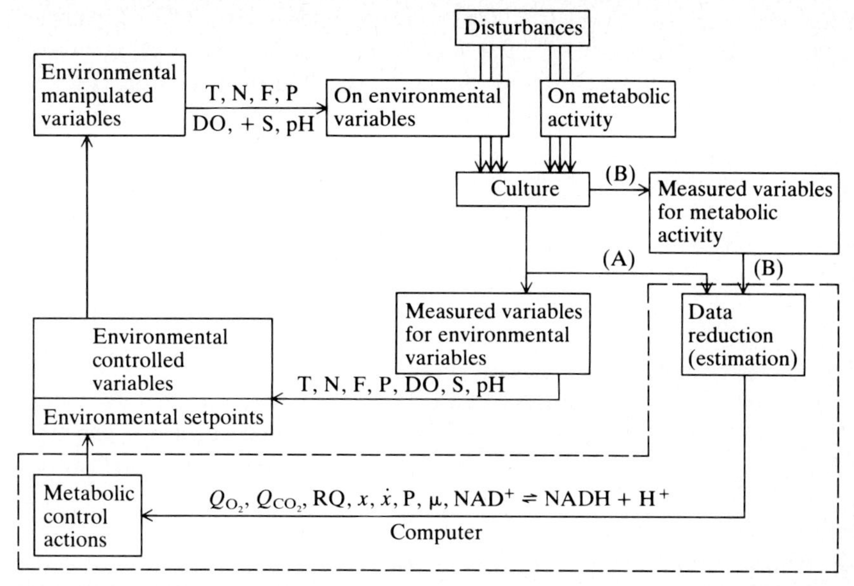

Figure 10.22 Information flow in a cascade control scheme in which deviations from desired metabolic states are used to modify environmental variables. *Legend:* DO, dissolved oxygen; F, gas flow rate; S, substrate; +S, substrate addition rate; P, pressure or product; N, agitation speed; Q_{O_2}, Q_{CO_2}, gas utilization rate. *(Reprinted by permission from L. P. Tannen and L. K. Nyiri, "Instrumentation of Fermentation Systems," p. 331 in "Microbial Technology," 2d ed., Vol. II [H. J. Peppler and D. Perlman (eds.)], Academic Press, New York, 1979.)*

growth rate or respiratory quotient is kept at a desired value. This is feasible since, as summarized in Sec. 10.5, we can estimate some metabolic properties of the culture based upon available measurements. Then, as illustrated in the information flowchart in Fig. 10.22, the estimated metabolic property may be compared with its set point value. Error here determines the "metabolic control action" in Fig. 10.22, perhaps using one of the feedback controller algorithms just described.

The output of the metabolic controller may be used directly to alter a process input such as a pumping rate. Alternatively, the metabolic controller output may be used to change the set point of an "environmental controller," say for pH or DO. The pH or DO controller will then alter the pH or DO which, in a good control system, will change the metabolic variable to reduce its deviation from the metabolic variable set point.

When the "environmental" variables are controlled by local single-loop controllers and the environmental controller set points come from a digital computer, the scheme is called *supervisory control* or *digital set point control* (DSC). Of course, everything may be done by digital computer/controllers. We have already seen such a system in Fig. 10.14, in which the first-level computer carries out the regulatory control of environmental variables, driven by set points provided by a higher level computer.

Computer control of fed-batch cultivation of baker's yeast (*Saccharomyces cerevisiae*) affords an interesting example of metabolic manipulation by proper environmental regulation. Regulation of glucose catabolism in this organism is quite complicated. At suitable values of glucose and dissolved oxygen concentration, glucose is utilized for respiration, providing maximal cell yields per unit amount of glucose consumed (recall Sec. 5.3). If glucose concentration increases above a certain level, metabolism switches to fermentation even in the presence of oxygen. This condition is termed *aerobic fermentation* and is the result of metabolic regulation known as the *Crabtree effect*. If aerobic fermentation occurs, cell yield on glucose is reduced, and ethanol and CO_2 are formed as end-products. As noted earlier (Sec. 7.2.3), ethanol inhibits yeast cell growth.

Consequently, aerobic fermentation should be avoided if production of yeast is the process objective (as is often the case). This can be accomplished by feeding glucose during the batch fermentation. The program of glucose feeding can be a preset schedule based on previous experience with the fermentation. However, due to batch-to-batch variability in the inoculum and medium (in practice, molasses rather than pure glucose), such a fixed feeding schedule may not match the glucose requirements of the culture over time, resulting in high glucose concentrations, aerobic fermentation, and yield reductions.

In order to adapt such a *feed-on-demand* strategy to the requirements for a particular batch, Wang, Cooney, and Wang [39, 40] used on-line material balancing to estimate the progress of the fermentation and to adjust the glucose feed rate accordingly. Respiratory quotient [RQ; Eq. (5.51)] was found to be a useful indicator of glucose utilization pathway, with RQ values greater than unity indicative of ethanol formation. RQ values in the ranges below 0.6, 0.7–0.8 and 0.9–1.0 signal ethanol utilization, endogeneous metabolism and oxidative growth, respectively.

Wang et al. [39, 40] described the baker's yeast fermentation using the following stoichiometric representation:

$$aC_6H_{12}O_6 + bO_2 + cNH_3 \longrightarrow C_6H_{10.9}N_{1.03}O_{3.06} + eH_2O + fCO_2 \tag{10.12}$$

$$C_6H_{12}O_6 \longrightarrow 2C_2H_5OH + 2CO_2. \tag{10.13}$$

The empirical cell formula [with unity coefficient in Eq. (10.12)] is based on an average of elemental analyses conducted at different stages of the batch fermentation. There are seven unknowns in describing this reaction system: the five stoichiometric coefficients in Eq. (10.12), and the extents to which these two reactions have occurred. Based upon elemental balances on C, H, O, and N in Eq. (10.12) and measurements of O_2 utilization, CO_2 evolution, and NH_3 addition, these unknowns may be determined on-line.

Controlling glucose addition based on the strategy of maintaining RQ less than unity produced the results shown in Fig. 10.23. Here, the control and associated estimation method performed extremely well. Glucose and ethanol levels both remain low throughout the fermentation, with the exception of a brief pulse

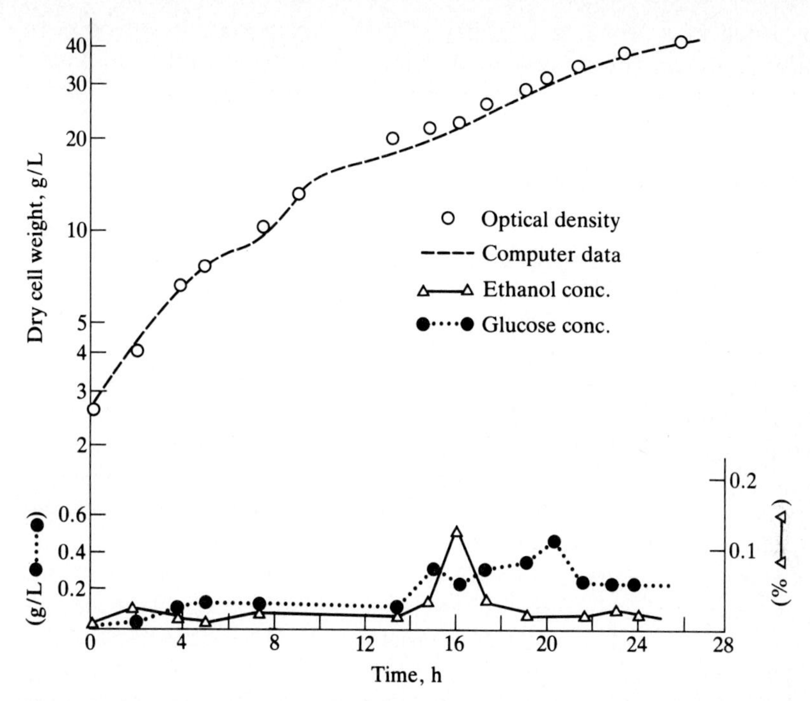

Figure 10.23 Results of a fed-batch baker's yeast fermentation using computer-controlled glucose feeding based on attaining desired respiratory quotient values. Measured and estimated biomass concentrations during the batch are shown for comparison. *(Reprinted by permission from H. Y. Wang, C. L. Cooney, and D. I. C. Wang, "Computer-Aided Baker's Yeast Fermentations," Biotech. Bioeng., vol. 19, p. 69, 1977.)*

in ethanol production around hour 16. The cell density trajectory obtained from material balancing computer estimates closely tracks the experimentally measured information.

Such close agreement between estimated and measured biomass concentrations was not obtained in other fed-batch experiments in which greater ethanol production occurred. In these cases, the cells consumed ethanol for growth, a reaction not considered in the biomass estimation procedure. These examples illustrate possible pitfalls in such direct material balancing approaches: errors in process state estimates, once made, tend to propagate, and existence of overall conversions not in the presumed stoichiometry can throw off the whole scheme.

10.7 ADVANCED CONTROL STRATEGIES

To conclude our overview of process instrumentation and control, we shall examine some of the strategies used to maximize product yield in batch reactors and to regulate and stabilize continuous reactors. In addition, we shall examine briefly some of the interesting scheduling and design problems which arise in a process consisting of a sequence of different batch operations. In all of these cases, application of computers is essential to implement the control or to do the calculations necessary to determine the desired control strategy.

10.7.1 Programmed Batch Bioreaction

Given a particular organism, maximizing production from batch bioreactions requires determination of the environmental conditions during the batch which drive the cells to their best possible performance—or, stated differently—which maximize the genetic potential of the organism. There are several different ways in which this environmental optimization problem can be defined, depending on the instrumentation and control provided on the reactor. In the simplest case, exemplified by a shake flask in which there is usually no on-line measurement or control capability beyond operating temperature, we seek the best temperature and initial medium composition. In a bioreactor with only direct environmental controls, we can look for the pH, agitation intensity, and other environmental parameters which optimize performance. Here, the environmental variables are maintained constant throughout the batch reaction—to the extent feasible for the reactor. (For example, dissolved oxygen level will not be controllable if the culture oxygen demand exceeds the oxygen transfer capacity of the bioreactor.)

However, we know in many cases that a constant environment is not optimal for many fermentations. For example, secondary metabolites are not actively synthesized during rapid growth, but, on the other hand, slow growth is undesirable during the initial stages of a batch fermentation in which cell density is low. Accordingly, we should operate the reactor initially under conditions which maximize growth. Later, when cell density is high, we utilize conditions which stimulate maximum net product formation. In this production stage of the batch, consideration of the rate of product inactivation or decomposition as well as the rate of product synthesis is essential.

In a similar fashion, when manufacturing a protein using a recombinant cell, it is usually best to avoid expression of the product early in the batch because this often inhibits cell growth and may accentuate any genetic instability problems which exist. By adding an inducer (or depleting an inhibitor of gene expression) later in the batch, product formation can be switched on after suitable culture growth has occurred. Thus, operating strategies for recombinant batch fermentations may closely resemble those for secondary metabolites although the biochemical processes involved and their regulation are much different.

The activity of a batch culture at any point in time is a function of both the environmental state at that time and the previous history of culture environments. A particular time sequence of pH, dissolved oxygen level, and other variables may be required in order to develop the culture over time in a way that provides the greatest productivity. Often this is accomplished empirically. However, a sufficiently structured process model makes possible production maximization through computer simulation and optimization. In the remainder of this section we shall examine different examples showing the benefits of programmed batch operation and the methods used to determine the batch programming strategy.

Production of the enzyme β-galactosidase by the mold *Aspergillus niger* in batch culture follows the pattern typical of secondary metabolites. A computer-controlled operating strategy for this process described by Lundell [41] is based upon the following guidelines:

Growth phase

1. Use carbon source feeding to extend the growth phase and increase the cell mass concentration.
2. Add additional carbon source intermittently as required, as determined by decrease of both the CO_2 evolution rate (CER) and respiratory quotient (RQ) to 20 percent of their maximum values.
3. Use the pH and temperature which maximizes cell growth.
4. To conserve energy, use the lowest possible air-feed rate and agitator rotation speed, as determined by the fermentor system's oxygen transfer rate capacity and the culture CER and RQ.

Enzyme formation phase

1. Switch to enzyme formation operating conditions when growth slows (dropping CER and RQ).
2. Use optimum temperature and pH for enzyme production.
3. Adjust agitator rotation and feed air-flow to meet process requirements (enzyme formation phase is characterized by lower oxygen requirement and decreased broth viscosity).
4. When enzyme formation rate slows, add more enzyme inducer.
5. Add additional nutrient to extend growth.
6. Add surfactant, then terminate the batch when enzyme formation rate declines rapidly.

Table 10.4 summarizes the different types of operating strategies which were examined experimentally. The continued batch (CB) mode employs nutrient feed during the enzyme formation phase. In the fed-batch (FB) mode, extra nutrient feed is provided during the growth phase. The FB/CB strategy combines both of these features. The enzyme production trajectories obtained using the conventional batch and continued batch operating actions are illustrated in Fig. 10.24. Performance characteristics of all four operating strategies are listed in Table 10.5. Here, careful selection of programmed batch conditions combined with computerized data analysis and control gives a 70 percent increase in enzyme production and simultaneously a 50 percent reduction in energy use. Notice that, although the basic features of the operating sequence are programmed in advance, on-line measurements and derived quantities are used to determine the particular points of switching from one operating region to another. This is very important in practice because of batch-to-batch variations in inocula and media.

It has been known for some time that high temperatures (30°C) maximize growth rate of the *Penicillium* mold while lower temperatures (20°C) are more favorable for high rates of penicillin synthesis. Although operation at a temperature between these extremes (24 to 25°C) has predominated in past commercial practice, intentional variation of the temperature during the batch, i.e., temperature programming, can conceivably give larger pencillin yields than any constant temperature. Standard mathematical procedures for obtaining the best temperature program have been applied to this problem [42, 43] with interesting results which are summarized next.

In order to apply optimal control theory, a mathematical model is necessary. The crosses in Fig. 10.25 are experimental data from commercial fermentors operated at 25°C. To avoid revelation of proprietary-process characteristics, these data were reported only in the nondimensional form shown. These plots do not show the lag phase: the first data point is about 50 h into the fermentation. The trends in these data suggest the following general forms of growth and product-formation kinetics:

$$\frac{dx_1}{dt} = b_1 x_1 \left(1 - \frac{x_1}{b_2}\right) \tag{10.14}$$

$$\frac{dx_2}{dt} = b_3 x_1 - b_4 x_2 + b_5 \frac{dx_1}{dt} \tag{10.15}$$

Table 10.4 Summary of different programmed batch operating strategies considered for optimization of β-galactosidase production by a strain of *Aspergillus niger*. In each case the total pressure was 130 kPa and the dissolved oxygen level was 20% of saturation

(Reprinted by permission from R. Lundell, "Practical Implementation of Basic Computer Control Strategies for Enzyme Production," p. 181 in "Computer Applications in Fermentation Technology," Society of Chemical Industry, London, 1982.)

1. Batch (B) fermentations used as reference.

temperature	30°C
pH	4.5
stirring speed	200 rpm
air flow rate	30 m^3/h

2. Continued-batch (CB) fermentations with addition of extra organic nutrient to yield extended enzyme formation.

temperature	30°C
pH	4.5
stirring speed	175 rpm/200 rpm/150 rpm
air flow rate	15 m^3/h–30 m^3/h–20 m^3/h

addition of extra organic nutrient

3. Fed-batch (FB) fermentation with addition of extra carbon source to provide extended cell formation.

temperature	35°C/30°C
pH	4.8/4.5
stirring speed	175 rpm/250 rpm/200 rpm
air flow rate	15 m^3/h–30 m^3/h–20 m^3/h

addition of starch and $(NH_4)_2HPO_4$

4. Fed-batch/continued-batch (FB/CB) fermentations with addition of extra organic nutrient during the enzyme formation phase.

temperature	35°C/30°C
pH	4.8/4.5
stirring speed	175 rpm/250 rpm/200 rpm
air flow rate	15 m^3/h–30 m^3/h–20 m^3/h

addition of starch and $(NH_4)_2HPO_4$ during the cell formation phase
addition of extra organic nutrient during the enzyme formation phase

where x_1 and x_2 are the dimensionless cell and pencillin concentrations, respectively. Equation (10.14) is the logistic equation familiar from Chap. 7, while (10.15) is of the Luedeking-Piret type, discussed before, with the addition of a term reflecting penicillin destruction. The need for this term is indicated by the flattening of the penicillin curve in Fig. 10.25 for large times. Also, other data indicate that penicillin hydrolyzes in aqueous solution and that the penicillin synthesis system may also decay with time.

Assuming various sets of values for parameters b_1 through b_5 and comparing the computed x_1 and x_2 time histories with the data show that the sum-of-squares residual for all data points is minimized at $T = 25°C$ when $b_5 = 0$, and

$$b_1 = 13.099 \equiv b_{10} \qquad b_2 = 0.9426 \equiv b_{20}$$
$$b_3 = 4.6598 \equiv b_{30} \qquad b_4 = 4.4555 \equiv b_{40} \tag{10.16}$$

The temperature dependence of these parameters can be estimated by combining these values at 25°C with (1) the known optimal temperatures for cell growth rate (30°C) and penicillin synthesis rate

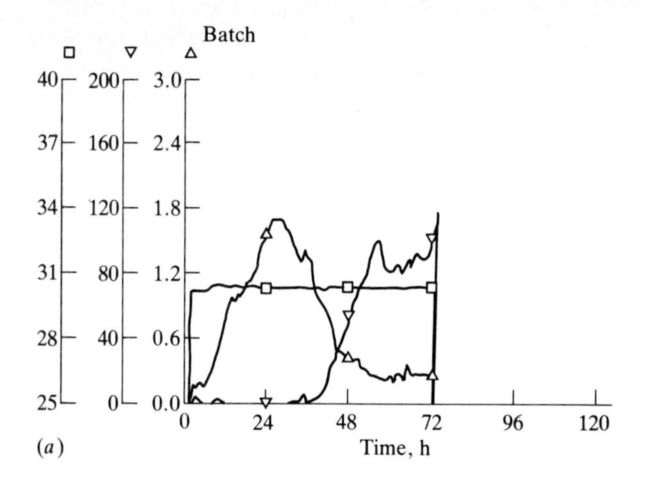

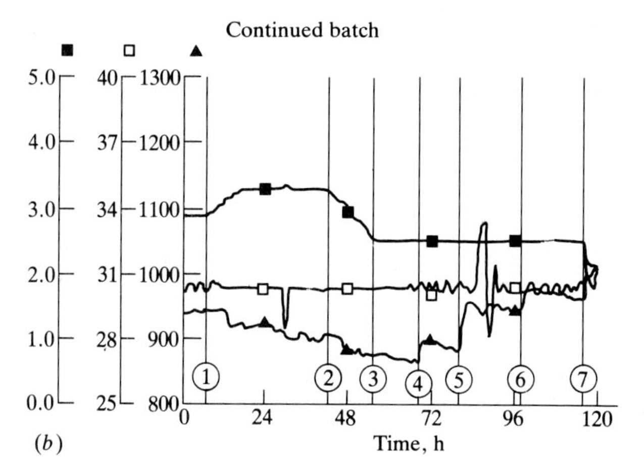

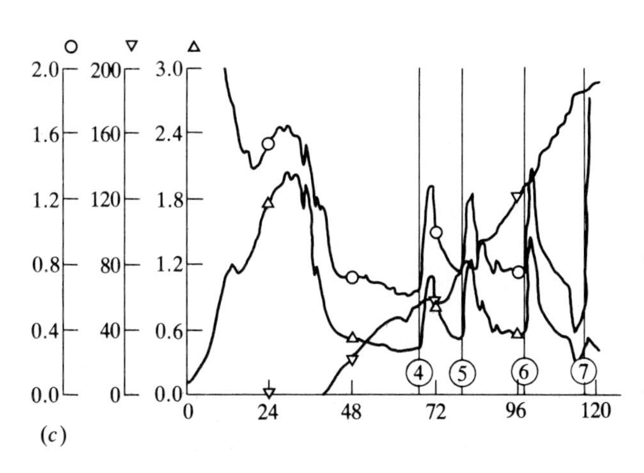

Figure 10.24 Data for β-galactosidase production by *Aspergillus niger* in (*a*) batch culture and (*b, c*) continued batch fermentation. The numbers in parts b and c designate computer-controlled changes in operating conditions during the batch as follows: 1, increase rotation speed; 2, decrease rotation speed; 3, constant rotation speed; 4, 5, 6, add organic nutrient; 7, add surfactant. *Legend:* □, temperature, °C; ▽, β-galactosidase activity; △, exhaust gas CO_2, %V; ▲, fermentor weight, kg; ■, agitator rotation speed, s^{-1}; ○, respiratory quotient (kg/kg). *(Reprinted by permission from R. Lundell, "Practical Implementation of Basic Computer Control Strategies for Enzyme Production," p. 181 in "Computer Applications in Fermentation Technology," Society of Chemical Industry, London, 1982.)*

Table 10.5 Summary of the results of different batch fermentation operating strategies for maximizing production of β-galactosidase by *Aspergillus niger*

(Reprinted by permission from R. Lundell, "Practical Implementation of Basic Computer Control Strategies for Enzyme Production," p. 181 in "Computer Applications in Fermentation Technology," Society of Chemical Industry, London, 1982.)

Fermentation batch	Fermentation time	Enzyme formed	Energy used	Energy/enzyme		Enzyme/time	
	h	rel. act.	kWh	kWh/rel. act.	% of B	rel. act./h	% of B
B Conventional batch	72	100	180	1.8	100	1.39	100
CB Continued batch	115	207	210	1.0	56	1.80	129
FB Fed-batch	100	163	195	1.2	67	1.63	117
FB/CB Fed-batch/continued batch	120	283	240	0.85	47	2.36	170

(20°C) and (2) experimental evidence showing Arrhenius dependence of the rate constant for penicillin decay with an activation energy of 12 to 15 kcal/mol. These facts suggest the forms

$$
\begin{aligned}
b_i(\theta) &= b_{i0}\,g(\theta) \qquad i = 1, 2 \\
g(\theta) &= 1.143[1 - 0.005(30 - \theta)^2] \\
b_3(\theta) &= 1.143 b_{30}[1 - 0.005(\theta - 20)^2] \\
b_4(\theta) &= b_{40} \exp\left[-6145\left(\frac{1}{273.1 + \theta} - \frac{1}{298}\right)\right]
\end{aligned}
\tag{10.17}
$$

where θ is temperature in degrees Celsius.

Utilizing all the above equations, we can now cast the fermentation model in the general form

$$
\frac{dx_i(t)}{dt} = f_i(x_1(t), x_2(t), \theta(t)) \qquad i = 1, 2 \tag{10.18}
$$

with initial values

$$
x_1(0) = 0.0294 \qquad x_2(0) = 0 \tag{10.19}
$$

The objective of maximizing penicillin yield can now be precisely stated mathematically. Find θ as a function of t from $t = 0$ to the final fermentation time $t = T$ such that $x_2(T)$ is maximized. The

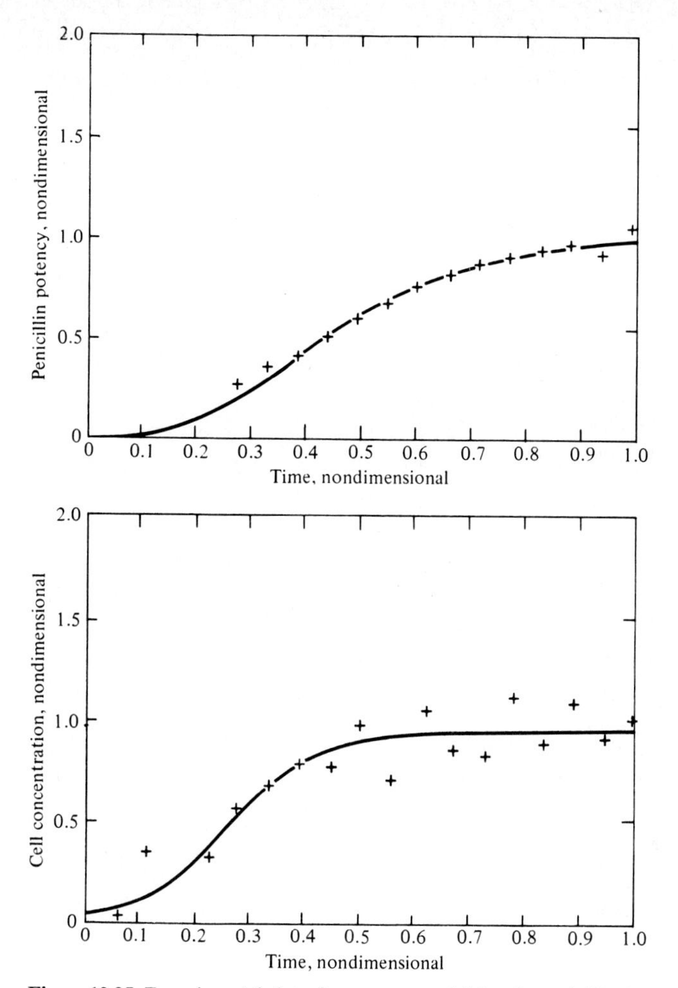

Figure 10.25 Experimental data for a commercial batch penicillin fermentation at 25°C. *(Reprinted by permission from A. Constantinides, J. L. Spencer, and E. L. Gaden, Jr., "Optimization of Batch Fermentation Processes. I. Development of Mathematical Models for Batch Penicillin Fermentations," Biotech. Bioeng., vol. 12, p. 803, 1970.)*

Maximum Principle [5] asserts that for the optimal temperature program, which we shall denote θ^*, it is necessary that the Hamiltonian function H

$$H(\theta) \equiv \sum_{i=1}^{2} \lambda_i^*(t) f_i[x_1^*(t), x_2^*(t), \theta] \tag{10.20}$$

be a maximum for $\theta = \theta^*(t)$ for all t between 0 and T. The superscript * in Eq. (10.20) denotes values obtained using θ^*, where the adjoint variables λ_i satisfy the differential equations

$$\frac{d\lambda_i}{dt} = -\sum_{j=1}^{2} \lambda_j(t) \frac{\partial f_j[x_1(t), x_2(t), \theta(t)]}{\partial x_i} \tag{10.21}$$

and the conditions

$$\lambda_1(T) = 0 \qquad \lambda_2(T) = 1 \tag{10.22}$$

These necessary conditions and associated theoretical results suggest the following iteration algorithm for computing the optimal temperature program:

1. Let $\theta^{(n)}(t)$ denote the nth guess for the program.
2. Solve Eqs. (10.14) and (10.15) for $x_1^{(n)}(t)$, $x_2^{(n)}(t)$ using $\theta = \theta^{(n)}$.
3. Using $x_1^{(n)}$, $x_2^{(n)}$, $\theta^{(n)}$, to evaluate the time-varying coefficients on the right-hand side of Eq. (10.21), integrate those equations backward numerically from $t = T$ to $t = 0$ to determine $\lambda_1^{(n)}(t)$, $\lambda_2^{(n)}(t)$ (this reverse integration is necessary to avoid numerical instabilities often encountered if this integration is begun at $t = 0$).
4. Compute

$$\frac{\partial H^{(n)}}{\partial \theta}(t) = \sum_{j=1}^{2} \lambda_j^{(n)}(t) \frac{\partial f_j(x_1^{(n)}(t), x_2^{(n)}(t), \theta^n(t))}{\partial \theta} \tag{10.23}$$

Determine the $(n + 1)$st guess for the optimal program using

$$\theta^{(n+1)}(t) = \theta^{(n)}(t) + \varepsilon \frac{\partial H^{(n)}}{\partial \theta}(t) \tag{10.24}$$

where ε is a small positive number.

Constantinides, Spencer, and Gaden [42, 43] employed this procedure with some adaptations to avoid decreasing cell concentrations. The final results are illustrated in Fig. 10.26. Notice that the temperature is initially large to maximize cell growth and then lower to optimize penicillin formation rate. The increase in penicillin yield provided by optimal temperature programming is quite dramatic: the yield is 76.6 percent larger than that obtained at the best constant temperature of 25°C.

Other models based on different data sets are formulated and optimized in the references mentioned earlier. In those cases, penicillin-yield improvements of about 15 percent result from programmed temperature. The temperature variations prescribed by these calculations can be approximated closely in commercial practice with little added cost. Consequently, the combined tools of mathematical modeling and optimization theory should also find fruitful application for other fermentations.

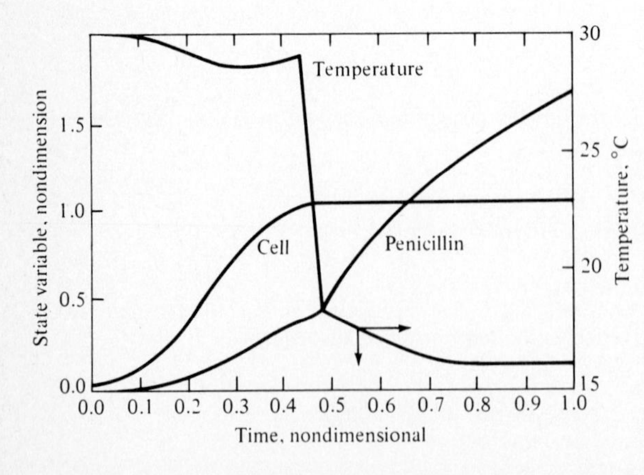

Figure 10.26 Computed optimal temperature policy and the corresponding growth and penicillin-production profiles. *(Reprinted by permission from A. Constantinides, J. L. Spencer, and E. L. Gaden, Jr., "Optimization of Batch Fermentation Processes. II. Optimum Temperature Profiles for Batch Penicillin Fermentations," Biotech. Bioeng., vol. 12, p. 1081, 1970.)*

10.7.2 Design and Operating Strategies for Batch Plants

An overall manufacturing process consisting of a series of different types of batch operations (e.g., substrate pretreatment, sterilization, fermentation, product recovery, packaging) poses a hierarchical series of problems from single units to overall process-to-process design and scheduling for manufacture of several different products. These may be summarized as follows [44]:

Characterize and optimize performance of individual process units.
Optimize the performance in a given sequence of batch process units producing a single product.
Specify equipment required for manufacture of one or several products.
Determine equipment interconnections to meet product requirements most efficiently.
Decide the operating strategies to be used in manufacturing several different products over a certain time span.

We have already considered the first of these problems in several contexts. To see how consideration of a sequence of batch processes can alter optimum design, we shall consider a simple example in which a function $f(t)$ describes the fraction of feed raw material converted during operating time t to useful products for the next stage. In addition, we assume the time to clean the process unit and charge it for the next batch is t_{cl}. The objective function F considered here is the amount of feed converted per unit time which may be written

$$F(t) = \frac{f(t - t_{cl})}{t} \qquad t \geq t_{cl} \tag{10.25}$$

Differentiating Eq. (10.25), equating the result to zero and rearranging this condition on the optimum cycle time (denoted t^*) gives the relationship

$$f'(t^* - t_{cl}) = \frac{f(t^* - t_{cl})}{t^*} \tag{10.26}$$

As sketched in Fig. 10.27, this condition has a simple graphical interpretation: the optimum batch cycle time is the point at which a straight line through the origin is tangent to the unit's performance function $f(t - t_{cl})$. This construction is shown also for a second batch process with performance function g.

Now consider two batch processes in series, the first characterized by f and the second by g. The fraction of raw material converted to desired product in the serial batch process is $f(t - t_{cl})g(t - t'_{cl})$ which we will call $h(t)$. Here, t'_{cl} is the restart time for the second unit. Maximizing feed conversion per unit time requires maximization of

$$F = \frac{f(t - t_{cl})g(t - t'_{cl})}{t} = \frac{h(t)}{t} \qquad t \geq \max\{t_{cl}, t'_{cl}\} \tag{10.27}$$

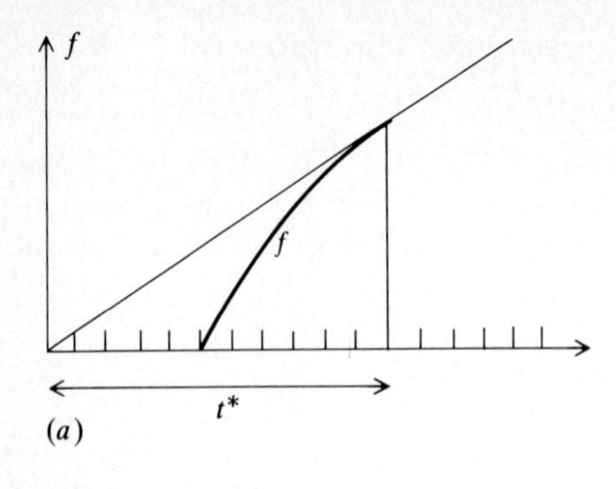

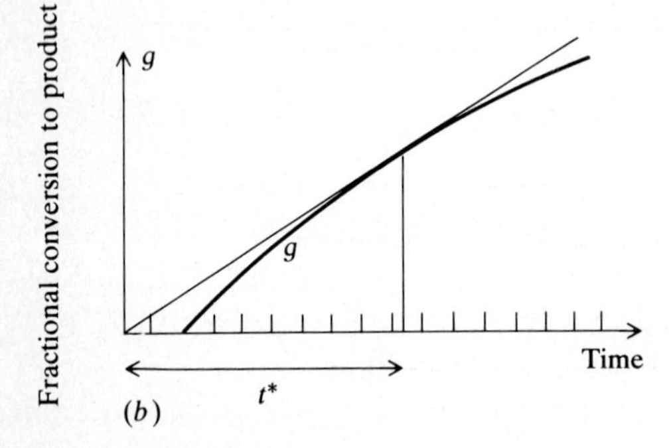

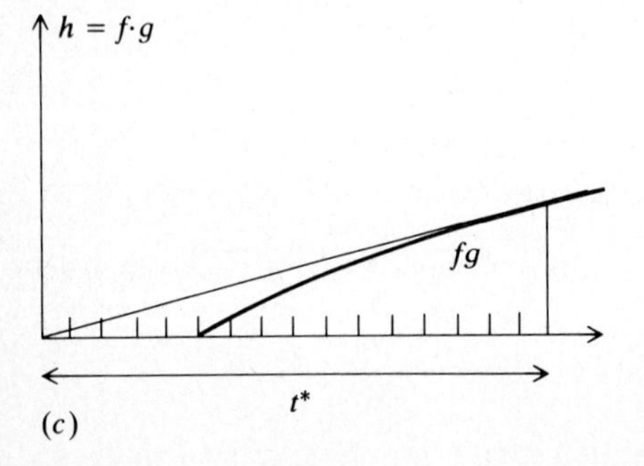

Figure 10.27 Performance curves and determination of optimum cycle times for single-batch units (a, b) and for two batch units in series (c).

Considering the same calculations as before, we arrive at the same condition for tangency, but now in terms of the composite performance function h (Fig. 10.27c). The following central point is illustrated by the example performance curves and graphical constructions in Fig. 10.27: the optimum cycle time for batch processes in series may be different from the optimum cycle time for any of the individual processes treated separately.

If the optimum cycle times for individual steps in a batch-processing sequence are significantly different, improved overall performance may be possible by allowing different cycle times for different units. The simplest case which allows different cycle times is selection of a cycle time for the slowest unit(s) which is an integer multiple (m) of the faster unit(s) cycle times. Then m of the slower units must be provided in parallel to keep up with process transitions from and to the faster unit(s).

Maximum flexibility in cycle time selection for batch-unit processes in series is afforded by including *intermediate storage* in the plant, so that products from one batch unit can be stored temporarily before entering the next batch unit in the processing sequence. Intermediate storage can also provide other useful functions in batch processing [45]. Just as storage or surge tanks in continuous processes can absorb transients, intermediate storage in batch processes helps to damp disturbances created by equipment failures, by fluctuations in unit start-up time, and by batch-to-batch variation in operating performance as is common, for example, in batch fermentation. Also, intermediate storage serves to moderate upsets which may arise in switchover from one product to another.

Several requirements must be kept in mind when intermediate storage is considered. First, enough units of each type must be provided in parallel so that the time-average processing capacity is equal for all steps in the overall process. Second, the stability of the raw materials and products under storage conditions must be considered. Intermediate storage is impractical for unstable materials. Finally, requirements for *batch integrity* may preempt any economic advantages for certain batch-process operating strategies. Batch integrity means that the material from one batch is processed separately from any other batch's material —with no intermixing at any point. In this way, a certain lot of product can be identified uniquely with a corresponding batch of production. This requirement is common in the pharmaceutical industry. Intermediate storage may still be desirable to address cycle time imbalances or to improve damping functions even when batch integrity must be maintained.

Many fascinating design and optimization problems and opportunities arise in consideration of equipment specification and in design and operation of multiproduct plants. Recent research on these and other aspects of batch (or semicontinuous) process design are reviewed in the chapter references.

10.7.3 Continuous Process Control

Different types of control problems and opportunities for use of advanced control strategies arise for continuous processes. Here, the system is usually designed for operation at some steady-state condition. One control objective in this case is minimizing start-up time or some function of start-up costs. This is essentially a batch-optimization problem of a somewhat different type from those just considered—now the goal is to move the system from its initial state to the desired operating point or some neighborhood of that point in order to minimize some objective function. Given a good process model, this problem can be solved by methods related closely to that described above for batch temperature programming for penicillin production.

After start-up, the usual control objective is keeping the process at the desired steady state. This can often be accomplished using several separate, single-loop controllers of the kind discussed in Sec. 10.6, and indeed this is current standard practice in biochemical processing. However, experience in system and process control in other contexts has shown that undesirable *interactions* often occur between different control loops. That is, when manipulating one process input to try to drive one process variable back toward its set-point value, other process variables are disturbed. Generally, many process variables are coupled with each other and with several process inputs, causing individual control loops to interact.

Several different approaches have been developed for dealing with control interactions. Most of these we must leave to the chapter references, but one, optimal multivariable control, will be considered briefly here.

First, we recall that a process system described by

$$\frac{d\mathbf{c}(t)}{dt} = \mathbf{f}[\mathbf{c}(t), \mathbf{d}(t)] \tag{10.28}$$

where $\mathbf{c}$ is the vector of process state variables and $\mathbf{d}$ is a vector of process inputs, can be approximated near a steady-state operating point ($\mathbf{c}_s$, $\mathbf{d}_s$) satisfying

$$\mathbf{f}(\mathbf{c}_s, \mathbf{d}_s) = \mathbf{0} \tag{10.29}$$

by the linearized differential equation

$$\frac{d\boldsymbol{\chi}(t)}{dt} = \mathbf{A}\boldsymbol{\chi}(t) + \mathbf{B}\mathbf{v}(t) \tag{10.30}$$

where $\boldsymbol{\chi}$ and $\mathbf{v}$ are the state and input deviations, respectively (see Sec. 9.2):

$$\boldsymbol{\chi}(t) = \mathbf{c}(t) - \mathbf{c}_s \tag{10.31}$$

$$\mathbf{v}(t) = \mathbf{d}(t) - \mathbf{d}_s \tag{10.32}$$

We base the multivariable control design on the local, linearized approximation of Eq. (10.30).

To formulate a mathematical optimization problem compatible with the goal of keeping $\mathbf{c}(t)$ near $\mathbf{c}_s$, or $\boldsymbol{\chi}(t)$ near zero, we shall seek to minimize the scalar objective function

$$J = \frac{1}{2}\int_0^\infty [\boldsymbol{\chi}^T(t)\mathbf{C}\boldsymbol{\chi}(t) + \mathbf{v}^T(t)\mathbf{R}\mathbf{v}(t)]\, dt \tag{10.33}$$

where $\mathbf{C}$ and $\mathbf{R}$ are positive definite matrices. The first term in the integrand in Eq. (10.33) has clear physical significance: this is the measure in some sense of the amount of deviation of the state from the set-point vector $\mathbf{c}_s$, and the integral adds together all of the instantaneous values of this measure of off-spec operation. How $\mathbf{C}$ is chosen is up to the control designer, who may wish to weight some variable deviations more than others based on the requirements or economics of that particular process. The simplest choice of $\mathbf{C}$ is the identity matrix, which makes the first term in the integrand simply the sum of the squares of all the state variable deviations.

The term $\mathbf{v}^T\mathbf{R}\mathbf{v}$ in the objective function is sometimes called the cost of control, but this interpretation usually does not make physical sense in the context of chemical process control. This term serves an essential mathematical and practical function: without it, the optimal control would be unbounded and thereby physically impossible. It is more useful to think of $\mathbf{R}$ as a matrix which scales the magnitude of the control action, with larger values of the norm of $\mathbf{R}$ corresponding to smaller deviations of the process inputs from their nominal values under steady-state design conditions.

Applying the Maximum Principle introduced in Sec. 10.7.1, we can show that the control which minimizes J in Eq. (10.33) subject to the state and control interactions described by Eq. (10.30) (often called the linear-quadratic optimal control problem) is given by

$$\mathbf{v}(t) = -\mathbf{R}^{-1}\mathbf{B}^T\mathbf{M}\boldsymbol{\chi}(t) \tag{10.34}$$

where the time-invariant matrix $\mathbf{M}$ satisfies the following nonlinear algebraic equation:

$$\mathbf{MA} + \mathbf{A}^T\mathbf{M} + \mathbf{C} - \mathbf{MBR}^{-1}\mathbf{B}^T\mathbf{M} = \mathbf{0} \tag{10.35}$$

This is an interesting result, since Eq. (10.34) has the form of a multivariable, feedback control. Thus, current control values are given explicitly in terms of the current values of the state variables.

If the state variable cannot be measured directly, we may be able to estimate these values based on one of the data analysis methods described earlier. Kalman filter methods are especially suitable here since filter calculations are formulated in a mathematical framework very similar to that used in calculating the optimal multivariable feedback control. Comparisons of controller performance with direct state measurements and results based on Kalman filter estimates for a simulated fermentation process are described in Ref. 46.

Fan, Erickson and coworkers [47] studied using mathematical models the response of a single biological CSTR in which cell growth was described by Monod kinetics with maintenance. An optimal controller of the design just described was used to manipulate flow rate through the vessel based upon measurements of effluent substrate and cell mass concentrations. The control design matrices $\mathbf{Q}$ and $\mathbf{R}$ were taken as

$$\mathbf{Q} = \begin{pmatrix} q_{11} & 0 \\ 0 & q_{22} \end{pmatrix} \qquad \mathbf{R} = 1 \text{ (scalar)} \tag{10.36}$$

With $\mathbf{R}$ fixed, increasing one or both nonzero components of $\mathbf{Q}$ has the effect of increasing the amount of control action.

Shown in Fig. 10.28 are the variations in volumetric flow rate (the manipulated input or control variable) and corresponding trajectories in dimensionless effluent cell density and substrate concentration following a 12.5 percent step increase in feed concentration from dimensionless time zero to dimensionless time = 2. The parameters on these trajectories correspond to different magnitudes of q_{11} with q_{22} fixed at 1000. Notice that as q_{11} increases, the control variable changes more, the deviations in dimensionless effluent substrate concentration (y_1) decrease, while larger fluctuations are obtained in dimensionless biomass concentration (y_2). Keeping q_{11} fixed and increasing q_{22} gives different control action during the disturbance which reduces y_2 deviations and gives greater y_1 deviations for increasing control action. This shows the ability to tune the controller response and effects according to the process requirements.

Another possible goal of continuous process control is stabilizing a steady state that is unstable in the absence of control. Two examples can be cited from the recent research literature. Growth in a CSTR of methanol-utilizing organisms which exhibit substrate-inhibited kinetics admits, according to reactor models, three steady states for some operating conditions. DiBiasio, Lim and Weigand [48, 49] showed theoretically and experimentally that stable operation at the intermediate, unstable steady state could be obtained by use of proportional control. The measured process output was culture turbidity and the manipulated variable was substrate feed rate.

Although operation at a desired steady state is usually considered the ideal for continuous reactor operation, a number of computational and experimental

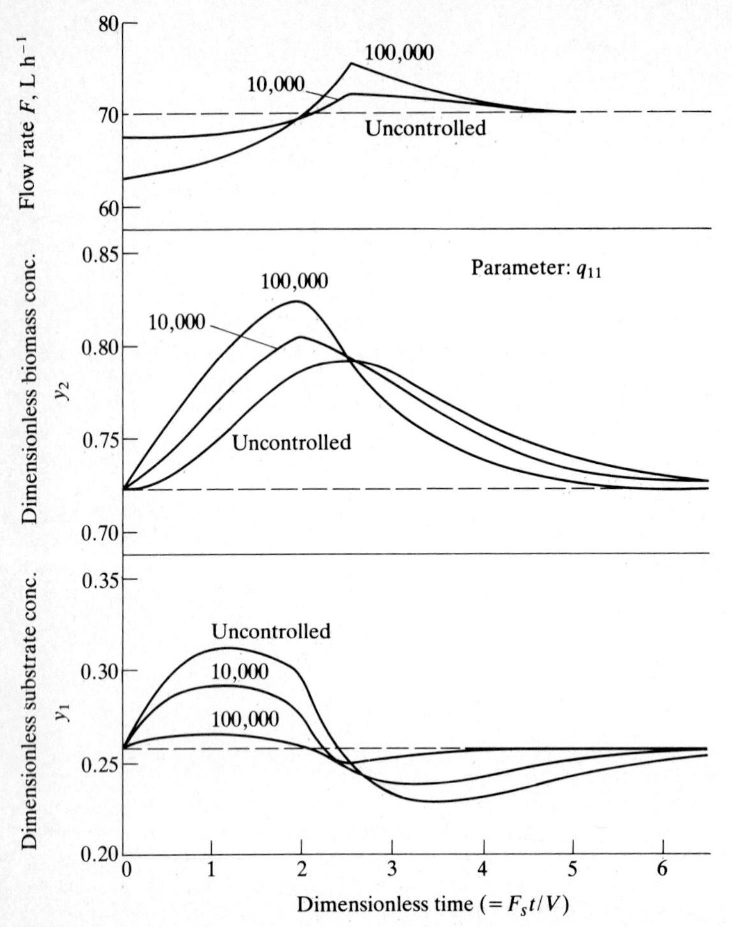

Figure 10.28 Simulation results showing application of an optimum linear-quadratic control to a microbial CSTR. The parameter on the different trajectories is q_{11}, and the dashed lines denote the reference steady-state values of dimensionless substrate concentration ($= y_1 \equiv s/s_{fs}$) and dimensionless biomass concentration ($= y_2 \equiv x/Y_{X/S}s_{fs}$). $q_{22} = 1000$ throughout. *(Reprinted by permission from L. T. Fan, P. S. Shah, N. C. Pereira, and L. E. Erickson, "Dynamic Analysis and Optimal Feedback Control Synthesis Applied to Biological Waste Treatment," Water Research, vol. 7, p. 1609, 1973.)*

studies indicate that, in some cases, improved reactor performance can be obtained in dynamic operation which is forced by time variations in feed or operating conditions [50]. Periodic operation, in which such forcing involves periodic functions of time and the reactor variables eventually become periodic also, is the most widely considered dynamic operating mode. The basic concept underlying the possibility of improved reactor operation by intentional transient operation is quite simple. In transient operation, medium and cellular compositions and reaction rates can achieve different interrelationships than any of those possible in steady-state operation, where steady-state stoichiometric constraints limit the available mixture compositions. Intentional periodic fluctuations in bioreactor operation have been shown, for example, to increase cytochrome production by *Candida utilis*, biophotolytic hydrogen yield by *Anabaena cylindrica*, and to modify macromolecular composition of *E. coli* [51].

As we will explore further in Chap. 13, growth of two different organisms competing for a common limiting substrate usually leads to washout of one species. However, Hatch, Cadman and Wilder [52] have shown in simulation studies that a stable steady state with both species coexisting can be maintained using a proportional control algorithm in which substrate feed rate and overall dilution rate are both manipulated in response to measurements of the cell densities of the two species. Further, they demonstrated experimentally the feasibility of rapidly monitoring the cell concentrations of both *Candida utilis* and *Corynebacterium glutamicum* using flow cytometer light-scattering measurements.

10.8 CONCLUDING REMARKS

Control of a biotechnological process, whether by human or computer, requires good information on process operating state. This in turn depends upon analytical instrumentation and rigorous, systematic analysis and interpretation of process data. Major advances should occur in coming years as more powerful analytical tools from chemistry, biochemistry, and cell biology are adapted and improved for process applications. Model-based data analysis methods will progress beyond applications of elementary stoichiometry to on-line parameter updating and more detailed metabolic state estimates based on robust, structured models.

Several major concepts and topics connected with process control which were not discussed above should be mentioned before turning to our next topic, separation processes. On a local level, we wish to conduct the process according to some design or operating specification. That specification arises in turn from a higher level optimization or control problem in which the goal is maximization in some sense, often profit, of the contribution of that process to the overall plant, thence the parent corporation, and finally to the society. The objective functions and the constraints which the process engineer considers are derived from a hierarchy of objectives and rules at larger levels of operation.

Often reliability and safety are important components of the objective function or the operating constraints. How should the process be controlled to maximize product uniformity, to minimize the chances of losing a batch, and to minimize the probability of worker exposure to potentially unhealthful conditions? In practice, these objectives may be much more important than a few percent more or less in product yield or deviation from set point. Control engineers have begun to explore these questions, and we can expect more advances in these areas and in their significance in bioprocess control design in the future.

PROBLEMS

10.1 On-line sensors Define and give, in one sentence, the operating principle of the following.

(*a*) Thermistor, thermocouple, diaphragm gauge, Hall effect wattmeter, torsion dynamometer, strain gauge, gas rotameter, thermal mass flowmeter, capacitance probe.

(*b*) pH electrode, galvanic and polarographic types of dissolved oxygen probe, p_{CO_2} electrochemical probe, immobilized enzyme electrode.

(*c*) Mass spectrometer, gas chromatograph, paramagnetic O_2 analyzer.

(*d*) Spectrometer, nephelometer, in situ fluorometry.

10.2 Computer and control functions Define (*a*) A/D converter, multiplexer, D/A converter, (*b*) hardware, software, (*c*) process, sensor, monitor, controller, final control element.

10.3 State and parameter estimation (*a*) Discuss *measured* vs *estimated* variables. (*b*) What hardware and/or software system is needed for state estimation?

10.4 Process control and Laplace transforms The Laplace transform has properties very convenient for evaluation of linear, or linearized, control systems. The definition of the transform of any function of time, $f(t)$ is $F(s)$ where

$$F(s) \equiv \int_0^\infty f(t)e^{-st}\,dt$$

which is often abbreviated as $F(s) \equiv L\{f(t)\}$.

(*a*) Establish the properties of transforms of a derivative and integral:

$$L\{df/dt\} = sF(s) \qquad \text{if } f(t=0) = 0$$

$$L\left\{\int_0^t f(t)\,dt\right\} = F(s)/s$$

(*b*) Show that the Laplace transform of the controller output deviation $o(t) - o_s$ (Eq. 10.10) is

$$O(s) = K_c\left\{E(s) + \frac{E(s)}{\tau_I s} + \tau_D E(s)\right\}$$

and thus that the transfer function, or ratio of output (correction) to input (error) transforms, is given in the s-domain by

$$\frac{O(s)}{E(s)} = K_c\left[1 + \frac{1}{\tau_I s} + \tau_D s\right] = G_c(s)$$

10.5 pH control The kinetics of product formation during the batch lactic acid fermentation was described by Luedeking and Piret as Eq. (7.93)

$$\frac{dp}{dt} = \beta x + \alpha\frac{dx}{dt}$$

(*a*) Suppose $x(t)$ is described, under approximately constant pH, by the logistic equation

$$dx/dt = \mu x(1 - x/x_{\max})$$

Obtain a solution for $p(t)$ assuming that $p(0) = 0$.

(*b*) Lactic acid ($pK_a = 3.88$) causes a pH shift if the fermentation is uncontrolled. Suppose we wish to control the pH at 6.5 (i.e., the desired "set-point"). Derive expressions for the error in an uncontrolled process

$$e(t) \equiv \mathrm{pH}(t) - 6.5$$

What time program for addition of a small stream of concentrated base will keep the error in pH no larger than 0.1 ($\equiv \mathrm{pH} - \mathrm{pH}_0$)?

(*c*) Derive an expression for the output $O(t)$ of a PID controller using Eq. (10.10) and the error $e(t)$ of the uncontrolled process.

(*d*) Consider a proportional-only controller (Eq. 10.10 with $\tau_D = 0$ and $\tau_I = \infty$). The largest error allowable ($\mathrm{pH} - \mathrm{pH}_0 = 0.1$) must generate a base addition rate $\dot{m}_B$ which just matches the maximum product generation rate $V_R(dp/dt)_{\max}$. Expressing the output $O(t)$ as equal to the base addition rate $\dot{m}_B$, what proportional gain K_c is needed for satisfactory control?

(*e*) For the simple block diagram shown in Fig. 10P5.1 derive an expression for pH(t) for the *controlled* process. Neglect the volume change caused by addition of concentrated base, $\dot{m}_B$. Assume that $\mathrm{pH}_m = \mathrm{pH}$, i.e., that the pH probe is both instantaneous and accurate. (For information on block diagram representations, see any process control text such as D. R. Coughanowr and L. B. Koppel, *Process Systems Analysis and Control*, McGraw-Hill, New York, 1965.)

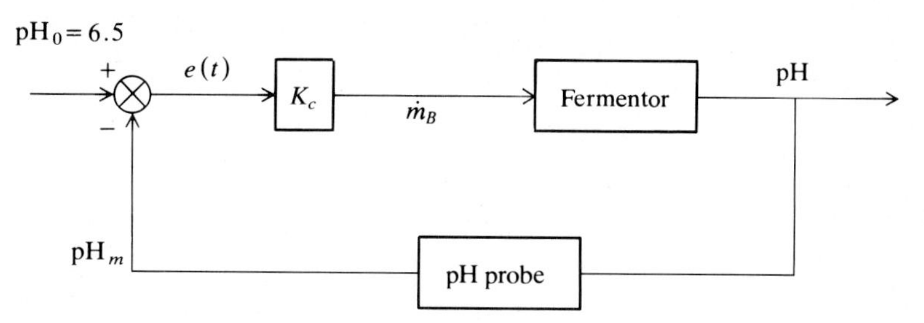

Fig. 10P5.1

10.6 Time lag in transient conditions In the presence of transients, the introduction of a time delay constant ($= 1/\gamma$) for the response of the instantaneous specific growth constant $\mu(t)$ to the changed substrate level has been proposed (T. B. Young, D. F. Bruley, and H. R. Bungay, "A Dynamic Mathematical Model of the Chemostat," *Biotect. Bioeng.*, **12**: 747, 1970). The movement of $\mu(t)$ toward the steady-state value μ_0 is given by

$$\mu_0 = \frac{\mu_{\max} s}{K_s + s} \qquad \frac{d\mu}{dt} = \frac{\mu_0 - \mu}{\gamma}$$

(*a*) If s is a function of time $s(t)$, show that the solution is given by

$$\mu(t) = \mu(t = 0)e^{-t/\gamma} + \frac{\mu_{\max}}{\gamma} \int_0^t \frac{e^{-(t-t')/\gamma} s(t')\, dt'}{K_s + s(t')}$$

(*b*) Suppose $s(t) = s_0(1 + \alpha \cos wt)$, $\alpha < 1$, and $K_s \gg s(t')$. Obtain the explicit solution for $\mu(t)$ above, plot the ratio $\mu(t)/\mu_{\max}$ vs. t for $\alpha = 0.5$, $\gamma = 1$, and $w = 0.1\gamma$, 1.0γ, and 10γ. Explain the differences between these curves. For what sorts of transients ought the time constant γ be included?

(*c*) If growth $r_x = \mu x s$, write down the equations needed to describe the behavior of a CSTR if $s_f = s_{f0}(1 + \cos wt)$.

10.7 Aerated chemostat monitoring When growing aerobic organisms in a chemostat, compressed air is passed through the fermentor at typically one volume of air per volume of liquid per minute (VVM). Compressed air is quite dry on entering the fermentor but saturated with water on leaving. As a consequence, the liquid flow rate in and out will be different.

How will this affect the interpretation of analytical data obtained by measurement on the fermentor effluent? If the fermentor volume is 10 liters, and the dilution rate is 0.1 h^{-1}, what will be the actual product concentration, corrected for evaporation, in the following system:

$$s_0 = 50 \text{ g glucose/liter}$$

$$K_s = 10 \text{ mg glucose/liter}$$

$$\mu_{max} = 0.7 \text{ h}^{-1}$$

$$Y_{X/S} = 0.5 \text{ g cell/g glucose}$$

$$m = 0.02 \text{ g glucose/g cell-h}$$

$$Y_{P/S} = 0.5$$

$$q_p = 0.2 \text{ g product/g cell-h}$$

$$T = 60°C$$

The vapor pressure of water at 60°C is 150 mmHg. Assume that Monod model is valid. Also you may assume that the concentration of CO_2 in the air is negligible and that the respiratory quotient for this fermentation is unity.

10.8 Feedforward control A batch fermentation is operated such that microbial growth is given by

$$\frac{dx}{dt} = \mu x$$

until a critical nutrient S is exhausted at $(x - x_0) = Ys_0$. The aerobic process has stoichiometric requirements for oxygen for biomass growth and for product formation, reflected in Y_{X/O_2} and Y_{P/O_2}.

(*a*) If product formation is nongrowth associated, so that $dp/dt = \beta x$, derive expressions for the instantaneous oxygen demand, Q_{O_2}, and for the specific oxygen demand (based on cell concentration), Q_{O_2}/x. At what condition do each of these have maximum values?

(*b*) Derive an expression for the $K_l a$ design value needed to meet the peak demand.

(*c*) If $K_l \sim$ constant and a varies as the gas inlet flow rate to the 3/4 power, how must F_g (inlet) be programmed to just achieve the necessary aeration rate with a minimum total gas utilization. Derive an expression for the ratio of air volume delivered, as cell mass grows from x_0 to x_{max}, divided by the volume delivered if F_g were held at the maximum value throughout this time.

(*d*) Feedforward control is often easier to implement than feedback control, but since the former involves no continued sensing of the process, it runs "blind." Discuss the consequences of having the programmed airflow meter for part (*c*) miscalibrated so that it consistently delivers (i) 10% below or (ii) 50% above the desired program rate. What modifications in the control system do you propose to address this potential problem?

10.9 State estimation via on-line mass balancing (*a*) Consider the rapid growth of cells during which the CO_2 production rate (CPR) is proportional to the biomass growth rate. Derive expressions for estimates of x and μ (i) if CPR is measured continuously and (ii) if CPR is measured at discrete intervals of length Δt.

(*b*) At low cell growth rates, corresponding to penicillin production, the simple CO_2 proportionality breaks down and a fuller carbon balance is needed. The overall reaction in a feed-batch system is given by

Substrate + penicillin precursor + seed biomass →
unconverted substrate + unconverted precursor + fermenter biomass
+ soluble unidentified products + penicillin + CO_2

If only substrate, precursor, CO_2, and penicillin provide appreciable carbon in addition to the biomass, express the biomass concentration through a carbon balance, in terms of measurable substrate, precursor, penicillin and CO_2 level, letting γ_i = fraction carbon in ith species. Ignore the volume change of fermentation fluid with time.

(*c*) Suppose that penicillin is not directly measurable, but that it does not contribute importantly to the carbon balance. If penicillin accumulation is assumed to have the kinetics given earlier, such that

$$\frac{dp}{dt} = nx + m\frac{dx}{dt} - kp$$

derive an expression from which you could estimate p concentration in terms of discrete measurements of substrate, precursor, and CO_2 levels. (see D.-G. Mou and C. L. Cooney, *Biotech. Bioeng.*, **25**, 225, 257, 1983, for a detailed example of on-line balancing for penicillin production).

10.10 Automatic supplementation of minerals in fed-batch culture *In situ* probes for NH_3 or volatile carbon substrate (e.g., ethanol) allow direct feedback control of feed addition rate of nitrogen or carbon. Many minerals necessary for growth and viability cannot be measured on-line. Full mineral addition at $t = 0$ would often be inhibitory to the inoculum. Accordingly, fed-batch programmed addition of minerals can be accomplished by use of stoichiometric relations between the measurable and nonmeasurable *in situ* levels.

Consider a simple balance of the following type:

$$\alpha(NH_3) + \beta(\text{carbon source}) + \gamma(\text{mineral source}) \rightarrow \text{biomass}$$

(*a*) For three feed reservoirs at concentrations n_0, c_0, and m_0 of nitrogen, carbon, and mineral, respectively, what ratio of volumetric feed rates will be stoichiometrically balanced for biomass production? How many different sensors are needed to set all feed rates?

(*b*) If both biomass and a product ($C_7NO_3H_{10}$) are produced, derive an expression for the proper volumetric mineral solution rate, $F_m(t)$, in terms of the feedback-controlled rates of nitrogen and carbon solutions, $F_n(t)$ and $F_c(t)$. How many different sensors are now needed? (See T. Suzuki et al. "Automatic supplementation of Minerals in Fed-Batch Culture to High Biomass Concentrations," *Biotech. Bioeng.*, **27**: 192, 1985.)

10.11 Enzyme assay for growth monitoring Studies on the microbial utilization of cellulose are difficult because of the water insolubility of the substrate. You are asked to examine the production of the enzyme glucose oxidase by a cellulolytic bacterium growing on cellulose. Glucose oxidase catalyses the reaction:

$$\text{Glucose} + O_2 \rightarrow \delta\text{-gluconolactone} + H_2O_2$$

Kinetics of this reaction may be measured with an enzyme-linked colorimetric assay or with a manometric apparatus (e.g. Warburg respirometer). You know that the organism of interest contains about one unit of glucose oxidase activity per gram of cell protein (1 unit = 1 μmole of glucose consumed/minute). The cellulase activity in this bacterium is associated with the cell wall, thus direct contact between the organism and substrate is required for growth.

(*a*) Design an experiment to study the kinetics of growth and glucose oxidase production by the bacterium growing on cellulose powder (particles are 100 μm in diameter). These studies should include both the exponential and stationary phases. In particular, you should describe: (1) the type of apparatus you would use, (2) the specific type of assays, and (3) the amount of cellulose and nitrogen source [$(NH_4)_2 SO_4$] to be added to the medium.

(*b*) How long would you expect the experiment to take and what would you expect the growth and enzyme specific activity curve to look like? You may assume that:

$$Y_{X/S} = 0.5 \qquad Y_{X/N} = 10$$

$$Y_{X/O} = 1.0$$

$$\text{Area of bacteria} \approx 5 \times 10^{-8} \text{cm}^2/\text{cell}$$

$$\text{Weight of bacteria} \approx 10^{12} \text{ g/cell}$$

$$\text{Desired final cell conc.} = 10 \text{ g/L}$$

$$\mu_{max} \approx 0.2\text{h}^{-1}$$

cells are 60% protein

REFERENCES

The following texts are useful summaries of chemical engineering process control:

1. G. Stephanopoulos, *Chemical Process Control. An Introduction to Theory and Practice*, Prentice-Hall, Inc., Englewood Cliffs, N.J., 1984.
2. W. L. Luyben, *Process Modeling, Simulation, and Control for Chemical Engineers*, McGraw-Hill, N.Y., 1973.
3. D. R. Coughanowr and L. B. Koppel, *Process Systems Analysis and Control*, McGraw-Hill, N.Y., 1965.

Kalman filtering and optimal control theory are summarized in many texts and monographs including the following:

4. A. H. Jazwinski, *Stochastic Processes and Filtering Theory*, Academic Press, New York, 1970.
5. M. Athans and P. L. Falb, *Optimal Control*, McGraw-Hill, N.Y., 1966.

Many recent review articles and monographs consider bioprocess instrumentation and control:

6. N. S. Wang and G. Stephanopoulos, "Computer Applications in Fermentation Processes," *CRC Critical Reviews in Biotechnology*, **2**: 1, 1984.
7. *Computer Applications in Fermentation Technology*, 3rd Int. Conf. on Computer Applications in Fermentation Technology, held in Manchester, England, 1981; Society of Chemical Industry, London, 1982.
8. W. B. Armiger (ed), *Computer Applications to Fermentation Processes* (Biotechnology and Bioengineering Symposium Series, No. 9), John Wiley, N.Y., 1979.
9. L. P. Tannen and L. K. Nyiri, "Instrumentation of Fermentation Systems," p. 331 in *Microbial Technology*, 2d ed., Vol. II, H. J. Peppler and D. Perlman (eds.), Academic Press, N.Y., 1979.
10. W. B. Armiger and A. E. Humphrey, "Computer Applications to Fermentation Technology," in *Microbial Technology*, 2d ed., Vol. II, H. J. Peppler and D. Perlman (eds.), Academic Press, N.Y., 1979.

References for specific topics cited in the chapter text are listed below.

Sensors

11. W. C. Wernan and C. R. Wilke, "New Method for Evaluation of Dissolved Oxygen Response for $k_L a$ Determination," *Biotech. Bioeng.*, **15**: 571 (1973).
12. I. Karube and S. Suzuki, "Application of Biosensor in Fermentation Processes," p. 203 in *Annual Reports on Fermentation Processes, Vol. 6*, G. T. Tsao, (ed.), Academic Press, N.Y., 1983.
13. J. S. Schultz and G. Sims, "Affinity Sensors for Individual Metabolites," p. 65, in *Biotech. Bioeng Symposium 9*, W. B. Armiger, (ed.), Wiley, N.Y., 1979.

Off-line analytical methods:

14. P. Gerhardt (ed.-in-chief), *Manual of Methods for General Bacteriology*, American Society for Microbiology, Washington, D.C., 1981.
15. D. H. Prescott, (ed.), *Methods in Cell Biology, Vol. XII, Yeast Cells*, Academic Press, N.Y., 1975.
16. R. L. Rodriguez and R. C. Tait, *Recombinant DNA Techniques: An Introduction*, Addison-Wesley Publishing Co., Reading, MA, 1983.
17. S. Pedersen, P. L. Bloch, S. Reeh, and F. C. Neidhardt, "Patterns of Protein Synthesis in *E. coli*: A Catalog of the Amount of 140 Individual Proteins at Different Growth Rates," *Cell* **14**: 179, 1978.
18. N. R. Rose and P. E. Bijayzi (eds.), *Methods in Immunodiagnosis*, 2d ed., Wiley, N.Y., 1980.
19. K. Ugurbil, H. Rottenberg, P. Glynn, and R. G. Shulman, "^{31}P Nuclear Magnetic Resonance Studies of Bioenergetics and Glycolysis in Anaerobic *Escherichia coli* Cells," *PNAS(US)*, **75**: 2244 (1978).
20. A. A. Herrero, R. F. Gomez, B. Snedecor, C. J. Tolman, and M. F. Roberts, "Growth Inhibition of *Clostridium thermocellum* by Carboxylic Acids: A Mechanism Based on Uncoupling by Weak Acids," *Appl. Microbiol. Biotechnol.*, **22**: 53, 1985.
21. J. A. den Hollander, K. Ugurbil, T. R. Brown, and R. G. Shulman, "Phosphorus-31 Nuclear Magnetic Resonance Studies of the Effect of Oxygen upon Glycolysis in Yeast," *Biochemistry*, **20**: 5871, 1981.
22. K. Nicolay, W. A. Sheffers, P. M. Bruinenberg, and R. Kaptein, "Phosphorus-31 Nuclear Magnetic Resonance Studies of Intracellular pH, Phosphate Compartmentation, and Phosphate Transport in Yeasts, "*Arch. Microbiol.* **133**: 83, 1982.
23. K. Ugurbil, T. R. Brown, J. A. den Hollander, P. Glynn, and R. G. Shulman, "High-Resolution ^{13}C Nuclear Magnetic Resonance Studies of Glucose Metabolism in *Escherichia coli*," *PNAS(US)*, **75**: 3742, 1978.
24. T. L. Legerton, K. Kanamori, R. L. Weiss, and J. D. Roberts "^{15}N NMR studies of Nitrogen Metabolism in Intact Mycelia of *Neurospora crassa*," *PNAS(US)*, **78**: 1495, 1981.
25. E. Nestaas, D. I. C. Wang, H. Suzuke, and L. B. Evans, "A New Sensor—the 'Filtration Probe' —for Quantitative Characterization of the Penicillin Fermentation. II. The Monitor of Mycelial Growth," *Biotech. Bioeng.*, **23**: 2815, 1981.
26. E. Nestaas, D. I. C. Wang, "A New Sensor—the 'Filtration Probe'—for Quantitative Characterization of the Penicillin Fermentation. III. An Automatically Operating Probe," *Biotech. Bioeng.*, **25**: 1981, 1983.
27. D. C. Thomas, V. K. Chittur, J. W. Cagney, and H. C. Lim, "On-Line Estimation of Mycelial Cell Mass Concentrations with a Computer-Interfaced Filtration Probe," *Biotech. Bioeng.*, **27**: 729, 1985.
28. F. Srienc, B. Arnold, and J. E. Bailey, "Characterization of Intracellular Accumulation of Poly-β-Hydroxybutyrate (PHB) in Individual cells of *Alcaligenes eutrophus* H16 by Flow Cytometry," *Biotech. Bioeng.*, **26**: 982, 1984.
29. M. R. Melamed, P. F. Mullaney, and M. L. Mendelsohn (ed.), *Flow Cytometry and Sorting*, John Wiley & Sons, N.Y., 1979.
30. J. E. Bailey, "Single-Cell Metabolic Model Determination by Analysis of Microbial Populations," p. 136 in *Foundations of Biochemical Engineering: Kinetics and Thermodynamics in Biological Systems* (H. W. Blanch, E. T. Papoutsakis, and G. Stephanopoulos, eds.), American Chemical Society, Washington, D.C., 1983.

31. F. Srienc, J. L. Campbell, and J. E. Bailey, "Detection of Bacterial β-Galactosidase Activity in Individual *Saccharomyces cerevisiae* Cells by Flow Cytometry," *Biotech. Ltrs.*, **5**: 43, 1983.
32. R. T. Hatch, C. Wilder, and T. W. Cadman, "Analysis and Control of Mixed Cultures," *Biotech. Bioeng. Symp. No. 9* [W. B. Armiger (ed.)], p. 25, John Wiley & Sons, 1979.
33. K.-J. Hutter, U. Punessen, and H. E. Eipel, "Flow Cytometric Determination of Microbial Contaminants," *Biotech. Ltrs.*, **1**: 35, 1979.

Microcomputers:

34. W. A. Hampel, "Application of Microcomputers in the Study of Microbial Processes," p. 1 in *Advances in Biochemical Engineering*, Vol. 13. [T. K. Ghose, A. Fiechter, and N. Blakebrough (eds.)], p. 1, Springer-Verlag, Berlin, 1979.

Filtering and recursive estimation:

35. R. P. Jefferis, H. Winter, and H. Vogelmann, "Digital Filtering for Automatic Analysis of Cell Density and Productivity," *Workshop Computer Applications in Fermentation Technology* [R. P. Jefferis III, (ed.)], Verlag Chemie, N.Y., 1977.

Material balancing estimation methods:

36. D. W. Zabriskie and A. E. Humphrey, "Real-Time Estimation of Aerobic Batch Fermentation Biomass Concentration by Component Balancing," *AIChE J.*, **24**: 138, 1978.

Kalman filter applications:

37. G. Stephanopoulos and K.-Y. San, "Studies on On-line Bioreactor Identification. Part I. Theory," *Biotech. Bioeng.*, **26**: 1176, 1984.
38. K.-Y. San and G. Stephanopoulos, "Studies on On-Line Bioreactor Identification. Part II. Numerical and Experimental Results," *Biotech. Bioeng.*, **26**: 1189, 1984.

Fed-batch fermentor control:

39. C. L. Cooney, H. Y. Wang, and D. I. C. Wang, "Computer-Aided Material Balancing for Prediction of Fermentation Parameters," *Biotech. Bioeng.*, **19**: 55, 1977.
40. H. Y. Wang, C. L. Cooney, and D. I. C. Wang, "Computer-Aided Baker's Yeast Fermentations," *Biotech. Bioeng.*, **19**: 69, 1977.

Batch fermentation control:

41. R. Lundell, "Practical Implementation of Basic Computer Control Strategies for Enzyme Production," p. 181 in *Computer Applications in Fermentation Technology*, Society of Chemical Industry, London, 1982.
42. A. Constantinides, J. L. Spencer, and E. L. Gaden, Jr., "Optimization of Batch Fermentation Processes. I. Development of Mathematical Models for Batch Penicillin Fermentations," *Biotech. Bioeng.*, **12**: 803, 1970.
43. A. Constantinides, J. L. Spencer, and E. L. Gaden, Jr., "Optimization of Batch Fermentation Processes. II. Optimum Temperature Profiles for Batch Penicillin Fermentations," *Biotech. Bioeng.*, **12**: 1081, 1970.

Batch-process design and operation:

44. D. W. T. Rippin, "Simulation of Single- and Multiproduct Batch Chemical Plants for Optimal Design and Operation," *Computers and Chem. Eng.*, **7**: 137, 1983.
45. I. A. Karimi and G. V. Reklaitis, "Intermediate Storage in Noncontinuous Processing," p. 425 in *Foundations of Computer Aided Process Design*, A. W. Westerberg and H. H. Chien, eds., CACHE Publications, Ann Arbor, Michigan, 1984.

Linear-quadratic optimal control:

46. A. S. Fawzy and O. R. Hinton, "Microprocessors Control of Fermentation Process," *J. Ferment. Technol.*, **58**: 61, 1980.

47. L. T. Fan, P. S. Shah, N. C. Pereira, and L. E. Erickson, "Dynamic Analysis and Optimal Feedback Control Synthesis Applied to Biological Waste Treatment," *Water Research*, **7**: 1609, 1973.

Multiple steady states and control:

48. D. DiBiasio, H. C. Lim, W. A. Weigand, and G. T. Tsao, "Phase-Plane Analysis of Feedback Control of Unstable Steady States in a Biological Reactor," *AIChE J.*, **24**: 686, 1978.
49. D. Dibiasio, H. C. Lim, and W. A. Weigand, "An Experimental Investigation of Stability and Multiplicity of Steady States in a Biological Reactor," *AIChE J.*, **27**: 284, 1981.

Forced periodic bioreactors:

50. J. E. Bailey, "Periodic Phenomena," p. 758, in *Chemical Reactor Theory: A Review* [L. Lapidus and N. R. Amundson (eds.)], Prentice-Hall, Englewood Cliffs, N.J., 1977.
51. A. M. Pickett, M. J. Bazin, and H. H. Topiwala, "Growth and Composition of *Escherichia coli* Subjected to Square-Wave Perturbation in Nutrient Supply: Effect of Varying Frequencies," *Biotech. Bioeng.*, **21**: 1043, 1979.

Mixed culture control:

52. C. T. Wilder, T. W. Cadman, and R. T. Hatch, "Feedback Control of a Competitive Mixed-Culture System," *Biotech. Bioeng.*, **22**: 89, 1980.

CHAPTER

ELEVEN

PRODUCT RECOVERY OPERATIONS

Fermentation broths are complex aqueous mixtures of cells, soluble extracellular products, intracellular products, and unconverted substrate or unconvertible components therein. Characteristically, a bioprocess includes, centrally, both the bioreactor and a subsequent product recovery section. The particular separation techniques useful for any given bioprocess depend not only on the location of the product (intracellular vs. extracellular) and size, charge, and solubility of the product but also on the size of the process itself and the product value. For example, various forms of chromatography are usefully applied to purification of high value pharmaceuticals or biologicals such as hormones, antibodies, and enzymes, but are both expensive and difficult to scale up beyond laboratory size.

The present chapter on separations pertinent to bioprocess operations is structured in the following sequence. We first examine individual recovery methods useful in whole or in part for recovery of microbial (or plant or animal) products. Next, we illustrate how these processes are arranged in series to produce the quality of separation needed for a given product (e.g., from filtered yeast cells, to crude enzyme extracts, to pyrogen-free insulin from recombinant *E. coli*). We also take note of the appreciable cost of many separation operations and reflect upon interactions between choice of bioreaction process and consequent required product recovery; for example it is often desirable from the viewpoint of product recovery to choose a strain which produces an extracellular, rather than intracellular, product. Also, impurities or residues in various possible substrates may affect overall process costs more in the separation section than in the fermentation itself. The processes and principles for product recovery apply to effluents from enzyme and cell culture systems, as well as the prototype process most often considered, a microbial fermentation process.

The ultimate challenge is to select the best combination of substrate, enzyme or organism, bioreactor, and separation for a given product. A common basis, the economics of bioprocess design, is the mechanism by which this choice is achieved and is considered in the next chapter.

Each separation needed depends on initial broth characteristics (viscosity, product concentration, impurities, undesired particulates, etc.) and final product concentration and form needed (crystallized product, concentrated liquid, crude solution, dried powder). For example, a crude product may be obtained as a dried residue of the fermentation broth. The operations *sequence* through which a bioreactor broth must pass for a highly purified product is typically as follows:

1. *Removal of particulates* (insolubles). Common operations here are filtration, centrifugation, and/or settling/sedimentation/decanting.
2. *Primary isolation.* Solvent extraction, sorption, precipitation, and ultrafiltration are best known. The latter offers filtration which discriminates at the level of molecular size. During primary isolation, desired product concentration increases considerably, and substances of widely differing polarities are separated from the product.
3. *Purification.* These operations often select for impurity removal as well as further product concentration. Approaches include fractional precipitation, many kinds of chromatography, and adsorption.
4. *Final product isolation.* The last step(s) must provide the desired product in a form suitable for final formulation and blending, or for direct shipping. Processes here include centrifugation and subsequent drying of a crystallized product, drum or spray drying, freeze drying (lyophilization), or organic solvent removal.

A characteristic processing profile for pharmaceuticals via bioprocessing is given in Table 11.1, which indicates both concentration and relative purity (%)

Table 11.1 Typical processing profile[†]

	Product	
Step	Concentration, g/L	Quality, %
Harvest broth	0.1–5	0.1–1.0
Filtration	0.1–5	0.1–2.0
Primary isolation	5.0–10	1.0–10
Purification	50–200	50–80
Crystallization		90–100

[†] *After P. Belter, "General Procedures for Isolation of Fermentation Products," p. 403 in Microbial Technology, 2d ed., vol. 2, H. J. Peppler and D. Perlman, eds., Academic Press, 1979.*

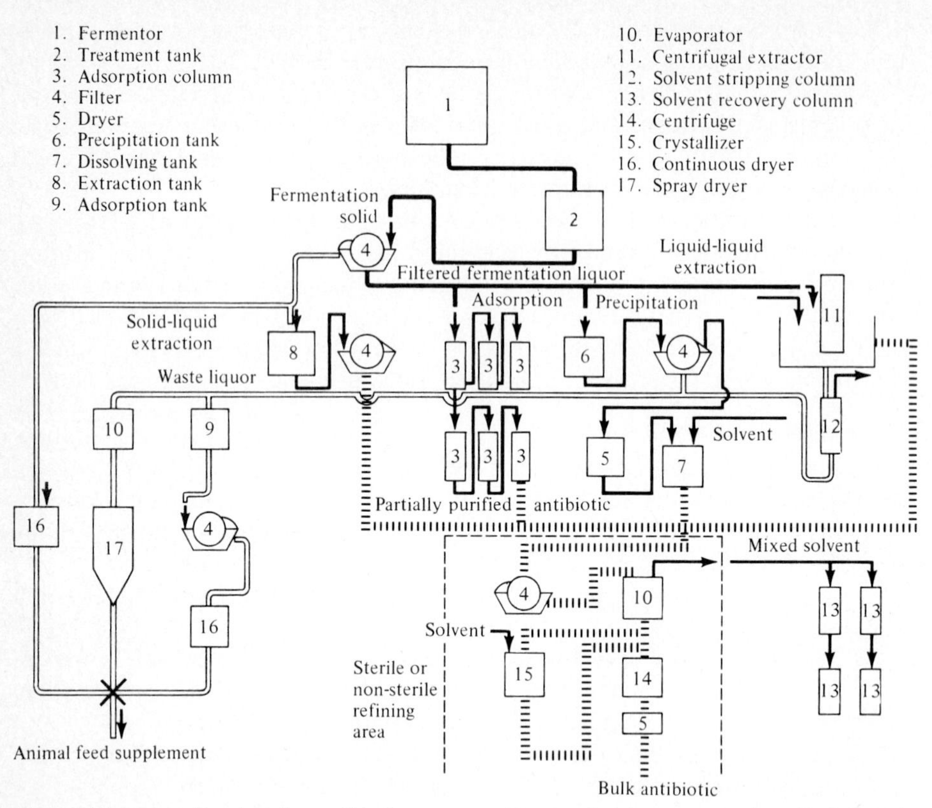

Figure 11.1 Process flowsheet for antibiotic recovery. *(Reprinted with permission from S. C. Beesch and G. M. Stull, "Fermentation," Ind. Eng. Chem., vol. 49, p. 1491, 1957. Copyright by the American Chemical Society.)*

through the product recovery process consisting of solids removal, primary isolation, purification, and final isolation (crystallization) steps.

An example of an overall separation system is shown in Figure 11.1. Here, fifteen different separation techniques, plus a fermentation broth pretreatment tank, are used to produce both a crude and a highly purified antibiotic product, the former as a spray or continuously dried crude solid, and the latter as a crystalline, essentially pure material.

11.1 RECOVERY OF PARTICULATES: CELLS AND SOLID PARTICLES

Particles, from cellular to molecular, may be separated from a solution based on their differences in key physical-chemical properties such as size, density, solubility and diffusivity. Figure 11.2, which summarizes applicable methods based on these characteristics vs. particle size, indicates that size and density are the features allowing separation of cell-sized (0.3 to 10 μm) particles by filtrations,

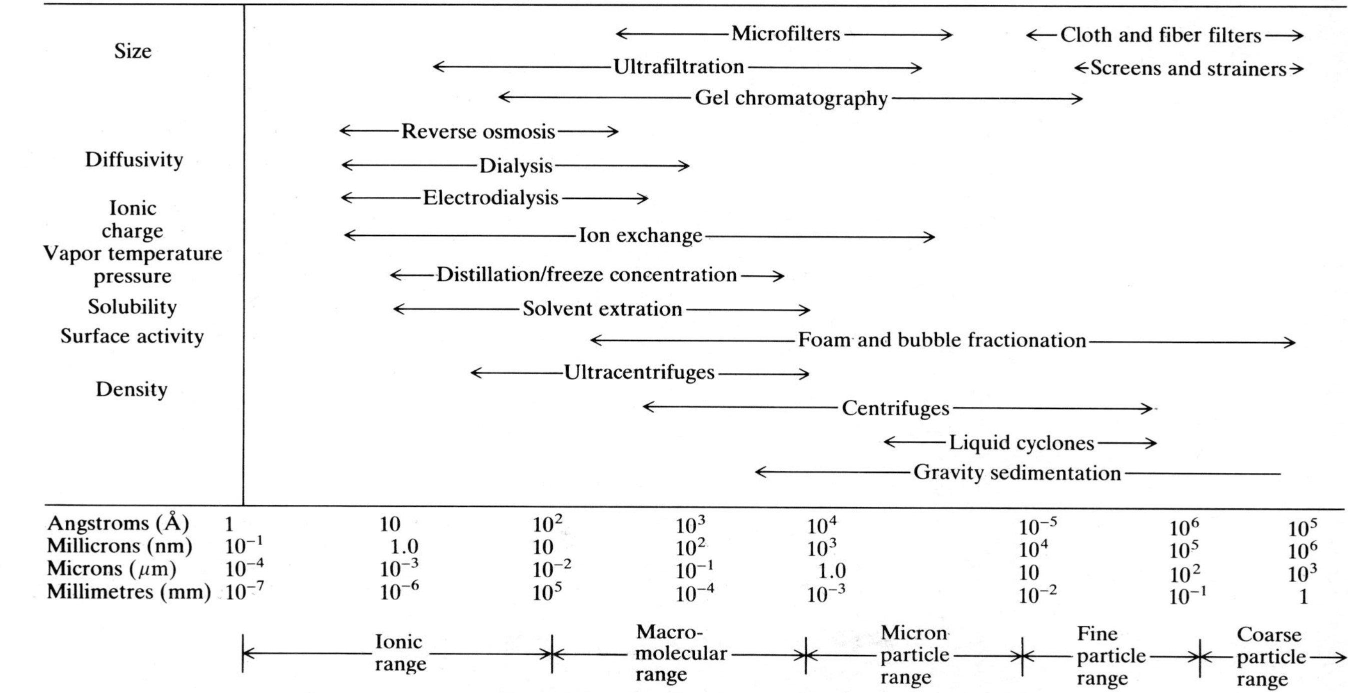

Figure 11.2 Primary factors affecting separation vs. particle size. *(After B. Atkinson and F. Mavituna, Biochemical and Biotechnology Handbook, Macmillan Publ. Ltd., Surrey, England, 1983, p. 935.)*

centrifugation, and/or settling. Sedimentation is settling in a simple gravitational field, whereas centrifugation requires production of enhanced settling velocities by centrifugal forces.

Bioprocess broths have characteristics which may render solid–liquid separation (and subsequent) processes difficult. These often include high viscosities, gelatinous broth materials, compressible filter cakes, particles with small density difference compared with water, high degree of initial dispersion, and diluteness of particulate suspension. In batch bioreactions, these characteristics change in time. Continuous processes also produce broths of variable character depending on flow rate and operating conditions: the successful continuous sedimentation of cells in domestic waste treatment will be considered in Chap. 14.

Broth pretreatments designed to increase ease of a subsequent cell separation may include aging (cells tend to clump together in later stationary phase), heat treatment, pH treatments, and/or addition of chemicals which aid cell flocculation (polyelectrolytes, calcium chloride, colloidal clay). The evident cost of this additional treatment is balanced by improvements in sedimentation or filtration rate, which may be spectacular: factors of ten are common, factors of 100–200 have occasionally been reported.

Selection of settler, filter, or centrifuge type depends upon both starting broth and final desired cell density:

Settling is used in large-scale waste treatment processes, as well as the traditional fermentation industry (why is good champagne finished in a nearly inverted bottle?). The resultant flocculent material may have several percent solids.

Centrifugation produces a cell-concentrated stream, often referred to as a cream because of its solids content (~15% w/v) yet appreciably fluid behavior.

Filtration produces yet more concentrated (dewatered) cell sludges (20–35% w/v) or cell solids (≥40% w/v). In some instances, the accumulating biomass filter cake has a modest permeability, allowing straightforward filtration. However, the cell biomass may often provide the major filtration resistance; use of a filter precoat diminishes this problem.

11.1.1 Filtration

Small fermentation batches can be handled in a plate-and-frame filter, which gradually accumulates biomass, then is opened and cleared of filter cake. Larger processes rely on continuous filters. One of these is the rotary vacuum filter which in some cases requires precoat. The type of continuous rotary vacuum filter schematically illustrated in Fig. 11.3 uses strings to lift off the rotating filter cake (a layer of concentrated solids) which has accumulated on the drum. While string discharge is satisfactory for removing *Penicillium* mycelia, *Streptomyces* mycelia are more difficult to process and require precoat of the filter cloth with filter aid, e.g., diatomaceous earth, and cake removal with a knife-blade to scrape the cake from the rotating drum.

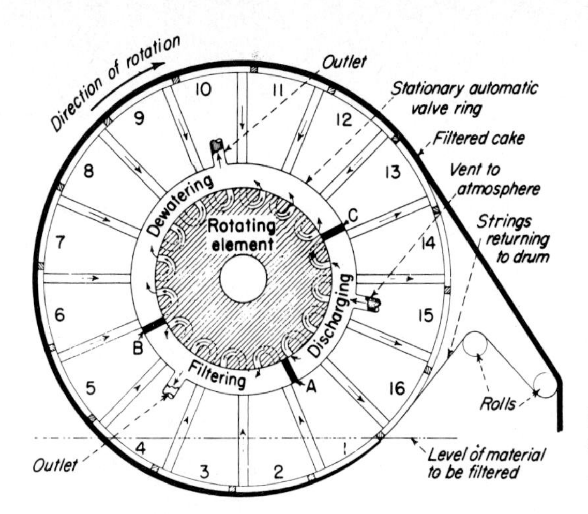

Figure 11.3 Schematic diagram of a string filter in operation. *(Courtesy of Ametek, Inc.)*

Filtering characteristics of the solid-liquid slurry are often described in terms of the elementary theory of filtration. Assuming laminar flow of filtrate liquid through the cake, we may write

$$\frac{1}{A}\frac{dV_f}{dt} = \frac{\Delta p}{\mu_c[\alpha(W/A) + r]} \tag{11.1}$$

where A = area of filtering surface
V_f = volume of filtrate collected
t = time
Δp = pressure drop across filter
μ_c = filtrate viscosity
α = average specific cake resistance
W = mass of accumulated dry cake solids = $[\rho w/(1 - mw)]V_f$, where ρ is filtrate density, w is the mass fraction of solids in the slurry, and m is the ratio of wet-cake to dry-cake mass
r = resistance coefficient of filter medium

Pressure drop across the filter is constant for the rotary vacuum filters commonly used in the fermentation industry. If we assume that the cake is incompressible, α is constant and Eq. (11.1) can be integrated to obtain

$$\frac{t}{V_f/A} = \frac{\mu_c \alpha \rho w}{2\Delta p(1 - mw)}\frac{V_f}{A} + \frac{\mu_c r}{\Delta p} \tag{11.2}$$

indicating that t/V_f is a linear function of V_f under the conditions assumed.

Figure 11.4*a* shows a plot of t/V_f against V_f for a *Streptomyces griseus* fermentation broth filtered at various pH values using a cotton cloth, diatomaceous-earth filter aid, and a Δp of 28.4 lb/in^2. Two important features are revealed by this data: (i) the cake is not incompressible since the data for each pH do not fall on a straight line. Generally cells and other organic material from

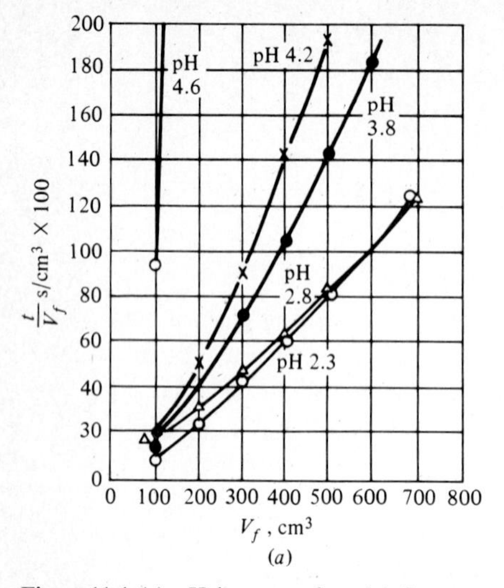

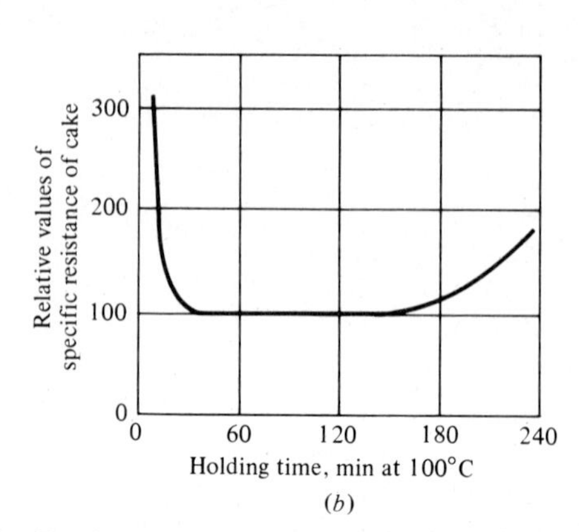

Figure 11.4 (*a*) pH has a profound influence on the filtration rate of *S. griseus* broth. (*b*) Heating pretreatment of *S. griseus* broth changes the specific resistance of the resulting cake. [*Reprinted from S. Shirato and S. Esumi, "Filtration of a Culture Broth of* Streptomyces griseus," *J. Ferment. Tech. (Japan), vol. 41, p. 87, 1963.*]

fermentations form compressible cakes. (ii) The data in Fig. 11.4 show a strong dependence of filtering properties on broth pretreatment and filtration conditions. Clearly the pH during filtration has a major influence on filtration rates. Figure 11.4*b* reveals that broth preheating can substantially lower the specific cake resistance, presumably by coagulating mycelial protein.

Cell recycle is receiving increasing emphasis in order to increase reactor volumetric productivity (recall Chap. 9, Sec. 9.1.3). In large-scale aerobic domestic waste treatment (Chap. 14), the flocculent multi-population biomass is conveniently settled and recycled as a settled cell sludge. For small processes involving nonflocculent organisms, a *cross-flow filtration* is possible. Here continuous fluid motion parallel to the filter surface continuously removes the accumulating cell mass, thereby allowing a nearly *steady* filtrate flow [vs. the time dependent behavior of a batch system, Eq. (11.2)]. The transmembrane flux of filtrate depends both on the applied transmembrane pressure Δp, and the steady resistance of the cellular layer, $\alpha(W/A)$. Accordingly, Eq. (11.1) is appropriate for cross-flow filtration where now the steady-state filter cake weight W and specific resistance α are functions of the operating fluid cross-flow (or tangential-flow) rate.

The cake composition may differ in conventional vs. cross-flow filtration, especially if the feed contains a variety of particles besides cells. Cell concentration factors of more than one order of magnitude are achievable; values reported for cross-flow with a 100,000 molecular weight cutoff membrane are 15–50 for harvesting of *E. coli*, mycoplasma (for veterinary vaccines), and influenza virus (whole virus vaccine). When a *batch* volume of cells is to be cross-flow filtered, the cell concentration in the continuously recycled cell stream builds up in time;

an analysis of this circumstance requires inclusion of an increasing feed concentration with time.

As with high-speed centrifugation, some prefiltering may be required prior to cross-flow filtration. For example, in antibiotic fermentations using soy grits and calcium carbonate, incomplete utilization of these nutrient components leaves particulates which are best removed by a coarse screen or filter prior to cross-flow filtration.

Where an extracellular product is extractable directly from a bioreactor broth by an immiscible solvent phase, the processes of cell elimination and product separation from the broth may be accomplished in a single operation. This *combined function* processing is considered in a later section.

11.1.2 Centrifugation

Centrifugation may be used to remove cells from fermentation broths; yeasts, for example, are sometimes harvested in this fashion. A schematic diagram of one type of continuous centrifuge is given in Fig. 11.5. For dilute suspensions, each cell may be treated as a single particle in an infinite fluid. In this case the analysis of Example 1.1 applies. Such an approach is not valid for concentrated slurries, in which a given particle's motion is influenced by neighboring particles.

Correlations of particle velocity u_h in such hindered-settling situations with the single-particle velocity u_0 and the volume fraction of particles ε_p have been developed. Possessing the general form

$$\frac{u_h}{u_0} = \frac{1}{1 + \beta\varepsilon_p^{1/3}} \tag{11.3}$$

the empirical relationships derived between β and ε_p are

$$\beta = \begin{cases} 1 + 3.05\varepsilon_p^{2.84} & 0.15 < \varepsilon_p < 0.5, \text{ irregular particles} \\ 1 + 2.29\varepsilon_p^{3.43} & 0.2 < \varepsilon_p < 0.5, \text{ spherical particles} \\ 1\text{–}2 & \text{dilute suspensions } (\varepsilon_p < 0.15) \end{cases} \tag{11.4}$$

Centrifuges may be classified by their internal structures, which have been developed to handle somewhat differing suspensions leading to different methods of solids discharge (Table 11.2). An important feature of the decanter or scroll-type centrifuge (Fig. 11.6) is that it can easily handle large solid particles, in contrast to other forms of continuous centrifuges. Accordingly, this type of centrifuge may be used in series with another, fine particle centrifuge, to allow full capacity utilization of the latter without clogging or overloading.

Example applications of these centrifuges appear in Table 11.3; note the scroll (decanter) centrifuge for recovery of large mold pellets, and the larger throughput capacity of the nozzle-type centrifuges.

Noncellular solids often occur in biological process fluids. For example, when enzymes are harvested from plant biomass, the latter is typically first

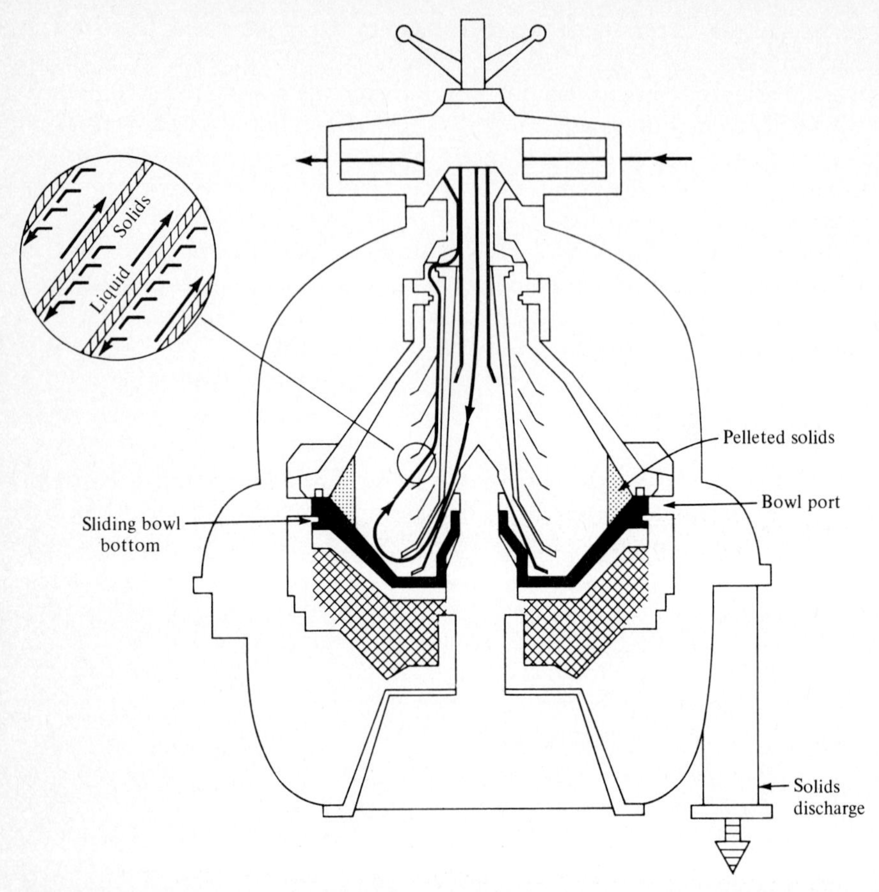

Figure 11.5 In this continuous centrifuge, solid particulates are removed in flow between closely stacked cones. Clarified effluent is withdrawn from the top of the unit.

crushed and extracted at cool temperatures into a high ionic strength solution, following which the major solid content must be removed. Similarly, partial utilization of solid substrates (e.g., cellulosics) may result in a fermentation broth containing macroscopic particles. Where an appreciable density difference exists, the scroll centrifuge (Fig. 11.6) is used first to remove large or easily settled solids.

11.1.3 Sedimentation

When cells have a high tendency to aggregate closely (coagulate) or to form multicelled flocs with the aid of polyvalent cations or extracellular polymers, the recovery of cell biomass by sedimentation is possible. Such aggregation provides an inexpensive cell recycle stream in activated sludge waste treatment (Chap. 14), and a number of highly flocculent yeast strains have been used (in brewing beer

Table 11.2 Types of centrifugal separators

	Transport of sediment	Solids content in feed material, % by vol	Maximum throughput in largest machine m^3/h (gpm)
Solids bowl separators	Stays in bowl	0–1	150 (650)
Solids ejecting nozzle separator	Intermittent discharge through axial channels	0.01–10	200 (880)
Solids ejecting separators	Intermittent discharge through radial slot	0.2–20	100 (440)
Nozzle separator	Continuous discharge through nozzles at or near bowl periphery	1–30	300 (1320)
Decanter (scroll)	Internal screw conveyor	5–80	200 (880)

After P. Belter, "Recovery Processes—Past, Present, and Future" 184th Amer. Chem. Soc. Mtg., Sept. 1982.

and in single-cell protein production) due to this desirable property which allows inexpensive cell separation. The mechanics of cell suspension settling and compaction are not well understood. In contrast to a single particle, which settles at a concentration independent rate, a suspension settles more slowly with increasing solids concentration [Eq. (11.3.)] For example, the interface dividing cell-free vs. cell-bearing liquid moves downward often according to Eq. (11.5):

$$\text{Settling velocity } u_s = kc^{-m} \qquad \text{where} \qquad m \approx 1.7\text{–}2.6 \tag{11.5}$$

Once the cell concentration achieves a large value, c_{max}, further concentration occurs at a negligible rate, as seen for an activated sludge example involving both

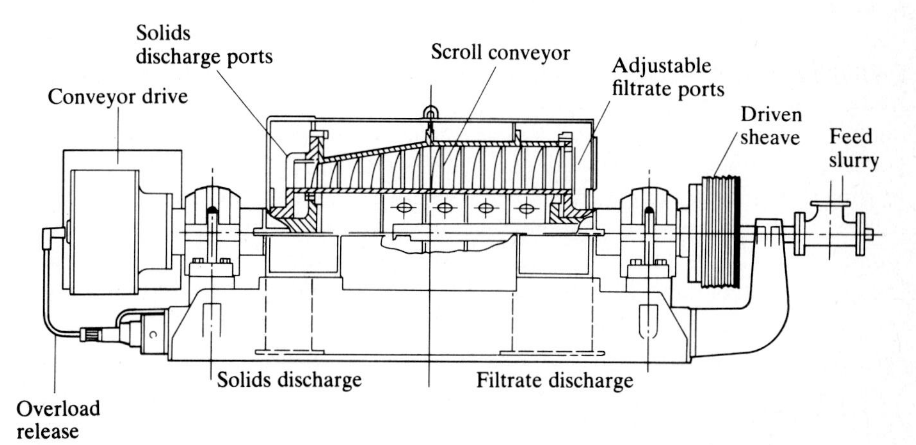

Figure 11.6 Configuration of a scroll conveyer centrifuge. *(After R. H. Perry and C. H. Chilton, Chemical Engineers Handbook, 5th ed., McGraw-Hill, New York, pp. 19–92, 1973.)*

Table 11.3 Centrifugal biomass separations

Product	Microorganism Type	Microorganism Size (microns)	Relative throughput in centrifuge	Type of separator
Bakers yeast	*Saccharomyces*	7–10	100	Nozzle
Brewers yeast	*Saccharomyces*	5–8	70	Nozzle,
Alcohol yeast	*Saccharomyces*	5–8	60	solids-ejecting
SCP	*Candida*	4–7	50	Nozzle, decanter
Antibiotics	Mold	....	10–20	Decanter
Antibiotics	*Actinomyces*	10–20	7	Solids-ejecting
Citric acid	Mold	....	20–30	Solids-ejecting decanter
Enzymes	*Bacillus*	1–3	7	Nozzle, solids-ejecting
Vaccines	*Clostridia*	1–3	5	Solid bowl, solids-ejecting

After P. Belter, "Recovery Processes—Past, Present, and Future," 184th Amer. Chem. Soc. Mtg., Sept., 1982.

mixed microbial cultures and extracellular polysaccharides (Fig. 11.7*a*). Settled sludge biomass may be further dewatered by centrifugation or filtration; in the former case there is again a (gravity dependent) maximum cell concentration achievable (Fig. 11.7*b*). The centrifugal acceleration number Z is defined as $r_0\omega^2/g$, where r_0 is the centrifuge drum radius, ω is the centrifuge angular velocity, and g is the acceleration of gravity.

The relative abilities of sedimentation, centrifugation, filtration, and drying to achieve dewatering up to a desired level are important in determining which processes are appropriate. As Fig. 11.8 indicates, the water content of a cell-containing suspension may be classified as free, floc, capillary, particle, and solid water. The free water is removable by sedimentation; further sedimentation (aging) of the initial layer floc allows some further water reduction. Centrifugation at $Z = 3000$ removes all floc water, giving a pellet of 5 percent water. Further dewatering is only achievable (to remove capillary water) in small lab centrifuges, and particle water is removable only by drying after cell lysis. Increased sedimentation rates may be possible in inclined tubes or narrow channels, because of a secondary flow phenomenon (Boycott effect) which results in rapid evolution of a clear fluid zone at the top of the channel or tube.

11.1.4 Emerging Technologies for Cell Recovery

Particulates can be removed from an aqueous suspension by attachment to rising air bubbles. This *flotation* method is widely used for recovery of small particles

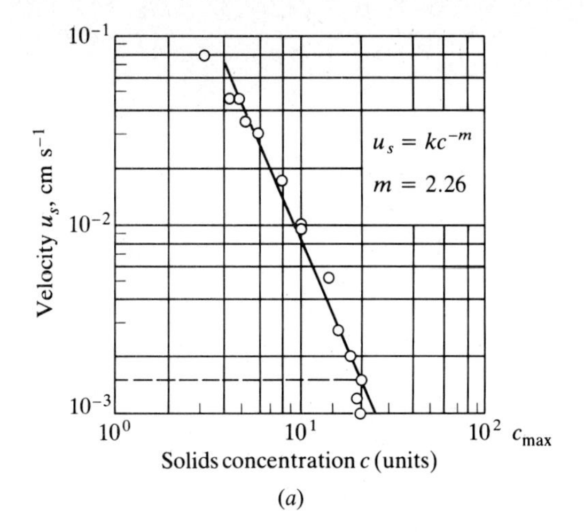

(a)

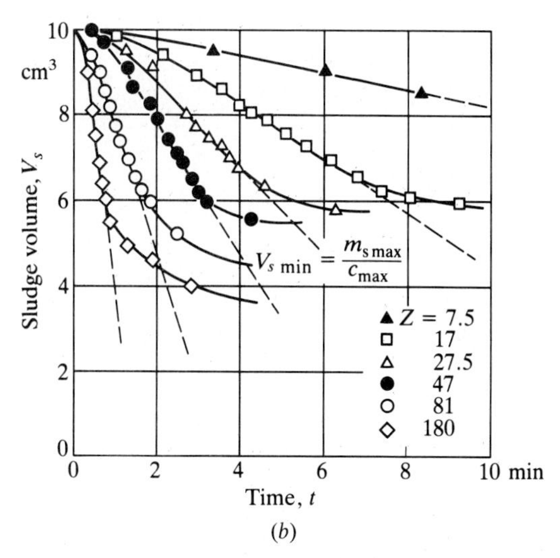

(b)

Figure 11.7 (*a*) Settling velocity (of interface) of oxygen-fed activated sludge. (*b*) Settled (mixed digested) sludge volume vs. time: test tube centrifuge. Parameter Z: centifugal acceleration number = $r_0\omega^2/g$. *(From U. Weissman and H. Binder, "Biomass Separation from Liquids by Sedimentation and Centrifugation," p. 119 in Adv. Biochem. Eng., vol. 24, A. Fiechter (ed.), Springer-Verlag, Berlin, 1982.)*

from aqueous suspensions of minerals. This method has also been used in beer processes. A related technique, *froth flotation*, uses the air–water surface tension to strip proteins from solution and accumulate them in a high protein, stable froth (a la European or English beer!). Flotation has been applied to concentrate *Acinetobacter cerificans* for SCP.

Electrokinetic deposition uses voltage gradients of 1050 volts cm^{-1} to produce solid biomass with densities up to 40% w/vol, thereby paying for power for deposition by savings in a subsequent dewatering or drying step. Current inefficiencies due to broth electrolysis diminish the fraction of total current carried by the desired cell deposition. This technique has also been tried in reverse to *prevent* solids deposition on filtration surfaces during conventional filtration. Electrochemically driven coagulation of cells has been used with SCP production of *Methylomonas clara* in the Hoechst process.

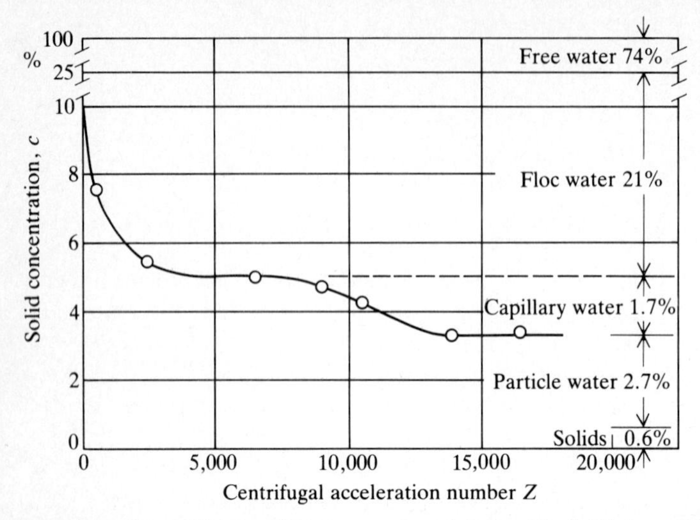

Figure 11.8 Water "fractions," in waste activated sludge: For 0.6% dry solids, the intracellular water is 2.7%, capillary water (external) is 1.7%, the interstitial water to complete a single floc particle is 21%, and, in a multifloc sample, the free water is 74%. Beginning with free water, each succeeding type is more difficult to remove. *(From U. Weissman and H. Binder, "Biomass Separation from Liquids by Sedimentation and Centrifugation," p. 119 in Adv. Biochem. Eng., vol. 24, A. Fiechter (ed.), Springer-Verlag, Berlin, 1982.)*

11.1.5 Summary

Cell or particle separation may be desired for any of the following reasons: recovery of whole cells per se, recovery of cells for subsequent lysis and intracellular product recovery, removal of major cell debris after extraction of lysed or ground cells, removal or recovery of trace unconverted soluble substrate, removal of biomass from product-extracted broth, and recovery and recycle of cells to the bioreactor. The choice of filtration (batch, continuous vacuum, cross-flow, etc.), centrifugation (vertical rotor, scroll (horizontal rotor)), or sedimentation/coagulation depends on broth conditions (T, pH, ionic strength), medium components (cells, polymers, polyvalent cations, presence of other particles), and final state of desired product.

11.2 PRIMARY ISOLATION

11.2.1 Extraction

Extraction requires the presence of two liquid phases. While aqueous–organic two-phase systems have been dominant in antibiotic recovery, a novel two-phase aqueous–aqueous system has recently been considered for protein recovery.

11.2.1.1 Solvent extraction Many antibiotics have excellent solubilities in organic solvents which are nearly water-immiscible. A multistep, alternating aqueous to organic, then organic-to-aqueous extraction can provide both *concentration* and subsequent *purification.*

A typical penicillin broth contains 20–35 g antibiotic/liter. Following filtration to remove mycelial biomass, the filtrate is cooled to 4°C or below, and may be subjected to a second, polymer-aided filtration. The pK_a values of penicillins lie in the range 2.5–3.1. Consequently, acidification of the broth to near pH 2.0–3.0 causes protonation (neutralization) of the penicillins, rendering them extractable by organic solvents. A countercurrent extraction with a solvent/broth ratio of 1/10 using amyl acetate or butyl acetate removes most of the product from the broth. A subsequent back extraction with an aqueous phosphate buffer (pH 5–7.5) increases penicillin concentration. The penicillin is finally recovered as sodium penicillin precipitate from a butanol–water mixture.

Operation of the extraction step as a continuous process is accomplished with a continuous flow, countercurrent device, the Podbielniak centrifugal extractor-separator (Fig. 11.9). The light (solvent) and heavy (broth) liquids enter through the rotating shaft, and leave the rotating contactor near the shaft and

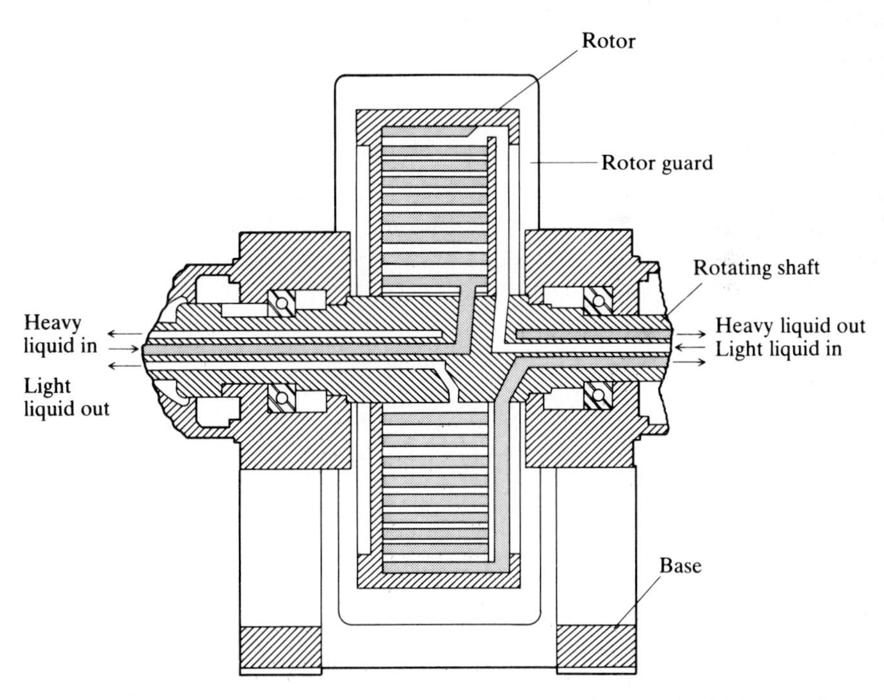

Figure 11.9 The Podbielniak centrifugal contactor provides for two liquid introduction points through the rotating shaft. The light liquid is forced to the outside first; the heavy liquid is introduced near the center. Rapid rotation allows centrifugal force to move the fluids past each other in a counter current contacting fashion. *[After W. J. Todd and D. B. Podbielniak, "Centrifugal Extraction," Chem. Eng. Progress, vol. 61(5), p. 69 (1965).]*

rotor periphery, respectively. Rapid rotation produces a centrifugal field which drives the two fluids countercurrent to each other.

Other important antibiotics have been concentrated and purified via similar aqueous-to-organic-to-aqueous extractions. Thus, erythromycin is treated with amyl acetate, and bacitracin, a polypeptide antibiotic, is directly extracted with *n*-butanol, then reextracted into aqueous buffer, concentrated by evaporation, and dried. We note again that separations needed depend on final use: bacitracin as an animal food supplement is obtained by simple evaporation of water from the whole fermentation broth (including cells) and a subsequent drying of the resultant product sludge.

Not all antibiotics are purified by extraction: streptomycin is stripped from the filtered fermentation broth and recovered by ionic adsorption involving ion exchange (see Secs. 11.2.2, 11.4, and 11.7.2).

In the original extraction steps, the fermentation broth is normally acidified to render the antibiotic neutral for removal to the organic phase. However, for a given solute, one particular pH range may give maximum selectivity for antibiotic vs. other extractables as well. As Fig. 11.10 indicates, choice of a pH < 2.0 will give maximum selectivity for penicillin extraction vs. extraction of other

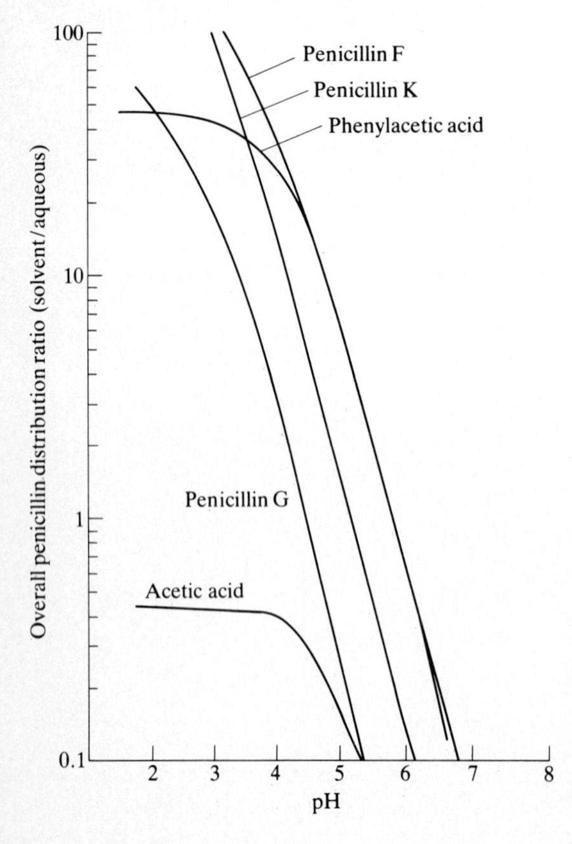

Figure 11.10 Distribution ratio (solvent/aqueous) for various penicillins and contaminants. *[After S. Queener and R. Swartz, "Secondary Products of Metabolism," Economic Microbiology, vol. 3, A. H. Rose (ed.), Academic Press, London, 1979.]*

solutes with a higher (less acidic) pK_a. However, the increased hydrolysis rate of penicillins with increased solution acidity makes the optimal extracting pH more basic than 2.0. The short residence time extraction allowed by the Podbielniak centrifuge also helps minimize this acid hydrolysis problem.

Direct extraction of the original microbial broth has also been considered for recovery of fine chemicals. These examples are treated later in a section titled *whole broth processing* (11.7.2).

11.2.1.2 Extraction using aqueous two-phase systems The need for an extracting phase which does not harm labile enzymes and other proteins, which can be readily scaled to process large throughputs of product mixtures, and which allows recovery of modestly stable whole organelles and other substructures, has motivated consideration of liquid–liquid extraction systems in which *each* phase is primarily aqueous. The production of separated aqueous phases is achieved by dissolution of two incompatible polymers, such as polyethylene glycol (PEG) and a dextran moiety (Fig. 11.11*a*). The resulting phases are >75% water, and either dextran or PEG rich; Fig. 11.11*a* demonstrates that the PEG rich phase may have almost no dextran. Similar systems with PEG-potassium phosphate are available (Fig. 11.11*b*) as are examples with dextran/ficoll/PEG, polyvinyl alcohol/dextran, and dextran/methylcellulose although those systems of Fig. 11.11*a* and *b* are the best examined to date.

The partition coefficient for eight enzymes in a PEG-Dextran system, $K = e(\text{PEG-rich})/e(\text{Dextran-rich})$, lies in the range 1.0 to 3.7, necessitating consideration of a derivatized polymer to produce a liquid ion-exchange, of a dissolved polysite affinity adsorber (see later section), or of salt addition to enhance one-stage partition coefficients. Ammonium sulfate and potassium phosphate provide 10-fold increases in K for pullulanase recovery at salt concentration of 12% and 0.3-M, respectively, in one PEG/ammonium sulfate system (see Ref. 16).

The polymer phases may be separated by decanting (moderate holding time, Fig. 11.12*a*) or centrifugation (short holding time, Fig. 11.12*b*), and then stripped of desired product (by, for example, ultrafiltration) giving polymer (PEG) recovery. As this contacting and separation process may be quickly accomplished at ambient temperature, the trade-off of polymer costs (investment, loss per separation cycle) against refrigeration needed in conventional (longer process time) protein solvent extraction or precipitation is evident. Dextran, even in crude form, dominates the medium makeup costs and is thus the key item for recovery. A summary of enzymes purified from microorganisms by aqueous two-phase extraction is provided in Table 11.4.

11.2.2 Sorption

Sorption involves the partitioning of a solute between a bulk solution phase and a typically porous or high surface area solid. A prime example is the sorption of

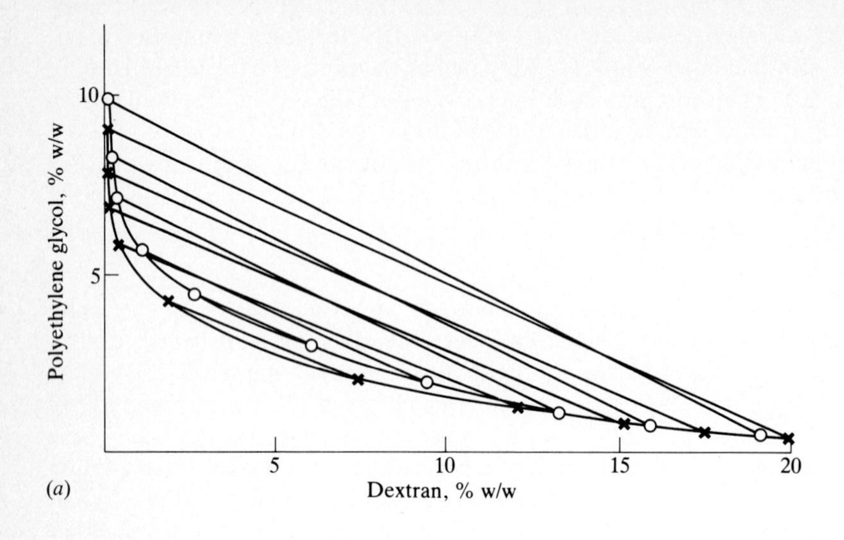

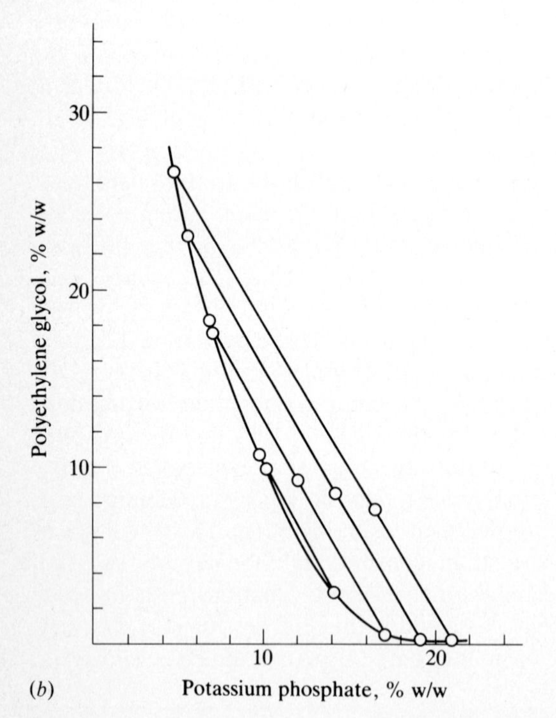

Figure 11.11 (*a*) Phase diagram of the dextran-polyethylene glycol system (D48-PEG 6000) at 20°C (○) and 0°C (×). (*b*) Phase diagram and phase compositions of the potassium phosphate—PEG 6000 system at 0°C. *(Reprinted by permission from P.-A. Albertsson, "Partition of Cell Particles and Macromolecules," 2d ed., p. 41, Wiley-Interscience, New York, 1971.)*

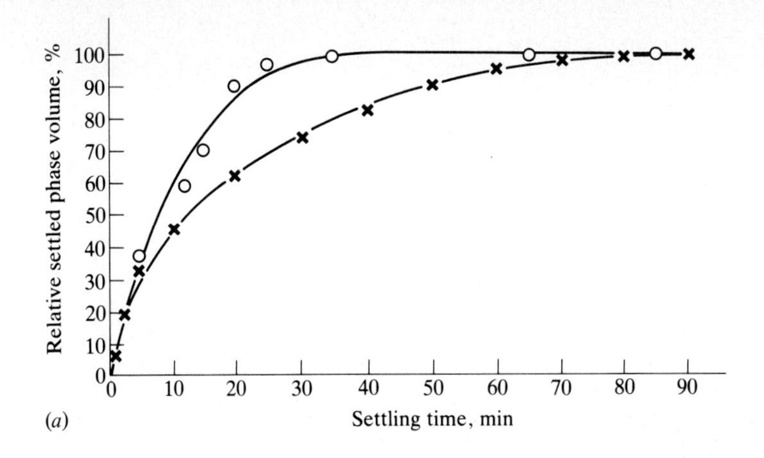

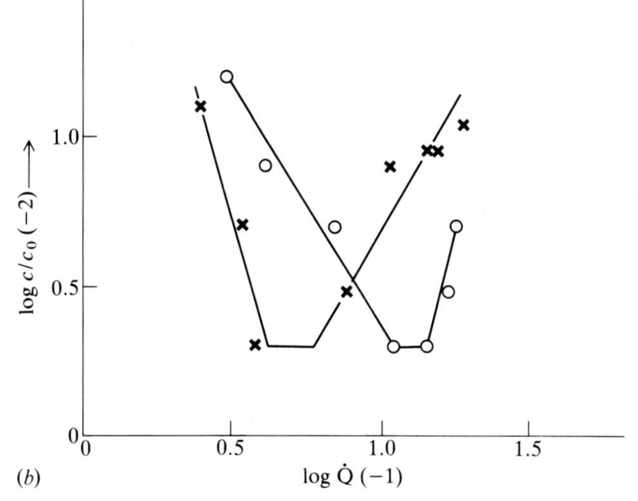

Figure 11.12 (*a*) Relative settled phase volume vs. time for gravity driven separation of two polyethylene glycol/salt systems used in FDH purification. Settling tank = 150 L glass tank. × = PEG potassium phosphate system: 6% PEG 1550, 9% PEG 400, 15% potassium phosphate, pH 7.8 including 2.5M KCl; top phase dispersed, H/D = 2.67. ○ = PEG potassium phosphate system (12% PEG 4000, 7% potassium phosphate, pH 7.8; bottom phase dispersed, H/D = 2.5. (*b*) Throughput characteristics of a centrifugal separator (α-Laval LAPX-202). Logarithmic ratio between dispersed phase effluent concentration, c, and feed concentration, c_0 vs. logarithm of throughput $\dot{Q}$(L/min). Feed; 14% PEG 4000, 11% potassium phosphate, 0.1% Blue-dextran as indicator. Rpm: 700(×), 9300(○). *(Reprinted by permission from M. R. Kula, K. H. Kroner, and H. Hustedt, "Purification of Enzymes by Liquid-Liquid Extraction," Adv. Biochem. Eng., vol. 24, A. Fiechter (ed.), p. 105, Springer-Verlag, Berlin, 1982.*

Table 11.4 Two-phase aqueous extractions of enzymes from microbial cells. Except for pullalanase recovery, cells were disrupted by high pressure homogenization or wet milling.
(Information courtesy of Dr. M.-R. Kula)

Enzyme	Organism	Biomas concentration, %	Phase system	Partition coefficient	Yield, %	Purification factor
Isoleucyl-tRNA synthetase	*Escherichia coli*	20	PEG/salt	3.6	93	2.3
Fumarase		25	PEG/salt	3.2	93	3.4
Aspartase		25	PEG/salt	5.7	96	6.6
Penicillin acylase		20	PEG/salt	2.5	90	8.2
α-Glucosidase	*Saccharomyces cerevisiae*	30	PEG/salt	2.5	95	3.2
Glucose-6-phosphate dehydrogenase		30	PEG/salt	4.1	91	1.8
Alcohol dehydrogenase		30	PEG/salt	8.2	96	2.5
Hexokinase		30	PEG/salt	....	92	1.6
Glucose isomerase	*Streptomyces* species	20	PEG/salt	3.0	86	2.5
Pullulanase	*Klebsiella pneumoniae*	25	PEG/dextran	3.0	91	2
Phosphorylase		16	PEG/dextran	1.4	85	1
Leucine dehydrogenase	*Bacillus sphaericus*	20	PEG/crude dextran	9.5	98	2.4
Diacetate dehydrogenase	*Lactobacillus* species	20	PEG/salt	4.8	95	1.5
L-2-hydroxyisocaproate dehydrogenase	*Lactobacillus confusus*	20	PEG/salt	10	94	16
D-2-hydroxyisocaproate dehydrogenase	*Lactobacillus casei*	20	PEG/salt	11	95	4.9

See also Partioning in Aqueous Two-Phase Systems: Theory, Methods, Uses and Applications to Biotechnology, H. Walter, D. E. Brooks, and D. Fisher (eds.). Academic Press, New York, 1985.

streptomycin by ion exchange. Ion exchange materials are simply dissociable ion pairs in which one type of charge is immobile; e.g.,

Cation exchanger

$$(\text{solid})^{-}\text{H}^{+} + \text{solute}^{+} \longrightarrow (\text{solid})^{-}\text{solute}^{+} + \text{H}^{+}$$

Anion exchanger

$$(\text{solid})^{+}\text{Cl}^{-} + \text{solute}^{-} \rightleftharpoons (\text{solid})^{+}\text{solute}^{-} + \text{Cl}^{-}$$

Streptomycin may be adsorbed onto a carboxylic acid cation exchange resin:

$$n(\text{Resin})^{-}\text{H}^{+} + \text{streptomycin}^{n+} \rightleftharpoons n(\text{Resin})^{-}\text{streptomycin}^{n+} + n\text{H}^{+}$$

The antibiotic-loaded resin is recovered from the fermentation broth, and subsequently the antibiotic is recovered by elution with acidulated water, reversing the original adsorption step and simultaneously regenerating the resin.

Ion-exchange materials may occur as soluble polymers giving rise to two-phase aqueous systems (see previous section). Among these, polyethylene glycol (PEG)-Dextran solutions are unstable and separate into two phases. As the PEG may be derivatized to yield cation or anion exchange properties, the value of a two-phase system with dielectric properties near that of water (to retain protein stability) is clear. With such two-phase systems, dramatic separations between beta interferon and contaminant protein are achievable [17].

As with other product concentrating operations, ion-exchange has been applied directly to whole broth processing, e.g., in the recovery of the antibiotic novobiocin (see Sec. 11.7.2).

11.3 PRECIPITATION

Organic solutes have solubilities dependent on solution temperature, pH, composition, ionic strength and dielectric constant. Precipitation may be brought about in many ways:

1. *Add precipitant* to react with solute and produce an insoluble product, often a salt. Examples here include antibiotic recovery operations:

$$\left.\begin{array}{l}\text{Procaine-hydrochloride}\\ \text{Sodium buffer salt}\\ \text{Potassium buffer salt}\end{array}\right\} + \text{penicillin} \longrightarrow \left\{\begin{array}{l}\text{procaine}\\ \text{sodium}\\ \text{potassium}\end{array}\right\} - \text{penicillin}$$

$$\begin{array}{c}\text{Organic solvent} +\\ \text{streptomycin} + H_2SO_4\end{array} \longrightarrow \begin{array}{c}\text{di-hydro-streptomycin}\\ \text{sulfate}\end{array}$$

$$\begin{array}{c}\text{Organic solvent} +\\ H_2O + \text{erythromycin}\end{array} \longrightarrow \begin{array}{c}\text{erythromycin}\\ \text{hydrate}\end{array}$$

Biopolymer recovery is also obtainable by salt addition. For example, xanthan gum is a polyanion. Divalent cation (calcium) addition can be used to form a gel precipitate. Alginate biopolymer is recoverable from algal biomass by cell removal (filtration), followed by a calcium chloride precipitation of the biopolymer. [Recall use of this method to form calcium alginate gels in common use for cell immobilization (Chap. 9).]

2. *Solvent driven precipitations* are useful in production of microbial biopolysaccharides including dextran and xanthan gum. These biogum fermentations are typically aerobic and produce a highly viscous final broth. With xanthan production, a final broth pasteurization kills the *Xanthamonas* cells. After potassium chloride addition (which diminishes subsequent solvent needed), the gum polysaccharide is directly precipitated by added methanol or isopropanol. The cell-bearing precipitate is then dewatered, washed, and dried, and contains no viable cells. Dextran recovery from broth is achieved by alcohol or acetone precipitation. In use of solvent-driven precipitation for production of bulk biopolysaccharide, the modest product value requires efficient recovery and reuse of solvent, as well as good solvent removal from food or pharmaceutical grade product.
3. *Protein precipitation techniques*, which result in a phase change to form a precipitate, require some alteration of protein solution conditions to render the original, thermodynamically stable one-phase system unstable with respect to precipitation. The various methods for causing the needed reduction in solubility of proteins include:

    (*a*) added high salt concentration, to give precipitate by "salting-out"
    (*b*) pH adjustment to a protein's pH of neutral charge, the isoelectric point (Table 11.5), at which point the protein has a minimum solubility
    (*c*) reduction of medium dielectric constant to enhance electrostatic interactions by, for example, addition of miscible organic solvent
    (*d*) addition of nonionic polymers, which reduce the amount of water available for protein solvation
    (*e*) addition of polyelectrolytes, which presumably act as flocculating agents
    (*f*) addition of polyvalent metal ions to form (reversibly) a protein precipitate

Table 11.5 Isoelectric points for several enzymes and other proteins[†]

	pI = isoelectric pH		
Pepsin	~1.0	Hemoglobin	6.8
Egg albumin	4.6	Myoglobin	7.0
Serum albumin	4.9	Chymotrypsinogen	9.5
Urease	5.0	Cytochrome *c*	10.65
β-Lactoglobulin	5.2	Lysozyme	11.0
γ_1-Globulin	6.6		

[†] Adapted from A. L. Lehninger, *Biochemistry* 2d ed., table 7.1, Worth Publishers, Inc., New York, 1975.

The method of choice includes consideration not only of the protein concentration needed and the cost of separation technique, but also the purity of the final product compared to precipitating agent.

It is not surprising that the solubility of a protein depends on the pH of the solution and is usually smallest at the isoelèctric pH. Since the protein molecules have no net charge at pI and some charge at different pH, electrostatic repulsive forces between solute molecules are minimized at pI. This behavior suggests a *fractional-precipitation* procedure for separating proteins with different isoelectric points: at a given pH the proteins with the nearest pI will tend to precipitate, other things (such as molecular weight) being equal. By varying pH, fractions containing different proteins are separated.

However, imposing wide pH variations on a protein solution runs the risk of denaturing many of its components. Consequently, another precipitation technique commonly called *salting out* is more prevalent. Protein solubility is markedly affected by the salt concentration of the solution. In 1889, Hofmeister observed that more negatively charged anions of salts were most effective in causing precipitation of proteins; a similar trend is evident with cations. The Hofmeister (or lyotropic) series of ions, in order of approximate diminishing effectiveness for salting out proteins is as follows:

Anions: citrate^{3-}, tartrate^{2-}, F^-, IO_3^-, $H_2PO_4^-$, acetate$^-$, $B_2O_3^-$, Cl^-, ClO_3^-, Br^-, NO_3^-, ClO_4^-, I^-, CNS^-

Cations: Th^{4+}, Al^{3+}, H^+, Ba^{2+}, Sr^{2+}, Ca^{2+}, Mg^{2+}, Cs^+, Rb^+, NH_4^+, K^+, Na^+, Li^+

As ammonium sulfate is very soluble in aqueous solutions, it is a common salt for biochemical isolations, although the Hofmeister series reveals the potential usefulness of many other salts.

The major variable governing salting out of proteins is the ionic strength μ_{is}, defined by

$$\mu_{is} = \frac{1}{2} \sum_{i=1}^{n} c_i Z_i^2 \tag{11.6}$$

where c_i and Z_i are the concentration and charge of the ith ion, respectively, and n is the total number of ionic species. It is believed that the addition of various salts tends to reduce the water concentration available for hydration of ionized or polarizable protein groups, thereby diminishing the protein solubility. As the data in Fig. 11.13 show, protein solubility at high ionic strength can often be correlated by the relation

$$\log(\text{solubility, g/L}) = \beta' - \kappa_s \mu_{is} \tag{11.7}$$

where κ_s is called the *salting-out constant.*

Proteins which denature at higher temperatures more easily than the component of interest can be removed from solution by "cooking" them. After denaturation, the protein is less soluble (remember the oil-drop model: unfolding a globular protein will contact hydrophobic groups with the aqueous phase) and it

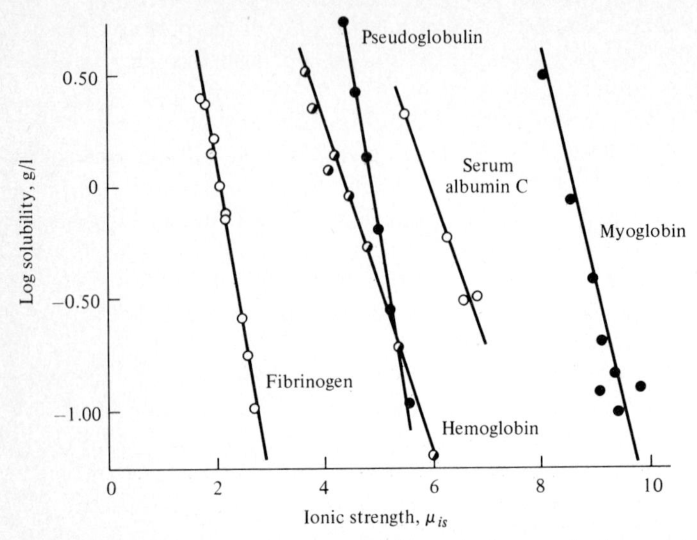

Figure 11.13 Dependence of protein solubility on ionic strength of ammonium sulfate solutions. *(Reprinted from E. Cohn and J. T. Edsall, "Proteins, Amino acids, and Peptides," p. 602, Academic Press, New York, 1943.)*

precipitates. Another precipitation mechanism involves use of casein, diatomaceous earth (containing diatoms, the porous skeletons of many algae), or gelatin: proteins settle out of solution by adsorption to these large particles. This method is advantageously applied, for example, to purify a thermostable protein expressed in a mesophilic host cell. Contaminant host proteins may be selectively precipitated while the thermostable protein remains in solution.

A combination of these precipitation techniques is evident in many protein procedures as shown in Example 11.1. The process begins with enzyme extraction in a buffered solution. This is followed by slight heating to remove, by precipitation, those proteins or components which denature more easily than the dehydrogenase of interest. Most of the enzyme is precipitated (along with other molecules of similar solubility) with acetone addition, which separates the enzyme from many of the low-molecular-weight extraction components: a subsequent resuspension and precipitation in ammonium sulfate effects a further purification in terms of enzyme activity per gram of precipitate. This purification is typically achieved, however, with loss of some of the total amount of enzyme in each step (bottom of Example 11.1).

A further purification to separate the dehydrogenase from other proteins of similar solubility is effected by suspending the precipitate from step 8 (Example 11.1) in progressively weaker solutions of ammonium sulfate. Since the solubility of a protein increases as salt concentration decreases below saturation values, resolution of the enzyme is accomplished with each progressively weaker contact. The supernatants from these resuspensions have specific activities (enzyme activity per total soluble protein) of the order of 100 times that of the first crude buffer extract. In addition, a great deal of extraneous material has been removed.

Example 11.1 Procedures for isolation of enzymes from isolated cells (yeast alcohol dehydrogenase from dried baker's yeast)[†]

1. Grind yeast in ball mill.
2. Mix 100 g powdered dried yeast with 100 mL 0.066 *M* disodium phosphate buffer which is 0.001 *M* in ethylene diamine tetraacetate (EDTA).
3. Add 200 mL buffer, stir 2 h at 37°C, and let stand at ambient 3 h.
4. Remove yeast-cell debris by centrifugation (28,000 g for 20 min) (assay).
5. Remove thermally labile protein by heating at ~55°C for 15 min (alcohol dehydrogenase relatively heat-stable)(assay).
6. Add 50 mL acetone per 100 mL heat-treated extract (mixing at −2 to −4°C) and discard precipitate. Add additional 55 mL acetone, centrifuge (28,000 g), and discard supernatant (assay).
7. Resuspend precipitate in 50 mL of 10^{-3} *M* phosphate buffer (pH 7.5) + 10^{-3} *M* EDTA buffer, dialyze twice with 3 L fresh phosphate buffer for 1 h.
8. At 0°C gradually add 36 g ammonium sulfate per 100 mL dialyzed solution and centrifuge at 0°C (assay).
9. Suspend the first $(NH_4)_2SO_4$ precipitate in 30 mL of 50% saturated $(NH_4)_2SO_4$ solution. Separate supernatant and remaining precipitate and repeat precipitate suspension with 45, 40, 35, and 25% saturated ammonium sulfate solutions (assay solutions).
10. Combine high-activity supernatants and add saturated ammonium sulfate solution until incipient turbidity is observed. Let crystallization proceed for several days at 0°C. Dissolve precipitate and recrystallize in 10 mL of 10^{-3} *M* EDTA (assay).

The results are

	Total units, $\times 10^{-6}$	Specific activity
First extract	25	4,000
After heat	30	8,000
Acetone precipitate	20	45,000
First $(NH_4)_2SO_4$ precipitate	14	70,000

	Protein in solution	
Saturation $(NH_4)_2SO_4$, %	Specific activity	Total protein, %
50	180,000	14.9
45	200,000	12.5
40	650,000	12.0
35	440,000	6.6
25	430,000	19.5

11.3.1 Kinetics of Precipitate Formation

Protein precipitation arises when a change in solution condition so reduces the solubility that it falls well below the actual protein (or protein-reagent) concentration. In favorable cases, such as for α-casein, following an initial lag, the

[†] From S. Chaykin, *Biochemistry Laboratory Techniques*, pp. 22–32, John Wiley & Sons, Inc., New York, 1966.

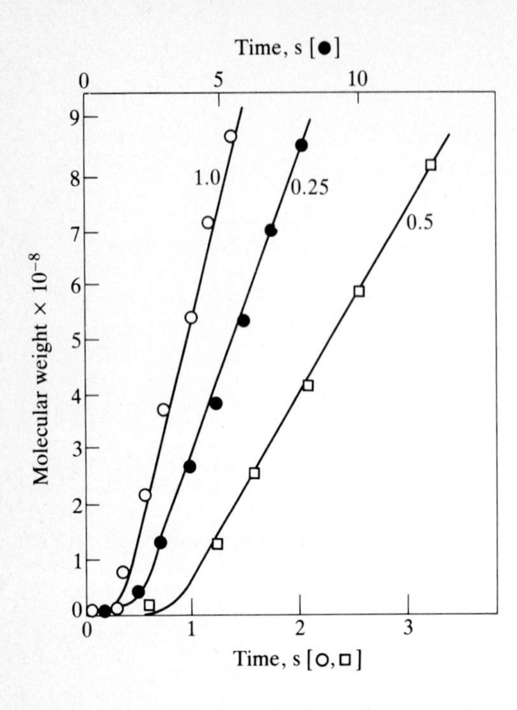

Figure 11.14 Molecular weight vs. time for α-casein precipitate at three concentrations (indicated near lines; kg/m^3) formed in presence of 0.008 M $CaCl_2$. Lines are predictions of Smoluchowski's perikinetic aggregation theory. *(After D. J. Bell, M. Hoare, and P. Dunnill, Adv. Biochem. Eng., vol. 26, p. 19, Springer-Verlag, Berlin, 1983.)*

precipitate particle average molecular weight, $\overline{MW}(t)$, increases linearly and rapidly with time (Fig. 11.14), consistent with a growth kinetics form derived by Smoluchoski:

$$\overline{MW}(t) = MW(0)[1 + K_A m_0 t] \tag{11.8}$$

where m_0 = molar concentration of the aggregating species.

The theory which predicts such a linear increase in a stagnant fluid of Brownian particles is a diffusion driven, or *perikinetic*, flocculation approach in which the diminution of species concentration, N, follows a second-order law,

$$-\frac{dN}{dt} = \frac{K_A}{W} N^2 \tag{11.9}$$

where $K_A = 8\pi \mathscr{D}_{\text{molecule}} \cdot d_{\text{molecule}}$, and W is a colloid stability parameter which includes accounting for the electric field barrier around each identical molecule.

While $\overline{MW}(t)$ vs. t is linear for short times, later, slower growth of particles in a sedimenting precipitate appears to follow a form closer to Eq. (11.10) below

$$\bar{d}(t) = \bar{d}(0)[1 + a \ln (t)] \tag{11.10}$$

where $\bar{d}$ denotes the number-averaged precipitate particle diameter as illustrated by casein precipitates in an ammonium sulfate solution (Fig. 11.15).

Fluid shear can increase flocculation rate and also effects compaction of the resulting floc to give a stable particle which will survive intact in subsequent passage through a centrifuge. By increasing the frequency of particle-particle

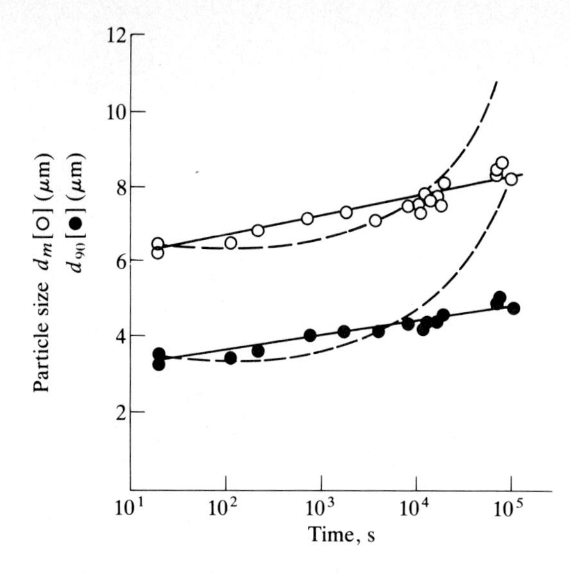

Figure 11.15 Subsequent aging: particle size vs. time in hindered settling of casein precipitate produced by salting out in 1.8 *M* ammonium surfate. Dashed line is perikinetic theory. *(Reprinted by permission from M. Hoare, "Protein Precipitation and Precipitate Ageing. Part I. Salting Out and Ageing of Case in Precipitates" Trans. I., Chem. E., vol. 60(2), p. 79, 1982.)*

contacts, it can enhance the rate of particle flocculation. For example, in a turbulent field generated by a power input per unit volume of P/V, a mean shear rate $\bar{G}$ is defined by:

$$\bar{G} \propto (P/V)^{1/2} \tag{11.11}$$

Since the particle–particle collision frequency increases as $\bar{G}$, the form of Eq. (11.9) appropriate to a shear-driven or *orthokinetic* flocculation is given by:

$$-\frac{dN}{dt} = \tfrac{2}{3}\alpha d^3_{\text{particle}} \cdot \bar{G} \cdot N^2 \tag{11.12}$$

where α is a collision effectiveness factor (fraction of collisions which lead to a permanent aggregate). The term $\bar{G}t$ is a dimensionless quantity, known as the Camp number, Ca, in water treatment flocculation; the precipitate growth behavior of the system can be correlated with this variable, which is sometimes called the *aging parameter* (Fig. 11.16).

This equation is usefully recast by assuming a constant volume fraction of particles, $\phi_v = \pi d^3_{\text{particle}} N/6$, to give a first-order disappearance:

$$-\frac{dN}{dt} = \frac{4}{\pi}\alpha\phi_v \bar{G} N \tag{11.13}$$

Ultimately, the precipitation operation must accomplish both *nucleation* and *growth* of individual particles as well as their subsequent *aggregation* to give a floc size convenient for subsequent particulate recovery (centrifugation, filtration, etc.), with an appropriate efficiency of product removal to the particulate phase. While the initial quiescent (perikinetic) or shear-induced (orthokinetic) nucleation and particle growth rates are described by Eqs. (11.9) and (11.12) above,

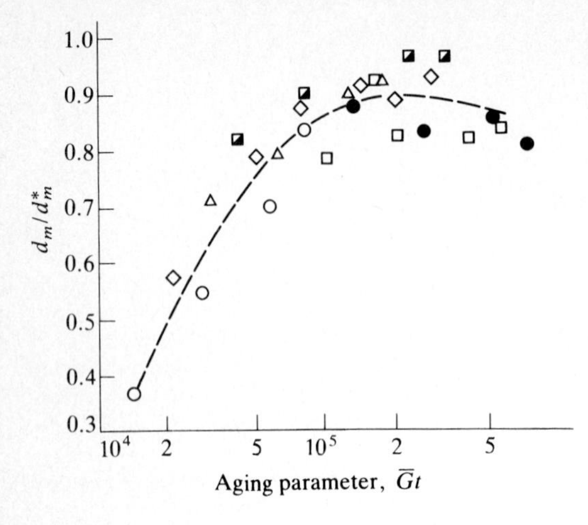

Figure 11.16 Ratio of mean particle size after and before exposure to shear in a capillary vs. aging parameter $\bar{G}$ (shear rate) × t(time). Average shear rate = $1.7 \times 10^4\ s^{-1}$; average shear time = 0.065 s, protein concentration = 30 kg/m^3. Mean initial diameters (μm) 53.5(○), 23.4(△), 19.5(◇), 15.2(□), 10.2(◪), and 8.8(●). *(Reprinted by permission from D. J. Bell and P. Dunhill, "Shear Disruption of Soya Protein Precipitate Particles and the Effect of Aging in a Stirred Tank." Biotech. Bioeng., vol. 24, p. 1271, 1982.)*

the slower subsequent precipitate flocculation rate, often achieved in a separate mixer device, may follow a first-order removal of individual precipitate particles into large aggregates. The relation of mixer residence time to precipitate particle concentration diminution has been given for both plug flow and m-CSTRs in series:

Plug flow:

$$\ln\left(\frac{N_0}{N(t)}\right) = \frac{4}{\pi}\alpha\phi_v \bar{G} t_{PFR} \tag{11.14}$$

m-CSTRs in series:

$$t_{\text{CSTR}} = \frac{\pi}{4}\frac{m}{\alpha\phi_v \bar{G}}\left[\left(\frac{N_0}{N(t)}\right)^{1/m} - 1\right] \tag{11.15}$$

where the overbar refers to a mean value. The utility of these simple equations for protein precipitates hinges on values of α and ϕ_v which are independent of particle size.

Shear forces also allow compaction of flocs to occur; Fig. 11.16 indicates that for fresh precipitates subjected to an exposure $\bar{G}t$(Ca) of at least 10^5, the most compact precipitate particles are obtained. Excessive shear forces can break up weak flocs by turbulent mechanisms analogous to those considered in the critical Weber number discussion of Chap. 8. The dependence of maximum stable floc diameter on mean shear rate is nearly inverse order, $d_{\max} \propto \bar{G}^{-1}$, when $d_{\max}$ is larger than the turbulence scale, whereas when $d_{\max}$ is considerably smaller, it varies only slightly with shear rate.

The use of centrifugation for recovery of precipitates requires sufficient time *prior to* centrifugation to allow for both floc formation and stabilization. The

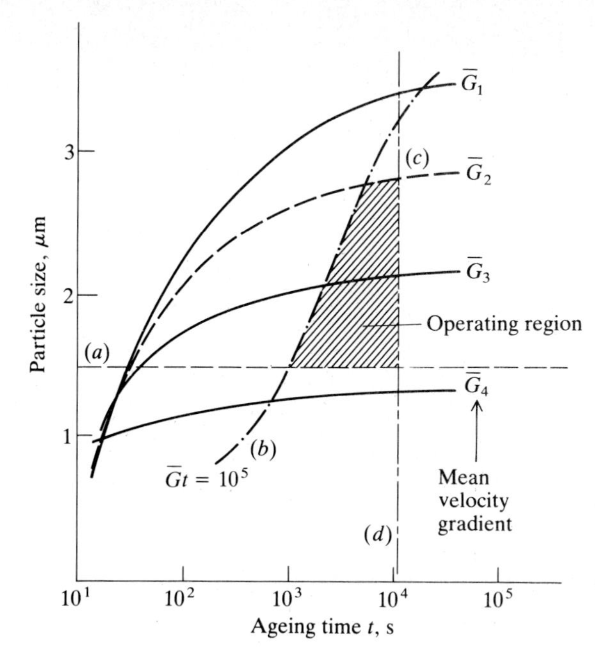

Figure 11.17 Precipitate preparation (floc) size vs. aging time, t(s) for batch precipitation. The shaded operating area provides a satisfactory precipitate for subsequent centrifugal recovery. Boundaries dictated by (*a*) minimum size for centrifugal recovery, (*b*) optimum strength parameter ($\bar{G}t = 10^5$), (*c*) minimum stirring speed to provide uniform mixing, and (*d*) maximum acceptable aging time. *[Reprinted by permission from D. J. Bell, M. Hoare, and P. Dunnill, "The Formation of Protein Precipitates and their Centrifugal Recovery," p. 65 in Adv. Biochem. Eng., vol. 2, A. Fiechter (ed.), Springer-Verlag, Berlin 1983]*

requirements for time and shear rate (stirring, etc.) as shown in Fig. 11.17 may be summarized as follows:

1. Centrifugation requires a precipitate of at least a certain (minimum) diameter.
2. A combined (shear rate × time) product sufficient to provide a strong floc.
3. Sufficient stirring to allow good bulk mixing.
4. Not excessive holding time.

These constraints for precipitate formation and aging are indicated in Fig. 11.17; the enclosed operating region is the feasible domain, i.e., that area of operating parameter values for which the system satisfies all the requirements.

11.4 CHROMATOGRAPHY AND FIXED-BED ADSORPTION: BATCH PROCESSING WITH SELECTIVE ADSORBATES

Chromatography involves the column resolution of a pulse (or small batch) of flowing, multicomponent fluid into separate fluid volumes of (nearly) pure solute solutions (Fig. 11.18). Biological molecules such as proteins have varying tendencies to adsorb on such materials as starch, diatomaceous earth, and polyacrylamide gel. As a result, when a solution containing proteins, for example, is passed over such a phase, each protein moves at an effective velocity which

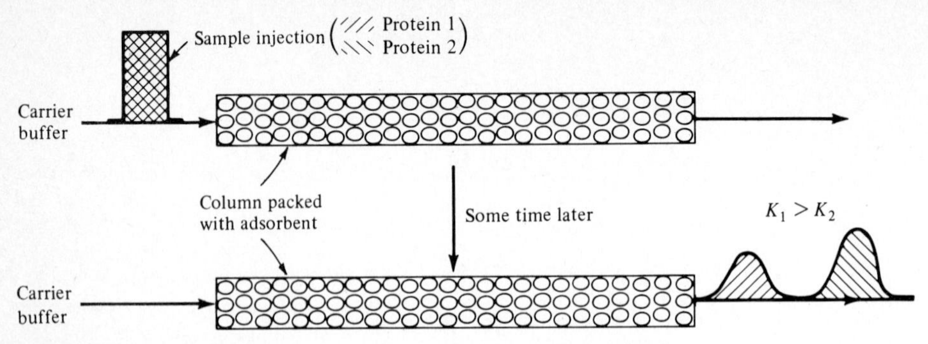

Figure 11.18 Schematic diagram of chromatographic separation of proteins.

decreases as the protein's propensity to adsorb increases. The following analogy is often used to explain the basic principle of chromatography. Imagine a number of coachmen who start together down a road lined with pubs. Obviously those with the greatest thirst will complete the journey last, while those with no taste for ale will progress rapidly! In case of extreme affinity for the fruit of the vine, the coachman will stay in one of the pubs until later coaxed out and given a ride by a later driver. This corresponds to fixed-bed adsorption and subsequent elution (e.g., see Fig. 11.19) which is also often called chromatography in the biochemistry and bioprocess literature.

To formulate an elementary quantitative description of chromatography, assume that fluid containing solutes $S_1, S_2, \ldots, S_N$ moves in plug flow through a packed column in which a fraction ε of the total volume is occupied by interstitial fluid. Solute concentrations in the bulk-fluid phase, s_i, are presumed to be sufficiently small so that solute exchange with the stationary packing phase does not alter total bulk-fluid concentration c_T (and thus bulk-fluid superficial velocity through the column). Writing an unsteady differential material balance on the two-phase column (recall Figs. 9.5 and 9.30, for example) gives

$$u\varepsilon \frac{\partial s_i}{\partial z} + \varepsilon \frac{\partial s_i}{\partial t} + (1 - \varepsilon) \frac{\partial w_i}{\partial t} = 0 \tag{11.16}$$

where u is the interstitial fluid velocity (=superficial velocity$\cdot\varepsilon^{-1}$ where superficial velocity is the total volumetric flow rate divided by the total column cross section). The variable w_i in Eq. (11.16) denotes the moles of component i in the solid per unit volume of solid.

The rate of solute i uptake by the stationary phase, $\partial w_i/\partial t$, is in general a complex function of flow conditions, solute diffusion in the fixed phase, adsorption kinetics and capacity, and other parameters. In the simplest case, solute in the column packing is in equilibrium locally with solute in the interstitial fluid, giving

$$\frac{w_i}{s_i} = K_i \tag{11.17}$$

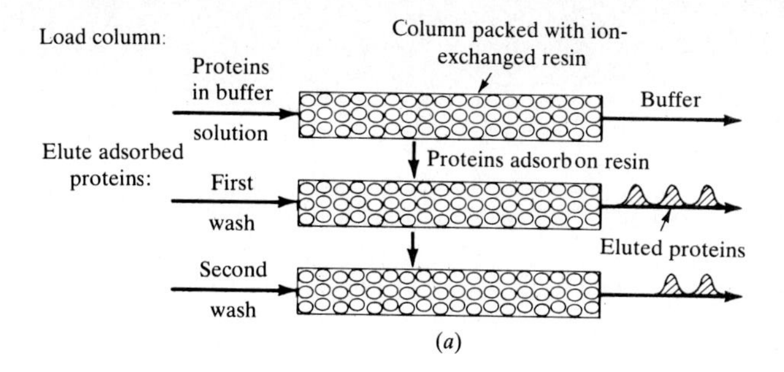

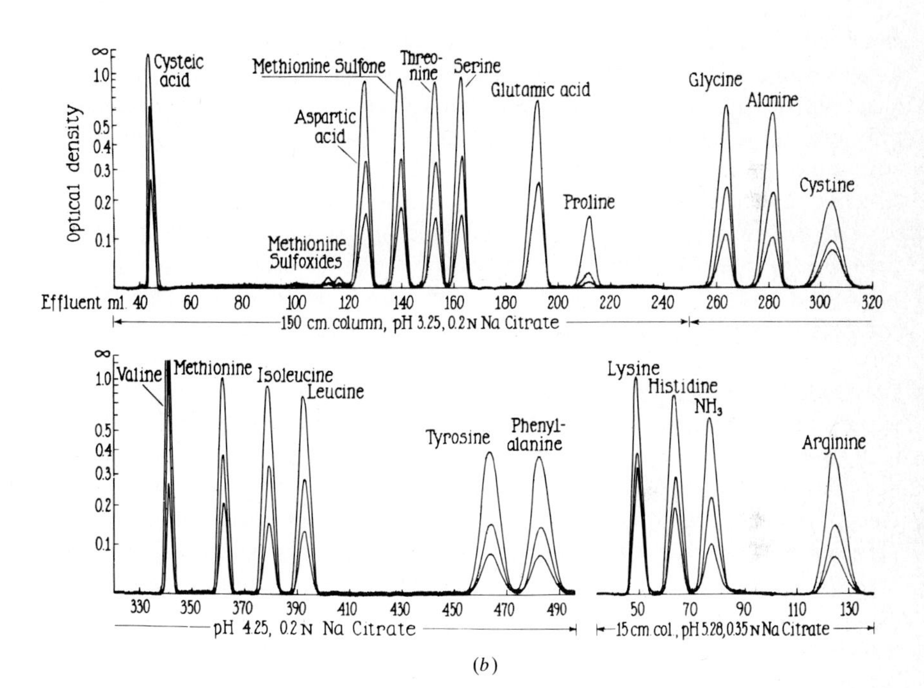

Figure 11.19 Ion exchange chromatography: (*a*) the procedure; (*b*) the results of an application to amino acid analysis. Labels above the chromatogram identify the feed conditions of the wash stream. [(*b*) *Reprinted with permission from D. H. Spackman, W. H. Stein, and S. Moore, "Automatic Recording Apparatus for Use in the Chromatography of Amino Acids," Anal. Chem., vol. 30, p. 1190, 1958. Copyright by the American Chemical Society.*]

The equilibrium relationship is in general dependent on all fixed-phase concentrations $w_1, \ldots, w_n$. In a dilute system, however, K_i may be assumed to be a constant. Using Eq. (11.17) to evaluate w_i and substituting in Eq. (11.16) and rearranging gives

$$u_i \frac{\partial s_i}{\partial z} + \frac{\partial s_i}{\partial t} = 0 \tag{11.18a}$$

where

$$u_i = u \cdot \left(1 + \frac{1-\varepsilon}{\varepsilon} K_i\right)^{-1} \tag{11.18b}$$

Equation (11.18*a*) is identical in form to the material balance for solute i moving at velocity u_i through a tube with no packing. Therefore u_i given by Eq. (11.18*b*) gives the effective velocity of component i movement through the packed chromatography column. In accordance with the coachman analogy, Eq. (11.18*b*) indicates that components with greater affinities for the second phase (large K_i's) will move through the column relatively slowly whereas compounds with very small tendencies to partition into or onto the second phase (small K_i) will move at velocities approaching that of the solvent. Extensions of this elementary theory (see references) encompass axial dispersion effects, nonequilibrium binding, and diffusive transport into the stationary phase. Consideration of irreversible adsorption or desorption is pertinent to fixed-bed adsorber loading and subsequent selective solvent elution.

An isolation technique known as molecular-sieve (or gel-filtration or gel-permeation) chromatography separates molecules of different sizes. In this instance the column is packed with gel particles with pores of a characteristic size. Molecules larger than this pore size cannot diffuse into the gel and pass rapidly through the column, while smaller species penetrate the gel and move more slowly (Fig. 11.20). We should be aware that the separation may not be governed entirely by molecular exclusion according to the pore diameter (or fiber diameter if the gel is regarded as a tangled collection of solid fibers). Other possible influential parameters include the effective solute diffusivity within the material, the tendency of the solute to adsorb on the internal material surface, and the osmotic pressure of the material. Currently the exact relative importance of these factors appears to be unknown. However, molecular exclusion often predominates. In this case, an equivalent equilibrium constant K_{av} for use in Eq. (11.18*b*) can be calculated from

$$K_{av,i} = \exp\left[-\pi L(r_g + r_i)^2\right] \tag{11.19}$$

where $K_{av,i}$ = fraction of total internal material volume available for spherical molecule of radius r_i

L = concentration of gel fiber, expressed in length per volume, cm/cm^3

r_g = radius of gel fiber

r_i = radius of spherical molecule of ith species

Some typical r_g and r_i values are listed in Table 11.6. By varying the degree of cross-linking between chains in the gel, the value of the effective gel-fiber radius can be altered. A new gel support can be characterized by plotting $-(\ln K_{av})^{0.5}$ vs. r_i. From Eq. (11.19), the slope is $(\pi L)^{0.5}$, and the intercept is $(\pi L)^{0.5}\, r_g$. The K_{av} values for a Sephadex and several Sephadex-agarose (Sepharose) gels are

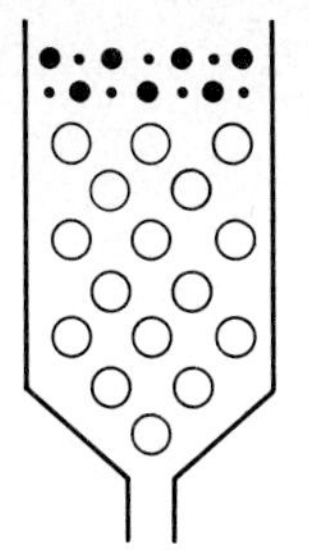

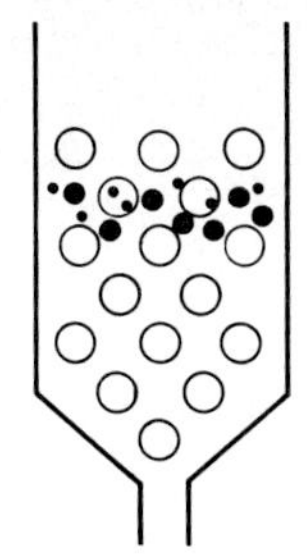

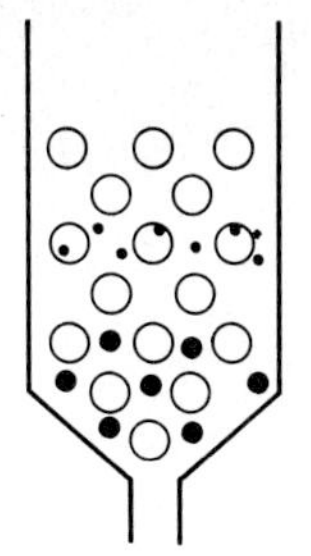

Figure 11.20 In molecular-sieve chromatography, larger molecules pass more rapidly through the column, while smaller molecules are retarded by occasional excursions into the gel particles.

shown in Fig. 11.21. Such plots using K_{av} vs. molecular weight rather than radius r_i may be used when the radii of interest are not available.

Figure 11.22 illustrates the results of a protein separation achieved by molecular-sieve chromatography. Notice that the various proteins do not appear all at once but instead emerge as diffuse peaks. Several factors can contribute to peak spreading, including molecular diffusion, mass-transfer resistances, and variation in axial velocity over the column cross section. The response in Fig. 11.22 is typical of chromatographic methods in general. The fractionation of very large proteins, viruses, cell fragments, and even different cell types in a mixture of cells can be carried out in an analogous chromatographic fashion provided the support material has a tendency to discriminate between various mixture components.

In ion-exchange chromatography the protein solution is passed over a fixed-bed column containing ion-exchange resin (Fig. 11.19). One common resin for protein purification is CM-cellulose, a cation-exchange resin obtained by linking negatively charged carboxymethyl groups to a cellulose backbone. Cationic (positively charged) proteins will bind to this resin by electrostatic forces, the

Table 11.6 Some radii estimated from diffusion studies for several molecules†

Protein	Mol wt	Diffusity, $D \times 10^7$, cm^2/s	r_i, Å
Ribonuclease	13,683	11.9	18.0
Lysozyme	14,100	10.4	20.6
Chymotrypsinogen	23,200	9.5	22.5
Serum albumin	65,000	5.94	36.1
Catalase	250,000	4.1	52.2
Urease	480,000	3.46	61.9

Typical fiber radii in gel	r_g, Å
Sephadex	7
Agarose	25

† Selected from summary in C. Tanford, *Physical Chemistry of Macromolecules*, table 21.1, John Wiley & Sons, Inc., New York, 1961.

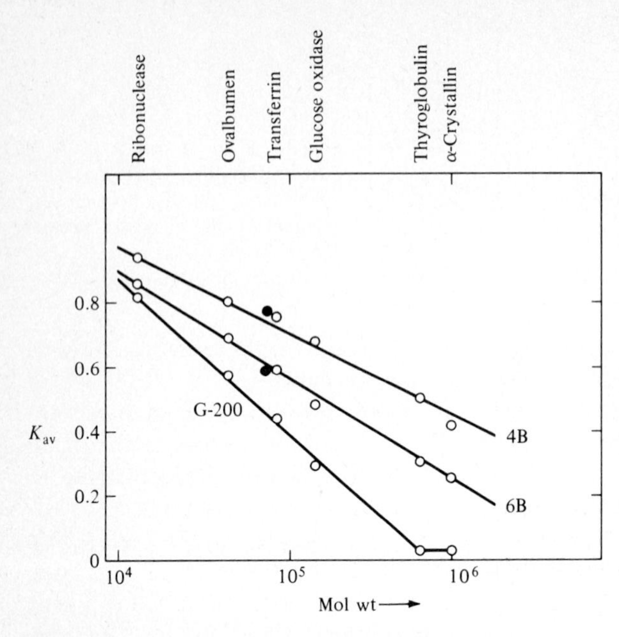

Figure 11.21 Protein vs. molecular-weight selectivity curves of Sephadex G-200, and Sepharose 6B (~6 percent agarose gel, experimental) and Sepharose 4B (~4 percent gel). *(Reprinted from M. K. Joustra, "Gel Filtration on Agarose Gels," in Progress in Separation and Purification, vol. 2, p. 183, Wiley-Interscience, New York, 1969.)*

strength of the attachment depending on the net positive charge at the pH of the column feed. After protein has been deposited in this manner, the column is developed by washing it with buffers of increasing pH and/or ionic strength. Such changes in the carrier solution cause weakly bound proteins to detach from the resin first, followed by more tightly attached molecules as conditions become far removed from those in the deposition step. Consequently the wash, which is protein-free when fed to the ion-exchange column, picks up a certain characteristic portion of the bound protein. The effluents (eluate) obtained by washing the column are then collected in small fractions under different conditions. Similar procedures are employed with anionic exchangers, which typically employ diethylaminoethyl (DEAE) cellulose ion-exchange resins.

Ion-exchange chromatography, like affinity chromatography to be discussed next, is somewhat different from gel permeation and other types of liquid and gas chromatography based on different yet finite effective velocities of different solutes. In ion exchange chromatography, one or more components adsorb essentially irreversibly ($K_i \rightarrow \infty$) under the conditions used for column loading. Thus, this is a selective, fixed-bed adsorption process. If multiple components are adsorbed, as is the case in the example in Figure 11.22, they are separated in the *elution* step. The feed is replaced by a series of elution fluids to achieve (ideally) sequential desorption of bound species.

An intriguing chromatographic technique based on the natural specificity of some biopolymers is affinity chromatography (Fig. 11.23). A number of proteins and other biological macromolecules (A) complex with some other molecular entity (B) with a high degree of specificity: if species B is attached to a solid which is then used to pack a column, the K for component A is very large while

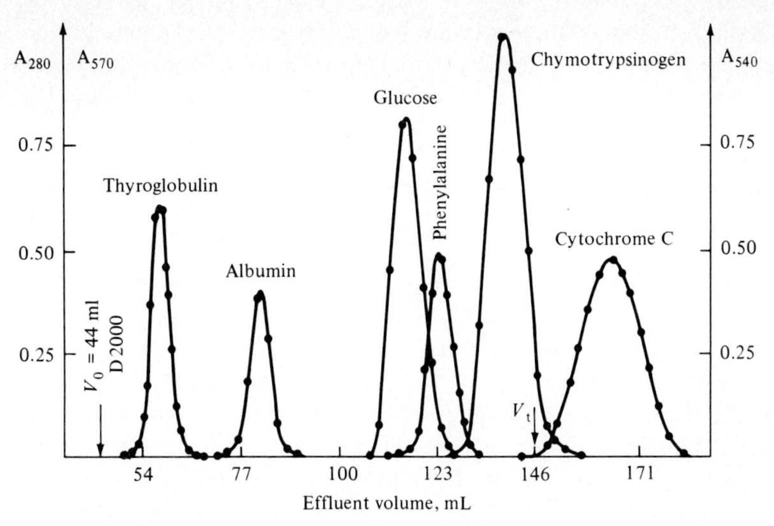

Figure 11.22 Chromatogram resulting from molecular sieve chromatography of a solute mixture (packing is 6 percent cross-linked desulfated agar. Buffer is pH 7.5 0.05 M tris-HCl). *[Reprinted by permission from J. Porath, "Chromatographic Methods in Fractionation of Enzymes," in L. B. Wingard, Jr. (ed.), Enzyme Engineering, p. 154, John Wiley and Sons, New York, 1972.]*

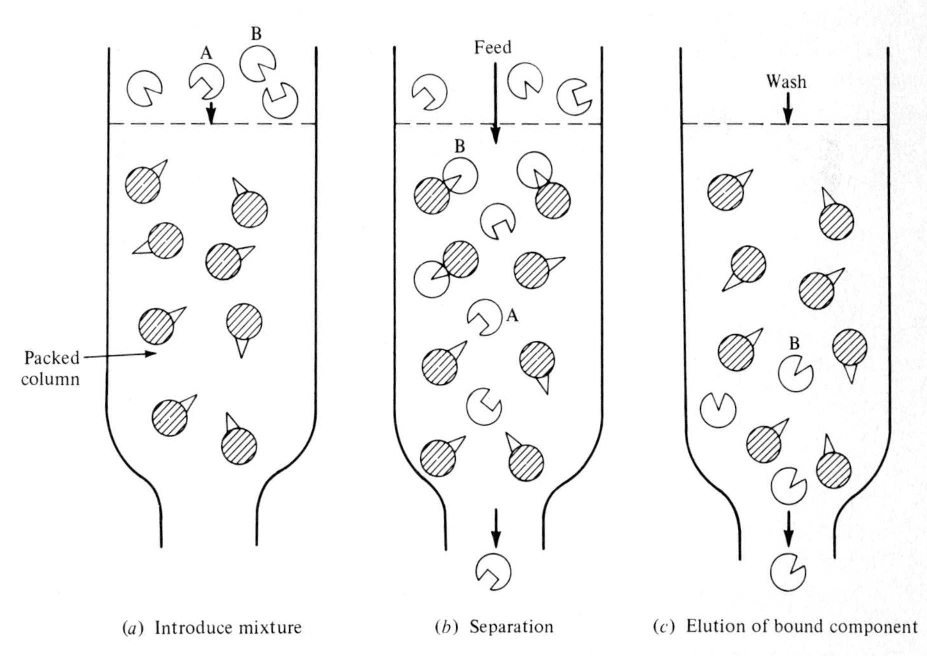

Figure 11.23 Schematic diagram of affinity chromatography, where fine separations are achieved by specific interactions between an immobilized agent and a component in the liquid phase. Three steps are shown: (*a*) introduction of mixture, (*b*) separation, and (*c*) elution of bound component.

the K's for other solution components are essentially zero. Examples of very specific pair interactions include enzyme-inhibitor, antigen-antibody, and lectin-cell:

$$\text{Enzyme} + \text{inhibitor} \rightleftharpoons \text{enzyme-inhibitor complex}$$

$$\text{Antibody} + \text{antigen} \longrightarrow \text{antibody-antigen precipitate}$$

$$\text{Lectin} + \text{cell-wall} \longrightarrow \text{lectin–cell-wall complex}$$

For example, an affinity-chromatography column for isolation of DNA depolymerase can be prepared by chemically binding DNA depolymerase inhibitor to the surface of some convenient bead material, e.g., agarose or polyacrylamide. Flow of a lysed-cell centrifugate (such as that obtained after step 4 of Example 11.1) through such a column could result in an essentially quantitative binding of the enzyme DNA depolymerase and passage of nearly all other materials. From Eq. (11.18*b*), $u_i \approx u$ for all other solution components. Subsequently, a solution containing DNA polymerase inhibitor or some other appropriate eluant can be passed through the column to release and recover the bound enzyme for further use.

With the advent of *monoclonal antibody* production, which allows the synthesis of a single type of antibody with a very high specific binding constant to its corresponding antigen (a particular protein or other molecule), the preparation of affinity columns has become not only routine, but commercial: an example is cited later involving human leukocyte interferon purification from an *E. coli* lysate using an immobilized-monoclonal antibody column. With such *immunosorbent* column separation, as this particular form of affinity chromatography has been denoted, some important practical and theoretical differences arise compared to the previously described more conventional forms of chromatographic resolution:

1. The antibody needed to make the immunosorbent column is a dominant cost, and may be much more costly than the antigen-containing broth itself, hence
2. A small column capable of repeated, high capacity use is required.
3. Elution of adsorbed product requires breaking the antigen-antibody complex which often involves denaturing conditions. Loss of some antibody binding affinity typically occurs, with gradual loss of column capacity.
4. Agarose appears to be a most suitable column support; the surface density of antibody is a column parameter which can be varied. Lack of mechanical strength (high compressibility) is a disadvantage of this and other polysaccharide support materials, limiting column depth which can be used without excessive pressure drop (recall Example 4.2).
5. Precolumn purification may not be necessary, but should be studied with each case (as a lack of long experience exists).
6. A first cycle on a new column gives poorer recovery than successive operations, apparently due to some irreversible binding.

7. Determination of optimal elution buffer wash volumes and concentrations is a key economic goal.

Other clear differences exist between affinity chromatography and more conventional adsorption or ion-exchange chromatography. The latter frequently gives a multi-component adsorption which is *resolved in the desorption step*, by, for example, a continuously increasing pH or salt gradient elution profile (Fig. 11.19). The former relies primarily on specificity in the adsorption step (Fig. 11.23). As a result, the affinity column can accept a continuous feed input until saturation is nearly obtained, since virtually everything adsorbed is the desired molecule. Thus, the operation strategies here are geared more toward a typical fixed-bed sorption operation (continuous feed until bed saturation achieved) rather than the conventional resolution of a single multicomponent pulse into its many constituents. The obvious trade-off is increased column specificity, allowing much larger volumes to be processed per cycle, vs. a much larger column preparation cost.

As a result of highly specific binding, a sharp adsorption front develops in the immunosorbent bed and moves through the bed with a constant concentration profile. In consequence, it is convenient to feed an affinity column until near saturation is achieved as noted by appearance of desired product in column effluent. Excessive loading gives rise to loss of some product in the effluent; insufficient loading leaves the downstream column end of this typically expensive column under utilized (Fig. 11.24).

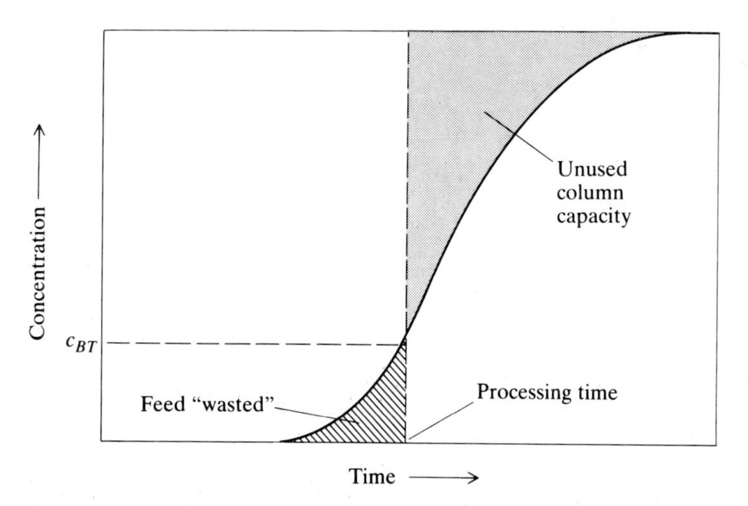

Figure 11.24 Breakthrough curve (effluent concentration of the adsorbed solute vs. time of solution feed to an initially fresh column) for affinity adsorption. If feed is stopped at effluent concentration c_{BT}, a portion of the column capacity has not been used. A small amount of solute is lost in the effluent. *(Reprinted by permission from F. H. Arnold, H. W. Blanch, and C. R. Wilke, "A Rational Approach to Affinity Chromatography," p. 113 in Purification of Fermentation Products, D. LeRoith et al. (eds), ACS Symposium Series* **271**, *1985. Copyright by the American Chemical Society.)*

Analyses of uptake processes for typical antibody-antigen binding and molecular weight indicates that, with porous column packing, internal mass transfer provides an important mass transfer resistance. Hence a simple equilibrium model, e.g., Eq. (11.18) above, is inappropriate. Also, recent experiments indicate that the *rate* of antigen-antibody complex formation can significantly influence the overall antigen uptake rate.

In mass transfer operations, it is convenient to replace column length by NTU's, defined as the *N*umber of *T*ransfer *U*nits needed to achieve a desired separation. For example, consider a simple external mass transfer-limited situation. If $k_l a(\Delta c)$ is the available transfer rate per volume, to achieve a given desired transfer rate (moles/time) requires $Vk_l a(\Delta c)$. Rendering this moles/time quantity dimensionless by division by $V\Delta c$ and multiplication by L/u_0 (u_0 is the superficial velocity) gives the requisite number of transfer units:

$$\mathrm{NTU} = \frac{k_l a(\Delta c)L}{\Delta c u_0} = \frac{k_l a L}{u_0}$$

In immunosorbent columns, internal transport is slow. If such internal transport and strong binding provide the transport rate limitation, then [29, 39]

$$\mathrm{NTU}_{\mathrm{pore}} = \frac{k_{\mathrm{pore}} a_p L}{u_0} = \frac{60(1-\varepsilon)\mathscr{D}_{\mathrm{eff}}}{d_p^2}\left(\frac{L}{u_0}\right) \tag{11.20}$$

The relationship between NTU, dimensionless effluent concentration $\bar{c}_e$ ($\equiv c/c_f$) and dimensionless time τ for the case of irreversible adsorption is given by [38, 39]

$$\mathrm{NTU}_{\mathrm{pore}}(\tau - 1) = 2.44 - 0.273[1 - \bar{c}_e]^{1/2} \tag{11.21a}$$

where

$$\tau = \left(\frac{u_0 t}{\varepsilon L} - 1\right)\left(\frac{\varepsilon}{1-\varepsilon}\right)\frac{1}{K_e} \tag{11.21b}$$

and

$$K_e = \frac{\rho_p q_f^*}{c_f} \tag{11.21c}$$

The parameter q_f^* in Eq. (11.21*c*) denotes the solid-phase adsorbate concentration in equilibrium with c_f.

For a pseudo-homogeneous diffusion model (e.g., in a gel phase where pore diameter is not meaningful) [38]

$$\frac{\mathrm{NTU}_{\mathrm{PH}}}{L} = \psi_{\mathrm{PH}}(1-\varepsilon)K_e\frac{k_p a_p}{u_0} \tag{11.22}$$

where ψ_{PH} is a correction factor of order unity. The time-effluent concentration relation for irreversible adsorption is for this system

$$\mathrm{NTU}_{\mathrm{PH}}(\tau - 1) = -1.69(\ln[1 - \bar{c}_e^2] + 0.61) \tag{11.23}$$

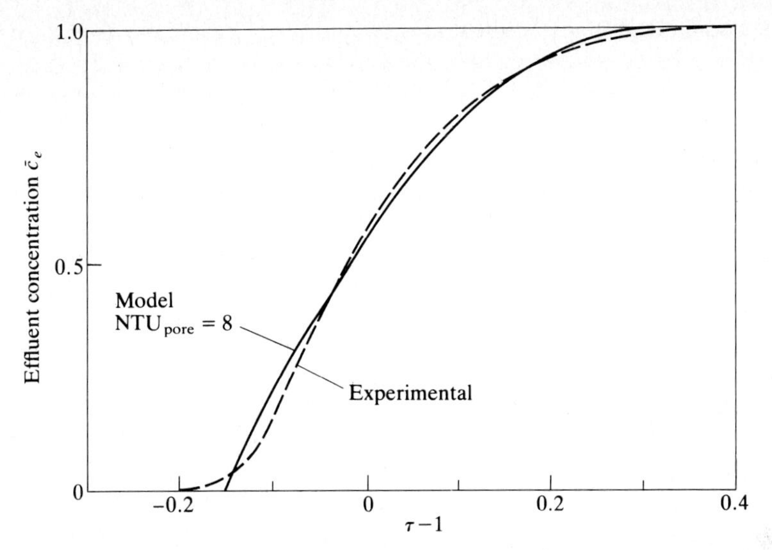

Figure 11.25 Experimental (---) and model (——) breakthrough curves for a 1.5 × 16.5 cm. Controlled pore glass (CPG)-monoclonal antibody column, $u_0 = 0.011$ cm/s, $\varepsilon = 0.6$, $d_p = 0.01$ cm, and $\tau = 1$ at $V_L - \varepsilon V_C = 211$ mL. *(Reprinted by permission from F. H. Arnold, H. W. Blanch, and C. R. Wilke, "A Rational Approach to Affinity Chromatography," p. 113 in Purification of Fermentation Products, D. LeRoith et. al. (eds.), ACS Symposium Series* **271**, *1985. Copyright by the American Chemical Society.)*

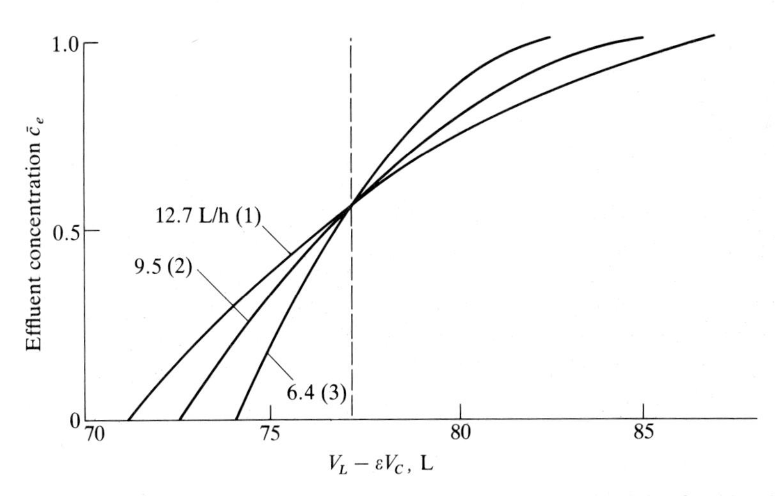

Figure 11.26 Breakthrough curve predicted by Eqs. (11.20) and (11.21) for 15 × 60 cm column of CPG-monoclonal antibody. Equilibrium capacity is the solute equivalent of 77.1 L of feed. V_L is volume of fluid fed to the column, V_C is the column volume, and ε is the interparticle void fraction. (1) $u_0 = 0.020$ cm/s, (2) $u_0 = 0.015$ cm/s, (3), $u_0 = 0.10$ cm/s. *(Reprinted by permission from F. H. Arnold, H. W. Blanch, and C. R. Wilke, "A Rational Approach to Affinity Chromatography," p. 113 in Purification of Fermentation Products, D. LeRoith et. al. (eds.), ACS Symposium Series* **271**, *1985. Copyright by the American Chemical Society.)*

Results of a small column experiment are given in Fig. 11.25. Controlled pore glass (CPG) was used as a support for monoclonal antibody against arsanilic acid (ARS). The solute separated was a conjugate of ARS and bovine serum albumin. Equation (11.21) gives a reasonable fit to the effluent concentration profile for this system and operating conditions using $NTU_{pore} = 8$. If pore diffusion is rate limiting, the performance of a scaled-up column can then be predicted using Eq. (11.21) and the definition of NTU_{pore} (Eq. (11.20)). The predicted breakthrough curves for a large column for different feed velocities are shown in Fig. 11.26. A kinetic resistance to antigen binding may also be important; see Ref. 39 for further details.

11.5 MEMBRANE SEPARATIONS

The principal advantage of membrane separations is that operation is achieved without change of phase or interphase transfer; thus any desired product is continually maintained in an aqueous environment. Size differences provide one basis on which membrane separations occur (Fig. 11.2). Given the following characteristic dimensions,

	Diameter, nm
Yeast and fungi	10^3–10^4
Bacteria	300–10^3
Colloidal solids	100–1000
Macromolecules (proteins, polysaccharides)(10^4–10^6 MW)	2–10
Antibiotics (300–1000 MW)	0.6–1.2
Mono-, disachaccarides (200–400 MW)	0.8–1.0
Organic acids (100–500 MW)	0.4–0.8
Inorganic ions (10–100 MW)	0.2–0.4
Water (18 MW)	0.2

two appropriate membrane separations are available, reverse osmosis and ultrafiltration.

11.5.1 Reverse Osmosis

Osmosis occurs when a solution and a volume of pure solvent are separated by a solute-impermeable membrane; this circumstance leads to diffusion of pure solvent through the membrane, into the solution phase, in order to equalize solvent chemical potentials in each phase. If an increasing pressure Δp is applied to the *solution* phase, osmosis halts when the applied pressure Δp equals the osmotic pressure, π, of the solution, where

$$\pi = cRT[1 + B_2[c] + B_3[c]^2 + \cdots] \tag{11.24}$$

where B_2, B_3 are virial coefficients for the solute in solution.

Thus, at low concentrations and zero solvent flux, $\Delta p = \pi \cong cRT$. When Δp exceeds π, a solvent flux occurs from the dilute solution into the pure solvent, giving rise to *reverse osmosis*, a process which produces a more concentrated solution.

In reality, membranes are not perfectly size selective, and it is convenient to consider both a passive solute permeability P as well as a flow-related *reflection coefficient* σ for each solute. The latter represents the fraction of solute molecules which are not passed through the membrane; thus $\sigma = 1.0$ is perfect reflection and $\sigma = 0$ is complete solute passage. These reflection coefficients are membrane dependent; the apparent pore size of the membrane provides an indication of the size-selective nature of this process (Table 11.7).

Under an applied pressure, the transmembrane solvent flux rate and solute transfer rate are represented by Eqs. (11.25) and (11.26):

$$N_1(\text{solvent}) = L_p(\Delta p - \pi) \tag{11.25}$$

$$N_2(\text{solute}) = \bar{c}_2(1 - \sigma)N_1 + P\Delta c_2 \tag{11.26}$$

where L_p and P are membrane permeabilities for solvent and solute, respectively, Δc_2 is the solute concentration difference across the membrane, and $\bar{c}_2$ is the average solute concentration in solution. The corresponding solute concentration in the liquid exiting the membrane is thus

$$\frac{N_2}{N_1} = \left[\bar{c}_2(1 - \sigma) + \frac{P\Delta c_2}{N_1}\right]$$

which, for σ near 1.0 and small P and/or large N_1, will be much smaller than the upstream solute concentration.

Table 11.7 Flow characteristics of three types of membranes[†]

			Reflection coefficient		
Substance	Mol wt	Molecular radius, Å	Dialysis tubing[‡]	Cellophane[§]	Wet gel[¶]
D_2O	20	1.9	0.002	....	0.001
Urea	60	2.7	0.024	0.006	0.004
Glucose	180	4.4	0.20	0.044	0.016
Sucrose	342	5.3	0.37	0.074	0.028
Raffinose	595	6.1	0.44	0.089	0.035
Inulin	991	12	0.76	0.43	0.23
Bovine serum albumin	66,000	37	1.02	1.03	0.73
Calculated pore radius, Å	....	....	23	41	82
Membrane constant, L_p 10^{-5} g·(cm^2·s·atm)$^{-1}$	....	....	1.7	6.5	25

[†] Data from R. P. Durbin, *J. Gen. Physiol.*, **44:** 315 (1960).
[‡] Visking cellulose.
[§] DuPont 450-PT-62 cellophane.
[¶] Sylvania 300 viscose wet gel.

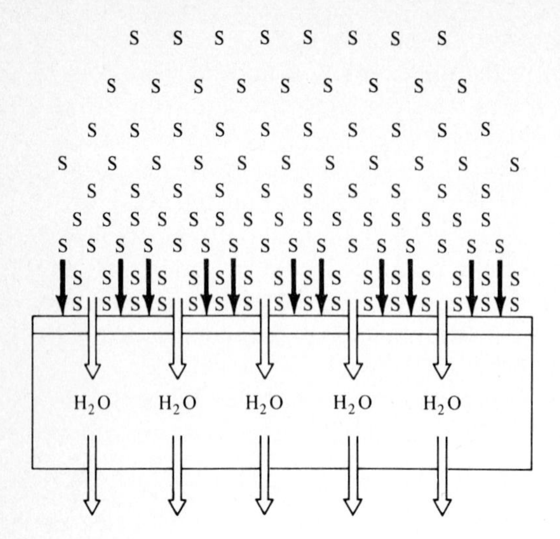

Figure 11.27 Concentration polarization caused by buildup of solute(s) near the upstream membrane surface.

Liquids are nearly incompressible, thus provision of appreciable pressures is not costly per se. However, as the flux increases, a layer of solute-rich fluid builds up at the interface (Fig. 11.27), giving rise to a greater solute permeation rate [Eq. (11.26)], as well as a diminished solvent flux by virtue of the locally increased osmotic pressure [Eq. (11.25)]. This *concentration polarization* provides, effectively, an upper limit to membrane flux rate: upstream stirring near the membrane surface can reduce this ultimate resistance, as indicated in Fig. 11.28 (880 vs. 1830 rpm stirring with 6.5 percent protein), in a manner analogous to the fluid motion in cross-flow filtration which diminishes the filtration cake resistance near the upstream membrane face.

With a multicomponent salt solution, the total osmotic pressure of the solution, ignoring any interaction tendencies, would be a simple sum:

$$\pi_t = \sum_i \pi_i = \sum_i c_i RT[1 + B_{2i}[c]_i + B_{3i}[c]_i^2 \cdots] \tag{11.27}$$

The solvent flux is again proportional to $(\Delta p - \pi_t)$, so now each *solute* flux depends on the *total* osmotic pressure as well as its own permeability and rejection coefficients: hence concentration polarization of any solute decreases the flux of all solutes.

In the absence of any appreciable solvent flux, if the *downstream* fluid is circulated and continuously replaced, *dialysis* occurs in which small solutes are continuously diffused into the circulating fluid and large solutes retained. This technique underlies operation of artificial kidney machines, which can remove salts and small waste organics (e.g., urea, Table 11.7), while retaining large molecules such as proteins and sugars. If removal of one low molecular weight species is *not* desired, it must be provided in the circulating dialysate fluid; thus the use of 10^{-3} *M* phosphate in the dialysis purification (for removal of other small solutes) in the enzyme purification example (step 7, Example 11.1).

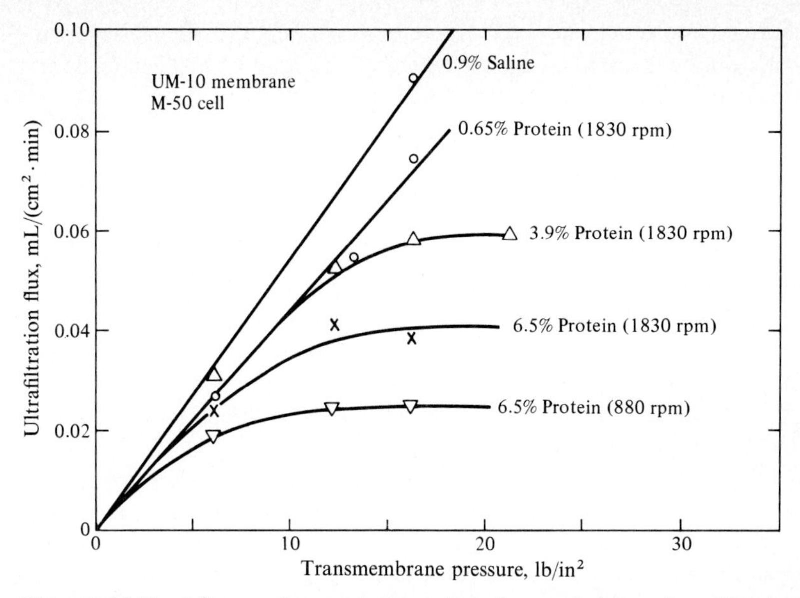

Figure 11.28 The influence of concentration polarization on deviations from linearity and approach to a constant flux as transmembrane pressure is increased. *[Reprinted by permission from M. C. Porter, "Applications of Membranes to Enzyme Isolation and Purification," in L. B. Wingard, Jr. (ed.), Enzyme Engineering, p. 119, John Wiley and Sons, Inc., New York, 1972.]*

11.5.2 Ultrafiltration

The use of larger pore membranes than found in reverse osmosis allows flow passage of 1–10 Å molecules and retention of proteins or other macromolecules. For molecules from 10 Å to 500–1000 Å diameter, ultrafiltration is useful both for product *concentration* (by solvent removal) and *purification* (by removal of low molecular weight impurities).

When macromolecules account for the concentration polarization layer, the osmotic pressure is typically negligible, as Eq. (11.24) suggests for MW large. However, the accumulating macromolecular solute polarization layer can create an appreciable mass-flow resistance.

Several analytical approaches are available, depending on the physical situation.

1. For very dilute feeds, membrane intrinsic resistance sets the solvent flux according to a simplified form of the reverse osmosis equations which neglect osmotic pressure, thus

$$N_1(\text{solvent}) = L_p \Delta p$$

$$N_2(\text{solute}) \simeq \bar{c}_2(1 - \sigma) L_p \,\Delta p + P\,\Delta c_2$$

2. The accumulating macromolecular polarization layer may, in its more concentrated volume, form a *gel phase* which adds an appreciable, perhaps even dominant, resistance to the ultrafiltration process.

Defining a molecular sieving parameter $\phi = c_p/c_w$ and the observed rejection of a solute R as $(c_f - c_p)/c_f$, where c_w, c_f, and c_p are gel, feed, and permeate concentrations, respectively, then

$$N_1 = \rho_M k \ln\left[\left(\frac{1-\phi}{\phi}\right)\left(\frac{1-R}{R}\right)\right] \tag{11.28}$$

where k = liquid phase mass transfer coefficient and ρ_M = molar density of solvent. The amylase data of Fig. 11.29 support the linearity of $\ln[(1 - R)/R]$ vs. N_1. Addition of a second polymer, β-lactoglobulin, again provides a linear plot of N_1 vs. $(1 - R)/R$, but with larger value of $(1 - \phi)/\phi$. Thus, for a given solvent flux, the rejection factor R of amylase is larger for the mixture. Note that the slopes vary somewhat as well, reflecting a composition-dependent gel layer resistance.

Further evidence of gel resistance is given in Fig. 11.30; here operation at the β-lactoglobulin pK value of pH = 5.2 gives maximum gel concentration. Intuitively, a denser gel would be expected from these neutral (isoelectric) molecules than from solutes of net charge other than zero; this notion is born out by the enormously increased rejection of α-amylase with increasing β-lactalbumin concentration at pH = pK for β-lactalbumin (Fig. 11.31).

These results indicate that, for a given filtration material, solvent flux

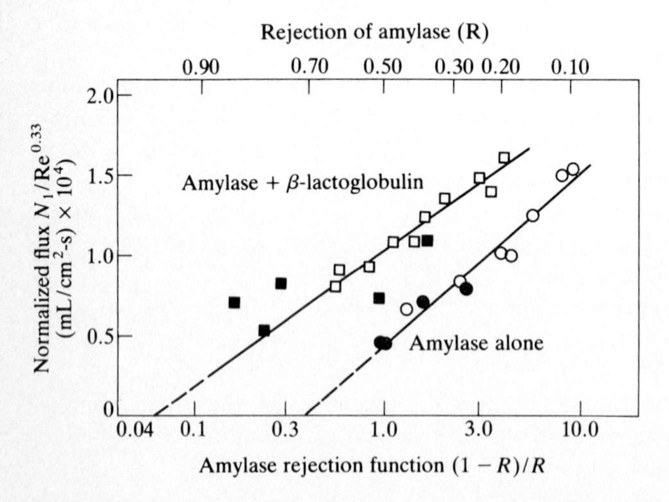

Figure 11.29 Variation of flux with amylase rejection (upper scale), pH and presence of β-lactoglobulin. Amylase only: pH = 6.0(○), 5.2(●). Amylase and β-lactoglobulin: pH = 5.2(■). HFA-300 membrane, T = 30°C, stirrer speed = 900 rpm (Re = 50,000), p = 5–20 psig (amylase only); 15 psig (β-lactoglobulin); 0.01 M acetate buffer, 0.002 M $CaCl_2$. *[After T. A. Butterworth and D. I. C. Wang, "Separation and Purification of Enzyme Mixtures by Ultrafiltration," p. 195 in Fermentation Technology Today, (G. Terui (ed.), Soc. Ferm. Technol., Japan, 1972.]*

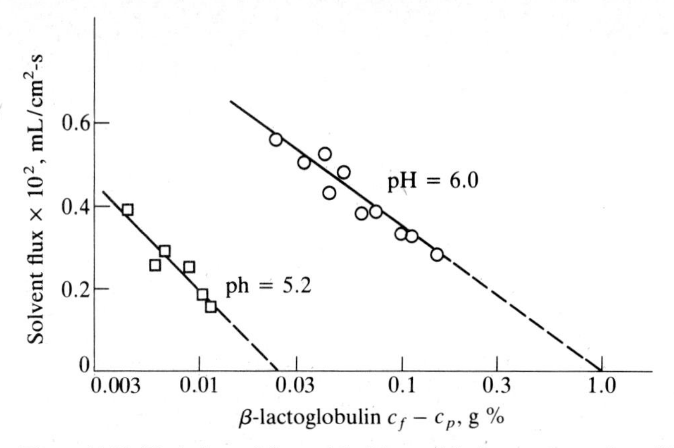

Figure 11.30 Variation of flux with β-lactoglobulin concentration difference $c_f - c_p$ (g %). 0.01 *M* acetate buffer, 0.002 *M* $CaCl_2$, 900 rpm, $p = 15$ psig. *[After T. A. Butterworth and D. I. C. Wang, "Separation and Purification of Enzyme Mixtures by Ultrafiltration," p. 195 in Fermentation Technology Today, G. Terui (ed.), Soc. Ferm. Technol., Japan, 1972.]*

and solute rejection may be correlated by Eq. (11.28). However, the slope $\rho_M k$ and intercept $\rho_M k \ln [(1 - \phi)/\phi]$ are concentration dependent, and the ultrafiltration efficiencies are not a function of molecular size alone: for the example above, the dilute, higher molecular weight α-amylase ($MW =$ 48,000) was passed through the membrane in preference to the more concentrated, gel-forming β-lactoglobulin ($MW = 36{,}000$). (The contradiction is only apparent, not real; what is the molecular weight of a gel?).

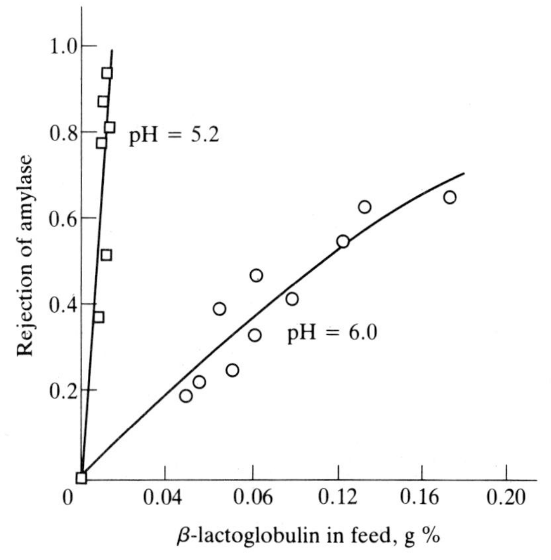

Figure 11.31 Amylase rejection R vs. β-lactoglobulin in feed: pH = 5.2(□), 6.0(○), 0.01 *M* acetate buffer, 0.002 *M* $CaCl_2$, rpm = 900, 15 psig = *p*. *[After T. A. Butterworth and D. I. C. Wang, "Separation and Purification of Enzyme Mixtures by Ultrafiltration," p. 195 in Fermentation Technology Today, G. Terui (ed.), Soc. Ferm. Technol., Japan, 1972.]*

11.6 ELECTROPHORESIS

Electrophoresis is the movement of a charged species in an electric field. The steady velocity u_E achieved by a particle of charge q in a fluid under the influence of an electric field of strength E is found from the momentum balance

$$-\text{Fluid drag on particle} = \text{electrostatic force} = qE \tag{11.29}$$

For globular proteins, use can be made of Stoke's law for the drag on a sphere of radius r_p moving in a Newtonian liquid of viscosity μ_c

$$-\text{Fluid drag on particle} = 6\pi\mu_c r_p u_E \tag{11.30}$$

so that

$$u_E = \frac{q_E}{6\pi\mu_c r_p} \tag{11.31}$$

Since in general each protein species will have a different net charge q, application of an electric field will result in different protein velocities. Thus a sample containing several proteins can be separated into its component parts (Fig. 11.32). By varying the pH of the electrophoretic medium, the velocity of a protein can be altered. If for a given protein the pI is smaller than the pH, its charge and velocity will be negative. Protein components with p$I >$ pH will move in the opposite direction. This technique can be utilized to measure pI: the pI of proteins or other molecules can easily be determined by placing the molecules in a pH gradient and noting where in the gradient the electrophoretic velocity u_E becomes zero.

In flow electrophoresis, bulk fluid flow occurs perpendicular to the applied field, allowing continuous processing. In paper and gel electrophoresis, particle motion is modified by adsorption/desorption and hindered diffusion interactions with the porous solid or gel.

A general problem in considering scale-up of electrophoresis techniques is ohmic heating of the solution. As solutions which stabilize proteins are often quite conductive, high current densities are typical and the achievement of large-scale heat-exchange capabilities for such systems is not yet at hand. Small-scale electrophoretic equipment has been developed for continuous operation.

11.7 COMBINED OPERATIONS

When simultaneous operations may be run in the same vessel, both processing time (operating costs) and equipment (capital costs) may be saved. The potential is often low, however, since the restrictions required for two different processes must be met simultaneously. Here, we illustrate several characteristic kinds of combined operation to illustrate both the advantages and requirements.

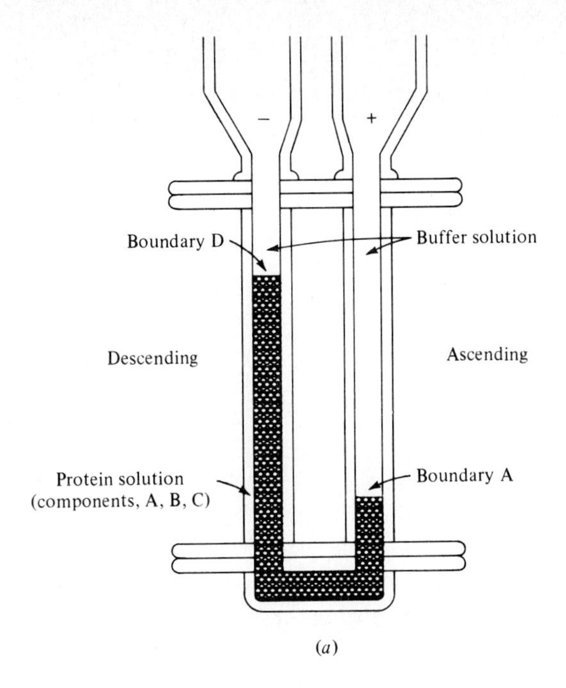

(a)

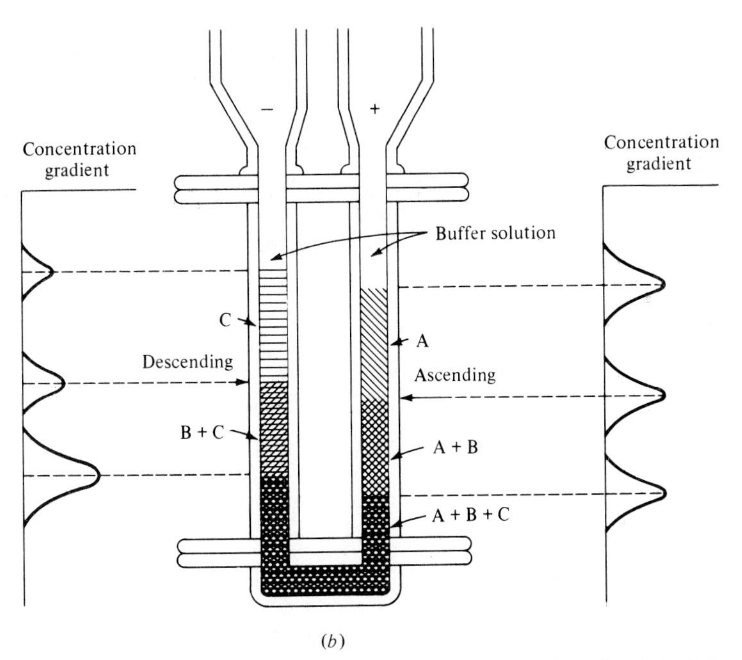

(b)

Figure 11.32 Location of protein components in an electrophoresis cell (a) before and (b) after application of an electric field. *(Courtesy of Beckman Instruments, Inc.)*

11.7.1 Immobilized Cells

Immobilized cells (Chap. 9) carry out simultaneous bioconversion (or biosynthesis) and biocatalyst recovery from the product broth. The cells may be operated in a fluidized or packed bed mode, and the trade-off includes the absence of a separate cell removal step compared to the cost of cell support and the immobilization process. Additionally, immobilized cells tend to grow, and eventually leak, into the medium, diminishing the original advantage if cell contamination of product is undesired. To avoid this, the cells may be starved of some crucial nutrient (e.g., nitrogen source) to prevent growth, and occasionally regenerated when catalyst activity declines appreciably.

11.7.2 Whole Broth Processing

Whole broth processing involves direct product removal without prior solids or cell removal. Example separations include extraction, ion-exchange, dialysis, adsorption (activated carbon) and membrane and/or filter operations.

As an illustration, antibiotic (cycloheximide) fermentation with and without added hydrophobic ion exchange resin is shown in Fig. 11.33. Over 98 percent of the cycloheximide was adsorbed on the resin, which must be regenerated periodically for commerical application. The case cited found, on subsequent elution from the resin into butyl acetate, that the final cycloheximide concentration was not greatly different from that found after direct broth solvent extraction, but product purity was improved.

Volatile inhibitory product ethanol may be continuously removed in a vacuum distillation process which feeds broth to a low temperature vacuum still allowing viable yeast (and broth) recycle with product removal.

Direct solvent extraction for penicillin and steroid recovery is practiced, and extraction into two-aqueous phase systems (PEG) has been demonstrated, although potentially expensive in terms of PEG cost.

Direct broth processing by adsorption on ion exchange resin has been realized at pilot and process levels for streptomycin and novobiocin recovery [37]. A simple set of fixed bed ion exchange adsorbers receives a continuous flow of fermentation whole broth (Fig. 11.34). This system has been described by a simple transient model of the form below:

$$V_c \frac{dc_n}{dt} = Fc_{n-1} - Fc_n - V_R \frac{dw_n}{dt} \tag{11.32}$$

where V_c = liquid volume in column n (L)

c_{n-1}, c_n = antibiotic concentration in bulk feed to or effluent from column n, respectively

F = volumetric flow rate (L/h)

V_R = resin volume in column n (L)

w_n = antibiotic concentration in resin phase, g/L-resin

t = time (h)

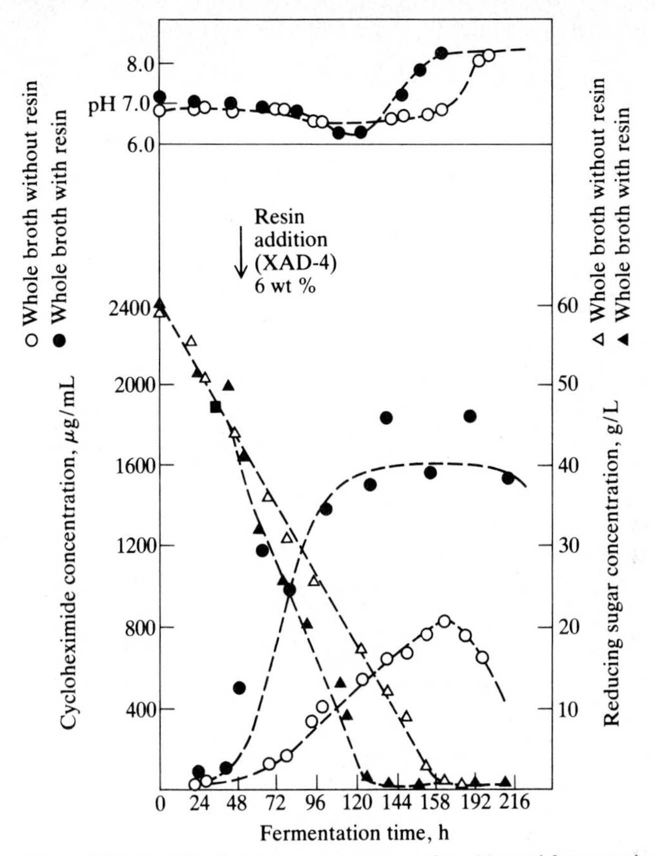

Figure 11.33 Cycloheximide concentration vs. time. Normal fermentation (○, △); fermentation modified at 48 hours by addition of 6 wt % resin (XAD-4) (•, ▲). *(Reprinted by permission from H. Wang, Ann. N.Y. Acad. Sciences, vol. 413, p. 313, 1983.)*

The resin binds the antibiotic strongly. In consequence the antibiotic distribution is nonuniform in a single resin particle, giving rise to a time dependent diffusional transport resistance to additional antibiotic uptake by the resin. This situation can be described conveniently using an uptake-dependent correlation for an apparent mass transfer coefficient. Hence the uptake rate is written:

$$\frac{dw_n}{dt} = k(c_n - c^*) \tag{11.33}$$

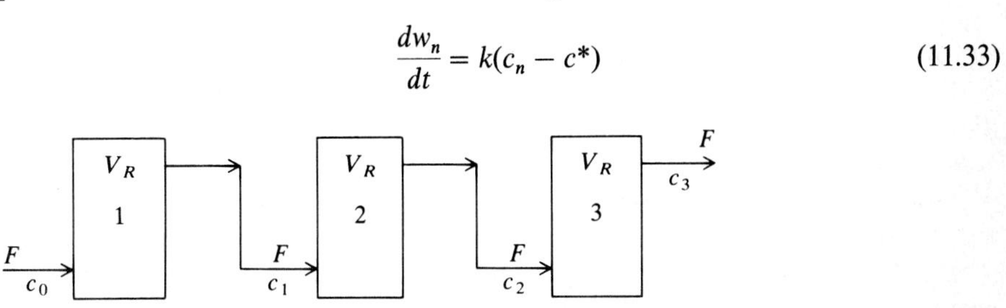

Figure 11.34 Multibed absorber for antibiotic recovery. *(Reprinted by permission from P. A. Belter, F. C. Cunningham, and J. W. Chen, "Development of a Recovery Process for Novobiocin," Biotech. Bioeng., vol. 15, 533, 1973.)*

where the apparent mass transfer coefficient k is given as $k = f(w/w_{max})$:

$$k = Ae^{-B(w/w_{max})} + De^{-E(w/w_{max})} \tag{11.34}$$

Finally, c^* follows a Freundlich isotherm:

$$c^* = b(w_n)^a \tag{11.35}$$

When the last-stage effluent antibiotic concentration reaches a particular value, the first adsorber is removed for product recovery and a fresh ion exchange adsorber added to the effluent end. The semiempirical model of Eqs. (11.33) to (11.35) provides a useful description of fed-batch adsorption for whole broth ion-exchange extraction. The process was developed in order to eliminate a costly mycelial filtration which had an accompanying appreciable antibiotic loss in filter cake materials.

Problem areas in whole broth extraction include:

Adsorption. Uptake of nonproduct solutes (other proteins, unconverted substrates, etc.) may diminish sorbent capacity and diminish purity of eluted product.

Membranes are susceptible to fouling when not regularly cleaned.

Solvent-broth systems may form difficult emulsions, allow solids deposition in recovery devices, or allow excessive solvent loss in discarded broth phase.

In general, a successful system should be selective, nontoxic to growing culture, capable of aseptic operation, have a high product-loading capacity, and be reusable if it is expensive.

11.7.3 Mass Recycle

Such recycle systems involve the combined operation of *separation* of particular mass fractions from a bioreactor or processed broth and its reintroduction upstream as an additional feed stream to the bioreactor. The three major possibilities are (i) gas phase recycle, (ii) water recycle, and (iii) biomass recycle (or retention). The first involves only bioreactor configurations allowing reintroduction of the freely separated gaseous effluent, and was discussed previously in Chap. 9. Biomass recycle is practiced advantageously primarily in aerobic waste treatment (Chap. 14) and was considered analytically in Chap. 9. Various membrane operations (cross-flow filtration, microfiltration) lead to cell *retention* in the bioreactor, providing advantages in specific rate similar to those of cell recycle.

The prospect of large-scale bioprocess plants for, for example, single-cell protein or ethanol production, requires consideration of water recycle. For example, a 50,000–70,000 dry ton/yr SCP process could require 2–2.8 × 10^6 tons/yr of process water. Discharge of this stream would give losses from incomplete nutrient conversion and useful (unrecovered) organic by-products, and could add a waste treatment cost, should such be needed to render the process effluent suitable for sewage or other discharge. Additionally, bioreaction processes in arid climates must address water as a critical resource; the above large process water requirement represents about one quarter of the fresh and distilled water annual usage in Kuwait's main industrial area.

Recycle brings several potential problems which could significantly alter reactor performance:

1. Buildup of inhibitory excreted metabolites or cell lysis components. (These may be produced in either the bioreactor or the product recovery section.)
2. Buildup of undesirable or unused feedstock constituents.
3. In mixed culture, selective cell recycle which alters culture makeup.

Analytical study of reactor performance predicted with models of the sort discussed previously in Chap. 9 indicate that, for systems where inhibition occurs:

1. Water recycle always lowers conversion vs. the same dilution rate in a nonrecycle system;
2. For a constant total nutrient input per unit volume, recycle again lowers conversion;
3. For every recycle system, a critical recycle ratio exists, below which the effect on conversion is mild, and above which the effect is considerable.

We note in passing that recycle provides a form of backmixing in the biological reactor; as with other areas of chemical kinetics, positive order conversions (such as given by Monod kinetics which are fractional order in substrates) are penalized by backmixing. Thus, water recycle will always lead to a larger (more expensive) reactor, and the additional economic gain due to decrease of disposal or water purchase costs must more than cover these additional bioreactor costs.

11.8 PRODUCT RECOVERY TRAINS

In this section, we examine integrated product separation and purification trains, indicating with several examples the general trend of solids removal, primary concentration, purification and final processing which characterize biological recovery schemes.

Prior to examining such detail, it is useful to reflect on interactions between the bioreaction and recovery section. The bioreactor feedstock may be a complex medium on one extreme, or a chemically defined one on the other. While the former often gives better cell or product yields, the resultant broth is more heterogeneous and may cause recovery difficulties. More costly defined media may be justified to facilitate recovery when the product has high value. Similarly, higher bioreactor productivity is a goal which conflicts with the notion of complete substrate utilization; the latter gives lower specific rates in the bioreactor but also provides a bioreaction broth for recovery which may be free of troublesome solutes (e.g., sugars) which interact unfavorably with a product recovery scheme, or which may impart instability to the stored, dried product.

Based on pharmaceutical industry experience, the following guidelines are profitably considered:

1. Antifoam should be minimal, as surfactants may stabilize emulsions in later extractions;
2. pH buffers are less desirable than use of simple acid/base additions;
3. Biocides are less desirable in sterilization than heat treatments;
4. Bioreactions which are clearly intracellular *or* extracellular yield product only in one phase and give an easier recovery train design than when the product is in *both* biomass and broth phases;
5. Substrate residues should be minimal. Thus, unconverted substrate, as well as unconvertible feed components and separate solids, are to be avoided if possible in order to minimize demand on the recovery train.

11.8.1 Commercial Enzymes

Enzyme products are available as crude, dried preparations, dilute or concentrated liquids, or purified (even crystallized) solids. Figure 11.35 provides a general process recovery scheme for enzyme derived from animal, plant, surface, or submerged fermentations. The former sources require immediate pretreatment to release enzyme into an extracting buffer, followed by the appropriate solids removal steps when liquid or purified products are required. A more detailed process recovery example for a plant enzyme (Fig. 11.36) shows the necessity of good mixing (flow in coil) at a low temperature (stabilize intracellular products) to maximize initial extraction. The two serial centrifuges perform a solids fractionation, removing large particles first by scroll centrifugation so that the more expensive, higher rpm bowl centrifuge is not clogged with large particles. Subsequent acidification shifts pH sufficiently to precipitate much originally soluble protein, provided a sufficient residence time is allowed in the cooled holding coil to form a centrifugal precipitate (recall Fig. 11.17). A disk centrifuge removes this (unwanted) protein precipitate; a second acidification and holding coil precipitate the desired protein, recovered as wet solid from the second disc centrifuge. Thus, this example contains two instances where similar or identical processes are placed serially to carry out a fractionation, first of solids by centrifugation and second by acidification/precipitation.

As subsequent smaller scale operations occur in protein purification (as processed volume diminishes), recovery steps may logically shift from continuous to batch as shown in Fig. 11.37 for protease production. Note additional steps to enhance enzyme yield: (1) repeated washing of biomass, (2) two-stage ultrafiltration to carry out sequential 5-fold volume reductions, (3) a switch from continuous to batch ultrafiltration to effect a further 40-fold volume reduction.

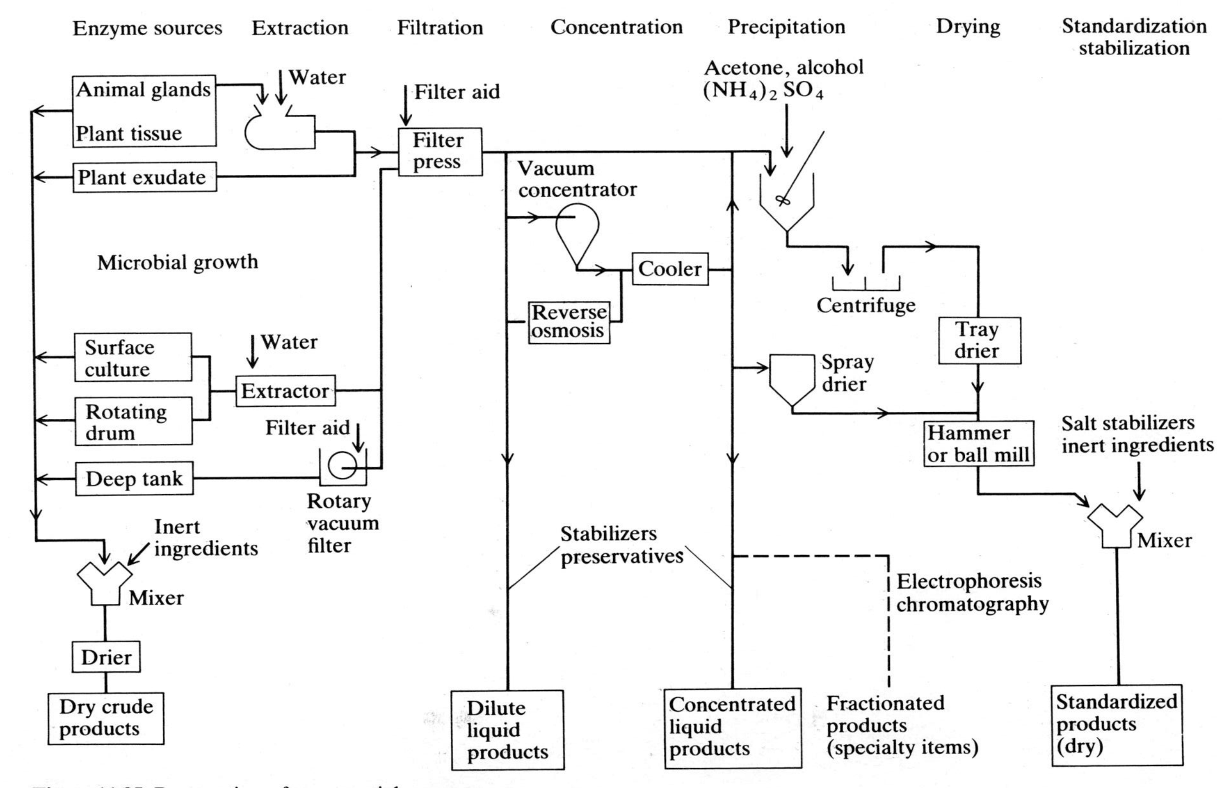

Figure 11.35 Preparation of commercial enzymes.

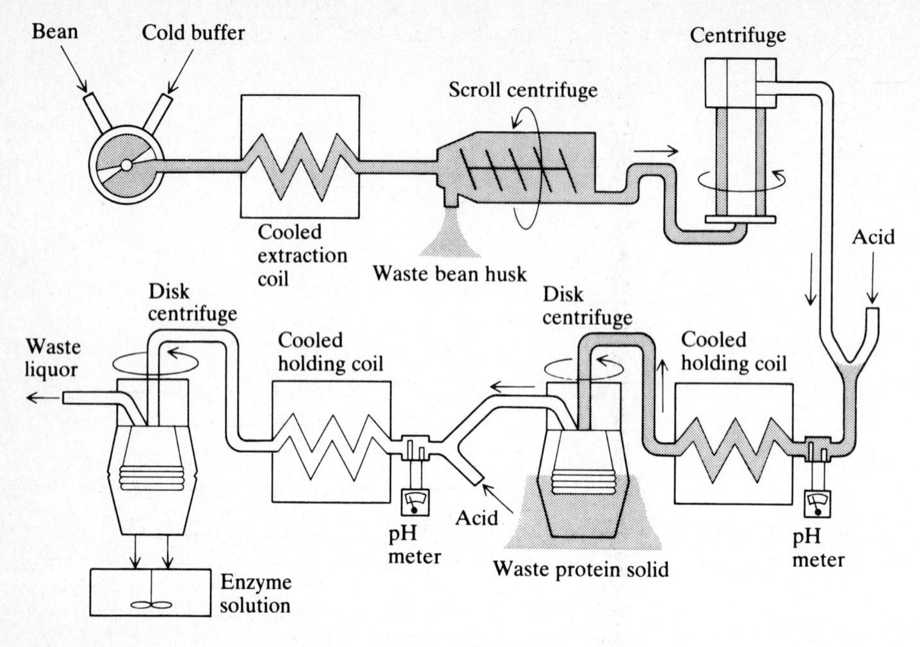

Figure 11.36 Continuous isolation of enzyme proly-tRNA synthetase from mung bean. *(Reprinted from M. D. Lilly and P. Dunnill, Science J., vol. 5, p. 59, 1969.)*

An important enzyme application occurs with protease (and other hydrolases) in synthetic detergents. *Bacillus* protease alone accounts for 40 percent of the world industrial enzyme sales, yielding roughly 500 tons of pure protein in 1978–1979. These proteases were first produced as dusty powders, including particles of less than 10 μm diameter. The resulting dust levels caused allergic reactions among plant workers, and led to temporary removal of enzyme detergents from the U.S. market. Subsequent development of a granulated, dust-free enzyme process (Fig. 11.38) allowed resumption of U.S. usage; the protease-containing granules are produced either as a wax-coated 500-μm sphere (Fig. 11.38) or, alternately, are directly embedded in a waxy particle matrix.

11.8.2 Intracellular Foreign Proteins From Recombinant *E. coli*

Proteins synthesized in genetically engineered organisms and intended for injection into animals must be stringently purified. Pyrogens from *E. coli* including the outer envelope lipopolysaccharide (LPS) must be removed or inactivated. (LPS causes fever response in humans at levels $\lesssim$0.5 ng/kg body weight.) The product must be protected from *E. coli* proteases. Often in purification, phenyl methyl sulfanyl fluoride, a serine protease inhibitor, is added to buffers for this purpose.

Figure 11.39 summarizes the sequence of processing steps which have been used to purify human insulin, human growth hormone, and human leukocyte

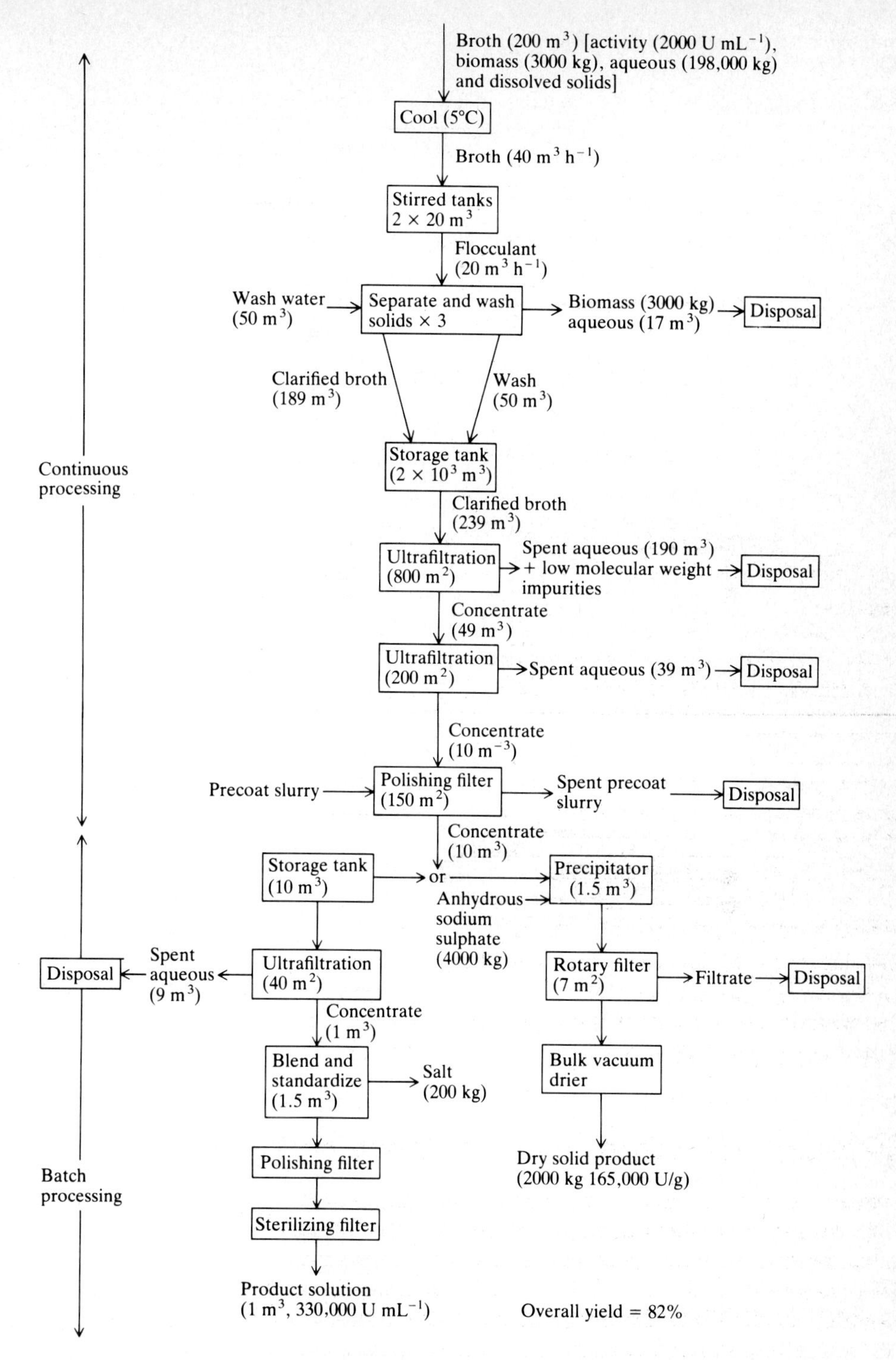

Figure 11.37 Extracellular enzyme (protease) recovery. Basis: 200 m^3 batch, 10-h operation with two hours' clean-down. *(Reprinted by permission from B. Atkinson and F. Mavituna, Biochemical Engineering and Biotechnology Handbook, Macmillan Publishers Ltd., Surrey, England, 1983, p. 918.)*

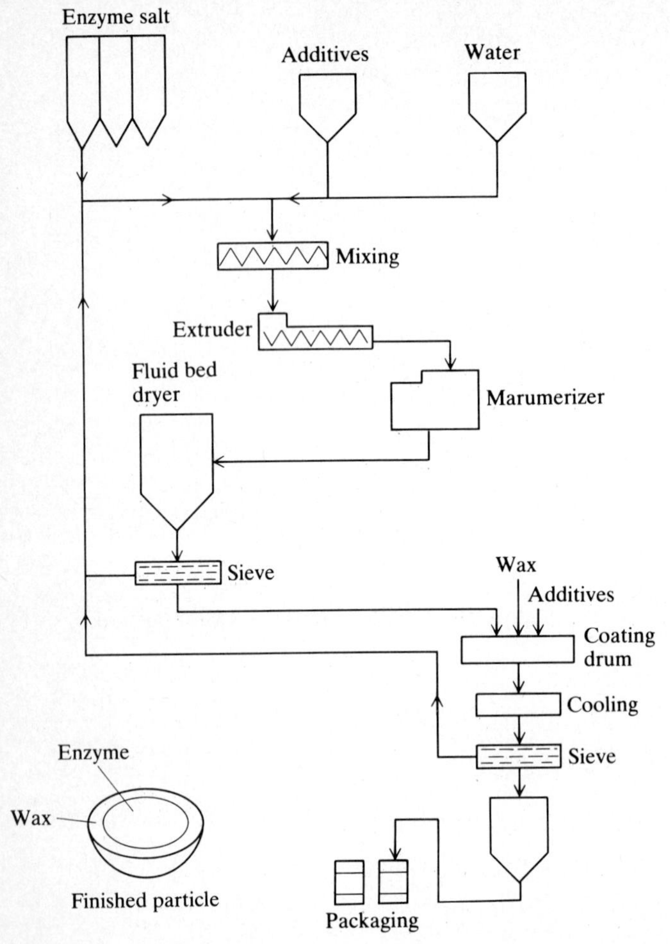

Figure 11.38 Flow process for dust-free enzyme product; the marumerizer provides wax-coated particles. (*Reprinted by permission from K. Austrup, O. Andresen, E. A. Falch, and T. K. Nielsen, "Production of Microbial Enzymes," p. 281 in Microbial Technology, 2d ed., Vol. I, H. J. Peppler and D. Perlman, eds., Academic Press, N.Y., 1979.)*

interferon made in recombinant *E. coli.* A series of precipitation and chromotography steps is applied in each case to achieve the required purity [22].

Proteins expressed at high levels in recombinant *E. coli* often accumulate in intracellular refractile bodies (Fig. 6.27). While easily separated from lysed cell solution by centrifugation, these crystalline agglomerates of highly cross-linked protein must be dissolved and renatured to obtain useful product. Summarized in Fig. 11.40 are an oxidative sulfitolysis protocol for dissolving refractile bodies and a refolding and disulfide bond formation procedure for renaturing soluble denatured protein. These methods were reported by scientists at Genencor, Inc. in connection with cloning and expression of rennin in *E. coli.*

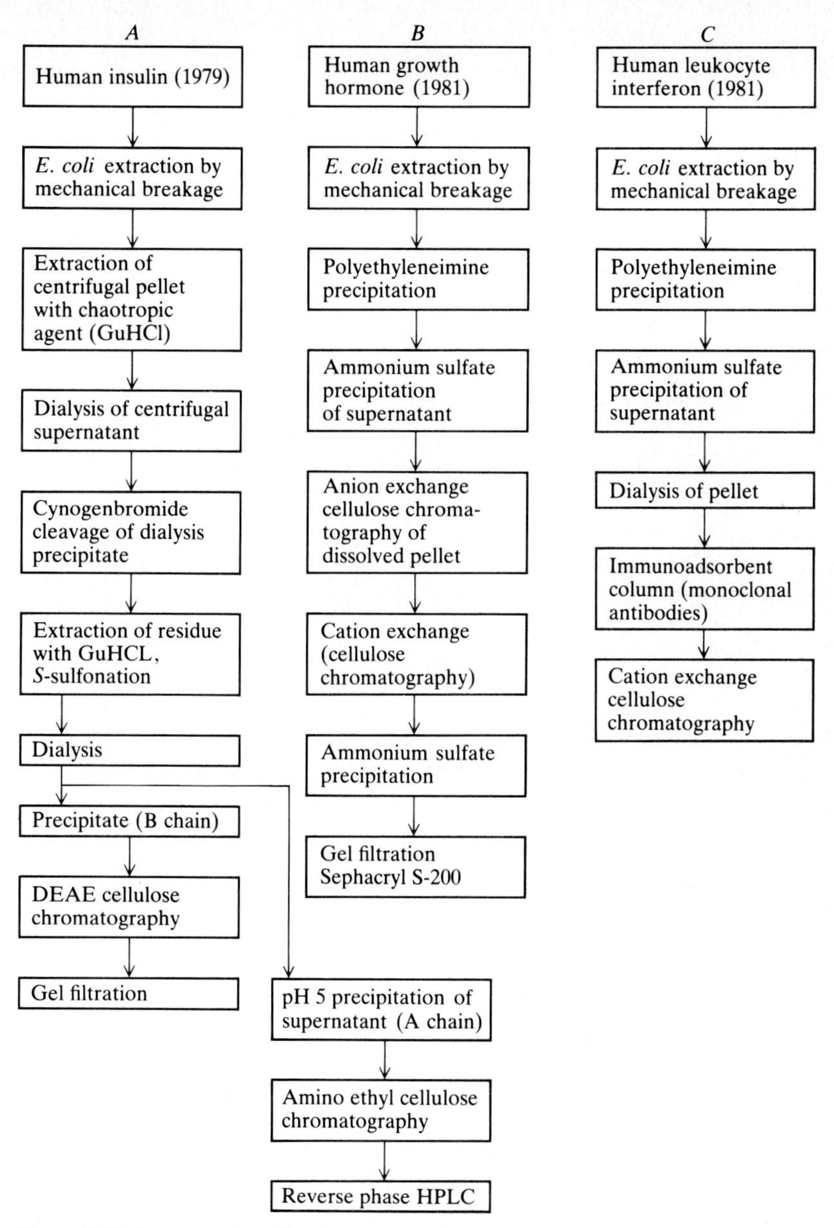

Figure 11.39 Summary of purification processes for three human proteins synthesized in recombinant *E. coli*. *(Reprinted by permission from W. C. McGregor, "Large-Scale Isolation and Purification of Proteins from Recombinant* E. coli," *Ann. N.Y. Acad. Sci., vol. 413, p. 231, 1983.)*

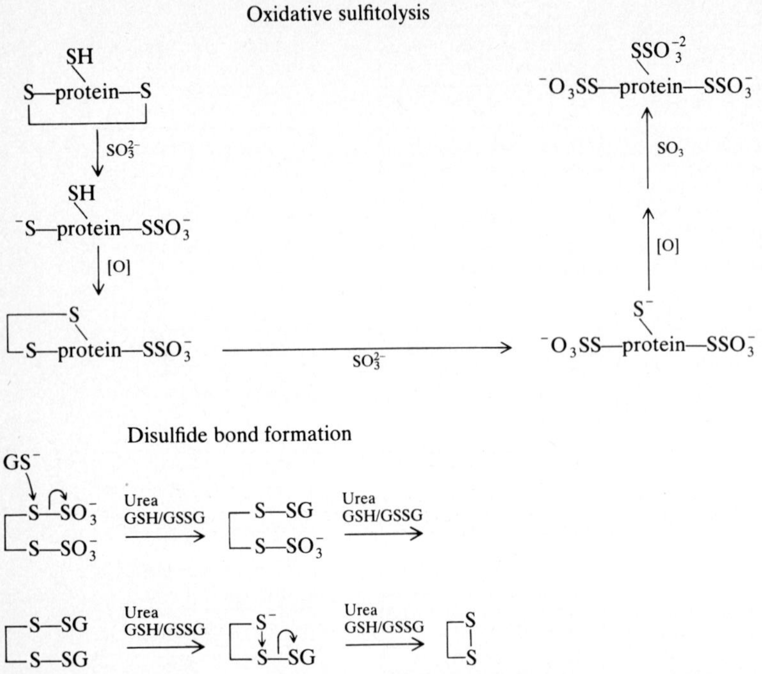

Oxidative sulfitolysis

SH, S—protein—S (cyclic) $\xrightarrow{SO_3^{2-}}$ SH, $^-$S—protein—SSO_3^- $\xrightarrow{[O]}$ S, S—protein—SSO_3^- (cyclic) $\xrightarrow{SO_3^{2-}}$ S^-, ^-O_3SS—protein—SSO_3^- $\xrightarrow{[O]}$ $\xrightarrow{SO_3}$ SSO_3^{-2}, ^-O_3SS—protein—SSO_3^-

Disulfide bond formation

GS^- + [S—SO_3^-, S—SO_3^-] $\xrightarrow{\text{Urea, GSH/GSSG}}$ [S—SG, S—SO_3^-] $\xrightarrow{\text{Urea, GSH/GSSG}}$

[S—SG, S—SG] $\xrightarrow{\text{Urea, GSH/GSSG}}$ [S^-, S—SG] $\xrightarrow{\text{Urea, GSH/GSSG}}$ [S—S]

Figure 11.40 Oxidative sulfitolysis and disulfide bond formation procedures used for conversion of refractile bodies to soluble, active protein. GSH and GSSG denote the reduced and oxidized forms, respectively, of glutathione. *(Courtesy of Bryan Lawlis and Kirk Hayenga, Genencor, Inc.)*

11.8.3 Polysaccharide and Biogum Recovery

Polysaccharides and closely related derivatives are precipitable by alcohol addition, forming the basis for recovery. Dextrans provide a high value polysaccharide product with uses which may include food or medicinal applications, hence the distinctive uses in its purification (Fig. 11.41) of alcohol precipitation and use of washes with pyrogen-free water. Partial hydrolysis of initial precipitate is followed by (unhydrolyzable) solids removal, and repeated fractional crystallizations, with final impurities removed by ion exchange. Xanthan gum is precipitated with isopropanol after broth pasteurization (to kill all cells), then spray-dried and milled for a food-grade product.

11.8.4 Antibiotics

Antibiotic production yields either a bulk salt form, e.g., sodium penicillin (Fig. 11.42), or a more purified precipitate (procaine penicillin) for clinical use. Note the additional use of adsorbent columns and line filters to remove dissolved and particulate contaminants, respectively. Final spray drying of the procaine-antibiotic precipitate is accomplished in a low temperature freeze dryer. Biomass may be fully recovered and sold as animal food supplement.

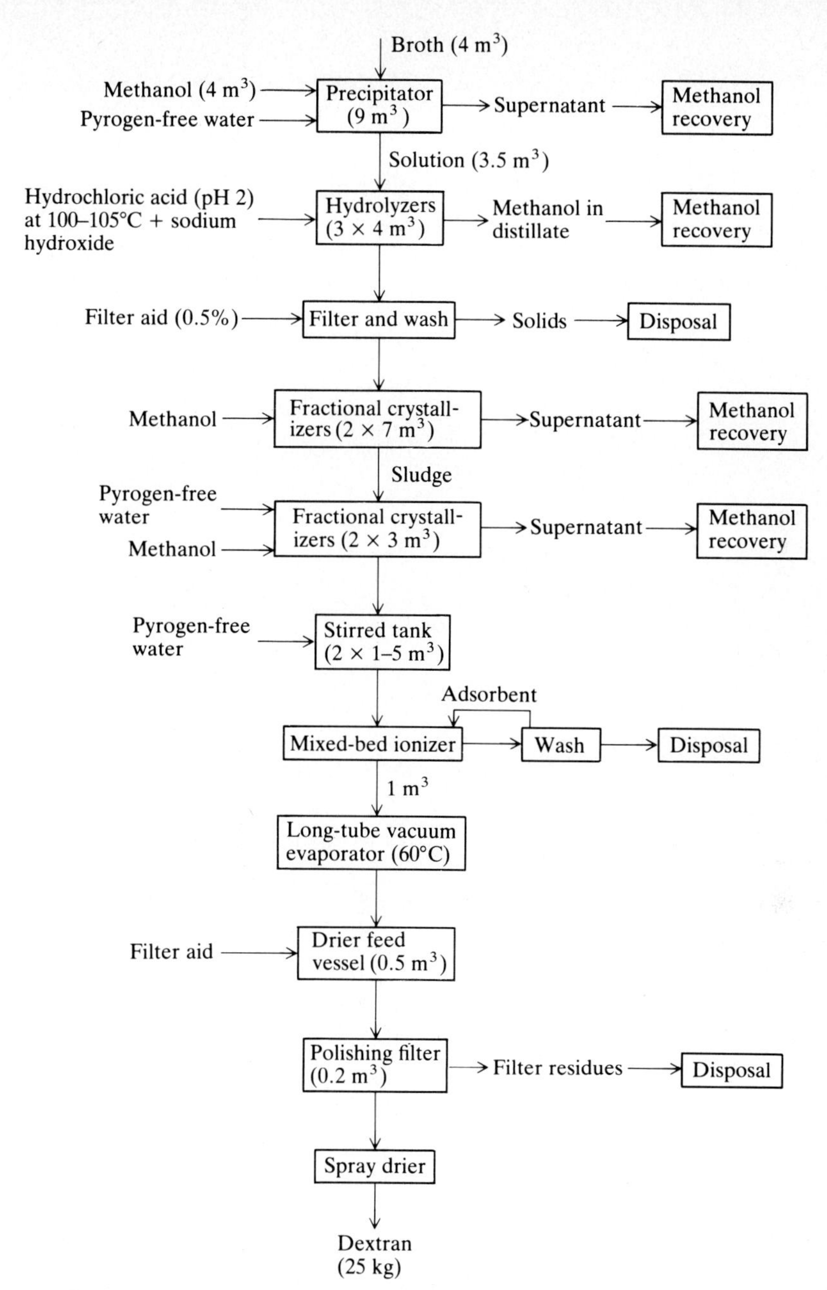

Figure 11.41 Processing of dextran from broth of low solids content. Basis: (4 batches of $4\ m^3$)/day. *(After B. Atkinson and F. Mavituna, Biochemical Engineering and Biotechnology Handbook, Macmillan Publishers Ltd. Surrey, England, 1983, p. 911.)*

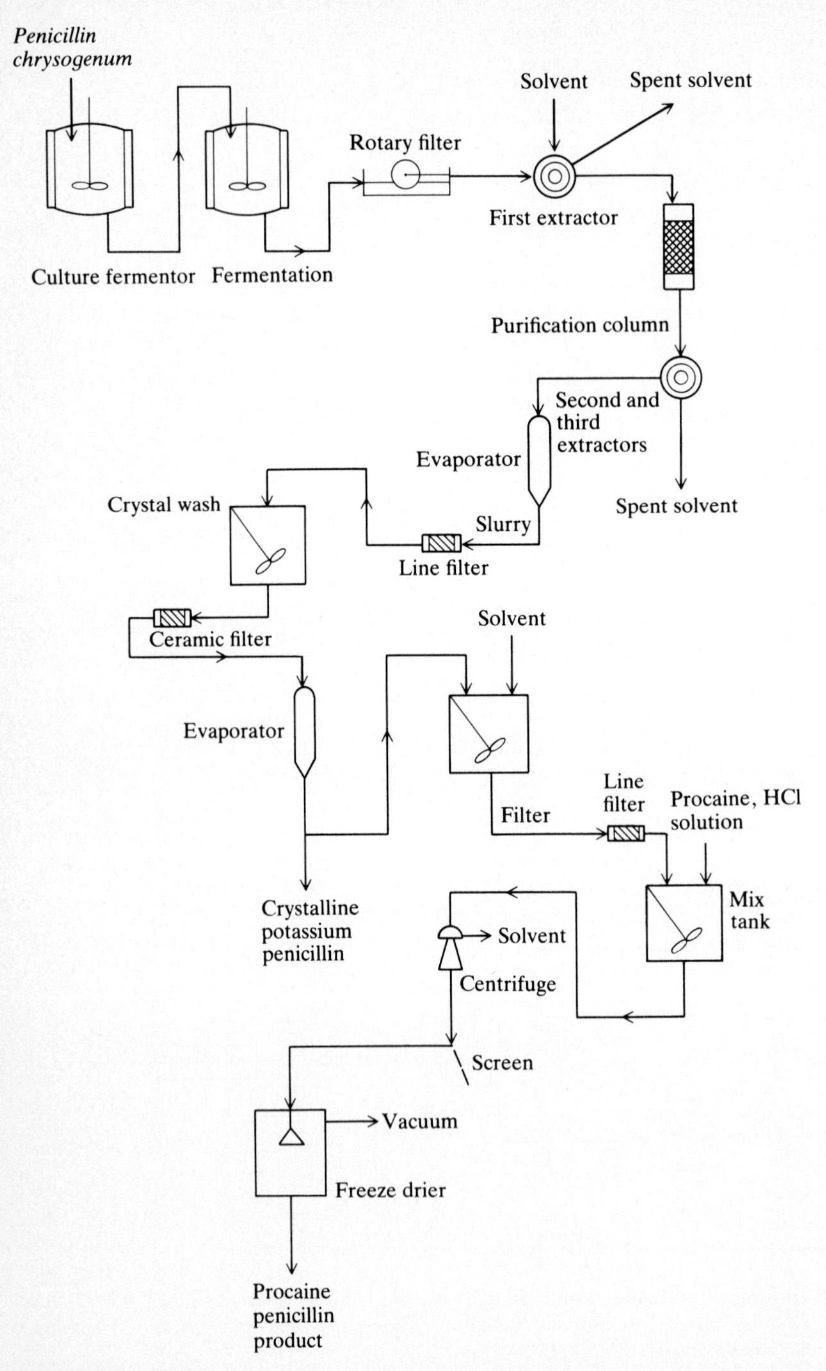

Figure 11.42 Penicillin production process. *(After B. Atkinson and F. Mavituna, Biochemical Engineering and Biotechnology Handbook, Macmillan Publishers Ltd., Surrey, England, 1983, p. 990.)*

11.8.5 Organic Acids

Citric, gluconic, and itaconic acids are all recovered by biomass removal by filtration followed by a precipitation (calcium citrate, calcium or sodium gluconate, itaconic acid). The citric acid requires calcium citrate dissolution with sulfuric acid (Fig. 11.43), discard of calcium sulfate product, and precipitation as citric acid (or sodium citrate) crystals with centrifugal recovery; subsequent drying depends on final product (anhydrous, monohydrate, sodium salt). Itaconic

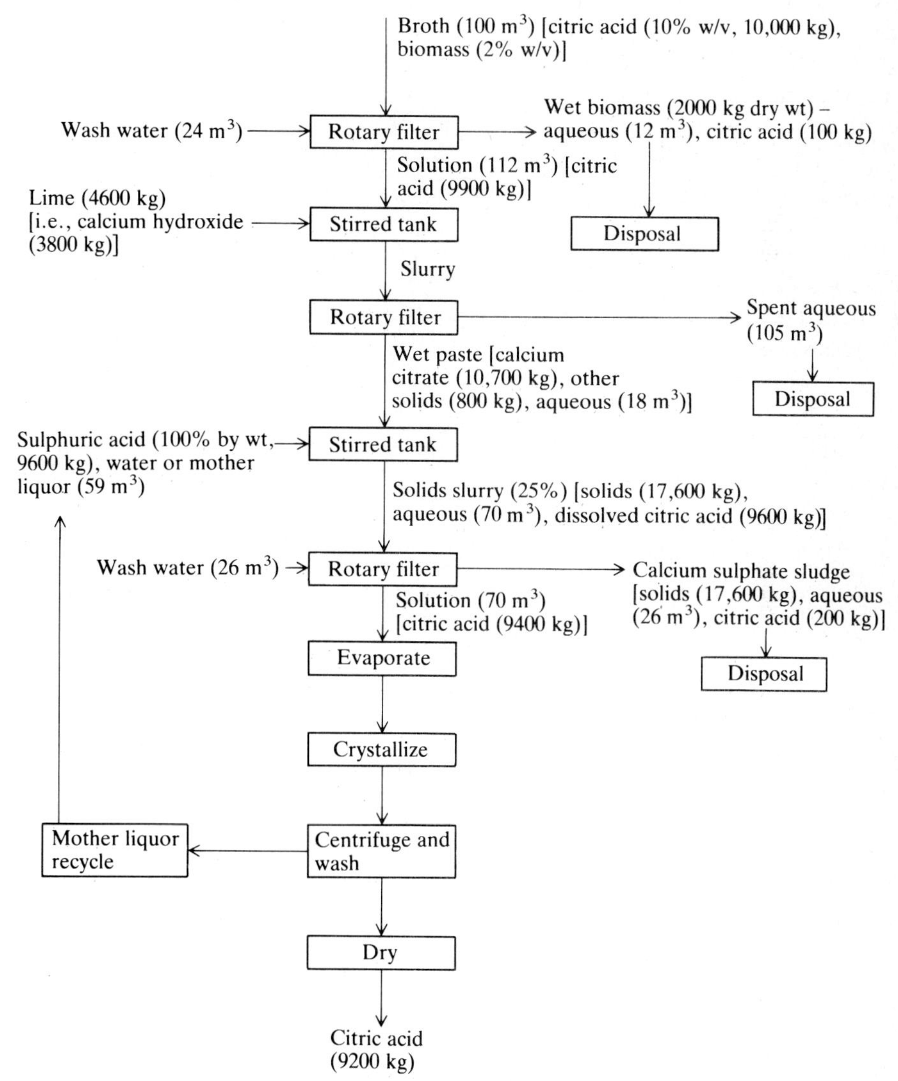

Figure 11.43 Organic (citric) acid recovery process. Basis: 100 m^3 batch yielding 9200 kg acid. *(After B. Atkinson and F. Mavituna, Biochemical Engineering and Biotechnology Handbook, Macmillan Publishers Ltd., Surrey, England, 1983, p. 910.)*

acid solutions are treated with activated carbon for high quality end use; this step is not needed for industrial grade product.

11.8.6 Ethanol

Ethanol forms an azeotropic mixture with water at 95.7 wt% (89 mol%) ethanol. Distillation recovery proceeds in two steps. First a three-column conventional distillation train produces a high alcohol product stream. For anhydrous alcohol production (as needed for motor fuel usage), this binary solution is then mixed with benzene, breaking the binary azeotrope, and allowing recovery of 100 percent ethanol by distillation (Fig. 11.44). Energy recovery costs depend on the ultimate ethanol grade needed; Table 11.8 summarizes columns useful for broth concentration, industrial and motor fuel production.

11.8.7 Single-Cell Protein

Biomass recovery, where biomass is the principle product rather than byproduct (as in glutamic acid or antibiotic production), requires a very inexpensive recovery step. Two recent bacterial processes, due to ICI and Hoechst, recover methanol-grown cell biomass in a simple preconcentration step. The ICI process

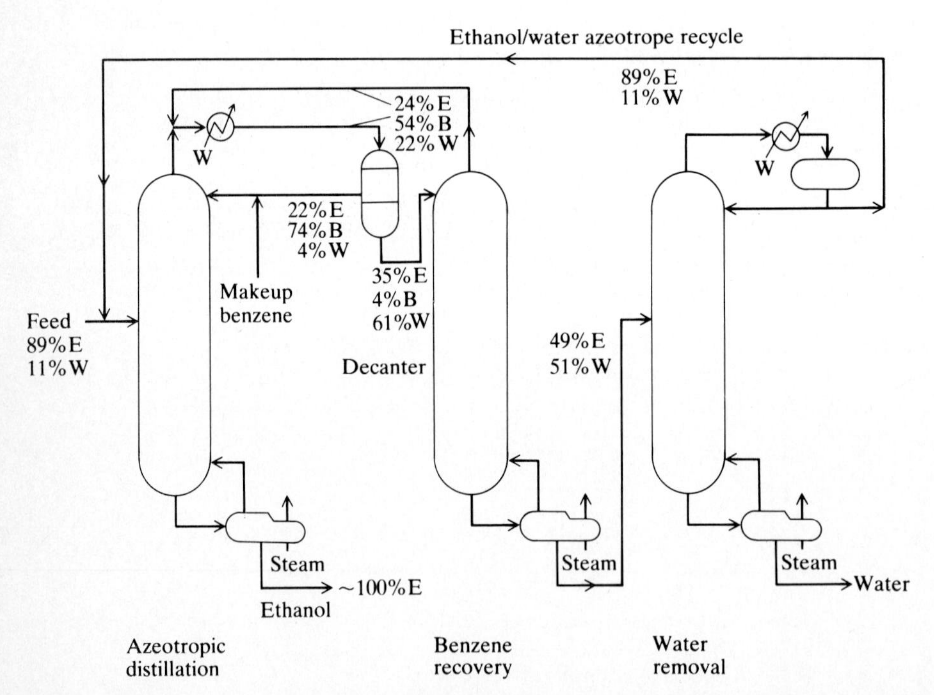

Figure 11.44 Pure ethanol via azeotropic distillation. *(After B. Maiorella, C. R. Wilke, and H. W. Blanch, "Alcohol Production and Recovery," Adv. Biochem. Eng., A. Fiechter (ed.), vol. 20, p. 43, Springer-Verlag, Berlin, 1981.)*

Table 11.8 Energy requirements of ethanol recovery[†]

Process	Energy consumption kg steam/L	Comments
Simple 1- or 2-column distillation to 96 vol % ethanol from 6.25 vol %	2.4	20-tray beer still, 30-tray rectifying column
3-column vapor reuse process for high-quality 96 wt % industrial alcohol from 6.25 vol %	2.4	20-tray beer still, 45-tray aldehyde column, 54-tray rectifier
4-column Barbet unit for 96 vol % industrial alcohol from 6.25 vol %	4.1	20-tray beer still, 30-try purifying column, 54-tray rectifying column
Benzene dehydration from 96 to 99.9 vol % (by vol.) ethanol	1.0	50-tray dehydrating column, 30-tray water-steam column, 45-tray supplementary rectifier
Combined 3-column vapor reuse and high-pressure ether distillation for 99.9% ethanol from 6.25%	2.5	20-tray beer still, 45-tray aldehyde column, 54-tray rectifier, 30-tray dehydrator, 20-tray water stripper
Pentane dehydration from 96 to 99.9 vol % ethanol	0.9	18-tray dehydration column, 14-tray water stripper column (supplementary rectifier also required)
Vacuum rectification to produce 95 wt % ethanol from 13 wt %	1.1	40-tray rectifier (assumes no aldehydes or fusel oil)
Extractive distillation using salts to produce 99.9 mole % ethanol from 5 wt% ethanol	1.1 (additional energy for salt recovery, depending on the process chosen)	Single 20-tray column (assumes no aldehydes or fusel oil) hinges on low-energy recovery of salts, not yet studied
Vapor-phase adsorption of water to produce 99.9% alcohol from 85% alcohol	Energy requirements for adsorbent regeneration have not been established	Only small laboratory-scale tests have been conducted
Extraction	Laboratory-scale process requires further study	
Membrane separation	Laboratory scale process	
Molecular sieve separation	Conceptual process requires further study	

[†] *After B. Maiorella, C. R. Wilke, and H. W. Blanch, "Alcohol Production and Recovery," p. 43 in Adv. Biochem. Eng., vol. 20, A. Fiechter, ed., Springer-Verlag, Berlin, 1981.*

employs coagulation and decanting to produce 25 g cell (dry weight) per liter, and cell recovery in the Hoechst process is based on electrocoagulation. The resulting cell fluid is sufficiently dense to be centrifuged directly, bypassing an intermediate filtration step. The final product is obtained by spray dryer operation. These filtration-free results arose from need: (i) large volumes must be processed on a continuous cell recovery basis (thus excluding plate-and-frame filter), (ii) the filter cakes formed by small (bacterial) particles would clog vacuum rotary filters, and (iii) use of any filter aid would compromise final product quality. These most recent processes use bacteria, which generally have higher specific growth rates than yeasts.

Yeast, while slower growing, provide easier recovery due to their larger size (typically $d = 5\text{–}8\ \mu\text{m}$) than bacteria ($d = 1\text{–}2\ \mu\text{m}$). For example, bakers' yeast (*S. cerevisiae*) ($\rho(\text{yeast}) = 1.133\ \text{g/cm}^3$) may be recovered from 3–5% broths by centrifuging with nozzle type, vertical continuous centrifuges. A single-pass triples the concentration, and subsequent passes produce 18–20% yeast solids as a pumpable cream product.

Molds and higher fungi are recoverable with basket centrifuge, rotary vacuum filter, or screens, yielding 22–45% solids product, which is then dried. The trade-offs are clearly slower growth rates vs. less expensive recovery schemes; recall that the doubling times are typically 20–30 minutes for bacteria, 2–3 hours for yeasts, and several hours for molds, algae, and higher fungi.

Final cell treatments prior to packaging must include removal of undesired substrate. For carbohydrates, no problem exists. Methane utilization is aided by recovery and recycle in the bioreactor, and trace methanol is easily removed during drying. Gas oil-based SCP processes repeatedly encountered difficulties due to unconverted substrate residue in the final SCP product, in spite of complex separation treatments which have included decantation, phase separation with solvents, surfactant-containing washes, and solvent extraction. Pure alkane grown SCP can receive a simple surfactant wash to remove residual substrate after the first centrifugation. The final drying treatments for SCP product must provide sufficient killing (5–8 log cycles) of desired as well as any contaminant strains, yet not be so harsh as to damage the product by, for example, excessive darkening with drum drying.

When usage as bulk human food is considered, a means of diminishing DNA content is required. Two approaches have been considered: reduction of DNA in whole cell material (by heat shock or alkaline extraction) or recovery of protein from broken cells, an approach which traditionally has had poor protein recovery yields.

11.9 SUMMARY

It is the challenge to a bioprocess separation system, whether a single-unit operation or a sequence of operations, to yield a product of the desired concentration, purity, and stability. This chapter has provided an introduction to the principles underlying each type of separation unit operation and has given several illustrative applications for bioproduct recovery.

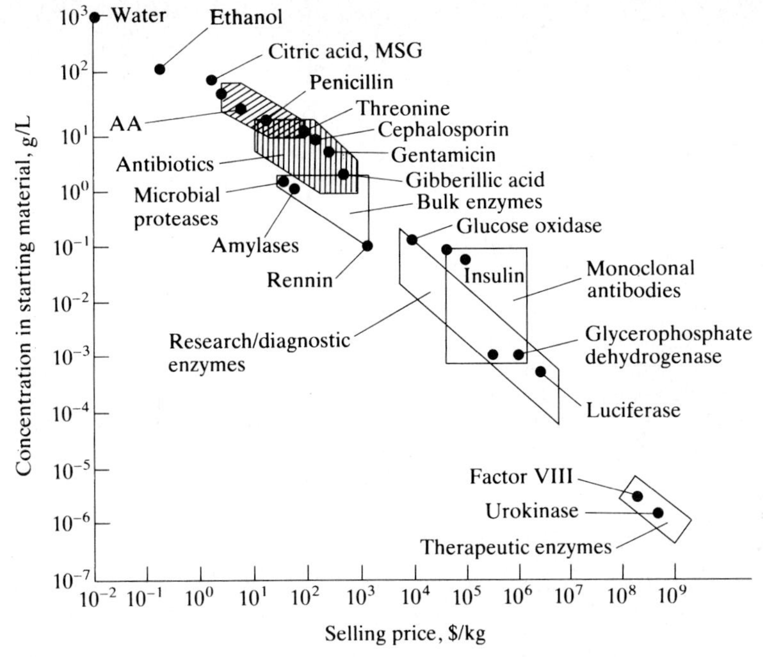

Figure 11.45 Relation between starting product concentration in completed broth or medium, and final selling price of prepared product. *(Reprinted by permission from J. L. Dwyer, "Scaling Up Bioproduct Separation with High Performance Liquid Chromatography," Bio/Technology, vol. 2, p. 957, 1984.)*

Economics drives the connection between separation costs and initial product concentration of the completed bioreaction broth as indicated in Fig. 11.45. The very clear trend from this important figure reminds us that a continual pressure for technical improvement exists on the upstream bioreactor section to yield a final broth of higher product concentration, so as to lower the costs for the subsequent separation and recovery sections. Ultimately, we must use a common denominator, namely costs, to compare the overall resources needed for each section of a bioprocess: raw materials handling and storage, bioreaction, separation, and recovery. A unifying basis for this cost engineering is the subject of the first half of the next chapter, followed by illustrative examples of bioprocess economics.

PROBLEMS

11.1 Cell settling (*a*) Calculate the apparent isolated particle size for the following biological particles given their densities (ρ) and settling velocities (u_s):

Primary sewage sludge (Chap. 14) ($\rho = 1.001$ g/cm^3; $u_s = 0.042$ cm/s)

Activated studge floc (Chap. 14) ($\rho = 1.005$; $u_s = 0.2$ cm/s)

Yeast ($\rho = 1.001$, $u_s = 0.0003$ cm/s) (Data: W. J. Weber (ed.), *Physicochemical Processes for Water Quality Control*, Wiley Interscience, 1972, p. 128).

(*b*) For bacteria ($\rho = 1.001$, effective diameter $= 1.0\ \mu\text{m}$), is settling a practical operation in less than 1 hour? Why (not)?

11.2 Cell filtration: Batch processing A shake flask broth is to be filtered at the following conditions: $A = 10\ \text{cm}^2$, $\mu_c = 2$ centipoise, $\Delta p = 0.3$ atm, $\rho = 1.1\ \text{g/cm}^3$, $w = 0.001$, and $m = 2.5$.

(*a*) Evaluate the specific cake resistance α and the resistance coefficient r of the filter if it is found at two different times that $(t\,(s), V(\text{mL})) = (30, 38)(60, 52)$.

(*b*) For scale-up purposes, what size filter A_f would be needed to process 10^5 liters of the same broth in 10 minutes?

11.3 Rotating vacuum filter: continuous filtration A rotating vacuum filter is partially submerged in a fermentation broth. As the fresh filter enters the broth, filtration and cake accumulation commences.

(*a*) Develop an expression for the accumulated local solids density (W/A) vs. time (t) as the filter rotates through the broth. What simpler forms result if filter resistance alone is dominant? If cake resistance alone is dominant? Sketch all three relations.

(*b*) Suppose that operation up to an accumulated solids density γ is practiced. If the filter operates at a rotation rate of ω radians per minute and has a radius R, derive an expression for the absolute broth processing rate of the filter. Note that you need to integrate the local flux over the entire submerged filter portion.

11.4 Centrifugation In centrifugation, a suspension of spherical particles of radius R and density ρ_p is placed in a centrifuge tube containing fluid medium of density ρ_f and viscosity μ_f. When the tube is spun at an angular velocity ω (rad/s), motion of an isolated particle in a dilute suspension is described by Eq. (1E1.4).

(*a*) At 10,000 rpm, a dilute suspension of yeast ($R = 5\mu\text{m}$, $\rho = 1.001$) is centrifuged in tubes filled with 4 cm of liquid. When the centrifuge is operating, the tubes are perpendicular to the rotation axis, and the bottom of the tube is 8 cm from the axis. What time is required for complete centrifugation? State any assumptions.

(*b*) A concentrated suspension of spherical particles, with volume fraction ε_ρ, is centrifuged in the same condition as part (*b*). Develop an expression for the times required for complete centrifugation assuming that ε_ρ is constant at all times. Is $\varepsilon_\rho =$ constant likely to be correct? Why (not)?

11.5 Filtration of mycelia The specific cake resistance in the filtration of mycelia (as in the penicillin producing fermentations) can be estimated from properties of fluid resistance to flow through a packed bed of cylinders. The specific cake resistance r is given by

$$r = K'' s_p^2 (1 - \varepsilon)^2/\varepsilon^3$$

where K'' = Kozeny shape factor, s_p = specific surface area of cake solids, ε = cake void volume. The parameter s_p is estimated from the hyphal (cell) diameter d_h according to

$$s_p = \pi d_h/(\pi d_h^2/4) = 4/d_h$$

and the void volume ε is given by

$$\varepsilon = 1 - 1/(\bar{v}\rho_h)$$

where ρ_h = hyphal density and $\bar{v}$ = specific cake volume.

(*a*) Show that the following equation arises from the filtration equation (11.2) and the defining equations above.

$$\bar{\tau} = \frac{t_F}{V_F V_C} = \frac{8K''\mu_F}{A\Delta p}\left[\frac{1}{A} + \frac{2L}{V_C}\right]\left(\frac{\rho_h \bar{v}}{(\rho_h \bar{v} - 1)^3\, d_h^2}\right)$$

where L= filter medium resistance.

(*b*) The mycelia form a *compressible* cake, thus causing a change in $\bar{v}$ as the filtration parameter $\bar{\tau}$ varies. This change, for given filtration conditions, has been correlated by the following equations:

$$\bar{v} = a - b(1/\bar{\tau}) \qquad \text{or} \qquad \bar{v} = a - b(1/\bar{\tau}) + c(1/\bar{\tau})^2$$

Derive, for each form, an expression for the change in the specific cake resistance, r, as a function of the modified filtration time, $\bar{\tau}$.

(*c*) The correlations in (*b*) involve dimensional variables, yet such relations enjoy maximum usage when recast in dimensionless form. Propose suitable dimensionless forms for the above equations for r and $\bar{\tau}$.

These equations can be used to design a filter, and have also been applied to design of an in situ biomass probe (next problem). (See E. Nestaas, PhD. thesis, MIT, 1980.)

11.6 Mycelial biomass filter probe The filter design equation (11.2) was modified to describe filtration of compressible mycelia in Prob. 11.5. An *in situ* fermentor probe for on-line biomass measurement has been designed based on determination of the time required to fill a small fixed volume reservoir located just above a filter. The filled reservoir is signalled by the movement of the cake-medium interface past a light beam, triggering determination of t_F and V_F (t_F).

(*a*) Show that the biomass concentration in the medium is given by $x = 1000\ V_C/[\bar{v}(V_C + V_F)]$.

(*b*) Show how you would calculate x from a single measure of t_F and $V_F(t_F)$ with this fixed filter cake volume probe.

In operation, x is measured by a single t_F, V_F, V_C determination. Then the filter is back flushed to dislodge the filter cake, allowing installation of the probe in the fermentor (no cake withdrawal is needed). (See D. C. Thomas, V. K. Chitter, J. W. Cagney, and H. C. Lim, *Biotech. Bioeng.*, **27**: 729, 1985.)

11.7 High-pressure homogenizer Dunnill and Lilly ("Protein Extraction from Microbial Cells," *Single Cell Protein II*, S. Tannenbaum and D. I. C. Wang (eds.) MIT Press, Cambridge, Mass., 1975) observed that protein recovery from passage of bakers'-yeast cell suspensions through a large pressure drop confined to a small volume (homogenizer) followed the equation

$$\ln \frac{R_m}{R_m - R} = KN$$

where K = first-order rate constant
N = number of passes through homogenizer
R_m = maximal achievable protein release
R = protein release after N passes
$K = \kappa p^{\alpha}$ where κ, α depend on microbe and p is operating pressure.

(*a*) Show that a first-order rate law gives the expression observed, treating N as a continuous variable.

(*b*) If the power input per pass W varies linearly with operating pressure p, show that a maximum exists in the curve of Q (percent protein released per kilowatt power input) vs. pressure.

(*c*) If $\alpha = 2.9$ for bakers' yeast and the maximum in Q is 7.0 at 570 kg_f/cm^2 for $N = 1$, evaluate κ and W/p and calculate the profiles of q vs. p for $N = 1, 2$.

11.8 Reactive extraction Solvent extraction of an unstable product can be hastened by addition of a reversibly complexing agent to the solvent phase. Solvent-soluble amines, for example, can be used to complex penicillin according to the reaction:

$$\text{Amine (organic)} + \text{penicillin}^- \text{ (aq)} + \text{H}^+ \text{ (aq)} \rightleftharpoons \text{amine} - \text{HP (organic)}$$

(*a*) Comment on the conceptual similarity of this process to facilitated transport (Chap. 5), exemplified by oxygen uptake in red blood cells by complexation with hemoglobin.

Full analysis of this mass transfer with chemical reaction system can be complex. The extractive reaction can aid the desired process in two ways: increase of the *rate* of extraction, and increase of the extent of product *equilibrium partitioning* into the solvent phase. Each of these can be analyzed simply for limiting behavior.

(*b*) For rate enhancement, consider the following network for penicillin (P):

$$P \xrightarrow{k_d} R \text{ (degradation product).}$$

$$P \underset{}{\overset{k_1}{\rightleftharpoons}} P_s \text{ (mass transfer to interface).}$$

$$P_s \xrightarrow{k_O} P_O \text{ (solubilization into organic phase)}$$

$$P_s \xrightarrow{k_2[A]} AHP \text{ (reactive extraction into organic phase)}$$

Write expressions for dp/dt and dp_s/dt. By applying the pseudo-steady state assumption to the intermediate, p_s (interface concentration), show that the p transfer rate (net) to the interface divided by the rate of p degradation is given by

$$\frac{\text{Net transfer rate}}{\text{Degradation rate}} = \frac{k_1}{k_d}\left(1 - \frac{k_d}{k_1 + k_O + k_2[A]}\right)$$

Sketch this function vs. amine concentration [A]. What transfer to degradation ratio obtains for conventional extraction ([A] = 0)? For dominant reactive extraction ([A] → large)?

(*c*) For equilibrium enhancement, only the network below needs consideration:

$$P \overset{K_O}{\rightleftharpoons} P_O$$

$$A + P_O \overset{K_R}{\rightleftharpoons} AP_O$$

Given volumes V_a and V_O for aqueous and solvent phases, respectively, show that the equilibrium ratio of total penicillin in aqueous phase to total penicillin in organic phase is given by $(V_a/V_O)/K_a(1 + K_R[A])$. Sketch this function vs. [A].

Data for degradation rates and equilibria for various extractive reactants for penicillin appear in M. Reschke and K. Schügerl, *Chem. Eng. Jl.*, **28**: B1, B11, 1984.

11.9 Recovery of intracellular vs. extracellular product For nongrowing cells, the time course of measured product concentration, intracellular and extracellular, are plausibly described by equations of the following form:

$$p_i(t) = p_{i\,\max}[1 - e^{-t/\tau}]$$

$$p_e(t) = \alpha[t - t_0(1 - e^{-t/t_0})]$$

(*a*) Sketch these functions.

(*b*) Suggest explanations, based on known microbial functions, for these time variations.

(*c*) Under what circumstances would you design a process to recover *intracellular* product or *extracellular* product?

(*d*) For each case in part (*c*), what kind of biocatalyst improvement research and development would you recommend?

(*e*) Sketch a process for each circumstance, showing principal operations needed to produce a solvent-extracted, crystalline product and a dried biomass by-product.

11.10 Whole broth antibiotic recovery Strong sorption of antibiotic (and perhaps other process solutes) changes the rate determining step in adsorption, resulting in an uptake-dependent mass transfer coefficient, k, given by a form of Eq. (11.34) for novobiocin:

$$k = 130\, e^{-23(q/q_{\max})} + 46\, e^{-3(q/q_{\max})} \quad (\text{h}^{-1})$$

(*a*) Consider a single tank contactor with continuous addition and withdrawal of antibiotic broth and continuous addition and withdrawal of resin beads. Here, the concentration driving force is a constant, $c - c^*$, and time is replaced by bead holding time τ in Eq. (11.33). For a single tank, integrate a form of (11.33) analytically to obtain $\tau = fcn(q)$. (This is most easily done by noting that

$dq/d\tau = (d\tau/dq)^{-1}$; i.e., let τ be the dependent variable. Integration via partial fractions is then straightforward.)

(*b*) In batch processing, a resin charge and broth volume would be added simultaneously, giving the conservation relation

$$V_r \frac{dq}{dt} = V_b \left(\frac{dc}{dt}\right) \tag{i}$$

and also a time changing value of c^*, given by $c^* = bq^a$ (Eq. 11.35). Using (i) and (11.35), repeat the procedure of part (*a*) to obtain the time required, t, as an integral of a function of the uptake achieved, q.

(*c*) Simple transient problems can often be conveniently solved by Laplace transform *if* the governing equations are linear. Is the set of Eqs. (11.32)–(11.35) for staged countercurrent processing amenable to solution by Laplace transform? Why (not)? (See P. A. Belter, F. L. Cunningham, and J. W. Chen, "Development of a Recovery Process for Novobiocin," *Biotech. Bioeng.*, **15**: 533, 1973.)

11.11 Capillary driven separations In paper and thin-layer chromatography, the carrier fluid velocity is not constant in time. The pressure differential due to the advancing solvent-solid interface in the paper capillaries is given by $\Delta p = 4\sigma \cos(\theta)/d_c$, where σ is surface tension, d_c is capillary diameter, and θ is fluid-solid contact angle. The viscous resistance to a flow rate m is proportional to μz (viscosity times length of liquid-filled portion of capillary) and inversely proportional to the capillary cross section ($\propto d_c^2$).

(*a*) Assuming m proportional to dz/dt, show that z^2 and t are proportional, where z is the position of solvent front.

(*b*) Prove that the distance between any two solute peaks also has the same $z{:}t$ dependence.

(*c*) To represent actual chromatographic supports more realistically, suppose that the normalized fraction of capillaries of diameter d varies about the size d_m of maximum frequency as $N(d) = A \exp[-(d - d_m)^2]$. Develop an expression giving the relative water content of the capillary material vs. time and distance.

(*d*) Ignoring other dispersion effects, derive an expression for the apparent peak-to-peak distance of any two solutes.

11.12 Chromatography In chromatography, each solute has an equilibrium partition coefficient $K_i = s_i/c_i$, where s_i is the adsorbed species concentration and c_i the bulk concentration. If a sample of width d at the injection point is to be separated into peaks of at least this peak-to-peak spacing, show that:

(*a*) The mean residence time of any single component i is given by

$$t_i = \frac{L}{u\varepsilon}(1 - \varepsilon)(1 + K_i)$$

where u is fluid velocity and L the column length.

(*b*) The required maximum column length L is the maximum value of

$$L = \frac{d(K_j - K_i)}{(1 + K_j)^2(1 + K_i)} \qquad \text{where } K_j > K_i$$

(*c*) In liquid chromatography, the diffusion coefficient is typically 10^{-6} cm^2/s for larger molecules (Table 11.5). If the chromatography column is packed with particles of 10^{-2} cm diameter, use the Einstein equation for the root-square (rms) distance traversed by a molecule in time t, $\langle z^2 \rangle^{1/2} = (Dt)^{1/2}$, to show that flow velocities larger than the order of 10^{-4} cm/s will require a longer column than indicated in part (*b*).

(*d*) Charm and Wong ("An Immunosorbent Process for Removing Hepatitis Antigen from Blood and Plasma," *Biotech. Bioeng.*, **16**: 539, 1974) have used an affinity-chromatography column to separate serum-hepatitis antigen from pooled blood plasma. For such specific separation $K_i \to \infty$ (i = antigen), $K_j \simeq 0$ ($j \neq i$). If the column is packed with nonporous beads of radius R and the antigen concentration is n particles per liter, show that the maximum number of blood volumes which can be cleared continuously of antigen by a single unit of column volume is given by

$$\frac{3(1-\varepsilon)}{n\pi r^2 R}$$

where ε is the column void volume and r the antigen radius.

11.13 Chromatography From Fig. 11.21 and a correlation from the data in Table 11.5, deduce the values of L and r_g for the three chromatographic materials of Fig. 11.21.

11.14 Protein precipitation via "salting-out" (*a*) For the solubility correlation of Eq. (11.7), determine the parameters β and K_s for each protein in Fig. 11.13.

(*b*) How would you use fractional precipitation to resolve a mixture of the proteins in this figure into nearly pure precipitates?

(*c*) The least effective ions for "salting-out" are actually useful for "salting-in" or dissolving a polymer, e.g., LiCNS solutions will dissolve silk stockings! Read a discussion of "salting-in" in your physical chemistry book, and briefly discuss the phenomena at work in "salting-in" and "salting-out."

11.15 Enzyme isolation efficiency Assuming a 50% yield in step 4 of Example 11.1, calculate the fraction of initial enzyme which is recovered in the final step.

11.16 Immunosorbent capacity The binding capacity C of some immunosorbent columns has been reported to diminish with adsorption/elution cycle number n according to $C(n) = C(0)e^{-\alpha n}$ with $n > 0$.

(*a*) Show that the total amount of protein which could be isolated by N cycles is

$$C(T) = \sum_{n=0}^{N-1} C(n)$$

(*b*) If α is small, the series in (*a*) can be approximated by an integral,

$$C(T) = \int_0^{(N-1)} C(0)e^{-\alpha n}\, dn$$

Show that the total amount of protein recoverable by the column is $C(0)/\alpha$.

(*c*) How would you decide how many cycles could be economically useful before the column was to be discarded?

11.17 Recycle of inhibitory feed component Recycle of water or of cells also brings recycle of solutes, some of which may be inhibitory. Suppose the specific growth rate is influenced by inhibitor I according to

$$\mu_i \propto \frac{1}{K_i + i}$$

(*a*) Consider a simple recycle reactor (Fig. 9.4) with dissolved feed inhibitor concentration $i_f \ll K_i$. Show that simple recycle has no effect on concentration of I in the reactor.

(*b*) Suppose now that the main product (e.g., ethanol) and water are removed continuously by distillation and that I is nonvolatile. What is the maximum allowable recycle ratio (F_r/F_0) if we cannot allow $\mu(i)/\mu(i=0) < 0.95$? (Example: A number of dissolved salts in cane molasses are inhibitory to *Saccharomyces cerevisiae* in ethanol production: B. L. Maiorella, H. W. Blanch, and C. R. Wilke, *Biotech. Bioeng.*, **26**: 1155, 1984.)

(*c*) Let I be an inhibitory product, produced by biomass at a rate equal to nx. Derive an expression for $(F_r/F_0)_{\max}$ in terms of n, α (cell enrichment ratio in separator), and the minimum allowable value $\mu(i)/\mu(i=0)$.

11.18 Kinetics of immobilized subtilisin in a two-phase reactor-recovery system Optically active D-arylglycines for use as side chains for semisynthetic penicillin and cephalosporins are preparable from a racemic mixture using immobilized subtilisin (protease). In reaction conditions, the racemic precursors are dissolved in an organic phase, which is then mixed with immobilized enzyme slurried previously in water. The sparingly soluble reactant is easily extracted and converted in the aqueous phase, liberating acid. Base titration thus provides a measure of reaction rate.

Conditions: 143.3 mmol DL-2-acetamido phenylacetic acid methylester dissolved in 250 mL methyl isobutyl ketone. 250 mL water added, along with immobilized catalyst.

Base added vs. time

mL 1N NaOH	0.	33	45	53	58	66	69	71	73
t (h)	0.	.5	1.0	1.5	2.0	3.	4.	5.	9.

(*a*) What rate form would you use to fit this data?

(*b*) For expensive precursor material, the yield from precursor is very important. Thus, the ability to predict time required for 95 vs. 90% yield may be crucial. How well does the above data (fail to) provide this information?

(*c*) Sketch a continuous process which would include high yield hydrolysis of racemate, recovery of product by aqueous crystallization, recycle of organic solvent, and evaporative recovery (for reracemization) from the solvent of the unhydrolyzed precursor. What fraction of the major equipment in your process is reactor- vs. separation-associated? (See H. Schutt et al., *Biotech. Bioeng.*, **27**:420, 1985.)

11.19 Cell affinity chromatography Bone marrow contains stem cells and mature T lymphocytes. The former can be constructively transplanted to a host with diseased marrow, in order to boost the number of healthy marrow cells with beneficial functions. The T lymphocytes may recognize the recipient host as "foreign," and thus initiate a harmful reaction against the host's tissues. As lymphocytes possess certain lectin-binding surface receptor sites, an immobilized lectin column may be used to provide T lymphocyte removal from bone marrow populations.

(*a*) Cell *adsorption* appears to depend on flow velocity, which determines both the frequency of cell-column particle collisions, v_p, and the average residence time of a cell, τ_p, if no binding occurs. Let v_p vary as u^{α}, and τ_p vary as $u^{-\beta}$, (α, $\beta > 0$). Show that the steady state coverage of cells θ_c is proportional to $u^{\alpha-\beta}$.

(*b*) Cell *binding* is suggested to depend on the formation of a critical number of cell-surface bonds during the cell's residence time τ on the surface. If bonds B are formed from lectin sites L and cell binding sites S according to the equation,

$$\frac{db}{dt} = k_f N_L N_S - k_r$$

where $L + S \rightleftharpoons B$ and, in the cell-surface contact area, there are N_{LO} and N_{SO} independent, diffusible sites of corresponding kind,

(i) Derive an expression for the equilibrium bond number b^*.

(ii) Integrate the rate equation for db/dt (use partial fractions method) to obtain $b(t)$ vs. t.

(iii) Let $b_{\text{crit}}(< b^*)$ be the number of bonds needed to irreversibly bind the T lymphocyte *cell* to the surface of the column. Derive a relationship between b_{crit} and the fastest fluid velocity which will lead to cell binding. (See C. M. Hertz, D. J. Graves, D. A. Lauffenberger, and F. T. Serota, "Cell Affinity Chromatography for Separation of Lymphocyte Subpopulations, *Bioch. Bioeng.*, **27**: 603, 1985.)

11.20 Membranes and maintenance for artificial organs The microencapsulation or entrapment in hollow fiber devices of mammalian cells has been proposed for artificial, implantable organs. The membrane porosity must allow outward diffusion of the desired polypeptides or protein product, yet must minimize outward diffusion of other nonhost proteins which could generate an antagonistic response in the host.

(*a*) Estimate the desired average pore size of membranes for artificial organs to release the following: insulin, human growth hormone.

(*b*) The encapsulated or fiber-entrapped cells are provided with nutrients from the host; these nutrients would diffuse into the immobilized cell mass to provide, ideally, sufficient food for maintenance and product formation. Taking the molecular diffusivity for oxygen to be $\mathscr{D}_{O_2} \sim 10^{-6}$ cm^2/s, estimate the immobilized cell particle radius which gives a zero order Thiele modulus of unity for immobilized growing yeast ($t_D = 2$ h); immobilized resting yeast (maintenance O_2 demand = 3 percent of growth demand), immobilized growing animal cells ($t_D = 40$ h), immobilized resting cell (maintenance O_2 demand = 10 percent of growth demand). Is immobilization of resting 10 μm mammalian cells feasible in 0.1 mm, 0.5 mm, and 2 mm microspheres?

REFERENCES

Downstream processing, product recovery

1. B. Atkinson and F. Mavituna, *Biochemical Engineering and Biotechnology Handbook*, Macmillan Publishers Ltd., Surrey, England, 1983 (chaps. 12–14 contain a wealth of specific examples).
2. P. Belter, "General Procedures for Isolation of Fermentation Products," p. 403 in *Microbial Technology, 2d ed., vol. 2*, H. J. Peppler and D. Perlman (eds.), Academic Press, New York, 1979.
3. V. Edwards, "Recovery and Purification of Biologicals," *Adv. Appl. Microbiology*, **11**: 159, 1969.

Cell and particle removal

4. S. Aiba and M. Nagatani, "Separation of Cells from Culture Media," *Adv. Biochem. Eng.*, **1**:3, 1971.
5. P. Belter, "Recovery Processes: Past, Present, Future," 184th Amer. Chem. Soc. Mtg., September, 1982.
6. *Fermentation and Biochemical Engineering Handbook*, H. C. Vogel (ed.), chaps. 5 (filtration) and 10 (centrifugation), Noyes Publications, Park Ridge, NJ, 1983.
7. U. Weissman and H. Binder, "Biomass Separation from Liquids by Sedimentation and Centrifugation," p. 119 in *Adv. Biochem. Eng.*, vol. 24, A. Fiechter, ed., Springer-Verlag, Berlin, 1982. (Sedimentation).
8. W. J. Weber, Jr., *Physicochemical Processes for Water Quality Control*, Wiley-Interscience, N.Y., 1972 (fundamental chapters on coagulation, sedimentation, and filtration).
9. S. Shirato and S. Esumi, "Filtration of a Culture Broth of *Streptomyces griseus*," *J. Ferment. Technol. (Japan)*, **41**: 87, 1963.
10. R. Gabler and M. Ryan, "Processing Cell Lysate with Tangential Flow Filtration," *Purification of Fermentation Products*, D. LeRoith et al. (eds.), *Amer. Chem. Soc., Symposium Ser.*, **271**: 1, 1985.
11. J. Zahka and T. J. Leahy, "Practical Aspects of Tangential Flow Filtration in Cell Separations," *Purification of Fermentation Products*, D. LeRoith et al. (eds.), *Amer. Chem. Soc. Symposium Ser.*, **271**: 51, 1985.
12. D. B. Todd and W. J. Podbielniak, "Centrifugal Extraction," *Chem. Eng. Progress*, **61**(5): 69, 1965.
13. S. Queener and R. Swartz, "Secondary Products of Metabolism," p. 35 in *Economic Microbiology, vol. 3*, A. H. Rose (ed.), Academic Press, London (1979) (solvent extraction of penicillin).
14. R. N. Shreve and J. Brink, *The Chemical Process Industries*, McGraw-Hill, New York, 1977 (chaps. on pharmaceuticals including adsorptive and extractive antibiotic recovery).
15. P. A. Albertsson, *Partition of Cell Particles and Macromolecules*, 2d ed., Wiley-Interscience, N.Y., 1960 (the first comprehensive presentation on 2-phase aqueous extraction systems).

Isolation and Purification

16. M. R. Kula, K. H. Kroner, and H. Hustedt, "Purification of Enzymes by Liquid-Liquid Extraction," *Adv. Biochemical. Eng.*, **24**: 103, 1982 (2-phase aqueous extraction of proteins).
17. U. Menge and M. R. Kula, "Purification Techniques for Human Interferons," *Enz. Microb. Technol.*, **6:** 101, 1984. (2-phase aqueous extraction and high pressure liquid chromatography).

18. J. Darbyshire, "Large Scale Enzyme Extraction and Recovery," *Topics in Enzyme and Fermentation Biotechnology*, **5**: 147, 1981, A. Wiseman (ed.), Halsted Press/J. Wiley, N.Y.
19. G. C. Grabender and C. E. Glatz, "Protein Precipitation: Analysis of Particle Size Distribution and Kinetics," *Chem. Eng. Commun.*, **12**: 203, 1981.
20. D. J. Bell, M. Hoare, and P. Dunnill, "Formation of Protein Precipitates and Their Centrifugal Recovery," *Adv. Biochem. Eng.* **26**: 1, 1983, A. Fiechter (ed.), Springer-Verlag, Berlin/N.Y.
21. E. Flaschel, Ch. Wandrey, and M. R. Kula, "Ultrafiltration for the Separation of Biocatalysts," *Adv. Biochem. Eng.*, **26**: 73, 1983, A. Fiechter (ed.), Springer-Verlag, Berlin, N.Y.
22. W. C. McGregor, "Large Scale Isolation and Purification of Proteins from *E. coli*," *Ann. N.Y. Acad. Sci.*, **413**: 231, 1983.
23. R. G. Schoner et al., "Isolation and Purification of Protein Granules from *E. coli* Cells Overproducing Bovine Growth Hormone," *Bio/Technology*, **3**: 151, 1985.
24. M. K. Joustra, "Gel Filtration on Agarose Gels," *Prog. in Separation and Purification*, vol. 2, p. 183, Wiley-Interscience, N.Y., 1969.
25. P. L. Dubin, "Aqueous Size Exclusion Chromatography," *Separation and Purification Methods*, **10**(2): 287, 1981.
26. D. J. Graves and Y. T. Wu, "The Rational Design of Affinity Chromatography Separation Processes," *Adv. Biochem. Eng.*, **12**: 219, 1979, A. Fiechter (ed.), Springer-Verlag, Berlin, N.Y.
27. J. W. Eveleigh, "Practical Considerations in the Use of Immunosorbents and Associated Instrumentation"; *Affinity Chromatography and Related Techniques*, T. C. J. Gribneau et al. (eds.), Elsevier, Amsterdam, 1982.
28. H. A. Chase, "Affinity Separations Using Immobilized Monoclonal Antibodies," *Chem. Eng. Sci.*, **39**: 1099, 1984.
29. F. H. Arnold, J. J. Chalmers, M. S. Saunders, M. S. Croughan, H. W. Blanch, and C. R. Wilke, "A Rational Approach to Scale-Up of Affinity Chromatography," *Purification of Fermentation Products*, D. LeRoith et al., (eds.), Amer. Chem. Soc. Symp. Ser. **271**: 113, 1985.
30. G. Sofer, "Chromatographic Removal of Pyrogenic Material," *Bio/Technology*, **2**: 1035, 1984.
31. J. C. Janson, "Large Scale Affinity Purification—State of the Art and Future Prospects," *Trends in Biotechnology*, **2**(2): 31, 1984.
32. B. Maiorella, C. R. Wilke, and H. Blanch, "Alcohol Production and Recovery," *Adv. Biochemical. Eng.*, **20**: 43, 1981 (ethanol recovery processes and economics).
33. G. M. Whitesides et al., "Magnetic Separations in Biotechnology," *Trends in Biotechnology*, **1**(5): 144, 1983.
34. N. C. Beaton and H. Steadly, "Industrial Ultrafiltration," *Recent Developments in Separation Science*, **VII**: 1, 1982.
35. I. H. Smith and G. H. Pace, "Recovery of Microbial Polysaccharides," *J. Chem. Tech. Biotechnol.*, **32**: 119, 1982.
36. N. M. Fish and M. D. Lilly, "The Interactions between Fermentation and Protein Recovery," *Bio/Technology*, **2**: 623, 1984
37. P. A. Belter, F. C. Cunningham, and J. W. Chen, "Development of a Recovery Process for Novobiocin," *Biotech. Bioeng.*, **15**: 533, 1973.
38. T. Vermeulen, "Adsorption and Ion Exchange," Section 16 in *Chemical Engineers' Handbook*, 5th ed., R. H. Perry and C. H. Chilton (eds.), McGraw-Hill, New York, 1973.
39. F. H. Arnold, H. W. Blanch, and C. R. Wilke, "Analysis of Affinity Separations," *The Chem. Eng. Jl.*, **30**: B9, B25, 1985.

CHAPTER

TWELVE

BIOPROCESS ECONOMICS

In this chapter, we examine the continuing role which economics plays in bioprocess research, development, and commercialization. The first section introduces general development phases, then illustrates them with a reasonably complete fermentation example which begins with a process flow concept, works through process equipment sizing, materials and utility needs, costs of initial plant and of operations, and an estimate of profitability (return on investment). Subsequently, characteristic features of particular fermentation processes are discussed, including fine chemicals, bulk chemicals, and single cell protein. In these examples, the relative cost importance of substrate feedstocks, equipment, utilities, and bioreactor vs. recovery sections will be examined, since identification of major cost areas frequently pinpoints process economic weaknesses and thereby indicates whether an engineering process improvement and/or strain development would be most logical to pursue in process optimization.

Circumstances peculiar to each general product area will also be noted. For example, the expenses for new drug development (fine chemical) must also include costs of obtaining clinical evidence for and sales approval by the Food and Drug Administration. Bulk chemicals from fermentation involve sufficiently large scale operation that a market for the incidental biomass must be found for by-product credit, or a waste disposal cost is incurred. Ethanol from biomass processes include not only fermentation and recovery sections but may also require a substantial pretreatment process to hydrolyze and solubilize the biomass components. Moreover, different countries (including the United States) have developed various forms of tax subsidies or credits for ethanol plants: such devices have a clear impact on process economics. Biological waste treatment provides the main process example for which substrate conversion (rather than biomass production or product formation) is the operating goal; here, biomass (cell sludge) disposal is responsible for a major fraction of plant operating costs.

12.1 PROCESS ECONOMICS

Economic considerations play a continuing role at nearly every stage of a plant design project (Table 12.1). For a project which eventually achieves commercialization, these stages include inception of process or product idea, a preliminary evaluation of economics and markets, the development of any additional data needed for final detailed design, a final complete economic evaluation in light of all data, creation of a detailed engineering design, procurement of site and equipment, construction of buildings and process, process start-up and trial runs, and regular production operation.

Inception may arise from any source: a sales department suggestion, emergence of a competing product, a customer request, an offshoot of a current process, or a new idea from research and development. The idea may come internally from within the (eventual) production company and may be patented by the producer, or it may arise through outside patents and achieve realization through royalty and licensing agreements.

In the long run, each project is expected to recover its costs and return an appropriate profit. Thus, an immediate need arises, given a process or product inception, to carry out a *preliminary evaluation of economics and market.* The latter determines the potential market size for various assumed product prices. The profitability, or lack thereof, is determined from the economics evaluation which includes estimates of costs incurred to meet the assumed market size and required price.

If the preliminary evaluations are promising, the *development of data necessary for final design* normally follows. This development includes both market

Table 12.1 Stages in plant design project

1. Inception
2. Preliminary evaluation of economics and market†
3. Development of data necessary for final design†
4. Final economic analysis
5. Detailed engineering design
6. Procurement
7. Construction
8. Start-up and trial runs
9. Production

† For products requiring FDA (or USDA) approval, clinical test and FDA (or USDA) application and approval are included here. (see Sec. 12.2)

and cost refinement. A full market analsyis is made, and prospective customers often receive final product samples to determine if the product is satisfactory and thus if the sales potential assumed is realistic. The cost studies include development of a complete plant process flowsheet, determination of corresponding capital and operating cost, and estimation of process profit each year over the assumed lifetime or pay-out period. These latter results would also include a simple sensitivity analysis to establish the key assumed cost and performance measures which most strongly affect the profitability calculations, since costs may change significantly before a process is realized.

Prior to a final detailed design, a management review provides a *final economic analysis*, in which the project is set within a larger picture for consideration. Questions asked here often include the following: How does the estimated profitability compare with similar estimates for other possible projects which are also competing for a limited pool of corporate investment capital? Does the proposed process build upon existing corporate technical manpower and marketing strengths, or is a new path indicated, requiring new personnel? Is the process in an area which is likely to develop related processes, or is the area singular? Will the proposed process have the flexibility to be used in other related projects being considered or anticipated, or is the process (and the personnel) singular? What are the anticipated growth directions for the company as a whole, and does the process continue an established strength, develop technological experience and products in an anticipated growth direction, or spread activities too thinly? If this review is positive, then funds are committed for the next several steps.

The *detailed engineering design* phase provides complete determinations of all information needed for subsequent plant construction: equipment sizing and specifications, controls, services, piping configurations, and final price quotations. The detailed design includes production of a complete construction design: elevation drawings, plant-layout arrangements, and any other information needed for actual plant construction.

Procurement includes purchase and receipt of land and equipment not previously on hand, and arrangement of permits for all anticipated activities: construction, operation, utility installment and hookups, discharges (air, water), sewer hookups, etc. Note that appreciable lead times will be necessary, especially for specialty or custom fabricated equipment.

The *construction* phase involves erection of the complete plant. Following this phase, *start-up and trial runs* are undertaken, including break-in and trials of the plant and its units, operator training, and establishment of start-up, operation, shutdown, and safety procedures. Full *production* is the final subsequent phase.

Repeatedly throughout all phases, economic assumptions are revisited and examined to see if the anticipated market, available raw materials, corporate goals, FDA or USDA approvals, or other important factors have shifted in a manner which significantly alters the economic prospects of the project. Since such evaluations may be demanding and complex, a standard objective method is typically used, e.g., a return-on-investment or related criterion. Objectivity aids in obtaining the clearest picture: a pessimistic or excessively conservative approach may cause abandonment of a project which may be the key to future corporate development; an overly optimistic approach can lead to expenditures which, in the form of a complete final plant, could be disastrous.

12.2 BIOPRODUCT REGULATION

Involvement of government in bioproduct regulation stems from legislated agency responsibility for worker health and safety, consumer safety, environmental statutes applicable to biotechnology, and specific regulations concerning newer aspects of biotechnology (recombinant DNA in particular). Appreciation of the timing needed for regulatory approvals allows proper inclusion of costs for such approvals (including clinical trials where necessary) in overall process economics.

The three most important U.S. agencies in regulation are the Food and Drug Administration (FDA), the U.S. Department of Agriculture (USDA), and the Environmental Protection Agency (EPA).

Regulation of drugs, biologics, food and food additives, and diagnostics fall in the FDA's domain. New drug approval, involving animal and human tests, may take several years. The "new drug" category is important because current (1984) interpretation regards each drug made by rDNA technology, even if apparently identical to existing drugs, as "new," and thus requiring approval as effective and safe. A human biologic is "a vaccine, therapeutic serum, toxin,

Table 12.2 U.S. Agencies involved in biotechnology regulation†

1. FDA
   (*a*) Office of New Drug Evaluation (all new drugs require animal and human testing; any rDNA drug products are all "new").
   (*b*) Office of Biologics: clinical trials required, followed by licensing of both product and producer.
   (*c*) National Center for Devices and Radiologic Health: regulates medical devices and medical diagnostics (including monoclonal antibody diagnostics, except those related to blood bank operation). Monoclonal antibody (MAb) systems currently must demonstrate performance equivalence to prior product, rather than safety and efficacy.
2. USDA
   Primary regulatory authority over animal biologics, with FDA secondary. MAb and rDNA open grey areas of agency responsibility uncertainty (e.g., USDA regulates interstate marketing, but FDA may regulate intrastate conditions.)
3. EPA
   Authority "to identify and evaluate potential hazards and then to regulate the production, use, distribution, and disposal of these substances, including chemicals, herbicides, and pesticides."
4. OSHA (Occupational Safety and Health Administration.)
   Responsible for worker health and safety regulation. As of 1984, no bioprocess standards issued, nor regulatory stance toward biotechnology declared.

† "Commercial Biotechnology: An International Analysis" Office of Technology Assessment, January 1984, pp. 359–365.

antitoxin, or analogous product for the prevention, treatment, or care of diseases or injuries." Both the product and producer must be FDA-licensed, and product approval may be on a lot-by-lot basis. Similar restrictions apply to other products. A brief agency summary is presented in Table 12.2.

12.3 GENERAL FERMENTATION PROCESS ECONOMICS

While considerable differences clearly exist between different fermentations, the plant design is most easily discussed from a common flow sheet, given in Fig. 12.1. This simple diagram indicates that the process is largely divisible into a *fermentation* (or bioreaction) section, and a *product recovery* section. Upstream is a section for receiving and storing raw materials, and downstream is final product preparation (formulation), packaging, and shipping. The individual section operations may each need one or more utilities (water, gas, electricity), and environmental circumstances including process effluent disposal, siting, and zoning must be addressed in a full process design.

Any bioprocess must be developed around a particular idea (the inception) for a product. For any given application, a more detailed process flow sheet may be prepared. To assure appropriate levels of completeness at such a stage, the specifications (if only for conceptual evaluation purposes) should be provided

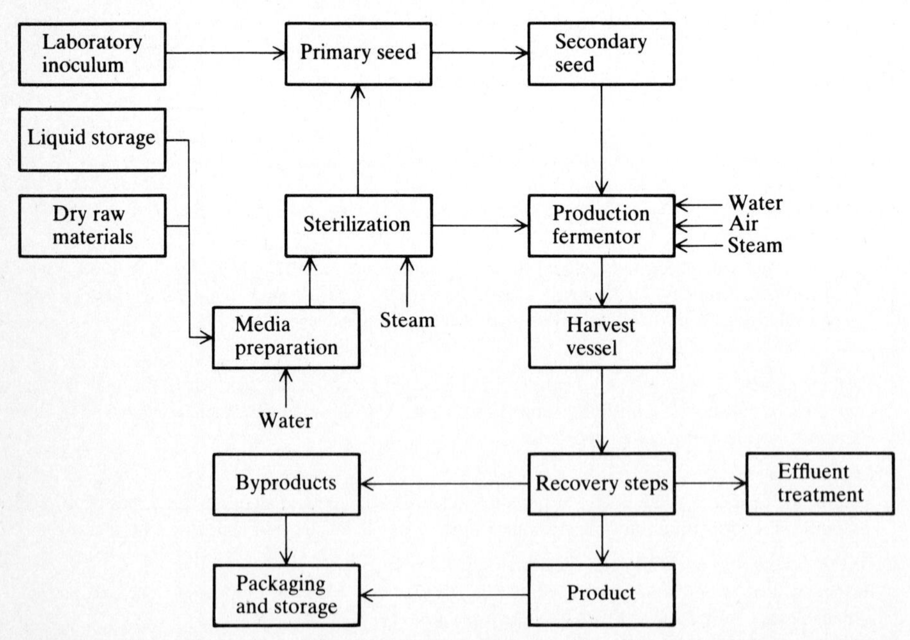

Figure 12.1 Generalized process flow sheet. *[By permission from W. H. Bartholomew and H. B. Reisman, "Economics of Fermentation Processes," Microbial Technology, 2d ed., vol. 2, H. J. Peppler and D. Perlman (eds.), Academic Press, 1978, p. 481.]*

for the following items: major substrate, other raw materials, type of reaction, principal products and by-products, organism utilized, agitation and aeration requirement, special operations desired or required for bioreactor operation or for separation processes, and process control requirements.

For preliminary economic analysis, it is convenient and fruitful to utilize costs from relatively standard sources; in a subsequent detailed cost exercise, all particulars must be evaluated as well. Cost analyses are typically performed on three bases: CAPITAL costs (bioreaction and product recovery *equipment*), PRODUCTION costs (salaries, supplies, maintenance, taxes, depreciation, etc.), and MATERIALS and UTILITIES costs (feedstock nutrients, additives, packaging agents and materials, electrical, gas, steam, and water needs).

Approximate costs for equipment commonly used in the fermentation industries are given in Table 12.3. Prices change with time; the Table 12.3 entries concern 1977 estimates obtained from appropriate manufacturers in the continental Unites States. The scaling of process equipment costs with time is conveniently accomplished (for preliminary cost estimation) with the aid of two cost indexes (the Engineering and News Record construction cost index, and the Marshall and Swift Equipment (MSE) cost index). These indices may be used to scale process equipment cost estimates made in a given year to other points in time. Given the size of process equipment needs for any specific process conception, 1977 costs may determined from Table 12.3, and the MSE index used to

Table 12.3 Typical equipment costs (1977)

Fermentors (stainless (304) with coils, baffles, 2.75 atm rating)	
15 m^3	\$35,000
113 m^3	\$95,000
225 m^3	\$165,000
Holding tank (304 stainless steel)	
190 m^3 silo	\$37,000
Installed, with instrumentation	\$60,000
Spray dryers (304 stainless steel)	
Nominal water removed, kg/h	Installed cost
1800	\$600,000
3200	\$1,250,000
4500	\$1,600,000

(Complete with fans, ducts, power lines, steam preheater, gas burner, dryer chamber, wet collection system, bagging system, buildings.)

[By permission, from W. H. Bartholomew and H. B. Reisman, "Economics of Fermentation Processes," Microbial Technology, 2d ed., vol. 2, H. J. Peppler and D. Perlman (eds.), Academic Press, New York, N.Y., 1978, p. 463.]

Table 12.4 Approximate production costs

Production costs

Depreciation: 6.7–10% of capital (10–15-year straight line)
Taxes and insurance: 2–6% of capital (varies with location)
All salaries except maintenance supervision: 70–100% of direct labor (administration, supervisory, clerical, technical, engineering).
Indirect labor: 30–50% of direct labor (includes control and quality labs, yard, materials handling, cleanup, etc.)
Associated payroll costs: 20–40% of payroll
Supplies: 20–30% of direct labor
Other plant overhead: 50–70% of direct labor (telephone, outside services, rentals, tools, travel, freight, dues, legal, medical, professional services, sewer charges, etc.)
Maintenance: 4–8% of capital (salaries, labor and associated payroll are 50% of total)

General cost factors

Sales/Administration/R & D	10–15% of net sales
Provision for taxes	50% of profit
Start-up expenses	5–10% of capital
Working capital	20–30% of net sales

[By permission, from W. H. Bartholomew and H. B. Reisman, "Economics of Fermentation Processes," Microbial Technology, 2d ed., vol. 2, H. J. Peppler and D. Perlman (eds.), Academic Press, New York, N.Y., 1978, p. 486.]

update cost estimates to current years. Current price data from manufacturers is preferred, however. A specific process example is considered shortly.

Production costs tend to be relatively independent of process details when the costs are expressed as percentages of appropriate costs. Table 12.4 presents typical production and general cost factors for cost estimation purposes. Given total capital requirements and associated direct labor costs (both specific to a given design), the corresponding production costs are available from Table 12.4 factors.

With the availability of capital, direct labor, production, and materials/utilities costs, a profitability analysis (return on investment) may be made. In the following section, we consider a complete specific example. A final subsequent section reviews specific economically important items related to general bioprocess classes: fine chemicals, bulk chemicals, single cell protein, and biological methane production.

12.4 A COMPLETE EXAMPLE

For illustration of a fermentation product which falls in an intermediate-sized fermentation chemical range, we consider a hypothetical example (of a microbial control agent for agricultural usage) (Ref. 1) which is characterized by the following:

1. Preliminary market research suggests that 15–25 percent of the product market could be available at a selling price of $2.50 per kilogram of product.
2. Given 4.5 to 7.5 million hectares of the assumed product-related crop, the potential market is 13–22 million kilograms of product each year.

The potential process size is now estimated:

A large, but convenient-sized fermentor is assumed: 225 m^3. The operating volume is 180 m^3 of liquid. If the fermentor is on-stream 97 percent (including turnaround) of the total available time, then the laboratory kinetics information is used to estimate production of 20,000 kg per batch (at a standardized activity per kilogram of product).

For a production per year of 17,000,000 kg and a production per batch of 20,000 kg (estimated from lab data), we require $(17 \times 10^6/20 \times 10^3) = 850$ batches per year. With a 40-h (1.67-day) cycle time (including a 4-h turnaround), a single fermentor can provide $365/(5/3)$ day $= 216$ batches per year. Thus, the number of 225 m^3 fermenters needed is $850/216 \approx 4$.

With this process basis established, a materials usage and cost table is constructed by scaling laboratory data and by considering prior or published experience with raw materials and product formulation ingredients. The results are illustrated in Table 12.5. Notice that these costs are determined by simple stoichiometry calculations, starting from lab data on fermentation product yield from a given quantity of nutrients, and from formulated product composition (active ingredient (fermentation product), diluent, stabilizer, sticker, anticaking agent, etc.); in other words, these costs are independent of the internal process structure chosen for the fermentation and recovery operations.

The development of a particular process design which can take the materials inputs of Table 12.5 and yield the assumed product rate [17,000,000 kg/year from 4 main fermentors] is now considered in three steps of economic interest:

1. A particular sequence of unit operations is set out to provide the plant *size* needed; from this process flow diagram the total *capital equipment costs* can be calculated from the data of Table 12.3. From literature guidelines, and from past operating experience, we calculate the manpower needed to operate the major equipment, leading to a *direct labor cost* calculation.
2. From the process flow sheet which indicates the explicit equipment sizes and number, the individual needs for steam, process water, wash water, tower (cooling) water, sterilized air, electricity, natural gas, or other fuel are identified. Their summation over all process components provides a *utilities* (steam, water, fuel, electricity) *cost* estimate. (From this same detailed flow sheet, a point-by-point indication of materials feed rates and product, by-product, and waste product flows is constructed; a final detailed scheme must indicate usage and/or disposal costs of all process exit streams.)

Table 12.5 Materials usage and cost

Basis

Production fermentor volume	225 m^3
Operating volume	180 m^3
Number of fermentors	4
Stream factor	97%
Cycle time (including 4-h turnaround)	40 h
Production: batch	20,000 kg (standard activity)
Batches per year	850
Production per year	17,000,000 kg

Raw materials

	kg/batch	Unit cost, $/kg		Cost, $1000
Fermentation				
Molasses	11,475	0.07		683
Adjuvant solids	3,668	0.30		935
Miscellaneous salts, etc.	922	0.70 (avg.)		549
			Subtotal:	2167
Formulation				
Diluent	9100	0.30		2320
Stabilizers	100	2.00		170
Sticker	400	0.50		170
Anticaking agent	400	0.24		82
			Subtotal:	2742
Packaging				
	25 kg/bag	0.40/bag		272
			Total:	$5181

[By permission, from W. H. Bartholomew and H. B. Reisman, "Economics of Fermentation Processes," Microbial Technology, 2d ed., vol 2, H. J. Peppler and D. Perlman (eds.), Academic Press, New York, N.Y., 1978, p. 463.]

3. With the *direct labor* costs and total *capital* costs (fermentation process equipment, utilities and materials and product storage as needed, engineering and construction costs, and contingency allocation) available, the appropriate *production cost estimate* is prepared from the guidelines of Table 12.4.

With the available costs for direct labor, equipment, materials, utilities, and production, a profitability analysis is then carried out for the base case assumed above, and for a group of related cases where the initial result is tested for sensitivity to major variables (selling price, market volume, capital cost, plant size).

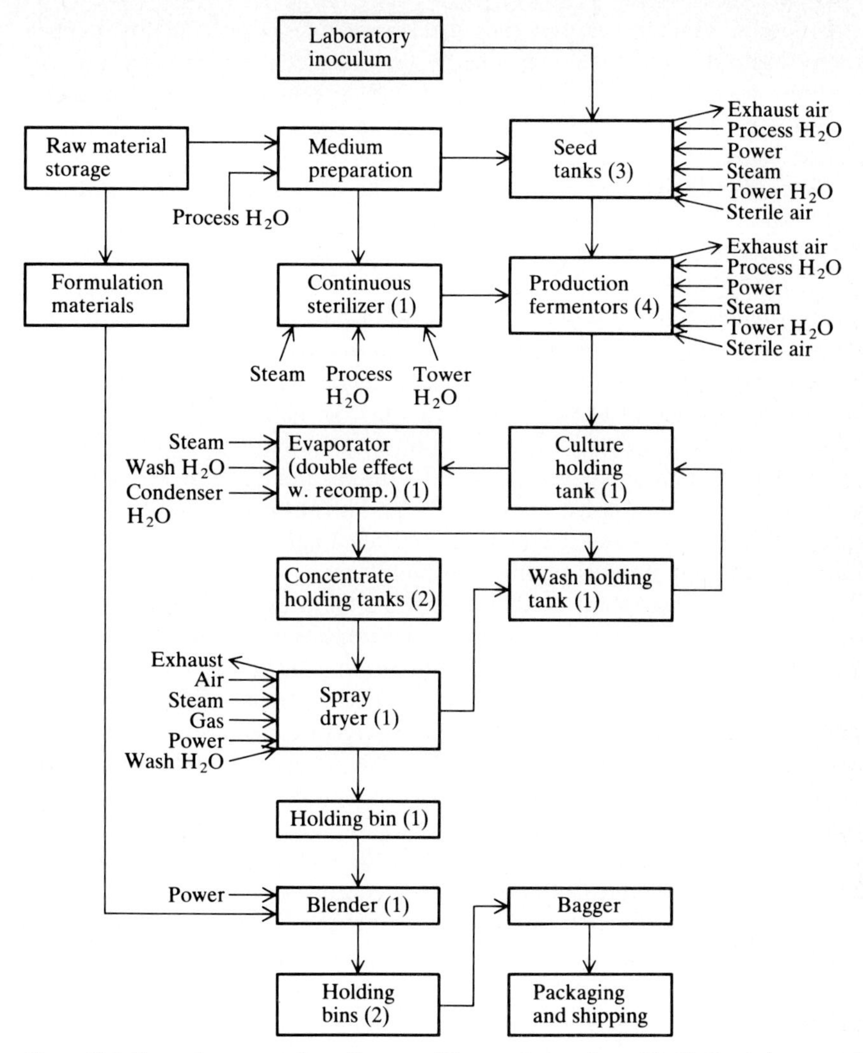

Figure 12.2 Example process flow diagram. *[By permission, from W. H. Bartholomew and H. B. Reisman, "Economics of Fermentation Processes," Microbial Technology, 2d ed., vol. 2., H. J. Peppler and D. Perlman (eds.), Academic Press, 1978, p. 488.]*

We now consider specific illustrations of a process flow sheet, and estimation of the required capital, direct labor, utilities, and production costs, and of return of investment.

A *process flow diagram* (Fig. 12.2) for the example under consideration is simply an appropriately detailed version of a general fermentation process (Fig. 12.1). This particular example must identify the flow sequence of operations, the number of identical units and types of utilities needed for each unit operation.

The steam, air, water, waste gases (CO_2, H_2O), and solid (biomass) waste rates are closely keyed to the materials processing rates. Consequently, explicit utility and disposal needs may conveniently be indicated on a detailed material balance flow sheet (Fig. 12.3). Here each individual unit operation has specific mass flow and compositions indicated for all streams, whether regarded as raw or formulation materials, utilities (water, steam, air, gas), product, or waste. Electricity needs are determined from power needs of the individual units, typically by direct consultation with equipment manufacturers.

The capital investment estimate for the major equipment of the process in Fig. 12.3 is presented in Table 12.6. The allowances for instrumentation, process piping, and buildings for the fermentation and processing (product recovery) sections are estimated as standard percentages of major capital equipment. Ultimately, all equipment costs are expressed on an installed basis. The total major equipment costs for the process in Fig. 12.3 may be described as approximate percentages of total fixed capital: fermentation capital (61 percent), product recovery and formulation (26 percent), and utilities and tankage (13 percent). Estimated engineering and construction costs, and a contingency fund, are taken as 20 percent each of the total fixed capital (installed equipment costs plus associated instrumentation, process piping, and buildings).

The size of the utility plant needed to support the proposed process is calculated from summing the individual utility consumption rates indicated in Fig. 12.3; the result is given in Table 12.7. The processing section consumes 84 percent of the steam, 73 percent of the electricity, and 100 percent of the natural gas, while the fermentation section accounts for 100 percent of the process water, and 84 percent of the tower (cooling) water. Patently, energy optimization of the overall process is keyed to the processing (fermentation product recovery) section. Since nearly all of the utility requirements of this process section are for removal of water (concentration, evaporation), the continual fermentation process concern with development of new microbial strains to *increase broth product concentrations* is easily understood. The economic implications of this central theme were cited earlier in the separations chapter. We note the ubiquitous miscellaneous and contingency percentage (33 percent) used. The total utility needs are summed for each type, and the cost of an installed utility plant of appropriate capacity is indicated on the bottom line. Note that the assumed size of the installed plants is larger than the estimated value (including contingency) by from 14 percent (tower water) to as much as approximately 50 percent (steam, electrical plants), presumably representing allowance for flexibility of process and/or of the size of available commercial utility units.

The total capital investment is estimated to be \$20,484,000 (exclusive of land) for a plant utilizing $4 \times 230 = 920\ m^3$ (9.2×10^5 liters) of installed fermentation capacity. With \$1,550,000 for plant utilities, these two costs correspond to approximately \$24 per liter of installed capacity. This result may be compared with an order of magnitude fermentation plant citation of \$20 to \$50 per liter of installed capacity at 1977 prices. Assuming an average cost increase of 10 percent per year since 1977, the 1986 cost range would be 135 percent higher, or \$47 to \$118 per liter of installed capacity.

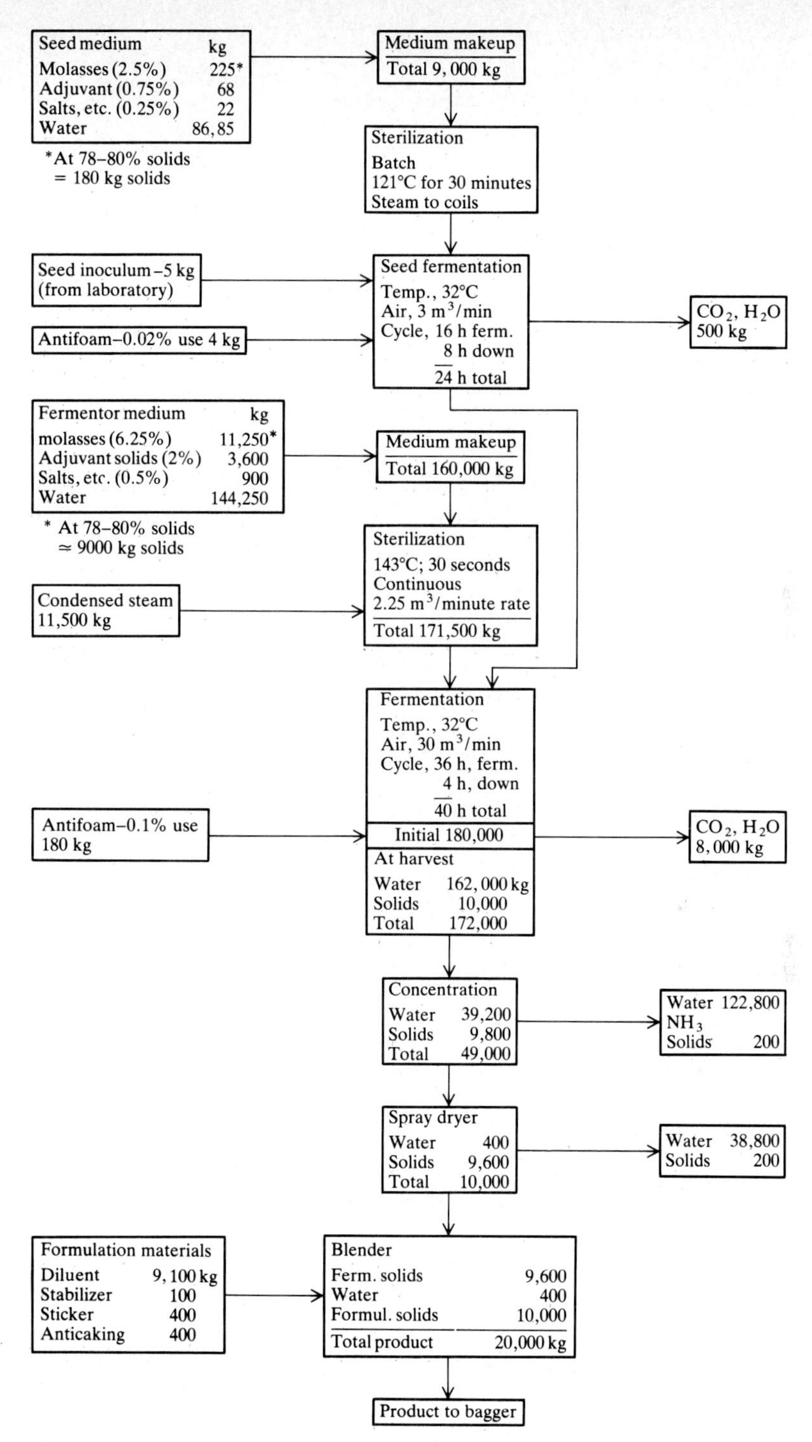

Figure 12.3 Material balance flow sheet. *[By permission, after W. H. Bartholomew and H. B. Reisman, "Economics of Fermentation Processes," Microbial Technology, 2d ed., vol. 2, H. J. Peppler and D. Perlman (eds.), Academic Press, 1978, p. 489.]*

Table 12.6 Capital investment estimate—cost (in thousands)

Fermentation	#	Size	Unit	Extended/Inst.
Formulation tanks	2	38 m^3	\$15	\$30
Agitator	2	3.75 kW	4	8
Continuous sterilizer	1	2.25 m^3/min	100	100
Compressors	2	85 m^3/min	225	450
Antifoam system	1		100	100
Seed tanks	3	15 m^3	35	105
Agitator	3	19 kW	14	42
Production fermentors	4	225 m^3	165	660
Agitator	4	150 kW	70	280
Holding tanks	2	190 m^3	40	80
Pumps				150
Separators				100
Special equipment				
		Major equipment subtotal		\$2,105
Installed costs: (2.5 × major equipment)				\$5,263
Instrumentation: (10% of major equipment))				526
Process piping: (20% of major equipment)				1,053
Buildings: mixed outdoor/indoor: (40% of major equipment)				2,105
		Fermentation capital total		\$8,947

Processing	#	Size	Unit	Extended	Installed
Evaporator	1	18,000 kg/h	\$400	\$400	\$600
Holding tanks	3	75 m^3	24	72	180
Holding bins	3	75 m^3	24	72	180
Blender	1	15 m^3			75
Vibrating screen	1	4 m^2, 3-deck			50
		Process equipment subtotal:			\$1,085
Instrumentation: 10% of process equipment					109
Process piping: 20% of process equipment					217
Buildings: indoor construction (75% of process equipment)					814
		Subtotal			\$2,225
Spray dryer	1	4,500 kg/h	1070	1070	\$1,600
		Process capital total			\$3,825

Table 12.6 *(continued)* **Capital investment estimate—cost (in thousands)**

	#	Unit	Extended	Installed
Utilities: (major addition at site at 20% of installed major equipment)				\$1,590
Molasses storage tanks	2	2800 m³		270
		Subtotal:		\$1,860
Total fixed capital				\$14,632
Engineering and construction (20% of total fixed capital)				\$2,926
Contingency: average, 20% (process subject to change)				\$2,926
Total capital investment (exclusive of land)				\$20,484

Notes: Instruments, piping, buildings, utilities basis Chilton factors (1949). Engineering and contingency: Chilton (1949)

[By permission, from W. H. Bartholomew and H. B. Reisman, "Economics of Fermentation Processes," Microbial Technology, 2d ed., vol. 2, H. J. Peppler and D. Perlman (eds.), Academic Press, New York, N.Y., 1978, p. 463.]

Table 12.7 Fermentation plant utilities

	Steam, kg	Process water, m^3	Tower water, m^3	Electrical, kWh	Gas, m^3
Seed	7,000/batch	9	300	300	
Fermentor	25,000	147	6,000	5,700	
Concentration	53,000	....	1,200	650	
Dryer	120,000	....	......	5,400	1800
Compressed Air	....	....	......	9,250	
Blender	....	....	......	900	
Subtotal	205,000	156	7,500	22,200	1800
Miscellaneous and contingency (33%)	68,000	52	2,500	7,325	600
Total/batch	273,000	208	10,000	29,525	2400
Per day	636,000	484	23,300	68,800	5600
Installed	1,000,000	....	27,000	100,000	
Installed† costs (\$1000)	600	100	450	373	25

† (estimated).

[By permission from W. H. Bartholomew and H. B. Reisman, "Economics of Fermentation Processes," Microbial Technology, 2d ed., vol. 2, H. J. Peppler and D. Perlman (eds.), Academic Press, New York, N.Y., 1978, p. 463.]

Table 12.8 Direct labor requirements

	Per shift
Fermentation	3
Evaporator	1
Dryer	1
Blending	1
Packing and shipping	2
Total/shift	8
Total/4 shifts†	32
Wages/operator	\$16,000/year†
Total direct labor	\$512,000

† 168 hours/week ÷ 4 = 42 hours/week/operator. Each operator thus has the equivalent of 42 hours straight time pay built into the schedule. (1978 rates.)

[Reprinted by permission from W. H. Bartholomew and H. B. Reisman, "Economics of Fermentation Processes," Microbial Technology, H. J. Peppler and D. Perlman (eds.), 2d ed., vol. 2, Academic Press, Inc., New York, N.Y., 1979, p. 463.]

The *direct labor* requirements are estimated from plant operating experience (Table 12.8). The *production cost estimate* for the specific example of Fig. 12.3 and 12.4 is presented in Table 12.9. The relative contributions to production costs are seen to be raw materials (37.0 percent), payroll (15 percent), supplies and expenses (5.3 percent), utilities (22.3 percent), and other costs (20.3 percent); the total production cost estimate is \$0.82 per kilogram for a product with an expected selling price of \$2.50/kg to the buyer.

The final data needed is a simple *return on investment* (ROI) estimate (or any equivalent economic figure-of-merit analysis). The straightforward 14-line accounting scheme of Table 12.10 indicates that the 10-year average net earnings after taxes (weighted sums of lines 10 and 11) divided by the sum of working capital (line 12) and original fixed capital investment (line 13) gives a 21.2 percent return on investment. The *sensitivity* of the calculated ROI to central assumptions used in developing the estimated costs is tested by 20 percent variation of selling price (downward), market volume (reduced), capital costs (overrun = upward), and plant size (downward). Given a particular ROI which is regarded as a typical dividing line between an acceptable case vs. a result which is marginal or worse, the worth of both the original ROI base case analysis (column *A*, Table 12.10) and of the simple sensitivity analysis indicated in columns *B*–*E* is evident.

The end results of these estimation schemes are eight simple tables and two diagrams which constitute the completion of data necessary for a final design

Table 12.9 Production cost estimate

Raw materials	Cost, in thousands
Fermentation	\$2167
Formulation	2742
Packaging	272
Subtotal	\$ 5,181
Payroll charges	
Direct labor	\$ 512
Indirect labor (40% of direct labor)	205
Salaried payroll (85% of direct labor)	435
Associated payroll costs (30% of payroll)	346
Maintenance salaries, labor and associated (3% of capital)	615
Subtotal	\$ 2,113
Supplies and expenses	
Maintenance (3% of capital)	\$ 615
Supplies (25% of direct labor)	128
Subtotal	\$ 743
Utilities	
Steam: \$8.50/100 kg × 273 × 850	\$1942
Electricity: \$0.035/kWh × 30,000 × 850	893
Water: \$0.008/m^3 × 10,000 × 850	68
Gas: 2400 m^3 × \$0.09/m^3 × 850	184
Subtotal	\$ 3,117
Other costs	
Depreciation (12 years)	\$1708
Taxes and insurance (4% of capital)	820
Other plant overhead (60% of direct labor)	307
Subtotal	\$ 2,835
Total	\$13,989
	\$0.82 per kg

[Reprinted by permission from W. H. Bartholomew and H. B. Reisman, "Economics of Fermentation Processes," Microbial Technology, H. J. Peppler and D. Perlman (eds.), 2d ed., vol. 2, Academic Press, Inc., New York, N.Y., 1979, p. 463.]

(stage 3, Table 12.1). At this point, the results obtained are typically moved to management consideration for a final economic analysis (stage 4, Table 12.1).

A brief aside is appropriate here regarding the interaction between management and engineering. The nature of management planning (or, for that matter, any human essay on long- or short-range planning) is illustrated by a summary in Fig. 12.4. The economic and process analysis materials presented in this chapter are part of the information flow which management must consider. Telling features in Fig. 12.4, however, are the *two* feedback loops on management review of processes. While the first (short-range) will be dominated by quantitative analyses of the type in Tables 12.4–10 and Figs. 12.1–4, the second is strongly

Table 12.10 Return on investment—10-year basis

	Value (dollars in thousands)				
	A Base case 17 million kg/year at $2,50/kg	*B* 20% reduced selling price at $2.00/kg	*C* 20% reduced volume at 13.6×10^6 kg/year	*D* 20% capital overrun	*E* 13.6×10^6 kg/year plant design
1. Sales to growers	42,500	34,000	34,000	42,500	34,000
2. Dealer discounts, distribution, freight (30%)	(12,750)	(10,200)	(10,200)	(12,750)	(10,200)
3. Net sales	29,750	23,800	23,800	29,750	23,800
4. Cost of goods sold	(13,989)	(13,989)	(12,210)‡	(14,743)‡	(11,818)‡
5. Gross profit	15,761	9,811	11,590	15,007	11,982
6. Sales and administration plus research and development (12.5% of net sales)	(3,719)	(2,975)	(2,975)	(3,719)	(2,975)
7. Profit from operations (years 2–10)	12,042	6,836	8,615	11,288	9,007
8. Startup costs, first year (10% of capital)	(2,050)	(2,050)	(2,050)	(2,460)	(1,770)
9. Profit from operations (first year)	9,992	4,786	6,565	8,828	7,237
10. Net earnings after taxes (50%)—first year	4,996	2,393	3,283	4,414	3,619
11. Net earnings after taxes—2–10 years	6,021	3,418	4,308	5,644	4,504
12. Working capital (25% of net sales)	7,438	5,950	5,950	7,438	5,950
13. Original fixed capital investment	20,500	20,500	20,500	24,600	17,000§
14. Return on investment (%)†	21.2	12.5	15.9	17.2	18.7

† $\text{ROI} = \dfrac{(\text{line } 10 + 9 \times \text{line } 11)/10}{\text{line } 13 + \text{line } 12}$

‡ Variable costs of materials and utilities × 0.8. Semifixed costs of labor related items × 0.94 (direct labor to 30 men from 32). Capital related costs follow their percentage of capital investment.

§ Calculated based on $(17/13.6)^{0.67}$ ratio of plant size to obtain new capital cost. The 0.67 factor is a balance between the 0.75 for fermentation section scale factor and the often used 0.6 for process plant scale factor.

[Reprinted by permission from W. H. Bartholomew and H. B. Reisman, "Economics of Fermentation Processes," Microbial Technology, H. J. Peppler and D. Perlman (eds.), 2d. ed., vol. 2, Academic Press, Inc., New York, N.Y., 1979, p. 463.]

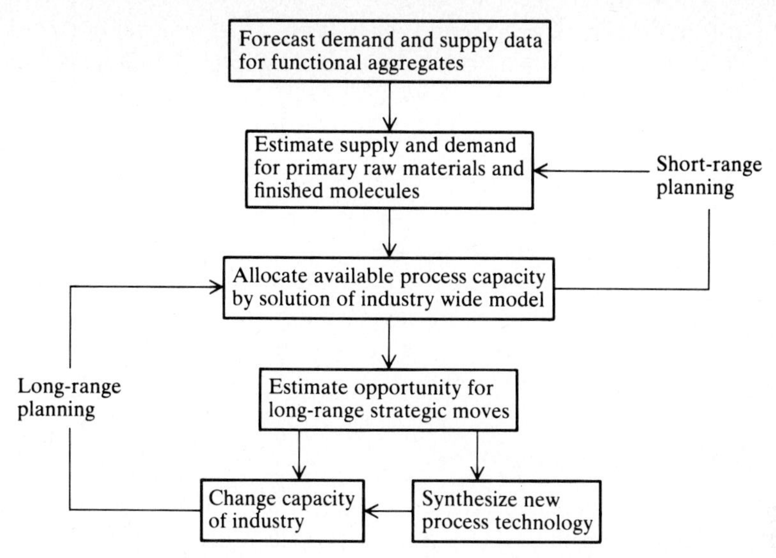

Figure 12.4 Flow of information in industrial development model. *(By permission, D. F. Rudd, "Modelling the Development of the Intermediate Chemicals Industry," The Chemical Engineering Journal, vol. 9, p. 11, 1975.)*

guided by larger strategic and tactical questions as mentioned at the outset of this chapter: Is the *area* of the proposed process one which may expand or contract in time? Is it one in which the company has *prior experience* and *market outlets*? Is it most interesting to *produce* the product internally, or to *license* the concept to another company? How stable are the raw materials and energy sources? What is the price of money needed for construction? While many of these questions lack clear answers, it is the responsibility of management to make the proper decision regarding process development and construction in the face of present and anticipated conditions.

With a fully described process example in hand, we can now sensibly consider particular examples of extant (proven) commercial processes and of emerging or potential processes.

12.5 FINE CHEMICALS

The category *fine chemicals* within the context of biotechnology is broadly interpreted to include *bioproduction* of high value molecules (vitamins, hormones, enzymes, antibiotics, monoclonal antibodies, etc.) and *bioconversion* of high value starting materials (antibiotics, steroids, etc.) to yet more valuable products. Process characteristics vary commonly from one example to the next. Some products occur at trace levels, others such as penicillin may exceed 50 g/L in the final broth. If the final product is to be administered medically or ingested with food, then product purification may be crucial and often costly, as in removal of pyrogen from *E. coli* molecular products, or removal of impurities to allow clean crystallization of, e.g., penicillin.

12.5.1 Enzymes

Production of enzymes for food, medical, diagnostic, and pharmaceutical applications is extensive as we have discussed previously in Chap. 4. The total U.S. industrial enzyme market in 1977–1978 was 50–55 million dollars. A recent estimate of world industrial enzyme market appears in Table 12.11. Note that detergent protease, two amylases, pectinase, and glucose isomerase together claimed 88 percent of the total market in 1977–1978.

12.5.2 Proteins Via Recombinant DNA

The three earliest products available via recombinant *E. coli* in clinically significant amounts are human insulin (1979), human growth hormone (1981), and human leukocyte interferon (1981). Human insulin and a vaccine against *E. coli* scours disease in newborn piglets were on the market in 1984, with numerous other products in various stages of commercial development. The economic assessment of these various products must include:

1. Initial research and development costs: these products are among the first from a new generation of biotechnology industries. Early products must pay for recovery of a (considerable) initial investment in laboratory, personnel, and production facility.
2. Conventional bioreaction and product recovery costs of the actual production process.

Table 12.11 World industrial enzyme production (1977–1978)

Enzyme	Amounts produced per year, tons of pure enzyme protein	Relative sales value, %
Bacillus protease	500	40
Glucoamylase	300	14
Bacillus amylase	300	12
Glucose isomerase	50	12
Microbial rennet	10	7
Fungal amylase	10	3
Pectinases	10	10
Fungal protease	10	1
Other	....	1

[After K. Aunstrup, O. Andresen, E. A. Falch, and T. K. Nielsen, "Production of Microbiol Enzymes," Microbiol Technology, 2d. ed., vol. 1, H. Peppler and D. Perlman (eds.), Academic Press, Inc., 1979, p. 283. Reprinted by permission of Academic Press, New York, N.Y.]

3. Recovery of costs for clinical testing and other procedures required by the U.S. Food and Drug Administration, U.S. Department of Agriculture, or other corresponding agency.

The bioreaction step is a relatively standard process, though exceptional interest arises with respect to contamination and problems associated with stability of the recombinant plasmid (Chap. 6). In the recombinant DNA approach (Chap. 6), a host-vector system is chosen. Only a small number of hosts are currently used or under serious consideration for industrial application (*E. coli*, *Saccharomyces cerevisiae*, several mammalian cell lines, perhaps *Bacillus subtilis*, and several *Pseudomonas* species). The host species imparts its character to the overall bioreactor and recovery design. For *E. coli*, for example, the simple genetics, good characterization, and high growth rate are advantages. Known bioreaction and recovery problems associated with *E. coli* hosts are:

1. *E. coli* contains at least eight soluble proteases: these may have played a role in observed degradation of the expressed recombinant proteins somatostatin and human growth hormone. Their influence is countered during product recovery by use of the protease inhibitor phenyl methyl sulfonyl fluoride which is effective against serine proteases but not metalloenzymes. Minimal iron levels diminish metalloenzyme protease activity, but induce a metal complexation agent which may hinder product purification.
2. Cell envelope lipopolysaccharide is a very powerful pyrogen, which provides fever responses even at 0.5 ng/kg body weight (one-half part per trillion). Previous use of *E. coli for* L-asparaginase production revealed product contamination by such pyrogens.

The separation processes for these early products were given in Fig. 11.39. Since large quantities of individual proteins expressed in *E. coli* tend to accumulate in refractile body precipitates of partially denatured products, special solubilization and refolding processing steps are needed. Subsequent purification involves chromatographic methods familiar at least on laboratory scale. The important immunosorbent synthesis and its use in purification for interferon and other protein production is novel; this technique was discussed in Chap. 11.

The financial impact of recombinant DNA (rDNA) products will be spread across many areas (pharmaceuticals, agriculture, specialty chemicals, etc.). The pace of development in each area may be generally keyed to the product selling price and technical complexity of the rDNA project, as is suggested by Fig. 12.5. A high-value pharmaceutical example is insulin, for which the U.S. and European market sales may double by 1985 (from 1981) to two-thirds of a billion dollars. Eli Lilly used Genentech's rDNA work and Lilly's own thorough familiarity with insulin markets and regulatory procedures to achieve FDA approval for human insulin (hI) in four years. The extent of hI replacement of pig and cow insulin is only now being determined.

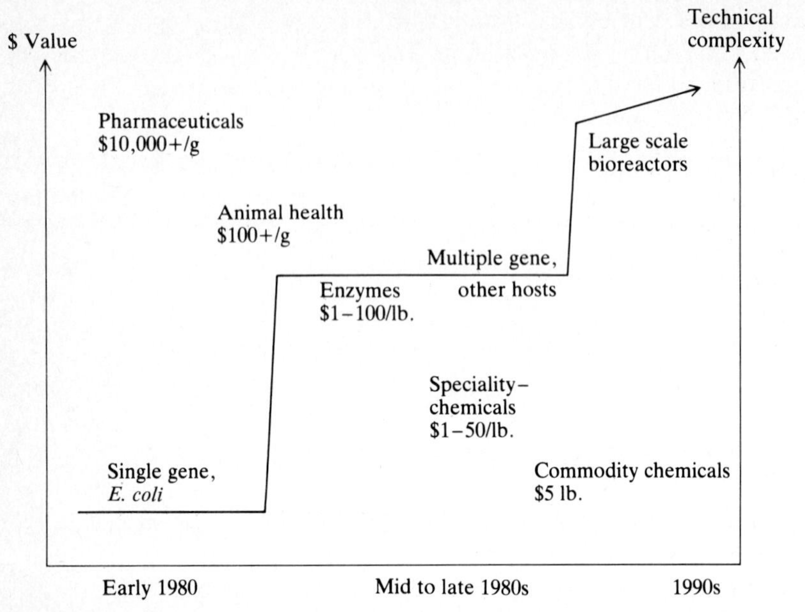

Figure 12.5 Approximate timing of development of rDNA products. *(After W. Hustedter, Genentech, Inc., by permission, 1985.)*

Other high value products undergoing rDNA research and development include regulatory proteins (interferons, human and animal growth hormones, neuroactive peptides, and lymphokines), blood factors (human serum albumin, antihemophilic factor, thrombolytic and fibrinolytic enzymes), vaccines (viral and bacterial), and antibiotics. Even high-value items have an uncertain rDNA future: a typical U.S. animal vaccine market is \$5–10 $\times 10^6$ per year which may be an insufficient single product to maintain a company; consequently development of several animal vaccines could be necessary.

The overall potential impact of rDNA technology is staggering, as virtually all areas of biotechnology may be strongly affected. The pace of R & D efforts in the early 1980s is extraordinary; a complete survey of present and potential technology and markets is provided in the 1984 OTA report; *Commercial Biotechnology: An International Analysis.* (See Ref. 2).

12.5.3 Antibiotics

Secondary metabolism products which inhibit growth of other microbial species, even at low levels, are called antibiotics. Uses for antibiotics are found in antimicrobials (human disease), antitumor agents, fungicides and pesticides for plant protection, animal disease and animal growth products, and research. Total 1980 estimated world annual production and sales were above 10^5 tons and 4 billion dollars, respectively. Screening, mutation, and other microbial research activity in antibiotic production continues to be a major area of industrial microbiology.

More than 6000 antibiotics are known, of which nearly 100 are produced commercially via fermentation.

Antibiotic classifications are based on breadth of antimicrobial action (*broad* vs. *narrow* range), basis of activity (mechanisms), source (producer strain), biosynthetic pathway, and molecular structure. Table 12.12 summarizes the major structural groups.

The full interaction between medium formulation, microbial activity, and synthetic organic chemistry is well illustrated in commercial penicillin production. All penicillins (e.g., Table 12.13) have the same general 6-aminopenicillanic acid (6-APA) structure, consisting of a 5-member thiazolidine ring, an associated 4-member β-lactam ring, and an acyl (R—CO—) side chain on the lactam ring.

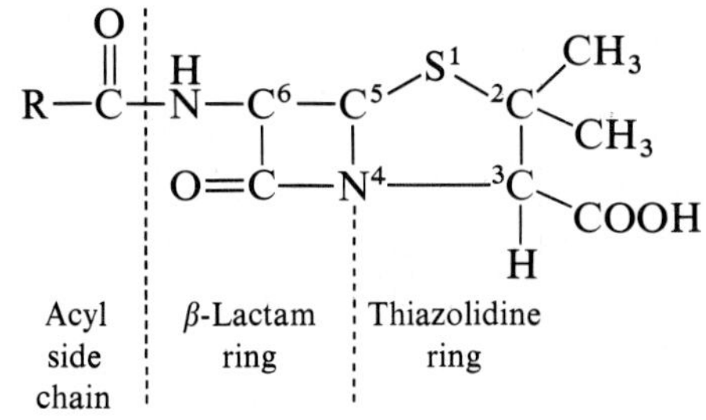

Acyl side chain | β-Lactam ring | Thiazolidine ring

The *natural* penicillins are produced microbially from a useful medium formulation; Penicillin G (Table 12.13) is a major example. Addition of side-chain precursors to the medium formulation allows change in the dominant acyl group substituted on the 6-APA structure; penicillins V and O are some of the consequent *biosynthetic* pencillins. Enzymatic removal of the acyl group, followed by a synthetic organic chemical substitution yields a *semisynthetic* penicillin. Such synthetic substitution has improved the breadth of antimicrobial effectiveness, resistance to natural degradations of β-lactamases, and acid stability of the antibiotics.

Cephalosporins are β-lactam antibiotics with a dihydrothiazine ring and 3 substituent groups R_1, R_2, and R_3; advantages of these antibiotics include a broad spectrum of antimicrobial activity, low toxicity, and resistance to natural penicillinases.

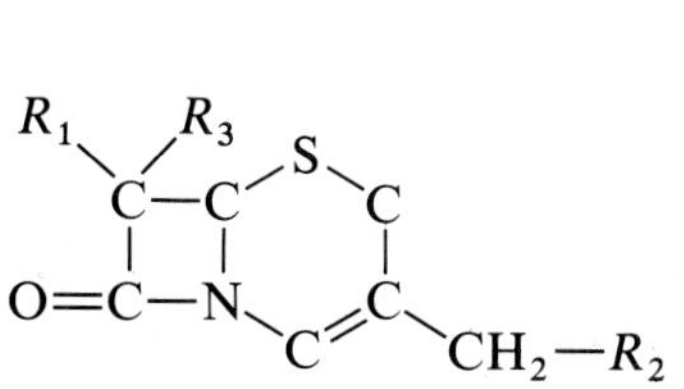

Cephalosporin structure

Semisynthetic cephalosporins are produced in two ways: (1) side-chain cleavages to give the precursor 7-aminocephalosporanic acid (7-ACA) with

Table 12.12 Classification of antibiotics according to their chemical structure. An example of each is given in parentheses

Class	Example
1. Carbohydrate-containing antibiotics	
Pure sugars	(Nojirimycin)
Aminoglycosides	(Streptomycin)
Orthosomycins	(Everninomicin)
N-Glycosides	(Streptothricin)
C-Glycosides	(Vancomycin)
Glycolipids	(Moenomycin)
2. Macrocyclic lactones	
Macrolide antibiotics	(Erythromycin)
Polyene antibiotics	(Candicidin)
Ansamycins	(Rifamycin)
Macrotetrolides	(Tetranactin)
3. Quinones and related antibiotics	
Tetracyclines	(Tetracycline)
Anthracyclines	(Adriamycin)
Naphthoquinones	(Actinorhodin)
Benzoquinones	(Mitomycin)
4. Amino acid and peptide antibiotics	
Amino acid derivatives	(Cycloserine)
β-Lactam antibiotics	(Penicillin)
Peptide antibiotics	(Bacitracin)
Chromopeptides	(Actinomycins)
Depsipeptides	(Valinomycin)
Chelate-forming peptides	(Bleomycins)
5. Heterocyclic antibiotics containing nitrogen	
Nucleoside antibiotics	(Polyoxins)
6. Heterocyclic antibiotics containing oxygen	
Polyether antibiotics	(Monensin)
7. Alicyclic derivatives	
Cycloalkane derivatives	(Cycloheximide)
Steroid antibiotics	(Fusidic acid)
8. Aromatic antibiotics	
Benzene derivatives	(Chloramphenicol)
Condensed aromatic antibiotics	(Griseofulvin)
Aromatic ether	(Novobiocin)
9. Aliphatic antibiotics	
Compounds containing phosphorus	(Fosfomycins)

[By permission from W. Crueger and A. Crueger, "Biotechnology: A Textbook of Industrial Microbiology," (English edition), Science Tech, Inc., Madison, WI, 1984, p. 198. (After J. Berdy, Adv. Appl. Microbiol., vol. 18, p. 309, 1979.)]

Table 12.13 Chemical structures and properties of three different types of penicillin[†]

R group	Chemical name	Trivial name	Oral absorption	Activity against: *Staphylococcus*	Activity against: Penicillinase-producing *Staphylococcus*	Activity against: Other cocci	Activity against: Gram-negative bacilli	Limitations
$-CH_2-$	6-(Phenylacetamido) penicillin	Penicillin G	Poor	High	Nil	High	Nil	Variable absorption
C—C—, N, O, N, CH_3	6-(5-Methyl-3-phenyl-4-isoxazole-carboxamido) penicillin	Oxacillin	High	High	High	Variable	Nil	Variable absorption; narrow spectrum
H, C—, NH_2	6-[D(−)-α-Aminophenyl-acetamido]penicillin	Ampicillin	High	High	Nil	High	Medium	Gram-negative bacilli may become resistant during therapy

[†] From G. T. Stewart, *The Penicillin Group of Drugs*, Elsevier Publishing Company, Amsterdam, 1965.

$R_1 = NH_3$, $R_2 =$ acetate, $R_3 = H$, and subsequent acylation, and (2) by ring expansion of penicillin (a metabolic pathway precursor) to 7-ACA and subsequent modification.

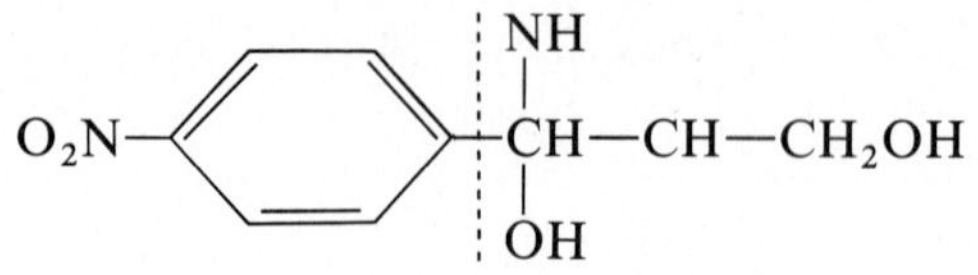

$$O_2N-C_6H_4-\underset{OH}{\underset{|}{CH}}-\overset{NH}{\overset{|}{CH}}-CH_2OH$$

P-Nitrophenol Propandiol

Chloramphenicol structure

Fully synthetic products exist: the antibiotics chloramphenicol (above) and pyrrolnitrin were discovered as biological metabolites but are now wholly chemical in production. This trend is expected to increase: A University of Pennsylvania pair of researchers recently completed synthesis of efrotomycin, a molecule with 21 stereocenters. Thus, antibiotic process economics may be guided by a purely biological process, a mixed biological-synthetic approach or a wholly synthetic, nonbiological sequence.

Production methods for antibiotics were reviewed in Chap. 11 because extensive product recovery and purification trains were at times used. Processes remain typically batch because of strain genetic instability, but periodic feed/withdrawal operations on a batch vessel have been effected without penicillin yield loss. Penicillin production now yields broth titres in the tens of grams per liter; the resultant low price and ease of manipulation (to semisynthetic penicillins, cephalosporin 7-ACA structure, etc.) renders penicillin production in the intermediate, rather than fine, chemical level.

12.5.4 Vitamins, Alkaloids, Nucleosides, Steroids

In addition to proteins (enzymes) and antibiotics, many other complex metabolites are produced by microbes; a few processes are of both commercial interest and competitive or better than straight chemical synthetic routes.

Of the microbially synthesized vitamins [B_{12}, riboflavin (B_2), thiamine, folic acid, pantothenic acid, pyridoxal], only B_{12} and B_2 are produced microbially, and the latter have been progressively displaced by synthetic routes. Vitamin B_{12}, also called cyanocobalamin, is synthesized exclusively by microbes in nature, yet is required by all animals. While microbial flora of the large intestine can synthesize B_{12}, its assimilation by humans does not occur, hence it must be obtained from food. Early 1980s world production was roughly ten tons, of which the pharmaceutical industries consumed about two-thirds (as cyanocobalamin, hydroxocobalamin, coenzyme B_{12}, and methylcobalamin), and the balance went to animal food (to supplement vegetable protein source B_{12} content).

Propionibacteria (freudenreichii and *shermanii*) are used primarily (Table 12.14) in a batch process employing two phases: a 2–4 day anaerobic operation

Table 12.14 Characteristics of microbial processes for vitamin B_{12} production†

Microorganism	Ingredient of medium	B_{12} yield, mg/L	Comments
Bacillus megaterium	Beet molasses; ammonium phosphate; cobalt salt; inorganic salts	0.45	18-h fermentation (aerated)
Propionibacterium freudenreichii	Corn-steep liquor; glucose cobalt salt; maintained at pH 7 with NH_4OH	19	6-day fermentation (3 days aerobic + 3 days anaerobic)
P. freudenreichii	Corn-steep liquor (or autolyzed penicillium mycelium); glucose; cobalt salt; maintained at pH 7 with NH_4OH	8	Continuous two-stage fermentation; 33-h retention time
Propionibacterium shermanii	Corn-steep liquor; glucose; cobalt salt; maintained at pH 7 with NH_4OH	23	7-day batch fermentation (3 days anaerobic + 4 days aerobic)
Streptomyces olivaceus	Glucose; soybean meal; distillers' solubles; cobalt salt; inorganic salts	3.3	6-day fermentation (aerated)
Streptomyces sp.	Soybean meal; glucose; cobalt salt; K_2HPO_4	5.7	6-day fermentation (aerated)

† From H. J. Peppler (ed)., *Microbial Technology*, p. 286, Reinhold Publishing Corporation, New York, 1967.

followed by a 3–4 day aerobic stage. The anaerobic product (5′-deoxyadenosylcobinamide guanosine phosphate) is coupled with the aerobic product (5,6-dimethyl benzimidazole) to give, ultimately, the desired vitamin B_{12}. Recovery involves heating (to release cell-bound vitamin B_{12}), and chemical conversion to stable cyanocobalamin. A continuous, two-stage analog has also been effected. *Pseudomonas denitrificans* also produces B_{12} in a one stage process; cobalt and 5,6-dimethyl benzimidazole are also added. A dozen years of strain improvement have moved yields from 0.6 to 60 mg/L. Sugar beet molasses, providing inexpensive betaine, is used. Activated sludge in waste treatment is also a B_{12} source (4–10 mg/L), but recovery is difficult due to cost of resolution of many B_{12} analogs.

Pharmacologically active ergot alkaloids are produced by the fungi *Claviceps*. While a complex chemical path is known, and propagation by rye infection is also possible, submerged fermentation cultures can achieve yields in excess of 5 g/L. Strain degradation is a common problem. Because of mixing sensitivity and high oxygen demand, scaleup is not straightforward.

Nucleosides and nucleotides are used as flavor enhancers in food. The most effective substances are the purine 5′-monophosphates: guanylic acid (5′-GMP), inosinic acid (5′-IMP) and xanthylic acid (5′-XMP). Sodium glutamate has a synergistic flavor-enhancing action with these 5′-monophosphates (see organic acids, Sec. 12.6.3). Approximately 3000 tons per year of 5′-IMP (66%) and 5′-GMP (34%) are produced annually in Japan. Since nucleoside synthesis is normally feedback regulated, overproduction requires either auxotrophic mutants with addition of growth-limiting end product or development of purine-analog resistance to regulation. The commercial products are ultimately produced either microbially (as discussed above) or by hydrolysis of cellular RNA (either *in vitro* using yeast RNA and microbial enzymes or *in vivo* using endogenous enzymes followed by mononucleotide excretion).

Biochemical conversion effected by microbes or enzymes is commonly called *bioconversion* or *biotransformation.* Microbial catalysis provides four advantages vs. nonbiological conversions: specificity of substrate, regiospecificity (selectivity among groups on same molecule), stereospecificity (racemate resolution; recall immobilized L-aminoacylase process, Fig. 4.12), and mildness of reaction conditions. The disadvantages, where competitive possibilities exist, are requirement of cell viability during conversion and cost of product recovery from a complex mixture. Steroid conversions typically involve very selective partial oxidation at one position on the typical basic steroid structure (e.g., progesterone).

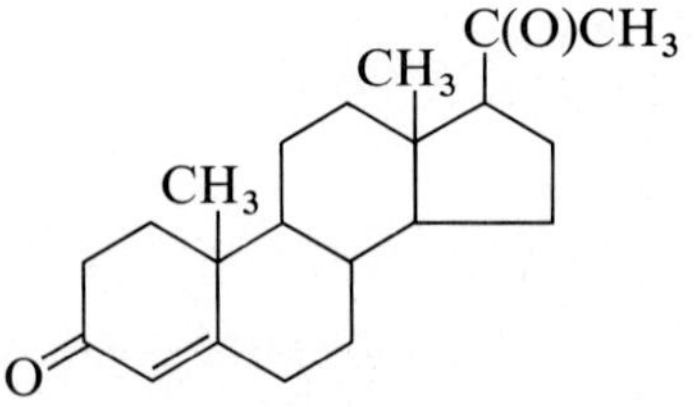

$C(O)CH_3$ CH_3 CH_3 O

Progesterone

For example, achievement of the very selective microbial 11-β-hydroxylation of progesterone allowed ultimate diminution of cortisone cost from \$200/g (1949) to \$1/g (1979).

$$\text{Progesterone} \xrightarrow[\text{11-}\beta\text{-Hydroxylation}]{} \text{cortisol} \longrightarrow \text{cortisone}$$

Medium and propagation conditions for such bioconversions appear in Table 12.15. Noteworthy items are the timing of precursor addition and the long conversion times. Applications for natural and derivatized steroid and sterols include therapeutic uses (estrogens, progesterone, and androgens); contraceptives (derivatized estrogen and progesterone), sedatives, antitumor therapy, veterinary products, anti-inflammatories for skin diseases and arthritis (cortisone), and medication for controlling sodium retention/potassium excretion. Use of immobilized cell and immobilized enzymes, as well as hybrid biological/synthetic routes, have characterized recent essays involving thousands of steroid-sterol conversions.

Table 12.15 Conditions for operation of several steroid and sterol transformations

	Substrate	Product	Yield, %	Microorganism	Medium	Conditions used
1	Progesterone	1-Dehydrotestololactone	50	*Cylindrocarpon radicicola*	†	72 h, 25°C
2	Progesterone	1,4-Androstadiene-3,17-dione	85	*Fusarium solani*	‡	96 h, 25°C
3	Progesterone	15α-Hydroxy-4-pregnene-3,20-dione	11	*Streptomyces aureus*	§	72 h, 25°C
4	4-Androstene-3,17-dione	11α-Hydroxy-4-androstene 3,17-dione	25	*Rhizopus arrhizus*	¶	96 h, 28°C
5	Progesterone	11-α-Hydroxyprogesterone	90	*Aspergillus ochraceus*	††	72 h, 28°C
6	Hydrocortisone	Prednisolone	93	*Arthrobacter simplex*	‡‡	120 h, 28°C Pseudocrystalline fermentation
7	Cholesterol	1,4-Androstadiene-3,17-dione	90	*Arthrobacter simplex*	§§	44 h, 30°C + chelating agents
8	β-Sitosterol, Cholesterol Stigmasterol Campesterol	9α-Hydroxy-4-androstene-3,17-dione	Data unavailable	*Mycobacterium fortuitum*	¶¶	336 h, 30°C mutant with block in breakdown of steroid nucleus

† 3 g corn steep liquor (dry weight), 3 g $NH_4H_2PO_4$, 2.5 g $CaCO_3$, 2.2 g soy bean oil, 0.5 g progesterone, distilled H_2O to 1 L, pH 7.0.

‡ 15 g peptone, 6 mL corn steep liquor, 50 g glucose, distilled H_2O to 1 L, pH 6.0; 0.25 g progesterone, added after 48 h.

§ 2.2 g soy bean oil, 15 g soy meal, 10 g glucose, 2.5 g $CaCO_3$, 0.25 g progesterone, distilled H_2O to 1 L.

¶ 20 g peptone, 5 mL corn steep liquor, 50 g glucose, tap water to 1 L, pH 5.5–5.9; 0.25 g androstendione, added after 27 h.

†† Edamin, glucose, corn steep liquor; 20 g progesterone, finely ground, suspended in 0.01% Tween 80, added after 18–24 h.

‡‡ 5 g peptone, 5 g corn steep liquor, 5 g glucose; distilled H_2O 1 L; pH 7.0; 1–50% finely ground hydrocortisone, suspended in ethanol, added to a 24-hour-old culture.

§§ 10 g corn steep liquor, 2 g meat extract, 5 g glucose, 0.5 g K_2HPO_4; distilled H_2O 1 L; pH 7.0; cholesterol (1 g) added after 20 h (dispersed in water); 0.8 mM α,α′-dipyridyl (an iron chelating agent) in ethanol added after 26 h.

¶¶ 10 g Glycerol, 8.4 g Na_2HPO_4, 4.5 g KH_2PO_4, 2 g NH_4Cl, trace elements; distilled H_2O 1 L; 1 g soy meal and 10 g sitosterol are added.

[Reprinted by permission, O. K. Sebek and D. Perlman, "Microbial Transformation of Steroids and Sterols," Microbiol Technology, 2d. ed., vol. 1, H. J. Peppler and D. Perlman (eds.), Academic Press, New York, N.Y., 1979 p. 483.]

12.5.5 Monoclonal Antibodies (MAb)

When a foreign molecule (*antigen*) is injected into an animal such as a mouse, the animal's immune system will often recognize the foreign molecule and produce specific antibodies which complex very selectively with the antigen, thereby deactivating it and/or tagging it so that other body functions will recognize the complex and remove it from circulation. Typically, antibodies are produced in specialized cells known as B lymphocytes (spleen, blood, lymph); a characteristic immune response will produce a number of antibodies (i.e., a mixture) to a given antigen. Each B lymphocyte produces (only) one kind of antibody. These cells (e.g., spleen) can be removed and cultured, external to the mouse, to produce antibody. The resultant multicelled propagation has two disadvantages: the cells produce a number of different antibodies since the culture is polyclonal; i.e., derived from a number of spleen B cells, and the cells tend to dedifferentiate or regress and lose their specific antibody production facility. These two problems are overcome by the technique of cell fusion (Sec. 6.2.3) whereby a productive lymphocyte cell may be joined to a rapidly growing myeloma (tumor) cell to produce a *hybridoma* cell. The hybrid cell can grow rapidly; this fact allows subsequent cloning to isolate individual cultures derived from a single lymphocyte cell. The result is a continuously growing or "immortalized" propagation, producing only a single kind of antibody, hence the term monoclonal antibody (MAb).

MAb technical applications have appeared rapidly in commercial use in the 1980s, primarily because the earliest major products, *in vitro* medical diagnostic kits, did not require the level of testing needed for *in vivo* applications. Potential use for *in vivo* diagnosis, prophylaxis, and therapy will be subjected to the same tighter agency requirements pertinent to biologics and drugs (Sec. 12.2).

The driving forces for MAb-based diagnostics are several. (1) Time savings: a number of diseases were previously diagnosed by time-demanding sample cultures; MAb kits provide rapid determinations. Examples here include human venereal diseases (gonorrhea, chlamydia, and herpes simplex virus), and several common bacterial infections found in hospitalized patients (e.g., *Pseudomonas aeruginosa* and group B *streptococci*). Potentially, cancer detections are also possible provided a tumor-specific marker is identifiable; a market example is an acid phosphatase released by a cancerous prostate gland. (2) Increased sensitivity: one MAb hepatitis B detection system is claimed to be 100 times more sensitive than a polyclonal antibody determination. Such sensitivity is of great interest to blood banks which must routinely screen for hepatitis.

As noted earlier in this chapter, development of new products which include appreciable new R & D components is necessarily more costly than subsequent similar products arising from the maturing knowledge base. One estimate suggests (1983) that development of a new MAb supplied to *in vitro* kit manufacturers could cost \$3.5 million over 3 years, with ultimate kit-production facilities costing 5–10 times as much.

In addition to diagnostics for hepatitis B, prostatic tumor, and the human

Table 12.16 Estimates: U.S. Monoclonal Antibody Market (1981 dollars in millions)

	Market size (estimated)	
Application	1982	1990
Diagnostics		
1. In vitro kits	\$5 to \$6	\$300 to \$500 (\$40)†‡
2. Immunohistochemical kits (biopsies, smears, etc).	Nil	\$25
3. In vivo diagnostics (primarily imaging)	Nil	Small to \$100§¶
Therapeutics (includes radiolabeled and toxin-labeled reagents)	Nil	\$500 to \$1,000‡§
Other		
Research	Small	\$10
Purification	Small	\$10

† High numbers: total kit value; (value): value of antibody alone.
‡ Variation depending on industrial source.
§ Could rise or fall depending on regulatory process.
¶ Basis: 1981 price for diagnostic kits of same type.
(From "*Commercial Biotechnology, An International Analysis*," Office of Technology Assessment, Washington, D.C., 20510, 1984.)

venereal and hospital bacterial infections, MAb products marketed in the early 1980s include diagnostics for pregnancy [human chorionic gonadotropin (HCG)], rubella, and rabies, among others.

The potential growth rate of MAb products for human and animal health products is summarized in Table 12.16. Note again that *in vitro* applications exist in 1982, and the *in vivo* markets (drug delivery, tumor imaging, and others) should make a strong contribution before 1990.

12.6 BULK OXYGENATES

Both anaerobic and aerobic processes are extant for bulk oxygenate production. We first review the traditional fermented beverage market, then consider use of simple oxygenates (alcohols, ketones, acids) and amino acids. A listing of such fermentation oxychemicals, and the corresponding microbial strain and substrate, appears in Table 12.17. The total U.S. production figures for simple oxygenates (Table 12.18), with market values and applications, indicate the enormous potential size of the oxychemical market via fermentation. The oldest oxychemical is ethanol, produced in traditional fermented beverage manufacture, which is considered in the next section. Subsequently, we discuss ethanol production for fuel, and microbial processes for acetone, butanol, organic acids, and amino acids.

Table 12.17 Microbial production of chemicals

Chemical	Process	Microorganism
Ethanol	$C_6H_{12}O_6 \xrightarrow{M^\dagger} C_2H_5OH$	*Saccharomyces cerevisiae*
Ethylene	$C_2H_5OH \xrightarrow{C} CH_2{=}CH_2$	*Zymomonas mobilis*
1,3-Butadiene	$C_2H_5OH \xrightarrow{C} CH_2{=}CH\text{-}CH{=}CH_2$	
Ethylene glycol	$C_2H_5OH \xrightarrow{C} CH_2OH\text{-}CH_2OH$	
Acetic acid	$C_6H_{12}O_6 \xrightarrow{M} CH_3COOH$	*Clostridium thermoaceticum*
	$C_5H_5OH \xrightarrow{M} CH_3COOH$	*Acetobacter aceti*
Acetone	$C_6H_{12}O_6 \xrightarrow{M} CH_3COCH_3$	*Clostridium acetobutylicum*
Butanol	$\rightarrow CH_3(CH_2)_2CH_2OH$	
Isopropyl alcohol	$C_6H_{12}O_6 \xrightarrow{M} (CH_3)_2CHOH$	*Clostridium aurianticum*
	$\rightarrow CH_3(CH_2)_2CH_2OH$	
Adipic acid	$CH_3(CH_2)_4CH_3 \xrightarrow{M} HOOC(CH_2)_4COOH$	*Pseudomonas species*
Acrylic	$C_6H_{12}O_6 \xrightarrow{M_1} CH_3CH(OH)COOH \xrightarrow{M_2/C} CH_2{=}CHCOOH$	1. (M_1) *Lactobacillus bulgarius* 2. (M_2/C) *Clostridium propionium*
	$CH_2OHCH(OH)CH_2OH \xrightarrow{M/C} CH_2{=}CHCOOH$	*Klebsiella pneumoniae* (*Aerobacter aerogenes*)
Methyl ethyl ketone	$C_6H_{12}O_6 \xrightarrow{M} CH_3CH(OH)CH(OH)CH_3 \xrightarrow{C} CH_3COCH_2CH_3$	*Klebsiella pneumoniae*
Propylene glycol	$CH_2OHCH(OH)CH_2OH \xrightarrow{C} CH_3CH(OH)CH_2OH$	
Glycerol	$C_6H_{12}O_6 \xrightarrow{M} CH_2OHCH(OH)CH_2OH$	*Saccharomyces cerevisiae*
	$H_2O + CO_2 \xrightarrow{M} CH_2OHCH(OH)CH_2OH$	*Dunaliella* sp.
Citric acid	$C_6H_{12}O_6 \xrightarrow{M} CH_2(COOH)(OH)C(COOH)CH_2(COOH)$	*Aspergillus niger*

† M, microbial fermentation; C, chemical processing.

(Reprinted by permission from T. K. Ng, R. M. Busche, C. C. McDonald. "Production of Feedstock Chemicals," Science, vol. 219, p. 736. Copyright 1983 by the American Association for the Advancement of Science.)

Table 12.18 Oxychemicals from renewable resources

Chemical	1981 U.S. production, million pounds	1983 price, cents per pound	1981 commercial value, million dollars	Major use or derivative
Ethanol				
Ethylene	28,867[†]	25 to $25\frac{1}{2}$[‡]	8,169[§]	Polyethylene, ethylene oxides
Butadiene	3,046	34	1,234	Styrene-butadiene rubber, polybutadiene rubber
Industrial	1,157	\$1.70 to \$1.82 per gallon (27.5 cents per pound)	359	Solvents, ethyl acetate and other esters
Ethylene glycol	4,055	$27\frac{1}{2}$ to $28\frac{1}{2}$	1,281	Polyethylene terephthalate, antifreeze
Acetic acid	2,706	$26\frac{1}{2}$	511	Vinyl acetate, cellulose acetate
Acetone	2,167	31	483	Solvents, methyl, and other methacrylates
Isopropyl alcohol	1,644	\$2.05 per gallon (31 cents per pound)	507	Acetone, solvents
Adipic acid	1,210	57	653	Nylon 66
Butanol	823	$33\frac{1}{2}$	251	Solvents, butyl acrylate
Acrylic acid	691	58	276	Polymers
Methyl ethyl ketone	626	37	260	Solvents
Propylene glycol	480	44	208	Unsaturated polyester resin
Glycerol	370	$80\frac{1}{2}$	259	Drugs, cosmetics
Citric acid	235	71 to $77\frac{1}{2}$	192	Food, drugs

[†] Values in this column were collected from the *Chemical Marketing Economics Handbook* and U.S. International Trade Commision data on organic chemicals.

[‡] From *Chemical Marketing Reporter* (20 Sept. 1982); actual prices depend on quality, quantity, and location.

[§] From *Chemical Marketing Economics Handbook*, U.S. Bureau of Census, and U.S. International Trade Commission. Total values do not reflect actual price of transaction or current price.

(Reprinted by permission from T. K. Ng, R. M. Busche, C. C. McDonald. "Production of Feedstock Chemicals," Science, vol. 219, p. 737. Copyright 1983 by the American Association for the Advancement of Science.)

12.6.1 Brewing and Wine Making

Beer production illustrates several principles of enzyme technology already discussed. To prepare barley for processing, it is incubated for 2 to 6 days so that germination occurs. This malting step promotes the formation of active α- and β-amylases as well as protease enzymes. The resulting grains are then carefully dried.

Preparation of a nutrient medium called beer wort for yeast growth requires mashing, which is carefully controlled warming of an aqueous mixture of malt and starch. Success of the mashing step requires artful exploitation of a complex mixture of substrates and enzymes. Proteins and carbohydrates in the malt-starch mixture must be hydrolyzed, since the yeast involved in the fermentation step can utilize only simple sugars and amino acids as nutrients. On the other hand, total hydrolysis is not desirable because dextrins, peptides, and peptones contribute flavor and body to the beer. Complicating the situation are the characteristics of the hydrolytic enzymes which participate in mashing. Elaborate cooking recipes with temperatures varying with time over a range from about 40 to 100°C are employed to optimize the complex mashing process (Table 12.19). The end product contains a suitable mixture of yeast nutrients and flavor components.

Next the beer wort is clarified and then boiled for about 2 h with periodic addition of hops (ripened, dried cones of the hop vine), which contribute flavor, aroma, and color to the beer and also exert an antibacterial action. The sterilized wort is then cooled and inoculated with a proprietary strain of brewer's yeast (*Saccharomyces cerevisiae*). The yeast is first cultivated under aerobic conditions followed by a switch to an anaerobic environment to cause ethanol and CO_2 production. Finally the product is clarified, pasteurized, aged, and packaged.

The magnitude of the brewing industry is impressive. The U.S. industry had 1975 sales of 160 million barrels (31 gallons per barrel). The approximate cost distribution, in millions, for this production was \$610 (malt), other grains (\$188), hops (\$49), directy salaries (\$860), packaging and containers (\$2500), fuel, power,

Table 12.19 Three enzyme-catalyzed hydrolyses used in mashing with different components and different rate characteristics†

Enzyme	Substrate (or bond) attacked	Hydrolysis product(s)	Temperature of maximum activity, °C
α-Amylase	Starch (all bonds)	Large-fragment dextrins, amylose	70–75
β-Amylase	Amylose (β-1,4 linkage)	Maltose	57–65
Proteases	Proteins	Amino acids, small peptones	50
		Larger peptides, peptones	60

† Data from Ref. 23.

and water (\$80), transportation (\$360), equipment and improvements (\$600), and advertising (\$230).

A flow diagram of a typical winery process is shown in Fig. 12.6. The feed for wine production may be nearly any ripe fruit or vegetable extract which contains 12 to 30 percent sugar in the juices. The fermentation is accomplished either with wild yeast, present on the grape or fruit skins, or with inoculated selected yeast cultures and pasteurized fruit juices.

Characteristic of many anaerobic fermentation processes, the yeast inoculum is first grown to desired size in the juice under aerobic conditions, generating more cells as well as carbon dioxide; subsequent fermentation to yield 7 to 15 percent ethanol solutions and further carbon dioxide is carried out anaerobically. Some acids produced during fermentation are degraded by bacteria in postfermentation stages, creating the characteristic bouquet of the wine. The type of wine produced is determined by the fruit used (grape variety, apricots, peaches, etc.), the yeast and bacterial strains involved, possible subsequent fortification (i.e., adding alcohol) and the series of mechanical and heating, cooling, and aging processes before bottling.

World wine production is concentrated in Europe (almost 80 percent), with 15 percent in the Americas. Production facilities vary from large-scale bulk producers to very small "family" operations. Wine evaluation is complex; procedures include estimation of taste character (acidity, sugar, bitterness, and astringency (due to tannin)), appearance (clarity, lack of sediment), odor, aroma and bouquet, and flavor (a composite measure including after-taste).

12.6.2 Fuel Alcohol Production

Ethanol production for use as a fuel has been widely considered again in the last decade, following appreciable usage in pre-World War II Europe when fuel ethanol production from agricultural crops was practiced. Inexpensive petroleum and new petrochemical processes developed after 1945 provided a feedstock source which then displaced fermentation-derived products, including ethanol and virtually all other simple oxychemicals (acetone, butanol, acetic acid, etc.). Petroleum price rises over the last decade have caused reconsideration of fermentation routes. The attractiveness of ethanol as a motor fuel derives from four characteristics (i) it is liquid, easily transported, (ii) its heating value per gallon is high (about 2/3 that of gasoline), (iii) it can be blended up to 10 percent with gasolene with no change of engine tuning or increase in emission, and (iv) it can enhance the octane rating of unleaded gasoline. Additionally, in many countries, ethanol fuel may serve as a partial replacement for imported oil.

Raw materials for ethanol fermentation process have included sugars (sugar cane juices, cassava, molasses), starches (grains), wood and agricultural residues (from wheat straw, corn stalks) and forestry, urban, and industrial wastes (newspaper, spent sulfite liquor, cheese whey, vegetable and fruit industry wastes, etc.). The trade-off in fermentation feedstocks is clearly a more fermentable, purer substrate vs. cheaper but more heterogeneous starting material. Use of the latter

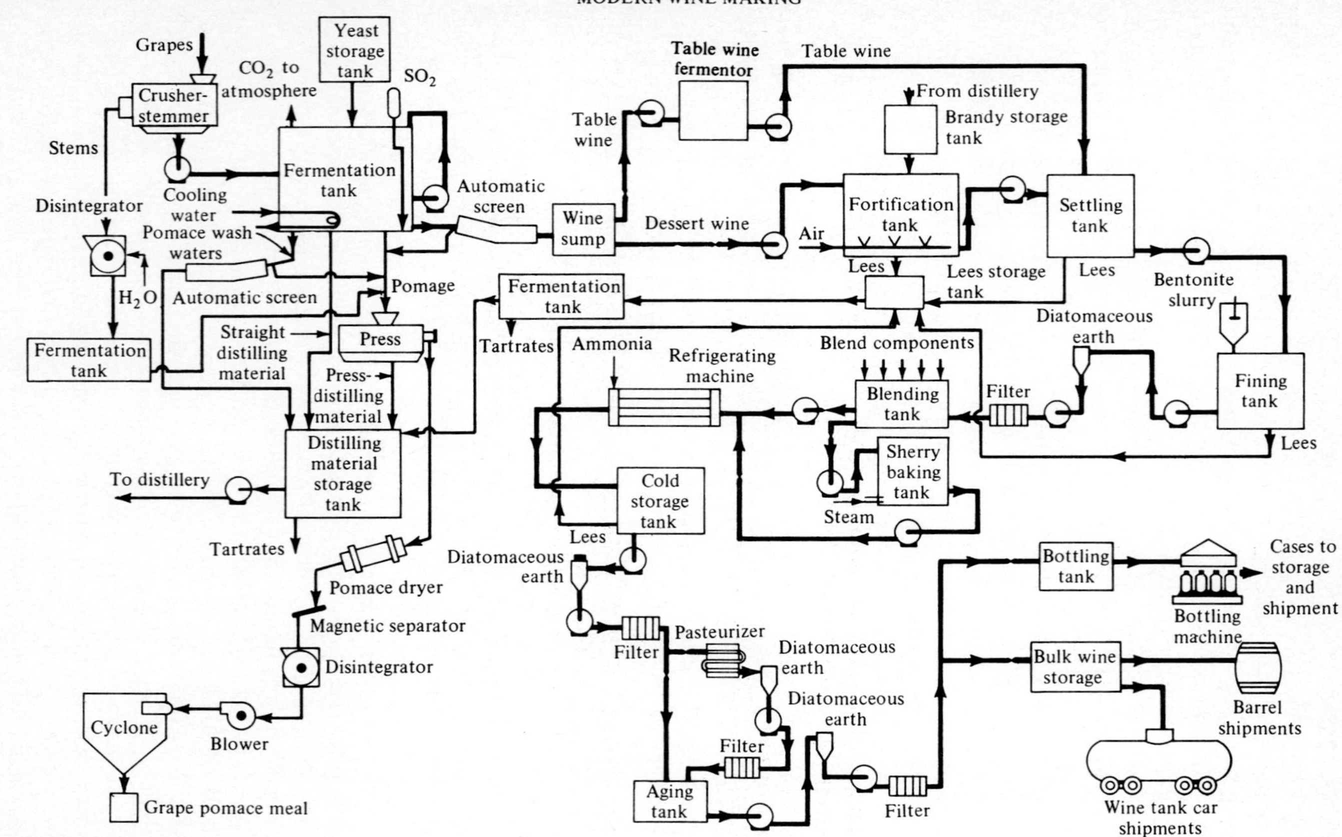

Figure 12.6 A process flowsheet illustrating the major operations in modern winemaking. *(Reprinted from W. Q. Hull, W. E. Kite, and R. C. Auerbach, "Modern Winemaking," Ind. Eng. Chem., vol. 43, p. 2182, 1951, copyright by American Chemical Society).*

adds considerably to pretreatment costs: one estimate for fuel alcohol production from corn stover indicates that the pretreatment equipment required for the pre-fermentation section shown in Fig. 12.7 would account for over 50 percent of the total process capital cost; the breakdown of the latter for a large ethanol-from-cellulosics plant is feedstock preparation (5 percent), hydrolysis (19 percent), acid recovery (29.4 percent), fermentation and purification (19.2 percent), and utilities/off-sites (27.4 percent).

The typical industrial operations are batch fermentations with fermentation times of about 50 hours, initial pH = 4.5, and $T = 20\text{–}30°C$ to give an ethanol yield of about 90 percent of the sugar theoretical value. Final ethanol levels are 10–16% v/v, and ethanol must subsequently be recovered from the final broth.

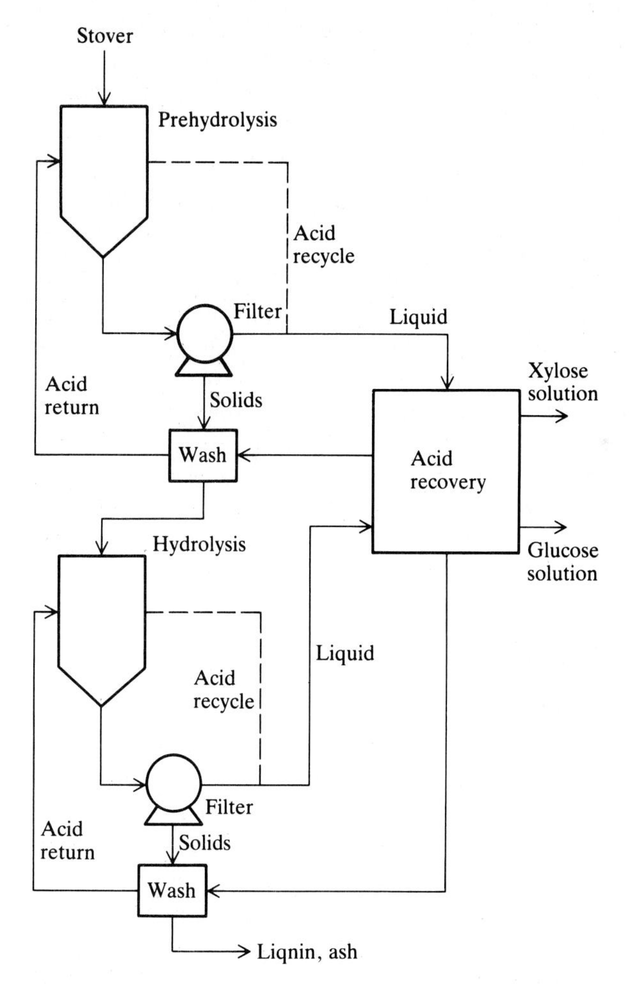

Figure 12.7 Corn stover pretreatment process for fuel ethanol feedstock production. *(Reprinted from E. C. Clausey and J. L. Gaddy, "Production of Ethanol from Biomass," Ann. N.Y. Acad. Sci., vol. 413, pp. 435–447, 1983.)*

The key features of the traditional batch fermentation process, which any replacement must improve upon, are:

1. Raw material costs are a predominant portion of the total alcohol costs.
2. The slow reaction rate leads to high fermentor volume requirements, contributing about 10 cents per gallon (1981) to the alcohol cost.
3. Valuable by-products need to be recovered: yeast sales (as feed supplement) can diminish alcohol costs by up to 22 cents per gallon.
4. Distillation is energy-intensive, but efficient: a standard "old" plant in Peoria, IL uses less than 30 percent of the ethanol heating value in the distillation process.

The revived interest in ethanol as fuel has included many novel process and microbial developments which have improved commercial prospects considerably; alcohol-containing motor fuels are available in a number of U.S. locations.

Important modifications arising in the last decade include:

1. Reduced pressure operation, allowing alcohol volatilization during fermentation, thereby reducing alcohol product inhibition, extending fermentation time length, and increasing sugar utilization.
2. Continuous process configurations which include cell recycle and vacuum, allowing a twelvefold increase of ethanol productivity vs. conventional continuous process. At 50 mmHg pressure and glucose feed of 33% (w/v), ethanol productivity of 40 g L^{-1} h^{-1} was achievable.
3. Bacteria (*Zymomonas mobilis*) which have higher glucose uptake and ethanol production specific rates than yeast.
4. Immobilized cells which exhibit much higher volumetric alcohol productivities than conventional suspension culture.
5. Tall tower fermentors which involve highly flocculent yeast which form, at steady-state throughput, a fixed plug in the base of the fermentor (cell densities to 50–80 g L^{-1} are obtained). A large-diameter baffled top allows retention (via sedimentation) of virtually all cells without need for an auxiliary mechanical separation device. Following a several-week start up, year-long continuous operation has been achieved.
7. Modified ethanol recovery processes to decrease high separation operating costs have been proposed which include vapor compression, recuperation of heat derived from wastewater treatment (corn alcohol plant), and dehydration of water–ethanol mixture by use of various synthetic or natural sorbents.

Many of the above developments have been tested, but others have appreciable development yet needed before the commercial prospects can be confidently assessed. A comparison of some of these alternate fermentation processes is given in Table 12.20.

In view of the many potential changes available for future ethanol plants, a comparative economic evaluation with four better studied configurations is instructive. Table 12.21 illustrates that for a 295 $m^3 d^{-1}$ plant producing 95% ethanol from 50% cane molasses sugar solution, the investment related portion of operating costs diminish in the sequence: batch > continuous > continuous (cell recycle) > (vacuum + cell recycle), indicating that the progressively higher volumetric productivity more than offsets the cost of increasing plant complexity. Lower labor costs are evident for all three continuous processes, and the steam utility savings in the (vacuum + cell recycle) system is striking.

Uncertainties responsible for major government policy shifts (U.S. and elsewhere) regarding ethanol fuel production include the future prices and supply of petroleum (and gasoline), the (un)desirability of using a major food product (corn) as a fuel feedstock, and availability (or lack thereof) of appreciable government guaranteed prices and/or subsidies. For no other single fermentation product since World War II has the potential impact been sufficient to require consideration on the part of many countries of national resources and security.

The acetone–butanol producing fermentation has been reconsidered in light of butanol as a diesel fuel extender ("diesohol"). Calculations for a 45×10^6 kg solvents per year plant indicated that a molasses-based process was uneconomical, with this feedstock contributing about 60 percent of total annual production costs. Liquid dairy whey feedstock provided a positive economic picture (Table 12.22) if the whey was free, but whey transportation costs were included. In this latter case, the original molasses-fed plant design was modified to include whey ultrafiltration (for whey protein product credit) and anaerobic waste digestion (methane source, Chap. 14).

With these circumstances, a net return on investment of 30.8 percent was calculated, basis the considerable number of products included. The gross plant profit was $\$15 \times 10^6$/year; a value equal to several by-product credits alone (dry slops + hydrogen + carbon dioxide + ethanol). Areas identified as particularly important: low butanol yields (1.9% w/v, 1980), strict anaerobic nature, and tendency toward phage and *Lactobocilli* infection.

12.6.3 Organic and Amino Acid Manufacture

Organic acid production examples illustrate the important effect of yield on cost. A submerged fermentation process with high citric acid yields of 70 kg acid from 100 kg sugar has fermentation capital costs which are 15–25 percent less than a surface fermentation, but requires 16–26 percent greater operating costs. Inclusion of product recovery costs would diminish these percentages by the order of half when the total processes are then compared. Here the recovery process is straightforward, given a high product concentration, and the process questions center on the fermentation section.

A more modest yield of 40 g/L itaconic acid led to consideration of recovery by evaporation, ion exchange, or electrodialysis based on a continuous fermentation. The fraction of total process construction costs for these three continuous recovery process options were estimated to be 92 percent (distillation), 96 percent (ion exchange), and 56.7 percent (electrodialysis). Raw materials and total utility

Table 12.20 Alternative fermentation processes

Process	Ethanol productivity	Comments
Batch fermentation	Very low (1.8–2.5 g L^{-1} h^{-1})	Very high capital and operating cost
Simple CSTR fermentor	Low (6 g L^{-1} h^{-1})	Mechanically simple equipment, simple continuous operation
Series CSTR fermentors	2–3 times rate for simple CSTR	Simple continuous operation
Perforated plate	Not reported	Mechanically complex
CSTR with centrifuge cell recycle	High (30–40 g L^{-1} h^{-1})	Added energy requirements and added operator attention required for centrifuge
CSTR with alternative recycle scheme	High	Settlers appear to be too large for industrial application. Whirlpool separators may be attractive
Tower fermentors	High	Mechanically simple with simple continuous operation, but start up period is very long
Slant tube fermentors	High	Mechanically complex
Packed bed, tower fermentor	High	Plugging and by-passing are major problems
Dialysis fermentor	Not reported	Reaction rate limited by substrate diffusion through membrane Membrane fouling may require frequent shutdowns
Pressure dialysis fermentor	Not reported but potentially high	Membrane fouling will be a major problem
Rotor fermentor	High (36 g L^{-1} h^{-1})	Membrane fouling problem overcome but the fermentor device is mechanically quite complex Membrane destruction by mechanical shear may be a problem

Hollow fiber	Not reported but potentially high	Expensive units. Mass transfer limitation may limit productivity
Plug fermentor	High (72 times greater than for similar batch fermentation)	Frequent shutdowns for yeast rejuvenation in an aerobic environment will be required. High power feed pumps are required
Gel entrapment	90% yield from 10% glucose solution in 10 h	Very simple system with no agitation or recycle equipment required. Half life of entrapped cells must be increased to make this process attractive
Chemically bound yeast cells	Not reported	The cell viabilities do remain high after covalent bonding. If productivities also remain high, this could yield a very simple, high productivity system
Direct extractive fermentation	Potentially very high	No suitable extractant has been found
Membrane extractive fermentation	Not reported, but potentially very high	Membrane fouling may be a problem. Process is otherwise simple
Selective membrane fermentation	Not reported but potentially very high	Membranes with sufficiently high throughput rates have not yet been developed
Vacuum fermentation	Very high (80 g L^{-1} h^{-1})	Mechanically complicated equipment requiring constant monitoring. Small added energy requirements for vapor compression. Contamination of the vacuum vessel may be a problem. Pure oxygen must be sparged
Flash fermentation	Very high	Mechanically very complicated. Energy requirements only slightly increased over conventional processes. Contamination problem greatly reduced as compared to direct vacuum process and pure oxygen is not required

(Reprinted by permission from B. Maiorella, C. R. Wilke, and H. W. Blanch, "Alcohol Production and Recovery" Adv. Biochem. Eng., vol. 20, p. 72, 1981.)

Table 12.21 Operating costs for different processes for ethanol production†

	Production cost, \$ m^{-3}			
	Batch	Continuous	Continuous cell recycle	Vacuum cell recycle
Investment related costs	27.21	12.94	10.57	9.25
Operating labor	8.45	2.38	1.32	1.06
Supervision and clerical	0.53	0.26	0.26	0.26
Utilities				
Water	1.58	1.58	1.58	1.06
Power	3.17	1.58	2.38	1.58
Steam	26.68	25.10	25.10	17.96
Oxygen	......	....	....	1.32
Laboratory changes	0.26	0.26	0.26	0.26
Plant overhead	4.75	1.58	1.06	0.79
Total	72.63	45.68	42.53	33.54

† Plant capacity was 295 m^3 d^{-1} 95% ethanol from 50% "cane" molasses sugar solution.

(Reprinted by permission, from G. R. Cysewski and C. R. Wilke, "Process Design and Economic Studies of Alternative Fermentation Methods for the Production of Ethanol," Biotech. Bioeng., vol. 20, p. 1421, 1978.)

Table 12.22 Income Summary for 45 × 10^6 kg solvents per year. (Acetone/butanol fermentation facility)

Item	Annual quantity, kg	Value, \$/kg	Income, \$/year
1-Butanol	26.4 × 10^6	0.53	13.9 × 10^6
Acetone	13.2 × 10^6	0.44	5.8 × 10^6
Ethanol	4.5 × 10^6	0.40	1.8 × 10^6
Dry slops	64.3 × 10^6	0.094	4.7 × 10^6
Hydrogen	1.8 × 10^6	0.29	0.52 × 10^6
Carbon dioxide	72.0 × 10^6	0.11	7.9 × 10^6
Whey protein	10 × 10^6	\$0.46/kg	9.3 × 10^6
Methane			0.46 × 10^6
		Total	\$44.4 × 10^6

Modified, from T. Lenz and A. Moreira, "Economic Evaluation of the Acetone-Butanol Fermentation," *Ind. Eng. Chem. Prod. Res. Dev.*, **19**: 478, 1980.

costs were estimated at 49 and 13 percent of operating costs for the least expensive recovery system (electrodialysis). As the relative advantages of different recovery processes are concentration dependent, it would be interesting to determine the sensitivity of these calculations to fermentation improvement leading to 60, 80, and 100 g/L itaconic acid yields in the fermentation section.

The economic aspects of amino acid production are now considered. Of the twenty amino acids, only two (lysine and glutamic acid) were produced primarily by fermentation in 1980 (in Japan), although more are potentially producible (arginine, glutamine, histidine, leucine, ornithine, phenylalanine, proline, threonine, and valine) (Table 12.23). The simultaneous existence of both chemical and fermentative synthesis activities for L-lysine (Table 12.23) indicates that other factors are evident: as of 1980, 80 percent of lysine production occurred in Japan and S. Korea, whereas the U.S. production was chemically based. A cost comparison for lysine via fermentation or chemical synthesis is illustrated in Table 12.24.

The new sweetener aspartame is L-aspartyl-L-methionine methyl ester. Its recent adoption in foods and especially in soft drinks has driven a major increase in the market for L-phenylalanine, moving from 50 tons (1981) to over 3000 tons (1984) to nearly 8000 tons (estimated) in 1990. An early $50/kilo price has dropped to $35/kilo (1984) and may be less than $20/kilo eventually. Current phenylalanine synthesis routes include fermentation (from sugars) and enzymatic conversions from cinnamic acid or hydantoin.

12.7 SINGLE-CELL PROTEIN (SCP)

Single-cell protein or SCP refers to proteinaceous materials which are dried cells of microorganisms; example species which have been cultivated for use in animal or human foods include algae, actinomycetes, bacteria, yeasts, molds, and high fungi. Human consumption of microbial protein is ancient in origin and includes yeast in leavened bread, lactic acid bacteria in fermented dairy products (milks, sausages, cheeses), and algae (*Spirulina*) harvested from ponds and lakes.

More recent studies have established microbial growth in aerated fermentors, using substrates such as gas oil, paraffins, natural gas, molasses, and cellulosics. The operating characteristics of some SCP processes appear in Table 12.25. The process variability includes factors such as aseptic vs. nonaseptic, open vs. enclosed growth vessels, full vs. partial substrate utilization, optimal pH values from acidic to alkaline, and product recovery by filtration vs. centrifugation. Not surprisingly, bulk biomass production is keyed to availability of bulk feedstocks which include hydrocarbons as well as waste streams from paper, agriculture, dairy, and industrial processes (Table 12.26). Virtually all major SCP plants sell their product as an animal food supplement. Human consumption occurs at much smaller levels, where production of baker's yeast continues to be the prime example. Specific growth rate ranges for many organisms considered for SCP production are 0.11 to 0.7 hr^{-1}. SCP production is widespread in the world; the largest total capacity is found in the Soviet Union.

Table 12.23 Data for commercially produced amino acids[†]

Amino acid	Price March 1980, per kg pure L	Present source	Production 1978, tonnes	Potential for application of biotechnology (de novo synthesis or bioconversion using organisms and enzymes)
Alanine	$ 80	Hydrolysis of protein; chemical synthesis	10–50(J)[‡]	
Arginine	28	Gelatin hydrolysis	200–300(J)	Fermentation in Japan
Asparagine	50	Extraction	10–50(J)	
Aspartic acid	12	Bioconversion of fumaric acid	500–1000(J)	Bioconversion
Citruline	250	....	10–90(J)	Fermentation in Japan
Cysteine	50	Extraction	100–200(J)	
Cystine	60	Extraction	100–200(J)	
DOPA (dihydrophenyl-alanine)	750	Chemical	100–200(J)	
Glutamic	4	Fermentation	10,000–100,000(J)	De novo: *Micrococcus glutamicus*
Glutamine	55	Extraction	200–300(J)	Fermentation in Japan
Histidine	160	....	100–200	Fermentation in Japan
Hydroxyproline	280	Extraction from collagen	10–50	

Isoleucine	350	Extraction	10–50(J)	
Leucine	55	....	50–100(J)	Fermentation in Japan
Lysine	350	Fermentation (80%) Chemical (20%)	10,000(J)	(80% by fermentation) De novo: *Corynebacterium glutamicum* and *Brevibacterium flavum*
Methionine	265	Chemical from acrolein	17,000(D, L)[§] 20,000(D, L)(J)	
Ornithine	60	....	10–50(J)	Fermentation in Japan
Phenylalanine	55	Chemical from benzaldehyde	50–100(J)	Fermentation in Japan
Proline	125	Hydrolysis of gelatin	10–50(J)	Fermentation in Japan
Serine	320	....	10–50(J)	Bioconversion in Japan
Threonine	150	....	50–10(J)	Fermentation in Japan
Tryptophan	110	Chemical from indole	55(J)	
Tyrosine	13	Extraction	50–100(J)	
Valine	60	....	50–100(J)	Fermentation in Japan

[†] Production data largely from Japan because of relative small U.S. production.

[‡] Japan.

[§] D and L forms.

(Reprinted from p. 88 of "Impacts of Applied Genetics: Microorganisms, Plants and Animals," Office of Technology Assessment, Washington, D.C., 1981.)

Table 12.24 Summary of recent estimates of primary U.S. cost factors in the production of L-lysine monohydrochloride by fermentation and chemical synthesis

Cost factors in production of 98% L-lysine monohydrochloride	By fermentation[†]			By chemical synthesis[‡]		
	Requirement (units per unit product)	Estimated 1976 cost per unit product		Requirement (units per unit product)	Estimated 1976 cost per unit product	
		cents/lb	cents/kg		cents/lb	cents/kg
Total labor[§]	....	8	18	....	9	20
Materials						
Molasses	44	7	16	....	....	
Soybean meal, hydrolyzed	0.462	4	9	....		
Cyclohexanol	....	....	....	0.595	17	37
Anhydrous ammonia	....	....	....	0.645	6	14
Other chemicals[¶]	....	7	15	....	4	10
Nutrients and solvents	....	....	....	....	4	8
Packaging, operating, and maintenance materials	....	10	22	....	9	21
Total materials	....	28	62	....	45	90

Total utilities[††]	6	12	7	16
Total dir. oper. cost	42	92	56	126
Plant overhead, taxes and insurance	10	21	10	21
Total cash cost	52	11	66	147
Depreciation[‡‡]	16	35	13	28
Interest on working capital	1	3	1	3
Total cost[§§]	69	151	80	178

[†] Assumes a 23 percent yield on molasses.

[‡] Assumes a 65 percent yield on cyclohexanol.

[§] Includes operating, maintenance, and control laboratory labor.

[¶] For both the process of fermentation and chemical synthesis, assumed use of hydrochloric acid (36 percent) and ammonia (29 percent). For fermentation includes also potassium diphosphate, urea, ammonium sulfate, calcium carbonate, and magnesium sulfate. For chemical synthesis also includes nitrosyl chloride, sulfuric acid, and a credit for ammonium sulfate byproduct.

[††] Total utilities for both processes include cooling water, steam process water, and electricity. For chemical synthesis, natural gas is also included.

[‡‡] Ten percent per year of fixed capital costs for a new 20-million-lb-per-year U.S. plant built in 1975 at assumed capital cost of $\$38.6 \times 10^8$ for fermentation and $\$32.5 \times 10^6$ for chemical synthesis exclusive of land costs.

Source: Stanford Research Institute, *Chemical Economics Handbook* **583**:3401, May 1979.

(Reprinted from p. 89 of "Impacts of Applied Genetics: Microorganisms, Plants and Animals," Office of Technology Assessment, Washington, DC., 1981.)

Table 12.25 Operating characteristics of selected single-cell protein processes

	Process			
Item	Algal *Spirulina maxima*	Bacterial *Methylophilus methylotrophus* (methanol)	Yeast *Candida utilis* (ethanol)	Mold *Paecilomyces varioti* (sulfite waste liquor)
Type of process	Batch or semicontinuous	Continuous	Continuous or batch	Continuous
Sterility	Nonaseptic	Aseptic	Aseptic	Nonaseptic
Fermentor	Ponds	Airlift	Agitated	Agitated
Feedstock utilization	Partially or fully utilized	Fully utilized	Fully utilized	Partially utilized
Temperature, °C	Ambient	35 to 42	30 to 40	38 to 39
pH	9 to 11	6.0 to 7.0	4.6	4.5 to 4.7
Product recovery	Filtration	Agglomeration and centrifugation	Centrifugation	Filtration

(From J. Litchfield, "Single Cell Protein," Science, vol. 219, p. 740. Copyright 1983 by the American Association for the Advancement of Science.)

Table 12.26 Selected raw materials used as carbon and energy sources in single-cell protein processes

Raw material	Process type and scale[†]	Organism	Producer or developer
	Algal, 2 metric tons per day[‡]	*Chlorella* sp.	Taiwan Chlorella Manufacture Co. Ltd., Taipei
CO_2	Photosynthetic		
Cane syrup, molasses (sucrose)	Nonphotosynthetic		
CO_2 or $NaHCO_3$, Na_2CO_3	Algal, 320 metric tons per year;[§] photosynthetic	*Spirulina maxima*	Sosa Texcoco, S.A., Mexico City
Methanol	Bacterial		
	70,000 metric tons per year	*Methylophilus methylotrophus*	Imperial Chemical Industries, Billingham
	1000 metric tons per year	*Methylomonas clara*	Hoechst-Uhde, Frankfurt, West Germany
Ethanol	Yeast, 7500 short tons per year	*Candida utilis* (Torula)	Pure Culture Products, Hutchinson, Minnesota
n-Alkanes, wood hydrolyzates	Yeast (several plants), 20,000 to 40,000 metric tons per year	*Candida* sp.	All-Union Research Institute of Protein Biosynthesis, U.S.S.R.
Sulfite waste liquor	Yeast, 15 short tons per day	*Candida utilis*	Rhinelander Paper Corp., Rhinelander, Wisconsin
	Mold, 10,000 metric tons per year	*Paecilomyces varioti*	Pekilo Process, Finnish Pulp and Paper Research Institute, Jamsankoski, Finland
Glucose	Mold, 50 to 100 metric tons per year	*Fusarium graminearum*	Rank Hovis MacDougall Research Limited, High Wycombe, U.K.
Cheese whey (lactose)	Yeast, 5000 short tons per year	*Kluyveromyces fragilis*	Amber Laboratories, Juneau, Wisconsin
	Mold, 300 metric tons per year	*Penicillium cyclopium*	Heurty, S. A., France

[†] Plant capacity, metric tons (1000 kilograms), or short tons (2000 pounds) per unit of time indicated.
[‡] Total pond area, 83,400 square meters.
[§] Pond area, 900 hectares.
(From J. Litchfield, "Single Cell Protein," Science, vol. 219, p. 740. Copyright 1983 by the American Association for the Advancement of Science.)

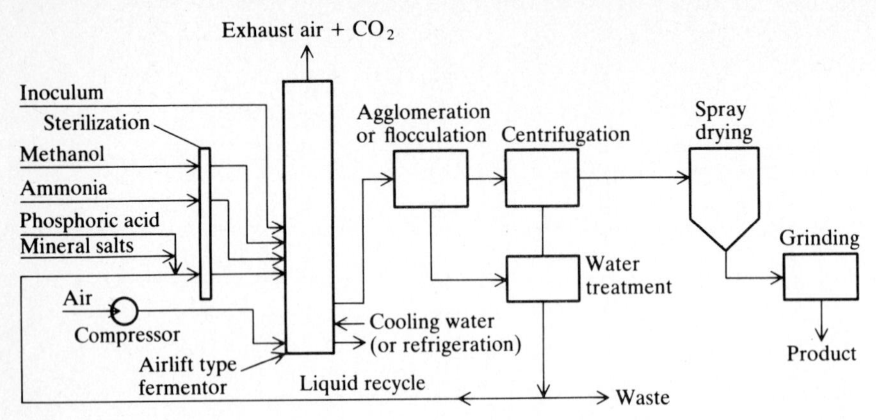

Figure 12.8 Schematic diagram of a recent process for producing single-cell protein. *(Reprinted by permission from J. Lichfield, "Single Cell Protein," Science, vol. 219, p. 740. Copyright 1983 by the American Association for the Advancement of Science.)*

A recent process for SCP production with a methanol-utilizing bacterium (*Methylophilus methylotrophus*) (Figure 12.8) provides the overall stoichiometry:

$$1.72\,CH_3OH + 0.23\,NH_3 + 1.51\,O_2 \longrightarrow 1.0\,CH_{1.68}O_{0.36}N_{0.22} + 0.72\,CO_2 + 2.94\,H_2O$$

The process provides rapid growth ($\mu = 0.5\ h^{-1}$), dense final broths (achieved by agglomeration in the fermentation vessel) which can be fed directly to centrifuges, CO_2 by-product recovery, and a growth yield above 0.50 g cell/g substrate.

Major improvements in SCP production have included strain improvement (large *S. cerevisiae* cells leading to easier recovery), genetic engineering of methanol utilizers for higher protein content and of *Methylophilus methylotrophus* for improved NH_3 utilization efficiency, development of protein-excreting *Bacillus brevis*, improved cell harvesting by agglomeration and electrocoagulation, and process automation.

Economically, SCP must always compete with other protein sources (Table 12.27). Soybeans in particular have provided a large reservoir of inexpensive protein; as market expansion of this legume is expected in the 1980s, SCP may be expected to retain by-product status in most locations. Prospects for markets for SCP depend on usage. Feed grade SCP competes with conventional feed sources and growth performance of the appropriate animal (e.g., chicken, turkey, swine). Food grade SCP provides particular functions (typically flavor, but also texture and nutritional value); Torula yeasts are commonly used here. The greater capital costs are associated with SCP for food grade processes: 1975–76 estimates required selling prices of \$660–\$1000/metric ton of annual capacity.

For large-scale aerobic fermentations, major inputs include raw materials (substrate), utilities (power for aeration, mixing and mass transfer), and cooling

Table 12.27 Comparison of selling price ranges for selected microbial, plant, and animal protein products

Product, substrate, and quality	Crude protein content	Price range 1979 U.S. dollars/kg
Single-cell proteins		
Candida utilis, ethanol, food grade	52	1.32–1.35
Kluyveromyces fragilis, cheese whey, food grade	54	1.32
Saccharomyces cerevisiae:		
Brewer's, debittered, food grade	52	1.00–1.20
Feed grade	52	0.39–0.50
Plant proteins		
Alfalfa (dehydrated)	17	0.12–0.13
Soybean meal, defatted	49	0.20–0.22
Soy protein concentrate	70–72	0.90–1.14
Soy protein isolate	90–92	1.96–2.20
Animal proteins		
Fishmeal (Peruvian)	65	0.41–0.45
Meat and bonemeal	50	0.24–0.25
Dry skim milk	37	0.88–1.00

Source: Office of Technology Assessment.

(Reprinted from p. 108 of "Impacts of Applied Genetics: Microorganisms, Plants, and Animals," Office of Technology Assessment, Washington, DC, 1981.)

(heat transfer). Inclusion of cellulosics as an SCP feedstock would have to include utilities for the very appreciable pretreatment section which would be required to convert cellulosics to a fermentable form.

A comparative estimate which includes all raw materials (Table 12.28) indicates that their costs are dominant, being 44–77 percent of the total production costs.

12.8 ANAEROBIC METHANE PRODUCTION

Anaerobic conversion of carbohydrate/cellulosics, especially of agricultural wastes, has been considered for biogas (methane) production, which is typically 50–65% CH_4, 35–50% CO_2, 30–160 g/m^3 H_2O, and 1.5–12.5 g/m^3 H_2S. Costs of producing biogas (option *A*) vs. methane (CO_2 scrubbed biogas) (option *B*) are summarized in Table 12.29. The largest plants have production costs of \$2.61–\$2.74/GJ (gigajoules) vs. 1980 natural gas prices of \$3.00/GJ. By-product utilization of fermentation residue at \$60/ton as animal feed (protein) supplement brings the production costs of even the smallest plant size (1 Mg TS/day) into a competitive range.

Table 12.28 Comparative production costs of biomass production

Production Item	Yeast paraffin, % (Italy)	Bacteria methanol, % (England)	Yeast ethanol, % (Czechoslovakia)	Fungi sulfite liquor, % (Finland)	Bacteria bagasse, % (United States)
Depreciation	9.3	5.8	5.8	9.1	11.5
Raw materials	58.5	73.8	77.1	55.1	43.6
Substrate	29.4	47.4	63.9	17.0	25.7
Phosphoric acid	11.1	11.8	3.2	16.2	5.7
Ammonia	9.9	12.0	4.8	13.3	3.6(?)
Mineral salts	2.9	2.6	1.9	4.2	4.2
Miscellaneous	5.2	....	3.3	4.4	4.4
Utilities	23.8	14.2	12.0	24.8	36.6
Labor, etc.	8.4	6.2	5.1	11.0	8.3
Total	100%	100%	100%	100%	100%

Note: Substrate cost assumption: paraffin = 1.3 methanol: ethanol = 2.6 methanol.
After M. Moo-Young, "Economics of SCP Production," *Process Biochem.*, **12** (4): 6–10 (1977).

Table 12.29 Costs for producing biogas (option A)† and methane (option B)‡ from anaerobic fermentation of beef cattle manure at 55°C

	Plant size, Mg TS/day			
Parameter	1	10	100	1000
Capital costs, $1000				
Option *A*	70.7	354	2185	13,490
Option *B*	89.6	449	2769	17,090
Fixed costs, $1000/year				
Option *A*	17.0	85.0	525	3,240
Option *B*	21.5	107.7	665	4,100
Utility costs, $1000/year				
Option *A*	2.24	21.9	213	2,126
Option *B*	2.86	28.2	276	2,760
Labor costs, $1000/year	11.8	22.9	44.3	85.9
Total annual costs, $1000/year				
Option *A*	31.0	130.1	782	5,450
Option *B*	36.2	158.8	985	6,942
Production cost, $/GJ				
Option *A*	13.94	5.72	3.41	2.61
Option *B*	16.28	6.98	4.29	2.74

† Biogas is compressed to 860 kPa, 12 h of mixing fermentor liquor and influent slurry.

‡ CO_2 is scrubbed and methane is compressed to 860 kPa, 12 h of mixing fermentor liquor and influent slurry.

12.9 OVERVIEW

Successful operation of microbial processes depends upon both technical and economic considerations. A detailed example presented in this chapter provides a model approach by which many aspects of any particular economic analysis may be carried out. Section discussions concerning fine chemicals, bulk chemicals and beverages, single-cell protein, and anaerobic methane production indicate the strong variations which occur in microbial process economics. Further examples appear in the problem section.

PROBLEMS

12.1 Plant design stages Define, from memory, the stages of a plant design: inception, preliminary evaluation, development of final design data, final economic analysis, detailed engineering design, procurement, construction, start-up and trial runs, production.

12.2 Sensitivity analysis Table 12.10 suggests strongly that, for major (20 percent unfavorable) deviations from the assumed base case, the order of greatest to least sensitive items was selling price, selling volume, and capital overrun. A better estimate of the sensitivity, or percent change of the ROI with respect to a percent change in each variable, is available from calculations of ± 5 and ± 10 percent variations. Carry out such a calculation for the same three variables, and plot ROI(%) vs. percent deviation from base case value of selling price, selling volume, and capital overrun. Is the order of sensitivities the same or different for small deviations (several percent) as for the 20 percent example of Table 12.10? (Consider the use of spreadsheet software for this type of calculation.)

12.3 Yield factor sensitivity Column B of Table 12.10 contains figures for a 20 percent reduction in selling price, without mention of cause.

(*a*) Assuming that selling price is directly related to product percentage content of active ingredient and that the latter is directly related to reciprocal of yield factor $(Y_{P/S})^{-1}$, plot ROI vs. percent deviation of $Y_{P/S}$ for 0, $\pm 5\%$, $\pm 10\%$, $\pm 20\%$.

(*b*) Scan a number of articles in the journal *Biotechnology and Bioengineering* for a data plot from which a product yield factor has been determined. From the original data, confirm the original yield value, and calculate a standard deviation, $\sigma^2 = (\mu_i - \bar{\mu})^{2/n}$, for the data. What is σ as a percentage of $Y_{P/S}$? How confidently would you predict $Y_{P/S}$ for a single fermentation run? How would this result impact your confidence in ROI estimates?

(*c*) Suppose the run-to-run variation of $Y_{P/S}$ under "identical" operating conditions is a distribution given by $N(Y_{P/S}) = e^{-(Y_{P/S} - \bar{Y}_{P/S})^{2}/\sigma}$. Plot this normalized distribution function, and discuss what additional costs would be incurred if 10 successive batches had routinely to be stored in order to have blending sufficiently consistent to maintain an average blend value deviating 5 percent or less from $\bar{Y}_{P/S}$.

12.4 Ethanol process alternatives An exhaustive comparison of ethanol process/recovery alternatives is recently available (B. L. Maiorella, H. W. Blanch, and C. R. Wilke, *Biotech. Bioeng.*, **26**: 1003, 1984). Read this review, choose one particular configuration, and prepare the equivalent of Tables 12.2 through 12.9. Summarize the tables in a memo of length less than two pages (attach all tables as an appendix).

12.5 Production competition avenues Your company has discovered a new microbial product which has appreciable pharmacological activity and must be produced at large scale to achieve a 20 percent share of the potential market. Your director of research, trained in organic chemistry, gives the synthetic organic chemists and the microbiologists 3 months to propose "ideal" wholly synthetic and microbial production processes, respectively. The results received 3 months later lead to an economic analysis which suggests that, costwise, the two routes are equal. What do you recommend to management, and why?

12.6 Amino acid production Read the section on amino acids in the Kirk-Othmer Encyclopedia of Chemical Technology. Rationalize the last column in Table 12.23 which indicates that only one country in the world has serious interests in amino acid production via fermentation.

12.7 L-lysine process economics (*a*) Calculate the sensitivity of the total costs (Table 12.24) for L-lysine production for each item which contributes. (Express sensitivity as percent change in total cost per 5 percent change in item cost.)

(*b*) Suppose it takes 5 years to complete construction and operation of a plant if a decision to build is made today. From Table 12.24, would you build a fermentation plant or a synthetic plant for L-lysine production if molasses prices are expected to rise 10 percent a year, relative to all else, and cyclohexanol prices are expected to rise at 0, 10, 20, or 30 percent per year?

12.8 Process modification: soluble to immobilized enzyme Consider an enzyme process involving N_b batches/year and E_b mass of enzyme/batch. It is compared to an immobilized enzyme configuration, also at N_b batches/year. Immobilized enzyme preparation involves a mass M_c of carrier, a specific catalyst loading C (mass E/mass carrier), an immobilization retention, R, of enzyme (immobilized/used), and an immobilized enzyme lifetime of N_i batches.

(*a*) If X_e and X_c are costs/mass of enzyme and carrier, respectively, and if E_b is lost after each soluble enzyme batch, and E_c and M_c lost after N_i immobilized batches (no recovery), show that the cost difference between the batch soluble enzyme and the batch immobilized enzyme processes is

$$X_e E_b N_b - M_c C(1 - R) X_e N_i - M_c C R X_e N_i - M_c X_c N_i$$

(*b*) If η = fraction of total immobilized enzyme activity which is utilized (effectiveness factor, Chapter 4), and product conversion and substrate costs are the same, show from part (*a*) that the processes are economically equal when the following equation holds:

$$1 - \left(\frac{N_i}{N_b}\right)\cdot\frac{1}{\eta R}\left[1 + \frac{X_r}{C}\right] = 0$$

Thus, an economic advantage for use of immobilized enzyme exists only if the catalyst lifetime satisfies

$$N_i > N_b(1 + X_r/C)/\eta R$$

Similar criteria including soluble batch vs. continuous immobilized operation, catalyst deactivation, and substrate costs are found in M. M. Domach, "Specifying the Values of Immobilized Enzyme Performance Parameters in Terms of Process Economic Feasibility," *The Chem. Eng. Jl.*, **29**: B1, 1984; also presented are effectiveness factor, η, and immobilization retention, R, target values for β-amylase, glucose isomerase, and lactase.

12.9 Production of biohazardous compounds The building, instrumentation and overhead costs associated with biochemical plant design in this chapter apply to conventional production of reasonably innocuous materials. Additional costs are incurred in manufacture of potentially biohazardous materials, such as cytotoxic antitumor compounds or mammalian virus-derived materials. Cost estimations must include the following additional National Cancer Institute design concepts:

1. Facility: controlled access (totally enclosed, fenced).
2. Air pressure: negative (no outflow of air from door, window or any unfiltered vent).
3. Air exhaust: always through HEPA filters (complete particle removal).
4. Local, complete containment of all potential aerosol generating equipment (fermentor, centrifuge, filters, etc).
5. Personnel protection: garments, masks, boots, gloves (purchase and daily laundry).
6. Personnel medical monitoring.
7. Environmental monitoring.

Suppose that a fermentation plant one-fourth the size of the main chapter example was designed to meet the annual demand for a number of different cytotoxic compounds (different fermentations; same average length as example). Using the fermentation section design in the chapter as the process concept, develop all tables needed to estimate a cost comparison of a conventional plant vs. a cytotoxin production facility. (Equipment, materials, utilities, labor, operating costs; do not do return-on-investment calculation). (M. C. Flickinger and E. B. Sansome, "Pilot and Production Scale Containment of Cytotoxic and Oncogenic Fermentation Processes," *Biotech. Bioeng.*, **26**: 860, 1984.)

12.10 Economics of activated-sludge process A detailed analysis of an aerated reactor with enriched biomass stream recycle, centrifugal recovery of final biomass, (for disposal) and chlorination of aqueous effluent was reported by R. B. Paterson and M. M. Denn, "Computer-aided Design and Control of Activated Sludge Process," *The Chemical Engineering Journal*, **27**: B13, 1983. They found that total plant costs were very insensitive to two key variables over appreciable ranges: reactor biomass levels (vary inversely with reactor volume) and average biomass residence times (differs from liquid residence time in cell-rich recycle systems).

(*a*) Read Sec. 14.4 through Eq. (14.5).

(*b*) Do Prob. 14.7 to understand the kinetics of the operation.

(*c*) Read the above reference and comment on why these intriguing conclusions were reached. Note that the performance criteria, met by all designs, are based on fixed (99.25%) reduction of influent carbon substrate.

REFERENCES

General

See Ref. 5 of Chap. 5 and

1. W. H. Bartholomew and H. B. Reisman, "Economics of Fermentation Process," *Microbial Technology, 2d ed.*, vol. 2, H. J. Peppler and D. Perlman (eds.), Academic Press, New York, 1978, pp. 463–496 (a complete bioprocess example, from inception to return on investment discussion).
2. *Commercial Biotechnology: An International Analysis*, Office of Technology Assessment, Washington, DC, January, 1984.
3. J. D. Stowell and J. B. Bateson, "Economic Aspects of Industrial Fermentation," *Biooactive Microbial Products 2: Development and Production*, L. J. Nisbet and D. J. Winstanley (eds.), Academic Press, New York, 1983, p. 117.
4. A. H. Rose, *Economic Microbiology*, vols. 1–5, Academic Press, New York, 1977–1981.
5. M. S. Peters and K. D. Timmerhaus, *Plant Design and Economics for Chemical Engineers*, McGraw-Hill, 3d ed., N.Y., 1980. (A complete text including general design considerations, cost estimating, depreciation, taxes, investment, profitability, optimal design, material selection, equipment for material handling, heat transfer, mass transfer, and reactors.)
6. G. Ulrich, *A Guide to Chemical Engineering Process Design and Economics*, John Wiley and Sons, New York, 1984. [A two-section text structured much like the Bartholomew/Reisman example of this chapter; Process Design (conception, flow sheet preparation, specification and design of equipment), Economic Analysis (capital and manufacturing cost estimation, optimization, profitability analysis), Technical reporting.]
7. W. Crueger and A. Crueger, *A Textbook of Industrial Microbiology*, (English ed.), Science Tech., Inc., Madison, WI, 1984. (Many chapters on a different classes of biochemical conversions contain useful economic information as well.)
8. *Impact of Applied Genetics: Microorganisms, Plants, and Animals*, Office of Technology Assessment, Washington, D.C., 1981. (Survey of potential technical and economic impact of the "new" biotechnologies.)
9. H. J. Peppler and D. Perlman, Eds., *Microbial Technology, 2d ed., vol. 1: Microbial Processes; vol. II: Fermentation Technology*, Academic Press, New York, 1979 (a wealth of process, economic, and technical information).

Particular references

10. R. W. Swartz, "The Use of Economic Analysis of Penicillin G Manufacturing Cost in Establishing Priorities for Fermentation Process Improvement," *Ann. Reports on Fermentation Processes*, **3**: 75 (1979).
11. G. C. Avgerinos and D. I. C. Wang, "Direct Microbiological Conversion of Cellulosics to Ethanol," *Ann. Reports on Fermentation Processes*, **4**: 165, 1979. (pp. 180–188 review ethanol economics).
12. C. Ratledge, "Fermentation Substrates," *Ann. Reports on Fermentation Processes*, **1**: 49, 1977. (Substrate costs.)
13. R. S. Roberts et al., "Process Optimization for Saccharification of Cellulose by Acid Hydrolysis," *Biotech. Bioeng. Symp.*, **10**: 125, 1980. (Example: hexose from cellulose at minimum cost.)
14. M. Ackerson et al., "Two Stage Acid Hydrolysis of Biomass," *Biotech. Bioeng. Symp.*, **10**: 103, 1980. (Example: corn stover to fermentable sugars.)
15. A. Constantinides, "Application of Rigorous Optimization Methods to the Control and Operation of Fermentation Processes," *Ann. N.Y. Acad. Sci.*, **326**: 193, 1979.
16. E. C. Clausen and J. L. Gaddy, "Production of Ethanol from Biomass," *Ann. N. Y. Acad. Sci.*, **413**: 435, 1983.
17. J. Litchfield, "Single Cell Protein," *Science*, **219**: 740, 1983.
18. K. Aunstrup, O. Andresen, E. A. Falch, and T. K. Nielsen, "Production of Microbial Enzymes," p. 283 in *Microbial Technology, 2d ed.*, vol 1, H. J. Peppler and D. Perlman (eds.), Academic Press, New York, 1979.

19. T. K. Ng, R. M. Busche, and C. C. McDonald, "Production of Feedstock Chemicals," *Science*, **219**: 737, 1983. (Emphasis on renewable resources.)
20. B. Maiorella, C. R. Wilke, and H. W. Blanch, "Alcohol Production and Recovery," *Adv. Biochem. Eng.*, **20**: 72, 1981. (Comparative process and economic summary.)
21. B. L. Maiorella, H. W. Blanch, and C. R. Wilke, "Economic Evaluation of Alternative Ethanol Fermentation Processes," *Biotech. Bioeng.*, **26**:1003, 1984.
22. T. Lenz and A. Moreira, "Economic Evaluation of the Acetone-Butanol Fermentation," *Ind. Eng. Chem. Proc. Des. Dev.*, **19**: 478, 1980.
23. M. Moo-Young, "Economics of SCP Production," *Process Biochemistry*, **12**: 6, 1977.
24. A. Klausner, "Building for Success in Phenylalanine," *Biotechnology*, **3** (April), 301 (1985).
25. L. E. Casida, *Industrial Microbiology*, John Wiley and Sons, Inc., New York, 1968.

CHAPTER

THIRTEEN

ANALYSIS OF MULTIPLE INTERACTING MICROBIAL POPULATIONS

Until now we have concentrated on systems dominated by a single type of microorganism. Untouched so far are the myriad situations where several different microbial species are important. Among commercial processes, we can cite biological waste-water treatment and cheese manufacture as examples where multiple microbial species are required. Moreover, mixed populations of microorganisms are the rule rather than the exception in natural systems. The natural cycles of carbon, nitrogen, oxygen, and numerous other elements on our planet all require the active participation of many different microorganisms. These applications and others will be pursued more thoroughly in the next chapter.

For the moment, we shall concentrate on the analysis of microbial interactions. We shall seek first to characterize two-species interactions and then extend our analysis to more complex populations. In the following section, four of the six basic types of microbial interactions are considered. Subsequently, the last two pairwise interactions are analyzed in detail mathematically. The final sections consider general and large systems and the development of spatial patterns in mixed populations.

13.1 NEUTRALISM, MUTUALISM, COMMENSALISM, AND AMENSALISM

The first two of these relationships are among the extreme cases possible when two microbial species interact. *Neutralism* means that there is no change in the growth rate of either microorganism due to the presence of the other. Thus, so far as growth rates are concerned, there is no observable interaction. On one extreme from this bland situation is *mutualism*, where both species grow faster together than they do separately. The other extreme, to be considered in Sec. 13.3, is *competition*, where each species exerts a negative influence on the growth rate of the other.

Very few instances of neutralism have been studied. One of these is growth of yogurt starter strains of *Streptococcus* and *Lactobacillus* in a chemostat. The total counts of these two species at a dilution rate of $0.4\ h^{-1}$ were quite similar whether the populations were cultured separately or together. Indeed, it is difficult to imagine many situations in which consumption of nutrients and evolution of products by each species has absolutely no effect on its neighbor. Neutralism can only occur, it would seem, in special environment-microorganism scenarios where each species consumes different limiting substrates and where end products are effectively neutralized or diluted.

Viewed from a different perspective, neutralism implies that the pure-culture behavior of both species is identical to their behavior in mixed culture. The apparent rarity of neutralism casts some shadows on the value of many pure-culture data in describing how mixed populations behave. Prediction of mixed-culture performance from pure-culture studies will be possible only when these studies characterize the relationship between a microorganism and its environment in great detail. Such information on each species can then be hooked together to describe the mixed-population situation.

Mutualism is much more common than neutralism and involves several different mechanisms. One of these is the exchange of growth factors. Such an interaction can be beautifully illustrated by growing a phenylalanine-requiring strain of *Lactobacillis* and a folic acid-requiring strain of *Streptococcus* in a mixed batch culture. Figure 13.1 shows the results with a synthetic medium lacking both phenylalanine and folic acid. The mixed culture grows well, while separate pure cultures exhibit almost no growth.

Exchange of nutrients may also be involved in mutualistic relationships. Numerous instances are known where mutually beneficial associations exist between aerobic bacteria and photosynthetic algae. While the bacteria use oxygen and carbohydrate, they produce CO_2 and growth factors. The algae, using sunlight as an energy source, convert CO_2 to carbohydrate and also liberate O_2. This system illustrates on a microscopic scale some features of the carbon and oxygen cycles considered further in Chap. 14.

Very close mutualistic ties, such that the partnership is necessary for the survival of one or both species, are often termed *symbiosis*. Microbes are found in many symbiotic relationships with each other as well as with higher organisms. Characteristics of several systems exhibiting symbiosis, some of them quite fascinating, are given in Table 13.1. We shall consider here one example in more detail since it reveals another mode of multualistic interaction.

Methanobacillus omelianskii, a "bacterium" abundant in anaerobic sludge (see Chap. 14), has been discovered to be a mixture of two species. The first

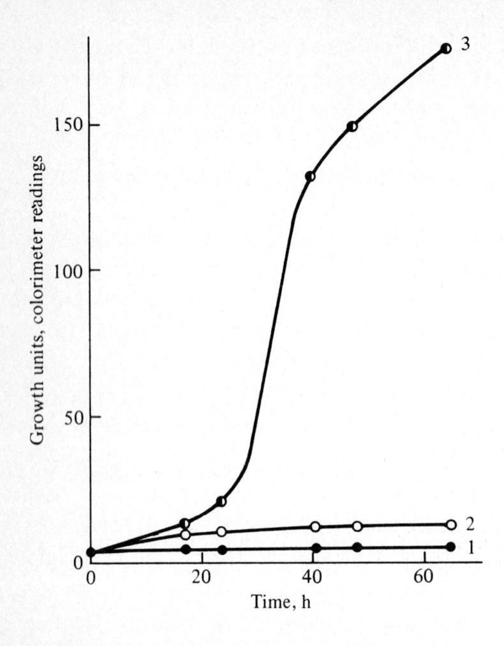

Figure 13.1 Batch growth of pure and mixed cultures of a phenylalanine requiring strain of *Lactobacillus arabinosas* (curve 2 = pure culture) and a folic acid-requiring strain of *Streptococcus faecalis* (curve 1 = pure culture) in a synthetic medium containing neither phenylalanine nor folic acid. Curve 3 shows enhanced growth as a result of mutualism in mixed culture of these two organisms. *(Reprinted from V. Nurmikko, "Biochemical Factors Affecting Symbiosis Among Bacteria," Experientia, vol. 12, p. 245, 1956.)*

converts ethanol to hydrogen and acetate

$$CH_3CH_2OH + H_2O \longrightarrow CH_3COO^- + H^+ + 2H_2$$

but is inhibited by the hydrogen it produces. The second species of the "bacterium" cannot grow on ethanol but consumes hydrogen, yielding methane

$$4H_2 + CO_2 \longrightarrow CH_4 + 2H_2O$$

Thus we have a situation in which one species destroys a toxin for its associate, which in turn provides a nutrient for the first. This detoxification type of mutualism may also arise when an aerobe shields an obligate anaerobe from too much free oxygen. Another possibility is maintenance of an advantageous pH by two organisms, one which tends to decrease pH and another which provides the opposite effect.

The final two classes of interactions to be examined in this section involve no significant effect on the first species. In *commensalism*, the second microbe enjoys benefits. The opposite occurs in *amensalism*, where the second species suffers as a result of its interaction with the first.

Several instances of commensalism are similar to the last kind of mutualism we considered. In the commensal version, one species removes a toxin for the second species, but, in contrast to mutualism, the latter organism provides no special benefits for the detoxifier. This type of commensalism is common, as suggested by Table 13.2.

Still more widespread, however, are commensal relationships where one species produces compounds which accelerate growth of another species. The end

Table 13.1 Examples of symbiotic relationships involving at least one microorganism

Microorganism	Other organism or site	Comments
Flagellated protozoa	Termites	Protozoa hydrolyze cellulose for termites in exchange for supply of this material, which termites alone cannot digest; flagellates are in turn hosts to bacteria which provide cellulase enzymes
Luminous bacteria	Squid, some fishes	Glands of squid house luminous bacteria, which provide the squid with a recognition device; bacteria accorded protection and nutrients
Rumen microorganisms	Cattle, sheep, goats	First two stomachs of the cow, the rumen, contain many microbial species, which in exchange for food supply aid the cow in digesting plant material including cellulose, starch, and lipids
Normal microbial flora	Skin, throat, mouth, intestines	Normal flora play an important although ill-defined role in preventing many diseases, as has been shown with studies on germ-free animals
Bacteria	Protozoa	Bacteria live inside the protozoa (an endosymbiosis) and derive nutrients; in one case, at least, bacteria provide their host with needed amino acids and growth factors
Algae	Protozoa	Each protozoan holds 50 to a few hundred algae; algae use light to fix CO_2 and free O_2, which in turn is used by the protozoan to oxidize nutrients, liberating CO_2
Algae	Fungus	Together these form an intimate association called a *lichen*; association is of benefit mostly in very wet or dry environments with scarce nutrients; the alga provides the fungus with organic nutrients; the fungal role not well understood
Rhizobium bacteria	Leguminous plants	Bacteria live in nodules formed in plant roots, where they enjoy nutrients provided by the plant; bacteria fix atmospheric nitrogen so that it becomes accessible to plant

products serving as bases for commensalism are numerous (see Table 13.3); depending on the particular commensal situation, the produced compound might serve as an energy and/or carbon source for the second species. Such commensal relationships are often strung together in a chain so that over time a succession of commensal pairs appears. In a batch system, for example, yeast can convert glucose to alcohol, which serves as a nutrient for *Acetobacter*. This species then produces acetic acid, which in turn is consumed by other microorganisms, and so on.

Also shown in Table 13.3 are several cases where a vitamin or some other growth factor is passed from one species to another. Such situations can readily

Table 13.2 Commensal relations: compound removed[†]

Toxic compound	Details of interrelationship
Concentrated sugar solutions	Osmophilic yeasts metabolize the sugar and thereby reduce the osmolarity, allowing the growth of species sensitive to high osmotic pressures
Oxygen	Aerobic organisms may reduce the oxygen tension, thus allowing anaerobes to grow
Hydrogen sulfide	Toxic H_2S is oxidized by photosynthetic sulfur bacteria, and the growth of other species is then possible
Food preservatives	The growth inhibitors benzoate and sulfur dioxide are destroyed biologically
Lactic acid	The fungus *Geotrichum candidum* metabolizes the lactic acid produced by *Streptococcus lactis*; the acid would otherwise accumulate and inhibit the growth of the bacteria
Mercury-containing germicides	*Desulfovibrio* sp. form H_2S from sulfate, and the sulfide combines with mercury-containing germicides and permits bacterial growth
Antibiotics	Enzymes are produced by some species of bacteria which break down antibiotics; thus the growth of antibiotic-sensitive species is allowed
Phenols	Some bacteria can oxidize phenols, thereby permitting other species to grow
Trichlorophenol	A number of gram-negative bacteria can absorb trichlorophenol in their cell wall lipids and thereby protect *Staphylococcus aureus* from its action

† From J. L. Meers, "Growth of Bacteria in Mixed Cultures," p. 158 in A. J. Laskin and H. Lechevalier (eds.), *Microbial Ecology*, CRC Press, Cleveland, 1974.

Table 13.3 Commensal relations: compound supplied[†]

Compound	Species producing compound	Species requiring compound
Purine	*Bacillus subtilis*	*B. subtilis* auxotrophs
Organic acid	*Aerobacter cloacae*	Unnamed bacterium
Isobutyrate	*Corynebacterium diphtheriae*	*Treponema microdentium*
Nicotinic acid	*Saccharomyces cerevisiae*	*Proteus vulgaris*
Vitamin K	*Staphylococcus aureus*	*Bacteroides melaninogenicus*
Nitrite ions	*Nitrosomonas*	*Nitrobacter*
Hydrogen sulfide	*Desulfovibrio*	Sulfur bacteria
Water	*Bacillus mesentericus*	*Clostridium botulinum*
Polysaccharides	Algae	Bacteria
Hydrogen	Rumen bacteria	*Methanobacterium ruminatium*
Methane	Anaerobic methane bacteria	Methane-oxidizing bacteria
Ammonium ions	Many heterotrophs	*Nitrosomonas*
Nitrite	*Nitrosomonas*	*Nitrobacter*
Nitrate	*Nitrobacter*	Denitrifying bacteria
Acetyl phosphate	*Corynebacterium diphtheriae*	*Borrelia vincenti*
Fructose	*Acetobacter suboxydans*	*Saccharomyces carlsbergensis*

† From J. L. Meers, "Growth of Bacteria in Mixed Cultures," p. 156 in A. J. Laskin and H. Lechevalier (eds.), *Microbial Ecology*, CRC Press, Cleveland, 1974.

be constructed by using an auxotrophic mutant as one of the species. Any other species which produces the necessary metabolic intermediate for the first then completes the commensal pair.

Most reported examples of commensalism have been based on batch-culture studies. Indeed, realization of a strictly commensal relationship in continuous culture or other open system is difficult. The problem is avoiding competition: although one organism may aid the other, they both may compete for a nutrient which eventually becomes a limiting nutrient. Therefore, we expect commensalism in open systems only when the species involved differ widely in their nutritional requirements.

Amensalism is the opposite of commensalism: in an amensal relationship, the growth of one species is inhibited by the presence of another. The harmful effects of the offensive species usually are due either to synthesis of toxic products or removal of essential nutrients. Most reported examples of amenalism involve the first mechanism, where the second organism creates an environment within which other species can survive only to a limited degree, if at all.

We have already seen that several antibiotics are produced by microorganisms. One standard laboratory test for microbial antibiotic synthesis is essentially a demonstration of amensalism: an antibiotic-synthesizing and an antibiotic-sensitive species are grown together on an agar surface. The presence of clear zones around the antibiotic-producing colonies is evidence of antibiotic activity (see Fig. 13.2). (Penicillin was discovered by just such an observation.) These clear zones result from diffusion of the antibiotic away from the synthesizing colonies, with resulting inhibition of the susceptible strain.

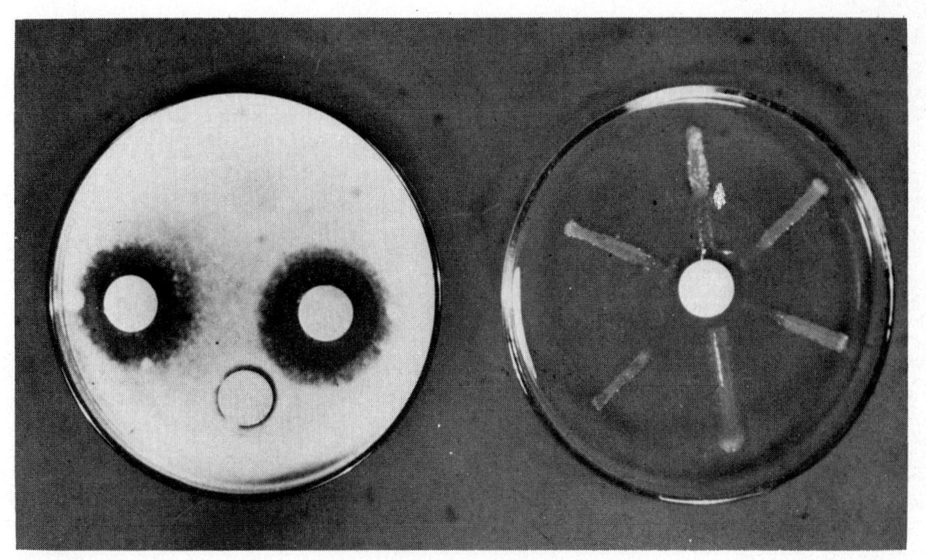

Figure 13.2 Amensalism: diffusion of antibiotic away from the colonies of antibiotic-producing fungus inhibits growth of other organisms on the agar medium. *(Photo courtesy of Charles Pfizer and Co., Inc.)*

Antibiotic synthesis by molds and actinomycetes has been emphasized earlier. To appreciate the possible roles of antibiotics in microbial ecology, we must also recognize that algae and other bacteria can also produce antibiotics. Most of these additional bacterial antibiotics are polypeptides. Like penicillin, some of them inhibit cell-wall synthesis while others serve to destroy the permeability barrier provided by the cell membrane.

A related kind of amensalism results because some microbes excrete enzymes which decompose cell-wall polymers. Such organisms derive two different benefits when the lytic enzymes depolymerize cell walls of other species. Possible competitors are destroyed, and the lysed cells release nutrients which can be used by the enzyme-producing microbes. Actually, if the nutrients so released constitute a significant resource for the enzyme-producing species, they benefit from the association. Thus, the ecological relationship ceases to be amensalistic and becomes parasitism (see Sec. 13.4).

Inhibition milder than the above cases may result from microbial synthesis of organic acids. This often lowers pH and inhibits growth of other organisms. More specific inhibitory effects of organic acids are also possible. *Shigella flexneri* is inhibited by formic and acetic acids, and propionate and acetate inhibit growth of the bacteria *Propionibacterium shermanii.* Another example of amensalism involves inhibition of chemoautotrophs by small amounts of ammonia produced by other species, especially those which derive energy from amino acids.

13.2 CLASSIFICATION OF INTERACTIONS BETWEEN TWO SPECIES

The classification of pairwise interactions between different species which we will employ is based upon the dynamic properties of the mixed culture. These dynamic properties are characterized by the linearized system description introduced in Sec. 9.2.1. Letting $\mathbf{n}$ denote the vector with components n_i equal to the number (or number concentration = number of cells per unit volume) of species i organisms in the system, the number balances for a well-mixed system can usually be written in the form

$$\frac{d\mathbf{n}(t)}{dt} = \mathbf{f}(\mathbf{n}(t), \mathbf{m}) \tag{13.1}$$

where $\mathbf{m}$ is a vector of parameters which characterize the organisms' environment. In the context of mixed culture ecology, the coefficient matrix $\mathbf{A}$ of the corresponding linearized model with entries

$$a_{ij} = \frac{\partial f_i(\mathbf{n}_s, \mathbf{m})}{\partial n_j} \tag{13.2}$$

is called the *community matrix*. As discussed in Chap. 9, $\mathbf{n}_s$ in Eq. (13.2) denotes a time-invariant vector $\mathbf{n}$ which is a steady state of Eq. (13.1).

The variables used to describe species' concentrations in mixed culture (and other) systems deserve careful attention. Studies of transient growth of pure cultures (e.g., batch cultivation) show changes in average single-cell mass (recall Sec. 7.3.1) which imply that cell *mass* concentration and cell *number* concentration follow different trajectories. Stated differently, the ratio x_i/n_i changes with time, where x_i and n_i are the mass and number concentrations of species i, respectively. Kinetic modeling studies indicate that growth kinetics descriptions are substantially simpler, possessing relatively straightforward functional descriptions considered in Chap. 7, when they are based on mass rather than number concentration. For example, based on established cell-cycle features of *E. coli*, first-order mass synthesis kinetics imply first-order number synthesis kinetics *with a time lag*, creating significantly more complicated dynamic behavior and much more difficult analyses of transient characteristics (recall Sec. 9.2.2). Consequently, use of total cell mass concentration is generally preferred.

Unfortunately, experimental assays for densities of different species in mixed cultures usually involve counting colonies grown on plates. Such data provides cell number but not mass concentrations. Simultaneous measurements of species mean cell size by microscopy or particle sizing (Coulter principle, light scattering) would permit approximate conversion to mass concentrations, but such data are rarely available for mixed cultures. Accordingly, since confrontation with experiment is essential, kinetic models considered in this chapter will, as above, usually be posed in terms of cell number concentration. It should be noted that some difficulties in representing cell number dynamics in terms of these kinetic models may be due to failure to use cell mass variables and to account for transient changes in cell size.

The interaction between two species near a steady state is characterized by two entries in the community matrix **A**. Considering for the moment species i and j, we notice that a small increase in the jth population from its steady-state value $[\chi_j(0) > 0]$ contributes a term $a_{ij}\chi_j(0)$ to the initial derivative of χ_i. (To avoid confusion with cell mass densities, the deviation $[n_i(t) - n_{is}]$ will be denoted $\chi_i(t)$.) Thus, if all other populations are initially at their steady-state values $[\chi_k(0) = 0,\ k \neq j]$, the sign of $d\chi_i(0)/dt$ depends on the sign of a_{ij}. We can say that if a_{ij} is positive, species j has a positive effect on growth of species i, and if a_{ij} is negative, that an inhibiting effect is evident. No interaction occurs if a_{ij} is zero. The same argument applies for determining the influence of species i on species j: the effect is stimulatory for positive a_{ji}, inhibitory for negative a_{ji}, and neutral for zero a_{ji}.

All possible combinations of interactions defined in this manner are shown in Table 13.4. We see that the four kinds of pairwise interactions considered in the previous section fit nicely into this scheme. Also, two new types of interactions, competition and predation, appear in the table. Before considering these interactions in detail, we shall explore some implications of the interaction type on population dynamics near a steady state. For the moment, we shall limit this inquiry to two interacting populations.

Table 13.4 Classification of pairwise interactions based on the signs of the entries a_{ji} and a_{ij} from the community matrix $\mathbf{A}(i \neq j)$†

		Effect of species j on species i (sign of a_{ij})		
		−	0	+
	−	− − Competition	−0 Amensalism	− + Predation
Effect of species i on species j (sign of a_{ji})	0	0− Amensalism	00 Neutralism	0+ Commensalism
	+	+ − Predation	+0 Commensalism	+ + Mutualism

† Adopted from R. M. May, *Stability and Complexity in Model Ecosystems*, p. 25, Princeton University Press, Princeton, N.J., 1973.

Example 13.1: Two-species dynamics near a steady state When only two species interact, the community matrix $\mathbf{A}$ is 2 × 2, as seen from the appropriate form of (9.30):

$$\frac{d(n_1 - n_{1s})}{dt} \equiv \frac{d\chi_1}{dt} = a_{11}\chi_1 + a_{12}\chi_2$$
$$\frac{d(n_2 - n_{2s})}{dt} \equiv \frac{d\chi_2}{dt} = a_{21}\chi_1 + a_{22}\chi_2 \tag{13E1.1}$$

or

$$\frac{d}{dt}\begin{bmatrix}\chi_1\\ \chi_2\end{bmatrix} = \begin{bmatrix}a_{11} & a_{12}\\ a_{21} & a_{22}\end{bmatrix}\begin{bmatrix}\chi_1\\ \chi_2\end{bmatrix} \tag{13E1.2}$$

The characteristic equation (9.34) can readily be expanded into the quadratic equation

$$\lambda^2 - (a_{11} + a_{22})\lambda + (a_{11}a_{22} - a_{12}a_{21}) = \lambda^2 - (\text{tr }\mathbf{A})\lambda + \det \mathbf{A} = 0 \tag{13E1.3}$$

where tr $\mathbf{A}$ and det $\mathbf{A}$ signify the trace and determinant of $\mathbf{A}$, respectively. Applying the Hurwitz criterion (9.41) to Eq. (13E1.3), we readily conclude that the linearized system is asymptotically stable if and only if

$$\det \mathbf{A} > 0 \tag{13E1.4}$$

$$\text{tr }\mathbf{A} < 0 \tag{13E1.5}$$

Additional features of the system dynamics are revealed by the solutions to Eq. (13E1.3), which are

$$\lambda_{1,2} = \frac{\text{tr }\mathbf{A}}{2} \pm \frac{1}{2}\sqrt{(\text{tr }\mathbf{A})^2 - 4 \det \mathbf{A}} \tag{13E1.6}$$

At least one eigenvalue will have zero real part if either tr $\mathbf{A} = 0$ (and det $\mathbf{A} > 0$) or det $\mathbf{A}$ is zero. Let us exclude this critical case from our considerations and proceed to other possibilities. (Bifurcations arising from critical case situations are discussed in Sec. 13.5.5.) If

$$\left(\frac{\text{tr }\mathbf{A}}{2}\right)^2 > \det \mathbf{A} > 0 \tag{13E1.7}$$

both eigenvalues are real and have the same sign: the steady state $\mathbf{n}_s$ in question is called a *node*. If det **A** is negative, there is one positive and one negative eigenvalue and the steady state is a *saddle point*. Complex eigenvalues occur when 4 det **A** is larger than the square of tr **A**. In this case the steady state is called a *focus*.

The motivation for these terms, which derive originally from mathematical studies in mechanics, becomes clearer by examining the population trajectories in the phase plane. By definition, the coordinates of the phase plane are χ_1 and χ_2. A specified initial condition $(\chi_1(0), \chi_2(0))$ can be represented by a point in this plane. Likewise, if we plot for all future t points $(\chi_1(t), \chi_2(t))$ in the phase plane, the result will be a continuous curve called a *trajectory*. Thus, plotting trajectories in the phase plane is a convenient device for representing the dynamic behavior of two variables in a single diagram.

Figure 13E1.1 shows typical trajectories for three common cases. Each trajectory shown in the figure corresponds to a different set of initial conditions, and the arrows point in the direction of increasing time. The typical phase behavior of unstable nodes and foci can be obtained simply by reversing the arrows on the trajectories in Fig. 13E1.1*a* and *b*. (Why does this work?)

Returning now to the classification scheme summarized in Table 13.4, we see that this scheme refers only to off-diagonal entries of the **A** matrix. Consequently, without further specifications we can say nothing about tr **A** and very little about det **A**.

Let us now consider reasonable assumptions concerning the autonomous specific growth rates a_{11} and a_{22} for some of the interaction types already defined. Mutualism may give rise to larger steady-state populations than possible without the beneficial interaction, so that $a_{11} < 0$, $a_{22} < 0$ obtains. Then condition (13E1.5) certainly holds, but satisfaction of (13E1.4) requires

$$a_{11}a_{22} > a_{12}a_{21} \tag{13E1.8}$$

If this inequality is not fulfilled, this mutualistic steady state is unstable.

For either commensalism or amensalism in a two-population system, the stability characteristics depend entirely on the autonomous growth rates, for $\det \mathbf{A} = a_{11}a_{22}$ and $\operatorname{tr} \mathbf{A} = a_{11} + a_{22}$. This reveals immediately that asymptotic stability appears if and only if both a_{11} and a_{22} are negative.

Several of the results just outlined can be generalized to more complex populations, as we shall see in Sec. 13.5.

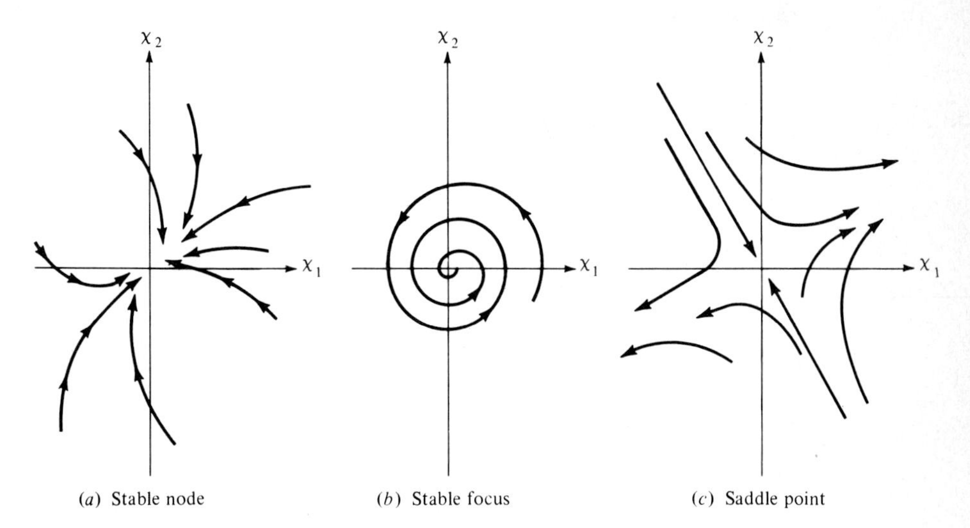

Figure 13E1.1 Some characteristic phase-plane portraits for the linearized dynamics of two-species ecosystems: (*a*) stable node, (*b*) stable focus, and (*c*) saddle point.

13.3 COMPETITION: SURVIVAL OF THE FITTEST

Darwin's work on natural selection emphasized the importance of competition between different species. As we shall use it here, the term *competition* refers to dependence of two species on a common factor such as food supply, light, space, or some other limiting resource. Consumption of this common factor by each species limits its availability to the other, so that the growth rates of both organisms are affected negatively.

In a competitive situation, we are interested in discovering whether either species enjoys a natural advantage. A little reflection suggests that the species with the fastest growth rate should do better, since by virtue of its rapid growth it will be able to utilize more of the limiting factor than the slower-growing organism. As we shall see, experimental results as well as mathematical analyses of reasonable models strongly support this idea.

Consider the example shown in Fig. 13.3. Grown separately (open symbols), both *E. coli* and *Staphylococcus aureus* reached similar maximum populations although *E. coli* grew faster. When the two bacteria were cultivated in a batch mixed culture, the *E. coli* growth pattern was identical to that found in pure culture. The *S. aureus* population, on the other hand, was restricted to much

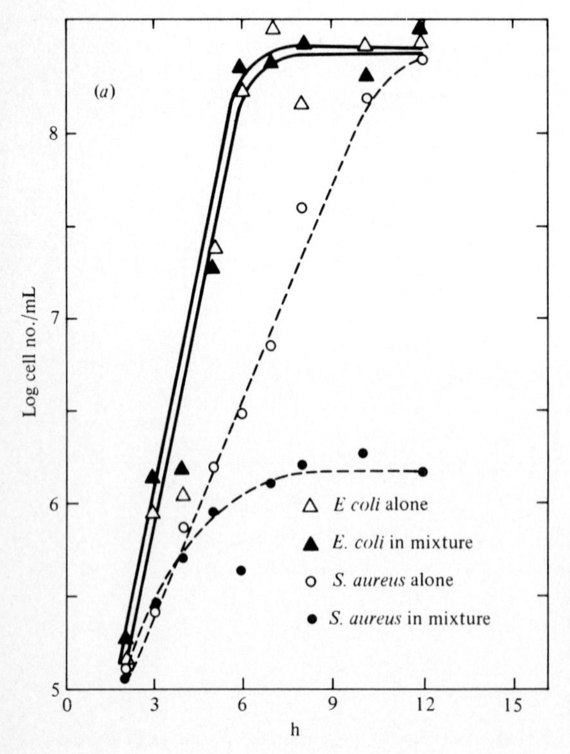

Figure 13.3 Growth of *S. aureus* is reduced when the organism grows in competition with *E. coli* in mixed culture. *(Reprinted from T. R. Oberhofer and W. C. Frazier, "Competition of Staphylococcus aureus with Other Organisms," J. Milk Food Technol., vol. 24, p. 172, 1961.)*

smaller maximal levels in the mixed-culture case. Since *E. coli* produces no inhibitors of the *Staphylococcus* organism, this effect can be attributed to the preferential nutrient uptake rate of *E. coli.* (What implicit assumption is involved here?)

13.3.1 Volterra's Analysis of Competition

Growth of two competing species in a closed environment has been analyzed by Vito Volterra, a famous Italian mathematician who established many of the foundations of mathematical ecology. In this nonlinear model, the autonomous specific growth rates μ_1 and μ_2 are both presumed positive. Furthermore, the first species utilizes the limiting nutrient to an extent given by $h_1 n_1$, and $h_2 n_2$ is the corresponding nutrient depletion for species 2, where h_1 and h_2 are both positive. Such consumption reduces the specific growth rates of the two species so that

$$\frac{dn_1}{dt} = [\mu_1 - \gamma_1(h_1 n_1 + h_2 n_2)]n_1 \tag{13.3}$$

$$\frac{dn_2}{dt} = [\mu_2 - \gamma_2(h_1 n_1 + h_2 n_2)]n_2 \tag{13.4}$$

The constants $\gamma_1 > 0$ and $\gamma_2 > 0$ in these equations reflect the different effects of nutrient depletion on species 1 and 2, respectively. Turning back to Sec. 7.3.2, we note that this model is simply a two-population generalization of the Verlhurst and Pearl model already discussed.

The ultimate effects of competition can be deduced by multiplying Eq. (13.3) by γ_2/n_1, multiplying Eq. (13.4) by γ_1/n_2, and then adding the resulting equations to obtain

$$\gamma_2 \frac{d \ln n_1}{dt} - \gamma_1 \frac{d \ln n_2}{dt} = \gamma_2 \mu_1 - \gamma_1 \mu_2 \tag{13.5}$$

Integration of this equation yields

$$\frac{n_1^{\gamma_2}}{n_2^{\gamma_1}} = C \exp{(\gamma_2 \mu_1 - \gamma_1 \mu_2)t} \tag{13.6}$$

where C is a constant of integration which depends on $n_1(0)$ and $n_2(0)$.

Let us suppose now that $\gamma_2 \mu_1 - \gamma_1 \mu_2$ is negative. (Does this involve any loss of generality?) Then Eq. (13.6) reveals that the ratio on the left-hand side approaches zero as time increases. We can readily rule out the possibility that this happens because n_2 grows without bound. Turning to Eq. (13.4), we can see immediately that dn_2/dt is negative for $n_2 > \mu_2/\gamma_2 h_2$, proving that n_2 is bounded. Then, the only conclusion consistent with the limiting behavior demonstrated by Eq. (13.6) is that n_1 approaches zero as time passes.

The net result of this analysis is known as the *competitive exclusion principle.* When two species compete in a *common* environment and $\gamma_1/\mu_1 > \gamma_2/\mu_2$, the

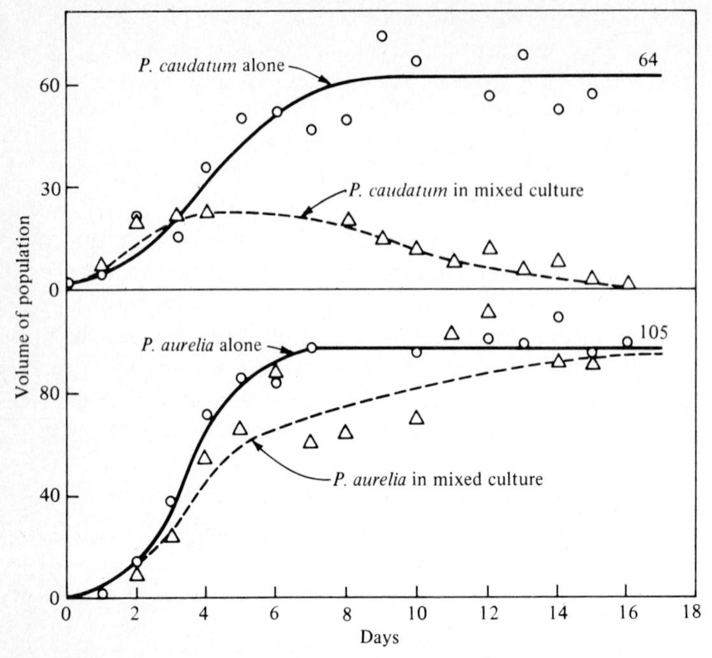

Figure 13.4 Demonstration of the exclusion principle in competitive growth of two protozoa. The slower-growing microbe, *Paramecium candatum*, is eliminated in mixed culture. *(Redrawn from G. F. Gause, "Experimental Analysis of Vito Volterra's Mathematical Theory of the Struggle for Existence," Science, vol. 79, p. 16, 1934.)*

second species population reaches a limiting size given by $\mu_2/\gamma_2 h_2$ while the first species eventually disappears. This principle can be extended to multiple-species situations described by

$$\frac{dn_i}{dt} = [\mu_i - \gamma_i F(n_1, n_2, \ldots, n_p)]n_i \qquad i = 1, 2, \ldots, p \tag{13.7}$$

by arguments identical to the above. Again, the conclusion is extinction of all species but one. Note that, because μ_i values are presumed independent of population densities, nutrient supplies are presumed unbounded, or continuously provided. Strictly speaking, the competitive exclusion principle applies only in *open* ecosystems.

Dramatic experimental evidence illustrating the exclusion principle is shown in Fig. 13.4. Here, in a batch system controlled to maintain constant food supply, a protozoan species which follows the logistic curve [Eq. (7.52)] when grown alone becomes extinct in competition with a faster-growing organism.

The success of the exclusion principle in explaining the outcome of this and numerous other actual instances of competition is somewhat surprising in view of the assumptions behind Eqs. (13.3) and (13.4). It is important to realize, however, that a model of this *form* may be applicable in physical circumstances very different from those assumed in our derivation. In fact, from a strictly mathematical viewpoint, we may view Eqs. (13.3) and (13.4) as approximate representations obtained by expanding a general model with a Taylor's series and retaining only terms through the second order.

13.3.2 Competition and Selection in a Chemostat

Next we shall consider the effects of competition in open systems, using the chemostat as a prototype representative. Considering the specific growth rate μ_i of species i to be a constant, the dynamic behavior of the species concentration is given by

$$\chi_i(t) = \chi_i(0)e^{(\mu_i - D)t} \qquad i = 1, 2, \ldots, p \tag{13.8}$$

Thus, whether χ_i increases, decreases, or stays constant with time depends on the relative values of μ_i and D. What happens in the two-species case can be deduced using a rather heuristic argument based on the Monod model.

It is reasonable to assume that the dependence of growth rate on limiting-substrate concentration will rarely be identical for two competing microorganisms. This leaves two possibilities: either the growth rates are related as shown in Fig. 13.5, or the growth-rate curves cross. For the moment, let us assume that the situation depicted in Fig. 13.5 obtains.

Suppose that organism 1 is growing at steady state in a chemostat with dilution rate D. Then, as indicated in Fig. 13.5, the limiting-substrate concentration within the vessel will be s_1. Now, if organism 2 is introduced into the chemostat, it will grow initially at rate μ_2. Since $\mu_2 > D$, the population of species 2 will increase with time, causing the substrate concentration to fall. With $s < s_1$, the specific growth rate of species 1 is smaller than D, so that organism 1 begins to wash out. This trend continues until the substrate concentration reaches s_2, at which point species 2 attains steady state while species 1 continues to wash out.

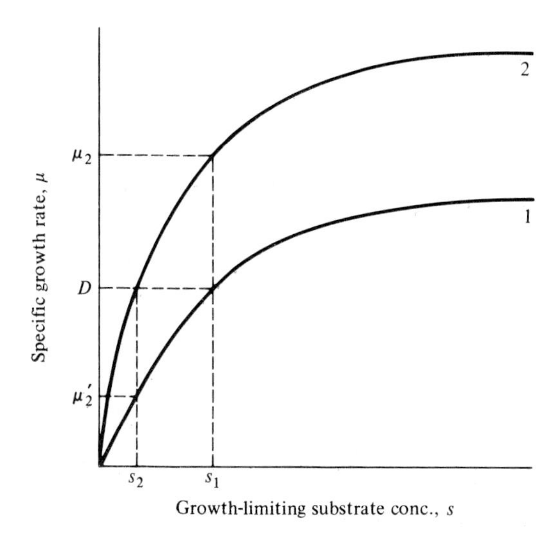

Figure 13.5 Hypothetical specific growth rates of organisms 1 and 2. The text explains why organism 1 will wash out of continuous culture. *(Reprinted from J. L. Meers and D. W. Tempest, "The Influence of Extra-Cellular Products on Mixed Microbial Populations in Magnesium-Limited Chemostat Cultures," J. Gen. Microbiol, vol. 52, p. 309, 1968.)*

This pattern of events has been observed in numerous experimental studies. In the simultaneous growth of the bacterium *Aerobacter aerogenes* and the yeast *Torula utilis* in a chemostat, the washout rate of the slower-growing yeast appears to be a function of dilution rate (see Fig. 13.6), as expected from Eq. (13.8). While Fig. 13.6 shows some overshoots and oscillations not included in our simple scenario, other data, such as those plotted in Fig. 13.7, follow our hypothesized chain of events very closely.

If the two functions $\mu_1(s)$ and $\mu_2(s)$ cross, so that which is larger depends on the particular substrate concentration involved, we can conclude by analogy with the argument above that the dilution rate used will determine which species dominates. A dramatic demonstration of such dependence of competitive advantage on dilution rate is revealed in Fig. 13.8. Initially the yeast dominates, but it begins to wash out as soon as the dilution rate is increased. Although the bacterium gains ascendency at the higher dilution rate, the roles of the two species are reversed as soon as the dilution rate returns to its starting level.

Before turning to the fascinating class of predation and parasitism interactions, we should point out that the previous discussion also applies to organisms of the same type which have different characteristics. For example, mutant strains especially suited for growth in a particular chemostat environment may develop. These strains may then be viewed as organism 2 in our account of chemostat competition: their population may increase dramatically, eventually completely

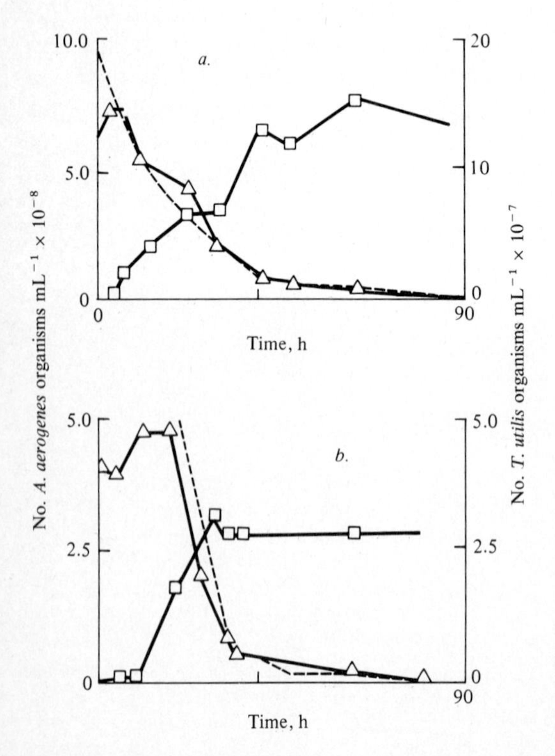

Figure 13.6 Data on competition between *A. aerogenes* (□) and *T. utilis* (△) in a chemostat: (*a*) $D = 0.05\ h^{-1}$, (*b*) $D = 0.3\ h^{-1}$; $T = 33°C$, pH 6.4, potassium-limited simple-salts medium. Dashed lines are the calculated washout curves for nongrowing organisms. *[Reprinted from J. L. Meers, "Growth of Bacteria in Mixed Cultures," in A. I. Laskin and H. Lechevalier (eds.), "Microbial Ecology," p. 142. © CRC Press, Inc., 1974. Used by permission of CRC Press, Inc.]*

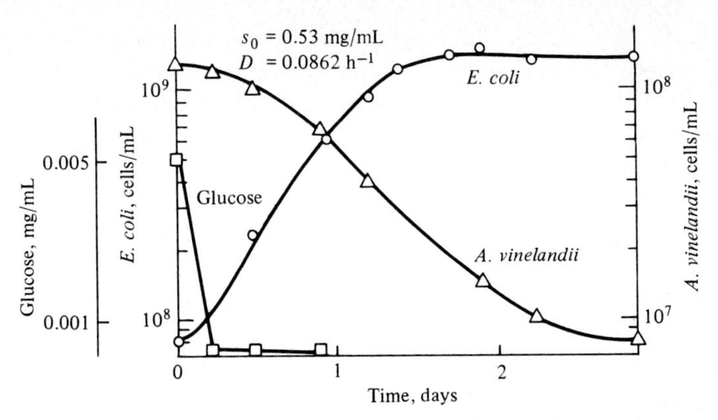

Figure 13.7 Competition between two bacteria (*E. coli* and *Azotobacter vinelandii*) in a chemostat. *(Reprinted from J. L. Jost et al., "Interactions of* Tetrahymena pyriformis, Escherichia coli, Azotobacter vinelandii *and Glucose in a Minimal Medium," J. Bacteriol., vol.* 113, *p.* 834, 1973.)

displacing the original species. Similarly, mutants competing in a batch reactor should be considered a mixed culture subject to the types of behavior discussed in the previous section.

This manifestation of exclusion caused by competition has very important practical implications. Many fermentations have maximal product formation rather than biomass production as their objective. Consequently, strains selected for desirable product yields may well be slower-growing than related mutant strains. Under these circumstances, it becomes increasingly probable that relatively fast-growing mutants will evolve in a chemostat, leading to the washout of

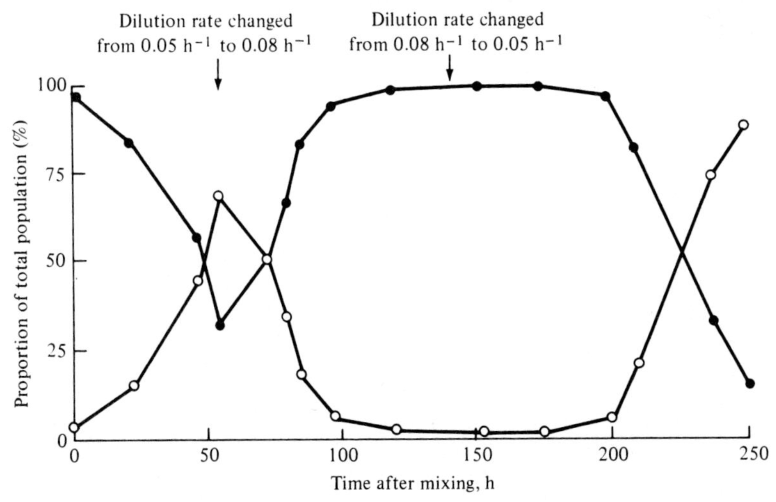

Figure 13.8 Effects of dilution-rate changes on continuous mixed culture of *Bacillus subtilis* (●) and *T. utilis* (○) ($T = 33°C$, pH = 6.4, magnesium-limited simple-salts medium). *(Reprinted from J. L. Meers, J. Gen. Microbiol., vol. 67, p. 359, 1967.)*

the productive strains. As Figs 13.6 to 13.8 amply demonstrate, any initial advantage in population density is lost in continuous culture and the contaminant or mutant strain with a larger μ value than the original culture will evidently prevail. As mutants with desirable properties have been produced for batch cultures, we can expect similar efforts in the future to develop productive strains able to compete successfully with useless mutants in continuous reactors.

The effects of wall growth, considered in various lights in Chap. 9, should also be mentioned in the context of microbial competition in open environments. We have already mentioned (Sec. 9.1.3) that a population attached to a surface can survive in a chemostat even when D exceeds the maximal specific growth rate. Similarly, a slower-growing organism, if present in an ecologically advantageous niche such as a film, has been observed experimentally to persist in competition with faster-growing species.

Example 13.2: Competitive growth in unstable recombinant cultures In growth of cells containing plasmids which do not stably replicate or segregate, many different "species," characterized by the number of plasmids contained in the cell, can be identified. Here we shall lump these various species into two groups: plasmid-containing cells, denoted X^+, and cells without plasmids, denoted X^-. Describing growth of these two populations in overall terms, we may write

$$X^+ \rightarrow (2-p)X^+ + pX^- \qquad \text{Rate} = \mu^+ x^+ \tag{13E2.1}$$

$$X^- \rightarrow 2X^- \qquad \text{Rate} = \mu^- x^- \tag{13E2.2}$$

The parameter p, which denotes the number of plasmid-free cells formed per generation of plasmid-containing cells, is a measure of the degree of plasmid segregational instability. The two strains are presumed to have different specific growth rates since, in nonselective media, plasmid-containing cells typically grow more slowly than plasmid-free cells. (Why?) The reverse situation applies in selective medium.

Considering batch cultivation of the unstable strain with constant specific growth rates, the fraction F of cells with plasmids after N generations of plasmid-containing cells is easily shown to be

$$F = \frac{1-\alpha-p}{1-\alpha-2^{N(\alpha+p-1)}p} \tag{13E2.3}$$

The parameter α is the ratio of specific growth rates

$$\alpha = \mu^-/\mu^+ \tag{13E2.4}$$

Propagation of a population from stock to full-scale bioreactor typically requires 25 generations or more. Figure 13E2.1 shows the fraction of plasmid-containing cells after 25 doublings (F_{25}) as a function of the parameters α and p. These calculations indicate the clear possibility for substantial loss of the productive, plasmid-containing population, as indeed has been observed in some cases in practice.

This type of genetic instability also has major implications for continuous reactor operation. Presuming a growth situation with $\mu^+ = \mu^- = \mu$ in a CSTR, the mass balances on the two strains are

$$\frac{dx^+}{dt} = (1-p)\mu x^+ - Dx^+ \tag{13E2.5}$$

$$\frac{dx^-}{dt} = \mu x^- + p\mu x^+ - Dx^- \tag{13E2.6}$$

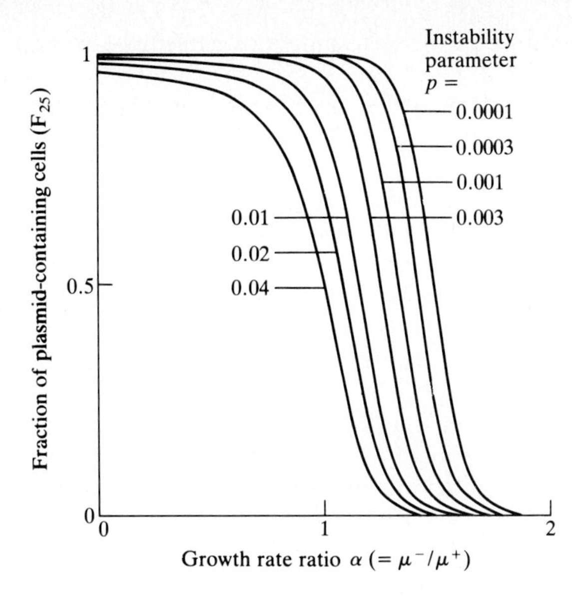

Figure 13E2.1 Fraction of plasmid-containing cells after 25 generations of batch cultivation (After Ref. 12).

Letting x denote total biomass concentration ($=x^+ + x^-$), it is easy to show by adding the last two equations that x has a nonzero steady-state value only for $D = \mu$. Under these conditions, however, the solution of Eq. (13E2.5) is

$$x^+(t) = x^+(0)e^{-\mu pt} \to 0 \tag{13E2.7}$$

Consequently, the concentration of plasmid-containing cells declines exponentially, approaching the value of zero at steady state. The rate of washout of plasmid-containing cells is scaled by the instability parameter p, which indicates that, as a practical matter, the continuous reactor could operate productively for a useful time interval if p is sufficiently small.

Extensions of these models which include substrate utilization, product formation, and Monod dependence of μ on substrate level have been presented [13, 14]. Alternative treatments based on population-balance models have evaluated errors introduced by the approximations in the above class of models (generally small) and connected single-cell growth cycle, plasmid replication, and plasmid segregation properties with overall parameters [15, 16]. We should note that these models apply equally well to nonrecombinant strains in which recurrent mutations occur, and that some of the mathematical treatments of unstable recombinant populations are special cases of population genetics theory [17, 18].

13.4 PREDATION AND PARASITISM

In both predation and parasitism, one species benefits at the expense of the other. We usually distinguish between these two types of interactions by the relative size of the organisms and the mechanism by which one species destroys the other. Predation involves the ingestion of one organism, the prey, by the predator organism. This mode of interaction is relatively common among microbes, with consumption of bacteria by protozoa a prime example. In parasitism, the host, which is usually the larger organism, is damaged by the parasite, which benefits from use of nutrients from the host. Attack of microorganisms by phages, already considered in some detail in Chap. 6, is the best example of parasitism in the microbial world.

Although the physical situations in predation and parasitism differ, their conceptual and mathematical descriptions share many common features. Consequently, for simplicity in our discussion of these forms of interaction, we shall use the generic terms predator and prey in the rest of this section to denote the respective species which benefit and which are damaged by their relationship.

In many instances of predator-prey interactions, the populations of predator and prey do not reach steady-state values but oscillate, as shown in Fig. 13.9 for a bacterium-protozoan system. Other examples of predator-prey oscillations beyond the microbial ones considered in this chapter include the lynx-hare, owl-lemming, and moth–conifer-tree system [Ref. 1, p. 188]. We can explain such regular fluctuations qualitatively as follows. Starting with a high prey and low predator concentration, the predators ravenously consume the prey, so that the number of predators increases while the prey population declines. These trends continue until there is an overshoot, where the small prey population is inadequate to support the large predator population. Then the predator population declines due to insufficient food supply while the prey population rebounds. This leads back to the start of the cycle, which may then repeat again and again.

13.4.1 The Lotka-Volterra Model of Predator-Prey Oscillations

A mathematical model for predator-prey interaction that produces such cycles in population sizes was developed by Lotka and Volterra in the late 1920s [8]. In this model, it is assumed that the prey species 1 can multiply autonomously and is consumed by predator at a rate proportional to the product of the numbers of predator and prey:

$$\frac{dn_1}{dt} = an_1 - \gamma n_1 n_2 \tag{13.9}$$

The product $n_1 n_2$ may be viewed as the frequency of encounter between the prey and predator species, and the coefficient γ attached to this term is a measure of how often such encounters lead to death of the prey. It is known, for example, that some bacteria are more resistant to attack by protozoa than others.

The predator species 2 benefits from encounters with prey, and the proportionality constant ε will denote the amount by which the predator population increases per kill of prey. Since the prey is presumed to be the only source of food for predators, the predator population declines in the absence of prey. Thus, the population balance for predators takes the form

$$\frac{dn_2}{dt} = -bn_2 + \varepsilon\gamma n_1 n_2 \tag{13.10}$$

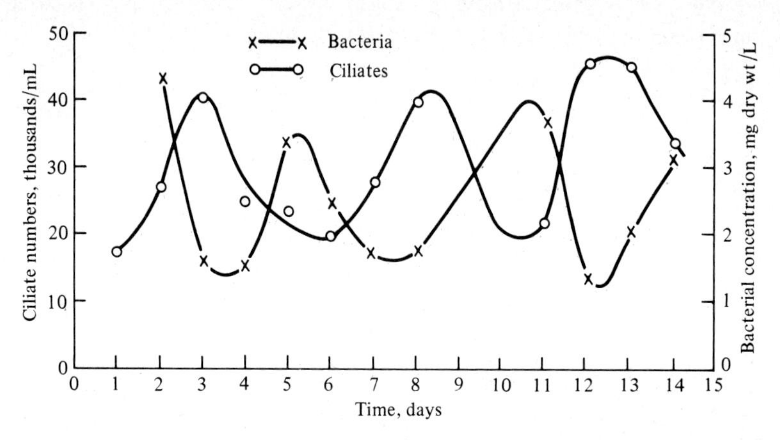

Figure 13.9 Oscillations in chemostat growth of a mixed population of the bacterium *A. aerogenes* and the protozoan *Tetrahymena pyriformis. [Reprinted from J. L. Meers, "Growth of Bacteria in Mixed Cultures," in A. I. Laskin and H. Lechevalier, (eds.), "Microbial Ecology," p. 151. © CRC Press, Inc., 1974. Used by permission of CRC Press, Inc.]*

Equations (13.9) and (13.10) constitute the basic Lotka-Volterra model. The nontrivial steady state admitted by the model is

$$n_{1s} = \frac{b}{\varepsilon\gamma} \qquad n_{2s} = \frac{a}{\gamma} \tag{13.11}$$

We can simplify the notation somewhat for our subsequent analysis by introducing new population variables y_1 and y_2, which are scaled by the steady-state values from Eq. (13.11)

$$y_1 = \frac{n_1}{n_{1s}} \qquad y_2 = \frac{n_2}{n_{2s}} \tag{13.12}$$

In terms of these variables, the Lotka-Volterra model becomes

$$\frac{dy_1}{dt} = a(1 - y_2)y_1 \tag{13.13}$$

$$\frac{dy_2}{dt} = -b(1 - y_1)y_2 \tag{13.14}$$

For this model, the phase-plane behavior is readily determined by eliminating time from the pair of Eqs. (13.13) and (13.14) to obtain a single differential equation relating y_1 and y_2. We accomplish this by dividing (13.14) by (13.13) to obtain

$$\frac{dy_2/dt}{dy_1/dt} = \frac{-by_2 + by_1y_2}{ay_1 - ay_1y_2} \tag{13.15}$$

then multiplying by $a(1 - y_2)(dy_1/dt)/y_2$, which yields

$$\frac{a}{y_2}\frac{dy_2}{dt} - a\frac{dy_2}{dt} + \frac{b}{y_1}\frac{dy_1}{dt} - b\frac{dy_1}{dt} = 0 \tag{13.16}$$

Integration of this equation gives

$$a \ln y_2 - ay_2 + b \ln y_1 - by_1 = C \tag{13.17}$$

or

$$\left(\frac{y_1}{e^{y_1}}\right)^b \left(\frac{y_2}{e^{y_2}}\right)^a = e^C \tag{13.18}$$

where C is an integration constant dependent on the initial population sizes.

Paths in the phase plane can now be computed from Eq. (13.18) by assuming a value for y_1 and computing compatible y_2 values. Care is necessary when doing this, as zero, one, or two positive solutions y_2 to this transcendental equation may occur for fixed y_1. In Fig. 13.10 several trajectories are illustrated. Immediately apparent are the existence of closed curves for all initial states except the steady state. (What does this tell us about asymptotic stability of the steady state?) If we transfer this representation to plots of populations vs. time, which requires solution of (13.13) and (13.14), we obtain prey and predator oscillations (see Fig. 13.11) quite reminiscent of data for numerous predator-prey systems, including the information shown earlier in Fig. 13.9

Before turning to extensions, criticisms, and modifications in the Lotka-Volterra model, let us explore some of its implications a little further. Integrating Eq. (13.13) over one period T of the oscillation, we find that

$$\ln \frac{y_1(T)}{y_1(0)} = aT - a\int_0^T y_2(t)\, dt \tag{13.19}$$

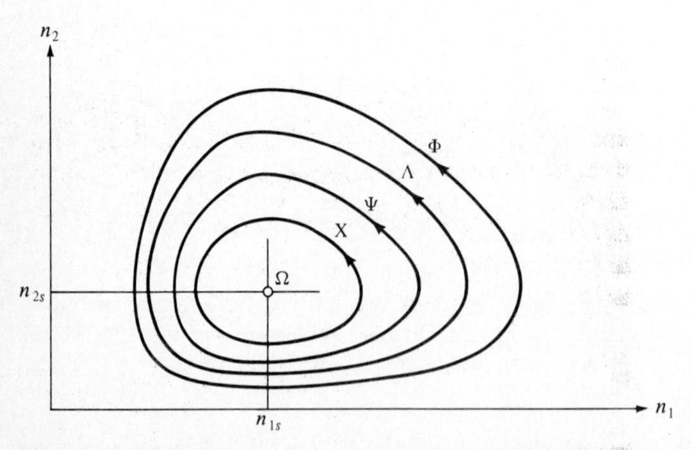

Figure 13.10 Motion in the phase plane as described by the Lotka-Volterra model. Point Ω is the steady state. Oscillatory trajectories labeled X, Ψ, Λ, and Φ correspond to different choices of initial population levels.

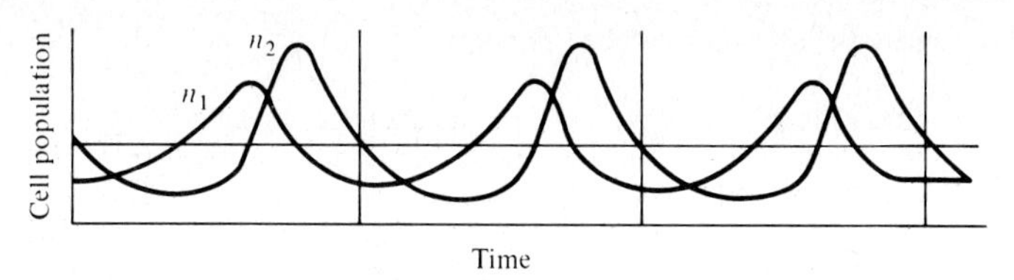

Figure 13.11 Periodic oscillations in the prey n_1 and predator n_2 population sizes from the Lotka-Volterra model.

Since y_1 is periodic with period T,

$$y_1(0) = y_1(T) \tag{13.20}$$

so that the left-hand side of Eq. (13.19) is zero. Rearranging what is left gives

$$\frac{1}{T}\int_0^T y_2(t)\,dt = 1 \tag{13.21}$$

or, in terms of the original variables,

$$\frac{1}{T}\int_0^T n_1(t)\,dt = n_{1s} \tag{13.22}$$

Thus, although the periods and the location of the population cycles in Fig. 13.10 depend on the initial conditions, the mean value of prey population, which is given by Eq. (13.22), is independent of the initial conditions and is equal to the steady-state value. Starting with Eq. (13.14) and applying exactly the same approach as that just described, we can easily prove the analogous relation for the mean predator population

$$\frac{1}{T}\int_0^T n_2(t)\,dt = n_{2s} \tag{13.23}$$

The Lotka-Volterra model has some interesting implications for control of microbial or other populations involving predator-prey interactions. Suppose that we introduce an agent which kills both predator and prey in proportion to their numbers. Mathematically, this corresponds to adding terms $-\delta_1 n_1$ and $-\delta_2 n_2 (\delta_1 > 0$ and $\delta_2 > 0)$ to Eqs. (13.9) and (13.10), respectively. The steady-state concentrations of prey and predator *with the control measure* now become

$$n_{1s} = \frac{b + \delta_2}{\varepsilon\gamma} \qquad n_{2s} = \frac{a - \delta_1}{\gamma} \tag{13.24}$$

Consequently, the effect of this attempt to control the interacting populations is an *increase* in the mean number of prey and a decrease in the predator average population.

13.4.2 A Multispecies Extension of the Lotka-Volterra Model

Ecologists are interested in the relationship between system complexity and its dynamic behavior. For example, is the system more or less likely to be stable or to exhibit oscillations as the number of different species is increased or as the number and intensity of interactions become larger? Although most of our considerations of such issues will be concentrated in Sec. 13.5, we shall consider now a generalized version of the Lotka-Volterra model following the argument of May [5].

The analogous forms of Eqs. (13.9) and (13.10) for a system with N predators and N preys is

$$\frac{dn_i}{dt} = n_i\left(a_i - \sum_{j=1}^{N} \alpha_{ij} m_j\right) \qquad i = 1, \ldots, N$$

$$\frac{dm_i}{dt} = m_i\left(-b_i + \sum_{j=1}^{N} \beta_{ij} n_j\right) \qquad i = 1, \ldots, N \tag{13.25}$$

where $n_1, \ldots, n_N$ are the numbers in the N prey populations and $m_1, \ldots, m_N$ are the predator population numbers. Letting $\boldsymbol{\alpha}$ denote the matrix with elements α_{ij}, $\boldsymbol{\beta}$ the matrix with elements β_{ij}, $\mathbf{a}$ the column vector with entries a_i, and $\mathbf{b}$ the column vector with elements b_i, it follows that the steady-state populations $n_{1s}, \ldots, n_{Ns}, m_{1s}, \ldots, m_{Ns}$ for the system described by Eqs. (13.25) satisfy

$$\boldsymbol{\beta}\mathbf{n}_s = \mathbf{b} \qquad \boldsymbol{\alpha}\mathbf{m}_s = \mathbf{a} \tag{13.26}$$

where $\mathbf{n}_s$ and $\mathbf{m}_s$ are vectors with elements n_{is} and m_{is}, respectively.

If we evaluate the $2N \times 2N$ community matrix $\mathbf{A}$ for this system, the result can be partitioned as follows:

$$\mathbf{A} = \left[\begin{array}{c:c} \mathbf{0} & -\boldsymbol{\alpha}^* \\ \hdashline \boldsymbol{\beta}^* & \mathbf{0} \end{array}\right] \tag{13.27}$$

where

$$\alpha_{ij}^* = n_{is}\alpha_{ij} \qquad \beta_{ij}^* = m_{is}\beta_{ij} \tag{13.28}$$

and the $\mathbf{0}$ entries in (13.27) denote $N \times N$ zero matrices. Significant information about the stability of this system can be obtained from the following general formula from matrix theory. For an arbitrary $q \times q$ matrix $\mathbf{C}$,

$$\operatorname{tr} \mathbf{C} = \sum_{i=1}^{q} \sigma_i \tag{13.29}$$

where the σ_i are the eigenvalues of $\mathbf{C}$. Since the trace of the matrix $\mathbf{A}$ in (11.40) is zero, we know from Eq. (13.29) that

$$\sum_{i=1}^{2N} \lambda_i = 0 \tag{13.30}$$

This result leaves two possibilities for the eigenvalues λ_i. First, they are zeros or pairs of conjugate imaginary numbers. Although this is the case for the original one-prey–one-predator Lotka-Volterra model, it seems unlikely when N is large. The other alternative is the occurrence of at least some of the eigenvalues of $\mathbf{A}$ in the form $c + id$, $-c - id$ with $c \neq 0$. In this instance, at least one eigenvalue has positive real part and the steady state is unstable. Consequently, this analysis suggests that the effect of additional complexity will be a destabilization of the prey-predator system.

13.4.3 Other One-Predator–One-Prey Models

Inspired by the work of Lotka and Volterra, subsequent investigators have proposed improvements to their model. One deficiency in the Lotka-Volterra approach can be seen by examining Eq. (13.9): in the absence of predator, the prey species n_1 will enjoy unbounded exponential growth. This unrealistic feature can be removed by explicitly accounting for the prey's utilization of the substrate which limits its growth.

Also, several experimental studies have revealed that predator specific growth rates do not vary linearly with prey concentration, as assumed in the original Lotka-Volterra analysis. As indicated by the Lineweaver-Burk plot for a protozoan-bacterium system in Fig. 13.12, an equation of the Monod form

$$\mu_p = \frac{\mu_{p,\max} n_1}{K_p + n_1} \tag{13.31}$$

will often provide a better representation of predator specific growth rates.

After incorporating both these refinements and assuming constant yield factors Y_s and Y_p for prey growth on substrate and predator growth on prey, respectively, the following populations balances are obtained for a chemostat:

$$\frac{ds}{dt} = D(s_0 - s) - \frac{1}{Y_s} \frac{\mu_{s,\max} s n_1}{K_s + s} \tag{13.32}$$

$$\frac{dn_1}{dt} = -Dn_1 + \frac{\mu_{s,\max} s n_1}{K_s + s} - \frac{1}{Y_p} \frac{\mu_{p,\max} n_1 n_2}{K_p + n_1} \tag{13.33}$$

$$\frac{dn_2}{dt} = -Dn_2 + \frac{\mu_{p,\max} n_1 n_2}{K_p + n_1} \tag{13.34}$$

This model was found by Tsuchiya et al. [19] to describe most of the important features of their experimental studies of predation by the amoeba *Dictyostelium discoideum* on *E. coli* bacteria. After determining kinetic constants for this system (Table 13.5), they computed the solid curves shown in Fig. 13.13. Not only does the model predict oscillations for these operating conditions ($D = 0.0625$/h, $s_0 = 0.5$ mg/mL), but it also does well in reproducing the experimental period and the amplitudes and phase relations of all three variables.

Another weakness in the Lotka-Volterra model can now be placed in perspective. We have already seen in Fig. 13.10 that Lotka-Volterra oscillations depend on initial conditions. Viewed differently, as in Fig. 13.14a, this means that the Lotka-Volterra oscillations will change period and amplitude following a

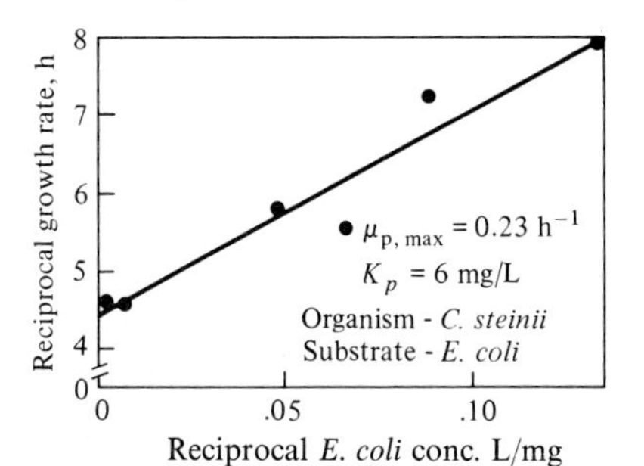

Figure 13.12 A double-reciprocal plot of predator (protozoan *Colpoda steinii*) growth rate and prey (bacterium *E. coli*) concentration. The approximately linear relationship in this plot suggests a growth-rate equation of Monod form. *(Reprinted from G. Praper and J. C. Garver, "Mass Culture of the Protozoa* Colpoda steinii," *Biotech. Bioeng., vol. 8, p. 287, 1966.)*

Table 13.5 Parameter values for the prey-predator model in Eqs. (13.32) to (13.34)†

Organism	Max specific growth rates μ_{max}, h^{-1}	Saturation constant K	Yield coefficient Y
D. discoideum	0.24	4×10^8 bacteria/mL	1.4×10^3 bacteria/amoeba
E. coli	0.25	5×10^{-4} mg glucose/mL	3.3×10^{-10} mg glucose/bacterium

† From H. M. Tsuchiya et al., *J. Bacteriol.*, **110**: 1151 (1972).

disturbance to the system. This behavior, sometimes termed a *soft oscillation* by mathematicians, is not very realistic from a physical viewpoint. It means that a small perturbation to the system can alter its dynamic characteristics for all future time.

The observance of regular sustained oscillations in nature, where small upsets occur frequently, suggests that natural predator-prey oscillations are more stable than those obtained from the Lotka-Volterra model. These stability characteristics are found, however, in the model described in Eqs. (13.32) to (13.34). Figure 13.14b indicates that oscillations predicted by this model are independent of initial conditions. Called *hard oscillations* or *limit cycles*, these oscillations are more reminiscent of actual predator-prey behavior than the soft Lotka-Volterra solutions.

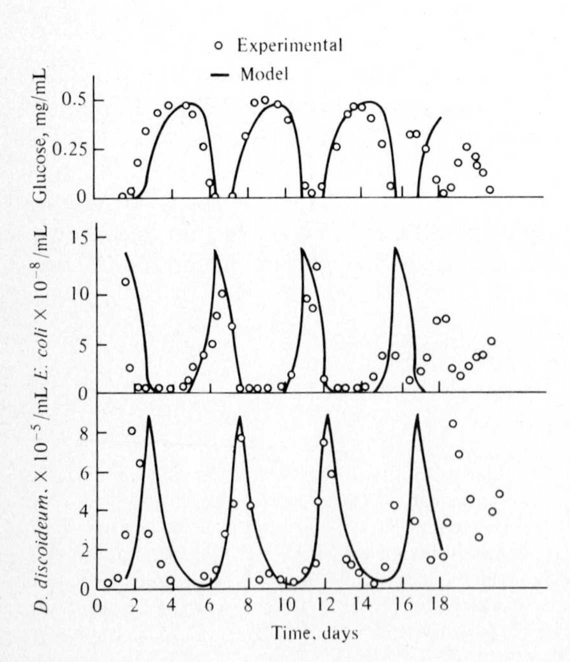

Figure 13.13 Model (—) and experimental (○) results for the prey-predator system *D. discoideum–E. coli* in continuous culture (25°C, $D = 0.0625\ h^{-1}$). *(Reprinted from H. M. Tsuchiya et al., "Predator-Prey Interactions of* Dictyostelium discoideum *and* Escherichia coli *in Continuous Culture," J. Bacteriol., vol. 110, p. 1147, 1972.)*

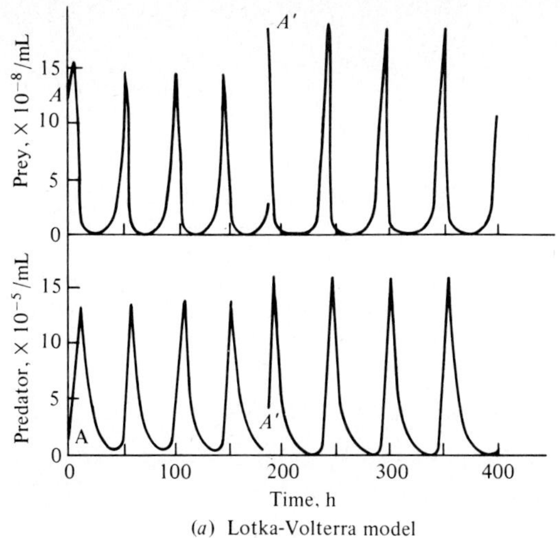

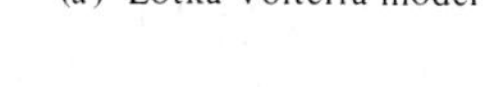

(*a*) Lotka-Volterra model

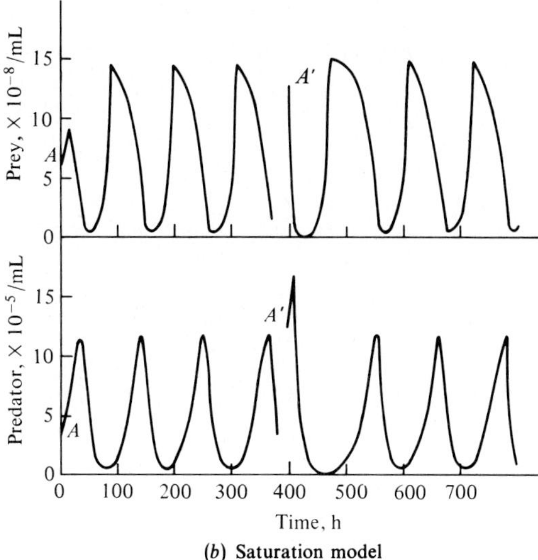

(*b*) Saturation model

Figure 13.14 Comparison of the characteristics of (*a*) the Lotka-Volterra model, which produces soft oscillations dependent on initial conditions, and (*b*) the model of Eq. (13.32) to (13.34), which gives hard oscillations independent of initial conditions. At point A' in these calculations, the numbers of predator and prey were shifted to new values. *(Reprinted from H. M. Tsuchiya et al., "Predator-Prey Interactions of* Dictyostelium discoideum *and* Escherichia coli *in Continuous Culture," J. Bacteriol., vol. 110, p. 1147, 1972.)*

Example 13.3: Model discrimination and development via stability analysis Tsuchiya, Frederickson, and coworkers [19–21] continued the above work and showed how stability analysis could be applied to differentiate between mathematical models of prey-predator systems. They built upon Canale's† results, which show that the chemostat described by Eqs. (13.32) to (13.34) can have three different steady states:

Total washout:

$$n_{1s} = n_{2s} = 0 \qquad s_s = s_0$$

† R. P. Canale: An Analysis of Models Describing Predator-Prey Interaction, *Biotech. Bioeng.*, **12:** 353, 1970.

Predator washout:

$$n_{1s} > 0 \qquad n_{2s} = 0 \qquad 0 < s_s < s_0$$

Prey-predator survival:

$$n_{1s} > 0 \qquad n_{2s} > 0 \qquad 0 < s_2 < s_0$$

Whether more than one of these steady states exists for given operating conditions D and s_0 depends upon the kinetic parameters of the system.

Likewise, stability properties depend upon all these system parameters. Before delving further into a discussion of stability analysis, we take advantage of the special structure of Eqs. (13.32) to (13.34) to reduce the number of dependent variables by one. Dividing Eq. (13.34) by Y_P, multiplying (13.32) by Y_s, and then adding the resulting two equations with (13.33) gives

$$\frac{d}{dt}\left[Y_s(s - s_0) + n_1 + \frac{n_2}{Y_p}\right] = -\frac{1}{D}\left[Y_s(s - s_0) + n_1 + \frac{n_2}{Y_p}\right] \tag{13E3.1}$$

where a term equal to zero $[d(-Y_s s_0)/dt]$ has been added to the left-hand side to make the expressions in brackets identical. Now, integrating Eq. (13E3.1) shows that

$$Y_s[s(t) - s_0] + n_1(t) + \frac{n_2(t)}{Y_p} = \left\{Y_s[s(0) - s_0] + n_1(0) + \frac{n_2(0)}{Y_p}\right\}\exp\left(\frac{-t}{D}\right) \tag{13E3.2}$$

Thus, after three to five chemostat holding times have elapsed, we can say to an excellent approximation that

$$Y_s[s(t) - s_0] + n_1(t) + \frac{n_2(t)}{Y_p} \approx 0 \tag{13E3.3}$$

Using (13E3.3), we can eliminate one of the dependent variables, say $s(t)$, from Eqs. (13.33) and (13.34). Henceforth we shall invoke (13E3.3) and view our problem as a two-dimensional one.

One way to represent the general dynamic features of our chemostat is an *operating diagram* like Fig. 13E3.1. Here for given values of the kinetic constants, regions with different characteristics are labeled in the s_0, $1/D$ plane. Thus, if the point corresponding to given operating conditions s_0 and D

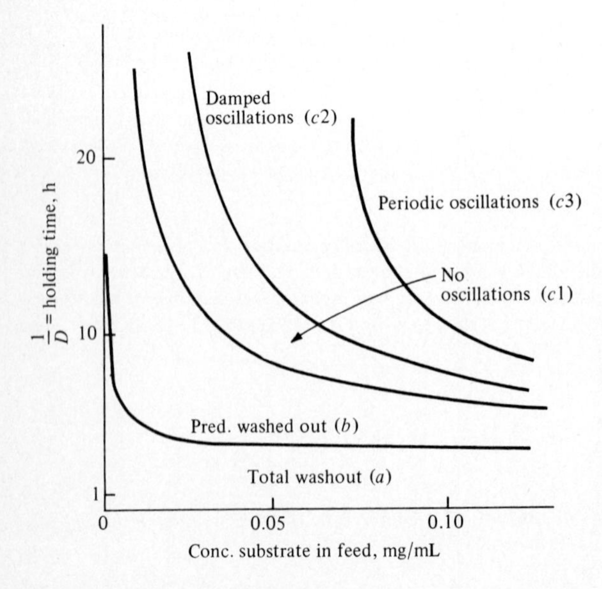

Figure 13E3.1 Qualitative dynamic characteristics of the prey-predator dynamic model given in Eqs. (13.32) to (13.34) in terms of chemostat operating conditions. *(Reprinted from H. M. Tsuchiya et al., "Predator-Prey Interactions of* Dictyostelium discoideum *and* Escherichia coli *in Continuous Culture," J. Bacteriol., vol. 110, p. 1147, 1972.)*

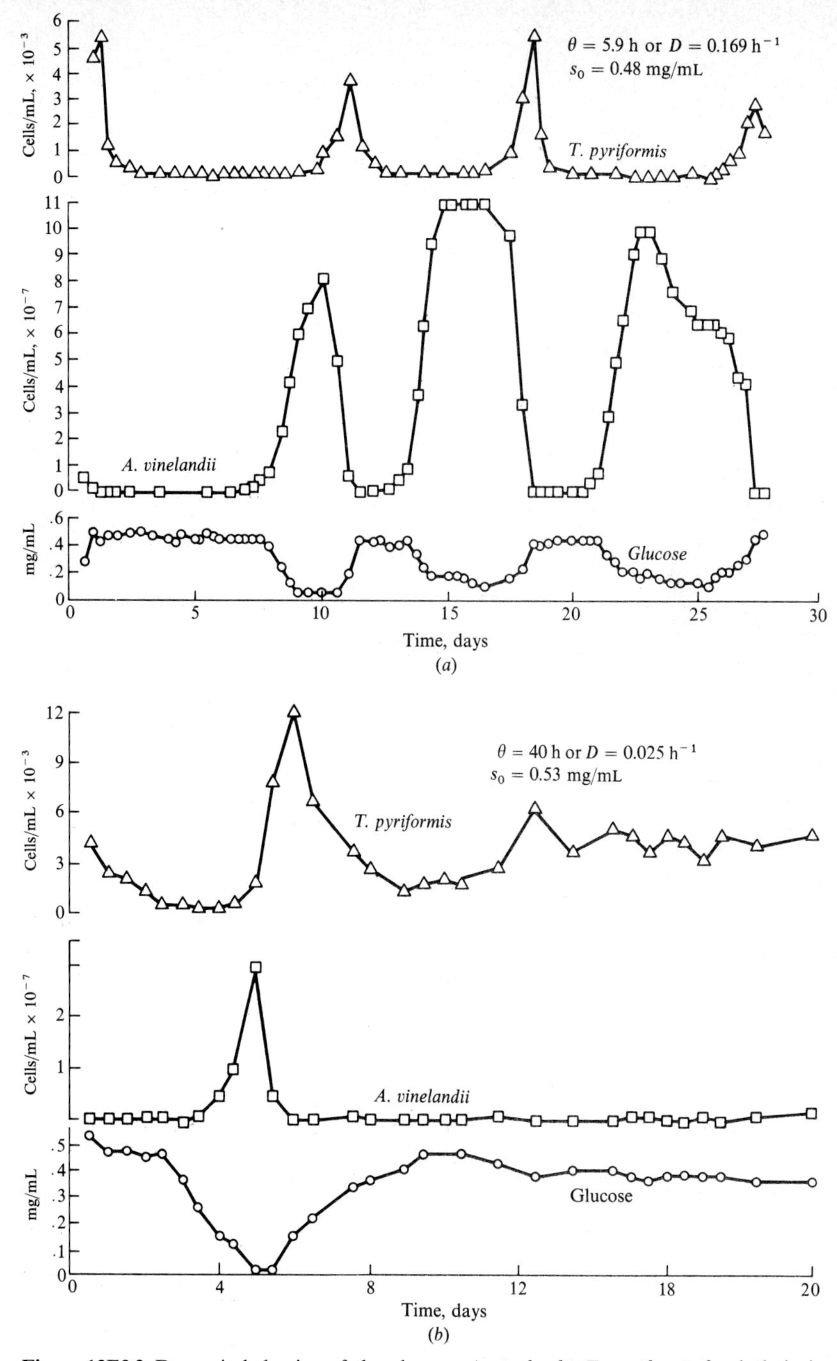

Figure 13E3.2 Dynamic behavior of the glucose–*A. vinelandii*–*T. pyriformis* food chain in a chemostat. (*a*) Sustained oscillations occur for $D = 0.169\ \text{h}^{-1}$. (*b*) The oscillations are damped when the dilution rate is reduced to $0.025\ \text{h}^{-1}$. *(Reprinted from J. L. Jost et al., "Interactions of* Tetrahymena pyriformis, Escherichia coli, Azobacter vinelandii, *and Glucose in a Minimal Medium," J. Bacteriol., vol. 113, p. 834, 1973.)*

lies in region *a*, for example, only the first of Canale's steady states exists and is asymptotically stable, so that washout of both predator and prey occurs.

Using the methods for local stability analysis already presented, we can show that when steady state *c* exists, the remaining steady states *a* and *b* are always unstable. The three uppermost regions in Fig. 13E3.1 are distinguished by the eigenvalues characterizing steady state *c*. In region *c*1, both eigenvalues are real and negative: here steady state *c* is a stable node (recall Example 13.1). Steady state *c* is a stable focus for operating conditions in region *c*2. Region *c*3 results in instability for steady state *c*.

This latter possibility apparently poses a dilemma since *none* of the system steady states are stable. This problem is resolved by the theory of Poincaré and Bendixson, which guarantees in this case that limit-cycle oscillations are obtained. Exploitation of this theorem requires that both independent variables have upper and lower bounds, as can be readily shown for this example, and that *the model have only two independent variables.* The latter important restriction has been overlooked in some of the literature on chemical and biological oscillations (see Ref. 10 for further details).

Although the location of the boundary lines in Fig. 13E3.1 will vary somewhat with the choice of the kinetic-parameter values, the general features shown there prevail for reasonable system parameters. One feature of this operating diagram is of great interest in the current context. Suppose that periodic oscillations result for a given feed concentration and holding time. Then sustained oscillations are also expected for all larger holding times and the same inlet concentration of substrate.

This behavior is at odds with the experimental results shown in Fig. 13E3.2 for simultaneous cultivation of *Azotobacter vinelandii* and protozoan *T. pyriformis* in a chemostat. Although oscillations appear for a 5.9-h holding time, they are damped when the holding time is 40 h. Obviously, the model proposed above is not adequate for this system.

In an attempt to improve the model, we shall add some structure to the description of predator physiology. Introducing two intermediate physiological states N_2' and N_2'' for the predator, we postulate a mechanism

$$N_2 + \alpha N_1 \longrightarrow N_2' \qquad N_2' + \beta N_1 \longrightarrow N_2'' \qquad N_2'' \longrightarrow 2N_2$$

Application of the quasi-steady-state approximation to N_2' and N_2'' gives the following expression for predator specific growth rate μ_p:

$$\mu_p = \frac{\mu_{p,\max}(n_1)^2}{(K_{p1} + n_1)(K_{p2} + n_1)} \tag{13E3.4}$$

According to this model, which reduces to the Monod form (13.31) for n_1 much larger than the smaller of K_{p1} and K_{p2}, the predator growth rate varies as n_1^2 when the prey concentration is small.

Exactly the same approach as described earlier can be employed to study the stability properties of the revised chemostat model incorporating Eq. (13E3.4). With kinetic parameters obtained from batch studies of *A. vinelandii* and *E. coli*, the operating diagram of Fig. 13E3.3 is obtained. This result

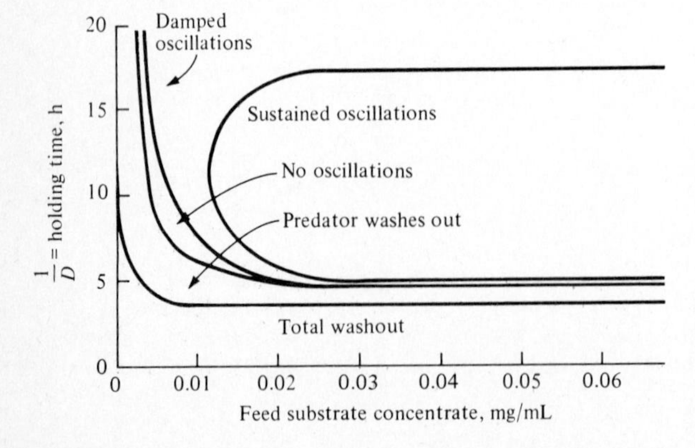

Figure 13E3.3 Operating diagram for the modified predator-prey chemostat model incorporating the multiple saturation predation rate of Eq. (13E2.4). *(Reprinted from J. L. Jost et al., "Interactions of* Tetrahymena pyriformis, Escherichia coli, Azotobacter vinelandii, *and Glucose in a Minimal Medium," J. Bacteriol., vol. 113, p. 834, 1973.)*

reveals the disappearance of sustained oscillations when D decreases in accordance with the experimental findings. Thus we have seen here how experimental data plus theoretical stability analysis can find weaknesses in a mathematical model and test an alternative formulation.

13.5 EFFECTS OF THE NUMBER OF SPECIES AND THEIR WEB OF INTERACTIONS

In this section we shall proceed from specific to rather abstract examinations of how mixed populations with more than two species behave. Before starting our investigation of these sometimes difficult matters, we require some additional terminology from the science of ecology.

13.5.1 Trophic Levels, Food Chains, and Food Webs: Definitions and an Example

Often in nature, there is a hierarchy of food-consumer relationships. For example, the bacteria in the previous section are the consumers of the food glucose, and these bacteria in turn are food for protozoan consumers, which are then consumed by larger organisms. Such hierarchies are called *food chains*, and the successive steps in the chain are designated *trophic levels*. Usually, food chains are not isolated from each other but merge and intertwine to form *food webs*. A simple food web, which has been studied in the laboratory, is illustrated in Fig. 13.15; here, two food chains (substrate → prey 1 → predator and substrate → prey 2 → predator) overlap at the first and third trophic levels.

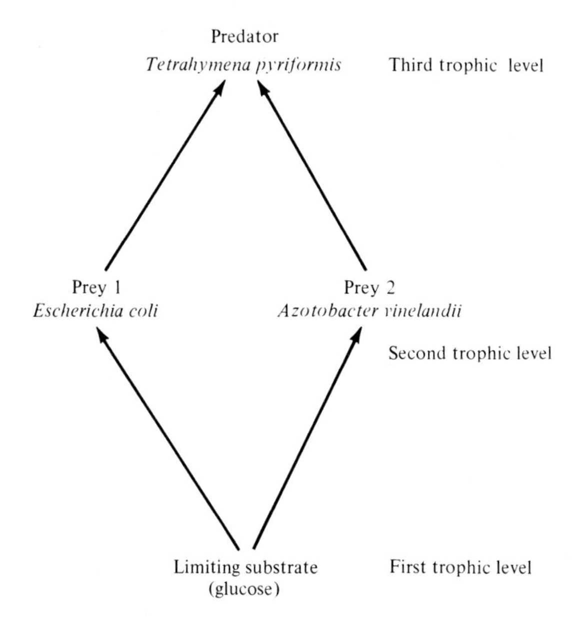

Figure 13.15 Schematic diagram of a simple food web containing three trophic levels. The two bacteria in the second level, which both consume the nutrient of the first level, are both consumed by the protozoan predator in the highest trophic level. *(Reprinted from J. L. Jost et al., "Interactions of* Tetrahymena pyriformis, Escherichia coli, Azotobacter vinelandii, *and Glucose in a Minimal Medium." J. Bacteriol., vol. 113, p. 834, 1973.)*

Before reviewing experimental results for the food web shown in Fig. 13.15 we should recall earlier discussions of its various components. Considering only the first and second trophic levels, we have a competitive situation and the exclusion principle applies. In fact, Fig. 13.7 illustrated the validity of the exclusion principle for the first two trophic levels in question. Also, the glucose → prey 2 → predator system of Fig. 13.15 has been studied experimentally and found to yield sustained oscillations under some conditions (see Fig. 13E3.2). Consequently, it will be quite interesting to see what types of dynamic behavior emerge when these and one more prey-predator pair coalesce into the food web of Fig. 13.15.

Results of an experimental study of this web in a chemostat with $D = 0.1\ \text{h}^{-1}$ are shown in Fig. 13.16. The most notable feature of these results is the *coexistence* of the two bacterial competitors when they share a common predator. Thus, the addition of a predator for both competitors has a stabilizing effect: washout of one of the competitor species is avoided.

Although oscillations are evident in the concentrations of all three populations, the fluctuations are smaller and smoother than in the simpler predator-prey food chain. Unfortunately, the data do not extend beyond 8 days, so that possible damping of the oscillations cannot be ruled out. The experiment was stopped at this point by rapid and extensive growth on the walls of the chemostat.

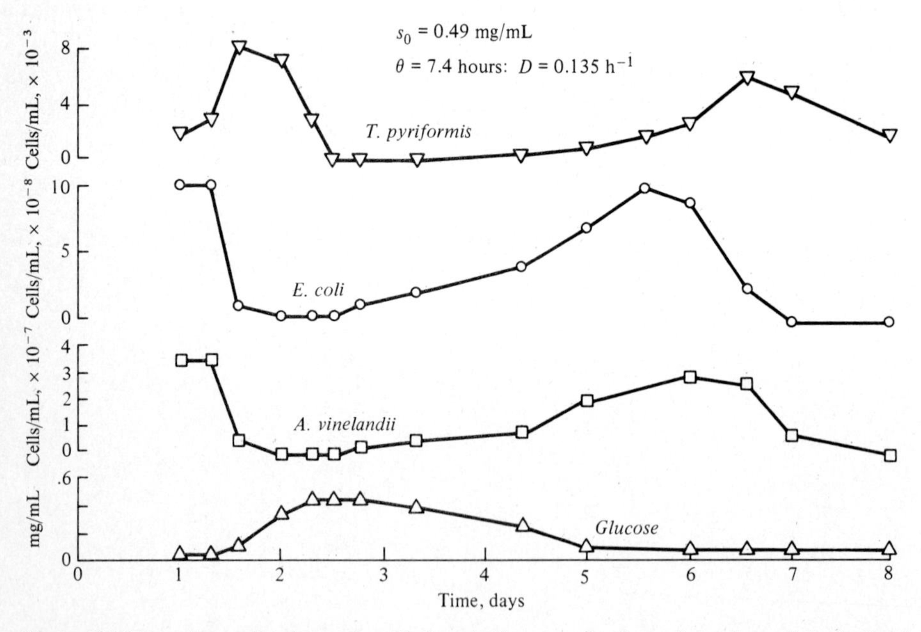

Figure 13.16 Experimentally observed oscillations in the concentrations of members of the food web shown in Fig. 13.15. *(Reprinted from J. L. Jost et al., "Interactions of* Tetrahymena pyriformis, Escherichia coli, Azotobacter vinelandii, *and Glucose in a Minimal Medium." J. Bacteriol., vol. 113, p. 834, 1973.)*

13.5.2 Population Dynamics in Models of Mass-Action Form

For the remainder of Sec. 13.5, we shall consider abstract methods for analyzing and characterizing complex interacting populations. Note that many of the results cited here will also apply to complex biological reaction systems. Indeed, a frequent starting point in formulating ecological models is the assumption of a set of microbial "reactions" such as those above Eq. (13E3.4). Since the theory outlined in this subsection relies heavily on a chemical-reaction orientation, let us consider an additional example of application of this approach to microbial growth models.

Suppose that the following reactions occur in an isothermal, homogeneous, constant-volume system to which component B is added continuously so that its concentration stays constant:

$$A_1 + B \xrightarrow{r_1} 2A_1 \qquad A_1 + A_2 \xrightarrow{r_2} 2A_2 \qquad A_2 \xrightarrow{r_3} C \tag{13.35}$$

Moreover, we shall assume that each reaction is elementary as written, so that the kinetics follows directly from the reaction stoichiometry. Consequently, for reaction system (13.35), we have

$$r_1 = k_1 a_1 b \qquad r_2 = k_2 a_1 a_2 \qquad r_3 = k_3 a_2 \tag{13.36}$$

As a matter of convention, we need not include in our analysis concentrations which are time-invariant. To this end, let us define

$$k^* = k_1 b \tag{13.37}$$

so that r_1 is

$$r_1 = k^* a_1 \tag{13.38}$$

The mass balances for A_1 and A_2 can now be written

$$\frac{da_1}{dt} = k^* a_1 - k_2 a_1 a_2 \tag{13.39}$$

$$\frac{da_2}{dt} = k_2 a_1 a_2 - k_3 a_2 \tag{13.40}$$

which, upon comparison with Eqs. (13.9) and (13.10), we recognize as the Lotka-Volterra equations.

Now let us progress toward presentation of some very general and powerful theorems which apply to such mass-action kinetic systems. These results often will permit us to conclude, relying only on the *algebraic structure* of the reaction network, that a given system has a single steady state which is globally asymptotically stable. Here the term algebraic structure refers only to the way species are transformed and connected in the reaction network: the results we shall reach apply independent of the specific values of the rate constants or other operating parameters.

Treasures like this theory, developed by Horn, Feinberg, and Jackson [9], rarely come free so we must learn several new concepts and definitions. Our understanding of them will be greatly facilitated by reference to several examples. Some hypothetical mechanisms, as well as two of interest in our biological studies, are listed in Table 13.6. We recognize (3) as an equivalent form of the Lotka-Volterra mechanism, and (6) is one of the proposed models for glycolysis. (Can you identify A_1 through A_7 in terms of some of the compounds in the glycolytic pathway?) The *zero species* 0 appearing in these two mechanisms has the following significance: $0 \rightarrow A_j$ means that component A_j is added to the system at a constant rate. Such a reaction would be used to designate, for example, the addition of substrate to a chemostat via the feed stream. Writing $A_k \rightarrow 0$ means that A_k is removed from the system at a rate proportional to its concentration, as in removal of a species from a chemostat in the effluent stream or the death processes of a species. Thus, the zero species is used to introduce interactions with the system environment: it is the use of this device which permits application of the mass-action theory to open systems.

Table 13.6 Various reaction mechanisms to illustrate definitions and conclusions of mass-action theory†

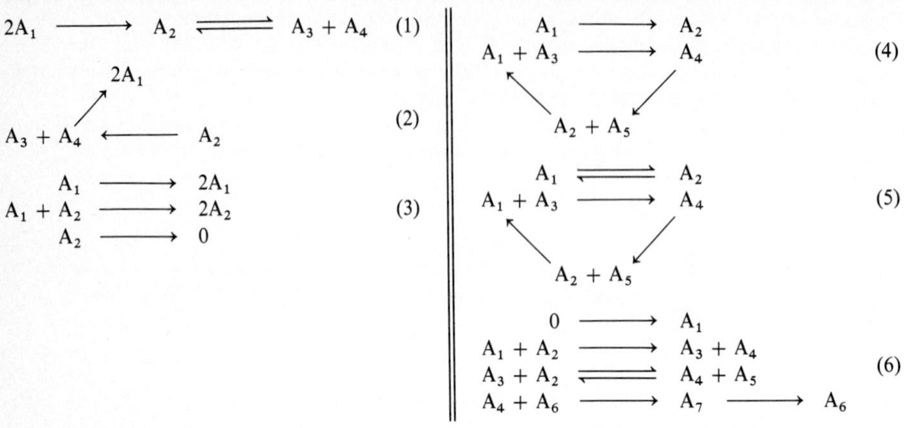

$$2A_1 \longrightarrow A_2 \rightleftharpoons A_3 + A_4 \qquad (1)$$

$$\begin{array}{ccc} & 2A_1 & \\ & \nearrow & \\ A_3 + A_4 & \longleftarrow & A_2 \end{array} \qquad (2)$$

$$\begin{array}{rcl} A_1 & \longrightarrow & 2A_1 \\ A_1 + A_2 & \longrightarrow & 2A_2 \\ A_2 & \longrightarrow & 0 \end{array} \qquad (3)$$

$$\begin{array}{rcl} A_1 & \longrightarrow & A_2 \\ A_1 + A_3 & \longrightarrow & A_4 \\ \nwarrow & & \swarrow \\ & A_2 + A_5 & \end{array} \qquad (4)$$

$$\begin{array}{rcl} A_1 & \rightleftharpoons & A_2 \\ A_1 + A_3 & \longrightarrow & A_4 \\ \nwarrow & & \swarrow \\ & A_2 + A_5 & \end{array} \qquad (5)$$

$$\begin{array}{rclcl} 0 & \longrightarrow & A_1 & & \\ A_1 + A_2 & \longrightarrow & A_3 + A_4 & & \\ A_3 + A_2 & \rightleftharpoons & A_4 + A_5 & & \\ A_4 + A_6 & \longrightarrow & A_7 & \longrightarrow & A_6 \end{array} \qquad (6)$$

† From M. Feinberg and F. J. M. Horn, *Chem. Eng. Sci.*, **29**: 775 (1974).

Next we need to define three integers for a reaction mechanism. These are n_c, the number of complexes in the mechanism; n_l, the number of linkage classes in the mechanism; and n_s, the dimension of the mechanism. Each of these in turn requires additional definitions. First, a *complex* is an entity which appears either at the head or the tail of a reaction arrow. For example, mechanisms (1) and (2) in Table 13.6 have three complexes: $2A_1$, A_2, and $A_3 + A_4$, so that here $n_c = 3$. The Lotka-Volterra mechanism (3) has $n_c = 6$ (A_1, $2A_1$, $A_1 + A_2$, $2A_2$, A_2, and 0), while there are five complexes ($n_c = 5$) for mechanisms (4) and (5): A_1, A_2, $A_1 + A_3$, A_4, and $A_2 + A_5$.

Turning now to linkage classes, let us disregard the direction of the reaction arrows and consider the complexes on either end of an arrow to be *linked* by that reaction. Considering mechanism (1), for example, we see that the complexes $2A_1$ and A_2 are linked directly in this manner. Also, while the complex A_2 is directly linked to the complex $A_3 + A_4$, we can say that the complexes $2A_1$ and $A_3 + A_4$ are linked indirectly: we can get from one complex to the other along some path consisting of direct linkage steps. A *linkage class* is a set of complexes which are all linked to one another either directly or indirectly such that no complex in the set is linked to any outside the set.

From what we have just said about mechanism (1), it has one linkage class $\{2A_1, A_2, A_3 + A_4\}$, so that $n_l = 1$ for this mechanism. Mechanism (2) has the same single linkage class. Mechanisms (4) and (5) have two ($n_l = 2$) linkage classes, which are $\{A_1, A_2\}$ and $\{A_1 + A_3, A_4, A_2 + A_5\}$, and $n_l = 3$ for mechanism (3): $\{A_1, 2A_1\}$, $\{A_1 + A_2, 2A_2\}$, $\{A_2, 0\}$.

The concept of *weak reversibility* is related to the linkage notion. Now we consider the direction of the reaction arrows and ask the following question: If there is a directed-arrow pathway leading from one complex to another, is there also a directed-arrow pathway leading from the second back to the first? If the answer is yes for any pair of complexes in the mechanism, we say that the mechanism is *weakly reversible*.

Mechanism (2) in Table 13.6, for example, is weakly reversible. We can move from complex $A_3 + A_4$ to complex $2A_1$ in a single step, and we can return from $2A_1$ to $A_3 + A_4$ by a two-step directed path through A_2. Moreover, similar arguments apply to every pair of complexes in the mechanism. Applying the test underlying the weak-reversibility concept to the other example mechanisms, we conclude that (1), (3), (4), and (6) are not weakly reversible and that (5) is.

One final preliminary matter must be considered before we reach the main theorems. To this end, we define a reaction vector for each reaction, i.e., each arrow, in the mechanism. If there are a total of M different chemical species $A_1, A_2, A_3, \ldots, A_M$ in the mechanism [$M = 4, 4, 2, 5, 5$ for mechanisms (1) to (5), respectively], each reaction vector will have M elements. These elements are determined as follows: the ith element is zero if A_i does not appear in the reaction. If A_i does appear, the ith entry is the stoichiometric coefficient of A_i. This coefficient is assigned a negative sign if A_i is on the tail side of the reaction arrow and positive if the reaction arrow points towards A_i.

For example, mechanism (1) of Table 13.6 has three arrows and four species. Each of these reactions is rewritten below along with the corresponding reaction vectors:

$$2A_1 \longrightarrow A_2 \qquad \mathbf{v}_1 = (-2, 1, 0, 0)$$

$$A_2 \longrightarrow A_3 + A_4 \qquad \mathbf{v}_2 = (0, -1, 1, 1)$$

$$A_3 + A_4 \longrightarrow A_2 \qquad \mathbf{v}_3 = (0, 1, -1, -1)$$

From the set of reaction vectors so obtained, we next determine the maximum number of these vectors which are linearly independent. This number we call n_s, the *dimension* of the mechanism. For instance, $n_s = 2$ for mechanism (1), since $\mathbf{v}_1$ and $\mathbf{v}_2$ above are linearly independent but $\mathbf{v}_3$ is not independent of these two ($\mathbf{v}_3 + \mathbf{v}_2 = \mathbf{0}$).

With the necessary foundation now established, we shall state the *zero-deficiency theorem* [9]. Suppose that for a given mechanism

$$n_c - n_l - n_s = 0 \tag{13.41}$$

Then:

1. For *any* kinetics, mass-action or otherwise, neither a steady state with all M concentrations positive nor periodic oscillations of the concentrations are possible if the mechanism is not weakly reversible.
2. If in addition to condition (13.41) the mechanism is weakly reversible and the kinetics are described by mass-action forms, for all stoichiometrically equivalent positive initial compositions, there is a unique steady state which is globally asymptotically stable. This conclusion applies regardless of the rate-constant values.

Two terms in the last sentence deserve further explanation. Positive means that each species has a positive concentration. Stoichiometrically equivalent compositions are those which can be transformed into each other by running one or several reactions in the mechanism in the forward or reverse direction. Considering first only $A_1 \rightleftharpoons 2A_2$, the initial concentrations (a_{10}, a_{20}) listed next are equivalent stoichiometrically: (2, 4), (3, 2), (1, 6), and so forth so long as $2a_{10} + a_{20} = 8$. Further, for mechanism (1), the following positive initial compositions $(a_{10}, a_{20}, a_{30}, a_{40})$ are stoichiometrically equivalent: (2, 1, 1, 1), (3, $\frac{1}{2}$, 1, 1), (2, $\frac{1}{2}$, $1\frac{1}{2}$, $1\frac{1}{2}$), (2, $1\frac{1}{2}$, $\frac{1}{2}$, $\frac{1}{2}$), etc.

The power of this theorem ultimately depends on how many reaction mechanisms satisfy condition (13.41). Horn has explored this question and found that the vast majority of possible mechanisms are consistent with (13.41). Our examples of Table 13.6 are therefore a somewhat atypical set, since only four [(1), (2), (3), (5)] of the six satisfy (13.41). Let us now consider further the potential applications of this theorem.

Suppose we wish to model an oscillating mixed population by postulating a population "reaction" mechanism involving several elementary steps with mass-action kinetics. The theorem then tells us that for any chance of success, the mechanism must violate either condition (13.41) or weak reversibility. Indeed, this is the case for mechanisms (3) and (6), both of which give rise to sustained oscillations. Since the vast majority of mechanisms we could choose will violate neither, we can use the theorem's conclusions to limit greatly the scope of our search. On the other hand, part 2 of the theorem tells us that for mass-action weakly reversible mechanisms which fulfill (13.41), asymptotic stability in the large is assured regardless of the exact location of the unique steady state or the values of the rate constants. Thus, in such cases, we need not bother with local-stability analyses or search for tricks to verify stability on a global scale.

Part 1 of the theorem also has important ecological consequences. It asserts that if (13.41) is satisfied while weak reversibility is not, some of the concentrations will be zero at steady state. From a population-dynamics viewpoint, then, this theory permits us to discover mechanisms for which at least one species becomes extinct.

Example 13.4: An application of the mass-action theory† In order to illustrate the usefulness of the zero-deficiency theorem, we shall use it to investigate the dynamic behavior of the following mechanism:

$$3A_1 \xrightarrow{k_a} A_1 + 2A_2 \qquad (a)$$

$$3A_2 \xrightarrow{k_b} 2A_1 + A_2 \qquad (b)$$

$$A_1 + 2A_2 \xrightarrow{k_c} 3A_2 \qquad (c)$$

$$2A_1 + A_2 \xrightarrow{k_d} 3A_1 \qquad (d)$$

which will be assumed to follow mass-action kinetics. The complexes for this mechanism are

$$3A_1 \qquad A_1 + 2A_2 \qquad 3A_2 \qquad \text{and} \qquad 2A_1 + A_2$$

so that $n_c = 4$. Next we note that the mechanism is weakly reversible. In fact, following the reaction arrows in the sequence (*a*) to (*d*) reveals a single directed path which connects all four complexes. Thus, there is only one linkage class and $n_l = 1$.

To find the mechanism dimension, we form four reaction vectors, one for each arrow in our mechanism. Each vector has two components since only two species, A_1 and A_2, are involved in this system. The reaction vectors are

$$\mathbf{v}_a = (-2, 2) \qquad \mathbf{v}_b = (2, -2) \qquad \mathbf{v}_c = (-1, 1) \qquad \mathbf{v}_d = (1, -1)$$

Since $-2\mathbf{v}_d = 2\mathbf{v}_c = -\mathbf{v}_b = \mathbf{v}_a$, there is only one linearly independent reaction: $n_s = 1$.

Testing condition (13.41) for our mechanism, we find

$$n_c - n_l - n_s = 4 - 1 - 1 = 2$$

so the condition is *not* fulfilled. Consequently, we cannot rule out the existence of oscillations or multiple steady states for this mechanism. Indeed, Feinberg and Horn show that for $k_c = k_d = 1$ and $k_a = k_b < \frac{1}{6}$, we shall observe multiple steady states for this example.

13.5.3 Qualitative Stability

In the preceding subsection, we have perused some of the strongest theoretical tools available for a class of nonlinear process models. Our concluding studies of the influences of ecosystem size and complexity will concentrate on the linearized model. First we shall consider what can be said about stability based only on the *signs* of the entries in the community matrix.

If all eigenvalues of **A** have negative real parts regardless of the magnitudes

† This example is based on a discussion in Feinberg and Horn [9] and is intended for illustrative purposes only. It does not necessarily correspond to any actual microbial or biochemical reaction system.

of the nonzero elements, we shall say that **A** is *qualitatively stable.* Necessary and sufficient conditions for qualitative stability of **A** are the following:

1. $a_{jj} \leq 0$ all j
2. $a_{ii} < 0$ for at least one i
3. $a_{ij}a_{ji} \leq 0$ for all $i \neq j$
4. For any sequence of three or more unequal indices $j, k, \ldots, s, t,$

$$a_{ij}a_{jk} \cdots a_{rs}a_{st} = 0$$

5. $\det \mathbf{A} \neq 0$

As discussed by May [5], each of these conditions has a direct physical interpretation. Condition 5 ensures that the linear system (9.30) has a unique steady state at $\chi = 0$. While condition 1 requires that the autonomous specific growth of every species be nonpositive, condition 2 strengthens this restriction by stipulating that at least one of the autonomous specific growth rates be negative. Condition 4 states that there are no closed loops of interactions containing more than two species.

Especially interesting in light of the interaction classifications in Table 13.4 is condition 3. Interpreted from an ecological outlook, it states that no mixed population exhibiting a mutualistic (a_{ij} and a_{ji} both positive) or a competitive (negative a_{ij} and a_{ji}) pairwise interaction can be qualitatively stable. On the other hand, all three remaining nontrivial interactions, namely commensalism, amensalism, and predation, satisfy condition 3.

> This mathematically rigorous statement may be plausibly extended into the broader, if rougher, statement that competition or mutualism between two species is less conducive to overall web stability than is a predator-prey relationship. It is tempting to speculate that stability considerations may make for communities in which strong predator-prey bonds are more common than symbiotic ones. This result is not intuitively obvious, yet it is a feature of many real world ecosystems ... [5, p. 73].

Example 13.5: Qualitative stability of a simple food web† Consider the simple food web sketched schematically in Fig. 13E5.1α. The arrow from species 1 to species 2 indicates that species 2 feeds on species 1, so that $a_{21} > 0$ and $a_{12} < 0$. All the other arrows have analogous interpretations. Consequently, we can determine from the figure that the community matrix $\mathbf{A}_\alpha$ corresponding to this web has entries with signs as indicated:

$$\mathbf{A}_\alpha = \begin{bmatrix} - & - & - \\ + & - & - \\ + & + & - \end{bmatrix} \tag{13E5.1}$$

where it has been assumed that all species have negative autonomous specific growth rates. Checking this sign structure against the qualitative stability conditions, we see that, for instance,

$$a_{12}a_{23}a_{31} > 0 \tag{13E5.2}$$

† This example is based on a discussion in May [5].

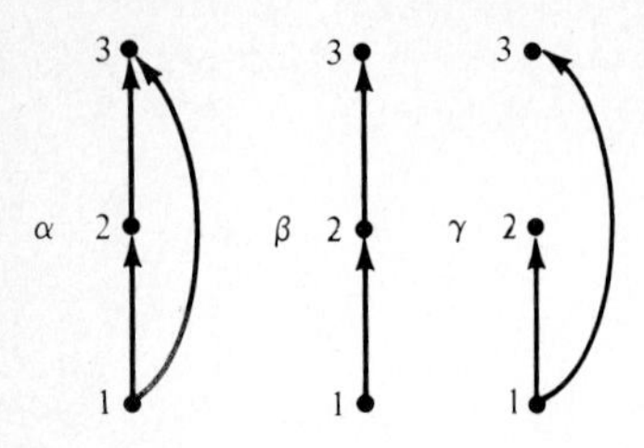

Figure 13E5.1 Three simple food webs, each with three trophic levels.

so that condition 4 is violated. Hence there exists some choice of magnitudes for the a_{ij} for which the system is unstable, and we must check the real part of the matrix eigenvalues or an equivalent condition such as (9.41), employing the specific numerical values for the a_{ij}.

In cases β and γ (Fig. 13E5.1), however, the sign structures of the corresponding community matrices take the respective forms

$$\mathbf{A}_\beta = \begin{bmatrix} - & - & 0 \\ + & - & - \\ 0 & + & - \end{bmatrix} \quad \text{and} \quad \mathbf{A}_\gamma = \begin{bmatrix} - & - & - \\ + & - & 0 \\ + & 0 & - \end{bmatrix} \tag{13E5.3}$$

Obviously conditions 1 to 3 are fulfilled for both these matrices. Checking condition 4, we see that det $\mathbf{A}_\beta < 0$ and det $\mathbf{A}_\gamma < 0$, so that singularity is no problem. Finally, the physical interpretation of condition 4 and examination of Fig. 13E5.1 immediately reveal that no closed loops of more than two species appear in webs β or γ. This could also be checked by applying condition 4 directly to (13E5.3).

Therefore, food webs β and γ, both obtained by removing one trophic link from web α, are stable regardless of the numerical values of the a_{ij}.

13.5.4 Stability of Large, Randomly Constructed Food Webs

Next we shall review some fascinating work by Gardner and Ashby,† May [5], and McMurtrie‡ on the relationship between stability and ecosystem size and complexity. Suppose that the system contains p different species which if left alone would all exhibit negative autonomous specific growth rates of similar magnitudes. By suitably scaling time, we can then assume that each species contributes an element -1 to the main diagonal of the community matrix.

The effects of interactions will be represented by a matrix **B**, so that the community matrix **A** takes the form

$$\mathbf{A} = \mathbf{B} - \mathbf{I} \tag{13.42}$$

where **I** is the $p \times p$ identity matrix. The extent of interactions and their magnitudes will be specified by two parameters, the connectance C $(0 < C < 1)$ and the strength σ^2, respectively.

† M. R. Gardner and W. R. Ashby, Connectance of Large Dynamical (Cybernetic) Systems: Critical Values for Stability, *Nature*, **228:** 784, 1970.

‡ Paper in preparation cited in Ref. 5.

B matrices will now be selected at random (we can use a random-number generator to pick each element b_{ij}) with the following constraints:

Of the p^2 matrix elements b_{ij}, a fraction C will be nonzero. Stated more precisely, the probability that any b_{ij} is zero is $1 - C$.
Nonzero elements have an equal probability of being positive or negative. Thus the mean of these b_{ij} is zero.
The mean square value of the nonzero b_{ij} is σ^2.

For each **B** matrix so selected, we can evaluate **A** with (13.42) and then pursue standard stability tests. The question of major interest here is this: What is the probability that a particular **B** matrix chosen from the randomly selected set gives a stable model? From our formulation it is evident that this probability will depend on the three parameters p, C, and σ^2. May has shown that if $p \gg 1$, the probability that the system is stable approaches unity when

$$\sigma\sqrt{pC} < 1 \tag{13.43}$$

On the contrary, if $p \gg 1$ and

$$\sigma\sqrt{pC} > 1 \tag{13.44}$$

the system is almost certain to be unstable. Figure 13.17 shows results of Monte Carlo simulations which are very consistent with Eqs. (13.43) and (13.44).

Consequently, we have seen that the more species which are present (larger p), and the greater the number which interact (larger C) and the more intense the effect of interaction on growth (larger σ), the more likely it is that the ecosystem will be unstable. While this conclusion is certainly consistent with stability studies of many other types of dynamic systems, we should emphasize here that the current thinking of many ecologists runs in the opposite track. Since the

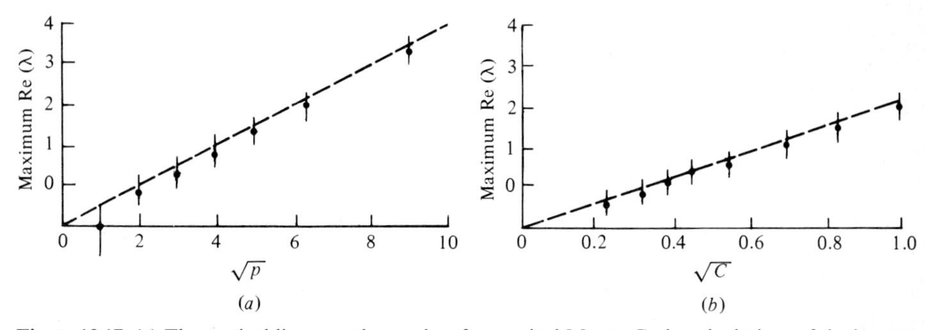

Figure 13.17 (*a*) The vertical lines are the results of numerical Monte Carlo calculations of the largest real-part eigenvalue of random matrices with $\sigma = 0.5$, $C = 1$. These results and the dashed line, which corresponds to Eqs. (13.43) and (13.44), show that instability becomes more likely as the number of species p increases. (*b*) Similar results with $\sigma = 0.5$, $p = 40$. Here the destabilizing effect of population interactions is evident. *(Reprinted from R. M. May, "Stability and Complexity in Model Ecosystems," Princeton University Press, Princeton, New Jersey, 1973.)*

highly stable ecological systems which occur in nature tend to involve many species with a very complex web of interactions, it is often assumed that diversity and complexity in an ecological system make for increased stability. Although empirical observations of natural systems support this view, mathematical analyses like those outlined above do not.

May [5] suggests an idea for resolving this apparent dilemma. Although, on the average, diversity and complexity tend to destabilize a system, the natural world is anything but an "average" system. It has evolved over millennia to its present state, presumably selecting stable interrelationships among populations in preference to unstable ones. This thesis suggests an interesting avenue for future studies in mathematical ecology: find what *special* characteristics a large-scale system should have to be stable. In a sense, the results summarized in the two previous subsections are steps in this direction.

13.5.5 Bifurcation and Complicated Dynamics

One useful way to think about steady-state and dynamic behavior of complex, interacting multiple species systems is to consider the changes which occur as a system parameter (e.g., nutrient concentration, dilution rate) shifts smoothly. Substantial, *qualitative* changes frequently arise including appearance of additional steady states or of limit cycles. Also, stable limit cycles may become unstable, leading to more complicated, often nonperiodic, transient behavior. In the extreme, trajectories may assume an almost random nature and are then termed *chaotic*.

Mathematical analysis of these qualitative changes in system dynamics arising from variation of one or more system parameters is the domain of *bifurcation theory*. This theory is important because in many cases the genesis of nonlinear features can be determined from relatively simple, linearized calculations. Only the beginning elements of this theory can be summarized here: additional background information and guidance to more advanced literature can be found in the chapter references.

We have already discussed local stability analysis for two-species systems in Example 13.1. Recalling Sec. 9.2.2, the steady state $\mathbf{n}_s$ is locally asymptotically stable for the multiple species case if and only if all eigenvalues of the community matrix [Eq. (13.2)] have negative real parts. Any eigenvalue with positive real part implies instability of the steady state $\mathbf{n}_s$.

Now suppose that, starting with a system with a single, locally asymptotically stable steady state, conditions (i.e., model parameters) change and the eigenvalues of $\mathbf{A}$ shift as well. A number of nonlinear phenomena can appear if the steady state loses stability, meaning one or more eigenvalues moves from the left- to the right-hand complex plane. These are characterized by the eigenvalue(s) position(s) when crossing the imaginary axis.

If a single eigenvalue crosses with zero imaginary part, steady-state bifurcation, meaning appearance of additional steady-state solutions of the system equations, occurs. Oscillations are unlikely. On the other hand, if a conjugate pair of eigenvalues crosses the imaginary axis at $\pm i\omega$, an oscillation appears. This is called a *Hopf bifurcation*. The steady states or oscillation which appear at bifurcation may or may not be stable; further analysis is needed to resolve this question. More complicated bifurcations, in which other eigenvalue configurations transit the imaginary axis, and bifurcations resulting from loss of stability of limit cycles are introduced in the references.

13.6 SPATIAL PATTERNS

Up until now, we have assumed that all populations are uniformly dispersed, or well mixed, within the volume of interest. While convenient for modeling and analysis, this assumption of spatial uniformity is not necessarily correct. Many systems have internal gradients of environmental conditions which tend to force inhomogeneities in population characteristics with position.

Even more interesting are the numerous cases where spatial differences in one or several microbial populations arise spontaneously in the absence of any environmental gradients. Although much additional research is required to explore the detailed mechanisms whereby such conditions arise and are maintained, we can safely assume that spatial differentiation is caused by interactions between the individuals involved. These individuals may be members of two different species or cells of a single species at different stages of development.

A vivid example of this phenomenon is presented in Fig. 13.18, which shows regular sporulation of the fungus *Nectria cinnabarina* at 1-mm intervals. Evident here is a spatial periodicity: as we move away from the colony's center, we periodically encounter bands. It is intriguing to speculate that the physicochemical mechanism underlying development of spatial periodicity is the same in concept as that causing periodicities in time. Some mathematical studies point to this conclusion, although it is far from proved in general. Also, there is an actual chemical-reaction system which supports this hypothesis. Left unstirred, the Belousov-Zhabotinskii reaction generates spatial patterns which can also fluctuate with time. Well mixed, this system exhibits periodic oscillations in time.

While much more could be said on this fascinating subject, we have space to consider here only one simple model of spatial differentiation. Access to other work in this field is available in the references.

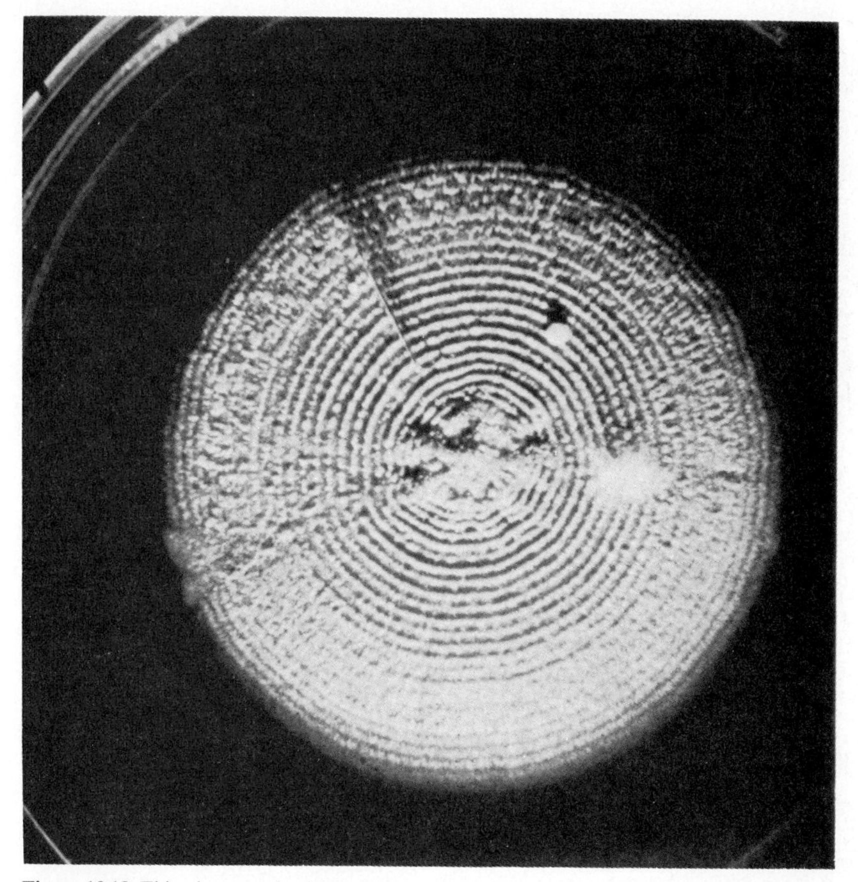

Figure 13.18 This photograph of a 6-cm-diameter mycelial colony of *N. cinnabarina* shows a regular spatial pattern of sporulation which appears here as spore ridges about 1 mm apart. *(Reprinted from A. T. Winfree, "Polymorphic Pattern Formation in the Fungus* Nectria,*" J. Theor. Biol., vol. 38, p. 363, 1973.)*

Following the work of M. E. Gurtin [11], consider two species, 1 and 2, which are confined to a one-dimensional strip from $z = 0$ *to* $z = L$. The confinement condition means that

$$\frac{\partial n_1}{\partial z} = \frac{\partial n_2}{\partial z} = 0 \qquad \text{at } z = 0 \text{ and } z = L \tag{13.45}$$

We shall also assume that neither species multiplies or dies. As will be proved shortly, this assumption, combined with condition (13.45), implies that the total number of n_1 and n_2 within the region is constant.

Several arguments exists for supposing that the fluxes (numbers per unit time crossing a given position z) $J_1(z)$ and $J_2(z)$ are given by

$$J_i(z) = -D_{i1}\frac{\partial n_1(z)}{\partial z} - D_{i2}\frac{\partial n_2(z)}{\partial z} \qquad i = 1, 2 \tag{13.46}$$

Of course we recognize (13.46) as a general Fick's law relationship for movement of populations. We shall assume for the remainder of this analysis that the flux of species 1 is much more sensitive to the gradient of species 2 than to its own gradient and that species 1 will move in the direction of decreasing density of species 2. This suggests assuming that

$$D_{11} = 0 \qquad D_{12} = D_p > 0 \tag{13.47}$$

A little reflection upon this assumption, perhaps after this analysis is completed, will reveal that it is reasonable in many instances. Making a similar assumption for species 2 gives

$$D_{21} = D_p > 0 \qquad D_{22} = 0 \tag{13.48}$$

Under these conditions, the unsteady-state population balances for organisms 1 and 2 are

$$\frac{\partial n_1}{\partial t} = D_p \frac{\partial^2 n_2}{\partial z^2} \tag{13.49}$$

$$\frac{\partial n_2}{\partial t} = D_p \frac{\partial^2 n_1}{\partial z^2} \tag{13.50}$$

The requisite boundary conditions are already indicated in (13.45), and initial conditions are the starting distributions $n_{10}(z)$, $n_{20}(z)$ of the two populations:

$$n_i(z, 0) = n_{i0}(z) \qquad i = 1, 2 \tag{13.51}$$

Adding Eqs. (13.49) and (13.50) gives

$$\frac{\partial n}{\partial t} = D_p \frac{\partial^2 n}{\partial z^2} \tag{13.52}$$

where n is the total population at a point

$$n(z, t) = n_1(z, t) + n_2(z, t) \tag{13.53}$$

From conditions (13.45) and (13.51), we can readily show that

$$\frac{\partial n}{\partial z} = 0 \qquad \text{at } z = 0 \text{ and } z = L \tag{13.54}$$

and

$$n(z, 0) = n_{10}(z) + n_{20}(z) \tag{13.55}$$

This boundary-value problem for n is of the standard form amenable to solution by separation of variables. Without writing the transient response, we need only recall for the moment that the solution approaches a constant E as time increases:

$$\lim_{t \to \infty} n(z, t) = E \tag{13.56}$$

We can evaluate E by integrating Eq. (13.52) from $z = 0$ to $z = L$ to find

$$\frac{\partial}{\partial t}\int_0^L n(z, t)\,dz = D_p \int_0^L \frac{\partial^2 n(z, t)}{\partial z^2}\,dz = D_p\left[\frac{\partial n(L, t)}{\partial z} - \frac{\partial n(0, t)}{\partial z}\right] \tag{13.57}$$

Condition (13.54) reveals that the right-hand side of (13.57) is zero, from which we deduce that

$$\int_0^L n(z, t)\,dz = \text{const} \tag{13.58}$$

Next, we apply Eq. (13.58) at large times [Eq. (13.56)] and the initial time $t = 0$ to assert that

$$\int_0^L E\,dz = \int_0^L [n_{10}(z) + n_{20}(z)]\,dz \tag{13.59}$$

The left-hand side of Eq. (13.59) is simply EL, and so E follows immediately, permitting us to restate Eq. (13.56) in the form

$$\lim_{t\to\infty} n(z, t) = \frac{1}{L}\int_0^L [n_{10}(z) + n_{20}(z)]\,dz \tag{13.60}$$

Consequently, we have shown that as time increases, the *total* population density approaches a constant equal to the average density.

More interesting results are available if we consider the difference w between the two population densities

$$w = n_1 - n_2 \tag{13.61}$$

Combining the equations and conditions on n_1 and n_2 in a manner parallel to the previous development gives

$$\frac{\partial w}{\partial t} = -D_p \frac{\partial^2 w}{\partial z^2} \tag{13.62}$$

$$\frac{\partial w}{\partial z} = 0 \qquad \text{at } z = 0 \text{ and } z = L \tag{13.63}$$

and

$$w(0, z) = w_0(z) = n_{10}(z) - n_{20}(z) \tag{13.64}$$

Here, however, the similarity with the n problem ceases. Those who have studied parabolic partial differential equations will immediately recognize Eqs. (13.62) to (13.64) as an ill-posed problem.

At any point z where $w_0(z)$ is nonzero, the magnitude of w will grow without bound as time advances. We can illustrate this general conclusion with a specific example. When we take

$$w_0(z) = \varepsilon \cos \frac{k\pi z}{L} \tag{13.65}$$

a solution of Eqs. (13.62) to (13.65) is

$$w(z, t) = \varepsilon \cos \frac{k\pi z}{L} \exp \frac{D_p k^2 \pi^2}{L^2} t \tag{13.66}$$

Clearly, $w(z, t)$ approaches either plus or minus infinity as $t \to \infty$ except at the points where $w_0(z)$ is zero. Of course, this result is physically meaningless once either n_1 or n_2 reaches zero, and we must assume that once this has occurred our model no longer applies.

This result plus Eq. (13.60) indicates that unless $n_{10}(z) = n_{20}(z)$ at a point z, one of the species will vanish at that point as time progresses while the other will approach a constant value. The species which survives in a given spatial region will be the one initially present in greatest density. Thus, the end result is total segregation of the two interacting populations.

This previous example, although amenable to complete analysis, is rather atypical of the types of interactions which describe and result in spatially heterogeneous multiple populations in practice. Other important situations include the interplay between environment composition gradients and cell chemotaxis (movement towards nutrients or away from inhibitors; see, for example, Ref. 29 and Prob. 13.15). Also, nonlinear interactions in a spatially uniform situation can result in instability, leading to "symmetry breaking" and "pattern formation." There is an extensive and rapidly growing literature in this area. Interestingly, systems which exhibit oscillations in well-stirred reactors are also frequently the ones which form patterns in space in stagnant systems with diffusive transport of solutes (and perhaps of cells). Spatial patterns which oscillate in time are observed, for example, in slime molds. A beautiful summary of mathematical aspects of spatial and temporal order arising from kinetic instabilities in chemical reaction systems is provided in Pismen's review [30].

In this chapter we have explored some of the biological and mathematical principles which underly the behavior of multiple interacting populations. We have seen that experimental studies of these systems are possible under carefully controlled conditions and that mathematical models can be formulated which are consistent with the general features observed in the laboratory. Next, in Chap. 14, we leave the sheltered environment of laboratory, analysis, and computer to consider some of the extremely complicated mixed-population systems which occur in nature and which are exploited industrially. We shall discover that our fundamental studies just concluded are very valuable in these contexts, even though we still lack the necessary insight to accomplish predictive analyses of complex ecosystems.

PROBLEMS

13.1 Competition in batch growth Two competing species A and B are placed in a batch fermentor containing a substrate concentration $s_0 \gg K_A, K_B$, where

$$\mu_i = \frac{\mu_{\max,i} s}{K_i + s} \qquad i = \text{A, B}$$

Show that at substrate exhaustion,

$$(n_{Af})^{\mu_{\max,A}} (n_{Bf})^{\mu_{\max,B}} = \text{const}$$

where n_{Af}, n_{Bf} = cell populations when $s = 0$. (This result was observed by Talling [31] using mixed diatom cultures.)

13.2 Simple mutualism model Meyer, Tsuchiya, and Frederickson [32] have examined the stability of two mutualistic populations with the forms

$$\frac{dn_i}{dt} = -Dn_i + \mu_i n_i \qquad i = 1, 2$$

$$\frac{ds_i}{dt} = -Ds_i + \alpha_i \mu_j n_j - \beta_i \mu_i n_i \qquad i, j = 1, 2; i \neq j$$

Here it is assumed that the medium contains all necessary nutrients except s_1 (produced by n_2) and s_2 (produced by n_1).

(*a*) For each of the growth forms $\mu_i = \mu_{\max,i} s_i$ and $\mu_i = \mu_{\max,i} s_i/(K_i + s_i)$, show that only two steady-state solutions exist and that the nonwashout solution exists only if $\alpha_1 \alpha_2 > \beta_1 \beta_2$ and, additionally, for the Monod substrate dependence, if $D <$ minimum of $(\mu_{\max,i}, i = 1, 2)$.

(*b*) Show by a Taylor's series expansion about the nonwashout solution of part (*a*) that the solution is unstable and hence that a purely mutualistic interaction of the sort considered above cannot exist.

13.3 Mutualism plus competition Meyer et al. [32] suggest that one resolution to the result of Prob. 13.2 is to add a dependence of the two species on a common substrate, s_3, which appears as a separate factor of the Monod form in each growth equation, for example, Eq. (7.34).

(*a*) Write out the five equations now needed to describe transient chemostat behavior.

(*b*) Discover (by graphical or computer means) the solutions when $\mu_{\max,1} = 0.3\ \mathrm{h}^{-1}$, $\mu_{\max,2} = 0.2\ \mathrm{h}^{-1}$, $K_1 = 10^{-3}$ g/L, $K_2 = 0.002$ g/L, $K_{13} = 0.02$ g/L, $\alpha_1 = 4.0$, $\beta_1 = 2.0$, $\alpha_2 = 10.0$, $\beta_2 = 10.0$, and the yield factors for s_3 are $Y_{13} = 2.5$ and $Y_{23} = 5$.

(*c*) Expand the original set of equations in a Taylor's series again, and establish the (in)stability to small perturbations in each variable.

13.4 Mutualism with substrate inhibition A second modification of the conditions of Prob. 13.2 which leads to some stable nonwashout solutions (coexistence) in a chemostat arises when the equation of Andrews (7.32) is assumed to govern species growth:

$$\mu_i = \frac{\mu_{\max,i} s_i}{K_i + s_i + s_i^2/K_i'}$$

(*a*) Write out the system equations for the time evolution of a chemostat.

(*b*) Show that four coexistence solutions (not including washout) may exist at steady state provided $\alpha_1\alpha_2 > \beta_1\beta_2$ and $D <$ minimum of the maximum absolute values which $\mu_i (i = 1, 2)$ above can have.

(*c*) Show by linearized stability analysis that the steady state will be stable if the following conditions are satisfied:

$$\beta_1 A_1 + \beta_2 A_2 > 0 \qquad \text{and} \qquad \bar{s}_1 < \sqrt{K_1 K_1'} \text{ and } \bar{s}_2 > \sqrt{K_2 K_2'}$$

or the reverse for *both* inequalities.

$$A_i = \frac{\mu_{\max,i}\bar{n}_i(K_i K_i' - \bar{s}_i^2)}{K_i'[K_i + \bar{s}_i + \bar{s}_i^2/K_i']^2} \qquad i = 1, 2$$

(The bar over a variable indicates the steady-state value.)

(*d*) The meaning of the inequalities in the stable solutions is that stability arises only when one species is substrate-rate-limited and the other is substrate-inhibited. Indicate by sketches why such a system is stable to perturbations in s_1, s_2 and why simultaneous limitation or inhibition for two species is unstable.

13.5 Methanobacillus omelianskii In the text discussion of mutualism, an anaerobic example was mentioned involving the two species now known to make up "*M. omelianskii*" (Sec. 13.1):

$$CH_3CH_2OH + H_2O \xrightarrow{\text{Specie 1}} CH_3COO^- + H^+ + 2H_2$$

$$4H_2 + CO_2 \xrightarrow{\text{Specie 2}} CH_4 + 2H_2O$$

Thus, the first species produces substrate for the second but is now *inhibited* by its own product, H_2.

(*a*) Why is this interaction *mutualism* rather than a form of *commensalism*?

(*b*) Write out appropriate transient system equations assuming reasonable forms for inhibition by H_2.

(*c*) Discuss the system stability using a Taylor's series expansion about a steady state.

13.6 Particles with aerobic and anaerobic zones In soil aggregates, obligate anaerobes may exist in the center of the moist aggregate, and their propagation toward the periphery depends on the reduction of dissolved oxygen to a subcritical level

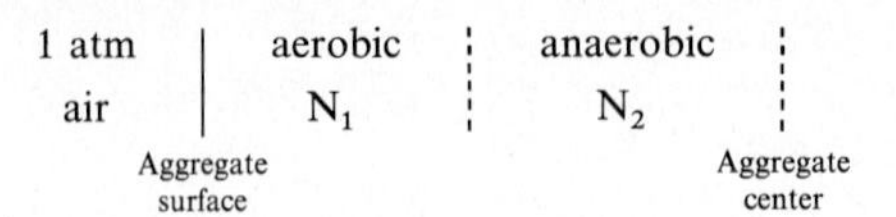

(*a*) Write down the equations describing the system behavior for the following cases where carbon nutrient S_1 is continuously trickled over the outer soil-particle surface:

1. N_1 requires both nutrient S_1 and oxygen, and N_2 consumes product from N_1 and returns nothing.
2. N_1 and N_2 compete for the nutrient S_1; no other interaction.

(*b*) How does the nonuniformity of the oxygen distribution, giving rise to aerobic and anaerobic *niches*, change stability concepts from those in Probs. 13.2, 13.3 and 13.4?

13.7 Niches from imperfect mixing Figure 13.8 indicates that the dilution rate will determine the dominant population for a two-species system consuming a common substrate if growth functions μ_1 and μ_2 cross when plotted vs. substrate. Figure 9.11 and the associated analysis for an imperfectly mixed reactor consider a division of the reactor into two well-mixed systems, each with its own dilution streams.

(*a*) Write the system equations for two species in such an imperfectly mixed system.

(*b*) Establish under what conditions one population dominates one vessel and the other the second, i.e., under what dilution rate, etc., a spatial niche stabilizes the otherwise unstable coexistence of the two species shown in Fig. 13.8.

(*c*) Why do students and faculty live in separate housing?

13.8 Feinberg-Horn-Jackson analysis For Probs. 13.2 through 13.5.

(*a*) Write down the microbial system description in the terms of the Feinberg-Horn-Jackson analysis (Table 13.6).

(*b*) Identify each complex, each linkage class, and the presence or absence of weak reversibility.

(*c*) Apply the zero-deficiency theorem (13.41) and discuss each result in terms of nontrivial existence and system stability.

(*d*) How would you apply this analysis to a tanks-in-series reactor configuration?

13.9 Intracellular network stability The Feinberg-Horn-Jackson approach applies to any system exhibiting mass-action kinetics, thus to intracellular reaction networks.

(*a*) Write out in appropriate notation the reaction steps involved in nucleotide biosynthesis (Fig. 6.20).

(*b*) Determine from the zero-deficiency theorem whether the normal network (Fig. 6.20*a*) or mutant network (Fig. 6.20*b*) is stable or may exhibit oscillatory behavior.

13.10 Leslie's equations Leslie [29] suggested a predator-prey formulation of the form

$$\frac{dx}{dt} = ax - bx^2 - cxy \qquad \frac{dy}{dt} = ey - \frac{fy^2}{x}$$

The last term in the second equation introduces a ratio of prey to predator.

(*a*) Plot the equations resulting when dx/dt and $dy/dt = 0$ on a y-vs.-x plot. Choose some simple positive values for a, b, c, e, f, and by occasional evaluation of dy/dx sketch the behavior of the system starting at other than the equilibrium point.

(*b*) Is the form of the apparent *specific* growth rate for y reasonable as y/x is held constant while x is decreased to small values? Why (not)?

(*c*) Prove that this system exhibits damped oscillations.

13.11 Statistical mechanics of populations As our essays include mathematical modeling of progressively more complex systems, we often are tempted to draw analogies with apparently similar systems. Write a short summary discussing the similarities and differences between the apparent behavior of molecular vs. cellular populations. Include considerations of open vs. closed systems, equilibrium vs. steady state, equations of motion and their invariants (the conserved quantities).

13.12 Food-web kinetics Consider the sequential food chain $1 \to 2 \to 3 \to 4$, where organisms 2, 3, and 4 are the predators of 1, 2, and 3 respectively.

(*a*) Assuming a logistic growth rate for each species, and including a predation term, derive the steady-state solutions for this system, showing in particular that the level of 1 depends on *all* the kinetic constants of the system while the level of 4 depends only on the logistic equation for 4 [35].

(*b*) Since carbon, nitrogen, etc., are obviously recycled, include predation of organism 4 by 1 so that the web is now closed. Show that the steady-state solutions now have identical forms for each species (each is dependent on *all* the web kinetic constants). Discuss quantitatively the stability of these two systems, i.e., the linear vs. cyclic food web.

13.13 Slime-mold aggregation [36–38] When a growing slime-mold amoeba population exhausts its food supply, a relatively quiescent period is followed by an aggregation of the originally uniformly dispersed amoeba into groups spaced ~0.1 mm apart. These aggregates eventually produce spores. The aggregation appears to be due to an attractive amoeba response to a gradient in cyclic AMP secreted by the starving amoeba themselves. Let a = amoeba density and ρ = attractant (cyclic AMP) density; then a model can be developed yielding

$$\frac{\partial a}{\partial t} = \frac{\partial}{\partial x}\left(\mu \frac{\partial a}{\partial x} - \chi a \frac{\partial \rho}{\partial x}\right) \qquad \frac{\partial \rho}{\partial t} = fa - k\rho + D\frac{\partial^2 \rho}{\partial x^2}$$

(*a*) Show from linearization of these equations that assumed solutions of the form $(\sin qx)e^{\sigma t}$ for a and ρ lead to stability if $\chi a_0 f < \mu(k + Dq^2)$, $q \neq 0$.

(*b*) Which wavelength perturbations lead to the earliest instability?

(*c*) It is argued that the initial system is stable but that starvation changes the system parameter values and produces a situation unstable for a = const. Identify the meaning of each parameter in the model. Which of these might reasonably change in a way to bring the system into instability?

(*d*) Why is aggregation necessary for spore formation?

13.14 Competition in continuous culture Experiments in continuous culture with lactate as the growth-limiting nutrient have revealed that the K_s for a pseudomonad is approximately 6 mg/L while that for a spirillium is approximately 12 mg/L. If the maximum growth rate for pseudomonas is $0.6\ h^{-1}$ and that for spirillium is $0.8\ h^{-1}$, discuss the effect of lactate concentration in a mixed culture containing both bacteria.

13.15 Chemotaxis and competitor coexistence The competitive exclusion principle discussed in the text does not always apply in environments with spatial gradients in nutrients or inhibitors or with time-varying conditions. The former situation has been modeled and analyzed by D. A. Lauffenberger ("Effects of Cell Motility Properties on Cell Populations in Ecosystems," p. 265 in *Foundations of Biochemical Engineering: Kinetics and Thermodynamics in Biological Systems*, H. W. Blanch, E. T. Papoutsakis, and G. N. Stephanopoulos (eds.), American Chemical Society, Washington, D.C., 1983). When nutrient gradients exist, *chemotactic* motion of the organisms must be considered. In such a situation, the cell flux J_x in the $+z$ direction may be written as

$$J_x = -\theta \frac{dx}{dz} + \chi x \frac{ds}{dz}$$

where x and s are cell and limiting nutrient concentrations, respectively. The parameters θ and χ are the random motility and chemotaxis coefficients, respectively. Assuming that substrate flux is described by Fick's law and that cell specific growth rate can be approximated by the step function

$$\mu = \begin{cases} \mu^{\circ} - k_e & \text{for } s > s_c \\ -k_e & \text{for } s \leq s_c \end{cases}$$

(*a*) Determine the steady-state distribution of cells of a single species and nutrient in a confined region between $z = 0$ and $z = L$. Cells cannot penetrate these boundaries, and no substrate diffuses through the boundary at $z = 0$. The substrate concentration at $z = L$ is s_0.

(*b*) Considering the important dimensionless groups and phenomena noted in the analysis of part (*a*), speculate on possible scenarios for two-species competition in such a confined domain. Are there situations in which two competitors can coexist? Test your intuition by reading Lauffenberger's analysis. How do these modeling results compare with your observations of competition in human endeavors?

REFERENCES

Some information on mixed population situations is available in the general microbiology references given in Chap. 1. Reference 4 of that chapter has three chapters on symbiosis. See also:

1. E. P. Odum, *Fundamentals of Ecology*, 3d ed., W. B. Saunders Company, Philadelphia, 1971. An extensive and popular textbook dealing with the entire spectrum of ecological science.
2. G. F. Gause, *The Struggle for Existence*, Williams & Wilkins Company, Baltimore, 1934. The work that set the stage for most of the future experimental and theoretical studies in microbial interactions. Many of the concepts, definitions, and outlooks presented in this chapter can be traced back directly to the influence of this classic monograph.
3. A. I. Laskin and H. Lechevalier (eds.), *Microbial Ecology*, CRC Press, Cleveland, 1974. The first two chapters deal with microbe-pesticide interactions, and the third chapter reviews the ecology of soil microorganisms. The final chapter, by J. L. Meers, "Growth of Bacteria in Mixed Cultures," is outstanding and highly recommended.
4. T. Hattori, *Microbial Life in the Soil: An Introduction*, Marcel Dekker, Inc., New York, 1973. A good overview of soil ecosystems, which are very complex and provide many interesting examples of microbial interactions.
5. R. M. May, *Stability and Complexity in Model Ecosystems*, Princeton University Press, Princeton, N.J., 1973. A beautifully written monograph on the mathematics of interacting populations which served as the basis for much of Secs. 13.4 and 13.5. The idea that complexity breeds instability is presented in compelling fashion.
6. H. R. Bungay and M. L. Bungay, "Microbial Interactions in Continuous Culture," *Adv. Appl. Microbiol.*, **10**: 269, 1968. Although now somewhat dated, still a useful source of ideas and additional reading.
7. A. Rescigno and I. W. Richardson, "The Deterministic Theory of Population Dynamics," p. 283 in R. Rosen (ed.), *Foundations of Mathematical Biology*, vol. 3, *Supercellular Systems*, Academic Press, Inc., New York, 1973. Contains an excellent presentation of Volterra's work on competition and predation and reviews some later developments in population dynamics, especially the work of Kostitzin.
8. A. J. Lotka, "Undamped Oscillations Derived from the Law of Mass Action," *J. Am. Chem. Soc.*, **42**: 1595, 1920. The first work leading to the famous Lotka-Volterra model for predator-prey interactions.
9. M. Feinberg and F. J. M. Horn, "Dynamics of Open Chemical Systems and the Algebraic Structure of the Underlying Reaction Network," *Chem. Eng. Sci.*, **29**: 775, 1974. A review paper of the mass-action theory developed by the authors and R. Jackson, summarized in Sec. 13.5.2. (A more recent account is given by M. Feinberg, same volume as Ref. 10, p. 1.)
10. J. E. Bailey, "Periodic Phenomena," p. 758 in L. Lapidus and N. R. Amundson (eds.), *The R. H. Wilhelm Memorial Volume on Chemical Reactor Theory*, Prentice-Hall, Inc., Englewood Cliffs, N.J., 1977. An introduction to the mathematical theory of spontaneous and forced oscillations set in the context of chemical reaction systems.
11. M. E. Gurtin, "Some Mathematical Models for Population Dynamics That Lead to Segregation," *Q. Appl. Math.*, **32**: 1, 1974. For further details on the example presented in Sec. 13.6 as well as related problems.

Growth of unstable recombinant populations and related systems:

12. T. Imanaka and S. Aiba, "A Perspective on the Application of Genetic Engineering: Stability of Recombinant Plasmid," *Ann. N.Y. Acad. Sci.*, **369**: 1, 1981.

13. D. F. Ollis and H.-T. Chang, "Batch Fermentation Kinetics with (Unstable) Recombinant Cultures," *Biotech. Bioeng.*, **24**: 2583, 1982.
14. D. F. Ollis, "Industrial Fermentations with (Unstable) Recombinant Cultures," *Phil. Trans. Roy. Soc. London B* **297**: 617, 1982.
15. M. A. Hjortso and J. E. Bailey, "Plasmid Stability in Budding Yeast Populations: Steady-State Growth with Selection Pressure," *Biotech. Bioeng.*, **26**: 528, 1984.
16. J.-H. Seo and J. E. Bailey, "A Segregated Model for Plasmid Content and Product Synthesis in Unstable Binary Fission Recombinant Organisms," *Biotech. Bioeng.*, **27**: 156, 1985.
17. M. Kimura and T. Ohta, *Theoretical Aspects of Population Genetics*, Princeton University Press, Princeton, NJ., 1971.
18. F. M. Stewart and B. R. Levin, "The Population Biology of Bacterial Plasmids: *A priori* Conditions for the Existence of Conjugationally Transmitted Factors," *Genetics*, **87**: 209, 1977.

Papers on predator-prey interactions:

19. H. M. Tsuchiya et al., "Predator-Prey Interactions of *Dictyostelium discoideum* and *Escherichia coli* in Continuous Culture," *J. Bacteriol.*, **110**: 1147, 1972.
20. J. L. Jost et al., "Interactions of *Tetrahymena pyriformis*, *Escherichia coli*, and *Azotobacter vinelandii* and Glucose in a Minimal Medium," *J. Bacteriol.*, **113**: 834, 1973.
21. J. L. Jost et al., "Microbial Food Chains and Food Webs," *J. Theoret. Biol.*, **41**: 461, 1973.

Further information on spatial patterns:

22. A. M. Turing, "The Chemical Basis of Morphogenesis," *Trans. Roy. Soc. Lond.*, **B237**: 337, 1952.
23. P. Glansdorff and I. Prigogine, *Thermodynamic Theory of Structure, Stability, and Fluctuations*, Wiley-Interscience, London, 1971.
24. H. G. Othmer and L. E. Scriven, "Instability and Dynamic Pattern in Cellular Networks," *J. Theoret. Biol.*, **32**: 507, 1971.
25. R. J. Feld, "A Reaction Periodic in Time and Space," *J. Chem. Educ.*, **49**: 308, 1972.
26. A. T. Winfree, "Rotating Chemical Reactions," *Sci. Am.*, June, 1974, p. 82.

Bifurcation phenomena:

See [10] and
27. C. B. Smith, B. Kuszta, G. Lyberatos, and J. E. Bailey, "Period Doubling and Complex Dynamics in an Isothermal Reaction System," *Chem. Eng. Sci.*, **38**: 425, 1983.
28. G. Lyberatos, B. Kuszta, and J. E. Bailey, "Versal Matrix Families, Normal Forms, and Higher Order Bifurcations in Dynamic Chemical Systems," *Chem. Eng. Sci.*, **40**: 1177 (1985).

Spatial patterns

29. D. A. Lauffenberger, "Effects of Cell Motility Properties on Cell Populations in Ecosystems," *ACS Symp. Ser.*, **207**: 265, 1983.
30. L. M. Pismen, "Kinetic Instabilities in Man-Made and Natural Reactors," *Chem. Eng. Sci.*, **35**: 1950, 1980.

Problems

31. J. F. Talling, "The Growth of Two Plankton Diatoms in Mixed Culture," *Physiol. Plant.* **10**: 215, 1957.
32. J. M. Meyer, H. M. Tsuchiya, and A. G. Frederickson, "Dynamics of Mixed Populations Having Complementary Metabolism," *Biotech. Bioeng.*, **17**: 1065, 1975.
33. P. H. Leslie, "Some Further Notes on the Use of Matrices in Population Mathematics," *Biometrika*, **35**: 213, 1948.
34. J. M. Smith, *Models in Ecology*, Cambridge University Press, London, 1974.

35. D. J. Rapport and J. Turner, "Predator-Prey Interactions in Natural Communities," *J. Theoret. Biol.*, **51**: 169, 1975.
36. J. T. Bonner, *The Cellular Slime Molds*, Princeton University Press, Princeton, N.J., 1967.
37. E. F. Keller and L. A. Segel, "Initiation of Slime Mold Aggregation Viewed as an Instability," *J. Theoret. Biol.*, **26**: 399, 1970.
38. C. C. Lin and L. A. Segel, *Mathematics Applied to Deterministic Problems in the Natural Sciences*, pp. 22–31, The Macmillan Company, New York, 1974.

CHAPTER

FOURTEEN

MIXED MICROBIAL POPULATIONS IN APPLICATIONS AND NATURAL SYSTEMS

The importance of mixed populations of microorganisms can be traced to the beginnings of life on earth. Since then, different species of microbes have played an integral role in the operation of the biosphere and in its evolution. For example, some time between 500 million and 2 billion years ago, the development of primitive algae capable of photosynthesis had reached the point where a significant amount of oxygen (about 1 percent of the current level) was present in the atmosphere. Although microbial life before that time was limited to strictly anaerobic forms, aerobes and facultative anaerobes emerged afterward. Thus the generation of an oxygen atmosphere and subsequent emergence of aerobic life forms on earth can be attributed to primeval photosynthetic microorganisms. In Sec. 14.3, we shall briefly survey contributions of the microbial world to the contemporary global ecosystem.

Mixed microbial populations abound in nature—in the air, soil, and bodies of water. They also grow on and inside higher organisms. Among the most interesting examples are the symbioses with the ruminant animals such as cattle, sheep, and goats (see Table 13.1). In the rumen, which comprises the first two of at least four stomachs, a dense mixed culture of bacteria (about 10^{10} cells per milliliter) and protozoa thrives. In this complicated and extremely diverse population, mutualistic, competitive, amensalistic, and prey-predatory interactions have been observed.

The overall effect of microbial activity in the rumen is decomposition of plant material, including cellulose and other complex carbohydrates, into simpler substances which can be absorbed in the animal's bloodstream. Digestive enzymes are secreted by the ruminant only in stomachs following the rumen. This permits unhindered microbial activity in the rumen and provides for lysis and digestion of ruminant microorganisms in later stomachs. Such a novel design allows ruminants to ingest and utilize much simpler nitrogen nutrients than man and other mammals. Ammonia or urea, for example, are incorporated into combined organic forms such as proteins by rumen microorganisms, and it is the microbially produced nitrogen compounds which are absorbed by the cow in stomachs following the rumen.

Other familiar examples of mixed microbial populations include the natural flora of microorganisms which inhabit the human body. Proper digestion requires the assistance of many bacteria which reside in the intestinal tract. Commensalism and amensalism among these organisms have been observed, and substantial populations of protozoa feed on the intestinal bacteria. Dental plaque consists of several microorganisms, facultative anaerobes being among the most common species. (Why would these be well suited to survive common dental hygienic practices?) The skin is populated by many different protists, with bacteria well entrenched in hair follicles, and yeasts and fungi growing on moist regions. Investigation of how the human body protects itself and often benefits from these numerous cohabitants is a fascinating study but would take us too far afield from our major theme. Consequently, we shall turn now to commercial exploitation of mixed microbial populations.

Here too there is a long history. Ancient artifacts reveal that wine and beer were made by fermenting fruits and grains before 2000 B.C. These processes, at least as conducted then, are excellent examples of applications of *naturally occurring mixed cultures.* In natural mixed-culture processes, inoculation with specific organisms is not practiced. The organisms naturally present when the fermentation begins are responsible for the desired changes. As we shall explore in greater detail below, spontaneous mixed populations are especially efficient in utilizing substrate mixtures.

These examples drawn from naturally arising populations suggest that simultaneous growth of several species offers special advantages and characteristics unattainable in pure cultures. In the next section, we shall briefly review the status of defined mixed-population technology. The remainder of this chapter is devoted to applications of naturally occurring mixed cultures, with heavy emphasis on biological wastewater treatment (Sec. 14.4).

14.1 USES OF WELL-DEFINED MIXED POPULATIONS

The foremost illustration of defined mixed-culture application is cheese manufacture. While the gastronomical benefits of such activities are well known, the economic importance of the cheese and dairy industries is not widely appreciated.

According to one estimate, roughly 4 percent of the United States' Gross National Product is contributed by the dairy industry, with about 15 percent of that figure due to cheese production.

Although indigenous mixed cultures were originally used in making cheese, special starter cultures are now employed to ensure reproducibility of product quality. Cheeses of various types are produced by inoculating pasteurized fresh milk with appropriate lactic acid–producing organisms. The resulting proteinaceous curd, precipitated by the acidity of the medium, is drained of liquid (*whey*) and allowed to age or ripen by action of bacteria or mold. With hard-curd cheese, an enzyme mixture of rennet (containing the proteolytic enzyme rennin) is added to the innoculated milk after slightly acid conditions are produced. A rubberlike curd eventually results, which is cut into small pieces, heated, drained of whey, and finally milled into shavings and pressed to remove further whey (the cheddaring process). Finally, a second milling, salting, draining, and pressing into molds yields the cheese to be cured. In curing, the slow fermentative (anaerobic) action partially breaks down lipids and proteins to produce additional partially oxidized products such as lactic, butyric, and acetic acids, which, with the continued increase in cheese age, contribute to the sharp taste of the resultant cheese.

Many natural cheeses are similar in initial stages; the final product consistency and taste are predominantly determined in the later curing stages by the organisms used, salt and humidity levels, and the curing temperature. A representative listing of common cheeses and organisms is given in Table 14.1. Many of these organisms synthesize vitamins which increase the nutritional value of the cheese during curing.

Lactic acid bacteria also participate in other defined mixed cultures used in food production. In whiskey manufacture, for instance, a *Lactobacillus* added to the yeast lowers pH to reduce contamination and also contributes to a desirable flavor and aroma. Another example of favorable interaction between a yeast and lactic acid bacterium is ginger-beer fermentation. Also, use of two *Lactobacillus* species increases yield in the lactic acid fermentation.

Consecutive transformation of a nutrient into a desired final product can sometimes be accomplished effectively by a tactic called *dual fermentation*. One such process makes L-lysine from glycerol, with α,ε-diaminopimelic acid (DAP) as an intermediate. In one fermentation, the DAP intermediate is accumulated using an *E. coli* auxotroph. Separately, an *Aerobacter aerogenes* population is grown. This organism synthesizes DAP decarboxylase, an enzyme which acts on DAP to yield L-lysine. Combining the two cultures and adding toluene liberates both DAP and its decarboxylase, so that L-lysine is produced.

There are few cases in which multiple defined cultures are grown simultaneously for commercial use. One is manufacture of β-carotene, where different mating types of the same organism grown together provide about 20 times more product than either type by itself. In the following example, we shall explore how methane utilization for single-cell protein production is improved by using a mixed culture.

Table 14.1 Some cultures used in manufacture of cheese†

a. Cultures used in cheese production

I. Bacteria
  A. Used for lactic acid production primarily:
    1. *Streptococcus lactis* "lactic"
    2. *Streptococcus cremoris* "lactic"
    3. *Streptococcus thermophilus* "coccus"
    4. *Streptococcus durans* (USDA modified cheddar "make" procedure)
    5. *Lactobacillus bulgaricus* "rod"
  B. Used to develop flavor and aroma with or without an effect upon the body and texture:
    6. *Brevibacterium linens* "red smear"
    7. *Propionibacterium shermanii* "props"
    8. *Leuconostoc* sp. (flavor associates or citric acid fermenters—"CAFs")
    9. *Streptococcus diacetilactis* (flavor and special uses)

II. Yeasts
  A. Used for maintenance and enhancement of bacterial growth in cultures and cheese (*not essential*)
    10. *Candida krusei*
    11. *Mycoderma* sp.

III. Molds (to enhance flavor, aroma, body, texture, composition and appearance)
  A. Externally grown
    12. *Penicillium camemberti*
    13. Miscellaneous species
  B. Internally grown
    14. *Penicillium roqueforti* and variants thereof

b. Use relationship between starter culture types and characteristic cheese varieties

Cheese variety	Type of cultures in use: Class I		Type of cultures in use: Classes II and III	How used‡
Brick	A.1–3	B.6, 8, 9	II.A	*x–z*
Camembert	A.1, 2	B.8, 9	III.A.12	*x, z*
Cheddar	A.1, 2, 4	B.8, 9	....	*x–z*
Cottage	A.1, 2	B.8, 9	....	*x, z*
Cream	A.1, 2	B.8, 9	....	*x, z*
Edam	A.1, 2	B.8, 9	....	*x, z*
Gouda	A.1, 2	B.8, 9	....	*x, z*
Limburger	A.1–3	B.6, 8, 9	II.A	*x–z*
Neufchatel	A.1, 2	B.8, 9	....	*x, z*
Parmesan	A.1–3, 5	B.8,9	....	*x–z*
Provolone	A.1–3, 5	B.8, 9	....	*x–z*
Romano	A.1–3, 5	B.8, 9	....	*x–z*
Roquefort	A.1, 2	B.8, 9	III.B.14	*x, z*
Swiss	A.1–3, 5	B.7–9	....	*x–z*
Trappist	A.1–3	B.6, 8, 9	II.A	*x–z*

† Courtesy of G. W. Reinbold, Iowa State University, Ames, Iowa.

‡ Key:
x = Single-strain cultures; *y* = multiple-strain cultures; *z* = combined mixed-strain cultures (dual purpose, and/or single or multiple strains).

Example 14.1: Enhanced growth of methane-utilizing *Pseudomonas* sp. due to mutualistic interactions in a chemostat[†] In Sec. 12.7 and Example 9.2, we considered some aspects of single-cell protein production. Substantial recent interest has been focused on gaseous hydrocarbon substrates, particularly methane. Several experimental studies have shown that mixed cultures often have the ability to grow more rapidly on methane than pure cultures of methane-utilizing microorganisms.

A tentative explanation for this observation is a mutualistic interaction like the one shown schematically in Fig. 14E1.1. In this mixed population, studied experimentally by Wilkinson, Topiwala, and Hamer, there are two major classes of bacteria: methane-utilizing *Pseudomonas* sp. and methanol-utilizing *Hyphomicrobium* sp. Because methanol is a metabolic end product of the *Pseudomonas* sp. and also inhibits the growth of those organisms, they benefit greatly from the second population, which removes methanol. Clearly, the *Hyphomicrobium* sp. also enjoy the interaction,

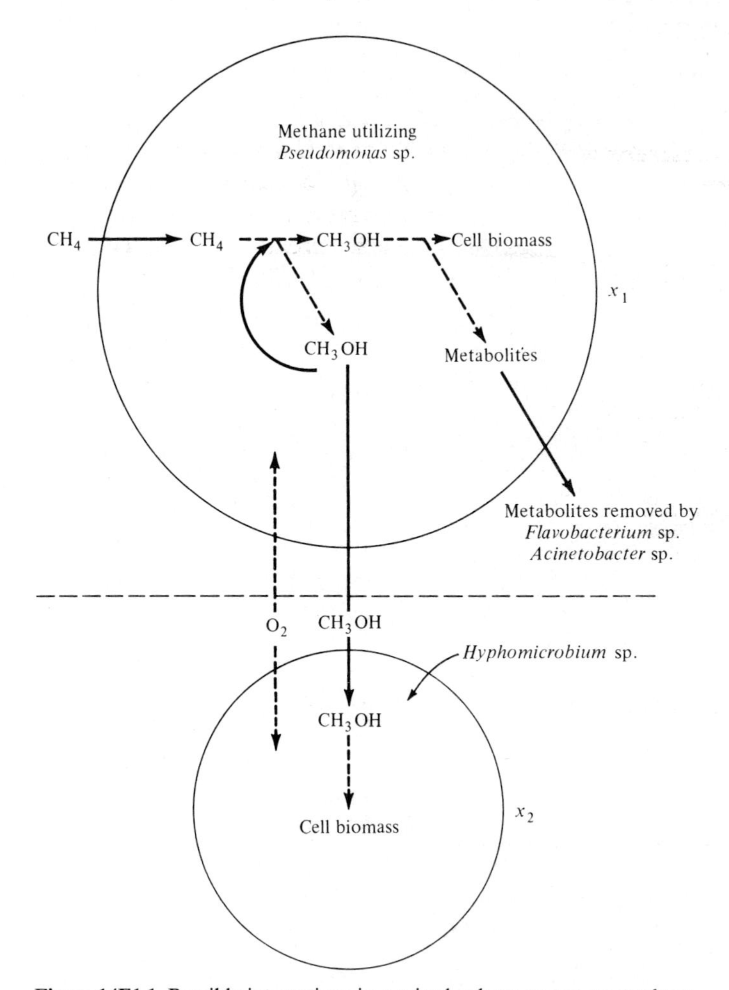

Figure 14E1.1 Possible interactions in a mixed culture grown on methane.

[†] This example is drawn from T. G. Wilkinson, H. H. Topiwala, and G. Hamer, "Interactions in a Mixed Population Growing on Methane in Continuous Culture," *Biotechnol. Bioeng.*, **16**: 41, 1974.

since they are supplied with a carbon source. The two other types of bacterial species in the mixed population are present in relatively small concentrations and are believed to serve useful functions by utilizing other end metabolites of the *Pseudomonas* sp.

While we shall leave the complete argument for this interpretation to the original paper, one experiment which shows several effects consistent with the scheme is indicated in Fig. 14E1.2. The mixed population was grown in a chemostat, and at $t = 0$ methanol was added to the fermentor and medium reservoir to establish a methanol concentration of 1.6 g/L. As Fig. 14E1.2 shows, this shock causes an immediate and severe drop in methane and oxygen utilization and a gentler decrease in total dry weight of bacteria. Also, the percentage of *Hypomicrobium* sp. in the fermentor increases. After about 17h, the methanol concentration in the vessel is near zero, leading to a reversal of the above trends. Presumably at this point *Pseudomonas* activity resumes to a significant extent.

All the component parts for the mathematical model of this system should now be familiar from our previous studies of mass transfer and biological kinetics. Letting x_1 and x_2 denote the concentrations of *Pseudomonas* and *Hyphomicrobium* species, respectively, we see that their unsteady-state material balances for a chemostat are

$$\frac{dx_i}{dt} = -Dx_i + r_{fi} \qquad i = 1, 2 \tag{14E1.1}$$

The forms of the *Pseudomonas* and *Hyphomicrobium* growth rates are chosen to reflect the situation depicted in Fig. 14E1.1. In particular, for oxygen-limited growth of the *Pseudomonas* bacteria, we take

$$r_{f1} = \frac{\mu_{1,\max} c_{O_2}}{K_1 + c_{O_2}} \frac{1}{1 + s/K_i} x_1 \tag{14E1.2}$$

where c_{O_2} and s are dissolved oxygen and methanol concentrations, respectively. The inhibition function used here parallels those of Chap. 3 (noncompetitive inhibition of enzyme-catalyzed reactions) and Chap. 7 (ethanol inhibition of yeast fermentation). Since the *Hyphomicrobium* can readily use nitrate as an electron acceptor when dissolved-oxygen concentration is low, we assume that r_{f2} is independent of c_{O_2}, while adopting the Monod dependence on s:

$$r_{f2} = \frac{\mu_{2,\max} s}{K_2 + s} x_2 \tag{14E1.3}$$

Assuming that yield factors for both species relative to methanol remain constant and that most oxygen uptake is by the *Pseudomonas*, we have the following material balances for s and c_{O_2}:

$$\frac{ds}{dt} = -Ds + \frac{1}{Y_1} r_{f1} - \frac{1}{Y_2} r_{f2} \tag{14E1.4}$$

$$\frac{dc_{O_2}}{dt} = k_l a(c_{O_2s} - c_{O_2}) - \frac{1}{Y_3} r_{f1} - Dc_{O_2} \tag{14E1.5}$$

From batch experiments and measurement of oxygen uptake rates, Wilkinson, Topiwala, and Hamer estimated the value of all parameters in this model. Their suggested values are listed in Table 14E1.1.

The steady states computed using this model for various dilution rates are displayed in Fig. 14E1.3, along with some experimental data on this system. The agreement, while imperfect, is adequate. Transient simulations of the effect of methanol addition to the mixed culture produce responses qualitatively similar to those shown in Fig. 14E1.2. However, the real system dynamics are far more sluggish than the model. Apparently, more structure is needed in the model for an adequate reflection of the unsteady-state interactions which occur in the mixed culture.

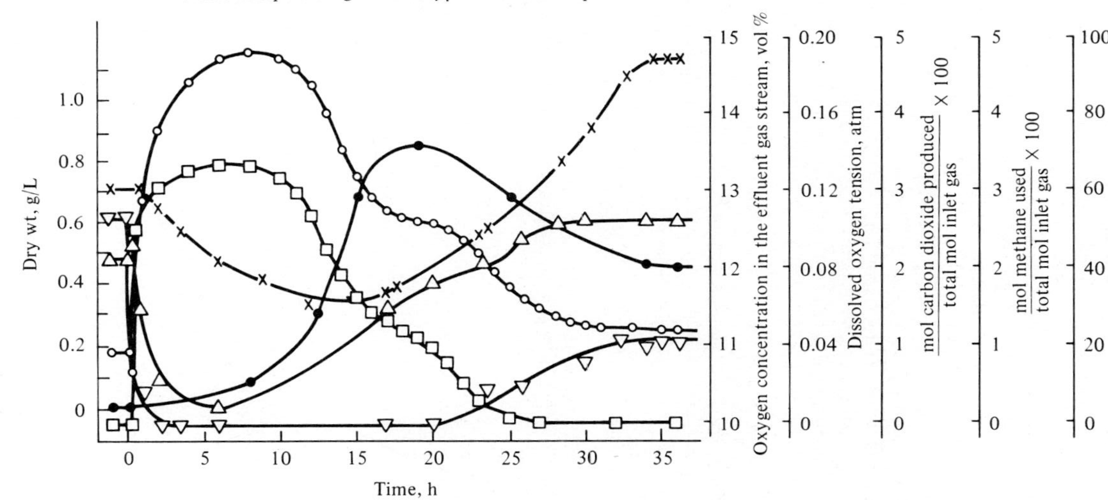

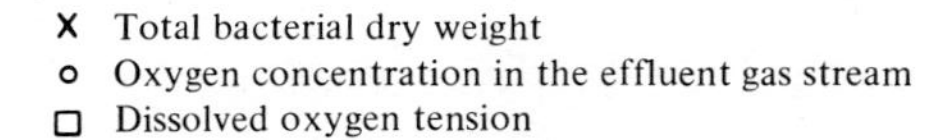

Figure 14E1.2 Transient behavior of a methane-utilizing mixed culture following a shock load to a chemostat. At $t = 0$, the methanol concentration in the fermentor and in the feed was raised to 1.6 g/L. ($D = 0.08\ h^{-1}$.) *(Reprinted from T. G. Wilkinson et al., "Interactions in a Mixed Bacterial Population Growing on Methane in Continuous Culture," Biotech. Bioeng., vol. 16, p. 41, 1974.)*

Table 14.E1.1 Kinetic parameters for a model of a methane-utilizing mixed population†

Parameter	Comment	Numerical value
$\mu_{1,\max}$	Maximum specific growth rate for methane-utilizing component	$0.185\ h^{-1}$
$\mu_{2,\max}$	Maximum specific growth rate for methanol-scavenging component	$0.185\ h^{-1}$
K_1	Michaelis constant for oxygen consumption by x_1	1×10^{-5} g/L
K_2	Michaelis constant for methanol consumption by x_2	5×10^{-6} g/L
K_i	Methanol-inhibition constant (hyperbolic) for x_1	1×10^{-4} g/L
Y_1	Stoichiometric yield constant for methanol production by x_1	5.0 g bacteria/g methanol
Y_2	Stoichiometric yield constant for methanol consumption by x_2	0.3 g bacteria/g methanol
Y_3	Stoichiometric yield constant for oxygen consumption by x_1	0.2 g bacteria/g oxygen
$k_l a$	Oxygen mass-transfer product (specific area · coefficient)	$42.0\ h^{-1}$
c_{O_2}	Dissolved-oxygen saturation level	0.128 atm (0.008 g/L)

† T. G Wilkinson, H. H. Topiwala, and G. Hamer, *Biotech. Bioeng.*, **16:** 56, 1974.

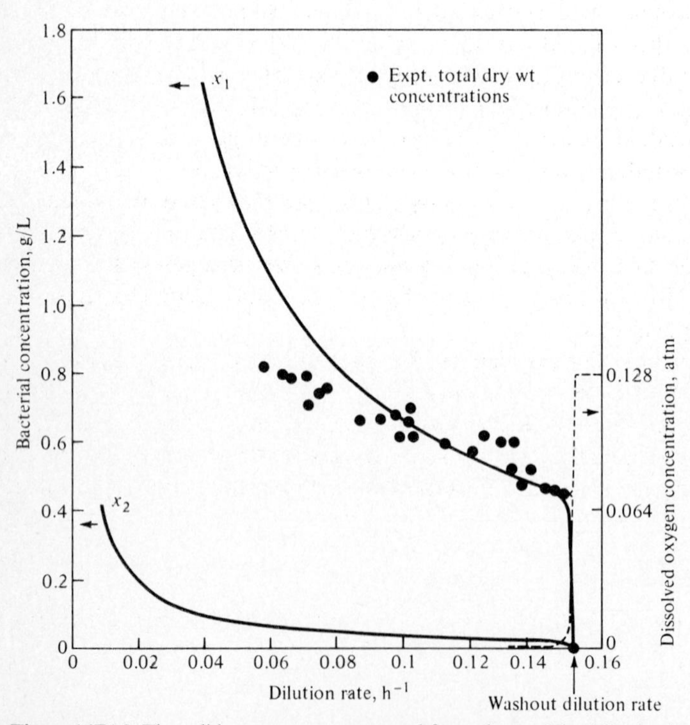

Figure 14E1.3 The solid curves were computed from the model, while the dots are experimental data. *(Reprinted from T. G. Wilkinson et al., "Interactions in a Mixed Bacterial Population Growing on Methane in Continuous Culture," Biotech Bioeng., vol. 16, p. 41, 1974.)*

14.2 SPOILAGE AND PRODUCT MANUFACTURE BY SPONTANEOUS MIXED CULTURES

We turn now to processes where inoculation takes place from natural sources. Under these conditions, the nutrient supply and other environment factors largely determine the resulting dominant mixed cultures. Perhaps the foremost intentional application of this strategy is biological wastewater treatment. To understand this and other natural mixed-culture systems, we must recognize that the microbial world is extremely diverse and dispersed. Thus, we may assume as a working rule of thumb that if there is an environment attractive for growth of a particular microorganism, the chances are high that that particular microbe is growing or will grow there. Consequently, the mixed population which arises in an aerated vessel containing wastewater is, by a type of natural selection, especially suited for growing in that environment.

Natural mixed populations are therefore particularly efficient means for utilization of substrate mixtures. While this characteristic is highly attractive in the wastewater-treatment context, it is troublesome when unwanted microbial attack occurs on "substrates" such as wood and food. We shall consider such undesirable activities next, saving discussion of biological waste-water treatment for Sec. 14.4.

Spoilage generally involves decomposition of organic molecules, including polymers such as proteins or carbohydrates. In some instances, e.g., wood rot, it is disappearance of the original substance which causes concern. On the other hand, most undesirable effects of food spoilage derive from the metabolic end products of the attacking microorganism. Both manifestations usually begin with the attack by extracellular enzymes produced by a microorganism.

In the case of wood rot, one or more of the three polymeric constituents of the wood are degraded by a cellulase enzyme system excreted by a fungus. Fungi as well as bacteria are implicated in attack on pectin in foodstuffs. This causes disintegration of canned fruits, softening of brined cuccumbers, and rotting of vegetables. However, as we saw earlier in Sec. 4.1.4, pectic enzymes also enjoy beneficial applications.

Protein spoilage, another problem in the food industry, can be viewed as a two-step sequence. In the first, called *proteolysis*, whole protein is hydrolyzed to yield peptides and amino acids. Liquefaction of gelatin is a common manifestation of this step. Subsequently, usually under anaerobic conditions, proteins are decomposed, and the amino acids are metabolized to yield foul-smelling products such as putrescine:

$$\underset{\text{Ornithine}}{H_2NCH_2(CH_2)_2CH(NH_2)COOH} \xrightarrow[\text{ornithine decarboxylase}]{-CO_2} \underset{\text{Putrescine}}{H_2NCH_2(CH_2)_2CH_2NH_2}$$

This process, called *putrefaction*, is evident in the vile odors of badly spoiled meats.

Perhaps the most familiar example of spoilage is sour milk. Although most milk is pasteurized before packaging, this achieves only disinfection, not sterilization. Many sporeforming bacteria survive the process and, with time, cause curdling or putrefaction. Especially interesting is the sequence of microbial populations which typically occur in raw milk at room temperature. Initially lactose (milk sugar) is fermented by streptococci, bacilli, and other bacteria. As a result of this activity, the pH of the milk drops (see Fig. 14.1). This inhibits the original population, and permits acid-tolerant species including *Lactobacilli* to gain ascendency. Further reduction in pH to below 4.7 causes curd (a rubbery material consisting primarily of casein, a protein) to form and precipitate.

Next yeasts and molds which can use lactic acid as a nutrient proliferate, increasing the pH. The preeminence of these populations eventually gives way to that of fungi and bacteria, which use fat and casein as nutrients. Eventually oxygen is depleted, and anaerobic bacteria cause putrefaction.

Such successions of microbial species, each enjoying an interlude of dominance during favorable conditions, occur frequently in indigenous mixed cultures. We shall see other examples of this phenomenon in our review below of soil microbiology and in biological trickling filters. Recall the immense practical importance of controlling the initial stages of spoilage: production of curd in milk by bacterial action is the starting point for cheese manufacture, as discussed in the last section.

We find natural fermentation by lactic acid–producing bacteria in other food processes. Pickles are made by a lactic acid fermentation of cucumbers using mixed populations. In the manufacture of sausage and other fermented meats, *Lactobacillus* sp. and other microorganisms produce lactic acid and also accomplish the reduction of nitrate to nitrite, a process that contributes significantly to the development of color and the production of tangy flavors. Batter for sourdough bread is allowed to ferment for 1 or 2 days so that ethanol and organic acids are produced. Sauerkraut is another food prepared using a mixed-culture fermentation.

Although we discussed production of wine, beer, and vinegar earlier, in some respects those processes belong here. In many cases, the microbial activity responsible for a successful final product arises from a spontaneous mixed culture.

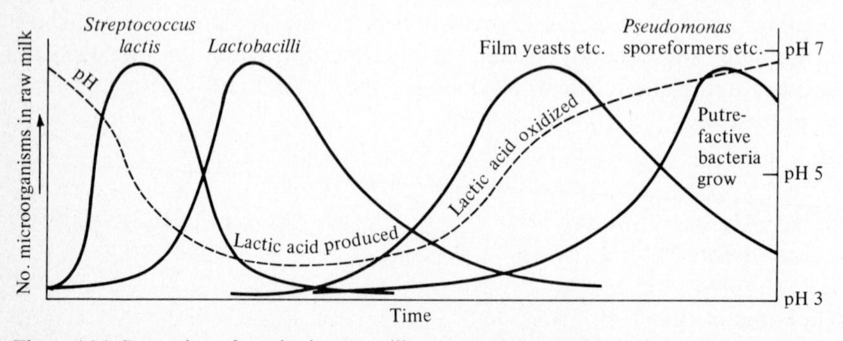

Figure 14.1 Succession of species in raw milk at room temperature.

Current technology in these areas, however, seems directed at better-defined, more reproducible operations. Consequently, use of carefully screened and preserved inoculum species is increasingly common.

14.3 MICROBIAL PARTICIPATION IN THE NATURAL CYCLES OF MATTER

Most of our previous discussions of examples of microbial utilization have dealt with processes largely under human operation and control. In order to preserve and improve our environment, it is also important to understand the basic features of natural microbial activities in the biosphere. With such knowledge we may be able to construct useful models for such natural processes as lake eutrophication, biodegradation and water repurification in soils, and stream and estuary ecosystem dynamics. Also, purposeful biological treatment of waste waters, one of the most important tasks microorganisms perform under human direction, in large part mimics components of natural ecological cycles. Thus, a study of microbial activities in nature aids understanding of how life on earth is sustained and also provides a valuable introduction to a critically important application of mixed populations.

Essential to proper function of the earth's ecosystem is cyclic turnover of the elements required for life. Organic matter derived from metabolic wastes or dead organisms must be broken down and converted into inorganic form. Microorganisms are especially suited to play vital roles in this process, which is called *mineralization.* Among the important attributes of the kingdom of protists in this regard are metabolic versatility, high rates of chemical activity, and natural abundance. Taken as a group, microorganisms have the power to decompose every naturally occurring organic compound.

In order to avoid excessive accumulation of any natural organic waste, it is of course not sufficient that degradation reactions occur: these reactions must occur with adequately large absolute rates (specific rate × population density). We have already noted that microorganisms grow at rates far in excess of those for higher organisms. As a consequence, specific rates of substrate utilization are also very large, so that the chemical conversions necessary for mineralization usually occur at rapid rates. The other important factor contributing to the natural importance of microorganisms is their density and widespread distribution. Microbes are extremely numerous in surface waters and topsoil; bacteria have even been discovered in soil samples from Antarctica. It is estimated that 1 acre of fertile soil contains 2 tons of bacteria and fungi in the top 6 in. Combining this density with high metabolic rates, the microorganisms in this amount of soil possess more potential metabolic activity than 10,000 human beings. As much as 90 percent of the CO_2 produced in the biosphere results from respiration in bacteria and fungi.

While microorganisms dominate the mineralization segment of most cycles of matter, they also participate significantly in other portions of these cycles. The particular involvements of protists will be explored in the next two sections. In the first, we consider a gross overview of elemental cycles from a global perspective. Next, some particular better-defined ecosystems are examined.

14.3.1 Overall Cycles of the Elements of Life

Cyclic turnover of the biologically significant elements is often accompanied by cyclic changes in their oxidation state. This is most important, since we are already aware that the biological suitability of an element is often directly linked to its chemical state (see Chap. 5). Although evident in all the element cycles considered below, the linked cycles of carbon and oxygen vividly illustrate the general principles of the cyclic transformation of matter.

The general features of the carbon and oxygen cycles are shown in Fig. 14.2. The major driving force underlying these and the other cycles is photosynthesis, which taps solar energy to reduce CO_2, bicarbonate, and carbonate, the oxidized forms of carbon, while simultaneously liberating molecular oxygen from water (Sec. 5.5). The amount of carbon fixed per year on land and in the oceans is roughly 1.6×10^{10} and 1.2×10^{10} tons, respectively. While green plants are the major contributors to photosynthetic activity on land, photosynthesis occurring in the oceans is almost entirely due to unicellular algae called phytoplankton. Although photosynthesis is the dominant means of CO_2 reduction, chemoautotrophs also reduce CO_2. As already noted, mineralization of organic carbon to CO_2 is primarily the consequence of bacterial and fungal metabolic activities.

Carbon is removed or sequestered from the life cycle just described by several mechanisms. Much of the CO_2 released into the atmosphere enters the oceans as bicarbonate ions. There, it can combine with calcium to form calcium carbonate, an insoluble compound which appears in coral shells and limestone. In this form carbon is relatively inaccessible, but much of it is ultimately made available by weathering or by attack of acids. Microorganisms participate in the latter process through synthesis of carbonic, sulfuric, nitric, and other acids.

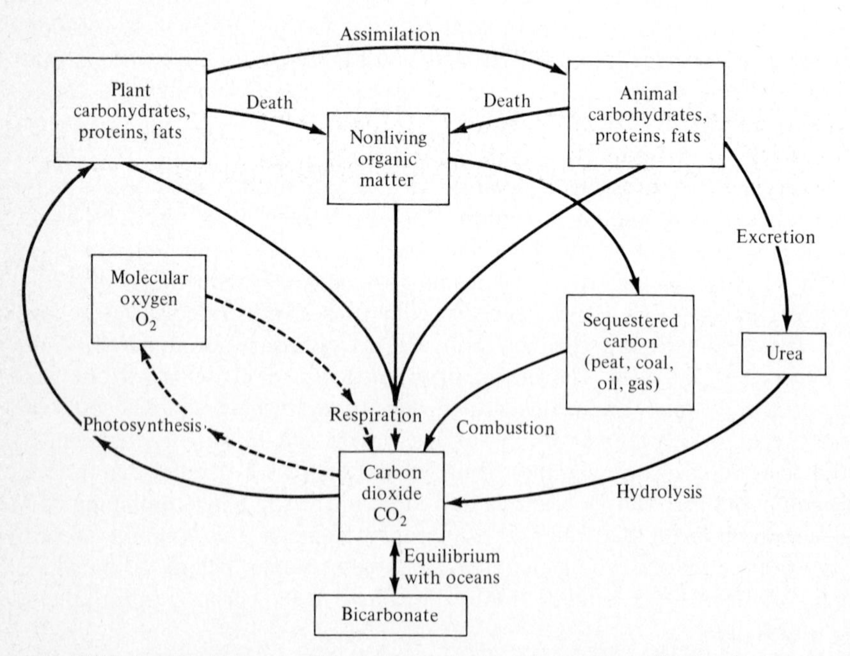

Figure 14.2 Simplified diagram of the carbon cycle. Also shown (dotted lines) is the major component of the oxygen cycle, which is closely linked to the cycle of carbon.

We are all familiar with sequestered carbon in organic form. *Humus*, an organic residue derived from microbial-resistant plant components, is an important constituent of rich soil. When conditions favor large accumulations of humus, deposits of *peat* are created which, on a geological time scale, can be transformed into coal. Oil and natural gas are other common forms of sequestered organic carbon. Carbon residing in these forms seems destined for eventual return to the biosphere due to man's apparently relentless demands.

Let us next consider how nitrogen cycles in the biosphere and how its chemical state alters in the process. Several basic principles provide useful guidelines for this study; molecular nitrogen in the atmosphere is quite inert; i.e., it is not an acceptable nutrient form for most organisms. Also, its chemical form in living organisms is primarily in a reduced state in proteins.

A general overview of the nitrogen cycle is provided in Fig. 14.3. Organic nitrogen is converted into ammonia by microbial action. Although some ammonia escapes into the atmosphere and some is utilized directly by plants and microorganisms, most is oxidized to nitrate (NO_3^-) in a two-step process called *nitrification*. Each half of the nitrification pathway is mediated by a special family of bacteria, primarily the *Nitrosomas* and the *Nitrobacter*. Both groups are chemoautotrophic and obligate aerobes.

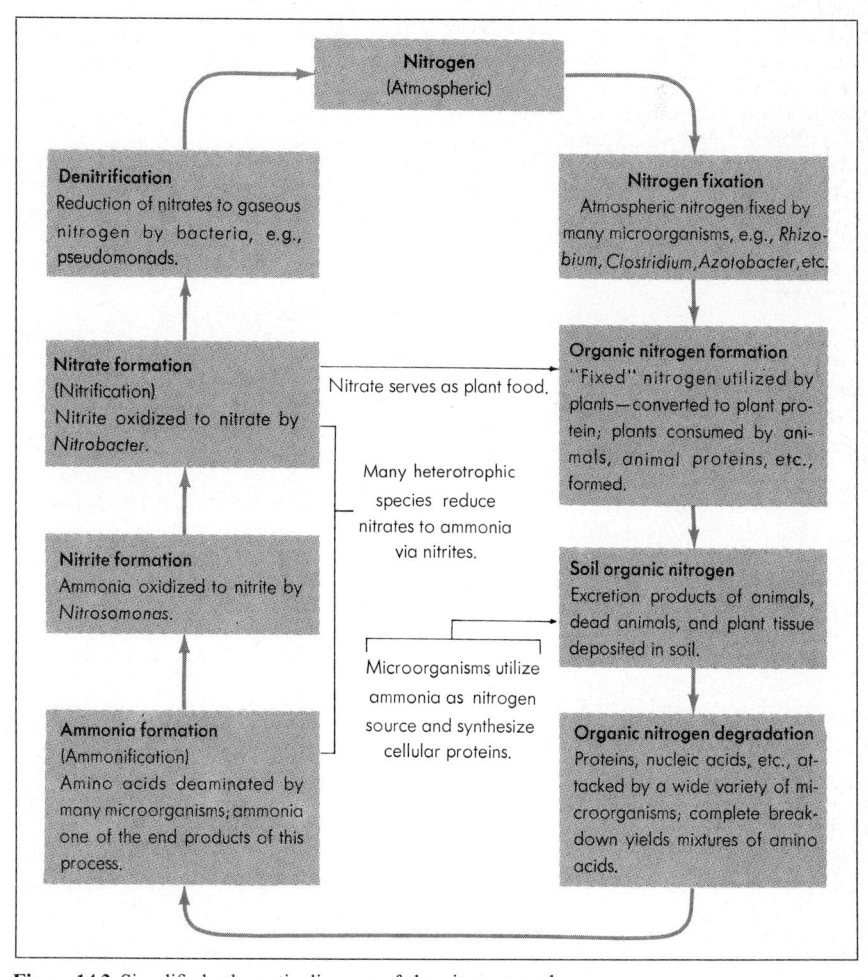

Figure 14.3 Simplified schematic diagram of the nitrogen cycle.

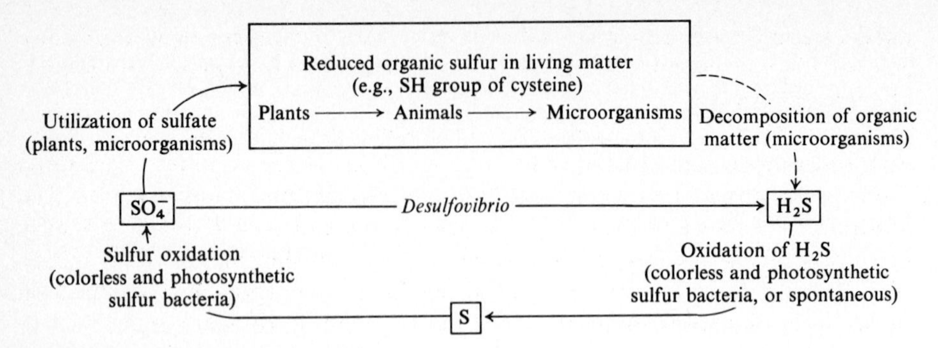

Figure 14.4 Major features of the sulfur cycle.

The nitrate thus formed is the best form of nitrogen for plant assimilation; consequently at this point much of the nitrogen flux reenters the pool of reduced organic nitrogen. After incorporation into organic compounds by algae and plants, the nitrogen is in a form suitable for animal utilization. However, microbial activities provide an alternate demand on nitrate. When oxygen is unavailable, as in a microenvironment where respiration has exhausted the supply, many bacteria possess sufficient metabolic dexterity to use nitrate as a hydrogen acceptor. The end result of their activities is denitrification: molecular nitrogen is formed.

Fortunately, several specialized microorganisms have the ability to utilize molecular nitrogen and to return it to the biosphere in combined form. Two general types of nitrogen fixation can be identified: in the symbiotic version, mutualistic relationships between *Rhizobium* bacteria and seed plants accomplish nitrogen fixation. The remaining nitrogen-fixing capcity derives from the nonsymbiotic activity of blue-green algae and a few aerobic (*Azotobacter*) and anaerobic (*Clostridium pasteurianum*) bacteria.

A major challenge to genetic engineering arises using either microbes or plants:

(*a*) bacteria which colonize other plant root systems could be altered to incorporate nitrogen-fixation (Nif) genes, or
(*b*) plants could be developed which contain Nif genes directly in the plant genetic makeup.

Either achievement would result in a new Nif-containing crop which would not need nitrogenous fertilizer, at least at current levels of application.

Sulfur also undergoes a cycle of oxidation, reduction, incorporation into, and liberation from, organic matter (see Fig. 14.4). Sulfate-reducing and sulfur-oxidizing bacteria play major roles in microbial corrosion. For additional information on the sulfur cycle and the organisms which participate in it, the references should be consulted. In the next section, we shall examine how these cycles or segments of them are manifested in particular environments.

14.3.2 Interrelationships of Microorganisms in the Soil and Other Natural Ecosystems

Soil provides a varied and complicated environmental system which is an excellent habitat for microorganisms. It consists of finely divided minerals (largely aluminum silicate compounds), decaying organic residues, and a living mixed microbial population. In addition, water is often present, as is a gaseous phase which may contain N_2, O_2, CO_2, H_2S, NH_3, and other gases. The extensive surface afforded by fine solid granules provides, through adsorption, concentration of certain nutrients and extracellular hydrolytic enzymes. From dissolved minerals and decaying organic material come ions, carbohydrates, nitrogeneous compounds, and vitamins. Hence a rich culture medium is available for support of microbial growth.

Syntrophism is a type of relationship in which organisms produce food for each other. Syntrophic relationships are ubiquitous in the soil, and indeed are essential for proper functioning of the cycles of elements. Some examples are illustrated schematically in Fig. 14.5. Notice that one organism (labeled A), which produces cellulolytic enzymes to provide its own nourishment, also feeds others (B, C, and D) from the simple sugars liberated. The metabolic end products of organism A are used by other microbes with different nutritional needs.

Another view of syntrophism, more analogous to the milk-spoilage scenario reviewed in the last section, can be obtained by considering the time sequence of events which follow plowing under of a grass or clover crop. Besides providing many soluble nutrients from the plants, plowing aerates the soil. Conditions consequently favor rapid growth of heterotrophic organisms and facultative autotrophs. Due to the metabolic action of these microorganisms, the temperature within the soil rises, as does its acidity.

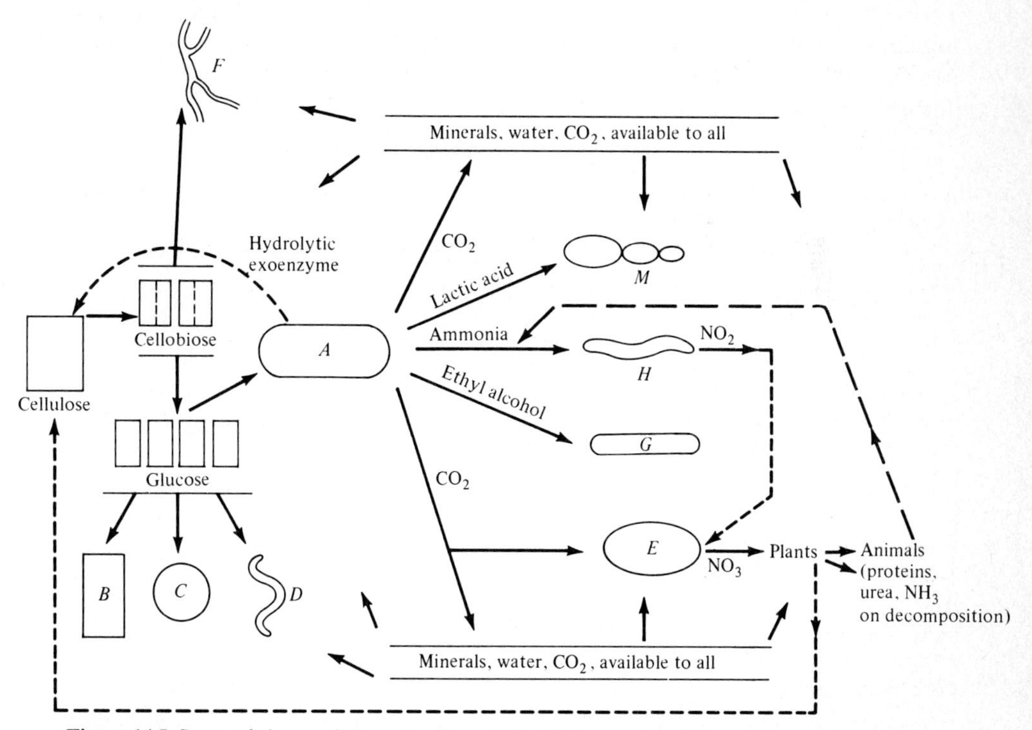

Figure 14.5 Some of the possible syntrophic relationships among microorganisms, plants, and animals in the soil. Metabolic end products of microbe A nourish organisms M, H, G, and E, and cellulolytic enzymes secreted by A hydrolyze cellulose. The resulting simple sugars are nutrients for microorganisms B, C, and D.

After all oxygen has been depleted, strict anaerobes appear, leading to further acid production. Eventually this explosive growth is arrested by nutrient depletion or toxin formation (recall Sec. 7.3.1), and many of the cells die, releasing compounds useful for plant growth. The residual, relatively low-level microbial population consists of species capable of attacking the resistant substances such as those found in humus.

We have only scratched the surface here in examining microbial activities in the soil. For example, almost complete cycles of the essential elements can and do occur under anaerobic conditions. The necessary transformations can all take place in a very small oxygen-depleted ecological niche since all of them are conducted by microorganisms. Figure 14.6 summarizes the close interrelationships which can exist between photosynthetic, fermentative, sulfate-reducing, and other bacteria in an anaerobic environment. Additional readings in soil microbiology are recommended in the references. We should note in closing this discussion that the vast majority of industrially important microorganisms have been isolated from the soil.

Microorganisms are the primary producers of organic matter in freshwater and marine environments. As illustrated schematically in Fig. 14.7, green plants and large multicellular algae (seaweed) can achieve photosynthesis only in shallow waters near shore. In open water, free-floating unicellular algae called phytoplankton conduct photosynthesis near the surface. Photosynthetic activity becomes difficult and eventually impossible with increasing depth because light is absorbed by water and intercepted by suspended solid materials.

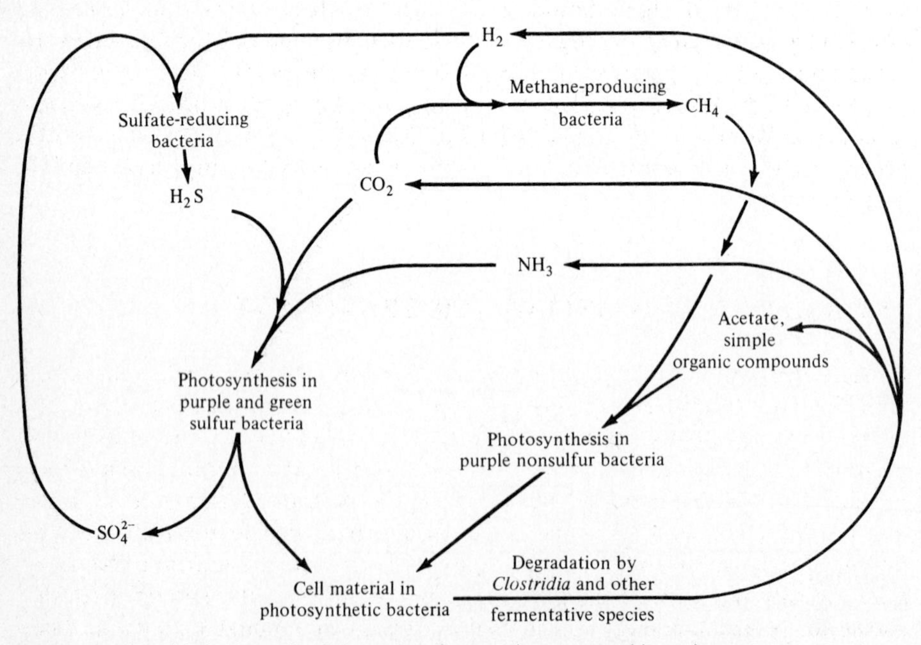

Figure 14.6 Simplified schematic of the cycles of matter in an anaerobic environment.

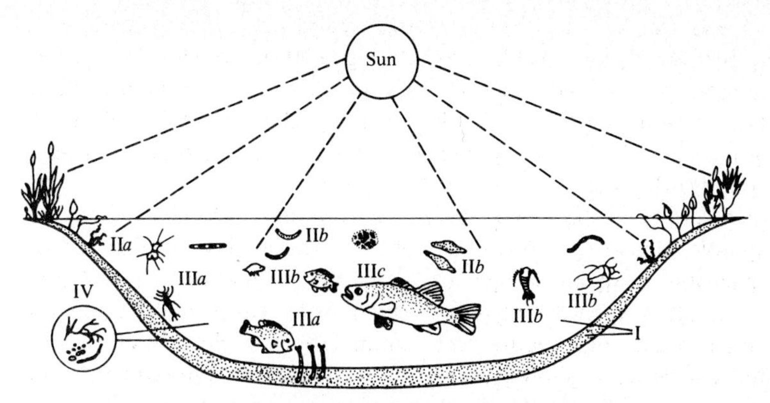

Figure 14.7 Members of the food chain in an aquatic ecosystem which relies on solar radiation as its major energy source. The labels I, II, III, and IV above denote the classes of abiotic components, producers, consumers, and microbial decomposers, respectively.

The phytoplankton produced in surface waters feeds an intricate chain of consumers, which begins with animal plankton (zooplankton) and ends with the fishes, whales, and other aquatic animals. The organic wastes resulting from metabolic activities and death in this chain are in turn decomposed by microorganisms. The simple compounds produced in decomposition can be utilized by the phytoplankton, thereby completing a food cycle.

In the damaging process called *eutrophication*, excess nutrient supplies cause explosive growth of algae. Some of the algae produce toxins, and unpleasant odors often accompany the algal bloom. Eventually much of the algae dies, releasing nutrients for heterotroph consumption. At this point, the net respiration rate exceeds the photosynthesis rate, so that the supply of dissolved oxygen is seriously reduced. This can cause extensive death of fish and many other aerobic organisms, totally upsetting the local balances necessary for survival of the lake or pond ecosystem.

14.4 BIOLOGICAL WASTEWATER TREATMENT

Wastewaters contain a complex mixture of solids and dissolved components, with the latter usually present in very small concentrations. In treatment plants, all these contaminants must be reduced to acceptably low concentrations or chemically transformed into inoffensive compounds. The overall system design used to accomplish this varies depending upon the type and amount of wastewater to be treated and economic and environmental considerations. Most of the alternatives, however, share enough common features to allow them to be shown schematically on a single diagram like Fig. 14.8. There, the parallel pathways shown for sludge handling and removal of various contaminants represent different options for accomplishing one treatment objective. A typical plant would

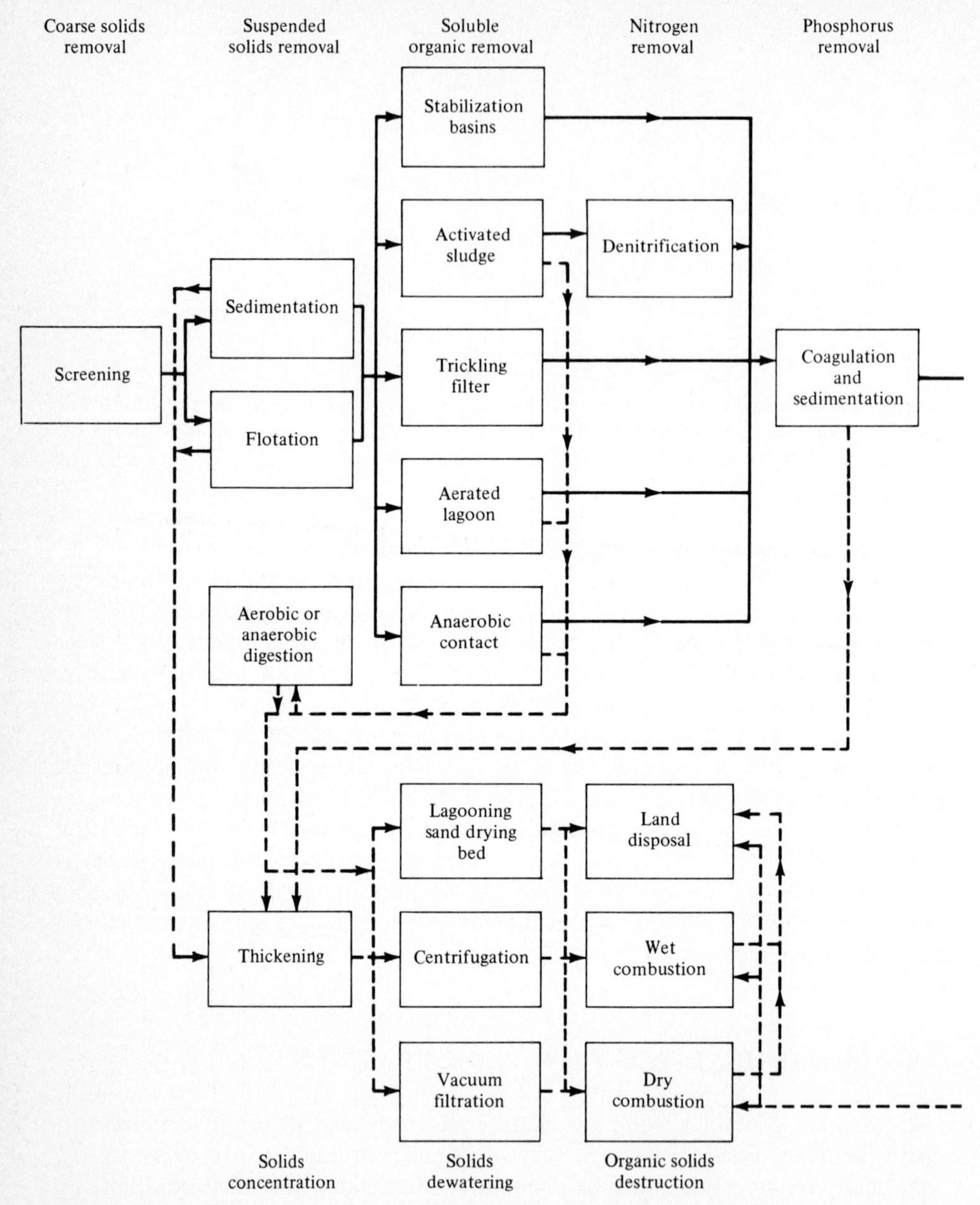

Figure 14.8 Available unit operations for primary, secondary, and tertiary wastewater treatment *(Reprinted from W. W. Eckenfelder, Jr., "Industrial Water Pollution Control," pp. 6–7, McGraw-Hill Book Company, Inc., New York, 1966.)*

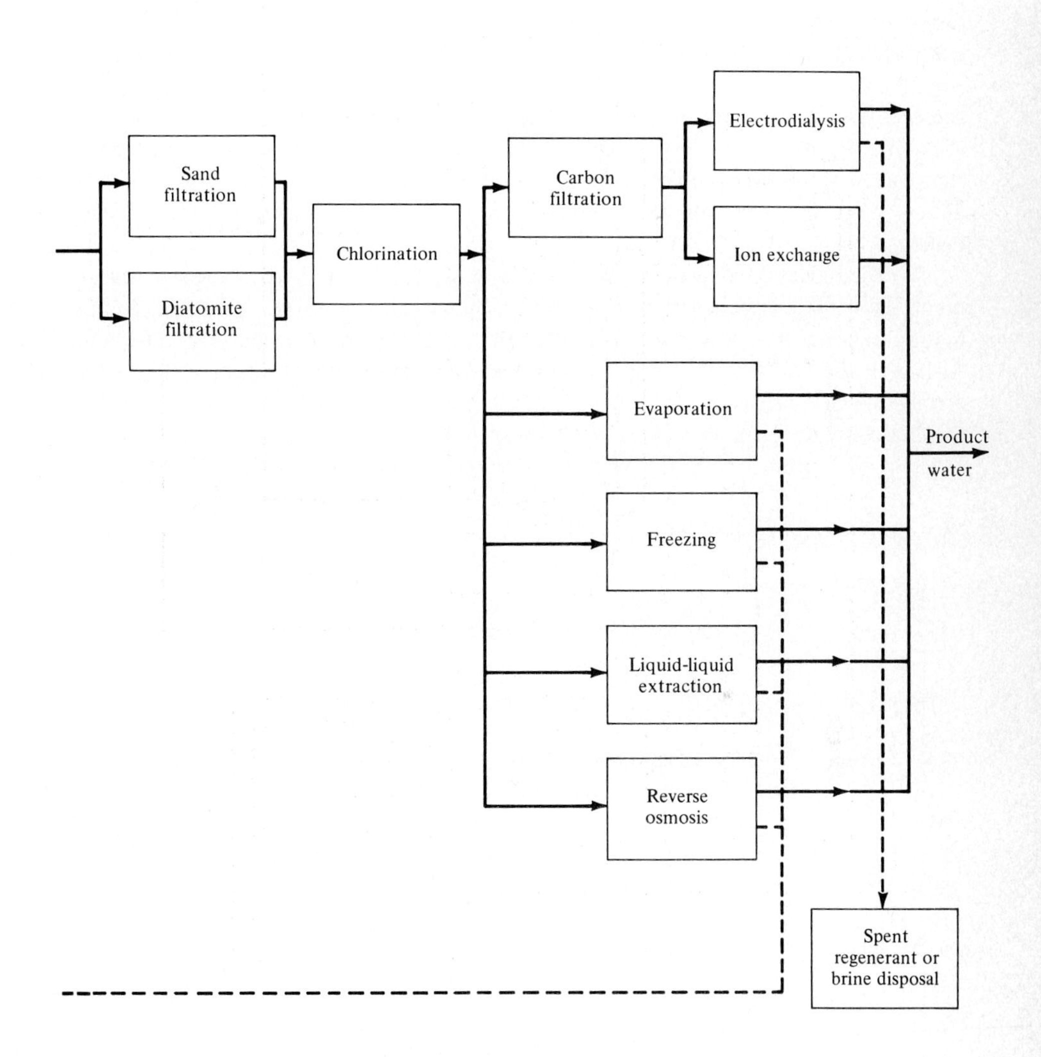
Fine suspended solids removal
Bacterial removal
Trace organics removal
Inorganic salt removal
Sand filtration
Diatomite filtration
Chlorination
Carbon filtration
Electrodialysis
Ion exchange
Evaporation
Freezing
Liquid-liquid extraction
Reverse osmosis
Product water
Spent regenerant or brine disposal

employ only a few of the many possible pathways from wastewater to final effluent.

We consider next the overall purpose of each of the major process trains. In *primary treatment*, the most easily separated contaminants are removed. Taken out here are readily settleable solids (see Fig. 14.9), oil films, and other "light" components. Suspended particles as well as soluble components are removed in *secondary treatment*. In many situations these waste materials are organic, and in these cases use of a biological oxidation process is common. We shall consider these biological processes in further detail later. *Tertiary treatment* is directed at removal of all or some of the remaining contaminants. Among the processes used at this stage are electrodialysis, reverse osmosis, deep-bed filtration, and adsorption.

Wet, concentrated solid wastes, called *sludge*, are removed in primary treatment; cell sludge is generated in secondary biological treatment. We have already mentioned the interplay between substrate utilization and biomass production. Although the secondary biological treatment processes, which involve many microbial species, are very efficient in attacking a dilute mixture of organic wastes, they also create biomass. Thus, very small particles and soluble components in

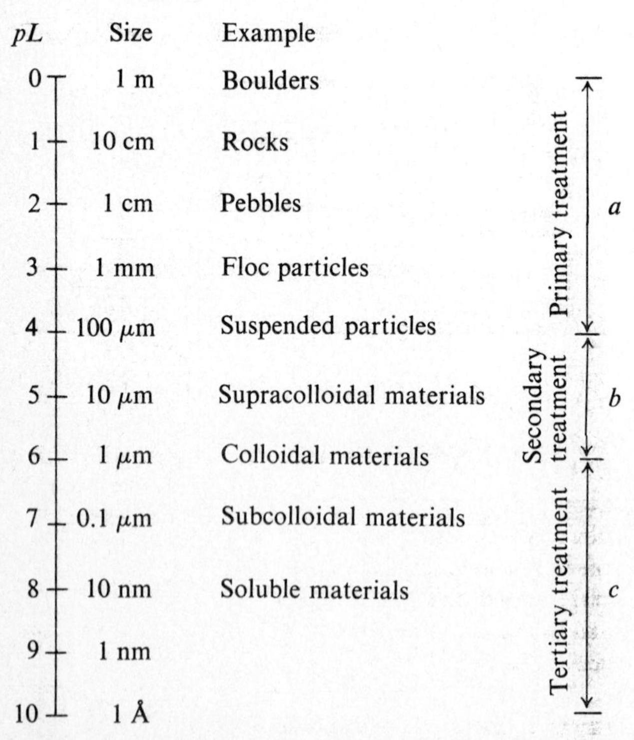

Figure 14.9 Different levels of treatment remove characteristic ranges of particulate sizes from the wastewater. *(Adapted from T. Helfgott et al., "Analytical and Process Characterization of Effluents," J. Sanit. Eng. Div. ASCE, vol. 96, p. 79, 1970.)*

liquid wastes are transformed in part into a sludge waste material which is easier to separate out than the original waste. Sludge handling and treatment is consequently an important part of the water-treatment plant. One popular process for sludge-volume reduction in sewage plants is *anaerobic digestion*, where organic material is biologically decomposed in an anaerobic environment.

We should not conclude from this discussion that all three levels of treatment and sludge digestion are always used. Some wastewaters are discharged into receiving waters such as streams, rivers, ponds, lakes, and oceans with no treatment whatsoever. In other situations, only primary treatment is applied. However, some form of secondary treatment and sludge processing is the norm for most municipal sewage-treatment plants in the United States, with tertiary treatment only infrequently used at present.

In keeping with the major theme of this text, we shall emphasize the biological components of wastewater-treatment processes. After reviewing the characteristics of typical wastewaters in Sec. 14.4.1, we shall consider analysis of activated sludge and related secondary biological treatment methods. Our rapid overview of these important applications of biochemical engineering will conclude with a discussion of anaerobic digestion. Additional information on these processes as well as the physical and chemical operations encountered in treatment plants is provided in the references.

14.4.1 Wastewater Characteristics

Naturally, the types and concentrations of contaminants in wastewater depend upon its source. There are two main classes of wastewaters to consider: industrial effluents and domestic wastes. The latter type is called *sewage*, and it consists of substances such as ground garbage, laundry water, and excrement.

More than 99 percent water, sewage typically contains about 300 ppm (mg/L) of suspended solids and about 500 mg/L volatile material. Much of the suspended solid component is cellulose, and the bulk of organic matter present is in the form of fatty acids, carbohydrates, and proteins in that order. As our earlier discussion of spoilage should suggest, the bad odor of sewage derives from protein decomposition under anaerobic conditions.

Because of its origins, it should be no surprise that sewage contains a varied population of soil and intestinal microorganisms. Included are aerobes, strict and facultative anaerobes, bacteria, yeasts, molds, and fungi. Since pathogenic organisms and numerous viruses including polioviruses and hepatitis viruses are often present in sewage, it is critically important to isolate drinking supplies and water lines from sewage contamination. The sewage microbial populations provide a continuous mixed-culture inoculum for the biological treatment processes and also supply the metabolic capacity used in the following standard analysis of wastewater composition.

Among the measures of sewage strength or concentration, perhaps the most common index is the *biochemical oxygen demand* (BOD). It is equal to the amount of dissolved oxygen which is consumed by a sewage incubated for a specified length of time at 20°C. The length of the incubation time is often shown as a subscript: thus the BOD determined from a 5-day incubation, which is one of the common intervals, is denoted BOD_5. The amount of dissolved oxygen consumed in an incubation which is continued until carbonaceous biological oxidation ceases is called the *ultimate* BOD (BOD_U). Originally devised in 1898 by the British Royal Commission on Sewage Disposal, this test was chosen to simulate the conditions of a stream and to provide a relatively direct measure of one of the most damaging effects of sewage discharge: depletion of dissolved oxygen in the receiving waters. A lowered dissolved-oxygen value can quickly lead to death of many aerobic organisms and animals; the result may be a murky, smelly river contaminated with pathogenic microbes.

The *chemical oxygen demand* (COD) is another indication of the overall oxygen load which a waste water will impose on the receiving water. It is equal to the number of milligrams of oxygen which a liter of sample will absorb from a hot, acidic solution of potassium dichromate. Generally, more components of the wastewater sample can be chemically oxidized in this manner than in the standard BOD test. Consequently, the COD value is usually greater than the BOD of the same sample. Although it is less directly related to the polluting effects of sewage than BOD, COD has the advantage of being measurable in about 2 h by conventional methods or in a few minutes using sophisticated instruments.

Both BOD and COD are gross, overall indicators of sewage composition. Nevertheless, they do provide a measure which relates to the environmental damage of the wastewater. Moreover, the necessary analyses can be performed with minimal equipment and training in analytical procedures.

Other parameters often used to characterize water quality are phosphorus, nitrogen, and suspended-solids concentrations. In Table 14.2, we see typical values of characteristic parameters for the influent and effluent streams of a sewage-treatment plant. Among the important contaminants which are not considered in the table are heavy metals and toxic organics such as pesticides, which are often present in small but significant quantities.

The composition of industrial wastes depends strongly on the source. As Table 14.3 reveals, many industrial wastes are far more concentrated than sewage. Also, those derived from processing hydrocarbon materials often contain

Table 14.2 Some characteristic parameters for water quality
Effluent standards for nitrogen and total phosphorus concentrations are not always applied

Parameter	Influent raw wastewater	Effluent in an acceptable plant
BOD, mg/L	100–250	5–15
COD, mg/L	200–700	15–75
Total phosphorus, mg/L	6–10	0.2–0.6
Nitrogen, mg/L	20–30	2–5
Suspended solids, mg/L	100–400	10–25

Table 14.3 Comparative strength of effluents[†]

Type of waste	Main pollutants	BOD_5	COD
Abattoir	Suspended solids, protein	2,600	4,150
Beet sugar	Suspended solids, carbohydrate	850	1,150
Board mill	Suspended solids, carbohydrate	430	1,400
Brewery (bottle washing)	Carbohydrate, protein	550	
Cannery (meat)	Suspended solids, fat, protein	8,000	17,940
Chemical plant	Suspended solids, extremes of acidity or alkalinity, organic chemicals	500	980
Coal carbonization:			
Coke ovens	Phenols, cyanide	780	1,670
Gas works	Thiocyanate, thiosulphate	6,500	16,400
Smokeless fuel	Ammonia	20,000	
Distillery	Suspended solids, carbohydrate, protein	7,000	10,000
Dairy	Carbohydrate, fat, protein	600	
Domestic sewage	Suspended solids, oil-grease, carbohydrate, protein	350	300
Grain-washing	Suspended solids, carbohydrate	1,500	1,800
Kier	Suspended solids, carbohydrate, lignin,	1,600	3,600
Laundry	Suspended solids, carbohydrate, soap	1,600	2,700
Maltings	Suspended solids, carbohydrates	1,240	1,480
Pulp mill	Suspended solids, carbohydrate, lignin, sulfate	25,000	76,000
Fermentation industry:			
Fermentation segment	....	4,560	4,120
Chemical-synthesis segment	....	960	1,580
Formulation, packaging segment	....	145	217
Petroleum refinery	Phenols, hydrocarbons, sulphur compounds	840	1,500
Resin manufacture	Phenol, formaldehyde, urea	7,400	12,900
Starch reduction of flour	Suspended solids, carbohydrate, protein	12,000	17,150
Tannery	Suspended solids, proteins, sulfide	2,300	5,100

[†] Adapted from J. W. Abson and K. H. Todhunter, pp. 318–319 in N. Blakebrough (ed.), *Biochemical and Biological Engineering Science*, vol. 1, Academic Press, London, 1967.

toxins such as formaldehyde, ammonia, or cyanide. We face two related problems with such wastes: (1) they are extremely damaging to living organisms of the receiving water, and (2) they may kill microorganisms utilized in aerobic and anaerobic waste treatment. Effective and reasonably economical methods for elimination of such toxic compounds from discharge waters still have not been perfected.

In the remainder of this chapter, we shall concentrate on processes and conditions typical of domestic sewage treatment. We should remember, however, that the same biological processes have important applications in treating industrial wastes. Moreover, these two problems often merge, since industrial effluents are in many cases discharged into domestic sewers.

14.4.2 The Activated-Sludge Process

The main component of the activated-sludge process is a continuous-flow aerated biological reactor. As indicated in Fig. 14.10, this aerobic reactor is closely tied to a sedimentation tank, in which the liquid is clarified. A portion of the sludge collected in the sedimentation tank is usually recycled to the biological reactor, providing a continuous sludge inoculation. This recycling extends the mean sludge residence time, giving the microorganisms present an opportunity to adapt to the available nutrients. Also, the sludge must reside in the aerobic reactor long enough for adsorbed organics to be oxidized. Other benefits of reactor recycle were discussed in Chap. 9.

To understand the basic mechanisms of substrate removal which operate in this unit, we must examine the nature and morphology of the community of mixed microbes which thrive in the aeration basin. A common bacterium in the activated-sludge population is *Zoogloea ramigera.* Perhaps the most important characteristic of this organism and others in the sludge is their propensity for synthesizing and secreting a polysaccharide gel. Because of this gel, the microbes tend to agglomerate into flocs, which are called activated sludge (see Fig. 14.11). A special property of activated sludge is its high affinity for suspended solids, including colloidal materials. Thus, the initial step in removing suspended solids from the wastewater is *attachment* to the floc. Following this, biodegradable components of the adsorbed particulates undergo *oxidation* by floc organisms, as indicated in Fig. 14.12.

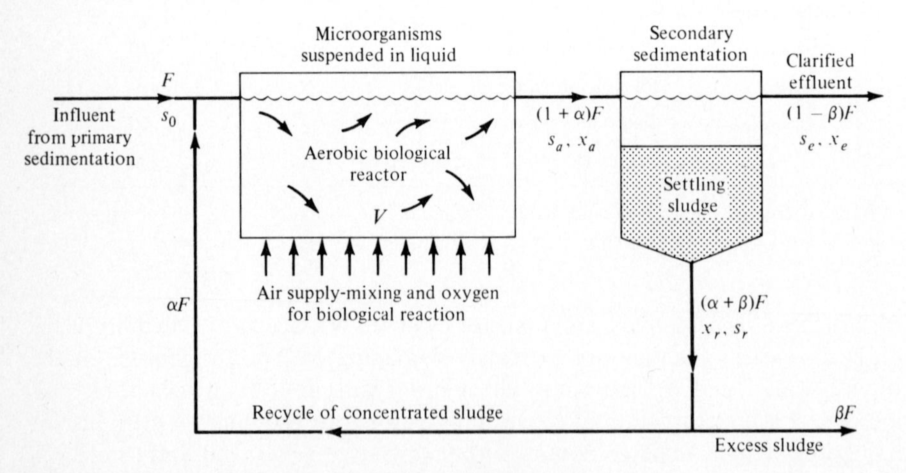

Figure 14.10 Schematic diagram of the activated-sludge process.

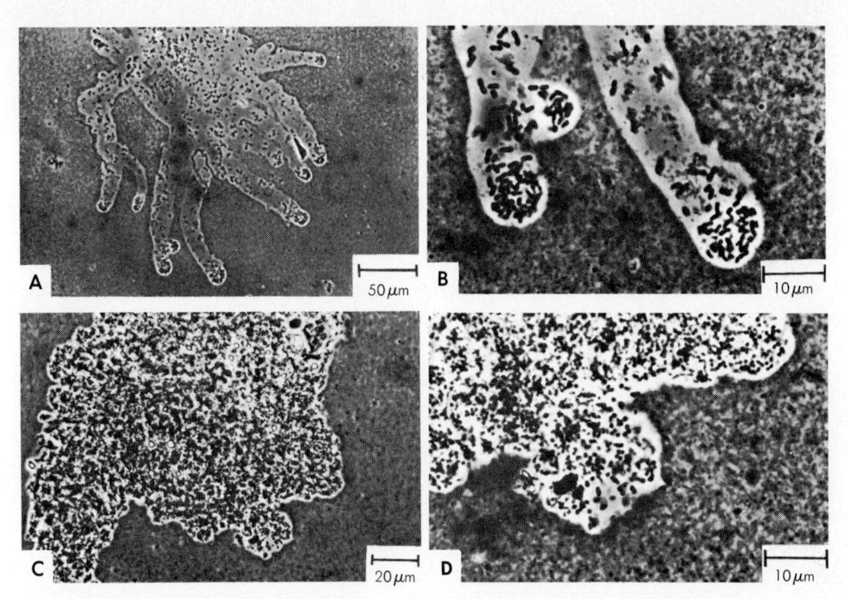

Figure 14.11 Photomicrograph of some of the microorganisms in activated sludge. *(Reprinted from R. F. Unz and N. C. Dondero, Water Research, vol. 4, p. 575, 1970.)*

In order to capitalize on the excellent adsorbent properties of activated sludge, a variation of the conventional process called *contact stabilization* has been devised. As shown in the process flow chart in Fig. 14.13, the recycle settled sludge is subjected to an additional aeration cycle before being mixed with the waste influent in an aerobic tank. In this latter contact basin, organics are removed almost entirely by physical attachment. Biological utilization of these

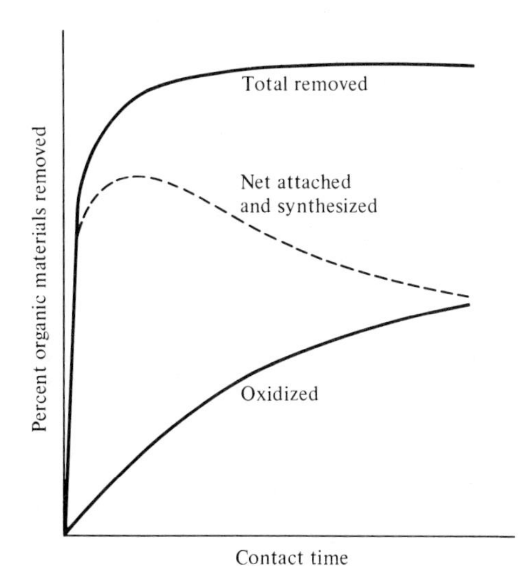

Figure 14.12 Removal of organic material in an aerated, batch activated-sludge system is believed to involve rapid initial physical capture of organics by the sludge, followed by slower biological oxidation.

stored organics occurs in the aerated recycle-sludge basin, which also serves the function of restoring the floc's adsorption capacity.

Other versions of the activated-sludge process differ from the conventional design primarily in the contacting pattern of wastewater, sludge, and air in the aeration basin. The flowsheets in Fig. 14.13 reveal that in the *step-feed process*, the influent stream is split and introduced at different points of the aeration basin. We appreciate the effects of this distributed feed from our reactor-analysis studies in Chap. 9: the conventional activated-sludge aeration basin is a long, narrow channel which behaves roughly like a tubular reactor with some dispersion. By distributing the feed along the reactor length, the basin is made to behave more like a well-mixed tank reactor.

Actually, a better approximation to a backmixed reactor is achieved by using a circular basin which is vigorously aerated to provide mass transfer and mixing. In such a system, gradients of dissolved oxygen and nutrient concentrations within the reactor are minimized. The activated-sludge population which develops under these conditions is often better suited for dealing with loading fluctuations or with shock loads of toxic material.

Although the underlying principles of aerator designs for activated-sludge processes are the same as considered in Chap. 8, the aeration systems used can vary widely. Besides the stirred-sparged combination familiar in fermentation applications, the air may be bubbled into the vessel through diffusers on the bottom or sides of the reactor. Alternatively, the surface of the basin may be brushed with rotating blades to create turbulence and promote gas absorption. Other possibilities include the simplex cone, which draws liquid from near the basin bottom and sprays it onto the tank surface. In all cases the aeration and agitation system serves to provide oxygen for microbial respiration, to suspend and mix the sludge and other particulates, and to strip out volatile metabolic products such as CO_2.

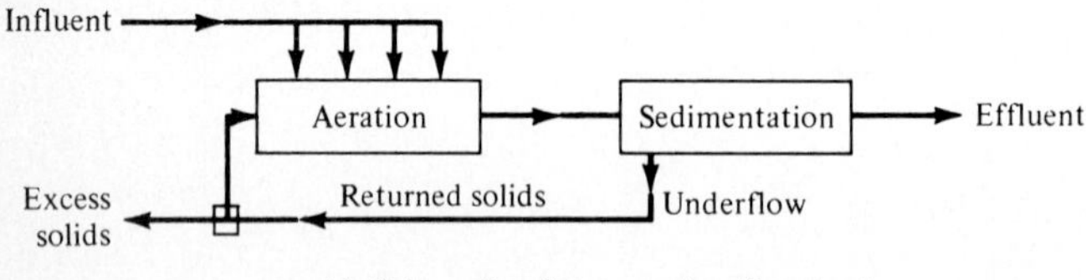

(*a*) Schematic of step aeration flowsheet

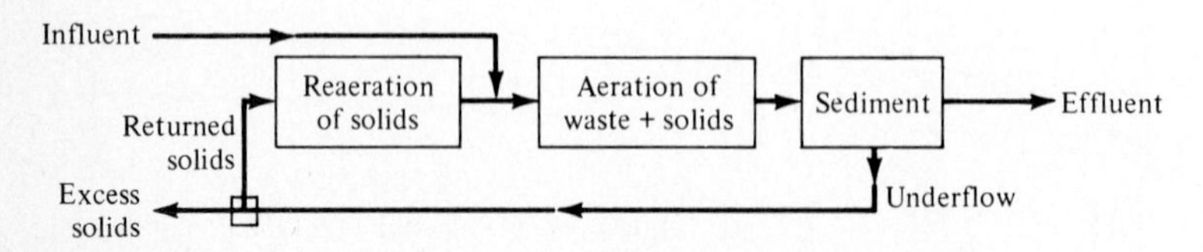

(*b*) Schematic of contact stabilization (solids reaeration) flowsheet

Figure 14.13 Two alternative flowsheets for biological oxidation processes. The conventional activated-sludge process was shown in Fig. 14.10.

Besides possessing the necessary adsorbent and metabolic qualities, a good sludge should settle rapidly. For example, after 30 min in a cylinder, the volume percent of settled sludge should be around 40 times the volume percentage of suspended solids. Much larger values, say 200, indicate a bad sludge, which will tend to overflow from the thickening tank. When the process *bulks*, as this condition is called, it has failed: the effluent will not meet the necessary standards.

While we do not yet understand the causes and mechanism of bulking, examination of the microbial constituents of poor sludge often reveals filamentous bacteria and flagellate protozoa. Healthy sludge, on the other hand, does not contain a significant population of filamentous organisms, and the protozoa present are mainly stalked ciliated species. The protozoa serve a valuable function in the overall process by preying on free, i.e., unflocculated, bacteria and thereby clarifying the effluent.

Normally, filamentous bacteria and fungi cannot compete with the heterotrophic bacteria found in healthy sludge. Large shocks in influent conditions or improper operation of the unit may, however, create conditions damaging to the desired population, permitting other microorganisms to invade the community. We could conjecture, therefore, that both normal operation and bulking are manifestations of the principles of competition in mixed populations.

14.4.3 Design and Modeling of Activated-Sludge Processes

Although a wastewater-treatment plant for a major city costs in excess of 100 million dollars, the biological reactors they contain are usually designed using extremely simplified and idealized models. Typically, the aeration basin is treated as a perfectly mixed vessel, and sludge is viewed as a single pseudo species whose growth rate follows Monod kinetics with an additional endogeneous metabolism decay term. As discussed in Sec. 14.4.2, the substrate or limiting nutrient concentration is usually expressed in terms of BOD.

With the nomenclature indicated in Fig. 14.10, which is consistent with that used in Chaps. 7 and 9, the steady-state mass balance on active solids in the process is

$$(1-\beta)Fx_e + \beta Fx_r = Vx_a\left(\frac{\mu_{\max} s_a}{s_a + K_s} - k_e\right) \tag{14.1}$$

Straightforward algebraic manipulation of this expression shows

$$\frac{1}{\theta_s} = \mu_{\max}\frac{s_a}{s_a + K_s} - k_e \tag{14.2}$$

where θ_s is the mean residence time of the activated sludge (ratio of active-solids retention to active-solids effluent rate), sometimes called the sludge age:

$$\theta_s = \frac{Vx_a}{F(1-\beta)x_e + \beta Fx_r} \tag{14.3}$$

In most sewage-treatment plants, the sludge age is of the order of 6 to 15 days.

Table 14.4 Monod model parameters for utilization of different substrates in aerobic mixed culture[†]

Organic substrate	Y	μ_{max}, day^{-1}	K_s, mg/L	k_e, day^{-1}	Coefficient basis	Temperature, °C
Domestic waste	0.5	....	....	0.055	BOD_5	
	0.67	....	....	0.048	BOD_5	20
	0.67	3.7	22	0.07	COD	
Skim milk	0.48	2.4	100	0.045	BOD_5	
Glucose	0.42	1.2	355	0.087	BOD_5	
Peptone	0.43	6.2	65		BOD_5	30

† A. W. Lawrence and P. L. McCarthy, Unfied Basis for Biological Treatment Design and Operation, *J. Sanit. Eng. Div. ASCE*, **96:** 768 (June 1970).

Using Eq. (14.2), we can compute the solids-retention time θ_s required to achieve a specified BOD level s_a from a given waste, providing the kinetic constants are known for that system. Typical values for these parameters are given in Table 14.4. We should recall, however, from our earlier discussions of lumping in Secs. 3.4.2 and 9.4.1, that the kinetic "constants" in such simplified models are actually dependent on composition and operating conditions. Consequently, appropriate values for μ_{max}, K_s, k_e, and Y should always be measured for the specific waste of interest.

Assuming a constant yield factor, we can now formulate a steady-state conservation equation for substrate in the aeration basin; assuming substrate is not separated in the clarifier ($s_a = s_e = s_r$) and recalling that the first term on the right-hand side of Eq. (14.2) is the rate of sludge synthesis, it follows that

$$Vx_a = \frac{YF\theta_s(s_0 - s_a)}{1 + k_e\theta_s} \tag{14.4}$$

This equation provides the active-solids weight Vx_a in the aeration basin in terms of quantities already specified. Next, sludge-settling characteristics must be determined experimentally to find x_r. With this known, the required basin volume V for a given recycle ratio α can be calculated from

$$V = F\theta_s\left(1 + \alpha - \alpha\frac{x_r}{x_a}\right) \tag{14.5}$$

which is obtained by writing an active-solids balance on the aeration basin and using Eq. (14.2).

Since the extreme simplicity of this design procedure differs so dramatically from the complex physical and biological processes which occur in an activated-sludge plant, it is reassuring to see some experimental data supporting this design

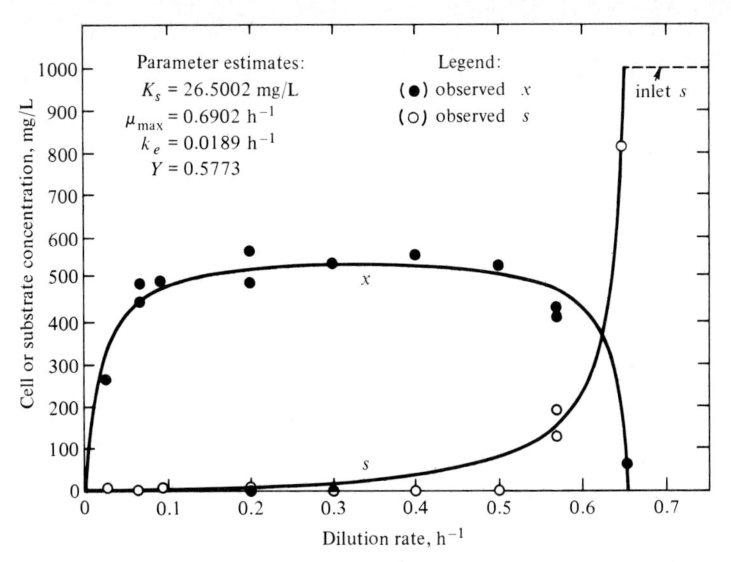

Figure 14.14 Comparison of Monod model calculations with experimental data for continuous mixed culture. *(Reprinted from S. Y. Chiu et al., "Kinetic Model Identification in Mixed Populations Using Continuous Culture Data," Biotech. Bioeng., vol. 14, p. 207, 1972.)*

approach. Figure 14.14 shows the results of cultivation of a mixed population of sewage microorganisms in a chemostat. The solid lines, computed using the same growth model as in Eq. (14.1) and the kinetic parameters shown in the figure, provide an excellent fit to the experimental data. Notice that although the kinetic parameters used here fall within the general ranges of magnitude suggested in Table 14.4, they differ considerably. This underscores the caveat just mentioned: the parameters must be determined for each waste stream, since they will vary with waste characteristics.

Our discussions in Chaps. 7 and 9 suggested that dependence of model parameters on input compositions and operating configuration can be reduced by introducing more structure into the model. Although much remains to be done, an excellent beginning on development of structured models for activated-sludge and other biological treatment processes has been made by Andrews and his colleagues [5–7]. Their work in this area has been motivated by the considerations we have summarized as well as a very important additional one: development of control strategies for wastewater-treatment plants. Every day most plants are subjected to large disturbances which require some adjustment in plant operating conditions. A model with inadequate structure would not accurately reflect the effects of process disturbances and would be useless for simulation studies of plant dynamics and control systems. We shall return to this point in Example 14.3; now let us examine Andrew's structured sludge model and its characteristics.

In this model, biomass is divided into three components, which are derived from substrate and interconverted according to the scheme

$$\text{Substrate} \xrightarrow[r_1]{\text{attachment}} \underset{\substack{\text{Stored}\\ \text{mass}}}{\overset{\overset{\text{respiration}}{\uparrow}}{X_s}} \xrightarrow[r_2]{\text{synthesis}} \underset{\substack{\text{Active}\\ \text{mass}}}{\overset{\overset{\text{respiration}}{\uparrow}}{X_A}} \xrightarrow[r_3]{\text{residue}} \underset{\substack{\text{Inert}\\ \text{mass}}}{X_i} \tag{14.6}$$

The rate of the attachment step is taken as

$$r_1 = k_s\left(x_T \frac{f_s s}{s + K_s} - x_s\right) \tag{14.7}$$

where k_s is a mass-transfer coefficient and x_s is the concentration of storage products in the floc phase. The total mixed-liquor volatile suspended-solids concentration (MLVSS) is denoted by x_T, which in turn is equal to the sum of stored-mass concentration x_s, the active-mass concentration x_A, and the inert-mass concentration x_i:

$$x_T = x_s + x_A + x_i \tag{14.8}$$

Finally, f_s is the maximum fraction of the MLVSS which can be storage products, s is substrate concentration, and K_s is a saturation constant. From literature reviews and their simulation work, Andrews' group suggests the parameter values listed in Table 14.5.

The specific rate r_2 of the active mass synthesis step is presumed to follow Monod kinetics so that

$$r_2 = \frac{\mu_A x_s}{K_A + x_s} x_A Y_A \tag{14.9}$$

Table 14.5 Parameter values suggested by Busby and Andrews for structured kinetic model of the activated-sludge process†

Term	Value	Definition
k_s	3.0	Substrate mass-transfer coefficient $[t^{-1}]$
f_s	0.45	Maximum fraction of MLVSS that can be storage products
K_s	150	Sorption coefficient $[\text{m/L}^3]$
μ_A	0.06	Maximum specific rate for conversion of stored mass to active mass $[t^{-1}]$
K_A	80.0	Saturation constant $[\text{m/L}^3]$
Y_A	0.66	Mass of active mass formed per unit mass of stored mass converted
Y_i	0.25	Mass of inert mass formed per unit mass of active mass converted
k_i	0.03	Specific decay rate of active mass $[t^{-1}]$

† Adapted from J. B. Busby and J. F. Andrews, *J. Water Pollut. Control Assoc.*, **47:** 1067, 1975.

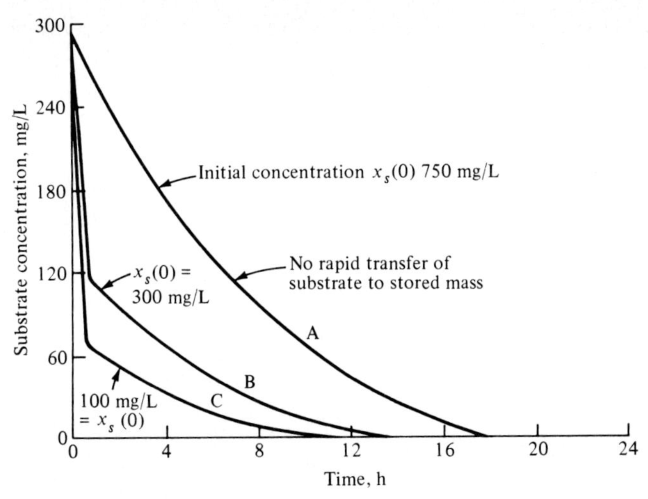

Figure 14.15 The initial stored-mass concentration has a major effect on batch substrate utilization in Busby and Andrews' structured model.

and the constant yield coefficient Y_A, defined as the active mass synthesized per stored mass utilized, characterizes the synthesis step's stoichiometry. Available data suggest that the inert-mass formation rate is first order in active mass:

$$r_3 = k_i x_A Y_i \tag{14.10}$$

Y_i denotes the constant-yield coefficient (mass i formed per mass of A consumed) for this step.

Figure 14.15 shows the results of simulated batch growth using this model. Alternatively these curves could of course be viewed as the concentration in a plug-flow aeration basin as a function of position. An especially interesting feature of these calculations is the great improvement in substrate uptake which occurs with decreasing initial stored-mass concentration. Such behavior is consistent with the increased efficiency provided by the contact-stabilization process, in which the stored mass is largely converted to other mass forms in the aeration tank which follows the clarifier.

Indeed, Andrews and his associates have simulated the steady-state behavior of most versions of the activated-sludge process using this kinetic model. In these calculations, mixing in the aeration basin was modeled using four ideal continuous-flow reactors in series. The primary and secondary settlers were represented by models which are presented in detail in the references. Using reasonable loading and sizing parameters, the model gave efficiencies, BOD removal rates, and operating characteristics in good agreement with experimental data on these processes. Some of these results are listed in Table 14.6.

Besides its value in simulating a wide variety of steady-state process configurations, the structured-sludge model just outlined is necessary for analyzing the

Table 14.6 Performance characteristics of several different biological oxidation-process configurations†
The average influent flow rate was 1000 m^3/h for all cases

Type of process	Removal BOD_U, %	Process loading intensity kg/(kg · day)	Active mass, %	Total tank volume, m^3	BOD_U removed, kg · (unit vol · day)$^{-1}$
Conventional	88	0.5	34	4,500	1.22
Extended air	96	0.15	16	16,000	0.392
Short term	80	0.75	40	2,200	2.21
Contact	88	0.29	28	3,000	1.8
Step feed	91	0.27	28	4,500	1.3

† From J. F. Busby and J. F. Andrews, *J. Water Pollut. Control Assoc.*, **47:** 1055, 1975.

dynamic behavior of activated-sludge plants. This should be evident from Fig. 14.15, in which the dynamic response of substrate concentration is shown to be highly sensitive to one segment of the total biomass. We discuss dynamics and control later, in connection with the anaerobic digestion process. Those interested in control of activated-sludge systems can gain a good entrée into the available literature from Andrews' review papers [5–7].

Consideration of even more detailed or structured models which explicitly consider population interactions between bacteria and protozoa in activated-sludge units has begun. In a series of papers considering models of activated sludge dynamics, Curds has investigated the food chain shown in Fig. 14.16. As indicated there, only the substrate and bacteria found in the sewage enter the plant in the influent stream. Reexamining Fig. 14.10, however, we recall that the feed to the aeration basin is a mixture of the influent waste and recycle-sludge streams, so that it is important to keep track of

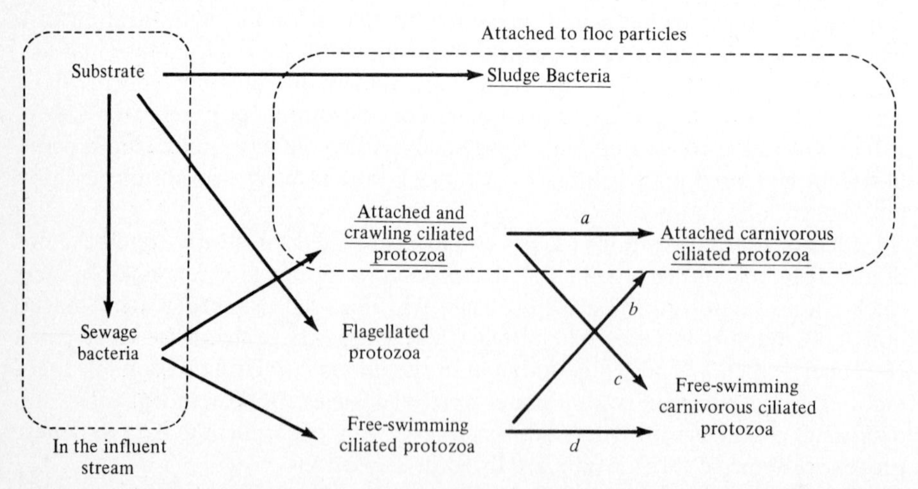

Figure 14.16 The food chain used in Curd's analysis of microbial interactions in an activated-sludge plant.

organisms which live on the floc phase and will be included in the recycled sludge. In Curds' model, these species, which are underlined in Fig. 14.16, are the sludge bacteria, the attached and crawling protozoa which prey on the suspended but not flocculated sewage bacteria, and finally the attached carnivorous protozoa, which prey on both free and attached protozoa.

Since by now the mechanics of writing the appropriate mass balances on the system should be familiar, we list only the assumptions needed to formulate Curds' equations:

1. The aeration basin is an ideal CSTR.
2. The specific growth rate of each species shown in Fig. 14.16 depends on the concentration of its nutrient in Monod fashion. The kinetic constants used in the simulation studies are given in Table 14.7; the yield factor for all steps is assumed to be 0.5.
3. The concentrations of floc organisms in the sludge recycle are directly proportional to their respective concentrations in the aeration-basin effluent. Thus, for example,

$$x_{ir} = b_i x_{ia} \tag{14.11}$$

where b_i is the *settler concentration factor* for species *i*. These constants are also listed in Table 14.7. We take $b = 1$ for substrate.

Assumption 3 makes it quite easy to include the sedimentation basin in the model thereby incorporating the effects of recycle in the simulations.

Table 14.7 Parameters for different microorganisms in Curds' model of activated sludge[†]

Organism	Concentration in nonvariable sewage, mg/L	Range in variable sewage, mg/L	Maximum specific growth rate, h^{-1}	Saturation constant, mg/L^{-1}	Food of organism	Settler concentration factor b
Sewage bacteria	30	15–45	0.5	10.0	Substrate	
Sludge bacteria	....	....	0.3	15.0	Substrate	1.90
Flagellates	....	....	0.4	12.0	Substrate	
Bacteria-consuming ciliates:						
Free-swimming	....	....	0.35	12.0	Sewage bacteria	
Crawling	....	....	0.35	12.0	Sewage bacteria	1.27
Attached	....	....	0.35	12.0	Sewage bacteria	1.90
Carnivorous ciliates:						
Free-swimming	....	....	0.35	12.0	Bacteria consuming ciliates	
Attached	....	....	0.35	12.0	Bacteria consuming ciliates	1.90

† From C. R. Curds, *Water Res.*, **7:** 1269, 1973.

The presence of free and attached forms of carnivorous protozoa and bacteria-consuming protozoa allows four different prey-predator pairs, labeled *a* to *d* in Fig. 14.16. Curds has considered each pair separately, assuming in the analysis that the other three prey-predator interactions at that trophic level are absent. In all cases, the following general trends are observed: in considering the carnivorous ciliates, which have been neglected in many similar analyses, it is found that their presence decreases the concentration of ciliate prey relative to the no-carnivore case. This in turn allows the population of sewage bacteria to increase. Because the sludge bacteria compete with sewage bacteria for a common nutrient, the sludge bacterial population is reduced.

For prey-predator pairs *a* and *b*, the ciliate prey is washed out of the system. Interaction *c* produces oscillations with mild damping which are displayed in Fig. 14.17. Only interaction *d*, which is relatively rare in activated-sludge populations, leads to stable, nonzero populations of all species at steady state. In all these calculations, the system parameter values assumed are $F = 100$ L/h, $\alpha = 1$, $\beta = 0.05264$, and $s_0 = 200$ mg/L; the influent concentration of sewage bacteria b_0 is 30 mg/L.

One shortcoming of these and many other simulations is the assumption of constant influent flow rate and concentrations. In actual practice the habits of the community lead to sewage variations with time which then appear in the plant influent. To a first approximation, these fluctuations are periodic, with a period of 24 h. The amplitude depends on the upstream system. In plants servicing large cities, for example, the sewage lengths between sewage source and treatment facility vary widely, so that substantial damping of influent load is achieved in the sewer network. Larger variations are expected in smaller plants. The ratio of maximum to minimum value of such oscillations is typically about 3, although this factor can be 10 or greater in some instances.

In one of the first attempts to examine the effects of periodic fluctuations in influent quantity or quality, Curds has considered the behavior of the system outlined above when the feed-stream conditions oscillate according to

$$F = 100 + 50 \sin 2\pi t \qquad \text{L/h}$$
$$s_0 = 200 + 60 \sin 2\pi t \qquad \text{mg/L}$$
$$b_0 = 30 + 15 \sin 2\pi t \qquad \text{mg/L} \tag{14.12}$$

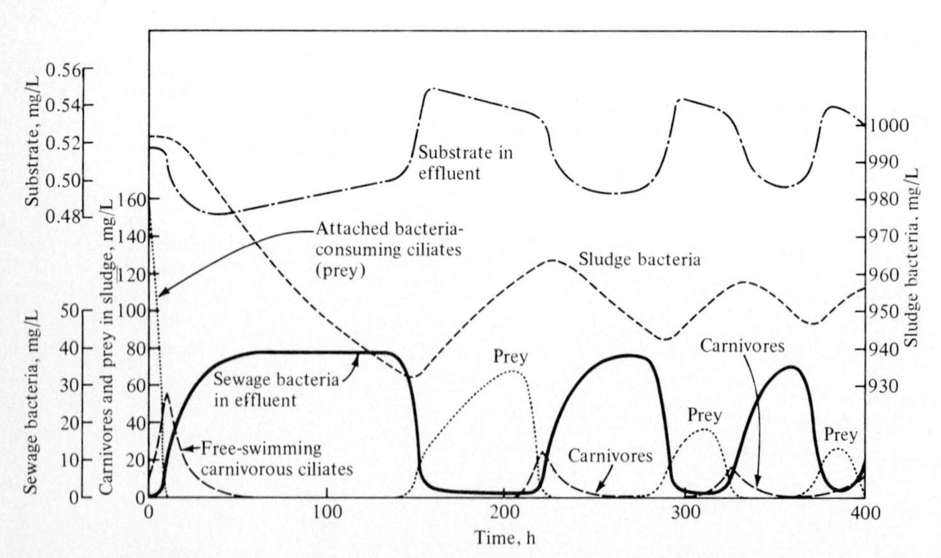

Figure 14.17 Simulation of an interacting activated-sludge population in which free-swimming ciliates prey on attached ciliates. *(Reprinted from C. R. Curds, "A Theoretical Study of Factors Influencing the Microbial Population Dynamics of the Activated-Sludge Process—I," Water Res., vol. 7, p. 1269, 1973.)*

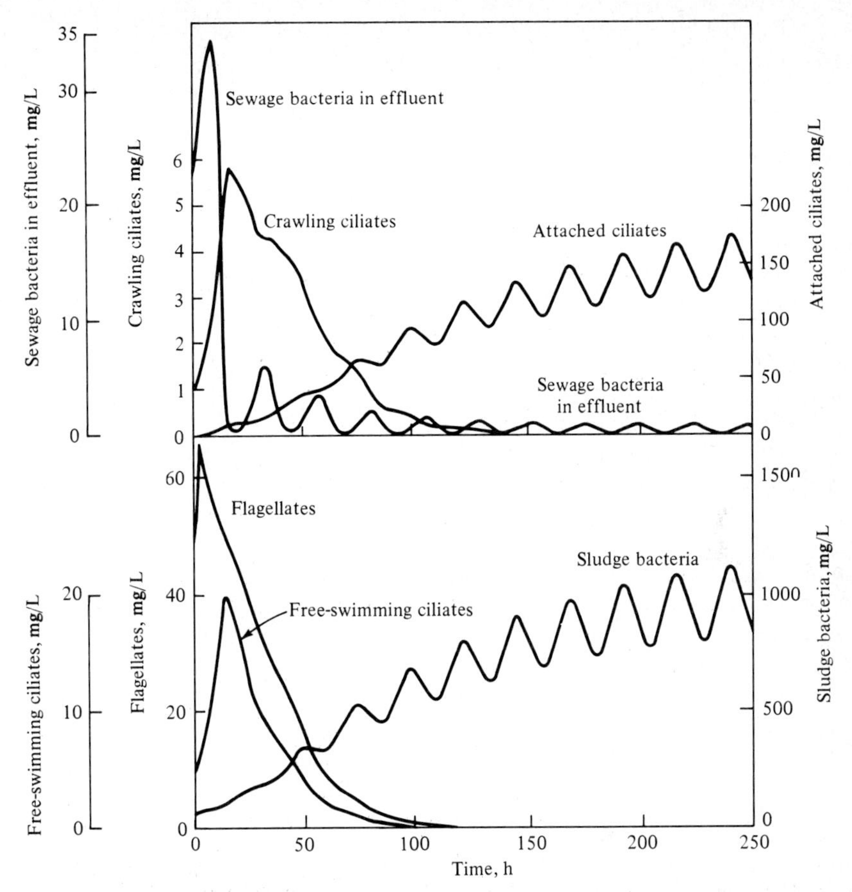

Figure 14.18 Simulation of start up of an activated-sludge unit in which the feed flow rate, feed substrate concentration, and feed sewage concentrations oscillate sinusoidally with a period of 1 day. *(Reprinted from C. R. Curds, "A Theoretical Study of Factors Influencing the Microbial Population Dynamics of the Activated-Sludge Process—I," Water Res. vol. 7, p. 1269, 1973.)*

where t is time in days and the carnivorous predators have been excluded from the calculations. The resulting behavior is illustrated in Fig. 14.18. This figure shows a succession of dominating populations like those we have seen before in spoilage and soil microbiology. Another interesting feature of these simulations is the survival of attached ciliates, while free-swimming and crawling ciliates are washed out. While free-swimming ciliates are normally not found in healthy sludge, crawling forms are, so that some further refinements in this model are desirable. We should point out in closing that the oscillations seen in Fig. 14.17 are very different from those of Fig. 14.18. In the latter case, the oscillations are *forced* by the cyclic variations in feed conditions. The former oscillations, which occur with constant feed conditions, are *autonomous* and reflect special nonlinear characteristics of the system. Bailey's review (Ref. 10 of Chap. 11) may be consulted for further comments on these matters.

While the structured models developed recently by Andrews, Curds, and others offer great promise for improved design methods and control systems, they have not yet enjoyed widespread application. With further testing and more careful experiments to check their validity, we can expect models of this type to gain wider acceptance in the future.

14.4.4 Aerobic Digestion

The high biological content sludge product from an activated sludge operation is frequently subjected to an additional aeration step which operates like an unfed activated sludge system. In this circumstance, the biomass utilizes its own carbon in endogenous respiration, and the net result is typically a 50 percent reduction of solids content. No cell recycle is used, and cell and hydraulic residence times are thus equal at 15 to 25 days. This treatment diminishes the total sludge mass to be disposed of by hauling (truck or barge).

14.4.5 Nitrification

In conventional aerated waste treatment processes, nitrogen-containing organics are among the substrates which are oxidized biologically. These oxidations often initially yield ammonia which must be oxidized to nitrite then nitrate in order to yield a final effluent of sufficiently low oxygen demand. As indicated earlier in Fig. 14.3, two microbial species are responsible for these conversions, *Nitrosomonas* and *Nitrobacter*. In unbalanced reaction forms

$$NH_3 + CO_2 + O_2 \xrightarrow{\text{Nitrosomonas}} \text{cells} + NO_2^-$$

$$NO_2^- + CO_2 + O_2 \xrightarrow{\text{Nitrobacter}} \text{cells} + NO_3^-$$

Specific growth rates for each of these species follow a simple Monod function form; data of Lawrence and McCarty† provide the following parameter values:

	Y_N g cells/gN	μ_{max} day^{-1}	K_S mg/L
Nitrosomonas	0.05	0.33	1.0
Nitrobacter	0.02	0.14	2.1

Material balances on *Nitrosomonas* and *Nitrobacter* species can be used to evaluate the effluent ammonia and nitrite concentrations from the activated sludge reactor, the throughput of which is generally based on BOD removal requirements. An example of such calculations follows.

† A. W. Lawrence and P. L. McCarty, "Unified Basis for Biological Treatment Design and Operation," *J. Sanitary Eng. Div., Proc. ASCE*, **96**: 757, 1970.

If cell residence time is too short in the activated sludge portion, a second aerated tank may be used subsequently to complete nitrification to the desired level.

Example 14.2: Nitrification design Schroeder (Ref. 10) provides the following case.

Industrial wastewater treatment:			
Waste characteristics:	Ultimate BOD:		400 mg/L (soluble)
	Organic N:		60 mg/L
	Phosphate (PO_4^{3-})		30 mg/L
	pH		7.0
	T		20°C
Process kinetics and stoichiometry	Substrate utilization	μ_s	2.0 d^{-1}
		K_m	5.0 mg/L
		Y_{obs}	$0.4/(1.0 + 0.06\theta_c)$
Sludge settling rate		u_H	$2.7 - 0.00026x$ cm/min

The maximum effluent levels are 10 and 0.5 mg/L for BOD and ammonia nitrogen, respectively. Design requires consideration of activated sludge tank nitrification performance to see if an additional nitrification tank is need.

If N and P are not growth limiting, but carbon C is, then the BOD removal specification implies

$$\text{Rate of BOD formation} = \frac{F \cdot (10 - 400)}{V_R} = -\frac{x}{Y_{obs}\theta_c} = -\frac{x\mu}{Y_{obs}} \tag{14E2.1}$$

A cell (biomass) balance provides a cell growth rate

$$\text{Rate of biomass formation} = \frac{x}{\theta_c} = \frac{Y_{obs}\mu_s s x}{K_m + s} \tag{14E2.2}$$

The minimum cell age which is needed to effect the required BOD conversion is obtained with $s = 10$ mg/L. Thus, with $Y_{obs} = 0.4/(1 + 0.06\theta_c)$, we find

$$\begin{aligned}\theta_c(\text{minimum}) &= \frac{(K_m + s)}{0.4\mu_g s - 0.06(K_m + s)}\\ &= \frac{5 + 10}{0.4(2)(10) - 0.06(5 + 10)}\\ &= 2.1 \text{ days}\end{aligned}$$

Thus, for operation at $\theta_c > 2.1$ days, a satisfactory BOD level is achieved.

For nitrification in the activated sludge plant, we have entering N at 60 mg/L and nitrogen assimilated at a concentration of $\frac{14}{113}(Y_{obs})(s_i - s)$. Thus, assimilated N is 16.6 and 14.3 mg/L at θ_c values of 3 and 6 days, respectively.

The material balance on *Nitrosomonas* biomass gives

$$\frac{1}{\theta_c} = \frac{\mu_{max, NH_3} s_{NH_3}}{K_{NH_3} + s_{NH_3}} \tag{14E2.3}$$

Evaluating the specific growth rate using the kinetic parameters from the table above and the specified effluent ammonia concentration gives

$$\frac{1}{\theta_c} = \frac{0.33(0.5)}{1. + 0.5} = 0.11 \text{ day}^{-1}$$

Thus, ammonia conversion meets effluent requirements if $\theta_c > 9$ days. Taking $\theta_c = 10$ days, the corresponding nitrite level is found from an equation of the same form with the nitrite conversion parameters:

$$\frac{1}{\theta_c} = \frac{\mu_{\max, NO_2} \cdot s_{NO_2}}{K_{NO_2} + s_{NO_2}} \tag{14E2.4}$$

or

$$s_{NO_2} = \frac{K_{NO_2}}{\theta_c \mu_{\max, NO_2} - 1}$$

Evaluating this formula for the *Nitrobacter* parameters above gives $s_{NO_2} = 5.25$ mg/L.

The preceding calculations show that in this case BOD removal is more rapid than nitrification. Hence either a larger single activated sludge tank, or an activated sludge BOD removal reactor followed by an aerated nitrifying system, is needed to obtain desired effluent nitrogen quality.

Nitrification adds to the overall oxygen demand of the activated sludge unit according to the following stoichiometry:

$$NH_3 + \tfrac{3}{2}O_2 \longrightarrow NO_2^- + H_2O + H^+$$

$$NO_2^- + \tfrac{1}{2}O_2 \longrightarrow NO_3$$

Fuller process design would include the contribution to nitrification made by the cell settling-sedimentation unit makes just down stream of the activated sludge reactor itself.

14.4.6 Secondary Treatment Using a Trickling Biological Filter

The so-called *trickling* or *percolating biological filter* is a popular alternative to the activated-sludge process. Here a film or slime of microorganisms lives on solid packing which loosely fills a vessel (void fraction about 0.5) designed to permit air to enter the lower portion of the bed. A typical biological-filter design is illustrated in Fig. 14.19.

The use of the term "filter" to describe this system is in a sense unfortunate, since the mechanism of waste removal is not due to straining but to the same attachment–biological-oxidation sequence which operates in the activated-sludge process. Before examining the microbial populations active in trickling filters, it is important to have a clear picture of the engineering and operational characteristics of the unit.

Liquid to be treated is fed to the top of the bed, which typically is 3 to 10 ft deep, either continuously, through fixed nozzles over the bed, or periodically, using a rotating distributor like that in Fig. 14.19. In both cases, the liquid rate must not be high enough to flood the bed. To ensure adequate oxygen supply, the liquid should trickle over the slime-covered packing in films sufficiently thin for the oxygen continuously to supply aerobic organisms in the outer surface of the microbial film. Unlike the activated-sludge process, which often receives forced aeration, air is circulated through the trickling filter by natural convection.

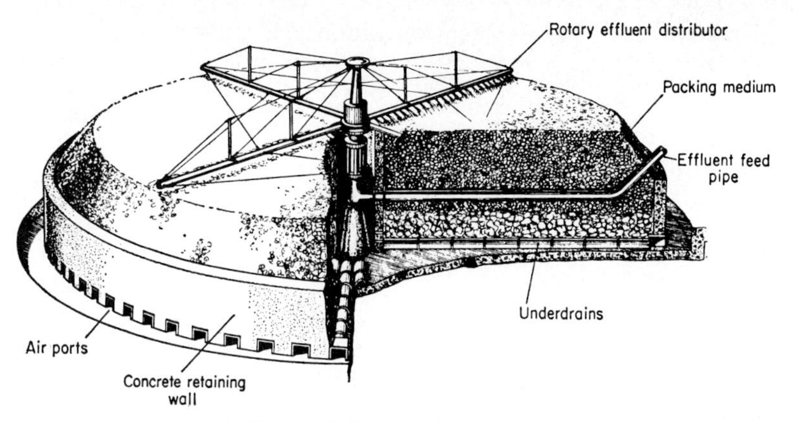

Figure 14.19 A trickling biological filter. *(From J. W. Abson and K. H. Todhunter, p. 326 in N. Blakebrough (ed.), "Biochemical and Biological Engineering Science," vol. 1, Academic Press, London 1967.)*

The driving force for this convection is the temperature difference generated in the trickling filter by biological oxidation of the sewage; air ports and accompanying ventilation pipes within the filter allow air to enter the bottom and intermediate portions of the bed.

The basic principles of substrate diffusion and reaction discussed in Sec. 4.4.2 also apply to biological filters. However, the complexities inherent in the mixed-substrate–mixed-population problem have prevented development of analytical design methods for complicated systems such as a biological filter, where there are gradations in concentrations and population densities both locally (within the films) and globally (within the bed from top to bottom). Still, the qualitative understanding of diffusion and reaction in films gained in Sec. 4.4.2 is useful. We can anticipate that development of anaerobic regions deep within the film will produce gases which cause portions of the slime to detach spontaneously from the packing. The organisms sloughed off of trickling filters in this fashion are often called *humus*, and this solid debris must be removed in a clarifier following the biological filter. The average film thickness achieved by this spontaneous regulation is a complex function of operating parameters. In a well-operated filter, a film thickness of 0.35 mm is typical.

The usual ranges of loadings and efficiencies of trickling biological filters are listed in Table 14.8. For the conventional process, the hydraulic loadings indicated result in liquid residence times in the filter of 20 to 60 min. The high-rate option indicated in Table 14.8 is sometimes called a *flushing trickle filter*; because of the higher liquid rate used, slime buildup is limited. On the other hand, this mode of operation flushes out more humus, which must be eliminated in a subsequent settler.

We can draw a useful space-time analogy by taking a Langrangian trip through the trickling filter; i.e., suppose that we ride through the filter, from top

Table 14.8 Characteristics of high-rate and low-rate trickling filters†

Feature	Low-rate	High-rate
Hydraulic loading, gal/(day · ft^2)	25–100	200–1000
Organic loading, (lb BOD_5)/ (1000 ft^3 · day)	5–25	25–300
Depth, ft		
Single-stage	5–8	3–6
Multistage	2.5–4	1.5–4
Dosing interval	Intermittent	Continuous
Recirculation	Generally not included	Always included
Effluent	Highly nitrified, 20 mg/L of BOD_5	Not fully nitrified, 30 mg/L or more of BOD_5

† From L. G. Rich, *Environmental Systems Engineering*, p. 370, McGraw-Hill Book Company, New York, 1973.

to bottom, in a liquid drop. Then, as we travel through the packed bed, we shall see changes with time in the liquid composition as different components are removed by different microorganisms. In a sense, these changes are similar conceptually to the course of events in milk spoilage (see Fig. 14.1) and in plowed soil. As conditions in the liquid medium change, different species of microorganisms gain ascendency, causing further changes in the liquid and continuing the succession of different microbial populations.

Now let us transfer this observation to a fixed, or Eulerian, frame of reference. What was seen as a sequence in time by a drop moving through the bed is, for a filter in steady state, a pattern in space. Organisms best suited to utilize the feed sewage as a nutrient predominate in the top of the bed, as do tenaciously holding fungi and free-swimming ciliated protozoa. In the lower portions of the filter live stalked ciliate protozoa and nitrifying bacteria. Higher animals are also among the inhabitants of biological filters, with worms and fly larvae the major populations. These animals graze on the slime film which grows on the filter packing, and control of their populations is an important factor in filter operation.

The spatial segregation of organisms in biological filters provides an opportunity for each species to adapt fully to its immediate environment. Because of this, low-rate biological filters usually provide clearer, more highly nitrified effluents than activated-sludge treatment does. Also, experience has shown that filters are less sensitive to shock loads of toxic substances than activated-sludge processes. As indicated in Table 14.9, however, activated-sludge units are in some respects superior to trickling biological filters. Choice between the two processes requires careful consideration of waste characteristics, costs, and environmental standards. In some cases, the optimum plant design involves application of both methods. Other options are provided by choice of clarifier and recycle arrangements, as discussed in further detail in the references.

Table 14.9 Comparison between trickling-filter and activated-sludge water treatment processes†

Item	Filter	Sludge tank
Capital costs	High	Low
Operating costs	Low	High
Space requirements	High	Low
Aeration control	Partial except in enclosed forced-draft types	Complete
Temperature control	Difficult due to large heat losses	Complete; heat losses small
Sensitivity to variations in applied feed concentrations	Fairly insensitive but slow to recover if upset	More sensitive but recovery quite rapid
Clarity of final effluent	Good	Not as good
Fly and odor nuisance	High	Low

† From J. W. Abson and K. H. Todhunter, p. 337 in N. Blakebrough (ed.), *Biochemical and Biological Engineering Science*, vol. 1, Academic Press, London, 1967.

Lagoon systems, while much more primitive than either the activated-sludge or trickling-biological-filter processes, provide another useful method for wastewater treatment. In *oxidation ponds*, which closely resemble natural aquatic-ecosystems, algae free oxygen through photosynthesis, thereby maintaining aerobic conditions for bacteria which consume organic wastes. Oxidation ponds are made quite shallow, typically 2 to 4 ft deep, to avoid establishment of anaerobic zones near the bottom. On the other hand, we find anaerobic conditions or an alternating temporal pattern of aerobic and anaerobic environment in *waste-stabilization lagoons*, which are used for wastes containing settleable solids. Additional data on these processes are provided in Rich [3].

14.4.7 Anaerobic Digestion†

Wastes containing substantial amounts of fermentable organic components can be treated biologically under anaerobic conditions. Although anaerobic treatment has broader applications, a major use arises in treatment of excess sludge solids (Figs. 14.10 and 14.13) produced in sewage-treatment processes. As we have discussed earlier, concentrated sludge is produced at several stages of these processes, including waste particulates removed in the screening and primary sedimentation units and also the sludge grown in the secondary biological oxidation process. This material is further concentrated, or thickened, often merely by settling, before disposal, frequently with anaerobic biological digestion as one of the steps in the process.

† This section is based on the discussion of mathematical modeling of anaerobic digestion in S. P. Graef and J. F. Andrews, *CEP Symp. Ser.*, [136] **70**: 101–127 (1974).

A simplified schematic of the overall mechanism of anaerobic digestion, which involves a multitude of microbial species, is

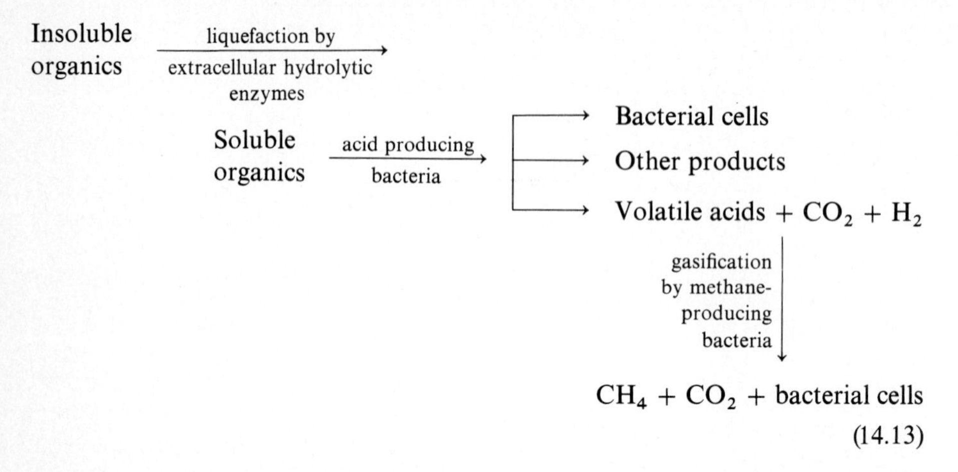

(14.13)

In the first step, large solid-sludge material is solubilized or dispersed by extracellular enzymes synthesized by a broad spectrum of bacteria. Among the enzymes found in anaerobic digesters are proteolytic, lipolytic, and several cellulolytic enzymes. Since solids do not build up in anaerobic digesters, these solubilization reactions apparently proceed fast enough to prevent this step from limiting the rate of the overall reaction sequence (14.13).

Experimental studies of the next portion of the digestion reaction, namely bacterial synthesis of short-chain fatty and volatile acids from soluble organic material, reveal that these steps also occur at a relatively rapid rate. The organisms responsible for these transformations, called *acid formers* for obvious reasons, are facultative anaerobic heterotrophs which function best in a range of pH from 4.0 to 6.5. While the major product of this step is acetic acid, propionic and butyric acid are also produced.

Acetic acid is the most important substrate for the final reaction of the sequence, since about 70 percent of the methane produced has been shown to derive from that component. This gasification step of the process involves methane bacteria, which are strict anaerobes. A narrower range of pH, from 7.0 to 7.8, is optimal for these organisms, which, although difficult to isolate in pure culture, thrive in mixed culture in a properly operated digester. Existing evidence suggests that this conversion of volatile acids to CH_4 and CO_2 is the rate-limiting step in the series of reactions shown in Eq. (14.13).

Figure 14.20 is a schematic diagram of an anaerobic digestion unit. Mixing is provided to prevent high local concentrations of acids from developing. In order to maintain a satisfactory environment for both acid formers and the methane bacteria, digesters are operated at a pH around 7. Also indicated in Fig. 14.20 is an external heat exchanger, which provides an above-ambient temperature in the vessel. At present, the usual practice is operation at the temperature in the mesophilic range which maximizes the rate of sludge digestion, about 90 to 100°F.

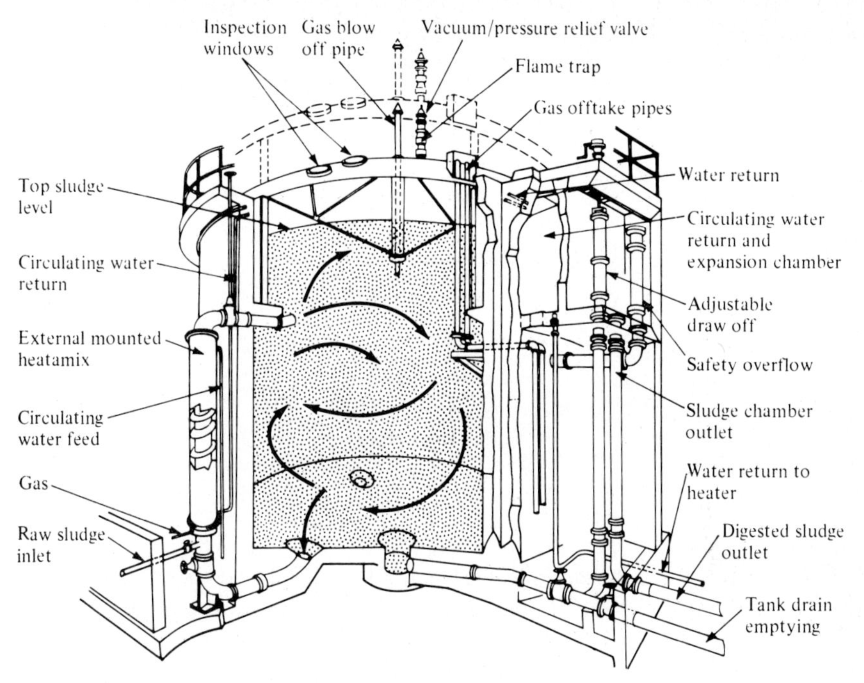

Figure 14.20 Schematic diagram of an anaerobic digestion unit. *(Reprinted from B. Atkinson, "Biochemical Reactors," p. 24, Pion Ltd., London, 1974.)*

There is limited evidence that more rapid digestion is possible in the thermophilic range, with largest rates at about 130°F. Operation at this temperature level is relatively rare: higher energy cost is one factor which weighs in favor of the mesophilic range of temperatures. The solids-residence time required for anaerobic sludge digestion at mesophilic temperatures is 10 to 30 days in a well-agitated unit.

Fortunately, the anaerobic digestion process produces a fuel which can be used to reduce energy costs for the wastewater-treatment plant. In some instances, the methane produced by anaerobic waste treatment is used outside the plant for heating and power. The gas mixture produced by anaerobic digestion, which is collected from the top of the unit as indicated in Fig. 14.20, is roughly 65 to 70 percent methane, with CO_2 comprising most of the remainder. Hydrogen sulfide, produced by sulfate-reducing bacteria, is present in small amounts, as are H_2 and CO. The digester off-gas has a heating value of 650 to 750 Btu/ft^3 and is produced with a yield of 12 to 18 std ft^3 per pound of organic matter decomposed. Since this gas has a substantially lower Btu value than natural gas (about 1000), it has not been such an attractive product in areas where natural-gas supplies are plentiful. With rising energy costs, however, increasing attention is being devoted to anaerobic digestion as a potential fuel source, albeit after the necessary H_2S removal.

Table 14.10 Effects of anaerobic digestion on sewage sludge†

Fraction	Raw sludge	Digested sludge
Ether-soluble	34.4	8.2
Water-soluble	9.5	5.5
Alcohol-soluble	2.5	1.6
Hemicellulose	3.2	1.6
Cellulose	3.8	0.6
Lignin	5.8	8.4
Protein	27.1	19.7
Ash	24.1	56.0

† From J. W. Abson and K. H. Todhunter, p. 339 in N. Blakebrough (ed.), *Biochemical and Biological Engineering Science*, vol. 1, Academic Press, London, 1967.

As a result of anaerobic digestion, the sludge is in much better condition for further treatment. First, the organic sludge solids are reduced by as much as 50 to 60 percent. Moreover, the composition of the sludge is profoundly changed (see Table 14.10). Because of these alterations, digested sludge is much less putrefactive than raw sludge, and it is also easier to dewater. After dewatering, which is often accomplished with rotary-drum vacuum filtration, the sludge is dried further, then spread on land as a fertilizer, dumped, or incinerated. Figure 14.8 indicates some of the other options for sludge treatment, and others are discussed in the references.

14.4.8 Mathematical Modeling of Anaerobic-Digester Dynamics

Despite production of a gaseous fuel and residual solids with fertilizer value, anaerobic digesters have a bad reputation because they are prone to operational problems. Many digester failures have been documented, with the major known causes classified as hydraulic, organic, and toxic overloading. In the first case, the dilution rate exceeds the growth rate of digester microbes, which are then washed out of the unit. High organic substrate concentrations, on the other hand, cause buildups of volatile acids. Methane bacteria are inhibited, and the digester "sours" as pH falls and failure ensues. When substances toxic to the methane bacteria enter the digester in sufficient amounts, washout of this population causes failure of the overall process.

Since improved operational practices could be of great benefit to enhanced success of anaerobic-treatment processes, there is an obvious incentive for studying the dynamic characteristics of these units and attempting to develop suitable control strategies. We shall review in this section a very interesting mathematical model of anaerobic digestion which was developed by Graef and Andrews. Also, some of the control schemes which they studied will be examined in Example 14.3. This model provides a fitting climax for this text because it involves intricate interplays between physical, chemical, and biological factors. It therefore exemplifies a synthesis of classical engineering skills with basic biological knowledge to achieve a biochemical engineering analysis.

We have already mentioned that conversion of volatile acids by the methane bacteria appears to be the rate-limiting step in the sequence of biological reactions (14.13). Assuming that all volatile acids can be represented as acetic acid and that the composition of methane bacteria can be approximated by the empirical formula $C_5H_7NO_2$, Graef and Andrews [7] develop the following stoichiometry for the gasification reaction:

$$CH_3COOH + 0.032\,NH_3 \longrightarrow 0.032\,C_5H_7NO_2 + 0.92\,CO_2 + 0.92\,CH_4 + 0.096\,H_2O \quad (14.14)$$

The limiting substrate for this reaction is presumed to be the nonionized volatile acids. The concentration of the nonionized form, HS, differs in general from the total concentration s due to the ionization reaction

$$HS \rightleftharpoons S^- + H^+ \qquad K_a \quad (14.15)$$

where S^- is used as a shorthand notation for ionized substrate. Since $-\log K_a \equiv pK_a$ is 4.5 and digesters operate at pH's above 6, we know that almost all the acid is in the ionized form, so that

$$s^- \approx s \quad (14.16)$$

and

$$(hs) \approx \frac{s(h^+)}{K_a} \quad (14.17)$$

In order to incorporate the known inhibitory effects of high substrate concentration in the model, Graef and Andrews [7] modified the Monod expression for specific growth rate to the form

$$\mu = \mu_{max}\left[\frac{1}{1 + K_s/(hs) + (hs)/K_i}\right] \quad (14.18)$$

which is familiar from our kinetic studies in Chaps. 3 and 7. Also included in the bacterial growth rate is a death rate, which is presumed to be first order in toxin concentration [tox][†]:

$$r_D = -k_T[\text{tox}] \quad (14.19)$$

Based on the available data and Graef and Andrews' estimates, the parameters appearing in these rate expressions have the following values: $\mu_{max} = 0.4\ \text{day}^{-1}$, $K_s = 0.0333$ mmol/L, $K_i = 0.667$ mmol/L, and $k_T = 2.0\ \text{day}^{-1}$.

If constant yield coefficients for the ratios (cell mass)/(limiting substrate), $Y_{X/S}$; CO_2/(cell mass), $Y_{CO_2/X}$; and CH_4/(cell mass), $Y_{CH_4/X}$, are assumed, the material balances on the substrate and biological phase in the digester take the form indicated at the bottom of Fig. 14.21. The quantities R_B and Q_{CH_4} denote the rate of CO_2 and methane formation, respectively, due to biological gasification. For

[†] To avoid confusion, the concentration of some components are indicated within brackets; others are denoted as usual by lower case italics.

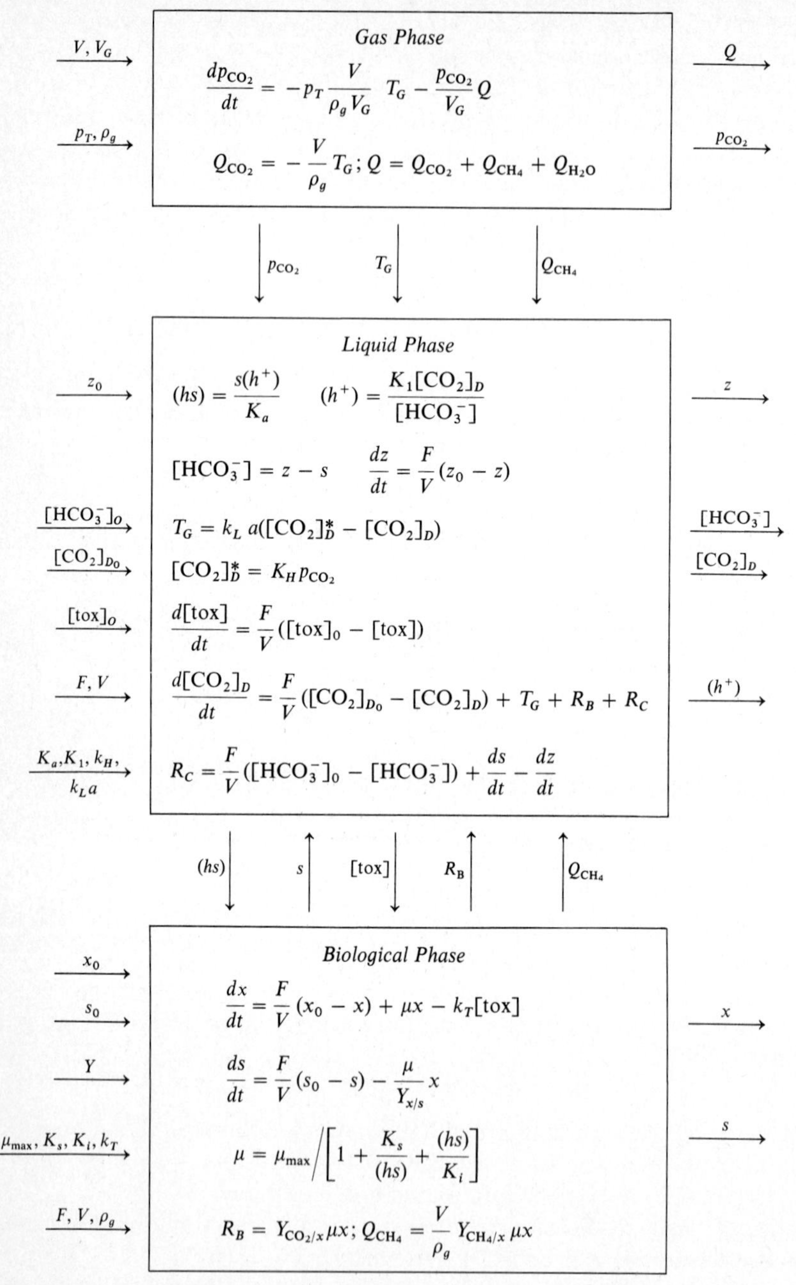

Figure 14.21 Summary diagram of the mathematic model of anaerobic digestion. The arrows indicate information flow between subsystems and interactions with the external environment. *(Reprinted from S. P. Graef and J. F. Andrews, "Mathematical Modeling and Control of Anaerobic Digestion" in G. F. Bennett (ed.), "Water–1973" CEP Symp. Ser. No. 136, vol. 76, p. 101, 1974.)*

conditions typical of anaerobic digesters, the gas density ρ_G is 0.0389 mol/L, and $Y_{CH_4/X}$ and $Y_{CO_2/X}$ are both 28.8 mol/mol. As Fig. 14.21 shows, methane is quite insoluble, so that all the methane formed enters the gas phase.

This is not the case for CO_2, however, which exists in the liquid phase in two forms as well as in the gas phase. The rate of mass transfer of CO_2 from the gas to liquid phases is given by the familiar form from Chap. 8:

$$T_G = k_L a([CO_2]_D^* - [CO_2]_D) \tag{14.20}$$

where $[CO_2]_D$ is the concentration of dissolved CO_2 and $[CO_2]_D^*$ is this concentration at equilibrium. From Henry's law

$$[CO_2]_D^* = K_H p_{CO_2} \tag{14.21}$$

where p_{CO_2} is the partial pressure of CO_2 in the gas phase. Graef and Andrews suggest values of 100 day^{-1} and 3.25×10^{-5} mol/(L/mmHg) for $k_L a$ and K_H, respectively.

Another pathway by which CO_2 can appear in the liquid phase is bicarbonate association, according to

$$HCO_3^- + H^+ \rightleftharpoons H_2O + CO_2 \qquad \text{equilibrium constant} = 1/K_1 \tag{14.22}$$

If we let R_C denote the rate of this reaction, the mass balance on bicarbonate in the liquid phase is

$$V\frac{d[HCO_3^-]}{dt} = F([HCO_3^-]_0 - [HCO_3^-]) - VR_C \tag{14.23}$$

We can obtain an independent expression for $d[HCO_3^-]/dt$ using the requirement of electroneutrality, which can be written

$$[H^+] + c = [HCO_3^-] + [S^-] + [OH^-] + a + 2[CO_3^{2-}] \tag{14.24}$$

where c is the total cation concentration, including contributions of calcium, sodium, magnesium, and ammonium, and a is the total anion (chlorides, phosphates, sulfide, etc.) concentration. For a digester operating in the normal pH range of 6 to 8, $[H^+]$, $[OH^-]$, and $[CO_3^{2-}]$ are negligible, and Eq. (14.24) becomes

$$z = [HCO_3^-] + s \tag{14.25}$$

where z, the net cation concentration, is defined by

$$z = c - a \tag{14.26}$$

and s^- has been replaced by s, as discussed earlier. If sulfide concentration is not too large, z corresponds approximately to ammonium-ion concentration. We shall suppose that the Z mass balance may be taken as

$$\frac{d(z)}{dt} = \frac{F}{V}(z_0 - z) \tag{14.27}$$

Now, differentiating Eq. (14.25) with respect to time gives

$$\frac{d[HCO_3^-]}{dt} = \frac{d(z)}{dt} - \frac{d(s)}{dt} \tag{14.28}$$

Eliminating $d[HCO_3^-]/dt$ from Eqs. (14.28) and (14.23) gives

$$R_c = \frac{F}{V}([HCO_3^-]_0 - [HCO_3]) + \frac{d(s)}{dt} - \frac{d(z)}{dt} \tag{14.29}$$

This rate is included in the material balance for liquid-phase CO_2, as indicated in Fig. 14.21.

The gas-phase mass balances are relatively straightforward, as Fig. 14.21 shows. Notice how the various gas-production rates computed elsewhere in the model are used to obtain the total effluent flow rate.

Next let us review some of the simulation results obtained using the model. Table 14.11 lists the parameter values and standard steady-state conditions employed in these calculations. Simulation of a batch digester, which is achieved simply by setting $F = 0$ in the continuous model, shows that increasing initial organism concentration, increasing initial pH, and decreasing initial substrate concentration all lead to smaller batch digestion times. The same trends have been observed in the field.

Another situation considered by Graef and Andrews is digester start-up. They showed that the model predicts: (1) a decreased time for start-up if initial pH is increased or the feed-sludge concentration is increased, (2) failure if initial pH or feed-sludge concentration is too low, and (3) alleviation of the possibility of digester failure during start-up by slowly raising the digester loading to its final value. Again, field units show similar characteristics.

The model summarized above also exhibits the three modes of failure discussed at the start of this section. Simulation results for two of these cases, organic and hydraulic overloading, are displayed in Figs. 14.22 and 14.23, respectively. In all these calculations, a step change in a process input occurs at $t = 1$ day. If the magnitude of this change is sufficiently small, say less than 35.7 g/L in feed-substrate concentration and less than 2.5 L/day in the case of a hydraulic-flow-rate disturbance, the digester attains a new, stable steady state in the vicinity of the original operating state. Substantially larger step changes in these inputs cause the process to run away: pH and methane product drop precipitously while the effluent substrate (volatile acids) concentration rapidly climbs. Computed results for a pulse of a toxic agent also reveal digester failure if the toxic overloading is too large.

Consequently, we see that the model outlined above characterizes the qualitatively dynamic features of anaerobic digestion quite well. Further study of this model and subsequent refinements will greatly aid our understanding of these complex processes and aid in development of design and operational strategies for improved performance. In fact, several suggestions for control design and possible flags for impending failure are discussed further in Andrews et al. [5–7].

Table 14.11 Standard steady-state conditions and parameter values used in simulation of anaerobic-digester dynamics[†]

Influent variables		Steady state conditions		Parameters and constants	
Term	Value	Term	Value	Term	Value
s_0	167 mM as acetic	s	2.0 mM as acetic	V	10 L
	10 g/L as acetic		120 mg/L as acetic	V_G	2 L
z_0	50 meq/L	(hs)	0.0112 mM as acetic	μ_{max}	0.4 day^{-1}
F	1.0 L/day		0.672 mg/L as acetic	p_T	760 mmHg
$[CO_2]_{D_0}$	0 mM	$[HCO_3^-]$	48.0 mM	D	25.7 L/mol
x_0	0 mM		24,000 mg/L as $CaCO_3$	I	0.1 mol/L
$[HCO_3^-]_0$	0 mM	x	5.28 mM as $C_5H_7NO_2$	k_La	100 day^{-1}
$[tox]_0$	0 mM		597 mg/L	K_s	0.0333 mM/L
		z	50 meq/L	K_i	0.667 mM/L
		$[CO_2]_D$	9.0 mM	$K_1^{‡}$	6.5×10^{-7}
		p_{CO_2}	273 mmHg	$Y_{X/S}$	0.032 mol organisms/mol substrate
		μ	0.1 day^{-1}		
		pH	6.91	$Y_{CO_2/X}$	28.8 mol CO_2/mol organisms produced
		Q_{dry}	6.35 L/day		
		Q_{CH_4}	3.91 L/day	$Y_{CH_4/X}$	28.8 mol CH_4/mol organisms produced

[†] From S. P. Graef and J. F. Andrews, *CEP Symp. Ser.*, [136] **70:** 130, 1974.
[‡] At 38°C; $I = 0.1$.

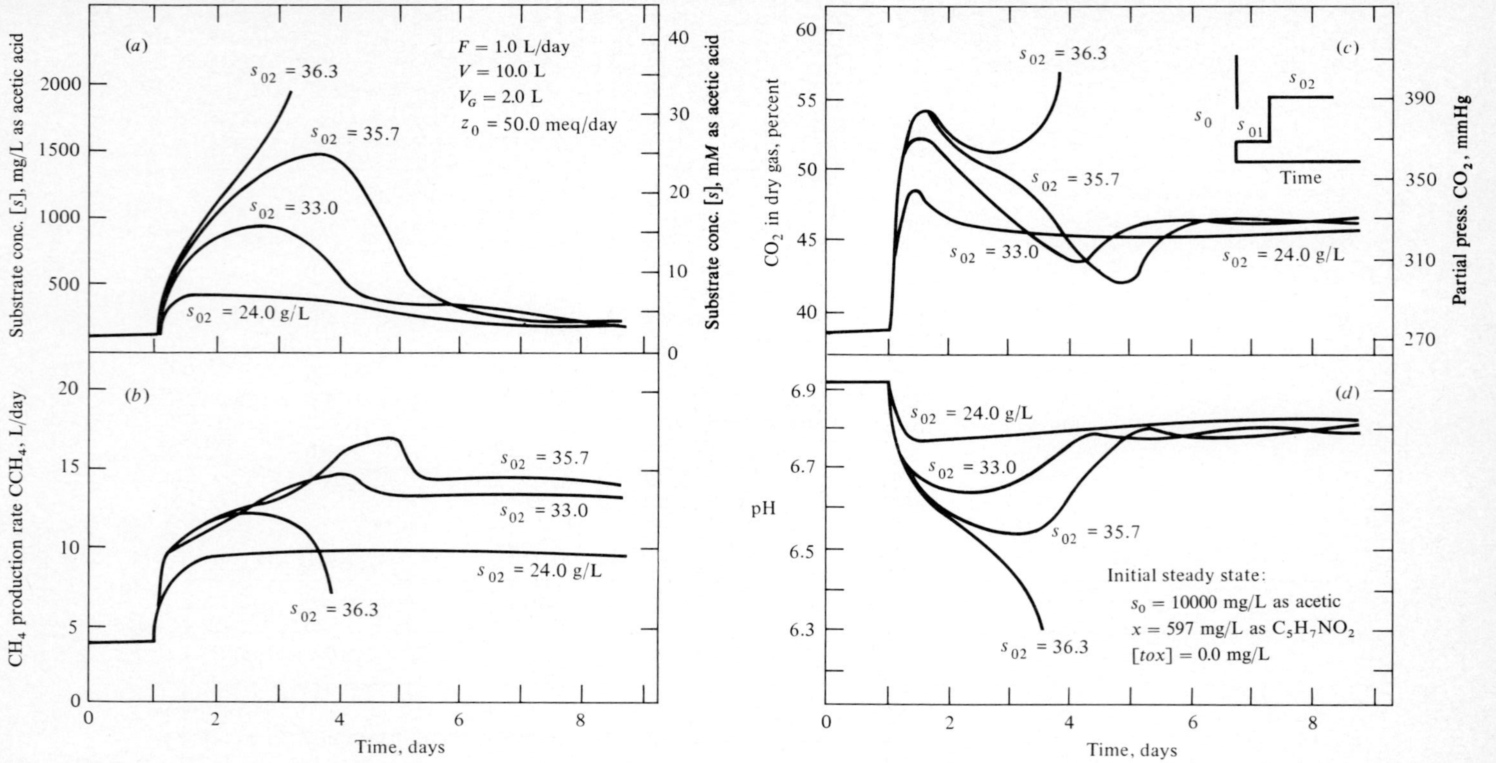

Figure 14.22 Different sized step changes in feed substrate concentrations s_0 to an anaerobic digester produce different patterns of dynamic response. For a step change to $s_0 = 36.3$ g/L., the process runs away. *(Reprinted from S. P. Graef and J. F. Andrews, "Mathematical Modeling and Control of Anaerobic Digestion" in G. F. Bennett (ed.), "Water–1973" CEP Symp. Ser. No. 136, vol. 76, p. 101, 1974.)*

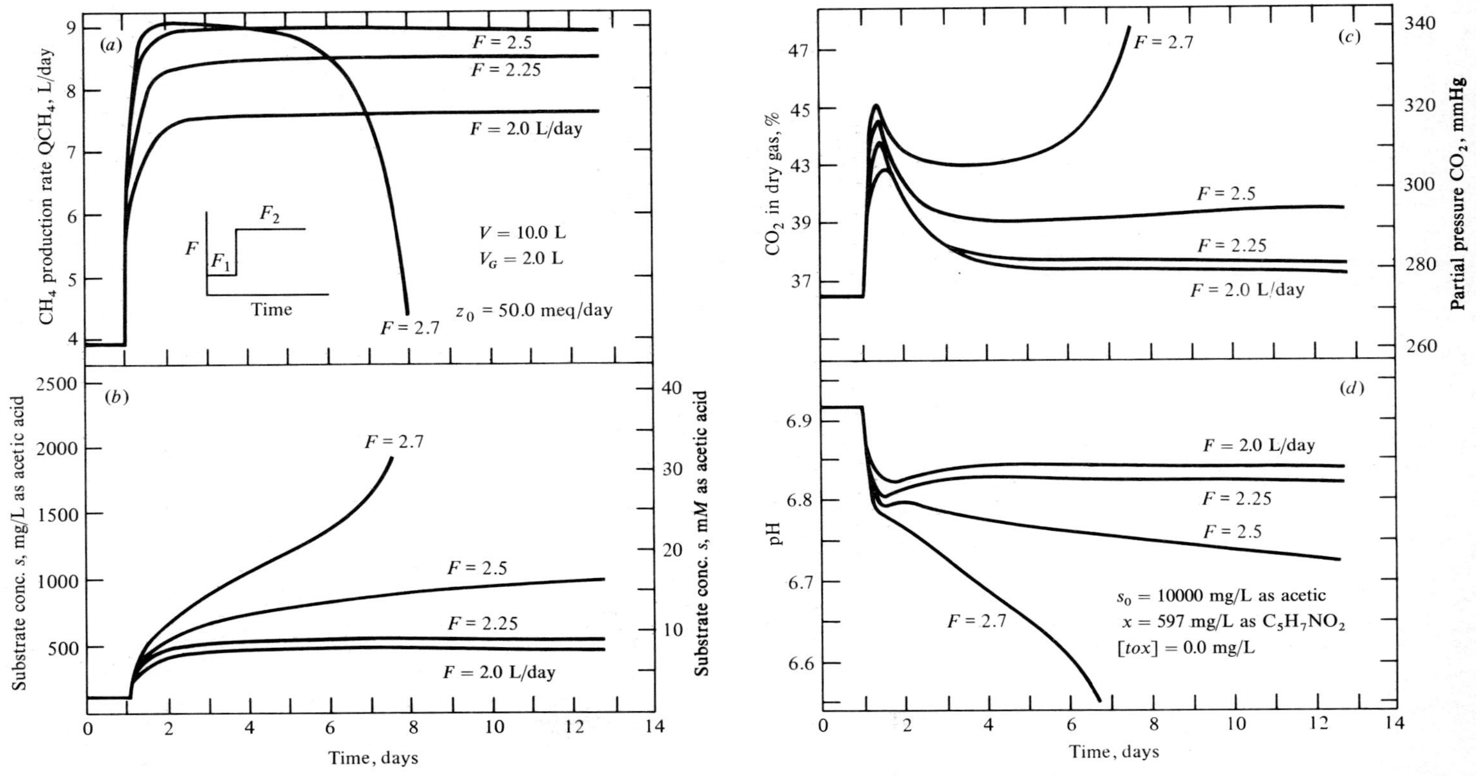

Figure 14.23 Calculated anaerobic digester response to step changes in hydraulic flow rate $F(F_1 = 1$ L/day). *(Reprinted from S. P. Graef and J. F. Andrews, "Mathematical Modeling and Control of Anaerobic Digestion," in G. F. Bennett (ed.), "Water–1973" CEP Symp. Ser. No. 136, vol. 76, p. 101, 1974.)*

One of the most interesting control schemes studied by these authors is reviewed in the following example. We should emphasize that a dynamic model which retains the known essential dynamic features of the real process is an indispensable ingredient for such investigations of controller design.

Example 14.3: Simulation studies of control strategies for anaerobic digesters The following four methods of feedback control of anaerobic digesters were considered by Graef and Andrews: (1) gas scrubbing and recycle, (2) base addition, (3) organism recycle, and (4) flow reduction. Since the first approach is the most unusual, and since it uses the ionic-equilibria portions of the model, we shall concentrate on control via gas scrubbing and recycle in this discussion. [See Prob. 14.9 for control via (3) or (4).]

A schematic diagram illustrating this control configuration is provided in Fig. 14E3.1. As indicated there, some of the effluent gas from the digester is scrubbed to remove CO_2, and then this gas is recycled to the digester. How much CO_2 is removed by passage through this loop is determined by

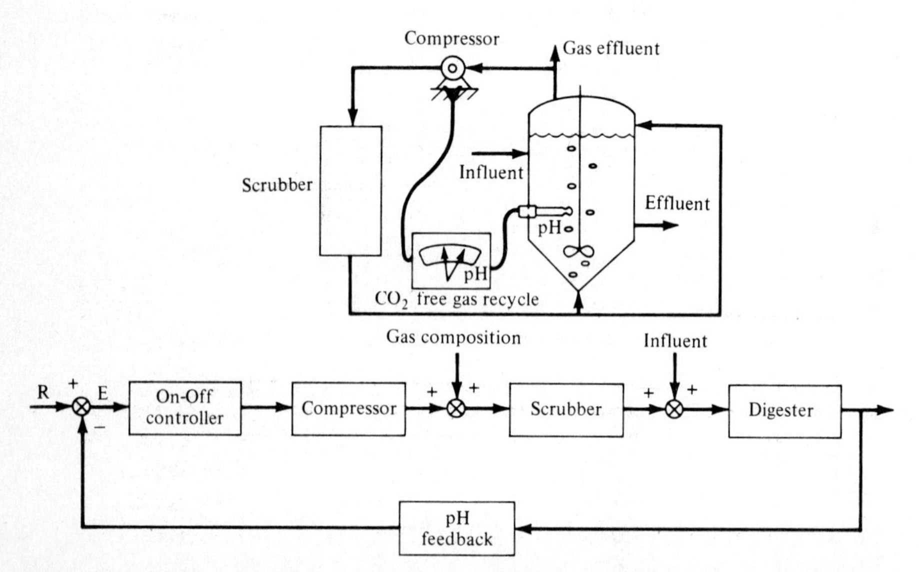

Figure 14E3.1 Off-gas scrubbing and recycle control system. *(Reprinted from S. P. Graef and J. F. Andrews, "Mathematical Modeling and Control of Anaerobic Digestion," in G. F. Bennett (ed.), "Water–1973" CEP Symp. Ser. No. 136, vol. 76, p. 101, 1974.)*

the pH within the digester. When digester pH falls below a threshold value, gas flow through the CO_2 scrubbing loop is increased. Removing CO_2 from the digester gas phase will cause the carbonic acid concentration in the liquid to drop, thereby creating an increase in pH.

This rather subtle approach to pH control has several potential advantages relative to conventional techniques, which require addition of a base. If strong alkali is added, it may create, at least temporarily, localized regions of very high pH without effectively raising pH in the total vessel. Moreover, metal cations in the alkali can be toxic to the microbial population of the digester. Lime addition, another possibility, has the problem of creating an insoluble carbonate precipitate.

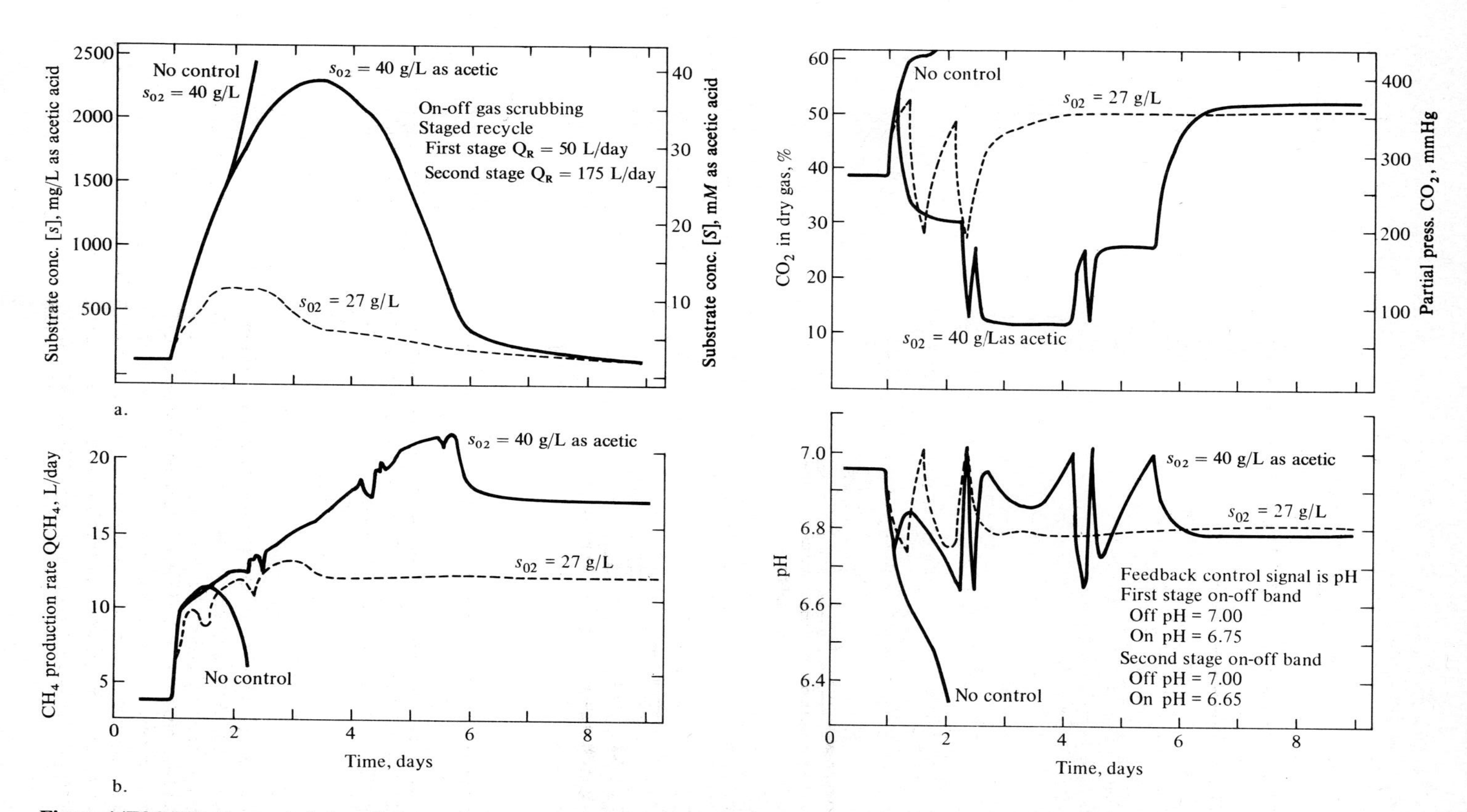

Figure 14E3.2 Effectiveness of the off-gas scrubbing control policy in counteracting an organic overload. *(Reprinted from S. P. Graef and J. F. Andrews, "Mathematical Modeling and Control of Anaerobic Digestion," in G. F. Bennett (ed.), "Water–1973" CEP Symp. Ser. No. 136, vol. 76, p. 101, 1974.)*

After gaining experience with this control system, Graef and Andrews concluded that a multilevel control provided best results. In this scheme, there are two overlapping on-off bands for the gas-recycle flow rate Q_R:

$$Q_{R1} = 0 \qquad \text{pH} < 6.75$$

$$Q_{R1} = Q_1 \qquad 6.75 \leq \text{pH} \leq 7.00$$

$$Q_{R1} = 0 \qquad \text{pH} > 7.00$$

Second stage Q_{R2}:

$$Q_{R2} = 0 \qquad \text{pH} < 6.65$$

$$Q_{R2} = Q_2 \qquad 6.65 \leq \text{pH} \leq 7.00$$

$$Q_{R2} = 0 \qquad \text{pH} > 7.00$$

with

$$Q_R = Q_{R1} + Q_{R2}$$

Figure 14E3.2 shows the response of the digester with gas-scrubber feedback control to organic overloads. Notice that the controlled system survives a step increase in feed substrate of 40 g/L, while this overload is more than sufficient to cause the uncontrolled digester to fail. Unfortunately, this mode of control could not prevent failure when toxic or hydraulic overloading occurred. The successes and shortcomings of other control strategies are summarized in Table 14E3.1. From the simulation results indicated there, it appears that a multivariable control scheme is necessary for stable digester operation in the face of all three kinds of overload. With such a control system, digester variables, e.g., pH and methane production rate, would be measured, and in response to all these measurements more than one variable (such as gas and sludge recycle and residence time) would be manipulated.

Table 14E3.1 Summary of control strategies for anaerobic digester control†
In each case, the manipulated variable was switched in an on-off fashion to maintain the measured variable near its set point value

		Control effective in preventing digester failure in the event of:		
Measured variable	Manipulated variable	Organic overloading	Toxic overloading	Hydraulic overloading
Digester pH	Flow rate through gas-recycle line with CO_2 scrubbing	Yes	No	No
	Flow rate of caustic added to digester	Yes	No	No
	Liquid flow rate in and out of digester	Uncertain	Uncertain	Yes
Rate of methane production	Rate of recycle of sludge collected from digester liquid effluent	No	Yes	No

† Adapted from S. P. Graef and J. F. Andrews, *CEP Symp. Ser.*, [136] **70:** 101, 1974.

14.4.9 Anaerobic Denitrification

Nitrogen reduction is accomplished under anaerobic circumstances by a variety of bacteria which can consume organics and utilize nitrate and nitrite as electron acceptors. Ultimately, two separate paths occur:

(*a*) In assimilatory nitrate reduction, some nitrogen becomes ammonia and is incorporated into cell biomass.
(*b*) In dissimilating reduction, molecular nitrogen is the final product.

As not all bacteria can effect both conversions, two independent reactions can be written:

$$NO_3^- + \text{organic} \longrightarrow \text{cells} + NO_2^- + CO_2$$

$$NO_2^- + \text{organic} \longrightarrow \text{cells} + N_2 + CO_2$$

Nitrite levels observed are very low, so a single combined expression yielding cells and N_2 may often suffice.

The organic may be added to give a level sufficient to effect the desired conversion of nitrate. Methanol has previously been used as a carbon and energy source with a relatively low cell yield, but cost increases have made this organic substrate less desirable.

Packed-bed lab denitrification studies have shown that the dissolved NO_3^- and NO_2^- levels vary with distance according to a simple sequential reaction pattern, leading to a high N removal efficiency as characterizes a plug flow operation. Cell growth eventually lead to reactor plugging; thus process design improvements are needed in this emerging area of denitrification conversions.

14.4.10 Phosphate Removal

Phosphorus is typically present in raw wastewaters at concentrations near 10 mgP/L, including orthophosphate, dehydrated orthophosphate (polyphosphate) and organic phosphorus. Biological treatment processes result in conversion of most P to the orthophosphate forms ($H_2PO_4^-$, HPO_4^{2-}, PO_4^{3-}). These latter forms may be removed by precipitation if effluent guidelines on phosphorus require low phosphate values relative to the influent. Precipitating agents may include calcium or aluminum:

$$3HPO_4^{2-} + 5Ca^{2+} + 4OH^- \longrightarrow Ca_5(OH)(PO_4)_3 + 3H_2O$$

$$HPO_4^{2-} + Al^{3+} \longrightarrow AlPO_4 + H^+$$

When lime is used as a Ca source, precipitation normally follows biological treatment. With alum (or iron) as precipitant, treatment may be effected in the activated sludge operation itself or in a primary settling basin (Fig. 14.8).

PROBLEMS

14.1 Nitrification design (*a*) For Schroeder's Example 14.2 (pp. 939–940), what is the effluent BOD if cell age is determined from ammonia effluent requirements, i.e., if $\theta_c = 9$ days? (*b*) The Monod form for carbon concerns the rate of substrate utilization, $r_s = \mu_s s/(K_m + s)$. Show, using Eq. (7.25), that $Y_{obs}\mu_s = \mu_m$ and that Y' (Eq. (7.25)) and Y_{obs} are the same variable. (*c*) How would you rationalize the behavior (Example 14.2) $Y'(=Y_{obs}) = a/(1.0 + b\theta_c)$?

14.2 Industrial vs. waste-treatment reactors A comparison between industrial and waste treatment microbial reactors is summarized in Table 14P2.1.

(*a*) Comment on the characteristic similarities and differences, including when the critical design criteria are likely to be similar or different.

(*b*) $k_l a$ has been observed to vary with $(P/V)^n$, where $n = 1.33$ (sewage). 0.72 (yeast broth), and 0.5 (endomyces or mycelia fermentation), and 0.4 (hydrocarbon fermentation). Attempt to rationalize these exponents. How do they affect the result of part (*a*)?

Table 14P2.1†

Property	Waste treatment	Industrial fermentation
Temp, °C	10–30	20–50
Variation	±5	±0.5
Rheology	Newtonian	Newtonian, non-Newtonian
Viscosity variation, cp	1–10	1–1000
Substrate concentration, g/L	0.1–5	5–40
Loading (BOD), ppm	100–5000	5000–40,000
Reactor size, gal	50,000–400,000	250–40,000
Power per unit volume, hp/10^3 gal	0.05–0.5	1–20
Growth rate, h^{-1}	0.05–0.1	0.1–1.0
Loading, lb BOD/1000 ft^3	10–1000	

† C. L. Cooney and D. I. C. Wang, *Biotech. Bioeng. Symp.* 2, p. 63, 1971.

14.3 Bifunctionalism in bacterial synergism When two distinct species carry out distinctly different chemical transformations, the result may be termed bifunctionalism, by analogy with bifunctional catalysis involving two distinct kinds of catalytic sites, e.g., dehydrogenation and isomerization. An example in the microbial context is

$$\begin{array}{ccc} \text{Arginine} & \xrightarrow{S.\ faecalis} & \text{Ornithine} \\ -CO_2 \Big\downarrow E.\ coli & & -CO_2 \Big\downarrow E.\ coli \\ \text{Agmatine} & & \text{Putrescine} \end{array}$$

(*a*) Assuming Monod forms for each of the three conversions, write down the system equations for a chemostat.

(*b*) At low arginine levels, sketch the system outlet behavior as the dilution rate is increased continuously.

(*c*) If μ_{max} for *E. coli* is the same for both steps but larger than μ_{max} for *S. faecalis*, what should be the relative size (V_1/V_2) to maximize putrescine concentration in two CSTRs in series? (fixed total volume). Repeat for μ_{max} (*E. coli*) < μ_{max} (*S. faecalis*).

(*d*) What design strategy, assuming a nonsterile feed, minimizes putrescine levels? Which strategy, (*c*) or (*d*), would please the neighbors?

14.4 Stability of activated-sludge interactions Apply the Feinberg-Horn-Jackson analysis to the calculated results for the activated-sludge pairwise species interactions *a*, *b*, *c*, and *d* in Fig. 14.16. Do the Feinberg-Horn-Jackson predictions agree with the explicit calculations from Curds' equations?

14.5 Activated-sludge reactor The inflow to an activated sludge reactor has a 5-day BOD of 220 mg/L; the outlet must be below 15 mg/L (Table 14.2). Given the following conditions, complete the design of an activated-sludge reactor as indicated below:

$$\text{Active solids concentration} = 3000 \text{ mg/L}$$

$$\text{Recycle ratio} = 0.46 \qquad F = 4.5 \times 10^6 \text{ gal/day}$$

Y, $\mu_{\max}$, K_s, k_d from Table 14.4. All nutrients except BOD are in excess.

(*a*) Calculate

(1) Sludge age. (2) reactor volume needed, (3) concentration of active solids in recirculation line, and (4) daily aeration rate, assuming 7.5 percent oxygen utilization. Equation (14.5) indicates that minimal V can be accomplished by increasing x_r/x_a. Clearly achievement of increased x_r demands a larger secondary sedimentation unit V_s. While settling of sludge is a complex process which proceeds in several stages (see, for example, Ref. 16 and Chapter 11), we can approximate the return active solids concentration by

$$x_r = x_{r,\max}(1 - e^{-\beta t}) + x_a e^{-\beta t} \qquad x_{r,\max} = x_r \text{ at } t = \infty$$

where $t(\equiv V_s/F)$ is the mean settling time elapsed and β the characteristic constant > 0.

(*b*) Evaluate x_r(optimum) if all operating costs are divided between the digester and the settler, and the ratio of such costs (1 ft^3 digester/1 ft^3 settler) is γ ($1.0 \le \gamma \le 10.0$). (Neglect construction costs.)

14.6 Substrate and biomass structure in activated sludge design Eqs. (14.1)–(14.4) present an analysis of activated sludge design which lumps all substrate and biomass into a single Monod-plus-endogenous metabolism model. R. B. Paterson and M. M. Denn ("Computer-aided Design and Control of an Activated Sludge Process," *The Chem. Eng. Jl.*, **27**: B13, 1983) presented a model in which substrates were carbonaceous, organic-N, and nitrite-N which were converted by, respectively, corresponding carbon, organic-N, and nitrite-N biomasses (x_C, x_O, x_N). Base operating plant conditions were specified as follows: $F = 2.0 \times 10^4$ m^3/day, $\alpha = 0.15$, $\beta = 0.0015$, $s_{Ca} = 16.8$ mg/L (5 day BOD), $s_{Co} = 1811$ mg/L, $s_{Oo} = 250$ mg/L, $s_{No} = 0.5$ mg/L. Data for Monod plus endogeneous metabolism kinetics:

$$[\mu_{\max}(\text{day}^{-1}), k_e\,(\text{day}^{-1}), K\,(\text{mg/L}), Y\,(\text{g/g})] = \text{carbonaceous}[5.0, 0.055, 100, 0.5],$$

$$\textit{Nitrosomonas}\ [0.33, 0.05, 1.0, 0.05]$$

$$\textit{Nitrobacter}\ [0.80, 0.05, 2.1, 0.02].$$

(*a*) Calculate the sludge age, θ_s, for the carbonaceous biomass.

(*b*) The secondary sedimentation unit is presumably equally selective on all biomass. Thus θ_s is the same for all conversions. Calculate the exit level concentrations for organic-N (s_{Oa}) and nitrite-N (s_{Na}).

(*c*) Calculate Vx and V from the data for carbonaceous biomass. Would calculation of V from similar data for the other biomass components give different V values? Why (not)? For the calculation, secondary sedimentation performance must be known. Plot your results as V (calc.) vs. (x_r/x_a) for the range $100 \le (x_r/x_a) \le 1000$.

14.7 Activated sludge plant controllability The ease of control of a waste treatment plant depends on the control objectives, the control and system dynamics, and the interactions between control objectives which may not be compatible. For example, sludge plant control may have as objectives (1) small variation in MLVSS [Eq. (14.8)] in the reactor and (2) maintenance of a relatively constant height of sludge "blanket" in the secondary clarifier (Fig. 14.10).

(*a*) Read the analysis of Paterson and Denn (see previous problem) and summarize the control strategies tested and the author's conclusions. Why are these two objectives "incompatible"?

(*b*) The first *control* objective was maintenance of an approximately constant MLVSS level, yet the *economic* analysis (same paper) based on steady-state operation found a very broad economic insensitivity to MLVSS variation. What important lesson for the engineer is apparent from these conclusions?

14.8 Two-tank anaerobic digestor Pohland and Ghosh [13] discuss the interaction between acid formers and methane formers in anaerobic digestors which is conceptually similar to that of Prob. 13.5. The detailed stoichiometry proposed is

Acid formation:

$$4C_3H_7O_2NS + 8H_2O \longrightarrow 4CH_3COOH + 4CO_2 + 4NH_3 + 4H_2S + 8H^+ + 8e^-$$

Methane formation:

$$8H^+ + 8e^- + 3CH_3COOH + CO_2 \longrightarrow 4CH_4 + 3CO_2 + 2H_2O$$

These workers suggest the possibility of an easier environmental control for the methane formers if two tanks in series are used. The growth parameters measured on a glucose (2 g/L) feed to the first reactor were:

Acid formers: μ_{max}(glucose) = 1.25 h^{-1} $\quad K_g = 22.5$ mg glucose/L $\quad Y_g = 0.2$

Methane formers: μ_{max}(acid) = 0.14 h^{-1} $\quad K_a = 600$ mg acetic acid/L $\quad Y_a = 0.05$

The glucose to acetic acid yield factor was 0.8, the inhibition constant for acid formers was 100 mg acetic acid/L, and endogeneous metabolism effects were negligible.

(*a*) Taking the kinetic forms for the two species to be those suggested in Prob. 13.5*b*, write out the steady-state equations governing two tanks.

(*b*) Prove that for dilution rates in the first tank in excess of 0.14 h^{-1}, the methane formers are washed out. What does this imply about using separate tanks to (largely) separate species in a food chain?

(*c*) Suppose the flow rate is greater than 0.14 h^{-1}. Solve for the outlet conditions in the first and second tanks when $V_2 = V_1$ and when $V_2 = 5V_1$ and sketch the outlet waste concentration (unconverted substrate) as a function of D. What trade-off is apparent in using increased D to favor species separation?

(*d*) In actual waste-treatment processes, settled cells and unconverted solid wastes would be recycled from each reactor exit to a settler-separator back to the same reactor inlet. Write out the substrates and species balances for such a two-tank system.

14.9 Anaerobic digester control Write out appropriate equations to describe anaerobic digester dynamics when (*a*) organism recycle or (*b*) flow reduction (increase) is used as the control method. Include a dynamic equation for the appropriate control variable, stating your rationale for its form.

14.10 Predator-prey kinetics A generalized predator-prey model formulated by Rozenweig and MacArthur [17] can be cast in the form

$$\frac{dx}{dt} = f(x) - y\phi(x) \qquad \frac{dy}{dt} = -ey + ky\phi(x)$$

where x and y are prey and predator densities, respectively.

(*a*) Compare this form with that of Lotka-Volterra [Eqs. (13.9), and (13.10)], and list as many physical or biological circumstances as you can conceive where the forms of $\phi(x)$ and $f(x)$ would differ from those implied by the Lotka-Volterra model. Be specific. What is the meaning of $\phi(x)$? Of $f(x)$?

(*b*) In experiments with the predator-prey system, *Paramecium* and *Didinium*, some 1973 results of Luckinbill are described by Smith (Ref. 34 of Chapter 13, p. 33): (in each case, the recorded densities for prey never approached their self-limiting density):

1. "... there is first a rapid increase of prey, followed by an increase in the predators, which capture all the prey and then starve."
2. "Prolonged coexistence was achieved by adding methyl cellulose to the medium, this renders the medium viscous and slows down the swimming of both species. However, there was still an oscillation of increased amplitude, ending in the extinction of the predator."
3. "Persistent coexistence was achieved by adding methyl cellulose and at the same time halving the concentration of food for the prey species."

Rationalize these observations. What further experiments would you perform to confirm or disprove any assumptions which you have made?

14.11 Nitrification in soil McLaren [14] has suggested that a first approximation to nitrification by soil microbes is given by considering a serial reaction:

$$NH_4^+ \xrightarrow{k_1} NO_2^- \xrightarrow{k_2} NO_3^-$$

In laboratory soil-enrichment cultures, the first and second conversions can each be associated with a single species, *Nitrosomonas* and *Nitrobacter*, respectively.

(*a*) At steady state, suppose that maintenance and cell replacement consume these nutrients by reactions which are zero order in NH_4^+ and NO_2^- with the rate constants k_1, k_2 above proportional to local microbial concentrations. If the respective biomass concentrations x_1, x_2 are independent of *depth* z, evaluate the vertical concentration profiles of NH_4^+, NO_2, and NO_3^- [normalized to NH_4^+ $(z = 0)$] if the downward fluid velocity is u and ion-exchange effects of nutrients with the soil are neglected. Plot the results for the cases $k_1/k_2 = 0.1$, 1.0, 10.0 using dimensionless distance (zk_1/u). Include the two situations: (1) nitrogen uptake for new cellular material can be neglected; (2) a separate zero-order reaction with rate constant βx_1 or βx_2, respectively, allows for NH_4^+ utilization by *each* species for maintenance and replacement.

(*b*) If NH_4^+ is indeed the limiting nutrient for *Nitrosomonas* maintenance and replacement, evaluate from part (*a*) the dimensionless depth at which the assumption of $x_1 = \text{const}$ everywhere fails. Experimentally, the NH_4^+ oxidizer profile has been measured in one instance to have the following values: $2 \times 10^5/\text{cm}^3$ (surface water), 2×10^5 per gram of soil (0 to 1 cm depth), 2×10^3 per gram of soil (1 to 3 cm depth), 2×10^2 per gram of soil (3 to 5 cm depth) [1, and references therein].

(*c*) The assumption of constant biomass level is an obvious convenience. Show that the model above does not allow determination of depth variations of x_1 and x_2.

(*d*) Discuss how you would devise a model for unambiguous prediction of *Nitrosomonas* and *Nitrobacter* profiles, keeping in mind the previous chapter and a quotation from 1923:

> [It] is shown that soil is normally inhabited by a very mixed population of organisms, varying in size from the smallest bacteria up to nematodes and others just visible to the unaided eye, on to larger animals, and finally earthworms, which can be readily seen and handled. These organisms all live in the soil, and therefore must find in it the conditions necessary for their growth [15].

14.12 Aerobic pipeline reactors "Sewerage systems of many cities comprise both pressure and gravity lines. It has been suggested that these lines may be used as aerobic biological reactors to reduce the biochemical oxygen demand (BOD) on treatment facilities" (C. M. Koch and I. Zandi, "Use of Pipelines as Aerobic Biological Reactors," *J. Water Pollut. Control Fed.*, **45**, 2537, 1973).

(*a*) Considering the moving liquid phase only, Powell and Lowe derive the result for a plug-flow tubular reactor:

$$\mu_{\max} t = \frac{1+\gamma}{\gamma} \ln \frac{\sigma - \gamma}{\sigma_0 - \gamma} - \frac{1}{\gamma} \ln \frac{\sigma}{\sigma_0}$$

where σ_0 = dimensionless substrate level at time 0 (entrance)
σ = dimensionless substrate level at time t (exit)
$\gamma = \sigma_0 + x_0$
x_0 = dimensionless entering cell concentration
μ_{max} = maximum specific growth rate
t = detention time

Find the assumed growth law and carefully define each variable in terms of dimensional variables.

(*b*) Koch and Zandi suggest that an aerobic pipeline reactor may be described as a two-phase flowing system with initial air- and liquid-phase (slurry) volumetric flow rates of Q_a and Q_l. Assuming (1) gas and liquid phases to each be in plug flow with identical velocities and (2) d(total pressure)/$dz = \lambda$ = const, where z = distance in pipeline, write out the two equations describing variation of oxygen concentrations with distance. [Assume that the slurry-phase oxygen consumption rate is linear in cell density, i.e., that the dissolved oxygen level c_{O_2} is always $> c_{cr}$ (Table 8.2).] State any assumptions.

(*c*) Integrate these equations to find the length of pipeline reactor at which $c_0 = c_{cr}$, that is, the point at which the gas phase should be replenished to maintain aerobic microbial activity. Take $Q_a = 5\ \text{ft}^3/\text{s}$, $Q_l = 500\ \text{ft}^3/\text{s}$, $x_0 = 10$ mg/L, pipe ID = 120 in, initial $c_{O_2} = 8$ mg/L, $c_{cr} = 0.5$ mg/L, s_0 = initial BOD = 150 mg O_2/L, $T = 300$ K, $K_s = 100$ mg O_2/L, $\mu_{max} = 0.3$/h, $\bar{Y}_s = 0.4$ g cell/g BOD, respiration rate = 0.375 g O_2/(h·g cells), $k_l a = 0.4$/min, $H = 4 \times 10^4$ atm/mol fraction, $\lambda = 0.005$ ft water pressure/ft pipe.

14.13 Oscillatory feed In waste-treatment plants, the feed rate may be periodic with a period length of hours to days. Using the Monod equation for single-substrate limitation, and taking s_f = const and $D = D_0(1 + \alpha \sin wt)$, $\alpha > 0$:

(*a*) Show that if x and its derivatives are periodic in time,

$$\int_0^{2\pi/w} \left\{ \frac{\mu_{max} s(t')}{K_s + s(t')} - D(t') \right\} x(t')\, dt' = 0$$

The solution $x(t') = 0$ for all t' is the washout result. Indicate the behavior of the quantity in brackets within the integral when a nontrivial solution exists, that is, $x(t') > 0$ for $0 \le t' \le 2\pi/w$.

(*b*) If $s_f \gg K$ and α is small, near the washout region $s(t')$ will always be close to s_f. Under this condition, show that the nontrivial solution requires $D_0 = \mu_{max}$ and that the variation of x for $0 \le t \le 2\pi/w$ is given by

$$x = x(0) \exp\left[\frac{D\alpha}{w}(1 - \cos wt)\right]$$

(*c*) If $s_f \ll K$, indicate explicitly what equations would have to be solved (by computer) in order to find $x(t)$ for the nontrivial case.

14.14 Detergent biodegradation Final biodegradation of long-lived substrates such as a number of detergent molecules appears to occur under substrate-limited conditions, thus the cell growth rate is linear in substrate, and the cell disappearance rate due to maintenance and death is appreciable. Under this circumstance, C. H. Wayman (*Prog. Ind. Microbiol.*, **10**: 219, 1971) proposed a relation between biomass x and detergent substrate s to be

$$\frac{dx}{dt} + Rx = \mu_m xs = -Y\frac{ds}{dt}$$

where R = respiration rate for no growth, that is, $s(\text{no growth}) = R/\mu_m$.

(*a*) Obtain the explicit solution $s = f(x)$ from the second equality.

(*b*) Using the approximation that $\int_0^{t'} x\, dt \sim xt'/2$ if t' represents the time at which x has achieved its maximum value, show that the biomass term x now follows Bernoulli's equation:

$$\frac{dx}{dt} + \phi_1 x = \phi_2 x^n$$

(*c*) By the change of variable $\psi = 1/x$, obtain a solution for x valid for a time period just somewhat beyond t'. [The solution becomes less accurate at longer times since the error in the approximation grows, but since the time (place) until nearly total substrate consumption has occurred is of main interest, this restriction seems less important.]

(*d*) Plot the shape of s for the case $t' \sim 1$ when $\phi_1/\phi_2 = 0.1, 1.0, 1.0$. How important are maintenance kinetics when slowly metabolized substrates are of interest?

14.15 Ultraviolet sterilizers In homes utilizing well water, microorganism contaminants may be reduced by UV irradiation accomplished by a bank of UV lamps arranged in hexagonal fashion parallel to a glass section of water pipe. Suppose the microbes die according to a first-order rate law, the rate "constant" being proportional to the local UV intensity I. Assume that the UV illumination is radially symmetric, and that the UV intensity varies radially according to Beer's law ($I(r) = I(R)e^{-\alpha(R-r)}$ where R = inner glass tube radius.

(*a*) For laminar flow, $u = u_{max}(1 - (r/R)^2)$. For a tube of length L, evaluate the ratio of viable organisms averaged over the entire exit section. What exit average is most meaningful?

(*b*) Repeat (*a*) for turbulent flow in the same tube (assume $L/D \gg 1$).

(*c*) If the drinking water standard is $\bar{c}$ viable organisms per liter, what is the upper limit of contamination of inlet water for laminar flow? For turbulent flow?

(*d*) Evaluate Re_{pipe} for $\frac{1}{2}$-in diameter pipe under expected volumetric flow rates in your residence.

REFERENCES

Substantial material on mixed microbial populations, including additional detail on the natural cycles of matter, can be found in Refs. 1 to 3 of Chap. 1. Also useful are most references in Chap. 13. Other readings on applications of mixed populations:

1. T. Hattori, *Microbial Life in the Soil: An Introduction*, Marcel Dekker, Inc., New York, 1973. A broad-based introduction to soil microbiology which emphasizes quantitative treatment wherever possible. Besides chapters on soil microbes, including their physiology, interactions, and roles in plant growth and geochemical transformations, this text offers substantial readings on the soil environment.
2. R. Mitchell, *Introduction to Environmental Microbiology*, Prentice-Hall, Inc., Englewood Cliffs, N.J., 1974. A broad survey of microbial action in the biosphere. Nutrient cycles, eutrophication, community ecology, and waste treatment are among the many examples discussed.
3. L. G. Rich, *Environmental Systems Engineering*. McGraw-Hill Book Company, New York, 1973. This environmental-science text emphasizes the application of modern modeling and systems methods to processes, including waste treatment.
4. J. W. Abson and K. H. Todhunter, "Effluent Disposal," chap. 9 in N. Blakebrough (ed.), *Biochemical and Biological Engineering Science*, vol. 1, Academic Press, London, 1967. A good brief review of industrial waste-water treatment. Chapter 10 by R. F. Wills, dealing with sedimentation and flocculation, provides an excellent complement to wastewater-treatment aspects emphasized here.
5. John F. Andrews: Review Paper: Dynamic Models and Control Strategies for Wastewater Treatment Processes, *Water Res.*, **8**: 261–289, 1974. An excellent summary of problems and approaches for describing and controlling treatment plant dynamics.

Detailed information on topics of special interest:

6. J. B. Busby and J. F. Andrews, "A Dynamic Model and Control Strategies for the Activated Sludge Process," *J. Water Pollut. Control Fed.*, **47**: 1055, 1975.
7. S. P. Graef and J. F. Andrews, "Mathematical Modeling and Control of Anaerobic Digestion," *CEP Symp. Ser.* [136] **70**: 101–127, 1974.

8. C. R. Curds, "A Theoretical Study of Factors Influencing the Microbial Population Dynamics of the Activated-Sludge Process, I: The Effects of Diurnal Variations of Sewage and Carnivorous Ciliated Protozoa," *Water Res.*, **7**: 1269–1284, 1973.
9. S. Y. Chiu, L. E. Erickson, L. T. Fan, and I. C. Kao, "Kinetic Model Identification in Mixed Populations Using Continuous Culture Data," *Biotech. Bioeng.*, **14**: 207–231, 1972.
10. E. R. Schroeder, *Water and Waste Water Treatment*, McGraw-Hill, New York, 1977.

Problems

11. C. T. Calam and E. W. Russell, "Microbial Aspects of Fermentation Process Development," *J. Appl. Chem. Biotech.*, **23**: 225, 1973.
12. H. W. Blanch and I. J. Dunn, "Modelling and Simulation in Biochemical Engineering," *Adv. Biochem. Eng.*, **3**: 159–162, 1974.
13. F. G. Pohland and S. Ghosh, "Developments in Anaerobic Treatment Processes, in Biological Waste Treatment," *Biotech. Bioeng. Symp.* 2, p. 85, 1971.
14. A. D. McLaren, *Soil Sci. Soc. Am. Proc.*, **33**: 55, 1969.
15. E. J. Russel, *Microorganisms of the Soil*, Longmans, Green and Co., London, 1923.
16. R. P. Canale and J. A. Borchardt, pp. 120–121 in W. J. Weber, Jr. (ed.), *Physiocochemical Processes for Water Quality Control*, Wiley-Interscience, New York, 1972.
17. M. L. Rozenweig and R. H. MacArthur, "Graphical Representation and Stability Conditions of Predator-Prey Interactions," *Am. Nat.*, **97**: 209, 1963; **103**: 81, 1969.

INDEX